Mastering Algebra

Intermediate Level

Book Title: Mastering Algebra - Intermediate Level

Author: Said Hamilton

Editor: Pat Eblen

Cover design by: Kathleen Myers

First published in 1998

Hamilton Education Guides
P.O. Box 681
Vienna, Va. 22183

Library of Congress Catalog Card Number 98-74114
Library of Congress Cataloging-in-Publication Data

ISBN 0-9649954-2-5

This book is dedicated to my wife and children for their support and understanding.

General Contents

Detailed Contents

Chapter 1 - Review of Introductory Algebra

Chapter 3 - Factoring Polynomials

Chapter 4 - Quadratic Equations and Factoring

Chapter 5 - Algebraic Fractions

Chapter 6 - Logarithms

Appendix - Exercise Solutions

Acknowledgments

I would like to acknowledge my wife and children for giving me inspiration and for their understanding and patience in allowing me to take on the task of writing this book. I am grateful to Pat Eblen for his editorial comments. As always, his constructive comments and suggestions on clearer presentation of topics truly elevated the usefulness of this book. His devotion to perfection is commendable. I would also like to acknowledge and give my thanks to numerous education professionals who reviewed and provided comments to further enhance this book. Finally, my special thanks to Kathleen Myers for her outstanding cover design.

Introduction and Overview

As we approach the twenty first century in our ever increasing technological society it is becoming more and more essential for individuals to obtain a thorough knowledge of technological concepts. One key to achieving this is through quality educational materials that can provide detailed instructions for solving mathematical problems.

My approach in teaching mathematics and other technical subjects has always been to present any topic in the simplest way possible. I have always believed that many of our academic texts, particularly those that lay the foundation for learning math and science, are either written in an abstract and difficult to follow language, lack sufficient number of detailed sample problems, or are not explained adequately for a student to become interested in the subject.

It is my belief that the key to learning mathematics is through stimulating student interest. Students can be greatly motivated if subjects are presented concisely and problems are solved in a detailed step-by-step approach. This provides a great deal of encouragement for students to learn the next subject or to solve the next problem. During my teaching career, I found this approach to be an effective way of teaching. I believe by presenting subjects with the methods used in this book, more students will become interested in the subject of mathematics and can carry this approach and philosophy to future generations.

The scope of this book is intended for educational levels ranging from the 9th grade to adult. The book can also be used by students in home study programs, parents, teachers, special education programs, tutors, high schools, preparatory schools, and adult educational programs including colleges and universities as a main text, a thorough reference, or a supplementary book. A basic knowledge of algebraic concepts such as the use of sign numbers, parentheses and brackets, fractional operations, exponents, radicals, and polynomials is required.

"Mastering Algebra: Intermediate Level" is the second book in a series of three books on algebra. It addresses subjects such as linear and non linear equations, inequalities, factoring, quadratic equations, algebraic fractions, and logarithms. In the first book *"Mastering Algebra: An Introduction"* students are introduced to topics such as integer fractions, exponents, radicals, fractional exponents, and polynomials. The current plan for the third book, *"Mastering Algebra: Advanced Level"* is to address areas such as solving and graphing linear equations, quadratic equations, parabolas, circles, ellipses, hyperbolas, and vector analysis.

"Mastering Algebra - Intermediate Level" is divided into six chapters. Chapter 1 reviews in some detail selected subjects addressed in the *"Mastering Algebra - An Introduction"* book. Topics such as parentheses and brackets, operations involving with integer fractions, exponents, radicals, and polynomials are reviewed. One variable linear equations and inequalities are addressed in Chapter 2. In this chapter students learn how to solve various math operations involving linear equations and linear inequalities. In addition, linear equations containing parentheses and brackets, fractions, and decimals including solution to formulas are also addressed in this chapter. Factoring polynomials using various methods such as the Greatest Common Factoring method, the Grouping method, the Trial and Error method, as well as factoring methods for polynomials with square and cubed terms are addressed in Chapter 3. Quadratic equations and quadratic formula are introduced in Chapter 4. Solving quadratic equations using methods such as

the Quadratic Formula method, the Square Root Property method, and Completing-the-Square method are addressed in this chapter. Additionally, examples for choosing the best factoring or solution method are also discussed in Chapter 4. Students are encouraged to gain a thorough understanding of the various factoring and solution methods introduced in Chapters 3 and 4. A strong knowledge of factoring methods will greatly simplify solving algebraic fraction problems introduced in Chapter 5. The concept of algebraic fractions and how algebraic expressions are simplified to lower terms are addressed in Chapter 5. Math operations involving algebraic fractions, including complex algebraic fractions, are also discussed in this chapter. (A review of fraction techniques introduced in the *"Mastering Fractions"* book, mainly chapters 3, 6, and 9, will greatly simplify solving algebraic fraction problems.) Logarithms are introduced in Chapter 6. Computations involving common (base 10), natural (base e), as well as other than base 10 or e logarithms are covered in this chapter. In addition, computing antilogarithms and the steps as to how math operations are simplified using logarithms are discussed in Chapter 6. How to solve one variable logarithmic and exponential equations using the laws of logarithms are also addressed in this chapter. Finally, detailed solution to the exercises are provided in the Appendix. Students are encouraged to solve each problem in the same detail and step-by-step format as shown in the text.

In keeping with our commitment of excellence in providing clear, easy to follow, and concise educational materials to our readers, I believe this book will add value to the Hamilton Education Guides series for its clarity and special attention to detail. I hope readers of this book will find it valuable as a learning tool and a reference. Any comments or suggestions for improvement of this book will be appreciated.

With best wishes,

Said Hamilton

Chapter 1
Review of Introductory Algebra

Quick Reference to Chapter 1 Case Problems

$\boxed{\frac{15}{60}} = ;\quad \boxed{\frac{25}{10}} = ;\quad \boxed{-\frac{327}{24}} =$

Case I - Adding Integer Fractions with Common Denominators, *p. 25*

$\boxed{\frac{2}{3}+\frac{8}{3}} = ;\quad \boxed{\frac{15}{4}+\frac{9}{4}} = ;\quad \boxed{\frac{12}{5}+\frac{33}{5}} =$

Case II - Adding Integer Fractions without a Common Denominator, *p. 26*

$\boxed{\frac{2}{5}+\frac{3}{4}} = ;\quad \boxed{\frac{3}{5}+\frac{2}{7}} = ;\quad \boxed{\frac{5}{6}+3} =$

Case III - Subtracting Integer Fractions with Common Denominators, *p. 28*

$\boxed{\frac{25}{3}-\frac{2}{3}} = ;\quad \boxed{\frac{5}{10}-\frac{14}{10}} = ;\quad \boxed{\frac{15}{6}-\frac{53}{6}} =$

Case IV - Subtracting Integer Fractions without a Common Denominator, *p. 29*

$\boxed{\frac{4}{5}-\frac{3}{8}} = ;\quad \boxed{\frac{9}{8}-\frac{3}{4}} = ;\quad \boxed{\frac{10}{6}-35} =$

Case V - Multiplying Integer Fractions with or without a Common Denominator, *p. 31*

$\boxed{\frac{4}{5}\times\frac{3}{8}} = ;\quad \boxed{25\times\frac{5}{8}} = ;\quad \boxed{\frac{140}{3}\times\frac{1}{5}} =$

Case VI - Dividing Integer Fractions with or without a Common Denominator, *p. 33*

$\boxed{\frac{3}{5}\div\frac{8}{15}} = ;\quad \boxed{\frac{125}{65}\div 230} = ;\quad \boxed{\frac{32}{18}\div\frac{50}{12}} =$

$\boxed{\frac{86}{5}} = ;\quad \boxed{\frac{506}{3}} = ;\quad \boxed{-\frac{597}{10}} =$

1.3 Exponents

Case I - Real Numbers Raised to Positive Integer Exponents, *p. 38*

$\boxed{2^3} = ;\quad \boxed{(-3)^5} = ;\quad \boxed{-8^3} =$

Case II - Real Numbers Raised to Negative Integer Exponents, *p. 40*

$\boxed{4^{-3}} = ;\quad \boxed{(-8)^{-3}} = ;\quad \boxed{-2^{-4}} =$

Case I - Multiplying Positive Integer Exponents, *p. 42*

$\boxed{\left(x^3y^2\right)\cdot\left(x^2y\right)\cdot y^3} = ;\quad \boxed{\left(e^3e^5e\right)\cdot\left(-\frac{4}{32}e^2\right)} = ;\quad \boxed{(2-3)^2\cdot\left(5x^3y^2\right)\cdot(-2xy)} =$

Case II - Dividing Positive Integer Exponents, *p. 44*

$\boxed{\dfrac{2ab}{-4a^3b^4}} = ;\quad \boxed{\left(\dfrac{u^2v^3w^2}{8u^7v^5}\right)\cdot\left(\dfrac{u}{v^2}\right)} = ;\quad \boxed{\dfrac{4k^3lm^2}{kl^2m^3}} =$

Case III - Adding and Subtracting Positive Integer Exponents, *p. 46*

$\boxed{x^3 + 3y^2 + 2x^3 - y^2 + 5} = ;\quad \boxed{\left(a^3 + 2a^2 + 4^3\right) - \left(4a^3 + 20\right)} = ;\quad \boxed{a^{3b} + 2a^{2b} - 4a^{3b} + 5 + 3a^{2b}} =$

Case I - Multiplying Negative Integer Exponents, *p. 48*

$\boxed{5^{-2}\cdot 5\cdot a^{-3}\cdot b^{-3}\cdot a^{-1}\cdot b} = ;\quad \boxed{\left(x^{-2}y^{-2}z^2\right)\cdot\left(x^{-1}y^3z^{-4}\right)} = ;\quad \boxed{\left(3^{-2}\cdot 2^{-1}\right)\cdot\left(3^{-1}\cdot 2^{-2}\right)\cdot 2} =$

Case II - Dividing Negative Integer Exponents, *p. 51*

$\boxed{\dfrac{2^{-3}u^{-1}v^{-3}}{2^{-1}u^{-2}v}} = ;\quad \boxed{\dfrac{e^{-5}f^{-2}e^0}{2^{-3}f^5e^{-4}}} = ;\quad \boxed{\dfrac{2^{-3}\cdot a^{-1}}{(-2)^{-2}a^{-3}}} =$

Case III - Adding and Subtracting Negative Integer Exponents, *p. 53*

$\boxed{x^{-1} + x^{-2} + 2x^{-1} - 4x^{-2} + 5^{-2}} = ;\quad \boxed{a^{-1} - b^{-1} + 2a^{-1} + 3b^{-1}} = ;\quad \boxed{k^{-2n} + k^{-3n} - 3k^{-2n} + 2^{-2}} =$

1.4 Radicals

Case I - Roots and Radical Expressions, *p. 56*

$\boxed{\sqrt{64}} = ;\quad \boxed{\sqrt[3]{375}} = ;\quad \boxed{-\sqrt[2]{324}} =$

Case II - Rational, Irrational, Real, and Imaginary Numbers, *p. 58*

$\boxed{\sqrt{25}\ \textit{is a rational number}}$; $\boxed{\sqrt{7}\ \textit{is an irrational number}}$; $\boxed{-\sqrt{-3}\ \textit{is not a real number}}$.

Case III - Simplifying Radical Expressions with Real Numbers as a Radicand, *p. 59*

$\boxed{\dfrac{-3}{-2}\sqrt{400}} = ;\quad \boxed{\dfrac{\sqrt[3]{162}}{9}} = ;\quad \boxed{\dfrac{1}{-5}\sqrt[2]{1000}} =$

Case I - Multiplying Monomial Expressions in Radical Form, with Real Numbers, *p. 61*

$\boxed{\sqrt{5}\cdot\sqrt{15}} = ;\quad \boxed{\left(-2\sqrt[3]{512}\right)\cdot\left(-5\sqrt[3]{108}\right)} = ;\quad \boxed{\sqrt{6}\cdot\sqrt{48}\cdot\sqrt{45}} =$

Case II - Multiplying Binomial Expressions in Radical Form, with Real Numbers, *p. 63*

$\boxed{\left(2+\sqrt{2}\right)\cdot\left(5-\sqrt{8}\right)} = ;\quad \boxed{\left(2\sqrt[4]{162}+3\right)\cdot\left(3\sqrt[4]{2}+5\right)} = ;\quad \boxed{\left(\sqrt{24}+3\sqrt{60}\right)\cdot\left(\sqrt{25}-\sqrt{72}\right)} =$

Case III - Multiplying Monomial and Binomial Expressions in Radical Form, with Real Numbers, *p. 66*

$\boxed{\sqrt{5}\cdot\left(\sqrt{50}+2\sqrt{27}\right)} = ;\quad \boxed{-2\sqrt{24}\cdot\left(\sqrt{36}-\sqrt{125}\right)} = ;\quad \boxed{-2\sqrt[4]{4}\cdot\left(\sqrt[4]{64}-\sqrt[4]{162}\right)} =$

Case IV - Rationalizing Radical Expressions - Monomial Denominators with Real Numbers, *p. 68*

$$\boxed{\frac{-8\sqrt{3}}{32\sqrt{45}}} = \;; \quad \boxed{\frac{5\sqrt{(-2)^2 \cdot 3}}{-\sqrt{7}}} = \;; \quad \boxed{\frac{3\sqrt[5]{8}}{\sqrt[5]{81}}} =$$

Case V - Rationalizing Radical Expressions - Binomial Denominators with Real Numbers, *p. 71*

$$\boxed{\frac{8}{2-\sqrt{2}}} = \;; \quad \boxed{\frac{\sqrt{125}}{\sqrt{3}-\sqrt{5}}} = \;; \quad \boxed{\frac{3+\sqrt{5}}{\sqrt{3}+\sqrt{5}}} =$$

Case VI - Adding and Subtracting Radical Terms, *p. 74*

$$\boxed{6\sqrt{2}+4\sqrt{2}} = \;; \quad \boxed{20\sqrt[5]{3}-8\sqrt[5]{3}+5\sqrt[5]{3}} = \;; \quad \boxed{a\sqrt{xy}+b\sqrt[3]{xy}-c^2\sqrt{xy}-d} =$$

1.5 Polynomials

1.5a Introduction to Polynomials..76

Case I - Classification of Polynomials, *p. 76*

$$\boxed{x^2y^2z} = \;; \quad \boxed{x^3-2x^2} = \;; \quad \boxed{w^5-2w^3+4w^2+7} =$$

Case II - Simplifying Polynomials, *p. 78*

$$\boxed{18x^3+2x^2-5x^3-2x-x^2} = \;; \quad \boxed{-8y^3+4y^5-5y^3-2y^5+3} = \;; \quad \boxed{2w^4+4w^3-w^4-8+2w-w^3+4} =$$

1.5b Operations with Polynomials..81

Case I - Multiplying Monomials, *p. 81*

Case I a - Multiplying Monomials by Monomials, p. 81

$$\boxed{\left(5x^3y^2\right)\cdot\left(3x^3y^2z\right)} = \;; \quad \boxed{\left(3a^2b^3c^5\right)\cdot\left(5b^2c^4\right)\cdot\left(4a^3b^0c^3\right)} = \;; \quad \boxed{\left(3x^2\right)\cdot(5xy)\cdot\left(2x^2y^3\right)} =$$

Case I b - Multiplying Polynomials by Monomials, p. 82

$$\boxed{\left(2x^4+3x^2+5x-x^4+x^2-3\right)\cdot\left(3x^2\right)} = \;; \quad \boxed{\left(5m^3n^3+2m^2n^2-3mn+mn+2\right)\cdot(5mn)} = \;;$$

$$\boxed{\left(\sqrt{27}x^2-2+\sqrt{8}x+\sqrt{36}\right)\cdot\sqrt{125}x} =$$

Case II - Multiplying Binomials by Binomials, *p. 85*

$$\boxed{\left(x^2+3x\right)(x+8)} = \;; \quad \boxed{\left(\sqrt{225}x+2\right)\left(5x-\sqrt{81}\right)} = \;; \quad \boxed{\left(\frac{2}{3}x^3+\frac{1}{3}x\right)\left(\frac{1}{2}x^2-\frac{3}{8}\right)} =$$

Case III - Dividing Polynomials by Monomials and Polynomials, *p. 87*

Case III a - Dividing Monomials by Monomials, p. 87

$$\boxed{\frac{\sqrt{8x^3y^2}}{\sqrt{243xy^3}}} = \;; \quad \boxed{\frac{-\sqrt{12a^2b^2c}}{\sqrt{225abc^4}}} = \;; \quad \boxed{\frac{\sqrt[3]{16u^2v^2}}{-\sqrt[3]{27uv^3}}} =$$

Case III b - Dividing Binomials by Monomials, p. 90

$$\boxed{\frac{8x^3-16x^2}{-8x}} = \;; \quad \boxed{\frac{-15\sqrt{a^3}+10\sqrt{a^2}}{-5a^2}} = \;; \quad \boxed{\frac{\sqrt[3]{x^3y^5}-4\sqrt[4]{x^8y^4}}{2\sqrt[3]{x^3y^6}}} =$$

Case III c - Dividing Polynomials by Polynomials, p. 92

$\boxed{Divide\ x^4 + 8x^3 + 16x^2 + 5x \ \ by\ \ x^2 + 3x + 1} = ;$ $\boxed{Divide\ 6x^2 + 19x + 18 \ \ by\ \ 3x + 5} = ;$

$\boxed{Divide\ x^4 + 2x^3 + 2x^2 + 2x + 6 \ \ by\ \ x + 1} =$

Case IV - Adding and Subtracting Polynomials Horizontally, *p. 95*

$\boxed{\left(x^2 + 3x^3 + 5\right) + \left(x^3 + 8x + 2x^2\right)} = ;$ $\boxed{\left(y + y^2 + 3y^3 + 3\right) - \left(3y^2 + 2y - y^3\right)} = ;$

$\boxed{\left(k^2 l + kl^2 + k + 2k^2 l\right) - \left(3k - 2kl^2 - k^2 l\right)} =$

Case V - Adding and Subtracting Polynomials Vertically, *p. 97*

$\boxed{\left(x^4 + x + 3x^3 + 4x\right) + \left(x^2 + 3x^4 - x^3 + x\right)} = ;$ $\boxed{\left(w^4 + 2w^3 + w + 2w^2\right) + \left(2w + 4w^2 + 6w\right)} = ;$

$\boxed{\left(a^2 + 3a + 2a^2 - a - 2\right) - \left(4 - 4a^2 - 3a - 6\right)} =$

Chapter 1 - Review of Introductory Algebra

The objective of this chapter is to review in some detail the subjects that were addressed in the "*Mastering Algebra - An Introduction*" book. Students are expected to be familiar with the subjects presented in the introductory book (Parentheses and Brackets, Integer Fractions, Exponents, Radicals, Fractional Exponents, Scientific Notation, and Polynomials) before proceeding with the subjects presented in this book. Students who may need to review additional examples need to study the "*Mastering Algebra - An Introduction*" book.
A summary of the math operations involving sign numbers including how Parentheses and Brackets are used in addition, subtraction, multiplication, and division of numbers is addressed in Section 1.1. Mathematical operations involving integer fractions which includes addition, subtraction, multiplication, and division of integer fractions is discussed in Section 1.2. A review of positive and negative integer exponents, including operations involving with positive and negative exponents is provided in Section 1.3. An introduction to radical expressions, the steps in simplifying radicals, as well as how rational, irrational, real, and imaginary numbers are identified is discussed in Section 1.4. In addition, multiplication, division, addition, and subtraction of radical expressions is addressed in this section. Classification of polynomials, how polynomials are simplified, including multiplying, dividing, adding, and subtracting different types of polynomials is reviewed in Section 1.5. Additional examples, followed by practice problems, are provided in the cases presented in each section to help meet the objective of this chapter.

1.1 Parentheses and Brackets

In this section addition, subtraction, multiplication, and division of signed numbers are reviewed in section 1.1a, Cases I through IV. In addition, the use of parentheses and brackets in solving math operations is addressed in section 1.1b, Cases I and II.

1.1a Signed Numbers

In mathematics, " + " and " − " symbols are used to indicate the use of positive and negative numbers, respectively. If a signed number has no symbol it is understood to be a positive number. Signed numbers are added, subtracted, multiplied, and divided as exemplified in the following cases:

Case I - Addition of Signed Numbers

When two numbers are added, the numbers are called **addends** and the result is called a **sum.** The sign of the sum dependents on the sign of the numbers. This is shown in the following cases with the sign change of two real numbers a and b:

Case I a.

$\boxed{a+b} = \boxed{A}$

For example,

1. $\boxed{5+6} = \boxed{\mathbf{11}}$ 2. $\boxed{7+8} = \boxed{\mathbf{15}}$ 3. $\boxed{1+0} = \boxed{\mathbf{1}}$

Case I b.

$\boxed{-a+b} = \boxed{B}$

For example,

1. $\boxed{-7+3} = \boxed{\mathbf{-4}}$ 2. $\boxed{-9+0} = \boxed{\mathbf{-9}}$ 3. $\boxed{-15+40} = \boxed{\mathbf{25}}$

Case I c.

$\boxed{a+(-b)} = \boxed{a-b} = \boxed{C}$

For example,

1. $\boxed{2+(-5)} = \boxed{2-5} = \boxed{\mathbf{-3}}$ 2. $\boxed{7+(-9)} = \boxed{7-9} = \boxed{\mathbf{-2}}$

Case I d.

$\boxed{(-a)+b} = \boxed{-a+b} = \boxed{D}$ *Note:* $(-a) = -a$

For example,

1. $\boxed{(-3)+9} = \boxed{-3+9} = \boxed{\mathbf{6}}$ 2. $\boxed{(-12)+8} = \boxed{-12+8} = \boxed{\mathbf{-4}}$

Case I e.

$\boxed{(-a)+(-b)} = \boxed{-a-b} = \boxed{E}$

For example,

1. $\boxed{(-6)+(-9)} = \boxed{-6-9} = \boxed{\mathbf{-15}}$ 2. $\boxed{(-45)+(-6)} = \boxed{-45-6} = \boxed{\mathbf{-51}}$

Case II - Subtraction of Signed Numbers

When two numbers are subtracted the result is called the **difference**. The sign of the difference depends on the sign of the numbers. This is shown in the following cases with the sign change of two real numbers a and b:

Case II a.

$\boxed{a-b} = \boxed{A}$

For example,

1. $\boxed{15-6} = \boxed{\mathbf{9}}$ 2. $\boxed{17-47} = \boxed{\mathbf{-30}}$ 3. $\boxed{1-0} = \boxed{\mathbf{1}}$

Case II b.

$\boxed{-a-b} = \boxed{B}$

For example,

1. $\boxed{-7-3} = \boxed{\mathbf{-10}}$ 2. $\boxed{-1+0} = \boxed{\mathbf{-1}}$ 3. $\boxed{-15\mp 45} = \boxed{\mathbf{-60}}$

Case II c.

$\boxed{a-(-b)} = \boxed{a+(b)} = \boxed{a+b} = \boxed{C}$

For example,

1. $\boxed{12-(-5)} = \boxed{12+(5)} = \boxed{12+5} = \boxed{\mathbf{17}}$
2. $\boxed{7-(-9)} = \boxed{7+(9)} = \boxed{7+9} = \boxed{\mathbf{16}}$

Case II d.

$\boxed{(-a)-(-b)} = \boxed{(-a)+(b)} = \boxed{-a+b} = \boxed{D}$

For example,

1. $\boxed{(-3)-(-9)} = \boxed{(-3)+(9)} = \boxed{-3+9} = \boxed{\mathbf{6}}$
2. $\boxed{(-32)-(-8)} = \boxed{(-32)+(8)} = \boxed{-32+8} = \boxed{\mathbf{-24}}$

Case III - Multiplication of Signed Numbers

When two numbers are multiplied, the numbers are called **factors** and the result is called a **product**. For example, when 12 is multiplied by 2 the result is 24.

$$\boxed{12\,(factor)\times 2\,(factor)} = \boxed{\mathbf{24}\,(product)}$$

Thus, 12 and 2 are the factors, and 24 is the product.

The sign of the product is positive if the factors have the same sign and is negative if the factors have different signs. This is shown in the following cases with the sign change of two real numbers a and b:

Case III a.

$\boxed{a\times b} = \boxed{ab}$

For example,

1. $\boxed{5\times 6} = \boxed{\mathbf{30}}$
2. $\boxed{7\times 8} = \boxed{\mathbf{56}}$
3. $\boxed{1\times 0} = \boxed{\mathbf{0}}$

Case III b.

$\boxed{(-a)\times b} = \boxed{-a\times b} = \boxed{-ab}$

For example,

1. $\boxed{(-7)\times 3} = \boxed{-7\times 3} = \boxed{\mathbf{-21}}$
2. $\boxed{(-1)\times 0} = \boxed{-1\times 0} = \boxed{\mathbf{0}}$

Case III c.

$\boxed{a\times(-b)} = \boxed{-a\times b} = \boxed{-ab}$

For example,

1. $\boxed{2\times(-5)} = \boxed{-2\times 5} = \boxed{\mathbf{-10}}$
2. $\boxed{7\times(-9)} = \boxed{-7\times 9} = \boxed{\mathbf{-63}}$

Case III d.

$\boxed{(-a)\times(-b)} = \boxed{+ab} = \boxed{ab}$

For example,

1. $\boxed{(-3)\times(-9)} = \boxed{+27} = \boxed{\mathbf{27}}$
2. $\boxed{(-12)\times(-4)} = \boxed{+48} = \boxed{\mathbf{48}}$

Case IV - Division of Signed Numbers

When one number is divided by another, the first number is called the **dividend**, the second number the **divisor**, and the result a **quotient**. For example, when 12 is divided by 2 the result is 6.

$$\boxed{\frac{12\,(dividend)}{2\,(divisor)}} = \boxed{6\,(quotient)}$$

Thus, 12 is the dividend, 2 is the divisor, and 6 is the quotient.

The sign of the quotient is positive if the divisor and the dividend have the same sign and is negative if the divisor and the dividend have different signs. This is shown in the following cases with the sign change of two real numbers a and b:

Case IV a.

$$\boxed{\frac{a}{b}} = \boxed{A}$$

For example,

1. $\boxed{\frac{9}{3}} = \boxed{3}$ 2. $\boxed{\frac{27}{3}} = \boxed{9}$ 3. $\boxed{\frac{75}{5}} = \boxed{15}$

Case IV b.

$$\boxed{\frac{-a}{b}} = \boxed{-\frac{a}{b}} = \boxed{B}$$

For example,

1. $\boxed{\frac{-10}{2}} = \boxed{-\frac{10}{2}} = \boxed{-5}$ 2. $\boxed{\frac{-66}{3}} = \boxed{-\frac{66}{3}} = \boxed{-22}$ 3. $\boxed{\frac{-75}{5}} = \boxed{-\frac{75}{5}} = \boxed{-15}$

Case IV c.

$$\boxed{\frac{a}{-b}} = \boxed{-\frac{a}{b}} = \boxed{C}$$

For example,

1. $\boxed{\frac{30}{-2}} = \boxed{-\frac{30}{2}} = \boxed{-15}$ 2. $\boxed{\frac{88}{-8}} = \boxed{-\frac{88}{8}} = \boxed{-11}$ 3. $\boxed{\frac{45}{-9}} = \boxed{-\frac{45}{9}} = \boxed{-5}$

Case IV d.

$$\boxed{\frac{-a}{-b}} = \boxed{+\frac{a}{b}} = \boxed{\frac{a}{b}} = \boxed{D}$$

For example,

1. $\boxed{\frac{-4}{-2}} = \boxed{+\frac{4}{2}} = \boxed{\frac{4}{2}} = \boxed{2}$ 2. $\boxed{\frac{-8}{-3}} = \boxed{+\frac{8}{3}} = \boxed{\frac{8}{3}} = \boxed{2.67}$ 3. $\boxed{\frac{-7}{-7}} = \boxed{+\frac{7}{7}} = \boxed{\frac{7}{7}} = \boxed{1}$

Signed Numbers - General Rules

Addition: $\boxed{(-)+(-)=(-)}$; $\boxed{(-)+(+)=(-) \text{ if negative No. is} \rangle \text{ positive No.}}$; $\boxed{(+)+(+)=(+)}$;

$\boxed{(+)+(-)=(+) \text{ if positive No. is} \rangle \text{ negative No.}}$ *Note: The symbol " $\rangle$ " means greater than.*

Subtraction: $\boxed{(-)-(-)=(-)+(+)=(-) \text{ if the 1st. negative No. is} \rangle \text{ the 2nd. negative No.}}$;

$\boxed{(+)-(+)=(+)+(-)=(+) \text{ if the 1st. positive No. is} \rangle \text{ the 2nd. positive No.}}$;

$\boxed{(+)-(-)=(+)+(+)=(+)}$; $\boxed{(-)-(+)=(-)+(-)=(-)}$

Multiplication: $\boxed{(-)\times(-)=(+)}$; $\boxed{(-)\times(+)=(-)}$; $\boxed{(+)\times(+)=(+)}$, $\boxed{(+)\times(-)=(-)}$

Division: $\boxed{\frac{(-)}{(-)}=(+)}$; $\boxed{\frac{(-)}{(+)}=(-)}$; $\boxed{\frac{(+)}{(+)}=(+)}$; $\boxed{\frac{(+)}{(-)}=(-)}$

Signed Numbers - Summary of Cases

1. Addition and Subtraction:

I a. $\boxed{a+b}=\boxed{A}$

I b. $\boxed{-a+b}=\boxed{B}$

I c. $\boxed{a+(-b)}=\boxed{a-b}=\boxed{C}$

I d. $\boxed{(-a)+b}=\boxed{-a+b}=\boxed{D}$

I e. $\boxed{(-a)+(-b)}=\boxed{-a-b}=\boxed{E}$

II a. $\boxed{a-b}=\boxed{A}$

II b. $\boxed{-a-b}=\boxed{B}$

II c. $\boxed{a-(-b)}=\boxed{a+(b)}=\boxed{a+b}=\boxed{C}$

II d. $\boxed{(-a)-(-b)}=\boxed{(-a)+(b)}=\boxed{-a+b}=\boxed{D}$

2. Multiplication and Division:

III a. $\boxed{a\times b}=\boxed{ab}$

III b. $\boxed{(-a)\times b}=\boxed{-a\times b}=\boxed{-ab}$

III c. $\boxed{a\times(-b)}=\boxed{-a\times b}=\boxed{-ab}$

III d. $\boxed{(-a)\times(-b)}=\boxed{+ab}=\boxed{ab}$

IV a. $\boxed{\frac{a}{b}}=\boxed{A}$

IV b. $\boxed{\frac{-a}{b}}=\boxed{-\frac{a}{b}}=\boxed{B}$

IV c. $\boxed{\frac{a}{-b}}=\boxed{-\frac{a}{b}}=\boxed{C}$

IV d. $\boxed{\frac{-a}{-b}}=\boxed{+\frac{a}{b}}=\boxed{\frac{a}{b}}=\boxed{D}$

Practice Problems - Signed Numbers

Section 1.1a Practice Problems - Show the correct sign by performing the following operations:

1. $\frac{-95}{-5}=$
2. $(-20)\times(-8)=$
3. $(-33)+(-14)=$
4. $(-18)-(+5)=$
5. $(-20)+8=$
6. $\frac{48}{-4}=$

1.1b Using Parentheses and Brackets in Mixed Operations

Parentheses and brackets are used to group numbers as a means to minimize mistakes in solving mathematical operations. It is important to solve math operations in the exact order in which parentheses or brackets are given. In this section the use of parentheses and brackets as applied to addition, subtraction, multiplication, and division, using integer numbers, is discussed. However, the properties associated with math operations are summarized first:

Commutative and Associative Property of Addition

1. Changing the order in which two numbers are added does not change the final answer. This property of real numbers is called the **Commutative Property of Addition**, e.g., for any two real numbers a and b

   $\boxed{a+b} = \boxed{b+a}$

   For example, $\boxed{9+7} = \boxed{\mathbf{16}}$ and $\boxed{7+9} = \boxed{\mathbf{16}}$

2. Re-grouping numbers does not change the final answer. This property of real numbers is called the **Associative Property of Addition**, e.g., for any real numbers a, b, and c

   $\boxed{(a+b)+c} = \boxed{a+(b+c)}$

   For example,

   $\boxed{(5+4)+7} = \boxed{(9)+7} = \boxed{9+7} = \boxed{\mathbf{16}}$

   $\boxed{5+(4+7)} = \boxed{5+(11)} = \boxed{5+11} = \boxed{\mathbf{16}}$

Properties Associated with Subtraction

1. Changing the order in which two numbers are subtracted does change the final answer. For example, for any two real numbers a and b

   $\boxed{a-b \neq b-a}$ *Note:* The symbol "$\neq$" means not equal.

   For example, $\boxed{20-8} = \boxed{\mathbf{12}}$, but $\boxed{8-20} = \boxed{\mathbf{-12}}$

2. Re-grouping numbers does change the final answer. For example, for any real numbers a, b, and c

   $\boxed{(a-b)-c \neq a-(b-c)}$

   For example,

   $\boxed{(25-6)-8} = \boxed{(19)-8} = \boxed{19-8} = \boxed{\mathbf{11}}$, however

   $\boxed{25-(6-8)} = \boxed{25-(-2)} = \boxed{25+(2)} = \boxed{25+2} = \boxed{\mathbf{27}}$

Commutative, Associative, and Distributive Property of Multiplication

1. Changing the order in which two numbers are multiplied does not change the final answer. This property of real numbers is called the **Commutative Property of Multiplication**, e.g., for any two real numbers a and b

   $\boxed{a \times b} = \boxed{b \times a}$

   For example, $\boxed{3 \times 15} = \boxed{\mathbf{45}}$ and $\boxed{15 \times 3} = \boxed{\mathbf{45}}$

2. Re-grouping numbers does not change the final answer. This property of real numbers is called the **Associative Property of Multiplication**, e.g., for any real numbers a, b, and c

   $\boxed{(a \times b) \times c} = \boxed{a \times (b \times c)}$

   For example,

   $\boxed{(4 \times 8) \times 5} = \boxed{(32) \times 5} = \boxed{32 \times 5} = \boxed{\mathbf{160}}$

   $\boxed{4 \times (8 \times 5)} = \boxed{4 \times (40)} = \boxed{4 \times 40} = \boxed{\mathbf{160}}$

3. Multiplication can be distributed over addition. This property is called the **Distributive Property of Multiplication**, e.g., for any real numbers a, b, and c

   $\boxed{a \times (b + c)} = \boxed{ab + ac}$

   For example,

   $\boxed{9 \times (4 + 5)} = \boxed{(9 \times 4) + (9 \times 5)} = \boxed{36 + 45} = \boxed{\mathbf{81}}$

Similar to addition, changing the order in which numbers are multiplied or grouped does not affect the final answer. However, it is important to learn how to solve math operations in the exact order in which parentheses or brackets are used in grouping numbers.

Properties Associated with Division

1. Changing the order in which two numbers are divided does change the final answer. For example, for any two real numbers a and b

   $\boxed{a \div b \neq b \div a}$ *Note 1:* $\frac{a}{b}$, $b \neq 0$ *and* $\frac{b}{a}$, $a \neq 0$ *Note 2:* $\frac{a}{0}$ *is not defined.*

   For example, $\boxed{15 \div 5} = \boxed{\mathbf{3}}$, but $\boxed{5 \div 15} = \boxed{\mathbf{0.33}}$

2. Re-grouping numbers does change the final answer. For example, for any real numbers a, b, and c

   $\boxed{(a \div b) \div c \neq a \div (b \div c)}$

   For example,

   $\boxed{(28 \div 4) \div 2} = \boxed{(7) \div 2} = \boxed{7 \div 2} = \boxed{\mathbf{3.5}}$, however

   $\boxed{28 \div (4 \div 2)} = \boxed{28 \div (2)} = \boxed{28 \div 2} = \boxed{\mathbf{14}}$

Case I - Use of Parentheses in Addition, Subtraction, Multiplication, and Division

In mixed mathematical operations, parentheses can be grouped in different ways, as shown in the following example cases:

Case I-1.

$\boxed{a+(b\div c)} =$

Let $\boxed{b\div c = k_1}$ and $\boxed{a+k_1 = A}$, then

$\boxed{a+(b\div c)} = \boxed{a+(k_1)} = \boxed{a+k_1} = \boxed{A}$

Example 1.1-1

$\boxed{30+(50\div 5)} =$

Solution:

$\boxed{30+(50\div 5)} = \boxed{30+(10)} = \boxed{30+10} = \boxed{\mathbf{40}}$

Case I-2.

$\boxed{a\div(b\times c)} =$

Let $\boxed{b\times c = k_1}$ and $\boxed{a\div k_1 = B}$, then

$\boxed{a\div(b\times c)} = \boxed{a\div(k_1)} = \boxed{a\div k_1} = \boxed{B}$

Example 1.1-2

$\boxed{18\div(4\times 2)} =$

Solution:

$\boxed{18\div(4\times 2)} = \boxed{18\div(8)} = \boxed{18\div 8} = \boxed{\mathbf{2.25}}$

Case I-3.

$\boxed{(a\times b)\div c} =$

Let $\boxed{a\times b = k_1}$ and $\boxed{k_1\div c = C}$, then

$\boxed{(a\times b)\div c} = \boxed{(k_1)\div c} = \boxed{k_1\div c} = \boxed{C}$

Example 1.1-3

$\boxed{(20\times 5)\div 8} =$

Solution:

$\boxed{(20\times 5)\div 8} = \boxed{(100)\div 8} = \boxed{100\div 8} = \boxed{\mathbf{12.5}}$

Case I-4.

$\boxed{(a \div b) + c} =$

Let $\boxed{a \div b = k_1}$ and $\boxed{k_1 + c = D}$, then

$\boxed{(a \div b) + c} = \boxed{(k_1) + c} = \boxed{k_1 + c} = \boxed{D}$

Example 1.1-4

$\boxed{(45 \div 5) + 25} =$

Solution:

$\boxed{(45 \div 5) + 25} = \boxed{(9) + 25} = \boxed{9 + 25} = \boxed{\mathbf{34}}$

Case I-5.

$\boxed{(a + b) \div (c - d)} =$

Let $\boxed{a + b = k_1}$, $\boxed{c - d = k_2}$, and $\boxed{k_1 \div k_2 = E}$, then

$\boxed{(a + b) \div (c - d)} = \boxed{(k_1) \div (k_2)} = \boxed{k_1 \div k_2} = \boxed{E}$

Example 1.1-5

$\boxed{(23 + 5) \div (20 - 8)} =$

Solution:

$\boxed{(23 + 5) \div (20 - 8)} = \boxed{(28) \div (12)} = \boxed{28 \div 12} = \boxed{\mathbf{2.33}}$

Case I-6.

$\boxed{(a \div b) - (c \times d)} =$

Let $\boxed{a \div b = k_1}$, $\boxed{c \times d = k_2}$, and $\boxed{k_1 - k_2 = F}$, then

$\boxed{(a \div b) - (c \times d)} = \boxed{(k_1) - (k_2)} = \boxed{k_1 - k_2} = \boxed{F}$

Example 1.1-6

$\boxed{(49 \div 5) - (12 \times 4)} =$

Solution:

$\boxed{(49 \div 5) - (12 \times 4)} = \boxed{(9.8) - (48)} = \boxed{9.8 - 48} = \boxed{\mathbf{-38.2}}$

Case II - Use of Brackets in Addition, Subtraction, Multiplication, and Division

In mixed operations, brackets are used in a similar way as parentheses. However, brackets are used to separate mathematical operations that contain integer numbers already grouped by parentheses. Brackets are used to group numbers in different ways, as shown in the following general and specific example cases:

Case II-1.

$\boxed{[a \div (b+c)] \div d} =$

Let $\boxed{b+c=k_1}$, $\boxed{a \div k_1 = k_2}$, and $\boxed{k_2 \div d = A}$, then

$\boxed{[a \div (b+c)] \div d} = \boxed{[a \div (k_1)] \div d} = \boxed{[a \div k_1] \div d} = \boxed{[k_2] \div d} = \boxed{k_2 \div d} = \boxed{A}$

Example 1.1-7

$\boxed{[350 \div (12+8)] \div 4} =$

Solution:

$\boxed{[350 \div (12+8)] \div 4} = \boxed{[350 \div (20)] \div 4} = \boxed{[350 \div 20] \div 4} = \boxed{[17.5] \div 4} = \boxed{17.5 \div 4} = \boxed{\mathbf{4.38}}$

Case II-2.

$\boxed{[(a \times b) \div c] + d} =$

Let $\boxed{a \times b = k_1}$, $\boxed{k_1 \div c = k_2}$, and $\boxed{k_2 + d = B}$, then

$\boxed{[(a \times b) \div c] + d} = \boxed{[(k_1) \div c] + d} = \boxed{[k_1 \div c] + d} = \boxed{[k_2] + d} = \boxed{k_2 + d} = \boxed{B}$

Example 1.1-8

$\boxed{[(12 \times 4) \div 2] + 46} =$

Solution:

$\boxed{[(12 \times 4) \div 2] + 46} = \boxed{[(48) \div 2] + 46} = \boxed{[48 \div 2] + 46} = \boxed{[24] + 46} = \boxed{24 + 46} = \boxed{\mathbf{70}}$

Case II-3.

$\boxed{a \times [b - (c+d)]} =$

Let $\boxed{c+d=k_1}$, $\boxed{b - k_1 = k_2}$, and $\boxed{ak_2 = C}$, then

$\boxed{a \times [b - (c+d)]} = \boxed{a \times [b - (k_1)]} = \boxed{a \times [b - k_1]} = \boxed{a \times [k_2]} = \boxed{a \times k_2} = \boxed{ak_2} = \boxed{C}$

Example 1.1-9

$\boxed{8 \times [10 - (5+9)]} =$

Solution:

$\boxed{8\times[10-(5+9)]} = \boxed{8\times[10-(14)]} = \boxed{8\times[10-14]} = \boxed{8\times[-4]} = \boxed{8\times -4} = \boxed{\mathbf{-32}}$

Case II-4.

$\boxed{[(a\times b)\div(c+d)]\div e} =$

Let $\boxed{a\times b = k_1}$, $\boxed{c+d = k_2}$, $\boxed{k_1 \div k_2 = k_3}$, and $\boxed{k_3 \div e = D}$, then

$\boxed{[(a\times b)\div(c+d)]\div e} = \boxed{[(k_1)\div(k_2)]\div e} = \boxed{[k_1 \div k_2]\div e} = \boxed{[k_3]\div e} = \boxed{k_3 \div e} = \boxed{D}$

Example 1.1-10

$\boxed{[(4\times 5)\div(28+9)]\div 5} =$

Solution:

$\boxed{[(4\times 5)\div(28+9)]\div 5} = \boxed{[(20)\div(37)]\div 5} = \boxed{[20\div 37]\div 5} = \boxed{[0.54]\div 5} = \boxed{0.54\div 5} = \boxed{\mathbf{0.108}}$

Case II-5.

$\boxed{[(a-b)-c]+(d+e)} =$

Let $\boxed{a-b = k_1}$, $\boxed{k_1 - c = k_2}$, $\boxed{d+e = k_3}$, and $\boxed{k_2 + k_3 = E}$, then

$\boxed{[(a-b)-c]+(d+e)} = \boxed{[(k_1)-c]+(k_3)} = \boxed{[k_1 - c]+k_3} = \boxed{[k_2]+k_3} = \boxed{k_2 + k_3} = \boxed{E}$

Example 1.1-11

$\boxed{[(23-6)-8]+(12+7)} =$

Solution:

$\boxed{[(23-6)-8]+(12+7)} = \boxed{[(17)-8]+(19)} = \boxed{[17-8]+19} = \boxed{[9]+19} = \boxed{9+19} = \boxed{\mathbf{28}}$

Case II-6.

$\boxed{a+[(b+c)-(d\times e)]} =$

Let $\boxed{b+c = k_1}$, $\boxed{d\times e = k_2}$, $\boxed{k_1 - k_2 = k_3}$, and $\boxed{a + k_3 = F}$, then

$\boxed{a+[(b+c)-(d\times e)]} = \boxed{a+[(k_1)-(k_2)]} = \boxed{a+[k_1 - k_2]} = \boxed{a+[k_3]} = \boxed{a+k_3} = \boxed{F}$

Example 1.1-12

$\boxed{35+[(12+5)-(4\times 2)]} =$

Solution:

$\boxed{35+[(12+5)-(4\times 2)]} = \boxed{35+[(17)-(8)]} = \boxed{35+[17-8]} = \boxed{35+[9]} = \boxed{35+9} = \boxed{\mathbf{44}}$

Case II-7.

$\boxed{[(a \div b)+(c \div d)] \times (e+f)} =$

Let $\boxed{a \div b = k_1}$, $\boxed{c \div d = k_2}$, $\boxed{e+f = k_3}$, $\boxed{k_1 + k_2 = k_4}$, and $\boxed{k_4 k_3 = G}$, then

$\boxed{[(a \div b)+(c \div d)] \times (e+f)} = \boxed{[(k_1)+(k_2)] \times (k_3)} = \boxed{[k_1 + k_2] \times k_3} = \boxed{[k_4] \times k_3} = \boxed{k_4 \times k_3} = \boxed{k_4 k_3} = \boxed{G}$

Example 1.1-13

$\boxed{[(45 \div 9)+(12 \div 4)] \times (10+5)} =$

Solution:

$\boxed{[(45 \div 9)+(12 \div 4)] \times (10+5)} = \boxed{[(5)+(3)] \times (15)} = \boxed{[5+3] \times 15} = \boxed{[8] \times 15} = \boxed{8 \times 15} = \boxed{\mathbf{120}}$

Case II-8.

$\boxed{(a-b)+[(c \div d) \times (e \div f)]} =$

Let $\boxed{a - b = k_1}$, $\boxed{c \div d = k_2}$, $\boxed{e \div f = k_3}$, $\boxed{k_2 k_3 = k_4}$, and $\boxed{k_1 + k_4 = H}$, then

$\boxed{(a-b)+[(c \div d) \times (e \div f)]} = \boxed{(k_1)+[(k_2) \times (k_3)]} = \boxed{k_1 + [k_2 \times k_3]} = \boxed{k_1 + [k_2 k_3]} = \boxed{k_1 + [k_4]} = \boxed{k_1 + k_4} = \boxed{H}$

Example 1.1-14

$\boxed{(45-6)+[(12 \div 4) \times (34 \div 4)]} =$

Solution:

$\boxed{(45-6)+[(12 \div 4) \times (34 \div 4)]} = \boxed{(39)+[(3) \times (8.5)]} = \boxed{39+[3 \times 8.5]} = \boxed{39+[25.5]} = \boxed{39+25.5} = \boxed{\mathbf{64.5}}$

Case II-9.

$\boxed{(a+b+c) \div [d \times (e-f)]} =$

Let $\boxed{a+b+c = k_1}$, $\boxed{e-f = k_2}$, $\boxed{dk_2 = k_3}$, and $\boxed{k_1 \div k_3 = I}$, then

$\boxed{(a+b+c) \div [d \times (e-f)]} = \boxed{(k_1) \div [d \times (k_2)]} = \boxed{k_1 \div [d \times k_2]} = \boxed{k_1 \div [dk_2]} = \boxed{k_1 \div dk_2} = \boxed{k_1 \div k_3} = \boxed{I}$

Example 1.1-15

$\boxed{(8+50+5) \div [3 \times (25-12)]} =$

Solution:

$\boxed{(8+50+5) \div [3 \times (25-12)]} = \boxed{(63) \div [3 \times (13)]} = \boxed{63 \div [3 \times 13]} = \boxed{63 \div [39]} = \boxed{63 \div 39} = \boxed{\mathbf{1.62}}$

Additional Examples - Use of Parentheses and Brackets in Addition, Subtraction, Multiplication, and Division

The following examples further illustrate how to use parentheses and brackets in mixed operations:

Example 1.1-16

$\boxed{(39+5)\div 4} = \boxed{(44)\div 4} = \boxed{44\div 4} = \boxed{\mathbf{11}}$

Example 1.1-17

$\boxed{36\times(12+3)} = \boxed{36\times(15)} = \boxed{36\times 15} = \boxed{\mathbf{540}}$

Example 1.1-18

$\boxed{(23+5)\div(8\times 2)} = \boxed{(28)\div(16)} = \boxed{28\div 16} = \boxed{\mathbf{1.75}}$

Example 1.1-19

$\boxed{38+[15\times(20\div 2)]} = \boxed{38+[15\times(10)]} = \boxed{38+[15\times 10]} = \boxed{38+[150]} = \boxed{38+150} = \boxed{\mathbf{188}}$

Example 1.1-20

$\boxed{[(35\times 2)+5]\div 3} = \boxed{[(70)+5]\div 3} = \boxed{[70+5]\div 3} = \boxed{[75]\div 3} = \boxed{75\div 3} = \boxed{\mathbf{25}}$

Example 1.1-21

$\boxed{(28-18)\times[16-(8-3)]} = \boxed{(10)\times[16-(5)]} = \boxed{10\times[16-5]} = \boxed{10\times[11]} = \boxed{10\times 11} = \boxed{\mathbf{110}}$

Example 1.1-22

$\boxed{[(20-4)+(15-5)]\div 2} = \boxed{[(16)+(10)]\div 2} = \boxed{[16+10]\div 2} = \boxed{[26]\div 2} = \boxed{26\div 2} = \boxed{\mathbf{13}}$

Example 1.1-23

$\boxed{[(15+6)\div 3]\times(8\div 2)} = \boxed{[(21)\div 3]\times(4)} = \boxed{[21\div 3]\times 4} = \boxed{[7]\times 4} = \boxed{7\times 4} = \boxed{\mathbf{28}}$

Practice Problems - Use of Parentheses and Brackets in Addition, Subtraction, Multiplication, and Division

Section 1.1b Practice Problems - Perform the indicated operations in the order grouped:

1. $(28\div 4)\times 3 =$
2. $250+(15\div 3) =$
3. $28\div[(23+5)\times 8] =$
4. $[(255-15)\div 20]+8 =$
5. $[230\div(15\times 2)]+12 =$
6. $55\times[(28+2)\div 3] =$

1.2 Integer Fractions

In this section the steps as to how integer fractions are simplified are addressed in section 1.2a, Cases I through IV. In addition, math operations involving integer fractions which include addition, subtraction, multiplication, and division of two or more integer fractions with or without common denominators are reviewed in section 1.2b, Cases I through VI.

1.2a Simplifying Integer Fractions

Integer fractions of the form $\frac{a}{b}$, where both the numerator a and the denominator b are integer numbers, are simplified as in the following cases:

Case I - The Numerator and the Denominator are Even Numbers

Use the following steps to simplify the integer fractions if the numerator and the denominator are even numbers:

Step 1 Check the numerator and the denominator of the integer fraction to see if it is an $\frac{even}{even}$ type of fraction.

Step 2 Simplify the fraction to its lowest term by dividing the numerator and the denominator by their Greatest Common Factor (G.C.F.) which is an even number, i.e., $(2, 4, 6, 8, 10, 12, 14, \ldots)$. See the methods introduced in finding G.C.F. at the end of this section.

Step 3 Change the improper fraction to a mixed fraction if the fraction obtained from Step 2 is an improper fraction (see Section 1.2 Appendix).

Examples with Steps

The following examples show the steps as to how integer fractions with even numerator and denominator are simplified:

Example 1.2-1

$$\boxed{-\frac{366}{64}} =$$

Solution:

Step 1 $$\boxed{-\frac{366}{64}} = \boxed{-\frac{366\,(is\ an\ even\ No.)}{64\,(is\ an\ even\ No.)}}$$

Step 2 $$\boxed{-\frac{366\,(is\ an\ even\ No.)}{64\,(is\ an\ even\ No.)}} = \boxed{-\frac{366 \div 2}{64 \div 2}} = \boxed{-\frac{183}{32}}$$

Step 3 $$\boxed{-\frac{183}{32}} = \boxed{-\left(5\frac{23}{32}\right)}$$

Example 1.2-2

$$\boxed{\frac{400}{350}} =$$

Solution:

Step 1 $\boxed{\dfrac{400}{350}} = \boxed{\dfrac{400\,(\textit{is an even No.})}{350\,(\textit{is an even No.})}}$

Step 2 $\boxed{\dfrac{400\,(\textit{is an even No.})}{350\,(\textit{is an even No.})}} = \boxed{\dfrac{400 \div 50}{350 \div 50}} = \boxed{\dfrac{8}{7}}$

Step 3 $\boxed{\dfrac{8}{7}} = \boxed{1\dfrac{1}{7}}$

Case II - The Numerator and the Denominator are Odd Numbers

Use the following steps to simplify the integer fractions if the numerator and the denominator are odd numbers:

Step 1 Check the numerator and the denominator of the integer fraction to see if it is an $\dfrac{odd}{odd}$ type of fraction.

Step 2 Simplify the fraction to its lowest term by dividing the numerator and the denominator by their Greatest Common Factor (G.C.F.) which is an odd number, i.e., $(3, 5, 7, 9, 11, 13, 15, \ldots)$. See the methods introduced in finding G.C.F. at the end of this section.

Step 3 Change the improper fraction to a mixed fraction if the fraction obtained from Step 2 is an improper fraction (see Section 1.2 Appendix).

Examples with Steps

The following examples show the steps as to how integer fractions with odd numerator and denominator are simplified:

Example 1.2-3

$\boxed{-\dfrac{3}{15}} =$

Solution:

Step 1 $\boxed{-\dfrac{3}{15}} = \boxed{-\dfrac{3\,(\textit{is an odd No.})}{15\,(\textit{is an odd No.})}}$

Step 2 $\boxed{-\dfrac{3\,(\textit{is an odd No.})}{15\,(\textit{is an odd No.})}} = \boxed{-\dfrac{3 \div 3}{15 \div 3}} = \boxed{-\dfrac{1}{5}}$

Step 3 *Not Applicable*

Example 1.2-4

$\boxed{\dfrac{17}{21}} =$

Solution:

Step 1 $\boxed{\frac{17}{21}} = \boxed{\frac{\mathbf{17}\,(is\ an\ odd\ No.)}{\mathbf{21}\,(is\ an\ odd\ No.)}}$

Step 2 $\boxed{Not\ Applicable}$

Step 3 $\boxed{Not\ Applicable}$

Note - In cases where the answer to Steps 2 and 3 are stated as "Not Applicable" this indicates that the fraction is in its lowest term and can not be simplified any further.

Case III - The Numerator is an Even Number and the Denominator is an Odd Number

Use the following steps to simplify the integer fractions if the numerator is an even number and the denominator is an odd number:

Step 1 Check the numerator and the denominator of the integer fraction to see if it is an $\frac{even}{odd}$ type of fraction.

Step 2 Simplify the fraction to its lowest term by dividing the numerator and the denominator by their Greatest Common Factor (G.C.F.) which is an odd number, i.e., $(3, 5, 7, 9, 11, 13, 15, ...)$. See the methods introduced in finding G.C.F. at the end of this section.

Step 3 Change the improper fraction to a mixed fraction if the fraction obtained from Step 2 is an improper fraction (see Section 1.2 Appendix).

Examples with Steps

The following examples show the steps as to how integer fractions with an even numerator and an odd denominator are simplified:

Example 1.2-5

$\boxed{\frac{18}{27}} =$

Solution:

Step 1 $\boxed{\frac{18}{27}} = \boxed{\frac{18\,(is\ an\ even\ No.)}{27\,(is\ an\ odd\ No.)}}$

Step 2 $\boxed{\frac{18\,(is\ an\ even\ No.)}{27\,(is\ an\ odd\ No.)}} = \boxed{\frac{18 \div 9}{27 \div 9}} = \boxed{\frac{2}{3}}$

Step 3 $\boxed{Not\ Applicable}$

Example 1.2-6

$\boxed{-\frac{108}{27}} =$

Solution:

Step 1 $\boxed{-\frac{108}{27}} = \boxed{-\frac{108\,(is\ an\ even\ No.)}{27\,(is\ an\ odd\ No.)}}$

Step 2 $\boxed{-\dfrac{108\,(is\ an\ even\ No.)}{27\,(is\ an\ odd\ No.)}} = \boxed{-\dfrac{108 \div 27}{27 \div 27}} = \boxed{-\dfrac{4}{1}} = \boxed{-4}$

Step 3 $\boxed{Not\ Applicable}$

Case IV - The Numerator is an Odd Number and the Denominator is an Even Number

Use the following steps to simplify the integer fractions if the numerator is an odd number and the denominator is an even number:

Step 1 Check the numerator and the denominator of the integer fraction to see if it is an $\dfrac{odd}{even}$ type of fraction.

Step 2 Simplify the fraction to its lowest term by dividing the numerator and the denominator by their Greatest Common Factor (G.C.F.) which is an odd number, i.e., $(3, 5, 7, 9, 11, 13, 15, ...)$. See the methods introduced in finding G.C.F. at the end of this section.

Step 3 Change the improper fraction to a mixed fraction if the fraction obtained from Step 2 is an improper fraction (see Section 1.2 Appendix).

Examples with Steps

The following examples show the steps as to how integer fractions with an odd numerator and an even denominator are simplified:

Example 1.2-7

$\boxed{\dfrac{15}{60}} =$

Solution:

Step 1 $\boxed{\dfrac{15}{60}} = \boxed{\dfrac{15\,(is\ an\ odd\ No.)}{60\,(is\ an\ even\ No.)}}$

Step 2 $\boxed{\dfrac{15\,(is\ an\ odd\ No.)}{60\,(is\ an\ even\ No.)}} = \boxed{\dfrac{15 \div 15}{60 \div 15}} = \boxed{\dfrac{\mathbf{1}}{\mathbf{4}}}$

Step 3 $\boxed{Not\ Applicable}$

Example 1.2-8

$\boxed{-\dfrac{327}{24}} =$

Solution:

Step 1 $\boxed{-\dfrac{327}{24}} = \boxed{-\dfrac{327\,(is\ an\ odd\ No.)}{24\,(is\ an\ even\ No.)}}$

Step 2 $\boxed{-\dfrac{327\,(is\ an\ odd\ No.)}{24\,(is\ an\ even\ No.)}} = \boxed{-\dfrac{327 \div 3}{24 \div 3}} = \boxed{-\dfrac{109}{8}}$

Step 3 $\boxed{-\dfrac{109}{8}} = \boxed{-\left(13\dfrac{5}{8}\right)}$

Note that in Cases II, III, and IV where the integer fractions are $\dfrac{odd}{odd}$, $\dfrac{even}{odd}$, and $\dfrac{odd}{even}$ respectively, odd numbers are always used to simplify the fractions.

Additional Examples - Simplifying Integer Fractions

The following examples further illustrate how to simplify integer fractions:

Example 1.2-9

$$\boxed{\frac{15}{3}} = \boxed{\frac{15\,(is\ an\ odd\ No.)}{3\,(is\ an\ odd\ No.)}} = \boxed{\frac{15 \div 3}{3 \div 3}} = \boxed{\frac{5}{1}} = \boxed{5}$$

Example 1.2-10

$$\boxed{-\frac{6}{8}} = \boxed{\frac{6\,(is\ an\ even\ No.)}{8\,(is\ an\ even\ No.)}} = \boxed{-\frac{6 \div 2}{8 \div 2}} = \boxed{-\frac{3}{4}}$$

Example 1.2-11

$$\boxed{\frac{12}{3}} = \boxed{\frac{12\,(is\ an\ even\ No.)}{3\,(is\ an\ odd\ No.)}} = \boxed{\frac{12 \div 3}{3 \div 3}} = \boxed{\frac{4}{1}} = \boxed{4}$$

Example 1.2-12

$$\boxed{\frac{35}{7}} = \boxed{\frac{35\,(is\ an\ odd\ No.)}{7\,(is\ an\ odd\ No.)}} = \boxed{\frac{35 \div 7}{7 \div 7}} = \boxed{\frac{5}{1}} = \boxed{5}$$

Example 1.2-13

$$\boxed{\frac{100}{3}} = \boxed{\frac{100\,(is\ an\ even\ No.)}{3\,(is\ an\ odd\ No.)}} = \boxed{33\frac{1}{3}}$$

Example 1.2-14

$$\boxed{-\frac{325}{40}} = \boxed{-\frac{325\,(is\ an\ odd\ No.)}{40\,(is\ an\ even\ No.)}} = \boxed{-\frac{325 \div 5}{40 \div 5}} = \boxed{-\frac{65}{8}} = \boxed{-\left(8\frac{1}{8}\right)}$$

Greatest Common Factor

Greatest Common Factor (G.C.F.) can be found in two ways: 1. Trial and error method, and 2. Prime factoring method.

1. ***Trial and Error Method***: In the trial and error method the numerator and the denominator are divided by odd or even numbers until the largest divisor for both the numerator and the denominator is found.
2. ***Prime Factoring Method***: The steps in using the prime factoring method are:
   a. Rewrite both the numerator and the denominator by their equivalent prime number products.
   b. Identify the prime numbers that are common in both the numerator and the denominator.
   c. Multiply the common prime numbers to obtain the G.C.F.

The following are examples of how G.C.F. can be found using the prime factoring method:

1. $\frac{24}{45}=\frac{8\times3}{9\times5}=\frac{4\times2\times3}{3\times3\times5}=\frac{2\times2\times2\times3}{3\times3\times5}$. The common prime number in both the numerator and the denominator is 3. Therefore, $G.C.F.=3$.

2. $\frac{400}{350}=\frac{4\times100}{35\times10}=\frac{2\times2\times4\times25}{7\times5\times5\times2}=\frac{2\times2\times2\times2\times5\times5}{7\times5\times5\times2}$. The common prime numbers in both the numerator and the denominator are 2, 5, and 5. Therefore, $G.C.F.=2\times5\times5=50$.

3. $\frac{15}{60}=\frac{5\times3}{6\times10}=\frac{5\times3}{2\times3\times5\times2}$. The common prime numbers in both the numerator and the denominator are 3 and 5. Therefore, $G.C.F.=3\times5=15$.

4. $\frac{108}{27}=\frac{12\times9}{9\times3}=\frac{6\times2\times3\times3}{3\times3\times3}=\frac{2\times3\times2\times3\times3}{3\times3\times3}$. The common prime numbers in both the numerator and the denominator are 3, 3, and 3. Therefore, $G.C.F.=3\times3\times3=27$.

Practice Problems - Simplifying Integer Fractions

Section 1.2a Practice Problems - Simplify the following integer fractions:

1. $\frac{60}{150}=$
2. $\frac{8}{18}=$
3. $\frac{355}{15}=$
4. $\frac{3}{8}=$
5. $\frac{27}{6}=$
6. $\frac{33}{6}=$

1.2b Operations Involving Integer Fractions

Integer fractions, i.e., fractions where both the numerator and the denominator are integers, are added, subtracted, multiplied, and divided as in the following cases:

Case I Adding Integer Fractions with Common Denominators

Add two integer fractions with common denominators using the following steps:

Step 1 a. Use the common denominator between the first and second fractions as the new denominator.

b. Add the numerators of the first and second fractions to obtain the new numerator.

Step 2 Simplify the fraction to its lowest term (see Section 1.2a).

Step 3 Change the improper fraction to a mixed fraction if the fraction obtained from Step 2 is an improper fraction (see Section 1.2 Appendix).

Examples with Steps

The following examples show the steps as to how two integer fractions with common denominators are added:

Example 1.2-15

$$\boxed{\frac{2}{3}+\frac{8}{3}}=$$

Solution:

Step 1 $\boxed{\frac{2}{3}+\frac{8}{3}}=\boxed{\frac{2+8}{3}}=\boxed{\frac{10}{3}}$

Step 2 *Not Applicable*

Step 3 $\boxed{\frac{10}{3}}=\boxed{3\frac{1}{3}}$

Example 1.2-16

$$\boxed{\frac{15}{4}+\frac{9}{4}}=$$

Solution:

Step 1 $\boxed{\frac{15}{4}+\frac{9}{4}}=\boxed{\frac{15+9}{4}}=\boxed{\frac{24}{4}}$

Step 2 $\boxed{\frac{24}{4}}=\boxed{\frac{24\div 4}{4\div 4}}=\frac{6}{1}=\boxed{6}$

Step 3 *Not Applicable*

In general, two integer fractions with a common denominator are added in the following way:

$$\boxed{\frac{a}{d}+\frac{b}{d}}=\boxed{\frac{a+b}{d}}$$

For example, $\boxed{\frac{5}{3}+\frac{13}{3}}=\boxed{\frac{5+13}{3}}=\boxed{\frac{\overset{6}{\cancel{18}}}{\underset{1}{\cancel{3}}}}=\boxed{\frac{6}{1}}=\boxed{6}$

Case II Adding Integer Fractions without a Common Denominator

Add two integer fractions without a common denominator using the following steps:

Step 1 Change the integer number a to an integer fraction of the form $\frac{a}{1}$, i.e., change 5 to $\frac{5}{1}$.

Step 2 a. Multiply the denominators of the first and second fractions to obtain the new denominator.

b. Cross multiply the numerator of the first fraction with the denominator of the second fraction.

c. Cross multiply the numerator of the second fraction with the denominator of the first fraction.

d. Add the results from the steps 2b and 2c above to obtain the new numerator.

Step 3 Simplify the fraction to its lowest term (see Section 1.2a).

Step 4 Change the improper fraction to a mixed fraction if the fraction obtained from Step 3 is an improper fraction (see Section 1.2 Appendix).

Examples with Steps

The following examples show the steps as to how two integer fractions without a common denominator are added:

Example 1.2-17

$$\boxed{40+\frac{4}{3}} =$$

Solution:

Step 1 $$\boxed{40+\frac{4}{3}} = \boxed{\frac{40}{1}+\frac{4}{3}}$$

Step 2 $$\boxed{\frac{40}{1}+\frac{4}{3}} = \boxed{\frac{(40\times 3)+(4\times 1)}{1\times 3}} = \boxed{\frac{120+4}{3}} = \boxed{\frac{124}{3}}$$

Step 3 *Not Applicable*

Step 4 $$\boxed{\frac{124}{3}} = \boxed{\mathbf{41\frac{1}{3}}}$$

Example 1.2-18

$$\boxed{\frac{5}{6}+3} =$$

Solution:

Step 1 $$\boxed{\frac{5}{6}+3} = \boxed{\frac{5}{6}+\frac{3}{1}}$$

Step 2 $$\boxed{\frac{5}{6}+\frac{3}{1}} = \boxed{\frac{(5\times 1)+(3\times 6)}{6\times 1}} = \boxed{\frac{5+18}{6}} = \boxed{\frac{23}{6}}$$

Step 3 *Not Applicable*

Step 4 $\boxed{\frac{23}{6}} = \boxed{\mathbf{3\frac{5}{6}}}$

In general, two integer fractions without a common denominator are added in the following way:

$$\boxed{\frac{a}{b}+\frac{c}{d}} = \boxed{\frac{(a\times d)+(c\times b)}{(b\times d)}} = \boxed{\frac{ad+cb}{bd}}$$

For example,

$$\boxed{\frac{6}{3}+\frac{9}{4}} = \boxed{\frac{(6\times 4)+(3\times 9)}{3\times 4}} = \boxed{\frac{24+27}{12}} = \boxed{\frac{\overset{17}{\cancel{51}}}{\underset{4}{\cancel{12}}}} = \boxed{\frac{17}{4}} = \boxed{\mathbf{4\frac{1}{4}}}$$

Additional Examples - Adding Integer Fractions with or without a Common Denominator

The following examples further illustrate how to add integer fractions:

Example 1.2-19

$$\boxed{\frac{3}{6}+\frac{4}{5}} = \boxed{\frac{(3\times 5)+(4\times 6)}{6\times 5}} = \boxed{\frac{15+24}{30}} = \boxed{\frac{\overset{13}{\cancel{39}}}{\underset{10}{\cancel{30}}}} = \boxed{\frac{13}{10}} = \boxed{\mathbf{1\frac{3}{10}}}$$

Example 1.2-20

$$\boxed{\frac{2}{5}+\frac{1}{4}+\frac{4}{3}} = \boxed{\left(\frac{2}{5}+\frac{1}{4}\right)+\frac{4}{3}} = \boxed{\left(\frac{(2\times 4)+(1\times 5)}{5\times 4}\right)+\frac{4}{3}} = \boxed{\left(\frac{8+5}{20}\right)+\frac{4}{3}} = \boxed{\left(\frac{13}{20}\right)+\frac{4}{3}} = \boxed{\frac{13}{20}+\frac{4}{3}}$$

$$= \boxed{\frac{(13\times 3)+(4\times 20)}{20\times 3}} = \boxed{\frac{39+80}{60}} = \boxed{\frac{119}{60}} = \boxed{\mathbf{1\frac{59}{60}}}$$

Example 1.2-21

$$\boxed{\frac{1}{2}+\left(\frac{2}{3}+\frac{1}{5}\right)} = \boxed{\frac{1}{2}+\left(\frac{(2\times 5)+(1\times 3)}{3\times 5}\right)} = \boxed{\frac{1}{2}+\left(\frac{10+3}{15}\right)} = \boxed{\frac{1}{2}+\left(\frac{13}{15}\right)} = \boxed{\frac{1}{2}+\frac{13}{15}} = \boxed{\frac{(1\times 15)+(13\times 2)}{2\times 15}}$$

$$= \boxed{\frac{15+26}{30}} = \boxed{\frac{41}{30}} = \boxed{\mathbf{1\frac{11}{30}}}$$

Example 1.2-22

$$\boxed{\left(\frac{2}{3}+\frac{6}{3}\right)+\left(\frac{8}{6}+\frac{2}{6}+\frac{1}{6}\right)} = \boxed{\left(\frac{2+6}{3}\right)+\left(\frac{8+2+1}{6}\right)} = \boxed{\frac{8}{3}+\frac{11}{6}} = \boxed{\frac{(8\times 6)+(11\times 3)}{3\times 6}} = \boxed{\frac{48+33}{18}} = \boxed{\frac{\overset{9}{\cancel{81}}}{\underset{2}{\cancel{18}}}} = \boxed{\frac{9}{2}}$$

$$= \boxed{\mathbf{4\frac{1}{2}}}$$

Practice Problems - Adding Integer Fractions with or without a Common Denominator

Section 1.2b Cases I and II Practice Problems - Add the following integer fractions:

1. $\frac{4}{9}+\frac{2}{9}=$
2. $\frac{3}{8}+\frac{2}{5}=$
3. $\left(\frac{3}{8}+\frac{2}{4}\right)+\frac{5}{6}=$
4. $\frac{4}{5}+\frac{2}{5}+\frac{3}{5}=$
5. $5+\frac{0}{10}+\frac{6}{1}+\frac{4}{8}=$
6. $\left(\frac{3}{16}+\frac{1}{8}\right)+\frac{1}{6}=$

Case III Subtracting Integer Fractions with Common Denominators

Subtract two integer fractions with common denominators using the following steps:

Step 1 a. Use the common denominator between the first and second fractions as the new denominator.

b. Subtract the numerators of the first and second fractions to obtain the new numerator.

Step 2 Simplify the fraction to its lowest term (see Section 1.2a).

Step 3 Change the improper fraction to a mixed fraction if the fraction obtained from Step 2 is an improper fraction (see Section 1.2 Appendix).

Examples with Steps

The following examples show the steps as to how two integer fractions with common denominators are subtracted:

Example 1.2-23

$$\boxed{\frac{25}{3}-\frac{2}{3}} =$$

Solution:

Step 1 $\boxed{\frac{25}{3}-\frac{2}{3}} = \boxed{\frac{25-2}{3}} = \boxed{\frac{23}{3}}$

Step 2 $\boxed{\textit{Not Applicable}}$

Step 3 $\boxed{\frac{23}{3}} = \boxed{7\frac{2}{3}}$

Example 1.2-24

$$\boxed{\frac{40}{4}-\frac{10}{4}} =$$

Solution:

Step 1 $\boxed{\frac{40}{4}-\frac{10}{4}} = \boxed{\frac{40-10}{4}} = \boxed{\frac{30}{4}}$

Step 2 $\boxed{\frac{30}{4}} = \boxed{\frac{30\div 2}{4\div 2}} = \boxed{\frac{15}{2}}$

Step 3 $\boxed{\frac{15}{2}} = \boxed{7\frac{1}{2}}$

In general, two integer fractions with a common denominator are subtracted in the following way:

$\boxed{\frac{a}{d}-\frac{b}{d}} = \boxed{\frac{a-b}{d}}$ For example, $\boxed{\frac{6}{8}-\frac{4}{8}} = \boxed{\frac{6-4}{8}} = \boxed{\frac{\frac{1}{2}}{\frac{8}{4}}} = \boxed{\frac{1}{4}}$

Case IV Subtracting Integer Fractions without a Common Denominator

Subtract two integer fractions without a common denominator using the following steps:

Step 1 Change the integer number a to an integer fraction of the form $\frac{a}{1}$.

Step 2 a. Multiply the denominators of the first and second fractions to obtain the new denominator.

b. Cross multiply the numerator of the first fraction with the denominator of the second fraction.

c. Cross multiply the numerator of the second fraction with the denominator of the first fraction.

d. Subtract the results from steps 2b and 2c above to obtain the new numerator.

Step 3 Simplify the fraction to its lowest term (see Section 1.2a).

Step 4 Change the improper fraction to a mixed fraction if the fraction obtained from Step 3 is an improper fraction (see Section 1.2 Appendix).

Examples with Steps

The following examples show the steps as to how two integer fractions without a common denominator are subtracted:

Example 1.2-25

$$\boxed{5-\frac{12}{8}} =$$

Solution:

Step 1

$$\boxed{5-\frac{12}{8}} = \boxed{\frac{5}{1}-\frac{12}{8}}$$

Step 2

$$\boxed{\frac{5}{1}-\frac{12}{8}} = \boxed{\frac{(5\times 8)-(12\times 1)}{1\times 8}} = \boxed{\frac{40-12}{8}} = \boxed{\frac{28}{8}}$$

Step 3

$$\boxed{\frac{28}{8}} = \boxed{\frac{28\div 4}{8\div 4}} = \boxed{\frac{7}{2}}$$

Step 4

$$\boxed{\frac{7}{2}} = \boxed{\mathbf{3\frac{1}{2}}}$$

Example 1.2-26

$$\boxed{\frac{10}{6}-35} =$$

Solution:

Step 1

$$\boxed{\frac{10}{6}-35} = \boxed{\frac{10}{6}-\frac{35}{1}}$$

Step 2

$$\boxed{\frac{10}{6}-\frac{35}{1}} = \boxed{\frac{(10\times 1)-(35\times 6)}{6\times 1}} = \boxed{\frac{10-210}{6}} = \boxed{\frac{-200}{6}}$$

Step 3 $\boxed{\frac{-200}{6}} = \boxed{\frac{-200 \div 2}{6 \div 2}} = \boxed{\frac{-100}{3}}$

Step 4 $\boxed{\frac{-100}{3}} = \boxed{-\left(33\frac{1}{3}\right)}$

In general, two integer fractions without a common denominator are subtracted in the following way:

$$\boxed{\frac{a}{b} - \frac{c}{d}} = \boxed{\frac{(a \times d) - (c \times b)}{b \times d}} = \boxed{\frac{ad - cb}{bd}}$$

For example, $\boxed{\frac{3}{4} - \frac{1}{8}} = \boxed{\frac{(3 \times 8) - (1 \times 4)}{4 \times 8}} = \boxed{\frac{24 - 4}{32}} = \boxed{\frac{\overset{5}{\cancel{20}}}{\underset{8}{\cancel{32}}}} = \boxed{\frac{5}{8}}$

Additional Examples - Subtracting Integer Fractions with or without a Common Denominator

The following examples further illustrate how to subtract integer fractions:

Example 1.2-27

$$\boxed{\frac{8}{3} - \frac{1}{6} - \frac{2}{5}} = \boxed{\left(\frac{8}{3} - \frac{1}{6}\right) - \frac{2}{5}} = \boxed{\left(\frac{(8 \times 6) - (1 \times 3)}{3 \times 6}\right) - \frac{2}{5}} = \boxed{\left(\frac{48 - 3}{18}\right) - \frac{2}{5}} = \boxed{\left(\frac{45}{18}\right) - \frac{2}{5}} = \boxed{\frac{\overset{15}{\cancel{45}}}{\underset{6}{\cancel{18}}} - \frac{2}{5}} = \boxed{\frac{15}{6} - \frac{2}{5}}$$

$$= \boxed{\frac{(15 \times 5) - (2 \times 6)}{6 \times 5}} = \boxed{\frac{75 - 12}{30}} = \boxed{\frac{\overset{21}{\cancel{63}}}{\underset{10}{\cancel{30}}}} = \boxed{\frac{21}{10}} = \boxed{2\frac{1}{10}}$$

Example 1.2-28

$$\boxed{\frac{16}{4} - \frac{2}{4} - \frac{4}{4}} = \boxed{\frac{16 - 2 - 4}{4}} = \boxed{\frac{16 - 6}{4}} = \boxed{\frac{\overset{5}{\cancel{10}}}{\underset{2}{\cancel{4}}}} = \boxed{\frac{5}{2}} = \boxed{2\frac{1}{2}}$$

Example 1.2-29

$$\boxed{\frac{3}{5} - \frac{2}{3} - 9} = \boxed{\left(\frac{3}{5} - \frac{2}{3}\right) - \frac{9}{1}} = \boxed{\left(\frac{(3 \times 3) - (2 \times 5)}{5 \times 3}\right) - \frac{9}{1}} = \boxed{\left(\frac{9 - 10}{15}\right) - \frac{9}{1}} = \boxed{\left(\frac{-1}{15}\right) - \frac{9}{1}} = \boxed{\frac{-1}{15} - \frac{9}{1}} = \boxed{\frac{(-1 \times 1) - (9 \times 15)}{15 \times 1}}$$

$$= \boxed{\frac{-1 - 135}{15}} = \boxed{\frac{-136}{15}} = \boxed{-\left(9\frac{1}{15}\right)}$$

Example 1.2-30

$$\boxed{\frac{2}{4} - \left(\frac{1}{3} - \frac{1}{5}\right)} = \boxed{\frac{\cancel{2}}{\underset{2}{\cancel{4}}} - \left(\frac{(5 \times 1) - (1 \times 3)}{3 \times 5}\right)} = \boxed{\frac{1}{2} - \left(\frac{5 - 3}{15}\right)} = \boxed{\frac{1}{2} - \frac{2}{15}} = \boxed{\frac{(1 \times 15) - (2 \times 2)}{2 \times 15}} = \boxed{\frac{15 - 4}{30}} = \boxed{\frac{11}{30}}$$

Practice Problems - Subtracting Integer Fractions with or without a Common Denominator

Section 1.2b Cases III and IV Practice Problems - Subtract the following integer fractions:

1. $\frac{3}{5} - \frac{2}{5} =$
2. $\frac{2}{5} - \frac{3}{4} =$
3. $\frac{12}{15} - \frac{3}{15} - \frac{6}{15} =$
4. $\frac{5}{8} - \frac{3}{4} - \frac{1}{3} =$
5. $\left(\frac{2}{8} - \frac{1}{6}\right) - \frac{2}{5} =$
6. $28 - \left(\frac{1}{8} - \frac{2}{3}\right) =$

Case V - Multiplying Integer Fractions with or without a Common Denominator

Multiply two integer fractions using the following steps:

Step 1 Change the integer number a to an integer fraction of the form $\frac{a}{1}$.

Step 2 a. Multiply the numerator of the first fraction with the numerator of the second fraction to obtain the new numerator.

b. Multiply the denominator of the first fraction with the denominator of the second fraction to obtain the new denominator.

Step 3 Simplify the fraction to its lowest term (see Section 1.2a).

Step 4 Change the improper fraction to a mixed fraction if the fraction obtained from Step 3 is an improper fraction (see Section 1.2 Appendix).

Examples with Steps

The following examples show the steps as to how two integer fractions with or without a common denominator are multiplied:

Example 1.2-31

$$\boxed{25\times\frac{5}{8}} =$$

Solution:

Step 1 $\boxed{25\times\frac{5}{8}} = \boxed{\frac{25}{1}\times\frac{5}{8}}$

Step 2 $\boxed{\frac{25}{1}\times\frac{5}{8}} = \boxed{\frac{25\times5}{1\times8}} = \boxed{\frac{125}{8}}$

Step 3 *Not Applicable*

Step 4 $\boxed{\frac{125}{8}} = \boxed{\mathbf{15\frac{5}{8}}}$

Example 1.2-32

$$\boxed{36\times\frac{4}{28}} =$$

Solution:

Step 1 $\boxed{36\times\frac{4}{28}} = \boxed{\frac{36}{1}\times\frac{4}{28}}$

Step 2 $\boxed{\frac{36}{1}\times\frac{4}{28}} = \boxed{\frac{36\times4}{1\times28}} = \boxed{\frac{144}{28}}$

Step 3 $\boxed{\frac{144}{28}} = \boxed{\frac{144\div4}{28\div4}} = \boxed{\frac{36}{7}}$

Step 4 $\boxed{\dfrac{36}{7}} = \boxed{5\dfrac{1}{7}}$

In general, two integer fractions with or without a common denominator are multiplied in the following way:

$$\boxed{\frac{a}{b}\times\frac{c}{d}} = \boxed{\frac{a\times c}{b\times d}} = \boxed{\frac{ac}{bd}}$$

For example,

$$\boxed{\frac{2}{5}\times\frac{3}{4}} = \boxed{\frac{2\times 3}{5\times 4}} = \boxed{\frac{\overset{3}{\cancel{6}}}{\underset{10}{\cancel{20}}}} = \boxed{\frac{3}{10}}$$

Additional Examples - Multiplying Integer Fractions with or without a Common Denominator

The following examples further illustrate how to multiply integer fractions by one another:

Example 1.2-33

$$\boxed{\frac{3}{5}\times\frac{2}{6}} = \boxed{\frac{\overset{1}{\cancel{3}}\times 2}{5\times\underset{2}{\cancel{6}}}} = \boxed{\frac{1\times\overset{1}{\cancel{2}}}{5\times\underset{1}{\cancel{2}}}} = \boxed{\frac{1\times 1}{5\times 1}} = \boxed{\frac{1}{5}}$$

Example 1.2-34

$$\boxed{\frac{2}{5}\times\frac{4}{5}\times\frac{25}{8}} = \boxed{\frac{2\times\overset{1}{\cancel{4}}\times\overset{5}{\cancel{25}}}{\underset{1}{\cancel{5}}\times 5\times\underset{2}{\cancel{8}}}} = \boxed{\frac{\overset{1}{\cancel{2}}\times 1\times\overset{1}{\cancel{5}}}{1\times\underset{1}{\cancel{5}}\times\underset{1}{\cancel{2}}}} = \boxed{\frac{1\times 1\times 1}{1\times 1\times 1}} = \boxed{\frac{1}{1}} = \boxed{1}$$

Example 1.2-35

$$\boxed{1000\times\frac{2}{100}\times\frac{1}{10}\times\frac{1}{2}} = \boxed{\frac{1000}{1}\times\frac{2}{100}\times\frac{1}{10}\times\frac{1}{2}} = \boxed{\frac{\overset{10}{\cancel{1000}}\times\overset{1}{\cancel{2}}\times 1\times 1}{1\times\underset{1}{\cancel{100}}\times 10\times\underset{1}{\cancel{2}}}} = \boxed{\frac{\overset{1}{\cancel{10}}\times 1\times 1\times 1}{1\times 1\times\underset{1}{\cancel{10}}\times 1}} = \boxed{\frac{1\times 1\times 1\times 1}{1\times 1\times 1\times 1}} = \boxed{\frac{1}{1}} = \boxed{1}$$

Example 1.2-36

$$\boxed{\left(\frac{3}{10}\times 24\right)\times\left(\frac{3}{8}\times\frac{25}{6}\right)} = \boxed{\left(\frac{3}{10}\times\frac{24}{1}\right)\times\left(\frac{3}{8}\times\frac{25}{6}\right)} = \boxed{\left(\frac{3\times\overset{12}{\cancel{24}}}{\underset{5}{\cancel{10}}\times 1}\right)\times\left(\frac{\overset{1}{\cancel{3}}\times 25}{8\times\underset{2}{\cancel{6}}}\right)} = \boxed{\left(\frac{3\times 12}{5\times 1}\right)\times\left(\frac{1\times 25}{8\times 2}\right)} = \boxed{\left(\frac{36}{5}\right)\times\left(\frac{25}{16}\right)}$$

$$= \boxed{\frac{36}{5}\times\frac{25}{16}} = \boxed{\frac{\overset{9}{\cancel{36}}\times\overset{5}{\cancel{25}}}{\underset{1}{\cancel{5}}\times\underset{4}{\cancel{16}}}} = \boxed{\frac{9\times 5}{1\times 4}} = \boxed{\frac{45}{4}} = \boxed{11\frac{1}{4}}$$

Practice Problems - Multiplying Integer Fractions with or without a Common Denominator

Section 1.2b Case V Practice Problems - Multiply the following integer fractions:

1. $\dfrac{4}{8}\times\dfrac{3}{5} =$

2. $\dfrac{4}{8}\times\dfrac{5}{6}\times 100 =$

3. $\dfrac{7}{3}\times\dfrac{9}{4}\times\dfrac{6}{3} =$

4. $34\times\dfrac{1}{5}\times\dfrac{3}{17}\times\dfrac{1}{8}\times 20 =$

5. $\left(\dfrac{2}{55}\times 3\right)\times\left(\dfrac{4}{5}\times\dfrac{25}{8}\right) =$

6. $\left(1000\times\dfrac{1}{5}\right)\times\left(\dfrac{25}{5}\times\dfrac{1}{8}\right)\times\dfrac{0}{100} =$

Case VI - Dividing Integer Fractions with or without a Common Denominator

Divide two integer fractions using the following steps:

Step 1 Change the integer number a to an integer fraction of the form $\frac{a}{1}$, i.e., change 9 to $\frac{9}{1}$.

Step 2
a. Change the division sign to a multiplication sign.
b. Replace the numerator of the second fraction with its denominator.
c. Replace the denominator of the second fraction with its numerator.
d. Multiply the numerator of the first fraction with the numerator of the second fraction to obtain the new numerator.
e. Multiply the denominator of the first fraction with the denominator of the second fraction to obtain the new denominator.

Step 3 Simplify the fraction to its lowest term (see Section 1.2a).

Step 4 Change the improper fraction to a mixed fraction if the fraction obtained from Step 3 is an improper fraction (see Section 1.2 Appendix).

Examples with Steps

The following examples show the steps as to how two integer fractions with or without a common denominator are divided:

Example 1.2-37

$$\boxed{\frac{3}{5} \div \frac{8}{15}} =$$

Solution:

Step 1 *Not Applicable*

Step 2 $$\boxed{\frac{3}{5} \div \frac{8}{15}} = \boxed{\frac{3}{5} \times \frac{15}{8}} = \boxed{\frac{3 \times 15}{5 \times 8}} = \boxed{\frac{45}{40}}$$

Step 3 $$\boxed{\frac{45}{40}} = \boxed{\frac{45 \div 5}{40 \div 5}} = \boxed{\frac{9}{8}}$$

Step 4 $$\boxed{\frac{9}{8}} = \boxed{\mathbf{1\frac{1}{8}}}$$

Example 1.2-38

$$\boxed{9 \div \frac{6}{12}} =$$

Solution:

Step 1 $$\boxed{9 \div \frac{6}{12}} = \boxed{\frac{9}{1} \div \frac{6}{12}}$$

Step 2 $$\boxed{\frac{9}{1} \div \frac{6}{12}} = \boxed{\frac{9}{1} \times \frac{12}{6}} = \boxed{\frac{9 \times 12}{1 \times 6}} = \boxed{\frac{108}{6}}$$

Step 3 $\boxed{\frac{108}{6}} = \boxed{\frac{108 \div 6}{6 \div 6}} = \boxed{\frac{18}{1}} = \boxed{\mathbf{18}}$

Step 4 $\boxed{\textit{Not Applicable}}$

In general, two integer fractions with or without a common denominator are divided in the following way:

$$\boxed{\frac{a}{b} \div \frac{c}{d}} = \boxed{\frac{a}{b} \times \frac{d}{c}} = \boxed{\frac{a \times d}{b \times c}} = \boxed{\frac{ad}{bc}}$$

For example,

$$\boxed{\frac{3}{5} \div \frac{2}{15}} = \boxed{\frac{3}{5} \times \frac{15}{2}} = \boxed{\frac{3 \times \overset{3}{\cancel{15}}}{\cancel{5} \times 2}} = \boxed{\frac{3 \times 3}{1 \times 2}} = \boxed{\frac{9}{2}} = \boxed{\mathbf{4\frac{1}{2}}}$$

Additional Examples - Dividing Integer Fractions with or without a Common Denominator

The following examples further illustrate how to divide integer fractions:

Example 1.2-39

$$\boxed{\frac{4}{5} \div \frac{2}{15}} = \boxed{\frac{4}{5} \times \frac{15}{2}} = \boxed{\frac{\overset{2}{\cancel{4}} \times \overset{3}{\cancel{15}}}{\cancel{5} \times \cancel{2}}} = \boxed{\frac{2 \times 3}{1 \times 1}} = \boxed{\frac{6}{1}} = \boxed{\mathbf{6}}$$

Example 1.2-40

$$\boxed{\left(\frac{3}{5} \div \frac{1}{5}\right) \div \frac{4}{15}} = \boxed{\left(\frac{3}{5} \times \frac{5}{1}\right) \div \frac{4}{15}} = \boxed{\left(\frac{3 \times 5}{5 \times 1}\right) \div \frac{4}{15}} = \boxed{\left(\frac{15}{5}\right) \div \frac{4}{15}} = \boxed{\frac{15}{5} \div \frac{4}{15}} = \boxed{\frac{15}{5} \times \frac{15}{4}} = \boxed{\frac{15 \times \overset{3}{\cancel{15}}}{\cancel{5} \times 4}} = \boxed{\frac{15 \times 3}{1 \times 4}}$$

$$= \boxed{\frac{45}{4}} = \boxed{\mathbf{11\frac{1}{4}}}$$

Example 1.2-41

$$\boxed{25 \div \left(\frac{2}{8} \div \frac{4}{3}\right)} = \boxed{\frac{25}{1} \div \left(\frac{2}{8} \times \frac{3}{4}\right)} = \boxed{\frac{25}{1} \div \left(\frac{\cancel{2} \times 3}{8 \times \underset{2}{\cancel{4}}}\right)} = \boxed{\frac{25}{1} \div \left(\frac{1 \times 3}{8 \times 2}\right)} = \boxed{\frac{25}{1} \div \left(\frac{3}{16}\right)} = \boxed{\frac{25}{1} \div \frac{3}{16}} = \boxed{\frac{25}{1} \times \frac{16}{3}}$$

$$= \boxed{\frac{25 \times 16}{1 \times 3}} = \boxed{\frac{400}{3}} = \boxed{\mathbf{133\frac{1}{3}}}$$

Example 1.2-42

$$\boxed{\left(\frac{9}{16} \div \frac{3}{32}\right) \div \left(\frac{4}{8} \div \frac{8}{1}\right)} = \boxed{\left(\frac{9}{16} \times \frac{32}{3}\right) \div \left(\frac{4}{8} \times \frac{1}{8}\right)} = \boxed{\left(\frac{\overset{3}{\cancel{9}} \times \overset{2}{\cancel{32}}}{\cancel{16} \times \cancel{3}}\right) \div \left(\frac{\cancel{4} \times 1}{8 \times \underset{2}{\cancel{8}}}\right)} = \boxed{\left(\frac{3 \times 2}{1 \times 1}\right) \div \left(\frac{1 \times 1}{8 \times 2}\right)} = \boxed{\left(\frac{6}{1}\right) \div \left(\frac{1}{16}\right)}$$

$$= \boxed{\frac{6}{1} \div \frac{1}{16}} = \boxed{\frac{6}{1} \times \frac{16}{1}} = \boxed{\frac{6 \times 16}{1 \times 1}} = \boxed{\frac{96}{1}} = \boxed{\mathbf{96}}$$

Practice Problems - Dividing Integer Fractions with or without a Common Denominator

Section 1.2b Case VI Practice Problems - Divide the following integer fractions:

1. $\frac{8}{10} \div \frac{4}{30} =$
2. $\left(\frac{3}{8} \div \frac{12}{16}\right) \div \frac{4}{8} =$
3. $\left(\frac{4}{16} \div \frac{1}{32}\right) \div 8 =$
4. $12 \div \left(\frac{9}{8} \div \frac{27}{16}\right) =$
5. $\left(\frac{2}{20} \div \frac{4}{5}\right) \div 2 =$
6. $\left(\frac{4}{15} \div \frac{8}{30}\right) \div \left(\frac{1}{5} \div \frac{4}{35}\right) =$

1.2 Appendix: Changing Improper Fractions to Mixed Fractions

Improper fractions of the form $\frac{c}{b}$ with absolute values of greater than one are changed to mixed fractions of the form $k\frac{a}{b}$, where k is a positive or negative whole number and $\frac{a}{b}$ is an integer fraction with value of less than one, using the following steps:

Step 1 Divide the dividend, i.e., the numerator of the improper fraction by the divisor, i.e., the denominator of the improper fraction using the general division process.

Step 2 a. Use the whole number portion of the quotient as the whole number portion of the mixed fraction.

b. Use the dividend of the remainder as the dividend (numerator) in the remainder portion of the quotient.

c. Use the divisor of the improper fraction as the divisor (denominator) in the remainder portion of the quotient.

Examples with Steps

The following examples show the steps as to how improper fractions are changed to integer fractions:

Example 1.2A-1

$$\boxed{\frac{86}{5}} =$$

Solution:

Step 1

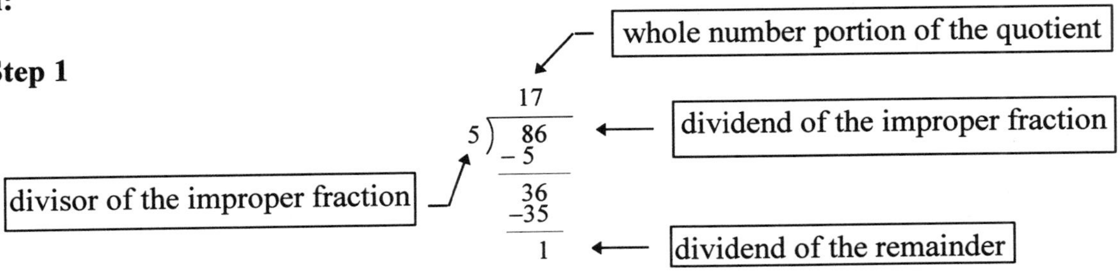

Step 2

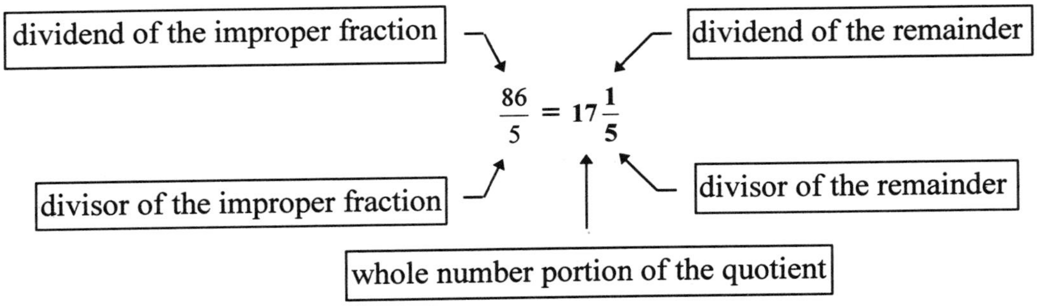

Example 1.2A-2

$$\boxed{\frac{1428}{45}} =$$

Solution:

Step 1

$$45\overline{)\,1428\,}$$ quotient: 31

$$\begin{array}{r} 1428 \\ -135 \\ \hline 78 \\ -45 \\ \hline 33 \end{array}$$

- 31 ← whole number portion of the quotient
- 1428 ← dividend of the improper fraction
- 45 ← divisor of the improper fraction
- 33 ← dividend of the remainder

Step 2

$$\frac{1428}{45} = 31\frac{33}{45}$$

- 1428: dividend of the improper fraction
- 45 (left side): divisor of the improper fraction
- 31: whole number portion of the quotient
- 33: dividend of the remainder
- 45 (right side): divisor of the remainder

Example 1.2A-3

$$\boxed{-\frac{38}{3}} =$$

Solution:

Step 1

$$3\overline{)\,38\,}$$ quotient: 12

$$\begin{array}{r} 38 \\ -\ 3 \\ \hline 08 \\ -\ 6 \\ \hline 2 \end{array}$$

- 12 ← whole number portion of the quotient
- 38 ← dividend of the improper fraction
- 3 ← divisor of the improper fraction
- 2 ← dividend of the remainder

Step 2

$$-\frac{38}{3} = -\left(12\frac{2}{3}\right)$$

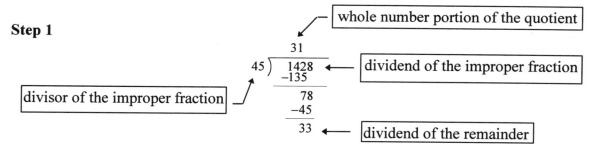

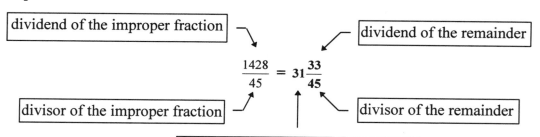

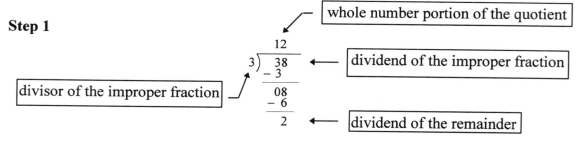

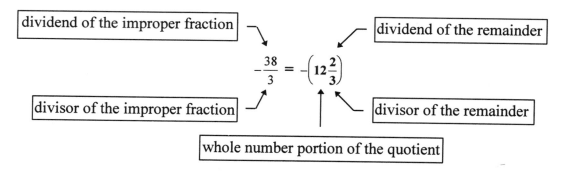

- 38: dividend of the improper fraction
- 3 (left side): divisor of the improper fraction
- 12: whole number portion of the quotient
- 2: dividend of the remainder
- 3 (right side): divisor of the remainder

In general, an improper integer fraction $\frac{c}{b}$, where c is bigger than b, is changed to a mixed fraction in the following way:

1. divide the numerator c by its denominator b using the general division process.

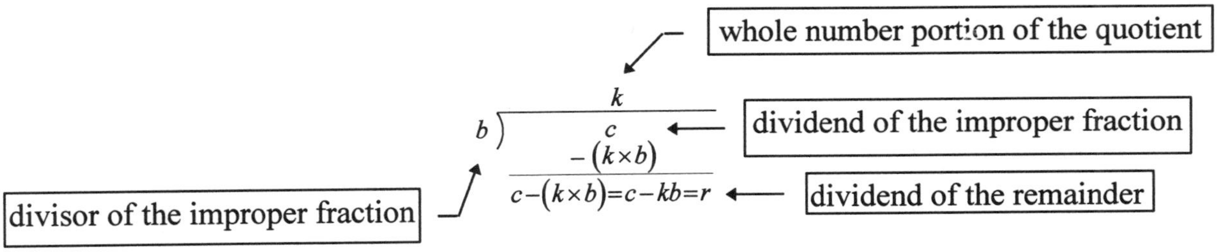

2. Use the whole number portion of the quotient k, the dividend of the remainder r, and the divisor of the improper fraction b to represent the mixed fraction as:

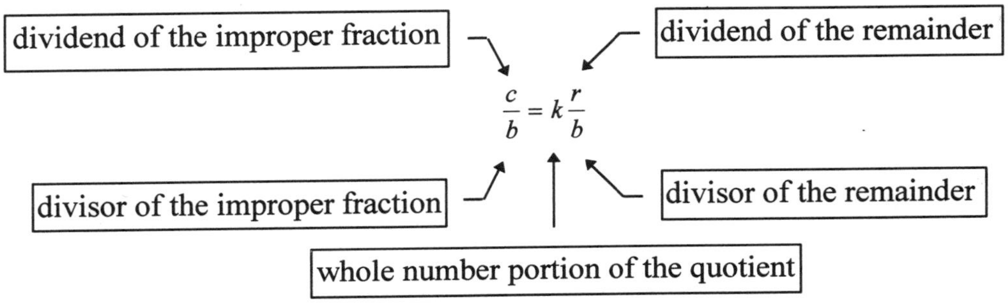

Note 1 - In the general equation $\left(\frac{c}{b} = k\frac{r}{b}\right)$; $\frac{c}{b}$ is the improper fraction, $k\frac{r}{b}$ is the quotient, k is the whole number portion of the quotient, and $\frac{r}{b}$ is the remainder portion of the quotient.

Note 2 - The divisor of the improper fraction is always used as the divisor of the remainder. This is shown in Step 2 of examples above.

1.2 Appendix Practice Problems - Changing Improper Fractions to Mixed Fractions

1.2 Appendix Practice Problems - Change the following improper fractions to mixed fractions:

1. $\frac{83}{4} =$
2. $\frac{13}{3} =$
3. $-\frac{26}{5} =$
4. $\frac{67}{10} =$
5. $\frac{9}{2} =$
6. $-\frac{332}{113} =$

1.3 Exponents

In this section an introduction to exponents and how real numbers are raised to positive or negative integer exponents are addressed in section 1.3a, Cases I and II. In addition, operations involving positive and negative integer exponents are reviewed in sections 1.3b, Cases I through III and 1.3c, Cases I through III, respectively.

1.3a Introduction to Integer Exponents

Integer exponents are defined as a^n where a is referred to as the **base**, and n is the **integer exponent**. Note that the base a can be a real number or a variable. The integer exponent n can be a positive or a negative integer. Real numbers raised to positive and negative integer exponents (Cases I and II) are addressed below:

Case I Real Numbers Raised to Positive Integer Exponents

In general, real numbers raised to positive integer exponents are shown as:

$$a^{+n} = a^n = a \cdot a \cdot a \cdot a \ldots a \qquad \textit{where } n \textit{ is a positive integer and } a \neq 0$$

For example,

$$8^{+4} = 8^4 = 8 \cdot 8 \cdot 8 \cdot 8 = 4096$$

Real numbers raised to a positive integer exponent are solved using the following steps:

Step 1 Multiply the base a by itself as many times as the number specified in the exponent. For example, 2^5 implies that multiply 2 by itself 5 times, i.e., $2^5 = 2 \cdot 2 \cdot 2 \cdot 2 \cdot 2$.

Step 2 Multiply the real numbers to obtain the product, i.e., $2 \cdot 2 \cdot 2 \cdot 2 \cdot 2 = 32$.

Examples with Steps

The following examples show the steps as to how real numbers raised to positive integer exponents are solved:

Example 1.3-1

$$\boxed{2^3} =$$

Solution:

Step 1 $\boxed{2^3} = \boxed{2 \cdot 2 \cdot 2}$

Step 2 $\boxed{2 \cdot 2 \cdot 2} = \boxed{\mathbf{8}}$

Example 1.3-2

$$\boxed{1.2^4} =$$

Solution:

Step 1 $\boxed{1.2^4} = \boxed{(1.2) \cdot (1.2) \cdot (1.2) \cdot (1.2)}$

Step 2 $\boxed{(1.2) \cdot (1.2) \cdot (1.2) \cdot (1.2)} = \boxed{\mathbf{2.074}}$

Example 1.3-3

$$\boxed{(-3)^5} =$$

Solution:

Step 1 $\boxed{(-3)^5} = \boxed{(-3)\cdot(-3)\cdot(-3)\cdot(-3)\cdot(-3)}$

Step 2 $\boxed{(-3)\cdot(-3)\cdot(-3)\cdot(-3)\cdot(-3)} = \boxed{\mathbf{-243}}$

Note that:

- A negative number raised to an even integer exponent such as 2, 4, 6, 8, 10, 12, etc. is always positive. For example,

$$(-3)^6 = (+3)^6 = +729 = 729 \qquad (-2)^2 = (+2)^2 = +4 = 4 \qquad (-5)^4 = (+5)^4 = +625 = 625$$

- A negative number raised to an odd integer exponent such as 1, 3, 5, 7, 9, 11, etc. is always negative. For example,

$$(-3)^5 = -243 \qquad (-2)^3 = -8 \qquad (-3)^7 = -2187$$

Additional Examples - Real Numbers Raised to Positive Integer Exponents

The following examples further illustrate how to solve real numbers raised to positive integer exponents:

Example 1.3-4

$\boxed{(-10)^0} = \boxed{\mathbf{1}}$ (See the note on page 41 on numbers raised to the zero power.)

Example 1.3-5

$\boxed{-(6)^5} = \boxed{-(6\cdot6\cdot6\cdot6\cdot6)} = \boxed{-(7776)} = \boxed{\mathbf{-7776}}$

Example 1.3-6

$\boxed{(-4.25)^3} = \boxed{(-4.25)\cdot(-4.25)\cdot(-4.25)} = \boxed{\mathbf{-76.77}}$

Example 1.3-7

$\boxed{(10.45)^4} = \boxed{(10.45)\cdot(10.45)\cdot(10.45)\cdot(10.45)} = \boxed{\mathbf{11925.19}}$

Example 1.3-8

$\boxed{-(-20)^3} = \boxed{-[(-20)\cdot(-20)\cdot(-20)]} = \boxed{-[-8000]} = \boxed{+8000} = \boxed{\mathbf{8000}}$

Practice Problems - Real Numbers Raised to Positive Integer Exponents

Section 1.3a Case I Practice Problems - Solve the following exponential expressions with real numbers raised to positive integer exponents:

1. $4^3 =$
2. $(-10)^4 =$
3. $0.25^3 =$
4. $12^5 =$
5. $-(3)^5 =$
6. $489^0 =$

Case II Real Numbers Raised to Negative Integer Exponents

Negative integer exponents are defined as a^{-n} where a is referred to as the **base**, and n is the **integer exponent**. Again, note that the base a can be a real number or a variable. The integer exponent n can be a positive or a negative integer. In this section, real numbers raised to negative integer exponents are addressed.
In general, real numbers raised to negative integer exponents are shown as:

$$a^{-n} = \frac{1}{a^{+n}} = \frac{1}{a^n} = \frac{1}{a \cdot a \cdot a \cdot a \ldots \cdot a} \qquad \textit{where } n \textit{ is a positive integer and } a \neq 0$$

For example,

$$5^{-4} = \frac{1}{5^4} = \frac{1}{5 \cdot 5 \cdot 5 \cdot 5} = \frac{1}{625}$$

Real numbers raised to a negative integer exponent are solved using the following steps:

Step 1 Change the negative integer exponent a^{-n} to a positive integer exponent of the form $\frac{1}{a^n}$. For example, change 3^{-4} to $\frac{1}{3^4}$.

Step 2 Multiply the base a in the denominator by itself as many times as the number specified in the exponent. For example, rewrite $\frac{1}{3^4}$ as $\frac{1}{3 \cdot 3 \cdot 3 \cdot 3}$.

Step 3 Multiply the real numbers in the denominator to obtain the answer, i.e., $\frac{1}{3 \cdot 3 \cdot 3 \cdot 3} = \frac{1}{81}$.

Examples with Steps

The following examples show the steps as to how real numbers raised to negative integer exponents are solved:

Example 1.3-9

$$\boxed{4^{-3}} =$$

Solution:

Step 1 $\boxed{4^{-3}} = \boxed{\frac{1}{4^3}}$

Step 2 $\boxed{\frac{1}{4^3}} = \boxed{\frac{1}{4 \cdot 4 \cdot 4}}$

Step 3 $\boxed{\frac{1}{4 \cdot 4 \cdot 4}} = \boxed{\mathbf{\frac{1}{64}}}$

Example 1.3-10

$$\boxed{(-8)^{-3}} =$$

Solution:

Step 1 $\boxed{(-8)^{-3}} = \boxed{\frac{1}{(-8)^3}}$

Step 2 $\boxed{\frac{1}{(-8)^3}} = \boxed{\frac{1}{(-8) \cdot (-8) \cdot (-8)}}$

Step 3 $\frac{1}{(-8)\cdot(-8)\cdot(-8)} = \frac{1}{-512} = -\frac{1}{512}$

Additional Examples - Real Numbers Raised to Negative Integer Exponents

The following examples further illustrate how to solve real numbers raised to negative integer exponents:

Example 1.3-11

$$2^{-3} = \frac{1}{2^3} = \frac{1}{2\cdot 2\cdot 2} = \mathbf{\frac{1}{8}}$$

Example 1.3-12

$$-(6)^{-4} = -\frac{1}{(6)^4} = -\frac{1}{6^4} = -\frac{1}{6\cdot 6\cdot 6\cdot 6} = -\mathbf{\frac{1}{1296}}$$

Example 1.3-13

$$(5.2)^{-4} = \frac{1}{(5.2)^4} = \frac{1}{5.2^4} = \frac{1}{(5.2)\cdot(5.2)\cdot(5.2)\cdot(5.2)} = \mathbf{\frac{1}{731.16}}$$

Example 1.3-14

$$(-9)^{-4} = \frac{1}{(-9)^4} = \frac{1}{(-9)\cdot(-9)\cdot(-9)\cdot(-9)} = \frac{1}{+6561} = \mathbf{\frac{1}{6561}}$$

Example 1.3-15

$$-(-4.5)^{-3} = -\frac{1}{(-4.5)^3} = -\frac{1}{(-4.5)\cdot(-4.5)\cdot(-4.5)} = -\frac{1}{-91.125} = +\frac{1}{91.125} = \mathbf{\frac{1}{91.125}}$$

Note 1: Any number or variable raised to the zero power is always equal to 1. For example,

$$55^0 = 1,\ (-15)^0 = 1,\ (5{,}689{,}763)^0 = 1,\ [(5x+2)-8]^0 = 1,\ \left[(axy)^3\right]^0 = 1, \text{ and } \left(\sqrt{x^3y^2z}\right)^0 = 1$$

Note 2: Zero raised to the zero power is not defined, i.e., 0^0 is undefined.

Note 3: Any number or variable divided by zero is not defined, i.e., $\frac{1}{0}$, $\frac{x}{0}$, $\frac{355}{0}$, etc. are undefined.

Note 4: Zero divided by any number or variable is always equal to zero. For example, $\frac{0}{1} = 0$,

$$\frac{0}{2{,}560} = 0,\ \frac{0}{\sqrt{10}} = 0,\ \frac{0}{x^2} = 0, \text{ and } \frac{0}{\sqrt[3]{a^2}} = 0.$$

Practice Problems - Real Numbers Raised to Negative Integer Exponents

Section 1.3a Case II Practice Problems - Solve the following exponential expressions with real numbers raised to negative integer exponents:

1. $4^{-3} =$
2. $(-5)^{-4} =$
3. $0.25^{-3} =$
4. $12^{-5} =$
5. $-(3)^{-4} =$
6. $48^{-2} =$

1.3b Operations with Positive Integer Exponents

To multiply, divide, add, and subtract integer exponents, we need to know the following laws of exponents (shown in Table 1.3-1). These laws are used to simplify the work in solving problems with exponential expressions and should be memorized.

Table 1.3-1: Exponent Laws 1 through 7 (Positive Integer Exponents)

I. **Multiplication**	$a^m \cdot a^n = a^{m+n}$	When multiplying positive exponential terms, if bases a are the same, add the exponents m and n.
II. **Power of a Power**	$(a^m)^n = a^{mn}$	When raising an exponential term to a power, multiply the powers (exponents) m and n.
III. **Power of a Product**	$(a \cdot b)^m = a^m \cdot b^m$	When raising a product to a power, raise each factor a and b to the power m.
IV. **Power of a Fraction**	$\left(\frac{a}{b}\right)^m = \frac{a^m}{b^m}$	When raising a fraction to a power, raise the numerator and the denominator to the power m.
V. **Division**	$\frac{a^m}{a^n} = a^m \cdot a^{-n} = a^{m-n}$	When dividing exponential terms, if the bases a are the same, subtract exponents m and n.
VI. **Negative Power**	$a^{-n} = \frac{1}{a^n}$	A non-zero base a raised to the $-n$ power equals 1 divided by the base a to the n power.
VII. **Zero Power**	$a^0 = 1$	A non-zero base a raised to the zero power is always equal to 1.

In this section students learn how to multiply (Case I), divide (Case II), and add or subtract (Case III) positive integer exponents by one another.

Case I Multiplying Positive Integer Exponents

Positive integer exponents are multiplied by each other using the following steps and the exponent laws I through III shown in Table 1.3-1.

Step 1 Group the exponential terms with similar bases.

Step 2 Apply the Multiplication Law (Law I) from Table 1.3-1 and simplify the exponential expressions by adding the exponents with similar bases.

Examples with Steps

The following examples show the steps as to how positive integer exponents are multiplied by one another:

Example 1.3-16

$$\boxed{\left(x^3y^2\right)\cdot\left(x^2y\right)\cdot y^3} =$$

Solution:

Step 1 $\boxed{\left(x^3y^2\right)\cdot\left(x^2y\right)\cdot y^3} = \boxed{\left(x^3x^2\right)\cdot\left(y^3y^2y\right)} = \boxed{\left(x^3x^2\right)\cdot\left(y^3y^2y^1\right)}$

Step 2 $\boxed{\left(x^3x^2\right)\cdot\left(y^3y^2y^1\right)} = \boxed{\left(x^{3+2}\right)\cdot\left(y^{3+2+1}\right)} = \boxed{x^5\cdot y^6} = \boxed{\mathbf{x^5y^6}}$

Example 1.3-17

$$\boxed{\left(-\frac{1}{5}a^2\right)\cdot\left(10\,a\,b\right)\cdot\left(-\frac{1}{4}a\,b^2\right)} =$$

Solution:

Step 1 $\boxed{\left(-\frac{1}{5}a^2\right)\cdot(10ab)\cdot\left(-\frac{1}{4}ab^2\right)} = \boxed{\left(-\frac{1}{5}\times-\frac{1}{4}\times 10\right)\cdot\left(a^2aa\right)\cdot\left(b^2b\right)} = \boxed{\left(\frac{10}{20}\right)\cdot\left(a^2a^1a^1\right)\cdot\left(b^2b^1\right)}$

Step 2 $\boxed{\left(\frac{10}{20}\right)\cdot\left(a^2a^1a^1\right)\cdot\left(b^2b^1\right)} = \boxed{\frac{1\not{0}}{2\not{0}}\cdot\left(a^{2+1+1}\right)\cdot\left(b^{2+1}\right)} = \boxed{\frac{1}{2}\cdot a^4\cdot b^3} = \boxed{\mathbf{\frac{1}{2}a^4b^3}}$

Note - Non zero numbers or variables raised to the zero power are always equal to 1, i.e., $10^0=1$, $(23456)^0=1$, $a^0=1\ for\ a\neq 0$, $(a\cdot b)^0=1\ for\ a\cdot b\neq 0$, $(x\cdot y\cdot z)^0=1\ for\ x\cdot y\cdot z\neq 0$, etc.

Additional Examples - Multiplying Positive Integer Exponents

The following examples further illustrate how positive exponential terms are multiplied:

Example 1.3-18

$\boxed{2^3\cdot 2^5} = \boxed{2^{3+5}} = \boxed{2^8} = \boxed{\mathbf{256}}$

Example 1.3-19

$\boxed{3^0\cdot 3\cdot 3^2\cdot 3^3} = \boxed{3^0\cdot 3^1\cdot 3^2\cdot 3^3} = \boxed{3^{0+1+2+3}} = \boxed{3^6} = \boxed{\mathbf{729}}$

Example 1.3-20

$\boxed{x^2\cdot y^2\cdot z^3\cdot x^2\cdot y^2\cdot z^4\cdot x} = \boxed{\left(x^2\cdot x^2\cdot x\right)\cdot\left(y^2\cdot y^2\right)\cdot\left(z^3\cdot z^4\right)} = \boxed{\left(x^{2+2+1}\right)\cdot\left(y^{2+2}\right)\cdot\left(z^{3+4}\right)} = \boxed{\mathbf{x^5y^4z^7}}$

Example 1.3-21

$\boxed{(-3-2)^2\cdot\left(-4k^2p^2\right)\cdot(-5\,k\,p)} = \boxed{\left[(-5)^2\cdot(-4\times -5)\right]\cdot\left(k^2k\right)\cdot\left(p^2p\right)} = \boxed{\left[(+25)\cdot(+20)\right]\cdot\left(k^2k^1\right)\cdot\left(p^2p^1\right)}$

$= \boxed{500\cdot\left(k^{2+1}\right)\cdot\left(p^{2+1}\right)} = \boxed{\mathbf{500\,k^3p^3}}$

Practice Problems - Multiplying Positive Integer Exponents

Section 1.3b Case I Practice Problems - Multiply the following positive integer exponents:

1. $x^2\cdot x^3\cdot x =$
2. $2\cdot a^2\cdot b^0\cdot a^3\cdot b^2 =$
3. $\frac{4}{-6}a^2b^3ab^4b^5 =$
4. $2^3\cdot 2^2\cdot x^{2a}\cdot x^{3a}\cdot x^a =$
5. $\left(x\cdot y^2\cdot z^3\right)^0\cdot w^2z^3zw^4z^2 =$
6. $2^0\cdot 4^2\cdot 4^2\cdot 2^2\cdot 4^1 =$

Case II Dividing Positive Integer Exponents

Positive integer exponents are divided by one another using the exponent laws I through VI shown in Table 1.3-1.

Case II Dividing Positive Integer Exponents

Positive integer exponents are divided by one another using the following steps:

Step 1 a. Apply the Division and/or the Negative Power Laws (Laws V and VI) from Table 1.3-1.

b. Group the exponential terms with similar bases.

Step 2 Apply the Multiplication Law (Law I) from Table 1.3-1 and simplify the exponential expressions by adding the exponents with similar bases.

Examples with Steps

The following examples show the steps as to how positive integer exponents are divided by one another:

Example 1.3-22

$$\frac{2ab}{-4a^3b^4} =$$

Solution:

Step 1

$$\frac{2ab}{-4a^3b^4} = -\frac{2}{4}\frac{a^1b^1}{a^3b^4} = -\frac{2}{4}\frac{1}{\left(a^3a^{-1}\right)\cdot\left(b^4b^{-1}\right)}$$

Step 2

$$-\frac{2}{4}\frac{1}{\left(a^3a^{-1}\right)\cdot\left(b^4b^{-1}\right)} = -\frac{\overset{1}{\cancel{2}}}{\underset{2}{\cancel{4}}}\frac{1}{\left(a^{3-1}\right)\cdot\left(b^{4-1}\right)} = -\frac{1}{2}\frac{1}{a^2\cdot b^3} = \mathbf{-\frac{1}{2}\left(\frac{1}{a^2b^3}\right)}$$

Example 1.3-23

$$\frac{-3xy^4z^3y}{-15x^2z^2} =$$

Solution:

Step 1

$$\frac{-3xy^4z^3y}{-15x^2z^2} = \frac{-3x^1y^4z^3y^1}{-15x^2z^2} = +\frac{3}{15}\frac{x^1y^4z^3y^1}{x^2z^2} = \frac{3}{15}\frac{\left(y^4y^1\right)\cdot\left(z^3z^{-2}\right)}{x^2x^{-1}}$$

Step 2

$$\frac{3}{15}\frac{\left(y^4y^1\right)\cdot\left(z^3z^{-2}\right)}{x^2x^{-1}} = \frac{\overset{1}{\cancel{3}}}{\underset{5}{\cancel{15}}}\frac{\left(y^{4+1}\right)\cdot\left(z^{3-2}\right)}{x^{2-1}} = \frac{1}{5}\frac{y^5\cdot z^1}{x^1} = \mathbf{\frac{1}{5}\left(\frac{y^5z}{x}\right)}$$

Additional Examples - Dividing Positive Integer Exponents

The following examples further illustrate how to divide positive integer exponential terms by one another:

Example 1.3-24

$$\frac{x^4y^3z}{x^2y^2} = \frac{\left(x^4x^{-2}\right)\cdot\left(y^3y^{-2}\right)\cdot z}{1} = \frac{\left(x^{4-2}\right)\cdot\left(y^{3-2}\right)\cdot z}{1} = \frac{x^2\cdot y^1\cdot z}{1} = \frac{x^2yz}{1} = \mathbf{x^2yz}$$

Example 1.3-25

$$\frac{5a^3b^5}{15a^2b^0} = \frac{5}{15}\,\frac{a^3b^5}{a^2\cdot 1} = \frac{\overset{1}{\not{5}}}{\underset{3}{\not{15}}}\,\frac{\left(a^3a^{-2}\right)\cdot b^5}{1} = \frac{1}{3}\,\frac{\left(a^{3-2}\right)\cdot b^5}{1} = \frac{1}{3}\,\frac{a^1\cdot b^5}{1} = \mathbf{\frac{1}{3}ab^5}$$

Example 1.3-26

$$\frac{8u^3w^3z^2}{2u^3w^2z} = \frac{8u^3w^3z^2}{2u^3w^2z^1} = \frac{\overset{4}{\not{8}}\cdot\left(w^3w^{-2}\right)\cdot\left(z^2\cdot z^{-1}\right)}{\underset{1}{\not{2}}\cdot\left(u^3u^{-3}\right)} = \frac{4\cdot\left(w^{3-2}\right)\cdot\left(z^{2-1}\right)}{1\cdot\left(u^{3-3}\right)} = \frac{4\cdot w^1\cdot z^1}{u^0} = \frac{4wz}{1} = \mathbf{4wz}$$

Example 1.3-27

$$\frac{100p^2t^2u}{5pt^4u^5} = \frac{100}{5}\,\frac{p^2t^2u^1}{p^1t^4u^5} = \frac{\overset{20}{\not{100}}}{\underset{1}{\not{5}}}\,\frac{p^2p^{-1}}{\left(t^4t^{-2}\right)\cdot\left(u^5u^{-1}\right)} = \frac{20}{1}\,\frac{p^{2-1}}{\left(t^{4-2}\right)\cdot\left(u^{5-1}\right)} = \frac{20}{1}\left(\frac{p^1}{t^2\cdot u^4}\right) = \mathbf{\frac{20p}{t^2u^4}}$$

Example 1.3-28

$$\left(\frac{w^2z^2}{z}\right)\cdot\left(\frac{w}{z^3}\right) = \left(\frac{w^2z^2}{z^1}\right)\cdot\left(\frac{w^1}{z^3}\right) = \frac{w^2z^2\cdot w^1}{z^1\cdot z^3} = \frac{w^2w^1}{z^1z^3z^{-2}} = \frac{w^{2+1}}{z^{1+3-2}} = \mathbf{\frac{w^3}{z^2}}$$

Practice Problems - Dividing Positive Integer Exponents

Section 1.3b Case II Practice Problems - Divide the following positive integer exponents:

1. $\dfrac{x^5}{x^3} =$
2. $\dfrac{a^2b^3}{a} =$
3. $\dfrac{a^3b^3c^2}{a^2b^6c} =$
4. $\dfrac{3^2\cdot\left(rs^2\right)}{(2rs)\cdot r^3} =$
5. $\dfrac{2p^2q^3pr^4}{-6p^4q^2r} =$
6. $\dfrac{\left(k^2l^3\right)\cdot\left(kl^2m^0\right)}{k^4l^3m^5} =$

$$\frac{9\cdot(rs^2)}{(2rs)\cdot r^3} \to \frac{9rs^2}{2rs\cdot r^3} \to \frac{9rs^2}{2r^4s} \to 4.5r^{-3}s \to \frac{9}{2}\frac{s}{r^3}$$

Case III Adding and Subtracting Positive Integer Exponents

A common source of mistakes among students is in dealing with addition and subtraction of exponential expressions. In this section positive integer exponents addressing addition and subtraction of numbers that are raised to positive exponents is introduced. Positive exponential expressions are added and subtracted using the following steps:

Step 1 Group the exponential terms with similar bases.

Step 2 Simplify the exponential expressions by adding or subtracting the like terms.

Note that **like terms** are defined as terms having the same variables raised to the same power. For example, x^3 *and* $2x^3$; y^2 *and* $4y^2$ are like terms of one another.

Examples with Steps

The following examples show the steps as to how exponential expressions having positive integer exponents are added or subtracted:

Example 1.3-29

$$\boxed{x^3+3y^2+2x^3-y^2+5}=$$

Solution:

Step 1

$$\boxed{x^3+3y^2+2x^3-y^2+5}=\boxed{\left(x^3+2x^3\right)+\left(3y^2-y^2\right)+5}$$

Step 2

$$\boxed{\left(x^3+2x^3\right)+\left(3y^2-y^2\right)+5}=\boxed{(1+2)x^3+(3-1)y^2+5}=\boxed{\mathbf{3x^3+2y^2+5}}$$

Example 1.3-30

$$\boxed{\left(2^3+x^2+4y\right)-\left(3x^2+y\right)+2x^2}=$$

Solution:

Step 1

$$\boxed{\left(2^3+x^2+4y\right)-\left(3x^2+y\right)+2x^2}=\boxed{8+x^2+4y-3x^2-y+2x^2}$$

$$=\boxed{\left(x^2+2x^2-3x^2\right)+(4y-y)+8}$$

Step 2

$$\boxed{\left(x^2+2x^2-3x^2\right)+(4y-y)+8}=\boxed{(1+2-3)x^2+(4-1)y+8}=\boxed{0x^2+3y+8}=\boxed{\mathbf{3y+8}}$$

Additional Examples - Adding and Subtracting Positive Integer Exponents

The following examples further illustrate addition and subtraction of exponential terms:

Example 1.3-31

$$\boxed{5x^3+3x^2+2x^3-x^2+5}=\boxed{\left(5x^3+2x^3\right)+\left(3x^2-x^2\right)+5}=\boxed{(5+2)x^3+(3-1)x^2+5}=\boxed{\mathbf{7x^3+2x^2+5}}$$

Example 1.3-32

$$\boxed{\left(-2m^4-3m^2+2m^4+3m-10\right)-\left(5m^2+2m+3\right)}=\boxed{-2m^4-3m^2+2m^4+3m-10-5m^2-2m-3}$$

$$=\boxed{\left(-2m^4+2m^4\right)+\left(-3m^2-5m^2\right)+(3m-2m)+(-10-3)}=\boxed{(-2+2)m^4+(-3-5)m^2+(3-2)m-13}$$

$$=\boxed{0m^4-8m^2+m-13}=\boxed{\mathbf{8m^2+m-13}}$$

Example 1.3-33

$$\boxed{x^2+3x^2+y^2+x-4y^2-5^2+2x^2+6x}=\boxed{\left(x^2+3x^2+2x^2\right)+\left(y^2-4y^2\right)+(x+6x)-25}$$

$$=\boxed{(1+3+2)x^2+(1-4)y^2+(1+6)x-25}=\boxed{\mathbf{6x^2-3y^2+7x-25}}$$

Example 1.3-34

$$\boxed{\left(-5w^3-3w-5\right)-\left(3w^3-w-4\right)+5w+2}=\boxed{\left(-5w^3-3w-5\right)+\left(-3w^3+w+4\right)+5w+2}$$

$$=\boxed{\left(-5w^3-3w^3\right)+(-3w+5w+w)+(-5+4+2)}=\boxed{(-5-3)w^3+(-3+5+1)w+1}=\boxed{\mathbf{-8w^3+3w+1}}$$

Example 1.3-35

$$\boxed{\left(a^2b^3+3a^2-b+2^4\right)+\left(2a^2b^3+2a^2+3^0\right)-3^3}=\boxed{a^2b^3+3a^2-b+16+2a^2b^3+2a^2+1-27}$$

$$=\boxed{\left(a^2b^3+2a^2b^3\right)+\left(3a^2+2a^2\right)-b+(16+1-27)}=\boxed{(1+2)a^2b^3+(3+2)a^2-b-10}=\boxed{\mathbf{3a^2b^3+5a^2-b-10}}$$

Practice Problems - Adding and Subtracting Positive Integer Exponents

Section 1.3b Case III Practice Problems - Add or subtract the following positive integer exponential expressions:

1. $x^2+4xy-2x^2-2xy+z^3=$
2. $\left(a^3+2a^2+4^3\right)-\left(4a^3+20\right)=$
3. $3x^4+2x^2+2x^4-\left(x^4-2x^2+3\right)=$
4. $-\left(-2l^3a^3+2l^2a^2-5^3\right)-\left(4l^3a^3-20\right)=$
5. $\left(m^{3n}-4m^{2n}\right)-\left(2m^{3n}+3m^{2n}\right)+5m=$
6. $\left(-7z^3+3z-5\right)-\left(-3z^3+z-4\right)+5z+20=$

1.3c Operations with Negative Integer Exponents

To proceed with simplification of negative exponents, we need to know the Negative Power Law in addition to the other exponent laws (shown in Table 1.3-2). The Negative Power Law states that a base raised to a negative exponent is equal to one divided by the same base raised to the positive exponent, or vice versa, i.e.,

$$a^{-n} = \frac{1}{a^n}$$

and

$$a^n = \frac{1}{a^{-n}} \qquad \text{since} \qquad a^n = \frac{1}{a^{-n}} = \frac{1}{\frac{1}{a^n}} = \frac{\frac{1}{1}}{\frac{1}{a^n}} = \frac{1 \times a^n}{1 \times 1} = \frac{a^n}{1} = a^n .$$

Note that the objective is to write the final answer without a negative exponent. To achieve that the exponent laws are used when simplifying expressions having negative integer exponents. These laws are used to simplify the work in solving exponential expressions and should be memorized.

Table 1.3-2: Exponent Laws 1 through 6 (Negative Integer Exponents)

I. **Multiplication**	$a^{-m} \cdot a^{-n} = a^{-m-n}$	When multiplying negative exponential terms, if bases a are the same, add the negative exponents $-m$ and $-n$.
II. **Power of a Power**	$\left(a^{-m}\right)^{-n} = a^{-m \times -n}$	When raising a negative exponential term to a negative power, multiply the negative powers (exponents) $-m$ and $-n$.
III. **Power of a Product**	$(a \cdot b)^{-m} = a^{-m} \cdot b^{-m}$	When raising a product to a negative power, raise each factor a and b to the negative power $-m$.
IV. **Power of a Fraction**	$\left(\frac{a}{b}\right)^{-m} = \frac{a^{-m}}{b^{-m}}$	When raising a fraction to a negative power, raise the numerator and the denominator to the negative power $-m$.
V. **Division**	$\frac{a^{-m}}{a^{-n}} = a^{-m} \cdot a^n = a^{-m+n}$	When dividing negative exponential terms, if the bases a are the same, add exponents $-m$ and n.
VI. **Negative Power**	$a^{-n} = \frac{1}{a^n}$	A non-zero base a raised to the $-n$ power equals 1 divided by the base a to the n power

In this section students learn how to multiply (Case I), divide (Case II), and add or subtract (Case III) negative integer exponents by one another.

Case I Multiplying Negative Integer Exponents

Negative integer exponents are multiplied by each other using the following steps and the exponent laws I through III shown in Table 1.3-2.

Step 1 Group the exponential terms with similar bases.

Step 2 Apply the Multiplication Law (Law I) from Table 1.3-2 and simplify the exponential expressions by adding the exponents with similar bases.

Step 3 Change the negative integer exponents to positive integer exponents.

Examples with Steps

The following examples show the steps as to how negative integer exponents are multiplied by one another:

Example 1.3-36

$$\boxed{3^{-2}\cdot 2\cdot 3^{-1}\cdot 2^{-3}} =$$

Solution:

Step 1 $$\boxed{3^{-2}\cdot 2\cdot 3^{-1}\cdot 2^{-3}} = \boxed{3^{-2}\cdot 2^{1}\cdot 3^{-1}\cdot 2^{-3}} = \boxed{\left(3^{-2}\cdot 3^{-1}\right)\cdot\left(2^{-3}\cdot 2^{1}\right)}$$

Step 2 $$\boxed{\left(3^{-2}\cdot 3^{-1}\right)\cdot\left(2^{-3}\cdot 2^{1}\right)} = \boxed{\left(3^{-2-1}\right)\cdot\left(2^{-3+1}\right)} = \boxed{3^{-3}\cdot 2^{-2}}$$

Step 3 $$\boxed{3^{-3}\cdot 2^{-2}} = \boxed{\frac{1}{3^{3}\cdot 2^{2}}} = \boxed{\frac{1}{(3\cdot 3\cdot 3)\cdot(2\cdot 2)}} = \boxed{\mathbf{\frac{1}{108}}}$$

Example 1.3-37

$$\boxed{5^{-2}\cdot 5\cdot a^{-3}\cdot b^{-3}\cdot a^{-1}\cdot b} =$$

Solution:

Step 1 $$\boxed{5^{-2}\cdot 5\cdot a^{-3}\cdot b^{-3}\cdot a^{-1}\cdot b} = \boxed{5^{-2}\cdot 5^{1}\cdot a^{-3}\cdot b^{-3}\cdot a^{-1}\cdot b^{1}} = \boxed{\left(5^{-2}5^{1}\right)\cdot\left(a^{-3}a^{-1}\right)\cdot\left(b^{1}b^{-3}\right)}$$

Step 2 $$\boxed{\left(5^{-2}5^{1}\right)\cdot\left(a^{-3}a^{-1}\right)\cdot\left(b^{1}b^{-3}\right)} = \boxed{\left(5^{-2+1}\right)\cdot\left(a^{-3-1}\right)\cdot\left(b^{1-3}\right)} = \boxed{5^{-1}\cdot a^{-4}\cdot b^{-2}} = \boxed{5^{-1}a^{-4}b^{-2}}$$

Step 3 $$\boxed{5^{-1}a^{-4}b^{-2}} = \boxed{\frac{1}{5^{1}}\cdot\frac{1}{a^{4}}\cdot\frac{1}{b^{2}}} = \boxed{\frac{1\cdot 1\cdot 1}{5\cdot a^{4}\cdot b^{2}}} = \boxed{\mathbf{\frac{1}{5a^{4}b^{2}}}}$$

Additional Examples - Multiplying Negative Integer Exponents

The following examples further illustrate how to multiply negative exponential terms by one another:

Example 1.3-38

$$\boxed{a^{-2}\cdot a^{3}\cdot a^{-1}} = \boxed{a^{-2+3-1}} = \boxed{a^{0}} = \boxed{\mathbf{1}}$$

Example 1.3-39

$$\boxed{3^{-2}\cdot 3^{-1}\cdot 2^{-2}\cdot 3^{-4}\cdot 2} = \boxed{3^{-2}\cdot 3^{-1}\cdot 2^{-2}\cdot 3^{-4}\cdot 2^{1}} = \boxed{\left(3^{-2}\cdot 3^{-1}\cdot 3^{-4}\right)\cdot\left(2^{1}\cdot 2^{-2}\right)} = \boxed{\left(3^{-2-1-4}\right)\cdot\left(2^{1-2}\right)} = \boxed{3^{-7}\cdot 2^{-1}}$$

$$= \boxed{\frac{1}{3^7} \cdot \frac{1}{2^1}} = \boxed{\frac{1}{3 \cdot 3 \cdot 3 \cdot 3 \cdot 3 \cdot 3 \cdot 3} \cdot \frac{1}{2}} = \boxed{\frac{1}{2187} \cdot \frac{1}{2}} = \boxed{\frac{1 \cdot 1}{2187 \cdot 2}} = \boxed{\mathbf{\frac{1}{4374}}}$$

Example 1.3-40

$$\boxed{a^{-5} \cdot b^{-3} \cdot a^{-1} \cdot b^{-2}} = \boxed{\left(a^{-5} \cdot a^{-1}\right) \cdot \left(b^{-3} \cdot b^{-2}\right)} = \boxed{\left(a^{-5-1}\right) \cdot \left(b^{-3-2}\right)} = \boxed{a^{-6} \cdot b^{-5}} = \boxed{a^{-6} b^{-5}} = \boxed{\mathbf{\frac{1}{a^6 b^5}}}$$

Example 1.3-41

$$\boxed{\left(5^{-3} \cdot 5^{-1}\right) \cdot \left(2^{-3} \cdot 5^{-2}\right)} = \boxed{5^{-3} \cdot 5^{-1} \cdot 2^{-3} \cdot 5^{-2}} = \boxed{\left(5^{-3} \cdot 5^{-1} \cdot 5^{-2}\right) \cdot 2^{-3}} = \boxed{\left(5^{-3-1-2}\right) \cdot 2^{-3}} = \boxed{5^{-6} \cdot 2^{-3}}$$

$$= \boxed{\frac{1}{5^6} \cdot \frac{1}{2^3}} = \boxed{\frac{1}{(5 \cdot 5 \cdot 5 \cdot 5 \cdot 5 \cdot 5)} \cdot \frac{1}{(2 \cdot 2 \cdot 2)}} = \boxed{\frac{1}{15625} \cdot \frac{1}{8}} = \boxed{\frac{1 \cdot 1}{(15625 \cdot 8)}} = \boxed{\mathbf{\frac{1}{125000}}}$$

Example 1.3-42

$$\boxed{(-1+3)^{-2}\left(r^{-2} s^2 t\right) \cdot \left(r^3 s^{-2} t^{-3} s\right)} = \boxed{(2)^{-2} r^{-2} s^2 t^1 \cdot r^3 s^{-2} t^{-3} s^1} = \boxed{2^{-2}\left(r^{-2} r^3\right) \cdot \left(s^2 s^{-2} s^1\right) \cdot \left(t^1 t^{-3}\right)}$$

$$= \boxed{2^{-2}\left(r^{-2+3}\right) \cdot \left(s^{2-2+1}\right) \cdot \left(t^{1-3}\right)} = \boxed{2^{-2} r^1 \cdot s^1 \cdot t^{-2}} = \boxed{2^{-2} \cdot r \cdot s \cdot t^{-2}} = \boxed{\frac{1}{2^2} \cdot \frac{r}{1} \cdot \frac{s}{1} \cdot \frac{1}{t^2}} = \boxed{\frac{1 \cdot r \cdot s \cdot 1}{2 \cdot 2 \cdot 1 \cdot 1 \cdot t^2}} = \boxed{\mathbf{\frac{rs}{4t^2}}}$$

Practice Problems - Multiplying Negative Integer Exponents

Section 1.3c Case I Practice Problems - Multiply the following exponential expressions by one another:

1. $\left(3^{-3} \cdot 2^{-1}\right) \cdot \left(2^{-3} \cdot 3^{-2} \cdot 2\right) =$
2. $a^{-6} \cdot b^{-4} \cdot a^{-1} \cdot b^{-2} \cdot a^0 =$
3. $\left(a^{-2} \cdot b^{-3}\right)^2 \cdot \left(a \cdot b^{-2}\right) =$
4. $(-2)^{-4}\left(r^{-2} s^2 t\right) \cdot \left(r^3 s t^{-2} s^{-1}\right) =$
5. $\left(\frac{4}{5}\right)^{-4} 2^2 v^{-5} 2^{-4} v^3 v^{-2} =$
6. $2^{-1} \cdot 3^2 \cdot 3^{-5} \cdot 2^2 \cdot 2^0 =$

Case II Dividing Negative Integer Exponents

Negative integer exponents are divided by one another using the exponent laws I through VI shown in Table 1.3-2. These laws are used in order to simplify division of negative fractional exponents by each other. Negative integer exponents are divided by one another using the following steps:

Step 1 a. Apply the Division and/or the Negative Power Laws (Laws V and VI) from Table 1.3-2.
b. Group the exponential terms with similar bases.

Step 2 Apply the Multiplication Law (Law I) from Table 1.3-2 and simplify the exponential expressions by adding the exponents with similar bases.

Examples with Steps

The following examples show the steps as to how negative integer exponents are divided by one another:

Example 1.3-43

$$\boxed{\frac{5^{-2}}{5^{-3}}} =$$

Solution:

Step 1

$$\boxed{\frac{5^{-2}}{5^{-3}}} = \boxed{\frac{5^{3} \cdot 5^{-2}}{1}} =$$

Step 2

$$\boxed{\frac{5^{3} \cdot 5^{-2}}{1}} = \boxed{\frac{5^{3-2}}{1}} = \boxed{\frac{5^{1}}{1}} = \boxed{\mathbf{5}}$$

Example 1.3-44

$$\boxed{\frac{a^{-2}b^{-3}}{a^{-1}b^{-2}}} =$$

Solution:

Step 1

$$\boxed{\frac{a^{-2}b^{-3}}{a^{-1}b^{-2}}} = \boxed{\frac{1}{\left(a^{2}a^{-1}\right)\cdot\left(b^{3}b^{-2}\right)}}$$

Step 2

$$\boxed{\frac{1}{\left(a^{2}a^{-1}\right)\cdot\left(b^{3}b^{-2}\right)}} = \boxed{\frac{1}{\left(a^{2-1}\right)\cdot\left(b^{3-2}\right)}} = \boxed{\frac{1}{a^{1} \cdot b^{1}}} = \boxed{\mathbf{\frac{1}{ab}}}$$

Additional Examples - Dividing Negative Integer Exponents

The following examples further illustrate how to divide negative integer exponents by one another:

Example 1.3-45

$$\boxed{\frac{a^{-2}c^{-3}}{-ac^{4}}} = \boxed{-\frac{a^{-2}c^{-3}}{a^{1}c^{4}}} = \boxed{-\frac{1}{\left(a^{1}a^{2}\right)\cdot\left(c^{4}c^{3}\right)}} = \boxed{-\frac{1}{\left(a^{1+2}\right)\cdot\left(c^{4+3}\right)}} = \boxed{-\mathbf{\frac{1}{a^{3}c^{7}}}}$$

Example 1.3-46

$$\boxed{\frac{(-3)^{-3}}{-(3)^{-3}}}=\boxed{-\frac{(-3)^{-3}}{(3)^{-3}}}=\boxed{-\frac{(3)^{3}}{(-3)^{3}}}=\boxed{-\frac{3\cdot3\cdot3}{-3\cdot-3\cdot-3}}=\boxed{-\frac{27}{-27}}=\boxed{+\frac{\overset{1}{\cancel{27}}}{\underset{1}{\cancel{27}}}}=\boxed{\frac{1}{1}}=\boxed{\mathbf{1}}$$

Example 1.3-47

$$\boxed{\frac{(-2)^{-4}}{-(-2)^{-3}}}=\boxed{-\frac{(-2)^{-4}}{(-2)^{-3}}}=\boxed{-\frac{(-2)^{3}}{(-2)^{4}}}=\boxed{-\frac{-2\cdot-2\cdot-2}{-2\cdot-2\cdot-2\cdot-2}}=\boxed{-\frac{-8}{16}}=\boxed{+\frac{\overset{1}{\cancel{8}}}{\underset{2}{\cancel{16}}}}=\boxed{\frac{\mathbf{1}}{\mathbf{2}}}$$

Example 1.3-48

$$\boxed{\frac{-2^{-3}\cdot a^{-1}}{(-2)^{-2}a^{-3}}}=\boxed{-\frac{2^{-3}\cdot a^{-1}}{(-2)^{-2}a^{-3}}}=\boxed{-\frac{(-2)^{2}\cdot\left(a^{3}\cdot a^{-1}\right)}{2^{3}}}=\boxed{-\frac{(-2\cdot-2)\cdot\left(a^{3-1}\right)}{2\cdot2\cdot2}}=\boxed{-\frac{4\cdot a^{2}}{8}}=\boxed{-\frac{\overset{1}{\cancel{4}}}{\underset{2}{\cancel{8}}}a^{2}}=\boxed{-\frac{\mathbf{1}}{\mathbf{2}}\boldsymbol{a}^{\mathbf{2}}}$$

Example 1.3-49

$$\boxed{\frac{a^{2}}{2^{-3}\cdot a^{-2}}}=\boxed{\frac{2^{3}\cdot a^{2}\cdot a^{2}}{1}}=\boxed{\frac{8\cdot a^{2+2}}{1}}=\boxed{\frac{8a^{4}}{1}}=\boxed{\mathbf{8a^{4}}}$$

Practice Problems - Dividing Negative Integer Exponents

Section 1.3c Case II Practice Problems - Divide the following negative integer exponents:

1. $\dfrac{x^{-2}x}{x^{3}x^{0}}=$

2. $\dfrac{-2a^{-2}b^{3}}{-6a^{-1}b^{-2}}=$

3. $-\dfrac{-(-3)^{-4}}{3\cdot(-3)^{-3}}=$

4. $\dfrac{-3^{3}y^{-3}yw}{(-3)^{-2}y^{2}w^{-3}}=$

5. $\dfrac{a^{-2}b^{2}a^{-5}y^{-2}}{a^{-3}y}=$

6. $\dfrac{(x\cdot y\cdot z)^{0}\cdot yx^{-2}}{x^{-4}y^{-1}}=$

Case III Adding and Subtracting Negative Integer Exponents

Negative exponential expressions are added and subtracted using the following steps:

Step 1 Group the exponential terms with similar bases.

Step 2 Apply the Negative Power Law (Law VI) from Table 1.3-2, i.e., change a^{-n} to $\frac{1}{a^n}$.

Step 3 Simplify the exponential expression by:

a. Using the fraction techniques learned in Section 1.2, and

b. Using appropriate exponent laws such as the Multiplication Law (Law I) from Table 1.3-2.

Examples with Steps

The following examples show the steps as to how exponential expressions having negative integer exponents are added or subtracted:

Example 1.3-50

$$\boxed{3^{-3}+3^{-2}} =$$

Solution:

Step 1 $\boxed{\textit{Not Applicable}}$

Step 2 $$\boxed{3^{-3}+3^{-2}} = \boxed{\frac{1}{3^3}+\frac{1}{3^2}} = \boxed{\frac{1}{27}+\frac{1}{9}}$$

Step 3 $$\boxed{\frac{1}{27}+\frac{1}{9}} = \boxed{\frac{(1\cdot 9)+(1\cdot 27)}{27\cdot 9}} = \boxed{\frac{9+27}{243}} = \boxed{\frac{\overset{4}{\cancel{36}}}{\underset{27}{\cancel{243}}}} = \boxed{\mathbf{\frac{4}{27}}}$$

Example 1.3-51

$$\boxed{x^{-1}+x^{-2}+2x^{-1}-4x^{-2}+5^{-2}} =$$

Solution:

Step 1 $$\boxed{x^{-1}+x^{-2}+2x^{-1}-4x^{-2}+5^{-2}} = \boxed{\left(x^{-1}+2x^{-1}\right)+\left(x^{-2}-4x^{-2}\right)+5^{-2}}$$

$$= \boxed{(1+2)x^{-1}+(1-4)x^{-2}+5^{-2}} = \boxed{3x^{-1}-3x^{-2}+5^{-2}}$$

Step 2 $$\boxed{3x^{-1}-3x^{-2}+5^{-2}} = \boxed{\frac{3}{x^1}-\frac{3}{x^2}+\frac{1}{5^2}} = \boxed{\frac{3}{x}-\frac{3}{x^2}+\frac{1}{5^2}}$$

Step 3 $$\boxed{\frac{3}{x}-\frac{3}{x^2}+\frac{1}{5^2}} = \boxed{\left(\frac{3}{x}-\frac{3}{x^2}\right)+\frac{1}{25}} = \boxed{\left(\frac{\left(3\cdot x^2\right)-(3\cdot x)}{x\cdot x^2}\right)+\frac{1}{25}} = \boxed{\left(\frac{3x^2-3x}{x^{1+2}}\right)+\frac{1}{25}}$$

$$= \frac{3x^2 - 3x}{x^3} + \frac{1}{25} = \frac{25 \cdot (3x^2 - 3x) + 1 \cdot x^3}{x^3 \cdot 25} = \frac{75x^2 - 75x + x^3}{25x^3} = \frac{x^3 + 75x^2 - 75x}{25x^3}$$

$$= \frac{x(x^2 + 75x - 75)}{25x^3} = \frac{x^2 + 75x - 75}{25(x^3 x^{-1})} = \frac{x^2 + 75x - 75}{25x^{3-1}} = \mathbf{\frac{x^2 + 75x - 75}{25x^2}}$$

Additional Examples - Adding and Subtracting Negative Integer Exponents

The following examples further illustrate addition and subtraction of negative exponential terms:

Example 1.3-52

$$(x+y)^{-5} = \mathbf{\frac{1}{(x+y)^5}}$$

Note: $(x \cdot y)^{-5} = \frac{1}{(x \cdot y)^5} = \mathbf{\frac{1}{x^5 y^5}}$

Example 1.3-53

$$a^{-1} - b^{-1} + 2a^{-1} + 3b^{-1} = (a^{-1} + 2a^{-1}) + (3b^{-1} - b^{-1}) = (1+2)a^{-1} + (3-1)b^{-1} = 3a^{-1} + 2b^{-1} = \frac{3}{a} + \frac{2}{b}$$

$$= \frac{(3 \cdot b) + (2 \cdot a)}{a \cdot b} = \mathbf{\frac{2a + 3b}{ab}}$$

Example 1.3-54

$$a^{-1} + b^{-2} + c^{-3} + 3b^{-2} = a^{-1} + (b^{-2} + 3b^{-2}) + c^{-3} = a^{-1} + (1+3)b^{-2} + c^{-3} = a^{-1} + 4b^{-2} + c^{-3}$$

$$= (a^{-1} + 4b^{-2}) + c^{-3} = \left(\frac{1}{a} + \frac{4}{b^2}\right) + \frac{1}{c^3} = \left(\frac{(1 \cdot b^2) + (4 \cdot a)}{a \cdot b^2}\right) + \frac{1}{c^3} = \left(\frac{b^2 + 4a}{ab^2}\right) + \frac{1}{c^3}$$

$$= \frac{[(b^2 + 4a) \cdot c^3] + (1 \cdot ab^2)}{ab^2 \cdot c^3} = \frac{b^2 \cdot c^3 + 4a \cdot c^3 + ab^2}{ab^2 c^3} = \mathbf{\frac{b^2c^3 + 4ac^3 + ab^2}{ab^2c^3}}$$

Example 1.3-55

$$5x^{-3} + 3x^{-2} + 2x^{-3} - x^{-2} + 5^{-2} = (5x^{-3} + 2x^{-3}) + (3x^{-2} - x^{-2}) + 5^{-2} = (5+2)x^{-3} + (3-1)x^{-2} + 5^{-2}$$

$$= 7x^{-3} + 2x^{-2} + 5^{-2} = \frac{7}{x^3} + \frac{2}{x^2} + \frac{1}{5^2} = \left(\frac{7}{x^3} + \frac{2}{x^2}\right) + \frac{1}{25} = \left(\frac{7 \cdot x^2 + 2 \cdot x^3}{x^3 \cdot x^2}\right) + \frac{1}{25} = \frac{7x^2 + 2x^3}{x^{3+2}} + \frac{1}{25}$$

$$= \frac{x^2(7+2x)}{x^5} + \frac{1}{25} = \frac{7+2x}{x^5 \cdot x^{-2}} + \frac{1}{25} = \frac{7+2x}{x^{5-2}} + \frac{1}{25} = \frac{7+2x}{x^3} + \frac{1}{25} = \frac{[(7+2x) \cdot 25] + (1 \cdot x^3)}{25 \cdot x^3}$$

$$= \boxed{\frac{175+50x+x^3}{25x^3}} = \boxed{\mathbf{\frac{x^3+50x+175}{25x^3}}}$$

Example 1.3-56

$$\boxed{x^{-2}+3x^{-2}+y^{-2}+x^2-4y^{-2}-5^2+2x^2} = \boxed{\left(x^{-2}+3x^{-2}\right)+\left(y^{-2}-4y^{-2}\right)+\left(x^2+2x^2\right)-25}$$

$$= \boxed{(1+3)x^{-2}+(1-4)y^{-2}+(1+2)x^2-25} = \boxed{4x^{-2}-3y^{-2}+3x^2-25} = \boxed{\left(\frac{4}{x^2}-\frac{3}{y^2}\right)+3x^2-25}$$

$$= \boxed{\left(\frac{4\cdot y^2-3\cdot x^2}{x^2\cdot y^2}\right)+\frac{3x^2-25}{1}} = \boxed{\frac{4y^2-3x^2}{x^2y^2}+\frac{3x^2-25}{1}} = \boxed{\frac{\left[\left(4y^2-3x^2\right)\cdot 1\right]+\left[x^2y^2\left(3x^2-25\right)\right]}{x^2y^2}}$$

$$= \boxed{\mathbf{\frac{4y^2-3x^2+3x^4y^2-25x^2y^2}{x^2y^2}}}$$

Practice Problems - Adding and Subtracting Negative Integer Exponents

Section 1.3c Case III Practice Problems - Simplify the following negative integer exponential expressions:

1. $x^{-1}+2x^{-2}+3x^{-1}-6x^{-2} =$
2. $\left(3a^{-4}-b^{-2}\right)+\left(-2a^{-4}+3b^{-2}\right) =$
3. $(xy)^{-1}+y^{-2}+4(xy)^{-1}-3y^{-2}+2^{-3} =$
4. $4x^{-1}+y^{-3}+5y^{-3} =$
5. $m^{-5}-\left(m^{-2}-3m^{-5}+m^0\right)+3m^{-2} =$
6. $\left(a^3\right)^{-2}+\left(a^{-2}b\right)^2-6a^{-6}+3a^{-4}b^2 =$

1.4 Radicals

In this section radical expressions and the steps as to how they are simplified are reviewed in section 1.4a, Cases I through III. In addition, operations involving radical expressions which include multiplication, division (rationalization), addition, and subtraction of radicals are addressed in section 1.4b, Cases I through VI.

1.4a Introduction to Radicals

A description of roots and radicals (Case I), classification of numbers (Case II), and simplification of radical expressions with real numbers as a radicand (Case III) are discussed below:

Case I Roots and Radical Expressions

In the general radical expression $\sqrt[a]{b} = c$, the symbol $\sqrt{\;}$ is called a **radical sign.** The expression under the radical b is called the **radicand,** a is called the **index**, and the positive square root of the number c is called the **principal square root.**
Exponents are a kind of shorthand for multiplication. For example, $5 \times 5 = 25$ can be expressed in exponential form as $5^2 = 25$. Radical signs are used to reverse this process. For example, to write the reverse of $5^2 = 25$ we take the square root of the terms on both sides of the equal sign, i.e., we write $\sqrt{25} = \sqrt{5^2} = 5$. Note that since $5^2 = 25$ and $(-5)^2 = 25$, we use $\sqrt{25}$ to indicate the positive square root of 25 is equal to 5 and $-\sqrt{25}$ to indicate the negative square root of 25 is equal to -5. Table 1.4-1 provides square roots, cube roots, fourth roots, and fifth roots of some common numbers used in solving radical expressions. This table should be used as a reference when simplifying radical terms. The students **are not** encouraged to memorize this table. Following are a few examples on simplifying radical expressions using Table 1.4-1:

a. $\sqrt{64} = \sqrt{8^2} = 8$

b. $-2\sqrt{25} = -2\sqrt{5^2} = -(2 \cdot 5) = -10$

c. $5\sqrt[5]{32} = 5\sqrt[5]{2^5} = (5 \cdot 2) = 10$

d. $\sqrt{125} = \sqrt{25 \cdot 5} = \sqrt{5^2 \cdot 5} = 5\sqrt{5}$

e. $\sqrt[2]{147} = \sqrt{49 \cdot 3} = \sqrt{7^2 \cdot 3} = 7\sqrt{3}$

f. $2\sqrt{32} = 2\sqrt{16 \cdot 2} = 2\sqrt{4^2 \cdot 2} = (2 \cdot 4)\sqrt{2} = 8\sqrt{2}$

g. $\sqrt[5]{2048} = \sqrt[5]{1024 \cdot 2} = \sqrt[5]{4^5 \cdot 2} = 4\sqrt[5]{2}$

h. $\sqrt[3]{375} = \sqrt[3]{125 \cdot 3} = \sqrt[3]{5^3 \cdot 3} = 5\sqrt[3]{3}$

i. $2\sqrt{250} = 2\sqrt{25 \cdot 10} = 2\sqrt{5^2 \cdot 10} = (2 \cdot 5)\sqrt{10} = 10\sqrt{10}$

j. $\sqrt[4]{324} = \sqrt[4]{81 \cdot 4} = \sqrt[4]{3^4 \cdot 4} = 3\sqrt[4]{4}$

k. $\sqrt[3]{648} = \sqrt[3]{216 \cdot 3} = \sqrt[3]{6^3 \cdot 3} = 6\sqrt[3]{3}$

l. $-\sqrt[2]{324} = -\sqrt[2]{81 \cdot 4} = -\sqrt{9^2 \cdot 2^2} = -(9 \cdot 2) = -18$

Practice Problems - Roots and Radical Expressions

Section 1.4a Case I Practice Problems - Simplify the following radical expressions by using Table 1.4-1:

1. $\sqrt[2]{98} =$
2. $3\sqrt{75} =$
3. $\sqrt[3]{125} =$
4. $\sqrt[5]{3125} =$
5. $\sqrt[4]{162} =$
6. $\sqrt[2]{192} =$

Table 1.4-1: Square roots, cube roots, fourth roots, and fifth roots

Square Roots	***Cube Roots***
$\sqrt{1}=\sqrt{1^2}=(1)^{\frac{1}{2}}=\left(1^2\right)^{\frac{1}{2}}=1$ *Note:* $\sqrt[2]{a}=\sqrt{a}$	$\sqrt[3]{1}=\sqrt[3]{1^3}=(1)^{\frac{1}{3}}=\left(1^3\right)^{\frac{1}{3}}=1$
$\sqrt{4}=\sqrt{2^2}=(4)^{\frac{1}{2}}=\left(2^2\right)^{\frac{1}{2}}=2$	$\sqrt[3]{8}=\sqrt[3]{2^3}=(8)^{\frac{1}{3}}=\left(2^3\right)^{\frac{1}{3}}=2$
$\sqrt{9}=\sqrt{3^2}=(9)^{\frac{1}{2}}=\left(3^2\right)^{\frac{1}{2}}=3$	$\sqrt[3]{27}=\sqrt[3]{3^3}=(27)^{\frac{1}{3}}=\left(3^3\right)^{\frac{1}{3}}=3$
$\sqrt{16}=\sqrt{4^2}=(16)^{\frac{1}{2}}=\left(4^2\right)^{\frac{1}{2}}=4$	$\sqrt[3]{64}=\sqrt[3]{4^3}=(64)^{\frac{1}{3}}=\left(4^3\right)^{\frac{1}{3}}=4$
$\sqrt{25}=\sqrt{5^2}=(25)^{\frac{1}{2}}=\left(5^2\right)^{\frac{1}{2}}=5$	$\sqrt[3]{125}=\sqrt[3]{5^3}=(125)^{\frac{1}{3}}=\left(5^3\right)^{\frac{1}{3}}=5$
$\sqrt{36}=\sqrt{6^2}=(36)^{\frac{1}{2}}=\left(6^2\right)^{\frac{1}{2}}=6$	$\sqrt[3]{216}=\sqrt[3]{6^3}=(216)^{\frac{1}{3}}=\left(6^3\right)^{\frac{1}{3}}=6$
$\sqrt{49}=\sqrt{7^2}=(49)^{\frac{1}{2}}=\left(7^2\right)^{\frac{1}{2}}=7$	$\sqrt[3]{343}=\sqrt[3]{7^3}=(343)^{\frac{1}{3}}=\left(7^3\right)^{\frac{1}{3}}=7$
$\sqrt{64}=\sqrt{8^2}=(64)^{\frac{1}{2}}=\left(8^2\right)^{\frac{1}{2}}=8$	$\sqrt[3]{512}=\sqrt[3]{8^3}=(512)^{\frac{1}{3}}=\left(8^3\right)^{\frac{1}{3}}=8$
$\sqrt{81}=\sqrt{9^2}=(81)^{\frac{1}{2}}=\left(9^2\right)^{\frac{1}{2}}=9$	$\sqrt[3]{729}=\sqrt[3]{9^3}=(729)^{\frac{1}{3}}=\left(9^3\right)^{\frac{1}{3}}=9$
$\sqrt{100}=\sqrt{10^2}=(100)^{\frac{1}{2}}=\left(10^2\right)^{\frac{1}{2}}=10$	$\sqrt[3]{1000}=\sqrt[3]{10^3}=(1000)^{\frac{1}{3}}=\left(10^3\right)^{\frac{1}{3}}=10$
Fourth Roots	***Fifth Roots***
$\sqrt[4]{1}=\sqrt[4]{1^4}=(1)^{\frac{1}{4}}=\left(1^4\right)^{\frac{1}{4}}=1$	$\sqrt[5]{1}=\sqrt[5]{1^5}=(1)^{\frac{1}{5}}=\left(1^5\right)^{\frac{1}{5}}=1$
$\sqrt[4]{16}=\sqrt[4]{2^4}=(16)^{\frac{1}{4}}=\left(2^4\right)^{\frac{1}{4}}=2$	$\sqrt[5]{32}=\sqrt[5]{2^5}=(32)^{\frac{1}{5}}=\left(2^5\right)^{\frac{1}{5}}=2$
$\sqrt[4]{81}=\sqrt[4]{3^4}=(81)^{\frac{1}{4}}=\left(3^4\right)^{\frac{1}{4}}=3$	$\sqrt[5]{243}=\sqrt[5]{3^5}=(243)^{\frac{1}{5}}=\left(3^5\right)^{\frac{1}{5}}=3$
$\sqrt[4]{256}=\sqrt[4]{4^4}=(256)^{\frac{1}{4}}=\left(4^4\right)^{\frac{1}{4}}=4$	$\sqrt[5]{1024}=\sqrt[5]{4^5}=(1024)^{\frac{1}{5}}=\left(4^5\right)^{\frac{1}{5}}=4$
$\sqrt[4]{625}=\sqrt[4]{5^4}=(625)^{\frac{1}{4}}=\left(5^4\right)^{\frac{1}{4}}=5$	$\sqrt[5]{3125}=\sqrt[5]{5^5}=(3125)^{\frac{1}{5}}=\left(5^5\right)^{\frac{1}{5}}=5$
$\sqrt[4]{1296}=\sqrt[4]{6^4}=(1296)^{\frac{1}{4}}=\left(6^4\right)^{\frac{1}{4}}=6$	$\sqrt[5]{7776}=\sqrt[5]{6^5}=(7776)^{\frac{1}{5}}=\left(6^5\right)^{\frac{1}{5}}=6$
$\sqrt[4]{2401}=\sqrt[4]{7^4}=(2401)^{\frac{1}{4}}=\left(7^4\right)^{\frac{1}{4}}=7$	$\sqrt[5]{16807}=\sqrt[5]{7^5}=(16807)^{\frac{1}{5}}=\left(7^5\right)^{\frac{1}{5}}=7$
$\sqrt[4]{4096}=\sqrt[4]{8^4}=(4096)^{\frac{1}{4}}=\left(8^4\right)^{\frac{1}{4}}=8$	$\sqrt[5]{32768}=\sqrt[5]{8^5}=(32768)^{\frac{1}{5}}=\left(8^5\right)^{\frac{1}{5}}=8$
$\sqrt[4]{6561}=\sqrt[4]{9^4}=(6561)^{\frac{1}{4}}=\left(9^4\right)^{\frac{1}{4}}=9$	$\sqrt[5]{59049}=\sqrt[5]{9^5}=(59049)^{\frac{1}{5}}=\left(9^5\right)^{\frac{1}{5}}=9$
$\sqrt[4]{10000}=\sqrt[4]{10^4}=(10000)^{\frac{1}{4}}=\left(10^4\right)^{\frac{1}{4}}=10$	$\sqrt[5]{100000}=\sqrt[5]{10^5}=(100000)^{\frac{1}{5}}=\left(10^5\right)^{\frac{1}{5}}=10$

Case II Rational, Irrational, Real, and Imaginary Numbers

A **rational number** is a number that **can** be expressed as:

1. An integer fraction $\frac{a}{b}$, where a and b are integer numbers and $b \neq 0$. For example: $\frac{3}{8}$, $-\frac{4}{5}$, $\frac{25}{100}$, and $\frac{2}{7}$ are rational numbers.
2. The square root of a perfect square, the cube root of a perfect cube, etc. For example: $\sqrt{36} = \sqrt{6^2} = 6$, $\sqrt{49} = \sqrt{7^2} = 7$, $-\sqrt[3]{125} = -\sqrt[3]{5^3} = -5$, $\sqrt[4]{81} = \sqrt[4]{3^4} = 3$, and $-\sqrt[5]{1024} = -\sqrt[5]{4^5} = -4$ are rational numbers.
3. An integer (a whole number). For example: $5 = \frac{5}{1}$, 0, $\frac{0}{28} = 0$, and 10 are rational numbers.
4. A terminating decimal. For example: $0.25 = \frac{25}{100}$, -0.75, and $5.5 = -5\frac{1}{2}$ are rational numbers.
5. A repeating decimal. For example: $0.3333333... = \frac{1}{3}$, $0.45454545...$, and are rational numbers.

An **irrational number** is a number that:

1. Can not be expressed as an integer fraction $\frac{a}{b}$, where a and b are integer numbers and $b \neq 0$. For example: π, $\frac{2}{\sqrt{2}}$, and $-\frac{5}{\sqrt{3}}$ are irrational numbers.
2. Can not be expressed as the square root of a perfect square, the cube root of a perfect cube, etc. For example: $\sqrt{5}$, $-\sqrt{7}$, $\sqrt{12}$, $\sqrt[3]{4}$, $-\sqrt[5]{6}$, and $\sqrt{3}$ are irrational numbers.
3. Is not a terminating or repeating decimal. For example: $0.432643...$, $-8.346723...$, and $3.14159...$ are irrational numbers.

The **real numbers** consist of all the rational and irrational numbers. For example: π, $\frac{2}{\sqrt{2}}$, $-\sqrt{7}$, $\sqrt[3]{4}$, $-\sqrt[5]{6}$, $\sqrt{3}$, $\sqrt{36} = \sqrt{6^2} = 6$, $0.25 = \frac{25}{100}$, -0.75, $-5.5 = -5\frac{1}{2}$, $-3.8 = -3\frac{4}{5}$, $5 = \frac{5}{1}$, 0, and $-25 = -\frac{25}{1}$ are real numbers.

The **not real numbers** or **imaginary numbers** are square root of any negative real number. For example: $\sqrt{-15}$, $\sqrt{-9}$, $\sqrt{-45}$, and $\sqrt{-36}$ are imaginary numbers. Note that imaginary numbers are also shown as $\sqrt{-15} = \sqrt{15}i$, $\sqrt{-9} = \sqrt{9}i$, $\sqrt{-45} = \sqrt{45}i$, and $\sqrt{-36} = \sqrt{36}i$ in more advanced math books.

Practice Problems - Rational, Irrational, Real, and Imaginary Numbers

Section 1.4a Case II Practice Problems - Identify which one of the following numbers are rational, irrational, real, or not real:

1. $\frac{5}{8} =$
2. $\sqrt{45} =$
3. $450 =$
4. $-\frac{2}{\sqrt{10}} =$
5. $-\sqrt{-5} =$
6. $\frac{\sqrt{5}}{-2} =$

Case III Simplifying Radical Expressions with Real Numbers as a Radicand

Radical expressions with a real number as radicand are simplified using the following general rule:

$$\sqrt[n]{a^n} = a^{\frac{n}{n}} = a \qquad \text{The } n^{th} \text{ root of } a^n \text{ is } a$$

Where a is a positive real number and n is an integer.

Radicals of the form $\sqrt[n]{a^n} = a$ are simplified using the following steps:

Step 1 Factor out the radicand a^n to a perfect square, cube, fourth, fifth, etc. term (use Table 1.4-1). Write any term under the radical that exceeds the index n as multiple sum of the index.

Step 2 Use the Multiplication Law for exponents (see Section 1.3) by writing a^{m+n} in the form of $a^m \cdot a^n$.

Step 3 Simplify the radical expression by using the general rule $\sqrt[n]{a^n} = a$. Note that any term under the radical which is less than the index n stays inside the radical.

Examples with Steps

The following examples show the steps as to how radical expressions with real terms are simplified:

Example 1.4-1

$$\boxed{\frac{1}{-8}\sqrt[2]{72}} =$$

Solution:

Step 1

$$\boxed{\frac{1}{-8}\sqrt[2]{72}} = \boxed{-\frac{1}{8}\sqrt{72}} = \boxed{-\frac{1}{8}\sqrt{36\cdot 2}} = \boxed{-\frac{1}{8}\sqrt{(6\cdot 6)\cdot 2}} = \boxed{-\frac{1}{8}\sqrt{\left(6^1\cdot 6^1\right)\cdot 2}}$$

$$= \boxed{-\frac{1}{8}\sqrt{\left(6^{1+1}\right)\cdot 2}} = \boxed{-\frac{1}{8}\sqrt{6^2\cdot 2}}$$

Step 2 *Not Applicable*

Step 3

$$\boxed{-\frac{1}{8}\sqrt{6^2\cdot 2}} = \boxed{-\frac{1}{8}\cdot 6\sqrt{2}} = \boxed{-\frac{\overset{3}{\cancel{6}}}{\underset{4}{\cancel{8}}}\sqrt{2}} = \boxed{-\frac{3}{4}\sqrt{2}}$$

Example 1.4-2

$$\boxed{\frac{-3}{10}\sqrt[4]{5^7\cdot 4}} =$$

Solution:

Step 1

$$\boxed{\frac{-3}{10}\sqrt[4]{5^7\cdot 4}} = \boxed{-\frac{3}{10}\sqrt[4]{5^{4+3}\cdot 4}}$$

Step 2　$\boxed{-\frac{3}{10}\sqrt[4]{5^{4+3}\cdot 4}} = \boxed{-\frac{3}{10}\sqrt[4]{5^4\cdot 5^3\cdot 4}}$

Step 3　$\boxed{-\frac{3}{10}\sqrt[4]{5^4\cdot 5^3\cdot 4}} = \boxed{-\frac{3}{10}\cdot 5\sqrt[4]{5^3\cdot 4}} = \boxed{-\frac{\overset{3}{\cancel{15}}}{\underset{2}{\cancel{10}}}\sqrt[4]{5^3\cdot 4}} = \boxed{-\frac{3}{2}\sqrt[4]{125\cdot 4}} = \boxed{\mathbf{-\frac{3}{2}\sqrt[4]{500}}}$

Additional Examples - Simplifying Radical Expressions with Real Numbers as a Radicand

The following examples further illustrate how to solve radical expressions with real numbers as radicand:

Example 1.4-3

$\boxed{\sqrt[2]{25}} = \boxed{\sqrt{25}} = \boxed{\sqrt{5\cdot 5}} = \boxed{\sqrt{5^1\cdot 5^1}} = \boxed{\sqrt{5^{1+1}}} = \boxed{\sqrt{5^2}} = \boxed{\mathbf{5}}$

Example 1.4-4

$\boxed{\sqrt[2]{18}} = \boxed{\sqrt{18}} = \boxed{\sqrt{9\cdot 2}} = \boxed{\sqrt{(3\cdot 3)\cdot 2}} = \boxed{\sqrt{\left(3^1\cdot 3^1\right)\cdot 2}} = \boxed{\sqrt{\left(3^{1+1}\right)\cdot 2}} = \boxed{\sqrt{3^2\cdot 2}} = \boxed{\mathbf{3\sqrt{2}}}$

Example 1.4-5

$\boxed{\sqrt[2]{125}} = \boxed{\sqrt{125}} = \boxed{\sqrt{25\cdot 5}} = \boxed{\sqrt{(5\cdot 5)\cdot 5}} = \boxed{\sqrt{\left(5^1\cdot 5^1\right)\cdot 5}} = \boxed{\sqrt{\left(5^{1+1}\right)\cdot 5}} = \boxed{\sqrt{5^2\cdot 5}} = \boxed{\mathbf{5\sqrt{5}}}$

Example 1.4-6

$\boxed{\sqrt[2]{2400}} = \boxed{\sqrt{2400}} = \boxed{\sqrt{400\cdot 6}} = \boxed{\sqrt{(20\cdot 20)\cdot 6}} = \boxed{\sqrt{\left(20^1\cdot 20^1\right)\cdot 6}} = \boxed{\sqrt{\left(20^{1+1}\right)\cdot 6}} = \boxed{\sqrt{20^2\cdot 6}} = \boxed{\mathbf{20\sqrt{6}}}$

Example 1.4-7

$\boxed{\sqrt[2]{120}} = \boxed{\sqrt{120}} = \boxed{\sqrt{4\cdot 30}} = \boxed{\sqrt{(2\cdot 2)\cdot 30}} = \boxed{\sqrt{\left(2^1\cdot 2^1\right)\cdot 30}} = \boxed{\sqrt{\left(2^{1+1}\right)\cdot 30}} = \boxed{\sqrt{2^2\cdot 30}} = \boxed{\mathbf{2\sqrt{30}}}$

Example 1.4-8

$\boxed{\frac{2}{15}\sqrt[2]{50}} = \boxed{\frac{2}{15}\sqrt{50}} = \boxed{\frac{2}{15}\sqrt{25\cdot 2}} = \boxed{\frac{2}{15}\sqrt{(5\cdot 5)\cdot 2}} = \boxed{\frac{2}{15}\sqrt{\left(5^1\cdot 5^1\right)\cdot 2}} = \boxed{\frac{2}{15}\sqrt{\left(5^{1+1}\right)\cdot 2}} = \boxed{\frac{2}{15}\sqrt{5^2\cdot 2}}$

$= \boxed{\frac{2}{15}\cdot 5\sqrt{2}} = \boxed{\frac{\overset{2}{\cancel{10}}}{\underset{3}{\cancel{15}}}\sqrt{2}} = \boxed{\mathbf{\frac{2}{3}\sqrt{2}}}$

Practice Problems - Simplifying Radical Expressions with Real Numbers as a Radicand

Section 1.4a Case III Practice Problems - Simplify the following radical expressions:

1. $-\sqrt{49} =$
2. $\sqrt{54} =$
3. $-\sqrt{500} =$
4. $\sqrt[5]{3^5\cdot 5} =$
5. $\sqrt[2]{216} =$
6. $-\frac{1}{4}\sqrt[4]{4^5\cdot 2} =$

1.4b Operations Involving Radical Expressions

In this section multiplication, division, addition, and subtraction of monomial and binomial radical expressions, with real numbers, is addressed in Cases I through VI.

Case I Multiplying Monomial Expressions in Radical Form, with Real Numbers

Radicals are multiplied by each other by using the following general product rule:

$$a\sqrt[n]{x}\cdot b\sqrt[n]{y}\cdot c\sqrt[n]{z} = (a\cdot b\cdot c)\sqrt[n]{x\cdot y\cdot z} = abc\sqrt[n]{xyz}$$

Note that radicals can only be multiplied by each other if they have the same index n.

A monomial expression in radical form is defined as:

$$\sqrt{8x^5}\,,\ \sqrt{y}\,,\ \sqrt{27}\,,\ -3\sqrt[5]{x^6y^7}\,,\ \sqrt[3]{x^2y^5}\,,\ 2\sqrt{125}\,,\ etc.$$

A binomial expression in radical form is defined as:

$$\sqrt{x}+\sqrt{y}\,,\ 1+\sqrt{8x}\,,\ xy+\sqrt{y^3}\,,\ x^3y^3-\sqrt{x^2y}\,,\ 9-\sqrt[5]{x^5y^6}\,,\ \sqrt[3]{64}+\sqrt[3]{x^2y^5}\,,\ etc.$$

Monomial expressions in radical form are multiplied by each other using the above general product rule. Radical expressions with real numbers as radicands are multiplied by each other using the following steps:

Step 1 Simplify the radical terms (see Section 1.4a, Case III).

Step 2 Multiply the radical terms by using the product rule. Repeat Step 1, if necessary.

$$k_1\sqrt[n]{a}\cdot k_2\sqrt[n]{b}\cdot k_3\sqrt[n]{c} = (k_1\cdot k_2\cdot k_3)\sqrt[n]{a\cdot b\cdot c} = k_1k_2k_3\sqrt[n]{abc} \qquad a,\ b,\ and\ c \ge 0$$

Examples with Steps

The following examples show the steps as to how radical expressions in monomial form are multiplied by one another:

Example 1.4-9

$$\boxed{\sqrt{5}\cdot\sqrt{15}} =$$

Solution:

Step 1 $\boxed{\textit{Not Applicable}}$

Step 2 $\boxed{\sqrt{5}\cdot\sqrt{15}} = \boxed{\sqrt{5\cdot 15}} = \boxed{\sqrt{75}} = \boxed{\sqrt{25\cdot 3}} = \boxed{\sqrt{5^2\cdot 3}} = \boxed{\mathbf{5\sqrt{3}}}$

Example 1.4-10

$$\boxed{\sqrt{98}\cdot\sqrt{48}\cdot\sqrt{108}} =$$

Solution:

Step 1 $\boxed{\sqrt{98}\cdot\sqrt{48}\cdot\sqrt{108}} = \boxed{\sqrt{49\cdot 2}\cdot\sqrt{16\cdot 3}\cdot\sqrt{36\cdot 3}} = \boxed{\sqrt{7^2\cdot 2}\cdot\sqrt{4^2\cdot 3}\cdot\sqrt{6^2\cdot 3}}$

$= \boxed{7\sqrt{2}\cdot 4\sqrt{3}\cdot 6\sqrt{3}}$

Step 2 $\boxed{7\sqrt{2}\cdot 4\sqrt{3}\cdot 6\sqrt{3}} = \boxed{(7\cdot 4\cdot 6)\cdot\left(\sqrt{2}\cdot\sqrt{3}\cdot\sqrt{3}\right)} = \boxed{168\left(\sqrt{2\cdot 3\cdot 3}\right)} = \boxed{168\sqrt{2\cdot 3^2}}$

$$= \boxed{(168 \cdot 3)\sqrt{2}} = \boxed{\mathbf{504\sqrt{2}}}$$

Additional Examples - Multiplying Monomial Expressions in Radical Form, with Real Numbers

The following examples further illustrate how to multiply radical terms by one another:

Example 1.4-11

$$\boxed{\sqrt{12} \cdot \sqrt{9}} = \boxed{\left(\sqrt{4 \cdot 3}\right) \cdot \sqrt{3^2}} = \boxed{\left(\sqrt{2^2 \cdot 3}\right) \cdot 3} = \boxed{2\sqrt{3} \cdot 3} = \boxed{(2 \cdot 3)\sqrt{3}} = \boxed{\mathbf{6\sqrt{3}}}$$

Example 1.4-12

$$\boxed{\sqrt{20} \cdot \sqrt{50}} = \boxed{\sqrt{4 \cdot 5} \cdot \sqrt{25 \cdot 2}} = \boxed{\sqrt{2^2 \cdot 5} \cdot \sqrt{5^2 \cdot 2}} = \boxed{2\sqrt{5} \cdot 5\sqrt{2}} = \boxed{(2 \cdot 5) \cdot \left(\sqrt{5 \cdot 2}\right)} = \boxed{\mathbf{10\sqrt{10}}}$$

Note that we can also simplify the radical terms in the following way:

$$\boxed{\sqrt{20} \cdot \sqrt{50}} = \boxed{\sqrt{20 \cdot 50}} = \boxed{\sqrt{1000}} = \boxed{\sqrt{100 \cdot 10}} = \boxed{\sqrt{10^2 \cdot 10}} = \boxed{\mathbf{10\sqrt{10}}}$$

Example 1.4-13

$$\boxed{\sqrt{50} \cdot \sqrt{32} \cdot \sqrt{3}} = \boxed{\sqrt{25 \cdot 2} \cdot \sqrt{16 \cdot 2} \cdot \sqrt{3}} = \boxed{\sqrt{5^2 \cdot 2} \cdot \sqrt{4^2 \cdot 2} \cdot \sqrt{3}} = \boxed{5\sqrt{2} \cdot 4\sqrt{2} \cdot \sqrt{3}} = \boxed{(5 \cdot 4) \cdot \left(\sqrt{2} \cdot \sqrt{2} \cdot \sqrt{3}\right)}$$

$$= \boxed{20 \cdot \left(\sqrt{2 \cdot 2 \cdot 3}\right)} = \boxed{20 \cdot \left(\sqrt{2^1 \cdot 2^1 \cdot 3}\right)} = \boxed{20 \cdot \left(\sqrt{2^{1+1} \cdot 3}\right)} = \boxed{20\sqrt{2^2 \cdot 3}} = \boxed{(20 \cdot 2)\sqrt{3}} = \boxed{\mathbf{40\sqrt{3}}}$$

Example 1.4-14

$$\boxed{\sqrt[3]{128} \cdot \sqrt[3]{500}} = \boxed{\sqrt[3]{2 \cdot 64} \cdot \sqrt[3]{4 \cdot 125}} = \boxed{\sqrt[3]{2 \cdot 4^3} \cdot \sqrt[3]{4 \cdot 5^3}} = \boxed{4\sqrt[3]{2} \cdot 5\sqrt[3]{4}} = \boxed{(4 \cdot 5) \cdot \left(\sqrt[3]{2} \cdot \sqrt[3]{4}\right)} = \boxed{20 \cdot \left(\sqrt[3]{2 \cdot 4}\right)}$$

$$= \boxed{20 \cdot \sqrt[3]{8}} = \boxed{20 \cdot \sqrt[3]{2^3}} = \boxed{20 \cdot 2} = \boxed{\mathbf{40}}$$

Example 1.4-15

$$\boxed{\sqrt[4]{54} \cdot \sqrt[4]{48}} = \boxed{\sqrt[4]{2 \cdot 27} \cdot \sqrt[4]{3 \cdot 16}} = \boxed{\sqrt[4]{2 \cdot 3^3} \cdot \sqrt[4]{3 \cdot 2^4}} = \boxed{\sqrt[4]{2 \cdot 3^3} \cdot 2\sqrt[4]{3^1}} = \boxed{2\sqrt[4]{2 \cdot 3^3 \cdot 3^1}} = \boxed{2\sqrt[4]{2 \cdot 3^{3+1}}} = \boxed{2\sqrt[4]{2 \cdot 3^4}}$$

$$= \boxed{(2 \cdot 3) \cdot \sqrt[4]{2}} = \boxed{\mathbf{6\sqrt[4]{2}}}$$

Practice Problems - Multiplying Monomial Expressions in Radical Form, with Real Numbers

Section 1.4b Case I Practice Problems - Multiply the following radical expressions:

1. $\sqrt{72} \cdot \sqrt{75} =$
2. $-3\sqrt{20} \cdot 2\sqrt{32} =$
3. $\sqrt[3]{16} \cdot \sqrt[3]{27} =$
4. $\sqrt{64} \cdot \sqrt{100} \cdot \sqrt{54} =$
5. $-\sqrt{125} \cdot -2\sqrt{98} =$
6. $\sqrt[4]{625} \cdot \sqrt[4]{324} \cdot \sqrt[4]{48} =$

Case II Multiplying Binomial Expressions in Radical Form, with Real Numbers

To multiply two binomial radical expressions the following multiplication method known as the **FOIL** method needs to be memorized:

$$(a+b)\cdot(c+d)=(a\cdot c)+(a\cdot d)+(b\cdot c)+(b\cdot d)$$

Multiply the **F**irst two terms, i.e., $(a\cdot c)$.
Multiply the **O**uter two terms, i.e., $(a\cdot d)$.
Multiply the **I**nner two terms, i.e., $(b\cdot c)$.
Multiply the **L**ast two terms, i.e., $(b\cdot d)$.

Examples:

1. $\boxed{\left(\sqrt{u}+\sqrt{v}\right)\cdot\left(\sqrt{u}-\sqrt{v}\right)}=\boxed{\left(\sqrt{u}\cdot\sqrt{u}\right)-\left(\sqrt{u}\sqrt{v}\right)+\left(\sqrt{v}\cdot\sqrt{u}\right)-\left(\sqrt{v}\cdot\sqrt{v}\right)}=\boxed{\left(\sqrt{u\cdot u}\right)-\left(\sqrt{u\cdot v}\right)+\left(\sqrt{u\cdot v}\right)-\left(\sqrt{v\cdot v}\right)}$

   $=\boxed{\sqrt{u^2}-\sqrt{v^2}}=\boxed{\mathbf{u-v}}$

2. $\boxed{\left(3-\sqrt{5}\right)\cdot\left(5+\sqrt{7}\right)}=\boxed{(3\cdot5)+\left(3\cdot\sqrt{7}\right)-\left(5\cdot\sqrt{5}\right)-\left(\sqrt{5}\cdot\sqrt{7}\right)}=\boxed{15+3\sqrt{7}-5\sqrt{5}-\sqrt{5\cdot7}}=\boxed{\mathbf{15+3\sqrt{7}-5\sqrt{5}-\sqrt{35}}}$

Binomial radical expressions are multiplied by each other using the following steps:

Step 1 Simplify the radical terms (see Section 1.4a, Case III).

Step 2 Use the FOIL method to multiply each term. Repeat Step 1, if necessary.

$$(a+b)\cdot(c+d)=(a\cdot c)+(a\cdot d)+(b\cdot c)+(b\cdot d)$$

Examples with Steps

The following examples show the steps as to how binomial radical expressions with real numbers as radicands are multiplied by one another:

Example 1.4-16

$$\boxed{\left(2+\sqrt{2}\right)\cdot\left(5-\sqrt{8}\right)}=$$

Solution:

Step 1 $\boxed{\left(2+\sqrt{2}\right)\cdot\left(5-\sqrt{8}\right)}=\boxed{\left(2+\sqrt{2}\right)\cdot\left(5-\sqrt{4\cdot2}\right)}=\boxed{\left(2+\sqrt{2}\right)\cdot\left(5-\sqrt{2^2\cdot2}\right)}$

$=\boxed{\left(2+\sqrt{2}\right)\cdot\left(5-2\sqrt{2}\right)}$

Step 2 $\boxed{\left(2+\sqrt{2}\right)\cdot\left(5-2\sqrt{2}\right)}=\boxed{(2\cdot5)-\left(2\cdot2\sqrt{2}\right)+\left(5\cdot\sqrt{2}\right)-\left(2\sqrt{2}\cdot\sqrt{2}\right)}$

$=\boxed{10-4\sqrt{2}+5\sqrt{2}-2\sqrt{2\cdot2}}=\boxed{10-4\sqrt{2}+5\sqrt{2}-2\sqrt{2^2}}=\boxed{10+(-4+5)\sqrt{2}-(2\cdot2)}$

$$= \boxed{10+\sqrt{2}-4} = \boxed{(10-4)+\sqrt{2}} = \boxed{\mathbf{6+\sqrt{2}}}$$

Example 1.4-17

$$\boxed{\left(\sqrt{24}+3\sqrt{60}\right)\cdot\left(\sqrt{25}-\sqrt{72}\right)} =$$

Solution:

Step 1

$$\boxed{\left(\sqrt{24}+3\sqrt{60}\right)\cdot\left(\sqrt{25}-\sqrt{72}\right)} = \boxed{\left(\sqrt{4\cdot 6}+3\sqrt{4\cdot 15}\right)\cdot\left(\sqrt{5\cdot 5}-\sqrt{36\cdot 2}\right)}$$

$$= \boxed{\left(\sqrt{2^2\cdot 6}+3\sqrt{2^2\cdot 15}\right)\cdot\left(\sqrt{5^2}-\sqrt{6^2\cdot 2}\right)} = \boxed{\left(2\sqrt{6}+(3\cdot 2)\sqrt{15}\right)\cdot\left(5-6\sqrt{2}\right)}$$

$$= \boxed{\left(2\sqrt{6}+6\sqrt{15}\right)\cdot\left(5-6\sqrt{2}\right)}$$

Step 2

$$\boxed{\left(2\sqrt{6}+6\sqrt{15}\right)\cdot\left(5-6\sqrt{2}\right)} = \boxed{\left(5\cdot 2\sqrt{6}\right)-\left(2\sqrt{6}\cdot 6\sqrt{2}\right)+\left(5\cdot 6\sqrt{15}\right)-\left(6\sqrt{15}\cdot 6\sqrt{2}\right)}$$

$$= \boxed{10\sqrt{6}-(2\cdot 6)\sqrt{6\cdot 2}+30\sqrt{15}-(6\cdot 6)\sqrt{15\cdot 2}} = \boxed{10\sqrt{6}-12\sqrt{12}+30\sqrt{15}-36\sqrt{30}}$$

$$= \boxed{10\sqrt{6}-12\sqrt{4\cdot 3}+30\sqrt{15}-36\sqrt{30}} = \boxed{10\sqrt{6}-12\sqrt{2^2\cdot 3}+30\sqrt{15}-36\sqrt{30}}$$

$$= \boxed{10\sqrt{6}-(12\cdot 2)\sqrt{3}+30\sqrt{15}-36\sqrt{30}} = \boxed{\mathbf{10\sqrt{6}-24\sqrt{3}+30\sqrt{15}-36\sqrt{30}}}$$

Additional Examples - Multiplying Binomial Expressions in Radical Form, with Real Numbers

The following examples further illustrate how to multiply radical expressions by one another:

Example 1.4-18

$$\boxed{\left(3+\sqrt{300}\right)\cdot\left(8-\sqrt{50}\right)} = \boxed{\left(3+\sqrt{100\cdot 3}\right)\cdot\left(8-\sqrt{25\cdot 2}\right)} = \boxed{\left(3+\sqrt{10^2\cdot 3}\right)\cdot\left(8-\sqrt{5^2\cdot 2}\right)} = \boxed{\left(3+10\sqrt{3}\right)\cdot\left(8-5\sqrt{2}\right)}$$

$$= \boxed{(3\cdot 8)-(3\cdot 5)\sqrt{2}+(8\cdot 10)\sqrt{3}-(10\cdot 5)\sqrt{3}\cdot\sqrt{2}} = \boxed{24-15\sqrt{2}+80\sqrt{3}-50\sqrt{3\cdot 2}} = \boxed{\mathbf{-15\sqrt{2}+80\sqrt{3}-50\sqrt{6}+24}}$$

Example 1.4-19

$$\boxed{\left(3+\sqrt{12}\right)\cdot\left(\sqrt{75}-\sqrt{2}\right)} = \boxed{\left(3+\sqrt{4\cdot 3}\right)\cdot\left(\sqrt{25\cdot 3}-\sqrt{2}\right)} = \boxed{\left(3+\sqrt{2^2\cdot 3}\right)\cdot\left(\sqrt{5^2\cdot 3}-\sqrt{2}\right)}$$

$$= \boxed{\left(3+2\sqrt{3}\right)\cdot\left(5\sqrt{3}-\sqrt{2}\right)} = \boxed{(3\cdot 5)\sqrt{3}-\left(3\cdot\sqrt{2}\right)+(2\cdot 5)\left(\sqrt{3}\cdot\sqrt{3}\right)-2\left(\sqrt{3}\cdot\sqrt{2}\right)} = \boxed{15\sqrt{3}-3\sqrt{2}+10\sqrt{3\cdot 3}-2\sqrt{3\cdot 2}}$$

$$= \boxed{15\sqrt{3}-3\sqrt{2}+10\sqrt{3^2}-2\sqrt{6}} = \boxed{15\sqrt{3}-3\sqrt{2}+(10\cdot 3)-2\sqrt{6}} = \boxed{\mathbf{15\sqrt{3}-3\sqrt{2}-2\sqrt{6}+30}}$$

Example 1.4-20

$$\boxed{\left(\sqrt[3]{3\cdot 5^3}-\sqrt[3]{4\cdot 2^3}\right)\cdot\left(\sqrt[3]{3}-\sqrt[3]{2\cdot 2^3}\right)} = \boxed{\left(5\sqrt[3]{3}-2\sqrt[3]{4}\right)\cdot\left(\sqrt[3]{3}-2\sqrt[3]{2}\right)}$$

$= \boxed{\left(5\sqrt[3]{3}\cdot\sqrt[3]{3}\right)-(5\cdot 2)\left(\sqrt[3]{3}\cdot\sqrt[3]{2}\right)-\left(2\sqrt[3]{4}\cdot\sqrt[3]{3}\right)+(2\cdot 2)\left(\sqrt[3]{4}\cdot\sqrt[3]{2}\right)} = \boxed{\left(5\sqrt[3]{3\cdot 3}\right)-10\left(\sqrt[3]{3\cdot 2}\right)-\left(2\sqrt[3]{4\cdot 3}\right)+4\left(\sqrt[3]{4\cdot 2}\right)}$

$= \boxed{5\sqrt[3]{9}-10\sqrt[3]{6}-2\sqrt[3]{12}+4\sqrt[3]{8}} = \boxed{5\sqrt[3]{9}-10\sqrt[3]{6}-2\sqrt[3]{12}+4\sqrt[3]{2^3}} = \boxed{5\sqrt[3]{9}-10\sqrt[3]{6}-2\sqrt[3]{12}+(4\cdot 2)}$

$= \boxed{\mathbf{5\sqrt[3]{9}-10\sqrt[3]{6}-2\sqrt[3]{12}+8}}$

Example 1.4-21

$\boxed{\left(6\sqrt{48}+2\right)\cdot\left(2\sqrt{18}-4\right)} = \boxed{\left(6\sqrt{3\cdot 16}+2\right)\cdot\left(2\sqrt{2\cdot 9}-4\right)} = \boxed{\left(6\sqrt{3\cdot 4^2}+2\right)\cdot\left(2\sqrt{2\cdot 3^2}-4\right)}$

$= \boxed{\left(6\cdot 4\sqrt{3}+2\right)\cdot\left(2\cdot 3\sqrt{2}-4\right)} = \boxed{\left(24\sqrt{3}+2\right)\cdot\left(6\sqrt{2}-4\right)} = \boxed{\left(24\sqrt{3}\cdot 6\sqrt{2}\right)-\left(4\cdot 24\sqrt{3}\right)+\left(2\cdot 6\sqrt{2}\right)-(2\cdot 4)}$

$= \boxed{\left(24\cdot 6\sqrt{3\cdot 2}\right)-\left(96\sqrt{3}\right)+\left(12\sqrt{2}\right)-8} = \boxed{\mathbf{144\sqrt{6}-96\sqrt{3}+12\sqrt{2}-8}}$

Example 1.4-22

$\boxed{\left[\left(-\sqrt{3}+2\right)\cdot\left(3-\sqrt{3}\right)\right]\cdot\left(\sqrt{3}-4\right)} = \boxed{\left[\left(-3\cdot\sqrt{3}\right)+\left(\sqrt{3}\cdot\sqrt{3}\right)+(2\cdot 3)-\left(2\cdot\sqrt{3}\right)\right]\cdot\left(\sqrt{3}-4\right)}$

$= \boxed{\left[-3\sqrt{3}+\sqrt{3\cdot 3}+6-2\sqrt{3}\right]\cdot\left(\sqrt{3}-4\right)} = \boxed{\left[\left(-3\sqrt{3}-2\sqrt{3}\right)+\sqrt{3^2}+6\right]\cdot\left(\sqrt{3}-4\right)} = \boxed{\left[-5\sqrt{3}+3+6\right]\cdot\left(\sqrt{3}-4\right)}$

$= \boxed{\left[-5\sqrt{3}+9\right]\cdot\left(\sqrt{3}-4\right)} = \boxed{\left(-5\sqrt{3}\cdot\sqrt{3}\right)+(5\cdot 4)\sqrt{3}+9\cdot\sqrt{3}-(9\cdot 4)} = \boxed{-5\sqrt{3\cdot 3}+20\sqrt{3}+9\sqrt{3}-36}$

$= \boxed{-5\sqrt{3^2}+(20+9)\sqrt{3}-36} = \boxed{-(5\cdot 3)+29\sqrt{3}-36} = \boxed{-15-36+29\sqrt{3}} = \boxed{\mathbf{-51+29\sqrt{3}}}$

Practice Problems - Multiplying Binomial Expressions in Radical Form, with Real Numbers

Section 1.4b Case II Practice Problems - Multiply the following radical expressions:

1. $\left(2\sqrt{3}+1\right)\cdot\left(2+\sqrt{2}\right) =$
2. $\left(1+\sqrt{5}\right)\cdot\left(\sqrt{8}+\sqrt{5}\right) =$
3. $\left(2-\sqrt{2}\right)\cdot\left(3+\sqrt{2}\right) =$
4. $\left(5+\sqrt{5}\right)\cdot\left(5-\sqrt{5^3}\right) =$
5. $\left(2+\sqrt{6}\right)\cdot\left(\sqrt[4]{16}-\sqrt{18}\right) =$
6. $\left(2-\sqrt{5}\right)\cdot\left(\sqrt{45}+\sqrt[4]{81}\right) =$

Case III Multiplying Monomial and Binomial Expressions in Radical Form, with Real Numbers

To multiply monomial and binomial expressions in radical form the following general multiplication rule is used:

$$a\cdot(b+c)=a\cdot b+a\cdot c$$

Monomial and binomial expressions in radical form are multiplied by each other using the following steps:

Step 1 Simplify the radical terms (see Section 1.4a, Case III).

Step 2 Multiply each term using the general multiplication rule, i.e., $a\cdot(b+c)=a\cdot b+a\cdot c$. Repeat Step 1, if necessary.

Examples with Steps

The following examples show the steps as to how monomial and binomial expressions in radical form are multiplied by one another:

Example 1.4-23

$$\sqrt{5}\cdot\left(\sqrt{50}+2\sqrt{27}\right)=$$

Solution:

Step 1

$$\sqrt{5}\cdot\left(\sqrt{50}+2\sqrt{27}\right)=\sqrt{5}\cdot\left(\sqrt{25\cdot2}+2\sqrt{9\cdot3}\right)=\sqrt{5}\cdot\left(\sqrt{5^2\cdot2}+2\sqrt{3^2\cdot3}\right)$$

$$=\sqrt{5}\cdot\left[5\sqrt{2}+(2\cdot3)\sqrt{3}\right]=\sqrt{5}\cdot\left[5\sqrt{2}+6\sqrt{3}\right]$$

Step 2

$$\sqrt{5}\cdot\left[5\sqrt{2}+6\sqrt{3}\right]=5\left(\sqrt{2}\cdot\sqrt{5}\right)+6\left(\sqrt{5}\cdot\sqrt{3}\right)=5\sqrt{2\cdot5}+6\sqrt{5\cdot3}=\mathbf{5\sqrt{10}+6\sqrt{15}}$$

Example 1.4-24

$$-2\sqrt[4]{4}\cdot\left(\sqrt[4]{64}-\sqrt[4]{162}\right)=$$

Solution:

Step 1

$$-2\sqrt[4]{4}\cdot\left(\sqrt[4]{64}-\sqrt[4]{162}\right)=-2\sqrt[4]{4}\cdot\left(\sqrt[4]{16\cdot4}-\sqrt[4]{81\cdot2}\right)=-2\sqrt[4]{4}\cdot\left(\sqrt[4]{2^4\cdot4}-\sqrt[4]{3^4\cdot2}\right)$$

$$=-2\sqrt[4]{4}\cdot\left(2\sqrt[4]{4}-3\sqrt[4]{2}\right)$$

Step 2

$$-2\sqrt[4]{4}\cdot\left(2\sqrt[4]{4}-3\sqrt[4]{2}\right)=-(2\cdot2)\cdot\left(\sqrt[4]{4}\cdot\sqrt[4]{4}\right)+(2\cdot3)\cdot\left(\sqrt[4]{4}\cdot\sqrt[4]{2}\right)=-4\left(\sqrt[4]{4\cdot4}\right)+6\left(\sqrt[4]{4\cdot2}\right)$$

$$=-4\sqrt[4]{16}+6\sqrt[4]{8}=-4\sqrt[4]{2^4}+6\sqrt[4]{8}=-(4\cdot2)+6\sqrt[4]{8}=-8+6\sqrt[4]{8}=\mathbf{2\left(3\sqrt[4]{8}-4\right)}$$

Additional Examples - Multiplying Monomial and Binomial Expressions in Radical Form, with Real Numbers

The following examples further illustrate how to multiply radical terms by one another:

Example 1.4-25

$$3\sqrt{5}\cdot\left(\sqrt{5}+6\sqrt{10}\right) = \left(3\sqrt{5}\cdot\sqrt{5}\right)+\left(3\sqrt{5}\cdot 6\sqrt{10}\right) = \left(3\sqrt{5\cdot 5}\right)+(3\cdot 6)\sqrt{5\cdot 10} = 3\sqrt{5^2}+18\sqrt{50}$$

$$= (3\cdot 5)+18\sqrt{25\cdot 2} = 15+18\sqrt{5^2\cdot 2} = 15+(18\cdot 5)\sqrt{2} = 15+90\sqrt{2} = \mathbf{15\left(1+6\sqrt{2}\right)}$$

Example 1.4-26

$$-2\sqrt{6}\cdot\left(-\sqrt{5}+\sqrt{50}\right) = \left(+2\sqrt{6}\cdot\sqrt{5}\right)-\left(2\sqrt{6}\cdot\sqrt{50}\right) = 2\sqrt{6\cdot 5}-2\sqrt{6\cdot 50} = 2\sqrt{30}-2\sqrt{300} = 2\sqrt{30}-2\sqrt{3\cdot 100}$$

$$= 2\sqrt{30}-2\sqrt{10^2\cdot 3} = 2\sqrt{30}-(2\cdot 10)\sqrt{3} = 2\sqrt{30}-20\sqrt{3} = \mathbf{2\left(\sqrt{30}-10\sqrt{3}\right)}$$

Example 1.4-27

$$\sqrt{3}\cdot\left(2\sqrt{3}+\sqrt{6}\right) = \left(2\sqrt{3}\cdot\sqrt{3}\right)+\left(\sqrt{3}\cdot\sqrt{6}\right) = \left(2\sqrt{3\cdot 3}\right)+\left(\sqrt{3\cdot 6}\right) = 2\sqrt{3^1\cdot 3^1}+\sqrt{18} = 2\sqrt{3^{1+1}}+\sqrt{9\cdot 2}$$

$$= 2\sqrt{3^2}+\sqrt{3^2\cdot 2} = (2\cdot 3)+3\sqrt{2} = 6+3\sqrt{2} = \mathbf{3\left(2+\sqrt{2}\right)}$$

Example 1.4-28

$$3\sqrt{2}\cdot\left(\sqrt{10}+4\sqrt{20}\right) = \left(3\sqrt{2}\cdot\sqrt{10}\right)+\left(3\sqrt{2}\cdot 4\sqrt{20}\right) = \left(3\sqrt{2\cdot 10}\right)+(3\cdot 4)\sqrt{2\cdot 20} = 3\sqrt{20}+12\sqrt{40}$$

$$= 3\sqrt{4\cdot 5}+12\sqrt{4\cdot 10} = 3\sqrt{2^2\cdot 5}+12\sqrt{2^2\cdot 10} = (3\cdot 2)\sqrt{5}+(12\cdot 2)\sqrt{10} = 6\sqrt{5}+24\sqrt{10} = \mathbf{6\left(\sqrt{5}+4\sqrt{10}\right)}$$

Example 1.4-29

$$\sqrt[3]{5}\cdot\left(\sqrt[3]{25}-\sqrt[3]{216}\right) = \sqrt[3]{5}\cdot\left(\sqrt[3]{5^2}-\sqrt[3]{6^3}\right) = \left(\sqrt[3]{5}\cdot\sqrt[3]{5^2}\right)-\left(\sqrt[3]{5}\cdot\sqrt[3]{6^3}\right) = \left(\sqrt[3]{5\cdot 5^2}\right)-\left(\sqrt[3]{5}\cdot 6\right) = \sqrt[3]{5^1\cdot 5^2}-6\sqrt[3]{5}$$

$$= \sqrt[3]{5^{1+2}}-6\sqrt[3]{5} = \sqrt[3]{5^3}-6\sqrt[3]{5} = \mathbf{5-6\sqrt[3]{5}}$$

Practice Problems - Multiplying Monomial and Binomial Expressions in Radical Form, with Real Numbers

Section 1.4b Case III Practice Problems - Multiply the following radical expressions:

1. $2\sqrt{3}\cdot\left(2+\sqrt{2}\right) =$
2. $\sqrt{5}\cdot\left(\sqrt{8}+\sqrt{5}\right) =$
3. $-\sqrt{8}\cdot\left(3-\sqrt{3}\right) =$
4. $4\sqrt{98}\cdot\left(3-\sqrt{2^3}\right) =$
5. $\sqrt[4]{48}\cdot\left(\sqrt[4]{324}+\sqrt[4]{32}\right) =$
6. $2\sqrt{5}\cdot\left(\sqrt{45}+\sqrt[4]{81}\right) =$

Case IV Rationalizing Radical Expressions - Monomial Denominators with Real Numbers

Radicals are divided by each other using the following general rule:

$$\sqrt[n]{\frac{x}{y}} = \frac{\sqrt[n]{x}}{\sqrt[n]{y}} \qquad x \ge 0,\ y > 0$$

In section 1.4a the difference between rational and irrational numbers was discussed. We learned that the square root of non perfect squares, the cube root of non perfect cubes, etc. are irrational numbers. For example, $\sqrt{3}, \sqrt{7}, \sqrt{10}, \sqrt[3]{4}, \sqrt[5]{7}$, *etc.* are classified as irrational numbers. In division of radicals, if the denominator of a fractional radical expression is not a rational number, we rationalize the denominator by changing the radicand of the denominator to a perfect square, a perfect cube, etc.

Simplification of radical expressions being divided requires rationalization of the denominator. A monomial and irrational denominator is rationalized by multiplying the numerator and the denominator by the irrational denominator. This change the radicand of the denominator to a perfect square.

Examples:

1. $\frac{\sqrt{1}}{\sqrt{7}} = \frac{\sqrt{1}}{\sqrt{7}} \times \frac{\sqrt{7}}{\sqrt{7}} = \frac{\sqrt{1 \cdot 7}}{\sqrt{7 \cdot 7}} = \frac{\sqrt{7}}{\sqrt{7^2}} = \frac{\sqrt{7}}{7}$

   Note that $\sqrt{7}$ is an irrational number. By multiplying $\sqrt{7}$ by itself the denominator is changed to a rational number, i.e., 7.

2. $\sqrt{\frac{20}{3}} = \frac{\sqrt{20}}{\sqrt{3}} = \frac{\sqrt{4 \cdot 5}}{\sqrt{3}} = \frac{\sqrt{2^2 \cdot 5}}{\sqrt{3}} = \frac{2\sqrt{5}}{\sqrt{3}} = \frac{2\sqrt{5}}{\sqrt{3}} \times \frac{\sqrt{3}}{\sqrt{3}} = \frac{2\sqrt{5 \cdot 3}}{\sqrt{3} \cdot \sqrt{3}} = \frac{2\sqrt{15}}{\sqrt{3^2}} = \mathbf{\frac{2\sqrt{15}}{3}}$

   Again, note that $\sqrt{3}$ is an irrational number. By multiplying $\sqrt{3}$ by itself the denominator is changed to a rational number, i.e., 3.

Radical expressions with monomial denominators are simplified using the following steps:

Step 1 Change the radical expression $\sqrt{\frac{a}{b}}$ to $\frac{\sqrt{a}}{\sqrt{b}}$ and simplify.

Step 2 Rationalize the denominator by multiplying the numerator and the denominator of the radical expression $\frac{\sqrt{a}}{\sqrt{b}}$ by $\sqrt{b}$.

Step 3 Simplify the radical expression (see Section 1.4a, Case III).

Examples with Steps

The following examples show the steps as to how radical expressions with monomial denominators are simplified:

Example 1.4-30

$$\boxed{\frac{-8\sqrt{3}}{32\sqrt{45}}} =$$

Solution:

Step 1

$$\boxed{\frac{-8\sqrt{3}}{32\sqrt{45}}} = \boxed{-\frac{\overset{1}{\cancel{8}}\sqrt{3}}{\underset{4}{\cancel{32}}\sqrt{45}}} = \boxed{-\frac{\sqrt{3}}{4\sqrt{9\cdot 5}}} = \boxed{-\frac{\sqrt{3}}{4\sqrt{3^2\cdot 5}}} = \boxed{-\frac{\sqrt{3}}{4\cdot 3\sqrt{5}}} = \boxed{-\frac{\sqrt{3}}{12\sqrt{5}}}$$

Step 2

$$\boxed{-\frac{\sqrt{3}}{12\sqrt{5}}} = \boxed{-\frac{\sqrt{3}}{12\sqrt{5}}\times\frac{\sqrt{5}}{\sqrt{5}}}$$

Step 3

$$\boxed{-\frac{\sqrt{3}}{12\sqrt{5}}\times\frac{\sqrt{5}}{\sqrt{5}}} = \boxed{-\frac{\sqrt{3}\times\sqrt{5}}{12\sqrt{5}\times\sqrt{5}}} = \boxed{-\frac{\sqrt{3\cdot 5}}{12\sqrt{5\cdot 5}}} = \boxed{-\frac{\sqrt{15}}{12\sqrt{5^1\cdot 5^1}}} = \boxed{-\frac{\sqrt{15}}{12\sqrt{5^{1+1}}}}$$

$$= \boxed{-\frac{\sqrt{15}}{12\sqrt{5^2}}} = \boxed{-\frac{\sqrt{15}}{12\cdot 5}} = \boxed{-\frac{\sqrt{15}}{60}}$$

Example 1.4-31

$$\boxed{\frac{3\sqrt[5]{8}}{\sqrt[5]{81}}} =$$

Solution:

Step 1

$$\boxed{\frac{3\sqrt[5]{8}}{\sqrt[5]{81}}} = \boxed{\frac{3\sqrt[5]{8}}{\sqrt[5]{3^4}}}$$

Step 2

$$\boxed{\frac{3\sqrt[5]{8}}{\sqrt[5]{3^4}}} = \boxed{\frac{3\sqrt[5]{8}}{\sqrt[5]{3^4}}\times\frac{\sqrt[5]{3^1}}{\sqrt[5]{3^1}}}$$

Note that radical expressions with third, fourth, or higher root in the denominator can also be rationalized by changing the denominator to a perfect third, fourth, or higher power.

Step 3

$$\boxed{\frac{3\sqrt[5]{8}}{\sqrt[5]{3^4}}\times\frac{\sqrt[5]{3^1}}{\sqrt[5]{3^1}}} = \boxed{\frac{3\sqrt[5]{8}\times\sqrt[5]{3^1}}{\sqrt[5]{3^4}\times\sqrt[5]{3^1}}} = \boxed{\frac{3\sqrt[5]{8\cdot 3}}{\sqrt[5]{3^4\cdot 3^1}}} = \boxed{\frac{3\sqrt[5]{24}}{\sqrt[5]{3^{4+1}}}} = \boxed{\frac{3\sqrt[5]{24}}{\sqrt[5]{3^5}}} = \boxed{\frac{\overset{1}{\cancel{3}}\cdot\sqrt[5]{24}}{\underset{1}{\cancel{3}}}} = \boxed{\frac{1\cdot\sqrt[5]{24}}{1}}$$

$$= \boxed{\sqrt[5]{24}}$$

Additional Examples: Rationalizing Radical Expressions - Monomial Denominators with Real Numbers

The following examples further illustrate how to solve radical expressions with monomial denominators:

Example 1.4-32

$$\boxed{\frac{8\sqrt{3}}{\sqrt{2}}} = \boxed{\frac{8\sqrt{3}}{\sqrt{2}}\times\frac{\sqrt{2}}{\sqrt{2}}} = \boxed{\frac{8\sqrt{3}\times\sqrt{2}}{\sqrt{2}\times\sqrt{2}}} = \boxed{\frac{8\sqrt{3\cdot 2}}{\sqrt{2\cdot 2}}} = \boxed{\frac{8\sqrt{6}}{\sqrt{2^1\cdot 2^1}}} = \boxed{\frac{8\sqrt{6}}{\sqrt{2^{1+1}}}} = \boxed{\frac{8\sqrt{6}}{\sqrt{2^2}}} = \boxed{\frac{\overset{4}{\cancel{8}}\sqrt{6}}{\underset{1}{\cancel{2}}}} = \boxed{\frac{4\sqrt{6}}{1}} = \boxed{4\sqrt{6}}$$

Example 1.4-33

$$\boxed{\sqrt[2]{\frac{5}{7}}} = \boxed{\sqrt{\frac{5}{7}}} = \boxed{\frac{\sqrt{5}}{\sqrt{7}}} = \boxed{\frac{\sqrt{5}}{\sqrt{7}} \times \frac{\sqrt{7}}{\sqrt{7}}} = \boxed{\frac{\sqrt{5} \times \sqrt{7}}{\sqrt{7} \times \sqrt{7}}} = \boxed{\frac{\sqrt{5 \cdot 7}}{\sqrt{7 \cdot 7}}} = \boxed{\frac{\sqrt{35}}{\sqrt{7^1 \cdot 7^1}}} = \boxed{\frac{\sqrt{35}}{\sqrt{7^{1+1}}}} = \boxed{\frac{\sqrt{35}}{\sqrt{7^2}}} = \boxed{\frac{\mathbf{\sqrt{35}}}{\mathbf{7}}}$$

Example 1.4-34

$$\boxed{\frac{1}{\sqrt[2]{5}}} = \boxed{\frac{1}{\sqrt{5}}} = \boxed{\frac{1}{\sqrt{5}} \times \frac{\sqrt{5}}{\sqrt{5}}} = \boxed{\frac{1 \times \sqrt{5}}{\sqrt{5} \times \sqrt{5}}} = \boxed{\frac{\sqrt{5}}{\sqrt{5 \cdot 5}}} = \boxed{\frac{\sqrt{5}}{\sqrt{5^1 \cdot 5^1}}} = \boxed{\frac{\sqrt{5}}{\sqrt{5^{1+1}}}} = \boxed{\frac{\sqrt{5}}{\sqrt{5^2}}} = \boxed{\frac{\mathbf{\sqrt{5}}}{\mathbf{5}}}$$

Example 1.4-35

$$\boxed{\frac{40\sqrt{12}}{5\sqrt{6}}} = \boxed{\frac{\overset{8}{\cancel{40}}\sqrt{12}}{\underset{1}{\cancel{5}}\sqrt{6}}} = \boxed{\frac{8\sqrt{12}}{\sqrt{6}}} = \boxed{\frac{8\sqrt{4 \cdot 3}}{\sqrt{6}}} = \boxed{\frac{8\sqrt{2^2 \cdot 3}}{\sqrt{6}}} = \boxed{\frac{(8 \cdot 2)\sqrt{3}}{\sqrt{6}}} = \boxed{\frac{16\sqrt{3}}{\sqrt{6}}} = \boxed{\frac{16\sqrt{3}}{\sqrt{6}} \times \frac{\sqrt{6}}{\sqrt{6}}} = \boxed{\frac{16\sqrt{3} \times \sqrt{6}}{\sqrt{6} \times \sqrt{6}}}$$

$$= \boxed{\frac{16\sqrt{3 \cdot 6}}{\sqrt{6 \cdot 6}}} = \boxed{\frac{16\sqrt{18}}{\sqrt{6^1 \cdot 6^1}}} = \boxed{\frac{16\sqrt{9 \cdot 2}}{\sqrt{6^{1+1}}}} = \boxed{\frac{16\sqrt{3^2 \cdot 2}}{\sqrt{6^2}}} = \boxed{\frac{(16 \cdot 3)\sqrt{2}}{6}} = \boxed{\frac{\overset{8}{\cancel{48}}\sqrt{2}}{\underset{1}{\cancel{6}}}} = \boxed{\frac{8\sqrt{2}}{1}} = \boxed{\mathbf{8\sqrt{2}}}$$

Example 1.4-36

$$\boxed{\sqrt{\frac{1000}{36}}} = \boxed{\sqrt{\frac{\overset{250}{\cancel{1000}}}{\underset{9}{\cancel{36}}}}} = \boxed{\sqrt{\frac{250}{9}}} = \boxed{\frac{\sqrt{250}}{\sqrt{9}}} = \boxed{\frac{\sqrt{25 \cdot 10}}{\sqrt{3^2}}} = \boxed{\frac{\sqrt{5^2 \times 10}}{3}} = \boxed{\frac{\mathbf{5\sqrt{10}}}{\mathbf{3}}}$$

Practice Problems: Rationalizing Radical Expressions - Monomial Denominators with Real Numbers

Section 1.4b Case IV Practice Problems - Solve the following radical expressions:

1. $\sqrt{\frac{1}{8}} =$
2. $\sqrt[2]{\frac{50}{7}} =$
3. $\frac{\sqrt{75}}{-5} =$
4. $\sqrt[3]{\frac{25}{16}} =$
5. $\sqrt[5]{\frac{32}{8}} =$
6. $\frac{-3\sqrt{100}}{-5\sqrt{3000}} =$

Case V Rationalizing Radical Expressions - Binomial Denominators with Real Numbers

Simplification of fractional radical expressions with binomial denominators requires rationalization of the denominator. A binomial denominator is rationalized by multiplying the numerator and the denominator by its conjugate. Two binomials that differ only by the sign between them are called **conjugates** of each other. Note that whenever conjugates are multiplied by each other, the two similar but opposite in sign middle terms drop out.

Examples:

1. The conjugate of $2+\sqrt{3}$ is $2-\sqrt{3}$.
2. The conjugate of $\sqrt{6}-10$ is $\sqrt{6}+10$.
3. The conjugate of $\sqrt{3}-\sqrt{5}$ is $\sqrt{3}+\sqrt{5}$.
4. The conjugate of $\sqrt{7}+\sqrt{2}$ is $\sqrt{7}-\sqrt{2}$.

Radical expressions with binomial denominators are simplified using the following steps:

Step 1 Simplify the radical terms in the numerator and the denominator (see Section 1.4a, Case III).

Step 2 Rationalize the denominator by multiplying the numerator and the denominator by its conjugate.

Step 3 Simplify the radical expression using the FOIL method (see Section 1.4b, Case II).

Examples with Steps

The following examples show the steps as to how radical expressions with two terms in the denominator are simplified:

Example 1.4-37

$$\frac{8}{2-\sqrt{2}}$$

Solution:

Step 1 *Not Applicable*

Step 2

$$\frac{8}{2-\sqrt{2}} = \frac{8}{2-\sqrt{2}} \times \frac{2+\sqrt{2}}{2+\sqrt{2}}$$

Step 3

$$\frac{8}{2-\sqrt{2}} \times \frac{2+\sqrt{2}}{2+\sqrt{2}} = \frac{8\times\left(2+\sqrt{2}\right)}{\left(2-\sqrt{2}\right)\times\left(2+\sqrt{2}\right)} = \frac{8\cdot\left(2+\sqrt{2}\right)}{(2\cdot 2)+\left(2\cdot\sqrt{2}\right)-\left(2\cdot\sqrt{2}\right)-\left(\sqrt{2}\cdot\sqrt{2}\right)}$$

$$= \frac{8\left(2+\sqrt{2}\right)}{4+2\sqrt{2}-2\sqrt{2}-\sqrt{2\cdot 2}} = \frac{8\left(2+\sqrt{2}\right)}{4-\sqrt{2^1\cdot 2^1}} = \frac{8\left(2+\sqrt{2}\right)}{4-\sqrt{2^{1+1}}} = \frac{8\left(2+\sqrt{2}\right)}{4-\sqrt{2^2}} = \frac{8\left(2+\sqrt{2}\right)}{4-2}$$

$$= \boxed{\frac{\overset{4}{\cancel{8}}\left(2+\sqrt{2}\right)}{\cancel{2}}} = \boxed{\frac{4\left(2+\sqrt{2}\right)}{1}} = \boxed{\mathbf{4\left(2+\sqrt{2}\right)}}$$

Example 1.4-38

$$\boxed{\frac{\sqrt{8}+\sqrt{4}}{4-\sqrt{2}}} =$$

Solution:

Step 1

$$\boxed{\frac{\sqrt{8}+\sqrt{4}}{4-\sqrt{2}}} = \boxed{\frac{\sqrt{2^2\cdot 2}+\sqrt{2^2}}{4-\sqrt{2}}} = \boxed{\frac{2\sqrt{2}+2}{4-\sqrt{2}}}$$

Step 2

$$\boxed{\frac{2\sqrt{2}+2}{4-\sqrt{2}}} = \boxed{\frac{2\sqrt{2}+2}{4-\sqrt{2}} \times \frac{4+\sqrt{2}}{4+\sqrt{2}}}$$

Step 3

$$\boxed{\frac{2\sqrt{2}+2}{4-\sqrt{2}} \times \frac{4+\sqrt{2}}{4+\sqrt{2}}} = \boxed{\frac{\left(2\sqrt{2}+2\right)\times\left(4+\sqrt{2}\right)}{\left(4-\sqrt{2}\right)\times\left(4+\sqrt{2}\right)}} = \boxed{\frac{\left(4\cdot 2\sqrt{2}\right)+\left(2\sqrt{2}\cdot\sqrt{2}\right)+\left(2\cdot 4\right)+\left(2\cdot\sqrt{2}\right)}{\left(4\cdot 4\right)+\left(4\cdot\sqrt{2}\right)-\left(4\cdot\sqrt{2}\right)-\left(\sqrt{2}\cdot\sqrt{2}\right)}}$$

$$= \boxed{\frac{8\sqrt{2}+2\sqrt{2\cdot 2}+8+2\sqrt{2}}{16+4\sqrt{2}-4\sqrt{2}-\sqrt{2\cdot 2}}} = \boxed{\frac{8\sqrt{2}+2\sqrt{2^2}+8+2\sqrt{2}}{16-\sqrt{2^2}}} = \boxed{\frac{8\sqrt{2}+(2\cdot 2)+8+2\sqrt{2}}{16-2}}$$

$$= \boxed{\frac{8\sqrt{2}+4+8+2\sqrt{2}}{14}} = \boxed{\frac{(8+2)\sqrt{2}+12}{14}} = \boxed{\frac{10\sqrt{2}+12}{14}} = \boxed{\frac{\overset{1}{\cancel{2}}\cdot\left(5\sqrt{2}+6\right)}{\underset{7}{\cancel{14}}}} = \boxed{\frac{1\cdot\left(5\sqrt{2}+6\right)}{7}}$$

$$= \boxed{\mathbf{\frac{5\sqrt{2}+6}{7}}}$$

Additional Examples: Rationalizing Radical Expressions - Binomial Denominators with Real Numbers

The following examples further illustrate how to rationalize radical expressions with binomial denominators:

Example 1.4-39

$$\boxed{\frac{\sqrt{5}}{3+\sqrt{5}}} = \boxed{\frac{\sqrt{5}}{3+\sqrt{5}} \times \frac{3-\sqrt{5}}{3-\sqrt{5}}} = \boxed{\frac{\sqrt{5}\times\left(3-\sqrt{5}\right)}{\left(3+\sqrt{5}\right)\times\left(3-\sqrt{5}\right)}} = \boxed{\frac{\left(3\cdot\sqrt{5}\right)-\left(\sqrt{5\cdot 5}\right)}{\left(3\cdot 3\right)-\left(3\cdot\sqrt{5}\right)+\left(3\cdot\sqrt{5}\right)-\left(\sqrt{5}\cdot\sqrt{5}\right)}}$$

$$= \boxed{\frac{3\sqrt{5}-\sqrt{5^2}}{9-3\sqrt{5}+3\sqrt{5}-\sqrt{5^2}}} = \boxed{\frac{3\sqrt{5}-5}{9-\sqrt{5^2}}} = \boxed{\frac{3\sqrt{5}-5}{9-5}} = \boxed{\mathbf{\frac{3\sqrt{5}-5}{4}}}$$

Example 1.4-40

$$\boxed{\frac{\sqrt{5}+\sqrt{3}}{\sqrt{7}+\sqrt{2}}} = \boxed{\frac{\sqrt{5}+\sqrt{3}}{\sqrt{7}+\sqrt{2}} \times \frac{\sqrt{7}-\sqrt{2}}{\sqrt{7}-\sqrt{2}}} = \boxed{\frac{\left(\sqrt{5}+\sqrt{3}\right)\times\left(\sqrt{7}-\sqrt{2}\right)}{\left(\sqrt{7}+\sqrt{2}\right)\times\left(\sqrt{7}-\sqrt{2}\right)}} = \boxed{\frac{\left(\sqrt{5}\cdot\sqrt{7}\right)-\left(\sqrt{5}\cdot\sqrt{2}\right)+\left(\sqrt{3}\cdot\sqrt{7}\right)-\left(\sqrt{3}\cdot\sqrt{2}\right)}{\left(\sqrt{7}\cdot\sqrt{7}\right)-\left(\sqrt{7}\cdot\sqrt{2}\right)+\left(\sqrt{2}\cdot\sqrt{7}\right)-\left(\sqrt{2}\cdot\sqrt{2}\right)}}$$

$$= \frac{\sqrt{5\cdot 7}-\sqrt{5\cdot 2}+\sqrt{3\cdot 7}-\sqrt{3\cdot 2}}{\sqrt{7\cdot 7}-\sqrt{7\cdot 2}+\sqrt{2\cdot 7}-\sqrt{2\cdot 2}} = \frac{\sqrt{35}-\sqrt{10}+\sqrt{21}-\sqrt{6}}{\sqrt{7^2}-\sqrt{14}+\sqrt{14}-\sqrt{2^2}} = \frac{\sqrt{35}-\sqrt{10}+\sqrt{21}-\sqrt{6}}{7-2}$$

$$= \boxed{\frac{\sqrt{35}-\sqrt{10}+\sqrt{21}-\sqrt{6}}{5}}$$

Example 1.4-41

$$\frac{3+\sqrt{27}}{3-\sqrt{3}} = \frac{3+\sqrt{9\cdot 3}}{3-\sqrt{3}} = \frac{3+\sqrt{3^2\cdot 3}}{3-\sqrt{3}} = \frac{3+3\sqrt{3}}{3-\sqrt{3}} = \frac{3+3\sqrt{3}}{3-\sqrt{3}}\times\frac{3+\sqrt{3}}{3+\sqrt{3}} = \frac{\left(3+3\sqrt{3}\right)\times\left(3+\sqrt{3}\right)}{\left(3-\sqrt{3}\right)\times\left(3+\sqrt{3}\right)}$$

$$= \frac{(3\cdot 3)+\left(3\cdot\sqrt{3}\right)+(3\cdot 3)\sqrt{3}+\left(3\sqrt{3}\cdot\sqrt{3}\right)}{(3\cdot 3)+\left(3\cdot\sqrt{3}\right)-\left(3\cdot\sqrt{3}\right)-\left(\sqrt{3}\cdot\sqrt{3}\right)} = \frac{9+3\sqrt{3}+9\sqrt{3}+3\sqrt{3\cdot 3}}{9+3\sqrt{3}-3\sqrt{3}-\sqrt{3\cdot 3}} = \frac{9+3\sqrt{3}+9\sqrt{3}+3\sqrt{3^2}}{9-\sqrt{3^2}}$$

$$= \frac{9+3\sqrt{3}+9\sqrt{3}+3\cdot 3}{9-3} = \frac{(9+9)+(3+9)\sqrt{3}}{9-3} = \frac{18+12\sqrt{3}}{6} = \frac{\not{6}\left(3+2\sqrt{3}\right)}{\not{6}} = \frac{3+2\sqrt{3}}{1} = \boxed{3+2\sqrt{3}}$$

Example 1.4-42

$$\frac{\sqrt{5}-1}{\sqrt{5}+1} = \frac{\sqrt{5}-1}{\sqrt{5}+1}\times\frac{\sqrt{5}-1}{\sqrt{5}-1} = \frac{\left(\sqrt{5}-1\right)\times\left(\sqrt{5}-1\right)}{\left(\sqrt{5}+1\right)\times\left(\sqrt{5}-1\right)} = \frac{\left(\sqrt{5}\cdot\sqrt{5}\right)-\left(1\cdot\sqrt{5}\right)-\left(1\cdot\sqrt{5}\right)+(1\cdot 1)}{\left(\sqrt{5}\cdot\sqrt{5}\right)-\left(1\cdot\sqrt{5}\right)+\left(1\cdot\sqrt{5}\right)-(1\cdot 1)} = \frac{\sqrt{5\cdot 5}-\sqrt{5}-\sqrt{5}+1}{\sqrt{5\cdot 5}-\sqrt{5}+\sqrt{5}-1}$$

$$= \frac{\sqrt{5^2}-\sqrt{5}-\sqrt{5}+1}{\sqrt{5^2}-1} = \frac{5-\sqrt{5}-\sqrt{5}+1}{5-1} = \frac{(5+1)+(-1-1)\sqrt{5}}{5-1} = \frac{6-2\sqrt{5}}{4} = \frac{2\left(3-\sqrt{5}\right)}{\not{4}\,2} = \boxed{\frac{3-\sqrt{5}}{2}}$$

Practice Problems: Rationalizing Radical Expressions - Binomial Denominators with Real Numbers

Section 1.4b Case V Practice Problems - Solve the following radical expressions:

1. $\frac{7}{1+\sqrt{7}} =$
2. $\frac{1-\sqrt{18}}{2+\sqrt{18}} =$
3. $\frac{\sqrt{5}}{\sqrt{5}+\sqrt{2}} =$
4. $\frac{3-\sqrt{5}}{\sqrt{7}-\sqrt{4}} =$
5. $\frac{-3+\sqrt{3}}{4+\sqrt{5}} =$
6. $\frac{3-\sqrt{3}}{3+\sqrt{3}} =$

Case VI Adding and Subtracting Radical Terms

Radicals are added and subtracted using the following general rule:

$$k_1\sqrt[n]{a} + k_2\sqrt[n]{a} + k_3\sqrt[n]{a} = (k_1 + k_2 + k_3)\sqrt[n]{a}$$

Only similar radicals can be added and subtracted. **Similar radicals** are defined as radical expressions with the same index n and the same radicand a. Note that the distributive property of multiplication (see Section 1.1) is used to group the numbers in front of the similar radical terms. Radicals are added and subtracted using the following steps:

Step 1 Group similar radicals.

Step 2 Simplify the radical expression.

Examples with Steps

The following examples show the steps as to how radical expressions are added and subtracted:

Example 1.4-43

$$\boxed{6\sqrt{2} + 4\sqrt{2}} =$$

Solution:

Step 1 $\boxed{6\sqrt{2} + 4\sqrt{2}} = \boxed{(6+4)\sqrt{2}}$

Step 2 $\boxed{(6+4)\sqrt{2}} = \boxed{\mathbf{10\sqrt{2}}}$

Example 1.4-44

$$\boxed{20\sqrt[5]{3} - 8\sqrt[5]{3} + 5\sqrt[5]{3}} =$$

Solution:

Step 1 $\boxed{20\sqrt[5]{3} - 8\sqrt[5]{3} + 5\sqrt[5]{3}} = \boxed{(20-8+5)\sqrt[5]{3}}$

Step 2 $\boxed{(20-8+5)\sqrt[5]{3}} = \boxed{\mathbf{17\sqrt[5]{3}}}$

Example 1.4-45

$$\boxed{\left(6\sqrt{7} + 2\sqrt{7}\right) - 2\sqrt[3]{7}} =$$

Solution:

Step 1 $\boxed{\left(6\sqrt{7} + 2\sqrt{7}\right) - 2\sqrt[3]{7}} = \boxed{(6+2)\sqrt{7} - 2\sqrt[3]{7}}$

Step 2 $\boxed{(6+2)\sqrt{7} - 2\sqrt[3]{7}} = \boxed{\mathbf{8\sqrt{7} - 2\sqrt[3]{7}}}$

Additional Examples - Adding and Subtracting Radical Terms

The following examples further illustrate how to add and subtract radical terms:

Example 1.4-46

$$2\sqrt{5}+3\sqrt{5}+6 = (2+3)\sqrt{5}+6 = \mathbf{5\sqrt{5}+6}$$

Example 1.4-47

$$8\sqrt[3]{4}+2\sqrt[3]{4}+5 = (8+2)\sqrt[3]{4}+5 = 10\sqrt[3]{4}+5 = \mathbf{5\left(2\sqrt[3]{4}+1\right)}$$

Example 1.4-48

$$2\sqrt[4]{3}+4\sqrt[4]{3}-3\sqrt[4]{3}+\sqrt[4]{5} = (2+4-3)\sqrt[4]{3}+\sqrt[4]{5} = \mathbf{3\sqrt[4]{3}+\sqrt[4]{5}}$$

Note that the two radical terms have the same index (4) but have different radicands (3 and 5). Therefore, they can not be combined.

Example 1.4-49

$$\sqrt[5]{5}+3\sqrt[5]{5}+a\sqrt[5]{5}-(4+a)\sqrt{2} = (1+3+a)\sqrt[5]{5}-(4+a)\sqrt{2} = (4+a)\sqrt[5]{5}-(4+a)\sqrt{2} = \mathbf{(4+a)\left[\sqrt[5]{5}-\sqrt{2}\right]}$$

Example 1.4-50

$$5\sqrt[3]{2x}+8\sqrt[3]{2x}-2c\sqrt[3]{2x}+4\sqrt{2x}-8\sqrt{2x} = (5+8-2c)\sqrt[3]{2x}+(4-8)\sqrt{2x} = \mathbf{(13-2c)\sqrt[3]{2x}-4\sqrt{2x}}$$

Example 1.4-51

$$a\sqrt{xy}+b\sqrt[3]{xy}-c^2\sqrt{xy}-d = a\sqrt{xy}-c^2\sqrt{xy}+b\sqrt[3]{xy}-d = \mathbf{\left(a-c^2\right)\sqrt{xy}+b\sqrt[3]{xy}-d}$$

Example 1.4-52

$$2\sqrt{75}+3\sqrt{125}+\sqrt{20}+3\sqrt{10}-4\sqrt{10} = 2\sqrt{25\cdot 3}+3\sqrt{25\cdot 5}+\sqrt{4\cdot 5}+(3-4)\sqrt{10}$$

$$= 2\sqrt{5^2\cdot 3}+3\sqrt{5^2\cdot 5}+\sqrt{2^2\cdot 5}-\sqrt{10} = (2\cdot 5)\sqrt{3}+(3\cdot 5)\sqrt{5}+2\sqrt{5}-\sqrt{10} = 10\sqrt{3}+15\sqrt{5}+2\sqrt{5}-\sqrt{10}$$

$$= 10\sqrt{3}+(15+2)\sqrt{5}-\sqrt{10} = \mathbf{10\sqrt{3}+17\sqrt{5}-\sqrt{10}}$$

Example 1.4-53

$$8\sqrt[3]{6}+4\sqrt[3]{6}+a\sqrt[3]{6}-\sqrt{5}-4\sqrt{5} = (8+4+a)\sqrt[3]{6}+(-1-4)\sqrt{5} = (12+a)\sqrt[3]{6}+(-5)\sqrt{5} = \mathbf{(12+a)\sqrt[3]{6}-5\sqrt{5}}$$

Practice Problems - Adding and Subtracting Radical Terms

Section 1.4b Case VI Practice Problems - Simplify the following radical expressions:

1. $5\sqrt{3}+8\sqrt{3} =$
2. $2\sqrt[3]{3}-4\sqrt[3]{3} =$
3. $12\sqrt[4]{5}+8\sqrt[4]{5}+2\sqrt[4]{3} =$
4. $a\sqrt{ab}-b\sqrt{ab}+c\sqrt{ab} =$
5. $3x\sqrt[3]{x}-2x\sqrt[3]{x}+4x\sqrt[3]{x^2} =$
6. $5\sqrt[3]{2}+8\sqrt[3]{5} =$

1.5 Polynomials

In this section classification of polynomials and how they are simplified are addressed in section 1.5a, Cases I and II. In addition, math operations involving polynomials which include multiplication, division, addition, and subtraction of polynomials by polynomials are reviewed in section 1.5b, Cases I through V.

1.5a Introduction to Polynomials

A polynomial is an algebraic expression that can be expressed in the following general form:

$$P(x) = a_n x^n + a_{n-1} x^{n-1} + a_{n-2} x^{n-2} + \cdots + a_0$$

where a_n, a_{n-1}, a_{n-2}, ..., and a_0 are real numbers, n is a positive integer number, and x is a variable. Note that in the above algebraic expression the $+$ or $-$ signs separate the polynomial to **terms**, i.e., $a_n x^n$, $a_{n-1} x^{n-1}$, $a_{n-2} x^{n-2}$, and a_0 are each referred to as a polynomial term. Classification of polynomials and how polynomials are simplified is discussed in the following two cases.

Case I Classification of Polynomials

Polynomials are usually named by their number of terms and are stated by the degree of the highest power of the variable in the polynomial. A polynomial is defined in the following way:

1. Definition of a Polynomial

A polynomial is a variable expression consisting of one or more terms. Note that in a polynomial the variable in each term has positive integer exponent. For example,

$x^3 + 5x$, $x^2 + 2x + 5$, $\frac{4}{5}x^2 + \frac{2}{3}x + \frac{1}{6}$, $3x^2$,

$9u^5 + 8u^3 - 6u - 5$, $x^3 - 1$, and $y^3 + 2y^2$

are polynomials. However,

$2x^4 - 5x^3 + 2x^{-2} + 5$, $6x^4 + \frac{2}{x^3} + 2x^2 - 5x$, $6w^6 + \frac{2}{w^4} - \frac{5}{w} + 3$, $\frac{5}{x^2} - 3$,

$2m^{-8} - \frac{5}{6}m^{-4} + \frac{2}{3}m^{-1}$, and $y^{-5} + 3y^{-2} - 2y + 6$

are not polynomials since the variable in one or more terms of the polynomials contain negative integer exponents.

Note that polynomial terms can have one or more variables. For example,

$x^4y^3 + 2x^3y^3 + 3x^2y^2 + 2xy - 5$, $\sqrt{5}a^4b^3 + \sqrt{3}a^2b^2 - \sqrt{2}ab + 12$,

$8x^4y^2z^3 + 3x^2y^2z - 2xyz + 1$, and $r^5s^4t^3 + 3rs^2t + 2r^2st - 4rst + 3$

are polynomials with two and three variable terms. In these instances a polynomial can be written in standard form in different ways depending on the variable selected.

For example, the polynomial $x^4y^3 + 2xy^2 - x^5 - 3x^2y^4 + 5$ is written in standard form as:

- $-x^5+\left(y^3\right)x^4+\left(-3y^4\right)x^2+\left(2y^2\right)x+5$ for the variable x, and
- $\left(-3x^2\right)y^4+\left(x^4\right)y^3+(2x)y^2-x^5+5$ for the variable y

(See the solutions to example 1.5-4 in Section 1.5a, Case II for the variables x, and y).

2. *Classification of Polynomials*

Polynomials are named by their number of terms. For example, a polynomial with one term only is called a **monomial**. A polynomial with two terms is called a **binomial** and a polynomial with three terms is called a **trinomial**. A polynomial with more than three terms is simply called a **polynomial**. For example,

- $5x^2$, 50 , $\sqrt{2}y^3$, x^3y^3 , x^2y^2z , $\frac{2}{3}y^2z^3$, and $8w$ are referred to as monomial expressions.
- x^3-2x^2 , y^5+1 , $2a^3b^4+\sqrt{5}$, $\frac{2}{3}a^4+\frac{1}{4}$, $\frac{1}{2}u^2w^4+4uw$, and $x+2$ are referred to as binomial expressions.
- x^5-2x^2+6x , y^7+4y^3+2y , $x^4y^3-2x^2y+\sqrt{3}$, a^8-4a^3+6a , and $-\frac{1}{3}m^4-\frac{4}{5}m^3+6m$ are referred to as trinomial expressions.
- x^6-4x^2+6x+1 , $w^5-2w^3+4w^2+7$, $x^5y^6-2x^2y^3+6xy+1$, and $-x^4-2x^3+6x^2+7x-5$ are referred to as polynomial expressions.

3. *Degree of Polynomials*

The degree of a polynomial is determined by the highest power of the variable in the polynomial. For example,

- $25x^0=25$ is a zero degree polynomial.
- $2x^1+1=2x+1$ is a first degree polynomial.
- $3z^2+6z-4$ is a second degree polynomial.
- $-3+5n^3$ is a third degree polynomial.
- $-4a^4+2a^3+2a^2-6a+2$ is a fourth degree polynomial.
- $2u-\sqrt{3}u^6-3u^2+2$ is a sixth degree polynomial.
- $m^4+2m^5+3m^8-m+2$ is an eighth degree polynomial.

In general, the degree of a polynomial is an indication of the number of roots that polynomial has. (In Chapter 3 students learn how to factor second or higher degree polynomials.)

4. *Polynomials in Standard Form*

A polynomial in **standard form** is defined as a polynomial in which the terms of the polynomial are written in order from the highest to the lowest power of the variable. For example,

$y^6+3y^5-2y^3+6$, x^5-2x^2+6x+1, x^4+2x-1, and a^3+a+1

are polynomials written in standard form. Note that the powers in a polynomial written in standard form decreases as we go from left to right.

In general, when a polynomial is written with the highest power of the variable first, followed by the second, third, fourth, fifth, etc. highest power of the variable, the polynomial is said to be in **descending order**.

Table 1.5-1 show examples of polynomials indicating their type, degree, and number of terms.

Table 1.5-1: Polynomials

Polynomial in Standard Form	Type	Degree	Number of Terms
x^3+6x^2	binomial	3	2
x^3+2x+5	trinomial	3	3
$x+1$	binomial	1	2
$5y^5$	monomial	5	1
$35=35x^0$	monomial	0	1
$5x^5-2x^4-8x^2+3$	polynomial	5	4
x^4+2x^3+5	trinomial	4	3
$x^{-4}+2x^2+5x-3$	not a polynomial		
$w^6-4w^5+2w^3+5w^2-1$	polynomial	6	5
x^2+3x-2	trinomial	2	3
$x^4+\frac{1}{x^3}-5$	not a polynomial		

Practice Problems - Classification of Polynomials

Section 1.5a Case I Practice Problems - Write the following polynomials in standard form and identify each polynomial type, its degree, and number of terms.

1. $3x+2x^3-6$
2. $-6y^8+2$
3. $2w+6w^2+8w^5$
4. $6y$
5. $\sqrt{72}$
6. $-16+2x^4$

Case II Simplifying Polynomials

Polynomials are simplified using the following steps:

Step 1 Group like terms.

Step 2 Combine like terms and write the polynomial in standard form.

Note that **like terms** are defined as polynomial terms having the same variables raised to the same power. For example, in the polynomial expression:

$8y^3+5y^2-2y^3-y+5y^3-20+y^2-3y+4$

$8y^3$, $-2y^3$, *and* $5y^3$; $5y^2$ *and* y^2; $-y$ *and* $-3y$; -20 *and* $+4$;

are like terms of one another.

Examples with Steps

The following examples show the steps as to how polynomials are simplified:

Example 1.5-1

$$\boxed{18x^3+2x^2-5x^3-2x-x^2}=$$

Solution:

Step 1 $\boxed{18x^3+2x^2-5x^3-2x-x^2}=\boxed{\left(18x^3-5x^3\right)+\left(2x^2-x^2\right)-2x}$

Step 2 $\boxed{\left(18x^3-5x^3\right)+\left(2x^2-x^2\right)-2x}=\boxed{(18-5)x^3+(2-1)x^2-2x}=\boxed{\mathbf{13x^3+x^2-2x}}$

Example 1.5-2

$$\boxed{2w^4+4w^3-w^4-8+2w-w^3+4}=$$

Solution:

Step 1 $\boxed{2w^4+4w^3-w^4-8+2w-w^3+4}=\boxed{\left(2w^4-w^4\right)+\left(4w^3-w^3\right)+(-8+4)+2w}$

Step 2 $\boxed{\left(2w^4-w^4\right)+\left(4w^3-w^3\right)+(-8+4)+2w}=\boxed{(2-1)w^4+(4-1)w^3-4+2w}$

$$=\boxed{w^4+3w^3-4+2w}=\boxed{\mathbf{w^4+3w^3+2w-4}}$$

Additional Examples - Simplifying Polynomials

The following examples further illustrate how to simplify and write polynomials in standard form:

Example 1.5-3

$$\boxed{-4w^7+3w^3-5+2w^7+2w^3-5w^7+w^3-3}=\boxed{\left(-4w^7+2w^7-5w^7\right)+\left(3w^3+2w^3+w^3\right)+(-5-3)}$$

$$=\boxed{(-4+2-5)w^7+(3+2+1)w^3-8}=\boxed{\mathbf{-7w^7+6w^3-8}}$$

Example 1.5-4

$$\boxed{-2x^4y^3+x^2y+5x^3y^5-8xy^2+3x^3y^5-6x^2y}=\boxed{-2x^4y^3+\left(5x^3y^5+3x^3y^5\right)-8xy^2+\left(-6x^2y+x^2y\right)}$$

$$=\boxed{-2x^4y^3+(5+3)x^3y^5-8xy^2+(-6+1)x^2y}=\boxed{-2x^4y^3+8x^3y^5-8xy^2-5x^2y}$$

$$=\boxed{\mathbf{\left(-2y^3\right)x^4+\left(8y^5\right)x^3+(-5y)x^2+\left(-8y^2\right)x}\quad \textit{in standard form for the variable x}}$$

$$=\boxed{\mathbf{\left(8x^3\right)y^5+\left(-2x^4\right)y^3+(-8x)y^2+\left(-5x^2\right)y}\quad \textit{in standard form for the variable y}}$$

Example 1.5-5

$$\boxed{-5y^7+y+5y^5+12y^7-5y^4+y^5-3y^4-5y}=\boxed{\left(-5y^7+12y^7\right)+(y-5y)+\left(5y^5+y^5\right)+\left(-5y^4-3y^4\right)}$$

$$= \boxed{(-5+12)y^7 + (1-5)y + (5+1)y^5 + (-5-3)y^4} = \boxed{7y^7 - 4y + 6y^5 - 8y^4} = \boxed{\mathbf{7y^7 + 6y^5 - 8y^4 - 4y}}$$

Example 1.5-6

$$\boxed{5m^5 + 5m^3 + 10m^5 - 6 - m^5 + 3m^3 + 4m + 9} = \boxed{\left(5m^5 + 10m^5 - m^5\right) + \left(5m^3 + 3m^3\right) + (-6+9) + 4m}$$

$$= \boxed{(5+10-1)m^5 + (5+3)m^3 + 3 + 4m} = \boxed{14m^5 + 8m^3 + 3 + 4m} = \boxed{\mathbf{14m^5 + 8m^3 + 4m + 3}}$$

Example 1.5-7

$$\boxed{\frac{2}{3}x + \frac{1}{2} + \frac{1}{4}x^2 + \frac{1}{3}x + \frac{3}{4}x^2 - \frac{2}{3}} = \boxed{\left(\frac{2}{3}x + \frac{1}{3}x\right) + \left(\frac{1}{2} - \frac{2}{3}\right) + \left(\frac{1}{4}x^2 + \frac{3}{4}x^2\right)}$$

$$= \boxed{\left(\frac{2}{3} + \frac{1}{3}\right)x + \left(\frac{1}{2} - \frac{2}{3}\right) + \left(\frac{1}{4} + \frac{3}{4}\right)x^2} = \boxed{\left(\frac{2+1}{3}\right)x + \left(\frac{(1\cdot 3)-(2\cdot 2)}{2\cdot 3}\right) + \left(\frac{1+3}{4}\right)x^2} = \boxed{\frac{\cancel{3}}{\cancel{3}}x + \frac{3-4}{6} + \frac{\cancel{4}}{\cancel{4}}x^2}$$

$$= \boxed{\frac{1}{1}x - \frac{1}{6} + \frac{1}{1}x^2} = \boxed{x - \frac{1}{6} + x^2} = \boxed{\mathbf{x^2 + x - \frac{1}{6}}}$$

Practice Problems - Simplifying Polynomials

Section 1.5a Case II Practice Problems - Simplify the following polynomial expressions. Write the answer in standard form.

1. $-x^3 + 4x - 8x^2 + 3x - 5x^3 - 5x =$
2. $2y + 2y^3 - 5 + 4y - 5y^3 + 1 + y =$
3. $2a^5 + 2a^2 - 3 + 4a^5 + a^2 =$
4. $3x + 2x^4 + 2x^3 - 7x - 5x^4 =$
5. $2rs + 4r^3s^3 - 20 + 2rs - 5r^3s^3 - 3 =$
6. $2xyz + 2x^3y^3z^3 + 10 - 4xyz - 4 =$

1.5b Operations with Polynomials

In this section multiplication, division, addition, and subtraction of polynomial expressions is addressed in Cases I through V.

Case I Multiplying Monomials

Polynomials are multiplied by each other using the general product rule (see Section 1.1). Note that polynomials with like terms are multiplied by one another using the multiplication law for exponents (see Section 1.3). Monomial expressions are multiplied by each other using the general exponent rule, i.e.,

$$\left(a_0x^m\right)\cdot\left(a_1x^n\right) = \left(a_0a_1\right)\cdot\left(x^mx^n\right) = \left(a_0a_1\right)\cdot\left(x^{m+n}\right)$$ When monomial terms have the same variable.

or,

$$\left(a_0x^m\right)\cdot\left(a_1y^n\right) = \left(a_0a_1\right)\cdot\left(x^my^n\right)$$ When monomial terms have different variables.

where a_0, and a_1 are real numbers, x and y are variables, and m and n are integer numbers. Multiplication of monomial expressions is divided to two cases. Case Ia - multiplication of monomials by monomials, and Case Ib - multiplication of polynomials by monomials.

Case Ia Multiplying Monomials by Monomials

Monomials are multiplied by one another using the following steps:

Step 1 Group the like terms with each other.

Step 2 a. Multiply the numerical coefficients (see Section 1.1).

b. Multiply the variables using the exponent rule $x^n \cdot x^m = x^{n+m}$ (see Section 1.3).

Examples with Steps

The following examples show the steps as to how monomials are multiplied by one another:

Example 1.5-8

$$\boxed{\left(3x^3\right)\cdot\left(2x^2y\right)} =$$

Solution:

Step 1

$$\boxed{\left(3x^3\right)\cdot\left(2x^2y\right)} = \boxed{(3\cdot 2)\cdot\left(x^3\cdot x^2\right)\cdot y}$$

Step 2

$$\boxed{(3\cdot 2)\cdot\left(x^3\cdot x^2\right)\cdot y} = \boxed{6\cdot x^{3+2}\cdot y} = \boxed{\mathbf{6x^5y}}$$

Example 1.5-9

$$\boxed{\left(3a^2b^3c^5\right)\cdot\left(5b^2c^4\right)\cdot\left(4a^3b^0c^3\right)} =$$

$15a^2b^5c^9\left(4a^3b^0c^3\right) = 60a^5b^5c^{12}$

Solution:

Step 1

$$\boxed{\left(3a^2b^3c^5\right)\cdot\left(5b^2c^4\right)\cdot\left(4a^3b^0c^3\right)} = \boxed{(3\cdot 5\cdot 4)\cdot\left(a^3\cdot a^2\right)\cdot\left(b^3\cdot b^2\cdot b^0\right)\cdot\left(c^5\cdot c^4\cdot c^3\right)}$$

Step 2

$$\boxed{(3\cdot 5\cdot 4)\cdot\left(a^3\cdot a^2\right)\cdot\left(b^3\cdot b^2\cdot b^0\right)\cdot\left(c^5\cdot c^4\cdot c^3\right)} = \boxed{60\cdot a^{3+2}\cdot b^{3+2+0}\cdot c^{5+4+3}} = \boxed{\mathbf{60a^5b^5c^{12}}}$$

Additional Examples - Multiplying Monomials by Monomials

The following examples further illustrate how to multiply monomials by monomials:

Example 1.5-10

$$\boxed{\left(8x^3y^2\right)\cdot\left(2x^2y\right)} = \boxed{(8\cdot 2)\cdot\left(x^3\cdot x^2\right)\cdot\left(y^2\cdot y\right)} = \boxed{16\cdot x^{3+2}\cdot y^{2+1}} = \boxed{\mathbf{16x^5y^3}}$$

Example 1.5-11

$$\boxed{\left(3x^2\right)\cdot(5xy)\cdot\left(2x^2y^3\right)} = \boxed{(3\cdot 5\cdot 2)\cdot\left(x^2\cdot x\cdot x^2\right)\cdot\left(y^3\cdot y\right)} = \boxed{30\cdot x^{2+1+2}\cdot y^{3+1}} = \boxed{\mathbf{30x^5y^4}}$$

Example 1.5-12

$$\boxed{(4xy)\cdot\left(5x^2y\right)\cdot\left(3x^0\right)} = \boxed{(4\cdot 5\cdot 3)\cdot\left(x\cdot x^2\cdot x^0\right)\cdot(y\cdot y)} = \boxed{60\cdot x^{1+2+0}\cdot y^{1+1}} = \boxed{\mathbf{60x^3y^2}}$$

Example 1.5-13

$$\boxed{\left(5x^2y\right)^0\cdot(3xy)\cdot x^2y^3} = \boxed{1\cdot(3xy)\cdot x^2y^3} = \boxed{3\cdot\left(x\cdot x^2\right)\cdot\left(y\cdot y^3\right)} = \boxed{3\cdot x^{1+2}\cdot y^{1+3}} = \boxed{\mathbf{3x^3y^4}}$$

Example 1.5-14

$$\boxed{\left(8x^2y\right)\cdot(2x)\cdot(4z)} = \boxed{(8\cdot 2\cdot 4)\cdot\left(x^2\cdot x\right)\cdot yz} = \boxed{64\cdot x^{2+1}\cdot yz} = \boxed{\mathbf{64x^3yz}}$$

Example 1.5-15

$$\boxed{\left(3a^2b^2\right)\cdot(2ab)\cdot\left(3a^0\right)} = \boxed{(3\cdot 2\cdot 3)\cdot\left(a^2\cdot a\cdot a^0\right)\cdot\left(b^2\cdot b\right)} = \boxed{18\cdot a^{2+1+0}\cdot b^{2+1}} = \boxed{\mathbf{18a^3b^3}}$$

Practice Problems - Multiplying Monomials by Monomials

Section 1.5b Case Ia Practice Problems - Multiply the following monomials by each other:

1. $(2ax)\cdot\left(3a^2x^2\right) =$
2. $\left(5x^2y^2\right)\cdot(2x)\cdot(4y) =$
3. $\left(6x^2\right)^0\cdot\left(3x^2\right)\cdot(-2x) =$
4. $\left(x^2y\right)\cdot(3xy)\cdot\left(4x^3y^2\right) =$
5. $\left(3x^2y^2\right)\cdot\left(2xy^0\right)\cdot\left(5x^0y\right) =$
6. $\left(8a^2b^2\right)\cdot(2a)\cdot\left(3a^2b^3\right) =$

Case Ib **Multiplying Polynomials by Monomials**

Polynomials are multiplied by monomials using the following steps:

Step 1 Group like terms with each other.

Step 2 Multiply each term of the polynomial by the monomial by:

a. Multiplying the numerical coefficients (see Section 1.1).

b. Multiplying the variables using the exponent rule $x^n\cdot x^m = x^{n+m}$ (see Section 1.3)

Examples with Steps

The following examples show the steps as to how polynomials are multiplied by monomials:

Example 1.5-16

$$\left(2x^4+3x^2+5x-x^4+x^2-3\right)\cdot\left(3x^2\right)=$$

Solution:

Step 1

$$\left(2x^4+3x^2+5x-x^4+x^2-3\right)\cdot\left(3x^2\right)=\left(2x^4-x^4+3x^2+x^2+5x-3\right)\cdot\left(3x^2\right)$$

$$=\left[\left(2x^4-x^4\right)+\left(3x^2+x^2\right)+5x-3\right]\cdot\left(3x^2\right)=\left[(2-1)x^4+(3+1)x^2+5x-3\right]\cdot\left(3x^2\right)$$

$$=\left[x^4+4x^2+5x-3\right]\cdot\left(3x^2\right)$$

Step 2

$$\left[x^4+4x^2+5x-3\right]\cdot\left(3x^2\right)=3\left(x^4\cdot x^2\right)+(4\cdot 3)\left(x^2\cdot x^2\right)+(5\cdot 3)\left(x\cdot x^2\right)-(3\cdot 3)x^2$$

$$=3x^{4+2}+12x^{2+2}+15x^{1+2}-9x^2=\mathbf{3x^6+12x^4+15x^3-9x^2}$$

Example 1.5-17

$$\left(\sqrt{27}x^2-2+\sqrt{8}x+\sqrt{36}\right)\cdot\sqrt{125x}=$$

Solution:

Step 1

$$\left(\sqrt{27}x^2-2+\sqrt{8}x+\sqrt{36}\right)\cdot\sqrt{125x}=\left(\sqrt{9\cdot 3}x^2-2+\sqrt{4\cdot 2}x+\sqrt{6^2}\right)\cdot\sqrt{25\cdot 5x}$$

$$=\left[\sqrt{3^2\cdot 3}x^2+\sqrt{2^2\cdot 2}x+(-2+6)\right]\cdot\sqrt{5^2\cdot 5x}=\left(3\sqrt{3}x^2+2\sqrt{2}x+4\right)\cdot 5\sqrt{5x}$$

Step 2

$$\left(3\sqrt{3}x^2+2\sqrt{2}x+4\right)\cdot 5\sqrt{5x}=(3\cdot 5)\left(\sqrt{3}\cdot\sqrt{5}\right)\left(x^2\cdot x\right)+(2\cdot 5)\left(\sqrt{2}\cdot\sqrt{5}\right)(x\cdot x)+(4\cdot 5)\sqrt{5x}$$

$$=15\left(\sqrt{3\cdot 5}\right)x^{2+1}+10\left(\sqrt{2\cdot 5}\right)x^{1+1}+20\sqrt{5x}=15\left(\sqrt{3\cdot 5}\right)x^{2+1}+10\left(\sqrt{2\cdot 5}\right)x^{1+1}+20\sqrt{5x}$$

$$=\mathbf{15\sqrt{15}x^3+10\sqrt{10}x^2+20\sqrt{5x}}$$

Additional Examples - Multiplying Polynomials by Monomials

The following examples further illustrate how to multiply polynomial expressions by monomials:

Example 1.5-18

$$\left(5a^2b^2+3ab-2a^2b^2-ab+1\right)\cdot(3ab)=\left[\left(5a^2b^2-2a^2b^2\right)+(3ab-ab)+1\right]\cdot(3ab)$$

$$=\boxed{\left[(5-2)a^2b^2+(3-1)ab+1\right]\cdot(3ab)}=\boxed{\left[3a^2b^2+2ab+1\right]\cdot(3ab)}=\boxed{(3\cdot3)\left(a^2\cdot a\right)\left(b^2\cdot b\right)+(2\cdot3)(a\cdot a)(b\cdot b)+3ab}$$

$$=\boxed{9a^{2+1}b^{2+1}+6a^{1+1}b^{1+1}+3ab}=\boxed{\mathbf{9a^3b^3+6a^2b^2+3ab}}$$

Example 1.5-19

$$\boxed{\left(-3x^2+4x^3+x-5+2x^3\right)\cdot\left(-2x^2\right)}=\boxed{\left[\left(4x^3+2x^3\right)-3x^2+x-5\right]\cdot\left(-2x^2\right)}=\boxed{\left[(4+2)x^3-3x^2+x-5\right]\cdot\left(-2x^2\right)}$$

$$=\boxed{\left(6x^3-3x^2+x-5\right)\cdot\left(-2x^2\right)}=\boxed{-(6\cdot2)\left(x^3\cdot x^2\right)+(3\cdot2)\left(x^2\cdot x^2\right)-2\left(x\cdot x^2\right)+(5\cdot2)x^2}$$

$$=\boxed{-12x^{3+2}+6x^{2+2}-2x^{1+2}+10x^2}=\boxed{\mathbf{-12x^5+6x^4-2x^3+10x^2}}$$

Example 1.5-20

$$\boxed{\left(4x^3+3x^2-3+3x^5+(3x)^0\right)\cdot\left(2x^3\right)}=\boxed{\left(4x^3+3x^2-3+3x^5+1\right)\cdot\left(2x^3\right)}=\boxed{\left[3x^5+4x^3+3x^2(-3+1)\right]\cdot\left(2x^3\right)}$$

$$=\boxed{\left[3x^5+4x^3+3x^2-2\right]\cdot\left(2x^3\right)}=\boxed{(3\cdot2)\left(x^5\cdot x^3\right)+(4\cdot2)\left(x^3\cdot x^3\right)+(3\cdot2)\left(x^2\cdot x^3\right)-(2\cdot2)x^3}$$

$$=\boxed{6x^{5+3}+8x^{3+3}+6x^{2+3}-4x^3}=\boxed{\mathbf{6x^8+8x^6+6x^5-4x^3}}$$

Example 1.5-21

$$\boxed{\left(3l^2+3l^3-5l+2l^4+2l-2\right)\cdot l^2}=\boxed{\left[2l^4+3l^3+3l^2+(2l-5l)-2\right]\cdot l^2}=\boxed{\left[2l^4+3l^3+3l^2+(2-5)l-2\right]\cdot l^2}$$

$$=\boxed{\left(2l^4+3l^3+3l^2-3l-2\right)\cdot l^2}=\boxed{2\left(l^4\cdot l^2\right)+3\left(l^3\cdot l^2\right)+3\left(l^2\cdot l^2\right)-3\left(l\cdot l^2\right)-2l^2}$$

$$=\boxed{2l^{4+2}+3l^{3+2}+3l^{2+2}-3l^{1+2}-2l^2}=\boxed{\mathbf{2l^6+3l^5+3l^4-3l^3-2l^2}}$$

Practice Problems - Multiplying Polynomials by Monomials

Section 1.5b Case Ib Practice Problems - Multiply the following polynomial expressions by monomials:

1. $2\cdot\left(5x^2+6x-2x^2-x+5\right)=$
2. $\left(2x^2y-5y^2+3x^2y-2y^2+3\right)\cdot\left(3x^2y^2\right)=$
3. $\left(5x^3+2x^2-5+3x-2x^3\right)\cdot(-2x)^2=$
4. $6w\cdot\left(4w+2w^2+2-3w+w^2\right)=$
5. $2x\cdot\left(2x^2\right)^2\cdot\left(5x^2+3x-2x^2+x-2\right)=$
6. $\left(\sqrt{162}+\sqrt{9}x-2x^2+\sqrt{16}x^3\right)\cdot\left(2x^3\right)=$

Case II Multiplying Binomials by Binomials

Binomials are multiplied by one another using the multiplication method known as the FOIL method (see Section 1.4b, Case II). In general, binomials are multiplied by each other in the following way:

$$\left(a_0x^n + a_1x^{n-m}\right)\left(b_0x^n + b_1x^{n-m}\right)$$

$$= \left(a_0 \cdot b_0\right)\cdot\left(x^n \cdot x^n\right) + \left(a_0 \cdot b_1\right)\cdot\left(x^n \cdot x^{n-m}\right) + \left(a_1 \cdot b_0\right)\cdot\left(x^{n-m} \cdot x^n\right) + \left(a_1 \cdot b_1\right)\cdot\left(x^{n-m} \cdot x^{n-m}\right)$$

$$= a_0b_0\left(x^{n+n}\right) + a_0b_1\left(x^{n+n-m}\right) + a_1b_0\left(x^{n-m+n}\right) + a_1b_1\left(x^{n-m+n-m}\right)$$

$$= a_0b_0\left(x^{2n}\right) + a_0b_1\left(x^{2n-m}\right) + a_1b_0\left(x^{2n-m}\right) + a_1b_1\left(x^{2n-2m}\right) = \mathbf{a_0b_0x^{2n} + (a_0b_1 + a_1b_0)x^{2n-m} + a_1b_1x^{2n-2m}}$$

where n and m are positive integer numbers and $n \geq m$.

Binomials are multiplied by one another using the following steps:

Step 1 a. Simplify each binomial term, if possible.

b. Multiply the terms of the first binomial by each term of the second binomial using the FOIL method.

Step 2 Group like terms with each other.

Examples with Steps

The following examples show the steps as to how binomials are multiplied by each other:

Example 1.5-22

$$\boxed{\left(x^2 + 3x\right)(x+8)} =$$

Solution:

Step 1 $\boxed{\left(x^2 + 3x\right)(x+8)} = \boxed{\left(x^2 \cdot x\right) + \left(8 \cdot x^2\right) + 3(x \cdot x) + (3 \cdot 8)x} = \boxed{x^3 + 8x^2 + 3x^2 + 24x}$

Step 2 $\boxed{x^3 + 8x^2 + 3x^2 + 24x} = \boxed{x^3 + (8+3)x^2 + 24x} = \boxed{\mathbf{x^3 + 11x^2 + 24x}}$

Example 1.5-23

$$\boxed{\left(\sqrt{225}x + 2\right)\left(5x - \sqrt{81}\right)} =$$

Solution:

Step 1 $\boxed{\left(\sqrt{225}x + 2\right)\left(5x - \sqrt{81}\right)} = \boxed{\left(\sqrt{15^2}x + 2\right)\left(5x - \sqrt{9^2}\right)} = \boxed{(15x+2)(5x-9)}$

$= \boxed{(15 \cdot 5)(x \cdot x) - (15 \cdot 9)x + (2 \cdot 5)x - (2 \cdot 9)} = \boxed{75x^2 - 135x + 10x - 18}$

Step 2 $\boxed{75x^2 - 135x + 10x - 18} = \boxed{75x^2 + (-135 + 10)x - 18} = \boxed{\mathbf{75x^2 - 125x - 18}}$

Additional Examples - Multiplying Binomials by Binomials

The following examples further illustrate how to multiply binomials by one another:

Example 1.5-24

$$\boxed{(x+\sqrt{98})(x-2\sqrt{162})}=\boxed{(x+\sqrt{49\cdot 2})(x-2\sqrt{81\cdot 2})}=\boxed{\left(x+\sqrt{7^2\cdot 2}\right)\left(x-2\sqrt{9^2\cdot 2}\right)}=\boxed{(x+7\sqrt{2})(x-(2\cdot 9)\sqrt{2})}$$

$$=\boxed{(x+7\sqrt{2})(x-18\sqrt{2})}=\boxed{(x\cdot x)-(18\sqrt{2}\cdot x)+(7\sqrt{2}\cdot x)-(7\cdot 18)\cdot(\sqrt{2}\cdot\sqrt{2})}=\boxed{x^2-18\sqrt{2}x+7\sqrt{2}x-126(\sqrt{2\cdot 2})}$$

$$=\boxed{x^2+(-18+7)\sqrt{2}x-126\sqrt{2^2}}=\boxed{x^2-11\sqrt{2}x-(126\cdot 2)}=\boxed{\mathbf{x^2-11\sqrt{2}x-252}}$$

Example 1.5-25

$$\boxed{(b^3-\sqrt{8})(b+\sqrt{50b^2})}=\boxed{(b^3-\sqrt{4\cdot 2})(b+\sqrt{25\cdot 2b^2})}=\boxed{\left(b^3-\sqrt{2^2\cdot 2}\right)\left(b+\sqrt{5^2\cdot 2b^2}\right)}=\boxed{(b^3-2\sqrt{2})(b+5\sqrt{2b^2})}$$

$$=\boxed{(b^3\cdot b)+5\sqrt{2}(b^3\cdot b^2)-2\sqrt{2}\cdot b-(2\sqrt{2}\cdot 5\sqrt{2})b^2}=\boxed{b^4+5\sqrt{2}b^5-2\sqrt{2}b-(2\cdot 5)\sqrt{2\cdot 2}b^2}$$

$$=\boxed{b^4+5\sqrt{2}b^5-2\sqrt{2}b-10\sqrt{2^2}b^2}=\boxed{b^4+5\sqrt{2}b^5-2\sqrt{2}b-(10\cdot 2)b^2}=\boxed{\mathbf{5\sqrt{2}b^5+b^4-20b^2-2\sqrt{2}b}}$$

Example 1.5-26

$$\boxed{(u^2+3u)(-u+u^2)}=\boxed{-(u^2\cdot u)+(u^2\cdot u^2)-3(u\cdot u)+3(u\cdot u^2)}=\boxed{-u^3+u^4-3u^2+3u^3}=\boxed{u^4+3u^3-u^3-3u^2}$$

$$=\boxed{u^4+(3-1)u^3-3u^2}=\boxed{\mathbf{u^4+2u^3-3u^2}}$$

Example 1.5-27

$$\boxed{\left(\sqrt[5]{q^{10}}+3q\right)\left(4q^2-3\sqrt[5]{q^5}\right)}=\boxed{\left(\sqrt[5]{q^{5+5}}+3q\right)(4q^2-3q)}=\boxed{\left(\sqrt[5]{q^5\cdot q^5}+3q\right)(4q^2-3q)}$$

$$=\boxed{[(q\cdot q)+3q](4q^2-3q)}=\boxed{(q^2+3q)(4q^2-3q)}=\boxed{4(q^2\cdot q^2)-3(q^2\cdot q)+(3\cdot 4)(q\cdot q^2)-(3\cdot 3)(q\cdot q)}$$

$$=\boxed{4q^4-3q^3+12q^3-9q^2}=\boxed{4q^4+(-3+12)q^3-9q^2}=\boxed{\mathbf{4q^4+9q^3-9q^2}}$$

Practice Problems - Multiplying Binomials by Binomials

Section 1.5b Case II Practice Problems - Multiply the following binomial expressions:

1. $(x+3)(x-2)=$
2. $(-y+8)(y-6)=$
3. $(x^2-2xy)(-y^2+2xy)=$
4. $(a^3-a^2)(a-6)=$
5. $\left(\sqrt{x^3}-2x\sqrt{x^5}\right)(\sqrt{x}-4)=$
6. $\left(\sqrt[3]{y^5}-\sqrt[3]{y^2}\right)\left(\sqrt[3]{y^7}-\sqrt[3]{y}\right)=$

Case III Dividing Polynomials by Monomials and Polynomials

Polynomials are divided by one another using a similar method like the long division used in arithmetic operations. To divide a polynomial, a trinomial, or a binomial by a monomial we **divide each term** in the numerator which is separated by a + or a − sign by the denominator. In general, polynomials are divided by monomials in the following way:

$$\frac{a_n x^n + a_{n-1}x^{n-1} + a_{n-2}x^{n-2} + \ldots + a_0}{bx^m} = \frac{a_n x^n}{bx^m} + \frac{a_{n-1}x^{n-1}}{bx^m} + \frac{a_{n-2}x^{n-2}}{bx^m} + \ldots + \frac{a_0}{bx^m}$$

where a_n, a_{n-1}, a_{n-2}, ..., a_0, and b are real numbers, n and m are positive integer numbers, and x is a variable. For example,

$$\frac{16y^4 + 5y^3 + 4y^2 + 8y + 20}{4} = \frac{16y^4}{4} + \frac{5y^3}{4} + \frac{4y^2}{4} + \frac{8y}{4} + \frac{20}{4}$$

Note that we **can not divide out only one term** of the polynomial in the numerator by the denominator, i.e., we can not do the following:

$$\frac{4x^3 - 5x^2 + 6x + 12}{2} \neq \frac{\overset{2}{\cancel{4}}x^3 - 5x^2 + 6x + 12}{\underset{1}{\cancel{2}}}$$

instead,

$$\frac{4x^3 - 5x^2 + 6x + 12}{2} = \frac{\overset{2}{\cancel{4}}x^3}{\underset{1}{\cancel{2}}} - \frac{5x^2}{2} + \frac{\overset{3}{\cancel{6}}x}{\underset{1}{\cancel{2}}} + \frac{\overset{6}{\cancel{12}}}{\underset{1}{\cancel{2}}} = 2x^3 - \frac{5}{2}x^2 + 3x + 6$$

Division by monomial and polynomial expressions is divided to three cases. Case III a - dividing monomial by monomials, Case III b - dividing binomials by monomials, and Case III c - dividing polynomials by polynomials.

Case IIIa - Dividing Monomials by Monomials

Monomials are divided by one another using the following steps:

Step 1 Simplify the monomials in both the numerator and the denominator.

Step 2 Divide the numerator by the denominator using the exponent rules (see Section 1.3) for dividing a variable, i.e.,

$$\frac{x^n}{x^m} = \frac{x^{n-m}}{1} = x^{n-m} \qquad \text{if } n > m$$

or,

$$\frac{x^n}{x^m} = \frac{1}{x^{m-n}} \qquad \text{if } n < m$$

where n and m are positive integer numbers and x is a variable.

Examples with Steps

The following examples show the steps as to how monomials are divided by each other:

Example 1.5-28

$$\boxed{\frac{\sqrt{8x^3y^2}}{\sqrt{243xy^3}}} =$$

Solution:

Step 1

$$\boxed{\frac{\sqrt{8x^3y^2}}{\sqrt{243xy^3}}} = \boxed{\frac{\sqrt{4\cdot 2x^3y^2}}{\sqrt{81\cdot 3xy^3}}} = \boxed{\frac{\sqrt{2^2\cdot 2x^3y^2}}{\sqrt{9^2\cdot 3xy^3}}} = \boxed{\frac{2\sqrt{2x^3y^2}}{9\sqrt{3xy^3}}}$$

Step 2

$$\boxed{\frac{2\sqrt{2x^3y^2}}{9\sqrt{3xy^3}}} = \boxed{\frac{2\sqrt{2}\ x^3y^2}{9\sqrt{3}\ x^1y^3}} = \boxed{\frac{2\sqrt{2}\ x^3x^{-1}}{9\sqrt{3}\ y^3y^{-2}}} = \boxed{\frac{2\sqrt{2}\ x^{3-1}}{9\sqrt{3}\ y^{3-2}}} = \boxed{\frac{2\sqrt{2}\ x^2}{9\sqrt{3}\ y^1}} = \boxed{\mathbf{\frac{2\sqrt{2}x^2}{9\sqrt{3}y}}}$$

Example 1.5-29

$$\boxed{\frac{-\sqrt{12a^2b^2c}}{\sqrt{225abc^4}}} =$$

Solution:

Step 1

$$\boxed{\frac{-\sqrt{12a^2b^2c}}{\sqrt{225abc^4}}} = \boxed{-\frac{\sqrt{4\cdot 3a^2b^2c}}{\sqrt{15\cdot 15abc^4}}} = \boxed{-\frac{\sqrt{2^2\cdot 3a^2b^2c}}{\sqrt{15^2\,abc^4}}} = \boxed{-\frac{2\sqrt{3a^2b^2c}}{15abc^4}}$$

Step 2

$$\boxed{-\frac{2\sqrt{3a^2b^2c}}{15abc^4}} = \boxed{-\frac{2\sqrt{3}}{15}\frac{a^2b^2c^1}{a^1b^1c^4}} = \boxed{-\frac{2\sqrt{3}}{15}\frac{\left(a^2a^{-1}\right)\cdot\left(b^2b^{-1}\right)}{c^4c^{-1}}} = \boxed{-\frac{2\sqrt{3}}{15}\frac{\left(a^{2-1}\right)\cdot\left(b^{2-1}\right)}{c^{4-1}}}$$

$$= \boxed{-\frac{2\sqrt{3}}{15}\frac{a^1b^1}{c^3}} = \boxed{\mathbf{-\frac{2\sqrt{3}ab}{15c^3}}}$$

Additional Examples - Dividing Monomials by Monomials

The following examples further illustrate how to divide monomials by each other:

Example 1.5-30

$$\boxed{\frac{x^6y^5}{x^4y}} = \boxed{\frac{x^6y^5}{x^4y^1}} = \boxed{\frac{\left(x^6x^{-4}\right)\cdot\left(y^5y^{-1}\right)}{1}} = \boxed{\frac{x^{6-4}y^{5-1}}{1}} = \boxed{\frac{x^2y^4}{1}} = \boxed{\mathbf{x^2y^4}}$$

Example 1.5-31

$$\boxed{\frac{4xy^2z^3}{-16x^2yz^2}} = \boxed{-\frac{\cancel{4}\ x^1y^2z^3}{\underset{4}{\cancel{16}}\ x^2y^1z^2}} = \boxed{-\frac{1}{4}\frac{\left(y^2y^{-1}\right)\cdot\left(z^3z^{-2}\right)}{x^2x^{-1}}} = \boxed{-\frac{1}{4}\frac{y^{2-1}z^{3-2}}{x^{2-1}}} = \boxed{-\frac{1}{4}\frac{y^1z^1}{x^1}} = \boxed{-\frac{1}{4}\frac{yz}{x}} = \boxed{\mathbf{-\frac{yz}{4x}}}$$

Example 1.5-32

$$\boxed{\frac{-5a^2b}{15a^2b^3}} = \boxed{-\frac{\cancel{5}\ a^2b^1}{\underset{3}{\cancel{15}}\ a^2b^3}} = \boxed{-\frac{1}{3}\frac{a^2a^{-2}}{b^3b^{-1}}} = \boxed{-\frac{1}{3}\frac{a^{2-2}}{b^{3-1}}} = \boxed{-\frac{1}{3}\frac{a^0}{b^2}} = \boxed{-\frac{1}{3}\frac{1}{b^2}} = \boxed{\mathbf{-\frac{1}{3b^2}}}$$

Example 1.5-33

$$\boxed{\frac{-4xy^4}{-\sqrt{64x^3y^2}}} = \boxed{+\frac{4}{\sqrt{64}}\frac{xy^4}{x^3y^2}} = \boxed{\frac{4}{\sqrt{8^2}}\frac{xy^4}{x^3y^2}} = \boxed{\frac{\cancel{4}\ x^1y^4}{\underset{2}{\cancel{8}}\ x^3y^2}} = \boxed{\frac{1}{2}\frac{y^4y^{-2}}{x^3x^{-1}}} = \boxed{\frac{1}{2}\frac{y^{4-2}}{x^{3-1}}} = \boxed{\frac{1}{2}\frac{y^2}{x^2}} = \boxed{\mathbf{\frac{y^2}{2x^2}}}$$

Example 1.5-34

$$\boxed{\frac{56}{-\sqrt{36}}} = \boxed{-\frac{56}{\sqrt{6^2}}} = \boxed{-\frac{\overset{28}{\cancel{56}}}{\underset{3}{\cancel{6}}}} = \boxed{-\frac{28}{3}} = \boxed{-\left(9\frac{1}{3}\right)}$$

Example 1.5-35

$$\boxed{\frac{49\sqrt{x^2y^3z^4}}{-7\sqrt{x^2y^4z^6}}} = \boxed{-\frac{49}{7}\frac{x\sqrt{y^{2+1}z^{2+2}}}{x\sqrt{y^{2+2}z^{2+2+2}}}} = \boxed{-\frac{\overset{7}{\cancel{49}}}{\cancel{7}}\frac{x\sqrt{\left(y^2\cdot y^1\right)\cdot\left(z^2\cdot z^2\right)}}{x\sqrt{\left(y^2\cdot y^2\right)\cdot\left(z^2\cdot z^2\cdot z^2\right)}}} = \boxed{-\frac{7}{1}\frac{x\cdot y(z\cdot z)\sqrt{y}}{x\cdot(y\cdot y)\cdot(z\cdot z\cdot z)\sqrt{1}}}$$

$$= \boxed{-\frac{7}{1}\frac{xyz^2\sqrt{y}}{xy^2z^3}} = \boxed{-\frac{7}{1}\frac{x^1y^1z^2\sqrt{y}}{x^1y^2z^3}} = \boxed{-\frac{7}{1}\frac{\left(x^1x^{-1}\right)\sqrt{y}}{\left(y^2y^{-1}\right)\cdot\left(z^3z^{-2}\right)}} = \boxed{-\frac{7}{1}\frac{x^{1-1}\sqrt{y}}{y^{2-1}z^{3-2}}} = \boxed{-\frac{7}{1}\frac{x^0\sqrt{y}}{y^1z^1}} = \boxed{-\frac{7\sqrt{y}}{yz}}$$

Practice Problems - Dividing Monomials by Monomials

Section 1.5b Case IIIa Practice Problems - Divide the following monomial expressions:

1. $\frac{-4xyz}{-8xyz} =$
2. $\frac{u^2v^3w}{-uw^4} =$
3. $\frac{\sqrt{72x^2y^4}}{-12xy^2} =$
4. $\frac{-36x^3y^3z^4}{-\sqrt{25}xyz^3} =$
5. $\frac{-9a^2b^2c^3}{\sqrt[3]{27a^6b^3c^3}} =$
6. $\frac{-24lm^3n^2}{12l^2mn} =$

Case IIIb Dividing Binomials by Monomials

Binomials are divided by monomial expressions using the following steps:

Step 1 Simplify each term in the numerator and the denominator.

Step 2 Divide each binomial term by the denominator using the exponent rules (see Section 1.3) for dividing a variable, i.e.,

$$\frac{x^n}{x^m} = \frac{x^{n-m}}{1} = x^{n-m} \qquad \text{if } n \rangle m$$

or,

$$\frac{x^n}{x^m} = \frac{1}{x^{m-n}} \qquad \text{if } n \langle m$$

where n and m are positive integer numbers and x is a variable.

Examples with Steps

The following examples show the steps as to how binomials are divided by monomials:

Example 1.5-36

$$\boxed{\frac{-15\sqrt{a^3}+10\sqrt{a^2}}{-5a^2}} =$$

Solution:

Step 1

$$\boxed{\frac{-15\sqrt{a^3}+10\sqrt{a^2}}{-5a^2}} = \boxed{\frac{-15\sqrt{a^{2+1}}+10a}{-5a^2}} = \boxed{\frac{-15\sqrt{a^2a^1}+10a}{-5a^2}} = \boxed{\frac{-15a\sqrt{a}+10a}{-5a^2}}$$

Step 2

$$\boxed{\frac{-15a\sqrt{a}+10a}{-5a^2}} = \boxed{\frac{-15a\sqrt{a}}{-5a^2}+\frac{10a}{-5a^2}} = \boxed{\frac{\overset{3}{\cancel{15}}a^1\sqrt{a}}{\underset{1}{\cancel{5}}a^2}-\frac{\overset{2}{\cancel{10}}a^1}{\underset{1}{\cancel{5}}a^2}} = \boxed{\frac{3\sqrt{a}}{a^2a^{-1}}-\frac{2}{a^2a^{-1}}}$$

$$= \boxed{\frac{3\sqrt{a}}{a^{2-1}}-\frac{2}{a^{2-1}}} = \boxed{\frac{3\sqrt{a}}{a^1}-\frac{2}{a^1}} = \boxed{\frac{3\sqrt{a}}{a}-\frac{2}{a}} = \boxed{\mathbf{\frac{3\sqrt{a}-2}{a}}}$$

Example 1.5-37

$$\boxed{\frac{\sqrt[3]{x^3y^5}-4\sqrt[4]{x^8y^4}}{2\sqrt[3]{x^3y^6}}} =$$

Solution:

Step 1

$$\boxed{\frac{\sqrt[3]{x^3y^5}-4\sqrt[4]{x^8y^4}}{2\sqrt[3]{x^3y^6}}} = \boxed{\frac{x\sqrt[3]{y^{3+2}}-4y\sqrt[4]{x^{4+4}}}{2x\sqrt[3]{y^{3+3}}}} = \boxed{\frac{x\sqrt[3]{y^3\cdot y^2}-4y\sqrt[4]{x^4\cdot x^4}}{2x\sqrt[3]{y^3\cdot y^3}}}$$

$$= \boxed{\frac{xy\sqrt[3]{y^2}-4(x\cdot x)y}{2x(y\cdot y)}} = \boxed{\frac{xy\sqrt[3]{y^2}-4x^2y}{2xy^2}}$$

Step 2

$$\frac{xy\sqrt[3]{y^2}-4x^2y}{2xy^2} = \frac{xy\sqrt[3]{y^2}}{2xy^2}+\frac{-4x^2y}{2xy^2} = \frac{x^1y^1\sqrt[3]{y^2}}{2x^1y^2}-\frac{\overset{2}{\cancel{4}}x^2y^1}{\underset{1}{\cancel{2}}x^1y^2} = \frac{x^1x^{-1}\sqrt[3]{y^2}}{2y^2y^{-1}}-\frac{2x^2x^{-1}}{y^2y^{-1}}$$

$$= \frac{x^{1-1}\sqrt[3]{y^2}}{2y^{2-1}}-\frac{2x^{2-1}}{y^{2-1}} = \frac{x^0\sqrt[3]{y^2}}{2y^1}-\frac{2x^1}{y^1} = \boxed{\frac{\sqrt[3]{y^2}}{2y}-\frac{2x}{y}}$$

Additional Examples - Dividing Binomials by Monomials

The following examples further illustrate how binomials are divided by monomials:

Example 1.5-38

$$\frac{10x^3-5x^2}{5x^2} = \frac{10x^3}{5x^2}+\frac{-5x^2}{5x^2} = \frac{10x^3x^{-2}}{5}-\frac{5x^2x^{-2}}{5} = \frac{\overset{2}{\cancel{10}}x^{3-2}}{\underset{1}{\cancel{5}}}-\frac{\overset{1}{\cancel{5}}x^{2-2}}{\underset{1}{\cancel{5}}} = \frac{2x^1}{1}-\frac{x^0}{1} = \frac{2x}{1}-\frac{1}{1} = \boxed{2x-1}$$

Example 1.5-39

$$\frac{m^3+n^4}{m^2n^2} = \frac{m^3}{m^2n^2}+\frac{n^4}{m^2n^2} = \frac{m^3m^{-2}}{n^2}+\frac{n^4n^{-2}}{m^2} = \frac{m^{3-2}}{n^2}+\frac{n^{4-2}}{m^2} = \frac{m^1}{n^2}+\frac{n^2}{m^2} = \boxed{\frac{m}{n^2}+\frac{n^2}{m^2}}$$

Example 1.5-40

$$\frac{x^3y^2z+x^4yz^2}{x^2y^3z^5} = \frac{x^3y^2z}{x^2y^3z^5}+\frac{x^4yz^2}{x^2y^3z^5} = \frac{x^3y^2z^1}{x^2y^3z^5}+\frac{x^4y^1z^2}{x^2y^3z^5} = \frac{x^3x^{-2}}{\left(y^3y^{-2}\right)\cdot\left(z^5z^{-1}\right)}+\frac{x^4x^{-2}}{\left(y^3y^{-1}\right)\cdot\left(z^5z^{-2}\right)}$$

$$= \frac{x^{3-2}}{y^{3-2}z^{5-1}}+\frac{x^{4-2}}{y^{3-1}z^{5-2}} = \frac{x^1}{y^1z^4}+\frac{x^2}{y^2z^3} = \boxed{\frac{x}{yz^4}+\frac{x^2}{y^2z^3}}$$

Example 1.5-41

$$\frac{\sqrt[3]{a^3b^4}+\sqrt[2]{a^2b^4}}{a^2b^2} = \frac{a\sqrt[3]{b^{3+1}}+a\sqrt{b^{2+2}}}{a^2b^2} = \frac{a\sqrt[3]{b^3\cdot b^1}+a\sqrt{b^2\cdot b^2}}{a^2b^2} = \frac{ab\sqrt[3]{b}+a(b\cdot b)}{a^2b^2} = \frac{ab\sqrt[3]{b}+ab^2}{a^2b^2}$$

$$= \frac{ab\sqrt[3]{b}}{a^2b^2}+\frac{ab^2}{a^2b^2} = \frac{a^1b^1\sqrt[3]{b}}{a^2b^2}+\frac{a^1b^2}{a^2b^2} = \frac{\sqrt[3]{b}}{a^2a^{-1}\cdot b^2b^{-1}}+\frac{b^2b^{-2}}{a^2a^{-1}} = \frac{\sqrt[3]{b}}{a^{2-1}b^{2-1}}+\frac{b^{2-2}}{a^{2-1}} = \boxed{\frac{\sqrt[3]{b}}{ab}+\frac{1}{a}}$$

Practice Problems - Dividing Binomials by Monomials

Section 1.5b Case IIIb Practice Problems - Divide the following binomial expressions by monomials:

1. $\frac{98-46}{-12} =$
2. $\frac{x^3y^3z+4x^2y^2}{-2xy^2z} =$
3. $\frac{-a^3b^3c+a^2bc^2}{-a^2b^2c^2} =$
4. $\frac{\sqrt[4]{a^5b^4c^3}-\sqrt[3]{a^3b^6c}}{\sqrt{a^2b^4c^6}} =$
5. $\frac{m^3n^2l+ml^2}{mnl} =$
6. $\frac{36y^2-18y^3}{-9y} =$

Case IIIc Dividing Polynomials by Polynomials

Whole numbers are divided by one another using the long division method which can be summarized as: selecting a quotient, multiplying the quotient by the divisor to obtain a product, subtracting the product from the dividend, and bringing down the next digit/dividend term. Polynomials are divided by one another in a similar way as the long division method used for whole numbers. The following are the steps for dividing two polynomials by each other:

Step 1 a. Select the first term for the quotient which divides the first term of the dividend by the first term of the divisor.

b. Multiply the selected first term of the quotient by the divisor.

c. Write the product under the dividend.

Step 1 a d. Change the signs of the product written under the dividend.

e. Subtract the product from the dividend.

f. Bring down the next term from the dividend to obtain a new dividend.

Step 2 a. Select the second term for the quotient which divides the first term of the new dividend by the first term of the divisor.

b. Multiply the selected second term of the quotient by the divisor.

c. Write the product under the new dividend.

Step 2 a d. Change the signs of the product written under the new dividend.

e. Subtract the product from the new dividend to obtain a remainder. If a remainder is not obtained, proceed with the next step.

f. Bring down the next term from the dividend to obtain another new dividend.

g. Repeat Steps $2a.$ through $2f.$ until a remainder is obtained.

To **check** the answer multiply the quotient by the divisor and add in the remainder. The result should match the dividend.

Examples with Steps

The following examples show the steps as to how polynomials are divided by one another:

Example 1.5-42: Divide $x^4+8x^3+16x^2+5x$ by x^2+3x+1.

Solution:

Step 1

$$\begin{array}{r|l} & x^2 \quad \textit{first term of the quotient} \\ \hline x^2+3x+1 & +x^4+8x^3+16x^2+5x \quad \textit{dividend} \\ \textit{divisor} & +x^4+3x^3+\;x^2 \end{array}$$

Step 1 a

$$\begin{array}{r|l} & x^2 \\ \hline x^2+3x+1 & +x^4+8x^3+16x^2+5x \\ & \mp x^4 \mp 3x^3 \mp \; x^2 \\ \hline & \qquad +5x^3+15x^2+5x \quad \textit{new dividend} \end{array}$$

Step 2

$$\begin{array}{r|l} & x^2+5x \\ \hline x^2+3x+1 & +x^4+8x^3+16x^2+5x \\ & \overset{-}{+}x^4 \overset{-}{+}3x^3 \overset{-}{+} x^2 \\ \hline & +5x^3+15x^2+5x \\ & +5x^3+15x^2+5x \end{array}$$

first and final term of the quotient

Step 2 a

$$\begin{array}{r|ll} & x^2+5x & \textit{quotient} \\ \hline x^2+3x+1 & +x^4+8x^3+16x^2+5x & \textit{dividend} \\ \textit{divisor} & \overset{-}{+}x^4 \overset{-}{+}3x^3 \overset{-}{+} x^2 & \\ \hline & +5x^3+15x^2+5x & \\ & \overset{-}{+}5x^3 \overset{-}{+}15x^2 \overset{-}{+}5x & \\ \hline & 0 & \textit{remainder} \end{array}$$

The answer is x^2+5x with remainder of zero.

Check: $(x^2+5x)(x^2+3x+1) = (x^2\cdot x^2)+3(x^2\cdot x)+(x^2\cdot 1)+5(x\cdot x^2)+(5\cdot 3)(x\cdot x)+(5\cdot 1)x$

$= x^4+3x^3+x^2+5x^3+15x^2+5x = x^4+(3x^3+5x^3)+(x^2+15x^2)+5x$

$= x^4+(3+5)x^3+(1+15)x^2+5x = x^4+8x^3+16x^2+5x$ *which is the same as the dividend*

Example 1.5-43: Divide $6x^2+19x+18$ by $3x+5$.

Solution:

Step 1

$$\begin{array}{r|ll} & 2x & \textit{first term of the quotient} \\ \hline 3x+5 & +6x^2+19x+18 & \textit{dividend} \\ \textit{divisor} & +6x^2+10x & \end{array}$$

Step 1 a

$$\begin{array}{r|ll} & 2x & \\ \hline 3x+5 & +6x^2+19x+18 & \\ & \overset{-}{+}6x^2 \overset{-}{+}10x & \\ \hline & +9x+18 & \textit{new dividend} \end{array}$$

Step 2

$$\begin{array}{r|ll} & 2x+3 & \textit{first and final term of the quotient} \\ \hline 3x+5 & +6x^2+19x+18 & \\ & \overset{-}{+}6x^2 \overset{-}{+}10x & \\ \hline & +9x+18 & \\ & +9x+15 & \end{array}$$

Step 2 a

$$\begin{array}{r|ll} & 2x+3 & \textit{quotient} \\ \hline 3x+5 & +6x^2+19x+18 & \textit{dividend} \\ \textit{divisor} & \overset{-}{+}6x^2 \overset{-}{+}10x & \\ \hline & +9x+18 & \\ & \overset{-}{+}9x \overset{-}{+}15 & \\ \hline & +3 & \textit{remainder} \end{array}$$

The answer is $2x+3$ with remainder of $+3$, or $2x+3+\dfrac{3}{3x+5}$.

Check: $(2x+3)(3x+5)+3 = (2\cdot 3)(x\cdot x)+(2\cdot 5)x+(3\cdot 3)x+(3\cdot 5)+3 = 6x^2+10x+9x+15+3$

$= 6x^2+(10x+9x)+(15+3) = 6x^2+19x+18$ *which is the same as the dividend*

Additional Examples - Dividing Polynomials by Polynomials

The following examples further illustrate how to divide two polynomials by each other:

Example 1.5-44: Divide $x^3+6x^2+14x+20$ by $x+3$.

Solution:

$$
\begin{array}{r|l}
 & x^2+3x+5 \\
\hline
x+3 & +x^3+6x^2+14x+20 \\
 & \underline{\mp x^3 \mp 3x^2} \\
 & \quad +3x^2+14x \\
 & \quad \underline{\mp 3x^2 \mp 9x} \\
 & \qquad +5x+20 \\
 & \qquad \underline{\mp 5x \mp 15} \\
 & \qquad\qquad +5
\end{array}
$$

The answer is x^2+3x+5 with remainder of $+5$, or $x^2+3x+5+\dfrac{5}{x+3}$.

Example 1.5-45: Divide $2x^5+3x^4-9x^3+12x-18$ by $2x-3$.

Solution:

$$
\begin{array}{r|l}
 & x^4+3x^3+6 \\
\hline
2x-3 & +2x^5+3x^4-9x^3+12x-18 \\
 & \underline{\mp 2x^5 \pm 3x^4} \\
 & \quad +6x^4-9x^3 \\
 & \quad \underline{+6x^4-9x^3} \\
 & \qquad\qquad +12x-18 \\
 & \qquad\qquad \underline{\mp 12x \pm 18} \\
 & \qquad\qquad\qquad 0
\end{array}
$$

The answer is x^4+3x^3+6 with remainder of zero.

Practice Problems - Dividing Polynomials by Polynomials

Section 1.5b Case IIIc Practice Problems - Divide the following polynomial expressions:

1. $3x^2+10x+3$ by $x+3$
2. $x^4+7x^3+13x^2+17x+10$ by $x+5$
3. $x^6-x^5-2x^4-x^3+2x^2+5x-10$ by $x-2$
4. $-2x^4+5x^3-4x^2+16x-15$ by $-2x+5$
5. $2x^4-13x^3+13x^2+15x-25$ by $x-5$
6. $-2x^4+7x^3-6x^2-2x+3$ by $-2x+3$

Case IV Adding and Subtracting Polynomials Horizontally

Polynomials are added and subtracted by combining their numerical coefficients while keeping the like terms. Polynomials can be added horizontally or vertically. Polynomials are horizontally added and subtracted using the following steps:

Step 1 Write the polynomial in descending order.

Step 2 Group the like terms. (Note: In the case of subtraction, change the sign in each term of the polynomial being subtracted before grouping the like terms.)

Step 3 Add or subtract the like terms.

Examples with Steps

The following examples show the steps as to how polynomials are added and subtracted horizontally:

Example 1.5-46

$$\left(x^2+3x^3+5\right)+\left(x^3+8x+2x^2\right)=$$

Solution:

Step 1 $\left(x^2+3x^3+5\right)+\left(x^3+8x+2x^2\right)=\left(3x^3+x^2+5\right)+\left(x^3+2x^2+8x\right)$

Step 2 $\left(3x^3+x^2+5\right)+\left(x^3+2x^2+8x\right)=\left(3x^3+x^3\right)+\left(x^2+2x^2\right)+8x+5$

Step 3 $\left(3x^3+x^3\right)+\left(x^2+2x^2\right)+8x+5=(3+1)x^3+(1+2)x^2+8x+5=\mathbf{4x^3+3x^2+8x+5}$

Example 1.5-47

$$\left(y+y^2+3y^3+3\right)-\left(3y^2+2y-y^3\right)=$$

Solution:

Step 1 $\left(y+y^2+3y^3+3\right)-\left(3y^2+2y-y^3\right)=\left(3y^3+y^2+y+3\right)-\left(-y^3+3y^2+2y\right)$

Step 2 $\left(3y^3+y^2+y+3\right)-\left(-y^3+3y^2+2y\right)=\left(3y^3+y^2+y+3\right)+\left(+y^3-3y^2-2y\right)$

$=\left(3y^3+y^3\right)+\left(y^2-3y^2\right)+(y-2y)+3$

Step 3 $\left(3y^3+y^3\right)+\left(y^2-3y^2\right)+(y-2y)+3=(3+1)y^3+(1-3)y^2+(1-2)y+3$

$=\mathbf{4y^3-2y^2-y+3}$

Additional Examples - Adding and Subtracting Polynomials Horizontally

The following examples further illustrate how to add and subtract polynomials horizontally:

Example 1.5-48

$$\left(-3x^3-5x-6x+7x^0\right)-\left(12x^3+5x^2-3x\right)=\left(-3x^3-5x-6x+(7\cdot 1)\right)+\left(-12x^3-5x^2+3x\right)$$

$$=\left(-3x^3-12x^3\right)-5x^2+(-5x-6x+3x)+7=(-3-12)x^3-5x^2+(-5-6+3)x+7=\mathbf{-15x^3-5x^2-8x+7}$$

Example 1.5-49

$$-\left(x^3+3x^2+6x^4-5\right)+\left(5x-3x^3-2x^2+2x^0\right)=-\left(6x^4+x^3+3x^2-5\right)+\left(-3x^3-2x^2+5x+2x^0\right)$$

$$=\left(-6x^4-x^3-3x^2+5\right)+\left(-3x^3-2x^2+5x+(2\cdot 1)\right)=-6x^4+\left(-x^3-3x^3\right)+\left(-3x^2-2x^2\right)+5x+(5+2)$$

$$=-6x^4+(-1-3)x^3+(-3-2)x^2+5x+7=\mathbf{-6x^4-4x^3-5x^2+5x+7}$$

Example 1.5-50

$$\left(3x^5+4x^0+2x^4+2x^2\right)+\left(5x-3x^2-3x^4+3\right)=\left(3x^5+2x^4+2x^2+4x^0\right)+\left(-3x^4-3x^2+5x+3\right)$$

$$=\left(3x^5+2x^4+2x^2+4\right)+\left(-3x^4-3x^2+5x+3\right)=3x^5+\left(2x^4-3x^4\right)+\left(2x^2-3x^2\right)+5x+(4+3)$$

$$=3x^5+(2-3)x^4+(2-3)x^2+5x+7=\mathbf{3x^5-x^4-x^2+5x+7}$$

Example 1.5-51

$$\left(7x+3x^3-2x^2+5\right)+\left(2x-3x^2+x^0\right)-\left(x^2+4x^3\right)=\left(3x^3-2x^2+7x+5\right)+\left(-3x^2+2x+x^0\right)-\left(4x^3+x^2\right)$$

$$=\left(3x^3-2x^2+7x+5\right)+\left(-3x^2+2x+1\right)+\left(-4x^3-x^2\right)=\left(3x^3-4x^3\right)+\left(-2x^2-3x^2-x^2\right)+(7x+2x)+(5+1)$$

$$=(3-4)x^3+(-2-3-1)x^2+(7+2)x+6=\mathbf{-x^3-6x^2+9x+6}$$

Practice Problems - Adding and Subtracting Polynomials Horizontally

Section 1.5b Case IV Practice Problems - Add or subtract the following polynomials horizontally:

1. $\left(x^3+2x^5-3x+2\right)+\left(3x^3+x-x^5\right)=$
2. $\left(y-y^2+2y^4+3y^2-3\right)+\left(2y^4+y^3+5-y^2\right)=$
3. $\left(3x-3x^2+5x-3\right)-\left(-2x+5-x^2+2\right)=$
4. $\left(xyz+2x^2yz+4xyz\right)+\left(4x^2yz-x^2yz+2xyz\right)=$
5. $\left(-2ab-3+2a^2b^2\right)+\left(-3ab+a^2b^2+2(ab)^0\right)=$
6. $\left(5x^6-x^5-4x^4+3x+x^2\right)-\left(x-3x^2+x^4-3x^6\right)=$

Case V Adding and Subtracting Polynomials Vertically

Polynomials are vertically added and subtracted using the following steps:

Step 1 Write the polynomials in descending order.

Step 2 Group the like terms in each polynomial separately. (Note: In the case of subtraction, change the sign in each term of the polynomial being subtracted before grouping the like terms.)

Step 3 Write the like terms under one another.

Step 4 Add or subtract the like terms.

Examples with Steps

The following examples show the steps as to how polynomials are added and subtracted vertically:

Example 1.5-52

$$\boxed{\left(x^4+x+3x^3+4x\right)+\left(x^2+3x^4-x^3+x\right)}=$$

Solution:

Step 1

$$\boxed{\left(x^4+x+3x^3+4x\right)+\left(x^2+3x^4-x^3+x\right)}=\boxed{\left(x^4+3x^3+4x+x\right)+\left(3x^4-x^3+x^2+x\right)}$$

Step 2

$$\boxed{\left(x^4+3x^3+4x+x\right)+\left(3x^4-x^3+x^2+x\right)}=\boxed{\left[x^4+3x^3+(4+1)x\right]+\left(3x^4-x^3+x^2+x\right)}$$

$$=\boxed{\left(x^4+3x^3+5x\right)+\left(3x^4-x^3+x^2+x\right)}$$

Step 3

$$\boxed{\left(x^4+3x^3+5x\right)+\left(3x^4-x^3+x^2+x\right)}=\boxed{\begin{array}{l} x^4+3x^3\qquad\quad+5x \\ 3x^4-x^3+x^2+x \end{array}}$$

Step 4

$$\boxed{\begin{array}{l} x^4+3x^3\qquad\quad+5x \\ 3x^4-x^3+x^2+x \end{array}}=\begin{array}{l} \quad x^4+3x^3\qquad\quad+5x \\ \underline{\quad 3x^4-x^3+x^2+x} \\ \mathbf{4x^4+2x^3+x^2+6x} \end{array}$$

Example 1.5-53

$$\boxed{\left(x^5+x^4+2x+5\right)-\left(5x-3x^4-x+3x+6\right)}=$$

Solution:

Step 1

$$\boxed{\left(x^5+x^4+2x+5\right)-\left(5x-3x^4-x+3x+6\right)}=\boxed{\left(x^5+x^4+2x+5\right)-\left(-3x^4+5x+3x-x+6\right)}$$

Step 2

$$\boxed{\left(x^5+x^4+2x+5\right)-\left(-3x^4+5x+3x-x+6\right)}=\boxed{\left(x^5+x^4+2x+5\right)+\left(3x^4-5x-3x+x-6\right)}$$

$$=\boxed{\left(x^5+x^4+2x+5\right)+\left[3x^4+(-5-3+1)x-6\right]}=\boxed{\left(x^5+x^4+2x+5\right)+\left(3x^4-7x-6\right)}$$

Step 3 $\boxed{\left(x^5 + x^4 + 2x + 5\right) + \left(3x^4 - 7x - 6\right)} = \boxed{\begin{array}{r} x^5 + \; x^4 + 2x + 5 \\ +3x^4 - 7x - 6 \end{array}}$

Step 4 $\boxed{\begin{array}{r} x^5 + \; x^4 + 2x + 5 \\ +3x^4 - 7x - 6 \end{array}} = \begin{array}{r} x^5 + \; x^4 + 2x + 5 \\ +3x^4 - 7x - 6 \\ \hline \mathbf{x^5 + 4x^4 - 5x - 1} \end{array}$

Additional Examples - Adding and Subtracting Polynomials Vertically

The following examples further illustrate how to add and subtract polynomials vertically:

Example 1.5-54

$$\boxed{\left(y^0 + 2y + 4y^3 + 3y^2\right) - \left(y^3 - 2y^2 - 4y + 3y + 4\right)} = \boxed{\left(4y^3 + 3y^2 + 2y + y^0\right) - \left(y^3 - 2y^2 - 4y + 3y + 4\right)}$$

$$= \boxed{\left(4y^3 + 3y^2 + 2y + y^0\right) + \left(-y^3 + 2y^2 + 4y - 3y - 4\right)} = \boxed{\left(4y^3 + 3y^2 + 2y + 1\right) + \left(-y^3 + 2y^2 + (4-3)y - 4\right)}$$

$$= \boxed{\left(4y^3 + 3y^2 + 2y + 1\right) + \left(-y^3 + 2y^2 + y - 4\right)} = \begin{array}{r} 4y^3 + 3y^2 + 2y + 1 \\ -y^3 + 2y^2 + \; y - 4 \\ \hline \mathbf{3y^3 + 5y^2 + 3y - 3} \end{array}$$

Example 1.5-55

$$\boxed{\left(y^2z^2 + 5y^2z^2 - 15 + 2yz\right) - \left(30 + 4y^2z^2 - 2yz + 3\right)} = \boxed{\left(y^2z^2 + 5y^2z^2 + 2yz - 15\right) - \left(4y^2z^2 - 2yz + 30 + 3\right)}$$

$$= \boxed{\left(y^2z^2 + 5y^2z^2 + 2yz - 15\right) + \left(-4y^2z^2 + 2yz - 30 - 3\right)} = \boxed{\left((1+5)y^2z^2 + 2yz - 15\right) + \left(-4y^2z^2 + 2yz - (30+3)\right)}$$

$$= \boxed{\left(6y^2z^2 + 2yz - 15\right) + \left(-4y^2z^2 + 2yz - 33\right)} = \begin{array}{r} 6y^2z^2 + 2yz - 15 \\ -4y^2z^2 + 2yz - 33 \\ \hline \mathbf{2y^2z^2 + 4yz - 48} \end{array}$$

Example 1.5-56

$$\boxed{\left(x^2y^2 + 2x^2y^2 + 2x + 5\right) + \left(3x^2y^2 - 3 - 6x + 5\right)} = \boxed{\left(x^2y^2 + 2x^2y^2 + 2x + 5\right) + \left(3x^2y^2 - 6x - 3 + 5\right)}$$

$$= \boxed{\left((1+2)x^2y^2 + 2x + 5\right) + \left(3x^2y^2 - 6x + (-3+5)\right)} = \boxed{\left(3x^2y^2 + 2x + 5\right) + \left(3x^2y^2 - 6x + 2\right)} = \begin{array}{r} 3x^2y^2 + 2x + 5 \\ 3x^2y^2 - 6x + 2 \\ \hline \mathbf{6x^2y^2 - 4x + 7} \end{array}$$

Practice Problems - Adding and Subtracting Polynomials Vertically

Section 1.5b Case V Practice Problems - Add or subtract the following polynomials vertically:

1. $\left(x^2 + 2x + x^3\right) + \left(3x - 2x^3\right) =$
2. $\left(y + y^2 + 3y^3 + 4\right) + \left(-2 + y^2 + 3y^2 + 2y\right) =$
3. $\left(x^3 + x^2 - 3 + 3x^2\right) - \left(-2x^3 - 5x + 5\right) =$
4. $\left(z^5 + 3z^2 + z - 2z^2 - 4z + 2\right) + \left(z^2 + 4z^5 + z^0\right) =$
5. $-\left(a^3 - 2a + a + 2 - 3a^3\right) + \left(-2a^3 - 4a - 3\right) =$
6. $\left(u^2 + 2u + u + 5\right) + \left(-2u^2 - 3 - 5u - 8\right) =$

Chapter 2
One Variable Linear Equations and Inequalities

Quick Reference to Chapter 2 Case Problems

Chapter 2 - One Variable Linear Equations and Inequalities

The objective of this chapter is to improve the student's ability to solve operations involving linear equations and linear inequalities. One variable linear equations and the process of determining the solution to an algebraic equation as well as how the solution set to a linear equation is verified is addressed in Section 2.1. Math operations involving addition, subtraction, multiplication, and division of linear equations are addressed in Section 2.2. In Section 2.3 linear equations containing parentheses and brackets, integer fractions, and decimals are introduced. Formulas, its definition, and the steps as to how they are solved for a specific variable is addressed in Section 2.4. Math operations involving linear inequalities are discussed in Section 2.5. Cases presented in each section are concluded by solving additional examples with practice problems to further enhance the student ability.

2.1 Introduction to Linear Equations

A numerical statement consists of two expressions which are separated by an equal sign "=". The symbol "=" implies that the left and the right hand side of the numerical statement must equal to each other in order for the equal sign to hold true. For example, $3+5=8$, $6-4=2$,and $1+5=6$ are true statements where as $7-3\neq 9$, $8+5\neq 23$, and $1+0\neq 2$ are false.

An algebraic equation consists of two algebraic expressions which are separated by an equal sign "=". For example, using y as variable, the statement $y+5=6$; $3y-2=7$; $y^2+3y=5$ are called algebraic equations. A solution to an equation is a value that when substituted for the variable, make the equation a true numerical equation. For example, if 1 is substituted for y in the equation $y+5=6$, we obtain $1+5=6$ or $6=6$ which is a true numerical statement. Therefore, we say that $y=1$ is the solution to the equation $y+5=6$.

The process of determining the solution to an algebraic equation is referred to as solving an equation. The set of all solutions to an algebraic equation is called its **solution set**. For example, the solution set of $y+5=6$ is 1, which is expressed as $\{1\}$, and the solution set of $3y-5=7$ is $\frac{2}{3}$, which is expressed as $\left\{\frac{2}{3}\right\}$.

In the following sections we will learn how to solve an algebraic equation. However, we first need to learn the process as to how a solution to a linear equation is verified. To check a solution to an equation we need to use the following steps:

Step 1: Substitute the solution into the original equation in place of the variable.

Step 2: Solve the equation.

Step 3: If both sides of the equation become equal to each other then the solution satisfies the original equation. Otherwise, the solution does not satisfy the original equation.

The following examples show how the solution to an algebraic linear equation is verified:

Example 2.1-1: Given the algebraic equation $5x-3=3x+5$, does $x=3$, $x=4$, and $x=5$ satisfy the original equation?

1. Substitute $x=3$ into the original equation and see if both sides of the equation become equal to each other.

$$(5\cdot 3)-3\overset{?}{=}(3\cdot 3)+5 \ ; \ 15-3\overset{?}{=}9+5 \ ; \ 12\neq 14$$

Since the left hand side of the equation is not equal to the right hand side of the equation, therefore $x=3$ does not satisfy the original equation. This implies that the two sides are not equal to each other.

2. Substitute $x=4$ into the original equation and see if both sides of the equation become equal to each other.

$$(5\cdot 4)-3\overset{?}{=}(3\cdot 4)+5 \ ; \ 20-3\overset{?}{=}12+5 \ ; \ \mathbf{17=17}$$

Since the left hand side of the equation is equal to the right hand side of the equation, therefore $x=4$ does satisfy the original equation. This implies that the two sides are equal to each other. **Therefore, $x=4$ is the solution to the equation $5x-3=3x+5$.**

3. Substitute $x=5$ into the original equation and see if both sides of the equation become equal to each other.

$$(5\cdot 5)-3\overset{?}{=}(3\cdot 5)+5 \ ; \ 25-3\overset{?}{=}15+5 \ ; \ 22\neq 20$$

Since the left hand side of the equation is not equal to the right hand side of the equation, therefore $x=5$ does not satisfy the original equation. This implies that the two sides are not equal to each other.

Example 2.1-2: Given the algebraic equation $y=3(y-2)+4$, does $y=-1$, $y=0$, and $y=1$ satisfy the original equation?

1. Substitute $y=-1$ into the original equation and see if both sides of the equation become equal to each other.

$$-1\overset{?}{=}3(-1-2)+4 \ ; \ -1\overset{?}{=}3\cdot -3+4 \ ; \ -1\overset{?}{=}-9+4 \ ; \ -1\neq -5$$

Since the left hand side of the equation is not equal to the right hand side of the equation, therefore $y=-1$ does not satisfy the original equation. This implies that the two sides are not equal to each other.

2. Substitute $y=0$ into the original equation and see if both sides of the equation become equal to each other.

$$0\overset{?}{=}3(0-2)+4 \ ; \ 0\overset{?}{=}3\cdot -2+4 \ ; \ 0\overset{?}{=}-6+4 \ ; \ 0\neq -2$$

Since the left hand side of the equation is not equal to the right hand side of the equation, therefore $y=0$ does not satisfy the original equation. This implies that the two sides are not equal to each other.

3. Substitute $y=1$ into the original equation and see if both sides of the equation become equal to each other.

$$1 \stackrel{?}{=} 3(1-2)+4 \;;\; 1 \stackrel{?}{=} 3\cdot -1+4 \;;\; 1 \stackrel{?}{=} -3+4 \;;\; \mathbf{1=1}$$

Since the left hand side of the equation is equal to the right hand side of the equation, therefore $y=1$ does satisfy the original equation. This implies that the two sides are equal to each other. **Therefore, $y=1$ is the solution to the equation $y=3(y-2)+4$.**

Example 2.1-3: Determine if $z=2$ is the solution to each of the following equations.

a. $3z-1=2z$ b. $6z-2=4+2z$ c. $z+3=9z-13$ d. $-3z+13=10$

Solution:

a. Let $z=2$ in the equation $3z-1=2z$, i.e., $3\cdot 2-1 \stackrel{?}{=} 2\cdot 2$; $6-1 \stackrel{?}{=} 4$; $5 \neq 4$. Therefore, $z=2$ is not the solution to $3z-1=2z$.

b. Let $z=2$ in the equation $6z-2=4+2z$, i.e., $6\cdot 2-2 \stackrel{?}{=} 4+2\cdot 2$; $12-2 \stackrel{?}{=} 4+4$; $10 \neq 8$. Therefore, $z=2$ is not the solution to $6z-2=4+2z$.

c. Let $z=2$ in the equation $z+3=9z-13$, i.e., $2+3 \stackrel{?}{=} 9\cdot 2-13$; $5 \stackrel{?}{=} 18-13$; $\mathbf{5=5}$. **Therefore, $z=2$ is the solution to $z+3=9z-13$.**

d. Let $z=2$ in the equation $-3z+13=10$, i.e., $-3\cdot 2+13 \stackrel{?}{=} 10$; $-6+13 \stackrel{?}{=} 10$; $7 \neq 10$. Therefore, $z=2$ is not the solution to $-3z+13=10$.

Practice Problems - Introduction to Linear Equations

Section 2.1 Practice Problems - Solve the following linear equations:

1. Determine whether 2 is the solution to each of the following equations:

   a. $3x-2=10$ b. $-2x+3=x$ c. $6-x=2x+1$ d. $2x-8=-3x+2$

2. Determine if $y=-2$ is the solution to the following equations:

   a. $y+3=-2y$ b. $6y+y=8y+2$ c. $6+3y=0$ d. $3y=5-y$

3. Given the algebraic equation $2x-8=(x-5)+3$, does $x=0$, $x=-1$, and $x=6$ satisfy the original equation?

4. Does $a=2$ satisfy any of the following equations?

   a. $3a+2=4a$ b. $3+7a=18$ c. $-5a+3=-3a-1$ d. $8=a+3$

2.2 Math Operations Involving Linear Equations

First degree equations of one variable, which are also referred to as linear equations, are solved by the proper use of addition, subtraction, multiplication, and division rules. It is important to learn how to apply these rules to linear equations in order to find the solution to the unknown variable. In this section, solving linear equations using either the addition and subtraction rules (Case I) or the multiplication and division rules (Case II) are discussed. Students are encouraged to learn how to solve first degree equations by properly organizing and applying the rules that are stated below in order to minimize mistakes.

Case I Addition and Subtraction of Linear Equations

To add or subtract the same positive or negative number to linear equations the following rules should be used:

Addition and Subtraction Rules: *The same positive or negative number can be added or subtracted to both sides of an equation without changing the solution: for all real numbers* a, b, *and* c,

1. $a = b$ *if and only if* $a + c = b + c$
2. $a = b$ *if and only if* $a - c = b - c$

The steps as to how linear equations are solved, using the addition and subtraction rules, are as follows:

Step 1 Isolate the variable to the left hand side of the equation by applying the addition and subtraction rules.

Step 2 Find the solution by simplifying the equation. Check the answer by substituting the solution into the original equation.

Examples with Steps

The following examples show the steps as to how linear equations are solved using the addition and subtraction rules:

Example 2.2-1

$\boxed{x - 3 = 5}$

Solution:

Step 1 $\boxed{x - 3 = 5}$; $\boxed{x - 3 + 3 = 5 + 3}$; $\boxed{x + 0 = 8}$; $\boxed{\mathbf{x = 8}}$

Step 2 $\boxed{\textit{Not Applicable}}$

The solution set is $\{\mathbf{8}\}$.

Check: $8 - 3 \stackrel{?}{=} 5$; $5 = 5$

Example 2.2-2

$\boxed{1\frac{2}{3} + w = 2\frac{3}{5}}$

Solution:

Step 1 $\boxed{1\frac{2}{3} + w = 2\frac{3}{5}}$; $\boxed{\frac{(1 \cdot 3) + 2}{3} + w = \frac{(2 \cdot 5) + 3}{5}}$; $\boxed{\frac{3 + 2}{3} + w = \frac{10 + 3}{5}}$; $\boxed{\frac{5}{3} + w = \frac{13}{5}}$

; $\boxed{\frac{5}{3}-\frac{5}{3}+w=\frac{13}{5}-\frac{5}{3}}$; $\boxed{0+w=\frac{13}{5}-\frac{5}{3}}$; $\boxed{w=\frac{13}{5}-\frac{5}{3}}$

Step 2 $\boxed{w=\frac{13}{5}-\frac{5}{3}}$; $\boxed{w=\frac{(13\cdot 3)-(5\cdot 5)}{5\cdot 3}}$; $\boxed{w=\frac{39-25}{15}}$; $\boxed{w=\frac{14}{15}}$; $\boxed{\mathbf{w=0.93}}$

The solution set is $\{\mathbf{0.93}\}$.

Check: $1\frac{2}{3}+0.93\overset{?}{=}2\frac{3}{5}$; $\frac{(1\cdot 3)+2}{3}+0.93\overset{?}{=}\frac{(2\cdot 5)+3}{5}$; $\frac{3+2}{3}+0.93\overset{?}{=}\frac{10+3}{5}$; $\frac{5}{3}+0.93\overset{?}{=}\frac{13}{5}$; $1.67+0.93\overset{?}{=}2.6$; $2.6=2.6$

Example 2.2-3

$$\boxed{u-0.48=\left(1\frac{3}{5}+2\frac{4}{5}\right)}$$

Solution:

Step 1 $\boxed{u-0.48=\left(1\frac{3}{5}+2\frac{4}{5}\right)}$; $\boxed{u-0.48+0.48=\left(1\frac{3}{5}+2\frac{4}{5}\right)+0.48}$; $\boxed{u+0=\left(1\frac{3}{5}+2\frac{4}{5}\right)+0.48}$

; $\boxed{u=\left(1\frac{3}{5}+2\frac{4}{5}\right)+0.48}$

Step 2 $\boxed{u=\left(1\frac{3}{5}+2\frac{4}{5}\right)+0.48}$; $\boxed{u=\left(\frac{(1\cdot 5)+3}{5}+\frac{(2\cdot 5)+4}{5}\right)+0.48}$; $\boxed{u=\left(\frac{5+3}{5}+\frac{10+4}{5}\right)+0.48}$

; $\boxed{u=\left(\frac{8}{5}+\frac{14}{5}\right)+0.48}$; $\boxed{u=\left(\frac{8+14}{5}\right)+0.48}$; $\boxed{u=\frac{22}{5}+0.48}$; $\boxed{u=4.4+0.48}$; $\boxed{\mathbf{u=4.88}}$

The solution set is $\{\mathbf{4.88}\}$.

Check: $4.88-0.48\overset{?}{=}1\frac{3}{5}+2\frac{4}{5}$; $4.4\overset{?}{=}\frac{(1\cdot 5)+3}{5}+\frac{(2\cdot 5)+4}{5}$; $4.4\overset{?}{=}\frac{5+3}{5}+\frac{10+4}{5}$; $4.4\overset{?}{=}\frac{8}{5}+\frac{14}{5}$; $4.4\overset{?}{=}\frac{8+14}{5}$; $4.4\overset{?}{=}\frac{22}{5}$; $4.4=4.4$

Example 2.2-4

$$\boxed{-\frac{1}{5}=-y+\left(3\frac{2}{3}-2\frac{1}{2}\right)}$$

Solution:

Step 1 $\boxed{-\frac{1}{5}=-y+\left(3\frac{2}{3}-2\frac{1}{2}\right)}$; $\boxed{y-\frac{1}{5}=y-y+\left(3\frac{2}{3}-2\frac{1}{2}\right)}$; $\boxed{y-\frac{1}{5}=0+\left(3\frac{2}{3}-2\frac{1}{2}\right)}$

; $\boxed{y-\frac{1}{5}=\left(3\frac{2}{3}-2\frac{1}{2}\right)}$; $\boxed{y-\frac{1}{5}+\frac{1}{5}=\left(3\frac{2}{3}-2\frac{1}{2}\right)+\frac{1}{5}}$; $\boxed{y+0=\left(3\frac{2}{3}-2\frac{1}{2}\right)+\frac{1}{5}}$

$; \boxed{y = \left(3\frac{2}{3} - 2\frac{1}{2}\right) + \frac{1}{5}}$

Step 2 $\boxed{y = \left(3\frac{2}{3} - 2\frac{1}{2}\right) + \frac{1}{5}}$; $\boxed{y = \left(\frac{(3\cdot3)+2}{3} - \frac{(2\cdot2)+1}{2}\right) + \frac{1}{5}}$; $\boxed{y = \left(\frac{9+2}{3} - \frac{4+1}{2}\right) + \frac{1}{5}}$

$; \boxed{y = \left(\frac{11}{3} - \frac{5}{2}\right) + \frac{1}{5}}$; $\boxed{y = \left(\frac{(11\cdot2)-(5\cdot3)}{3\cdot2}\right) + \frac{1}{5}}$; $\boxed{y = \left(\frac{22-15}{6}\right) + \frac{1}{5}}$; $\boxed{y = \frac{7}{6} + \frac{1}{5}}$

$; \boxed{y = \frac{(7\cdot5)+(1\cdot6)}{6\cdot5}}$; $\boxed{y = \frac{35+6}{30}}$; $\boxed{y = \frac{41}{30}}$; $\boxed{y = 1\frac{11}{30}}$; $\boxed{\mathbf{y = 1.37}}$

The solution set is $\{\mathbf{1.37}\}$.

Check: $-\frac{1}{5} \stackrel{?}{=} -1.37 + \left(3\frac{2}{3} - 2\frac{1}{2}\right)$; $-0.2 \stackrel{?}{=} -1.37 + \left(\frac{(3\cdot3)+2}{3} - \frac{(2\cdot2)+1}{2}\right)$; $-0.2 \stackrel{?}{=} -1.37 + \left(\frac{9+2}{3} - \frac{4+1}{2}\right)$

$; -0.2 \stackrel{?}{=} -1.37 + \left(\frac{11}{3} - \frac{5}{2}\right)$; $-0.2 \stackrel{?}{=} -1.37 + \left(\frac{(11\cdot2)-(5\cdot3)}{3\cdot2}\right)$; $-0.2 \stackrel{?}{=} -1.37 + \left(\frac{22-15}{3\cdot2}\right)$

$; -0.2 \stackrel{?}{=} -1.37 + \frac{7}{6}$; $-0.2 \stackrel{?}{=} -1.37 + 1.17$; $-0.2 = -0.2$

Example 2.2-5

$\boxed{\frac{2}{5} + x = \left(3 - 4\frac{3}{7}\right)}$

Solution:

Step 1 $\boxed{\frac{2}{5} + x = \left(3 - 4\frac{3}{7}\right)}$; $\boxed{\frac{2}{5} - \frac{2}{5} + x = -\frac{2}{5} + \left(3 - 4\frac{3}{7}\right)}$; $\boxed{0 + x = -\frac{2}{5} + \left(3 - 4\frac{3}{7}\right)}$

$; \boxed{x = -\frac{2}{5} + \left(3 - 4\frac{3}{7}\right)}$

Step 2 $\boxed{x = -\frac{2}{5} + \left(3 - 4\frac{3}{7}\right)}$; $\boxed{x = -\frac{2}{5} + \left(\frac{3}{1} - \frac{(4\cdot7)+3}{7}\right)}$; $\boxed{x = -\frac{2}{5} + \left(\frac{3}{1} - \frac{28+3}{7}\right)}$

$; \boxed{x = -\frac{2}{5} + \left(\frac{3}{1} - \frac{31}{7}\right)}$; $\boxed{x = -\frac{2}{5} + \left(\frac{(3\cdot7)-(31\cdot1)}{1\cdot7}\right)}$; $\boxed{x = -\frac{2}{5} + \left(\frac{21-31}{7}\right)}$

$; \boxed{x = -\frac{2}{5} - \frac{10}{7}}$; $\boxed{x = \frac{(-2\cdot7)+(-10\cdot5)}{5\cdot7}}$; $\boxed{x = \frac{-14-50}{35}}$; $\boxed{x = \frac{-64}{35}}$; $\boxed{\mathbf{x = -1.83}}$

The solution set is $\{\mathbf{-1.83}\}$.

Check: $\frac{2}{5} + (-1.83) \stackrel{?}{=} 3 - 4\frac{3}{7}$; $0.4 - 1.83 \stackrel{?}{=} 3 - \frac{(4\cdot7)+3}{7}$; $-1.43 \stackrel{?}{=} 3 - \frac{28+3}{7}$; $-1.43 \stackrel{?}{=} 3 - \frac{31}{7}$

$; -1.43 \stackrel{?}{=} 3 - 4.43$; $-1.43 = -1.43$

Additional Examples - Addition and Subtraction of Linear Equations

The following examples further illustrate how linear equations are solved using the addition and subtraction rules:

Example 2.2-6

$\boxed{x+6=8}$; $\boxed{x+6-6=8-6}$; $\boxed{x+0=2}$; $\boxed{\mathbf{x=2}}$ The solution set is $\{\mathbf{2}\}$.

Check: $2+6\stackrel{?}{=}8$; $8=8$

Example 2.2-7

$\boxed{-3+y=-5}$; $\boxed{-3+3+y=-5+3}$; $\boxed{0+y=-2}$; $\boxed{\mathbf{y=-2}}$ The solution set is $\{\mathbf{-2}\}$.

Check: $-3+(-2)\stackrel{?}{=}-5$; $-3-2\stackrel{?}{=}-5$; $-5=-5$

Example 2.2-8

$\boxed{2=u-5}$; $\boxed{2+5=u-5+5}$; $\boxed{7=u+0}$; $\boxed{7=u}$; $\boxed{\mathbf{u=7}}$ The solution set is $\{\mathbf{7}\}$.

Check: $2\stackrel{?}{=}7-5$; $2=2$

Example 2.2-9

$\boxed{5+x=-10}$; $\boxed{-5+5+x=-10-5}$; $\boxed{0+x=-15}$; $\boxed{\mathbf{x=-15}}$ The solution set is $\{\mathbf{-15}\}$.

Check: $5+(-15)\stackrel{?}{=}-10$; $5-15\stackrel{?}{=}-10$; $-10=-10$

Example 2.2-10

$\boxed{-24=-10+h}$; $\boxed{-24+10=-10+10+h}$; $\boxed{-14=0+h}$; $\boxed{-14=h}$; $\boxed{\mathbf{h=-14}}$ The solution set is $\{\mathbf{-14}\}$.

Check: $-24\stackrel{?}{=}-10+(-14)$; $-24\stackrel{?}{=}-10-14$; $-24=-24$

Example 2.2-11

$\boxed{5=y+20}$; $\boxed{5-20=y+20-20}$; $\boxed{-15=y+0}$; $\boxed{-15=y}$; $\boxed{\mathbf{y=-15}}$ The solution set is $\{\mathbf{-15}\}$.

Check: $5\stackrel{?}{=}-15+20$; $5=5$

Example 2.2-12

$\boxed{s-12=15}$; $\boxed{s-12+12=15+12}$; $\boxed{s+0=27}$; $\boxed{\mathbf{s=27}}$ The solution set is $\{\mathbf{27}\}$.

Check: $27-12\stackrel{?}{=}15$; $15=15$

Example 2.2-13

$\boxed{8.5+x=-2.4}$; $\boxed{8.5-8.5+x=-2.4-8.5}$; $\boxed{0+x=-10.9}$; $\boxed{\mathbf{x=-10.9}}$ The solution set is $\{\mathbf{-10.9}\}$.

Check: $8.5+(-10.9)\stackrel{?}{=}-2.4$; $8.5-10.9\stackrel{?}{=}-2.4$; $-2.4=-2.4$

Example 2.2-14

$\boxed{-9=w+8}$; $\boxed{-9-8=w+8-8}$; $\boxed{-17=w+0}$; $\boxed{-17=w}$; $\boxed{\mathbf{w=-17}}$ The solution set is $\{\mathbf{-17}\}$.

Check: $-9\stackrel{?}{=}-17+8$; $-9=-9$

Example 2.2-15

$\boxed{-1.2+y=-2.8}$; $\boxed{-1.2+1.2+y=-2.8+1.2}$; $\boxed{0+y=-1.6}$; $\boxed{\mathbf{y=-1.6}}$ The solution set is $\{\mathbf{-1.6}\}$.

Check: $-1.2+(-1.6)\stackrel{?}{=}-2.8$; $-1.2-1.6\stackrel{?}{=}-2.8$; $-2.8=-2.8$

Example 2.2-16

$\boxed{x+2\frac{3}{5}=1\frac{1}{8}}$; $\boxed{x+2\frac{3}{5}-2\frac{3}{5}=1\frac{1}{8}-2\frac{3}{5}}$; $\boxed{x+0=1\frac{1}{8}-2\frac{3}{5}}$; $\boxed{x=\frac{(1\times 8)+1}{8}-\frac{(2\times 5)+3}{5}}$; $\boxed{x=\frac{8+1}{8}-\frac{10+3}{5}}$

; $\boxed{x=\frac{9}{8}-\frac{13}{5}}$; $\boxed{x=\frac{(9\cdot 5)-(13\cdot 8)}{8\cdot 5}}$; $\boxed{x=\frac{45-104}{40}}$; $\boxed{x=\frac{-59}{40}}$; $\boxed{x=-\frac{59}{40}}$; $\boxed{x=-1\frac{19}{40}}$; $\boxed{\mathbf{x=-1.475}}$

The solution set is $\{\mathbf{-1.475}\}$.

Check: $-1.475+2\frac{3}{5}\overset{?}{=}1\frac{1}{8}$; $-1.475+\frac{(2\cdot 5)+3}{5}\overset{?}{=}\frac{(1\cdot 8)+1}{8}$; $-1.475+\frac{10+3}{5}\overset{?}{=}\frac{8+1}{8}$; $-1.475+\frac{13}{5}\overset{?}{=}\frac{9}{8}$

; $-1.475+2.6\overset{?}{=}1.125$; $1.125=1.125$

Example 2.2-17

$\boxed{2\frac{3}{4}=y+3\frac{1}{4}}$; $\boxed{2\frac{3}{4}-3\frac{1}{4}=y+3\frac{1}{4}-3\frac{1}{4}}$; $\boxed{2\frac{3}{4}-3\frac{1}{4}=y+0}$; $\boxed{2\frac{3}{4}-3\frac{1}{4}=y}$; $\boxed{y=\frac{(2\times 4)+3}{4}-\frac{(3\times 4)+1}{4}}$

; $\boxed{y=\frac{11}{4}-\frac{13}{4}}$; $\boxed{y=\frac{11-13}{4}}$; $\boxed{y=-\frac{\overset{1}{\cancel{2}}}{\underset{2}{\cancel{4}}}}$; $\boxed{y=-\frac{1}{2}}$; $\boxed{\mathbf{y=-0.5}}$

The solution set is $\{\mathbf{-0.5}\}$.

Check: $2\frac{3}{4}\overset{?}{=}-0.5+3\frac{1}{4}$; $\frac{(2\cdot 4)+3}{4}\overset{?}{=}-0.5+\frac{(3\cdot 4)+1}{4}$; $\frac{8+3}{4}\overset{?}{=}-0.5+\frac{12+1}{4}$; $\frac{11}{4}\overset{?}{=}-0.5+\frac{13}{4}$

; $2.75\overset{?}{=}-0.5+3.25$; $2.75=2.75$

Example 2.2-18

$\boxed{y-2.35=2\frac{3}{8}}$; $\boxed{y-2.35+2.35=2\frac{3}{8}+2.35}$; $\boxed{y+0=\frac{(2\times 8)+3}{8}+2.35}$; $\boxed{y=\frac{16+3}{8}+2.35}$; $\boxed{y=\frac{19}{8}+2.35}$

; $\boxed{y=2.375+2.35}$; $\boxed{\mathbf{y=4.725}}$

The solution set is $\{\mathbf{4.725}\}$.

Check: $4.725-2.35\overset{?}{=}2\frac{3}{8}$; $2.375\overset{?}{=}\frac{(2\cdot 8)+3}{8}$; $2.375\overset{?}{=}\frac{16+3}{8}$; $2.375\overset{?}{=}\frac{19}{8}$; $2.375=2.375$

Example 2.2-19

$\boxed{-2.5=u+1\frac{3}{5}}$; $\boxed{-2.5-1\frac{3}{5}=u+1\frac{3}{5}-1\frac{3}{5}}$; $\boxed{-2.5-1\frac{3}{5}=u+0}$; $\boxed{-2.5-\frac{(1\times 5)+3}{5}=u}$; $\boxed{-2.5-\frac{5+3}{5}=u}$

; $\boxed{-2.5-\frac{8}{5}=u}$; $\boxed{-2.5-1.6=u}$; $\boxed{-4.1=u}$; $\boxed{\mathbf{u=-4.1}}$

The solution set is $\{\mathbf{-4.1}\}$.

Check: $-2.5\overset{?}{=}-4.1+1\frac{3}{5}$; $-2.5\overset{?}{=}-4.1+\frac{(1\cdot 5)+3}{5}$; $-2.5\overset{?}{=}-4.1+\frac{5+3}{5}$; $-2.5\overset{?}{=}-4.1+\frac{8}{5}$; $-2.5\overset{?}{=}-4.1+1.6$

; $-2.5=-2.5$

Example 2.2-20

$\boxed{5\frac{1}{5}-x=4\frac{2}{5}-1\frac{1}{5}}$; $\boxed{\frac{(5\cdot 5)+1}{5}-x=\frac{(4\cdot 5)+2}{5}-\frac{(1\cdot 5)+1}{5}}$; $\boxed{\frac{25+1}{5}-x=\frac{20+2}{5}-\frac{5+1}{5}}$; $\boxed{\frac{26}{5}-x=\frac{22}{5}-\frac{6}{5}}$

; $\boxed{\frac{26}{5}-x=\frac{22-6}{5}}$; $\boxed{\frac{26}{5}-x=\frac{16}{5}}$; $\boxed{\frac{26}{5}-\frac{26}{5}-x=\frac{16}{5}-\frac{26}{5}}$; $\boxed{0-x=\frac{16-26}{5}}$; $\boxed{-x=\frac{-10}{5}}$; $\boxed{x=\frac{10}{5}}$; $\boxed{\mathbf{x=2}}$

The solution set is $\{\mathbf{2}\}$.

Check: $5\frac{1}{5}-2\overset{?}{=}4\frac{2}{5}-1\frac{1}{5}$; $\frac{(5\cdot5)+1}{5}-2\overset{?}{=}\frac{(4\cdot5)+2}{5}-\frac{(1\cdot5)+1}{5}$; $\frac{25+1}{5}-2\overset{?}{=}\frac{20+2}{5}-\frac{5+1}{5}$; $\frac{26}{5}-2\overset{?}{=}\frac{22}{5}-\frac{6}{5}$

; $5.2-2\overset{?}{=}4.4-1.2$; $3.2=3.2$

Example 2.2-21

$\boxed{x+1\frac{1}{3}=\frac{2}{3}-1\frac{2}{5}}$; $\boxed{x+\frac{(1\cdot3)+1}{3}=\frac{2}{3}-\frac{(1\cdot5)+2}{5}}$; $\boxed{x+\frac{3+1}{3}=\frac{2}{3}-\frac{5+2}{5}}$; $\boxed{x+\frac{4}{3}=\frac{2}{3}-\frac{7}{5}}$; $\boxed{x+\frac{4}{3}=\frac{(2\cdot5)-(7\cdot3)}{3\cdot5}}$

; $\boxed{x+\frac{4}{3}=\frac{10-21}{15}}$; $\boxed{x+\frac{4}{3}=\frac{-11}{15}}$; $\boxed{x+\frac{4}{3}-\frac{4}{3}=-\frac{11}{15}-\frac{4}{3}}$; $\boxed{x+0=\frac{(-11\cdot3)+(-4\cdot15)}{15\cdot3}}$; $\boxed{x=\frac{-33-60}{45}}$

; $\boxed{x=\frac{-93}{45}}$; $\boxed{x=-\frac{93}{45}}$; $\boxed{x=-2\frac{3}{45}}$; $\boxed{\mathbf{x=-2.07}}$ The solution set is $\{\mathbf{-2.07}\}$.

Check: $-2.07+1\frac{1}{3}\overset{?}{=}\frac{2}{3}-1\frac{2}{5}$; $-2.07+\frac{(1\cdot3)+1}{3}\overset{?}{=}\frac{2}{3}-\frac{(1\cdot5)+2}{5}$; $-2.07+\frac{3+1}{3}\overset{?}{=}\frac{2}{3}-\frac{5+2}{5}$; $-2.07+\frac{4}{3}\overset{?}{=}\frac{2}{3}-\frac{7}{5}$

; $-2.07+1.33\overset{?}{=}0.66-1.4$; $-0.74=-0.74$

Example 2.2-22

$\boxed{0.45+w=2\frac{2}{3}+6}$; $\boxed{0.45+w=\frac{(2\cdot3)+2}{3}+6}$; $\boxed{0.45+w=\frac{6+2}{3}+6}$; $\boxed{0.45+w=\frac{8}{3}+6}$; $\boxed{0.45+w=2.67+6}$

; $\boxed{0.45+w=8.67}$; $\boxed{0.45-0.45+w=8.67-0.45}$; $\boxed{0+w=8.67-0.45}$; $\boxed{w=8.67-0.45}$; $\boxed{\mathbf{w=8.22}}$

The solution set is $\{\mathbf{8.22}\}$.

Check: $0.45+8.22\overset{?}{=}2\frac{2}{3}+6$; $0.45+8.22\overset{?}{=}\frac{(2\cdot3)+2}{3}+6$; $0.45+8.22\overset{?}{=}\frac{6+2}{3}+6$; $0.45+8.22\overset{?}{=}\frac{8}{3}+6$

; $0.45+8.22\overset{?}{=}2.67+6$; $8.67=8.67$

Example 2.2-23

$\boxed{\sqrt{20}-x=\sqrt{80}+1\frac{2}{3}}$; $\boxed{\sqrt{4\cdot5}-x=\sqrt{16\cdot5}+\frac{(1\cdot3)+2}{3}}$; $\boxed{\sqrt{2^2\cdot5}-x=\sqrt{4^2\cdot5}+\frac{3+2}{3}}$; $\boxed{2\sqrt{5}-x=4\sqrt{5}+\frac{5}{3}}$

; $\boxed{(2\cdot2.24)-x=(4\cdot2.24)+1.67}$; $\boxed{4.48-x=8.96+1.67}$; $\boxed{4.48-x=10.63}$; $\boxed{4.48-4.48-x=10.63-4.48}$

; $\boxed{0-x=6.15}$; $\boxed{-x=6.15}$; $\boxed{\mathbf{x=-6.15}}$ The solution set is $\{\mathbf{-6.15}\}$.

Check: $\sqrt{20}-(-6.15)\overset{?}{=}\sqrt{80}+1\frac{2}{3}$; $\sqrt{4\cdot5}-(-6.15)\overset{?}{=}\sqrt{16\cdot5}+\frac{(1\cdot3)+2}{3}$; $2\sqrt{5}+6.15\overset{?}{=}4\sqrt{5}+\frac{3+2}{3}$

; $(2\cdot2.4)+6.15\overset{?}{=}(4\cdot2.24)+\frac{5}{3}$; $4.48+6.15\overset{?}{=}8.96+1.67$; $10.63=10.63$

Example 2.2-24

$\boxed{-1\frac{3}{5}=h+2\frac{3}{4}}$; $\boxed{-\frac{(1\cdot5)+3}{5}=h+\frac{(2\cdot4)+3}{4}}$; $\boxed{-\frac{5+3}{5}=h+\frac{8+3}{4}}$; $\boxed{-\frac{8}{5}=h+\frac{11}{4}}$; $\boxed{-\frac{8}{5}-\frac{11}{4}=h+\frac{11}{4}-\frac{11}{4}}$

; $\boxed{-\frac{8}{5}-\frac{11}{4}=h+0}$; $\boxed{\frac{(-8\cdot4)+(-11\cdot5)}{5\cdot4}=h}$; $\boxed{\frac{-32-55}{20}=h}$; $\boxed{\frac{-87}{20}=h}$; $\boxed{h=-\frac{87}{20}}$; $\boxed{h=-4\frac{7}{20}}$; $\boxed{\mathbf{h=-4.35}}$

The solution set is $\{-4.35\}$.

Check: $-1\frac{3}{5}\overset{?}{=}-4.35+2\frac{3}{4}$; $-\frac{(1\cdot5)+3}{5}\overset{?}{=}-4.35+\frac{(2\cdot4)+3}{4}$; $-\frac{5+3}{5}\overset{?}{=}-4.35+\frac{8+3}{4}$; $-\frac{8}{5}\overset{?}{=}-4.35+\frac{11}{4}$

; $-1.6\overset{?}{=}-4.35+2.75$; $-1.6=-1.6$

Example 2.2-25

$\boxed{x-3\frac{2}{5}=\frac{2}{3}-1\frac{1}{8}}$; $\boxed{x-\frac{(3\cdot5)+2}{5}=\frac{2}{3}-\frac{(1\cdot8)+1}{8}}$; $\boxed{x-\frac{15+2}{5}=\frac{2}{3}-\frac{8+1}{8}}$; $\boxed{x-\frac{17}{5}=\frac{2}{3}-\frac{9}{8}}$

; $\boxed{x-\frac{17}{5}=\frac{(2\cdot8)-(9\cdot3)}{3\cdot8}}$; $\boxed{x-\frac{17}{5}=\frac{16-27}{24}}$; $\boxed{x-\frac{17}{5}=\frac{-11}{24}}$; $\boxed{x-\frac{17}{5}+\frac{17}{5}=-\frac{11}{24}+\frac{17}{5}}$; $\boxed{x+0=-\frac{11}{24}+\frac{17}{5}}$

; $\boxed{x=\frac{(-11\cdot5)+(17\cdot24)}{24\cdot5}}$; $\boxed{x=\frac{-55+408}{120}}$; $\boxed{x=\frac{353}{120}}$; $\boxed{x=2\frac{113}{120}}$; $\boxed{\mathbf{x=2.942}}$

The solution set is $\{\mathbf{2.942}\}$.

Check: $2.94-3\frac{2}{5}\overset{?}{=}\frac{2}{3}-1\frac{1}{8}$; $2.94-\frac{(3\cdot5)+2}{5}\overset{?}{=}\frac{2}{3}-\frac{(1\cdot8)+1}{8}$; $2.94-\frac{15+2}{5}\overset{?}{=}\frac{2}{3}-\frac{8+1}{8}$; $2.94-\frac{17}{5}\overset{?}{=}\frac{2}{3}-\frac{9}{8}$

; $2.94-3.4\overset{?}{=}0.67-1.13$; $-0.46=-0.46$

Practice Problems - Addition and Subtraction of Linear Equations

Section 2.2 Case I Practice Problems - Solve the following linear equations by adding or subtracting the same positive or negative number to both sides of the equation:

1. $x-13=12$
2. $8+h=20$
3. $5=x-3$
4. $-3=u-5$
5. $2.8+x=-3.7$
6. $x-\frac{3}{8}=2\frac{3}{8}$
7. $4.9+x=1\frac{2}{3}$
8. $u+2\frac{1}{3}=-2\frac{3}{5}$
9. $6\frac{2}{3}=y-2\frac{4}{5}$
10. $y-2.38=-3\frac{2}{5}$

Case II Multiplication and Division of Linear Equations

To multiply or divide linear equations by the same positive or negative number the following rules should be used:

Multiplication Rule: *The same positive or negative number can be multiplied by both sides of an equation without changing the solution: for all real numbers* a, b, *and* c,

$$a = b \text{ if and only if } a \cdot c = b \cdot c .$$

Division Rule: *The same positive or negative number (except zero) can be divided by both sides an equation without changing the solution: for all real numbers* a, b, *and* c, *where* $c \neq 0$

$$a = b \text{ if and only if } \frac{a}{c} = \frac{b}{c} .$$

The steps as to how linear equations are solved, using the multiplication and division rules, are as follows:

Step 1 Isolate the variable to the left hand side of the equation by applying the addition and subtraction rules.

Step 2 Find the solution by applying the multiplication or division rules. Check the answer by substituting the solution in the original equation.

Examples with Steps

The following examples show the steps as to how linear equations are solved using the multiplication or division rules:

Example 2.2-26

$$\boxed{2x = -\frac{1}{5}}$$

Solution:

Step 1 *Not Applicable*

Step 2 $\boxed{2x \cdot \frac{1}{2} = -\frac{1}{5} \cdot \frac{1}{2}}$; $\boxed{x = -\frac{1}{5} \cdot \frac{1}{2}}$; $\boxed{x = -\frac{1}{5} \cdot \frac{1}{2}}$; $\boxed{x = -\frac{1 \cdot 1}{5 \cdot 2}}$; $\boxed{x = -\frac{\mathbf{1}}{\mathbf{10}}}$

The solution set is $\left\{-\frac{\mathbf{1}}{\mathbf{10}}\right\}$.

Check: $2 \cdot -\frac{1}{10} \stackrel{?}{=} -\frac{1}{5}$; $\frac{2}{1} \cdot -\frac{1}{10} \stackrel{?}{=} -\frac{1}{5}$; $-\frac{2 \cdot 1}{1 \cdot 10} \stackrel{?}{=} -\frac{1}{5}$; $-\frac{\cancel{2}}{\cancel{10}_{5}} \stackrel{?}{=} -\frac{1}{5}$; $-\frac{1}{5} = -\frac{1}{5}$

Example 2.2-27

$$\boxed{\frac{3}{8} = -3w}$$

Solution:

Step 1 $\boxed{\frac{3}{8} = -3w}$; $\boxed{\frac{3}{8} + 3w = -3w + 3w}$; $\boxed{\frac{3}{8} + 3w = 0}$; $\boxed{\frac{3}{8} - \frac{3}{8} + 3w = 0 - \frac{3}{8}}$; $\boxed{0 + 3w = -\frac{3}{8}}$

; $\boxed{3w = -\frac{3}{8}}$

Step 2 $\boxed{3w=-\frac{3}{8}}$; $\boxed{\cancel{3}w\cdot\frac{1}{\cancel{3}}=-\frac{\cancel{3}}{8}\cdot\frac{1}{\cancel{3}}}$; $\boxed{\boldsymbol{w=-\frac{1}{8}}}$ The solution set is $\left\{-\frac{\mathbf{1}}{\mathbf{8}}\right\}$.

Check: $\frac{3}{8}\overset{?}{=}-3\cdot-\frac{1}{8}$; $\frac{3}{8}=\frac{3}{8}$

Example 2.2-28

$\boxed{1\frac{1}{3}x=-2\frac{3}{5}}$

Solution:

Step 1 $\boxed{\textit{Not Applicable}}$

Step 2 $\boxed{1\frac{1}{3}x=-2\frac{3}{5}}$; $\boxed{\frac{(1\cdot 3)+1}{3}x=-\frac{(2\cdot 5)+3}{5}}$; $\boxed{\frac{3+1}{3}x=-\frac{10+3}{5}}$; $\boxed{\frac{4}{3}x=-\frac{13}{5}}$

; $\boxed{\frac{\cancel{4}}{\cancel{3}}x\cdot\frac{\cancel{3}}{\cancel{4}}=-\frac{13}{5}\cdot\frac{3}{4}}$; $\boxed{x=-\frac{13}{5}\cdot\frac{3}{4}}$; $\boxed{x=-\frac{13}{5}\cdot\frac{3}{4}}$; $\boxed{x=-\frac{13\cdot 3}{5\cdot 4}}$; $\boxed{x=-\frac{39}{20}}$; $\boxed{x=-1\frac{19}{20}}$

; $\boxed{\boldsymbol{x=-1.95}}$ The solution set is $\{\mathbf{-1.95}\}$.

Check: $1\frac{1}{3}\cdot(-1.95)\overset{?}{=}-2\frac{3}{5}$; $\frac{4}{3}\cdot(-1.95)\overset{?}{=}-\frac{13}{5}$; $\frac{4\cdot(-1.95)}{3}\overset{?}{=}-\frac{13}{5}$; $-\frac{7.8}{3}\overset{?}{=}-\frac{13}{5}$; $-2.6=-2.6$

Example 2.2-29

$\boxed{-26=-12y}$

Solution:

Step 1 $\boxed{-26=-12y}$; $\boxed{-26+12y=-12y+12y}$; $\boxed{-26+12y=0}$; $\boxed{-26+26+12y=0+26}$

; $\boxed{0+12y=26}$; $\boxed{12y=26}$

Step 2 $\boxed{12y=26}$; $\boxed{\frac{-\cancel{12}y}{-\cancel{12}}=\frac{-26}{-12}}$; $\boxed{y=\frac{26}{12}}$; $\boxed{y=\frac{\overset{13}{\cancel{26}}}{\underset{6}{\cancel{12}}}}$; $\boxed{y=\frac{13}{6}}$; $\boxed{y=2\frac{1}{6}}$; $\boxed{\boldsymbol{y=2.166}}$

The solution set is $\{\mathbf{2.166}\}$.

Check: $-26\overset{?}{=}-12\cdot 2.166$; $-26=-26$

Example 2.2-30

$\boxed{-3.8h=2\frac{3}{8}}$

Solution:

Step 1 $\boxed{\textit{Not Applicable}}$

Step 2 $\boxed{-3.8h=2\frac{3}{8}}$; $\boxed{-3.8h=\frac{(2\cdot 8)+3}{8}}$; $\boxed{-3.8h=\frac{16+3}{8}}$; $\boxed{-3.8h=\frac{19}{8}}$; $\boxed{-3.8h=2.375}$

; $\boxed{\frac{-\cancel{3.8}h}{-\cancel{3.8}}=\frac{2.375}{-3.8}}$; $\boxed{\boldsymbol{h=-0.625}}$ The solution set is $\{\mathbf{-0.625}\}$.

Check: $-3.8\cdot(-0.625)\overset{?}{=}2\frac{3}{8}$; $2.375\overset{?}{=}\frac{19}{8}$; $2.375=2.375$

Additional Examples - Multiplication and Division of Linear Equations

The following examples further illustrate how linear equations are simplified using the above multiplication or division rules:

Example 2.2-31

$\boxed{3y=60}$; $\boxed{\frac{\cancel{3}y}{\cancel{3}}=\frac{\overset{20}{\cancel{60}}}{\cancel{3}}}$; $\boxed{y=\frac{20}{1}}$; $\boxed{\mathbf{y=20}}$

The solution set is $\{\mathbf{20}\}$.

Check: $3\cdot 20\overset{?}{=}60$; $60=60$

Example 2.2-32

$\boxed{\frac{1}{5}x=8}$; $\boxed{\frac{1}{\cancel{5}}x\cdot\cancel{5}=8\cdot 5}$; $\boxed{\mathbf{x=40}}$

The solution set is $\{\mathbf{40}\}$.

Check: $\frac{1}{5}\cdot 40\overset{?}{=}8$; $\frac{\overset{8}{\cancel{40}}}{\cancel{5}}\overset{?}{=}8$; $\frac{8}{1}\overset{?}{=}8$; $8=8$

Example 2.2-33

$\boxed{\frac{5}{7}=-3h}$; $\boxed{\frac{5}{7}+3h=-3h+3h}$; $\boxed{\frac{5}{7}+3h=0}$; $\boxed{\frac{5}{7}-\frac{5}{7}+3h=0-\frac{5}{7}}$; $\boxed{0+3h=-\frac{5}{7}}$; $\boxed{3h=-\frac{5}{7}}$; $\boxed{\frac{1}{\cancel{3}}\cdot\cancel{3}h=\frac{1}{3}\cdot-\frac{5}{7}}$

; $\boxed{h=\frac{1\cdot -5}{3\cdot 7}}$; $\boxed{\mathbf{h=-\frac{5}{21}}}$

The solution set is $\left\{-\frac{\mathbf{5}}{\mathbf{21}}\right\}$.

Note that another way of solving for h is by not isolating the variable to the left hand side of the equation. However, in the very last step, we should write the variable to the left hand side of the equation and the solution to the right hand side of the equation as shown below.

$\boxed{\frac{5}{7}=-3h}$; $\boxed{\frac{1}{-3}\cdot\frac{5}{7}=\frac{1}{-\cancel{3}}\cdot-\cancel{3}h}$; $\boxed{-\frac{1\cdot 5}{3\cdot 7}=h}$; $\boxed{-\frac{5}{21}=h}$; $\boxed{\mathbf{h=-\frac{5}{21}}}$

Check: $\frac{5}{7}\overset{?}{=}-3\cdot-\frac{5}{21}$; $\frac{5}{7}\overset{?}{=}+\frac{3}{1}\cdot\frac{5}{21}$; $\frac{5}{7}\overset{?}{=}\frac{3\cdot 5}{1\cdot 21}$; $\frac{5}{7}\overset{?}{=}\frac{15}{21}$; $\frac{5}{7}\overset{?}{=}\frac{\overset{5}{\cancel{15}}}{\underset{7}{\cancel{21}}}$; $\frac{5}{7}=\frac{5}{7}$

Example 2.2-34

$\boxed{\frac{u}{5}=-3}$; $\boxed{\frac{u}{\cancel{5}}\cdot\cancel{5}=-3\cdot 5}$; $\boxed{\mathbf{u=-15}}$

The solution set is $\{\mathbf{-15}\}$.

Check: $\frac{-15}{5}\overset{?}{=}-3$; $-\frac{\overset{3}{\cancel{15}}}{\cancel{5}}\overset{?}{=}-3$; $-\frac{3}{1}\overset{?}{=}-3$; $-3=-3$

Example 2.2-35

$\boxed{-2w=-28}$; $\boxed{\frac{\cancel{-2}w}{\cancel{-2}}=\frac{-28}{-2}}$; $\boxed{w=+\frac{\overset{14}{\cancel{28}}}{\cancel{2}}}$; $\boxed{w=\frac{14}{1}}$; $\boxed{\mathbf{w=14}}$

The solution set is $\{\mathbf{14}\}$.

Check: $-2\cdot 14\overset{?}{=}-28$; $-28=-28$

Example 2.2-36

$\boxed{-y=15}$; $\boxed{\frac{-y}{-1}=\frac{15}{-1}}$; $\boxed{+y=-\frac{15}{1}}$; $\boxed{\mathbf{y=-15}}$ The solution set is $\{\mathbf{-15}\}$.

Check: $-(-15)\overset{?}{=}15$; $15=15$

Example 2.2-37

$\boxed{-3=-\frac{2}{5}x}$; $\boxed{-3+\frac{2}{5}x=-\frac{2}{5}x+\frac{2}{5}x}$; $\boxed{-3+\frac{2}{5}x=0}$; $\boxed{-3+3+\frac{2}{5}x=0+3}$; $\boxed{0+\frac{2}{5}x=3}$; $\boxed{\frac{2}{5}x=3}$

; $\boxed{\frac{\cancel{2}}{\cancel{5}}x\cdot\frac{\cancel{5}}{\cancel{2}}=3\cdot\frac{5}{2}}$; $\boxed{x=\frac{3}{1}\cdot\frac{5}{2}}$; $\boxed{x=\frac{3\cdot 5}{1\cdot 2}}$; $\boxed{x=\frac{15}{2}}$; $\boxed{x=7\frac{1}{2}}$; $\boxed{\mathbf{x=7.5}}$ The solution set is $\{\mathbf{7.5}\}$.

Second Approach: Keep the variable x to the right hand side of the equation.

$\boxed{-3=-\frac{2}{5}x}$; $\boxed{-3\cdot-\frac{5}{2}=-\frac{\cancel{2}}{\cancel{5}}x\cdot-\frac{\cancel{5}}{\cancel{2}}}$; $\boxed{+\frac{3}{1}\cdot\frac{5}{2}=x}$; $\boxed{\frac{3\cdot 5}{1\cdot 2}=x}$; $\boxed{\frac{15}{2}=x}$; $\boxed{x=\frac{15}{2}}$; $\boxed{x=7\frac{1}{2}}$; $\boxed{\mathbf{x=7.5}}$

Check: $-3\overset{?}{=}-\frac{2}{5}\cdot 7.5$; $-3\overset{?}{=}-\frac{2}{5}\cdot\frac{7.5}{1}$; $-3\overset{?}{=}-\frac{2\cdot 7.5}{5\cdot 1}$; $-3\overset{?}{=}-\frac{\overset{3}{\cancel{15}}}{\cancel{5}}$; $-3\overset{?}{=}-\frac{3}{1}$; $-3=-3$

Example 2.2-38

$\boxed{\frac{1}{8}x=-3}$; $\boxed{\frac{1}{\cancel{8}}x\cdot\cancel{8}=-3\cdot 8}$; $\boxed{\mathbf{x=-24}}$ The solution set is $\{\mathbf{24}\}$.

Check: $\frac{1}{8}\cdot-24\overset{?}{=}-3$; $-\frac{\overset{3}{\cancel{24}}}{\cancel{8}}\overset{?}{=}-3$; $-\frac{3}{1}\overset{?}{=}-3$; $-3=-3$

Example 2.2-39

$\boxed{\frac{h}{-5}=7}$; $\boxed{-\cancel{5}\cdot\frac{h}{-\cancel{5}}=-5\cdot 7}$; $\boxed{\mathbf{h=-35}}$ The solution set is $\{\mathbf{-35}\}$.

Check: $\frac{-35}{-5}\overset{?}{=}7$; $+\frac{\overset{7}{\cancel{35}}}{\cancel{5}}\overset{?}{=}7$; $\frac{7}{1}\overset{?}{=}7$; $7=7$

Example 2.2-40

$\boxed{20x=-30}$; $\boxed{\frac{\cancel{20}x}{\cancel{20}}=-\frac{30}{20}}$; $\boxed{x=-\frac{\overset{3}{\cancel{30}}}{\underset{2}{\cancel{20}}}}$; $\boxed{x=-\frac{3}{2}}$; $\boxed{x=-1\frac{1}{2}}$; $\boxed{\mathbf{x=-1.5}}$ The solution set is $\{\mathbf{-1.5}\}$.

Check: $20\cdot(-1.5)\overset{?}{=}-30$; $-30=-30$

Example 2.2-41

$\boxed{-5=\frac{w}{-3}}$; $\boxed{-5=-\frac{w}{3}}$; $\boxed{-5+\frac{w}{3}=-\frac{w}{3}+\frac{w}{3}}$; $\boxed{-5+\frac{w}{3}=0}$; $\boxed{-5+5+\frac{w}{3}=0+5}$; $\boxed{0+\frac{w}{3}=5}$; $\boxed{\frac{w}{3}=5}$

; $\boxed{\cancel{3}\cdot\frac{w}{\cancel{3}}=3\cdot 5}$; $\boxed{\mathbf{w=15}}$ The solution set is $\{\mathbf{15}\}$.

Second Approach: Keep the variable w to the right hand side of the equation.

$\boxed{-5=\frac{w}{-3}}$; $\boxed{-5\cdot-3=\frac{w}{-\cancel{3}}\cdot-\cancel{3}}$; $\boxed{15=w}$; $\boxed{\mathbf{w=15}}$

Check: $-5 \overset{?}{=} \frac{15}{-3}$; $-5 \overset{?}{=} -\frac{\overset{5}{\cancel{15}}}{\cancel{3}}$; $-5 \overset{?}{=} -\frac{5}{1}$; $-5 = -5$

Example 2.2-42

$\boxed{-12 = -6y}$; $\boxed{-12 + 6y = -6y + 6y}$; $\boxed{-12 + 6y = 0}$; $\boxed{-12 + 12 + 6y = 0 + 12}$; $\boxed{0 + 6y = 12}$; $\boxed{6y = 12}$

; $\boxed{\cancel{6}y \cdot \frac{1}{\cancel{6}} = 12 \cdot \frac{1}{6}}$; $\boxed{y = \frac{\overset{2}{\cancel{12}}}{\cancel{6}}}$; $\boxed{y = \frac{2}{1}}$; $\boxed{\mathbf{y = 2}}$ The solution set is $\{\mathbf{2}\}$.

Second Approach: Keep the variable y to the right hand side of the equation.

$\boxed{-12 = -6y}$; $\boxed{-12 \cdot \frac{1}{-6} = -\cancel{6}y \cdot \frac{1}{-\cancel{6}}}$; $\boxed{\frac{\overset{2}{\cancel{12}}}{\cancel{6}} = y}$; $\boxed{\frac{2}{1} = y}$; $\boxed{2 = y}$; $\boxed{\mathbf{y = 2}}$

Check: $-12 \overset{?}{=} -6 \cdot 2$; $-12 = -12$

Example 2.2-43

$\boxed{-3y = 18}$; $\boxed{-\cancel{3}y \cdot \frac{1}{-\cancel{3}} = 18 \cdot \frac{1}{-3}}$; $\boxed{y = \frac{18}{1} \cdot \frac{1}{-3}}$; $\boxed{y = \frac{18 \cdot 1}{1 \cdot -3}}$; $\boxed{y = -\frac{\overset{6}{\cancel{18}}}{\cancel{3}}}$; $\boxed{y = -\frac{6}{1}}$; $\boxed{\mathbf{y = -6}}$

The solution set is $\{\mathbf{-6}\}$.

Check: $-3 \cdot (-6) \overset{?}{=} 18$; $18 = 18$

Example 2.2-44

$\boxed{\frac{2}{3}x = -\frac{5}{8}}$; $\boxed{\frac{\cancel{2}}{\cancel{3}}x \cdot \frac{\cancel{3}}{\cancel{2}} = -\frac{5}{8} \cdot \frac{3}{2}}$; $\boxed{x = -\frac{5 \cdot 3}{8 \cdot 2}}$; $\boxed{x = -\frac{15}{16}}$; $\boxed{\mathbf{x = -0.94}}$ The solution set is $\{\mathbf{-0.94}\}$.

Check: $\frac{2}{3} \cdot (-0.94) \overset{?}{=} -\frac{5}{8}$; $\frac{2 \cdot (-0.94)}{3} \overset{?}{=} -0.625$; $-\frac{1.88}{3} \overset{?}{=} -0.625$; $-0.625 = -0.625$

Example 2.2-45

$\boxed{3.6x = -0.22}$; $\boxed{\frac{\cancel{3.6}x}{\cancel{3.6}} = -\frac{0.22}{3.6}}$; $\boxed{x = -\frac{\frac{22}{100}}{\frac{36}{10}}}$; $\boxed{x = -\frac{22 \cdot 10}{100 \cdot 36}}$; $\boxed{x = -\frac{\overset{11}{\cancel{220}}}{\underset{180}{\cancel{3600}}}}$; $\boxed{x = -\frac{11}{180}}$; $\boxed{\mathbf{x = -0.06}}$

The solution set is $\{\mathbf{-0.06}\}$.

Check: $3.6 \cdot (-0.06) \overset{?}{=} -2.2$; $-0.22 = -0.22$

Example 2.2-46

$\boxed{-\frac{1}{3}y = -23}$; $\boxed{-\frac{1}{\cancel{3}}y \cdot -\cancel{3} = -23 \cdot -3}$; $\boxed{\mathbf{y = 69}}$ The solution set is $\{\mathbf{69}\}$.

Check: $-\frac{1}{3} \cdot 69 \overset{?}{=} -23$; $-\frac{1}{3} \cdot \frac{69}{1} \overset{?}{=} -23$; $-\frac{1 \cdot 69}{3 \cdot 1} \overset{?}{=} -23$; $-\frac{\overset{23}{\cancel{69}}}{\cancel{3}} \overset{?}{=} -23$; $-\frac{23}{1} \overset{?}{=} -23$; $-23 = -23$

Example 2.2-47

$\boxed{2\frac{3}{5} = -1\frac{1}{3}x}$; $\boxed{\frac{(2 \cdot 5) + 3}{5} = -\frac{(1 \cdot 3) + 1}{3}x}$; $\boxed{\frac{10 + 3}{5} = -\frac{3 + 1}{3}x}$; $\boxed{\frac{13}{5} = -\frac{4}{3}x}$; $\boxed{\frac{13}{5} + \frac{4}{3}x = -\frac{4}{3}x + \frac{4}{3}x}$

; $\boxed{\frac{13}{5} + \frac{4}{3}x = 0}$; $\boxed{\frac{13}{5} - \frac{13}{5} + \frac{4}{3}x = 0 - \frac{13}{5}}$; $\boxed{0 + \frac{4}{3}x = -\frac{13}{5}}$; $\boxed{\frac{4}{3}x = -\frac{13}{5}}$; $\boxed{\frac{\cancel{4}}{\cancel{3}}x \cdot \frac{\cancel{3}}{\cancel{4}} = -\frac{13}{5} \cdot \frac{3}{4}}$; $\boxed{x = -\frac{13 \cdot 3}{5 \cdot 4}}$

$; \boxed{x = -\frac{39}{20}} ; \boxed{x = -1.95}$ The solution set is $\{-\mathbf{1.95}\}$.

Second Approach: Keep the variable x to the right hand side of the equation.

$\boxed{2\frac{3}{5} = -1\frac{1}{3}x} ; \boxed{\frac{(2\cdot 5)+3}{5} = -\frac{(1\cdot 3)+1}{3}x} ; \boxed{\frac{10+3}{5} = -\frac{3+1}{3}x} ; \boxed{\frac{13}{5} = -\frac{4}{3}x} ; \boxed{\frac{13}{5}\cdot -\frac{3}{4} = -\frac{\not 4}{\not 3}x\cdot -\frac{\not 3}{\not 4}} ; \boxed{-\frac{13\cdot 3}{5\cdot 4} = x}$

$; \boxed{-\frac{39}{20} = x} ; \boxed{-1.95 = x} ; \boxed{x = -1.95}$

Check: $2\frac{3}{5} \overset{?}{=} -1\frac{1}{3}\cdot(-1.95)$; $\frac{13}{5} \overset{?}{=} -\frac{4}{3}\cdot(-1.95)$; $\frac{13}{5} \overset{?}{=} -\frac{4\cdot(-1.95)}{3}$; $\frac{13}{5} \overset{?}{=} \frac{7.8}{3}$; $2.6 = 2.6$

Example 2.2-48

$\boxed{2y = -\frac{3}{8}} ; \boxed{\not 2y\cdot\frac{1}{\not 2} = -\frac{3}{8}\cdot\frac{1}{2}} ; \boxed{y = -\frac{3\cdot 1}{8\cdot 2}} ; \boxed{y = -\frac{3}{16}} ; \boxed{y = -0.1875}$ The solution set is $\{-\mathbf{0.1875}\}$.

Check: $2\cdot(-0.1875) \overset{?}{=} -\frac{3}{8}$; $-0.375 \overset{?}{=} -\frac{3}{8}$; $-0.375 = -0.375$

Example 2.2-49

$\boxed{-x = \frac{2}{3}} ; \boxed{-x = 0.67} ; \boxed{\frac{-x}{-1} = \frac{0.67}{-1}} ; \boxed{+x = -\frac{0.67}{1}} ; \boxed{x = -0.67}$ The solution set is $\{-\mathbf{0.67}\}$.

Check: $-(-0.67) \overset{?}{=} \frac{2}{3}$; $0.67 = 0.67$

Example 2.2-50

$\boxed{2\frac{3}{5}x = -1\frac{1}{5}} ; \boxed{\frac{(2\cdot 5)+3}{5}x = -\frac{(1\cdot 5)+1}{5}} ; \boxed{\frac{10+3}{5}x = -\frac{5+1}{5}} ; \boxed{\frac{13}{5}x = -\frac{6}{5}} ; \boxed{\frac{\not{13}}{\not 5}x\cdot\frac{\not 5}{\not{13}} = -\frac{6}{\not 5}\cdot\frac{\not 5}{13}} ; \boxed{x = -\frac{6}{13}}$

$; \boxed{x = -0.462}$ The solution set is $\{-\mathbf{0.462}\}$.

Check: $2\frac{3}{5}\cdot(-0.462) \overset{?}{=} -1\frac{1}{5}$; $\frac{13}{5}\cdot(-0.462) \overset{?}{=} -\frac{6}{5}$; $\frac{13\cdot(-0.462)}{5} \overset{?}{=} -\frac{6}{5}$; $-\frac{6}{5} = -\frac{6}{5}$

Practice Problems - Multiplication and Division of Linear Equations

Section 2.2 Case II Practice Problems - Solve the following linear equations by applying the multiplication or division rules:

1. $3y = -\frac{2}{3}$
2. $-\frac{1}{2}x = 1\frac{2}{3}$
3. $\frac{3}{8} = -2h$
4. $\frac{x}{8} = -2$
5. $-x = -35$
6. $2\frac{1}{8}u = -1\frac{1}{2}$
7. $-w = 1\frac{4}{5}$
8. $-\frac{1}{2}y = -12$
9. $2.8x = -1.4$
10. $-2\frac{3}{5}x = -4.3$

Case III Mixed Operations Involving Linear Equations

In Cases I and II we learned how to solve linear equations by either applying: 1. The addition and subtraction rules or, 2. The multiplication or division rules. In this section, solution to linear equations which may involve using all four rules is discussed.

Addition and Subtraction Rules: *The same positive or negative number can be added or subtracted to both sides of an equation without changing the solution: for all real numbers* a, b, *and* c,

1. $a = b$ *if and only if* $a + c = b + c$
2. $a = b$ *if and only if* $a - c = b - c$.

Multiplication Rule: *The same positive or negative number can be multiplied by both sides of an equation without changing the solution: for all real numbers* a, b, *and* c,

$$a = b \text{ if and only if } a \cdot c = b \cdot c.$$

Division Rule: *The same positive or negative number (except zero) can be divided by both sides an equation without changing the solution: for all real numbers* a, b, *and* c, *where* $c \neq 0$

$$a = b \text{ if and only if } \frac{a}{c} = \frac{b}{c}.$$

The steps as to how linear equations are solved using the addition and subtraction, multiplication, and division rules are as follows:

Step 1 Isolate the variable to the left hand side of the equation by applying the addition and subtraction rules.

Step 2 Find the solution by applying the multiplication or division rules. Check the answer by substituting the solution into the original equation.

Examples with Steps

The following examples show the steps as to how linear equations are solved using the addition, subtraction, multiplication, and division rules:

Example 2.2-51

$$\boxed{5x = 20 + 3x}$$

Solution:

Step 1 $\boxed{5x = 20 + 3x}$; $\boxed{5x - 3x = 20 + 3x - 3x}$; $\boxed{2x = 20 + 0}$; $\boxed{2x = 20}$

Step 2 $\boxed{2x = 20}$; $\boxed{\frac{2x}{2} = \frac{\overset{10}{20}}{2}}$; $\boxed{x = \frac{10}{1}}$; $\boxed{\mathbf{x = 10}}$ The solution set is $\{\mathbf{10}\}$.

Check: $5 \cdot 10 \overset{?}{=} 20 + 3 \cdot 10$; $50 \overset{?}{=} 20 + 30$; $50 = 50$

Example 2.2-52

$$\boxed{4y - 2 = 3y + 8}$$

Solution:

Step 1 $\boxed{4y - 2 = 3y + 8}$; $\boxed{4y - 3y - 2 = 3y - 3y + 8}$; $\boxed{y - 2 = 0 + 8}$; $\boxed{y - 2 = 8}$

$\boxed{y-2+2=8+2}$; $\boxed{y+0=10}$; $\boxed{\mathbf{y=10}}$

Step 2 $\boxed{\textit{Not Applicable}}$ The solution set is $\{\mathbf{10}\}$.

Check: $4\cdot10-2\overset{?}{=}3\cdot10+8$; $40-2\overset{?}{=}30+8$; $38=38$

Example 2.2-53

$\boxed{\frac{u}{5}+3=-2}$

Solution:

Step 1 $\boxed{\frac{u}{5}+3=-2}$; $\boxed{\frac{u}{5}+3-3=-2-3}$; $\boxed{\frac{u}{5}+0=-5}$; $\boxed{\frac{u}{5}=-5}$

Step 2 $\boxed{\frac{u}{5}=-5}$; $\boxed{\not{5}\cdot\frac{u}{\not{5}}=5\cdot-5}$; $\boxed{\mathbf{u=-25}}$ The solution set is $\{\mathbf{-25}\}$.

Check: $\frac{-25}{5}+3\overset{?}{=}-2$; $-\frac{\overset{5}{\not{25}}}{\not{5}}+3\overset{?}{=}-2$; $-\frac{5}{1}+3\overset{?}{=}-2$; $-5+3\overset{?}{=}-2$; $-2=-2$

Example 2.2-54

$\boxed{-5m+5=-3m+2}$

Solution:

Step 1 $\boxed{-5m+5=-3m+2}$; $\boxed{-5m+3m+5=-3m+3m+2}$; $\boxed{-2m+5=0+2}$; $\boxed{-2m+5=2}$

; $\boxed{-2m+5-5=2-5}$; $\boxed{-2m+0=-3}$; $\boxed{-2m=-3}$

Step 2 $\boxed{-2m=-3}$; $\boxed{\frac{-\not{2}m}{-\not{2}}=\frac{-3}{-2}}$; $\boxed{\mathbf{m=\frac{3}{2}}}$ The solution set is $\left\{\mathbf{\frac{3}{2}}\right\}$.

Check: $-5\cdot\frac{3}{2}+5\overset{?}{=}-3\cdot\frac{3}{2}+2$; $-\frac{15}{2}+\frac{5}{1}\overset{?}{=}-\frac{9}{2}+\frac{2}{1}$; $\frac{(-15\cdot1)+(5\cdot2)}{2\cdot1}\overset{?}{=}\frac{(-9\cdot1)+(2\cdot2)}{2\cdot1}$; $\frac{-15+10}{2}\overset{?}{=}\frac{-9+4}{2}$

; $-\frac{5}{2}=-\frac{5}{2}$

Example 2.2-55

$\boxed{15x-5=-3x+2}$

Solution:

Step 1 $\boxed{15x-5=-3x+2}$; $\boxed{15x+3x-5=-3x+3x+2}$; $\boxed{18x-5=0+2}$; $\boxed{18x-5=2}$

; $\boxed{18x-5+5=2+5}$; $\boxed{18x+0=7}$; $\boxed{18x=7}$

Step 2 $\boxed{18x=7}$; $\boxed{\frac{\not{18}x}{\not{18}}=\frac{7}{18}}$; $\boxed{x=\frac{7}{18}}$; $\boxed{\mathbf{x=0.389}}$ The solution set is $\{\mathbf{0.389}\}$.

Check: $(15\cdot0.389)-5\overset{?}{=}(-3\cdot0.389)+2$; $5.84-5\overset{?}{=}-1.16+2$; $0.84=0.84$

Additional Examples - Mixed Operations Involving Linear Equations

The following examples further illustrate how to solve linear equations using the addition, subtraction, multiplication, and division rules:

Example 2.2-56

$$\boxed{3x-5=10}\;;\;\boxed{3x-5+5=10+5}\;;\;\boxed{3x+0=15}\;;\;\boxed{3x=15}\;;\;\boxed{\frac{\cancel{3}x}{\cancel{3}}=\frac{\overset{5}{\cancel{15}}}{\cancel{3}}}\;;\;\boxed{x=\frac{5}{1}}\;;\;\boxed{\mathbf{x=5}}$$

The solution set is $\{\mathbf{5}\}$.

Check: $(3\cdot5)-5\overset{?}{=}10\;;\;15-5\overset{?}{=}10\;;\;10=10$

Example 2.2-57

$$\boxed{4y-2=3y+5}\;;\;\boxed{4y-3y-2=3y-3y+5}\;;\;\boxed{y-2=0+5}\;;\;\boxed{y-2=5}\;;\;\boxed{y-2+2=5+2}\;;\;\boxed{y+0=7}\;;\;\boxed{\mathbf{y=7}}$$

The solution set is $\{\mathbf{7}\}$.

Check: $(4\cdot7)-2\overset{?}{=}(3\cdot7)+5\;;\;28-2\overset{?}{=}21+5\;;\;26=26$

Example 2.2-58

$$\boxed{\frac{2}{3}x+5=3}\;;\;\boxed{\frac{2}{3}x+5-5=3-5}\;;\;\boxed{\frac{2}{3}x+0=-2}\;;\;\boxed{\frac{2}{3}x=-2}\;;\;\boxed{\frac{\cancel{3}}{\cancel{2}}\cdot\frac{\cancel{2}}{\cancel{3}}x=\frac{3}{\cancel{2}}\cdot-\cancel{2}}\;;\;\boxed{x=-\frac{3}{1}}\;;\;\boxed{\mathbf{x=-3}}$$

The solution set is $\{\mathbf{-3}\}$.

Check: $\frac{2}{3}\cdot(-3)+5\overset{?}{=}3\;;\;\frac{2}{\cancel{3}}\cdot(-\cancel{3})+5\overset{?}{=}3\;;\;\frac{2}{1}\cdot(-1)+5\overset{?}{=}3\;;\;-2+5\overset{?}{=}3\;;\;3=3$

Example 2.2-59

$$\boxed{4+\frac{u}{2}=5-u}\;;\;\boxed{4-4+\frac{u}{2}=5-4-u}\;;\;\boxed{0+\frac{u}{2}=1-u}\;;\;\boxed{\frac{u}{2}=1-u}\;;\;\boxed{\frac{u}{2}+u=1-u+u}\;;\;\boxed{\frac{u}{2}+\frac{u}{1}=1+0}$$

$$;\;\boxed{\frac{(1\cdot u)+(2\cdot u)}{2\cdot1}=1}\;;\;\boxed{\frac{u+2u}{2}=1}\;;\;\boxed{\frac{3u}{2}=1}\;;\;\boxed{\frac{3}{2}u=1}\;;\;\boxed{\frac{\cancel{2}}{\cancel{3}}\cdot\frac{\cancel{3}}{\cancel{2}}u=1\cdot\frac{2}{3}}\;;\;\boxed{u=\frac{2}{3}}\;;\;\boxed{\mathbf{u=0.67}}$$

The solution set is $\{\mathbf{0.67}\}$.

Check: $4+\frac{0.67}{2}\overset{?}{=}5-0.67\;;\;4+0.33\overset{?}{=}4.33\;;\;4.33=4.33$

Example 2.2-60

$$\boxed{-5x+3=10}\;;\;\boxed{-5x+3-3=10-3}\;;\;\boxed{-5x+0=7}\;;\;\boxed{-5x=7}\;;\;\boxed{\frac{-\cancel{5}x}{-\cancel{5}}=\frac{7}{-5}}\;;\;\boxed{x=-\frac{7}{5}}\;;\;\boxed{\mathbf{x=-1.4}}$$

The solution set is $\{\mathbf{-1.4}\}$.

Check: $(-5\cdot-1.4)+3\overset{?}{=}10\;;\;7+3\overset{?}{=}10\;;\;10=10$

Example 2.2-61

$$\boxed{-8t+5=2t-3}\;;\;\boxed{-8t-2t+5=2t-2t-3}\;;\;\boxed{-10t+5=0-3}\;;\;\boxed{-10t+5=-3}\;;\;\boxed{-10t+5-5=-3-5}$$

$$;\;\boxed{-10t+0=-8}\;;\;\boxed{-10t=-8}\;;\;\boxed{\frac{-\cancel{10}t}{-\cancel{10}}=\frac{-8}{-10}}\;;\;\boxed{t=\frac{8}{10}}\;;\;\boxed{\mathbf{t=0.8}}$$

The solution set is $\{\mathbf{0.8}\}$.

Check: $(-8\cdot0.8)+5\overset{?}{=}(2\cdot0.8)-3\;;\;-6.4+5\overset{?}{=}1.6-3\;;\;-1.4=-1.4$

Example 2.2-62

$\boxed{5+3y=5y}$; $\boxed{5-5+3y=5y-5}$; $\boxed{0+3y=5y-5}$; $\boxed{3y=5y-5}$; $\boxed{3y-5y=5y-5y-5}$; $\boxed{-2y=0-5}$

; $\boxed{-2y=-5}$; $\boxed{\frac{-2y}{-2}=\frac{-5}{-2}}$; $\boxed{y=\frac{5}{2}}$; $\boxed{y=2\frac{1}{2}}$; $\boxed{\mathbf{y=2.5}}$ The solution set is $\{\mathbf{2.5}\}$.

Check: $5+(3\cdot 2.5)\stackrel{?}{=}5\cdot 2.5$; $5+7.5\stackrel{?}{=}12.5$; $12.5=12.5$

Example 2.2-63

$\boxed{6x+10=8x+4}$; $\boxed{6x+10-10=8x+4-10}$; $\boxed{6x+0=8x-6}$; $\boxed{6x=8x-6}$; $\boxed{6x-8x=8x-8x-6}$

; $\boxed{-2x=0-6}$; $\boxed{-2x=-6}$; $\boxed{\frac{-2x}{-2}=\frac{-6}{-2}}$; $\boxed{x=\frac{\overset{3}{\cancel{6}}}{\cancel{2}}}$; $\boxed{x=\frac{3}{1}}$; $\boxed{\mathbf{x=3}}$ The solution set is $\{\mathbf{3}\}$.

Check: $(6\cdot 3)+10\stackrel{?}{=}(8\cdot 3)+4$; $18+10\stackrel{?}{=}24+4$; $28=28$

Example 2.2-64

$\boxed{\frac{w}{5}-10=4}$; $\boxed{\frac{w}{5}-10+10=4+10}$; $\boxed{\frac{w}{5}+0=4+10}$; $\boxed{\frac{w}{5}=14}$; $\boxed{\cancel{5}\times\frac{w}{\cancel{5}}=5\times 14}$; $\boxed{\mathbf{w=70}}$

The solution set is $\{\mathbf{70}\}$.

Check: $\frac{70}{5}-10\stackrel{?}{=}4$; $\frac{\overset{14}{\cancel{70}}}{\cancel{5}}-10\stackrel{?}{=}4$; $\frac{14}{1}-10\stackrel{?}{=}4$; $14-10\stackrel{?}{=}4$; $4=4$

Example 2.2-65

$\boxed{0.25x+3.5=1.2-0.5}$; $\boxed{0.25x+3.5=0.7}$; $\boxed{0.25x+3.5-3.5=0.7-3.5}$; $\boxed{0.25x+0=-2.8}$; $\boxed{0.25x=-2.8}$

; $\boxed{\frac{0.2\cancel{5}x}{0.2\cancel{5}}=\frac{-2.8}{0.25}}$; $\boxed{x=-\frac{2.8}{0.25}}$; $\boxed{x=-\frac{\frac{28}{10}}{\frac{25}{100}}}$; $\boxed{x=-\frac{28\cdot\overset{10}{\cancel{100}}}{\cancel{10}\cdot 25}}$; $\boxed{x=-\frac{28\cdot 10}{1\cdot 25}}$ = $\boxed{x=-\frac{280}{25}}$ = $\boxed{\mathbf{x=-11.2}}$

The solution set is $\{\mathbf{-11.2}\}$.

Check: $0.25\cdot(-11.2)+3.5\stackrel{?}{=}1.2-0.5$; $-2.8+3.5\stackrel{?}{=}0.7$; $0.7=0.7$

Example 2.2-66

$\boxed{5m+6=-3m-1}$; $\boxed{5m+3m+6=-3m+3m-1}$; $\boxed{8m+6=0-1}$; $\boxed{8m+6=-1}$; $\boxed{8m+6-6=-1-6}$

; $\boxed{8m+0=-7}$; $\boxed{8m=-7}$; $\boxed{\frac{\mathbf{8m}}{\mathbf{8}}=-\frac{7}{8}}$; $\boxed{\mathbf{m=-\frac{7}{8}}}$ The solution set is $\left\{-\frac{\mathbf{7}}{\mathbf{8}}\right\}$.

Check: $5\times-\frac{7}{8}+6\stackrel{?}{=}-3\times-\frac{7}{8}-1$; $-\frac{35}{8}+6\stackrel{?}{=}\frac{21}{8}-1$; $-\frac{35}{8}+\frac{6}{1}\stackrel{?}{=}\frac{21}{8}-\frac{1}{1}$; $\frac{(-35\cdot 1)+(6\cdot 8)}{8\cdot 1}\stackrel{?}{=}\frac{(21\cdot 1)+(-1\cdot 8)}{8\cdot 1}$

; $\frac{-35+48}{8}\stackrel{?}{=}\frac{21-8}{8}$; $\frac{13}{8}=\frac{13}{8}$

Example 2.2-67

$\boxed{x-\frac{1}{2}=3x-\frac{2}{5}}$; $\boxed{x-3x-\frac{1}{2}=3x-3x-\frac{2}{5}}$; $\boxed{-2x-\frac{1}{2}=0-\frac{2}{5}}$; $\boxed{-2x-\frac{1}{2}=-\frac{2}{5}}$; $\boxed{-2x-\frac{1}{2}+\frac{1}{2}=-\frac{2}{5}+\frac{1}{2}}$

$; \boxed{-2x+0=\frac{(-2\cdot 2)+(1\cdot 5)}{5\cdot 2}} ; \boxed{-2x=\frac{-4+5}{10}} ; \boxed{-2x=\frac{1}{10}} ; \boxed{\frac{-2x}{-2}=\frac{\frac{1}{10}}{-2}} ; \boxed{x=\frac{\frac{1}{10}}{\frac{-2}{1}}} ; \boxed{x=\frac{1\cdot 1}{10\cdot -2}} ; \boxed{\mathbf{x=-\frac{1}{20}}}$

The solution set is $\left\{-\frac{\mathbf{1}}{\mathbf{20}}\right\}$.

Check: $-\frac{1}{20}-\frac{1}{2}\overset{?}{=}3\cdot\left(-\frac{1}{20}\right)-\frac{2}{5}$; $\frac{(-1\cdot 2)+(-1\cdot 20)}{20\cdot 2}\overset{?}{=}-\frac{3}{20}-\frac{2}{5}$; $\frac{-2-20}{40}\overset{?}{=}-\frac{3}{20}-\frac{2}{5}$; $-\frac{22}{40}\overset{?}{=}\frac{(-3\cdot 5)+(-2\cdot 20)}{20\cdot 5}$

$; -\frac{22}{40}\overset{?}{=}\frac{-15-40}{100} ; -\frac{\overset{11}{\cancel{22}}}{\underset{20}{\cancel{40}}}\overset{?}{=}-\frac{\overset{11}{\cancel{55}}}{\underset{20}{\cancel{100}}} ; -\frac{11}{20}=-\frac{11}{20}$

Example 2.2-68

$\boxed{6a-3=4a+4} ; \boxed{6a-4a-3=4a-4a+4} ; \boxed{2a-3=0+4} ; \boxed{2a-3=4} ; \boxed{2a-3+3=4+3} ; \boxed{2a+0=7}$

$; \boxed{2a=7} ; \boxed{\frac{\cancel{2}a}{\cancel{2}}=\frac{7}{2}} ; \boxed{a=\frac{7}{2}} ; \boxed{a=3\frac{1}{2}} ; \boxed{\mathbf{a=3.5}}$

The solution set is $\{\mathbf{3.5}\}$.

Check: $6\cdot 3.5-3\overset{?}{=}4\cdot 3.5+4$; $21-3\overset{?}{=}14+4$; $18=18$

Example 2.2-69

$\boxed{0.4m+5=0.6m} ; \boxed{0.4m-0.6m+5=0.6m-0.6m} ; \boxed{-0.2m+5=0} ; \boxed{-0.2m+5-5=0-5} ; \boxed{-0.2m+0=-5}$

$; \boxed{-0.2m=-5} ; \boxed{\frac{-\cancel{0.2}m}{-\cancel{0.2}}=\frac{-5}{-0.2}} ; \boxed{m=\frac{5}{0.2}} ; \boxed{m=\frac{\frac{5}{1}}{\frac{2}{10}}} ; \boxed{m=\frac{5\cdot 10}{1\cdot 2}} ; \boxed{m=\frac{\overset{25}{\cancel{50}}}{\cancel{2}}} ; \boxed{m=\frac{25}{1}} ; \boxed{\mathbf{m=25}}$

The solution set is $\{\mathbf{25}\}$.

Check: $(0.4\cdot 25)+5\overset{?}{=}0.6\cdot 25$; $10+5\overset{?}{=}15$; $15=15$

Example 2.2-70

$\boxed{8x=20+4x} ; \boxed{8x-4x=20+4x-4x} ; \boxed{4x=20+0} ; \boxed{4x=20} ; \boxed{\frac{\cancel{4}x}{\cancel{4}}=\frac{\overset{5}{\cancel{20}}}{\cancel{4}}} ; \boxed{x=\frac{5}{1}} ; \boxed{\mathbf{x=5}}$

The solution set is $\{\mathbf{5}\}$.

Check: $8\cdot 5\overset{?}{=}20+(4\cdot 5)$; $40\overset{?}{=}20+20$; $40=40$

Example 2.2-71

$\boxed{5z-3=2z-8} ; \boxed{5z-2z-3=2z-2z-8} ; \boxed{3z-3=0-8} ; \boxed{3z-3=-8} ; \boxed{3z-3+3=-8+3} ; \boxed{3z+0=-5}$

$; \boxed{3z=-5} ; \boxed{\frac{\cancel{3}z}{\cancel{3}}=\frac{-5}{3}} ; \boxed{z=-\frac{5}{3}} ; \boxed{z=-1\frac{2}{3}} ; \boxed{\mathbf{z=-1.667}}$

The solution set is $\{\mathbf{-1.667}\}$.

Check: $(5\cdot -1.667)-3\overset{?}{=}(2\cdot -1.667)-8$; $-8.335-3\overset{?}{=}-3.335-8$; $-11.335=-11.335$

Example 2.2-72

$\boxed{7y+1=1\frac{2}{3}} ; \boxed{7y+1-1=1\frac{2}{3}-1} ; \boxed{7y+0=\frac{(1\cdot 3)+2}{3}-1} ; \boxed{7y=\frac{5}{3}-\frac{1}{1}} ; \boxed{7y=\frac{(5\cdot 1)-(1\cdot 3)}{3\cdot 1}} ; \boxed{7y=\frac{5-3}{3}}$

$$; \boxed{7y = \frac{2}{3}} ; \boxed{\frac{7y}{7} = \frac{\frac{2}{3}}{7}} ; \boxed{y = \frac{\frac{2}{3}}{\frac{7}{1}}} ; \boxed{y = \frac{2 \cdot 1}{3 \cdot 7}} ; \boxed{y = \frac{\mathbf{2}}{\mathbf{21}}}$$

The solution set is $\left\{\frac{\mathbf{2}}{\mathbf{21}}\right\}$.

Check: $7 \cdot \frac{2}{21} + 1 \overset{?}{=} 1\frac{2}{3}$; $\frac{14}{21} + \frac{1}{1} \overset{?}{=} \frac{(1 \cdot 3) + 2}{3}$; $\frac{(14 \cdot 1) + (1 \cdot 21)}{21 \cdot 1} \overset{?}{=} \frac{3+2}{3}$; $\frac{14+21}{21} \overset{?}{=} \frac{5}{3}$; $\frac{35}{21} \overset{?}{=} \frac{5}{3}$; $\frac{\overset{5}{\cancel{35}}}{\underset{3}{\cancel{21}}} \overset{?}{=} \frac{5}{3}$; $\frac{5}{3} = \frac{5}{3}$

Example 2.2-73

$$\boxed{-2z + 1 = 12} ; \boxed{-2z + 1 - 1 = 12 - 1} ; \boxed{-2z + 0 = 11} ; \boxed{-2z = 11} ; \boxed{\frac{-2z}{-2} = \frac{11}{-2}} ; \boxed{z = -\frac{11}{2}} ; \boxed{z = -5\frac{1}{2}} ; \boxed{\mathbf{z = -5.5}}$$

The solution set is $\{\mathbf{-5.5}\}$.

Check: $-2 \cdot (-5.5) + 1 \overset{?}{=} 12$; $11 + 1 \overset{?}{=} 12$; $12 = 12$

Example 2.2-74

$$\boxed{6x + 3 = 5x} ; \boxed{6x - 5x + 3 = 5x - 5x} ; \boxed{x + 3 = 0} ; \boxed{x + 3 - 3 = 0 - 3} ; \boxed{x + 0 = -3} ; \boxed{\mathbf{x = -3}}$$

The solution set is $\{\mathbf{-3}\}$.

Check: $6 \cdot -3 + 3 \overset{?}{=} 5 \cdot -3$; $-18 + 3 \overset{?}{=} -15$; $-15 = -15$

Example 2.2-75

$$\boxed{-2y - 8 = 5y + 13} ; \boxed{-2y - 5y - 8 = 5y - 5y + 13} ; \boxed{-7y - 8 = 0 + 13} ; \boxed{-7y - 8 = 13} ; \boxed{-7y - 8 + 8 = 13 + 8}$$

$$; \boxed{-7y + 0 = 21} ; \boxed{-7y = 21} ; \boxed{\frac{-7y}{-7} = \frac{21}{-7}} ; \boxed{y = -\frac{\overset{3}{\cancel{21}}}{\cancel{7}}} ; \boxed{y = -\frac{3}{1}} ; \boxed{\mathbf{y = -3}}$$

The solution set is $\{\mathbf{-3}\}$.

Check: $(-2 \cdot -3) - 8 \overset{?}{=} (5 \cdot -3) + 13$; $6 - 8 \overset{?}{=} -15 + 13$; $-2 = -2$

Practice Problems - Mixed Operations Involving Linear Equations

Section 2.2 Case III Practice Problems - Solve the following linear equations by applying the addition, subtraction, multiplication, and division rules:

1. $3x - 20 = 5x - 8$
2. $-6y + 2 = -3 + 10y$
3. $\frac{x}{-2} + 3 = 5$
4. $5x - 3 = -15$
5. $\frac{y}{4} + 4 = -3$
6. $5 + \frac{w}{2} = 10$
7. $25 - 3y = 2y$
8. $10y + 2 = 8y$
9. $\frac{2}{3}x + 5 = 12$
10. $m + \frac{1}{2} = 4m - \frac{2}{3}$

2.3 Solving Other Classes of Linear Equations

In many instances linear equations contain parentheses and brackets, fractions, or decimals. To simplify and solve these classes of equations students should be familiar with the rules that are applicable to each class. For example, solving linear equations that contain either parentheses and brackets or integer fractions require familiarity with the commutative, associative, and distributive rules (see Section 1.1) as well as the fractions rules (see Section 1.2). In this section students learn how to solve linear equations containing parentheses and brackets (Case I), fractions (Case II), and decimals (Case III).

Case I Solving Linear Equations Containing Parentheses and Brackets

To solve linear equations containing parentheses and brackets students need to be familiar with the concepts of signed numbers and the proper use of parentheses and brackets (review section 1.1). Note that the method used in solving this class of linear equations is similar with what we have already learned in section 2.2. Linear equations containing parentheses and brackets are solved using the following steps:

Step 1 Simplify the linear equation by properly multiplying the negative sign inside the parentheses or brackets (see Section 1.1b, Cases I and II).

Step 2 Isolate the variable to the left hand side of the equation by applying the addition and subtraction rules (see Section 2.2, Case I).

Step 3 Find the solution by applying the multiplication or division rules. Check the answer by substituting the solution into the original equation (see Section 2.2, Case II).

Examples with Steps

The following examples show the steps as to how linear equations containing parentheses and brackets are solved using the addition, subtraction, multiplication, and division rules:

Example 2.3-1

$$\boxed{2-(3+x)=5x}$$

Solution:

Step 1 $\boxed{2-(3+x)=5x}$; $\boxed{2-3-x=5x}$; $\boxed{-1-x=5x}$

Step 2 $\boxed{-1-x=5x}$; $\boxed{-1+(-x-5x)=5x-5x}$; $\boxed{-1-6x=0}$; $\boxed{(-1+1)-6x=0+1}$; $\boxed{0-6x=1}$; $\boxed{-6x=1}$

Step 3 $\boxed{-6x=1}$; $\boxed{\frac{-6x}{-6}=\frac{1}{-6}}$; $\boxed{x=-\frac{1}{6}}$; $\boxed{\mathbf{x=-0.166}}$

Check: $2-(3-0.166)\overset{?}{=}5\cdot-0.166$; $2-3+0.166\overset{?}{=}-0.83$; $-0.83=-0.83$

Example 2.3-2

$$\boxed{-(x-2)+3x-1=-4}$$

Solution:

Step 1 $\boxed{-(x-2)+3x-1=-4}$; $\boxed{-x+2+3x-1=-4}$; $\boxed{(-x+3x)+(-1+2)=-4}$; $\boxed{2x+1=-4}$

Step 2 $\boxed{2x+1=-4}$; $\boxed{2x+1-1=-4-1}$; $\boxed{2x+0=-5}$; $\boxed{2x=-5}$

Step 3 $\boxed{2x=-5}$; $\boxed{\frac{\cancel{2}x}{\cancel{2}}=\frac{-5}{2}}$; $\boxed{x=-\frac{5}{2}}$; $\boxed{x=-2\frac{1}{2}}$; $\boxed{\mathbf{x=-2.5}}$

Check: $-(-2.5-2)+(3\cdot-2.5)-1\overset{?}{=}-4$; $-(-4.5)-7.5-1\overset{?}{=}-4$; $4.5-8.5\overset{?}{=}-4$; $-4=-4$

Example 2.3-3

$\boxed{2x-5-(3x-8)=0}$

Solution:

Step 1 $\boxed{2x-5-(3x-8)=0}$; $\boxed{2x-5-3x+8=0}$; $\boxed{(2x-3x)+(8-5)=0}$; $\boxed{-x+3=0}$

Step 2 $\boxed{-x+3=0}$; $\boxed{-x+3-3=0-3}$; $\boxed{-x+0=-3}$; $\boxed{-x=-3}$

Step 3 $\boxed{-x=-3}$; $\boxed{\frac{-x}{-1}=\frac{-3}{-1}}$; $\boxed{\mathbf{x=3}}$

Check: $2\cdot3-5-(3\cdot3-8)\overset{?}{=}0$; $6-5-(9-8)\overset{?}{=}0$; $6-5-1\overset{?}{=}0$; $1-1\overset{?}{=}0$; $0=0$

Example 2.3-4

$\boxed{-[(x-5)+3]=(x-2)+3}$

Solution:

Step 1 $\boxed{-[(x-5)+3]=(x-2)+3}$; $\boxed{-[x-5+3]=x-2+3}$; $\boxed{-[x-2]=x+1}$; $\boxed{-x+2=x+1}$

Step 2 $\boxed{-x+2=x+1}$; $\boxed{(-x-x)+2=(x-x)+1}$; $\boxed{-2x+2=0+1}$; $\boxed{-2x+2=1}$

; $\boxed{-2x+2-2=1-2}$; $\boxed{-2x+0=-1}$; $\boxed{-2x=-1}$

Step 3 $\boxed{-2x=-1}$; $\boxed{\frac{-\cancel{2}x}{-\cancel{2}}=\frac{-1}{-2}}$; $\boxed{x=\frac{1}{2}}$; $\boxed{\mathbf{x=0.5}}$

Check: $-[(0.5-5)+3]\overset{?}{=}(0.5-2)+3$; $-(-4.5+3)\overset{?}{=}-1.5+3$; $-(-1.5)\overset{?}{=}1.5$; $1.5=1.5$

Example 2.3-5

$\boxed{-[(x-1)-(x-5)]=2x}$

Solution:

Step 1 $\boxed{-[(x-1)-(x-5)]=2x}$; $\boxed{-[x-1-x+5]=2x}$; $\boxed{-[(x-x)+(5-1)]=2x}$; $\boxed{-[0+4]=2x}$

; $\boxed{-4=2x}$

Step 2 ; $\boxed{-4=2x}$; $\boxed{-4-2x=2x-2x}$; $\boxed{-4-2x=0}$; $\boxed{-4+4-2x=0+4}$; $\boxed{0-2x=4}$

; $\boxed{-2x=4}$

Step 3 $\boxed{-2x = 4}$; $\boxed{\frac{-2x}{-2} = \frac{4}{-2}}$; $\boxed{x = -\frac{4}{2}}$; $\boxed{x = -\frac{2}{1}}$; $\boxed{x = -2}$

Check: $-[(-2-1)-(-2-5)] \stackrel{?}{=} 2\cdot -2$; $-[-3-(-7)] \stackrel{?}{=} -4$; $-(-3+7) \stackrel{?}{=} -4$; $-4 = -4$

Additional Examples - Solving Linear Equations Containing Parentheses and Brackets

The following examples further illustrate how to solve linear equations containing parentheses and brackets using the addition, subtraction, multiplication, and division rules:

Example 2.3-6

$\boxed{2(x-4)=5}$; $\boxed{2x-8=5}$; $\boxed{2x-8+8=5+8}$; $\boxed{2x+0=13}$; $\boxed{2x=13}$; $\boxed{\frac{2x}{2}=\frac{13}{2}}$; $\boxed{x=\frac{13}{2}}$; $\boxed{x=6\frac{1}{2}}$; $\boxed{x=6.5}$

Check: $2(6.5-4) \stackrel{?}{=} 5$; $2\cdot 2.5 \stackrel{?}{=} 5$; $5=5$

Example 2.3-7

$\boxed{3+4(x+1)=5(x-2)}$; $\boxed{3+4x+4=5x-10}$; $\boxed{4x+(3+4)=5x-10}$; $\boxed{4x+7=5x-10}$; $\boxed{4x+7-7=5x-10-7}$

; $\boxed{4x+0=5x-17}$; $\boxed{4x=5x-17}$; $\boxed{4x-5x=5x-5x-17}$; $\boxed{-x=0-17}$; $\boxed{-x=-17}$; $\boxed{\frac{-x}{-1}=\frac{-17}{-1}}$; $\boxed{x=17}$

Check: $3+4(17+1) \stackrel{?}{=} 5(17-2)$; $3+4\cdot 18 \stackrel{?}{=} 5\cdot 15$; $3+72 \stackrel{?}{=} 75$; $75=75$

Example 2.3-8

$\boxed{2-(x-1)+2x=5}$; $\boxed{2-x+1+2x=5}$; $\boxed{(2+1)+(2x-x)=5}$; $\boxed{3+x=5}$; $\boxed{3-3+x=5-3}$; $\boxed{0+x=2}$; $\boxed{x=2}$

Check: $2-(2-1)+2\cdot 2 \stackrel{?}{=} 5$; $2-1+4 \stackrel{?}{=} 5$; $1+4 \stackrel{?}{=} 5$; $5=5$

Example 2.3-9

$\boxed{3y-(y+5)+10=0}$; $\boxed{3y-y-5+10=0}$; $\boxed{(3y-y)+(-5+10)=0}$; $\boxed{2y+5=0}$; $\boxed{2y+5-5=0-5}$

; $\boxed{2y+0=-5}$; $\boxed{2y=-5}$; $\boxed{\frac{2y}{2}=\frac{-5}{2}}$; $\boxed{y=-\frac{5}{2}}$; $\boxed{y=-2\frac{1}{2}}$; $\boxed{y=-2.5}$

Check: $(3\cdot -2.5)-(-2.5+5)+10 \stackrel{?}{=} 0$; $-7.5-(+2.5)+10 \stackrel{?}{=} 0$; $-7.5-2.5+10 \stackrel{?}{=} 0$; $-10+10 \stackrel{?}{=} 0$; $0=0$

Example 2.3-10

$\boxed{-2[3(6x-2)+5]+3=0}$; $\boxed{-2[18x-6+5]+3=0}$; $\boxed{-2[18x-1]+3=0}$; $\boxed{-36x+2+3=0}$; $\boxed{-36x+5=0}$

; $\boxed{-36x+5-5=0-5}$; $\boxed{-36x+0=-5}$; $\boxed{-36x=-5}$; $\boxed{\frac{-36x}{-36}=\frac{-5}{-36}}$; $\boxed{x=\frac{5}{36}}$; $\boxed{x=0.139}$

Check: $-2[3(6\cdot 0.139-2)+5]+3 \stackrel{?}{=} 0$; $-2[3(0.834-2)+5]+3 \stackrel{?}{=} 0$; $-2[3(-1.166)+5]+3 \stackrel{?}{=} 0$

; $-2[-3.5+5]+3 \stackrel{?}{=} 0$; $-2\cdot 1.5+3 \stackrel{?}{=} 0$; $-3+3 \stackrel{?}{=} 0$; $0=0$

Example 2.3-11

$\boxed{5-4[(3x+5)-2x]=0}$; $\boxed{5-4[3x+5-2x]=0}$; $\boxed{5-4[(3x-2x)+5]=0}$; $\boxed{5-4[x+5]=0}$; $\boxed{5-4x-20=0}$

$5-12x-20+8x=0$
$5-4x-20=0$

; $\boxed{-4x+(-20+5)=0}$; $\boxed{-4x-15=0}$; $\boxed{-4x-15+15=0+15}$; $\boxed{-4x+0=15}$; $\boxed{-4x=15}$; $\boxed{\frac{\not{-4}x}{\not{-4}}=\frac{15}{-4}}$

; $\boxed{x=-\frac{15}{4}}$; $\boxed{x=-3\frac{3}{4}}$; $\boxed{\mathbf{x=-3.75}}$

Check: $5-4[(3\cdot-3.75+5)-(2\cdot-3.75)]\overset{?}{=}0$; $5-4[(-11.25+5)-(-7.5)]\overset{?}{=}0$; $5-4[-6.25+7.5]\overset{?}{=}0$

; $5-4\cdot1.25\overset{?}{=}0$; $5-5\overset{?}{=}0$; $0=0$

Example 2.3-12

$\boxed{3-2(x-2)+5=3x-4}$; $\boxed{3-2x+4+5=3x-4}$; $\boxed{-2x+(3+4+5)=3x-4}$; $\boxed{-2x+12=3x-4}$

; $\boxed{-2x+12-12=3x-4-12}$; $\boxed{-2x+0=3x-16}$; $\boxed{-2x=3x-16}$; $\boxed{-2x-3x=3x-3x-16}$; $\boxed{-5x=0-16}$

; $\boxed{-5x=-16}$; $\boxed{\frac{\not{-5}x}{\not{-5}}=\frac{-16}{-5}}$; $\boxed{x=\frac{16}{5}}$; $\boxed{x=3\frac{1}{5}}$; $\boxed{\mathbf{x=3.2}}$

Check: $3-2(3.2-2)+5\overset{?}{=}3\cdot3.2-4$; $3-(2\cdot1.2)+5\overset{?}{=}9.6-4$; $3-2.4+5\overset{?}{=}5.6$; $0.6+5\overset{?}{=}5.6$; $5.6=5.6$

Example 2.3-13

$\boxed{10+[2x-(4-x)]=-3x+1}$; $\boxed{10+[2x-4+x]=-3x+1}$; $\boxed{10+[(2x+x)-4]=-3x+1}$; $\boxed{10+[3x-4]=-3x+1}$

; $\boxed{10+3x-4=-3x+1}$; $\boxed{(10-4)+3x=-3x+1}$; $\boxed{6+3x=-3x+1}$; $\boxed{6-6+3x=-3x+1-6}$; $\boxed{0+3x=-3x-5}$

; $\boxed{3x=-3x-5}$; $\boxed{3x+3x=-3x+3x-5}$; $\boxed{6x=0-5}$; $\boxed{6x=-5}$; $\boxed{\frac{\not{6}x}{\not{6}}=\frac{-5}{6}}$; $\boxed{x=-\frac{5}{6}}$; $\boxed{\mathbf{x=-0.83}}$

Check: $10+[(2\cdot-0.83)-(4+0.83)]\overset{?}{=}(-3\cdot-0.83)+1$; $10+[-1.66-4.83]\overset{?}{=}2.5+1$; $10-6.49\overset{?}{=}3.5$; $3.5=3.5$

Example 2.3-14

$\boxed{5x+(2x-1)=-3(x-5)}$; $\boxed{5x+2x-1=-3x+15}$; $\boxed{7x-1=-3x+15}$; $\boxed{7x-1+1=-3x+15+1}$

; $\boxed{7x+0=-3x+16}$; $\boxed{7x=-3x+16}$; $\boxed{7x+3x=-3x+3x+16}$; $\boxed{10x=0+16}$; $\boxed{10x=16}$; $\boxed{\frac{\not{10}x}{\not{10}}=\frac{16}{10}}$

; $\boxed{x=\frac{16}{10}}$; $\boxed{x=1\frac{6}{10}}$; $\boxed{\mathbf{x=1.6}}$

Check: $5\cdot1.6+(2\cdot1.6-1)\overset{?}{=}-3(1.6-5)$; $8+(3.2-1)\overset{?}{=}-3(-3.4)$; $8+2.2\overset{?}{=}10.2$; $10.2=10.2$

Example 2.3-15

$\boxed{8-2[3x-(2+x)]=-5(x-1)+2}$; $\boxed{8-2[3x-2-x]=-5x+5+2}$; $\boxed{8-2[(3x-x)-2]=-5x+(5+2)}$

; $\boxed{8-2[(3x-x)-2]=-5x+7}$; $\boxed{8-2[2x-2]=-5x+7}$; $\boxed{8-4x+4=-5x+7}$; $\boxed{(8+4)-4x=-5x+7}$

; $\boxed{12-4x=-5x+7}$; $\boxed{12-12-4x=-5x+7-12}$; $\boxed{0-4x=-5x-5}$; $\boxed{-4x=-5x-5}$

; $\boxed{-4x+5x=-5x+5x-5}$; $\boxed{x=0-5}$; $\boxed{\mathbf{x=-5}}$

Check: $8-2[(3\cdot -5)-(2-5)]\stackrel{?}{=}-5(-5-1)+2$; $8-2[-15-(-3)]\stackrel{?}{=}-5(-6)+2$; $8-2(-15+3)\stackrel{?}{=}30+2$; $8-2(-12)\stackrel{?}{=}30+2$; $8+24\stackrel{?}{=}32$; $32=32$

Practice Problems - Solving Linear Equations Containing Parentheses and Brackets

Section 2.3 Case I Practice Problems - Solve the following linear equations by applying the addition, subtraction, multiplication, and division rules:

1. $x-(2x+3)=3$
2. $2+3(x-1)=-3-(x+5)$
3. $2-3(x-1)+5x=0$
4. $-4(-x+1)-3x=2(x-1)$
5. $2[5-(x-2)]-(x-3)=0$
6. $(x-5)-[3(x-1)+2]=2$
7. $3-[(-x+1)+2]=3x-5$
8. $-[(5-x)+(3-4x)]=8$
9. $3+(2x+5)-4x=3-3x$
10. $6(x-2)-2(x+1)=3(x+2)$

Case II Solving Linear Equations Containing Integer Fractions

A class of linear equations contains integer fractions. To solve these type of problems students need to be familiar with the fraction rules (review fraction concepts discussed in Section 1.2). Note that the method used in solving linear equations with integer fractions is similar with what we have already learned in section 2.2. However, these type of problems require more attention due to computations involving with integer fractions. Linear equations containing fractions are solved using the following steps:

Step 1 Isolate the variable to the left hand side of the equation by applying the addition and subtraction rules (see Section 2.2, Case I).

Step 2 Find the solution by applying the multiplication or division rules. Check the answer by substituting the solution into the original equation (see Section 2.2, Case II).

Examples with Steps

The following examples show the steps as to how linear equations containing integer fractions are solved using the addition, subtraction, multiplication, and division rules:

Example 2.3-16

$$\boxed{x-\frac{1}{3}=\frac{2}{3}x}$$

Solution:

Step 1 $\boxed{x-\frac{1}{3}=\frac{2}{3}x}$; $\boxed{x-\frac{2}{3}x-\frac{1}{3}=\frac{2}{3}x-\frac{2}{3}x}$; $\boxed{\left(1-\frac{2}{3}\right)x-\frac{1}{3}=0}$; $\boxed{\left(\frac{1}{1}-\frac{2}{3}\right)x-\frac{1}{3}=0}$

; $\boxed{\left(\frac{(1\cdot 3)-(2\cdot 1)}{1\cdot 3}\right)x-\frac{1}{3}=0}$; $\boxed{\left(\frac{3-2}{3}\right)x-\frac{1}{3}=0}$; $\boxed{\frac{1}{3}x-\frac{1}{3}=0}$; $\boxed{\frac{1}{3}x-\frac{1}{3}+\frac{1}{3}=0+\frac{1}{3}}$

; $\boxed{\frac{1}{3}x+0=\frac{1}{3}}$; $\boxed{\frac{1}{3}x=\frac{1}{3}}$

Step 2 $\boxed{\frac{1}{3}x=\frac{1}{3}}$; $\boxed{\cancel{3}\cdot\frac{1}{\cancel{3}}x=\cancel{3}\cdot\frac{1}{\cancel{3}}}$; $\boxed{\mathbf{x=1}}$

Check: $1-\frac{1}{3}\overset{?}{=}\frac{2}{3}\cdot 1$; $\frac{1}{1}-\frac{1}{3}\overset{?}{=}\frac{2}{3}$; $\frac{(1\cdot 3)-(1\cdot 1)}{1\cdot 3}\overset{?}{=}\frac{2}{3}$; $\frac{3-1}{3}\overset{?}{=}\frac{2}{3}$; $\frac{2}{3}=\frac{2}{3}$

Example 2.3-17

$$\boxed{u-\frac{1}{4}u=\frac{2}{5}}$$

Solution:

Step 1 $\boxed{u-\frac{1}{4}u=\frac{2}{5}}$; $\boxed{\left(1-\frac{1}{4}\right)u=\frac{2}{5}}$; $\boxed{\left(\frac{1}{1}-\frac{1}{4}\right)u=\frac{2}{5}}$; $\boxed{\left(\frac{(1\cdot 4)-(1\cdot 1)}{1\cdot 4}\right)u=\frac{2}{5}}$; $\boxed{\left(\frac{4-1}{4}\right)u=\frac{2}{5}}$

; $\boxed{\frac{3}{4}u=\frac{2}{5}}$

Step 2
$$\boxed{\frac{3}{4}u=\frac{2}{5}} \;;\; \boxed{\frac{4}{3}\cdot\frac{3}{4}u=\frac{2}{5}\cdot\frac{4}{3}} \;;\; \boxed{u=\frac{2\cdot 4}{5\cdot 3}} \;;\; \boxed{u=\frac{8}{15}} \;;\; \boxed{\mathbf{u=0.53}}$$

Check: $0.53-\frac{1}{4}\cdot 0.53\overset{?}{=}\frac{2}{5}$; $0.53-\frac{0.53}{4}\overset{?}{=}\frac{2}{5}$; $0.53-0.13\overset{?}{=}0.4$; $0.4=0.4$

Example 2.3-18

$$\boxed{4y-2=1\frac{2}{3}y}$$

Solution:

Step 1
$$\boxed{4y-2=1\frac{2}{3}y} \;;\; \boxed{4y-2=\frac{(1\cdot 3)+2}{3}y} \;;\; \boxed{4y-2=\frac{3+2}{3}y} \;;\; \boxed{4y-2=\frac{5}{3}y}$$

$$;\; \boxed{4y-\frac{5}{3}y-2=\frac{5}{3}y-\frac{5}{3}y} \;;\; \boxed{\left(4-\frac{5}{3}\right)y-2=0} \;;\; \boxed{\left(\frac{4}{1}-\frac{5}{3}\right)y-2=0}$$

$$;\; \boxed{\left(\frac{(4\cdot 3)-(5\cdot 1)}{1\cdot 3}\right)y-2=0} \;;\; \boxed{\left(\frac{12-5}{3}\right)y-2=0} \;;\; \boxed{\frac{7}{3}y-2=0} \;;\; \boxed{\frac{7}{3}y-2+2=0+2}$$

$$;\; \boxed{\frac{7}{3}y+0=2} \;;\; \boxed{\frac{7}{3}y=2}$$

Step 2
$$\boxed{\frac{7}{3}y=2} \;;\; \boxed{\frac{3}{7}\cdot\frac{7}{3}y=\frac{3}{7}\cdot 2} \;;\; \boxed{y=\frac{3\cdot 2}{7}} \;;\; \boxed{y=\frac{6}{7}} \;;\; \boxed{\mathbf{y=0.857}}$$

Check: $4\cdot 0.857-2\overset{?}{=}1\frac{2}{3}\cdot 0.857$; $3.43-2\overset{?}{=}\frac{(1\times 3)+2}{3}\cdot 0.857$; $1.43\overset{?}{=}\frac{3+2}{3}\cdot 0.857$; $1.43\overset{?}{=}\frac{5}{3}\cdot 0.857$

; $1.43\overset{?}{=}\frac{5\cdot 0.857}{3}$; $1.43\overset{?}{=}\frac{4.285}{3}$; $1.43=1.43$

Example 2.3-19

$$\boxed{\frac{1}{4}m=\frac{1}{3}-\frac{1}{2}m}$$

Solution:

Step 1
$$\boxed{\frac{1}{4}m=\frac{1}{3}-\frac{1}{2}m} \;;\; \boxed{\frac{1}{4}m+\frac{1}{2}m=\frac{1}{3}-\frac{1}{2}m+\frac{1}{2}m} \;;\; \boxed{\frac{1}{4}m+\frac{1}{2}m=\frac{1}{3}+0} \;;\; \boxed{\left(\frac{1}{4}+\frac{1}{2}\right)m=\frac{1}{3}}$$

$$;\; \boxed{\left(\frac{(1\cdot 2)+(1\cdot 4)}{4\cdot 2}\right)m=\frac{1}{3}} \;;\; \boxed{\left(\frac{2+4}{8}\right)m=\frac{1}{3}} \;;\; \boxed{\frac{\overset{3}{6}}{\underset{4}{8}}m=\frac{1}{3}} \;;\; \boxed{\frac{3}{4}m=\frac{1}{3}}$$

Step 2
$$\boxed{\frac{3}{4}m=\frac{1}{3}} \;;\; \boxed{\frac{4}{3}\cdot\frac{3}{4}m=\frac{4}{3}\cdot\frac{1}{3}} \;;\; \boxed{m=\frac{4\cdot 1}{3\cdot 3}} \;;\; \boxed{m=\frac{4}{9}} \;;\; \boxed{\mathbf{m=0.44}}$$

Check: $\frac{1}{4}\cdot 0.44\overset{?}{=}\frac{1}{3}-\frac{1}{2}\cdot 0.44$; $\frac{0.44}{4}\overset{?}{=}\frac{1}{3}-\frac{0.44}{2}$; $0.11\overset{?}{=}0.33-0.22$; $0.11=0.11$

Example 2.3-20

$$\boxed{x+5=4x-2\frac{3}{5}}$$

Solution:

Step 1 $\boxed{x+5=4x-2\frac{3}{5}}$; $\boxed{x-4x+5=4x-4x-2\frac{3}{5}}$; $\boxed{-3x+5=0-\frac{(2\cdot5)+3}{5}}$

; $\boxed{-3x+5=-\frac{13}{5}}$; $\boxed{-3x+5-5=-\frac{13}{5}-5}$; $\boxed{-3x+0=-\frac{13}{5}-\frac{5}{1}}$; $\boxed{-3x=\frac{(-13\cdot1)+(-5\cdot5)}{5\cdot1}}$

; $\boxed{-3x=\frac{-13-25}{5}}$; $\boxed{-3x=-\frac{38}{5}}$

Step 2 $\boxed{-3x=-\frac{38}{5}}$; $\boxed{\frac{-3x}{-3}=\frac{-\frac{38}{5}}{-3}}$; $\boxed{\frac{\cancel{3}x}{\cancel{3}}=\frac{\frac{38}{5}}{\frac{3}{1}}}$; $\boxed{x=\frac{38\cdot1}{5\cdot3}}$; $\boxed{x=\frac{38}{15}}$; $\boxed{x=2\frac{8}{15}}$; $\boxed{x=2.53}$

Check: $2.53+5\overset{?}{=}(4\cdot2.53)-2\frac{3}{5}$; $7.53\overset{?}{=}10.12-2.6$; $7.53\approx7.52$

Additional Examples - Solving Linear Equations Containing Integer Fractions

The following examples further illustrate how to solve linear equations containing integer fractions using the addition, subtraction, multiplication, and division rules:

Example 2.3-21

$\boxed{\frac{2}{3}x+5=\frac{1}{2}x+\frac{1}{5}}$; $\boxed{\frac{2}{3}x-\frac{1}{2}x+5=\frac{1}{2}x-\frac{1}{2}x+\frac{1}{5}}$; $\boxed{\left(\frac{2}{3}-\frac{1}{2}\right)x+5=0+\frac{1}{5}}$; $\boxed{\left(\frac{(2\cdot2)-(1\cdot3)}{3\cdot2}\right)x+5=\frac{1}{5}}$

; $\boxed{\left(\frac{4-3}{6}\right)x+5=\frac{1}{5}}$; $\boxed{\frac{1}{6}x+5=\frac{1}{5}}$; $\boxed{\frac{1}{6}x+5-5=\frac{1}{5}-5}$; $\boxed{\frac{1}{6}x+0=\frac{1}{5}-\frac{5}{1}}$; $\boxed{\frac{1}{6}x=\frac{(1\cdot1)-(5\cdot5)}{5\cdot1}}$; $\boxed{\frac{1}{6}x=\frac{1-25}{5}}$

; $\boxed{\frac{1}{6}x=-\frac{24}{5}}$; $\boxed{\frac{1}{\cancel{6}}x\cdot\cancel{6}=-\frac{24}{5}\cdot6}$; $\boxed{x=-\frac{24}{5}\cdot\frac{6}{1}}$; $\boxed{x=-\frac{24\cdot6}{5\cdot1}}$; $\boxed{x=-\frac{144}{5}}$; $\boxed{x=-28\frac{4}{5}}$; $\boxed{x=-28.8}$

Check: $\frac{2}{3}\cdot-28.8+5\overset{?}{=}\frac{1}{2}\cdot-28.8+\frac{1}{5}$; $\frac{2\cdot-28.8}{3}+5\overset{?}{=}\frac{1\cdot-28.8}{2}+\frac{1}{5}$; $-\frac{57.6}{3}+5\overset{?}{=}-\frac{28.8}{2}+\frac{1}{5}$

; $-19.2+5\overset{?}{=}-14.4+0.2$; $-14.2=-14.2$

Example 2.3-22

$\boxed{\frac{1}{3}u-\frac{1}{5}u=4}$; $\boxed{\left(\frac{1}{3}-\frac{1}{5}\right)u=4}$; $\boxed{\left(\frac{(1\cdot5)-(1\cdot3)}{3\cdot5}\right)u=4}$; $\boxed{\left(\frac{5-3}{15}\right)u=4}$; $\boxed{\frac{2}{15}u=4}$; $\boxed{\frac{2u}{15}=\frac{4}{1}}$; $\boxed{2u\cdot1=15\cdot4}$

; $\boxed{2u=60}$; $\boxed{\frac{\cancel{2}u}{\cancel{2}}=\frac{60}{2}}$; $\boxed{u=\frac{\overset{30}{\cancel{60}}}{\cancel{2}}}$; $\boxed{u=\frac{30}{1}}$; $\boxed{u=30}$

Check: $\frac{1}{3}\cdot30-\frac{1}{5}\cdot30\overset{?}{=}4$; $\frac{30}{3}-\frac{30}{5}\overset{?}{=}4$; $\frac{(30\cdot5)-(30\cdot3)}{3\cdot5}\overset{?}{=}4$; $\frac{150-90}{15}\overset{?}{=}4$; $\frac{\overset{4}{\cancel{60}}}{\cancel{15}}\overset{?}{=}4$; $\frac{4}{1}\overset{?}{=}4$; $4=4$

Example 2.3-23

$\boxed{\frac{1}{5}y+9=2y}$; $\boxed{\frac{1}{5}y-2y+9=2y-2y}$; $\boxed{\left(\frac{1}{5}-2\right)y+9=0}$; $\boxed{\left(\frac{1}{5}-\frac{2}{1}\right)y+9=0}$; $\boxed{\left(\frac{(1\cdot1)-(2\cdot5)}{5\cdot1}\right)y+9=0}$

$$;\ \boxed{\left(\frac{1-10}{5}\right)y+9=0}\ ;\ \boxed{-\frac{9}{5}y+9=0}\ ;\ \boxed{-\frac{9}{5}y+9-9=0-9}\ ;\ \boxed{-\frac{9}{5}y+0=-9}\ ;\ \boxed{-\frac{9}{5}y=-9}$$

$$;\ \boxed{-\frac{\cancel{5}}{\cancel{9}}\cdot-\frac{\cancel{9}}{\cancel{5}}y=-\cancel{9}\cdot-\frac{5}{\cancel{9}}}\ ;\ \boxed{y=\frac{5}{1}}\ ;\ \boxed{\mathbf{y=5}}$$

Check: $\frac{1}{5}\cdot 5+9\overset{?}{=}2\cdot 5$; $\frac{\cancel{5}}{\cancel{5}}+9\overset{?}{=}10$; $1+9\overset{?}{=}10$; $10=10$

Example 2.3-24

$$\boxed{\frac{1}{3}m-\frac{1}{5}m=\frac{1}{2}}\ ;\ \boxed{\left(\frac{1}{3}-\frac{1}{5}\right)m=\frac{1}{2}}\ ;\ \boxed{\left(\frac{(1\cdot 5)-(1\cdot 3)}{3\cdot 5}\right)m=\frac{1}{2}}\ ;\ \boxed{\left(\frac{5-3}{15}\right)m=\frac{1}{2}}\ ;\ \boxed{\frac{2}{15}m=\frac{1}{2}}\ ;\ \boxed{\frac{2m}{15}=\frac{1}{2}}$$

$$;\ \boxed{2m\cdot 2=1\cdot 15}\ ;\ \boxed{4m=15}\ ;\ \boxed{\frac{\cancel{4}m}{\cancel{4}}=\frac{15}{4}}\ ;\ \boxed{m=\frac{15}{4}}\ ;\ \boxed{m=3\frac{3}{4}}\ ;\ \boxed{\mathbf{m=3.75}}$$

Check: $\frac{1}{3}\cdot 3.75-\frac{1}{5}\cdot 3.75\overset{?}{=}\frac{1}{2}$; $\frac{3.75}{3}-\frac{3.75}{5}\overset{?}{=}\frac{1}{2}$; $1.25-0.75\overset{?}{=}0.5$; $0.5=0.5$

Example 2.3-25

$$\boxed{x-\frac{1}{3}=6+\frac{2}{3}x}\ ;\ \boxed{x-\frac{2}{3}x-\frac{1}{3}=6+\frac{2}{3}x-\frac{2}{3}x}\ ;\ \boxed{x-\frac{2}{3}x-\frac{1}{3}=6+0}\ ;\ \boxed{\left(1-\frac{2}{3}\right)x-\frac{1}{3}=6}\ ;\ \boxed{\left(\frac{1}{1}-\frac{2}{3}\right)x-\frac{1}{3}=6}$$

$$;\ \boxed{\left(\frac{(1\cdot 3)-(1\cdot 2)}{1\cdot 3}\right)x-\frac{1}{3}=6}\ ;\ \boxed{\left(\frac{3-2}{3}\right)x-\frac{1}{3}=6}\ ;\ \boxed{\frac{1}{3}x-\frac{1}{3}=6}\ ;\ \boxed{\frac{1}{3}x-\frac{1}{3}+\frac{1}{3}=6+\frac{1}{3}}\ ;\ \boxed{\frac{1}{3}x+0=\frac{6}{1}+\frac{1}{3}}$$

$$;\ \boxed{\frac{1}{3}x=\frac{(6\cdot 3)+(1\cdot 1)}{1\cdot 3}}\ ;\ \boxed{\frac{1}{3}x=\frac{18+1}{3}}\ ;\ \boxed{\frac{x}{3}=\frac{19}{3}}\ ;\ \boxed{\cancel{3}\cdot\frac{x}{\cancel{3}}=\cancel{3}\cdot\frac{19}{\cancel{3}}}\ ;\ \boxed{\mathbf{x=19}}$$

Check: $19-\frac{1}{3}\overset{?}{=}6+\frac{2}{3}\cdot 19$; $19-\frac{1}{3}\overset{?}{=}6+\frac{2\cdot 19}{3}$; $19-\frac{1}{3}\overset{?}{=}6+\frac{38}{3}$; $19-0.33\overset{?}{=}6+12.67$; $18.67=18.67$

Example 2.3-26

$$\boxed{\frac{3}{10}x+\frac{1}{3}=\frac{2}{5}}\ ;\ \boxed{\frac{3}{10}x+\frac{1}{3}-\frac{1}{3}=\frac{2}{5}-\frac{1}{3}}\ ;\ \boxed{\frac{3}{10}x+0=\frac{2}{5}-\frac{1}{3}}\ ;\ \boxed{\frac{3}{10}x=\frac{(2\cdot 3)-(1\cdot 5)}{5\cdot 3}}\ ;\ \boxed{\frac{3}{10}x=\frac{6-5}{15}}\ ;\ \boxed{\frac{3}{10}x=\frac{1}{15}}$$

$$;\ \boxed{\frac{3x}{10}=\frac{1}{15}}\ ;\ \boxed{3x\cdot 15=1\cdot 10}\ ;\ \boxed{45x=10}\ ;\ \boxed{\frac{\cancel{45}x}{\cancel{45}}=\frac{10}{45}}\ ;\ \boxed{x=\frac{10}{45}}\ ;\ \boxed{\mathbf{x=0.22}}$$

Check: $\frac{3}{10}\cdot 0.22+\frac{1}{3}\overset{?}{=}\frac{2}{5}$; $\frac{3\cdot 0.22}{10}+\frac{1}{3}\overset{?}{=}\frac{2}{5}$; $\frac{0.66}{10}+\frac{1}{3}\overset{?}{=}\frac{2}{5}$; $0.066+0.333\overset{?}{=}0.4$; $0.4=0.4$

Example 2.3-27

$$\boxed{\frac{1}{5}w+8=2w-\frac{1}{3}}\ ;\ \boxed{\frac{1}{5}w-2w+8=2w-2w-\frac{1}{3}}\ ;\ \boxed{\frac{1}{5}w-\frac{2}{1}w+8=0-\frac{1}{3}}\ ;\ \boxed{\left(\frac{1}{5}-\frac{2}{1}\right)w+8=-\frac{1}{3}}$$

$$;\ \boxed{\left(\frac{(1\cdot 1)-(2\cdot 5)}{5\cdot 1}\right)w+8=-\frac{1}{3}}\ ;\ \boxed{\left(\frac{1-10}{5}\right)w+8=-\frac{1}{3}}\ ;\ \boxed{-\frac{9}{5}w+8=-\frac{1}{3}}\ ;\ \boxed{-\frac{9}{5}w+8-8=-\frac{1}{3}-8}$$

$$;\ \boxed{-\frac{9}{5}w+0=-\frac{1}{3}-\frac{8}{1}}\ ;\ \boxed{-\frac{9}{5}w=\frac{(-1\cdot 1)+(-8\cdot 3)}{3\cdot 1}}\ ;\ \boxed{-\frac{9}{5}w=\frac{-1-24}{3}}\ ;\ \boxed{-\frac{9}{5}w=-\frac{25}{3}}\ ;\ \boxed{-\frac{\cancel{9}}{\cancel{5}}w\cdot-\frac{\cancel{5}}{\cancel{9}}=-\frac{25}{3}\cdot-\frac{5}{9}}$$

$$; \boxed{w=\frac{25\cdot 5}{3\cdot 9}} ; \boxed{w=\frac{125}{27}} ; \boxed{w=4\frac{17}{27}} ; \boxed{\mathbf{w=4.629}}$$

Check: $\frac{1}{5}\cdot 4.629+8\stackrel{?}{=}2\cdot 4.629-\frac{1}{3}$; $\frac{4.629}{5}+8\stackrel{?}{=}9.258-\frac{1}{3}$; $0.93+8\stackrel{?}{=}9.258-0.333$; $8.93=8.93$

Example 2.3-28

$$\boxed{x+\frac{2}{5}=\frac{1}{4}x} ; \boxed{x-\frac{1}{4}x+\frac{2}{5}=\frac{1}{4}x-\frac{1}{4}x} ; \boxed{\left(1-\frac{1}{4}\right)x+\frac{2}{5}=0} ; \boxed{\left(\frac{1}{1}-\frac{1}{4}\right)x+\frac{2}{5}=0} ; \boxed{\left(\frac{(1\cdot 4)-(1\cdot 1)}{1\cdot 4}\right)x+\frac{2}{5}=0}$$

$$; \boxed{\left(\frac{4-1}{4}\right)x+\frac{2}{5}=0} ; \boxed{\frac{3}{4}x+\frac{2}{5}=0} ; \boxed{\frac{3}{4}x+\frac{2}{5}-\frac{2}{5}=0-\frac{2}{5}} ; \boxed{\frac{3}{4}x+0=-\frac{2}{5}} ; \boxed{\frac{3}{4}x=-\frac{2}{5}} ; \boxed{\frac{\cancel{3}}{\cancel{4}}x\cdot\frac{\cancel{4}}{\cancel{3}}=-\frac{2}{5}\cdot\frac{4}{3}}$$

$$; \boxed{x=-\frac{2\cdot 4}{5\cdot 3}} ; \boxed{x=-\frac{8}{15}} ; \boxed{\mathbf{x=-0.533}}$$

Check: $-0.533+\frac{2}{5}\stackrel{?}{=}\frac{1}{4}\cdot -0.533$; $-0.533+0.4\stackrel{?}{=}-\frac{0.533}{4}$; $-0.133=-0.133$

Example 2.3-29

$$\boxed{\frac{2}{3}x-\frac{1}{4}=\frac{1}{2}x+\frac{2}{3}} ; \boxed{\frac{2}{3}x-\frac{1}{2}x-\frac{1}{4}=\frac{1}{2}x-\frac{1}{2}x+\frac{2}{3}} ; \boxed{\left(\frac{2}{3}-\frac{1}{2}\right)x-\frac{1}{4}=0+\frac{2}{3}} ; \boxed{\left(\frac{(2\cdot 2)-(1\cdot 3)}{3\cdot 2}\right)x-\frac{1}{4}=\frac{2}{3}}$$

$$; \boxed{\left(\frac{4-3}{6}\right)x-\frac{1}{4}=\frac{2}{3}} ; \boxed{\frac{1}{6}x-\frac{1}{4}=\frac{2}{3}} ; \boxed{\frac{1}{6}x-\frac{1}{4}+\frac{1}{4}=\frac{2}{3}+\frac{1}{4}} ; \boxed{\frac{1}{6}x=\frac{(2\cdot 4)+(1\cdot 3)}{3\cdot 4}} ; \boxed{\frac{1}{6}x=\frac{8+3}{12}} ; \boxed{\frac{1}{6}x=\frac{11}{12}}$$

$$; \boxed{\cancel{6}\cdot\frac{1}{\cancel{6}}x=\cancel{6}\cdot\frac{11}{\cancel{12}_{2}}} ; \boxed{x=\frac{11}{2}} ; \boxed{\mathbf{x=5.5}}$$

Check: $\frac{2}{3}\cdot 5.5-\frac{1}{4}\stackrel{?}{=}\frac{1}{2}\cdot 5.5+\frac{2}{3}$; $\frac{2\cdot 5.5}{3}-\frac{1}{4}\stackrel{?}{=}\frac{1\cdot 5.5}{2}+\frac{2}{3}$; $\frac{11}{3}-\frac{1}{4}\stackrel{?}{=}\frac{5.5}{2}+\frac{2}{3}$; $\frac{(11\cdot 4)-(1\cdot 3)}{3\cdot 4}\stackrel{?}{=}\frac{(5.5\cdot 3)+(2\cdot 2)}{2\cdot 3}$

; $\frac{44-3}{12}\stackrel{?}{=}\frac{16.5+4}{6}$; $\frac{41}{12}\stackrel{?}{=}\frac{20.5}{6}$; $3.42=3.42$

Example 2.3-30

$$\boxed{\frac{1}{2}t+\frac{2}{3}t=5-\frac{1}{4}t} ; \boxed{\left(\frac{1}{2}+\frac{2}{3}\right)t=5-\frac{1}{4}t} ; \boxed{\left(\frac{(1\cdot 3)+(2\cdot 2)}{2\cdot 3}\right)t=5-\frac{1}{4}t} ; \boxed{\left(\frac{3+4}{6}\right)t=5-\frac{1}{4}t} ; \boxed{\frac{7}{6}t=5-\frac{1}{4}t}$$

$$; \boxed{\frac{7}{6}t+\frac{1}{4}t=5-\frac{1}{4}t+\frac{1}{4}t} ; \boxed{\left(\frac{7}{6}+\frac{1}{4}\right)t=5+0} ; \boxed{\left(\frac{(7\cdot 4)+(1\cdot 6)}{6\cdot 4}\right)t=5} ; \boxed{\left(\frac{28+6}{24}\right)t=5} ; \boxed{\frac{\overset{17}{\cancel{34}}}{\underset{12}{\cancel{24}}}t=5} ; \boxed{\frac{17}{12}t=5}$$

$$; \boxed{\frac{\cancel{12}}{\cancel{17}}\cdot\frac{\cancel{17}}{\cancel{12}}t=\frac{12}{17}\cdot 5} ; \boxed{t=\frac{12\cdot 5}{17}} ; \boxed{t=\frac{60}{17}} ; \boxed{t=3\frac{9}{17}} ; \boxed{\mathbf{t=3.53}}$$

Check: $\frac{1}{2}\cdot 3.53+\frac{2}{3}\cdot 3.53\stackrel{?}{=}5-\frac{1}{4}\cdot 3.53$; $\frac{1\cdot 3.53}{2}+\frac{2\cdot 3.53}{3}\stackrel{?}{=}5-\frac{1\cdot 3.53}{4}$; $\frac{3.53}{2}+\frac{7.06}{3}\stackrel{?}{=}5-\frac{3.53}{4}$

; $1.77+2.35\stackrel{?}{=}5-0.88$; $4.12=4.12$

Practice Problems - Solving Linear Equations Containing Integer Fractions

Section 2.3 Case II Practice Problems - Solve the following linear equations by applying the addition, subtraction, multiplication, and division rules:

1. $\frac{1}{2}y = \frac{1}{5}y + 5$
2. $x = 3 - \frac{x}{2}$
3. $y + \frac{2}{3}y = 1\frac{2}{3}$
4. $u - \frac{1}{3}u = 6$
5. $s + 1\frac{2}{3} = 2\frac{3}{5}s$
6. $\frac{w}{3} - 1\frac{2}{3}w = 4$
7. $x - \frac{2}{3} = 1\frac{1}{4}x$
8. $t = 5 - \frac{2}{3}t$
9. $4\frac{2}{3}z - 2\frac{3}{5}z = 1\frac{1}{4}$
10. $6 + \frac{1}{2}t = t - 1\frac{2}{5}t$

Case III Solving Linear Equations Containing Decimals

Another class of linear equations contains decimals. To solve these type of problems students need to be familiar with conversion of decimal numbers to integer fraction form (review decimal fraction concepts discussed in Chapter 4 of the *Mastering Fractions* book). Note that the method used in solving linear equations with decimal numbers is similar with what we have already learned in section 2.2. However, these type of problems require more attention due to computations involving with decimals. Linear equations containing decimals are solved using the following steps:

Step 1 Isolate the variable to the left hand side of the equation by applying the addition and subtraction rules (see Section 2.2, Case I).

Step 2 Find the solution by applying the multiplication or division rules. Check the answer by substituting the solution into the original equation (see Section 2.2, Case II).

Examples with Steps

The following examples show the steps as to how linear equations containing decimals are solved using the addition, subtraction, multiplication, and division rules:

Example 2.3-31

$$\boxed{3.4x - 2.5 = -2.8x + 0.5}$$

Solution:

Step 1

$$\boxed{3.4x - 2.5 = -2.8x + 0.5}\;;\;\boxed{(3.4x + 2.8x) - 2.5 = (-2.8x + 2.8x) + 0.5}\;;\;\boxed{6.2x - 2.5 = 0 + 0.5}$$

$$;\;\boxed{6.2x - 2.5 = 0.5}\;;\;\boxed{6.2x + (-2.5 + 2.5) = 0.5 + 2.5}\;;\;\boxed{6.2x + 0 = 3}\;;\;\boxed{6.2x = 3}$$

Step 2

$$\boxed{6.2x = 3}\;;\;\boxed{\frac{6.2}{6.2}x = \frac{3}{6.2}}\;;\;\boxed{x = \frac{3}{6.2}}\;;\;\boxed{x = \frac{\frac{3}{1}}{\frac{62}{10}}}\;;\;\boxed{x = \frac{3 \cdot 10}{1 \cdot 62}}\;;\;\boxed{x = \frac{30}{62}}\;;\;\boxed{\mathbf{x = 0.484}}$$

Check: $(3.4 \cdot 0.484) - 2.5 \stackrel{?}{=} (-2.8 \cdot 0.484) + 0.5$; $1.65 - 2.5 \stackrel{?}{=} -1.35 + 0.5$; $-0.85 = -0.85$

Example 2.3-32

$$\boxed{1.25(x - 0.2) - (0.5x - 1) = 0}$$

Solution:

Step 1

$$\boxed{1.25(x - 0.2) - (0.5x - 1) = 0}\;;\;\boxed{1.25x - 0.25 - 0.5x + 1 = 0}\;;\;\boxed{(1.25x - 0.5x) + (-0.25 + 1) = 0}$$

$$;\;\boxed{0.75x + 0.75 = 0}\;;\;\boxed{0.75x + (0.75 - 0.75) = 0 - 0.75}\;;\;\boxed{0.75x + 0 = -0.75}\;;\;\boxed{0.75x = -0.75}$$

Step 2

$$\boxed{0.75x = -0.75}\;;\;\boxed{\frac{0.75}{0.75}x = \frac{-0.75}{0.75}}\;;\;\boxed{x = -\frac{0.75}{0.75}}\;;\;\boxed{x = -\frac{\frac{75}{100}}{\frac{75}{100}}}\;;\;\boxed{x = -\frac{75 \cdot 100}{100 \cdot 75}}$$

$$;\;\boxed{x = -\frac{7500}{7500}}\;;\;\boxed{\mathbf{x = -1}}$$

Check: $1.25(-1-0.2)-[(0.5\cdot -1)-1]\overset{?}{=}0$; $(1.25\cdot -1.2)-[-0.5-1]\overset{?}{=}0$; $-1.5-(-1.5)\overset{?}{=}0$; $-1.5+1.5\overset{?}{=}0$; $0=0$

Example 2.3-33

$$\boxed{8.4x-(0.5-0.2x)=1.25x}$$

Solution:

Step 1 $\boxed{8.4x-(0.5-0.2x)=1.25x}$; $\boxed{8.4x-0.5+0.2x=1.25x}$; $\boxed{(8.4x+0.2x)-0.5=1.25x}$

; $\boxed{8.6x-0.5=1.25x}$; $\boxed{(8.6x-1.25x)-0.5=1.25x-1.25x}$; $\boxed{7.35x-0.5=0}$

; $\boxed{7.35x+(-0.5+0.5)=0+0.5}$; $\boxed{7.35x+0=0.5}$; $\boxed{7.35x=0.5}$

Step 2 $\boxed{7.35x=0.5}$; $\boxed{\frac{7.35}{7.35}x=\frac{0.5}{7.35}}$; $\boxed{x=\frac{0.5}{7.35}}$; $\boxed{x=\frac{\frac{5}{10}}{\frac{735}{100}}}$; $\boxed{x=\frac{5\cdot 100}{10\cdot 735}}$; $\boxed{x=\frac{500}{7350}}$

; $\boxed{\mathbf{x=0.068}}$

Check: $(8.4\cdot 0.068)-[0.5+(-0.2\cdot 0.068)]\overset{?}{=}1.25\cdot 0.068$; $0.57-[0.5-0.014]\overset{?}{=}0.09$; $0.57-0.48\overset{?}{=}0.09$

; $0.09=0.09$

Example 2.3-34

$$\boxed{5.5-(x-0.2)-[(x-5)+0.45]=0}$$

Solution:

Step 1 $\boxed{5.5-(x-0.2)-[(x-5)+0.45]=0}$; $\boxed{5.5-x+0.2-[x-5+0.45]=0}$

; $\boxed{5.5-x+0.2-[x-4.55]=0}$; $\boxed{5.5-x+0.2-x+4.55=0}$

; $\boxed{(-x-x)+(5.5+4.55+0.2)=0}$; $\boxed{-2x+(5.5+4.55+0.2)=0}$; $\boxed{-2x+10.25=0}$

; $\boxed{-2x+(10.25-10.25)=0-10.25}$; $\boxed{-2x+0=-10.25}$; $\boxed{-2x=-10.25}$

Step 2 $\boxed{-2x=-10.25}$; $\boxed{\frac{-2}{-2}x=\frac{-10.25}{-2}}$; $\boxed{x=\frac{10.25}{2}}$; $\boxed{x=\frac{\frac{1025}{100}}{\frac{2}{1}}}$; $\boxed{x=\frac{1025\cdot 1}{100\cdot 2}}$; $\boxed{x=\frac{1025}{200}}$

; $\boxed{\mathbf{x=5.125}}$

Check: $5.5-(5.125-0.2)-[(5.125-5)+0.45]\overset{?}{=}0$; $5.5-4.925-[0.125+0.45]\overset{?}{=}0$; $(5.5-4.925)-0.575\overset{?}{=}0$

; $0.575-0.575\overset{?}{=}0$; $0=0$

Example 2.3-35

$$\boxed{0.5x=(x-2.5)-(0.45x-1)}$$

Solution:

Step 1 $\boxed{0.5x=(x-2.5)-(0.45x-1)}$; $\boxed{0.5x=x-2.5-0.45x+1}$; $\boxed{0.5x=(x-0.45x)+(1-2.5)}$

; $\boxed{0.5x = 0.55x - 1.5}$; $\boxed{0.5x - 0.55x = (0.55x - 0.55x) - 1.5}$; $\boxed{-0.05x = 0 - 1.5}$; $\boxed{-0.05x = -1.5}$

Step 2 $\boxed{-0.05x = -1.5}$; $\boxed{\frac{-0.05}{-0.05}x = \frac{-1.5}{-0.05}}$; $\boxed{x = \frac{1.5}{0.05}}$; $\boxed{x = \frac{\frac{15}{10}}{\frac{5}{100}}}$; $\boxed{x = \frac{15 \cdot 100}{10 \cdot 5}}$; $\boxed{x = \frac{1500}{50}}$

; $\boxed{\mathbf{x = 30}}$

Check: $0.5 \cdot 30 \stackrel{?}{=} (30 - 2.5) - [(0.45 \cdot 30) - 1]$; $0.5 \cdot 30 \stackrel{?}{=} 27.5 - [13.5 - 1]$; $15 \stackrel{?}{=} 27.5 - 12.5$; $15 = 15$

Additional Examples - Solving Linear Equations Containing Decimals

The following examples further illustrate how to solve linear equations containing decimals using the addition, subtraction, multiplication, and division rules:

Example 2.3-36

$\boxed{0.2x + 2.4 = 0.52x + 3.5}$; $\boxed{(0.2x - 0.52x) + 2.4 = (0.52x - 0.52x) + 3.5}$; $\boxed{-0.32x + 2.4 = 0 + 3.5}$; $\boxed{-0.32x + 2.4 = 3.5}$

; $\boxed{-0.32x + (2.4 - 2.4) = (3.5 - 2.4)}$; $\boxed{-0.32x + 0 = 1.1}$; $\boxed{-0.32x = 1.1}$; $\boxed{\frac{-0.32x}{-0.32} = \frac{1.1}{-0.32}}$; $\boxed{\frac{x}{1} = -\frac{1.1}{0.32}}$; $\boxed{x = -\frac{\frac{11}{10}}{\frac{32}{100}}}$

; $\boxed{x = -\frac{11 \cdot 100}{10 \cdot 32}}$; $\boxed{x = -\frac{1100}{320}}$; $\boxed{\mathbf{x = -3.44}}$

Check: $0.2 \cdot (-3.44) + 2.4 \stackrel{?}{=} 0.52 \cdot (-3.44) + 3.5$; $-0.69 + 2.4 \stackrel{?}{=} -1.79 + 3.5$; $1.71 = 1.71$

Example 2.3-37

$\boxed{-0.65(x + 0.2) = 0.25x}$; $\boxed{-0.65x - 0.13 = 0.25x}$; $\boxed{(-0.65x - 0.25x) - 0.13 = 0.25x - 0.25x}$; $\boxed{-0.9x - 0.13 = 0}$

; $\boxed{-0.9x - 0.13 + 0.13 = 0 + 0.13}$; $\boxed{-0.9x + 0 = 0.13}$; $\boxed{-0.9x = 0.13}$; $\boxed{x = \frac{0.13}{-0.9}}$; $\boxed{x = -\frac{0.13}{0.9}}$; $\boxed{x = -\frac{\frac{13}{100}}{\frac{9}{10}}}$

; $\boxed{x = -\frac{13 \cdot 10}{100 \cdot 9}}$; $\boxed{x = -\frac{130}{900}}$; $\boxed{\mathbf{x = -0.144}}$

Check: $-0.65(-0.144 + 0.2) \stackrel{?}{=} 0.25 \cdot (-0.144)$; $-0.65 \cdot 0.056 \stackrel{?}{=} -0.036$; $-0.036 = -0.036$

Example 2.3-38

$\boxed{4.5x - 0.2(x - 0.1) + 0.3 = 0}$; $\boxed{4.5x - 0.2x + 0.02 + 0.3 = 0}$; $\boxed{4.3x + 0.32 = 0}$; $\boxed{4.3x + 0.32 - 0.32 = 0 - 0.32}$

; $\boxed{4.3x + 0 = -0.32}$; $\boxed{4.3x = -0.32}$; $\boxed{x = \frac{-0.32}{4.3}}$; $\boxed{x = -\frac{\frac{32}{100}}{\frac{43}{10}}}$; $\boxed{x = -\frac{32 \cdot 10}{100 \cdot 43}}$; $\boxed{x = -\frac{320}{4300}}$; $\boxed{\mathbf{x = -0.074}}$

Check: $4.5 \cdot (-0.074) - 0.2(-0.074 - 0.1) + 0.3 \stackrel{?}{=} 0$; $-0.33 + 0.01 + 0.02 + 0.3 \stackrel{?}{=} 0$; $-0.33 + 0.33 \stackrel{?}{=} 0$; $0 = 0$

Example 2.3-39

$\boxed{4(x-0.45)=-3.9x+0.005}$; $\boxed{4x-1.8=-3.9x+0.005}$; $\boxed{(4x+3.9x)-1.8=(-3.9x+3.9x)+0.005}$

; $\boxed{7.9x-1.8=0+0.005}$; $\boxed{7.9x-1.8=0.005}$; $\boxed{7.9x+(-1.8+1.8)=0.005+1.8}$; $\boxed{7.9x+0=1.805}$; $\boxed{7.9x=1.805}$

; $\boxed{x=\frac{1.805}{7.9}}$; $\boxed{x=\frac{\frac{1805}{1000}}{\frac{79}{10}}}$; $\boxed{x=\frac{1805\cdot 10}{1000\cdot 79}}$; $\boxed{x=\frac{18050}{79000}}$; $\boxed{\mathbf{x=0.23}}$

Check: $4(0.23-0.45)\overset{?}{=}(-3.9\cdot 0.23)+0.005$; $0.92-1.8\overset{?}{=}-0.89+0.005$; $-0.88=-0.88$

Example 2.3-40

$\boxed{0.1(0.4x-1)=0.2x+0.28}$; $\boxed{0.04x-0.1=0.2x+0.28}$; $\boxed{(0.04x-0.2x)-0.1=(0.2x-0.2x)+0.28}$

; $\boxed{-0.16x-0.1=0+0.28}$; $\boxed{-0.16x-0.1=0.28}$; $\boxed{-0.16x+(-0.1+0.1)=0.28+0.1}$; $\boxed{-0.16x+0=0.38}$

; $\boxed{-0.16x=0.38}$; $\boxed{x=\frac{0.38}{-0.16}}$; $\boxed{x=-\frac{0.38}{0.16}}$; $\boxed{x=-\frac{\frac{38}{100}}{\frac{16}{100}}}$; $\boxed{x=-\frac{38\cdot 100}{100\cdot 16}}$; $\boxed{x=-\frac{3800}{1600}}$; $\boxed{\mathbf{x=-2.375}}$

Check: $0.1[0.4\cdot(-2.375)-1]\overset{?}{=}0.2\cdot(-2.375)+0.28$; $0.1(-0.95-1)\overset{?}{=}-0.475+0.28$; $0.1\cdot(-1.95)\overset{?}{=}-0.195$

; $-0.195=-0.195$

Example 2.3-41

$\boxed{0.5[(x-0.2)-0.45x]=2.5x}$; $\boxed{0.5[(x-0.45x)-0.2]=2.5x}$; $\boxed{0.5[0.55x-0.2]=2.5x}$; $\boxed{0.275x-0.1=2.5x}$

; $\boxed{(0.275x-2.5x)-0.1=2.5x-2.5x}$; $\boxed{-2.225x-0.1=0}$; $\boxed{-2.225x-0.1+0.1=0+0.1}$; $\boxed{-2.225x+0=0.1}$

; $\boxed{-2.225x=0.1}$; $\boxed{x=\frac{0.1}{-2.225}}$; $\boxed{x=-\frac{0.1}{2.225}}$; $\boxed{x=-\frac{\frac{1}{10}}{\frac{2225}{1000}}}$; $\boxed{x=-\frac{1\cdot 1000}{10\cdot 2225}}$; $\boxed{x=-\frac{1000}{22250}}$; $\boxed{\mathbf{x=-0.045}}$

Check: $0.5[(-0.045-0.2)-0.45\cdot(-0.045)]\overset{?}{=}2.5\cdot(-0.045)$; $0.5[-0.245+0.02]\overset{?}{=}-0.11$; $0.5\cdot(-0.25)\overset{?}{=}-0.11$

; $-0.11=-0.11$

Example 2.3-42

$\boxed{(0.2-x)+[1-0.2(x+2)]=2.4x}$; $\boxed{(0.2-x)+[1-0.2x-0.4]=2.4x}$; $\boxed{0.2-x+[(1-0.4)-0.2x]=2.4x}$

; $\boxed{0.2-x+[0.6-0.2x]=2.4x}$; $\boxed{(0.2+0.6)+(-x-0.2x)=2.4x}$; $\boxed{0.8-1.2x=2.4x}$

; $\boxed{0.8+(-1.2x-2.4x)=2.4x-2.4x}$; $\boxed{0.8-3.6x=0}$; $\boxed{(0.8-0.8)-3.6x=0-0.8}$; $\boxed{0-3.6x=-0.8}$

; $\boxed{-3.6x=-0.8}$; $\boxed{x=\frac{-0.8}{-3.6}}$; $\boxed{x=\frac{0.8}{3.6}}$; $\boxed{x=\frac{\frac{8}{10}}{\frac{36}{10}}}$; $\boxed{x=\frac{8\cdot 10}{36\cdot 10}}$; $\boxed{x=\frac{80}{360}}$; $\boxed{\mathbf{x=0.22}}$

Check: $(0.2-0.22)+[1-0.2(0.22+2)]\overset{?}{=}2.4\cdot 0.22$; $-0.02+[1-0.2\cdot 2.22]\overset{?}{=}0.53$; $-0.02+[1-0.444]\overset{?}{=}0.53$; $-0.02+0.556\overset{?}{=}0.53$; $0.54\approx 0.53$

Example 2.3-43

$\boxed{0.01x+0.25(x+0.1)=1}$; $\boxed{0.01x+0.25x+0.025=1}$; $\boxed{0.26x+0.025=1}$; $\boxed{0.26x+(0.025-0.025)=1-0.025}$; $\boxed{0.26x+0=0.975}$; $\boxed{0.26x=0.975}$; $\boxed{x=\frac{0.975}{0.26}}$; $\boxed{x=\frac{\frac{975}{1000}}{\frac{26}{100}}}$; $\boxed{x=\frac{975\cdot 100}{1000\cdot 26}}$; $\boxed{x=\frac{97500}{26000}}$; $\boxed{\mathbf{x=3.75}}$

Check: $(0.01\cdot 3.75)+0.25(3.75+0.1)\overset{?}{=}1$; $0.0375+(0.25\cdot 3.85)\overset{?}{=}1$; $0.0375+0.9625\overset{?}{=}1$; $1=1$

Example 2.3-44

$\boxed{0.3(x+0.1)-0.5x=2.35-1.2x}$; $\boxed{0.3x+0.03-0.5x=2.35-1.2x}$; $\boxed{(0.3x-0.5x)+0.03=2.35-1.2x}$; $\boxed{-0.2x+0.03=2.35-1.2x}$; $\boxed{(-0.2x+1.2x)+0.03=2.35+(-1.2x+1.2x)}$; $\boxed{x+0.03=2.35+0}$; $\boxed{x+0.03=2.35}$; $\boxed{x+(0.03-0.03)=2.35-0.03}$; $\boxed{x+0=2.32}$; $\boxed{\mathbf{x=2.32}}$

Check: $0.3(2.32+0.1)-(0.5\cdot 2.32)\overset{?}{=}2.35-(1.2\cdot 2.32)$; $(0.3\cdot 2.42)-1.16\overset{?}{=}2.35-2.784$; $0.726-1.16\overset{?}{=}-0.434$; $-0.434=-0.434$

Example 2.3-45

$\boxed{[(0.2x-2.3)+0.2(x-1)]-1.8x=0}$; $\boxed{[0.2x-2.3+0.2x-0.2]-1.8x=0}$; $\boxed{[(0.2x+0.2x)+(-0.2-2.3)]-1.8x=0}$; $\boxed{[0.4x-2.5]-1.8x=0}$; $\boxed{(0.4x-1.8x)-2.5=0}$; $\boxed{-1.4x-2.5=0}$; $\boxed{-1.4x+(-2.5+2.5)=0+2.5}$; $\boxed{-1.4x+0=2.5}$; $\boxed{-1.4x=2.5}$; $\boxed{x=\frac{2.5}{-1.4}}$; $\boxed{x=-\frac{\frac{25}{10}}{\frac{14}{10}}}$; $\boxed{x=-\frac{25\cdot 10}{10\cdot 14}}$; $\boxed{x=-\frac{250}{140}}$; $\boxed{\mathbf{x=-1.78}}$

Check: $[[(0.2\cdot -1.78)-2.3]+0.2(-1.78-1)]+(-1.8\cdot -1.78)\overset{?}{=}0$; $[[-0.35-2.3]+(0.2\cdot -2.78)]+3.21\overset{?}{=}0$; $[-2.65-0.56]+3.21\overset{?}{=}0$; $-3.21+3.21\overset{?}{=}0$; $0=0$

Practice Problems - Solving Linear Equations Containing Decimals

Section 2.3 Case III Practice Problems - Solve the following linear equations by applying the addition, subtraction, multiplication, and division rules:

1. $0.35-0.2x=0.5+0.1x$
2. $5.2x+0.1(x-0.25)=0.2x$
3. $0.4(x-2)-0.2(x-1)=0.25$
4. $1.2x+0.56-0.6x=1.25x$
5. $(x-0.5)-[(x+0.1)-3x]=-x$
6. $5(0.02x+0.002)-0.5x=1.25$
7. $0.5x=-(2-2.5x)+2.8$
8. $1.35-0.5(x+0.2)=0$
9. $0.5[-(0.8x-0.2)-5]=2.2x$
10. $0.25x-1.3+(1.2x-1.7)=-2.8$

2.4 Formulas

Formulas are rules that are stated using symbols, called variables, and are expressed as equations. For example, to find the area of a circle with a radius equal to $3\,cm$ we multiply the constant π "pronounced pi" by the square of the radius. Thus, the area is $\pi\cdot 3^2 = 9\pi\ cm^2$. Note that in this case, the stated rule can then be expressed as a formula $A = \pi\cdot r^2$ which is an equation involving two variables, A and r, and a constant, π. In this section we learn how to solve for a specific variable in a given formula by using the following steps:

Step 1: Isolate the variable either to the left or right hand side of the equation by applying the addition and subtraction rule (see Section 2.2, Case I).

Step 2: Solve for the variable by applying the multiplication or division rules (see Section 2.2, Case II).

Note - If the variable is isolated to the right hand side of the equation, to be consistent with the steps used in the previous sections, at the very last step move the variable to the left hand side of the equation (see examples 2.4-2, 2.4-6, and 2.4-7).

Examples with Steps

The following examples show the steps for solving a specific variable in a given formula:

Example 2.4-1

Solve $V = \frac{1}{3}bh$ for b.

Solution:

Step 1 $\boxed{\textit{Not Applicable}}$

Step 2 $\boxed{V = \frac{1}{3}bh}$; $\boxed{3\cdot V = \cancel{3}\cdot\frac{1}{\cancel{3}}bh}$; $\boxed{3V = bh}$; $\boxed{\frac{3V}{h} = \frac{b\cancel{h}}{\cancel{h}}}$; $\boxed{\frac{3V}{h} = b}$; $\boxed{\mathbf{b = \frac{3V}{h}}}$

Example 2.4-2

Solve $a = b + (c+2)d$ for c and d.

Solution:

I. Step 1 $\boxed{a = b+(c+2)d}$; $\boxed{a = b+cd+2d}$; $\boxed{a = b+2d+cd}$; $\boxed{a-b = (b-b)+2d+cd}$; $\boxed{a-b = 0+2d+cd}$; $\boxed{a-b = 2d+cd}$; $\boxed{a-b-2d = (2d-2d)+cd}$; $\boxed{a-b-2d = 0+cd}$; $\boxed{a-b-2d = cd}$

Step 2 $\boxed{a-b-2d = cd}$; $\boxed{\frac{a-b-2d}{d} = \frac{c\cancel{d}}{\cancel{d}}}$; $\boxed{\frac{a-b-2d}{d} = c}$; $\boxed{\mathbf{c = \frac{a-b-2d}{d}}}$

II. Step 1 $\boxed{a = b+(c+2)d}$; $\boxed{a-b = b-b+(c+2)d}$; $\boxed{a-b = 0+(c+2)d}$; $\boxed{a-b = (c+2)d}$

Step 2 $\boxed{a-b=(c+2)d}$; $\boxed{\dfrac{a-b}{(c+2)}=\dfrac{(\cancel{c+2})d}{(\cancel{c+2})}}$; $\boxed{\dfrac{a-b}{c+2}=d}$; $\boxed{\mathbf{d=\dfrac{a-b}{c+2}}}$

Example 2.4-3

Solve $A=2\pi r^2+2\pi rh$ for π and h.

Solution:

I. Step 1 $\boxed{\textit{Not Applicable}}$

Step 2 $\boxed{A=2\pi r^2+2\pi rh}$; $\boxed{A=\pi\left(2r^2+2rh\right)}$; $\boxed{\dfrac{A}{\left(2r^2+2rh\right)}=\dfrac{\pi\cancel{\left(2r^2+2rh\right)}}{\cancel{\left(2r^2+2rh\right)}}}$; $\boxed{\dfrac{A}{2r^2+2rh}=\pi}$

; $\boxed{\mathbf{\pi=\dfrac{A}{2r^2+2rh}}}$

II. Step 1 $\boxed{A=2\pi r^2+2\pi rh}$; $\boxed{A-2\pi r^2=2\pi r^2-2\pi r^2+2\pi rh}$; $\boxed{A-2\pi r^2=0+2\pi rh}$

; $\boxed{A-2\pi rr^2=2\pi rh}$

Step 2 $\boxed{A-2\pi r^2=2\pi rh}$; $\boxed{\dfrac{A-2\pi r^2}{2\pi r}=\dfrac{\cancel{2\pi r}h}{\cancel{2\pi r}}}$; $\boxed{\dfrac{A-2\pi r^2}{2\pi r}=h}$; $\boxed{\mathbf{h=\dfrac{A-2\pi r^2}{2\pi r}}}$

Example 2.4-4

Solve $A=p+prt$ for p and t.

Solution:

I. Step 1 $\boxed{\textit{Not Applicable}}$

Step 2 $\boxed{A=p+prt}$; $\boxed{A=p(1+rt)}$; $\boxed{\dfrac{A}{(1+rt)}=\dfrac{p\cancel{(1+rt)}}{\cancel{(1+rt)}}}$; $\boxed{\dfrac{A}{1+rt}=p}$; $\boxed{\mathbf{p=\dfrac{A}{1+rt}}}$

II. Step 1 $\boxed{A=p+prt}$; $\boxed{A-p=p-p+prt}$; $\boxed{A-p=0+prt}$; $\boxed{A-p=prt}$

Step 2 $\boxed{A-p=prt}$; $\boxed{\dfrac{A-p}{pr}=\dfrac{\cancel{pr}t}{\cancel{pr}}}$; $\boxed{\dfrac{A-p}{pr}=t}$; $\boxed{\mathbf{t=\dfrac{A-p}{pr}}}$

Example 2.4-5

Solve $\frac{1}{2}x+\frac{2}{3}y=\frac{1}{5}z$ for x, y, and z.

Solution:

I. Step 1 $\boxed{\dfrac{1}{2}x+\dfrac{2}{7}y=\dfrac{1}{5}z}$; $\boxed{\dfrac{1}{2}x+\dfrac{2}{7}y-\dfrac{2}{7}y=\dfrac{1}{5}z-\dfrac{2}{7}y}$; $\boxed{\dfrac{1}{2}x+0=\dfrac{1}{5}z-\dfrac{2}{7}y}$; $\boxed{\dfrac{1}{2}x=\dfrac{1}{5}z-\dfrac{2}{7}y}$

Step 2 $\boxed{\dfrac{1}{2}x=\dfrac{1}{5}z-\dfrac{2}{7}y}$; $\boxed{2\cdot\dfrac{1}{2}x=2\left(\dfrac{1}{5}z-\dfrac{2}{7}y\right)}$; $\boxed{\mathbf{x=\dfrac{2}{5}z-\dfrac{4}{7}y}}$

II. Step 1 $\boxed{\frac{1}{2}x+\frac{2}{7}y=\frac{1}{5}z}$; $\boxed{\frac{1}{2}x-\frac{1}{2}x+\frac{2}{7}y=\frac{1}{5}z-\frac{1}{2}x}$; $\boxed{0+\frac{2}{7}y=\frac{1}{5}z-\frac{1}{2}x}$; $\boxed{\frac{2}{7}y=\frac{1}{5}z-\frac{1}{2}x}$

Step 2 $\boxed{\frac{2}{7}y=\frac{1}{5}z-\frac{1}{2}x}$; $\boxed{\frac{7}{2}\cdot\frac{2}{7}y=\frac{7}{2}\left(\frac{1}{5}z-\frac{1}{2}x\right)}$; $\boxed{y=\frac{7}{2}\left(\frac{1}{5}z-\frac{1}{2}x\right)}$; $\boxed{\mathbf{y=\frac{7}{10}z-\frac{7}{4}x}}$

III. Step 1 *Not Applicable*

Step 2 $\boxed{\frac{1}{2}x+\frac{2}{7}y=\frac{1}{5}z}$; $\boxed{5\cdot\left(\frac{1}{2}x+\frac{2}{7}y\right)=5\cdot\frac{1}{5}z}$; $\boxed{\frac{5}{2}x+\frac{10}{7}y=z}$; $\boxed{\mathbf{z=\frac{5}{2}x+\frac{10}{7}y}}$

Additional Examples - Formulas

The following examples further illustrate how to solve for a specific variable in a given formula:

Example 2.4-6 Solve $y=mx+b$ for x and b.

Solution: I. $\boxed{y=mx+b}$; $\boxed{y-b=mx+b-b}$; $\boxed{y-b=mx+0}$; $\boxed{y-b=mx}$; $\boxed{\frac{y-b}{m}=\frac{mx}{m}}$; $\boxed{\frac{y-b}{m}=x}$; $\boxed{\mathbf{x=\frac{y-b}{m}}}$

II. $\boxed{y=mx+b}$; $\boxed{y-mx=mx-mx+b}$; $\boxed{y-mx=0+b}$; $\boxed{y-mx=b}$; $\boxed{\mathbf{b=y-mx}}$

Example 2.4-7 Solve $s=2t(a+b)$ for t and b.

Solution: I. $\boxed{s=2t(a+b)}$; $\boxed{\frac{s}{2(a+b)}=\frac{2t(a+b)}{2(a+b)}}$; $\boxed{\frac{s}{2(a+b)}=t}$; $\boxed{\mathbf{t=\frac{s}{2(a+b)}}}$

II. $\boxed{s=2t(a+b)}$; $\boxed{s=2at+2bt}$; $\boxed{s-2at=2at-2at+2bt}$; $\boxed{s-2at=0+2bt}$; $\boxed{s-2at=2bt}$; $\boxed{\frac{s-2at}{2t}=\frac{2bt}{2t}}$; $\boxed{\frac{s-2at}{2t}=b}$; $\boxed{\mathbf{b=\frac{s-2at}{2t}}}$

Example 2.4-8 Solve $2s-3t=4(s+2t)+5$ for s and t.

Solution: I. $\boxed{2s-3t=4(s+2t)+5}$; $\boxed{2s-3t=4s+8t+5}$; $\boxed{2s-3t+3t=4s+(8t+3t)+5}$; $\boxed{2s+0=4s+11t+5}$; $\boxed{2s=4s+11t+5}$; $\boxed{2s-4s=(4s-4s)+11t+5}$; $\boxed{-2s=0+11t+5}$; $\boxed{-2s=11t+5}$; $\boxed{\frac{-2s}{-2}=\frac{11t+5}{-2}}$; $\boxed{\mathbf{s=-\frac{11t+5}{2}}}$

II. $\boxed{2s-3t=4(s+2t)+5}$; $\boxed{2s-3t=4s+8t+5}$; $\boxed{2s-2s-3t=(4s-2s)+8t+5}$; $\boxed{0-3t=2s+8t+5}$; $\boxed{-3t=2s+8t+5}$; $\boxed{-3t-8t=2s+(8t-8t)+5}$; $\boxed{-11t=2s+0+5}$

; $\boxed{-11t = 2s+5}$; $\boxed{\frac{-11t}{-11} = \frac{2s+5}{-11}}$; $\boxed{\mathbf{t = -\frac{2s+5}{11}}}$

Example 2.4-9 Solve $I = prt$ for p, r, and t.

Solution: I. $\boxed{I = prt}$; $\boxed{\frac{I}{rt} = \frac{p\cancel{r}\cancel{t}}{\cancel{r}\cancel{t}}}$; $\boxed{\frac{I}{rt} = p}$; $\boxed{\mathbf{p = \frac{I}{rt}}}$

II. $\boxed{I = prt}$; $\boxed{\frac{I}{pt} = \frac{\cancel{p}r\cancel{t}}{\cancel{p}\cancel{t}}}$; $\boxed{\frac{I}{pt} = r}$; $\boxed{\mathbf{r = \frac{I}{pt}}}$

III. $\boxed{I = prt}$; $\boxed{\frac{I}{pr} = \frac{\cancel{p}\cancel{r}t}{\cancel{p}\cancel{r}}}$; $\boxed{\frac{I}{pr} = t}$; $\boxed{\mathbf{t = \frac{I}{pr}}}$

Example 2.4-10 Solve $A = \pi r^2$ for π and r.

Solution: I. $\boxed{A = \pi r^2}$; $\boxed{\frac{A}{r^2} = \frac{\pi \cancel{r^2}}{\cancel{r^2}}}$; $\boxed{\frac{A}{r^2} = \pi}$; $\boxed{\mathbf{\pi = \frac{A}{r^2}}}$

II. $\boxed{A = \pi r^2}$; $\boxed{\frac{A}{\pi} = \frac{\cancel{\pi} r^2}{\cancel{\pi}}}$; $\boxed{\frac{A}{\pi} = r^2}$; $\boxed{r^2 = \frac{A}{\pi}}$; $\boxed{\sqrt{r^2} = \pm\sqrt{\frac{A}{\pi}}}$; $\boxed{\mathbf{r = \pm\sqrt{\frac{A}{\pi}}}}$

Example 2.4-11 Solve $P = 2l + 2w$ for l and w.

Solution: I. $\boxed{P = 2l + 2w}$; $\boxed{P - 2w = 2l + 2w - 2w}$; $\boxed{P - 2w = 2l + o}$; $\boxed{P - 2w = 2l}$; $\boxed{\frac{P-2w}{2} = \frac{\cancel{2}l}{\cancel{2}}}$

; $\boxed{\frac{P-2w}{2} = l}$; $\boxed{\mathbf{l = \frac{P-2w}{2}}}$

II. $\boxed{P = 2l + 2w}$; $\boxed{P - 2l = 2l - 2l + 2w}$; $\boxed{P - 2l = 0 + 2w}$; $\boxed{P - 2l = 2w}$; $\boxed{\frac{P-2l}{2} = \frac{\cancel{2}w}{\cancel{2}}}$

; $\boxed{\frac{P-2l}{2} = w}$; $\boxed{\mathbf{w = \frac{P-2l}{2}}}$

Example 2.4-12 Solve $A = \frac{1}{2}bh$ for h.

Solution: $\boxed{A = \frac{1}{2}bh}$; $\boxed{A = \frac{bh}{2}}$; $\boxed{2 \cdot A = \cancel{2} \cdot \frac{bh}{\cancel{2}}}$; $\boxed{2A = bh}$; $\boxed{\frac{2A}{b} = \frac{\cancel{b}h}{\cancel{b}}}$; $\boxed{\frac{2A}{b} = h}$; $\boxed{\mathbf{h = \frac{2A}{b}}}$

Example 2.4-13 Solve $C = \frac{5}{9}(F - 32)$ for F.

Solution: $\boxed{C = \frac{5}{9}(F - 32)}$; $\boxed{\frac{9}{5} \cdot C = \frac{\cancel{9}}{\cancel{5}} \cdot \frac{\cancel{5}}{\cancel{9}}(F - 32)}$; $\boxed{\frac{9}{5}C = F - 32}$; $\boxed{\frac{9}{5}C + 32 = F - 32 + 32}$

; $\boxed{\frac{9}{5}C + 32 = F + 0}$; $\boxed{\frac{9}{5}C + 32 = F}$; $\boxed{\mathbf{F = \frac{9}{5}C + 32}}$

Example 2.4-14 Solve $A = \frac{1}{2}h(b_1 + b_2)$ for h and b_2.

Solution: I. $\boxed{A=\frac{1}{2}h(b_1+b_2)}$; $\boxed{2\cdot A=\cancel{2}\cdot\frac{1}{\cancel{2}}h(b_1+b_2)}$; $\boxed{2A=h(b_1+b_2)}$; $\boxed{\frac{2A}{b_1+b_2}=\frac{h(\cancel{b_1}+\cancel{b_2})}{(\cancel{b_1}+\cancel{b_2})}}$

; $\boxed{\frac{2A}{b_1+b_2}=h}$; $\boxed{\mathbf{h=\frac{2A}{b_1+b_2}}}$

II. $\boxed{A=\frac{1}{2}h(b_1+b_2)}$; $\boxed{2\cdot A=\cancel{2}\cdot\frac{1}{\cancel{2}}h(b_1+b_2)}$; $\boxed{2A=h(b_1+b_2)}$; $\boxed{2A=b_1h+b_2h}$

; $\boxed{2A-b_1h=(b_1h-b_1h)+b_2h}$; $\boxed{2A-b_1h=0+b_2h}$; $\boxed{2A-b_1h=b_2h}$; $\boxed{\frac{2A-b_1h}{h}=\frac{b_2\cancel{h}}{\cancel{h}}}$

; $\boxed{\frac{2A-b_1h}{h}=b_2}$; $\boxed{\mathbf{b_2=\frac{2A-b_1h}{h}}}$

Example 2.4-15 Solve $\frac{x-4}{8}=\frac{y-5}{10}$ for x and y.

Solution: I. $\boxed{\frac{x-4}{8}=\frac{y-5}{10}}$; $\boxed{\cancel{8}\cdot\frac{x-4}{\cancel{8}}=\overset{4}{\cancel{8}}\cdot\frac{y-5}{\underset{5}{\cancel{10}}}}$; $\boxed{x-4=\frac{4}{5}(y-5)}$; $\boxed{x-\cancel{4}+\cancel{4}=\frac{4}{5}(y-5)+4}$

; $\boxed{x+0=\frac{4}{5}(y-5)+4}$; $\boxed{x=\frac{4}{5}y-\frac{\overset{4}{\cancel{20}}}{\cancel{5}}+4}$; $\boxed{x=\frac{4}{5}y-\cancel{4}+\cancel{4}}$; $\boxed{\mathbf{x=\frac{4}{5}y}}$

II. $\boxed{\frac{x-4}{8}=\frac{y-5}{10}}$; $\boxed{\overset{5}{\cancel{10}}\cdot\frac{x-4}{\underset{4}{\cancel{8}}}=\cancel{10}\cdot\frac{y-5}{\cancel{10}}}$; $\boxed{\frac{5}{4}(x-4)=y-5}$; $\boxed{\frac{5}{4}(x-4)+5=y-\cancel{5}+\cancel{5}}$

; $\boxed{\frac{5}{4}(x-4)+5=y+0}$; $\boxed{\frac{5}{4}(x-4)+5=y}$; $\boxed{\frac{5}{4}x-\frac{\overset{5}{\cancel{20}}}{\cancel{4}}+5=y}$; $\boxed{\frac{5}{4}x-\cancel{5}+\cancel{5}=y}$; $\boxed{\mathbf{y=\frac{5}{4}x}}$

Note that in some instances formulas are solved for a specific variable while numerical values for the remaining variables are given. The following are few examples of this case:

Example 2.4-16 Solve $A=\frac{1}{2}h(b_1+b_2)$ for h, if $A=50$, $b_1=3$, and $b_2=5$.

Solution: $\boxed{A=\frac{1}{2}h(b_1+b_2)}$; $\boxed{2\cdot A=\cancel{2}\cdot\frac{1}{\cancel{2}}h(b_1+b_2)}$; $\boxed{2A=h(b_1+b_2)}$; $\boxed{\frac{2A}{(b_1+b_2)}=\frac{h(\cancel{b_1}+\cancel{b_2})}{(\cancel{b_1}+\cancel{b_2})}}$

; $\boxed{\frac{2A}{b_1+b_2}=h}$; $\boxed{h=\frac{2A}{b_1+b_2}}$

Substituting the given numerical values in the above equation we obtain a specific value for h.

$\boxed{h=\frac{2A}{b_1+b_2}}$; $\boxed{h=\frac{2\cdot 50}{3+5}}$; $\boxed{h=\frac{\overset{25}{\cancel{100}}}{\underset{2}{\cancel{8}}}}$; $\boxed{h=\frac{25}{2}}$; $\boxed{h=12\frac{1}{2}}$; $\boxed{\mathbf{h=12.5}}$

Example 2.4-17 Solve $F=\frac{9}{5}C+32$ for C, if $F=100$, $F=25$, and $F=-10$.

Solution: $\boxed{F=\frac{9}{5}C+32}$; $\boxed{F-32=\frac{9}{5}C+32-32}$; $\boxed{F-32=\frac{9}{5}C+0}$; $\boxed{F-32=\frac{9}{5}C}$

; $\boxed{\frac{5}{9}\cdot(F-32)=\frac{\cancel{5}}{\cancel{9}}\cdot\frac{\cancel{9}}{\cancel{5}}C}$; $\boxed{\frac{5}{9}(F-32)=C}$; $\boxed{C=\frac{5}{9}(F-32)}$

I. If $F=100$, then $\boxed{C=\frac{5}{9}(100-32)}$; $\boxed{C=\frac{5}{9}\cdot 68}$; $\boxed{C=\frac{340}{9}}$; $\boxed{C=37\frac{7}{9}}$; $\boxed{\mathbf{C=37.78}}$

Note that 100 degrees Fahrenheit corresponds with 27.78 degrees Centigrade.

II. If $F=25$, then $\boxed{C=\frac{5}{9}(25-32)}$; $\boxed{C=\frac{5}{9}\cdot(-7)}$; $\boxed{C=-\frac{35}{9}}$; $\boxed{C=-3\frac{8}{9}}$; $\boxed{\mathbf{C=-3.89}}$

III. If $F=-10$, then $\boxed{C=\frac{5}{9}(-10-32)}$; $\boxed{C=\frac{5}{9}\cdot(-42)}$; $\boxed{C=-\frac{210}{9}}$; $\boxed{C=-23\frac{3}{9}}$

; $\boxed{\mathbf{C=-23.33}}$

Example 2.4-18 Solve $V=\frac{1}{3}bh$ for h, if $V=150$ and $b=2$.

Solution: $\boxed{V=\frac{1}{3}bh}$; $\boxed{3\cdot V=\cancel{3}\cdot\frac{1}{\cancel{3}}bh}$; $\boxed{3V=bh}$; $\boxed{\frac{3V}{b}=\frac{\cancel{b}h}{\cancel{b}}}$; $\boxed{\frac{3V}{b}=h}$; $\boxed{h=\frac{3V}{b}}$

Substituting the given numerical values in the above equation we obtain a specific value for h.

$\boxed{h=\frac{3V}{b}}$; $\boxed{h=\frac{3\cdot 150}{2}}$; $\boxed{h=\frac{450}{2}}$; $\boxed{\mathbf{h=225}}$

Example 2.4-19 Solve $S=2\pi r^2+2\pi rh$ 1. For S, if $r=2$ and $h=5$. 2. For h, if $S=40$ and $r=2$.

Solution: I. If $r=2$ and $h=5$, then $\boxed{S=2\pi r^2+2\pi rh}$; $\boxed{S=2\pi\cdot 2^2+2\pi\cdot 2\cdot 5}$; $\boxed{S=2\pi\cdot 4+2\pi\cdot 10}$

; $\boxed{S=8\pi+20\pi}$; $\boxed{\mathbf{S=28\pi}}$

II. $\boxed{S=2\pi r^2+2\pi rh}$; $\boxed{S-2\pi r^2=2\pi r^2-2\pi r^2+2\pi rh}$; $\boxed{S-2\pi r^2=0+2\pi rh}$

; $\boxed{S-2\pi r^2=2\pi rh}$; $\boxed{\frac{S-2\pi r^2}{2\pi r}=\frac{\cancel{2\pi r}h}{\cancel{2\pi r}}}$; $\boxed{\frac{S-2\pi r^2}{2\pi r}=h}$; $\boxed{h=\frac{S-2\pi r^2}{2\pi r}}$

If $S=40$ and $r=2$, then $\boxed{h=\frac{40-2\pi\cdot 2^2}{2\pi\cdot 2}}$; $\boxed{h=\frac{40-8\pi}{4\pi}}$; $\boxed{h=\frac{4(10-2\pi)}{4\pi}}$

; $\boxed{\mathbf{h=\frac{10-2\pi}{\pi}}}$

Example 2.4-20 Solve $C = \frac{5}{9}(F - 32)$ for F, if $C = 37.78$, $C = 0$, and $C = -10$.

Solution: $\boxed{C = \frac{5}{9}(F - 32)}$; $\boxed{\frac{9}{5} \cdot C = \frac{\cancel{9}}{\cancel{5}} \cdot \frac{\cancel{5}}{\cancel{9}}(F - 32)}$; $\boxed{\frac{9}{5}C = F - 32}$; $\boxed{\frac{9}{5}C + 32 = F - 32 + 32}$

; $\boxed{\frac{9}{5}C + 32 = F + 0}$; $\boxed{\frac{9}{5}C + 32 = F}$; $\boxed{F = \frac{9}{5}C + 32}$

I. If $C = 37.78$, then $\boxed{F = \frac{9}{5} \cdot 37.78 + 32}$; $\boxed{F = \frac{340}{5} + 32}$; $\boxed{F = 68 + 32}$; $\boxed{\mathbf{F = 100}}$

II. If $C = 0$, then $\boxed{F = \frac{9}{5} \cdot 0 + 32}$; $\boxed{F = \frac{0}{5} + 32}$; $\boxed{F = 0 + 32}$; $\boxed{\mathbf{F = 32}}$

Note that zero degree Centigrade corresponds with 32 degrees Fahrenheit.

III. If $C = -10$, then $\boxed{F = \frac{9}{5} \cdot -10 + 32}$; $\boxed{F = -\frac{90}{5} + 32}$; $\boxed{F = -18 + 32}$; $\boxed{\mathbf{F = 14}}$

Practice Problems - Formulas

Section 2.4 Practice Problems - Solve each formula for the indicated variable.

1. $V = \pi r^2 h$ for r and h
2. $2x + 2y = 3(x + y) - 5$ for x and y
3. $C = 2\pi r$ for r
4. $d = rt$ for t and r
5. $y - b = \frac{1}{3}x + \frac{2}{3}b$ for x and b
6. $y = \frac{a - b - c}{3}$ for c
7. $m = \frac{y - b}{x}$ for y and b
8. $V = \frac{1}{3}\pi r^2 h$ for π, r, and h
9. $E = mc^2$ for c and m
10. $y - (2x - 3y) + 3 = 5y - x$ for x and y

2.5 Math Operations Involving Linear Inequalities

Just as the symbol " $=$ " represents *equality*, the symbols " $<$ " and " $>$ " represent *less than* or *greater than*, respectively. In general, an equation states that two algebraic expressions are equal. An inequality, in the other hand, states that an algebraic expression is either greater than or less than another one. Note that an inequality with numbers on both sides is a *numerical statement of inequality*. Numerical statements of inequality are either true or false. For example, $6+5>8$, $4\cdot 3>10$, and $15-3>2$ are true statements where as $5+3>10$, $8-2<2$, and $6\cdot 3>30$ are false statements. **Algebraic inequalities** are inequalities that contain one or more variables. For example, $x-3<5$, $2x+5<x-8$, and $3y<5y-2$ are referred to as algebraic inequalities.

An algebraic inequality becomes either a true or false numerical statement, each time a number is substituted for the variable. For example, the algebraic inequality $y+3>5$ is a false numerical statement if $y=-3$, because by substituting $y=-3$ in $y+3>5$ we obtain $-3+3>5$ which implies $0>5$. On the other hand, $y+3>5$ is a true numerical statement if $y=3$, because $y+3>5$ becomes $3+3>5$ which implies $6>5$. Note that the set of all solutions to an inequality that make an algebraic inequality to become a true numerical statement is referred to as its **solution set**. For example, the solution set for $y+3>5$ is the set of all real numbers that are greater than 2. This is expressed as $\{y \mid y>2\}$. The notation $\{y \mid y>2\}$ is read as "the set of all y such that y is greater than 2."

In this section students learn how to solve algebraic inequalities. The rules for solving inequalities, with only one exception, are the same ones we have learned for solving equations. Solving inequalities using addition and subtraction rules (Case I), and multiplication or division rules (Case II) are addressed below:

Case I Addition and Subtraction of Linear Inequalities

To add or subtract the same positive or negative number to linear inequalities the following rules should be used:

Addition and Subtraction Rules: *The same positive or negative number, or variable, can be added or subtracted to both sides of an inequality without changing the solution: for all real numbers a, b, and c,*

1. $a>b$ *if and only if* $a+c>b+c$
2. $a>b$ *if and only if* $a-c>b-c$.

Linear inequities are added or subtracted using the following steps:

Step 1 Isolate the variable to the left hand side of the inequality by applying the addition and subtraction rules.

Step 2 Find the solution by simplifying the inequality. Check the answer by substituting different values, that are greater than or less than the solution found for the inequality, into the original inequality.

Examples with Steps

The following examples show the steps as to how linear inequalities are solved using the addition and subtraction rules:

Example 2.5-1

$$\boxed{x-5<8}$$

Solution:

Step 1 $\boxed{x-5<8}$; $\boxed{x-5+5<8+5}$; $\boxed{x+0<13}$; $\boxed{\mathbf{x<13}}$

Step 2 $\boxed{\textit{Not Applicable}}$ The solution set is $\{\mathbf{x} \mid \mathbf{x<13}\}$.

Check 1: Let $x=-2$ which is less than 13. Then, is $-2-5\overset{?}{<}8$; $-7\overset{?}{<}8$? The answer is "yes" because $-7<8$.

Check 2: Let $x=15$ which is greater than 13. Then, is $15-5\overset{?}{<}8$; $10\overset{?}{<}8$? The answer is "no" because $10>8$.

Example 2.5-2

$$\boxed{2\frac{3}{4}+w\geq 1\frac{2}{3}}$$

Solution:

Step 1 $\boxed{2\frac{3}{4}+w\geq 1\frac{2}{3}}$; $\boxed{\frac{(2\cdot 4)+3}{4}+w\geq\frac{(1\cdot 3)+2}{3}}$; $\boxed{\frac{8+3}{4}+w\geq\frac{3+2}{3}}$; $\boxed{\frac{11}{4}+w\geq\frac{5}{3}}$

; $\boxed{\frac{11}{4}-\frac{11}{4}+w\geq\frac{5}{3}-\frac{11}{4}}$; $\boxed{0+w\geq\frac{5}{3}-\frac{11}{4}}$; $\boxed{w\geq\frac{5}{3}-\frac{11}{4}}$

Step 2 $\boxed{w\geq\frac{5}{3}-\frac{11}{4}}$; $\boxed{w\geq\frac{(5\cdot 4)-(3\cdot 11)}{3\cdot 4}}$; $\boxed{w\geq\frac{20-33}{12}}$; $\boxed{w\geq-\frac{13}{12}}$; $\boxed{\mathbf{w\geq-1.08}}$

The solution set is $\{\mathbf{w} \mid \mathbf{w\geq-1.08}\}$.

Check 1: Let $w=-1.08$. Then, is $2\frac{3}{4}-1.08\overset{?}{=}1\frac{2}{3}$; $\frac{11}{4}-1.08\overset{?}{=}\frac{5}{3}$; $2.75-1.08\overset{?}{=}1.67$? The answer is "yes" because $1.67=1.67$.

Check 2: Let $w=2$ which is greater than -1.08. Then, is $2\frac{3}{4}+2\overset{?}{>}1\frac{2}{3}$; $\frac{11}{4}+2\overset{?}{>}\frac{5}{3}$; $2.75+2\overset{?}{>}1.67$? The answer is "yes" because $4.75>1.67$.

Example 2.5-3

$$\boxed{5u-0.45\leq 4u+1\frac{1}{3}}$$

Solution:

Step 1 $\boxed{5u-0.45\leq 4u+1\frac{1}{3}}$; $\boxed{5u-4u-0.45\leq 4u-4u+\frac{(1\cdot 3)+1}{3}}$; $\boxed{u-0.45\leq 0+\frac{4}{3}}$

; $\boxed{u-0.45\leq 1.33}$; $\boxed{u-0.45+0.45\leq 1.33+0.45}$; $\boxed{u+0\leq 1.78}$; $\boxed{\mathbf{u\leq 1.78}}$

Step 2 $\boxed{\textit{Not Applicable}}$ The solution set is $\{\mathbf{u} \mid \mathbf{u\leq 1.78}\}$.

Check 1: Let $u = 1.78$. Then, is $(5 \cdot 1.78) - 0.45 \overset{?}{=} (4 \cdot 1.78) + 1\frac{1}{3}$; $8.9 - 0.45 \overset{?}{=} 7.12 + \frac{4}{3}$; $8.45 \overset{?}{=} 7.12 + 1.33$? The answer is "yes" because $8.45 = 8.45$.

Check 2: Let $u = 0$ which is less than 1.78. Then, is $(5 \cdot 0) - 0.45 \overset{?}{<} (4 \cdot 0) + 1\frac{1}{3}$; $0 - 0.45 \overset{?}{<} 0 + \frac{4}{3}$; $-0.45 \overset{?}{<} \frac{4}{3}$? The answer is "yes" because $-0.45 < 1.33$.

Check 3: Let $u = 5$ which is greater than 1.78. Then, is $(5 \cdot 5) - 0.45 \overset{?}{<} (4 \cdot 5) + 1\frac{1}{3}$; $25 - 0.45 \overset{?}{<} 20 + \frac{4}{3}$; $24.55 \overset{?}{<} 20 + 1.33$; $24.55 \overset{?}{<} 21.33$? The answer is "no" because we can not choose $u = 5$ since $5 > 1.78$.

Example 2.5-4

$$\boxed{-\frac{1}{4} > -y + \left(\frac{1}{3} + 1\frac{2}{3}\right)}$$

Solution:

Step 1

$$\boxed{-\frac{1}{4} > -y + \left(\frac{1}{3} + 1\frac{2}{3}\right)} ; \boxed{-\frac{1}{4} + y > -y + y + \left(\frac{1}{3} + \frac{(1 \cdot 3) + 2}{3}\right)} ; \boxed{-\frac{1}{4} + y > 0 + \left(\frac{1}{3} + \frac{3+2}{3}\right)}$$

$$; \boxed{-\frac{1}{4} + y > \frac{1}{3} + \frac{5}{3}} ; \boxed{-\frac{1}{4} + y > \frac{1+5}{3}} ; \boxed{-\frac{1}{4} + \frac{1}{4} + y > \frac{6}{3} + \frac{1}{4}} ; \boxed{0 + y > \frac{2}{1} + \frac{1}{4}} ; \boxed{y > \frac{2}{1} + \frac{1}{4}}$$

Step 2

$$\boxed{y > \frac{2}{1} + \frac{1}{4}} ; \boxed{y > \frac{(2 \cdot 4) + (1 \cdot 1)}{1 \cdot 4}} ; \boxed{y > \frac{8+1}{4}} ; \boxed{y > \frac{9}{4}} ; \boxed{y > 2\frac{1}{4}} ; \boxed{\mathbf{y > 2.25}}$$

The solution set is $\{y \mid y > \mathbf{2.25}\}$.

Check: Let $y = 10$ which is greater than 2.25. Then, is $-\frac{1}{4} \overset{?}{>} -10 + \left(\frac{1}{3} + 1\frac{2}{3}\right)$; $-\frac{1}{4} \overset{?}{>} -10 + \left(\frac{1}{3} + \frac{5}{3}\right)$; $-\frac{1}{4} \overset{?}{>} -10 + \frac{1+5}{3}$; $-\frac{1}{4} \overset{?}{>} -10 + \frac{6}{3}$; $-0.25 \overset{?}{>} -10 + 2$? The answer is "yes" because $-0.25 > -8$ or $0.25 < 8$.

Example 2.5-5

$$\boxed{\frac{2}{3} + y \ge 2 - 1\frac{2}{7}}$$

Solution:

Step 1

$$\boxed{\frac{2}{3} + y \ge 2 - 1\frac{2}{7}} ; \boxed{\frac{2}{3} - \frac{2}{3} + y \ge 2 - 1\frac{2}{7} - \frac{2}{3}} ; \boxed{0 + y \ge 2 - 1\frac{2}{7} - \frac{2}{3}} ; \boxed{y \ge 2 - 1\frac{2}{7} - \frac{2}{3}}$$

Step 2

$$\boxed{y \ge 2 - 1\frac{2}{7} - \frac{2}{3}} ; \boxed{y \ge \left(2 - 1\frac{2}{7}\right) - \frac{2}{3}} ; \boxed{y \ge \left(\frac{2}{1} - \frac{(1 \cdot 7) + 2}{7}\right) - \frac{2}{3}} ; \boxed{y \ge \left(\frac{2}{1} - \frac{7+2}{7}\right) - \frac{2}{3}}$$

$$; \boxed{y \ge \left(\frac{2}{1} - \frac{9}{7}\right) - \frac{2}{3}} ; \boxed{y \ge \left(\frac{(2 \cdot 7) - (9 \cdot 1)}{1 \cdot 7}\right) - \frac{2}{3}} ; \boxed{y \ge \left(\frac{14-9}{7}\right) - \frac{2}{3}} ; \boxed{y \ge \frac{5}{7} - \frac{2}{3}}$$

$$; \boxed{y \ge \frac{(5\cdot 3)-(2\cdot 7)}{7\cdot 3}} ; \boxed{y \ge \frac{15-14}{21}} ; \boxed{y \ge \frac{1}{21}} ; \boxed{\mathbf{y \ge 0.05}}$$

The solution set is $\{y \mid y \ge \mathbf{0.05}\}$.

Check 1: Let $y = 0.05$. Then, is $\frac{2}{3} + 0.05 \overset{?}{=} 2 - 1\frac{2}{7}$; $0.67 + 0.05 \overset{?}{=} 2 - \frac{9}{7}$; $0.72 \overset{?}{=} 2 - 1.28$? The answer is "yes" because $0.72 = 0.72$.

Check 2: Let $y = 5$ which is greater than 0.05. Then is $\frac{2}{3} + 5 \overset{?}{\rangle} 2 - 1\frac{2}{7}$; $0.67 + 5 \overset{?}{\rangle} 2 - \frac{9}{7}$; $5.67 \overset{?}{\rangle} 2 - 1.28$? The answer is "yes" because $5.67 \rangle 0.72$.

Additional Examples - Addition and Subtraction of Linear Inequalities

The following examples further illustrate how linear inequalities are solved using the addition and subtraction rules:

Example 2.5-6

$$\boxed{x + 8 \langle 12} ; \boxed{x + 8 - 8 \langle 12 - 8} ; \boxed{x + 0 \langle 4} ; \boxed{\mathbf{x \langle 4}}$$

The solution set is $\{x \mid x \langle \mathbf{4}\}$.

Check: Let $x = 3$ which is less than 4. Then, is $3 + 8 \overset{?}{\langle} 12$? The answer is "yes" because $11 \langle 12$.

Example 2.5-7

$$\boxed{-5 + y \rangle -8} ; \boxed{-5 + 5 + y \rangle -8 + 5} ; \boxed{0 + y \rangle -3} ; \boxed{\mathbf{y \rangle -3}}$$

The solution set is $\{y \mid y \rangle \mathbf{-3}\}$.

Check 1: Let $y = -1$ which is greater than -3. Then, is $-5 - 1 \overset{?}{\rangle} -8$; $-6 \overset{?}{\rangle} -8$; $\frac{-6}{-1} \overset{?}{\langle} \frac{-8}{-1}$? The answer is "yes" because $6 \langle 8$.

Check 2: Let $y = 5$ which is greater than -3. Then, is $-5 + 5 \overset{?}{\rangle} -8$? The answer is "yes" because $0 \rangle -3$.

Example 2.5-8

$$\boxed{5 \rangle -w + 9} ; \boxed{5 + w \rangle -w + w + 9} ; \boxed{5 + w \rangle 0 + 9} ; \boxed{5 + w \rangle 9} ; \boxed{5 - 5 + w \rangle 9 - 5} ; \boxed{0 + w \rangle 4} ; \boxed{\mathbf{w \rangle 4}}$$

The solution set is $\{w \mid w \rangle \mathbf{4}\}$.

Check: Let $w = 8$ which is greater than 4. Then, is $5 \overset{?}{\rangle} -8 + 9$? The answer is "yes" because $5 \rangle 1$.

Example 2.5-9

$$\boxed{6 + x \rangle -20} ; \boxed{6 - 6 + x \rangle -20 - 6} ; \boxed{0 + x \rangle -26} ; \boxed{\mathbf{x \rangle -26}}$$

The solution set is $\{x \mid x \rangle \mathbf{-26}\}$.

Check 1: Let $x = -20$ which is greater than -26. Then, is $6 - 20 \overset{?}{\rangle} -20$; $-14 \overset{?}{\rangle} -20$; $\frac{-14}{-1} \overset{?}{\langle} \frac{-20}{-1}$? The answer is "yes" because $14 \langle 20$.

Check 2: Let $x = 5$ which is greater than -26. Then, is $6 + 5 \overset{?}{\rangle} -20$? The answer is "yes" because $11 \rangle -20$.

Check 3: Let $x = -50$ which is less than -26. Then, is $6 - 50 \overset{?}{\rangle} -20$; $-44 \overset{?}{\rangle} -20$; $\frac{-44}{-1} \overset{?}{\langle} \frac{-20}{-1}$

; $44 \overset{?}{<} 20$? The answer is "no" because we can not choose $x = -50$ since $-50 < -26$.

Example 2.5-10

$\boxed{m - 12 < 15}$; $\boxed{m - 12 + 12 < 15 + 12}$; $\boxed{m + 0 < 27}$; $\boxed{\boldsymbol{m < 27}}$ The solution set is $\{\boldsymbol{m} \mid \boldsymbol{m < 27}\}$.

Check 1: Let $m = 5$ which is less than 27. Then, is $5 - 12 \overset{?}{<} 15$; $-7 \overset{?}{<} 15$; $\frac{-7}{-1} \overset{?}{>} \frac{15}{-1}$? The answer is "yes" because $7 > -15$.

Check 2: Let $m = 20$ which is less than 27. Then, is $20 - 12 \overset{?}{<} 15$? The answer is "yes" because $8 < 15$.

Example 2.5-11

$\boxed{9.2 + x < -2.4}$; $\boxed{9.2 - 9.2 + x < -2.4 - 9.2}$; $\boxed{0 + x < -11.6}$; $\boxed{\boldsymbol{x < -11.6}}$ The solution set is $\{\boldsymbol{x} \mid \boldsymbol{x < -11.6}\}$.

Check 1: Let $x = -15$ which is less than -11.6. Then, is $9.2 - 15 \overset{?}{<} -2.4$; $-5.8 \overset{?}{<} -2.4$; $\frac{-5.8}{-1} \overset{?}{>} \frac{-2.4}{-1}$? The answer is "yes" because $5.8 > 2.4$.

Check 2: Let $x = 5$ which is greater than -11.6. Then, is $9.2 + 5 \overset{?}{<} -2.4$; $14.2 \overset{?}{<} -2.4$? The answer is "no" because we can not choose $x = 5$ since $5 > -11.6$.

Example 2.5-12

$\boxed{-2.3 + w \ge -4.8}$; $\boxed{-2.3 + 2.3 + w \ge -4.8 + 2.3}$; $\boxed{0 + w \ge -2.5}$; $\boxed{\boldsymbol{w \ge -2.5}}$ The solution set is $\{\boldsymbol{w} \mid \boldsymbol{w \ge -2.5}\}$.

Check 1: Let $w = -2.5$. Then, is $-2.3 - 2.5 \overset{?}{=} -4.8$? The answer is "yes" because $-4.8 = -4.8$.

Check 2: Let $w = 10$ which is greater than -2.5. Then, is $-2.3 + 10 \overset{?}{>} -4.8$? The answer is "yes" because $7.7 > -4.8$.

Example 2.5-13

$\boxed{h + 1\frac{2}{3} < 2\frac{3}{7}}$; $\boxed{h + \frac{(1 \cdot 3) + 2}{3} < \frac{(2 \cdot 7) + 3}{7}}$; $\boxed{h + \frac{3 + 2}{3} < \frac{14 + 3}{7}}$; $\boxed{h + \frac{5}{3} < \frac{17}{7}}$; $\boxed{h + \frac{5}{3} - \frac{5}{3} < \frac{17}{7} - \frac{5}{3}}$; $\boxed{h + 0 < \frac{17}{7} - \frac{5}{3}}$

; $\boxed{h < \frac{(17 \cdot 3) - (5 \cdot 7)}{7 \cdot 3}}$; $\boxed{h < \frac{51 - 35}{21}}$; $\boxed{h < \frac{16}{21}}$; $\boxed{\boldsymbol{h < 0.76}}$ The solution set is $\{\boldsymbol{h} \mid \boldsymbol{h < 0.76}\}$.

Check: Let $h = 0$ which is less than 0.76. Then, is $0 + 1\frac{2}{3} \overset{?}{<} 2\frac{3}{7}$; $1\frac{2}{3} \overset{?}{<} 2\frac{3}{7}$; $\frac{(1 \cdot 3) + 2}{3} \overset{?}{<} \frac{(2 \cdot 7) + 3}{7}$; $\frac{3 + 2}{3} \overset{?}{<} \frac{14 + 3}{7}$; $\frac{5}{3} \overset{?}{<} \frac{17}{7}$? The answer is "yes" because $1.67 < 2.43$.

Example 2.5-14

$\boxed{y - 3.85 \le 1\frac{3}{8}}$; $\boxed{y - 3.85 \le \frac{(1 \cdot 8) + 3}{8}}$; $\boxed{y - 3.85 \le \frac{8 + 3}{8}}$; $\boxed{y - 3.85 \le \frac{11}{8}}$; $\boxed{y - 3.85 \le 1.38}$

; $\boxed{y - 3.85 + 3.85 \le 1.38 + 3.85}$; $\boxed{y + 0 \le 5.23}$; $\boxed{\boldsymbol{y \le 5.23}}$ The solution set is $\{\boldsymbol{y} \mid \boldsymbol{y \le 5.23}\}$.

Check 1: Let $y = 5.23$. Then, is $5.23 - 3.85 \stackrel{?}{=} 1\frac{3}{8}$; $1.38 \stackrel{?}{=} \frac{(1\cdot 8)+3}{8}$; $1.38 \stackrel{?}{=} \frac{8+3}{8}$; $1.38 \stackrel{?}{=} \frac{11}{8}$? The answer is "yes" because $1.38 = 1.38$.

Check 2: Let $y = 2$ which is greater than 1.78. Then, is $2 - 3.85 \stackrel{?}{<} 1\frac{3}{8}$; $-1.85 \stackrel{?}{<} \frac{(1\cdot 8)+3}{8}$; $-1.85 \stackrel{?}{<} \frac{8+3}{8}$; $-1.85 \stackrel{?}{<} \frac{11}{8}$? The answer is "yes" because $-1.85 < 1.38$.

Example 2.5-15

$\boxed{3\frac{1}{5} + w \le 2\frac{3}{5} + 1\frac{1}{5}}$; $\boxed{\frac{(3\cdot 5)+1}{5} + w \le \frac{(2\cdot 5)+3}{5} + \frac{(1\cdot 5)+1}{5}}$; $\boxed{\frac{15+1}{5} + w \le \frac{10+3}{5} + \frac{5+1}{5}}$; $\boxed{\frac{16}{5} + w \le \frac{13}{5} + \frac{6}{5}}$

; $\boxed{\frac{16}{5} + w \le \frac{13+6}{5}}$; $\boxed{\frac{16}{5} + w \le \frac{19}{5}}$; $\boxed{\frac{16}{5} - \frac{16}{5} + w \le \frac{19}{5} - \frac{16}{5}}$; $\boxed{0 + w \le \frac{19}{5} - \frac{16}{5}}$; $\boxed{w \le \frac{19-16}{5}}$; $\boxed{w \le \frac{3}{5}}$

; $\boxed{\mathbf{w \le 0.6}}$

The solution set is $\{\mathbf{w} \mid \mathbf{w \le 0.6}\}$.

Check 1: Let $w = 0.6$. Then, is $3\frac{1}{5} + 0.6 \stackrel{?}{=} 2\frac{3}{5} + 1\frac{1}{5}$; $\frac{16}{5} + 0.6 \stackrel{?}{=} \frac{13}{5} + \frac{6}{5}$; $3.2 + 0.6 \stackrel{?}{=} 2.6 + 1.2$? The answer is "yes" because $3.8 = 3.8$.

Check 2: Let $w = 0$ which is less than 0.6. Then, is $3\frac{1}{5} + 0 \stackrel{?}{<} 2\frac{3}{5} + 1\frac{1}{5}$; $\frac{16}{5} \stackrel{?}{<} \frac{13}{5} + \frac{6}{5}$; $3.2 \stackrel{?}{<} 2.6 + 1.2$? The answer is "yes" because $3.2 < 3.8$.

Check 3: Let $w = 5$ which is greater than 0.6. Then, is $3\frac{1}{5} + 5 \stackrel{?}{\le} 2\frac{3}{5} + 1\frac{1}{5}$; $\frac{16}{5} + 5 \stackrel{?}{\le} \frac{13}{5} + \frac{6}{5}$; $3.2 + 5 \stackrel{?}{\le} 2.6 + 1.2$; $8.2 \stackrel{?}{\le} 3.8$? The answer is "no" because we can not choose $w = 5$ since $5 > 0.6$.

Practice Problems - Addition and Subtraction of Linear Inequalities

Section 2.5 Case I Practice Problems - Solve the following linear inequalities by adding or subtracting the same positive or negative number to both sides of the inequality.

1. $x - 10 > 12$
2. $-3 \le -u + 8$
3. $8 < -x + 5$
4. $3.2 + w \ge -2.8$
5. $0.65 + t > 2\frac{2}{3}$
6. $s - \frac{3}{5} < 1\frac{3}{5}$
7. $0.8 + w \ge 1\frac{1}{3} + 0.9$
8. $-1\frac{2}{7} < -h + 2\frac{3}{8}$
9. $y - 1.25 \le -2\frac{3}{4}$
10. $x + 1\frac{2}{3} < 2\frac{2}{5} - \frac{2}{7}$

Case II Multiplication and Division of Linear Inequalities

To multiply or divide linear inequalities by the same positive or negative number the following rules should be used:

Multiplication Rule:

a. *The same positive number can be multiplied by both sides of an inequality without changing the solution: for all real numbers* a, b, *and* c, *with* $c > 0$ *(a positive number),*

$$a > b \text{ if and only if } a \cdot c > b \cdot c.$$

b. *The same negative number can be multiplied by both sides on an inequality, however, the direction of the inequality must be changed in order to keep the same inequality: for all real numbers* a, b, *and* c, *with* $c < 0$ *(a negative number),*

$$a > b \text{ if and only if } a \cdot c < b \cdot c.$$

Division Rule:

a. *The same positive number (except zero) can be divided by both sides of an inequality without changing the solution: for all real numbers* a, b, *and* c, *with* $c > 0$ *(a positive number),*

$$a > b \text{ if and only if } \frac{a}{c} > \frac{b}{c}.$$

b. *The same negative number can be divided by both sides of an inequality, however, the direction of the inequality must be changed in order to keep the same inequality: for all real numbers* a, b, *and* c, *with* $c < 0$ *(a negative number),*

$$a > b \text{ if and only if } \frac{a}{c} < \frac{b}{c}.$$

Linear inequalities are multiplied or divided using the following steps:

Step 1 Isolate the variable to the left hand side of the inequality by applying the addition and subtraction rules.

Step 2 Find the solution by applying the multiplication or division rules. Check the answer by substituting different values, that are greater than or less than the solution found for the inequality, into the original inequality.

Examples with Steps

The following examples show the steps as to how linear inequalities are solved using the multiplication or division rules:

Example 2.5-16

$$\boxed{3x < -\frac{1}{5}}$$

Solution:

Step 1 *Not Applicable*

Step 2 $\boxed{3x < -\frac{1}{5}}$; $\boxed{\frac{1}{\cancel{3}} \cdot \cancel{3}x < -\frac{1}{5} \cdot \frac{1}{3}}$; $\boxed{x < -\frac{1 \cdot 1}{5 \cdot 3}}$; $\boxed{x < -\frac{1}{15}}$; $\boxed{\mathbf{x < -0.06}}$

The solution set is $\{x \mid x < -0.06\}$.

Check 1: Let $x=-10$ which is less than -0.06. Then, is $3\cdot -10 \overset{?}{<} -\frac{1}{5}$; $-30 \overset{?}{<} -\frac{1}{5}$? The answer is "yes" because $-30 < -0.2$ or $\frac{-30}{-1} > \frac{-0.2}{-1}$; $30 > 0.2$.

Check 2: Let $x=0$ which is greater than -0.06. Then, is $3\cdot 0 \overset{?}{<} -\frac{1}{5}$; $0 \overset{?}{<} -0.2$? The answer is "no" because we can not choose $x=0$ since $0 > -0.06$.

Example 2.5-17

$$\boxed{-\frac{2}{3} \ge 4y}$$

Solution:

Step 1 $\boxed{-\frac{2}{3} \ge 4y}$; $\boxed{-\frac{2}{3} - 4y \ge 4y - 4y}$; $\boxed{-\frac{2}{3} - 4y \ge 0}$; $\boxed{-\frac{2}{3} - 4y \ge 0}$; $\boxed{-\frac{2}{3} + \frac{2}{3} - 4y \ge 0 + \frac{2}{3}}$; $\boxed{0 - 4y \ge \frac{2}{3}}$; $\boxed{-4y \ge \frac{2}{3}}$

Step 2 $\boxed{-4y \ge \frac{2}{3}}$; $\boxed{-4y \cdot \frac{1}{-4} \le \frac{2}{3} \cdot \frac{1}{-4}}$; $\boxed{y \le \frac{2\cdot 1}{3\cdot -4}}$; $\boxed{y \le \frac{2}{-12}}$; $\boxed{y \le -\frac{1}{6}}$; $\boxed{\mathbf{y \le -0.166}}$

The solution set is $\{y \mid y \le -0.166\}$.

Check 1: Let $y=-0.166$. Then, is $-\frac{2}{3} \overset{?}{=} 4\cdot -0.166$; $-0.66 \overset{?}{=} 4\cdot -0.166$? The answer is "yes" because $-0.66 = -0.66$.

Check 2: Let $y=-2$ which is less than -0.166. Then, is $-\frac{2}{3} \overset{?}{>} 4\cdot -2$; $-\frac{2}{3} \overset{?}{>} -8$? The answer is "yes" because $-0.66 > -8$ or $\frac{-0.66}{-1} < \frac{-8}{-1}$; $0.66 < 8$.

Check 3: Let $y=2$ which is greater than -0.166. Then, is $-\frac{2}{3} \overset{?}{\ge} 4\cdot 2$; $-\frac{2}{3} \overset{?}{\ge} 8$; $-0.66 \overset{?}{\ge} 8$? The answer is "no" because we can not choose $y=2$ since $2 > -0.166$.

Example 2.5-18

$$\boxed{1\frac{2}{3} w < -2\frac{3}{5}}$$

Solution:

Step 1 $\boxed{\textit{Not Applicable}}$

Step 2 $\boxed{1\frac{2}{3} w < -2\frac{3}{5}}$; $\boxed{\frac{(1\cdot 3)+2}{3} w < -\frac{(2\cdot 5)+3}{5}}$; $\boxed{\frac{3+2}{3} w < -\frac{10+3}{5}}$; $\boxed{\frac{5}{3} w < -\frac{13}{5}}$; $\boxed{\frac{\cancel{3}}{\cancel{5}} \cdot \frac{\cancel{5}}{\cancel{3}} w < -\frac{13}{5} \cdot \frac{3}{5}}$; $\boxed{w < -\frac{13\cdot 3}{5\cdot 5}}$; $\boxed{w < -\frac{39}{25}}$; $\boxed{\mathbf{w < -1.56}}$

The solution set is $\{w \mid w < -1.56\}$.

Check: Let $w=-10$ which is less than -1.56. Then, is $1\frac{2}{3}\cdot -10 \overset{?}{<} -2\frac{3}{5}$; $\frac{5}{3}\cdot -10 \overset{?}{<} -\frac{13}{5}$

; $-\frac{50}{3}\overset{?}{<}-\frac{13}{5}$? The answer is "yes" because $-16.6<-2.6$ or $\frac{-16.6}{-1}>\frac{-2.6}{-1}$; $16.6>2.6$.

Example 2.5-19

$\boxed{-2>-5t}$

Solution:

Step 1 $\boxed{-2>-5t}$; $\boxed{-2+5t>-5t+5t}$; $\boxed{-2+5t>0}$; $\boxed{-2+2+5t>0+2}$; $\boxed{0+5t>2}$; $\boxed{5t>2}$

Step 2 $\boxed{5t>2}$; $\boxed{\frac{\cancel{5}t}{\cancel{5}}>\frac{2}{5}}$; $\boxed{\mathbf{t>0.4}}$ The solution set is $\{t \mid t>0.4\}$.

Check: Let $t=5$ which is greater than 0.4. Then, is $-2\overset{?}{>}-5\cdot5$? The answer is "yes" because $-2>-25$ or $\frac{-2}{-1}<\frac{-25}{-1}$; $2<25$.

Example 2.5-20

$\boxed{2.6m<3\frac{2}{3}}$

Solution:

Step 1 *Not Applicable*

Step 2 $\boxed{2.6m<3\frac{2}{3}}$; $\boxed{2.6m<\frac{(3\cdot3)+2}{3}}$; $\boxed{2.6m<\frac{9+2}{3}}$; $\boxed{2.6m<\frac{11}{3}}$; $\boxed{2.6m<3.66}$; $\boxed{\frac{\cancel{2.6}}{\cancel{2.6}}m<\frac{3.66}{2.6}}$

; $\boxed{m<\frac{\frac{366}{100}}{\frac{26}{10}}}$; $\boxed{m<\frac{366\cdot10}{100\cdot26}}$; $\boxed{m<\frac{3660}{2600}}$; $\boxed{\mathbf{m<1.4}}$

The solution set is $\{m \mid m<1.4\}$.

Check 1: Let $m=1$ which is less than 1.4. Then, is $2.6\cdot1\overset{?}{<}3\frac{2}{3}$; $2.6\overset{?}{<}\frac{11}{3}$? The answer is "yes" because $2.6<3.7$.

Check 2: Let $m=2$ which is greater than 1.4. Then, is $2.6\cdot2\overset{?}{<}3\frac{2}{3}$; $5.2\overset{?}{<}\frac{11}{3}$; $5.2\overset{?}{<}3.7$? The answer is "no" because we can not choose $m=2$ since $2>1.4$.

Additional Examples - Multiplication and Division of Linear Inequalities

The following examples further illustrate how linear inequalities are simplified using the above multiplication or division rules:

Example 2.5-21

$\boxed{-5y>75}$; $\boxed{\frac{-\cancel{5}y}{-\cancel{5}}<\frac{75}{-5}}$; $\boxed{\mathbf{y<-15}}$ The solution set is $\{y \mid y<-15\}$.

Check: Let $y=-40$ which is less than -15. Then, is $-5\cdot(-40)\overset{?}{>}75$? The answer is "yes" because $200>75$.

Example 2.5-22

$$\boxed{\frac{2}{3}x\le -4} \;;\; \boxed{\frac{\cancel{3}}{\cancel{2}}\cdot\frac{\cancel{2}}{\cancel{3}}x\le -4\cdot\frac{3}{2}} \;;\; \boxed{x\le -\frac{4}{1}\cdot\frac{3}{2}} \;;\; \boxed{x\le -\frac{4\cdot 3}{1\cdot 2}} \;;\; \boxed{x\le -\frac{12}{2}} \;;\; \boxed{\mathbf{x\le -6}}$$

The solution set is $\{x \mid x\le -6\}$.

Check 1: Let $x=-6$. Then, is $\frac{2}{3}\cdot(-6)\overset{?}{=}-4$; $-\frac{12}{3}\overset{?}{=}-4$? The answer is "yes" because $-4=-4$.

Check 2: Let $x=-10$ which is less than -6. Then, is $\frac{2}{3}\cdot(-10)\overset{?}{<}-4$; $-\frac{20}{3}\overset{?}{<}-4$? The answer is "yes" because $-6.67<-4$ or $\frac{-6.67}{-1}>\frac{-4}{-1}$; $6.67>4$.

Example 2.5-23

$$\boxed{\frac{4}{7}\ge -2h} \;;\; \boxed{\frac{4}{7}+2h\ge -2h+2h} \;;\; \boxed{\frac{4}{7}+2h\ge 0} \;;\; \boxed{\frac{4}{7}-\frac{4}{7}+2h\ge 0-\frac{4}{7}} \;;\; \boxed{0+2h\ge -\frac{4}{7}} \;;\; \boxed{2h\ge -\frac{4}{7}} \;;\; \boxed{\frac{1}{\cancel{2}}\cdot\cancel{2}h\ge -\frac{4}{7}\cdot\frac{1}{2}}$$

$$;\; \boxed{h\ge -\frac{4}{14}} \;;\; \boxed{\mathbf{h\ge -0.28}}$$

The solution set is $\{h \mid h\ge -0.28\}$.

Note that another way of solving these type of inequalities is as shown below:

$$\boxed{\frac{4}{7}\ge -2h} \;;\; \boxed{\frac{4}{7}\cdot\frac{1}{-2}\le -\cancel{2}h\cdot\frac{1}{-\cancel{2}}} \;;\; \boxed{-\frac{4}{14}\le h} \;;\; \boxed{-0.28\le h} \text{ or } \boxed{\mathbf{h\ge -0.28}}$$

Check 1: Let $h=-0.28$. Then, is $\frac{4}{7}\overset{?}{=}-2\cdot -0.28$; $0.56\overset{?}{=}-2\cdot -0.28$? The answer is "yes" because $0.56=0.56$.

Check 2: Let $h=7$ which is greater than -0.28. Then, is $\frac{4}{7}\overset{?}{>}-2\cdot 7$? The answer is "yes" because $0.57>-14$.

Example 2.5-24

$$\boxed{-w>-9} \;;\; \boxed{\frac{-w}{-1}<\frac{-9}{-1}} \;;\; \boxed{\mathbf{w<9}}$$

The solution set is $\{w \mid w<9\}$.

Check: Let $w=2$ which is less than 9. Then, is $-2\overset{?}{>}-9$? The answer is "yes" because $\frac{-2}{-1}<\frac{-9}{-1}$; $2<9$.

Example 2.5-25

$$\boxed{-5<-\frac{2}{5}x} \;;\; \boxed{-5+\frac{2}{5}x<-\frac{2}{5}x+\frac{2}{5}x} \;;\; \boxed{-5+\frac{2}{5}x<0} \;;\; \boxed{-5+5+\frac{2}{5}x<0+5} \;;\; \boxed{0+\frac{2}{5}x<5} \;;\; \boxed{\frac{2}{5}x<5} \;;\; \boxed{\frac{\cancel{5}}{\cancel{2}}\cdot\frac{\cancel{2}}{\cancel{5}}x<5\cdot\frac{5}{2}}$$

$$;\; \boxed{x<\frac{25}{2}} \;;\; \boxed{\mathbf{x<12.5}}$$

The solution set is $\{x \mid x<12.5\}$.

Note that another way of solving these type of inequalities is as shown below:

$$\boxed{-5<-\frac{2}{5}x} \;;\; \boxed{-5\cdot -\frac{5}{2}>-\frac{\cancel{2}}{\cancel{5}}\cdot -\frac{\cancel{5}}{\cancel{2}}x} \;;\; \boxed{\frac{25}{2}>x} \;;\; \boxed{12.5>x} \text{ or } \boxed{\mathbf{x<12.5}}$$

Check: Let $x=5$ which is less than 12.5. Then, is $-5\overset{?}{<}-\frac{2}{5}\cdot 5$; $-5\overset{?}{<}-\frac{2}{\cancel{5}}\cdot\cancel{5}$? The answer is "yes" because $-5 < -2$ or $\frac{-5}{-1} > \frac{-2}{-1}$; $5 > 2$.

Example 2.5-26

$\boxed{\frac{1}{8}k \le 3}$; $\boxed{8\cdot\frac{1}{8}k \le 8\cdot 3}$; $\boxed{\mathbf{k \le 24}}$ The solution set is $\{k \mid k \le 24\}$.

Check 1: Let $k = 24$. Then, is $\frac{1}{8}\cdot 24 \stackrel{?}{=} 3$; $\frac{\overset{3}{\cancel{24}}}{\cancel{8}} \stackrel{?}{=} 3$; $\frac{3}{1} \stackrel{?}{=} 3$? The answer is "yes" because $3 = 3$.

Check 2: Let $k = 8$ which is less than 24. Then, is $\frac{1}{8}\cdot 8 \stackrel{?}{<} 3$; $\frac{1}{\cancel{8}}\cdot\cancel{8} \stackrel{?}{<} 3$? The answer is "yes" because $1 < 3$.

Example 2.5-27

$\boxed{-5 < \frac{w}{-6}}$; $\boxed{-5+\frac{w}{6} < -\frac{w}{6}+\frac{w}{6}}$; $\boxed{-5+\frac{w}{6} < 0}$; $\boxed{-5+5+\frac{w}{6} < 0+5}$; $\boxed{0+\frac{w}{6} < 5}$; $\boxed{\frac{w}{6} < 5}$; $\boxed{\cancel{6}\cdot\frac{w}{\cancel{6}} < 5\cdot 6}$; $\boxed{\mathbf{w < 30}}$

Or the inequality can be solved in the following way:

$\boxed{-5 < \frac{w}{-6}}$; $\boxed{-5\cdot -6 > \frac{w}{-6}\cdot -6}$; $\boxed{30 > \frac{w}{\cancel{-6}}\cdot \cancel{-6}}$; $\boxed{30 > w}$ or $\boxed{\mathbf{w < 30}}$ The solution set is $\{w \mid w < 30\}$.

Check: Let $w = 10$ which is less than 30. Then, is $-5 \stackrel{?}{<} \frac{10}{-6}$; $-5 \stackrel{?}{<} -\frac{10}{6}$? The answer is "yes" because $-5 < -1.66$ or $\frac{-5}{-1} > \frac{-1.66}{-1}$; $5 > 1.66$.

Example 2.5-28

$\boxed{\frac{1}{5}x \le -\frac{5}{7}}$; $\boxed{\cancel{5}\cdot\frac{1}{\cancel{5}}x \le -\frac{5}{7}\cdot 5}$; $\boxed{x \le -\frac{25}{7}}$; $\boxed{\mathbf{x \le -3.57}}$ The solution set is $\{x \mid x \le -3.57\}$.

Check 1: Let $x = -3.57$. Then, is $\frac{1}{5}\cdot -3.57 \stackrel{?}{=} -\frac{5}{7}$; $-\frac{3.57}{5} \stackrel{?}{=} -\frac{5}{7}$? The answer is "yes" because $-0.714 = -0.714$

Check 2: Let $x = -5$ which is less than -3.57. Then, is $\frac{1}{5}\cdot -5 \stackrel{?}{<} -\frac{5}{7}$; $\frac{1}{\cancel{5}}\cdot -\cancel{5} \stackrel{?}{<} -\frac{5}{7}$; $-1 \stackrel{?}{<} -\frac{5}{7}$? The answer is "yes" because $-1 < -0.71$ or $\frac{-1}{-1} > \frac{-0.7}{-1}$; $1 > 0.7$

Check 3: Let $x = 0$ which is greater than -3.57. Then, is $\frac{1}{5}\cdot 0 \stackrel{?}{\le} -\frac{5}{7}$; $0 \stackrel{?}{\le} -\frac{5}{7}$; $0 \stackrel{?}{\le} -0.71$? The answer is "no" because we can not choose $x = 0$ since $0 > -3.57$.

Example 2.5-29

$\boxed{-\frac{1}{4}y \le 5}$; $\boxed{-\frac{4}{1}\cdot -\frac{1}{4}y \ge 5\cdot -\frac{4}{1}}$; $\boxed{y \ge -\frac{20}{1}}$; $\boxed{\mathbf{y \ge -20}}$ The solution set is $\{y \mid y \ge -20\}$.

Check 1: Let $y = -20$. Then, is $-\frac{1}{4}\cdot -20 \stackrel{?}{=} 5$; $\frac{\overset{5}{\cancel{20}}}{\cancel{4}} \stackrel{?}{=} 5$; $\frac{5}{1} \stackrel{?}{=} 5$? The answer is "yes" because $5 = 5$.

Check 2: Let $y = -8$ which is greater than -20. Then, is $-\frac{1}{4}\cdot -8 \stackrel{?}{<} 5$; $\frac{\overset{2}{\cancel{8}}}{\cancel{4}} \stackrel{?}{<} 5$; $\frac{2}{1} \stackrel{?}{<} 5$? The answer is "yes" because $2 < 5$.

Check 3: Let $y = -40$ which is less than -20. Then, is $-\frac{1}{4}\cdot -40 \overset{?}{\le} 5$; $\frac{\overset{10}{\cancel{40}}}{\cancel{4}} \overset{?}{\le} 5$; $\frac{10}{1} \overset{?}{\le} 5$; $10 \overset{?}{\le} 5$?

The answer is "no" because we can not choose $y = -40$ since $-40 < -20$.

Example 2.5-30

$\boxed{2\frac{2}{3} \ge -1\frac{1}{5}x}$; $\boxed{\frac{(2\cdot 3)+2}{3} \ge -\frac{(1\cdot 5)+1}{5}x}$; $\boxed{\frac{6+2}{3} \ge -\frac{5+1}{5}x}$; $\boxed{\frac{8}{3} \ge -\frac{6}{5}x}$; $\boxed{\frac{8}{3}+\frac{6}{5}x \ge -\frac{6}{5}x+\frac{6}{5}x}$; $\boxed{\frac{8}{3}+\frac{6}{5}x \ge 0}$

; $\boxed{\frac{8}{3}-\frac{8}{3}+\frac{6}{5}x \ge 0-\frac{8}{3}}$; $\boxed{0+\frac{6}{5}x \ge -\frac{8}{3}}$; $\boxed{\frac{6}{5}x \ge -\frac{8}{3}}$; $\boxed{\frac{\cancel{5}}{\cancel{6}}\cdot\frac{\cancel{6}}{\cancel{5}}x \ge -\frac{8}{3}\cdot\frac{5}{6}}$; $\boxed{x \ge -\frac{40}{18}}$; $\boxed{\boldsymbol{x \ge -2.22}}$

Or the inequality can be solved in the following way:

$\boxed{2\frac{2}{3} \ge -1\frac{1}{5}x}$; $\boxed{\frac{(2\cdot 3)+2}{3} \ge -\frac{(1\cdot 5)+1}{5}x}$; $\boxed{\frac{6+2}{3} \ge -\frac{5+1}{5}x}$; $\boxed{\frac{8}{3} \ge -\frac{6}{5}x}$; $\boxed{-\frac{5}{6}\cdot\frac{8}{3} \le -\frac{\cancel{6}}{\cancel{5}}\cdot-\frac{\cancel{5}}{\cancel{6}}x}$; $\boxed{-\frac{40}{18} \le x}$

; $\boxed{-2.22 \le x}$ or $\boxed{\boldsymbol{x \ge -2.22}}$ The solution set is $\{\boldsymbol{x} \mid \boldsymbol{x \ge -2.22}\}$.

Check 1: Let $x = -2.22$. Then, is $2\frac{2}{3} \overset{?}{=} -1\frac{1}{5}\cdot(-2.22)$; $\frac{8}{3} \overset{?}{=} \frac{6}{5}\cdot 2.22$; $\frac{8}{3} \overset{?}{=} \frac{13.32}{5}$? The answer is "yes" because $2.66 = 2.66$.

Check 2: Let $x = 0$ which is greater than -2.22. Then, is $2\frac{2}{3} \overset{?}{>} -1\frac{1}{5}\cdot 0$; $\frac{8}{3} \overset{?}{>} 0$? The answer is "yes" because $2.66 > 0$.

Check 3: Let $x = -15$ which is less than -2.22. Then, is $2\frac{2}{3} \overset{?}{\ge} -1\frac{1}{5}\cdot(-15)$; $\frac{8}{3} \overset{?}{\ge} \frac{6}{5}\cdot 15$; $\frac{8}{3} \overset{?}{\ge} \frac{90}{5}$; $2.66 \overset{?}{\ge} 18$? The answer is "no" because we can not choose $x = -15$ since $-15 < -2.22$.

Practice Problems - Multiplication and Division of Linear Inequalities

Section 2.5 Case II Practice Problems - Solve the following linear inequalities by applying the multiplication or division rules:

1. $4y > -\frac{2}{3}$
2. $-\frac{2}{3}x \le 1\frac{2}{3}$
3. $\frac{2}{5} < -2h$
4. $\frac{w}{7} \ge -3$
5. $\frac{w}{-2} < -5$
6. $2\frac{1}{4}u > -1\frac{1}{5}$
7. $-2x \le 2\frac{3}{4}+1$
8. $3.28x \ge 2.4$
9. $-\frac{1}{4}y > -2$
10. $5\frac{1}{3} < -2\frac{3}{4}x$

Case III Mixed Operations Involving Linear Inequalities

In Cases I and II we learned how to solve linear inequalities by either applying: 1. The addition and subtraction rules, or 2. The multiplication or division rules. In this section, solution to linear inequalities which may involve using all four rules is discussed.

Addition and Subtraction Rules: *The same positive or negative number, or variable, can be added or subtracted to both sides of an inequality without changing the solution: for all real numbers* a, b, *and* c,

1. $a > b$ *if and only if* $a + c > b + c$
2. $a > b$ *if and only if* $a - c > b - c$.

Multiplication Rule:

a. *The same positive number can be multiplied by both sides of an inequality without changing the solution: for all real numbers* a, b, *and* c, *with* $c > 0$ *(a positive number),*

$$a > b \text{ if and only if } a \cdot c > b \cdot c.$$

b. *The same negative number can be multiplied by both sides on an inequality, however, the direction of the inequality must be changed in order to keep the same inequality: for all real numbers* a, b, *and* c, *with* $c < 0$ *(a negative number),*

$$a > b \text{ if and only if } a \cdot c < b \cdot c.$$

Division Rule:

a. *The same positive number (except zero) can be divided by both sides of an inequality without changing the solution: for all real numbers* a, b, *and* c, *with* $c > 0$ *(a positive number),*

$$a > b \text{ if and only if } \frac{a}{c} > \frac{b}{c}.$$

b. *The same negative number can be divided by both sides of an inequality, however, the direction of the inequality must be changed in order to keep the same inequality: for all real numbers* a, b, *and* c, *with* $c < 0$ *(a negative number),*

$$a > b \text{ if and only if } \frac{a}{c} < \frac{b}{c}.$$

Linear inequalities are solved using the following steps:

Step 1 Isolate the variable to the left hand side of the inequality by applying the addition and subtraction rules.

Step 2 Find the solution by applying the multiplication or division rules. Check the answer by substituting different values, that are greater than or less than the solution found for the inequality, into the original inequality.

Examples with Steps

The following examples show the steps as to how linear inequalities are solved using the addition, subtraction, multiplication, and division rules:

Example 2.5-31

$\boxed{6y+1 > 2\frac{3}{7}}$

Solution:

Step 1 $\boxed{6y+1 > 2\frac{3}{7}}$; $\boxed{6y+1-1 > 2\frac{3}{7}-1}$; $\boxed{6y+0 > 2\frac{3}{7}-1}$; $\boxed{6y > 2\frac{3}{7}-1}$

Step 2 $\boxed{6y > 2\frac{3}{7}-1}$; $\boxed{6y > \frac{(2\cdot 7)+3}{7}-1}$; $\boxed{6y > \frac{14+3}{7}-1}$; $\boxed{6y > \frac{17}{7}-\frac{1}{1}}$; $\boxed{6y > \frac{(17\cdot 1)-(1\cdot 7)}{7\cdot 1}}$

; $\boxed{6y > \frac{17-7}{7}}$; $\boxed{6y > \frac{10}{7}}$; $\boxed{6y > 1.43}$; $\boxed{\frac{\cancel{6}y}{\cancel{6}} > \frac{1.43}{6}}$; $\boxed{\mathbf{y > 0.24}}$

The solution set is $\{y \mid y > \mathbf{0.24}\}$.

Check 1: Let $y=4$ which is greater than 0.24. Then, is $6\cdot 4+1 \overset{?}{>} 2\frac{3}{7}$; $24+1 \overset{?}{>} \frac{17}{7}$; $25 \overset{?}{>} \frac{17}{7}$? The answer is "yes" because $25 > 2.43$.

Check 2: Let $y=0$ which is less than 0.24. Then, is $6\cdot 0+1 \overset{?}{>} 2\frac{3}{7}$; $0+1 \overset{?}{>} \frac{17}{7}$; $1 \overset{?}{>} 2.43$? The answer is "no" because we can not choose $y=0$ since $0 < 0.24$.

Example 2.5-32

$\boxed{6t+10 \geq 9t+5}$

Solution:

Step 1 $\boxed{6t+10 \geq 9t+5}$; $\boxed{6t-9t+10 \geq 9t-9t+5}$; $\boxed{-3t+10 \geq 0+5}$; $\boxed{-3t+10 \geq 5}$

; $\boxed{-3t+10-10 \geq 5-10}$; $\boxed{-3t+0 \geq -5}$; $\boxed{-3t \geq -5}$

Step 2 $\boxed{-3t \geq -5}$; $\boxed{\frac{-\cancel{3}t}{-\cancel{3}} \leq \frac{-5}{-3}}$; $\boxed{\mathbf{t \leq 1.66}}$ The solution set is $\{t \mid t \leq \mathbf{1.66}\}$.

Check 1: Let $t=1.66$. Then, is $6\cdot 1.66+10 \overset{?}{=} 9\cdot 1.66+5$; $9.9+10 \overset{?}{=} 14.9+5$? The answer is "yes" because $19.9 = 19.9$.

Check 2: Let $t=1$ which is less than 1.66. Then, is $6\cdot 1+10 \overset{?}{>} 9\cdot 1+5$; $6+10 \overset{?}{>} 9+5$? The answer is "yes" because $16 > 14$.

Check 3: Let $t=3$ which is greater than 1.66. Then, is $6\cdot 3+10 \overset{?}{\geq} 9\cdot 3+5$; $18+10 \overset{?}{\geq} 27+5$; $28 \overset{?}{\geq} 32$? The answer is "no" because we can not choose $t=3$ since $3 > 1.66$.

Example 2.5-33

$\boxed{3x \leq 25+8x}$

Solution:

Step 1 $\boxed{3x \leq 25+8x}$; $\boxed{3x-8x \leq 25+8x-8x}$; $\boxed{-5x \leq 25+0}$; $\boxed{-5x \leq 25}$

Step 2 $\boxed{-5x \leq 25}$; $\boxed{\frac{-\cancel{5}x}{-\cancel{5}} \geq \frac{25}{-5}}$; $\boxed{\mathbf{x \geq -5}}$ The solution set is $\{x \mid x \geq -5\}$.

Check 1: Let $x = -5$. Then, is $3 \cdot -5 \overset{?}{=} 25 + 8 \cdot -5$; $-15 \overset{?}{=} 25 - 40$? The answer is "yes" because $-15 = -15$.

Check 2: Let $x = -2$ which is greater than -5. Then, is $3 \cdot -2 \overset{?}{<} 25 + 8 \cdot -2$; $-6 \overset{?}{<} 25 - 16$? The answer is "yes" because $-6 < 9$.

Example 2.5-34

$$\boxed{0.8n + 10 \ge 1.2n}$$

Solution:

Step 1

$$\boxed{0.8n - 1.2n + 10 \ge 1.2n - 1.2n} ; \boxed{-0.4n + 10 \ge 0} ; \boxed{-0.4n + 10 \ge 0} ; \boxed{-0.4n + 10 - 10 \ge 0 - 10}$$

$$; \boxed{-0.4n + 0 \ge -10} ; \boxed{-0.4n \ge -10}$$

Step 2

$$\boxed{-0.4n \ge -10} ; \boxed{\frac{-0.4n}{-0.4} \le \frac{-10}{-0.4}} ; \boxed{n \le +\frac{10}{0.4}} ; \boxed{n \le \frac{\frac{10}{1}}{\frac{4}{10}}} ; \boxed{n \le \frac{10 \cdot 10}{1 \cdot 4}} ; \boxed{n \le \frac{100}{4}} ; \boxed{\mathbf{n \le 25}}$$

The solution set is $\{\mathbf{n} \mid \mathbf{n \le 25}\}$.

Check 1: Let $n = 25$. Then, is $0.8 \cdot 25 + 10 \overset{?}{=} 1.2 \cdot 25$; $20 + 10 \overset{?}{=} 30$? The answer is "yes" because $30 = 30$.

Check 2: Let $n = 20$ which is less than 25. Then, is $0.8 \cdot 20 + 10 \overset{?}{>} 1.2 \cdot 20$; $16 + 10 \overset{?}{>} 24$? The answer is "yes" because $26 > 24$.

Check 3: Let $n = 30$ which is greater than 25. Then, is $0.8 \cdot 30 + 10 \overset{?}{\ge} 1.2 \cdot 30$; $24 + 10 \overset{?}{\ge} 36$; $34 \overset{?}{\ge} 36$? The answer is "no" because we can not choose $n = 30$ since $30 > 25$.

Example 2.5-35

$$\boxed{z - \frac{2}{3} \le 4z - \frac{1}{5}}$$

Solution:

Step 1

$$\boxed{z - \frac{2}{3} \le 4z - \frac{1}{5}} ; \boxed{z - 4z - \frac{2}{3} \le 4z - 4z - \frac{1}{5}} ; \boxed{-3z - \frac{2}{3} \le 0 - \frac{1}{5}} ; \boxed{-3z - \frac{2}{3} \le -\frac{1}{5}}$$

$$; \boxed{-3z - \frac{2}{3} + \frac{2}{3} \le -\frac{1}{5} + \frac{2}{3}} ; \boxed{-3z + 0 \le -\frac{1}{5} + \frac{2}{3}} ; \boxed{-3z \le -\frac{1}{5} + \frac{2}{3}}$$

Step 2

$$\boxed{-3z \le -\frac{1}{5} + \frac{2}{3}} ; \boxed{-3z \le \frac{(-1 \cdot 3) + (2 \cdot 5)}{5 \cdot 3}} ; \boxed{-3z \le \frac{-3 + 10}{15}} ; \boxed{-3z \le \frac{7}{15}}$$

$$; \boxed{\frac{1}{\cancel{-3}} \cdot \cancel{-3}z \ge \frac{7}{15} \cdot \frac{1}{-3}} ; \boxed{z \ge \frac{7 \cdot 1}{15 \cdot -3}} ; \boxed{z \ge -\frac{7}{45}} ; \boxed{z \ge -0.155}$$

The solution set is $\{z \mid z \ge -\mathbf{0.155}\}$.

Check 1: Let $z = -0.155$. Then, is $-0.155 - \frac{2}{3} \overset{?}{=} (4 \cdot -0.155) - \frac{1}{5}$; $-0.155 - 0.66 \overset{?}{=} -0.62 - 0.2$? The answer is "yes" because $-0.82 = -0.82$.

Check 2: Let $z = 5$ which is greater than -0.155. Then, is $5 - \frac{2}{3} \overset{?}{<} (4 \cdot 5) - \frac{1}{5}$; $5 - 0.66 \overset{?}{<} 20 - 0.2$? The answer is "yes" because $4.34 < 19.8$.

Check 3: Let $z = -5$ which is less than -0.155. Then, is $-5 - \frac{2}{3} \overset{?}{\leq} (4 \cdot -5) - \frac{1}{5}$; $-5 - 0.66 \overset{?}{\leq} -20 - 0.2$; $-5.66 \overset{?}{\leq} -20.2$ or ; $\frac{-5.66}{-1} \overset{?}{\geq} \frac{-20.2}{-1}$; $5.66 \overset{?}{\geq} 20.2$? The answer is "no" because we can not choose $z = -5$ since $-5 < -0.155$.

Additional Examples - Mixed Operations Involving Linear Inequalities

The following examples further illustrate how to solve linear inequalities using the addition, subtraction, multiplication, and division rules:

Example 2.5-36

$\boxed{2x - 8 > 10}$; $\boxed{2x - 8 + 8 > 10 + 8}$; $\boxed{2x + 0 > 18}$; $\boxed{2x > 18}$; $\boxed{\frac{\not{2}x}{\not{2}} > \frac{18}{2}}$; $\boxed{\mathbf{x > 9}}$

The solution set is $\{x \mid x > 9\}$.

Check 1: Let $x = 20$ which is greater than 9. Then, is $(2 \cdot 20) - 8 \overset{?}{>} 10$; $40 - 8 \overset{?}{>} 10$?. The answer is "yes" because $32 > 10$.

Check 2: Let $x = 0$ which is less than 9. Then, is $(2 \cdot 0) - 8 \overset{?}{>} 10$; $0 - 8 \overset{?}{>} 10$; $-8 \overset{?}{>} 10$? The answer is "no" because we can not choose $x = 0$ since $0 < 9$.

Example 2.5-37

$\boxed{6w - 5 \leq 2w + 8}$; $\boxed{6w - 5 + 5 \leq 2w + 8 + 5}$; $\boxed{6w + 0 \leq 2w + 13}$; $\boxed{6w \leq 2w + 13}$; $\boxed{6w - 2w \leq 2w - 2w + 13}$

; $\boxed{4w \leq 0 + 13}$; $\boxed{4w \leq 13}$; $\boxed{\frac{4w}{4} \leq \frac{13}{4}}$; $\boxed{\mathbf{w \leq 3.25}}$ The solution set is $\{w \mid w \leq 3.25\}$.

Check 1: Let $w = 3.25$. Then, is $6 \cdot 3.25 - 5 \overset{?}{=} 2 \cdot 3.25 + 8$; $19.5 - 5 \overset{?}{=} 6.5 + 8$? The answer is "yes" because $14.5 = 14.5$.

Check 2: Let $w = 2$ which is less than 3.25. Then, is $6 \cdot 2 - 5 \overset{?}{<} 2 \cdot 2 + 8$; $12 - 5 \overset{?}{<} 4 + 8$?. The answer is "yes" because $7 < 12$.

Example 2.5-38

$\boxed{\frac{2}{5}x + 3 < 8}$; $\boxed{\frac{2}{5}x + 3 - 3 < 8 - 3}$; $\boxed{\frac{2}{5}x + 0 < 5}$; $\boxed{\frac{2}{5}x < 5}$; $\boxed{\frac{\not{5}}{\not{2}} \cdot \frac{\not{2}}{\not{5}}x < 5 \cdot \frac{5}{2}}$; $\boxed{x < \frac{25}{2}}$; $\boxed{\mathbf{x < 12.5}}$

The solution set is $\{x \mid x < 12.5\}$.

Check: Let $x = 5$ which is less than 12.5. Then, is $\frac{2}{5} \cdot 5 + 3 \overset{?}{<} 8$; $\frac{2}{\not{5}} \cdot \not{5} + 3 \overset{?}{<} 8$; $2 + 3 \overset{?}{<} 8$? The answer is "yes" because $5 < 8$.

Example 2.5-39

$$\boxed{4+\frac{u}{3}\ge 5-2u} \;;\; \boxed{4-4+\frac{u}{3}\ge 5-4-2u} \;;\; \boxed{0+\frac{u}{3}\ge 1-2u} \;;\; \boxed{\frac{u}{3}\ge 1-2u} \;;\; \boxed{\frac{u}{3}+2u\ge 1-2u+2u} \;;\; \boxed{\frac{u}{3}+\frac{2u}{1}\ge 1+0}$$

$$;\; \boxed{\frac{(1\cdot u)+(3\cdot 2u)}{3\cdot 1}\ge 1} \;;\; \boxed{\frac{u+6u}{3}\ge 1} \;;\; \boxed{\frac{7u}{3}\ge 1} \;;\; \boxed{\frac{7}{3}u\ge 1} \;;\; \boxed{\frac{\cancel{3}}{\cancel{7}}\cdot\frac{\cancel{7}}{\cancel{3}}u\ge 1\cdot\frac{3}{7}} \;;\; \boxed{u\ge\frac{3}{7}} \;;\; \boxed{\mathbf{u\ge 0.43}}$$

The solution set is $\{\boldsymbol{u} \mid \boldsymbol{u} \ge \mathbf{0.43}\}$.

Check 1: Let $u=0.43$. Then, is $4+\frac{0.43}{3}\overset{?}{=}5-(2\cdot 0.43)$; $4+0.14\overset{?}{=}5-0.86$? The answer is "yes" because $4.14=4.14$.

Check 2: Let $u=6$ which is greater than 0.43. Then, is $4+\frac{6}{3}\overset{?}{>}5-(2\cdot 6)$; $4+2\overset{?}{>}5-12$?. The answer is "yes" because $6>-7$.

Check 3: Let $u=0$ which is less than 0.43. Then, is $4+\frac{0}{3}\overset{?}{\ge}5-(2\cdot 0)$; $4+\frac{0}{3}\overset{?}{\ge}5-(2\cdot 0)$; $4\overset{?}{\ge}5$? The answer is "no" because we can not choose $u=0$ since $0<0.43$.

Example 2.5-40

$$\boxed{-25<-18+h} \;;\; \boxed{-25-h<-18+h-h} \;;\; \boxed{-25-h<-18+0} \;;\; \boxed{-25-h<-18} \;;\; \boxed{-25+25-h<-18+25}$$

$$;\; \boxed{0-h<7} \;;\; \boxed{-h<7} \;;\; \boxed{\frac{-h}{-1}>\frac{7}{-1}} \;;\; \boxed{\mathbf{h>-7}}$$

The solution set is $\{\boldsymbol{h} \mid \boldsymbol{h} > -\mathbf{7}\}$.

Check 1: Let $h=-1$ which is greater than -7. Then, is $-25\overset{?}{<}-18-1$; $-25\overset{?}{<}-19$; $\frac{-25}{-1}\overset{?}{>}\frac{-19}{-1}$? The answer is "yes" because $25>19$.

Check 2: Let $h=2$ which is greater than -7. Then, is $-25\overset{?}{<}-18+2$; $-25\overset{?}{<}-16$; $\frac{-25}{-1}\overset{?}{>}\frac{-16}{-1}$? The answer is "yes" because $25>16$.

Example 2.5-41

$$\boxed{5>y+35} \;;\; \boxed{5-y>y-y+35} \;;\; \boxed{5-y>0+35} \;;\; \boxed{5-y>35} \;;\; \boxed{5-5-y>35-5} \;;\; \boxed{0-y>30} \;;\; \boxed{-y>30}$$

$$;\; \boxed{\frac{-y}{-1}<\frac{30}{-1}} \;;\; \boxed{\mathbf{y<-30}}$$

The solution set is $\{\boldsymbol{y} \mid \boldsymbol{y} < -\mathbf{30}\}$.

Check 1: Let $y=-40$ which is less than -30. Then, is $5\overset{?}{>}-40+35$? The answer is "yes" because $5>-5$.

Check 2: Let $y=0$ which is greater than -30. Then, is $5\overset{?}{>}0+35$; $5\overset{?}{>}35$? The answer is "no" because we can not choose $y=0$ since $0>-30$.

Example 2.5-42

$$\boxed{-8>w+6} \;;\; \boxed{-8-w>w-w+6} \;;\; \boxed{-8-w>0+6} \;;\; \boxed{-8-w>6} \;;\; \boxed{-8+8-w>6+8} \;;\; \boxed{0-w>6+8} \;;\; \boxed{-w>14}$$

$$;\; \boxed{\frac{-w}{-1}<\frac{14}{-1}} \;;\; \boxed{\mathbf{w<-14}}$$

The solution set is $\{\boldsymbol{w} \mid \boldsymbol{w} < -\mathbf{14}\}$.

Check 1: Let $w=-20$ which is less than -14. Then, is $-8 \overset{?}{>} -20+6$; $-8 \overset{?}{>} -14$; $\frac{-8}{-1} \overset{?}{<} \frac{-14}{-1}$? The answer is "yes" because $8 < 14$.

Check 2: Let $w=10$ which is greater than -14. Then, is $-8 \overset{?}{>} 10+6$; $-8 \overset{?}{>} 16$? The answer is "no" because we can not choose $w=10$ since $10 > -14$.

Example 2.5-43

$\boxed{\frac{x}{4}-7>5}$; $\boxed{\frac{x}{4}-7+7>5+7}$; $\boxed{\frac{x}{4}+0>12}$; $\boxed{\frac{x}{4}>12}$; $\boxed{4\cdot\frac{x}{4}>12\cdot 4}$; $\boxed{\mathbf{x>48}}$

The solution set is $\{x \mid x > 48\}$.

Check 1: Let $x=50$ which is greater than 48. Then, is $\frac{50}{4}-7 \overset{?}{>} 5$; $12.5-7 \overset{?}{>} 5$? The answer is "yes" because $5.5 > 5$.

Check 2: Let $x=12$ which is less than 48. Then, is $\frac{12}{4}-7 \overset{?}{>} 5$; $3-7 \overset{?}{>} 5$; $-4 \overset{?}{>} 5$. The answer is "no" because we can not choose $x=12$ since $12 < 48$.

Example 2.5-44

$\boxed{2\frac{3}{5}>k+2\frac{1}{4}}$; $\boxed{\frac{(2\cdot 5)+3}{5}>k+\frac{(2\cdot 4)+1}{4}}$; $\boxed{\frac{10+3}{5}>k+\frac{8+1}{4}}$; $\boxed{\frac{13}{5}>k+\frac{9}{4}}$; $\boxed{2.6>k+2.3}$; $\boxed{2.6-k>k-k+2.3}$

; $\boxed{2.6-k>0+2.3}$; $\boxed{2.6-k>2.3}$; $\boxed{2.6-2.6-k>2.3-2.6}$; $\boxed{0-k>-0.3}$; $\boxed{-k>-0.3}$; $\boxed{\frac{-k}{-1}<\frac{-0.3}{-1}}$; $\boxed{\mathbf{k<0.3}}$

The solution set is $\{k \mid k < 0.3\}$.

Check: Let $k=0.1$ which is less than 0.3. Then, is $2\frac{3}{5} \overset{?}{>} 0.1+2\frac{1}{4}$; $\frac{(2\cdot 5)+3}{5} \overset{?}{>} 0.1+\frac{(2\cdot 4)+1}{4}$; $\frac{10+3}{5} \overset{?}{>} 0.1+\frac{8+1}{4}$; $\frac{13}{5} \overset{?}{>} 0.1+\frac{9}{4}$; $2.6 \overset{?}{>} 0.1+2.3$? The answer is "yes" because $2.6 > 2.4$.

Example 2.5-45

$\boxed{-2.8 \ge u+2\frac{3}{4}}$; $\boxed{-2.8-u \ge u-u+\frac{(2\cdot 4)+3}{4}}$; $\boxed{-2.8-u \ge 0+\frac{8+3}{4}}$; $\boxed{-2.8-u \ge \frac{11}{4}}$; $\boxed{-2.8-u \ge 2.75}$

; $\boxed{-2.8+2.8-u \ge 2.75+2.8}$; $\boxed{0-u \ge 5.55}$; $\boxed{-u \ge 5.55}$; $\boxed{\frac{-u}{-1} \le \frac{5.55}{-1}}$; $\boxed{\mathbf{u \le -5.55}}$

The solution set is $\{u \mid u \le -5.55\}$.

Check 1: Let $u=-5.55$. Then, is $-2.8 \overset{?}{=} -5.55+2\frac{3}{4}$; $-2.8 \overset{?}{=} -5.55+\frac{(2\cdot 4)+3}{4}$; $-2.8 \overset{?}{=} -5.55+\frac{8+3}{4}$; $-2.8 \overset{?}{=} -5.55+\frac{11}{4}$; $-2.8 \overset{?}{=} -5.55+2.75$? The answer is "yes" because $-2.8=-2.8$.

Check 2: Let $u=-10$ which is less than -5.55. Then, is $-2.8 \overset{?}{>} -10+2\frac{3}{4}$; $-2.8 \overset{?}{>} -10+\frac{(2\cdot 4)+3}{4}$; $-2.8 \overset{?}{>} -10+\frac{8+3}{4}$; $-2.8 \overset{?}{>} -10+\frac{11}{4}$; $-2.8 \overset{?}{>} -10+2.75$? The answer is "yes" because $-2.8 > -7.25$ or $\frac{-2.8}{-1} < \frac{-7.25}{-1}$; $2.8 < 7.25$.

Practice Problems - Mixed Operations Involving Linear Inequalities

Section 2.5 Case III Practice Problems - Solve the following linear inequalities by applying the addition, subtraction, multiplication, and division rules:

1. $-2x-9>9x-20$
2. $15x+3\leq 20x$
3. $-4x+5<10$
4. $-12t+4>4t-8$
5. $-4w-5\geq 8w+17$
6. $10y-4<4y-12$
7. $\frac{y}{3}-5\frac{2}{3}>1\frac{3}{5}$
8. $3\frac{4}{5}<t+2\frac{1}{3}$
9. $-3.4\geq w-2\frac{3}{5}$
10. $0.48x+2.5<1.5x-0.35$

Chapter 3
Factoring Polynomials

Quick Reference to Chapter 3 Case Problems

Chapter 3 - Factoring Polynomials

The primary objective of this chapter is to teach students different methods in factoring polynomials. (To review different classification of polynomials see Chapter 1, Section 1.5a, Case I.) In Section 3.1 monomial, binomial, and polynomial expressions are factored using the Greatest Common Factoring method. Factoring by using the Grouping method is addressed in Section 3.2. How polynomials are factored using a technique known as the Trial and Error method is discussed in Section 3.3. Section 3.4 introduces the factoring methods for polynomials with square and cubed terms. The steps as to how trinomials are factored using the Perfect Square Trinomial method as well as factorization of various other types of polynomials are discussed in Section 3.5. Cases presented in each section are concluded by solving additional examples with practice problems to further enhance the students ability. Additional methods for solving quadratic equations and presenting the solutions in factored form is discussed in Chapter 4.

3.1 Factoring Polynomials Using the Greatest Common Factoring Method

Solving algebraic fractions, which are introduced in Chapter 5, requires thorough knowledge of the factoring and solution methods that are introduced in this and the following chapter. Therefore, it is essential that students learn how to factor polynomials of second or higher degrees. For example, simplification of math operations such as:

$$\frac{x-y}{2x-y}\cdot\frac{2x^2+xy-y^2}{2y^2-3y+x^2}\;;\qquad \frac{x^2-y^2}{2x^2+5xy+3y^2}\cdot\frac{4x^2+4xy-3y^2}{5x-5y}\;;\qquad \frac{x^2-9}{x^3-5x^2+6x}\cdot\frac{16x^2}{8x+24}$$

require familiarization with various factoring methods. It is recommended that students spend adequate time to learn the different factoring and solution methods presented in this and the following chapter. In this section students are introduced to factoring the Greatest Common Factor to: monomial terms (Case I) and binomial and polynomial terms (Case II).

Case I Factoring the Greatest Common Factor to Monomial Terms

Factoring a polynomial means writing the polynomial as a product of two or more simpler polynomials. One method in factoring polynomials is by using the Greatest Common Factoring method where the Greatest Common Factor (G.C.F.) is factored out. The Greatest Common Factor to monomial terms is found using the following steps:

Step 1 a. Write the numbers and the variables in their prime factored form.

b. Identify the prime numbers and variables that are common in monomials.

Step 2 Multiply the common prime numbers and variables to obtain the G.C.F.

The following examples show the steps as to how monomial expressions are factored using the Greatest Common Factoring method:

Example 3.1-1

Find the G.C.F. to $6x^2$ and $8x^3$.

Solution:

Step 1 $\boxed{6x^2} = \boxed{2\cdot 3\cdot x\cdot x}$ and

$\boxed{8x^3} = \boxed{2 \cdot 4 \cdot x \cdot x^2} = \boxed{2 \cdot 2 \cdot 2 \cdot x \cdot x \cdot x}$

Therefore, the common terms are 2, x, and x.

Step 2 G.C.F. $= \boxed{2 \cdot x \cdot x} = \boxed{\mathbf{2x^2}}$

Example 3.1-2

Find the G.C.F. to $16y^4$ and $24y^3$.

Solution:

Step 1 $\boxed{16y^4} = \boxed{2 \cdot 8 \cdot y^2 \cdot y^2} = \boxed{2 \cdot 2 \cdot 4 \cdot y \cdot y \cdot y \cdot y} = \boxed{2 \cdot 2 \cdot 2 \cdot 2 \cdot y \cdot y \cdot y \cdot y}$ and

$\boxed{24y^3} = \boxed{3 \cdot 8 \cdot y \cdot y^2} = \boxed{3 \cdot 2 \cdot 4 \cdot y \cdot y \cdot y} = \boxed{3 \cdot 2 \cdot 2 \cdot 2 \cdot y \cdot y \cdot y}$

Therefore, the common terms are 2, 2, 2, y, y and y.

Step 2 G.C.F. $= \boxed{2 \cdot 2 \cdot 2 \cdot y \cdot y \cdot y} = \boxed{\mathbf{8y^3}}$

Example 3.1-3

Find the G.C.F. to $8x^2yz^3$, $3x^2y^2$ and $24xy^3z$.

Solution:

Step 1 $\boxed{8x^2yz^3} = \boxed{2 \cdot 4 \cdot x \cdot x \cdot y \cdot z \cdot z^2} = \boxed{2 \cdot 2 \cdot 2 \cdot x \cdot x \cdot y \cdot z \cdot z \cdot z}$

$\boxed{3x^2y^2} = \boxed{3 \cdot x \cdot x \cdot y \cdot y}$ and

$\boxed{24xy^3z} = \boxed{3 \cdot 8 \cdot x \cdot y \cdot y^2 \cdot z} = \boxed{3 \cdot 2 \cdot 4 \cdot x \cdot y \cdot y \cdot y \cdot z} = \boxed{3 \cdot 2 \cdot 2 \cdot 2 \cdot x \cdot y \cdot y \cdot y \cdot z}$

Therefore, the common terms are x and y.

Step 2 G.C.F. $= \boxed{x \cdot y} = \boxed{\mathbf{xy}}$

Example 3.1-4

Find the G.C.F. to $27x^2y^3$, $9x^3y$, and $15xy^2$.

Solution:

Step 1 $\boxed{27x^2y^3} = \boxed{3 \cdot 9 \cdot x \cdot x \cdot y \cdot y^2} = \boxed{3 \cdot 3 \cdot 3 \cdot x \cdot x \cdot y \cdot y \cdot y}$

$\boxed{9x^3y} = \boxed{3 \cdot 3 \cdot x \cdot x^2 \cdot y} = \boxed{3 \cdot 3 \cdot x \cdot x \cdot x \cdot y}$ and

$\boxed{15xy^2} = \boxed{3 \cdot 5 \cdot x \cdot y \cdot y}$

Therefore, the common terms are 3, x and y.

Step 2 G.C.F. $= \boxed{3 \cdot x \cdot y} = \boxed{\mathbf{3xy}}$

Example 3.1-5

Find the G.C.F. to $32a^3b^3$, $46ab^2$, and $56a^2b^4$.

Solution:

Step 1

$\boxed{32a^3b^3} = \boxed{4 \cdot 8 \cdot a \cdot a^2 \cdot b \cdot b^2} = \boxed{2 \cdot 2 \cdot 2 \cdot 4 \cdot a \cdot a \cdot a \cdot b \cdot b \cdot b} = \boxed{2 \cdot 2 \cdot 2 \cdot 2 \cdot 2 \cdot a \cdot a \cdot a \cdot b \cdot b \cdot b}$

$\boxed{46ab^2} = \boxed{3 \cdot 12 \cdot a \cdot b \cdot b} = \boxed{3 \cdot 3 \cdot 4 \cdot a \cdot b \cdot b} = \boxed{3 \cdot 3 \cdot 2 \cdot 2 \cdot a \cdot b \cdot b}$ and

$\boxed{56a^2b^4} = \boxed{7 \cdot 8 \cdot a \cdot a \cdot b^2 \cdot b^2} = \boxed{7 \cdot 2 \cdot 4 \cdot a \cdot a \cdot b \cdot b \cdot b \cdot b} = \boxed{7 \cdot 2 \cdot 2 \cdot 2 \cdot a \cdot a \cdot b \cdot b \cdot b \cdot b}$

Therefore, the common terms are 2, 2, a, b, and b.

Step 2

G.C.F. $= \boxed{2 \cdot 2 \cdot a \cdot b \cdot b} = \boxed{\mathbf{4ab^2}}$

Additional Examples - Factoring the Greatest Common Factor to Monomial Terms

The following examples further illustrate how to find the Greatest Common Factor to monomial terms:

Example 3.1-6

Find the G.C.F. to r^3s^3, $8rs^2$, and $9r^2s$.

1. $\boxed{r^3s^3} = \boxed{r \cdot r^2 \cdot s \cdot s^2} = \boxed{r \cdot r \cdot r \cdot s \cdot s \cdot s}$
2. $\boxed{8rs^2} = \boxed{2 \cdot 4 \cdot r \cdot s \cdot s} = \boxed{2 \cdot 2 \cdot 2 \cdot r \cdot s \cdot s}$
3. $\boxed{9r^2s} = \boxed{3 \cdot 3 \cdot r \cdot r \cdot s}$

Therefore, the common terms are r and s. Thus, G.C.F. $= \boxed{r \cdot s} = \boxed{\mathbf{rs}}$

Example 3.1-7

Find the G.C.F. to $48xy^2$, $16x^2y$, $4x^3y^2$, and $12xy$.

1. $\boxed{48xy^2} = \boxed{12 \cdot 4 \cdot x \cdot y \cdot y} = \boxed{6 \cdot 2 \cdot 2 \cdot 2 \cdot x \cdot y \cdot y} = \boxed{3 \cdot 2 \cdot 2 \cdot 2 \cdot 2 \cdot x \cdot y \cdot y}$
2. $\boxed{16x^2y} = \boxed{2 \cdot 8 \cdot x \cdot x \cdot y} = \boxed{2 \cdot 2 \cdot 4 \cdot x \cdot x \cdot y} = \boxed{2 \cdot 2 \cdot 2 \cdot 2 \cdot x \cdot x \cdot y}$
3. $\boxed{4x^3y^2} = \boxed{2 \cdot 2 \cdot x \cdot x^2 \cdot y \cdot y} = \boxed{2 \cdot 2 \cdot x \cdot x \cdot x \cdot y \cdot y}$
4. $\boxed{12xy} = \boxed{3 \cdot 4 \cdot x \cdot y} = \boxed{3 \cdot 2 \cdot 2 \cdot x \cdot y}$

Therefore, the common terms are 2, 2, x, and y. Thus, G.C.F. $= \boxed{2 \cdot 2 \cdot x \cdot y} = \boxed{\mathbf{4xy}}$

Example 3.1-8

Find the G.C.F. to $55u^2w^3z$, $50uw^2z^2$, and $15u^3w$.

1. $\boxed{55u^2w^3z} = \boxed{5 \cdot 11 \cdot u \cdot u \cdot w \cdot w^2 \cdot z} = \boxed{5 \cdot 11 \cdot u \cdot u \cdot w \cdot w \cdot w \cdot z}$

2. $\boxed{50uw^2z^2} = \boxed{5 \cdot 10 \cdot u \cdot w \cdot w \cdot z \cdot z} = \boxed{5 \cdot 5 \cdot 2 \cdot u \cdot w \cdot w \cdot z \cdot z}$

3. $\boxed{15u^3w} = \boxed{5 \cdot 3 \cdot u \cdot u^2 \cdot w} = \boxed{5 \cdot 3 \cdot u \cdot u \cdot u \cdot w}$

Therefore, the common terms are 5, u, and w. Thus, G.C.F. $= \boxed{5 \cdot u \cdot w} = \boxed{\mathbf{5uw}}$

Example 3.1-9

Find the G.C.F. to $27abc$, $36a^2b^2c^3$, and $24ac^2$.

1. $\boxed{27abc} = \boxed{9 \cdot 3 \cdot a \cdot b \cdot c} = \boxed{3 \cdot 3 \cdot 3 \cdot a \cdot b \cdot c}$

2. $\boxed{36a^2b^2c^3} = \boxed{2 \cdot 18 \cdot a \cdot a \cdot b \cdot b \cdot c \cdot c^2} = \boxed{2 \cdot 2 \cdot 9 \cdot a \cdot a \cdot b \cdot b \cdot c \cdot c \cdot c} = \boxed{2 \cdot 2 \cdot 3 \cdot 3 \cdot a \cdot a \cdot b \cdot b \cdot c \cdot c \cdot c}$

3. $\boxed{24ac^2} = \boxed{3 \cdot 8 \cdot a \cdot c \cdot c} = \boxed{3 \cdot 2 \cdot 4 \cdot a \cdot c \cdot c} = \boxed{3 \cdot 2 \cdot 2 \cdot 2 \cdot a \cdot c \cdot c}$

Therefore, the common terms are 3, a, and c. Thus, G.C.F. $= \boxed{3 \cdot a \cdot c} = \boxed{\mathbf{3ac}}$

Example 3.1-10

Find the G.C.F. to $12x$, $60xy^2$, and $63x^2$.

1. $\boxed{12x} = \boxed{3 \cdot 4 \cdot x} = \boxed{3 \cdot 2 \cdot 2 \cdot x}$

2. $\boxed{60xy^2} = \boxed{4 \cdot 15 \cdot x \cdot y \cdot y} = \boxed{2 \cdot 2 \cdot 5 \cdot 3 \cdot x \cdot y \cdot y}$

3. $\boxed{63x^2} = \boxed{3 \cdot 21 \cdot x \cdot x} = \boxed{3 \cdot 3 \cdot 7 \cdot x \cdot x}$

Therefore, the common terms are 3 and x. Thus, G.C.F. $= \boxed{3 \cdot x} = \boxed{\mathbf{3x}}$

Practice Problems - Factoring the Greatest Common Factor to Monomial Terms

Section 3.1 Case I Practice Problems - Find the Greatest Common Factor to the following monomial terms:

1. $5x^3$ and $15x$
2. $18x^2y^3z^4$ and $24xy^4z^5$
3. $16a^2bc^3$, $38ab^4c^2$, and $6a^3bc$
4. r^5s^4, $4r^3s^2$, and $3rs$
5. $10u^2vw^3$, $2uv^3w^2$, and uv^2
6. $19a^3b^3$, $12ab^2$, and $6ab$
7. $30xyz$, $2x$, and z
8. $25p^2q^7$, $5pq$, and p^3
9. a^3b^3c and a^7b^7
10. z^5, xy^3z, and $3x^2y$

Case II Factoring the Greatest Common Factor to Binomial and Polynomial Terms

The concept of obtaining Greatest Common Factor can be extended to binomial expressions by obtaining the greater common monomial factor which is found by using the following steps:

Step 1 a. Write each monomial term in its prime factored form.

b. Identify the prime numbers and variables that are common to monomials.

c. Multiply the common prime numbers and variables to obtain the greatest common monomial factor.

Step 2 Factor out the greatest common monomial factor from the binomial expression.

The following examples show the steps as to how binomial expressions are factored:

Example 3.1-11

Factor $4x^3 + 8x^2$.

Solution:

Step 1

$\boxed{4x^3} = \boxed{2 \cdot 2 \cdot x \cdot x \cdot x}$ and

$\boxed{8x^2} = \boxed{2 \cdot 4 \cdot x \cdot x} = \boxed{2 \cdot 2 \cdot 2 \cdot x \cdot x}$

Therefore, the common terms are 2, 2, x, and x which implies that the greatest common monomial factor is $2 \cdot 2 \cdot x \cdot x = 4x^2$. Thus,

Step 2

$\boxed{4x^3 + 8x^2} = \boxed{\mathbf{4x^2(x+2)}}$

Example 3.1-12

Factor $9u^3w^2 - 18uw$.

Solution:

Step 1

$\boxed{9u^3w^2} = \boxed{3 \cdot 3 \cdot u \cdot u^2 \cdot w \cdot w} = \boxed{3 \cdot 3 \cdot u \cdot u \cdot u \cdot w \cdot w}$ and

$\boxed{18uw} = \boxed{2 \cdot 9 \cdot u \cdot w} = \boxed{2 \cdot 3 \cdot 3 \cdot u \cdot w}$

Therefore, the common terms are 3, 3, u, and w which implies that the greatest common monomial factor is $3 \cdot 3 \cdot u \cdot w = 9uw$. Thus,

Step 2

$\boxed{9u^3w^2 - 18uw} = \boxed{\mathbf{9uw(u^2w - 2)}}$

Example 3.1-13

Factor $6x^2y + 12x$.

Solution:

Step 1

$\boxed{6x^2y} = \boxed{2 \cdot 3 \cdot x \cdot x \cdot y}$ and

$\boxed{12x} = \boxed{2 \cdot 6 \cdot x} = \boxed{2 \cdot 2 \cdot 3 \cdot x}$

Therefore, the common terms are 2, 3, and x which implies that the greatest common monomial factor is $2 \cdot 3 \cdot x = 6x$. Thus,

Step 2 $\boxed{6x^2y + 12x} = \boxed{\mathbf{6x(xy+2)}}$

Example 3.1-14

Factor $6a^3b^2c^2 - 2a^2bc^2$.

Solution:

Step 1 $\boxed{6a^3b^2c^2} = \boxed{2 \cdot 3 \cdot a \cdot a \cdot a \cdot b \cdot b \cdot c \cdot c}$ and

$\boxed{2a^2bc^2} = \boxed{2 \cdot a \cdot a \cdot b \cdot c \cdot c}$

Therefore, the common terms are 2, a, a, b, c, and c which implies that the greatest common monomial factor is $2 \cdot a \cdot a \cdot b \cdot c \cdot c = 2a^2bc^2$. Thus,

Step 2 $\boxed{6a^3b^2c^2 - 2a^2bc^2} = \boxed{\mathbf{2a^2bc^2(3ab-1)}}$

Example 3.1-15

Factor $12x^3y^2z + 36x^2z^2$.

Solution:

Step 1 $\boxed{12x^3y^2z} = \boxed{4 \cdot 3 \cdot x \cdot x^2 \cdot y \cdot y \cdot z} = \boxed{2 \cdot 2 \cdot 3 \cdot x \cdot x \cdot x \cdot y \cdot y \cdot z}$ and

$\boxed{36x^2z^2} = \boxed{2 \cdot 18 \cdot x \cdot x \cdot z \cdot z} = \boxed{2 \cdot 2 \cdot 9 \cdot x \cdot x \cdot z \cdot z} = \boxed{2 \cdot 2 \cdot 3 \cdot 3 \cdot x \cdot x \cdot z \cdot z}$

Therefore, the common terms are 2, 2, 3, x, x, and z which implies that the greatest common monomial factor is $2 \cdot 2 \cdot 3 \cdot x \cdot x \cdot z = 12x^2z$. Thus,

Step 2 $\boxed{12x^3y^2z + 36x^2z^2} = \boxed{\mathbf{12x^2z(xy^2+3z)}}$

Additional Examples - Factoring the Greatest Common Factor to Binomial and Polynomial Terms

The following examples further illustrate how to find the greatest common monomial factor to binomial terms:

Example 3.1-16

Find the greatest common monomial factor to $35m^2n^3 + 5mn^2$.

1. $\boxed{35m^2n^3} = \boxed{5 \cdot 7 \cdot m \cdot m \cdot n \cdot n^2} = \boxed{5 \cdot 7 \cdot m \cdot m \cdot n \cdot n \cdot n}$

2. $\boxed{5mn^2} = \boxed{5 \cdot m \cdot n \cdot n}$

Therefore, the common terms are 5, m, n, and n which implies that the greatest common monomial factor is $5 \cdot m \cdot n \cdot n = 5mn^2$. Thus, $\boxed{35m^2n^3 + 5mn^2} = \boxed{\mathbf{5mn^2(7mn+1)}}$

Example 3.1-17

Find the greatest common monomial factor to $6a^2b + 66ab^4$.

1. $\boxed{6a^2b} = \boxed{2\cdot 3\cdot a\cdot a\cdot b}$

2. $\boxed{66ab^4} = \boxed{6\cdot 11\cdot a\cdot b^2\cdot b^2} = \boxed{2\cdot 3\cdot 11\cdot a\cdot b\cdot b\cdot b\cdot b}$

Therefore, the common terms are 2, 3, a, and b which implies that the greatest common monomial factor is $2\cdot 3\cdot a\cdot b = 6ab$. Thus, $\boxed{6a^2b + 66ab^4} = \boxed{\mathbf{6ab(a + 11b^3)}}$

Example 3.1-18

Find the greatest common monomial factor to $7p^3q^6 - 49p^2q^5$.

1. $\boxed{7p^3q^6} = \boxed{7\cdot p\cdot p^2\cdot q^3\cdot q^3} = \boxed{7\cdot p\cdot p\cdot p\cdot q\cdot q^2\cdot q\cdot q^2} = \boxed{7\cdot p\cdot p\cdot p\cdot q\cdot q\cdot q\cdot q\cdot q\cdot q}$

2. $\boxed{49p^2q^5} = \boxed{7\cdot 7\cdot p\cdot p\cdot q\cdot q^4} = \boxed{7\cdot 7\cdot p\cdot p\cdot q\cdot q^2\cdot q^2} = \boxed{7\cdot 7\cdot p\cdot p\cdot q\cdot q\cdot q\cdot q\cdot q}$

Therefore, the common terms are 7, p, p, q, q, q, q, and q which implies that the greatest common monomial factor is $7\cdot p\cdot p\cdot q\cdot q\cdot q\cdot q\cdot q = 7p^2q^5$. Thus, $\boxed{7p^3q^6 - 49p^2q^5} = \boxed{\mathbf{7p^2q^5(pq - 7)}}$

Example 3.1-19

Find the greatest common monomial factor to $48x - 20xy$.

1. $\boxed{48x} = \boxed{4\cdot 12\cdot x} = \boxed{2\cdot 2\cdot 4\cdot 3\cdot x} = \boxed{2\cdot 2\cdot 2\cdot 2\cdot 3\cdot x}$

2. $\boxed{20xy} = \boxed{4\cdot 5\cdot x\cdot y} = \boxed{2\cdot 2\cdot 5\cdot x\cdot y}$

Therefore, the common terms are 2, 2, and x which implies that the greatest common monomial factor is $2\cdot 2\cdot x = 4x$. Thus, $\boxed{48x - 20xy} = \boxed{\mathbf{4x(12 - 5y)}}$

Example 3.1-20

Find the greatest common monomial factor to $24x^3y^3 + 12x^2y^4$.

1. $\boxed{24x^3y^3} = \boxed{2\cdot 12\cdot x\cdot x^2\cdot y\cdot y^2} = \boxed{2\cdot 4\cdot 3\cdot x\cdot x\cdot x\cdot y\cdot y\cdot y} = \boxed{2\cdot 2\cdot 2\cdot 3\cdot x\cdot x\cdot x\cdot y\cdot y\cdot y}$

2. $\boxed{12x^2y^4} = \boxed{2\cdot 6\cdot x\cdot x\cdot y^2\cdot y^2} = \boxed{2\cdot 2\cdot 3\cdot x\cdot x\cdot y\cdot y\cdot y\cdot y}$

Therefore, the common terms are 2, 2, 3, x, x, y, y and y which implies that the greatest common monomial factor is $2\cdot 2\cdot 3\cdot x\cdot x\cdot y\cdot y\cdot y = 12x^2y^3$. Thus, $\boxed{24x^3y^3 + 12x^2y^4}$

$= \boxed{\mathbf{12x^2y^3(2x + y)}}$

Note 1: As one gains more proficiency in solving this class of problems the need for factoring each monomial term to its prime factored form lessens. Therefore, students may simplify a given binomial expression by mentally factoring out the common terms. For example, we can quickly factor out the expression $24x^4y^2z + 12x^2y^3z^2$ by observing that its greatest common monomial factor is $12x^2y^2z$, e.g., $24x^4y^2z + 12x^2y^3z^2 = 12x^2y^2z\left(2x^2 + yz\right)$.

Note 2: The process of factoring binomial expressions can further be expanded to include trinomials and polynomials. Following are few additional examples indicating how the greatest common monomial factor to polynomials is obtained:

1. $5x + 15x^2 + 50x^3 = 5x\left(1 + 3x + 10x^2\right)$
2. $24xy^2 + 15xy + 12y = 3y(8xy + 5x + 4)$
3. $8a^2b^3 + 4ab^2 - 2a^2b = 2ab\left(4ab^2 + 2b - a\right)$
4. $20u^2w^2 + 15u^3w^2 + 5uw = 5uw\left(4uw + 3u^2w + 1\right)$
5. $4x^3 + 6x^2 + 2x = 2x\left(2x^2 + 3x + 1\right)$
6. $20x^2y^2 - 5x^3y + 15y = 5y\left(4x^2y - x^3 + 3\right)$
7. $40x^3y^4z^2 - 12x^2y^2z^2 + 8x^2yz = 4x^2yz\left(10xy^3z - 3yz + 2\right)$

Practice Problems - Factoring the Greatest Common Factor to Binomial and Polynomial Terms

Section 3.1 Case II Practice Problems - Find the Greatest Common Factor to the following binomial and polynomial terms:

1. $18x^3y^3 - 12x^2y =$
2. $3a^2b^3c + 15ab^2c^3 =$
3. $xyz^3 + 4x^2y^2z^5 =$
4. $25p^3 + 5p^2q^3 + pq =$
5. $r^2s^2t - 5rst^2 =$
6. $36x^3yz^3 + 4xy^2z^4 - 12x^3y^3z =$
7. $17ab^2c^2d^3 - 7a^3bc + 8a^2cd^2 =$
8. $5pq^2r + 30p^2q^2 - 20p^3qr^3 =$
9. $9x^2y^2z - 3xyz =$
10. $7abc^2d + 49a^3b^3 - 14a^2b^3d =$

3.2 Factoring Polynomials Using the Grouping Method

In many instances polynomials with four or more terms have common terms that can be grouped together. This process is called factoring by grouping. The steps as to how polynomials are grouped are shown below:

Step 1 Factor the common variables, or numbers, from the monomial terms.

Step 2 Factor the common binomial factor obtained in step 1 by grouping.

Examples with Steps

The following examples show the steps as to how polynomials are factored using the grouping method:

Example 3.2-1

$$\boxed{3x+3y+ax+ay}=$$

Solution:

Step 1 $\boxed{3x+3y+ax+ay}=\boxed{(3x+ax)+(3y+ay)}=\boxed{x(3+a)+y(3+a)}$

Step 2 $\boxed{x(3+a)+y(3+a)}=\boxed{\mathbf{(3+a)(x+y)}}$

Example 3.2-2

$$\boxed{x^2-4x-9x+36}=$$

Solution:

Step 1 $\boxed{x^2-4x-9x+36}=\boxed{x(x-4)+9(x-4)}$

Step 2 $\boxed{x(x-4)+9(x-4)}=\boxed{\mathbf{(x-4)(x+9)}}$

Example 3.2-3

$$\boxed{u^2-2u-16u+32}=$$

Solution:

Step 1 $\boxed{u^2-2u-16u+32}=\boxed{u(u-2)-16(u-2)}$

Step 2 $\boxed{u(u-2)-16(u-2)}=\boxed{\mathbf{(u-2)(u-16)}}$

Example 3.2-4

$$\boxed{60m^2+24m-15m-6}=$$

Solution:

Step 1 $\boxed{60m^2+24m-15m-6}=\boxed{12m(5m+2)-3(5m+2)}$

Step 2 $\boxed{12m(5m+2)-3(5m+2)}=\boxed{(5m+2)(12m-3)}=\boxed{\mathbf{4(5m+2)(3m-1)}}$

Example 3.2-5

$$\boxed{15y^3+25y^2+9y+15}=$$

Solution:

Step 1 $\boxed{15y^2+25y+9y+15}=\boxed{5y(3y+5)+3(3y+5)}$

Step 2 $\boxed{5y(3y+5)+3(3y+5)}=\boxed{(\mathbf{3y+5})(\mathbf{5y+3})}$

Additional Examples - Factoring Polynomials Using the Grouping Method

The following examples further illustrate how to factor polynomials using the grouping method:

Example 3.2-6

$$\boxed{x^2+3x-2x-6}=\boxed{x(x+3)-2(x+3)}=\boxed{(x+3)(x-2)}$$

Example 3.2-7

$$\boxed{ax^2+bx^2+axy+bxy}=\boxed{x^2(a+b)+xy(a+b)}=\boxed{(a+b)\left(x^2+xy\right)}=\boxed{x(a+b)(x+y)}$$

Example 3.2-8

$$\boxed{5(x+y)^2+15x+15y}=\boxed{5(x+y)^2+15(x+y)}=\boxed{5(x+y)[(x+y)+3]}$$

Example 3.2-9

$$\boxed{8(a+b)^2+4(a+b)^3+2(a+b)}=\boxed{2(a+b)\left[4(a+b)+2(a+b)^2+1\right]}=\boxed{2(a+b)\{2(a+b)[2+(a+b)]+1\}}$$

Example 3.2-10

$$\boxed{-x^2+2x-3x+6}=\boxed{x(-x+2)+3(-x+2)}=\boxed{(-x+2)(x+3)}$$

Example 3.2-11

$$\boxed{3ab-7b-3a+7}=\boxed{b(3a-7)-(3a+7)}=\boxed{(3a-7)(b-1)}$$

Example 3.2-12

$$\boxed{x^3+5x^2+x+5}=\boxed{x^2(x+5)+x+5}=\boxed{x^2(x+5)+(x+5)}=\boxed{(x+5)\left(x^2+1\right)}$$

Example 3.2-13

$$\boxed{y^3+3y^2+4y+12}=\boxed{y^2(y+3)+4(y+3)}=\boxed{(y+3)\left(y^2+4\right)}$$

Example 3.2-14

$$\boxed{24x^2y-12xy-36x+18}=\boxed{12xy(2x-1)-18(2x-1)}=\boxed{(2x-1)(12xy-18)}=\boxed{6(2x-1)(2xy-3)}$$

Example 3.2-15

$$\boxed{12r^3s-6r^2s+4r-2}=\boxed{6r^2s(2r-1)+2(2r-1)}=\boxed{(2r-1)\left(6r^2s+2\right)}=\boxed{2(2r-1)\left(3r^2s+1\right)}$$

In the following sections, additional factoring methods are introduced. These methods are used to present polynomials in their equivalent factored form. Students are encouraged to spend adequate time learning each method.

Practice Problems - Factoring Polynomials Using the Grouping Method

Section 3.2 Practice Problems - Factor the following polynomials using the grouping method:

1. $2ab - 5b - 6a + 15 =$
2. $y^3 + 4y^2 + y + 4 =$
3. $42x^2y + 21xy - 70x - 35 =$
4. $(x+y)^3 + (x+y)^2 + x + y =$
5. $4(a+b)^2 + 32a + 32b =$
6. $36r^3s - 6r^2s + 18r - 3 =$
7. $3u^2 + 7u + 2 =$
8. $25(x+y)^3 + 5(x+y) =$
9. $ax^3 + bx^3 - ax^2y^2 - bx^2y^2 =$
10. $6r^3s^2 + 6r + 9r^2s^2 + 9 =$

3.3 Factoring Polynomials Using the Trail and Error Method

Expressing trinomials as the product of two binomials is one of the most common ways of factoring. In this section, we will learn how to factor trinomials of the form $ax^2 + bx + c$, where $a = 1$ (Case I) and where $a \rangle 1$ (Case II), using a factoring method which in this book is referred to as the Trial and Error method.

Case I Factoring Trinomials of the Form $ax^2 + bx + c$ where $a = 1$

To express a trinomial of the form $ax^2 + bx + c$, where $a = 1$, in its factored form $(x+m)(x+n)$, let us consider the product $(x+m)(x+n)$ and use the FOIL method to see how each term of the resulting trinomial is formed , i.e.,

$$(x+m)(x+n) = x \cdot x + n \cdot x + m \cdot x + m \cdot n = x^2 + (m+n)x + mn$$

Note that the coefficient of the x term is the **sum** of m and n and the constant term is the **product** of m and n. We use this concept in order to express trinomials of the form $x^2 + bx + c$ in their equivalent factored form. In addition, in order to choose the right sign for the two integer numbers m and n, the knowledge of the following general sign rules for the indicated cases is needed:

General Sign Rules

When factoring a trinomial of the form $x^2 + ax + b$ to its equivalent factored form of $(x+m)(x+n)$, the sign of the two integer numbers m and n is determined based on the following cases:

Case I. If the sum of the two integer numbers $(a+b)$ is positive $(+)$ and the product of the two integer numbers $(a \cdot b)$ is negative $(-)$, then the two integer numbers m and n must have opposite signs. See examples 3.3-2, 3.3-7, 3.3-11, 3.3-12, 3.3-15, and 3.3-19.

Case II. If the sum of the two integer numbers $(a+b)$ is negative $(-)$ and the product of the two integer numbers $(a \cdot b)$ is positive $(+)$, then the two numbers must have the same sign. However, since the sum is negative, the two integer numbers m and n must both be negative. See examples 3.3-1, 3.3-14, and 3.3-18.

Case III. If the sum of the two integer numbers $(a+b)$ is positive $(+)$ and the product of the two integer numbers $(a \cdot b)$ is also positive $(+)$, then the two integer numbers m and n must both be positive. See examples 3.3-5, 3.3-6, 3.3-9, and 3.3-13.

Case IV. If the sum of the two integer numbers $(a+b)$ is negative $(-)$ and the product of the two integer numbers $(a \cdot b)$ is also negative $(-)$, then the two integer numbers m and n must have opposite signs. See examples 3.3-3, 3.3-8, 3.3-16, 3.3-17, and 3.3-20.

General Sign Rules - Summary

If sign of the sum $(a+b)$ is	**and sign of the product $(a \cdot b)$ is**	**then, the two integer numbers m and n must**
+	−	have opposite signs
−	+	have negative signs
+	+	have positive signs
−	−	have opposite signs

To factor a trinomial of the form $x^2 + ax + b$ to its equivalent factored form of $(x+m)(x+n)$ use the following steps:

Step 1 Obtain two numbers m and n whose sum equals to a and whose product equals to b.

Step 2 Write the trinomial in its factored form. Check the answer by using the FOIL method.

Examples with Steps

The following examples show the steps as to how trinomials of the form $x^2 + ax + b$ are factored:

Example 3.3-1

Factor $x^2 - 16x + 55$.

Solution:

Step 1 Obtain two numbers whose sum is -16 and whose product is 55. Note that since the sum is negative and the product is positive the two integer numbers must both be negative (Case II). Let's construct a table as follows:

Sum	***Product***
$-15-1=-16$	$(-15)\cdot(-1)=15$
$-14-2=-16$	$(-14)\cdot(-2)=28$
$-13-3=-16$	$(-13)\cdot(-3)=39$
$-12-4=-16$	$(-12)\cdot(-4)=48$
$\mathbf{-11-5=-16}$	$\mathbf{(-11)\cdot(-5)=55}$

Step 2 The last line contains the sum and the product of the two numbers that we need. Therefore, $x^2 - 16x + 55 = \mathbf{(x-11)(x-5)}$

Check: $(x-11)(x-5) = x \cdot x - 5 \cdot x - 11 \cdot x + (-11)(-5) = x^2 + (-5-11)x + 55 = x^2 - 16x + 55$

Example 3.3-2

Factor $x^2 + 2x - 48$.

Solution:

Step 1 Obtain two numbers whose sum is $+2$ and whose product is -48. Note that since the sum is positive and the product is negative the two integer numbers must have opposite signs (Case I). Let's construct a table as follows:

Sum	*Product*
$10-8=2$	$10\cdot(-8)=-80$
$9-7=2$	$9\cdot(-7)=-63$
$\mathbf{8-6=2}$	$\mathbf{8\cdot(-6)=-48}$

Step 2 The last line contains the sum and the product of the two numbers that we need. Therefore, $x^2+2x-48 = \mathbf{(x+8)(x-6)}$

Check: $(x+8)(x-6) = x\cdot x-6\cdot x+8\cdot x+8\cdot(-6) = x^2+(-6+8)x-48 = x^2+2x-48$

Example 3.3-3

Factor $x^2-6x-40$.

Solution:

Step 1 Obtain two numbers whose sum is -6 and whose product is -40. Note that since the sum and the product are both negative the two integer numbers must have opposite signs (Case IV). Let's construct a table as follows:

Sum	*Product*
$1-7=-6$	$1\cdot(-7)=-7$
$2-8=-6$	$2\cdot(-8)=-16$
$3-9=-6$	$3\cdot(-9)=-27$
$\mathbf{4-10=-6}$	$\mathbf{4\cdot(-10)=-40}$

Step 2 The last line contains the sum and the product of the two numbers that we need. Therefore, $x^2-6x-40 = \mathbf{(x+4)(x-10)}$

Check: $(x+4)(x-10) = x\cdot x-10\cdot x+4\cdot x+4\cdot(-10) = x^2+(4-10)x-40 = x^2-6x-40$

Example 3.3-4

Factor $y^2+4y+26$.

Solution:

Step 1 Obtain two numbers whose sum is 4 and whose product is 26. Note that since the sum and the product are both positive the two integer numbers must both be positive (Case III). Let's construct a table as follows:

Sum	*Product*
$1+3=4$	$1\cdot3=3$
$2+2=4$	$2\cdot2=4$

It is obvious that we can not find two positive integer numbers whose sum is 4 and whose product is 26. Therefore, we conclude that the trinomial $y^2+4y+26$ is ***not factorable*** using integers, or it is ***prime***.

Step 2 *Not Applicable*

Example 3.3-5

Factor $x^2+26x+169$.

Solution:

Step 1 Obtain two numbers whose sum is 26 and whose product is 169. Note that since the sum and the product are both positive the two integer numbers must both be positive (Case III). Let's construct a table as follows:

Sum	*Product*
$10+16=26$	$10\cdot 16=160$
$11+15=26$	$11\cdot 15=165$
$12+14=26$	$12\cdot 14=168$
$\mathbf{13+13=26}$	$\mathbf{13\cdot 13=169}$

Step 2 The last line contains the sum and the product of the two numbers that we need. Therefore, $x^2+26x+169 = \mathbf{(x+13)(x+13)}$

Check: $(x+13)(x+13) = x\cdot x+13\cdot x+13\cdot x+13\cdot 13 = x^2+(13+13)x+169 = x^2+26x+169$

Additional Examples - Factoring Trinomials of the Form ax^2+bx+c where $a=1$

The following examples further illustrate how to factor trinomials using the Trial and Error method:

Example 3.3-6: Factor $x^2+8x+15$.

Solution:

To factor the above trinomial we need to obtain two numbers whose sum is 8 and whose product is 15. Let's construct a table as follows:

Sum	*Product*
$1+7=8$	$1\cdot 7=7$
$2+6=8$	$2\cdot 6=12$
$\mathbf{3+5=8}$	$\mathbf{3\cdot 5=15}$

The last line contains the sum and the product of the two numbers that we need. Therefore, $x^2+8x+15 = \mathbf{(x+3)(x+5)}$

Check: $(x+3)(x+5) = x\cdot x+5\cdot x+3\cdot x+3\cdot 5 = x^2+5x+3x+15 = x^2+(5+3)x+15 = x^2+8x+15$

Example 3.3-7: Factor x^2+x-2.

Solution:

To factor the above trinomial we need to obtain two numbers whose sum is 1 and whose product is -2. Let's construct a table as follows:

Sum	*Product*
$0+1=1$	$0\cdot 1=0$
$\mathbf{1-2=-1}$	$\mathbf{1\cdot(-2)=-2}$

The last line contains the sum and the product of the two numbers that we need. Therefore, $x^2+x-2 = \mathbf{(x+1)(x-2)}$

Check: $(x+3)(x+5) = x\cdot x+5\cdot x+3\cdot x+3\cdot 5 = x^2+5x+3x+15 = x^2+(5+3)x+15 = x^2+8x+15$

Example 3.3-8: Factor $x^2 - x - 20$.

Solution:

To factor the above trinomial we need to obtain two numbers whose sum is -1 and whose product is -20. Let's construct a table as follows:

Sum	***Product***
$-2+1=-1$	$-2\cdot 1=-2$
$-3+2=-1$	$-3\cdot 2=-6$
$-4+3=-1$	$-4\cdot 3=-12$
$\mathbf{-5+4=-1}$	$\mathbf{-5\cdot 4=-20}$

The last line contains the sum and the product of the two numbers that we need. Thus, $x^2 - x - 20 = \mathbf{(x-5)(x+4)}$

Check: ~~$(x+3)(x+5) = x\cdot x + 5\cdot x + 3\cdot x + 3\cdot 5 = x^2 + 5x + 3x + 15 = x^2 + (5+3)x + 15 = x^2 + 8x + 15$~~

Example 3.3-9: Factor $w^2 + 9w + 20$.

Solution:

To factor the above trinomial we need to obtain two numbers whose sum is 9 and whose product is 20. Let's construct a table as follows:

Sum	***Product***
$1+8=9$	$1\cdot 8=8$
$2+7=9$	$2\cdot 7=14$
$3+6=9$	$3\cdot 6=18$
$\mathbf{4+5=9}$	$\mathbf{4\cdot 5=20}$

The last line contains the sum and the product of the two numbers that we need. Thus, $w^2 + 9w + 20 = \mathbf{(w+5)(w+4)}$

Check: $(w+5)(w+4) = w\cdot w + 4\cdot w + 5\cdot w + 4\cdot 5 = w^2 + 4w + 5w + 20 = w^2 + (4+5)w + 20 = w^2 + 9w + 20$

Example 3.3-10: Factor $x^2 - 10x + 12$.

Solution:

To factor the above trinomial we need to obtain two numbers whose sum is -10 and whose product is 12. Let's construct a table as follows:

Sum	***Product***
$-1-9=-10$	$(-1)\cdot(-9)=9$
$-2-8=-10$	$(-2)\cdot(-8)=16$
$-3-7=-10$	$(-3)\cdot(-7)=21$
$-4-6=-10$	$(-4)\cdot(-6)=24$
$-5-5=-10$	$(-5)\cdot(-5)=25$
$-1-12=-13$	$(-1)\cdot(-12)=12$
$-3-4=--7$	$(-3)\cdot(-4)=12$
$-2-6=-8$	$(-2)\cdot(-6)=12$

Since non of the numbers that add to the sum of -10, when multiplied, has a product of 12 and none of the factors of 12, when added, has a sum of -10. Therefore, we conclude that the trinomial $x^2 - 10x + 12$ is **not factorable** using integers, or it is **prime**.

Note: A prime polynomial is one that is not factorable using integers. For example, $4x^2 + 6x + 9$, $2y^2 + y + 7$, $6w^2 + 2w - 5$, $x^2 + 7y^2$, $y^2 - 6y + 2$, *and* $4x^2 + 9$ *are* ***prime*** *polynomials.*

Example 3.3-11: Factor $x^2 + x - 12$.

Solution:

To factor the above trinomial we need to obtain two numbers whose sum is 1 and whose product is -12. Let's construct a table as follows:

Sum	*Product*
$2-1=1$	$2\cdot(-1)=-2$
$3-2=1$	$3\cdot(-2)=-6$
$\mathbf{4-3=1}$	$\mathbf{4\cdot(-3)=-12}$

The last line contains the sum and the product of the two numbers that we need. Therefore, $x^2 + x - 12 = \mathbf{(x+4)(x-3)}$

Check: $(x+4)(x-3) = x \cdot x - 3 \cdot x + 4 \cdot x + 4 \cdot (-3) = x^2 - 3x + 4x - 12 = x^2 + (4-3)x - 12 = x^2 + x - 12$

Example 3.3-12: Factor $x^2 + 4x - 5$.

Solution:

To factor the above trinomial we need to obtain two numbers whose sum is 4 and whose product is -5. Let's construct a table as follows:

Sum	*Product*
$\mathbf{5-1=4}$	$\mathbf{5\cdot(-1)=-5}$

In this case, at first trial we obtained the sum and the product of the two numbers that we need. Thus, $x^2 + 4x - 5 = \mathbf{(x+5)(x-1)}$

Check: $(x+5)(x-1) = x \cdot x - 1 \cdot x + 5 \cdot x + 5 \cdot (-1) = x^2 - x + 5x - 5 = x^2 + (5-1)x - 5 = x^2 + 4x - 5$

Example 3.3-13: Factor $x^2 + 10x + 16$.

Solution:

To factor the above trinomial we need to obtain two numbers whose sum is 10 and whose product is 16. Let's construct a table as follows:

Sum	*Product*
$1+9=10$	$1\cdot 9=9$
$\mathbf{2+8=10}$	$\mathbf{2\cdot 8=16}$
$3+7=10$	$3\cdot 7=21$

The middle line contains the sum and the product of the two numbers that we need. Thus, $x^2 + 10x + 16 = \mathbf{(x+2)(x+8)}$

Check: $(x+2)(x+8) = x \cdot x + 8 \cdot x + 2 \cdot x + 2 \cdot 8 = x^2 + 8x + 2x + 16 = x^2 + (8+2)x + 16 = x^2 + 10x + 16$

Example 3.3-14: Factor $w^2 - 15w + 56$.

Solution:

To factor the above trinomial we need to obtain two numbers whose sum is -15 and whose product is 56. Let's construct a table as follows:

Sum	*Product*
$-2-13=-15$	$(-2)\cdot(-13)=26$
$-3-12=-15$	$(-3)\cdot(-12)=36$
$-4-11=-15$	$(-4)\cdot(-11)=44$
$-5-10=-15$	$(-5)\cdot(-10)=50$
$-6-9=-15$	$(-6)\cdot(-9)=54$
$\mathbf{-7-8=-15}$	$\mathbf{(-7)\cdot(-8)=56}$

The last line contains the sum and the product of the two numbers that we need. Thus, $w^2-15w+56 = \mathbf{(w-7)(w-8)}$

Check: $(w-7)(w-8) = w\cdot w-8\cdot w-7\cdot w+(-7)\cdot(-8) = w^2-8w-7w+56 = w^2+(-8-7)w+56$
$= w^2-15x+56$

Example 3.3-15: Factor $y^2+3y-108$.

Solution:

To factor the above trinomial we need to obtain two numbers whose sum is 3 and whose product is -108. Let's construct a table as follows:

Sum	*Product*
$8-5=3$	$8\cdot(-5)=-40$
$9-6=3$	$9\cdot(-6)=-54$
$10-7=3$	$10\cdot(-7)=-70$
$11-8=3$	$11\cdot(-8)=-88$
$\mathbf{12-9=3}$	$\mathbf{12\cdot(-9)=-108}$

The last line contains the sum and the product of the two numbers that we need. Thus, $y^2+3y-108 = \mathbf{(y+12)(y-9)}$

Check: $(y+12)(y-9) = y\cdot y+(-9)\cdot y+12\cdot y+12\cdot(-9) = y^2-9y+12y-108 = y^2+(-9+12)y-108$
$= y^2+3y-108$

Example 3.3-16: Factor $x^2-19x-66$.

Solution:

To factor the above trinomial we need to obtain two numbers whose sum is -19 and whose product is -66. Let's construct a table as follows:

Sum	*Product*
$10-29=-19$	$10\cdot(-29)=-290$
$9-28=-19$	$9\cdot(-28)=-252$
$8-27=-19$	$8\cdot(-27)=-216$
$7-26=-19$	$7\cdot(-26)=-182$
$6-25=-19$	$6\cdot(-25)=-150$
$5-24=-19$	$5\cdot(-24)=-120$
$4-23=-19$	$4\cdot(-23)=-92$
$\mathbf{3-22=-19}$	$\mathbf{3\cdot(-22)=-66}$

The last line contains the sum and the product of the two numbers that we need. Thus, $x^2 - 19x - 66 = \mathbf{(x+3)(x-22)}$

Check: $(x+3)(x-22) = x \cdot x + (-22) \cdot x + 3 \cdot x + 3 \cdot (-22) = x^2 - 22x + 3x - 66 = x^2 + (-22+3)x - 66$
$= x^2 - 19x - 66$

Example 3.3-17: Factor $x^2 - 16x - 80$.

Solution:

To factor the above trinomial we need to obtain two numbers whose sum is -16 and whose product is -80. Let's construct a table as follows:

Sum	*Product*
$1-17=-16$	$1 \cdot (-17) = -17$
$2-18=-16$	$2 \cdot (-18) = -36$
$3-19=-16$	$3 \cdot (-19) = -57$
$\mathbf{4-20=-16}$	$\mathbf{4 \cdot (-20) = -80}$

The last line contains the sum and the product of the two numbers that we need. Thus, $x^2 - 16x - 80 = \mathbf{(x+4)(x-20)}$

Check: $(x+4)(x-20) = x \cdot x + (-20) \cdot x + 4 \cdot x + 4 \cdot (-20) = x^2 - 20x + 4x - 80 = x^2 + (4-20)x - 80$
$= x^2 - 16x - 80$

Example 3.3-18: Factor $u^2 - 16u + 28$.

Solution:

To factor the above trinomial we need to obtain two numbers whose sum is -16 and whose product is 28. Let's construct a table as follows:

Sum	*Product*
$-10-6=-16$	$(-10) \cdot (-6) = 60$
$-11-5=-16$	$(-11) \cdot (-5) = 55$
$-12-4=-16$	$(-12) \cdot (-4) = 48$
$-13-3=-16$	$(-13) \cdot (-3) = 39$
$\mathbf{-14-2=-16}$	$\mathbf{(-14) \cdot (-2) = 28}$

The last line contains the sum and the product of the two numbers that we need. Thus, $u^2 - 16u + 28 = \mathbf{(u-14)(u-2)}$

Check: $(u-14)(u-2) = u \cdot u - 2 \cdot u - 14 \cdot u + (-14) \cdot (-2) = u^2 - 2u - 14u + 28 = u^2 + (-2-14)u + 28$
$= u^2 - 16u + 28$

Example 3.3-19: Factor $y^2 + 8y - 20$.

Solution:

To factor the above trinomial we need to obtain two numbers whose sum is 8 and whose product is -20. Let's construct a table as follows:

Sum	*Product*
$12-4=8$	$12\cdot-4=-48$
$11-3=8$	$11\cdot-3=-33$
$\mathbf{10-2=8}$	$\mathbf{10\cdot-2=-20}$
$9-1=8$	$9\cdot-1=-9$

The third line contains the sum and the product of the two numbers that we need. Thus, $y^2+8y-20 = (\mathbf{y+10})(\mathbf{y-2})$

Check: $(y+10)(y-2) = y\cdot y-2\cdot y+10\cdot y+10\cdot(-2) = y^2-2y+10y-20 = y^2+(-2+10)x-20$
$= y^2+8y-20$

Example 3.3-20: Factor x^2-x-6.

Solution:

To factor the above trinomial we need to obtain two numbers whose sum is -1 and whose product is -6. Let's construct a table as follows: $-4+3=-1$

Sum	*Product*
$-9+8=-1$	$1\cdot(-17)=-17$
$-8+7=-1$	$2\cdot(-18)=-36$
$-7+6=-1$	
$-6+5=-1$	
$-5+4=-1$	
$-4+3=-1$	$3\cdot(-19)=-57$
$\mathbf{4-20=-16}$	$\mathbf{4\cdot(-20)=-80}$

The last line contains the sum and the product of the two numbers that we need. Thus, $x^2-16x-80 = (\mathbf{x+4})(\mathbf{x-20})$

Check: $(x+4)(x-20) = x\cdot x+(-20)\cdot x+4\cdot x+4\cdot(-20) = x^2-20x+4x-80 = x^2+(4-20)x-80$
$= x^2-16x-80$

Practice Problems - Factoring Trinomials of the Form ax^2+bx+c where $a=1$

Section 3.3 Case I Practice Problems - Factor the following trinomials using the Trial and Error method:

1. $x^2-2x-15$
2. y^2-9y+8
3. $t^2+2t-15$
4. $y^2-2y+11$
5. $x^2+10x+21$
6. $u^2+4u-32$
7. $a^2+9a+18$
8. $w^2-11w+30$
9. $x^2-8x-20$
10. $v^2+120v+2000$

Case II Factoring Trinomials of the Form $ax^2 + bx + c$ where $a \rangle 1$

To express a trinomial of the form $ax^2 + bx + c$, where $a \rangle 1$, in its factored form $(lx + m)(kx + n)$, let us consider the product $(lx + m)(kx + n)$ and use the FOIL method to see how each term of the resulting trinomial is formed ,e.g.,

$$(lx + m)(kx + n) = (k \cdot l) \cdot x \cdot x + (l \cdot n) \cdot x + (k \cdot m) \cdot x + m \cdot n = (kl)x^2 + (ln + km)x + mn$$

Note that the product of the coefficient of the x^2 term and the constant term is $kl \cdot mn$. In addition, the product of the coefficients of x is also $kl \cdot mn$. We use this concept in order to express trinomials of the form $ax^2 + bx + c$, where $a \rangle 1$, in their equivalent factored form. The following show the steps in factoring this class of trinomials:

Step 1 Obtain two numbers m and n whose sum equals to b and whose product equals to $a \cdot c$.

Step 2 Rewrite the middle term of the trinomial as the sum of the two numbers found in Step 1.

Step 3 Write the trinomial in its factored form by grouping the first two terms and the last two terms (see Section 3.2). Check the answer by using the FOIL method.

Examples with Steps

The following examples further illustrate how to factor trinomials of the form $ax^2 + bx + c$, where $a \rangle 1$, using the Trial and Error method:

Example 3.3-21

Factor $6x^2 + 23x + 20$.

Solution:

Step 1 Obtain two numbers whose sum is 23 and whose product is $6 \cdot 20 = 120$. Let's construct a table as follows:

Sum	***Product***
$12 + 11 = 23$	$12 \cdot 11 = 132$
$13 + 10 = 23$	$13 \cdot 10 = 130$
$14 + 9 = 23$	$14 \cdot 9 = 126$
$\mathbf{15 + 8 = 23}$	$\mathbf{15 \cdot 8 = 120}$

The last line contains the sum and the product of the two numbers that we need. Therefore,

Step 2 $6x^2 + 23x + 20 = 6x^2 + (15 + 8)x + 20 = 6x^2 + 15x + 8x + 20 = 3x(2x + 5) + 4(2x + 5)$

Step 3 $3x(2x + 5) + 4(2x + 5) = \mathbf{(2x + 5)(3x + 4)}$

Check: $(2x + 5)(3x + 4) = (2 \cdot 3) \cdot x \cdot x + (2 \cdot 4) \cdot x + (5 \cdot 3) \cdot x + (5 \cdot 4) = 6x^2 + 8x + 15x + 20$
$= 6x^2 + (8 + 15)x + 20 = 6x^2 + 23x + 20$

Example 3.3-22

Factor $10x^2 - 9x - 91$.

Solution:

Step 1 Obtain two numbers whose sum is -9 and whose product is $10 \cdot (-91) = -910$. Let's construct a table as follows:

Sum	*Product*
$20 - 29 = -9$	$20 \cdot (-29) = -580$
$21 - 30 = -9$	$21 \cdot (-30) = -630$
$22 - 31 = -9$	$22 \cdot (-31) = -682$
$23 - 32 = -9$	$23 \cdot (-32) = -736$
$24 - 33 = -9$	$24 \cdot (-33) = -792$
$25 - 34 = -9$	$25 \cdot (-34) = -850$
$\mathbf{26 - 35 = -9}$	$\mathbf{26 \cdot (-35) = -910}$

The last line contains the sum and the product of the two numbers that we need. Therefore,

Step 2 $10x^2 - 9x - 91 = 10x^2 + (26 - 35)x - 91 = 10x^2 + 26x - 35x - 91 = 2x(5x + 13) - 7(5x + 13)$

Step 3 $2x(5x + 13) - 7(5x + 13) = \mathbf{(5x + 13)(2x - 7)}$

Check: $(5x + 13)(2x - 7) = (5 \cdot 2) \cdot x \cdot x + (5 \cdot -7) \cdot x + (13 \cdot 2) \cdot x + (13 \cdot -7) = 10x^2 - 35x + 26x - 91$
$= 10x^2 + (-35 + 26)x - 91 = 10x^2 - 9x - 91$

Example 3.3-23

Factor $36x^2 - 25$.

Solution:

Step 1 Write $36x^2 - 25$ in its standard form, i.e., $36x^2 + 0x - 25$. Obtain two numbers whose sum is 0 and whose product is $36 \cdot (-25) = -900$. Let's construct a table as follows:

Sum	*Product*
$10 - 10 = 0$	$10 \cdot (-10) = -100$
$20 - 20 = 0$	$20 \cdot (-20) = -400$
$\mathbf{30 - 30 = 0}$	$\mathbf{30 \cdot (-30) = -900}$
$40 - 40 = 0$	$40 \cdot (-40) = -1600$

The third line contains the sum and the product of the two numbers that we need. Therefore,

Step 2 $36x^2 - 25 = 36x^2 + 0x - 25 = 36x^2 + (30 - 30)x - 25 = 36x^2 + 30x - 30x - 25$
$= 6x(6x + 5) - 5(6x + 5)$

Step 3 $6x(6x + 5) - 5(6x + 5) = \mathbf{(6x + 5)(6x - 5)}$

Check: $(6x+5)(6x-5) = (6\cdot 6)\cdot x\cdot x+(6\cdot -5)\cdot x+(5\cdot 6)\cdot x+(5\cdot -5) = 36x^2-30x+30x-25$
$= 36x^2+(-30+30)x-25 = 36x^2+0x-25 = 36x^2-25$

Example 3.3-24

Factor $56x^2-13x-3$.

Solution:

Step 1 Obtain two numbers whose sum is -13 and whose product is $56\cdot(-3)=-168$. Let's construct a table as follows:

Sum	*Product*
$4-17=-13$	$4\cdot(-17)=-68$
$5-18=-13$	$5\cdot(-18)=-90$
$6-19=-13$	$6\cdot(-19)=-114$
$7-20=-13$	$7\cdot(-20)=-140$
$\mathbf{8-21=-13}$	$\mathbf{8\cdot(-21)=-168}$

The last line contains the sum and the product of the two numbers that we need. Therefore,

Step 2 $56x^2-13x-3 = 56x^2+(8-21)x-3 = 56x^2+8x-21x-3 = 8x(7x+1)-3(7x+1)$

Step 3 $8x(7x+1)-3(7x+1) = \mathbf{(7x+1)(8x-3)}$

Check: $(7x+1)(8x-3) = (7\cdot 8)\cdot x\cdot x+(7\cdot -3)\cdot x+(1\cdot 8)\cdot x+(1\cdot -3) = 56x^2-21x+8x-3$
$= 56x^2+(-21+8)x-3 = 56x^2-13x-3$

Example 3.3-25

Factor $18x^2+21x+5$.

Solution:

Step 1 Obtain two numbers whose sum is 21 and whose product is $18\cdot 5=90$. Let's construct a table as follows:

Sum	*Product*
$1+20=21$	$1\cdot 20=20$
$2+19=21$	$2\cdot 19=38$
$3+18=21$	$3\cdot 18=54$
$4+17=21$	$4\cdot 17=68$
$5+16=21$	$5\cdot 16=80$
$\mathbf{6+15=21}$	$\mathbf{6\cdot 15=90}$

The last line contains the sum and the product of the two numbers that we need. Therefore,

Step 2 $18x^2+21x+5 = 18x^2+(6+15)x+5 = 18x^2+6x+15x+5 = 6x(3x+1)+5(3x+1)$

Step 3 $6x(3x+1)+5(3x+1) = \mathbf{(6x+5)(3x+1)}$

Check: $(6x+5)(3x+1) = (6\cdot 3)\cdot x\cdot x+(6\cdot 1)\cdot x+(5\cdot 3)\cdot x+(5\cdot 1) = 18x^2+6x+15x+5$
$= 18x^2+(6+15)x+5 = 18x^2+21x+5$

Additional Examples - Factoring Trinomials of the Form ax^2+bx+c where $a \rangle 1$

The following examples further illustrate how to factor trinomials using the Trial and Error method:

Example 3.3-26: Factor $6x^2+16x+10$.

Solution:

To factor the above trinomial we need to obtain two numbers whose sum is 16 and whose product is $6\cdot 10=60$. Let's construct a table as follows:

Sum	*Product*
$8+8=16$	$8\cdot 8=64$
$9+7=16$	$9\cdot 7=63$
$\mathbf{10+6=16}$	$\mathbf{10\cdot 6=60}$

The last line contains the sum and the product of the two numbers that we need. Therefore,
$6x^2+16x+10 = 6x^2+(10+6)x+10 = 6x^2+10x+6x+10 = 2x(3x+5)+2(3x+5) = \mathbf{(3x+5)(2x+2)}$

Check: $(3x+5)(2x+2) = (3\cdot 2)\cdot x\cdot x+(3\cdot 2)\cdot x+(2\cdot 5)\cdot x+5\cdot 2 = 6x^2+6x+10x+10$
$= 6x^2+(6+10)x+10 = 6x^2+16x+10$

Example 3.3-27: Factor $5x^2+8x+3$.

Solution:

To factor the above trinomial we need to obtain two numbers whose sum is 8 and whose product is $5\cdot 3=15$. Let's construct a table as follows:

Sum	*Product*
$1+7=8$	$1\cdot 7=7$
$2+6=8$	$2\cdot 6=12$
$\mathbf{3+5=8}$	$\mathbf{3\cdot 5=15}$
$4+4=8$	$4\cdot 4=16$

The third line contains the sum and the product of the two numbers that we need. Therefore,
$5x^2+8x+3 = 5x^2+(3+5)x+3 = 5x^2+3x+5x+3 = x(5x+3)+(5x+3) = \mathbf{(5x+3)(x+1)}$

Check: $(5x+3)(x+1) = (5\cdot 1)\cdot x\cdot x+(5\cdot 1)\cdot x+(3\cdot 1)\cdot x+3\cdot 1 = 5x^2+5x+3x+3 = 5x^2+(5+3)x+3$
$= 5x^2+8x+3$

Example 3.3-28: Factor $6x^2+19x+10$.

Solution:

To factor the above trinomial we need to obtain two numbers whose sum is 19 and whose product is $6\cdot 10=60$. Let's construct a table as follows:

Sum	*Product*
$1+18=19$	$1\cdot 18=18$
$2+17=19$	$2\cdot 17=34$
$3+16=19$	$3\cdot 16=48$

$\mathbf{4+15=19}$	$\mathbf{4\cdot 15=60}$
$5+14=19$	$5\cdot 14=70$

The fourth line contains the sum and the product of the two numbers that we need. Thus,

$6x^2+19x+10 = 6x^2+(4+15)x+10 = 6x^2+4x+15x+10 = 2x(3x+2)+5(3x+2) = \mathbf{(2x+5)(3x+2)}$

Check: $(2x+5)(3x+2) = (2\cdot 3)\cdot x\cdot x+(2\cdot 2)\cdot x+(5\cdot 3)\cdot x+5\cdot 2 = 6x^2+4x+15x+10$

$= 6x^2+(4+15)x+10 = 6x^2+19x+10$

Example 3.3-29: Factor $2w^2-13w+15$.

Solution:

To factor the above trinomial we need to obtain two numbers whose sum is -13 and whose product is $2\cdot 15=30$. Let's construct a table as follows:

Sum	***Product***
$-1-12=-13$	$(-1)\cdot(-12)=12$
$-2-11=-13$	$(-2)\cdot(-11)=22$
$\mathbf{-3-10=-13}$	$\mathbf{(-3)\cdot(-10)=30}$
$-4-9=-13$	$(-4)\cdot(-9)=36$

The third line contains the sum and the product of the two numbers that we need. Thus,

$2w^2-13w+15 = 2w^2+(-3-10)w+15 = 2w^2-3w-10w+15 = w(2w-3)-5(2w-3) = \mathbf{(2w-3)(w-5)}$

Check: $(2w-3)(w-5) = (2\cdot 1)\cdot w\cdot w+(2\cdot -5)\cdot w+(-3\cdot 1)\cdot w+(-3\cdot -5) = 2w^2-10w-3w+15$

$= 2w^2+(-10-3)w+15 = 2w^2-13w+15$

Example 3.3-30: Factor $5y^2-16y+3$.

Solution:

To factor the above trinomial we need to obtain two numbers whose sum is -16 and whose product is $5\cdot 3=15$. Let's construct a table as follows:

Sum	***Product***
$-5-11=-16$	$(-5)\cdot(-11)=55$
$-4-12=-16$	$(-4)\cdot(-12)=48$
$-3-13=-16$	$(-3)\cdot(-13)=39$
$-2-14=-16$	$(-2)\cdot(-14)=28$
$\mathbf{-1-15=-16}$	$\mathbf{(-1)\cdot(-15)=15}$

The last line contains the sum and the product of the two numbers that we need. Thus,

$5y^2-16y+3 = 5y^2+(-1-15)y+3 = 5y^2-y-15y+3 = y(5y-1)-3(5y-1) = \mathbf{(5y-1)(y-3)}$

Check: $(5y-1)(y-3) = 5\cdot y\cdot y+(5\cdot -3)\cdot y-y+(-1)\cdot(-3) = 5y^2-15y-y+3 = 5y^2+(-15-1)y+3$

$= 5y^2-16y+3$

Example 3.3-31: Factor $5u^2+23u+12$.

Solution:

To factor the above trinomial we need to obtain two numbers whose sum is 23 and whose product is $5\cdot 12=60$. Let's construct a table as follows:

Sum	*Product*
$20+3=23$	**$20\cdot 3=60$**

In this case, at first trial we obtained the sum and the product of the two numbers that we need. Thus,

$5u^2+23u+12 = 5u^2+(20+3)u+12 = 5u^2+20u+3u+12 = 5u(u+4)+3(u+4) = \mathbf{(u+4)(5u+3)}$

Check: $(u+4)(5u+3) = 5\cdot u\cdot u+3\cdot u+(4\cdot 5)\cdot u+4\cdot 3 = 5u^2+3u+20u+12 = 5u^2+(20+3)u+12$

$= 5u^2+23u+12$

Example 3.3-32: Factor $2a^2+3a-2$.

Solution:

To factor the above trinomial we need to obtain two numbers whose sum is 3 and whose product is $2\cdot -2=-4$. Let's construct a table as follows:

Sum	*Product*
$6-3=3$	$6\cdot -3=-18$
$5-2=3$	$5\cdot -2=-10$
$4-1=3$	**$4\cdot -1=-4$**

The last line contains the sum and the product of the two numbers that we need. Therefore,

$2a^2+3a-2 = 2a^2+(4-1)a-2 = 2a^2+4a-a-2 = 2a(a+2)-(a+2) = \mathbf{(a+2)(2a-1)}$

Check: $(a+2)(2a-1) = 2\cdot a\cdot a-a+(2\cdot 2)\cdot a-2 = 2a^2-a+4a-2 = 2a^2+(-1+4)a-2 = 2a^2+3a-2$

Example 3.3-33: Factor $6b^2-14b+4$.

Solution:

To factor the above trinomial we need to obtain two numbers whose sum is -14 and whose product is $6\cdot 4=24$. Let's construct a table as follows:

Sum	*Product*
$-10-4=-14$	$-10\cdot -4=40$
$-11-3=-14$	$-11\cdot -3=33$
$-12-2=-14$	**$-12\cdot -2=24$**
$-13-1=-14$	$-13\cdot -1=13$

The third line contains the sum and the product of the two numbers that we need. Thus,

$6b^2-14b+4 = 6b^2+(-12-2)b+4 = 6b^2-12b-2b+4 = 6b(b-2)-2(b-2) = (b-2)(6b-2)$

$= (b-2)\cdot 2(3b-1) = \mathbf{2(b-2)(3b-1)}$

Check: $2(b-2)(3b-1) = 2[3\cdot b\cdot b-b+(-2\cdot 3)\cdot b+2] = 2[3b^2-b-6b+2] = 2[3b^2+(-1-6)b+2]$

$= 2[3b^2-7b+2] = 6b^2-14b+4$

Example 3.3-34: Factor $25w^2+5w-2$.

Solution:

To factor the above trinomial we need to obtain two numbers whose sum is 5 and whose product is $25\cdot -2=-50$. Let's construct a table as follows:

Sum	*Product*
$15-10=5$	$15\cdot(-10)=-150$
$14-9=5$	$14\cdot(-9)=-126$
$13-8=5$	$13\cdot(-8)=-104$
$12-7=5$	$12\cdot(-7)=-84$
$11-6=5$	$11\cdot(-6)=-66$
$\mathbf{10-5=5}$	$\mathbf{10\cdot(-5)=-50}$

The last line contains the sum and the product of the two numbers that we need. Thus, $25w^2+5w-2 = 25w^2+(10-5)w-2 = 25w^2+10w-5w-2 = 5w(5w+2)-(5w+2) = \mathbf{(5w+2)(5w-1)}$

Check: $(5w+2)(5w-1) = (5\cdot5)\cdot w\cdot w+(5\cdot-1)\cdot w+(2\cdot5)\cdot w+2\cdot-1 = 25w^2-5w+10w-2$
$= 25w^2+(-5+10)w-2 = 25w^2+5w-2$

Example 3.3-35: Factor $2y^2+11y+15$.

Solution:

To factor the above trinomial we need to obtain two numbers whose sum is 11 and whose product is $2\cdot15=30$. Let's construct a table as follows:

Sum	*Product*
$\mathbf{5+6=11}$	$\mathbf{5\cdot6=30}$
$4+7=11$	$4\cdot7=28$
$3+8=11$	$3\cdot8=24$
$2+9=11$	$2\cdot9=18$
$1+10=11$	$1\cdot10=10$

The first line contains the sum and the product of the two numbers that we need. Thus, $2y^2+11y+15 = 2y^2+(5+6)y+15 = 2y^2+5y+6y+15 = y(2y+5)+3(2y+5) = \mathbf{(2y+5)(y+3)}$

Check: $(2y+5)(y+3) = 2\cdot y\cdot y+(2\cdot3)\cdot y+5\cdot y+5\cdot3 = 2y^2+6y+5y+15 = 2y^2+(6+5)y+15$
$= 2y^2+11y+15$

Example 3.3-36: Factor $3m^2+17m+10$.

Solution:

To factor the above trinomial we need to obtain two numbers whose sum is 17 and whose product is $3\cdot10=30$. Let's construct a table as follows:

Sum	*Product*
$10+7=17$	$10\cdot7=70$
$11+6=17$	$11\cdot6=66$
$12+5=17$	$12\cdot5=60$
$13+4=17$	$13\cdot4=52$
$14+3=17$	$14\cdot3=42$
$\mathbf{15+2=17}$	$\mathbf{15\cdot2=30}$

The last line contains the sum and the product of the two numbers that we need. Therefore,

$3m^2+17m+10 = 3m^2+(15+2)m+10 = 3m^2+15m+2m+10 = 3m(m+5)+2(m+5) = \mathbf{(m+5)(3m+2)}$

Check: $(m+5)(3m+2) = 3\cdot m\cdot m+2\cdot m+(5\cdot 3)\cdot m+5\cdot 2 = 3m^2+2m+15m+10 = 3m^2+(2+15)m+10$
$= 3m^2+17m+10$

Example 3.3-37: Factor $12x^2+79x-35$.

Solution:

To factor the above trinomial we need to obtain two numbers whose sum is 79 and whose product is $12\cdot -35 = -420$. Let's construct a table as follows:

Sum	***Product***
$89-10=79$	$89\cdot(-10)=-890$
$88-9=79$	$88\cdot(-9)=-792$
$87-8=79$	$87\cdot(-8)=-696$
$86-7=79$	$86\cdot(-7)=-602$
$85-6=79$	$85\cdot(-6)=-510$
$\mathbf{84-5=79}$	$\mathbf{84\cdot(-5)=-420}$

The last line contains the sum and the product of the two numbers that we need. Thus,
$12x^2+79x-35 = 12x^2+(84-5)x-35 = 12x^2+84x-5x-35 = 12x(x+7)-5(x+7) = \mathbf{(x+7)(12x-5)}$

Check: $(x+7)(12x-5) = 12\cdot x\cdot x-5\cdot x+(7\cdot 12)\cdot x+7\cdot(-5) = 12x^2-5x+84x-35$
$= 12x^2+(-5+84)x-35 = 12x^2+79x-35$

Example 3.3-38: Factor $18a^2+55a-28$.

Solution:

To factor the above trinomial we need to obtain two numbers whose sum is 55 and whose product is $18\cdot -28 = -504$. Let's construct a table as follows:

Sum	***Product***
$65-10=55$	$65\cdot(-10)=-650$
$64-9=55$	$64\cdot(-9)=-576$
$\mathbf{63-8=55}$	$\mathbf{63\cdot(-8)=-504}$
$62-7=55$	$62\cdot(-7)=-434$

The third line contains the sum and the product of the two numbers that we need. Thus,
$18a^2+55a-28 = 18a^2+(63-8)a-28 = 18a^2+63a-8a-28 = 9a(2a+7)-4(2a+7) = \mathbf{(2a+7)(9a-4)}$

Check: $(2a+7)(9a-4) = (2\cdot 9)\cdot a\cdot a+(2\cdot -4)\cdot a+(7\cdot 9)\cdot a+7\cdot(-4) = 18a^2-8a+63a-28$
$= 18a^2+(-8+63)a-28 = 18a^2+55a-28$

Example 3.3-39: Factor $10x^2-27x+18$.

Solution:

To factor the above trinomial we need to obtain two numbers whose sum is -27 and whose product is $10\cdot 18 = 180$. Let's construct a table as follows:

Sum	*Product*
$-20-7=-27$	$(-20)\cdot(-7)=140$
$-19-8=-27$	$(-19)\cdot(-8)=152$
$-18-9=-27$	$(-18)\cdot(-9)=162$
$-17-10=-27$	$(-17)\cdot(-10)=170$
$-16-11=-27$	$(-16)\cdot(-11)=176$
$\mathbf{-15-12=-27}$	$\mathbf{(-15)\cdot(-12)=180}$

The last line contains the sum and the product of the two numbers that we need. Therefore,
$10x^2-27x+18 = 10x^2+(-15-12)x+18 = 10x^2-15x-12x+18 = 5x(2x-3)-6(2x-3) = \mathbf{(2x-3)(5x-6)}$

Check: $(2x-3)(5x-6) = (2\cdot5)\cdot x\cdot x+(2\cdot-6)\cdot x+(-3\cdot5)\cdot x+(-3\cdot-6) = 10x^2-12x-15x+18$
$= 10x^2+(-12-15)x+18 = 10x^2-27x+18$

Example 3.3-40: Factor $-15t^2+77t-10$.

Solution:

To factor the above trinomial we need to obtain two numbers whose sum is 77 and whose product is $-15\cdot-10=150$. Let's construct a table as follows:

Sum	*Product*
$76+1=77$	$76\cdot1=76$
$\mathbf{75+2=77}$	$\mathbf{75\cdot2=150}$
$74+3=77$	$74\cdot3=222$

The middle line contains the sum and the product of the two numbers that we need. Thus,
$-15t^2+77t-10 = -15t^2+(75+2)t-10 = -15t^2+75t+2t-10 = -15t(t-5)+2(t-5) = \mathbf{(t-5)(-15t+2)}$

Check: $(t-5)(-15t+2) = -15\cdot t\cdot t+2\cdot t+(-5\cdot-15)\cdot t+2\cdot-5 = -15t^2+2t+75t-10$
$= -15t^2+(2+75)t-10 = -15t^2+77t-10$

The steps in factoring the following class of trinomials is very similar, if not identical, to the previous problems solved in this section. However, in the following set of examples to ensure proper factorization, we need to accurately match the given coefficients of x^2, x, and the constant term with the coefficients and the constant term of the standard trinomial ax^2+bx+c. For example, given the trinomial $10x^2-14xy-12y^2$, where x is variable, we know that $a=10$, $b=-14y$, and $c=-12y^2$. Once this equality is established, then the remaining steps are identical to the steps used in factoring the previous problems. The following examples further illustrate this point:

Example 3.3-41: Factor $6x^2+10xy+4y^2$ (x is variable and y is constant).

Solution:

First - Write the equation in its standard form, i.e., write $6x^2+10xy+4y^2$ as $6x^2+(10y)x+4y^2$.

Second - Equate the coefficient of the standard trinomial with the given trinomial, i.e., let $a=6$, $b=10y$, and $c=4y^2$.

Third - Obtain two numbers whose sum is $10y$ and whose product is $6 \cdot 4y^2 = 24y^2$. Construct the following table:

Sum	***Product***
$1y + 9y = 10y$	$1y \cdot 9y = 9y^2$
$2y + 8y = 10y$	$2y \cdot 8y = 16y^2$
$3y + 7y = 10y$	$3y \cdot 7y = 21y^2$
$\mathbf{4y + 6y = 10y}$	$\mathbf{4y \cdot 6y = 24y^2}$

The last line contains the sum and the product of the two numbers that we need.

Fourth - Write the polynomial in its factored form by factoring the common binomial factor.

$$6x^2 + 10yx + 4y^2 = 6x^2 + (4y + 6y)x + 4y^2 = 6x^2 + 4yx + 6yx + 4y^2 = 2x(3x + 2y) + 2y(3x + 2y)$$
$$= \mathbf{(3x + 2y)(2x + 2y)}$$

Fifth - Check the answer using the FOIL method.

$$(3x + 2y)(2x + 2y) = (3 \cdot 2) \cdot x \cdot x + (3 \cdot 2y) \cdot x + (2y \cdot 2) \cdot x + 2y \cdot 2y = 6x^2 + 6yx + 4yx + 4y^2$$
$$= 6x^2 + (6y + 4y)x + 4y^2 = 6x^2 + 10yx + 4y^2$$

Example 3.3-42 A:

Factor $2x^2 - 19xy + 35y^2$ (x is variable and y is constant).

Solution:

First - Write the equation in its standard form, i.e., write $2x^2 - 19xy + 35y^2$ as $2x^2 + (-19y)x + 35y^2$

Second - Equate the coefficient of the standard trinomial with the given trinomial; i.e., let $a = 2$, $b = -19y$, and $c = 35y^2$.

Third - Obtain two numbers whose sum is $-19y$ and whose product is $2 \cdot 35y^2 = 70y^2$. Construct the following table:

Sum	***Product***
$-18y - y = -19y$	$(-18y) \cdot (-y) = 18y^2$
$-17y - 2y = -19y$	$(-17y) \cdot (-2y) = 34y^2$
$-16y - 3y = -19y$	$(-16y) \cdot (-3y) = 48y^2$
$-15y - 4y = -19y$	$(-15y) \cdot (-4y) = 60y^2$
$\mathbf{-14y - 5y = -19y}$	$\mathbf{(-14y) \cdot (-5y) = 70y^2}$

The last line contains the sum and the product of the two numbers that we need.

Fourth - Write the polynomial in its factored form by factoring the common binomial factor.

$$2x^2 - 19yx + 35y^2 = 2x^2 + (-14y - 5y)x + 35y^2 = 2x^2 - 14yx - 5yx + 35y^2$$
$$= 2x(x - 7y) - 5y(x - 7y) = \mathbf{(x - 7y)(2x - 5y)}$$

Fifth - Check the answer using the FOIL method.

$$(x - 7y)(2x - 5y) = 2 \cdot x \cdot x + (-5y) \cdot x + (-7y \cdot 2) \cdot x + (-7y) \cdot (-5y) = 2x^2 - 5yx - 14yx + 35y^2$$
$$= 2x^2 + (-5y - 14y)x + 35y^2 = 2x^2 - 19yx + 35y^2$$

Let's rework this problem. However, this time let y be the variable and x be the constant as follows:

Example 3.3-42 B:

Factor $2x^2 - 19xy + 35y^2$ (y is variable and x is constant).

Solution:

First - Write the equation in its standard form, i.e., write $2x^2 - 19xy + 35y^2$ as $35y^2 + (-19x)y + 2x^2$

Second - Equate the coefficient of the standard trinomial with the given trinomial, i.e., let $a = 35$, $b = -19x$, and $c = 2x^2$.

Third - Obtain two numbers whose sum is $-19x$ and whose product is $35 \cdot 2x^2 = 70x^2$. Construct the following table:

Sum	*Product*
$-18x - x = -19x$	$(-18x) \cdot (-x) = 18x^2$
$-17x - 2x = -19x$	$(-17x) \cdot (-2x) = 34x^2$
$-16x - 3x = -19x$	$(-16x) \cdot (-3x) = 48x^2$
$-15x - 4x = -19x$	$(-15x) \cdot (-4x) = 60x^2$
$\mathbf{-14x - 5x = -19x}$	$\mathbf{(-14x) \cdot (-5x) = 70x^2}$

The last line contains the sum and the product of the two numbers that we need.

Fourth - Write the polynomial in its factored form by factoring the common binomial factor.

$$35y^2 - 19xy + 2x^2 = 35y^2 + (-14x - 5x)y + 2x^2 = 35y^2 - 14xy - 5xy + 2x^2 = 35y^2 - 5xy - 14xy + 2x^2$$
$$= 5y(7y - x) - 2x(7y - x) = \mathbf{(7y - x)(5y - 2x)}$$

Fifth - Check the answer using the FOIL method.

$$(7y - x)(5y - 2x) = (7 \cdot 5) \cdot y \cdot y + (-2x \cdot 7) \cdot y - 5x \cdot y + (-x) \cdot (-2x) = 35y^2 - 14xy - 5xy + 2x^2$$
$$= 35y^2(-14x - 5x)y + 2x^2 = 35y^2 - 19xy + 2x^2$$

Example 3.3-43:

Factor $3r^2 + 11rs + 10s^2$ (r is variable and s is constant).

Solution:

First - Write the equation in its standard form, i.e., write $3r^2 + 11rs + 10s^2$ as $3r^2 + (11s)r + 10s^2$.

Second - Equate the coefficient of the standard trinomial with the given trinomial, i.e., let $a = 3$, $b = 11s$, and $c = 10s^2$.

Third - Obtain two numbers whose sum is $11s$ and whose product is $3 \cdot 10s^2 = 30s^2$. Construct the following table:

Sum	*Product*
$8s + 3s = 11s$	$8s \cdot 3s = 24s^2$
$7s + 4s = 11s$	$7s \cdot 4s = 28s^2$
$\mathbf{6s + 5s = 11s}$	$\mathbf{6s \cdot 5s = 30s^2}$

The last line contains the sum and the product of the two numbers that we need.

Fourth - Write the polynomial in its factored form by factoring the common binomial factor.

$$3r^2 + 11sr + 10s^2 = 3r^2 + (6 + 5)sr + 10s^2 = 3r^2 + 6sr + 5sr + 10s^2 = 3r(r + 2s) + 5s(r + 2s)$$
$$= \mathbf{(r + 2s)(3r + 5s)}$$

Fifth - Check the answer using the FOIL method.

$(r+2s)(3r+5s) = 3\cdot r\cdot r+5s\cdot r+(2s\cdot 3)\cdot r+2s\cdot 5s = 3r^2+5sr+6sr+10s^2 = 3r^2+(5+6)sr+10s^2$

$= 3r^2+11sr+10s^2$

Example 3.3-44:

Factor $21n^2+41mn+10m^2$ (n is variable and m is constant).

Solution:

First - Write the equation in its standard form, i.e., write $21n^2+41mn+10m^2$ as $21n^2+(41m)n+10m^2$

Second - Equate the coefficient of the standard trinomial with the given trinomial, i.e., let $a=21$, $b=41n$, and $c=10m^2$.

Third - Obtain two numbers whose sum is $41m$ and whose product is $21\cdot 10m^2=210m^2$.

Construct the following table:

Sum	***Product***
$40m+m=41m$	$40m\cdot m=40m^2$
$39m+2m=41m$	$39m\cdot 2m=78m^2$
$38m+3m=41m$	$38m\cdot 3m=114m^2$
$37m+4m=41m$	$37m\cdot 4m=148m^2$
$36m+5m=41m$	$36m\cdot 5m=180m^2$
$\mathbf{35m+6m=41m}$	$\mathbf{35m\cdot 6m=210m^2}$
$34m+7m=41m^2$	$34m\cdot 7m=238m^2$

The sixth line contains the sum and the product of the two numbers that we need.

Fourth - Write the polynomial in its factored form by factoring the common binomial factor.

$21n^2+41mn+10m^2 = 21n^2+(35m+6m)n+10m^2 = 21n^2+35mn+6mn+10m^2$

$= 21n^2+6mn+35mn+10m^2 = 3n(7n+2m)+5m(7n+2m) = \mathbf{(7n+2m)(3n+5m)}$

Fifth - Check the answer using the FOIL method.

$(7n+2m)(3n+5m) = (7\cdot 3)\cdot n\cdot n+(7\cdot 5m)\cdot n+(2m\cdot 3)\cdot n+2m\cdot 5m = 21n^2+35mn+6mn+10m^2$

$= 21n^2+(35+6)mn+10m^2 = 21n^2+41mn+10m^2$

Example 3.3-45:

Factor $6c^2-cd-d^2$ (c is variable and d is constant).

Solution:

First - Write the equation in its standard form, i.e., write $6c^2-cd-d^2$ as $6c^2+(-d)c+(-d^2)$.

Second - Equate the coefficient of the standard trinomial with the given trinomial, i.e., let $a=6$, $b=-d$, and $c=-d^2$.

Third - Obtain two numbers whose sum is $-d$ and whose product is $6\cdot(-d^2)=-6d^2$. Construct the following table:

Sum	***Product***
$-5d+4d=-d$	$-5d\cdot 4d=-20d^2$
$-4d+3d=-d$	$-4d\cdot 3d=-12d^2$
$\mathbf{-3d+2d=-d}$	$\mathbf{-3d\cdot 2d=-6d^2}$

The last line contains the sum and the product of the two numbers that we need.

Fourth - Write the polynomial in its factored form by factoring the common binomial factor.

$$6c^2 - dc - d^2 = 6c^2 + (-3d + 2d)c - d^2 = 6c^2 - 3dc + 2dc - d^2 = 3c(2c - d) + d(2c - d)$$
$$= \mathbf{(2c - d)(3c + d)}$$

Fifth - Check the answer using the FOIL method.

$$(2c - d)(3c + d) = (2 \cdot 3) \cdot c \cdot c + (2 \cdot d) \cdot c - 3d \cdot c - d \cdot d = 6c^2 + 2dc - 3dc - d^2 = 6c^2 + (2 - 3)dc - d^2$$
$$= 6c^2 - dc - d^2$$

Practice Problems - Factoring Trinomials of the Form $ax^2 + bx + c$ where $a \rangle 1$

Section 3.3 Case II Practice Problems - Factor the following trinomials using the Trial and Error method:

1. $10x^2 + 11x - 35$
2. $6x^2 - x - 12$
3. $-7x^2 + 46x + 21$
4. $6x^2 - 11xy + 3y^2$ y is variable
5. $6x^2 + x - 40$
6. $2x^2 + 3x - 27$
7. $12x^2 + 10y^2 - 23xy$ x is variable
8. $5x^2 - 17x + 14$
9. $18x^2 + 9x - 20$
10. $27x^2 + 42x + 16$

3.4 Factoring Methods for Polynomials with Square and Cubed Terms

The key to successful factorization of polynomials is recognition and use of the right factoring method. In this section we will learn how to factor binomials of the form $a^2 - b^2$ (Case I) and $a^3 \pm b^3$ (case II) by using formulas that reduce the binomials to lower product terms.

Case I Factoring Polynomials Using the Difference of Two Squares Method

Binomials of the form $a^2 - b^2$ are factored to product of two first degree binomials using the following factorization method:

$$a^2 - b^2 = (a-b)(a+b)$$

Note that $a^2 + b^2$ is a prime polynomial and can not be factored. The difference of two square terms can be factored using the following steps:

Step 1 Factor the common terms and write the binomial in the standard form of $a^2 - b^2$.

Step 2 Write the binomial in its equivalent factorable form. Check the answer using the FOIL method.

Examples with Steps

The following examples show the steps as to how binomials of the form $a^2 - b^2$ are factored:

Example 3.4-1 Factor $x^4y - x^2y^3$ completely.

Solution:

Step 1 $\boxed{x^4y - x^2y^3} = \boxed{x^2y(x^2 - y^2)}$

Step 2 $\boxed{x^2y(x^2 - y^2)} = \boxed{\mathbf{x^2y(x-y)(x+y)}}$

Check: $x^2y(x-y)(x+y) = x^2y(x \cdot x + x \cdot y - x \cdot y - y \cdot y) = x^2y(x^2 + \cancel{xy} - \cancel{xy} - y^2)$

$= x^2y(x^2 - y^2) = x^2 \cdot x^2y - y^2 \cdot x^2y = x^4y - x^2y^3$

Example 3.4-2 Factor $5k^4 - 3125$ completely.

Solution:

Step 1 $\boxed{5k^4 - 3125} = \boxed{5(k^4 - 625)} = \boxed{5\left(k^{2^2} - 25^2\right)}$

Step 2 $\boxed{5\left(k^{2^2} - 25^2\right)} = \boxed{5(k^2 - 25)(k^2 + 25)} = \boxed{5(k^2 - 5^2)(k^2 + 25)} = \boxed{\mathbf{5(k-5)(k+5)(k^2+25)}}$

Check: $5(k-5)(k+5)(k^2+25) = 5(k \cdot k + 5 \cdot k - 5 \cdot k - 5 \cdot 5)(k^2+25) = 5(k^2 + 5k - 5k - 25)(k^2 + 25)$

$= 5(k^2 - 25)(k^2 + 25) = 5(k^2 \cdot k^2 + 25 \cdot k^2 - 25 \cdot k^2 - 25 \cdot 25) = 5(k^4 + \cancel{25k^2} - \cancel{25k^2} - 625)$

$= 5(k^4 - 625) = 5k^4 - 3125$

Example 3.4-3 Factor $81m^4 - n^4$ completely.

Solution:

Step 1 $\boxed{81m^4 - n^4} = \boxed{\left(9m^2\right)^2 - n^{2^2}}$

Step 2 $\boxed{\left(9m^2\right)^2 - n^{2^2}} = \boxed{\left(9m^2 - n^2\right)\left(9m^2 + n^2\right)} = \boxed{\left(3^2 m^2 - n^2\right)\left(9m^2 + n^2\right)}$

$= \boxed{\left[(3m)^2 - n^2\right]\left(9m^2 + n^2\right)} = \boxed{\mathbf{(3m - n)(3m + n)\left(9m^2 + n^2\right)}}$

Check: $(3m - n)(3m + n)\left(9m^2 + n^2\right) = (3m \cdot 3m + 3m \cdot n - 3m \cdot n - n \cdot n)\left(9m^2 + n^2\right)$

$= \left(9m^2 + 3mn - 3mn - n^2\right)\left(9m^2 + n^2\right) = \left(9m^2 - n^2\right)\left(9m^2 + n^2\right)$

$= 9m^2 \cdot 9m^2 + 9m^2 \cdot n^2 - n^2 \cdot 9m^2 - n^2 \cdot n^2 = 81m^4 + 9m^2n^2 - 9m^2n^2 - n^4 = 81m^4 - n^4$

Example 3.4-4 Factor $(x + 5)^2 - y^2$ completely.

Solution:

Step 1 *Not Applicable*

Step 2 $\boxed{(x + 5)^2 - y^2} = \boxed{[(x + 5) - y][(x + 5) + y]} = \boxed{(x + 5 - y)\,(x + 5 + y)} = \boxed{\mathbf{(x - y + 5)\,(x + y + 5)}}$

Check: $(x - y + 5)\,(x + y + 5) = (x \cdot x + x \cdot y + 5 \cdot x - x \cdot y - y \cdot y - 5 \cdot y + 5 \cdot x + 5 \cdot y + 5 \cdot 5)$

$= \left(x^2 + xy + 5x - xy - y^2 - 5y + 5x + 5y + 25\right) = \left(x^2 - y^2 + 10x + 25\right) = \left(x^2 + 10x + 25\right) - y^2$

$= (x + 5)^2 - y^2$

Example 3.4-5 Factor $u^2 - (v + 1)^2$ completely.

Solution:

Step 1 *Not Applicable*

Step 2 $\boxed{u^2 - (v + 1)^2} = \boxed{[u - (v + 1)][u + (v + 1)]} = \boxed{\mathbf{(u - v - 1)\,(u + v + 1)}}$

Check: $(u - v - 1)\,(u + v + 1) = u \cdot u + u \cdot v + 1 \cdot u - u \cdot v - v \cdot v - 1 \cdot v - 1 \cdot u - 1 \cdot v - 1 \cdot 1$

$= u^2 + uv + u - uv - v^2 - v - u - v - 1 = u^2 - v^2 - v - v - 1 = u^2 - v^2 - 2v - 1$

$= u^2 - \left(v^2 + 2v + 1\right) = u^2 - (v + 1)^2$

Additional Examples - Factoring Polynomials Using the Difference of Two Squares Method

The following examples further illustrate how to factor binomials of the form $a^2 - b^2$:

Example 3.4-6:

$\boxed{6u^2 - 6v^2} = \boxed{6\left(u^2 - v^2\right)} = \boxed{\mathbf{6(u - v)(u + v)}}$

Check: $6(u - v)(u + v) = 6(u \cdot u + u \cdot v - u \cdot v - v \cdot v) = 6\left(u^2 + uv - uv - v^2\right) = 6\left(u^2 - v^2\right) = 6u^2 - 6v^2$

Example 3.4-7:

$$\boxed{b^3 - 16b} = \boxed{b\left(b^2 - 16\right)} = \boxed{b\left(b^2 - 4^2\right)} = \boxed{\mathbf{b(b-4)(b+4)}}$$

Check: $b(b-4)(b+4) = b(b \cdot b + 4 \cdot b - 4 \cdot b - 4 \cdot 4) = b\left(b^2 + \cancel{4b} - \cancel{4b} - 4^2\right) = b\left(b^2 - 16\right) = b^3 - 16b$

Example 3.4-8:

$$\boxed{x^5 - 16x} = \boxed{x\left(x^4 - 16\right)} = \boxed{x\left(x^{2^2} - 4^2\right)} = \boxed{x\left(x^2 - 4\right)\left(x^2 + 4\right)} = \boxed{x\left(x^2 - 2^2\right)\left(x^2 + 4\right)} = \boxed{\mathbf{x(x-2)(x+2)\left(x^2+4\right)}}$$

Check: $x(x-2)(x+2)\left(x^2+4\right) = x(x \cdot x + 2 \cdot x - 2 \cdot x - 2 \cdot 2)\left(x^2+4\right) = x\left(x^2 + 2x - 2x - 4\right)\left(x^2+4\right)$

$= x\left(x^2-4\right)\left(x^2+4\right) = x\left(x^2 \cdot x^2 + 4 \cdot x^2 - 4 \cdot x^2 - 4 \cdot 4\right) = x\left(x^4 + \cancel{4x^2} - \cancel{4x^2} - 16\right) = x\left(x^4 - 16\right)$

$= x^5 - 16x$

Example 3.4-9:

$$\boxed{3x^4 - 27x^2} = \boxed{3x^2\left(x^2 - 9\right)} = \boxed{3x^2\left(x^2 - 3^2\right)} = \boxed{\mathbf{3x^2(x-3)(x+3)}}$$

Check: $3x^2(x-3)(x+3) = 3x^2(x \cdot x + 3 \cdot x - 3 \cdot x - 3 \cdot 3) = 3x^2\left(x^2 + \cancel{3x} - \cancel{3x} - 9\right) = 3x^2\left(x^2 - 9\right) = 3x^4 - 27x^2$

Example 3.4-10:

$$\boxed{t^4 - 625} = \boxed{t^{2^2} - 25^2} = \boxed{\left(t^2 - 25\right)\left(t^2 + 25\right)} = \boxed{\left(t^2 - 5^2\right)\left(t^2 + 25\right)} = \boxed{\mathbf{(t-5)(t+5)\left(t^2+25\right)}}$$

Check: $(t-5)(t+5)\left(t^2+25\right) = (t \cdot t + 5 \cdot t - 5 \cdot t - 5 \cdot 5)\left(t^2+25\right) = \left(t^2 + 5t - 5t - 25\right)\left(t^2+25\right)$

$= \left(t^2-25\right)\left(t^2+25\right) = t^2 \cdot t^2 - 25 \cdot t^2 + 25 \cdot t^2 - 25 \cdot 25 = t^4 - \cancel{25t^2} + \cancel{25t^2} - 625 = t^4 - 625$

Example 3.4-11:

$$\boxed{4x^6 - 64x^2} = \boxed{4x^2\left(x^4 - 16\right)} = \boxed{4x^2\left(x^{2^2} - 4^2\right)} = \boxed{4x^2\left(x^2 - 4\right)\left(x^2 + 4\right)} = \boxed{4x^2\left(x^2 - 2^2\right)\left(x^2 + 4\right)}$$

$$= \boxed{\mathbf{4x^2(x-2)(x+2)\left(x^2+4\right)}}$$

Check: $4x^2(x-2)(x+2)\left(x^2+4\right) = 4x^2(x \cdot x + 2 \cdot x - 2 \cdot x - 2 \cdot 2)\left(x^2+4\right) = 4x^2\left(x^2 + 2x - 2x - 4\right)\left(x^2+4\right)$

$= 4x^2\left(x^2-4\right)\left(x^2+4\right) = 4x^2\left(x^2 \cdot x^2 + 4 \cdot x^2 - 4 \cdot x^2 - 4 \cdot 4\right) = 4x^2\left(x^4 + \cancel{4x^2} - \cancel{4x^2} - 16\right)$

$= 4x^2\left(x^4 - 16\right) = 4x^2 \cdot x^4 - 4x^2 \cdot 16 = 4x^6 - 64x^2$

Example 3.4-12:

$$\boxed{w^8 - 256} = \boxed{w^{4^2} - 16^2} = \boxed{\left(w^4 - 16\right)\left(w^4 + 16\right)} = \boxed{\left(w^{2^2} - 4^2\right)\left(w^4 + 16\right)} = \boxed{\left(w^2 - 4\right)\left(w^2 + 4\right)\left(w^4 + 16\right)}$$

$$= \boxed{\left(w^2 - 2^2\right)\left(w^2 + 4\right)\left(w^4 + 16\right)} = \boxed{\mathbf{(w-2)(w+2)\left(w^2+4\right)\left(w^4+16\right)}}$$

Check: $(w-2)(w+2)\left(w^2+4\right)\left(w^4+16\right) = (w \cdot w + 2 \cdot w - 2 \cdot w - 2 \cdot 2)\left(w^2+4\right)\left(w^4+16\right)$

$= \left(w^2 + 2w - 2w - 4\right)\left(w^2+4\right)\left(w^4+16\right) = \left(w^2-4\right)\left(w^2+4\right)\left(w^4+16\right)$

$$= \left(w^2 \cdot w^2 + 4 \cdot w^2 - 4 \cdot w^2 - 4 \cdot 4\right)\left(w^4 + 16\right) = \left(w^4 + 4w^2 - 4w^2 - 16\right)\left(w^4 + 16\right)$$

$$= \left(w^4 - 16\right)\left(w^4 + 16\right) = w^4 \cdot w^4 + 16 \cdot w^4 - 16 \cdot w^4 - 16 \cdot 16 = w^8 + \cancel{16w^4} - \cancel{16w^4} - 256 = w^8 - 256$$

Example 3.4-13:

$$\boxed{8d^4 - 200d^2} = \boxed{8d^2\left(d^2 - 25\right)} = \boxed{8d^2\left(d^2 - 5^2\right)} = \boxed{\mathbf{8d^2(d-5)(d+5)}}$$

Check: $8d^2(d-5)(d+5) = 8d^2(d \cdot d + 5 \cdot d - 5 \cdot d - 5 \cdot 5) = 8d^2\left(d^2 + \cancel{5d} - \cancel{5d} - 25\right) = 8d^2\left(d^2 - 25\right)$

$$= 8d^2 \cdot d^2 - 8 \cdot 25d^2 = 8d^4 - 200d^2$$

Example 3.4-14:

$$\boxed{x^2 - (y+5)^2} = \boxed{[x - (y+5)][x + (y+5)]} = \boxed{\mathbf{(x - y - 5)\,(x + y + 5)}}$$

Check: $(x - y - 5)\,(x + y + 5) = x \cdot x + x \cdot y + 5 \cdot x - x \cdot y - y \cdot y - 5 \cdot y - 5 \cdot x - 5 \cdot y - 5 \cdot 5$

$$= x^2 + \cancel{xy} + \cancel{5x} - \cancel{xy} - y^2 - 5y - \cancel{5x} - 5y - 25 = x^2 - y^2 - 5y - 5y - 25 = x^2 - y^2 - 10y - 25$$

$$= x^2 - \left(y^2 + 10y + 25\right) = x^2 - (y+5)^2$$

Example 3.4-15:

$$\boxed{(u+3)^2 - v^2} = \boxed{[(u+3) - v][(u+3) + v]} = \boxed{(u + 3 - v)\,(u + 3 + v)} = \boxed{\mathbf{(u - v + 3)\,(u + v + 3)}}$$

Check: $(u - v + 3)\,(u + v + 3) = u \cdot u + u \cdot v + 3 \cdot u - u \cdot v - v \cdot v - 3 \cdot v + 3 \cdot u + 3 \cdot v + 3 \cdot 3$

$$= u^2 + \cancel{uv} + 3u - \cancel{uv} - v^2 - \cancel{3v} + 3u + \cancel{3v} + 9 = u^2 - v^2 + 6u + 9 = \left(u^2 + 6u + 9\right) - v^2 = (u+3)^2 - v^2$$

Example 3.4-16:

$$\boxed{x^2 - y^2 - 2y - 1} = \boxed{x^2 - \left(y^2 + 2y + 1\right)} = \boxed{x^2 - (y+1)^2} = \boxed{[x - (y+1)][x + (y+1)]} = \boxed{\mathbf{(x - y - 1)(x + y + 1)}}$$

Check: $(x - y - 1)(x + y + 1) = x \cdot x + x \cdot y + 1 \cdot x - x \cdot y - y \cdot y - 1 \cdot y - 1 \cdot x - 1 \cdot y - 1 \cdot 1$

$$= x^2 + \cancel{xy} + \cancel{x} - \cancel{xy} - y^2 - y - \cancel{x} - y - 1 = x^2 - y^2 - y - y - 1 = x^2 - y^2 - 2y - 1$$

Example 3.4-17:

$$\boxed{x^2 + y^2 + 9y + 9} = \boxed{x^2 + \left(y^2 + 6y + 9\right)} = \boxed{x^2 + (y+3)^2}$$

Note that the answer is in the same form as $a^2 + b^2$ which, as we know, **is a prime polynomial** and can not be factored.

Example 3.4-18:

$$\boxed{(x+2)^2 - (y+4)^2} = \boxed{[(x+2) - (y+4)][(x+2) + (y+4)]} = \boxed{(x + 2 - y - 4)(x + 2 + y + 4)} = \boxed{\mathbf{(x - y - 2)(x + y + 6)}}$$

Check: $(x - y - 2)(x + y + 6) = x \cdot x + x \cdot y + 6 \cdot x - x \cdot y - y \cdot y - 6 \cdot y - 2 \cdot x - 2 \cdot y - 2 \cdot 6$

$$= x^2 + \cancel{xy} + 6x - \cancel{xy} - y^2 - 6y - 2x - 2y - 12 = x^2 + 6x - 2x - y^2 - 6y - 2y - 12$$

$$= x^2 + 4x - y^2 - 8y - 12 = x^2 + 4x - y^2 - 8y + (-16 + 4) = x^2 + 4x + 4 - y^2 - 8y - 16$$

$$= \left(x^2 + 4x + 4\right) - \left(y^2 + 8y + 16\right) = (x+2)^2 - (y+4)^2$$

Example 3.4-19:

$$\boxed{\left(s^2 + 12s + 36\right) - t^2} = \boxed{(s+6)^2 - t^2} = \boxed{[(s+6) - t][(s+6) + t]} = \boxed{(s + 6 - t)\,(s + 6 + t)} = \boxed{\mathbf{(s - t + 6)\,(s + t + 6)}}$$

Check: $(s-t+6)(s+t+6) = (s\cdot s + s\cdot t + 6\cdot s - s\cdot t - t\cdot t - 6\cdot t + 6\cdot s + 6\cdot t + 6\cdot 6)$

$= \left(s^2 + \cancel{st} + 6s - \cancel{st} - t^2 - \cancel{6t} + 6s + \cancel{6t} + 36\right) = \left(s^2 - t^2 + 12s + 36\right) = \left(s^2 + 12s + 36\right) - t^2$

$= (s+6)^2 - t^2$

Example 3.4-20:

$$\boxed{(x-5)^2 - y^2} = \boxed{[(x-5)-y][(x-5)+y]} = \boxed{(x-5-y)(x-5+y)} = \boxed{(\mathbf{x-y-5})(\mathbf{x+y-5})}$$

Check: $(x-y-5)(x+y-5) = x\cdot x + x\cdot y - 5\cdot x - x\cdot y - y\cdot y + 5\cdot y - 5\cdot x - 5\cdot y + 5\cdot 5$

$= x^2 + \cancel{xy} - 5x - \cancel{xy} - y^2 + \cancel{5y} - 5x - \cancel{5y} + 25 = x^2 - 5x - y^2 - 5x + 25 = x^2 - 5x - 5x + 25 - y^2$

$= \left(x^2 - 10x + 25\right) - y^2 = (x-5)^2 - y^2$

Practice Problems - Factoring Polynomials Using the Difference of Two Squares Method

Section 3.4 Case I Practice Problems - Use the Difference of Two Squares method to factor the following polynomials:

1. $x^3 - 16x =$
2. $(x+1)^2 - (y+3)^2 =$
3. $t^5 - 81t =$
4. $\left(x^2 + 10x + 25\right) - y^2 =$
5. $c^4 - 9c^2 =$
6. $p^2 - q^2 - 4q - 4 =$
7. $x^2 - y^2 + 6y - 9 =$
8. $p^2 + q^2 + 4q + 4 =$
9. $m^{16} - 256 =$
10. $r^2 - (s+7)^2 =$

Case II Factoring Polynomials Using the Sum and Difference of Two Cubes Method

To factor binomials of the form a^3+b^3 or a^3-b^3 we use the following formulas:

$$a^3+b^3=(a+b)\left(a^2-ab+b^2\right)$$
$$a^3-b^3=(a-b)\left(a^2+ab+b^2\right)$$

Students are encouraged to memorize these two formulas in order to successfully factor this class of polynomials. The sum and difference of two cubed binomial terms can be factored using the following steps:

Step 1 Write the binomial in the standard form of a^3+b^3 or a^3-b^3.

Step 2 Write the binomil in its equivalent factorable form. Check the answer by multiplication.

Examples with Steps

The following examples show the steps as to how binomials of the form a^3+b^3 and a^3-b^3 are factored:

Example 3.4-21 Factor $3x^7+81x^4$.

Solution:

Step 1 $\boxed{3x^7+81x^4}=\boxed{3x^4\left(x^3+27\right)}=\boxed{3x^4\left(x^3+3^3\right)}$

Step 2 $\boxed{3x^4\left(x^3+3^3\right)}=\boxed{3x^4\left[(x+3)\left(x^2-3\cdot x+3^2\right)\right]}=\boxed{\mathbf{3x^4\left[(x+3)\left(x^2-3x+9\right)\right]}}$

Check: $3x^4\left[(x+3)\left(x^2-3x+9\right)\right]=3x^4\left(x\cdot x^2-3x\cdot x+9\cdot x+3\cdot x^2-3\cdot 3x+3\cdot 9\right)$

$=3x^4\left(x^3-\cancel{3x^2}+\cancel{9x}+\cancel{3x^2}-\cancel{9x}+27\right)=3x^4\left(x^3+27\right)=3x^4\cdot x^3+27\cdot 3x^4=3x^{4+3}+81x^4$

$=3x^7+81x^4$

Example 3.4-22 Factor $2a^3-250$.

Solution:

Step 1 $\boxed{2a^3-250}=\boxed{2\left(a^3-125\right)}=\boxed{2\left(a^3-5^3\right)}$

Step 2 $\boxed{2\left(a^3-5^3\right)}=\boxed{2(a-5)\left(a^2+5\cdot a+5^2\right)}=\boxed{\mathbf{2(a-5)\left(a^2+5a+25\right)}}$

Check: $2(a-5)\left(a^2+5a+25\right)=2\left(a\cdot a^2+5a\cdot a+25\cdot a-5\cdot a^2-5\cdot 5a-5\cdot 25\right)$

$=2\left(a^3+\cancel{5a^2}+\cancel{25a}-\cancel{5a^2}-\cancel{25a}-125\right)=2\left(a^3-125\right)=2a^3-250$

Example 3.4-23 Factor $x^5y^5-64x^2y^2$.

Solution:

Step 1 $\boxed{x^5y^5-64x^2y^2}=\boxed{x^2y^2\left(x^3y^3-64\right)}=\boxed{x^2y^2\left[(xy)^3-4^3\right]}$

Step 2 $\boxed{x^2y^2\left[(xy)^3-4^3\right]}=\boxed{x^2y^2\left\{(xy-4)\left[(xy)^2+4\cdot(xy)+4^2\right]\right\}}$

$=\boxed{\mathbf{x^2y^2\left\{(xy-4)\left[(xy)^2+4xy+16\right]\right\}}}$

Check: $x^2y^2\left\{(xy-4)\left[(xy)^2+4xy+16\right]\right\}=x^2y^2\left[xy\cdot(xy)^2+4xy\cdot xy+16xy-4\cdot(xy)^2-4\cdot 4xy-4\cdot 16\right]$

$=x^2y^2\left[(xy)^3+4(xy)^2+16xy-4(xy)^2-16xy-64\right]=x^2y^2\left[(xy)^3-64\right]$

$=x^2y^2\cdot(xy)^3-64\cdot x^2y^2=x^2y^2\cdot x^3y^3-64\cdot x^2y^2=x^{2+3}y^{2+3}-64x^2y^2=x^5y^5-64x^2y^2$

Example 3.4-24 Factor $(x-1)^3+y^3$.

Solution:

Step 1 The binomial is already in the standard a^3+b^3 form.

Step 2 $\boxed{(x-1)^3+y^3}=\boxed{\left[(x-1)+y\right]\left[(x-1)^2-(x-1)\cdot y+y^2\right]}$

$=\boxed{\mathbf{(x-1+y)\left[(x-1)^2-(x-1)y+y^2\right]}}$

Check: $(x-1+y)\left[(x-1)^2-(x-1)y+y^2\right]=\left[(x-1)+y\right]\left[(x-1)^2-(x-1)y+y^2\right]$

$=(x-1)\cdot(x-1)^2-(x-1)\cdot(x-1)y+(x-1)\cdot y^2+y\cdot(x-1)^2-y\cdot(x-1)y+y\cdot y^2$

$=(x-1)^3-(x-1)^2y+(x-1)y^2+(x-1)^2y-(x-1)y^2+y^3=(x-1)^3+y^3$

Example 3.4-25 Factor $x^3-(y+5)^3$.

Solution:

Step 1 The binomial is already in the standard a^3-b^3 form.

Step 2 $\boxed{x^3-(y+5)^3}=\boxed{\left[x-(y+5)\right]\left[x^2+x\cdot(y+5)+(y+5)^2\right]}$

$=\boxed{\mathbf{(x-y-5)\left[x^2+x(y+5)+(y+5)^2\right]}}$

Check: $(x-y-5)\left[x^2+x(y+5)+(y+5)^2\right]=\left[x-(y+5)\right]\left[x^2+x(y+5)+(y+5)^2\right]$

$=x\cdot x^2+x\cdot x(y+5)+x\cdot(y+5)^2-x^2\cdot(y+5)-x(y+5)\cdot(y+5)-(y+5)\cdot(y+5)^2$

$=x^3+x^2(y+5)+x(y+5)^2-x^2(y+5)-x(y+5)^2-(y+5)^3=x^3-(y+5)^3$

Additional Examples - Factoring Polynomials Using the Sum and Difference of Two Cubes Method

The following examples further illustrate how to factor binomials of the form $a^3\pm b^3$ using the sum and difference of two cubes method:

Example 3.4-26:

$\boxed{x^3+1}=\boxed{x^3+1^3}=\boxed{(x+1)\left(x^2-1\cdot x+1^2\right)}=\boxed{\mathbf{(x+1)\left(x^2-x+1\right)}}$

Check: $(x+1)\left(x^2-x+1\right) = x\cdot x^2 - x\cdot x + 1\cdot x + 1\cdot x^2 - 1\cdot x + 1\cdot 1 = x^3 - x^2 + \cancel{x} + x^2 - \cancel{x} + 1 = x^3 + 1$

Example 3.4-27:

$$\boxed{x^3-1} = \boxed{x^3-1^3} = \boxed{(x-1)\left(x^2+1\cdot x+1^2\right)} = \boxed{\mathbf{(x-1)\left(x^2+x+1\right)}}$$

Check: $(x-1)\left(x^2+x+1\right) = x\cdot x^2 + x\cdot x + 1\cdot x - 1\cdot x^2 - 1\cdot x - 1\cdot 1 = x^3 + x^2 + \cancel{x} - x^2 - \cancel{x} - 1 = x^3 - 1$

Example 3.4-28:

$$\boxed{x^3y^3-27} = \boxed{x^3y^3-3^3} = \boxed{(xy)^3-3^3} = \boxed{(xy-3)\left[(xy)^2+3\cdot xy+3^2\right]} = \boxed{\mathbf{(xy-3)\left[(xy)^2+3xy+9\right]}}$$

Check: $(xy-3)\left[(xy)^2+3xy+9\right] = xy\cdot(xy)^2 + 3xy\cdot xy + 9\cdot xy - 3\cdot(xy)^2 - 3\cdot 3xy - 3\cdot 9$

$= (xy)^3 + \cancel{3}(xy)^2 + \cancel{9xy} - \cancel{3}(xy)^2 - \cancel{9xy} - 27 = (xy)^3 - 27 = x^3y^3 - 27$

Example 3.4-29:

$$\boxed{x^3y^3z^3+8} = \boxed{x^3y^3z^3+2^3} = \boxed{(xyz)^3+2^3} = \boxed{(xyz+2)\left[(xyz)^2-2\cdot xyz+2^2\right]} = \boxed{\mathbf{(xyz+2)\left[(xyz)^2-2xyz+4\right]}}$$

Check: $(xyz+2)\left[(xyz)^2-2xyz+4\right] = xyz\cdot(xyz)^2 - 2xyz\cdot xyz + 4\cdot xyz + 2\cdot(xyz)^2 - 2\cdot 2xyz + 2\cdot 4$

$= (xyz)^3 - \cancel{2}(xyz)^2 + \cancel{4xyz} + \cancel{2}(xyz)^2 - \cancel{4xyz} + 8 = (xyz)^3 + 8 = x^3y^3z^3 + 8$

Example 3.4-30:

$$\boxed{a^4-a} = \boxed{a\left(a^3-1\right)} = \boxed{a\left(a^3-1^3\right)} = \boxed{a\left[(a-1)\left(a^2+1\cdot a+1^2\right)\right]} = \boxed{\mathbf{a\left[(a-1)\left(a^2+a+1\right)\right]}}$$

Check: $a\left[(a-1)\left(a^2+a+1\right)\right] = a\left(a\cdot a^2 + a\cdot a + 1\cdot a - 1\cdot a^2 - 1\cdot a - 1\cdot 1\right) = a\left(a^3 + a^2 + \cancel{a} - a^2 - \cancel{a} - 1\right)$

$= a\left(a^3-1\right) = a^4 - a$

Example 3.4-31:

$$\boxed{a^5+a^2} = \boxed{a^2\left(a^3+1\right)} = \boxed{a^2\left(a^3+1^3\right)} = \boxed{a^2\left[(a+1)\left(a^2-1\cdot a+1^2\right)\right]} = \boxed{\mathbf{a^2\left[(a+1)\left(a^2-a+1\right)\right]}}$$

Check: $a^2\left[(a+1)\left(a^2-a+1\right)\right] = a^2\left(a\cdot a^2 - a\cdot a + 1\cdot a + 1\cdot a^2 - 1\cdot a + 1\cdot 1\right) = a^2\left(a^3 - a^2 + \cancel{a} + a^2 - \cancel{a} + 1\right)$

$= a^2\left(a^3+1\right) = a^5 + a^2$

Example 3.4-32:

$$\boxed{8p^5+125p^2q^3} = \boxed{p^2\left(8p^3+125q^3\right)} = \boxed{p^2\left(2^3p^3+5^3q^3\right)} = \boxed{p^2\left[(2p)^3+(5q)^3\right]}$$

$$= \boxed{p^2\left\{(2p+5q)\left[(2p)^2-2p\cdot 5q+(5q)^2\right]\right\}} = \boxed{\mathbf{p^2\left[(2p+5q)\left(4p^2-10pq+25q^2\right)\right]}}$$

Check: $p^2\left[(2p+5q)\left(4p^2-10pq+25q^2\right)\right] = p^2(2p\cdot 4p^2 - 2p\cdot 10pq + 2p\cdot 25q^2 + 5q\cdot 4p^2 - 5q\cdot 10pq$

$+5q\cdot 25q^2) = p^2\left(8p^3 - \cancel{20p^2q} + \cancel{50pq^2} + \cancel{20p^2q} - \cancel{50pq^2} + 125q^3\right) = p^2\left(8p^3+125q^3\right)$

$= 8p^5 + 125p^2q^3$

Example 3.4-33:

$$\boxed{64x^4-27xy^3}=\boxed{x\left(64x^3-27y^3\right)}=\boxed{x\left(4^3x^3-3^3y^3\right)}=\boxed{x\left[(4x)^3-(3y)^3\right]}=\boxed{x\left\{(4x-3y)\left[(4x)^2+4x\cdot 3y+(3y)^2\right]\right\}}$$

$$=\boxed{\mathbf{x(4x-3y)\left(16x^2+12xy+9y^2\right)}}$$

Check: $x(4x-3y)\left(16x^2+12xy+9y^2\right)=x\left(4x\cdot 16x^2+4x\cdot 12xy+4x\cdot 9y^2-3y\cdot 16x^2-3y\cdot 12xy-3y\cdot 9y^2\right)$

$=x\left(64x^3+48yx^2+36xy^2-48yx^2-36xy^2-27y^3\right)=x\left(64x^3-27y^3\right)=64x^4-27xy^3$

Example 3.4-34:

$$\boxed{343x^3+125y^3}=\boxed{7^3x^3+5^3y^3}=\boxed{(7x)^3+(5y)^3}=\boxed{(7x+5y)\left[(7x)^2-7x\cdot 5y+(5y)^2\right]}$$

$$=\boxed{\mathbf{(7x+5y)\left(49x^2-35xy+25y^2\right)}}$$

Check: $(7x+5y)\left(49x^2-35xy+25y^2\right)=7x\cdot 49x^2-7x\cdot 35xy+7x\cdot 25y^2+5y\cdot 49x^2-5y\cdot 35xy+5y\cdot 25y^2$

$=343x^3-245x^2y+175xy^2+245x^2y-175xy^2+125y^3=343x^3+125y^3$

Example 3.4-35:

$$\boxed{r^3-8s^3}=\boxed{r^3-2^3s^3}=\boxed{r^3-(2s)^3}=\boxed{(r-2s)\left[r^2+r\cdot 2s+(2s)^2\right]}=\boxed{\mathbf{(r-2s)\left(r^2+2rs+4s^2\right)}}$$

Check: $(r-2s)\left(r^2+2rs+4s^2\right)=r\cdot r^2+r\cdot 2rs+r\cdot 4s^2-2s\cdot r^2-2s\cdot 2rs-2s\cdot 4s^2$

$=r^3+2r^2s+4rs^2-2r^2s-4rs^2-8s^3=r^3-8s^3$

Example 3.4-36:

$$\boxed{x^3+27y^3}=\boxed{x^3+3^3y^3}=\boxed{x^3+(3y)^3}=\boxed{(x+3y)\left[x^2-x\cdot 3y+(3y)^2\right]}=\boxed{\mathbf{(x+3y)\left(x^2-3xy+9y^2\right)}}$$

Check: $(x+3y)\left(x^2-3xy+9y^2\right)=x\cdot x^2-x\cdot 3xy+x\cdot 9y^2+3y\cdot x^2-3y\cdot 3xy+3y\cdot 9y^2$

$=x^3-3x^2y+9xy^2+3x^2y-9xy^2+27y^3=x^3+27y^3$

Example 3.4-37:

$$\boxed{x^3-(y+5)^3}=\boxed{[x-(y+5)]\left[x^2+x\cdot(y+5)+(y+5)^2\right]}=\boxed{\mathbf{[x-(y+5)]\left[x^2+x(y+5)+(y+5)^2\right]}}$$

Check: $[x-(y+5)]\left[x^2+x(y+5)+(y+5)^2\right]=x\cdot x^2+x\cdot x(y+5)+x\cdot(y+5)^2$

$=-x^2\cdot(y+5)-x(y+5)\cdot(y+5)-(y+5)\cdot(y+5)^2=x^3+x^2(y+5)+x(y+5)^2$

$-x^2(y+5)-x(y+5)^2-(y+5)^3=x^3-(y+5)^3$

Example 3.4-38:

$$\boxed{c^3-27d^3}=\boxed{c^3-3^3d^3}=\boxed{c^3-(3d)^3}=\boxed{(c-3d)\left[c^2+c\cdot 3d+(3d)^2\right]}=\boxed{\mathbf{(c-3d)\left(c^2+3cd+9d^2\right)}}$$

Check: $(c-3d)\left(c^2+3cd+9d^2\right)=c\cdot c^2+c\cdot 3cd+c\cdot 9d^2-3d\cdot c^2-3d\cdot 3cd-3d\cdot 9d^2$

$=c^3+3c^2d+9cd^2-3c^2d-9cd^2-27d^3=c^3-27d^3$

Example 3.4-39:

$$\boxed{a^3+64b^3}=\boxed{a^3+4^3b^3}=\boxed{a^3+(4b)^3}=\boxed{(a+4b)\left[a^2-a\cdot 4b+(4b)^2\right]}=\boxed{\mathbf{(a+4b)\left[a^2-4ab+16b^2\right]}}$$

Check: $(a+4b)\left[a^2-4ab+16b^2\right] = a\cdot a^2 - a\cdot 4ab + a\cdot 16b^2 + 4b\cdot a^2 - 4b\cdot 4ab + 4b\cdot 16b^2$

$= a^3 - 4a^2b + 16ab^2 + 4a^2b - 16ab^2 + 64b^3 = a^3 + 64b^3$

Example 3.4-40:

$$\boxed{2x^5y-16x^2y^4}=\boxed{2x^2y\left(x^3-8y^3\right)}=\boxed{2x^2y\left(x^3-2^3y^3\right)}=\boxed{2x^2y\left[x^3-(2y)^3\right]}$$

$$=\boxed{2x^2y\left\{(x-2y)\left[x^2+x\cdot 2y+(2y)^2\right]\right\}}=\boxed{\mathbf{2x^2y\left\{(x-2y)\left(x^2+2xy+4y^2\right)\right\}}}$$

Check: $2x^2y\left\{(x-2y)\left(x^2+2xy+4y^2\right)\right\} = 2x^2y\left(x\cdot x^2 + x\cdot 2xy + x\cdot 4y^2 - 2y\cdot x^2 - 2y\cdot 2xy - 2y\cdot 4y^2\right)$

$= 2x^2y\left(x^3 + 2x^2y + 4xy^2 - 2x^2y - 4xy^2 - 8y^3\right) = 2x^2y\left(x^3-8y^3\right) = 2x^2y\cdot x^3 - 2x^2y\cdot 8y^3$

$= 2x^5y - 16x^2y^4$

Practice Problems - Factoring Polynomials Using the Sum and Difference of Two Cubes Method

Section 3.4 Case II Practice Problems - Use the sum and difference of two cubes method to factor the following polynomials:

1. $4x^6+4 =$
2. $x^6y^6+8 =$
3. $(x+2)^3-y^3 =$
4. $2r^6-128 =$
5. $(x-7)^3+y^3 =$
6. $x^6y^5+x^3y^2 =$
7. $27a^4-125a =$
8. $x^5y+64x^2y^4 =$
9. $u^3+(v+1)^3 =$
10. $a^3-(b+7)^3 =$

3.5 Factoring Various Types of Polynomials

In this section factoring perfect square trinomials as well as factoring various other types of polynomials that have not been addressed in the previous sections are discussed in Cases I and II, respectively.

Case I Factoring Perfect Square Trinomials

Trinomials of the form $a^2 + 2ab + b^2$ and $a^2 - 2ab + b^2$ are called perfect square trinomials. Note that these types of polynomials are easy to recognize because their first and last terms are always square and their middle term is twice the product of the quantities being squared in the first and last terms. Once perfect square trinomials are identified, they can then be represented in their equivalent factored form as shown below:

$$a^2 + 2ab + b^2 = (a+b)^2$$

$$a^2 - 2ab + b^2 = (a-b)^2$$

For example, $36x^2 - 24x + 4$, $1 + 8y + 16y^2$, $25x^2 + 30xy + 9y^2$, and $49m^2 - 70mn + 25n^2$ are perfect square trinomials because:

1. Their first term is a square, i.e., $(6x)^2$; $(1)^2$; $(5x)^2$; $(7m)^2$.
2. Their last term is a square, i. e., $(-2)^2$; $(4y)^2$; $(3y)^2$; $(-5n)^2$, and
3. Their middle term is twice the product of the quantities being squared in the first and last terms, i.e., $2 \cdot (6x \cdot -2)$; $2 \cdot (1 \cdot 4y)$; $2 \cdot (5x \cdot 3y)$; $2 \cdot (7m \cdot -5n)$.

Therefore, the above examples can be represented in their equivalent factored form as: $(6x-2)^2$; $(1+4y)^2$; $(5x+3y)^2$; and $(7m-5n)^2$, respectively.

The following show the steps as to how perfect square trinomials are represented in their equivalent factored form:

Step 1 Write the trinomial in descending order.

Step 2 Check and see if the trinomial match the general forms $a^2 + 2ab + b^2$ or $a^2 - 2ab + b^2$.

Step 3 Write the trinomial in its equivalent form, i.e., $(a+b)^2$ or $(a-b)^2$.

Examples with Steps

The following examples show the steps as to how perfect square trinomials are factored:

Example 3.5-1

Factor $x^2 + 49 + 14x$.

Solution:

Step 1 $\boxed{x^2 + 49 + 14x} = \boxed{x^2 + 14x + 49}$

Step 2 $\boxed{x^2 + 14x + 49} = \boxed{x^2 + 14x + 7^2} = \boxed{x^2 + 2 \cdot (x \cdot 7) + 7^2}$

Step 3 $\boxed{x^2 + 2 \cdot (x \cdot 7) + 7^2} = \boxed{(x+7)^2}$

Example 3.5-2

Factor $25y^2 + 16 + 40y$.

Solution:

Step 1 $\boxed{25y^2 + 16 + 40y} = \boxed{25y^2 + 40y + 16}$

Step 2 $\boxed{25y^2 + 40y + 16} = \boxed{5^2 y^2 + 40y + 4^2} = \boxed{(5y)^2 + 2\cdot(5y\cdot 4) + 4^2}$

Step 3 $\boxed{(5y)^2 + 2\cdot(5y\cdot 4) + 4^2} = \boxed{\mathbf{(5y+4)^2}}$

Example 3.5-3

Factor $16x^2 + 24xy + 9y^2$.

Solution:

Step 1 *Not Applicable*

Step 2 $\boxed{16x^2 + 24xy + 9y^2} = \boxed{4^2 x^2 + 24xy + 3^2 y^2} = \boxed{(4x)^2 + 2\cdot(4x\cdot 3y) + (3y)^2}$

Step 3 $\boxed{(4x)^2 + 2\cdot(4x\cdot 3y) + (3y)^2} = \boxed{\mathbf{(4x+3y)^2}}$

Example 3.5-4

Factor $25r^2 + 64s^2 - 80rs$.

Solution:

Step 1 $\boxed{25r^2 + 64s^2 - 80rs} = \boxed{25r^2 - 80rs + 64s^2}$

Step 2 $\boxed{25r^2 - 80rs + 64s^2} = \boxed{5^2 r^2 - 80rs + 8^2 s^2} = \boxed{(5r)^2 - 2\cdot(5r\cdot 8s) + (8s)^2}$

Step 3 $\boxed{(5r)^2 - 2\cdot(5r\cdot 8s) + (8s)^2} = \boxed{\mathbf{(5r-8s)^2}}$

Example 3.5-5

Factor $9x^4 - 42x^2y^2 + 49y^2$.

Solution:

Step 1 *Not Applicable*

Step 2 $\boxed{9x^4 - 42x^2y^2 + 49y^4} = \boxed{3^2 x^{2^2} - 42x^2y^2 + 7^2 y^{2^2}} = \boxed{\left(3x^2\right)^2 - 2\cdot\left(3x^2\cdot 7y^2\right) + (7y)^2}$

Step 3 $\boxed{\left(3x^2\right)^2 - 2\cdot\left(3x^2\cdot 7y^2\right) + (7y)^2} = \boxed{\mathbf{\left(3x^2 - 7y^2\right)^2}}$

Additional Examples - Factoring Perfect Square Trinomials

The following examples further illustrate how to factor perfect square trinomials:

Example 3.5-6:

$\boxed{x^2 + 4x + 4} = \boxed{x^2 + 4x + 2^2} = \boxed{x^2 + 2\cdot(x\cdot 2) + 2^2} = \boxed{\mathbf{(x+2)^2}}$

Example 3.5-7:

$$\boxed{9-6y+y^2}=\boxed{y^2-6y+9}=\boxed{y^2-2\cdot(y\cdot 3)+3^2}=\boxed{\mathbf{(y-3)^2}}$$

Example 3.5-8:

$$\boxed{12t+9+4t^2}=\boxed{4t^2+12t+9}=\boxed{2^2t^2+12t+3^2}=\boxed{(2t)^2+2\cdot(2t\cdot 3)+3^2}=\boxed{\mathbf{(2t+3)^2}}$$

Example 3.5-9:

$$\boxed{16x^2-40x+25}=\boxed{4^2x^2-40x+5^2}=\boxed{(4x)^2-2\cdot(4x\cdot 5)+5^2}=\boxed{\mathbf{(4x-5)^2}}$$

Example 3.5-10:

$$\boxed{10x+1+25x^2}=\boxed{25x^2+10x+1}=\boxed{5^2x^2+10x+1^2}=\boxed{(5x)^2+2\cdot(5x\cdot 1)+1^2}=\boxed{\mathbf{(5x+1)^2}}$$

Example 3.5-11:

$$\boxed{4+9t^2-12t}=\boxed{9t^2-12t+4}=\boxed{3^2t^2-12t+2^2}=\boxed{(3t)^2-2\cdot(3t\cdot 2)+2^2}=\boxed{\mathbf{(3t-2)^2}}$$

Example 3.5-12:

$$\boxed{9p^2-30pq+25q^2}=\boxed{3^2p^2-30pq+5^2q^2}=\boxed{(3p)^2-2\cdot(3p\cdot 5q)+(5q)^2}=\boxed{\mathbf{(3p-5q)^2}}$$

Example 3.5-13:

$$\boxed{100x^4-100x^2y^2+25y^4}=\boxed{10^2x^{2^2}-100x^2y^2+5^2y^{2^2}}=\boxed{\left(10x^2\right)^2-2\cdot\left(10x^2\cdot 5y^2\right)+\left(5y^2\right)^2}$$

$$=\boxed{\mathbf{\left(10x^2-5y^2\right)^2}}$$

Example 3.5-14:

$$\boxed{121u^4-88u^2v^2+16v^4}=\boxed{11^2u^{2^2}-88u^2v^2+4^2v^{2^2}}=\boxed{\left(11u^2\right)^2-2\cdot\left(11u^2\cdot 4v^2\right)+\left(4v^2\right)^2}=\boxed{\mathbf{\left(11u^2-4v^2\right)^2}}$$

Example 3.5-15:

$$\boxed{16x^2+40xy+25y^2}=\boxed{4^2x^2+40xy+5^2y^2}=\boxed{(4x)^2+2\cdot(4x\cdot 5y)+(5y)^2}=\boxed{\mathbf{(4x+5y)^2}}$$

Practice Problems - Factoring Perfect Square Trinomials

Section 3.5 Case I Practice Problems - Factor the following trinomials:

1. $x^2+18x+81=$
2. $9+64p^2-48p=$
3. $9w^2+25+30w=$
4. $25+k^2-10k=$
5. $49x^2-84x+36=$
6. $1+16z+64z^2=$
7. $100u^4-40u^2v^2+4v^4=$
8. $49p^2-126pq+81q^2=$
9. $25x^4-30x^2y^2+9y^4=$
10. $9x^2+12xy+4y^2=$

Case II Factoring Other Types of Polynomials

In many instances a polynomial need to be factored more than once. Hence, in factoring a polynomial we need to check and see if the polynomial terms have a greatest common factor among them. If so, we should always factor the greatest common factor in a polynomial first. Following are the steps in factoring polynomials which can be factored more than once.

Step 1 Factor the greatest common factor.

Step 2 Factor the polynomial further by choosing one of the factoring methods learned in this chapter. Check the answer using the FOIL method.

Examples with Steps

The following examples show the steps as to how various types of polynomials are factored:

Example 3.5-16

Factor $30t^2-5t-10$ completely.

Solution:

Step 1 $\boxed{30t^2-5t-10}=\boxed{5(6t^2-t-2)}$

Step 2 Obtain two numbers whose sum is -1 and whose product is $6\cdot -2=-12$. Let's construct a table as follows:

Sum	***Product***
$1-2=-1$	$1\cdot(-2)=-2$
$2-3=-1$	$2\cdot(-3)=-6$
$\mathbf{3-4=-1}$	$\mathbf{3\cdot(-4)=-12}$

The last line contains the sum and the product of the two numbers that we need. Therefore,

$$\boxed{5(6t^2-t-2)}=\boxed{5[6t^2+(3-4)t-2]}=\boxed{5[6t^2+3t-4t-2]}=\boxed{5[3t(2t+1)-2(2t+1)]}$$

$$=\boxed{5[(2t+1)(3t-2)]}=\boxed{\mathbf{5(2t+1)(3t-2)}}$$

Check: $5(2t+1)(3t-2)=5[2t\cdot 3t-2t\cdot 2+1\cdot 3t+1\cdot(-2)]=5[6t^2-4t+3t-2]=5[6t^2-t-2]$

$=30t^2-5t-10$

Example 3.5-17

Factor $y^5+3y^4-16y-48$ completely.

Solution:

Step 1 $\boxed{y^5+3y^4-16y-48}=\boxed{y^4(y+3)-16(y+3)}=\boxed{(y+3)(y^4-16)}$

Step 2 $\boxed{(y+3)(y^4-16)}=\boxed{(y+3)(y^{2^2}-4^2)}=\boxed{(y+3)[(y^2-4)(y^2+4)]}$

$$= \boxed{(y+3)\left[\left(y^2-2^2\right)\left(y^2+4\right)\right]} = \boxed{\mathbf{(y+3)\left[(y-2)(y+2)\left(y^2+4\right)\right]}}$$

Check: $(y+3)\left[(y-2)(y+2)\left(y^2+4\right)\right] = (y+3)\left[(y\cdot y+2\cdot y-2\cdot y-2\cdot 2)\left(y^2+4\right)\right]$

$= (y+3)\left[\left(y^2+2y-2y-4\right)\left(y^2+4\right)\right] = (y+3)\left[\left(y^2-4\right)\left(y^2+4\right)\right]$

$= (y+3)\left(y^2\cdot y^2+4\cdot y^2-4\cdot y^2-4\cdot 4\right) = (y+3)\left(y^4+4y^2-4y^2-16\right) = (y+3)\left(y^4-16\right)$

$= y\cdot y^4-16\cdot y+3\cdot y^4-3\cdot 16 = y^5-16y+3y^4-48 = y^5+3y^4-16y-48$

Example 3.5-18

Factor $48x^8-243x^4$ completely.

Solution:

Step 1 $\boxed{48x^8-243x^4} = \boxed{3x^4\left(16x^4-81\right)}$

Step 2 $\boxed{3x^4\left(16x^4-81\right)} = \boxed{3x^4\left(4^2x^{2^2}-9^2\right)} = \boxed{3x^4\left(4x^2-9\right)\left(4x^2+9\right)}$

$$= \boxed{3x^4\left(2^2x^2-3^2\right)\left(4x^2+9\right)} = \boxed{\mathbf{3x^4\left[(2x-3)(2x+3)\left(4x^2+9\right)\right]}}$$

Check: $3x^4\left[(2x-3)(2x+3)\left(4x^2+9\right)\right] = 3x^4\left[(2x\cdot 2x+2x\cdot 3-3\cdot 2x-3\cdot 3)\left(4x^2+9\right)\right]$

$= 3x^4\left[\left(4x^2+6x-6x-9\right)\left(4x^2+9\right)\right] = 3x^4\left[\left(4x^2-9\right)\left(4x^2+9\right)\right]$

$= 3x^4\left(4x^2\cdot 4x^2+4x^2\cdot 9-9\cdot 4x^2-9\cdot 9\right) = 3x^4\left(16x^4+36x^2-36x^2-81\right) = 3x^4\left(16x^4-81\right)$

$= 3x^4\cdot 16x^4-3x^4\cdot 81 = 48x^8-243x^4$

Example 3.5-19

Factor $12y^2+38y+30$ completely.

Solution:

Step 1 $\boxed{12y^2+38y+30} = \boxed{2\left(6y^2+19y+15\right)}$

Step 2 Obtain two numbers whose sum is 19 and whose product is $6\cdot 15=90$. Let's construct a table as follows:

Sum	*Product*
$\mathbf{10+9=19}$	$\mathbf{10\cdot 9=90}$
$11+8=19$	$11\cdot 8=88$
$12+7=19$	$12\cdot 7=84$
$13+6=19$	$13\cdot 6=78$

The first line contains the sum and the product of the two numbers that we need. Therefore,

$$\boxed{2\left(6y^2+19y+15\right)} = \boxed{2\left[6y^2+(10+9)y+15\right]} = \boxed{2\left[6y^2+10y+9y+15\right]}$$

$$= \boxed{2[2y(3y+5)+3(3y+5)]} = \boxed{\mathbf{2(3y+5)(2y+3)}}$$

Check: $2(3y+5)(2y+3) = 2[3y\cdot 2y+3y\cdot 3+5\cdot 2y+5\cdot 3] = 2\left[6y^2+9y+10y+15\right]$

$= 2\left[6y^2+19y+15\right] = 12y^2+38y+30$

Example 3.5-20

Factor $6y^3+33y^2+45y$ completely.

Solution:

Step 1 $\boxed{6y^3+33y^2+45y} = \boxed{3y\left(2y^2+11y+15\right)}$

Step 2 Obtain two numbers whose sum is 11 and whose product is $2\cdot 15=30$. Let's construct a table as follows:

Sum	***Product***
$\mathbf{5+6=11}$	$\mathbf{5\cdot 6=30}$
$4+7=11$	$4\cdot 7=28$
$3+8=11$	$3\cdot 8=24$
$2+9=11$	$2\cdot 9=18$
$1+10=11$	$1\cdot 10=10$

The first line contains the sum and the product of the two numbers that we need. Therefore,

$$\boxed{3y\left(2y^2+11y+15\right)} = \boxed{3y\left[2y^2+(5+6)y+15\right]} = \boxed{3y\left[2y^2+5y+6y+15\right]}$$

$$= \boxed{3y[y(2y+5)+3(2y+5)]} = \boxed{\mathbf{3y[(2y+5)(y+3)]}}$$

Check: $3y[(2y+5)(y+3)] = 3y[2\cdot y\cdot y+(2\cdot 3)\cdot y+5\cdot y+5\cdot 3] = 3y\left[2y^2+6y+5y+15\right]$

$= 3y\left[2y^2+(6+5)y+15\right] = 3y\left[2y^2+11y+15\right] = 6y^3+33y^2+45y$

Additional Examples - Factoring Other Types of Polynomials

The following examples further illustrate how to factor polynomials:

Example 3.5-21:

$$\boxed{2x^2+10x+8} = \boxed{2\left(x^2+5x+4\right)} = \boxed{\mathbf{2(x+1)(x+4)}}$$

Check: $2(x+1)(x+4) = 2(x\cdot x+4\cdot x+1\cdot x+1\cdot 4) = 2\left(x^2+4x+x+4\right) = 2\left(x^2+5x+4\right) = 2x^2+10x+8$

Example 3.5-22:

$$\boxed{7y^2+7y-42} = \boxed{7\left(y^2+y-6\right)} = \boxed{\mathbf{7(y+3)(y-2)}}$$

Check: $7(y+3)(y-2) = 7(y\cdot y-2\cdot y+3\cdot y-2\cdot 3) = 7\left(y^2-2y+3y-6\right) = 7\left(y^2+y-6\right) = 7y^2+7y-42$

Example 3.5-23:

$$\boxed{6x^3+21x^2+9x} = \boxed{3x\left(2x^2+7x+3\right)} = \boxed{3x\left[2x^2+(6+1)x+3\right]} = \boxed{3x\left[2x^2+6x+x+3\right]} = \boxed{3x[2x(x+3)+(x+3)]}$$

$$= \boxed{3x[(x+3)(2x+1)]} = \boxed{\mathbf{3x(x+3)(2x+1)}}$$

Check: $3x(x+3)(2x+1) = 3x(x\cdot 2x+1\cdot x+3\cdot 2x+3\cdot 1) = 3x\left(2x^2+x+6x+3\right) = 3x\left(2x^2+7x+3\right)$

$= 6x^3+21x^2+9x$

Example 3.5-24:

$$\boxed{-6x^2-11x-3} = \boxed{-\left(6x^2+11x+3\right)} = \boxed{-\left[6x^2+(9+2)x+3\right]} = \boxed{-\left[6x^2+9x+2x+3\right]} = \boxed{-[3x(2x+3)+(2x+3)]}$$

$$= \boxed{-[(2x+3)(3x+1)]} = \boxed{\mathbf{-(2x+3)(3x+1)}}$$

Check: $-(2x+3)(3x+1) = -(2x\cdot 3x+2x\cdot 1+3\cdot 3x+1\cdot 3) = -\left(6x^2+2x+9x+3\right) = -\left(6x^2+11x+3\right)$

$= -6x^2-11x-3$

Example 3.5-25:

$$\boxed{4x^2+20x+24} = \boxed{4\left(x^2+5x+6\right)} = \boxed{4\left[x^2+(2+3)x+6\right]} = \boxed{4\left[x^2+2x+3x+6\right]} = \boxed{4[x(x+2)+3(x+2)]}$$

$$= \boxed{4[(x+2)(x+3)]} = \boxed{\mathbf{4(x+2)(x+3)}}$$

Check: $4(x+2)(x+3) = 4(x\cdot x+3\cdot x+2\cdot x+2\cdot 3) = 4\left(x^2+3x+2x+6\right) = 4\left(x^2+5x+6\right) = 4x^2+20x+24$

Example 3.5-26:

$$\boxed{12y^3+46y^2+42y} = \boxed{2y\left(6y^2+23y+21\right)} = \boxed{2y\left[6y^2+(14+9)y+21\right]} = \boxed{2y\left[6y^2+14y+9y+21\right]}$$

$$= \boxed{2y[2y(3y+7)+3(3y+7)]} = \boxed{2y[(3y+7)(2y+3)]} = \boxed{\mathbf{2y(3y+7)(2y+3)}}$$

Check: $2y(3y+7)(2y+3) = 2y(3y\cdot 2y+3y\cdot 3+7\cdot 2y+7\cdot 3) = 2y\left(6y^2+9y+14y+21\right)$

$= 2y\left(6y^2+23y+21\right) = 12y^3+46y^2+42y$

Example 3.5-27:

$$\boxed{y^3+5y^2-4y-20} = \boxed{y^2(y+5)-4(y+5)} = \boxed{(y+5)\left(y^2-4\right)} = \boxed{(y+5)\left(y^2-2^2\right)} = \boxed{\mathbf{(y+5)(y-2)(y+2)}}$$

Check: $(y+5)(y-2)(y+2) = (y+5)(y\cdot y+2\cdot y-2\cdot y-2\cdot 2) = (y+5)\left(y^2+2y-2y-4\right)$

$= (y+5)\left(y^2-4\right) = y^2\cdot y-4\cdot y+5\cdot y^2-4\cdot 5 = y^3-4y+5y^2-20 = y^3+5y^2-4y-20$

Example 3.5-28:

$$\boxed{a^2-b^2+2a-2b} = \boxed{\left(a^2-b^2\right)+2(a-b)} = \boxed{(a-b)(a+b)+2(a-b)} = \boxed{(a-b)[(a+b)+2]} = \boxed{\mathbf{(a-b)[a+b+2]}}$$

Check: $(a-b)[a+b+2] = a\cdot a+a\cdot b+2\cdot a-a\cdot b-b\cdot b-2\cdot b = a^2+ab+2a-ab-b^2-2b$

$= a^2+2a-b^2-2b = a^2-b^2+2a-2b$

Example 3.5-29:

$\boxed{x^2 - y^2 + 7x + 7y} = \boxed{\left(x^2 - y^2\right) + 7(x + y)} = \boxed{(x - y)(x + y) + 7(x + y)} = \boxed{(x + y)[(x - y) + 7]}$

$= \boxed{(x + y)[x - y + 7]}$

Check: $(x + y)[x - y + 7] = x \cdot x - x \cdot y + 7 \cdot x + x \cdot y - y \cdot y + 7 \cdot y = x^2 - \cancel{xy} + 7x + \cancel{xy} - y^2 + 7y$

$= x^2 + 7x - y^2 + 7y = x^2 - y^2 + 7x + 7y$

Example 3.5-30:

$\boxed{5x^4 + 20x^3 - 25x^2 + 3} = \boxed{5x^2\left(x^2 + 4x - 5\right) + 3} = \boxed{5x^2(x + 5)(x - 1) + 3}$

Check: $5x^2(x + 5)(x - 1) + 3 = 5x^2(x \cdot x - x + 5 \cdot x - 5) + 3 = 5x^2\left(x^2 - x + 5x - 5\right) + 3 = 5x^2\left(x^2 + 4x - 5\right) + 3$

$= 5x^2 \cdot x^2 + 5x^2 \cdot 4x + 5x^2 \cdot -5 + 3 = 5x^4 + 20x^3 - 25x^2 + 3$

Practice Problems - Factoring Other Types of Polynomials

Section 3.5 Case II Practice Problems - Factor the following polynomials completely:

1. $a^3 + 7a^2 - 9a - 63 =$
2. $2x^2 + 16x - 40 =$
3. $a^2 - b^2 + 9a + 9b =$
4. $6y^2 + 39y + 60 =$
5. $2w^3 - 4w^2 - 16w =$
6. $-25x^4 + 70x^2y - 49y^2 =$
7. $12y^3 + 26y^2 + 10y =$
8. $24x^3 + 74x^2 - 35x =$
9. $24x^2 + 78x + 45 =$
10. $16u^8 - 256u^4 =$

Chapter 4
Quadratic Equations and Factoring

Quick Reference to Chapter 4 Case Problems

Chapter 4 - Quadratic Equations and Factoring

The objective of this chapter is to improve the student's ability to solve equations and provide additional factoring methods involving quadratic equations. Quadratic equations and the quadratic formula are introduced in Section 4.1. Solving different forms of quadratic equations using the quadratic formula is introduced in Section 4.2. Steps as to how quadratic equations are solved using the Square Root Property method are addressed in Section 4.3. Solving quadratic equations by Completing-the-Square method are addressed in Section 4.4. In Section 4.5, solving quadratic equations containing radicals and fractions are discussed. Choosing the most suitable method in factoring polynomials and solving second degree equations is discussed in Section 4.6. Cases presented in each section are concluded by solving additional examples with practice problems to further enhance the students ability. Students are encouraged to gain a thorough knowledge on the different factoring and solution methods introduced in Chapters 3 and 4 prior to studying Chapter 5. Knowing how to factor polynomials and solve quadratic equations will greatly improve the student's ability in solving algebraic fractions. A subject which is addressed in Chapter 5.

4.1 Quadratic Equations and the Quadratic Formula

A quadratic equation is an equation in which the highest power of the variable is 2. For example, $3x^2 - 16x + 5 = 0$, $x^2 = 16$, $w^2 + 9w = 0$, $x^2 - 4x + 3 = 0$, $x^2 = -11x - 24$, and $y^2 - 4 = 0$ are all examples of quadratic equations. Note that any equation that can be written in the form of $ax^2 + bx + c = 0$, where a, b, and c are real numbers and $a \neq 0$, is called a **quadratic equation**. A quadratic equation represented in the form of $ax^2 + bx + c = 0$ is said to be in its **standard form**. In the following sections we will learn how to solve and represent the solutions to quadratic equations in factored form. However, in order to solve any quadratic equation we first need to become familiar with the quadratic formula.

The Quadratic Formula

To derive the quadratic formula we start with the standard quadratic equation $ax^2 + bx + c = 0$, where a, b, and c are real numbers and use the method of completing the square to solve the equation as follows:

Step 1 Add $-c$ to both sides of the equation.

$$ax^2 + bx + c - c = -c \; ; \; ax^2 + bx = -c$$

Step 2 Divide both sides of the equation by a.

$$\frac{ax^2}{a} + \frac{bx}{a} = -\frac{c}{a} \; ; \; x^2 + \frac{bx}{a} = -\frac{c}{a}$$

Step 3 Divide $\frac{b}{a}$, the coefficient of x, by 2 and square the term to obtain $\left(\frac{b}{2a}\right)^2$. Add $\left(\frac{b}{2a}\right)^2$ to both sides of the equation.

$$x^2 + \frac{b}{a}x + \left(\frac{b}{2a}\right)^2 = -\frac{c}{a} + \left(\frac{b}{2a}\right)^2$$

Step 4 Write the left hand side of the equation , which is a perfect square trinomial, in its equivalent square form.

$$\left(x+\frac{b}{2a}\right)^2=-\frac{c}{a}+\left(\frac{b}{2a}\right)^2$$

Step 5 Simplify the right hand side of the equation using the fraction techniques.

$$\left(x+\frac{b}{2a}\right)^2=-\frac{c}{a}+\left(\frac{b}{2a}\right)^2 \; ; \; \left(x+\frac{b}{2a}\right)^2=-\frac{c}{a}+\frac{b^2}{4a^2} \; ; \; \left(x+\frac{b}{2a}\right)^2=\frac{\left(4a^2\cdot -c\right)+\left(a\cdot b^2\right)}{4a^2\cdot a}$$

$$; \; \left(x+\frac{b}{2a}\right)^2=\frac{ab^2-4a^2c}{4a^3} \; ; \; \left(x+\frac{b}{2a}\right)^2=\frac{a\left(b^2-4ac\right)}{4a^3} \; ; \; \left(x+\frac{b}{2a}\right)^2=\frac{b^2-4ac}{4a^2}$$

Step 6 Take the square root of both sides of the equation.

$$\sqrt{\left(x+\frac{b}{2a}\right)^2}=\pm\sqrt{\frac{b^2-4ac}{4a^2}} \; ; \; x+\frac{b}{2a}=\pm\frac{\sqrt{b^2-4ac}}{\sqrt{2^2a^2}} \; ; \; x+\frac{b}{2a}=\pm\frac{\sqrt{b^2-4ac}}{2a}$$

Step 7 Solve for x by adding $-\frac{b}{2a}$ to both sides of the equation.

$$x+\frac{b}{2a}-\frac{b}{2a}=-\frac{b}{2a}\pm\frac{\sqrt{b^2-4ac}}{2a} \; ; \; x=-\frac{b}{2a}\pm\frac{\sqrt{b^2-4ac}}{2a} \; ; \; x=\frac{-b\pm\sqrt{b^2-4ac}}{2a}$$

The equation $\boldsymbol{x=\frac{-b\pm\sqrt{b^2-4ac}}{2a}}$ is referred to as the **quadratic formula**. Note that the quadratic formula has two solutions $x=\frac{-b+\sqrt{b^2-4ac}}{2a}$ and $x=\frac{-b-\sqrt{b^2-4ac}}{2a}$. We use these solutions to write the quadratic equation $ax^2+bx+c=0$ in its equivalent factored form, i.e.,

$$\boldsymbol{ax^2+bx+c=0} \text{ is factorable to } \boldsymbol{\left(x-\frac{-b+\sqrt{b^2-4ac}}{2a}\right)\left(x+\frac{b+\sqrt{b^2-4ac}}{2a}\right)=0}$$

Let's check the above factored product using the FOIL method. The result should be equal to $ax^2+bx+c=0$.

Check:

$$\left(x-\frac{-b+\sqrt{b^2-4ac}}{2a}\right)\left(x+\frac{b+\sqrt{b^2-4ac}}{2a}\right)=0 \; ; \; \left(x+\frac{b-\sqrt{b^2-4ac}}{2a}\right)\left(x+\frac{b+\sqrt{b^2-4ac}}{2a}\right)=0$$

$$; \; x\cdot x+\left(\frac{b+\sqrt{b^2-4ac}}{2a}\right)\cdot x+\left(\frac{b-\sqrt{b^2-4ac}}{2a}\right)\cdot x+\left(\frac{b+\sqrt{b^2-4ac}}{2a}\right)\cdot\left(\frac{b-\sqrt{b^2-4ac}}{2a}\right)=0$$

$$; \; x^2+\left(\frac{b+\sqrt{b^2-4ac}}{2a}+\frac{b-\sqrt{b^2-4ac}}{2a}\right)x+\left(\frac{\left(b+\sqrt{b^2-4ac}\right)\left(b-\sqrt{b^2-4ac}\right)}{2a\cdot 2a}\right)=0$$

$$; \; x^2+\left(\frac{b+\sqrt{b^2-4ac}+b-\sqrt{b^2-4ac}}{2a}\right)x+\left(\frac{\left(b^2-b\sqrt{b^2-4ac}+b\sqrt{b^2-4ac}-\sqrt{b^2-4ac}\cdot\sqrt{b^2-4ac}\right)}{4a^2}\right)=0$$

$; \; x^2 + \left(\frac{b+b}{2a}\right)x + \left[\frac{b^2 - \left(b^2 - 4ac\right)}{4a^2}\right] = 0 \; ; \; x^2 + \left(\frac{2b}{2a}\right)x + \left(\frac{b^2 - b^2 + 4ac}{4a^2}\right) = 0 \; ; \; x^2 + \frac{b}{a}x + \frac{4ac}{4a^2} = 0$

$; \; x^2 + \frac{b}{a}x + \frac{c}{a} = 0 \; ; \; \frac{x^2}{1} + \frac{bx}{a} + \frac{c}{a} = 0 \; ; \; \frac{ax^2 + bx + c}{a} = 0 \; ; \; \frac{ax^2 + bx + c}{a} = \frac{0}{1} \; ; \; \left(ax^2 + bx + c\right) \cdot 1 = a \cdot 0$ which is the same as $ax^2 + bx + c = 0$.

The quadratic formula is a powerful formula and should be memorized. In the following sections we will use this formula to solve different types of quadratic equations.

Practice Problems - Quadratic Equations and Quadratic Formula

Section 4.1 Practice Problems - Given the following quadratic equations identify the coefficients a, b, and c.

1. $3x = -5 + 2x^2$
2. $2x^2 = 5$
3. $3w^2 - 5w = 2$
4. $15 = -y^2 - 3$
5. $x^2 + 3 = 5x$
6. $-u^2 + 2 = 3u$
7. $y^2 + 5y - 2 = 0$
8. $-3x^2 = 2x - 1$
9. $p^2 = p - 1$
10. $3x - 2 = x^2$

4.2 Solving Quadratic Equations Using the Quadratic Formula

As was stated earlier, the quadratic formula can be used to solve any quadratic equation by expressing the equation in the standard form of $ax^2 + bx + c = 0$ and by substituting the equivalent numbers for a, b, and c into the quadratic formula. In this section we will learn how to solve quadratic equations of the form $ax^2 + bx + c = 0$, where $a = 1$ (Case I) and where $a > 1$ (Case II), using the quadratic formula.

Case I Solving Quadratic Equations of the Form $ax^2 + bx + c = 0$, where $a = 1$, Using the Quadratic Formula

Quadratic equations of the form $ax^2 + bx + c = 0$, where $a = 1$, are solved using the following steps:

Step 1 Write the equation in standard form.

Step 2 Identify the coefficients a, b, and c.

Step 3 Substitute the values for a, b, and c into the quadratic equation $x = \frac{-b \pm \sqrt{b^2 - 4ac}}{2a}$. Simplify the equation.

Step 4 Solve for the values of x. Check the answers by either substituting the x values into the original equation or by multiplying the factored product using the FOIL method.

Step 5 Write the quadratic equation in its factored form.

Examples with Steps

The following examples show the steps as to how quadratic equations are solved using the quadratic formula:

Example 4.2-1

Solve the quadratic equation $x^2 + 5x = -4$.

Solution:

Step 1 $\boxed{x^2 + 5x = -4}$; $\boxed{x^2 + 5x + 4 = -4 + 4}$; $\boxed{x^2 + 5x + 4 = 0}$

Step 2 Let: $\boxed{a = 1}$, $\boxed{b = 5}$, and $\boxed{c = 4}$. Then,

Step 3 Given: $\boxed{x = \frac{-b \pm \sqrt{b^2 - 4ac}}{2a}}$; $\boxed{x = \frac{-5 \pm \sqrt{5^2 - 4 \times 1 \times 4}}{2 \times 1}}$; $\boxed{x = \frac{-5 \pm \sqrt{25 - 16}}{2}}$

; $\boxed{x = \frac{-5 \pm \sqrt{9}}{2}}$; $\boxed{x = \frac{-5 \pm \sqrt{3^2}}{2}}$; $\boxed{x = \frac{-5 \pm 3}{2}}$

Step 4 Separate $x = \frac{-5 \pm 3}{2}$ into two equations.

I. $\boxed{x = \frac{-5 + 3}{2}}$; $\boxed{x = -\frac{2}{2}}$; $\boxed{x = -\frac{1}{1}}$; $\boxed{\mathbf{x = -1}}$

II. $\boxed{x=\frac{-5-3}{2}}$; $\boxed{x=-\frac{\overset{4}{\cancel{8}}}{\cancel{2}}}$; $\boxed{x=-\frac{4}{1}}$; $\boxed{x=-4}$

Check No. 1: I. *Let* $x=-1$ *in* $x^2+5x=-4$; $(-1)^2+(5\times-1)\overset{?}{=}-4$; $1-5\overset{?}{=}-4$; $-4=-4$

II. *Let* $x=-4$ *in* $x^2+5x=-4$; $(-4)^2+(5\times-4)\overset{?}{=}-4$; $16-20\overset{?}{=}-4$; $-4=-4$

Check No. 2: $x^2+5x+4\overset{?}{=}(x+1)(x+4)$; $x^2+5x+4\overset{?}{=}(x\cdot x)+(4\cdot x)+(1\cdot x)+(1\cdot 4)$

; $x^2+5x+4\overset{?}{=}x^2+4x+x+4$; $x^2+5x+4\overset{?}{=}x^2+(4+1)x+4$; $x^2+5x+4=x^2+5x+4$

Step 5 Therefore, the equation $x^2+5x+4=0$ can be factored to $\mathbf{(x+1)(x+4)=0}$.

Example 4.2-2

Solve the quadratic equation $x^2=-12x-35$.

Solution:

Step 1 $\boxed{x^2=-12x-35}$; $\boxed{x^2+12x=-12x+12x-35}$; $\boxed{x^2+12x=0-35}$; $\boxed{x^2+12x=-35}$

; $\boxed{x^2+12x+35=-35+35}$; $\boxed{x^2+12x+35=0}$

Step 2 Let: $\boxed{a=1}$, $\boxed{b=12}$, and $\boxed{c=35}$. Then,

Step 3 Given: $\boxed{x=\frac{-b\pm\sqrt{b^2-4ac}}{2a}}$; $\boxed{x=\frac{-12\pm\sqrt{12^2-4\times1\times35}}{2\times1}}$; $\boxed{x=\frac{-12\pm\sqrt{144-140}}{2}}$

; $\boxed{x=\frac{-12\pm\sqrt{4}}{2}}$; $\boxed{x=\frac{-12\pm\sqrt{2^2}}{2}}$; $\boxed{x=\frac{-12\pm2}{2}}$

Step 4 Separate $x=\frac{-12\pm2}{2}$ into two equations.

I. $\boxed{x=\frac{-12+2}{2}}$; $\boxed{x=-\frac{\overset{5}{\cancel{10}}}{\cancel{2}}}$; $\boxed{x=-\frac{5}{1}}$; $\boxed{x=-5}$

II. $\boxed{x=\frac{-12-2}{2}}$; $\boxed{x=-\frac{\overset{7}{\cancel{14}}}{\cancel{2}}}$; $\boxed{x=-\frac{7}{1}}$; $\boxed{x=-7}$

Check No. 1: I. *Let* $x=-5$ *in* $x^2=-12x-35$; $(-5)^2\overset{?}{=}(-12\times-5)-35$; $25\overset{?}{=}60-35$; $25=25$

II. *Let* $x=-7$ *in* $x^2=-12x-35$; $(-7)^2\overset{?}{=}(-12\times-7)-35$; $49\overset{?}{=}84-35$; $49=49$

Check No. 2: $x^2+12x+35\overset{?}{=}(x+5)(x+7)$; $x^2+12x+35\overset{?}{=}(x\cdot x)+(7\cdot x)+(5\cdot x)+(5\cdot 7)$

; $x^2+12x+35\overset{?}{=}x^2+7x+5x+35$; $x^2+12x+35\overset{?}{=}x^2+(7+5)x+35$

; $x^2+12x+35=x^2+12x+35$

Step 5 Therefore, the equation $x^2+12x+35=0$ can be factored to $\mathbf{(x+5)(x+7)=0}$.

Example 4.2-3

Solve the quadratic equation $x^2 - 5x + 6 = 0$.

Solution:

Step 1 $\boxed{\textit{Not Applicable}}$

Step 2 Let: $\boxed{a = 1}$, $\boxed{b = -5}$, and $\boxed{c = 6}$. Then,

Step 3 Given: $\boxed{x = \frac{-b \pm \sqrt{b^2 - 4ac}}{2a}}$; $\boxed{x = \frac{-(-5) \pm \sqrt{(-5)^2 - 4 \times 1 \times 6}}{2 \times 1}}$; $\boxed{x = \frac{5 \pm \sqrt{25 - 24}}{2}}$

; $\boxed{x = \frac{5 \pm \sqrt{1}}{2}}$; $\boxed{x = \frac{5 \pm 1}{2}}$

Step 4 Separate $x = \frac{5 \pm 1}{2}$ into two equations.

I. $\boxed{x = \frac{5 + 1}{2}}$; $\boxed{x = \frac{\overset{3}{\cancel{6}}}{\cancel{2}}}$; $\boxed{x = \frac{3}{1}}$; $\boxed{\mathbf{x = 3}}$

II. $\boxed{x = \frac{5 - 1}{2}}$; $\boxed{x = \frac{\overset{2}{\cancel{4}}}{\cancel{2}}}$; $\boxed{x = \frac{2}{1}}$; $\boxed{\mathbf{x = 2}}$

Check No. 1: I. *Let* $x = 3$ *in* $x^2 - 5x + 6 = 0$; $(3)^2 + (-5 \times 3) + 6 \stackrel{?}{=} 0$; $9 - 15 + 6 \stackrel{?}{=} 0$; $15 - 15 \stackrel{?}{=} 0$; $0 = 0$

II. *Let* $x = 2$ *in* $x^2 - 5x + 6 = 0$; $(2)^2 + (-5 \times 2) + 6 \stackrel{?}{=} 0$; $4 - 10 + 6 \stackrel{?}{=} 0$; $4 - 4 \stackrel{?}{=} 0$; $0 = 0$

Check No. 2: $x^2 - 5x + 6 \stackrel{?}{=} (x - 3)(x - 2)$; $x^2 - 5x + 6 \stackrel{?}{=} (x \cdot x) + (-2 \cdot x) + (-3 \cdot x) + (-3 \cdot -2)$; $x^2 - 5x + 6 \stackrel{?}{=} x^2 - 2x - 3x + 6$; $x^2 - 5x + 6 \stackrel{?}{=} x^2 + (-2 - 3)x + 6$; $x^2 - 5x + 6 = x^2 - 5x + 6$

Step 5 Therefore, the equation $x^2 - 5x + 6 = 0$ can be factored to $\mathbf{(x - 3)(x - 2) = 0}$.

Example 4.2-4

Solve the quadratic equation $x^2 + 1 = -2x$.

Solution:

Step 1 $\boxed{x^2 + 1 = -2x}$; $\boxed{x^2 - 2x + 1 = -2x + 2x}$; $\boxed{x^2 + 2x + 1 = 0}$

Step 2 Let: $\boxed{a = 1}$, $\boxed{b = 2}$, and $\boxed{c = 1}$. Then,

Step 3 Given: $\boxed{x = \frac{-b \pm \sqrt{b^2 - 4ac}}{2a}}$; $\boxed{x = \frac{-2 \pm \sqrt{2^2 - 4 \times 1 \times 1}}{2 \times 1}}$; $\boxed{x = \frac{-2 \pm \sqrt{4 - 4}}{2}}$; $\boxed{x = \frac{-2 \pm 0}{2}}$

Step 4 Separate $x = \frac{-2 \pm 0}{2}$ into two equations.

I. $\boxed{x = \frac{-2+0}{2}}$; $\boxed{x = -\frac{\cancel{2}}{\cancel{2}}}$; $\boxed{x = -\frac{1}{1}}$; $\boxed{\mathbf{x = -1}}$

II. $\boxed{x = \frac{-2-0}{2}}$; $\boxed{x = -\frac{\cancel{2}}{\cancel{2}}}$; $\boxed{x = -\frac{1}{1}}$; $\boxed{\mathbf{x = -1}}$

Check No. 1: *Let* $x = -1$ *in* $x^2 + 1 = -2x$; $(-1)^2 + 1 \stackrel{?}{=} -2 \times -1$; $1 + 1 \stackrel{?}{=} 2$; $2 = 2$

Check No. 2: $x^2 + 2x + 1 \stackrel{?}{=} (x+1)(x+1)$; $x^2 + 2x + 1 \stackrel{?}{=} (x \cdot x) + (1 \cdot x) + (1 \cdot x) + (1 \cdot 1)$

; $x^2 + 2x + 1 \stackrel{?}{=} x^2 + x + x + 1$; $x^2 + 2x + 1 \stackrel{?}{=} x^2 + (1+1)x + 1$; $x^2 + 2x + 1 = x^2 + 2x + 1$

Step 5 Thus, the equation $x^2 + 2x + 1 = 0$ has two identical solutions and can be factored to $\mathbf{(x+1)(x+1) = 0}$

Example 4.2-5

Solve the quadratic equation $7x = -x^2 - 2$.

Solution:

Step 1 $\boxed{7x = -x^2 - 2}$; $\boxed{+x^2 + 7x = -x^2 + x^2 - 2}$; $\boxed{x^2 + 7x = 0 - 2}$; $\boxed{x^2 + 7x = -2}$

; $\boxed{x^2 + 7x + 2 = -2 + 2}$; $\boxed{x^2 + 7x + 2 = 0}$

Step 2 Let: $\boxed{a = 1}$, $\boxed{b = 7}$, and $\boxed{c = 2}$. Then,

Step 3 Given: $\boxed{x = \frac{-b \pm \sqrt{b^2 - 4ac}}{2a}}$; $\boxed{x = \frac{-7 \pm \sqrt{7^2 - 4 \times 1 \times 2}}{2 \times 1}}$; $\boxed{x = \frac{-7 \pm \sqrt{49 - 8}}{2}}$

; $\boxed{x = \frac{-7 \pm \sqrt{41}}{2}}$; $\boxed{x = \frac{-7 \pm 6.4}{2}}$

Step 4 Separate $x = \frac{-7 \pm 6.4}{2}$ into two equations.

I. $\boxed{x = \frac{-7 + 6.4}{2}}$; $\boxed{x = -\frac{0.6}{2}}$; $\boxed{\mathbf{x = -0.3}}$

II. $\boxed{x = \frac{-7 - 6.4}{2}}$; $\boxed{x = -\frac{13.4}{2}}$; $\boxed{\mathbf{x = -6.7}}$

Check No. 1: I. *Let* $x = -0.3$ *in* $7x = -x^2 - 2$; $7 \times -0.3 \stackrel{?}{=} -(-0.3)^2 - 2$; $-2.1 \stackrel{?}{=} -0.09 - 2$; $-2.1 = -2.1$

II. *Let* $x = -6.7$ *in* $7x = -x^2 - 2$; $7 \times -6.7 \stackrel{?}{=} -(-6.7)^2 - 2$; $-46.9 \stackrel{?}{=} -44.9 - 2$

; $-46.9 = -46.9$

Check No. 2: $x^2 + 7x + 2 \stackrel{?}{=} (x + 0.3)(x + 6.7)$; $x^2 + 7x + 2 \stackrel{?}{=} (x \cdot x) + (6.7 \cdot x) + (0.3 \cdot x) + (0.3 \cdot 6.7)$

; $x^2 + 7x + 2 \stackrel{?}{=} x^2 + 6.7x + 0.3x + 2$; $x^2 + 7x + 2 \stackrel{?}{=} x^2 + (6.7 + 0.3)x + 2$

$; \ x^2 + 7x + 2 = x^2 + 7x + 2$

Step 5 Thus, the equation $x^2 + 7x + 2 = 0$ can be factored to $\mathbf{(x + 0.3)(x + 6.7) = 0}$.

Additional Examples - Solving Quadratic Equations of the Form $ax^2 + bx + c = 0$, where $a = 1$, Using the Quadratic Formula

The following examples further illustrate how to solve quadratic equations using the quadratic formula:

Example 4.2-6

Solve the quadratic equation $x^2 = 16x - 55$.

Solution:

First, write the equation in standard form, i.e., $x^2 - 16x + 55 = 0$

Next, let: $\boxed{a = 1}$, $\boxed{b = -16}$, and $\boxed{c = 55}$. Then,

Given: $\boxed{x = \frac{-b \pm \sqrt{b^2 - 4ac}}{2a}}$; $\boxed{x = \frac{-(-16) \pm \sqrt{(-16)^2 - 4 \times 1 \times 55}}{2 \times 1}}$; $\boxed{x = \frac{16 \pm \sqrt{256 - 220}}{2}}$; $\boxed{x = \frac{16 \pm \sqrt{36}}{2}}$

; $\boxed{x = \frac{16 \pm \sqrt{6^2}}{2}}$; $\boxed{x = \frac{16 \pm 6}{2}}$ Therefore:

I. $\boxed{x = \frac{16 + 6}{2}}$; $\boxed{x = \frac{\overset{11}{\cancel{22}}}{\cancel{2}}}$; $\boxed{x = \frac{11}{1}}$; $\boxed{\mathbf{x = 11}}$ II. $\boxed{x = \frac{16 - 6}{2}}$; $\boxed{x = \frac{\overset{5}{\cancel{10}}}{\cancel{2}}}$; $\boxed{x = \frac{5}{1}}$; $\boxed{\mathbf{x = 5}}$

Check No. 1: I. *Let* $x = 11$ *in* $x^2 = 16x - 55$; $11^2 \overset{?}{=} 16 \times 11 - 55$; $121 \overset{?}{=} 176 - 55$; $121 = 121$

II. *Let* $x = 5$ *in* $x^2 = 16x - 55$; $5^2 \overset{?}{=} 16 \times 5 - 55$; $25 \overset{?}{=} 80 - 55$; $25 = 25$

Check No. 2: $x^2 - 16x + 55 \overset{?}{=} (x - 11)(x - 5)$; $x^2 - 16x + 55 \overset{?}{=} (x \cdot x) + (-5 \cdot x) + (-11 \cdot x) + (-11 \cdot -5)$

; $x^2 - 16x + 55 \overset{?}{=} x^2 - 5x - 11x + 55$; $x^2 - 16x + 55 \overset{?}{=} x^2 + (-5 - 11)x + 55$

; $x^2 - 16x + 55 = x^2 - 16x + 55$

Therefore, the equation $x^2 - 16x + 55 = 0$ can be factored to $\mathbf{(x - 11)(x - 5) = 0}$.

Example 4.2-7

Solve the quadratic equation $x^2 = -9x + 36$.

Solution:

First, write the equation in standard form, i.e., $x^2 + 9x - 36 = 0$

Next, let: $\boxed{a = 1}$, $\boxed{b = 9}$, and $\boxed{c = -36}$. Then,

Given: $\boxed{x = \frac{-b \pm \sqrt{b^2 - 4ac}}{2a}}$; $\boxed{x = \frac{-9 \pm \sqrt{9^2 - 4 \times 1 \times -36}}{2 \times 1}}$; $\boxed{x = \frac{-9 \pm \sqrt{81 + 144}}{2}}$; $\boxed{x = \frac{-9 \pm \sqrt{225}}{2}}$

; $\boxed{x=\frac{-9\pm\sqrt{15^2}}{2}}$; $\boxed{x=\frac{-9\pm 15}{2}}$ Therefore:

I. $\boxed{x=\frac{-9+15}{2}}$; $\boxed{x=\frac{\overset{3}{\cancel{6}}}{\cancel{2}}}$; $\boxed{x=\frac{3}{1}}$; $\boxed{\mathbf{x=3}}$ II. $\boxed{x=\frac{-9-15}{2}}$; $\boxed{x=-\frac{\overset{12}{\cancel{24}}}{\cancel{2}}}$; $\boxed{x=-\frac{12}{1}}$; $\boxed{\mathbf{x=-12}}$

and the solution set is $\{\mathbf{3, -12}\}$.

Check No. 1: I. *Let* $x=3$ *in* $x^2=-9x+36$; $3^2\overset{?}{=}-9\times3+36$; $9\overset{?}{=}-27+36$; $9=9$

II. *Let* $x=-12$ *in* $x^2=-9x+36$; $(-12)^2\overset{?}{=}-9\times-12+36$; $144\overset{?}{=}108+36$; $144=144$

Check No. 2: $x^2+9x-36\overset{?}{=}(x-3)(x+12)$; $x^2+9x-36\overset{?}{=}(x\cdot x)+(12\cdot x)+(-3\cdot x)+(-3\cdot 12)$

; $x^2+9x-36\overset{?}{=}x^2+12x-3x-36$; $x^2+9x-36\overset{?}{=}x^2+(12-3)x-36$

; $x^2+9x-36=x^2+9x-36$

Therefore, the equation $x^2+9x-36=0$ can be factored to $\mathbf{(x-3)(x+12)=0}$.

Example 4.2-8

Solve the quadratic equation $x^2+11x+24=0$.

Solution:

The equation is already in standard form.

Let: $\boxed{a=1}$, $\boxed{b=11}$, and $\boxed{c=24}$. Then,

Given: $\boxed{x=\frac{-b\pm\sqrt{b^2-4ac}}{2a}}$; $\boxed{x=\frac{-11\pm\sqrt{11^2-4\times1\times24}}{2\times1}}$; $\boxed{x=\frac{-11\pm\sqrt{121-96}}{2}}$; $\boxed{x=\frac{-11\pm\sqrt{25}}{2}}$

; $\boxed{x=\frac{-11\pm\sqrt{5^2}}{2}}$; $\boxed{x=\frac{-11\pm5}{2}}$ Therefore:

I. $\boxed{x=\frac{-11+5}{2}}$; $\boxed{x=-\frac{\overset{3}{\cancel{6}}}{\cancel{2}}}$; $\boxed{x=-\frac{3}{1}}$; $\boxed{\mathbf{x=-3}}$ II. $\boxed{x=\frac{-11-5}{2}}$; $\boxed{x=-\frac{\overset{8}{\cancel{16}}}{\cancel{2}}}$; $\boxed{x=-\frac{8}{1}}$; $\boxed{\mathbf{x=-8}}$

and the solution set is $\{\mathbf{-3, -8}\}$.

Check No. 1: I. *Let* $x=-3$ *in* $x^2+11x+24=0$; $(-3)^2+11\times-3+24\overset{?}{=}0$; $9-33+24\overset{?}{=}0$; $0=0$

II. *Let* $x=-8$ *in* $x^2+11x+24=0$; $(-8)^2+11\times-8+24\overset{?}{=}0$; $64-88+24\overset{?}{=}0$; $0=0$

Check No. 2: $x^2+11x+24\overset{?}{=}(x+3)(x+8)$; $x^2+11x+24\overset{?}{=}(x\cdot x)+(8\cdot x)+(3\cdot x)+(3\cdot 8)$

; $x^2+11x+24\overset{?}{=}x^2+8x+3x+24$; $x^2+11x+24\overset{?}{=}x^2+(8+3)x+24$

; $x^2+11x+24=x^2+11x+24$

Therefore, the equation $x^2+11x+24=0$ can be factored to $\mathbf{(x+3)(x+8)=0}$.

Example 4.2-9

Solve the quadratic equation $9 = -x^2 - 6x$.

Solution:

First, write the equation in standard form, i.e., $x^2 + 6x + 9 = 0$.

Next, let: $\boxed{a = 1}$, $\boxed{b = 6}$, and $\boxed{c = 9}$. Then,

Given: $\boxed{x = \frac{-b \pm \sqrt{b^2 - 4ac}}{2a}}$; $\boxed{x = \frac{-6 \pm \sqrt{6^2 - 4 \times 1 \times 9}}{2 \times 1}}$; $\boxed{x = \frac{-6 \pm \sqrt{36 - 36}}{2}}$; $\boxed{x = \frac{-6 \pm \sqrt{0}}{2}}$; $\boxed{x = \frac{-6 \pm 0}{2}}$

; $\boxed{x = -\frac{\overset{3}{\cancel{6}}}{\cancel{2}}}$; $\boxed{x = -\frac{3}{1}}$; $\boxed{x = -3}$

In this case the equation has one repeated solution, i.e., $\boxed{x = -3}$ and $\boxed{x = -3}$.

Thus, the solution set is $\{-3, -3\}$.

Check No. 1: *Let* $x = -3$ *in* $x^2 + 6x + 9 = 0$; $(-3)^2 + 6 \times -3 + 9 \overset{?}{=} 0$; $9 - 18 + 9 \overset{?}{=} 0$; $18 - 18 \overset{?}{=} 0$; $0 = 0$

Check No. 2: $x^2 + 6x + 9 \overset{?}{=} (x+3)(x+3)$; $x^2 + 6x + 9 \overset{?}{=} (x \cdot x) + (3 \cdot x) + (3 \cdot x) + (3 \cdot 3)$

; $x^2 + 6x + 9 \overset{?}{=} x^2 + 3x + 3x + 9$; $x^2 + 6x + 9 \overset{?}{=} x^2 + (3+3)x + 9$; $x^2 + 6x + 9 = x^2 + 6x + 9$

Therefore, the equation $x^2 + 6x + 9 = 0$ can be factored to $(x+3)(x+3) = 0$.

Example 4.2-10

Solve the quadratic equation $w^2 + 1 = -5w$.

Solution:

First, write the equation in standard form, i.e., $w^2 + 5w + 1 = 0$.

Next, let: $\boxed{a = 1}$, $\boxed{b = 5}$, and $\boxed{c = 1}$. Then,

Given: $\boxed{w = \frac{-b \pm \sqrt{b^2 - 4ac}}{2a}}$; $\boxed{w = \frac{-5 \pm \sqrt{5^2 - 4 \times 1 \times 1}}{2 \times 1}}$; $\boxed{w = \frac{-5 \pm \sqrt{25 - 4}}{2}}$; $\boxed{w = \frac{-5 \pm \sqrt{21}}{2}}$

; $\boxed{w = \frac{-5 \pm 4.58}{2}}$ Therefore:

I. $\boxed{w = \frac{-5 + 4.58}{2}}$; $\boxed{w = -\frac{0.42}{2}}$; $\boxed{w = -0.21}$

II. $\boxed{w = \frac{-5 - 4.58}{2}}$; $\boxed{x = -\frac{9.58}{2}}$; $\boxed{w = -4.79}$

and the solution set is $\{-0.21, -4.79\}$.

Check No. 1: I. *Let* $w = -0.21$ *in* $w^2 + 1 = -5w$; $(-0.21)^2 + 1 \overset{?}{=} -5 \times -0.21$; $0.05 + 1 \overset{?}{=} 1.05$; $1.05 = 1.05$

II. *Let* $w = -4.79$ *in* $w^2 + 1 = -5w$; $(-4.79)^2 + 1 \overset{?}{=} -5 \times -4.79$; $22.9 + 1 \overset{?}{=} 23.9$; $23.9 = 23.9$

Check No. 2: $w^2 + 5w + 1 \overset{?}{=} (w + 0.21)(w + 4.79)$; $w^2 + 5w + 1 \overset{?}{=} (w \cdot w) + (4.79 \cdot w) + (0.21 \cdot w) + (0.21 \cdot 4.79)$

$; \; w^2+5w+1\overset{?}{=}w^2+4.79w+0.21w+1 \; ; \; w^2+5w+1\overset{?}{=}w^2+(4.79+0.21)w+1$

$; \; w^2+5w+1=w^2+5w+1$

Therefore, the equation $w^2+5w+1=0$ can be factored to $\mathbf{(w+0.21)(w+4.79)=0}$.

Note that when $c=0$ the quadratic equation $ax^2+bx+c=0$ reduces to $ax^2+bx=0$. For cases where $a=1$, we can solve equations of the form $x^2+bx=0$ using the quadratic formula in the following way:

Example 4.2-11

Solve the quadratic equation $x^2+5x=0$.

Solution:

The equation is already in standard form.

Let: $\boxed{a=1}$, $\boxed{b=5}$, and $\boxed{c=0}$. Then,

Given: $\boxed{x=\frac{-b\pm\sqrt{b^2-4ac}}{2a}}$; $\boxed{x=\frac{-5\pm\sqrt{5^2-4\times1\times0}}{2\times1}}$; $\boxed{x=\frac{-5\pm\sqrt{25-0}}{2}}$; $\boxed{x=\frac{-5\pm\sqrt{25}}{2}}$; $\boxed{x=\frac{-5\pm\sqrt{5^2}}{2}}$

; $\boxed{x=\frac{-5\pm5}{2}}$ Therefore:

I. $\boxed{x=\frac{-5+5}{2}}$; $\boxed{x=\frac{0}{2}}$; $\boxed{\mathbf{x=0}}$ II. $\boxed{x=\frac{-5-5}{2}}$; $\boxed{x=-\frac{\overset{5}{\cancel{10}}}{\cancel{2}}}$; $\boxed{x=-\frac{5}{1}}$; $\boxed{x=-5}$; $\boxed{\mathbf{x=-5}}$

and the solution set is $\{\mathbf{0,-5}\}$.

Check No. 1: I. *Let* $x=0$ *in* $x^2+5x=0$; $0^2+5\cdot0\overset{?}{=}0$; $0+0\overset{?}{=}0$; $0=0$

II. *Let* $x=-5$ *in* $x^2+5x=0$; $(-5)^2+5\cdot-5\overset{?}{=}0$; $25-25\overset{?}{=}0$; $0=0$

Check No. 2: $x^2+5x\overset{?}{=}(x+0)(x+5)$; $x^2+5x\overset{?}{=}(x\cdot x)+(5\cdot x)+(0\cdot x)+(0\cdot5)$; $x^2+5x\overset{?}{=}x^2+5x+0+0$

; $x^2+5x=x^2+5x$

Therefore, the equation $x^2+5x=0$ can be factored to $(x+0)(x+5)=0$ which is the same as $\mathbf{x(x+5)=0}$.

Example 4.2-12

Solve the quadratic equation $x^2=9x$.

Solution:

First, write the equation in standard form, i.e., $x^2-9x=0$.

Next, let: $\boxed{a=1}$, $\boxed{b=-9}$, and $\boxed{c=0}$. Then,

Given: $\boxed{x=\frac{-b\pm\sqrt{b^2-4ac}}{2a}}$; $\boxed{x=\frac{-(-9)\pm\sqrt{(-9)^2-4\times1\times0}}{2\times1}}$; $\boxed{x=\frac{9\pm\sqrt{81-0}}{2}}$; $\boxed{x=\frac{9\pm\sqrt{81}}{2}}$

; $\boxed{x=\frac{9\pm\sqrt{9^2}}{2}}$; $\boxed{x=\frac{9\pm 9}{2}}$ Therefore:

I. $\boxed{x=\frac{9-9}{2}}$; $\boxed{x=\frac{0}{2}}$; $\boxed{x=0}$

II. $\boxed{x=\frac{9+9}{2}}$; $\boxed{x=\frac{\overset{9}{\cancel{18}}}{\cancel{2}}}$; $\boxed{x=\frac{9}{1}}$; $\boxed{x=9}$

and the solution set is $\{\mathbf{0, 9}\}$.

Check No. 1: I. *Let* $x=0$ *in* $x^2=9x$; $0^2\overset{?}{=}9\cdot 0$; $0=0$

II. *Let* $x=9$ *in* $x^2=9x$; $9^2\overset{?}{=}9\cdot 9$; $81=81$

Check No. 2: $x^2-9x\overset{?}{=}(x+0)(x-9)$; $x^2-9x\overset{?}{=}(x\cdot x)+(-9\cdot x)+(0\cdot x)+(0\cdot -9)$; $x^2-9x\overset{?}{=}x^2-9x+0+0$; $x^2-9x=x^2-9x$

Therefore, the equation $x^2-9x=0$ can be factored to $(x+0)(x-9)=0$ which is the same as $\mathbf{x(x-9)=0}$.

Practice Problems - Solving Quadratic Equations of the Form ax^2+bx+c, where $a=1$, Using the Quadratic Formula

Section 4.2 Case I Practice Problems - Use the quadratic formula to solve the following quadratic equations.

1. $x^2=-5x-6$
2. $y^2-40y=-300$
3. $-x=-x^2+20$
4. $x^2+3x+4=0$
5. $x^2-80-2x=0$
6. $x^2+4x+4=0$
7. $-6=-w^2+w$
8. $4x=x^2$
9. $z^2-37z-120=0$
10. $x^2-20=-8x$

Case II Solving Quadratic Equations of the Form $ax^2+bx+c=0$, where $a\rangle 1$, Using the Quadratic Equation

Trinomial equations of the form $ax^2+bx+c=0$, where $a\rangle 1$, are solved using the following steps:

Step 1 Write the equation in standard form.

Step 2 Identify the coefficients a, b, and c.

Step 3 Substitute the values for a, b, and c into the quadratic equation $x=\frac{-b\pm\sqrt{b^2-4ac}}{2a}$. Simplify the equation.

Step 4 Solve for the values of x. Check the answers by either substituting the x values into the original equation or by multiplying the factored product using the FOIL method.

Step 5 Write the quadratic equation in its factored form.

Examples with Steps

The following examples show the steps as to how second degree trinomial equations are solved using the quadratic formula:

Example 4.2-13

Solve the quadratic equation $2x^2+5x=-3$.

Solution:

Step 1 $\boxed{2x^2+5x=-3}$; $\boxed{2x^2+5x+3=0}$

Step 2 Let: $\boxed{a=2}$, $\boxed{b=5}$, and $\boxed{c=3}$. Then,

Step 3 Given: $\boxed{x=\frac{-b\pm\sqrt{b^2-4ac}}{2a}}$; $\boxed{x=\frac{-5\pm\sqrt{5^2-4\times2\times3}}{2\times2}}$; $\boxed{x=\frac{-5\pm\sqrt{25-24}}{4}}$

; $\boxed{x=\frac{-5\pm\sqrt{1}}{4}}$; $\boxed{x=\frac{-5\pm1}{4}}$

Step 4 Separate $x=\frac{-5\pm1}{4}$ into two equations:

I. $\boxed{x=\frac{-5+1}{4}}$; $\boxed{x=-\frac{4}{4}}$; $\boxed{x=-\frac{1}{1}}$; $\boxed{\mathbf{x=-1}}$ II. $\boxed{x=\frac{-5-1}{4}}$; $\boxed{x=-\frac{\overset{3}{\cancel{6}}}{\underset{2}{\cancel{4}}}}$; $\boxed{\mathbf{x=-\frac{3}{2}}}$

Thus, the solution set is $\left\{\mathbf{-1,\ -\frac{3}{2}}\right\}$.

Check No. 1: I. *Let* $x=-1$ *in* $2x^2+5x=-3$; $2(-1)^2+(5\times-1)\overset{?}{=}-3$; $2-5\overset{?}{=}-3$; $-3=-3$

II. *Let* $x=-\frac{3}{2}$ *in* $2x^2+5x=-3$; $2\left(-\frac{3}{2}\right)^2+\left(5\times-\frac{3}{2}\right)\overset{?}{=}-3$; $2\times\frac{9}{4}-\frac{15}{2}\overset{?}{=}-3$

; $\frac{18}{4}-\frac{15}{2}\overset{?}{=}-3$; $\frac{(2\times18)-(4\times15)}{4\times2}\overset{?}{=}-3$; $\frac{36-60}{8}\overset{?}{=}-3$; $-\frac{24}{8}\overset{?}{=}-3$; $-3=-3$

Check No. 2: $2x^2+5x+3\stackrel{?}{=}(x+1)(2x+3)$; $2x^2+5x+3\stackrel{?}{=}(2x\cdot x)+(3\cdot x)+(1\cdot 2x)+(1\cdot 3)$

; $2x^2+5x+3\stackrel{?}{=}2x^2+3x+2x+3$; $2x^2+5x+3\stackrel{?}{=}2x^2+(3+2)x+3$

; $2x^2+5x+3=2x^2+5x+3$

Step 5 Therefore, the equation $2x^2+5x+3=0$ can be factored to $(x+1)\left(x+\frac{3}{2}\right)=0$ which is the same as $\mathbf{(x+1)(2x+3)=0}$

Example 4.2-14

Solve the quadratic equation $15x^2=-7x+2$.

Solution:

Step 1 $\boxed{15x^2=-7x+2}$; $\boxed{15x^2+7x=-7x+7x+2}$; $\boxed{15x^2+7x=0+2}$; $\boxed{15x^2+7x=2}$

; $\boxed{15x^2+7x-2=2-2}$; $\boxed{15x^2+7x-2=0}$

Step 2 Let: $\boxed{a=15}$, $\boxed{b=7}$, and $\boxed{c=-2}$. Then,

Step 3 Given: $\boxed{x=\frac{-b\pm\sqrt{b^2-4ac}}{2a}}$; $\boxed{x=\frac{-7\pm\sqrt{7^2-4\times 15\times -2}}{2\times 15}}$; $\boxed{x=\frac{-7\pm\sqrt{49+120}}{30}}$

; $\boxed{x=\frac{-7\pm\sqrt{169}}{30}}$; $\boxed{x=\frac{-7\pm 13}{30}}$

Step 4 Separate $x=\frac{-7\pm 13}{30}$ into two equations:

I. $\boxed{x=\frac{-7+13}{30}}$; $\boxed{x=\frac{\overset{1}{\cancel{6}}}{\underset{5}{\cancel{30}}}}$; $\boxed{\mathbf{x=\frac{1}{5}}}$ II. $\boxed{x=\frac{-7-13}{30}}$; $\boxed{x=-\frac{\overset{2}{\cancel{20}}}{\underset{3}{\cancel{30}}}}$; $\boxed{\mathbf{x=-\frac{2}{3}}}$

Thus, the solution set is $\left\{\mathbf{-\frac{2}{3},\frac{1}{5}}\right\}$.

Check No. 1: I. *Let* $x=\frac{1}{5}$ *in* $15x^2=-7x+2$; $15\left(\frac{1}{5}\right)^2\stackrel{?}{=}\left(-7\times\frac{1}{5}\right)+2$; $15\times\frac{1}{25}\stackrel{?}{=}-\frac{7}{5}+2$

; $\frac{\overset{3}{\cancel{15}}}{\underset{5}{\cancel{25}}}\stackrel{?}{=}-\frac{7}{5}+\frac{2}{1}$; $\frac{3}{5}\stackrel{?}{=}\frac{(-7\times 1)+(2\times 5)}{5\times 1}$; $\frac{3}{5}\stackrel{?}{=}\frac{-7+10}{5}$; $\frac{3}{5}=\frac{3}{5}$

II. *Let* $x=-\frac{2}{3}$ *in* $15x^2=-7x+2$; $15\left(-\frac{2}{3}\right)^2\stackrel{?}{=}\left(-7\times-\frac{2}{3}\right)+2$; $15\times\frac{4}{9}\stackrel{?}{=}\frac{14}{3}+2$

; $\frac{\overset{20}{\cancel{60}}}{\underset{3}{\cancel{9}}}\stackrel{?}{=}\frac{14}{3}+\frac{2}{1}$; $\frac{20}{3}\stackrel{?}{=}\frac{(14\times 1)+(2\times 3)}{3\times 1}$; $\frac{20}{3}\stackrel{?}{=}\frac{14+6}{3}$; $\frac{20}{3}=\frac{20}{3}$

Check No. 2: $15x^2+7x-2\stackrel{?}{=}(5x-1)(3x+2)$; $15x^2+7x-2\stackrel{?}{=}(5x\cdot 3x)+(2\cdot 5x)+(-1\cdot 3x)+(-1\cdot 2)$

; $15x^2+7x-2\stackrel{?}{=}15x^2+10x-3x-2$; $15x^2+7x-2\stackrel{?}{=}15x^2+(10-3)x-2$

; $15x^2+7x-2=15x^2+7x-2$

Step 5 Therefore, the equation $15x^2+7x-2=0$ can be factored to $\left(x-\frac{1}{5}\right)\left(x+\frac{2}{3}\right)=0$ which is the same as $\mathbf{(5x-1)(3x+2)=0}$

Example 4.2-15

Solve the quadratic equation $4x^2+4xy=3y^2$. Let x be the variable.

Solution:

Step 1 $\boxed{4x^2+4xy=3y^2}$; $\boxed{4x^2+4xy-3y^2=3y^2-3y^2}$; $\boxed{4x^2+4yx-3y^2=0}$

Step 2 Let: $\boxed{a=4}$, $\boxed{b=4y}$, and $\boxed{c=-3y^2}$. Then,

Step 3 Given: $\boxed{x=\frac{-b\pm\sqrt{b^2-4ac}}{2a}}$; $\boxed{x=\frac{-(4y)\pm\sqrt{(4y)^2-4\times4\times-3y^2}}{2\times4}}$

; $\boxed{x=\frac{-4y\pm\sqrt{16y^2+48y^2}}{8}}$; $\boxed{x=\frac{-4y\pm\sqrt{64y^2}}{8}}$; $\boxed{x=\frac{-4y\pm8y}{8}}$

Step 4 Separate $x=\frac{-4y\pm8y}{8}$ into two equations:

I. $\boxed{x=\frac{-4y+8y}{8}}$; $\boxed{x=\frac{4y}{\underset{2}{\cancel{8}}}}$; $\boxed{\mathbf{x=\frac{y}{2}}}$ II. $\boxed{x=\frac{-4y-8y}{8}}$; $\boxed{x=-\frac{\overset{3}{\cancel{12}}y}{\underset{2}{\cancel{8}}}}$; $\boxed{\mathbf{x=-\frac{3y}{2}}}$

Thus, the solution set is $\left\{\mathbf{-\frac{3y}{2},\frac{y}{2}}\right\}$.

Check No. 1: I. *Let* $x=\frac{y}{2}$ *in* $4x^2+4xy=3y^2$; $4\left(\frac{y}{2}\right)^2+\left(4\times\frac{y}{2}\times y\right)\overset{?}{=}3y^2$; $4\times\frac{y^2}{4}+\frac{4y^2}{2}\overset{?}{=}3y^2$

; $y^2+2y^2\overset{?}{=}3y^2$; $3y^2=3y^2$

II. *Let* $x=-\frac{3y}{2}$ *in* $4x^2+4xy=3y^2$; $4\left(-\frac{3y}{2}\right)^2+\left(4\times-\frac{3y}{2}\times y\right)\overset{?}{=}3y^2$

; $4\times\frac{9y^2}{4}-\frac{\overset{6}{\cancel{12}}y^2}{\cancel{2}}\overset{?}{=}3y^2$; $9y^2-6y^2\overset{?}{=}3y^2$; $3y^2=3y^2$

Check No. 2: $4x^2+4yx-3y^2\overset{?}{=}(2x-y)(2x+3y)$; $4x^2+4yx-3y^2\overset{?}{=}(2x\cdot2x)+(2x\cdot3y)+(2x\cdot-y)$ $+(-y\cdot3y)$; $4x^2+4yx-3y^2\overset{?}{=}4x^2+6xy-2xy-3y^2$; $4x^2+4yx-3y^2\overset{?}{=}4x^2$ $+(6-2)xy-3y^2$; $4x^2+4yx-3y^2=4x^2+4xy-3y^2$

Step 5 Therefore, the equation $4x^2+4yx-3y^2=0$ can be factored to $\left(x-\frac{y}{2}\right)\left(x+\frac{3y}{2}\right)=0$ which is the same as $\mathbf{(2x-y)(2x+3y)=0}$

Example 4.2-16

Solve the quadratic equation $2x^2+15=13x$.

Solution:

Step 1 $\boxed{2x^2+15=13x}$; $\boxed{2x^2-13x+15=13x-13x}$; $\boxed{2x^2-13x+15=0}$

Step 2 Let: $\boxed{a=2}$, $\boxed{b=-13}$, and $\boxed{c=15}$. Then,

Step 3 Given: $\boxed{x=\frac{-b\pm\sqrt{b^2-4ac}}{2a}}$; $\boxed{x=\frac{-(-13)\pm\sqrt{(-13)^2-(4\times2\times15)}}{2\times2}}$

; $\boxed{x=\frac{13\pm\sqrt{169-120}}{4}}$; $\boxed{x=\frac{13\pm\sqrt{49}}{4}}$; $\boxed{x=\frac{13\pm7}{4}}$

Step 4 Separate $x=\frac{13\pm7}{4}$ into two equations:

I. $\boxed{x=\frac{13+7}{4}}$; $\boxed{x=\frac{\overset{5}{\cancel{20}}}{\cancel{4}}}$; $\boxed{x=\frac{5}{1}}$; $\boxed{\mathbf{x=5}}$ II. $\boxed{x=\frac{13-7}{4}}$; $\boxed{x=\frac{\overset{3}{\cancel{6}}}{\underset{2}{\cancel{4}}}}$; $\boxed{\mathbf{x=\frac{3}{2}}}$

Thus, the solution set is $\left\{\frac{3}{2},5\right\}$.

Check No. 1: I. *Let* $x=5$ *in* $2x^2+15=13x$; $2(5)^2+15\overset{?}{=}13\times5$; $2\times25+15\overset{?}{=}65$; $50+15\overset{?}{=}65$

; $65=65$

II. *Let* $x=\frac{3}{2}$ *in* $2x^2+15=13x$; $2\left(\frac{3}{2}\right)^2+15\overset{?}{=}13\times\frac{3}{2}$; $\cancel{2}\times\frac{9}{\cancel{4}}+15\overset{?}{=}\frac{39}{2}$; $\frac{9}{2}+15\overset{?}{=}\frac{39}{2}$

; $\frac{9}{2}+\frac{15}{1}\overset{?}{=}\frac{39}{2}$; $\frac{(1\times9)+(2\times15)}{1\times2}\overset{?}{=}\frac{39}{2}$; $\frac{9+30}{2}\overset{?}{=}\frac{39}{2}$; $\frac{39}{2}=\frac{39}{2}$

Check No. 2: $2x^2-13x+15\overset{?}{=}(x-5)(2x-3)$; $2x^2-13x+15\overset{?}{=}(x\cdot2x)+(x\cdot-3)+(-5\cdot2x)+(-5\cdot-3)$

; $2x^2-13x+15\overset{?}{=}2x^2-3x-10x+15$; $2x^2-13x+15\overset{?}{=}2x^2+(-3-10)x+15$

; $2x^2-13x+15=2x^2-13x+15$

Step 5 Thus, the equation $2x^2-13x+15=0$ can be factored to $(x-5)\left(x-\frac{3}{2}\right)=0$

which is the same as $\mathbf{(x-5)(2x-3)=0}$

Example 4.2-17

Solve the quadratic equation $4x^2-15x-4=0$.

Solution:

Step 1 $\boxed{\textit{Not Applicable}}$

Step 2 Let: $\boxed{a=4}$, $\boxed{b=-15}$, and $\boxed{c=-4}$. Then,

Step 3 Given: $\boxed{x=\frac{-b\pm\sqrt{b^2-4ac}}{2a}}$; $\boxed{x=\frac{-(-15)\pm\sqrt{(-15)^2-(4\times4\times-4)}}{2\times4}}$

; $\boxed{x=\frac{15\pm\sqrt{225+64}}{8}}$; $\boxed{x=\frac{15\pm\sqrt{289}}{8}}$; $\boxed{x=\frac{15\pm17}{8}}$

Step 4 Separate $x = \frac{15 \pm 17}{8}$ into two equations:

I. $\boxed{x = \frac{15+17}{8}}$; $\boxed{x = \frac{\overset{4}{\cancel{32}}}{\cancel{8}}}$; $\boxed{x = 4}$ II. $\boxed{x = \frac{15-17}{8}}$; $\boxed{x = -\frac{\cancel{2}}{\underset{4}{\cancel{8}}}}$; $\boxed{x = -\frac{1}{4}}$

Thus, the solution set is $\left\{-\frac{1}{4}, 4\right\}$.

Check No. 1: I. *Let* $x = 4$ *in* $4x^2 - 15x - 4 = 0$; $(4 \times 4^2) - (15 \times 4) - 4 \overset{?}{=} 0$; $(4 \times 16) - 60 - 4 \overset{?}{=} 0$

; $64 - 60 - 4 \overset{?}{=} 0$; $64 - 64 \overset{?}{=} 0$; $0 = 0$

II. *Let* $x = -\frac{1}{4}$ *in* $4x^2 - 15x - 4 = 0$; $4 \times \left(-\frac{1}{4}\right)^2 - \left(15 \times -\frac{1}{4}\right) - 4 \overset{?}{=} 0$; $\cancel{4} \times \frac{1}{\cancel{16}} + \frac{15}{4} - 4 \overset{?}{=} 0$

; $\frac{1}{4} + \frac{15}{4} - 4 \overset{?}{=} 0$; $\frac{1+15}{4} - 4 \overset{?}{=} 0$; $\frac{\cancel{16}}{\cancel{4}} - 4 \overset{?}{=} 0$; $\frac{4}{1} - 4 \overset{?}{=} 0$; $4 - 4 \overset{?}{=} 0$; $0 = 0$

Check No. 2: $4x^2 - 15x - 4 \overset{?}{=} (x-4)\left(x + \frac{1}{4}\right) = 0$; $4x^2 - 15x - 4 \overset{?}{=} (x \cdot x) + \left(\frac{1}{4} \cdot x\right) + (-4 \cdot x) + \left(-\cancel{4} \cdot \frac{1}{\cancel{4}}\right)$

; $4x^2 - 15x - 4 \overset{?}{=} x^2 + \frac{x}{4} - 4x - 1$; $4x^2 - 15x - 4 \overset{?}{=} x^2 + \left(\frac{x}{4} - \frac{4x}{1}\right) - 1$

; $4x^2 - 15x - 4 \overset{?}{=} x^2 + \left(\frac{(1 \cdot x) - (4x \cdot 4)}{1 \cdot 4}\right) - 1$; $4x^2 - 15x - 4 \overset{?}{=} x^2 + \left(\frac{x - 16x}{4}\right) - 1$

; $4x^2 - 15x - 4 \overset{?}{=} x^2 - \frac{15}{4}x - 1$; $4x^2 - 15x - 4 \overset{?}{=} 4 \cdot \left(x^2 - \frac{15}{4}x - 1\right)$

; $4x^2 - 15x - 4 = 4x^2 - 15x - 4$

Step 5 Thus, the equation $4x^2 - 15x - 4 = 0$ can be factored to $(x-4)\left(x + \frac{1}{4}\right) = 0$ which is the same as $\mathbf{(x-4)(4x+1) = 0}$.

Additional Examples - Solving Quadratic Equations of the Form $ax^2 + bx + c = 0$, where $a \rangle 1$, Using the Quadratic Formula

The following examples further illustrate how to solve quadratic equations:

Example 4.2-18

Solve the quadratic equation $3x^2 + 7x - 6 = 0$.

Solution:

The equation is already in standard form. Let: $\boxed{a = 3}$, $\boxed{b = 7}$, and $\boxed{c = -6}$. Then,

Given: $\boxed{x = \frac{-b \pm \sqrt{b^2 - 4ac}}{2a}}$; $\boxed{x = \frac{-7 \pm \sqrt{7^2 - 4 \times 3 \times -6}}{2 \times 3}}$; $\boxed{x = \frac{-7 \pm \sqrt{49 + 72}}{6}}$; $\boxed{x = \frac{-7 \pm \sqrt{121}}{6}}$

; $\boxed{x = \frac{-7 \pm \sqrt{11^2}}{6}}$; $\boxed{x = \frac{-7 \pm 11}{6}}$ Therefore:

I. $\boxed{x = \frac{-7+11}{6}}$; $\boxed{x = \frac{\overset{2}{\cancel{4}}}{\underset{3}{\cancel{6}}}}$; $\boxed{x = \frac{2}{3}}$ II. $\boxed{x = \frac{-7-11}{6}}$; $\boxed{x = -\frac{\overset{3}{\cancel{18}}}{\cancel{6}}}$; $\boxed{x = -\frac{3}{1}}$; $\boxed{x = -3}$

Thus, the solution set is $\left\{-3, \frac{2}{3}\right\}$.

Check No. 1: I. *Let* $x=-3$ *in* $3x^2+7x-6=0$; $3\cdot(-3)^2+7\cdot(-3)-6\overset{?}{=}0$; $3\cdot 9-21-6\overset{?}{=}0$; $27-21-6\overset{?}{=}0$; $27-27\overset{?}{=}0$; $0=0$

II. *Let* $x=\frac{2}{3}$ *in* $3x^2+7x-6=0$; $3\cdot\left(\frac{2}{3}\right)^2+7\cdot\left(\frac{2}{3}\right)-6\overset{?}{=}0$; $3\cdot\frac{4}{9}+\frac{14}{3}-6\overset{?}{=}0$; $\frac{\cancel{12}^{\,4}}{\cancel{9}_{\,3}}+\frac{14}{3}-6\overset{?}{=}0$; $\frac{4}{3}+\frac{14}{3}-6\overset{?}{=}0$; $\frac{4+14}{3}-6\overset{?}{=}0$; $\frac{\cancel{18}^{\,6}}{\cancel{3}}-6\overset{?}{=}0$; $\frac{6}{1}-6\overset{?}{=}0$; $6-6\overset{?}{=}0$; $0=0$

Check No. 2: $3x^2+7x-6\overset{?}{=}(x+3)(3x-2)$; $3x^2+7x-6\overset{?}{=}(x\cdot 3x)+(-2\cdot x)+(3\cdot 3x)+(3\cdot -2)$; $3x^2+7x-6\overset{?}{=}3x^2-2x+9x-6$; $3x^2+7x-6\overset{?}{=}3x^2+(-2+9)x-6$; $3x^2+7x-6=3x^2+7x-6$

Therefore, the equation $3x^2+7x-6=0$ can be factored to $(x+3)\left(x-\frac{2}{3}\right)=0$ which is the same as $(x+3)\left(\frac{x}{1}-\frac{2}{3}\right)=0$; $(x+3)\left(\frac{(3\cdot x)-(1\cdot 2)}{1\cdot 3}\right)=0$; $(x+3)\left(\frac{3x-2}{3}\right)=0$; $\left(\frac{x+3}{1}\right)\left(\frac{3x-2}{3}\right)=0$; $\frac{(x+3)\cdot(3x-2)}{1\cdot 3}=0$; $\frac{(x+3)\cdot(3x-2)}{1\cdot 3}=\frac{0}{1}$; $[(x+3)\cdot(3x-2)]\cdot 1=0\cdot 3$; $\mathbf{(x+3)(3x-2)=0}$

Example 4.2-19

Solve the quadratic equation $6x^2=-7x-2$.

Solution:

First, write the equation in standard form, i.e., $6x^2+7x+2=0$

Next, let: $\boxed{a=6}$, $\boxed{b=7}$, and $\boxed{c=2}$. Then,

Given: $\boxed{x=\frac{-b\pm\sqrt{b^2-4ac}}{2a}}$; $\boxed{x=\frac{-7\pm\sqrt{7^2-4\times 6\times 2}}{2\times 6}}$; $\boxed{x=\frac{-7\pm\sqrt{49-48}}{12}}$; $\boxed{x=\frac{-7\pm\sqrt{1}}{12}}$; $\boxed{x=\frac{-7\pm 1}{12}}$

Therefore: I. $\boxed{x=\frac{-7+1}{12}}$; $\boxed{x=-\frac{\cancel{6}^{\,1}}{\cancel{12}_{\,2}}}$; $\boxed{\mathbf{x=-\frac{1}{2}}}$ II. $\boxed{x=\frac{-7-1}{12}}$; $\boxed{x=-\frac{\cancel{8}^{\,2}}{\cancel{12}_{\,3}}}$; $\boxed{\mathbf{x=-\frac{2}{3}}}$

Thus, the solution set is $\left\{-\frac{1}{2}, -\frac{2}{3}\right\}$.

Check No. 1: I. *Let* $x=-\frac{1}{2}$ *in* $6x^2=-7x-2$; $6\left(-\frac{1}{2}\right)^2\overset{?}{=}\left(-7\times-\frac{1}{2}\right)-2$; $6\cdot\frac{1}{4}\overset{?}{=}\frac{7}{2}-2$; $\frac{\cancel{6}^{\,3}}{\cancel{4}_{\,2}}\overset{?}{=}\frac{7}{2}-\frac{2}{1}$; $\frac{3}{2}\overset{?}{=}\frac{(7\cdot 1)-(2\cdot 2)}{2\cdot 1}$; $\frac{3}{2}\overset{?}{=}\frac{7-4}{2}$; $\frac{3}{2}=\frac{3}{2}$

II. *Let* $x=-\frac{2}{3}$ *in* $6x^2=-7x-2$; $6\left(-\frac{2}{3}\right)^2\overset{?}{=}\left(-7\times-\frac{2}{3}\right)-2$; $6\cdot\frac{4}{9}\overset{?}{=}\frac{14}{3}-2$; $\frac{\overset{8}{\cancel{24}}}{\underset{3}{\cancel{9}}}\overset{?}{=}\frac{14}{3}-\frac{2}{1}$

; $\frac{8}{3}\overset{?}{=}\frac{(14\cdot1)-(2\cdot3)}{3\cdot1}$; $\frac{8}{3}\overset{?}{=}\frac{14-6}{3}$; $\frac{8}{3}=\frac{8}{3}$

Check No. 2: $6x^2+7x+2\overset{?}{=}(2x+1)(3x+2)$; $6x^2+7x+2\overset{?}{=}(2x\cdot3x)+(2\cdot2x)+(1\cdot3x)+(1\cdot2)$

; $6x^2+7x+2\overset{?}{=}6x^2+4x+3x+2$; $6x^2+7x+2\overset{?}{=}6x^2+(4+3)x+2$

; $6x^2+7x+2=6x^2+7x+2$

Therefore, the equation $6x^2+7x+2=0$ can be factored to $\left(x+\frac{1}{2}\right)\left(x+\frac{2}{3}\right)=0$ which is the same as $\left(\frac{x}{1}+\frac{1}{2}\right)\left(\frac{x}{1}+\frac{2}{3}\right)=0$; $\left(\frac{(2\cdot x)+(1\cdot1)}{1\cdot2}\right)\left(\frac{(3\cdot x)+(1\cdot2)}{1\cdot3}\right)=0$; $\left(\frac{2x+1}{2}\right)\left(\frac{3x+2}{3}\right)=0$; $\frac{(2x+1)\cdot(3x+2)}{2\cdot3}=0$

; $\frac{(2x+1)(3x+2)}{6}=\frac{0}{1}$; $[(2x+1)\cdot(3x+2)]\cdot1=0\cdot6$; $\mathbf{(2x+1)(3x+2)=0}$

Example 4.2-20

Solve the quadratic equation $-16x+5=-3x^2$.

Solution:

First, write the equation in standard form, i.e., $3x^2-16x+5=0$.

Next, let: $\boxed{a=3}$, $\boxed{b=-16}$, and $\boxed{c=5}$. Then,

Given: $\boxed{x=\frac{-b\pm\sqrt{b^2-4ac}}{2a}}$; $\boxed{x=\frac{-(-16)\pm\sqrt{(-16)^2-4\times3\times5}}{2\times3}}$; $\boxed{x=\frac{16\pm\sqrt{256-60}}{6}}$; $\boxed{x=\frac{16\pm\sqrt{196}}{6}}$

; $\boxed{x=\frac{16\pm\sqrt{14^2}}{6}}$; $\boxed{x=\frac{16\pm14}{6}}$ Therefore:

I. $\boxed{x=\frac{16+14}{6}}$; $\boxed{x=\frac{\overset{5}{\cancel{30}}}{\cancel{6}}}$; $\boxed{x=\frac{5}{1}}$; $\boxed{\mathbf{x=5}}$

II. $\boxed{x=\frac{16-14}{6}}$; $\boxed{x=\frac{\cancel{2}}{\underset{3}{\cancel{6}}}}$; $\boxed{\mathbf{x=\frac{1}{3}}}$

Thus, the solution set is $\left\{\mathbf{\frac{1}{3}, 5}\right\}$.

Check No. 1: I. *Let* $x=\frac{1}{3}$ *in* $-16x+5=-3x^2$; $-16\cdot\frac{1}{3}+5\overset{?}{=}-3\cdot\left(\frac{1}{3}\right)^2$; $-\frac{16}{3}+5\overset{?}{=}-3\cdot\frac{1}{9}$

; $-\frac{16}{3}+\frac{5}{1}\overset{?}{=}-\frac{\cancel{3}}{\underset{3}{\cancel{9}}}$; $\frac{(-16\cdot1)+(5\cdot3)}{3\cdot1}\overset{?}{=}-\frac{1}{3}$; $\frac{-16+15}{3}\overset{?}{=}-\frac{1}{3}$; $-\frac{1}{3}=-\frac{1}{3}$

II. *Let* $x=5$ *in* $-16x+5=-3x^2$; $-16\cdot5+5\overset{?}{=}-3\cdot5^2$; $-80+5\overset{?}{=}-3\cdot25$; $-75=-75$

Check No. 2: $3x^2-16x+5\overset{?}{=}\left(x-\frac{1}{3}\right)(x-5)$; $3x^2-16x+5\overset{?}{=}(x\cdot x)+(-5\cdot x)+\left(-\frac{1}{3}\cdot x\right)+\left(-\frac{1}{3}\cdot-5\right)$

; $3x^2-16x+5\overset{?}{=}x^2-5x-\frac{1}{3}x+\frac{5}{3}$; $3x^2-16x+5\overset{?}{=}x^2+\left(-5-\frac{1}{3}\right)x+\frac{5}{3}$

$; \; 3x^2 - 16x + 5 \stackrel{?}{=} x^2 + \left(-\frac{5}{1} - \frac{1}{3}\right)x + \frac{5}{3} \; ; \; 3x^2 - 16x + 5 \stackrel{?}{=} x^2 + \left(\frac{(-5 \cdot 3) - (1 \cdot 1)}{1 \cdot 3}\right)x + \frac{5}{3}$

$; \; 3x^2 - 16x + 5 \stackrel{?}{=} x^2 + \left(\frac{-15 - 1}{3}\right)x + \frac{5}{3} \; ; \; 3x^2 - 16x + 5 \stackrel{?}{=} x^2 - \frac{16}{3}x + \frac{5}{3}$

$; \; 3x^2 - 16x + 5 \stackrel{?}{=} 3 \cdot \left(x^2 - \frac{16}{3}x + \frac{5}{3}\right) \; ; \; 3x^2 - 16x + 5 = 3x^2 - 16x + 5$

Therefore, the equation $3x^2 - 16x + 5 = 0$ can be factored to $\left(x - \frac{1}{3}\right)(x - 5) = 0$ which is the same as $\mathbf{(3x - 1)(x - 5) = 0}$.

Example 4.2-21

Solve the quadratic equation $4x^2 + 9x = -6$.

Solution:

First, write the equation in standard form, i.e., $4x^2 + 9x + 6 = 0$.

Next, let: $\boxed{a = 4}$, $\boxed{b = 9}$, and $\boxed{c = 6}$. Then,

Given: $\boxed{x = \frac{-b \pm \sqrt{b^2 - 4ac}}{2a}}$; $\boxed{x = \frac{-9 \pm \sqrt{9^2 - 4 \times 4 \times 6}}{2 \times 4}}$; $\boxed{x = \frac{-9 \pm \sqrt{81 - 96}}{8}}$; $\boxed{x = \frac{-9 \pm \sqrt{-15}}{8}}$

Since the number under the radical is negative, therefore the quadratic equation does not have any real solutions. We state that **the equation is not factorable.**

Example 4.2-22

Solve the quadratic equation $3y^2 - 2y = 2$.

Solution:

First, write the equation in standard form, i.e., $3y^2 - 2y - 2 = 0$.

Next, let: $\boxed{a = 3}$, $\boxed{b = -2}$, and $\boxed{c = -2}$. Then,

Given: $\boxed{y = \frac{-b \pm \sqrt{b^2 - 4ac}}{2a}}$; $\boxed{y = \frac{-(-2) \pm \sqrt{(-2)^2 - 4 \times 3 \times -2}}{2 \times 3}}$; $\boxed{y = \frac{2 \pm \sqrt{4 + 24}}{6}}$; $\boxed{y = \frac{2 \pm \sqrt{28}}{6}}$

; $\boxed{y = \frac{2 \pm 5.3}{6}}$ Therefore:

I. $\boxed{y = \frac{2 + 5.3}{6}}$; $\boxed{y = \frac{7.3}{6}}$; $\boxed{\mathbf{y = 1.22}}$ II. $\boxed{y = \frac{2 - 5.3}{6}}$; $\boxed{y = -\frac{3.3}{6}}$; $\boxed{\mathbf{y = -0.55}}$

Thus, the solution set is $\{\mathbf{-0.55, 1.22}\}$.

Check No. 1: I. *Let* $y = 1.22$ *in* $3y^2 - 2y = 2$; $3 \cdot (1.22)^2 - 2 \cdot 1.22 \stackrel{?}{=} 2$; $3 \cdot 1.48 - 2.44 \stackrel{?}{=} 2$

; $4.44 - 2.44 \stackrel{?}{=} 2$; $2 = 2$

II. *Let* $y = -0.55$ *in* $3y^2 - 2y = 2$; $3 \cdot (-0.55)^2 - 2 \cdot (-0.55) \stackrel{?}{=} 2$; $3 \cdot 0.3 + 1.1 \stackrel{?}{=} 2$

; $0.9 + 1.1 \stackrel{?}{=} 2$; $2 = 2$

Check No. 2: $3y^2-2y-2\overset{?}{=}(y+0.55)(y-1.22)$; $3y^2-2y-2\overset{?}{=}(y\cdot y)+(-1.22\cdot y)+(0.55\cdot y)+(0.55\cdot -1.22)$

; $3y^2-2y-2\overset{?}{=}y^2-1.22y+0.55y-0.67$; $3y^2-2y-2\overset{?}{=}y^2+(-1.22+0.55)y-0.67$

; $3y^2-2y-2\overset{?}{=}y^2-0.67y-0.67$; $3y^2-2y-2\overset{?}{=}3\cdot\left(y^2-0.67y-0.67\right)$

; $3y^2-2y-2=3y^2-2y-2$

Therefore, the equation $3y^2-2y-2=0$ can be factored to $\mathbf{(y+0.55)(y-1.22)=0}$.

Note that when $c=0$ the quadratic equation $ax^2+bx+c=0$ reduces to $ax^2+bx=0$. For cases where $a\rangle 1$, we can solve equations of the form $ax^2+bx=0$ using the quadratic formula in the following way:

Example 4.2-23

Solve the quadratic equation $2x^2+5x=0$.

Solution:

First write the equation in standard form, i.e., $2x^2+5x+0=0$.

Next, let: $\boxed{a=2}$, $\boxed{b=5}$, and $\boxed{c=0}$. Then,

Given: $\boxed{x=\frac{-b\pm\sqrt{b^2-4ac}}{2a}}$; $\boxed{x=\frac{-5\pm\sqrt{5^2-4\times2\times0}}{2\times2}}$; $\boxed{x=\frac{-5\pm\sqrt{25-0}}{4}}$; $\boxed{x=\frac{-5\pm\sqrt{25}}{4}}$

; $\boxed{x=\frac{-5\pm\sqrt{5^2}}{4}}$; $\boxed{x=\frac{-5\pm5}{4}}$ Therefore:

I. $\boxed{x=\frac{-5+5}{4}}$; $\boxed{x=\frac{0}{4}}$; $\boxed{\mathbf{x=0}}$ II. $\boxed{x=\frac{-5-5}{4}}$; $\boxed{x=-\frac{\overset{5}{\cancel{10}}}{\underset{2}{\cancel{4}}}}$; $\boxed{x=-\frac{5}{2}}$; $\boxed{\mathbf{x=-2.5}}$

Thus, the solution set is $\{\mathbf{0,-2.5}\}$.

Check No. 1: I. *Let* $x=0$ *in* $2x^2+5x=0$; $2\cdot0^2+5\cdot0\overset{?}{=}0$; $0+0\overset{?}{=}0$; $0=0$

II. *Let* $x=-2.5$ *in* $2x^2+5x=0$; $2\cdot(-2.5)^2+5\cdot-2.5\overset{?}{=}0$; $2\cdot6.25-12.5\overset{?}{=}0$; $12.5=12.5$

Check No. 2: $2x^2+5x\overset{?}{=}(x+0)(x+2.5)$; $2x^2+5x\overset{?}{=}(x\cdot x)+(2.5\cdot x)+(0\cdot x)+(0\cdot2.5)$

; $2x^2+5x\overset{?}{=}x^2+2.5x+0+0$; $2x^2+5x\overset{?}{=}x^2+2.5x$; $2x^2+5x\overset{?}{=}2\left(x^2+2.5x\right)$

; $2x^2+5x=2x^2+5x$

Therefore, the equation $2x^2+5x=0$ can be factored to $(x+0)(x+2.5)=0$ which is the same as $\mathbf{x(x+2.5)=0}$.

Example 4.2-24

Solve the quadratic equation $3x^2=2x$.

Solution:

First, write the equation in standard form, i.e., $3x^2-2x+0=0$.

Next, let: $\boxed{a=3}$, $\boxed{b=-2}$, and $\boxed{c=0}$. Then,

Given: $\boxed{x = \frac{-b \pm \sqrt{b^2 - 4ac}}{2a}}$; $\boxed{x = \frac{-(-2) \pm \sqrt{(-2)^2 - 4 \times 3 \times 0}}{2 \times 3}}$; $\boxed{x = \frac{2 \pm \sqrt{4-0}}{6}}$; $\boxed{x = \frac{2 \pm \sqrt{4}}{6}}$

; $\boxed{x = \frac{2 \pm \sqrt{2^2}}{6}}$; $\boxed{x = \frac{2 \pm 2}{6}}$ Therefore:

I. $\boxed{x = \frac{2-2}{6}}$; $\boxed{x = \frac{0}{6}}$; $\boxed{x = 0}$ II. $\boxed{x = \frac{2+2}{6}}$; $\boxed{x = \frac{\overset{2}{\not 4}}{\underset{3}{\not 6}}}$; $\boxed{x = \frac{2}{3}}$; $\boxed{x = 0.67}$

Thus, the solution set is $\{\mathbf{0, 0.67}\}$.

Check No. 1: I. *Let* $x = 0$ *in* $3x^2 = 2x$; $3 \cdot 0^2 \overset{?}{=} 2 \cdot 0$; $0 = 0$

II. *Let* $x = 0.67$ *in* $3x^2 = 2x$; $3 \cdot 0.67^2 \overset{?}{=} 2 \cdot 0.67$; $3 \cdot 0.448 \overset{?}{=} 1.34$; $1.34 = 1.34$

Check No. 2: $3x^2 - 2x \overset{?}{=} (x+0)(x-0.67)$; $3x^2 - 2x \overset{?}{=} (x \cdot x) + (-0.67 \cdot x) + (0 \cdot x) + (0 \cdot -0.67)$

; $3x^2 - 2x \overset{?}{=} x^2 - 0.67x + 0 + 0$; $3x^2 - 2x \overset{?}{=} x^2 - 0.67x$; $3x^2 - 2x \overset{?}{=} 3\left(x^2 - 0.67x\right)$

; $3x^2 - 2x = 3x^2 - 2x$

Therefore, the equation $3x^2 - 2x = 0$ can be factored to $(x+0)(x-0.67) = 0$ which is the same as $\boldsymbol{x(x-0.67) = 0}$. Note that if both sides of the equation are multiplied by 3 we obtain $3 \cdot x(x-0.67) = 0 \cdot 3$; $3x^2 - 2x = 0$ which is the same as the original equation.

Similar to the examples presented in Section 3.3 Case II, the steps in solving the following class of quadratic equations is very similar, if not identical, to the previous problems solved in this section. However, in the following set of examples to ensure proper factorization, we need to accurately match the given coefficients of x^2, x, and the constant term with the coefficient and the constant term of the standard quadratic equation $ax^2 + bx + c = 0$. For example, given the quadratic equation $10x^2 - 14xy - 12y^2 = 0$ we know that $a = 10$, $b = -14y$, and $c = -12y^2$. Once this equality is established, then the remaining steps are identical to the steps used in solving the previous problems. To further illustrate this point the same examples that were used in Section 3.3 Case II, i.e., examples 3.3-41 through 3.3-44 are solved below. However, the method used here is the Quadratic Formula method as opposed to the Trail and Error method which was used in Section 3.3.

Example 4.2-25:

Solve $6x^2 + 10xy + 4y^2 = 0$ (x is variable and y is constant).

Solution:

First - Simplify the equation, i.e., $6x^2 + 10xy + 4y^2 = 0$; $2\left(3x^2 + 5xy + 2y^2\right) = 0$; $3x^2 + 5xy + 2y^2 = 0$

Second - Write the equation in standard form, i.e., write $6x^2 + 10xy + 4y^2 = 0$ as

$$6x^2 + (10y)x + 4y^2 = 0 .$$

Third - Equate the coefficient of the standard quadratic equation with the given equation, i.e., let $a = 6$, $b = 10y$, and $c = 4y^2$.

Fourth - Use the quadratic formula to solve the equation, i.e., given

$x=\frac{-b\pm\sqrt{b^2-4ac}}{2a}$; $x=\frac{-10y\pm\sqrt{(10y)^2-4\times6\times4y^2}}{2\times6}$; $x=\frac{-10y\pm\sqrt{100y^2-96y^2}}{12}$

; $x=\frac{-10y\pm\sqrt{4y^2}}{12}$; $x=\frac{-10y\pm\sqrt{2^2y^2}}{12}$; $x=\frac{-10y\pm2y}{12}$. Therefore:

I. $x=\frac{-10y+2y}{12}$; $x=-\frac{\overset{2}{\cancel{8}}}{\underset{3}{\cancel{12}}}y$; $x=-\frac{2}{3}y$; $\mathbf{x=-0.67y}$

II. $x=\frac{-10y-2y}{12}$; $x=-\frac{\cancel{12}}{\cancel{12}}y$; $\mathbf{x=-y}$

Thus, the solution set is $\{-y,-\mathbf{0.67y}\}$.

Fifth - Check the answer by substituting the solutions into the original equation.

I. *Let* $x=-0.67y$ *in* $3x^2+5xy+2y^2=0$; $3\cdot(-0.67y)^2+5\cdot(-0.67y)\cdot y+2y^2\overset{?}{=}0$

; $3\times0.45y^2-3.35y^2+2y^2\overset{?}{=}0$; $1.35y^2-3.35y^2+2y^2\overset{?}{=}0$; $(1.35+2)y^2-3.35y^2\overset{?}{=}0$

; $3.35y^2-3.35y^2\overset{?}{=}0$; $0=0$

II. *Let* $x=-y$ *in* $3x^2+5xy+2y^2=0$; $3\cdot(-y)^2+5\cdot(-y)\cdot y+2y^2\overset{?}{=}0$; $3y^2-5y^2+2y^2\overset{?}{=}0$

; $(3+2)y^2-5y^2\overset{?}{=}0$; $5y^2-5y^2\overset{?}{=}0$; $0=0$

Therefore, the equation $3x^2+5xy+2y^2=0$ can be factored to $\mathbf{(x+0.67y)(x+y)=0}$ which is the same as $\left(x+\frac{2}{3}y\right)(2x+2y)=0$; $(3x+2y)(2x+2y)=0$. (Compare this answer with the result obtained in example 3.3-41.)

Sixth - Check the answer using the FOIL method.

$(x+0.67y)(x+y)=0$; $x\cdot x+x\cdot y+0.67y\cdot x+0.67y\cdot y=0$; $x^2+xy+0.67xy+0.67y^2=0$; $x^2+(1+0.67)xy+0.67y^2=0$; $x^2+1.67xy+0.67y^2=0$. Let's multiply both sides of the equation by 6, i.e., $6\cdot\left(x^2+1.67xy+0.67y^2\right)=6\cdot0$; $6x^2+10xy+4y^2=0$ which is the same as the original equation.

Example 4.2-26 A:

Solve $2x^2-19xy+35y^2=0$ (x is variable and y is constant).

Solution:

First - The equation is already in its simplest from.

Second - Write the equation in standard form, i.e., write $2x^2-19xy+35y^2=0$ as $2x^2+(-19y)x+35y^2=0$.

Third - Equate the coefficient of the standard quadratic equation with the given equation, i.e., let $a=2$, $b=-19y$, and $c=35y^2$.

Fourth - Use the quadratic formula to solve the equation, i.e., given

$x=\frac{-b\pm\sqrt{b^2-4ac}}{2a}$; $x=\frac{-(-19y)\pm\sqrt{(-19y)^2-4\times2\times35y^2}}{2\times2}$; $x=\frac{19y\pm\sqrt{361y^2-280y^2}}{4}$

$; x = \frac{19y \pm \sqrt{81y^2}}{4} \; ; \; x = \frac{19y \pm \sqrt{9^2 y^2}}{4} \; ; \; x = \frac{19y \pm 9y}{4}$. Therefore:

I. $x = \frac{19y + 9y}{4} \; ; \; x = \frac{\mathbf{28}}{\mathbf{4}} y \; ; \; x = \frac{7}{1} y \; ; \; \mathbf{x = 7y}$ II. $x = \frac{19y - 9y}{4} \; ; \; x = \frac{\mathbf{10}}{\mathbf{4}} y \; ; \; \mathbf{x = \frac{5}{2} y}$

Thus, the solution set is $\left\{ \mathbf{\frac{5}{2} y, \, 7y} \right\}$.

Fifth - Check the answer by substituting the solutions into the original equation.

I. *Let* $x = -7y$ *in* $2x^2 - 19xy + 35y^2 = 0 \; ; \; 2 \cdot (7y)^2 - 19 \cdot 7y \cdot y + 35y^2 \stackrel{?}{=} 0 \; ; \; 2 \cdot 49y^2 - 133y^2 + 35y^2 \stackrel{?}{=} 0$

$; \; 98y^2 - 133y^2 + 35y^2 \stackrel{?}{=} 0 \; ; \; (98 - 133)y^2 + 35y^2 \stackrel{?}{=} 0 \; ; \; -35y^2 + 35y^2 \stackrel{?}{=} 0 \; ; \; 0 = 0$

II. *Let* $x = \frac{5}{2} y$ *in* $2x^2 - 19xy + 35y^2 = 0 \; ; \; 2 \cdot \left(\frac{5}{2} y \right)^2 - 19 \cdot \left(\frac{5}{2} y \right) \cdot y + 35y^2 \stackrel{?}{=} 0$

$; \; \mathbf{2} \cdot \frac{25}{\underset{2}{\mathbf{4}}} y^2 - 19 \cdot \frac{5}{2} y \cdot y + 35y^2 \stackrel{?}{=} 0 \; ; \; \frac{25}{2} y^2 - \frac{95}{2} y^2 + 35y^2 \stackrel{?}{=} 0 \; ; \; \left(\frac{25}{2} - \frac{95}{2} \right) y^2 + 35y^2 \stackrel{?}{=} 0$

$; \; \left(\frac{25 - 95}{2} \right) y^2 + 35y^2 \stackrel{?}{=} 0 \; ; \; -\frac{\overset{35}{\mathbf{70}}}{\mathbf{2}} y^2 + 35y^2 \stackrel{?}{=} 0 \; ; \; -35y^2 + 35y^2 \stackrel{?}{=} 0 \; ; \; 0 = 0$

Therefore, the equation $2x^2 - 19xy + 35y^2$ can be factored to $\mathbf{(x - 7y)\left(x - \frac{5}{2} y \right) = 0}$ which is the same as $(x - 7y)(2x - 5y) = 0$. (Compare this answer with the result obtained in example 3.3-42A.)

Sixth - Check the answer using the FOIL method.

$(x - 7y)(2x - 5y) = 0 \; ; \; x \cdot 2x - x \cdot 5y - 7y \cdot 2x - 7y \cdot (-5y) = 0 \; ; \; 2x^2 - 5xy - 14xy + 35y^2 = 0$

$; \; 2x^2 + (-5 - 14)xy + 35y^2 = 0 \; ; \; 2x^2 - 19xy + 35y^2 = 0$ which is the same as the original equation.

Let's rework this problem. However, this time let y be the variable and x be the constant as follows:

Example 4.2-26 B:

Solve $2x^2 - 19xy + 35y^2 = 0$ (y is variable and x is constant).

Solution:

First - Write the equation in standard form, i.e., write $2x^2 - 19xy + 35y^2 = 0$ as $35y^2 + (-19x)y + 2x^2 = 0$.

Second - Equate the coefficient of the standard quadratic equation with the given equation, i.e., let $a = 35$, $b = -19x$, and $c = 2x^2$.

Third - Use the quadratic formula to solve the equation, i.e., given

$y = \frac{-b \pm \sqrt{b^2 - 4ac}}{2a} \; ; \; y = \frac{-(-19x) \pm \sqrt{(-19x)^2 - 4 \times 35 \times 2x^2}}{2 \times 35} \; ; \; y = \frac{19x \pm \sqrt{361x^2 - 280x^2}}{70}$

$; \; y = \frac{19x \pm \sqrt{81x^2}}{70} \; ; \; y = \frac{19x \pm \sqrt{9^2 x^2}}{70} \; ; \; y = \frac{19x \pm 9x}{70}$. Therefore:

I. $y = \frac{19x + 9x}{70} \; ; \; y = \frac{\overset{14}{\mathbf{28}}}{\underset{35}{\mathbf{70}}} x \; ; \; y = \frac{14}{35} x \; ; \; \mathbf{y = 0.4x}$ II. $y = \frac{19x - 9x}{70} \; ; \; y = \frac{\mathbf{10}}{\underset{7}{\mathbf{70}}} x \; ; \; \mathbf{y = \frac{1}{7} x}$

Thus, the solution set is $\left\{\frac{1}{7}x, 0.4x\right\}$.

Fourth - Check the answer by substituting the solutions into the original equation.

I. *Let* $y = 0.4x$ *in* $2x^2 - 19xy + 35y^2 = 0$; $2x^2 - 19x\cdot(0.4x) + 35\cdot(0.4x)^2 \overset{?}{=} 0$

; $2x^2 - 7.6x^2 + 35\cdot 0.16x^2 \overset{?}{=} 0$; $2x^2 - 7.6x^2 + 5.6x^2 \overset{?}{=} 0$; $2x^2 + (-7.6 + 5.6)x^2 \overset{?}{=} 0$; $2x^2 - 2x^2 \overset{?}{=} 0$

; $0 = 0$

II. *Let* $y = \frac{1}{7}x$ *in* $2x^2 - 19xy + 35y^2 = 0$; $2x^2 - 19x\cdot\left(\frac{1}{7}x\right) + 35\cdot\left(\frac{1}{7}x\right)^2 \overset{?}{=} 0$

; $2\cdot\frac{25}{\underset{2}{4}}y^2 - 19\cdot\frac{5}{2}y\cdot y + 35y^2 \overset{?}{=} 0$; $\frac{25}{2}y^2 - \frac{95}{2}y^2 + 35y^2 \overset{?}{=} 0$; $\left(\frac{25}{2} - \frac{95}{2}\right)y^2 + 35y^2 \overset{?}{=} 0$

; $2x^2 - \frac{19}{7}x^2 + 35\cdot\frac{1}{49}x^2 \overset{?}{=} 0$; $2x^2 - 2.71x^2 + 0.71x^2 \overset{?}{=} 0$; $2x^2 + (-2.71 + 0.71)x^2 \overset{?}{=} 0$

; $2x^2 - 2x^2 \overset{?}{=} 0$; $0 = 0$

Therefore, the equation $35y^2 - 19xy + 2x^2 = 0$ can be factored to $(y - 0.4x)\left(y - \frac{1}{7}x\right) = 0$ which is the same as $\left(y - \frac{4}{10}x\right)\cdot\left(y - \frac{1}{7}x\right) = 0$; $\left(y - \frac{2}{5}x\right)\cdot\left(y - \frac{1}{7}x\right) = 0$; $(5y - 2x)\cdot(7y - x) = 0$.

(Compare this answer with the result obtained in example 3.3-42B.)

Fifth - Check the answer using the FOIL method.

$(y - 0.4x)\left(y - \frac{1}{7}x\right) = 0$; $\left(y - \frac{14}{35}x\right)\left(y - \frac{1}{7}x\right) = 0$; $\left(\frac{35y - 14x}{35}\right)\left(\frac{7y - x}{7}\right) = 0$; $\frac{(35y - 14x)\cdot(7y - x)}{35\cdot 7} = 0$

; $\frac{(35y - 14x)\cdot(7y - x)}{245} = \frac{0}{1}$; $[(35y - 14x)\cdot(7y - x)]\cdot 1 = 245\cdot 0$; $(35y - 14x)\cdot(7y - x) = 0$

; $245y^2 - 35xy - 98xy + 14x^2 = 0$; $245y^2 + (-35 - 98)xy + 14x^2 = 0$; $245y^2 - 133xy + 14x^2 = 0$

; $\frac{245y^2 - 133xy + 14x^2}{7} = \frac{0}{7}$; $35y^2 - 19xy + 2x^2 = 0$ which is the same as the original equation.

Example 4.2-27:

Solve $3r^2 + 11rs + 10s^2 = 0$ (r is variable and s is constant).

Solution:

First - Write the equation in standard form, i.e., write $3r^2 + 11rs + 10s^2 = 0$ as $3r^2 + (11s)r + 10s^2 = 0$.

Second - Equate the coefficient of the standard quadratic equation with the given equation, i.e., let $a = 3$, $b = 11s$, and $c = 10s^2$.

Third - Use the quadratic formula to solve the equation, i.e., given

$r = \frac{-b \pm \sqrt{b^2 - 4ac}}{2a}$; $r = \frac{-11s \pm \sqrt{(11s)^2 - 4\times 3\times 10s^2}}{2\times 3}$; $r = \frac{-11s \pm \sqrt{121s^2 - 120s^2}}{6}$; $r = \frac{-11s \pm \sqrt{s^2}}{6}$

; $r = \frac{-11s \pm s}{6}$. Therefore:

I. $r=\frac{-11s+s}{6}$; $r=-\frac{\overset{5}{\cancel{10}}s}{\underset{3}{\cancel{6}}}$; $r=-\frac{5}{3}s$; $\mathbf{r=-1.67s}$

II. $r=\frac{-11s-s}{6}$; $r=-\frac{\overset{2}{\cancel{12}}}{\cancel{6}}s$; $r=-\frac{2}{1}s$; $\mathbf{r=-2s}$

and the solution set is $\{\mathbf{-2s,\,-1.67s}\}$.

Fourth - Check the answer by substituting the solutions into the original equation.

I. *Let* $r=-1.67s$ *in* $3r^2+11rs+10s^2=0$; $3\cdot(-1.67s)^2+11\cdot(-1.67s)\cdot s+10s^2\stackrel{?}{=}0$

; $3\cdot\left(2.79s^2\right)-18.37s^2+10s^2\stackrel{?}{=}0$; $8.37s^2-18.37s^2+10s^2\stackrel{?}{=}0$; $(8.37+10)s^2-18.37s^2\stackrel{?}{=}0$

; $18.37s^2-18.37s^2\stackrel{?}{=}0$; $0=0$

II. *Let* $r=-2s$ *in* $3r^2+11rs+10s^2=0$; $3\cdot(-2s)^2+11\cdot(-2s)\cdot s+10s^2\stackrel{?}{=}0$; $3\cdot4s^2-22s^2+10s^2\stackrel{?}{=}0$

; $12s^2-22s^2+10s^2\stackrel{?}{=}0$; $(12+10)s^2-22s^2\stackrel{?}{=}0$; $22s^2-22s^2\stackrel{?}{=}0$; $0=0$

Therefore, the equation $3r^2+11rs+10s^2=0$ can be factored to $\mathbf{(r+1.67s)(r+2s)=0}$ which is the same as $\left(r+\frac{5}{3}s\right)(r+2s)=0$; $(3r+5s)(r+2s)=0$. (Compare this answer with the result obtained in example 3.3-43.)

Fifth - Check the answer using the FOIL method.

$(r+1.67s)(r+2s)=0$; $r\cdot r+r\cdot2s+1.67s\cdot r+1.67s\cdot2s=0$; $r^2+2rs+1.67rs+3.34s^2=0$

; $r^2+(2+1.67)rs+3.34s^2=0$; $r^2+3.67rs+3.34s^2=0$. Let's multiply both sides of the equation by 3, i.e., $3\cdot\left(r^2+3.67rs+3.34s^2\right)=3\cdot0$; $3r^2+11rs+10r^2=0$ which is the same as the original equation.

Example 4.2-28:

Solve $21n^2+41mn+10m^2=0$ (n is variable and m is constant).

Solution:

First - Write the equation in standard form, i.e., write $21n^2+41mn+10m^2=0$ as

$21n^2+(41m)n+10m^2=0$.

Second - Equate the coefficient of the standard quadratic equation with the given equation, i.e., let $a=21$, $b=41m$, and $c=10m^2$.

Third - Use the quadratic formula to solve the equation, i.e., given:

$n=\frac{-b\pm\sqrt{b^2-4ac}}{2a}$; $n=\frac{-41m\pm\sqrt{(41m)^2-4\times21\times10m^2}}{2\times21}$; $n=\frac{-41m\pm\sqrt{1681m^2-840m^2}}{42}$

; $n=\frac{-41m\pm\sqrt{841m^2}}{42}$; $n=\frac{-41m\pm\sqrt{29^2m^2}}{42}$; $n=\frac{-41m\pm29m}{42}$. Therefore:

I. $n=\frac{-41m+29m}{42}$; $n=-\frac{\overset{6}{\cancel{12}}}{\underset{21}{\cancel{42}}}m$; $n=-\frac{6}{21}m$; $\mathbf{n=-0.28m}$

II. $n=\frac{-41m-29m}{42}$; $n=-\frac{\overset{35}{\cancel{70}}}{\underset{21}{\cancel{42}}}m$; $n=-\frac{35}{21}m$; $\boldsymbol{n=-1.66m}$

Thus, the solution set is $\{\boldsymbol{-0.28m,\ -1.66m}\}$.

Fourth - Check the answer by substituting the solutions into the original equation.

I. *Let* $n=-0.28m$ *in* $21n^2+41mn+10m^2=0$; $21\cdot(-0.28m)^2+41m\cdot(-0.28m)+10m^2\overset{?}{=}0$

; $21\cdot\left(0.08m^2\right)-11.48m^2+10m^2\overset{?}{=}0$; $1.6m^2-11.6m^2+10m^2\overset{?}{=}0$; $(1.6+10)m^2-11.6m^2\overset{?}{=}0$

; $11.6m^2-11.6m^2\overset{?}{=}0$; $0=0$

II. *Let* $n=-1.66m$ *in* $21n^2+41mn+10m^2=0$; $21\cdot(-1.66m)^2+41m\cdot(-1.66m)+10m^2\overset{?}{=}0$

; $21\cdot\left(2.75m^2\right)-68m^2+10m^2\overset{?}{=}0$; $58m^2-68m^2+10m^2\overset{?}{=}0$; $(58+10)m^2-68m^2\overset{?}{=}0$

; $68m^2-68m^2\overset{?}{=}0$; $0=0$

Therefore, the equation $21n^2+41mn+10m^2=0$ can be factored to $\boldsymbol{(n+0.28m)(n+1.66m)=0}$ which is the same as $\left(n+\frac{2}{7}m\right)\left(n+\frac{5}{3}m\right)=0$; $(7n+2m)(3n+5m)=0$. (Compare this answer with the result obtained in example 3.3-44.)

Fifth - Check the answer using the FOIL method.

$(n+0.28m)(n+1.66m)=0$; $n\cdot n+n\cdot 1.66m+0.28m\cdot n+0.28m\cdot 1.66m=0$

; $n^2+1.66mn+0.28mn+0.46m^2=0$; $n^2+(1.66+0.28)mn+0.46m^2=0$; $n^2+1.94mn+0.46m^2=0$.

Let's multiply both sides of the equation by 21, i.e., $21\cdot\left(n^2+1.94mn+0.46m^2\right)=21\cdot 0$

; $21n^2+41mn+10m^2=0$ which is the same as the original equation.

Note That since the solutions are rounded off to the first two digits, in some instances, we do not obtain an exact match with the coefficients of the original equation.

Practice Problems - Solving Quadratic Equations of the Form $ax^2+bx+c=0$, where $a \rangle 1$, Using the Quadratic Formula

Section 4.2 Case II Practice Problems - Use the quadratic formula to solve the following quadratic equations.

1. $4u^2+6u+1=0$
2. $4w^2+10w=-3$
3. $6x^2+4x-2=0$
4. $15y^2+3=-14y$
5. $2x^2-5x+3=0$
6. $2x^2+xy-y^2=0$ x is variable
7. $6x^2+7x-3=0$
8. $5x^2=-3x$
9. $3x^2+4x+5=0$
10. $-3y^2+13y+10=0$

4.3 Solving Quadratic Equations Using the Square Root Property Method

Quadratic equations of the form $(ax+b)^2 = c$ are solved using a method known as the Square Root Property method where the square root of both sides of the equation are taken and the terms are simplified. Following show the steps as to how quadratic equations are solved using the Square Root property method:

Step 1 Take the square root of the left and the right hand side of the equation. Simplify the terms on both sides of the equation.

Step 2 Solve for the values of x. Check the answers by substituting the x values into the original equation.

Step 3 Write the equation in its factored form.

Examples with Steps

The following examples show the steps as to how equations of the form $(ax+b)^2 = c$ are solved using the Square Root Property method:

Example 4.3-1

Solve the quadratic equation $(x+4)^2 = 36$.

Solution:

Step 1 $\boxed{(x+4)^2 = 36}$; $\boxed{\sqrt{(x+4)^2} = \pm\sqrt{36}}$; $\boxed{\sqrt{(x+4)^2} = \pm\sqrt{6^2}}$; $\boxed{x+4 = \pm 6}$

Step 2 Separate $x+4 = \pm 6$ into two equations.

I. $\boxed{x+4 = +6}$; $\boxed{x = +6-4}$; $\boxed{\boldsymbol{x = 2}}$

II. $\boxed{x+4 = -6}$; $\boxed{x = -6-4}$; $\boxed{\boldsymbol{x = -10}}$

Thus, the solution set is $\{\mathbf{-10, 2}\}$.

Check: I. *Let* $x = 2$ *in* $(x+4)^2 = 36$; $(2+4)^2 \stackrel{?}{=} 36$; $6^2 \stackrel{?}{=} 36$; $36 = 36$

II. *Let* $x = -10$ *in* $(x+4)^2 = 36$; $(-10+4)^2 \stackrel{?}{=} 36$; $(-6)^2 \stackrel{?}{=} 36$; $36 = 36$

Step 3 Therefore, the equation $(x+4)^2 = 36$ can be factored to $\mathbf{(x-2)(x+10) = 0}$.

Example 4.3-2

Solve the quadratic equation $(x-2)^2 = 25$.

Solution:

Step 1 $\boxed{(x-2)^2 = 25}$; $\boxed{\sqrt{(x-2)^2} = \pm\sqrt{25}}$; $\boxed{\sqrt{(x-2)^2} = \pm\sqrt{5^2}}$; $\boxed{x-2 = \pm 5}$

Step 2 Separate $x-2 = \pm 5$ into two equations.

I. $\boxed{x-2 = +5}$; $\boxed{x = 5+2}$; $\boxed{\boldsymbol{x = 7}}$

II. $\boxed{x-2 = -5}$; $\boxed{x = -5+2}$; $\boxed{\boldsymbol{x = -3}}$

Thus, the solution set is $\{-3, 7\}$.

Check: I. $Let\ x = 7\ in\ \ (x-2)^2 = 25\ ;\ (7-2)^2 \stackrel{?}{=} 25\ ;\ 5^2 \stackrel{?}{=} 25\ ;\ 25 = 25$

II. $Let\ x = -3\ in\ \ (x-2)^2 = 25\ ;\ (-3-2)^2 \stackrel{?}{=} 25\ ;\ (-5)^2 \stackrel{?}{=} 25\ ;\ 25 = 25$

Step 3 Therefore, the equation $(x-2)^2 = 25$ can be factored to $(\mathbf{x-7})(\mathbf{x+3}) = \mathbf{0}$.

Example 4.3-3

Solve the quadratic equation $(x+2)^2 = 8$.

Solution:

Step 1 $\boxed{(x+2)^2 = 8}\ ;\ \boxed{\sqrt{(x+2)^2} = \pm\sqrt{8}}\ ;\ \boxed{\sqrt{(x+2)^2} = \pm\sqrt{4\cdot 2}}\ ;\ \boxed{x+2 = \pm 2\sqrt{2}}$

Step 2 Separate $x+2 = \pm 2\sqrt{2}$ into two equations.

I. $\boxed{x+2 = +2\sqrt{2}}\ ;\ \boxed{x+2-2 = -2+2\sqrt{2}}\ ;\ \boxed{x+0 = -2+2\sqrt{2}}\ ;\ \boxed{\mathbf{x = -2+2\sqrt{2}}}$

II. $\boxed{x+2 = -2\sqrt{2}}\ ;\ \boxed{x+2-2 = -2-2\sqrt{2}}\ ;\ \boxed{x+0 = -2-2\sqrt{2}}\ ;\ \boxed{\mathbf{x = -2-2\sqrt{2}}}$

Thus, the solution set is $\{\mathbf{-2-2\sqrt{2},\ -2+2\sqrt{2}}\}$.

Check: I. $Let\ x = -2+2\sqrt{2}\ in\ \ (x+2)^2 = 8\ ;\ \left(-2+2\sqrt{2}+2\right)^2 \stackrel{?}{=} 8\ ;\ \left(2\sqrt{2}\right)^2 \stackrel{?}{=} 8\ ;\ 4\cdot 2 \stackrel{?}{=} 8$

$;\ 8 = 8$

II. $Let\ x = -2-2\sqrt{2}\ in\ \ (x+2)^2 = 8\ ;\ \left(-2-2\sqrt{2}+2\right)^2 \stackrel{?}{=} 8\ ;\ \left(-2\sqrt{2}\right)^2 \stackrel{?}{=} 8\ ;\ 4\cdot 2 \stackrel{?}{=} 8$

$;\ 8 = 8$

Step 3 Thus, the equation $(x+2)^2 = 8$ can be factored to $\left(\mathbf{x+2-2\sqrt{2}}\right)\left(\mathbf{x+2+2\sqrt{2}}\right)$.

Example 4.3-4

Solve the quadratic equation $(2x-4)^2 = 16$.

Solution:

Step 1 $\boxed{(2x-4)^2 = 16}\ ;\ \boxed{\sqrt{(2x-4)^2} = \pm\sqrt{16}}\ ;\ \boxed{\sqrt{(2x-4)^2} = \pm\sqrt{4^2}}\ ;\ \boxed{2x-4 = \pm 4}$

Step 2 Separate $2x-4 = \pm 4$ into two equations.

I. $\boxed{2x-4 = +4}\ ;\ \boxed{2x = 4+4}\ ;\ \boxed{2x = 8}\ ;\ \boxed{x = \frac{\overset{4}{\cancel{8}}}{\cancel{2}}}\ ;\ \boxed{x = \frac{4}{1}}\ ;\ \boxed{\mathbf{x = 4}}$

II. $\boxed{2x-4 = -4}\ ;\ \boxed{2x = -4+4}\ ;\ \boxed{2x = 0}\ ;\ \boxed{x = \frac{0}{2}}\ ;\ \boxed{\mathbf{x = 0}}$

Thus, the solution set is $\{\mathbf{0, 4}\}$.

Check: I. $Let\ x = 4\ in\ \ (2x-4)^2 = 16\ ;\ (2\cdot 4-4)^2 \stackrel{?}{=} 16\ ;\ (8-4)^2 \stackrel{?}{=} 16\ ;\ 4^2 \stackrel{?}{=} 16\ ;\ 16 = 16$

II. $Let\ x = 0\ in\ \ (2x-4)^2 = 16\ ;\ (2\cdot 0-4)^2 \stackrel{?}{=} 16\ ;\ (0-4)^2 \stackrel{?}{=} 16\ ;\ (-4)^2 \stackrel{?}{=} 16$

$; \; 16 = 16$

Step 3 Thus, the equation $(2x-4)^2 = 16$ can be factored to $(x-4)(x+0) = 0$ which is the same as $\mathbf{x(x-4) = 0}$

Example 4.3-5

Solve the quadratic equation $\left(y+\frac{2}{3}\right)^2 = \frac{4}{9}$.

Solution:

Step 1 $\boxed{\left(y+\frac{2}{3}\right)^2 = \frac{4}{9}}$; $\boxed{\sqrt{\left(y+\frac{2}{3}\right)^2} = \pm\sqrt{\frac{4}{9}}}$; $\boxed{\sqrt{\left(y+\frac{2}{3}\right)^2} = \pm\sqrt{\frac{2^2}{3^2}}}$; $\boxed{y+\frac{2}{3} = \pm\frac{2}{3}}$

Step 2 Separate $y+\frac{2}{3} = \pm\frac{2}{3}$ into two equations

I. $\boxed{y+\frac{2}{3} = +\frac{2}{3}}$; $\boxed{y = -\frac{2}{3}-\frac{2}{3}}$; $\boxed{y = \frac{-2-2}{3}}$; $\boxed{\mathbf{y = -\frac{4}{3}}}$

II. $\boxed{y+\frac{2}{3} = -\frac{2}{3}}$; $\boxed{y = -\frac{2}{3}+\frac{2}{3}}$; $\boxed{y = \frac{-2+2}{3}}$; $\boxed{y = \frac{0}{3}}$; $\boxed{\mathbf{y = 0}}$

Thus, the solution set is $\left\{\mathbf{0, -\frac{4}{3}}\right\}$.

Check: I. *Let* $y = -\frac{4}{3}$ *in* $\left(y+\frac{2}{3}\right)^2 = \frac{4}{9}$; $\left(-\frac{4}{3}+\frac{2}{3}\right)^2 \stackrel{?}{=} \frac{4}{9}$; $\left(\frac{-4+2}{3}\right)^2 \stackrel{?}{=} \frac{4}{9}$; $\left(\frac{-2}{3}\right)^2 \stackrel{?}{=} \frac{4}{9}$

$; \; \frac{4}{9} = \frac{4}{9}$

II. *Let* $y = 0$ *in* $\left(y+\frac{2}{3}\right)^2 = \frac{4}{9}$; $\left(0+\frac{2}{3}\right)^2 \stackrel{?}{=} \frac{4}{9}$; $\left(\frac{2}{3}\right)^2 \stackrel{?}{=} \frac{4}{9}$; $\frac{4}{9} = \frac{4}{9}$

Step 3 Therefore, the equation $\left(y+\frac{2}{3}\right)^2 = \frac{4}{9}$ can be factored to $\left(y+\frac{4}{3}\right)(y+0) = 0$; $y\left(y+\frac{4}{3}\right) = 0$ which is the same as $\mathbf{y(3y+4) = 0}$.

Additional Examples - Solving Quadratic Equations Using the Square Root Property Method

The following examples further illustrate how to solve quadratic equations using the Square Root Property method:

Example 4.3-6

Solve the quadratic equation $(6u-3)^2 = 25$ using the Square Root Property method.

Solution:

$\boxed{(6u-3)^2 = 25}$; $\boxed{\sqrt{(6u-3)^2} = \pm\sqrt{25}}$; $\boxed{\sqrt{(6u-3)^2} = \pm\sqrt{5^2}}$; $\boxed{6u-3 = \pm 5}$

Therefore, the two solutions are:

I. $\boxed{6u-3 = +5}$; $\boxed{6u = 5+3}$; $\boxed{6u = 8}$; $\boxed{u = \frac{\overset{4}{\cancel{8}}}{\underset{3}{\cancel{6}}}}$; $\boxed{\mathbf{u = \frac{4}{3}}}$

II. $\boxed{6u-3=-5}$; $\boxed{6u=-5+3}$; $\boxed{6u=-2}$; $\boxed{u=-\frac{\overset{1}{\cancel{2}}}{\underset{3}{\cancel{6}}}}$; $\boxed{u=-\frac{1}{3}}$

Thus, the solution set is $\left\{-\frac{1}{3},\frac{4}{3}\right\}$.

Check: I. *Let* $u=\frac{4}{3}$ *in* $(6u-3)^2=25$; $\left(6\cdot\frac{4}{3}-3\right)^2\overset{?}{=}25$; $\left(\frac{\overset{8}{\cancel{24}}}{\cancel{3}}-3\right)^2\overset{?}{=}25$; $\left(\frac{8}{1}-3\right)^2\overset{?}{=}25$; $(8-3)^2\overset{?}{=}25$

; $5^2\overset{?}{=}25$; $25=25$

II. *Let* $u=-\frac{1}{3}$ *in* $(6u-3)^2=25$; $\left[\left(6\cdot-\frac{1}{3}\right)-3\right]^2\overset{?}{=}25$; $\left(-\frac{\overset{2}{\cancel{6}}}{\cancel{3}}-3\right)^2\overset{?}{=}25$; $\left(-\frac{2}{1}-3\right)^2\overset{?}{=}25$

; $(-2-3)^2\overset{?}{=}25$; $(-5)^2\overset{?}{=}25$; $25=25$

Therefore, the equation $(6u-3)^2=25$ can be factored to $\left(u-\frac{4}{3}\right)\left(u+\frac{1}{3}\right)=0$ which is the same as $(3u-4)(3u+1)=0$.

Example 4.3-7

Solve the quadratic equation $(5y+3)^2=15$ using the Square Root Property method.

Solution:

$\boxed{(5y+3)^2=15}$; $\boxed{\sqrt{(5y+3)^2}=\pm\sqrt{15}}$; $\boxed{5y+3=\pm\sqrt{15}}$ Therefore, the two solutions are:

I. $\boxed{5y+3=+\sqrt{15}}$; $\boxed{5y=\sqrt{15}-3}$; $\boxed{y=\frac{\sqrt{15}-3}{5}}$ II. $\boxed{5y+3=-\sqrt{15}}$; $\boxed{5y=-\sqrt{15}-3}$; $\boxed{y=-\frac{\sqrt{15}+3}{5}}$

Thus, the solution set is $\left\{-\frac{\sqrt{15}+3}{5},\frac{\sqrt{15}-3}{5}\right\}$.

Check: I. *Let* $y=\frac{\sqrt{15}-3}{5}$ *in* $(5y+3)^2=15$; $\left(\cancel{5}\cdot\frac{\sqrt{15}-3}{\cancel{5}}+3\right)^2\overset{?}{=}15$; $\left(\sqrt{15}-\cancel{3}+\cancel{3}\right)^2\overset{?}{=}15$; $\left(\sqrt{15}\right)^2\overset{?}{=}15$

; $15=15$

II. *Let* $y=-\frac{\sqrt{15}+3}{5}$ *in* $(5y+3)^2=15$; $\left[\left(\cancel{5}\cdot-\frac{\sqrt{15}+3}{\cancel{5}}\right)+3\right]^2\overset{?}{=}15$; $\left[\left(-\sqrt{15}-3\right)+3\right]^2\overset{?}{=}15$

; $\left(-\sqrt{15}-\cancel{3}+\cancel{3}\right)^2\overset{?}{=}15$; $\left(-\sqrt{15}\right)^2\overset{?}{=}15$; $15=15$

Therefore, the equation $(5y+3)^2=15$ can be factored to $\left(y-\frac{\sqrt{15}-3}{5}\right)\left(y+\frac{\sqrt{15}+3}{5}\right)=0$ which is the same as $(y-0.175)(y+1.375)=0$; $y^2+1.2y-0.24=0$; $25y^2+30y-6=0$, or $(5y+3)^2=15$.

Example 4.3-8

Solve the quadratic equation $(2w-4)^2=1$ using the Square Root Property method.

Solution:

$\boxed{(2w-4)^2=1}$; $\boxed{\sqrt{(2w-4)^2}=\pm\sqrt{1}}$; $\boxed{\sqrt{(2w-4)^2}=\pm 1}$; $\boxed{2w-4=\pm 1}$ Therefore, the two solutions are:

I. $\boxed{2w-4=+1}$; $\boxed{2w=4+1}$; $\boxed{2w=5}$; $\boxed{\mathbf{w=\frac{5}{2}}}$ II. $\boxed{2w-4=-1}$; $\boxed{2w=4-1}$; $\boxed{2w=3}$; $\boxed{\mathbf{w=\frac{3}{2}}}$

Thus, the solution set is $\left\{\mathbf{\frac{3}{2}, \frac{5}{2}}\right\}$.

Check: I. *Let* $w=\frac{5}{2}$ *in* $(2w-4)^2=1$; $\left(2\cdot\frac{5}{2}-4\right)^2\overset{?}{=}1$; $(5-4)^2\overset{?}{=}1$; $1^2\overset{?}{=}1$; $1=1$

II. *Let* $w=\frac{3}{2}$ *in* $(2w-4)^2=1$; $\left(2\cdot\frac{3}{2}-4\right)^2\overset{?}{=}1$; $(3-4)^2\overset{?}{=}1$; $(-1)^2\overset{?}{=}1$; $1=1$

Therefore, the equation $(2w-4)^2=1$ can be factored to $\left(w-\frac{5}{2}\right)\left(w-\frac{3}{2}\right)=0$ which is the same as $\mathbf{(2w-5)(2w-3)=0}$.

Example 4.3-9

Solve the quadratic equation $\left(x-\frac{1}{2}\right)^2=\frac{1}{16}$ using the Square Root Property method.

Solution:

$\boxed{\left(x-\frac{1}{2}\right)^2=\frac{1}{16}}$; $\boxed{\sqrt{\left(x-\frac{1}{2}\right)^2}=\pm\sqrt{\frac{1}{16}}}$; $\boxed{\sqrt{\left(x-\frac{1}{2}\right)^2}=\pm\sqrt{\frac{1}{4^2}}}$; $\boxed{x-\frac{1}{2}=\pm\frac{1}{4}}$

Therefore, the two solutions are:

I. $\boxed{x-\frac{1}{2}=+\frac{1}{4}}$; $\boxed{x=\frac{1}{4}+\frac{1}{2}}$; $\boxed{x=\frac{(1\cdot 2)+(1\cdot 4)}{2\cdot 4}}$; $\boxed{x=\frac{2+4}{8}}$; $\boxed{x=\frac{\overset{3}{\cancel{6}}}{\underset{4}{\cancel{8}}}}$; $\boxed{\mathbf{x=\frac{3}{4}}}$

II. $\boxed{x-\frac{1}{2}=-\frac{1}{4}}$; $\boxed{x=-\frac{1}{4}+\frac{1}{2}}$; $\boxed{x=\frac{(-1\cdot 2)+(1\cdot 4)}{2\cdot 4}}$; $\boxed{x=\frac{-2+4}{8}}$; $\boxed{x=\frac{\cancel{2}}{\underset{4}{\cancel{8}}}}$; $\boxed{\mathbf{x=\frac{1}{4}}}$

Thus, the solution set is $\left\{\mathbf{\frac{1}{4}, \frac{3}{4}}\right\}$.

Check: I. *Let* $x=\frac{3}{4}$ *in* $\left(x-\frac{1}{2}\right)^2=\frac{1}{16}$; $\left(\frac{3}{4}-\frac{1}{2}\right)^2\overset{?}{=}\frac{1}{16}$; $\left(\frac{(2\cdot 3)-(1\cdot 4)}{2\cdot 4}\right)^2\overset{?}{=}\frac{1}{16}$; $\left(\frac{6-4}{8}\right)^2\overset{?}{=}\frac{1}{16}$

; $\left(\frac{\cancel{2}}{\underset{4}{\cancel{8}}}\right)^2\overset{?}{=}\frac{1}{16}$; $\left(\frac{1}{4}\right)^2\overset{?}{=}\frac{1}{16}$; $\frac{1}{4^2}\overset{?}{=}\frac{1}{16}$; $\frac{1}{16}=\frac{1}{16}$

II. *Let* $x=\frac{1}{4}$ *in* $\left(x-\frac{1}{2}\right)^2=\frac{1}{16}$; $\left(\frac{1}{4}-\frac{1}{2}\right)^2\overset{?}{=}\frac{1}{16}$; $\left(\frac{(1\cdot 2)-(1\cdot 4)}{2\cdot 4}\right)^2\overset{?}{=}\frac{1}{16}$; $\left(\frac{2-4}{8}\right)^2\overset{?}{=}\frac{1}{16}$

; $\left(-\frac{\cancel{2}}{\underset{4}{\cancel{8}}}\right)^2\overset{?}{=}\frac{1}{16}$; $\left(-\frac{1}{4}\right)^2\overset{?}{=}\frac{1}{16}$; $\frac{1}{4^2}\overset{?}{=}\frac{1}{16}$; $\frac{1}{16}=\frac{1}{16}$

Therefore, the equation $\left(x-\frac{1}{2}\right)^2 = \frac{1}{16}$ can be factored to $\left(x-\frac{3}{4}\right)\left(x-\frac{1}{4}\right)=0$ which is the same as $\mathbf{(4x-3)(4x-1)=0}$.

Example 4.3-10

Solve the quadratic equation $(x+5)^2 = 49$ using the Square Root Property and the Quadratic Formula method.

Solution:

First Method - The Square Root Property method:

$\boxed{(x+5)^2 = 49}$; $\boxed{\sqrt{(x+5)^2} = \pm\sqrt{49}}$; $\boxed{\sqrt{(x+5)^2} = \pm\sqrt{7^2}}$; $\boxed{x+5=\pm 7}$ Therefore, the two solutions are:

I. $\boxed{x+5=+7}$; $\boxed{x=7-5}$; $\boxed{\mathbf{x=2}}$

II. $\boxed{x+5=-7}$; $\boxed{x=-7-5}$; $\boxed{\mathbf{x=-12}}$

Thus, the solution set is $\{\mathbf{-12, 2}\}$.

Check: I. *Let* $x=2$ *in* $(x+5)^2 = 49$; $(2+5)^2 \overset{?}{=} 49$; $7^2 \overset{?}{=} 49$; $49 = 49$

II. *Let* $x = $ ~~-13~~ -12 *in* $(x+5)^2 = 49$; $(-12+5)^2 \overset{?}{=} 49$; $(-7)^2 \overset{?}{=} 49$; $49 = 49$

Therefore, the equation $(x+5)^2 = 49$ can be factored to $\mathbf{(x-2)(x+12)=0}$.

Second Method - The Quadratic Formula method:

Given the expression $(x+5)^2 = 49$, expand the left hand side of the equation and write the quadratic equation in its standard form, i.e.,

$\boxed{(x+5)^2 = 49}$; $\boxed{x^2+25+10x=49}$; $\boxed{x^2+10x+(25-49)=0}$; $\boxed{x^2+10x-24=0}$

Let: $\boxed{a=1}$, $\boxed{b=10}$, and $\boxed{c=-24}$. Then,

Given: $\boxed{x=\frac{-b\pm\sqrt{b^2-4ac}}{2a}}$; $\boxed{x=\frac{-10\pm\sqrt{10^2-(4\times 1\times -24)}}{2\times 1}}$; $\boxed{x=\frac{-10\pm\sqrt{100+96}}{2}}$; $\boxed{x=\frac{-10\pm\sqrt{196}}{2}}$

; $\boxed{x=\frac{-10\pm\sqrt{14^2}}{2}}$; $\boxed{x=\frac{-10\pm 14}{2}}$ Therefore, we can separate x into two equations:

I. $\boxed{x=\frac{-10+14}{2}}$; $\boxed{x=\frac{\overset{2}{\cancel{4}}}{\cancel{2}}}$; $\boxed{x=\frac{2}{1}}$; $\boxed{\mathbf{x=2}}$ II. $\boxed{x=\frac{-10-14}{2}}$; $\boxed{x=-\frac{\overset{12}{\cancel{24}}}{\cancel{2}}}$; $\boxed{x=-\frac{12}{1}}$; $\boxed{\mathbf{x=-12}}$

Thus, the solution set is $\{\mathbf{-12, 2}\}$.

The equation $(x+5)^2 = 49$ can be factored to $\mathbf{(x-2)(x+12)=0}$.

Note: As you may have already noticed, using the quadratic formula may not be a good choice since it requires more work and takes longer to solve. The key to solving quadratic equations is selection of a method that is easiest to use. Further discussions on selection of a best method is addressed in Section 4.6.

Note that when $b=0$ the quadratic equation $(ax+b)^2=c$ reduces to $(ax)^2=c$. The following examples show the steps as to how quadratic equations of the form $(ax)^2=c$ are solved for cases where the coefficient of x is equal to or greater than one.

- For cases where $a=1$, we can solve equations of the form $x^2=c$ using the Square Root Property method in the following way:

Example 4.3-11

Solve $x^2=16$ using the Square Root Property method.

Solution:

First - Take the square root of both sides of the equation, i.e., $\sqrt{x^2}=\pm\sqrt{16}$

Second - Simplify the terms on both sides to obtain the solutions, i.e., $\mathbf{x=\pm 4}$. Therefore, the solution set is $\{-4, 4\}$ and the equation $x^2=16$ can be factored to $\mathbf{(x-4)(x+4)=0}$.

Check: I. $Let\ x=-4\ in\quad x^2=16\ ;\ (-4)^2\stackrel{?}{=}16\ ;\ 16=16$

II. $Let\ x=4\ in\quad x^2=16\ ;\ 4^2\stackrel{?}{=}16\ ;\ 16=16$

Example 4.3-12

Solve $w^2=5$ using the Square Root Property method.

Solution:

First - Take the square root of both sides of the equation, i.e., $\sqrt{w^2}=\pm\sqrt{5}$

Second - Simplify the terms on both sides to obtain the solutions, i.e., $\mathbf{w=\pm\sqrt{5}}$. Therefore, the solution set is $\{-\sqrt{5}, \sqrt{5}\}$ and the equation $w^2=5$ can be factored to $\mathbf{(w-\sqrt{5})(w+\sqrt{5})=0}$.

Check: I. $Let\ w=-\sqrt{5}\ in\quad w^2=5\ ;\ \left(-\sqrt{5}\right)^2\stackrel{?}{=}5\ ;\ \left(+5^{\frac{1}{2}}\right)^2\stackrel{?}{=}5\ ;\ 5^{\frac{1}{2}\times 2}\stackrel{?}{=}5\ ;\ 5=5$

II. $Let\ w=\sqrt{5}\ in\quad w^2=5\ ;\ \left(\sqrt{5}\right)^2\stackrel{?}{=}5\ ;\ \left(5^{\frac{1}{2}}\right)^2\stackrel{?}{=}5\ ;\ 5^{\frac{1}{2}\times 2}\stackrel{?}{=}5\ ;\ 5=5$

- For cases where $a\rangle 1$, we can solve equations of the form $(ax)^2=c$ (which is the same as $kx^2=c$, where $k=a^2$) using the Square Root Property method in the following way:

Example 4.3-13

Solve $3x^2=27$ using the Square Root Property method.

Solution:

First - Divide both sides of the equation by the coefficient x, i.e., $\frac{\cancel{3}x^2}{\cancel{3}}=\frac{\overset{9}{\cancel{27}}}{\cancel{3}}\ ;\ x^2=\frac{9}{1}\ ;\ x^2=9$

Second - Take the square root of both sides of the equation, i.e., $\sqrt{x^2}=\pm\sqrt{9}$

Third - Simplify the terms on both sides to obtain the solutions, i.e., $\mathbf{x=\pm 3}$

Therefore, the solution set is $\{-3, 3\}$ and the equation $3x^2=27$ can be factored to $\mathbf{(x-3)(x+3)=0}$.

Check: I. *Let* $x=-3$ *in* $3x^2=27$; $3\cdot(-3)^2\overset{?}{=}27$; $3\cdot 9\overset{?}{=}27$; $27=27$

II. *Let* $x=3$ *in* $3x^2=27$; $3\cdot 3^2\overset{?}{=}27$; $3\cdot 9\overset{?}{=}27$; $27=27$

Example 4.3-14

Solve $2y^2=9$ using the Square Root Property method.

Solution:

First - Divide both sides of the equation by the coefficient y, i.e., $\frac{2y^2}{2}=\frac{9}{2}$; $y^2=\frac{9}{2}$; $y^2=4.5$

Second - Take the square root of both sides of the equation, i.e., $\sqrt{y^2}=\pm\sqrt{4.5}$

Third - Simplify the terms on both sides to obtain the solutions, i.e., $\mathbf{y=\pm 2.12}$. Therefore, the solution set is $\{-2.12, 2.12\}$ and the equation $2y^2=9$ can be factored to $\mathbf{(y-2.12)(y+2.12)=0}$.

Check: I. *Let* $y=-2.12$ *in* $2y^2=9$; $2\cdot(-2.12)^2\overset{?}{=}9$; $2\cdot 4.5\overset{?}{=}9$; $9=9$

II. *Let* $y=-2.12$ *in* $2y^2=9$; $2\cdot 2.12^2\overset{?}{=}9$; $2\cdot 4.5\overset{?}{=}9$; $9=9$

Practice Problems - Solving Quadratic Equations Using the Square Root Property Method

Section 4.3 Practice Problems - Solve the following equations using the Square Root Property method:

1. $(2y+5)^2=36$
2. $(x+1)^2=7$
3. $(2x-3)^2=1$
4. $x^2+3=0$
5. $(y-5)^2=5$
6. $16x^2-25=0$
7. $x^2-49=0$
8. $(3x-1)^2=25$
9. $(x-2)^2=-7$
10. $\left(x-\frac{1}{3}\right)^2=\frac{1}{9}$

4.4 Solving Quadratic Equations Using Completing-the-Square Method

One of the methods used in solving quadratic equations is called Completing-the-Square method. Note that this method involves construction of perfect square trinomials which was addressed in Section 3.5, Case I. In this section we will learn how to solve quadratic equations of the form $ax^2 + bx + c = 0$, where $a = 1$ (case I) and where $a \rangle 1$ (Case II), using Completing-the-Square method.

Case I Solving Quadratic Equations of the Form $ax^2 + bx + c = 0$, where $a = 1$, by Completing the Square

The following show the steps as to how quadratic equations, where the coefficient of the squared term is equal to one, are solved using Completing-the-Square method:

Step 1 Write the equation in the form of $x^2 + bx = -c$.

Step 2 a. Divide the coefficient of x by 2, i.e., $\frac{b}{2}$.

b. Square half the coefficient of x obtained in step 2a, i.e., $\left(\frac{b}{2}\right)^2$.

c. Add the square of half the coefficient of x to both sides of the equation, i.e.,
$x^2 + bx + \left(\frac{b}{2}\right)^2 = -c + \left(\frac{b}{2}\right)^2$.

d. Simplify the equation.

Step 3 Factor the trinomial on the left hand side of the equation as the square of a binomial, i.e., $\left(x + \frac{b}{2}\right)^2 = -c + \left(\frac{b}{2}\right)^2$.

Step 4 Take the square root of both sides of the equation and solve for the x values, i.e.,
$\sqrt{\left(x + \frac{b}{2}\right)^2} = \pm\sqrt{-c + \left(\frac{b}{2}\right)^2}$; $x + \frac{b}{2} = \pm\sqrt{-c + \left(\frac{b}{2}\right)^2}$; $x = -\frac{b}{2} \pm \sqrt{-c + \left(\frac{b}{2}\right)^2}$.

Step 5 Check the answers by substituting the x values into the original equation.

Step 6 Write the quadratic equation in its factored form.

Examples with Steps

The following examples show the steps as to how quadratic equations, where the coefficient of the squared term is equal to one, are solved using Completing-the-Square method:

Example 4.4-1

Solve the quadratic equation $x^2 + 8x + 5 = 0$ by completing the square.

Solution:

Step 1 $\boxed{x^2 + 8x + 5 = 0}$; $\boxed{x^2 + 8x + 5 - 5 = -5}$; $\boxed{x^2 + 8x + 0 = -5}$; $\boxed{x^2 + 8x = -5}$

Step 2 $\boxed{x^2 + 8x = -5}$; $\boxed{x^2 + 8x + \left(\overset{4}{\frac{8}{2}}\right)^2 = -5 + \left(\overset{4}{\frac{8}{2}}\right)^2}$; $\boxed{x^2 + 8x + 4^2 = -5 + 4^2}$

; $\boxed{x^2+8x+16=-5+16}$; $\boxed{x^2+8x+16=11}$

Step 3 $\boxed{x^2+8x+16=11}$; $\boxed{(x+4)^2=11}$

Step 4 $\boxed{(x+4)^2=11}$; $\boxed{\sqrt{(x+4)^2}=\pm\sqrt{11}}$; $\boxed{x+4=\pm\sqrt{11}}$; $\boxed{x+4=\pm 3.3166}$ therefore:

I. $\boxed{x+4=+3.3166}$; $\boxed{x=3.3166-4}$; $\boxed{\mathbf{x=-0.6834}}$

II. $\boxed{x+4=-3.3166}$; $\boxed{x=-3.3166-4}$; $\boxed{\mathbf{x=-7.3166}}$

Thus, the solution set is $\{\mathbf{-7.3166,\ -0.6834}\}$.

Step 5 Check: Substitute $x=-0.6834$ and $x=-7.3166$ in $x^2+8x+5=0$

I. *Let* $x=-0.6834$ *in* $x^2+8x+5=0$; $(-0.6834)^2+(8\times-0.6834)+5\stackrel{?}{=}0$

; $0.467-5.467+5\stackrel{?}{=}0$; $-5+5\stackrel{?}{=}0$; $0=0$

II. *Let* $x=-7.3166$ *in* $x^2+8x+5=0$; $(-7.3166)^2+(8\times-7.3166)+5\stackrel{?}{=}0$

; $53.533-58.533+5\stackrel{?}{=}0$; $-5+5\stackrel{?}{=}0$; $0=0$

Step 6 Thus, the equation $x^2+8x+5=0$ can be factored to $\mathbf{(x+0.6834)(x+7.3166)=0}$

Example 4.4-2

Solve the quadratic equation $x^2-4x+3=0$ by completing the square.

Solution:

Step 1 $\boxed{x^2-4x+3=0}$; $\boxed{x^2-4x+3-3=-3}$; $\boxed{x^2-4x+0=-3}$; $\boxed{x^2-4x=-3}$

Step 2 $\boxed{x^2-4x=-3}$; $\boxed{x^2-4x+\left(-\frac{\overset{2}{\cancel{4}}}{2}\right)^2=-3+\left(-\frac{\overset{2}{\cancel{4}}}{2}\right)^2}$; $\boxed{x^2-4x+2^2=-3+2^2}$

; $\boxed{x^2-4x+4=-3+4}$; $\boxed{x^2-4x+4=1}$

Step 3 $\boxed{x^2-4x+4=1}$; $\boxed{(x-2)^2=1}$

Step 4 $\boxed{(x-2)^2=1}$; $\boxed{\sqrt{(x-2)^2}=\pm\sqrt{1}}$; $\boxed{x-2=\pm\sqrt{1}}$; $\boxed{x-2=\pm 1}$ therefore:

I. $\boxed{x-2=+1}$; $\boxed{x=2+1}$; $\boxed{\mathbf{x=3}}$

II. $\boxed{x-2=-1}$; $\boxed{x=2-1}$; $\boxed{\mathbf{x=1}}$

Thus, the solution set is $\{\mathbf{1,\ 3}\}$.

Step 5 Check: Substitute $x=3$ and $x=1$ in $x^2-4x+3=0$

I. *Let* $x = 3$ *in* $x^2 - 4x + 3 = 0$; $3^2 - (4 \times 3) + 3 \stackrel{?}{=} 0$; $9 - 12 + 3 \stackrel{?}{=} 0$; $12 - 12 \stackrel{?}{=} 0$; $0 = 0$

II. *Let* $x = 1$ *in* $x^2 - 4x + 3 = 0$; $1^2 - (4 \times 1) + 3 \stackrel{?}{=} 0$; $1 - 4 + 3 \stackrel{?}{=} 0$; $4 - 4 \stackrel{?}{=} 0$; $0 = 0$

Step 6 Thus, the equation $x^2 - 4x + 3 = 0$ can be factored to $\mathbf{(x-3)(x-1)=0}$

Example 4.4-3

Solve the quadratic equation $x^2 + x - 6 = 0$ by completing the square.

Solution:

Step 1 $\boxed{x^2 + x - 6 = 0}$; $\boxed{x^2 + x - 6 + 6 = +6}$; $\boxed{x^2 + x + 0 = 6}$; $\boxed{x^2 + x = 6}$

Step 2 $\boxed{x^2 + x = 6}$; $\boxed{x^2 + x + \left(\frac{1}{2}\right)^2 = 6 + \left(\frac{1}{2}\right)^2}$; $\boxed{x^2 + x + \frac{1}{4} = 6 + \frac{1}{4}}$; $\boxed{x^2 + x + \frac{1}{4} = \frac{6}{1} + \frac{1}{4}}$

; $\boxed{x^2 + x + \frac{1}{4} = \frac{(6 \cdot 4) + (1 \cdot 1)}{1 \cdot 4}}$; $\boxed{x^2 + x + \frac{1}{4} = \frac{24 + 1}{4}}$; $\boxed{x^2 + x + \frac{1}{4} = \frac{25}{4}}$

Step 3 $\boxed{x^2 + x + \frac{1}{4} = \frac{25}{4}}$; $\boxed{\left(x + \frac{1}{2}\right)^2 = \frac{25}{4}}$

Step 4 $\boxed{\left(x + \frac{1}{2}\right)^2 = \frac{25}{4}}$; $\boxed{\sqrt{\left(x + \frac{1}{2}\right)^2} = \pm\sqrt{\frac{25}{4}}}$; $\boxed{x + \frac{1}{2} = \pm\sqrt{\frac{5^2}{2^2}}}$; $\boxed{x + \frac{1}{2} = \pm\frac{5}{2}}$ therefore:

I. $\boxed{x + \frac{1}{2} = +\frac{5}{2}}$; $\boxed{x = -\frac{1}{2} + \frac{5}{2}}$; $\boxed{x = \frac{-1+5}{2}}$; $\boxed{x = \frac{\overset{2}{\cancel{4}}}{\cancel{2}}}$; $\boxed{\mathbf{x = 2}}$

II. $\boxed{x + \frac{1}{2} = -\frac{5}{2}}$; $\boxed{x = -\frac{1}{2} - \frac{5}{2}}$; $\boxed{x = \frac{-1-5}{2}}$; $\boxed{x = -\frac{\overset{3}{\cancel{6}}}{\cancel{2}}}$; $\boxed{\mathbf{x = -3}}$

Thus, the solution set is $\{\mathbf{-3, 2}\}$.

Step 5 Check: Substitute $x = 2$ and $x = -3$ in $x^2 + x - 6 = 0$

I. *Let* $x = 2$ *in* $x^2 + x - 6 = 0$; $2^2 + 2 - 6 \stackrel{?}{=} 0$; $4 + 2 - 6 \stackrel{?}{=} 0$; $6 - 6 \stackrel{?}{=} 0$; $0 = 0$

II. *Let* $x = -3$ *in* $x^2 + x - 6 = 0$; $(-3)^2 - 3 - 6 \stackrel{?}{=} 0$; $9 - 3 - 6 \stackrel{?}{=} 0$; $9 - 9 \stackrel{?}{=} 0$; $0 = 0$

Step 6 Thus, the equation $x^2 + x - 6 = 0$ can be factored to $\mathbf{(x-2)(x+3)=0}$

Example 4.4-4

Solve the quadratic equation $x^2 - 6x + 2 = 0$ by completing the square.

Solution:

Step 1 $\boxed{x^2 - 6x + 2 = 0}$; $\boxed{x^2 - 6x + 2 - 2 = -2}$; $\boxed{x^2 - 6x + 0 = -2}$; $\boxed{x^2 - 6x = -2}$

Step 2 $\boxed{x^2 - 6x = -2}$; $\boxed{x^2 - 6x + \left(-\frac{\overset{3}{\cancel{6}}}{\cancel{2}}\right)^2 = -2 + \left(-\frac{\overset{3}{\cancel{6}}}{\cancel{2}}\right)^2}$; $\boxed{x^2 - 6x + 3^2 = -2 + 3^2}$

; $\boxed{x^2 - 6x + 9 = -2 + 9}$; $\boxed{x^2 - 6x + 9 = 7}$

Step 3 $\boxed{x^2 - 6x + 9 = 7}$; $\boxed{(x-3)^2 = 7}$

Step 4 $\boxed{(x-3)^2 = 7}$; $\boxed{\sqrt{(x-3)^2} = \pm\sqrt{7}}$; $\boxed{x - 3 = \pm\sqrt{7}}$; $\boxed{x - 3 = \pm 2.646}$ therefore:

I. $\boxed{x - 3 = +2.646}$; $\boxed{x = 3 + 2.646}$; $\boxed{\mathbf{x = 5.646}}$

II. $\boxed{x - 3 = -2.646}$; $\boxed{x = 3 - 2.646}$; $\boxed{\mathbf{x = 0.354}}$

Thus, the solution set is $\{\mathbf{0.354, 5.646}\}$.

Step 5 Check: Substitute $x = 5.646$ and $x = 0.354$ in $x^2 - 6x + 2 = 0$

I. *Let* $x = 5.646$ *in* $x^2 - 6x + 2 = 0$; $5.646^2 - (6 \times 5.646) + 2 \overset{?}{=} 0$

; $31.877 - 33.877 + 2 \overset{?}{=} 0$; $33.877 - 33.877 \overset{?}{=} 0$; $0 = 0$

II. *Let* $x = 0.354$ *in* $x^2 - 6x + 2 = 0$; $0.354^2 - (6 \times 0.354) + 2 \overset{?}{=} 0$; $0.13 - 2.13 + 2 \overset{?}{=} 0$

; $2.13 - 2.13 \overset{?}{=} 0$; $0 = 0$

Step 6 Thus, the equation $x^2 - 6x + 2 = 0$ can be factored to $\mathbf{(x - 5.646)(x - 0.354) = 0}$

Example 4.4-5

Solve the quadratic equation $x^2 + 2x + 5 = 0$ by completing the square.

Solution:

Step 1 $\boxed{x^2 + 2x + 5 = 0}$; $\boxed{x^2 + 2x + 5 - 5 = -5}$; $\boxed{x^2 + 2x + 0 = -5}$; $\boxed{x^2 + 2x = -5}$

Step 2 $\boxed{x^2 + 2x = -5}$; $\boxed{x^2 + 2x + \left(\frac{2}{2}\right)^2 = -5 + \left(\frac{2}{2}\right)^2}$; $\boxed{x^2 + 2x + 1^2 = -5 + 1^2}$

; $\boxed{x^2 + 2x + 1 = -5 + 1}$; $\boxed{x^2 + 2x + 1 = -4}$

Step 3 $\boxed{x^2 + 2x + 1 = -4}$; $\boxed{(x+1)^2 = -4}$

Step 4 $\boxed{(x+1)^2 = -4}$; $\boxed{\sqrt{(x+1)^2} = \pm\sqrt{-4}}$; $\boxed{x + 1 = \pm\sqrt{-4}}$ $\sqrt{-4}$ is not a real number.

Therefore, the equation $x^2 + 2x + 5 = 0$ **does not have any real solutions.**

Step 5 $\boxed{\textit{Not Applicable}}$

Step 6 $\boxed{\textit{Not Applicable}}$

Additional Examples - Solving Quadratic Equations of the Form $ax^2 + bx + c = 0$, where $a = 1$, by Completing the Square

The following examples further illustrate how to solve quadratic equations using Completing the Square method:

Example 4.4-6

Solve the quadratic equation $x^2 + 3x - 7 = 0$ using Completing-the-Square method.

Solution:

$\boxed{x^2 + 3x - 7 = 0}$; $\boxed{x^2 + 3x = 7}$; $\boxed{x^2 + 3x + \left(\frac{3}{2}\right)^2 = 7 + \left(\frac{3}{2}\right)^2}$; $\boxed{x^2 + 3x + \frac{9}{4} = 7 + \frac{9}{4}}$; $\boxed{\left(x + \frac{3}{2}\right)^2 = \frac{7}{1} + \frac{9}{4}}$

; $\boxed{\left(x + \frac{3}{2}\right)^2 = \frac{(7 \cdot 4) + (1 \cdot 9)}{1 \cdot 4}}$; $\boxed{\left(x + \frac{3}{2}\right)^2 = \frac{28 + 9}{4}}$; $\boxed{\left(x + \frac{3}{2}\right)^2 = \frac{37}{4}}$; $\boxed{\sqrt{\left(x + \frac{3}{2}\right)^2} = \pm\sqrt{\frac{37}{4}}}$; $\boxed{x + \frac{3}{2} = \pm\frac{\sqrt{37}}{2}}$

therefore: I. $\boxed{x + \frac{3}{2} = +\frac{\sqrt{37}}{2}}$; $\boxed{x = \frac{\sqrt{37}}{2} - \frac{3}{2}}$; $\boxed{x = \frac{6.083 - 3}{2}}$; $\boxed{x = \frac{3.083}{2}}$; $\boxed{\mathbf{x = 1.541}}$

II. $\boxed{x + \frac{3}{2} = -\frac{\sqrt{37}}{2}}$; $\boxed{x = -\frac{\sqrt{37}}{2} - \frac{3}{2}}$; $\boxed{x = \frac{-6.083 - 3}{2}}$; $\boxed{x = -\frac{9.083}{2}}$; $\boxed{\mathbf{x = -4.541}}$

and the solution set is $\{\mathbf{-4.541, 1.541}\}$.

Check: I. *Let* $x = 1.541$ *in* $x^2 + 3x - 7 = 0$; $(1.541)^2 + (3 \times 1.541) - 7 \stackrel{?}{=} 0$; $2.38 + 4.62 - 7 \stackrel{?}{=} 0$; $0 = 0$

II. *Let* $x = -4.541$ *in* $x^2 + 3x - 7 = 0$; $(-4.541)^2 + (3 \times -4.541) - 7 \stackrel{?}{=} 0$; $20.62 - 13.62 - 7 \stackrel{?}{=} 0$; $0 = 0$

Therefore, the equation $x^2 + 3x - 7 = 0$ can be factored to $\mathbf{(x - 1.541)(x + 4.541) = 0}$.

Example 4.4-7

Solve the quadratic equation $x^2 - x - 20 = 0$ using Completing-the-Square method.

Solution:

$\boxed{x^2 - x - 20 = 0}$; $\boxed{x^2 - x = 20}$; $\boxed{x^2 - x + \left(-\frac{1}{2}\right)^2 = 20 + \left(-\frac{1}{2}\right)^2}$; $\boxed{x^2 - x + \frac{1}{4} = 20 + \frac{1}{4}}$; $\boxed{\left(x - \frac{1}{2}\right)^2 = \frac{20}{1} + \frac{1}{4}}$

; $\boxed{\left(x - \frac{1}{2}\right)^2 = \frac{(20 \cdot 4) + (1 \cdot 1)}{1 \cdot 4}}$; $\boxed{\left(x - \frac{1}{2}\right)^2 = \frac{80 + 1}{4}}$; $\boxed{\left(x - \frac{1}{2}\right)^2 = \frac{81}{4}}$; $\boxed{\sqrt{\left(x - \frac{1}{2}\right)^2} = \pm\sqrt{\frac{81}{4}}}$; $\boxed{x - \frac{1}{2} = \pm\frac{9}{2}}$

therefore: I. $\boxed{x - \frac{1}{2} = +\frac{9}{2}}$; $\boxed{x = \frac{9}{2} + \frac{1}{2}}$; $\boxed{x = \frac{9 + 1}{2}}$; $\boxed{x = \frac{\overset{5}{\cancel{10}}}{\cancel{2}}}$; $\boxed{\mathbf{x = 5}}$

II. $\boxed{x - \frac{1}{2} = -\frac{9}{2}}$; $\boxed{x = -\frac{9}{2} + \frac{1}{2}}$; $\boxed{x = \frac{-9 + 1}{2}}$; $\boxed{x = -\frac{\overset{4}{\cancel{8}}}{\cancel{2}}}$; $\boxed{\mathbf{x = -4}}$

and the solution set is $\{\mathbf{-4, 5}\}$.

Check: I. *Let* $x = 5$ *in* $x^2 - x - 20 = 0$; $5^2 - 5 - 20 \overset{?}{=} 0$; $25 - 25 \overset{?}{=} 0$; $0 = 0$

II. *Let* $x = -4$ *in* $x^2 - x - 20 = 0$; $(-4)^2 - (-4) - 20 \overset{?}{=} 0$; $16 + 4 - 20 \overset{?}{=} 0$; $20 - 20 \overset{?}{=} 0$; $0 = 0$

Therefore, the equation $x^2 - x - 20 = 0$ can be factored to $\mathbf{(x-5)(x+4)=0}$.

Example 4.4-8

Solve the quadratic equation $x^2 + 5x + 6 = 0$ using Completing-the-Square method.

Solution:

$\boxed{x^2 + 5x + 6 = 0}$; $\boxed{x^2 + 5x = -6}$; $\boxed{x^2 + 5x + \left(\frac{5}{2}\right)^2 = -6 + \left(\frac{5}{2}\right)^2}$; $\boxed{x^2 + 5x + \frac{25}{4} = -6 + \frac{25}{4}}$; $\boxed{\left(x + \frac{5}{2}\right)^2 = -\frac{6}{1} + \frac{25}{4}}$

; $\boxed{\left(x + \frac{5}{2}\right)^2 = \frac{(-6 \cdot 4) + (1 \cdot 25)}{1 \cdot 4}}$; $\boxed{\left(x + \frac{5}{2}\right)^2 = \frac{-24 + 25}{4}}$; $\boxed{\left(x + \frac{5}{2}\right)^2 = \frac{1}{4}}$; $\boxed{\sqrt{\left(x + \frac{5}{2}\right)^2} = \pm\sqrt{\frac{1}{4}}}$; $\boxed{x + \frac{5}{2} = \pm\frac{1}{2}}$

therefore: I. $\boxed{x + \frac{5}{2} = +\frac{1}{2}}$; $\boxed{x = \frac{1}{2} - \frac{5}{2}}$; $\boxed{x = \frac{1-5}{2}}$; $\boxed{x = -\frac{\overset{2}{\cancel{4}}}{\cancel{2}}}$; $\boxed{x = -\frac{2}{1}}$; $\boxed{\mathbf{x = -2}}$

II. $\boxed{x + \frac{5}{2} = -\frac{1}{2}}$; $\boxed{x = -\frac{1}{2} - \frac{5}{2}}$; $\boxed{x = \frac{-1-5}{2}}$; $\boxed{x = -\frac{\overset{3}{\cancel{6}}}{\cancel{2}}}$; $\boxed{x = -\frac{3}{1}}$; $\boxed{\mathbf{x = -3}}$

and the solution set is $\{\mathbf{-3, -2}\}$.

Check: I. *Let* $x = -2$ *in* $x^2 + 5x + 6 = 0$; $(-2)^2 + (5 \times -2) + 6 \overset{?}{=} 0$; $4 - 10 + 6 \overset{?}{=} 0$; $10 - 10 \overset{?}{=} 0$; $0 = 0$

II. *Let* $x = -3$ *in* $x^2 + 5x + 6 = 0$; $(-3)^2 + (5 \times -3) + 6 \overset{?}{=} 0$; $9 - 15 + 6 \overset{?}{=} 0$; $15 - 15 \overset{?}{=} 0$; $0 = 0$

Therefore, the equation $x^2 + 5x + 6 = 0$ can be factored to $\mathbf{(x+2)(x+3)=0}$.

Example 4.4-9

Solve the quadratic equation $y^2 - 9y + 11 = 0$ using Completing-the-Square method.

Solution:

$\boxed{y^2 - 9y + 11 = 0}$; $\boxed{y^2 - 9y = -11}$; $\boxed{y^2 - 9y + \left(-\frac{9}{2}\right)^2 = -11 + \left(-\frac{9}{2}\right)^2}$; $\boxed{y^2 - 9y + \frac{81}{4} = -11 + \frac{81}{4}}$

; $\boxed{\left(y - \frac{9}{2}\right)^2 = -\frac{11}{1} + \frac{81}{4}}$; $\boxed{\left(y - \frac{9}{2}\right)^2 = \frac{(-11 \cdot 4) + (1 \cdot 81)}{1 \cdot 4}}$; $\boxed{\left(y - \frac{9}{2}\right)^2 = \frac{-44 + 81}{4}}$; $\boxed{\left(y - \frac{9}{2}\right)^2 = \frac{37}{4}}$

; $\boxed{\sqrt{\left(y - \frac{9}{2}\right)^2} = \pm\sqrt{\frac{37}{4}}}$; $\boxed{y - \frac{9}{2} = \pm\frac{\sqrt{37}}{2}}$ therefore:

I. $\boxed{y - \frac{9}{2} = +\frac{\sqrt{37}}{2}}$; $\boxed{y = \frac{\sqrt{37}}{2} + \frac{9}{2}}$; $\boxed{y = \frac{6.083 + 9}{2}}$; $\boxed{y = \frac{15.083}{2}}$; $\boxed{\mathbf{y = 7.541}}$

II. $\boxed{y - \frac{9}{2} = -\frac{\sqrt{37}}{2}}$; $\boxed{y = -\frac{\sqrt{37}}{2} + \frac{9}{2}}$; $\boxed{y = \frac{-6.083 + 9}{2}}$; $\boxed{y = \frac{2.917}{2}}$; $\boxed{\mathbf{y = 1.459}}$

and the solution set is $\{\mathbf{1.459, 7.541}\}$.

Check: I. *Let* $y = 7.541$ *in* $y^2 - 9y + 11 = 0$; $(7.541)^2 + (-9 \times 7.541) + 11 \stackrel{?}{=} 0$; $56.87 - 67.87 + 11 \stackrel{?}{=} 0$; $67.87 - 67.87 \stackrel{?}{=} 0$; $0 = 0$

II. *Let* $y = 1.459$ *in* $y^2 - 9y + 11 = 0$; $(1.459)^2 + (-9 \times 1.459) + 11 \stackrel{?}{=} 0$; $2.13 - 13.13 + 11 \stackrel{?}{=} 0$; $13.13 - 13.13 \stackrel{?}{=} 0$; $0 = 0$

Therefore, the equation $y^2 - 9y + 11 = 0$ can be factored to $(\mathbf{y - 7.541})(\mathbf{y - 1.459}) = \mathbf{0}$.

Example 4.4-10

Solve the quadratic equation $x(x-2) - 24 = 0$ using Completing-the-Square method.

Solution:

$\boxed{x(x-2) - 24 = 0}$; $\boxed{x^2 - 2x - 24 = 0}$; $\boxed{x^2 - 2x = 24}$; $\boxed{x^2 - 2x + \left(\frac{2}{2}\right)^2 = 24 + \left(\frac{2}{2}\right)^2}$; $\boxed{x^2 - 2x + 1 = 24 + 1}$; $\boxed{(x-1)^2 = 25}$; $\boxed{\sqrt{(x-1)^2} = \pm\sqrt{25}}$; $\boxed{x - 1 = \pm 5}$ therefore:

I. $\boxed{x - 1 = +5}$; $\boxed{x = 5 + 1}$; $\boxed{\mathbf{x = 6}}$

II. $\boxed{x - 1 = -5}$; $\boxed{x = -5 + 1}$; $\boxed{\mathbf{x = -4}}$

and the solution set is $\{\mathbf{-4, 6}\}$.

Check: I. *Let* $x = 6$ *in* $x(x-2) - 24 = 0$; $6 \cdot (6-2) - 24 \stackrel{?}{=} 0$; $6 \cdot 4 - 24 \stackrel{?}{=} 0$; $24 - 24 \stackrel{?}{=} 0$; $0 = 0$

II. *Let* $x = -4$ *in* $x(x-2) - 24 = 0$; $-4 \cdot (-4-2) - 24 \stackrel{?}{=} 0$; $-4 \cdot (-6) - 24 \stackrel{?}{=} 0$; $24 - 24 \stackrel{?}{=} 0$; $0 = 0$

Therefore, the equation $x(x-2) - 24 = 0$ can be factored to $(\mathbf{x - 6})(\mathbf{x + 4}) = \mathbf{0}$.

Practice Problems - Solving Quadratic Equations of the Form $ax^2 + bx + c = 0$, where $a = 1$, by Completing the Square

Section 4.4 Case I Practice Problems - Solve the following quadratic equations using Completing-the-Square method:

1. $x^2 + 10x - 2 = 0$
2. $x^2 - x - 1 = 0$
3. $x(x+2) = 80$
4. $y^2 - 10y + 5 = 0$
5. $x^2 + 4x - 5 = 0$
6. $y^2 + 4y = 14$
7. $w^2 + \frac{1}{3}w - \frac{1}{2} = 0$
8. $z^2 + 3z = -\frac{1}{4}$
9. $z^2 + \frac{5}{3}z - \frac{1}{2} = 0$
10. $x^2 - 6x = -4$

Case II Solving Quadratic Equations of the Form $ax^2 + bx + c = 0$, where $a \rangle 1$, by Completing the Square

The following show the steps as to how quadratic equations, where the coefficient of the squared term is not equal to one, are solved using Completing-the-Square method:

Step 1 Write the equation in the form of $ax^2 + bx = -c$.

Step 2 Divide both sides of the equation by a, i.e., $\frac{\cancel{a}x^2}{\cancel{a}} + \frac{bx}{a} = -\frac{c}{a}$; $x^2 + \frac{b}{a}x = -\frac{c}{a}$.

Step 3 a. Divide the coefficient of x by 2, i.e., $\frac{1}{2} \cdot \frac{b}{a} = \frac{b}{2a}$.

b. Square half the coefficient of x obtained in step 3a, i.e., $\left(\frac{b}{2a}\right)^2$

c. Add the square of half the coefficient of x to both sides of the equation, i.e.,

$$x^2 + \frac{b}{a}x + \left(\frac{b}{2a}\right)^2 = -\frac{c}{a} + \left(\frac{b}{2a}\right)^2.$$

d. Simplify the equation.

Step 4 Factor the trinomial on the left hand side of the equation as the square of a binomial, i.e., $\left(x + \frac{b}{2a}\right)^2 = -\frac{c}{a} + \left(\frac{b}{2a}\right)^2$.

Step 5 Take the square root of both sides of the equation and solve for the x values, i.e.,

$$\sqrt{\left(x + \frac{b}{2a}\right)^2} = \pm\sqrt{-\frac{c}{a} + \left(\frac{b}{2a}\right)^2} \; ; \; x + \frac{b}{2a} = \pm\sqrt{-\frac{c}{a} + \left(\frac{b}{2a}\right)^2} \; ; \; x = -\frac{b}{2a} \pm \sqrt{-\frac{c}{a} + \left(\frac{b}{2a}\right)^2}.$$

Step 6 Check the answers by substituting the x values into the original equation.

Step 7 Write the quadratic equation in its factored form.

Examples with Steps

The following examples show the steps as to how quadratic equations, where the coefficient of the squared term is not equal to one, are solved using completing-the-square method:

Example 4.4-11

Solve the quadratic equation $3x^2 - 16x + 5 = 0$ by completing the square.

Solution:

Step 1 $\boxed{3x^2 - 16x + 5 = 0}$; $\boxed{3x^2 - 16x + 5 - 5 = -5}$; $\boxed{3x^2 - 16x + 0 = -5}$; $\boxed{3x^2 - 16x = -5}$

Step 2 $\boxed{3x^2 - 16x = -5}$; $\boxed{\frac{\cancel{3}}{\cancel{3}}x^2 - \frac{16}{3}x = -\frac{5}{3}}$; $\boxed{x^2 - \frac{16}{3}x = -\frac{5}{3}}$

Step 3 $\boxed{x^2 - \frac{16}{3}x = -\frac{5}{3}}$; $\boxed{x^2 - \frac{16}{3}x + \left(-\frac{\overset{8}{\cancel{16}}}{\underset{3}{\cancel{6}}}\right)^2 = -\frac{5}{3} + \left(-\frac{\overset{8}{\cancel{16}}}{\underset{3}{\cancel{6}}}\right)^2}$; $\boxed{x^2 - \frac{16}{3}x + \left(-\frac{8}{3}\right)^2 = -\frac{5}{3} + \left(-\frac{8}{3}\right)^2}$

; $\boxed{x^2 - \frac{16}{3}x + \frac{64}{9} = -\frac{5}{3} + \frac{64}{9}}$; $\boxed{x^2 - \frac{16}{3}x + \frac{64}{9} = \frac{(-5\cdot 9) + (3\cdot 64)}{3\cdot 9}}$

; $\boxed{x^2 - \frac{16}{3}x + \frac{64}{9} = \frac{-45+192}{27}}$; $\boxed{x^2 - \frac{16}{3}x + \frac{64}{9} = \frac{\overset{49}{\cancel{147}}}{\underset{9}{\cancel{27}}}}$; $\boxed{x^2 - \frac{16}{3}x + \frac{64}{9} = \frac{49}{9}}$

Step 4 $\boxed{x^2 - \frac{16}{3}x + \frac{64}{9} = \frac{49}{9}}$; $\boxed{\left(x - \frac{8}{3}\right)^2 = \frac{49}{9}}$

Step 5 $\boxed{\left(x - \frac{8}{3}\right)^2 = \frac{49}{9}}$; $\boxed{\sqrt{\left(x - \frac{8}{3}\right)^2} = \pm\sqrt{\frac{49}{9}}}$; $\boxed{x - \frac{8}{3} = \pm\sqrt{\frac{7^2}{3^2}}}$; $\boxed{x - \frac{8}{3} = \pm\frac{7}{3}}$ therefore:

I. $\boxed{x - \frac{8}{3} = +\frac{7}{3}}$; $\boxed{x = \frac{8}{3} + \frac{7}{3}}$; $\boxed{x = \frac{8+7}{3}}$; $\boxed{x = \frac{\overset{5}{\cancel{15}}}{\cancel{3}}}$; $\boxed{x = \frac{5}{1}}$; $\boxed{\mathbf{x = 5}}$

II. $\boxed{x - \frac{8}{3} = -\frac{7}{3}}$; $\boxed{x = \frac{8}{3} - \frac{7}{3}}$; $\boxed{x = \frac{8-7}{3}}$; $\boxed{\mathbf{x = \frac{1}{3}}}$

and the solution set is $\left\{\mathbf{\frac{1}{3}, 5}\right\}$.

Step 6 Check: Substitute $x = 5$ and $x = \frac{1}{3}$ in $3x^2 - 16x + 5 = 0$

I. *Let* $x = 5$ *in* $3x^2 - 16x + 5 = 0$; $3 \cdot 5^2 - (16 \times 5) + 5 \overset{?}{=} 0$; $75 - 80 + 5 \overset{?}{=} 0$; $80 - 80 + 5 \overset{?}{=} 0$; $0 = 0$

II. *Let* $x = \frac{1}{3}$ *in* $3x^2 - 16x + 5 = 0$; $3 \cdot \left(\frac{1}{3}\right)^2 - \left(16 \times \frac{1}{3}\right) + 5 \overset{?}{=} 0$; $\frac{\cancel{3}}{\underset{3}{\cancel{9}}} - \frac{16}{3} + 5 \overset{?}{=} 0$; $\frac{1}{3} - \frac{16}{3} + 5 \overset{?}{=} 0$; $\frac{1-16}{3} + 5 \overset{?}{=} 0$; $-\frac{\overset{5}{\cancel{15}}}{\cancel{3}} + 5 \overset{?}{=} 0$; $-5 + 5 \overset{?}{=} 0$; $0 = 0$

Step 7 Thus, the equation $3x^2 - 16x + 5 = 0$ can be factored to $(x-5)\left(x - \frac{1}{3}\right) = 0$ which is the same as $\mathbf{(x-5)(3x-1) = 0}$

Example 4.4-12

Solve the quadratic equation $2x^2 + 3x - 6 = 0$ by completing the square.

Solution:

Step 1 $\boxed{2x^2 + 3x - 6 = 0}$; $\boxed{2x^2 + 3x - 6 + 6 = +6}$; $\boxed{2x^2 + 3x + 0 = +6}$; $\boxed{2x^2 + 3x = 6}$

Step 2 $\boxed{2x^2 + 3x = 6}$; $\boxed{\frac{\cancel{2}}{\cancel{2}}x^2 + \frac{3}{2}x = \frac{\overset{3}{\cancel{6}}}{\cancel{2}}}$; $\boxed{x^2 + \frac{3}{2}x = \frac{3}{1}}$; $\boxed{x^2 + \frac{3}{2}x = 3}$

Step 3 $\boxed{x^2 + \frac{3}{2}x = 3}$; $\boxed{x^2 + \frac{3}{2}x + \left(\frac{3}{4}\right)^2 = 3 + \left(\frac{3}{4}\right)^2}$; $\boxed{x^2 + \frac{3}{2}x + \frac{9}{16} = 3 + \frac{9}{16}}$

$; \boxed{x^2 + \frac{3}{2}x + \frac{9}{16} = \frac{3}{1} + \frac{9}{16}}\ ; \boxed{x^2 + \frac{3}{2}x + \frac{9}{16} = \frac{48+9}{16}}\ ; \boxed{x^2 + \frac{3}{2}x + \frac{9}{16} = \frac{57}{16}}$

Step 4 $\boxed{x^2 + \frac{3}{2}x + \frac{9}{16} = \frac{57}{16}}\ ; \boxed{\left(x + \frac{3}{4}\right)^2 = \frac{57}{16}}$

Step 5 $\boxed{\left(x + \frac{3}{4}\right)^2 = \frac{57}{16}}\ ; \boxed{\sqrt{\left(x + \frac{3}{4}\right)^2} = \pm\sqrt{\frac{57}{16}}}\ ; \boxed{x + \frac{3}{4} = \pm\sqrt{\frac{57}{4^2}}}\ ; \boxed{x + \frac{3}{4} = \pm\frac{\sqrt{57}}{4}}$ therefore:

I. $\boxed{x + \frac{3}{4} = +\frac{\sqrt{57}}{4}}\ ; \boxed{x = \frac{7.55}{4} - \frac{3}{4}}\ ; \boxed{x = \frac{7.55-3}{4}}\ ; \boxed{x = \frac{4.55}{4}}\ ; \boxed{\mathbf{x = 1.138}}$

II. $\boxed{x + \frac{3}{4} = -\frac{\sqrt{57}}{4}}\ ; \boxed{x = -\frac{7.55}{4} - \frac{3}{4}}\ ; \boxed{x = \frac{-7.55-3}{4}}\ ; \boxed{x = \frac{-10.55}{4}}\ ; \boxed{\mathbf{x = -2.638}}$

and the solution set is $\{\mathbf{-2.638, 1.138}\}$.

Step 6 Check: Substitute $x = 1.138$ and $x = -2.638$ in $2x^2 + 3x - 6 = 0$

I. *Let* $x = 1.138$ *in* $2x^2 + 3x - 6 = 0\ ;\ 2\cdot(1.138)^2 + (3 \times 1.138) - 6 \stackrel{?}{=} 0$

$;\ 2.59 + 3.41 - 6 \stackrel{?}{=} 0\ ;\ 6 - 6 \stackrel{?}{=} 0\ ;\ 0 = 0$

II. *Let* $x = -2.638$ *in* $2x^2 + 3x - 6 = 0\ ;\ 2\cdot(-2.638)^2 + (3 \times -2.638) - 6 \stackrel{?}{=} 0$

$;\ 13.92 - 7.92 - 6 \stackrel{?}{=} 0\ ;\ 13.92 - 13.92 \stackrel{?}{=} 0\ ;\ 0 = 0$

Step 7 Thus, the equation $2x^2 + 3x - 6 = 0$ which is equal to $x^2 + 1.5x - 3 = 0$ can be factored to $\mathbf{(x - 1.138)(x + 2.638) = 0}$

Example 4.4-13

Solve the quadratic equation $8x^2 + 8x - 30 = 0$ by completing the square.

Solution:

Step 1 $\boxed{8x^2 + 8x - 30 = 0}\ ; \boxed{8x^2 + 8x - 30 + 30 = +30}\ ; \boxed{8x^2 + 8x + 0 = 30}\ ; \boxed{8x^2 + 8x = 30}$

Step 2 $\boxed{8x^2 + 8x = 30}\ ; \boxed{\frac{\cancel{8}}{\cancel{8}}x^2 + \frac{\cancel{8}}{\cancel{8}}x = \frac{\overset{15}{\cancel{30}}}{\underset{4}{\cancel{8}}}}\ ; \boxed{x^2 + x = \frac{15}{4}}$

Step 3 $\boxed{x^2 + x = \frac{15}{4}}\ ; \boxed{x^2 + x + \left(\frac{1}{2}\right)^2 = \frac{15}{4} + \left(\frac{1}{2}\right)^2}\ ; \boxed{x^2 + x + \frac{1}{4} = \frac{15}{4} + \frac{1}{4}}\ ; \boxed{x^2 + x + \frac{1}{4} = \frac{15+1}{4}}$

$; \boxed{x^2 + x + \frac{1}{4} = \frac{\overset{4}{\cancel{16}}}{\cancel{4}}}\ ; \boxed{x^2 + x + \frac{1}{4} = 4}$

Step 4 $\boxed{x^2 + x + \frac{1}{4} = 4}\ ; \boxed{\left(x + \frac{1}{2}\right)^2 = 4}$

Step 5 $\boxed{\left(x+\frac{1}{2}\right)^2=4}$; $\boxed{\sqrt{\left(x+\frac{1}{2}\right)^2}=\pm\sqrt{4}}$; $\boxed{x+\frac{1}{2}=\pm 2}$ therefore:

I. $\boxed{x+\frac{1}{2}=+2}$; $\boxed{x=-\frac{1}{2}+\frac{2}{1}}$; $\boxed{x=\frac{-(1\cdot 1)+(2\cdot 2)}{1\cdot 2}}$; $\boxed{x=\frac{-1+4}{2}}$; $\boxed{\mathbf{x=\frac{3}{2}}}$

II. $\boxed{x+\frac{1}{2}=-2}$; $\boxed{x=-\frac{1}{2}-\frac{2}{1}}$; $\boxed{x=\frac{-(1\cdot 1)-(2\cdot 2)}{1\cdot 2}}$; $\boxed{x=\frac{-1-4}{2}}$; $\boxed{\mathbf{x=-\frac{5}{2}}}$

and the solution set is $\left\{-\frac{\mathbf{5}}{\mathbf{2}},\frac{\mathbf{3}}{\mathbf{2}}\right\}$.

Step 6 Check: Substitute $x=\frac{3}{2}$ and $x=-\frac{5}{2}$ in $8x^2+8x-30=0$

I. *Let* $x=\frac{3}{2}$ *in* $8x^2+8x-30=0$; $8\cdot\left(\frac{3}{2}\right)^2+8\cdot\frac{3}{2}-30\overset{?}{=}0$; $8\cdot\frac{9}{4}+\frac{24}{2}-30\overset{?}{=}0$

; $\frac{\overset{18}{\cancel{72}}}{\cancel{4}}+\frac{\overset{12}{\cancel{24}}}{\cancel{2}}-30\overset{?}{=}0$; $18+12-30\overset{?}{=}0$; $30-30\overset{?}{=}0$; $0=0$

II. *Let* $x=-\frac{5}{2}$ *in* $8x^2+8x-30=0$; $8\cdot\left(-\frac{5}{2}\right)^2+8\cdot\left(-\frac{5}{2}\right)-30\overset{?}{=}0$

; $8\cdot\frac{25}{4}-\frac{40}{2}-30\overset{?}{=}0$; $\frac{\overset{50}{\cancel{200}}}{\cancel{4}}-\frac{\overset{20}{\cancel{40}}}{\cancel{2}}-30\overset{?}{=}0$; $50-20-30\overset{?}{=}0$; $50-50\overset{?}{=}0$; $0=0$

Step 7 Thus, the equation $8x^2+8x-30=0$ which is the same as $4x^2+4x-15=0$ can be factored to $\left(x-\frac{3}{2}\right)\left(x+\frac{5}{2}\right)=0$ which is the same as $\mathbf{(2x-3)(2x+5)=0}$

Example 4.4-14

Solve the quadratic equation $2u^2+6u-7=0$ by completing the square.

Solution:

Step 1 $\boxed{2u^2+6u-7=0}$; $\boxed{2u^2+6u-7+7=+7}$; $\boxed{2u^2+6u+0=7}$; $\boxed{2u^2+6u=7}$

Step 2 $\boxed{2u^2+6u=7}$; $\boxed{\frac{\cancel{2}}{\cancel{2}}u^2+\frac{\overset{3}{\cancel{6}}}{\cancel{2}}u=\frac{7}{2}}$; $\boxed{u^2+3u=\frac{7}{2}}$

Step 3 $\boxed{u^2+3u=\frac{7}{2}}$; $\boxed{u^2+3u+\left(\frac{3}{2}\right)^2=\frac{7}{2}+\left(\frac{3}{2}\right)^2}$; $\boxed{u^2+3u+\frac{9}{4}=\frac{7}{2}+\frac{9}{4}}$

; $\boxed{u^2+3u+\frac{9}{4}=\frac{(7\cdot 4)+(2\cdot 9)}{2\cdot 4}}$; $\boxed{u^2+3u+\frac{9}{4}=\frac{28+18}{8}}$; $\boxed{u^2+3u+\frac{9}{4}=\frac{46}{8}}$

Step 4 $\boxed{u^2+3u+\frac{9}{4}=\frac{46}{8}}$; $\boxed{\left(u+\frac{3}{2}\right)^2=\frac{\overset{23}{\cancel{46}}}{\underset{4}{\cancel{8}}}}$; $\boxed{\left(u+\frac{3}{2}\right)^2=\frac{23}{4}}$

Step 5 $\boxed{\left(u+\frac{3}{2}\right)^2=\frac{23}{4}}$; $\boxed{\sqrt{\left(u+\frac{3}{2}\right)^2}=\pm\sqrt{\frac{23}{4}}}$; $\boxed{u+\frac{3}{2}=\pm\sqrt{\frac{23}{2^2}}}$; $\boxed{u+\frac{3}{2}=\pm\frac{\sqrt{23}}{2}}$ therefore:

I. $\boxed{u+\frac{3}{2}=+\frac{\sqrt{23}}{2}}$; $\boxed{u=-\frac{3}{2}+\frac{\sqrt{23}}{2}}$; $\boxed{u=\frac{-3+\sqrt{23}}{2}}$; $\boxed{u=\frac{-3+4.8}{2}}$; $\boxed{u=\frac{1.8}{2}}$; $\boxed{\mathbf{u=0.9}}$

II. $\boxed{u+\frac{3}{2}=-\frac{\sqrt{23}}{2}}$; $\boxed{u=-\frac{3}{2}-\frac{\sqrt{23}}{2}}$; $\boxed{u=\frac{-3-\sqrt{23}}{2}}$; $\boxed{u=\frac{-3-4.8}{2}}$; $\boxed{u=\frac{-7.8}{2}}$; $\boxed{\mathbf{u=-3.9}}$

and the solution set is $\{\mathbf{-3.9,\ 0.9}\}$.

Step 6 Check: Substitute $u=0.9$ and $u=-3.9$ in $2u^2+6u-7=0$

I. *Let* $u=0.9$ *in* $2u^2+6u-7=0$; $2\cdot 0.9^2+(6\times 0.9)-7\overset{?}{=}0$; $1.6+5.4-7\overset{?}{=}0$; $7-7\overset{?}{=}0$; $0=0$

II. *Let* $u=-3.9$ *in* $2u^2+6u-7=0$; $2\cdot(-3.9)^2+(6\times -3.9)-7\overset{?}{=}0$; $2\times 15.2-23.4-7\overset{?}{=}0$; $30.4-23.4-7\overset{?}{=}0$; $7-7\overset{?}{=}0$; $0=0$

Step 7 Thus, the equation $2u^2+6u-7=0$, which is the same as $u^2+3u-3.5=0$, can be factored to $\mathbf{(u-0.9)(u+3.9)=0}$.

Example 4.4-15

Solve the quadratic equation $4a^2+24a-5=0$ by completing the square.

Solution:

Step 1 $\boxed{4a^2+24a-5=0}$; $\boxed{4a^2+24a-5+5=+5}$; $\boxed{4a^2+24a+0=5}$; $\boxed{4a^2+24a=5}$

Step 2 $\boxed{4a^2+24a=5}$; $\boxed{\frac{\cancel{4}}{\cancel{4}}a^2+\frac{\overset{6}{\cancel{24}}}{\cancel{4}}a=\frac{5}{4}}$; $\boxed{a^2+6a=\frac{5}{4}}$

Step 3 $\boxed{a^2+6a=\frac{5}{4}}$; $\boxed{a^2+6a+\left(\frac{\overset{3}{\cancel{6}}}{\cancel{2}}\right)^2=\frac{5}{4}+\left(\frac{\overset{3}{\cancel{6}}}{\cancel{2}}\right)^2}$; $\boxed{a^2+6a+\left(\frac{3}{1}\right)^2=\frac{5}{4}+\left(\frac{3}{1}\right)^2}$

; $\boxed{a^2+6a+3^2=\frac{5}{4}+3^2}$; $\boxed{a^2+6a+9=\frac{5}{4}+\frac{9}{1}}$; $\boxed{a^2+6a+9=\frac{(5\cdot 1)+(9\cdot 4)}{4\cdot 1}}$

; $\boxed{a^2+6a+9=\frac{5+36}{4}}$; $\boxed{a^2+6a+9=\frac{41}{4}}$

Step 4 $\boxed{a^2+6a+9=\frac{41}{4}}$; $\boxed{(a+3)^2=10.25}$

Step 5 $\boxed{(a+3)^2=10.25}$; $\boxed{\sqrt{(a+3)^2}=\pm\sqrt{10.25}}$; $\boxed{a+3=\pm 3.2}$ therefore:

I. $\boxed{a+3=+3.2}$; $\boxed{a=3.2-3}$; $\boxed{\mathbf{a=0.2}}$

II. $\boxed{a+3=-3.2}$; $\boxed{a=-3.2-3}$; $\boxed{\mathbf{a=-6.2}}$

and the solution set is $\{\mathbf{-6.2,\ 0.2}\}$.

Step 6 Check: Substitute $a = 0.2$ and $a = -6.2$ in $4a^2 + 24a - 5 = 0$

I. *Let* $a = 0.2$ *in* $4a^2 + 24a - 5 = 0$; $4 \cdot 0.2^2 + (24 \times 0.2) - 5 \stackrel{?}{=} 0$; $4 \times 0.04 + 4.8 - 5 \stackrel{?}{=} 0$; $0.16 + 4.8 - 5 \stackrel{?}{=} 0$; $5 - 5 \stackrel{?}{=} 0$; $0 = 0$

II. *Let* $a = -6.2$ *in* $4a^2 + 24a - 5 = 0$; $4 \cdot (-6.2)^2 + (24 \times -6.2) - 5 \stackrel{?}{=} 0$; ; $4 \times 38.44 - 148.8 - 5 \stackrel{?}{=} 0$; $153.8 - 148.8 - 5 \stackrel{?}{=} 0$; $153.8 - 153.8 \stackrel{?}{=} 0$; $0 = 0$

Step 7 Thus, the equation $4a^2 + 24a - 5 = 0$ can be factored to $\mathbf{(a - 0.2)(a + 6.2) = 0}$.

Additional Examples - Solving Quadratic Equations of the Form $ax^2 + bx + c = 0$, where $a \rangle 1$, by Completing the Square

The following examples further illustrate how to solve quadratic equations, where the coefficient of the squared term is not equal to one, using Completing-the-Square method:

Example 4.4-16

Solve the quadratic equation $3x^2 + 2x - 1 = 0$ using Completing-the-Square method.

Solution:

$\boxed{3x^2 + 2x - 1 = 0}$; $\boxed{3x^2 + 2x = 1}$; $\boxed{\frac{\cancel{3}}{\cancel{3}}x^2 + \frac{2}{3}x = \frac{1}{3}}$; $\boxed{x^2 + \frac{2}{3}x = \frac{1}{3}}$; $\boxed{x^2 + \frac{2}{3}x + \left(\frac{2}{\cancel{6}}\right)^2 = \frac{1}{3} + \left(\frac{2}{\cancel{6}}\right)^2}$

; $\boxed{x^2 + \frac{2}{3}x + \left(\frac{1}{3}\right)^2 = \frac{1}{3} + \left(\frac{1}{3}\right)^2}$; $\boxed{x^2 + \frac{2}{3}x + \frac{1}{9} = \frac{1}{3} + \frac{1}{9}}$; $\boxed{\left(x + \frac{1}{3}\right)^2 = \frac{1}{3} + \frac{1}{9}}$; $\boxed{\left(x + \frac{1}{3}\right)^2 = \frac{(1 \cdot 9) + (1 \cdot 3)}{3 \cdot 9}}$

; $\boxed{\left(x + \frac{1}{3}\right)^2 = \frac{9 + 3}{27}}$; $\boxed{\left(x + \frac{1}{3}\right)^2 = \frac{12}{27}}$; $\boxed{\sqrt{\left(x + \frac{1}{3}\right)^2} = \pm\sqrt{\frac{12}{27}}}$; $\boxed{x + \frac{1}{3} = \pm\sqrt{\frac{12}{27}}}$; $\boxed{x + \frac{1}{3} = \pm\sqrt{0.44}}$

; $\boxed{x + 0.33 = \pm 0.67}$ therefore:

I. $\boxed{x + 0.33 = +0.67}$; $\boxed{x = 0.67 - 0.33}$; $\boxed{\mathbf{x = 0.34}}$ II. $\boxed{x + 0.33 = -0.67}$; $\boxed{x = -0.67 - 0.33}$; $\boxed{\mathbf{x = -1}}$

and the solution set is $\{\mathbf{-1,\ 0.34}\}$.

Check: I. *Let* $x = -1$ *in* $3x^2 + 2x - 1 = 0$; $3 \cdot (-1)^2 + (2 \cdot -1) - 1 \stackrel{?}{=} 0$; $3 \cdot 1 - 2 - 1 \stackrel{?}{=} 0$; $3 - 3 \stackrel{?}{=} 0$; $0 = 0$

II. *Let* $x = 0.34$ *in* $3x^2 + 2x - 1 = 0$; $3 \cdot 0.34^2 + (2 \cdot 0.34) - 1 \stackrel{?}{=} 0$; $3 \cdot 0.11 + 0.68 - 1 \stackrel{?}{=} 0$; $0.33 + 0.68 - 1 \stackrel{?}{=} 0$; $1 - 1 \stackrel{?}{=} 0$; $0 = 0$

Therefore, the equation $3x^2 + 2x - 1 = 0$ can be factored to $\mathbf{(x - 0.34)(x + 1) = 0}$.

Example 4.4-17

Solve the quadratic equation $3y^2 - 8y + 2 = 0$ using Completing-the-Square method.

Solution:

$\boxed{3y^2 - 8y + 2 = 0}$; $\boxed{3y^2 - 8y = -2}$; $\boxed{\frac{\cancel{3}}{\cancel{3}}y^2 - \frac{8}{3}y = -\frac{2}{3}}$; $\boxed{y^2 - \frac{8}{3}y = -\frac{2}{3}}$; $\boxed{y^2 - \frac{8}{3}y + \left(-\frac{\frac{4}{\cancel{8}}}{\frac{\cancel{6}}{3}}\right)^2 = -\frac{2}{3} + \left(-\frac{\frac{4}{\cancel{8}}}{\frac{\cancel{6}}{3}}\right)^2}$

$; \boxed{y^2 - \frac{8}{3}y + \left(\frac{4}{3}\right)^2 = -\frac{2}{3} + \left(\frac{4}{3}\right)^2} ; \boxed{y^2 - \frac{8}{3}y + \frac{16}{9} = -\frac{2}{3} + \frac{16}{9}} ; \boxed{\left(y - \frac{4}{3}\right)^2 = -\frac{2}{3} + \frac{16}{9}}$

$; \boxed{\left(y - \frac{4}{3}\right)^2 = \frac{(-2 \cdot 9) + (16 \cdot 3)}{3 \cdot 9}} ; \boxed{\left(y - \frac{4}{3}\right)^2 = \frac{-18 + 48}{27}} ; \boxed{\left(y - \frac{4}{3}\right)^2 = \frac{30}{27}} ; \boxed{\sqrt{\left(y - \frac{4}{3}\right)^2} = \pm\sqrt{\frac{30}{27}}}$

$; \boxed{y - \frac{4}{3} = \pm\sqrt{\frac{30}{27}}} ; \boxed{y - \frac{4}{3} = \pm\sqrt{1.11}} ; \boxed{y - 1.33 = \pm 1.05}$ therefore:

I. $\boxed{y - 1.33 = +1.05} ; \boxed{y = 1.05 + 1.33} ; \boxed{\mathbf{y = 2.38}}$ II. $\boxed{y - 1.33 = -1.05} ; \boxed{y = -1.05 + 1.33} ; \boxed{\mathbf{y = 0.28}}$

and the solution set is $\{\mathbf{0.28, 2.38}\}$.

Check: I. *Let* $y = 0.28$ *in* $3y^2 - 8y + 2 = 0 ; 3 \cdot (0.28)^2 - 8 \cdot 0.28 + 2 \stackrel{?}{=} 0 ; 0.24 - 2.24 + 2 \stackrel{?}{=} 0$

$; 2.24 - 2.24 \stackrel{?}{=} 0 ; 0 = 0$

II. *Let* $y = 2.38$ *in* $3y^2 - 8y + 2 = 0 ; 3 \cdot (2.38)^2 - 8 \cdot 2.38 + 2 \stackrel{?}{=} 0 ; 17 - 19 + 2 \stackrel{?}{=} 0 ; 0 = 0$

Therefore, the equation $3y^2 - 8y + 2 = 0$ can be factored to $\mathbf{(y - 0.28)(y - 2.38) = 0}$.

Example 4.4-18

Solve the quadratic equation $3t^2 + 12t - 4 = 0$ using Completing-the-Square method.

Solution:

$\boxed{3t^2 + 12t - 4 = 0} ; \boxed{3t^2 + 12t = 4} ; \boxed{\frac{\cancel{3}}{\cancel{3}}t^2 + \frac{\overset{4}{\cancel{12}}}{\cancel{3}}x = \frac{4}{3}} ; \boxed{t^2 + 4t = \frac{4}{3}} ; \boxed{t^2 + 4t + \left(\frac{\overset{2}{\cancel{4}}}{2}\right)^2 = \frac{4}{3} + \left(\frac{\overset{2}{\cancel{4}}}{2}\right)^2}$

$; \boxed{t^2 + 4t + \left(\frac{2}{1}\right)^2 = \frac{4}{3} + \left(\frac{2}{1}\right)^2} ; \boxed{t^2 + 4t + 4 = \frac{4}{3} + 4} ; \boxed{(t + 2)^2 = \frac{4}{3} + \frac{4}{1}} ; \boxed{(t + 2)^2 = \frac{(4 \cdot 1) + (4 \cdot 3)}{3 \cdot 1}}$

$; \boxed{(t + 2)^2 = \frac{4 + 12}{3}} ; \boxed{(t + 2)^2 = \frac{16}{3}} ; \boxed{\sqrt{(t + 2)^2} = \pm\sqrt{\frac{16}{3}}} ; \boxed{t + 2 = \pm\sqrt{5.33}} ; \boxed{t + 2 = \pm 2.31}$ therefore:

I. $\boxed{t + 2 = +2.31} ; \boxed{t = 2.31 - 2} ; \boxed{\mathbf{t = 0.31}}$ II. $\boxed{t + 2 = -2.31} ; \boxed{t = -2.31 - 2} ; \boxed{\mathbf{t = -4.31}}$

and the solution set is $\{\mathbf{0.31, -4.31}\}$.

Check: I. *Let* $t = 0.31$ *in* $3t^2 + 12t - 4 = 0 ; 3 \cdot (0.31)^2 + (12 \cdot 0.31) - 4 \stackrel{?}{=} 0 ; 3 \cdot 0.096 - 3.72 - 4 \stackrel{?}{=} 0$

$; 0.288 + 3.72 - 4 \stackrel{?}{=} 0 ; 4 - 4 \stackrel{?}{=} 0 ; 0 = 0$

II. *Let* $t = -4.31$ *in* $3t^2 + 12t - 4 = 0 ; 3 \cdot (-4.31)^2 + (12 \cdot -4.31) - 4 \stackrel{?}{=} 0 ; 3 \cdot 18.57 - 51.72 - 4 \stackrel{?}{=} 0$

$; 55.72 - 51.72 - 4 \stackrel{?}{=} 0 ; 55.72 - 55.72 \stackrel{?}{=} 0 ; 0 = 0$

Therefore, the equation $3t^2 + 12t - 4 = 0$ can be factored to $\mathbf{(t - 0.31)(t + 4.31) = 0}$.

Example 4.4-19

Solve the quadratic equation $2a^2+16a-6=0$ using Completing-the-Square method.

Solution:

$\boxed{2a^2+16a-6=0}$; $\boxed{2a^2+16a=6}$; $\boxed{\frac{2}{2}a^2+\frac{16}{2}a=\frac{6}{2}}$; $\boxed{a^2+8a=3}$; $\boxed{a^2+8a+\left(\frac{8}{2}\right)^2=3+\left(\frac{8}{2}\right)^2}$

; $\boxed{a^2+8a+\left(\frac{4}{1}\right)^2=3+\left(\frac{4}{1}\right)^2}$; $\boxed{a^2+8a+16=3+16}$; $\boxed{(a+4)^2=19}$; $\boxed{\sqrt{(a+4)^2}=\pm\sqrt{19}}$; $\boxed{a+4=\pm 4.36}$

Therefore:

I. $\boxed{a+4=+4.36}$; $\boxed{a=4.36-4}$; $\boxed{\mathbf{a=0.36}}$　　II. $\boxed{a+4=-4.36}$; $\boxed{a=-4.36-4}$; $\boxed{\mathbf{a=-8.36}}$

and the solution set is $\{\mathbf{-8.36, 0.36}\}$.

Check: I. *Let* $a=-8.36$ *in* $2a^2+16a-6=0$; $2\cdot(-8.36)^2+(16\cdot -8.36)-6\overset{?}{=}0$; $2\cdot 69.9-133.8-6\overset{?}{=}0$

; $139.8-133.8-6\overset{?}{=}0$; $139.8-139.8\overset{?}{=}0$; $0=0$

II. *Let* $a=0.36$ *in* $2a^2+16a-6=0$; $2\cdot(0.36)^2+(16\cdot 0.36)-6\overset{?}{=}0$; $2\cdot 0.129+5.7-6\overset{?}{=}0$

; $0.3+5.7-6\overset{?}{=}0$; $6-6\overset{?}{=}0$; $0=0$

Therefore, the equation $2a^2+16a-6=0$ can be factored to $\mathbf{(a+8.36)(a-0.36)=0}$.

Example 4.4-20

Solve the quadratic equation $4n^2+5n-2=0$ using Completing-the-Square method.

Solution:

$\boxed{4n^2+5n-2=0}$; $\boxed{4n^2+5n=2}$; $\boxed{\frac{4}{4}n^2+\frac{5}{4}n=\frac{2}{4}}$; $\boxed{n^2+\frac{5}{4}n=\frac{1}{2}}$; $\boxed{n^2+\frac{5}{4}n+\left(\frac{5}{8}\right)^2=\frac{1}{2}+\left(\frac{5}{8}\right)^2}$

; $\boxed{n^2+\frac{5}{4}n+\frac{25}{64}=\frac{1}{2}+\frac{25}{64}}$; $\boxed{\left(n+\frac{5}{8}\right)^2=\frac{(1\cdot 64)+(25\cdot 2)}{2\cdot 64}}$; $\boxed{\left(n+\frac{5}{8}\right)^2=\frac{64+50}{128}}$; $\boxed{\left(n+\frac{5}{8}\right)^2=\frac{114}{128}}$

; $\boxed{\sqrt{\left(n+\frac{5}{8}\right)^2}=\pm\sqrt{\frac{114}{128}}}$; $\boxed{n+\frac{5}{8}=\pm\sqrt{0.89}}$; $\boxed{n+0.63=\pm 0.94}$ therefore:

I. $\boxed{n+0.63=+0.94}$; $\boxed{n=0.94-0.63}$; $\boxed{\mathbf{n=0.31}}$　　II. $\boxed{n+0.63=-0.94}$; $\boxed{n=-0.94-0.63}$; $\boxed{\mathbf{n=-1.6}}$

and the solution set is $\{\mathbf{-1.6, 0.31}\}$.

Check: I. *Let* $n=-1.6$ *in* $4n^2+5n-2=0$; $4\cdot(-1.6)^2+(5\cdot -1.6)-2\overset{?}{=}0$; $4\cdot 2.5-8-2\overset{?}{=}0$; $0=0$

II. *Let* $n=0.31$ *in* $4n^2+5n-2=0$; $4\cdot 0.31^2+(5\cdot 0.31)-2\overset{?}{=}0$; $4\cdot 0.1+1.6-2\overset{?}{=}0$

; $0.4+1.6-2\overset{?}{=}0$; $2-2\overset{?}{=}0$; $0=0$

Therefore, the equation $4n^2+5n-2=0$ can be factored to $\mathbf{(n-0.31)(n+1.6)=0}$.

Practice Problems - Solving Quadratic Equations of the Form $ax^2 + bx + c = 0$, where $a \rangle 1$, by Completing the Square

Section 4.4 Case II Practice Problems - Solve the following quadratic equations using Completing-the-Square method. (Note that these problems are identical to the exercises given in Section 4.2 Case II.)

1. $4u^2 + 6u + 1 = 0$
2. $4w^2 + 10w = -3$
3. $6x^2 + 4x - 2 = 0$
4. $15y^2 + 3 = -14y$
5. $2x^2 - 5x + 3 = 0$
6. $2x^2 + xy - y^2 = 0$ x is variable
7. $6x^2 + 7x - 3 = 0$
8. $5x^2 = -3x$
9. $3x^2 + 4x + 5 = 0$
10. $-3y^2 + 13y + 10 = 0$

4.5 Solving Other Types of Quadratic Equations

In this section two classes of quadratic equations are addressed: one containing radicals (case I) and the second containing fractions (Case II).

Case I - Solving Quadratic Equations Containing Radicals

In general, radical equations are solved by squaring both sides of the equation. This squaring process sometimes produces solutions that when substituted into the original equation do not produce equality in both sides of the equation. These solutions are called apparent solutions. For example, the equation $y = 7$ has only one solution, i.e., 7. Let's square both sides of the equation $y = 7$ to obtain $y^2 = 49$ and solve for y by taking square root of both sides, i.e., $\sqrt{y^2} = \pm\sqrt{49}$. Solving for y we obtain $y = \pm 7$. However, note that by substituting the two solutions into the original equation $y = 7$, it is clear that only $y = +7$ is the real solution and $y = -7$ is the apparent solution. Therefore, in order to identify the real solutions **we must check all solutions in the original equation**. The following show the steps as to how equations containing radical expressions are solved:

Step 1 Square both sides of the equation.

Step 2 Write the quadratic equation in standard form.

Step 3 Solve the quadratic equation by choosing a solution method.

Step 4 Check the answers by substituting the solutions into the original equation. Disregard any apparent solution.

Examples with Steps

The following examples show the steps as to how equations containing radicals are solved:

Example 4.5-1

Solve the radical equation $\sqrt{x^2 + 5} = 3$.

Solution:

Step 1 $\boxed{\sqrt{x^2 + 5} = 3}$; $\boxed{\left(\sqrt{x^2 + 5}\right)^2 = 3^2}$; $\boxed{x^2 + 5 = 9}$

Step 2 $\boxed{x^2 + 5 = 9}$; $\boxed{x^2 + 5 - 9 = 9 - 9}$; $\boxed{x^2 - 4 = 0}$

Step 3 $\boxed{x^2 - 4 = 0}$; $\boxed{x^2 = 4}$; $\boxed{\sqrt{x^2} = \pm\sqrt{4}}$; $\boxed{x = \pm 2}$ (Solve using the Square Root Property method)

Therefore, the two apparent solutions are $x = -2$ and $x = 2$.

Step 4 Check: Substitute $x = 2$ and $x = -2$ in $\sqrt{x^2 + 5} = 3$.

I. *Let* $x = 2$ *in* $\sqrt{x^2 + 5} = 3$; $\sqrt{2^2 + 5} \stackrel{?}{=} 3$; $\sqrt{4 + 5} \stackrel{?}{=} 3$; $\sqrt{9} \stackrel{?}{=} 3$; $3 = 3$

II. *Let* $x=-2$ *in* $\sqrt{x^2+5}=3$; $\sqrt{(-2)^2+5}\overset{?}{=}3$; $\sqrt{4+5}\overset{?}{=}3$; $\sqrt{9}\overset{?}{=}3$; $3=3$

Thus, $\mathbf{x=-2}$ and $\mathbf{x=2}$ are the real solutions to $\sqrt{x^2+5}=3$. Furthermore, the equation $\sqrt{x^2+5}=3$ can be factored to $(\mathbf{x-2})(\mathbf{x+2})=\mathbf{0}$.

Example 4.5-2

Solve the radical equation $\sqrt{-12x-4}=3x+1$.

Solution:

Step 1 $\boxed{\sqrt{-12x-4}=3x+1}$; $\boxed{\left(\sqrt{-12x-4}\right)^2=(3x+1)^2}$; $\boxed{-12x-4=(3x+1)^2}$

Step 2 $\boxed{-12x-4=(3x+1)^2}$; $\boxed{-12x-4=9x^2+1+6x}$; $\boxed{9x^2+1+6x+12x+4=0}$

; $\boxed{9x^2+(6x+12x)+(4+1)=0}$; $\boxed{9x^2+18x+5=0}$

Step 3 $\boxed{9x^2+18x+5=0}$; $\boxed{\left(x+\frac{1}{3}\right)\left(x+\frac{5}{3}\right)}$ (Solve using the Quadratic Formula)

Therefore, the two apparent solutions are $x=-\frac{1}{3}$ and $x=-\frac{5}{3}$.

Step 4 Check: Substitute $x=-\frac{1}{3}$ and $x=-\frac{5}{3}$ in $\sqrt{-12x-4}=3x+1$.

I. *Let* $x=-\frac{1}{3}$ *in* $\sqrt{-12x-4}=3x+1$; $\sqrt{-\overset{4}{\cancel{12}}\cdot\left(-\frac{1}{\cancel{3}}\right)-4}\overset{?}{=}\cancel{3}\cdot\left(-\frac{1}{\cancel{3}}\right)+1$; $\sqrt{4-4}\overset{?}{=}-1+1$

; $\sqrt{0}\overset{?}{=}0$; $0=0$

II. *Let* $x=-\frac{5}{3}$ *in* $\sqrt{-12x-4}=3x+1$; $\sqrt{-\overset{4}{\cancel{12}}\cdot\left(-\frac{5}{\cancel{3}}\right)-4}\overset{?}{=}\cancel{3}\cdot\left(-\frac{5}{\cancel{3}}\right)+1$

; $\sqrt{-4\cdot(-5)-4}\overset{?}{=}-5+1$; $\sqrt{20-4}\overset{?}{=}-4$; $\sqrt{16}\overset{?}{=}-4$; $\sqrt{4^2}\overset{?}{=}-4$; $4\neq-4$

Thus, the equation $\sqrt{-12x-4}=3x+1$ has one real solution, i.e., $\mathbf{x=-\frac{1}{3}}$.

Example 4.5-3

Solve the radical equation $x+1=\sqrt{x+1}$.

Solution:

Step 1 $\boxed{x+1=\sqrt{x+1}}$; $\boxed{(x+1)^2=\left(\sqrt{x+1}\right)^2}$; $\boxed{x^2+1+2x=x+1}$

Step 2 $\boxed{x^2+1+2x=x+1}$; $\boxed{x^2+(2x-x)+(1-1)=0}$; $\boxed{x^2+x+0=0}$; $\boxed{x^2+x=0}$

Step 3 $\boxed{x^2+x=0}$; $\boxed{x(x+1)=0}$

Therefore, the two apparent solutions are $x=0$ and $x=-1$.

Step 4 Check: Substitute $x=0$ and $x=-1$ in $x+1=\sqrt{x+1}$.

I. *Let* $x = 0$ *in* $x+1 = \sqrt{x+1}$; $0+1 \stackrel{?}{=} \sqrt{0+1}$; $1 \stackrel{?}{=} \sqrt{1}$; $1 = 1$

II. *Let* $x = -1$ *in* $x+1 = \sqrt{x+1}$; $-1+1 \stackrel{?}{=} \sqrt{-1+1}$; $0 \stackrel{?}{=} \sqrt{0}$; $0 = 0$

Thus, $\mathbf{x = 0}$ and $\mathbf{x = -1}$ are the real solutions to $x+1 = \sqrt{x+1}$. Furthermore, the equation $x+1 = \sqrt{x+1}$ can be factored to $\mathbf{x(x+1) = 0}$.

Example 4.5-4

Solve the radical equation $2t = \sqrt{11t-6}$.

Solution:

Step 1 $\boxed{2t = \sqrt{11t-6}}$; $\boxed{(2t)^2 = \left(\sqrt{11t-6}\right)^2}$; $\boxed{4t^2 = 11t-6}$

Step 2 $\boxed{4t^2 = 11t-6}$; $\boxed{4t^2 - 11t + 6 = 0}$

Step 3 $\boxed{4t^2 - 11t + 6 = 0}$; $\boxed{\left(t - \frac{3}{4}\right)(t-2)}$ (Solve Using the Quadratic Formula)

Therefore, the two apparent solutions are $t = \frac{3}{4}$ and $t = 2$.

Step 4 Check: Substitute $t = \frac{3}{4}$ and $t = 2$ in $2t = \sqrt{11t-6}$.

I. *Let* $t = \frac{3}{4}$ *in* $2t = \sqrt{11t-6}$; $\cancel{2} \cdot \frac{3}{\cancel{4}_{2}} \stackrel{?}{=} \sqrt{11 \cdot \frac{3}{4} - 6}$; $\frac{3}{2} \stackrel{?}{=} \sqrt{\frac{33}{4} - \frac{6}{1}}$; $\frac{3}{2} \stackrel{?}{=} \sqrt{\frac{(33 \cdot 1) - (6 \cdot 4)}{4 \cdot 1}}$

; $\frac{3}{2} \stackrel{?}{=} \sqrt{\frac{33-24}{4}}$; $\frac{3}{2} \stackrel{?}{=} \sqrt{\frac{9}{4}}$; $\frac{3}{2} \stackrel{?}{=} \sqrt{\frac{3^2}{2^2}}$; $\frac{3}{2} = \frac{3}{2}$

II. *Let* $t = 2$ *in* $2t = \sqrt{11t-6}$; $2 \cdot 2 \stackrel{?}{=} \sqrt{11 \cdot 2 - 6}$; $4 \stackrel{?}{=} \sqrt{22-6}$; $4 \stackrel{?}{=} \sqrt{16}$; $4 = 4$

Thus, $\mathbf{t = \frac{3}{4}}$ and $\mathbf{t = 2}$ are the real solutions to $2t = \sqrt{11t-6}$. Furthermore, the equation $2t = \sqrt{11t-6}$ can be factored to $\left(t - \frac{3}{4}\right)(t-2) = 0$ which is the same as $\mathbf{(4t-3)(t-2) = 0}$

Example 4.5-5

Solve the radical equation $\sqrt{2w} = \sqrt{3w-1}$.

Solution:

Step 1 $\boxed{\sqrt{2w} = \sqrt{3w-1}}$; $\boxed{\left(\sqrt{2w}\right)^2 = \left(\sqrt{3w-1}\right)^2}$; $\boxed{2w^2 = 3w-1}$

Step 2 $\boxed{2w^2 = 3w-1}$; $\boxed{2w^2 - 3w + 1 = 0}$

Step 3 $\boxed{2w^2-3w+1=0}$; $\boxed{\left(w-\frac{1}{2}\right)(w-1)=0}$ (Use Completing-the-Square method)

Therefore, the two apparent solutions are $w=\frac{1}{2}$ and $w=1$.

Step 4 Check: Substitute $w=\frac{1}{2}$ and $w=1$ in $\sqrt{2}w=\sqrt{3w-1}$.

I. *Let* $w=\frac{1}{2}$ *in* $\sqrt{2}w=\sqrt{3w-1}$; $\sqrt{2}\cdot\frac{1}{2}\overset{?}{=}\sqrt{3\cdot\frac{1}{2}-1}$; $\frac{\sqrt{2}}{2}\overset{?}{=}\sqrt{\frac{3}{2}-\frac{1}{1}}$

; $\frac{\sqrt{2}}{2}\overset{?}{=}\sqrt{\frac{(3\cdot1)-(1\cdot2)}{2\cdot1}}$; $\frac{\sqrt{2}}{2}\overset{?}{=}\sqrt{\frac{3-2}{2}}$; $\frac{\sqrt{2}}{2}\overset{?}{=}\sqrt{\frac{1}{2}}$; $\frac{\sqrt{2}}{2}\overset{?}{=}\frac{\sqrt{1}}{\sqrt{2}}$; $\frac{\sqrt{2}}{2}\overset{?}{=}\frac{\sqrt{1}\times\sqrt{2}}{\sqrt{2}\times\sqrt{2}}$

; $\frac{\sqrt{2}}{2}\overset{?}{=}\frac{\sqrt{1\cdot2}}{\sqrt{2\cdot2}}$; $\frac{\sqrt{2}}{2}\overset{?}{=}\frac{\sqrt{2}}{\sqrt{2^2}}$; $\frac{\sqrt{2}}{2}=\frac{\sqrt{2}}{2}$

II. *Let* $w=1$ *in* $\sqrt{2}w=\sqrt{3w-1}$; $\sqrt{2}\cdot1\overset{?}{=}\sqrt{3\cdot1-1}$; $\sqrt{2}\overset{?}{=}\sqrt{3-1}$; $\sqrt{2}=\sqrt{2}$

Thus, $\mathbf{w=\frac{1}{2}}$ and $\mathbf{w=1}$ are the real solutions to $\sqrt{2}w=\sqrt{3w-1}$. Furthermore, the equation $\sqrt{2}w=\sqrt{3w-1}$ can be factored to $(w-1)\left(w-\frac{1}{2}\right)=0$ which is the same as $\mathbf{(w-1)(2w-1)=0}$.

Additional Examples - Solving Quadratic Equations Containing Radicals

The following examples further illustrate how to solve quadratic equations that contain radical expressions:

Example 4.5-6

Solve the radical equation $x+4=\sqrt{x+4}$.

Solution:

First - Square both sides of the equation. $x+4=\sqrt{x+4}$; $(x+4)^2=\left(\sqrt{x+4}\right)^2$; $(x+4)^2=x+4$

Second - Complete the square on the left hand side of the equation and simplify.

$(x+4)^2=x+4$; $x^2+16+8x=x+4$; $x^2+16-4+8x-x=0$; $x^2+12+7x=0$

Third - Write the quadratic equation in standard form. $x^2+7x+12=0$

Fourth - Solve the quadratic equation by choosing a solution method.

$x^2+7x+12=0$; $(x+4)(x+3)=0$. Therefore, the two apparent solutions are:

$x+4=0$; $x=-4$ and $x+3=0$; $x=-3$

Fifth - Check the answers by substituting the x values into the original equation.

I. *Let* $x=-4$ *in* $x+4=\sqrt{x+4}$; $-4+4\overset{?}{=}\sqrt{-4+4}$; $0=0$

II. *Let* $x=-3$ *in* $x+4=\sqrt{x+4}$; $-3+4\overset{?}{=}\sqrt{-3+4}$; $1\overset{?}{=}\sqrt{1}$; $1=1$

Therefore, $\mathbf{x=-4}$ and $\mathbf{x=-3}$ are the real solutions to $x+4=\sqrt{x+4}$. Furthermore, the equation $x+4=\sqrt{x+4}$ can be factored to $(\mathbf{x+4})(\mathbf{x+3})=\mathbf{0}$.

Example 4.5-7

Solve the radical equation $\sqrt{3x+4}=x$.

Solution:

First - Square both sides of the equation. $\sqrt{3x+4}=x$; $\left(\sqrt{3x+4}\right)^2=x^2$; $3x+4=x^2$

Second - Write the quadratic equation in standard form. $x^2-3x-4=0$

Third - Solve the quadratic equation by choosing a solution method.

$x^2-3x-4=0$; $(x-4)(x+1)=0$. Therefore, the two apparent solutions are:

$x-4=0$; $x=4$ and $x+1=0$; $x=-1$

Fourth - Check the answers by substituting the x values into the original equation.

I. *Let* $x=4$ *in* $\sqrt{3x+4}=x$; $\sqrt{3\cdot 4+4}\overset{?}{=}4$; $\sqrt{12+4}\overset{?}{=}4$; $\sqrt{16}\overset{?}{=}4$; $\sqrt{4^2}\overset{?}{=}4$; $4=4$

II. *Let* $x=-1$ *in* $\sqrt{3x+4}=x$; $\sqrt{(3\cdot -1)+4}\overset{?}{=}-1$; $\sqrt{-3+4}\overset{?}{=}-1$; $\sqrt{1}\overset{?}{=}-1$; $1\neq -1$

Therefore, the equation $\sqrt{3x-4}=x$ has one real solution, i.e., $\mathbf{x=4}$.

Example 4.5-8

Solve the radical equation $\sqrt{u^2+5}=u+2$.

Solution:

First - Square both sides of the equation. $\sqrt{u^2+5}=u+2$; $\left(\sqrt{u^2+5}\right)^2=(u+2)^2$; $u^2+5=(u+2)^2$

Second - Complete the square on the right hand side of the equation and simplify.

$u^2+5=u^2+4+4u$; $u^2-u^2+4-5+4u=0$; $-1+4u=0$

Third - Solve for u, i.e., $-1+4u=0$; $4u=1$; $u=\frac{1}{4}$; $u=0.25$

Fourth - Check the answers by substituting the u solution into the original equation, i.e.,

Let $u=0.25$ *in* $\sqrt{u^2+5}=u+2$; $\sqrt{0.25^2+5}\overset{?}{=}0.25+2$; $\sqrt{0.0625+5}\overset{?}{=}2.25$; $\sqrt{5.0625}\overset{?}{=}2.25$

; $2.25=2.25$. Therefore, the equation $\sqrt{u^2+5}=u+2$ has one solution, i.e., $\mathbf{u=0.25}$.

Example 4.5-9

Solve the radical equation $\sqrt{2x+15}=x$.

Solution:

First - Square both sides of the equation. $\sqrt{2x+15}=x$; $\left(\sqrt{2x+15}\right)^2=x^2$; $2x+15=x^2$

Second - Write the quadratic equation in standard form. $x^2-2x-15=0$

Third - Solve the quadratic equation by choosing a solution method.

$x^2-2x-15=0$; $(x-5)(x+3)=0$. Therefore, the two apparent solutions are:

$x-5=0$; $x=5$ and $x+3=0$; $x=-3$

Fourth - Check the answers by substituting the x values into the original equation.

I. *Let* $x=5$ *in* $\sqrt{2x+15}=x$; $\sqrt{2\cdot 5+15}\overset{?}{=}5$; $\sqrt{10+15}\overset{?}{=}5$; $\sqrt{25}\overset{?}{=}5$; $5=5$

II. *Let* $x=-3$ *in* $\sqrt{2x+15}=x$; $\sqrt{2\cdot -3+15}\stackrel{?}{=}-3$; $\sqrt{-6+15}\stackrel{?}{=}-3$; $\sqrt{9}\stackrel{?}{=}-3$; $3\neq -3$

Therefore, the equation $\sqrt{2x+15}=x$ has one real solution, i.e., $\mathbf{x=5}$.

Example 4.5-10

Solve the radical equation $\sqrt{x+30}=x$.

Solution:

First - Square both sides of the equation. $\sqrt{x+30}=x$; $\left(\sqrt{x+30}\right)^2=x^2$; $x+30=x^2$

Second - Write the quadratic equation in standard form. $x^2-x-30=0$

Third - Solve the quadratic equation by choosing a solution method.

$x^2-x-30=0$; $(x-6)(x+5)=0$. Therefore, the two apparent solutions are:

$x-6=0$; $x=6$ and $x+5=0$; $x=-5$

Fourth - Check the answers by substituting the x values into the original equation.

I. *Let* $x=6$ *in* $\sqrt{x+30}=x$; $\sqrt{6+30}\stackrel{?}{=}6$; $\sqrt{36}\stackrel{?}{=}6$; $6=6$

II. *Let* $x=-5$ *in* $\sqrt{x+30}=x$; $\sqrt{-5+30}\stackrel{?}{=}-5$; $\sqrt{25}\stackrel{?}{=}-5$; $5\neq -5$

Therefore, the equation $\sqrt{x+30}=x$ has one real solution, i.e., $\mathbf{x=6}$.

Practice Problems - Solving Quadratic Equations Containing Radicals

Section 4.5 Case I Practice Problems - Solve the following equations. Check the answers by substituting the solutions into the original equation.

1. $\sqrt{-9y+28}-y+2=0$
2. $2x=\sqrt{9x+3}$
3. $t^2=-\sqrt{5}t$
4. $y^2-\sqrt{8}y=7$
5. $\sqrt{5}x=2x^2$
6. $\sqrt{x^2-12}=2$
7. $\sqrt{-8x-4}=2x+1$
8. $x=\sqrt{-x+2}$
9. $x=\sqrt{-2x+3}$
10. $\sqrt{x^2+3}=x+1$

Case II - Solving Quadratic Equations Containing Fractions

In this section solutions to quadratic equations with fractional coefficients are discussed. Note that in dealing with fractional equations not all solutions may satisfy the original equation. This is because fractions may encounter division by zero which is undefined. Therefore, it is essential that all solutions to a quadratic equation be checked by substitution into the original equation in order to ensure division by zero does not occur. Equations containing algebraic fractions are solved using the following steps:

Step 1 Write both sides of the equation in fraction form.

Step 2 Cross multiply the terms in both sides of the equation.

Step 3 Write the quadratic equation in standard form.

Step 4 Solve the quadratic equation by choosing a solution method.

Step 5 Check the answers by substituting the apparent solutions into the original equation. Disregard any apparent solution if equality on both sides of the equation is not obtained.

Examples with Steps

The following examples show the steps as to how equations containing fractions are solved:

Example 4.5-11

Solve the fractional equation $x-1=\frac{20}{x}$.

Solution:

First - Write the left hand side of the equation in fraction form. $x-1=\frac{20}{x}$; $\frac{x-1}{1}=\frac{20}{x}$

Second - Cross multiply the terms in both sides of the equation. $x\cdot(x-1)=1\cdot 20$; $x^2-x=20$

Third - Write the quadratic equation in standard form, i.e., $x^2-x-20=0$

Fourth - Solve the quadratic equation by choosing a method. $x^2-x-20=0$; $(x-5)(x+4)=0$.

Therefore, the two apparent solutions are: $x-5=0$; $x=5$ and $x+4=0$; $x=-4$

Fifth - Check the answers by substituting the x values into the original equation.

I. *Let* $x=5$ *in* $x-1=\frac{20}{x}$; $5-1\overset{?}{=}\frac{20}{5}$; $4=4$

II. *Let* $x=-4$ *in* $x-1=\frac{20}{x}$; $-4-1\overset{?}{=}\frac{20}{-4}$; $-5=-5$

Thus, $\mathbf{x=5}$ and $\mathbf{x=-4}$ are the real solutions to $x-1=\frac{20}{x}$. In addition, the equation $x-1=\frac{20}{x}$ can be factored to $\mathbf{(x-5)(x+4)=0}$.

Example 4.5-12

Solve the fractional equation $1+\frac{1}{x+1}=x+3$.

Solution:

First - use fraction techniques to rewrite the left hand side of the equation in a single fraction form.

$$1+\frac{1}{x+1}=x+3\ ;\ \frac{1}{1}+\frac{1}{x+1}=\frac{x+3}{1}\ ;\ \frac{[1\cdot(x+1)]+(1\cdot 1)}{1\cdot(x+1)}=\frac{x+3}{1}\ ;\ \frac{x+1+1}{x+1}=\frac{x+3}{1}\ ;\ \frac{x+2}{x+1}=\frac{x+3}{1}$$

Second - Cross multiply the terms in both sides of the equation.

$$\frac{x+2}{x+1}=\frac{x+3}{1}\ ;\ (x+2)\cdot 1=(x+3)\cdot(x+1)\ ;\ x+2=x^2+x+3x+3\ ;\ x+2=x^2+4x+3$$

Third - write the quadratic equation in standard form, i.e., $x^2+4x-x+3-2=0$; $x^2+3x+1=0$

Fourth - Solve the quadratic equation using the Quadratic Formula method.

$$x^2+3x+1=0\ ;\ \left(x+\frac{3-\sqrt{5}}{2}\right)\left(x+\frac{3+\sqrt{5}}{2}\right)=0\ ;\ \left(x+\frac{3-2.24}{2}\right)\left(x+\frac{3+2.24}{2}\right)=0$$

$$;\ (x+0.38)(x+2.62)=0\,.$$

Therefore, the two apparent solutions are: $x+0.38=0$; $x=-0.38$ and $x+2.62=0$; $x=-2.62$

Fifth - Check the answers by substituting the x values into the original equation.

I. *Let* $x=-0.38$ *in* $1+\frac{1}{x+1}=x+3\ ;\ 1+\frac{1}{-0.38+1}\overset{?}{=}-0.38+3\ ;\ 1+\frac{1}{0.62}\overset{?}{=}2.62\ ;\ 1+1.62\overset{?}{=}2.62$

$;\ 2.62=2.62$

II. *Let* $x=-2.62$ *in* $1+\frac{1}{x+1}=x+3\ ;\ 1+\frac{1}{-2.62+1}\overset{?}{=}-2.62+3\ ;\ 1-\frac{1}{1.62}\overset{?}{=}0.38\ ;\ 1-0.62\overset{?}{=}0.38$

$;\ 0.38=0.38$

Thus, $\mathbf{x=-0.38}$ and $\mathbf{x=-2.62}$ are the real solutions to $1+\frac{1}{x+1}=x+3$. In addition, the equation $1+\frac{1}{x+1}=x+3$ can be factored to $\mathbf{(x+0.38)(x+2.62)=0}$.

Example 4.5-13

Solve the fractional equation $\frac{1}{2y}=6y-\frac{1}{y}$.

Solution:

First - Write the left hand side of the equation in fraction form and simplify the right hand side of the equation.

$$\frac{1}{2y}=6y-\frac{1}{y}\ ;\ \frac{1}{2y}=\frac{6y}{1}-\frac{1}{y}\ ;\ \frac{1}{2y}=\frac{(6y\cdot y)-(1\cdot 1)}{1\cdot y}\ ;\ \frac{1}{2y}=\frac{6y^2-1}{y}$$

Second - Cross multiply the terms in both sides of the equation.

$$\frac{1}{2y}=\frac{6y^2-1}{y}\ ;\ 1\cdot y=2y\cdot\left(6y^2-1\right)\ ;\ y=12y^3-2y\ ;\ 12y^3-2y-y=0\ ;\ 12y^3-3y=0$$

$;\ 3y\left(4y^2-1\right)=0$. Thus, $y=0$ is an apparent solution.

Third - Solve the quadratic equation $4y^2-1=0$ by choosing a method.

$4y^2-1=0\ ;\ 4y^2=1\ ;\ \sqrt{4y^2}=\pm\sqrt{1}\ ;\ 2y=\pm 1\ ;\ y=\pm\frac{1}{2}$. Therefore, the other two apparent solutions are: $y=+\frac{1}{2}\ ;\ y=+0.5$ and $y=-\frac{1}{2}\ ;\ y=-0.5$.

Fourth - Check the answers by substituting the y values into the original equation.

I. *Let* $y=0$ *in* $\frac{1}{2y}=6y-\frac{1}{y}$. Since division by zero is encountered therefore, $y=0$ is not a real solution.

II. *Let* $y=0.5$ *in* $\frac{1}{2y}=6y-\frac{1}{y}$; $\frac{1}{2\times 0.5}\overset{?}{=}6\times 0.5-\frac{1}{0.5}$; $\frac{1}{1}\overset{?}{=}3-2$; $1=1$

III. *Let* $y=-0.5$ *in* $\frac{1}{2y}=6y-\frac{1}{y}$; $\frac{1}{(2\times -0.5)}\overset{?}{=}(6\times -0.5)-\frac{1}{(-0.5)}$; $\frac{1}{-1}\overset{?}{=}-3+2$; $-1=-1$

Thus, $\mathbf{y=0.5}$ and $\mathbf{y=-0.5}$ are the real solutions to $\frac{1}{2y}=6y-\frac{1}{y}$. In addition, the equation $\frac{1}{2y}=6y-\frac{1}{y}$ can be factored to $\mathbf{3y(y+0.5)(y-0.5)=0}$.

Example 4.5-14

Solve the fractional equation $\frac{2}{y}+\frac{3}{y^2}-1=0$.

Solution:

First - Use fraction techniques to simplify the left hand side of the equation.

$$\frac{2}{y}+\frac{3}{y^2}-1=0 \;;\; \frac{(2\cdot y^2)+(3\cdot y)}{y\cdot y^2}-1=0 \;;\; \frac{2y^2+3y}{y^3}-\frac{1}{1}=0 \;;\; \frac{1\cdot(2y^2+3y)-1\cdot y^3}{1\cdot y^3}=0$$

$$;\; \frac{2y^2+3y-y^3}{y^3}=0$$

Second - Cross multiply the terms in both sides of the equation.

$$\frac{2y^2+3y-y^3}{y^3}=\frac{0}{1} \;;\; 1\cdot(2y^2+3y-y^3)=0\cdot y^3 \;;\; 2y^2+3y-y^3=0 \;;\; -y(-2y-3+y^2)=0.$$ Thus $y=0$ is an apparent solution.

Third - Write the quadratic equation $-2y-3+y^2=0$ in standard form, i.e., $y^2-2y-3=0$

Fourth - Solve the quadratic equation by choosing a method. $y^2-2y-3=0$; $(y-3)(y+1)=0$.

Therefore, the other two apparent solutions are: $y-3=0$; $y=3$ and $y+1=0$; $y=-1$

Fifth - Check the answers by substituting the y values into the original equation.

I. *Let* $y=0$ *in* $\frac{2}{y}+\frac{3}{y^2}-1=0$. Since division by zero is encountered therefore, $y=0$ is not a real solution.

II. *Let* $y=3$ *in* $\frac{2}{y}+\frac{3}{y^2}-1=0$; $\frac{2}{3}+\frac{3}{3^2}-1\overset{?}{=}0$; $\frac{2}{3}+\frac{\not{3}}{\frac{\not{9}}{3}}-1\overset{?}{=}0$; $\frac{2}{3}+\frac{1}{3}-1\overset{?}{=}0$; $\frac{2+1}{3}-1\overset{?}{=}0$

; $\frac{\not{3}}{\not{3}}-1\overset{?}{=}0$; $\frac{1}{1}-1\overset{?}{=}0$; $1-1=0$; $0=0$

III. *Let* $y=-1$ *in* $\frac{2}{y}+\frac{3}{y^2}-1=0$; $\frac{2}{-1}+\frac{3}{(-1)^2}-1\overset{?}{=}0$; $-\frac{2}{1}+\frac{3}{1}-1\overset{?}{=}0$; $\frac{-2+3}{1}-1\overset{?}{=}0$; $1-1\overset{?}{=}0$

; $0=0$

Thus, $\mathbf{y=3}$ and $\mathbf{y=-1}$ are the real solutions to $\frac{2}{y}+\frac{3}{y^2}-1=0$. In addition, the equation

$\frac{2}{y}+\frac{3}{y^2}-1=0$ can be factored to $-y(y-3)(y+1)$.

Example 4.5-15

Solve the fractional equation $\frac{2y}{y-1}=\frac{1}{y^2-1}$.

Solution:

First - Write the denominator in the right hand side of the equation in its factored form.

$$\frac{2y}{y-1}=\frac{1}{y^2-1} \; ; \; \frac{2y}{y-1}=\frac{1}{(y-1)(y+1)}$$

Second - Simplify the equation and cross multiply the terms in both sides of the equation.

$$\frac{2y}{y-1}=\frac{1}{(y-1)(y+1)} \; ; \; \frac{2y}{1}=\frac{1}{(y+1)} \; ; \; 2y(y+1)=1\cdot 1 \; ; \; 2y^2+2y=1$$

Third - Write the quadratic equation in standard form, i.e., $2y^2+2y-1=0$

Fourth - Solve the quadratic equation by choosing a method.

$2y^2+2y-1=0 \; ; \; \left(y+\frac{1-\sqrt{3}}{2}\right)\left(y+\frac{1+\sqrt{3}}{2}\right)=0$. Therefore, the two apparent solutions are:

$y+\frac{1-\sqrt{3}}{2}=0 \; ; \; y+\frac{1-1.732}{2}=0 \; ; \; y-0.37=0 \; ; \; y=0.37$, and

$$y+\frac{1+\sqrt{3}}{2}=0 \; ; \; y+\frac{1+1.732}{2}=0 \; ; \; y+1.37=0 \; ; \; y=-1.37$$

Fifth - Check the answers by substituting the y values into the original equation.

I. *Let* $y=0.37$ *in* $\frac{2y}{y-1}=\frac{1}{y^2-1} \; ; \; \frac{2\times 0.37}{0.37-1}\stackrel{?}{=}\frac{1}{(0.37)^2-1} \; ; \; \frac{0.74}{-0.63}\stackrel{?}{=}\frac{1}{0.137-1} \; ; \; -1.17=-1.17$

II. *Let* $y=-1.37$ *in* $\frac{2y}{y-1}=\frac{1}{y^2-1} \; ; \; \frac{2\times -1.37}{-1.37-1}\stackrel{?}{=}\frac{1}{(-1.37)^2-1} \; ; \; \frac{-2.74}{-2.37}\stackrel{?}{=}\frac{1}{1.88-1} \; ; \; 1.15=1.15$

Thus, $\mathbf{y=0.37}$ and $\mathbf{y=-1.37}$ are the real solutions to $\frac{2y}{y-1}=\frac{1}{y^2-1}$. In addition, the equation $\frac{2y}{y-1}=\frac{1}{y^2-1}$ can be factored to $(\mathbf{y+1.37})(\mathbf{y-0.37})=\mathbf{0}$.

Additional Examples - Solving Quadratic Equations Containing Fractions

The following examples further illustrate how to solve quadratic equations with fractional coefficients:

Example 4.5-16

Solve the fractional equation $x+5=\frac{-4}{x}$.

Solution:

First - Write the left hand side of the equation in fraction form. $x+5=\frac{-4}{x} \; ; \; \frac{x+5}{1}=\frac{-4}{x}$

Second - Cross multiply the terms in both sides of the equation. $x\cdot(x+5)=1\cdot(-4) \; ; \; x^2+5x=-4$

Third - Write the quadratic equation in standard form, i.e., $x^2+5x+4=0$

Fourth - Solve the quadratic equation by choosing a method. $x^2+5x+4=0$; $(x+1)(x+4)=0$.

Therefore, the two apparent solutions are: $x+1=0$; $x=-1$ and $x+4=0$; $x=-4$

Fifth - Check the answers by substituting the x values into the original equation.

I. *Let* $x=-1$ *in* $x+5=\frac{-4}{x}$; $-1+5\overset{?}{=}\frac{-4}{-1}$; $4=4$

II. *Let* $x=-4$ *in* $x+5=\frac{-4}{x}$; $-4+5\overset{?}{=}\frac{-4}{-4}$; $1\overset{?}{=}\frac{1}{1}$; $1=1$

Therefore, $\mathbf{x=-1}$ and $\mathbf{x=-4}$ are the real solutions to $x+5=\frac{-4}{x}$. In addition, the equation $x+5=\frac{-4}{x}$ can be factored to $\mathbf{(x+1)(x+4)=0}$.

Example 4.5-17

Solve the fractional equation $6x+13=\frac{-5}{x}$.

Solution:

First - Write the left hand side of the equation in fraction form. $6x+13=\frac{-5}{x}$; $\frac{6x+13}{1}=\frac{-5}{x}$

Second - Cross multiply the terms in both sides of the equation.

$x\cdot(6x+13)=1\cdot(-5)$; $6x^2+13x=-5$

Third - Write the quadratic equation in standard form, i.e., $6x^2+13x+5=0$

Fourth - Solve the quadratic equation by choosing a method.

$6x^2+13x+5=0$; $(3x+5)(2x+1)=0$. Therefore, the two apparent solutions are:

$3x+5=0$; $3x=-5$; $x=-\frac{5}{3}$; $x=-1.67$ and $2x+1=0$; $2x=-1$; $x=-\frac{1}{2}$; $x=-0.5$

Fifth - Check the answers by substituting the x values into the original equation.

I. *Let* $x=-1.67$ *in* $6x+13=\frac{-5}{x}$; $6\cdot(-1.67)+13\overset{?}{=}\frac{-5}{-1.67}$; $-10+13\overset{?}{=}3$; $3=3$

II. *Let* $x=-0.5$ *in* $6x+13=\frac{-5}{x}$; $6\cdot(-0.5)+13\overset{?}{=}\frac{-5}{-0.5}$; $-3+13\overset{?}{=}10$; $10=10$

Therefore, $\mathbf{x=-1.67}$ and $\mathbf{x=-0.5}$ are the real solutions to $6x+13=\frac{-5}{x}$. In addition, equation $6x+13=\frac{-5}{x}$ can be factored to $\mathbf{(3x+5)(2x+1)=0}$.

Example 4.5-18

Solve the fractional equation $y=\frac{25}{y}$.

Solution:

First - Write the left hand side of the equation in fraction form. $y=\frac{25}{y}$; $\frac{y}{1}=\frac{25}{y}$

Second - Cross multiply the terms in both sides of the equation. $y\cdot y=1\cdot 25$; $y^2=25$

Third - Solve the quadratic equation by choosing the Square Root method.

$y^2=25$; $\sqrt{y^2}=\pm\sqrt{25}$; $y=\pm\sqrt{5^2}$; $y=\pm 5$. Therefore, the two apparent solutions are:

$y=+5$ and $y=-5$

Fourth - Check the answers by substituting the x values into the original equation.

I. *Let* $y = 5$ *in* $y = \frac{25}{y}$; $5 \stackrel{?}{=} \frac{\overset{5}{\cancel{25}}}{\cancel{5}}$; $5 \stackrel{?}{=} \frac{5}{1}$; $5 = 5$

II. *Let* $y = -5$ *in* $y = \frac{25}{y}$; $-5 \stackrel{?}{=} \frac{25}{-5}$; $-5 \stackrel{?}{=} -\frac{\overset{5}{\cancel{25}}}{\cancel{5}}$; $-5 \stackrel{?}{=} -\frac{5}{1}$; $-5 = -5$

Therefore, $\mathbf{y = 5}$ and $\mathbf{y = -5}$ are the real solutions to $y = \frac{25}{y}$. In addition, the equation $y = \frac{25}{y}$ can be factored to $\mathbf{(y-5)(y+5) = 0}$.

Example 4.5-19

Solve the fractional equation $\frac{2y-15}{y} + y = 0$.

Solution:

First - use fraction techniques to rewrite the equation in quadratic form.

$$\frac{2y-15}{y} + y = 0 \; ; \; \frac{2y-15}{y} + \frac{y}{1} = 0 \; ; \; \frac{[1\cdot(2y-15)] + (y\cdot y)}{1\cdot y} = 0 \; ; \; \frac{2y-15+y^2}{y} = 0 \; ; \; \frac{2y-15+y^2}{y} = \frac{0}{1}$$

$$; \; 1\cdot\left(2y-15+y^2\right) = y\cdot 0 \; ; \; 2y-15+y^2 = 0$$

Second - write the quadratic equation in standard form, i.e., $y^2 + 2y - 15 = 0$

Third - Solve the quadratic equation by choosing a method.

$y^2 + 2y - 15 = 0$; $(y+5)(y-3) = 0$.

Therefore, the two apparent solutions are: $y + 5 = 0$; $y = -5$ and $y - 3 = 0$; $y = 3$

Fourth - Check the answers by substituting the x values into the original equation.

I. *Let* $y = -5$ *in* $\frac{2y-15}{y} + y = 0$; $\frac{2\cdot(-5)-15}{-5} - 5 \stackrel{?}{=} 0$; $\frac{-10-15}{-5} - 5 \stackrel{?}{=} 0$; $\frac{-25}{-5} - 5 \stackrel{?}{=} 0$; $0 = 0$

II. *Let* $y = 3$ *in* $\frac{2y-15}{y} + y = 0$; $\frac{6-15}{3} + 3 \stackrel{?}{=} 0$; $\frac{-9}{3} + 3 \stackrel{?}{=} 0$; $-3 + 3 \stackrel{?}{=} 0$; $0 = 0$

Therefore, $\mathbf{y = -5}$ and $\mathbf{y = 3}$ are the real solutions to $\frac{2y-15}{y} + y = 0$. In addition, the equation $\frac{2y-15}{y} + y = 0$ can be factored to $\mathbf{(y+5)(y-3) = 0}$.

Example 4.5-20

Solve the fractional equation $\frac{x^2}{x+1} = \frac{4}{x+1}$.

Solution:

First - Cross multiply the terms in both sides of the equation.

$\frac{x^2}{x+1} = \frac{4}{x+1}$; $x^2\cdot(x+1) = 4\cdot(x+1)$.

Note that $x+1$ can be eliminated from both sides of the equation where we obtain $x^2 = 4$.

Second - Factor out the quadratic equation by choosing the Square Root factoring method.

$x^2 = 4$; $\sqrt{x^2} = \pm\sqrt{4}$; $x = \pm\sqrt{2^2}$; $x = \pm 2$.

Therefore, the two apparent solutions are: $x = +2$ and $x = -2$

Third - Check the answers by substituting the x values into the original equation.

I. *Let* $x = 2$ *in* $\frac{x^2}{x+1} = \frac{4}{x+1}$; $\frac{2^2}{2+1} \stackrel{?}{=} \frac{4}{2+1}$; $\frac{4}{3} = \frac{4}{3}$

II. *Let* $x = -2$ *in* $\frac{x^2}{x+1} = \frac{4}{x+1}$; $\frac{(-2)^2}{-2+1} \stackrel{?}{=} \frac{4}{-2+1}$; $\frac{4}{-1} = \frac{4}{-1}$; $-4 = -4$

Therefore, $\mathbf{x = 2}$ and $\mathbf{x = -2}$ are the real solutions to $\frac{x^2}{x+1} = \frac{4}{x+1}$. In addition, the equation $\frac{x^2}{x+1} = \frac{4}{x+1}$ can be factored to $\mathbf{(x+2)(x-2) = 0}$.

Practice Problems - Solving Quadratic Equations Containing Fractions

Section 4.5 Case II Practice Problems - Solve the following equations. Check the answers by substituting the solution into the original equation.

1. $\frac{8}{y+1} = y - 1$
2. $\frac{11x+15}{x} = -2x$
3. $\frac{x^2}{x+3} = \frac{1}{x+3}$
4. $\frac{1-2u}{u} = -u$
5. $x = \frac{3}{x} - 2$
6. $\frac{3x-10}{x} = -x$
7. $u = \frac{49}{u}$
8. $6x + 17 = \frac{-5}{x}$
9. $y + 4 = -\frac{3}{y}$
10. $3x = \frac{-5x-2}{x}$

4.6 How to Choose the Best Factoring or Solution Method

To factor polynomials and to solve quadratic equations a total of seven basic methods have been introduced in Chapters 3 and 4. Those methods are:

1. The Greatest Common Factoring method
2. The Grouping method
3. The Trial and Error method
4. Factoring methods for polynomials with square and cubed terms
5. The Quadratic Formula method
6. The Square Root Property method, and
7. Completing-the-Square method

The decision as to which one of the above methods is most suitable in factoring a polynomial or solving an equation is left to the student. For example, in some cases, using the Trial and Error method in solving a quadratic equation may be easier than using the Quadratic Formula or Completing-the-Square method. In certain cases, using the quadratic formula in solving a polynomial may be faster than the Grouping or the Trial and Error method. Note that the key in choosing the best and/or the easiest method is through solving many problems. After sufficient practice, students start to gain confidence on selection of one method over the other.

Assumption *- In many instances, the methods used in factoring polynomials (shown in Chapter 3) can also be used in solving quadratic equations (shown in Chapter 4) by recognizing that the left hand side of the equation* $ax^2+bx+c=0$, *namely* ax^2+bx+c *is a polynomial and can be factored as such, using polynomial factoring methods covered in Chapter 3.*

Note 1 - Any quadratic equation can be solved using the quadratic formula. Once the student has memorized the quadratic formula and has learned how to substitute the equivalent values of a, b, and c into the quadratic formula, then the next steps are merely the process of solving the quadratic equation using mathematical operations.

Note 2 - The quadratic formula can be used as an alternative method in factoring polynomials of the form ax^2+bx+c as is stated in the above assumption.

The following examples are solved using the seven factoring and solution methods shown above:

Example 4.6-1

Use different methods to solve the equation $x^2=25$.

Solution:

First Method: (The Trial and Error Method)

Write the equation in the standard quadratic equation form $ax^2+bx+c=0$, i.e., write $x^2=25$ as $x^2+0x-25=0$. To solve the given equation using the Trial and Error method we only consider the left hand side of the equation which is a second degree polynomial. Next, we need to obtain two numbers whose sum is 0 and whose product is -25 by constructing a table as shown below:

Sum	*Product*
$1-1=0$	$1\cdot(-1)=-1$
$2-2=0$	$2\cdot(-2)=-4$
$3-3=0$	$3\cdot(-3)=-9$
$4-4=0$	$4\cdot(-4)=-16$
$\mathbf{5-5=0}$	$\mathbf{5\cdot(-5)=-25}$

The last line contains the sum and the product of the two numbers that we need. Thus, $x^2=25$ or $x^2+0x-25=0$ can be factored to $\mathbf{(x-5)(x+5)=0}$.

Check: $(x-5)(x+5)=0$; $x\cdot x+5\cdot x-5\cdot x+5\cdot(-5)=0$; $x^2+5x-5x-25=0$; $x^2+(5-5)x-25=0$; $x^2+0x-25=0$

Second Method: (The Quadratic Formula Method)

First, write the equation in the standard quadratic equation form $ax^2+bx+c=0$, i.e., write $x^2=25$ as $x^2+0x-25=0$. Second, equate the coefficients of $x^2+0x-25=0$ with the standard quadratic equation by letting $a=1$, $b=0$, and $c=-25$. Then,

Given: $x=\frac{-b\pm\sqrt{b^2-4ac}}{2a}$; $x=\frac{-0\pm\sqrt{0^2-(4\times1\times-25)}}{2\times1}$; $x=\frac{\pm\sqrt{0+100}}{2}$; $x=\frac{\pm\sqrt{100}}{2}$; $x=\pm\frac{\sqrt{10^2}}{2}$; $x=\pm\frac{10}{2}$. Therefore:

I. $x=+\frac{\overset{5}{\cancel{10}}}{\cancel{2}}$; $x=\frac{5}{1}$; $\mathbf{x=5}$

II. $x=-\frac{\overset{5}{\cancel{10}}}{\cancel{2}}$; $x=-\frac{5}{1}$; $\mathbf{x=-5}$

Check: I. *Let* $x=5$ *in* $x^2=25$; $5^2\overset{?}{=}25$; $25=25$

II. *Let* $x=-5$ *in* $x^2=25$; $(-5)^2\overset{?}{=}25$; $25=25$

Therefore, the equation $x^2+0x-25=0$ can be factored to $\mathbf{(x+5)(x-5)=0}$.

Third Method: (The Square Root Property Method)

Take the square root of both sides of the equation, i.e., write $x^2=25$ as $\sqrt{x^2}=\pm\sqrt{25}$; $x=\pm\sqrt{5^2}$; $x=\pm5$. Thus, $\mathbf{x=+5}$ or $\mathbf{x=-5}$ are the solution sets to the equation $x^2=25$ which can be represented in its factorable form as $\mathbf{(x+5)(x-5)=0}$.

Fourth Method: (Completing-the-Square Method) - Is not applicable.

Note that from the above three methods using the Square Root Property method is the fastest and the easiest method to obtain the factored terms. The Trial and Error method is the second easiest method to use, followed by the Quadratic Formula method which is the most difficult way of obtaining the factored terms.

Example 4.6-2

Use different methods to solve the equation $x^2+11x+24=0$.

Solution:

First Method: (The Trial and Error Method)

To solve the given equation using the Trial and Error method we only consider the left hand side of the equation which is a second degree polynomial. Next, we need to obtain two numbers whose sum is 11 and whose product is 24 by constructing a table as shown below:

Sum	***Product***
$6+5=11$	$6\cdot 5=30$
$7+4=11$	$7\cdot 4=28$
$\mathbf{8+3=11}$	$\mathbf{8\cdot 3=24}$
$9+2=11$	$9\cdot 2=18$

The third line contains the sum and the product of the two numbers that we need. Thus, $x^2+11x+24=0$ can be factored to $\mathbf{(x+8)(x+3)=0}$.

Check: $(x+8)(x+3)=0$; $x\cdot x+3\cdot x+8\cdot x+8\cdot 3=0$; $x^2+3x+8x+24=0$; $x^2+(3+8)x+24=0$; $x^2+11x+24=0$

Second Method: (The Quadratic Formula Method)

Given the standard quadratic equation $ax^2+bx+c=0$, equate the coefficients of $x^2+11x+24=0$ with the standard quadratic equation by letting $a=1$, $b=11$, and $c=24$. Then,

Given: $x=\dfrac{-b\pm\sqrt{b^2-4ac}}{2a}$; $x=\dfrac{-11\pm\sqrt{11^2-(4\times 1\times 24)}}{2\times 1}$; $x=\dfrac{-11\pm\sqrt{121-96}}{2}$; $x=\dfrac{-11\pm\sqrt{25}}{2}$; $x=\dfrac{-11\pm\sqrt{5^2}}{2}$; $x=\dfrac{-11\pm 5}{2}$. Therefore:

I. $x=\dfrac{-11+5}{2}$; $x=-\dfrac{\overset{3}{\cancel{6}}}{\cancel{2}}$; $x=-\dfrac{3}{1}$; $\mathbf{x=-3}$ II. $x=\dfrac{-11-5}{2}$; $x=-\dfrac{\overset{8}{\cancel{16}}}{\cancel{2}}$; $x=-\dfrac{8}{1}$; $\mathbf{x=-8}$

Check: I. *Let* $x=-3$ *in* $x^2+11x+24=0$; $(-3)^2+11\cdot(-3)+24\overset{?}{=}0$; $9-33+24\overset{?}{=}0$; $33-33\overset{?}{=}0$; $0=0$

II. *Let* $x=-8$ *in* $x^2+11x+24=0$; $(-8)^2+11\cdot(-8)+24\overset{?}{=}0$; $64-88+24\overset{?}{=}0$; $88-88\overset{?}{=}0$; $0=0$

Therefore, the equation $x^2+11x+24=0$ can be factored to $\mathbf{(x+8)(x+3)=0}$.

Third Method: (Completing-the-Square Method)

$x^2+11x+24=0$; $x^2+11x=-24$; $x^2+11x+\left(\dfrac{11}{2}\right)^2=-24+\left(\dfrac{11}{2}\right)^2$; $x^2+11x+\dfrac{121}{4}=-24+\dfrac{121}{4}$; $\left(x+\dfrac{11}{2}\right)^2=-\dfrac{24}{1}+\dfrac{121}{4}$; $\left(x+\dfrac{11}{2}\right)^2=\dfrac{(-24\cdot 4)+(1\cdot 121)}{1\cdot 4}$; $\left(x+\dfrac{11}{2}\right)^2=\dfrac{-96+121}{4}$; $\left(x+\dfrac{11}{2}\right)^2=\dfrac{25}{4}$; $x+\dfrac{11}{2}=\pm\sqrt{\dfrac{25}{4}}$; $x+\dfrac{11}{2}=\pm\dfrac{5}{2}$

Therefore: I. $x+\dfrac{11}{2}=+\dfrac{5}{2}$; $x=\dfrac{5}{2}-\dfrac{11}{2}$; $x=\dfrac{5-11}{2}$; $x=-\dfrac{\overset{3}{\cancel{6}}}{\cancel{2}}$; $x=-\dfrac{3}{1}$; $\mathbf{x=-3}$

II. $x+\dfrac{11}{2}=-\dfrac{5}{2}$; $x=-\dfrac{5}{2}-\dfrac{11}{2}$; $x=\dfrac{-5-11}{2}$; $x=-\dfrac{\overset{8}{\cancel{16}}}{\cancel{2}}$; $x=-\dfrac{8}{1}$; $\mathbf{x=-8}$

Check: I. *Let* $x=-3$ *in* $x^2+11x+24=0$; $(-3)^2+(11\times -3)+24\overset{?}{=}0$; $9-33+24\overset{?}{=}0$; $0=0$

II. *Let* $x=-8$ *in* $x^2+11x+24=0$; $(-8)^2+(11\times -8)+24\overset{?}{=}0$; $64-88+24\overset{?}{=}0$; $0=0$

Therefore, the equation $x^2+11x+24=0$ can be factored to $\mathbf{(x+8)(x+3)=0}$.

Fourth Method: (The Square Root Property Method) - Is not applicable

Note that from the above three methods using the Trial and Error method is the fastest and the easiest method to obtain the factored terms. Completing-the-Square method is the second easiest method to use, followed by the Quadratic Formula method which is the longest and most difficult way of obtaining the factored terms.

Example 4.6-3

Use different methods to solve the equation $x^2+5x+2=0$.

Solution:

First Method: (The Trial and Error Method)

To solve the given equation using the Trial and Error method we only consider the left hand side of the equation which is a second degree polynomial. Next, we need to obtain two numbers whose sum is 5 and whose product is 2. However, after few trials, it becomes clear that such a combination of integer numbers is not possible to obtain. Hence, the Trial and Error method is not applicable to this particular example.

Second Method: (The Quadratic Formula Method)

Given the standard quadratic equation $ax^2+bx+c=0$, equate the coefficients of $x^2+5x+2=0$ with the standard quadratic equation by letting $a=1$, $b=5$, and $c=2$. Then,

Given: $x=\frac{-b\pm\sqrt{b^2-4ac}}{2a}$; $x=\frac{-5\pm\sqrt{5^2-(4\times1\times2)}}{2\times1}$; $x=\frac{-5\pm\sqrt{25-8}}{2}$; $x=\frac{-5\pm\sqrt{17}}{2}$. Therefore:

I. $x=\frac{-5+\sqrt{17}}{2}$; $x=\frac{-5+4.12}{2}$; $x=-\frac{0.88}{2}$; $\mathbf{x=-0.44}$

II. $x=\frac{-5-\sqrt{17}}{2}$; $x=\frac{-5-4.12}{2}$; $x=-\frac{9.12}{2}$; $\mathbf{x=-4.56}$

Check: I. *Let* $x=-0.44$ *in* $x^2+5x+2=0$; $(-0.44)^2+5\cdot(-0.44)+2\overset{?}{=}0$; $0.2-2.2+2\overset{?}{=}0$; $2.2-2.2\overset{?}{=}0$; $0=0$

II. *Let* $x=-4.56$ *in* $x^2+5x+2=0$; $(-4.56)^2+5\cdot(-4.56)+2\overset{?}{=}0$; $20.8-22.8+2\overset{?}{=}0$; $22.8-22.8\overset{?}{=}0$; $0=0$

Therefore, the equation $x^2+5x+2=0$ can be factored to $\mathbf{(x+0.44)(x+4.56)=0}$.

Third Method: (Completing-the-Square Method)

$x^2+5x+2=0$; $x^2+5x=-2$; $x^2+5x+\left(\frac{5}{2}\right)^2=-2+\left(\frac{5}{2}\right)^2$; $x^2+5x+\frac{25}{4}=-2+\frac{25}{4}$; $\left(x+\frac{5}{2}\right)^2=-\frac{2}{1}+\frac{25}{4}$; $\left(x+\frac{5}{2}\right)^2=\frac{(-2\cdot4)+(1\cdot25)}{1\cdot4}$; $\left(x+\frac{5}{2}\right)^2=\frac{-8+25}{4}$; $\left(x+\frac{5}{2}\right)^2=\frac{17}{4}$; $x+\frac{5}{2}=\pm\sqrt{\frac{17}{4}}$; $x+\frac{5}{2}=\pm\frac{\sqrt{17}}{2}$

Therefore: I. $x+\frac{5}{2}=+\frac{\sqrt{17}}{2}$; $x=\frac{\sqrt{17}}{2}-\frac{5}{2}$; $x=\frac{\sqrt{17}-5}{2}$; $x=\frac{4.12-5}{2}$; $x=-\frac{0.88}{2}$; $\mathbf{x=-0.44}$

II. $x+\frac{5}{2}=-\frac{\sqrt{17}}{2}$; $x=-\frac{\sqrt{17}}{2}-\frac{5}{2}$; $x=\frac{-\sqrt{17}-5}{2}$; $x=\frac{-4.12-5}{2}$; $x=-\frac{9.12}{2}$; $\mathbf{x=-4.56}$

Check: I. *Let* $x=-0.44$ *in* $x^2+5x+2=0$; $(-0.44)^2+5\cdot(-0.44)+2\overset{?}{=}0$; $0.2-2.2+2\overset{?}{=}0$; $2.2-2.2\overset{?}{=}0$; $0=0$

II. *Let* $x=-4.56$ *in* $x^2+5x+2=0$; $(-4.56)^2+5\cdot(-4.56)+2\overset{?}{=}0$; $20.8-22.8+2\overset{?}{=}0$; $22.8-22.8\overset{?}{=}0$; $0=0$

Therefore, the equation $x^2+5x+2=0$ can be factored to $\mathbf{(x+0.44)(x+4.56)=0}$.

Fourth Method: (The Square Root Property Method) - Is not applicable.

Note that from the above two methods using the Quadratic Formula method may be the faster method, for some, than Completing-the-Square method.

Example 4.6-4

Use different methods to solve the equation $6x^2+4x-2=0$.

Solution:

First Divide both sides of the equation by 2, i.e., $6x^2+4x-2=0$; $\frac{\overset{3}{\cancel{6}}}{\cancel{2}}x^2+\frac{\overset{2}{\cancel{4}}}{\cancel{2}}x-\frac{\cancel{2}}{\cancel{2}}=0$; $3x^2+2x-1=0$. Then consider other methods to solve the equation $3x^2+2x-1=0$.

First Method: (The Trial and Error Method)

To solve the given equation using the Trial and Error method we only consider the left hand side of the equation which is a second degree polynomial. Next, we need to obtain two numbers whose sum is 2 and whose product is $3\cdot-1=-3$ by constructing a table as shown below:

Sum	*Product*
$6-4=2$	$6\cdot(-4)=-24$
$5-3=2$	$5\cdot(-3)=-15$
$4-2=2$	$4\cdot(-2)=-8$
$\mathbf{3-1=2}$	$\mathbf{3\cdot(-1)=-3}$

The last line contains the sum and the product of the two numbers that we need. Therefore, $3x^2+2x-1=0$; $3x^2+(3-1)x-1=0$; $3x^2+3x-x-1=0$; $3x(x+1)-(x+1)=0$; $\mathbf{(x+1)(3x-1)=0}$.

Check: $(x+1)(3x-1)=0$; $3\cdot x\cdot x-1\cdot x+(1\cdot 3)\cdot x+1\cdot(-1)=0$; $3x^2-x+3x-1=0$; $3x^2+(3-1)x-1=0$; $3x^2+2x-1=0$

Second Method: (The Quadratic Formula Method)

Given the standard quadratic equation $ax^2+bx+c=0$, equate the coefficients of $3x^2+2x-1=0$ with the standard quadratic equation by letting $a=3$, $b=2$, and $c=-1$. Then,

Given: $x=\frac{-b\pm\sqrt{b^2-4ac}}{2a}$; $x=\frac{-2\pm\sqrt{2^2-(4\times 3\times -1)}}{2\times 3}$; $x=\frac{-2\pm\sqrt{4+12}}{6}$; $x=\frac{-2\pm\sqrt{16}}{6}$; $x=\frac{-2\pm\sqrt{4^2}}{6}$; $x=\frac{-2\pm 4}{6}$. Therefore:

I. $x=\frac{-2+4}{6}$; $x=\frac{\not 2}{\not 6_3}$; $\mathbf{x=\frac{1}{3}}$ II. $x=\frac{-2-4}{6}$; $x=-\frac{\not 6}{\not 6}$; $x=-\frac{1}{1}$; $\mathbf{x=-1}$

Check: I. *Let* $x=\frac{1}{3}$ *in* $3x^2+2x-1=0$; $3\cdot\left(\frac{1}{3}\right)^2+2\cdot\left(\frac{1}{3}\right)-1\overset{?}{=}0$; $\frac{\not 3}{\not 9_3}+\frac{2}{3}-1\overset{?}{=}0$; $\frac{1}{3}+\frac{2}{3}-1\overset{?}{=}0$

; $\frac{1+2}{3}-1\overset{?}{=}0$; $\frac{\not 3}{\not 3}-1\overset{?}{=}0$; $1-1\overset{?}{=}0$; $0=0$

II. *Let* $x=-1$ *in* $3x^2+2x-1=0$; $3\cdot(-1)^2+2\cdot(-1)-1\overset{?}{=}0$; $3-2-1\overset{?}{=}0$; $3-3\overset{?}{=}0$; $0=0$

Therefore, the equation $3x^2+2x-1=0$ can be factored to $\left(x-\frac{1}{3}\right)(x+1)=0$ which is the same as $\mathbf{(3x-1)(x+1)=0}$.

Third Method: (Completing-the-Square Method)

$3x^2+2x-1=0$; $3x^2+2x=1$; $\frac{\not 3}{\not 3}x^2+\frac{2}{3}x=\frac{1}{3}$; $x^2+\frac{2}{3}x=\frac{1}{3}$; $x^2+\frac{2}{3}x+\left(\frac{1}{2}\cdot\frac{2}{3}\right)^2=\frac{1}{3}+\left(\frac{1}{2}\cdot\frac{2}{3}\right)^2$

; $x^2+\frac{2}{3}x+\frac{1}{9}=\frac{1}{3}+\frac{1}{9}$; $\left(x+\frac{1}{3}\right)^2=\frac{(1\cdot 9)+(1\cdot 3)}{3\cdot 9}$; $\left(x+\frac{1}{3}\right)^2=\frac{9+3}{27}$; $\left(x+\frac{1}{3}\right)^2=\frac{12}{27}$; $x+\frac{1}{3}=\pm\sqrt{\frac{12}{27}}$

; $x+\frac{1}{3}=\pm\sqrt{\frac{4\cdot\not 3}{9\cdot\not 3}}$; $x+\frac{1}{3}=\pm\sqrt{\frac{2^2}{3^2}}$; $x+\frac{1}{3}=\pm\frac{2}{3}$

Therefore: I. $x+\frac{1}{3}=+\frac{2}{3}$; $x=\frac{2}{3}-\frac{1}{3}$; $x=\frac{2-1}{3}$; $\mathbf{x=\frac{1}{3}}$

II. $x+\frac{1}{3}=-\frac{2}{3}$; $x=-\frac{1}{3}-\frac{2}{3}$; $x=\frac{-1-2}{3}$; $x=-\frac{\not 3}{\not 3}$; $x=-\frac{1}{1}$; $\mathbf{x=-1}$

Check: I. *Let* $x=\frac{1}{3}$ *in* $3x^2+2x-1=0$; $3\cdot\left(\frac{1}{3}\right)^2+2\cdot\left(\frac{1}{3}\right)-1\overset{?}{=}0$; $\frac{\not 3}{\not 9_3}+\frac{2}{3}-1\overset{?}{=}0$; $\frac{1}{3}+\frac{2}{3}-1\overset{?}{=}0$

; $\frac{1+2}{3}-1\overset{?}{=}0$; $\frac{\not 3}{\not 3}-1\overset{?}{=}0$; $1-1\overset{?}{=}0$; $0=0$

II. *Let* $x=-1$ *in* $3x^2+2x-1=0$; $3\cdot(-1)^2+2\cdot(-1)-1\overset{?}{=}0$; $3-2-1\overset{?}{=}0$; $3-3\overset{?}{=}0$; $0=0$

Therefore, the equation $3x^2+2x-1=0$ can be factored to $\left(x-\frac{1}{3}\right)(x+1)=0$ which is the same as $\mathbf{(3x-1)(x+1)=0}$.

Fourth Method: (The Square Root Property Method) - Is not applicable.

Note that from the above three methods using the Trial and Error method is the easiest method to obtain the factored terms. The Quadratic Formula method is the second easiest method to use, followed by Completing-the-Square method.

Example 4.6-5

Use different methods to solve the equation $(2x+5)^2=25$.

Solution:

First Method: (The Trial and Error Method)

To apply the Trial and Error method to the equation $(2x+5)^2=25$ we need to complete and simplify the square in the left hand side of the equation, i.e., $(2x+5)^2=25$; $4x^2+25+20x=25$

; $4x^2+20x+25-25=25-25$; $4x^2+20x+0=0$; $\frac{\cancel{4}}{\cancel{4}}x^2+\frac{\overset{5}{\cancel{20}}}{\cancel{4}}x+0=0$; $x^2+5x+0=0$. Then, we only consider the left hand side of the equation which is a second degree polynomial. Next we need to obtain two numbers whose sum is 5 and whose product is $5\cdot 0=0$ by constructing a table as shown below:

Sum	***Product***
$3+2=5$	$3\cdot 2=6$
$4+1=5$	$4\cdot 1=4$
$\mathbf{5+0=5}$	$\mathbf{5\cdot 0=0}$

The last line contains the sum and the product of the two numbers that we need. Thus, $(2x+5)^2=25$ can be factored to $(x+0)(x+5)=0$ which is the same as $\mathbf{x(x+5)=0}$

Second Method: (The Square Root Property Method)

$(2x+5)^2=25$; $\sqrt{(2x+5)^2}=\pm\sqrt{25}$; $2x+5=\pm 5$. Therefore:

I. $2x+5=+5$; $2x=5-5$; $2x=0$; $\mathbf{x=0}$

II. $2x+5=-5$; $2x=-5-5$; $2x=-10$; $x=-\frac{\overset{5}{\cancel{10}}}{\cancel{2}}$; $\mathbf{x=-5}$

Check: I. *Let* $x=0$ *in* $(2x+5)^2=25$; $[(2\cdot 0)+5]^2\overset{?}{=}25$; $5^2\overset{?}{=}25$; $25=25$

II. *Let* $x=-5$ *in* $(2x+5)^2=25$; $[(2\cdot -5)+5]^2\overset{?}{=}25$; $(-10+5)^2\overset{?}{=}25$; $(-5)^2\overset{?}{=}25$; $25=25$

Therefore, the equation $(2x+5)^2=25$ can be factored to $(x-0)(x+5)=0$ which is the same as $\mathbf{x(x+5)=0}$.

Third Method: (The Quadratic Formula Method)

First complete the square term on the left hand side and simplify the equation:

$(2x+5)^2=25$; $4x^2+20x+25=25$; $4x^2+20x=25-25$; $4x^2+20x=0$; $\frac{\cancel{4}}{\cancel{4}}x^2+\frac{\overset{5}{\cancel{20}}}{\cancel{4}}x=\frac{0}{4}$; $x^2+5x=0$

Then, given the standard quadratic equation $ax^2+bx+c=0$, equate the coefficients of $x^2+5x=0$ with the standard quadratic equation by letting $a=1$, $b=5$, and $c=0$. Then,

Given: $x=\frac{-b\pm\sqrt{b^2-4ac}}{2a}$; $x=\frac{-5\pm\sqrt{5^2-(4\times 1\times 0)}}{2\times 1}$; $x=\frac{-5\pm\sqrt{25-0}}{2}$; $x=\frac{-5\pm 5}{2}$. Therefore:

I. $x=\frac{-5+5}{2}$; $x=\frac{0}{2}$; $\mathbf{x=0}$

II. $x=\frac{-5-5}{2}$; $x=-\frac{\overset{5}{\cancel{10}}}{\cancel{2}}$; $\mathbf{x=-5}$

Check: I. *Let* $x=0$ *in* $x^2+5x=0$; $0^2+5\times 0\overset{?}{=}0$; $0+0\overset{?}{=}0$; $0=0$

II. *Let* $x=-5$ *in* $x^2+5x=0$; $(-5)^2+5\cdot(-5)\overset{?}{=}0$; $25-25\overset{?}{=}0$; $0=0$

Therefore, the equation $(2x+5)^2=25$ can be factored to $(x-0)(x+5)=0$ which is the same as $\mathbf{x(x+5)=0}$.

Fourth Method: (The Greatest Common Factoring Method)

First complete the square term on the left hand side and simplify the equation:

$(2x+5)^2=25$; $4x^2+20x+25=25$; $4x^2+20x=25-25$; $4x^2+20x=0$; $\frac{\cancel{4}}{\cancel{4}}x^2+\frac{\overset{5}{\cancel{20}}}{\cancel{4}}x=\frac{0}{4}$; $x^2+5x=0$

Then, Factor out the greatest common monomial term x, i.e., $x^2+5x=0$; $x(x+5)=0$. Thus, the two solution to the equation are:

I. $x=0$ and II. $x+5=0$; $\mathbf{x=-5}$

Hence, the equation $(2x+5)^2=25$ can be factored to $(x-0)(x+5)=0$ which is the same as $\mathbf{x(x+5)=0}$.

Fifth Method: (Completing-the-Square Method)

First complete the square term on the left hand side and simplify the equation:

$(2x+5)^2=25$; $4x^2+20x+25=25$; $4x^2+20x=25-25$; $4x^2+20x=0$; $\frac{\cancel{4}}{\cancel{4}}x^2+\frac{\overset{5}{\cancel{20}}}{\cancel{4}}x=\frac{0}{4}$; $x^2+5x=0$

Then, complete the square.

$x^2+5x=0$; $x^2+5x+\left(\frac{5}{2}\right)^2=0+\left(\frac{5}{2}\right)^2$; $x^2+5x+\frac{25}{4}=\frac{25}{4}$; $\left(x+\frac{5}{2}\right)^2=\frac{25}{4}$; $x+\frac{5}{2}=\pm\sqrt{\frac{25}{4}}$; $x+\frac{5}{2}=\pm\frac{5}{2}$.

Therefore: I. $x+\frac{5}{2}=+\frac{5}{2}$; $x=\frac{5}{2}-\frac{5}{2}$; $x=\frac{5-5}{2}$; $x=\frac{0}{2}$; $\mathbf{x=0}$

II. $x+\frac{5}{2}=-\frac{5}{2}$; $x=-\frac{5}{2}-\frac{5}{2}$; $x=\frac{-5-5}{2}$; $x=-\frac{\overset{5}{\cancel{10}}}{\cancel{2}}$; $\mathbf{x=-5}$

Check: I. *Let* $x=0$ *in* $x^2+5x=0$; $0^2+5\times 0\overset{?}{=}0$; $0+0\overset{?}{=}0$; $0=0$

II. *Let* $x=-5$ *in* $x^2+5x=0$; $(-5)^2+5\cdot(-5)\overset{?}{=}0$; $25-25\overset{?}{=}0$; $0=0$

Therefore, the equation $(2x+5)^2=25$ can be factored to $(x-0)(x+5)=0$ which is the same as $\mathbf{x(x+5)=0}$.

Note that from the above five methods the Square Root Property and the Trial and Error methods are the easiest methods in solving the quadratic equation, followed by the Greatest Common Factoring method, Quadratic Formula method, and Completing-the-Square method.

Practice Problems - How to Choose the Best Factoring or Solution Method

Section 4.6 Practice Problems - Choose three methods to solve the following quadratic equations. State the degree of difficulty associated with each method you selected.

1. $x^2=16$
2. $x^2+7x+3=0$
3. $(3x+4)^2=36$
4. $x^2+11x+30=0$
5. $5t^2+4t-1=0$
6. $(2x+6)^2=36$
7. $y^2-8y+15=0$
8. $w^2=-7$
9. $6x^2+x-1=0$
10. x^2-4x+4

Chapter 5
Algebraic Fractions

Quick Reference to Chapter 5 Case Problems

$\left(\frac{x}{y}\cdot\frac{1}{x^2}\right)\div\frac{x}{y}=$; $\left(\frac{1}{x}+\frac{1}{y}\right)\cdot\frac{xy}{x^2-y^2}=$; $\left(\frac{xyz}{x^2y}\div\frac{xy}{y^2z}\right)+\frac{2yz^2}{x^2}=$

5.4 Math Operations Involving Complex Algebraic Fractions 330

Case I - Addition and Subtraction of Complex Algebraic Fractions, *p. 330*

$\dfrac{\frac{1}{2x}+\frac{2}{3y}}{\frac{3}{3y}-\frac{5}{2x}}=$; $\dfrac{\frac{3x^3y^2}{xy}-1}{\frac{x^2y}{xy^2}+1}=$; $\dfrac{\frac{1}{a}+\frac{1}{b}}{\frac{1}{a}-\frac{1}{b}}-\frac{a}{b}=$

Case II - Multiplication of Complex Algebraic Fractions, *p. 337*

$\dfrac{\frac{xy}{x^2}\cdot x}{x^2y^2\cdot\frac{1}{xy}}=$; $\dfrac{\frac{x^2y^2z}{z^2}\cdot\frac{z}{x^3y^3}}{\frac{1}{xy}}=$; $\dfrac{\frac{xyz^2}{xy}\cdot\frac{x^2}{y^3}}{\frac{x}{y}\cdot\frac{y^2}{x^3}}=$

Case III - Division of Complex Algebraic Fractions, *p. 342*

$\dfrac{\frac{a^3b^2c}{b^2c^3}\div a^3}{\frac{ab}{a^3}\div b^2}=$; $\dfrac{\frac{uv^2w}{w^3}\div\frac{u^2v^3}{u}}{\frac{u}{v}}=$; $3xyz\div\dfrac{\frac{xy}{z}\div z^2}{\frac{xy^2}{xyz^2}}=$

Case IV - Mixed Operations Involving Complex Algebraic Fractions, *p. 347*

$\dfrac{\frac{x}{y}-1}{\frac{x}{y}+1}\div\frac{1}{x+y}=$; $\dfrac{\frac{x+1}{x-1}-\frac{x}{x+1}}{x}\cdot\frac{x^2-x}{5}=$; $\dfrac{\frac{1}{2x}+\frac{2}{y}}{\frac{3}{2x}-\frac{1}{y}}-\dfrac{4}{\frac{3y-2x}{x}}=$

5.5 Solving One Variable Equations Containing Algebraic Fractions 353

$\frac{2}{2x+5}+\frac{3}{3x+2}=\frac{1}{2}$; $\frac{y-1}{y}=1+\frac{4}{y-1}$; $\frac{x+3}{x-1}-\frac{x}{x+2}=\frac{5}{x^2+x-2}$

Chapter 5 - Algebraic Fractions

The objective of this chapter is to teach students how to solve and simplify math operations involving algebraic fractions. An introduction to algebraic fractions and the difference between algebraic and arithmetic/integer fractions is discussed in Section 5.1. Simplification of algebraic fractions to lower terms is addressed in Section 5.2. The steps as to how algebraic fractions are added, subtracted, multiplied, divided, and mixed is discussed in Section 5.3. Section 5.4 describes the math operations involving complex algebraic fractions. Students are encouraged to spend adequate time learning how to solve complex algebraic fractions which is the most difficult form of an algebraic expression. Solving one variable equations containing algebraic fractions is introduced in Section 5.5. Cases presented in each section are concluded by solving additional examples with practice problems to further enhance the students ability.

5.1 Introduction to Algebraic Fractions

Arithmetic/integer fractions are fractions where the numerator and the denominator are integer numbers. For example, $\frac{2}{3}$, $\frac{1}{5}$, $-\frac{3}{8}$, and $\frac{2}{1}$ are examples of arithmetic fractions. **Algebraic fractions** are fractions where the numerator or the denominator (or both) are variables. For example, $\frac{a}{3}$, $\frac{x}{y}$, $\frac{3}{x+1}$, and $\frac{1}{x}$ are examples of algebraic fractions. The concepts and procedures learned in simplifying, adding, subtracting, multiplying, and dividing arithmetic fractions can directly be applied to algebraic fractions. In fact, with a good knowledge of operations involving arithmetic fractions students will find this chapter a "non-challenging" chapter. (The subject of arithmetic fractions has been addressed in detail in the "*Mastering Fractions*" book. Students are encouraged to review chapters 3 and 9 for an overall review of the fractional operations.) In this section students are introduced to sign rules for fractions; division of algebraic fractions by zero; and equivalent algebraic fractions.

A. Sign Rules For Fractions

In division, we need to consider two signs. The sign of the numerator and the sign of the denominator. Thus, the sign rules for division are:

1. $\frac{-a}{+b} = -\frac{a}{b}$ For example: $\frac{-8}{+2} = -\frac{8}{2} = -\frac{4}{1} = -4$

2. $\frac{+a}{-b} = -\frac{a}{b}$ For example: $\frac{+8}{-2} = -\frac{8}{2} = -\frac{4}{1} = -4$

3. $\frac{-a}{-b} = +\frac{a}{b}$ For example: $\frac{-8}{-2} = +\frac{8}{2} = +\frac{4}{1} = +4$

4. $\frac{+a}{+b} = +\frac{a}{b}$ For example: $\frac{+8}{+2} = +\frac{8}{2} = +\frac{4}{1} = +4$

In fractions, we need to consider three signs. The fractions sign itself, the numerator sign, and the denominator sign. Therefore, the sign rules for fractions are:

1. $+\dfrac{-a}{+b} = +\left(-\dfrac{a}{b}\right) = -\dfrac{a}{b}$

2. $+\dfrac{+a}{-b} = +\left(-\dfrac{a}{b}\right) = -\dfrac{a}{b}$

3. $+\dfrac{-a}{-b} = +\left(+\dfrac{a}{b}\right) = +\dfrac{a}{b}$

4. $+\dfrac{+a}{+b} = +\left(+\dfrac{a}{b}\right) = +\dfrac{a}{b}$

5. $-\dfrac{-a}{+b} = -\left(-\dfrac{a}{b}\right) = +\dfrac{a}{b}$

6. $-\dfrac{+a}{-b} = -\left(-\dfrac{a}{b}\right) = +\dfrac{a}{b}$

7. $-\dfrac{-a}{-b} = -\left(+\dfrac{a}{b}\right) = -\dfrac{a}{b}$

8. $-\dfrac{+a}{+b} = -\left(+\dfrac{a}{b}\right) = -\dfrac{a}{b}$

For example:

1. $+\dfrac{-8}{+2} = +\left(-\dfrac{8}{2}\right) = -\dfrac{8}{2} = -\dfrac{4}{1} = -4$

2. $+\dfrac{+8}{-2} = +\left(-\dfrac{8}{2}\right) = -\dfrac{8}{2} = -\dfrac{4}{1} = -4$

3. $+\dfrac{-8}{-2} = +\left(+\dfrac{8}{2}\right) = +\dfrac{8}{2} = +\dfrac{4}{1} = +4$

4. $+\dfrac{+8}{+2} = +\left(+\dfrac{8}{2}\right) = +\dfrac{8}{2} = +\dfrac{4}{1} = +4$

5. $-\dfrac{-8}{+2} = -\left(-\dfrac{8}{2}\right) = +\dfrac{8}{2} = +\dfrac{4}{1} = +4$

6. $-\dfrac{+8}{-2} = -\left(-\dfrac{8}{2}\right) = +\dfrac{8}{2} = +\dfrac{4}{1} = +4$

7. $-\dfrac{-8}{-2} = -\left(+\dfrac{8}{2}\right) = -\dfrac{8}{2} = -\dfrac{4}{1} = -4$

8. $-\dfrac{+8}{+2} = -\left(+\dfrac{8}{2}\right) = -\dfrac{8}{2} = -\dfrac{4}{1} = -4$

B. Division of Algebraic Fractions by Zero

An algebraic fraction is an expression of the form

$$\frac{A}{B} \quad \text{where } A \text{ and } B \text{ are polynomials}$$

Note that the denominator B in an algebraic fraction can not be equal to zero, since division by zero is not defined. For example,

a. $\dfrac{1}{x^2}$ is not defined when $x = 0$ because $\dfrac{1}{0^2} = \dfrac{1}{0}$ is not defined.

b. $\dfrac{2}{x+1}$ is not defined when $x = -1$ because $\dfrac{2}{-1+1} = \dfrac{2}{0}$ is not defined.

c. $\dfrac{a-2}{2-a}$ is not defined when $a = 2$ because $\dfrac{2-2}{2-2} = \dfrac{0}{0}$ is not defined.

d. $\dfrac{5x}{x-9}$ is not defined when $x = 9$ because $\dfrac{5 \times 9}{9-9} = \dfrac{45}{0}$ is not defined.

e. $\dfrac{x-1}{x^2+6x+5} = \dfrac{x-1}{(x+1)(x+5)}$ is not defined when $x = -1$ and $x = -5$ because

when $x = -1$ $\quad \dfrac{-1-1}{(-1+1)(-1+5)} = \dfrac{-2}{0 \times 4} = -\dfrac{2}{0}$ is not defined, and

when $x = -5$ $\quad \dfrac{-5-1}{(-5+1)(-5+5)} = \dfrac{-6}{-4 \times 0} = -\dfrac{6}{0}$ is not defined.

f. $\dfrac{x+5}{x^2+2x-15} = \dfrac{x+5}{(x-3)(x+5)} = \dfrac{(x+5)}{(x-3)(x+5)} = \dfrac{1}{x-3}$ is not defined when $x=3$ because $\dfrac{1}{3-3}$ $= \dfrac{1}{0}$ is not defined.

g. $\dfrac{3}{-x-3}$ is not defined when $x=-3$ because $\dfrac{3}{-(-3)-3} = \dfrac{3}{3-3} = \dfrac{3}{0}$ is not defined.

h. $\dfrac{2x+3}{2x-3}$ is not defined when $x=\dfrac{3}{2}$ because $\dfrac{2\cdot\frac{3}{2}+3}{2\cdot\frac{3}{2}-3} = \dfrac{\frac{3}{1}+3}{\frac{3}{1}-3} = \dfrac{3+3}{3-3} = \dfrac{6}{0}$ is not defined.

i. $\dfrac{x}{4x-3}$ is not defined when $x=\dfrac{3}{4}$ because $\dfrac{\frac{3}{4}}{4\cdot\frac{3}{4}-3} = \dfrac{\frac{3}{4}}{4\cdot\frac{3}{4}-3} = \dfrac{\frac{3}{4}}{\frac{3}{1}-3} = \dfrac{\frac{3}{4}}{3-3} = \dfrac{\frac{3}{4}}{0} = \dfrac{\frac{3}{4}}{\frac{0}{1}} = \dfrac{3\times1}{4\times0}$ $= \dfrac{3}{0}$ is not defined.

C. Equivalent Algebraic Fractions

When the numerator and the denominator of an algebraic fraction is multiplied by the same number, sign, or a variable the new algebraic fraction is said to be equivalent to the original algebraic fraction. For example, the following algebraic fractions are equivalent to one another:

a. $\dfrac{a}{b} = \dfrac{3a}{3b} = \dfrac{100a}{100b} = \dfrac{5a}{5b} = \dfrac{-a}{-b}$

b. $\dfrac{1+x}{x} = \dfrac{2+2x}{2x} = \dfrac{a+ax}{ax}$

c. $\dfrac{x-3}{4} = \dfrac{5x-15}{20} = \dfrac{3x-9}{12} = \dfrac{ax-3a}{4a} = \dfrac{-(x-3)}{-4} = \dfrac{3-x}{-4}$

d. $\dfrac{x^2}{1+x^2} = \dfrac{5x^2}{5+5x^2} = \dfrac{\frac{1}{2}x^2}{\frac{1}{2}+\frac{1}{2}x^2} = \dfrac{x^2y}{y+x^2y}$

e. $\dfrac{a-b}{3a} = \dfrac{2a-2b}{6a} = \dfrac{10a-10b}{30a} = \dfrac{ax-bx}{3ax}$

f. $\dfrac{5}{x-y} = \dfrac{-5}{-(x-y)} = \dfrac{-5}{y-x} = \dfrac{10}{2x-2y} = \dfrac{5xy}{x^2y-xy^2}$

g. $\dfrac{x-3}{2x-7} = \dfrac{-(x-3)}{-(2x-7)} = \dfrac{3-x}{7-2x} = \dfrac{2\cdot(x-3)}{2\cdot(2x-7)} = \dfrac{2x-6}{4x-14} = \dfrac{-3\cdot(x-3)}{-3\cdot(2x-7)} = \dfrac{9-3x}{21-6x}$

h. $-\dfrac{2}{x-1} = -\dfrac{-2}{-(x-1)} = \dfrac{2}{1-x} = -\dfrac{6}{3x-3} = \dfrac{6}{3-3x}$

i. $\dfrac{1-x}{1-y} = \dfrac{-(1-x)}{-(1-y)} = \dfrac{x-1}{y-1} = \dfrac{2\cdot(1-x)}{2\cdot(1-y)} = \dfrac{2-2x}{2-2y}$

Practice Problems - Introduction to Algebraic Fractions

A. Section 5.1 Practice Problems - Write the correct sign for the following fractions.

1. $-\dfrac{2}{-5} =$
2. $+\dfrac{-3}{-6} =$
3. $-\dfrac{-8}{-4} =$
4. $\dfrac{10}{-2} =$
5. $-\dfrac{5}{-15} =$
6. $+\dfrac{-8}{6} =$
7. $-\dfrac{+3}{+15} =$
8. $-\dfrac{2}{+12} =$
9. $-\dfrac{-6}{3} =$
10. $+\dfrac{2}{-3} =$

B. Section 5.1 Practice Problems - State the value(s) of the variable for which the following fractions are not defined.

1. $\dfrac{3}{x-1}$
2. $\dfrac{x-5}{5-x}$
3. $-\dfrac{1}{x}$
4. $\dfrac{x}{x+10}$
5. $\dfrac{2x}{3x-5}$
6. $\dfrac{5x-2}{x-7}$
7. $\dfrac{4}{x^3-x}$
8. $\dfrac{x+1}{x^2+5x+4}$
9. $\dfrac{3x}{2x^2-5x-3}$
10. $\dfrac{x-3}{x^2-2x-3}$

C. Section 5.1 Practice Problems - State which of the following algebraic fractions are equivalent.

1. $\dfrac{2x}{3y} \stackrel{?}{=} \dfrac{4x}{9y}$
2. $\dfrac{3x+1}{2x} \stackrel{?}{=} \dfrac{9x+3}{6x}$
3. $\dfrac{2}{a-b} \stackrel{?}{=} -\dfrac{2}{b-a}$
4. $\dfrac{x-5}{x+1} \stackrel{?}{=} \dfrac{5-x}{x-1}$
5. $-\dfrac{a}{a-1} \stackrel{?}{=} \dfrac{a}{1-a}$
6. $\dfrac{3-x}{-x} \stackrel{?}{=} \dfrac{x+3}{x}$
7. $\dfrac{u-w}{w} \stackrel{?}{=} \dfrac{-2u+2w}{2w}$
8. $\dfrac{a^2}{1-a^2} \stackrel{?}{=} \dfrac{-a^2}{a^2-1}$
9. $\dfrac{2-x}{-3} \stackrel{?}{=} \dfrac{x}{3} - \dfrac{2}{3}$
10. $-\dfrac{x+2}{-2} \stackrel{?}{=} \dfrac{-x-2}{2}$

5.2 Simplifying Algebraic Fractions to Lower Terms

In dealing with integer fractions we learned that integer (arithmetic) fractions are reduced to their lowest terms by dividing both the numerator and the denominator by their common terms. For example, the integer fraction $\frac{14}{7}$ is simplified to its lowest term by dividing both the numerator and the denominator by 7, which is common to both, i.e., $\frac{14}{7} = \frac{\cancel{7}\cdot 2}{\cancel{7}} = \frac{2}{1} = 2$. The same principle holds true when simplifying algebraic fractions. Algebraic fractions are simplified using the following steps:

Step 1 Factor both the numerator and the denominator completely (see Chapters 3 and 4).

Step 2 Simplify the algebraic fraction by eliminating the common terms in both the numerator and the denominator.

Examples with Steps

The following examples show the steps as to how algebraic fractions are simplified to their lowest terms:

Example 5.2-1

$$\boxed{\frac{3a^2b^2c}{15ab^3c^2}} =$$

Solution:

Step 1 $\boxed{\textit{Not Applicable}}$

Step 2 $\boxed{\frac{3a^2b^2c}{15ab^3c^2}} = \boxed{\frac{\overset{a}{\cancel{3}\cancel{a^2}}\,\cancel{b^2}\cancel{c}}{\underset{5}{\cancel{15}}\cancel{a}\underset{b}{\cancel{b^3}}\,\underset{c}{\cancel{c^2}}}} = \boxed{\frac{\mathbf{a}}{\mathbf{5bc}}}$

Example 5.2-2

$$\boxed{\frac{-xyz}{-3x^2z^2}} =$$

Solution:

Step 1 $\boxed{\textit{Not Applicable}}$

Step 2 $\boxed{\frac{-xyz}{-3x^2z^2}} = \boxed{+\frac{xyz}{3x^2z^2}} = \boxed{\frac{\cancel{x}y\cancel{z}}{3\underset{x}{\cancel{x^2}}\,\underset{z}{\cancel{z^2}}}} = \boxed{\frac{\mathbf{y}}{\mathbf{3xz}}}$

Example 5.2-3

$$\boxed{\frac{3x+6}{x^2-x-6}} =$$

Solution:

Step 1 $\boxed{\frac{3x+6}{x^2-x-6}} = \boxed{\frac{3(x+2)}{(x+2)(x-3)}}$

Step 2 $\boxed{\frac{3(x+2)}{(x+2)(x-3)}} = \boxed{\frac{3(\not{x}+\not{2})}{(\not{x}+\not{2})(x-3)}} = \boxed{\mathbf{\frac{3}{x-3}}}$

Example 5.2-4

$\boxed{\frac{2y^2-7y-15}{y^2-25}} =$

Solution:

Step 1 $\boxed{\frac{2y^2-7y-15}{y^2-25}} = \boxed{\frac{(2y+3)(y-5)}{y^2-5^2}} = \boxed{\frac{(2y+3)(y-5)}{(y-5)(y+5)}}$

Step 2 $\boxed{\frac{(2y+3)(y-5)}{(y-5)(y+5)}} = \boxed{\frac{(2y+3)(\not{y}-\not{5})}{(\not{y}-\not{5})(y+5)}} = \boxed{\mathbf{\frac{2y+3}{y+5}}}$

Example 5.2-5

$\boxed{\frac{x^3-x}{x^3-2x^2-3x}} =$

Solution:

Step 1 $\boxed{\frac{x^3-x}{x^3-2x^2-3x}} = \boxed{\frac{x(x^2-1)}{x(x^2-2x-3)}} = \boxed{\frac{x(x-1)(x+1)}{x(x+1)(x-3)}}$

Step 2 $\boxed{\frac{x(x-1)(x+1)}{x(x+1)(x-3)}} = \boxed{\frac{\not{x}(x-1)(\not{x}+\not{1})}{\not{x}(\not{x}+\not{1})(x-3)}} = \boxed{\mathbf{\frac{x-1}{x-3}}}$

Additional Examples - Simplifying Algebraic Fractions to Lower Terms

The following examples further illustrate how to simplify algebraic fractions to their lowest terms:

Example 5.2-6

$\boxed{\frac{5x^2}{15x}} = \boxed{\frac{\overset{x}{\not{5}x^{\not{2}}}}{\underset{3}{\not{15}\not{x}}}} = \boxed{\mathbf{\frac{x}{3}}}$

Example 5.2-7

$\boxed{\frac{x^2y^2z^3}{xy^3z}} = \boxed{\frac{\overset{x}{x^{\not{2}}}\,\not{y^2}\,\overset{z^2}{z^{\not{3}}}}{\not{x}\,\underset{y}{y^{\not{3}}}\,\not{z}}} = \boxed{\mathbf{\frac{xz^2}{y}}}$

Example 5.2-8

$\boxed{\frac{6a^2b^2}{30a^3b}} = \boxed{\frac{\not{6}\not{a^2}\,\overset{b}{b^{\not{2}}}}{\underset{5}{\not{30}}\,\underset{a}{a^{\not{3}}}\,\not{b}}} = \boxed{\mathbf{\frac{b}{5a}}}$

Example 5.2-9

$$\boxed{\frac{2u^2v^3w}{-4uvw^3}} = \boxed{-\frac{2u^2v^3w}{4uvw^3}} = \boxed{-\frac{\overset{u\ \ v^2}{\cancel{2}\cancel{u^2}\ \cancel{v^3}\ \cancel{w}}}{\underset{2\qquad w^2}{\cancel{4}\ \cancel{uv}\ \cancel{w^3}}}} = \boxed{-\frac{\mathbf{uv^2}}{\mathbf{2w^2}}}$$

Example 5.2-10

$$\boxed{\frac{-48x^3yz^2}{6x^4z^3}} = \boxed{-\frac{48x^3yz^2}{6x^4z^3}} = \boxed{-\frac{\overset{8}{\cancel{48}}\cancel{x^3}y\cancel{z^2}}{\underset{\ \ x\quad z}{\cancel{6}\cancel{x^4}\ \cancel{z^3}}}} = \boxed{-\frac{\mathbf{8y}}{\mathbf{xz}}}$$

Example 5.2-11

$$\boxed{\frac{-5xyz}{-15xy^2z}} = \boxed{+\frac{5xyz}{15xy^2z}} = \boxed{\frac{\cancel{5}\cancel{x}\cancel{y}\cancel{z}}{\underset{3\qquad y}{\cancel{15}\cancel{x}\ \cancel{y^2}\ \cancel{z}}}} = \boxed{\frac{\mathbf{1}}{\mathbf{3y}}}$$

Example 5.2-12

$\boxed{\dfrac{x^2y^2z}{3}}$ can not be simplified

Example 5.2-13

$$\boxed{\frac{x^2y^2}{-x^3y^4}} = \boxed{-\frac{x^2y^2}{x^3y^4}} = \boxed{-\frac{\cancel{x^2}\cancel{y^2}}{\underset{x\quad y^2}{\cancel{x^3}\ \cancel{y^4}}}} = \boxed{-\frac{\mathbf{1}}{\mathbf{xy^2}}}$$

Example 5.2-14

$$\boxed{\frac{x^2+5x}{x^2+2x-15}} = \boxed{\frac{x(x+5)}{(x-3)(x+5)}} = \boxed{\frac{x(\cancel{x+5})}{(x-3)(\cancel{x+5})}} = \boxed{\frac{\mathbf{x}}{\mathbf{x-3}}}$$

Example 5.2-15

$\boxed{\dfrac{5-3n}{3-5n}}$ can not be simplified

Example 5.2-16

$$\boxed{\frac{x^2-1}{4x+4}} = \boxed{\frac{(x-1)(x+1)}{4(x+1)}} = \boxed{\frac{(x-1)(\cancel{x+1})}{4(\cancel{x+1})}} = \boxed{\frac{\mathbf{x-1}}{\mathbf{4}}}$$

Example 5.2-17

$$\boxed{\frac{5}{5x+15}} = \boxed{\frac{5}{5(x+3)}} = \boxed{\frac{\cancel{5}}{\cancel{5}(x+3)}} = \boxed{\frac{\mathbf{1}}{\mathbf{x+3}}}$$

Example 5.2-18

$$\boxed{\frac{x^2+y^2}{(x+y)^2}} = \boxed{\frac{\mathbf{x^2+y^2}}{\mathbf{(x+y)(x+y)}}}$$

Example 5.2-19

$$\boxed{\frac{7x}{-14x^2+7x}}=\boxed{\frac{7x}{7x(-2x+1)}}=\boxed{\frac{\cancel{7x}}{\cancel{7x}(-2x+1)}}=\boxed{\frac{1}{-2x+1}}=\boxed{-\frac{\mathbf{1}}{\mathbf{2x-1}}}$$

Example 5.2-20

$\boxed{\frac{x+9}{7}}$ can not be simplified

Example 5.2-21

$$\boxed{-\frac{5x-3}{3-5x}}=\boxed{\frac{-(5x-3)}{3-5x}}=\boxed{\frac{3-5x}{3-5x}}=\boxed{\frac{\cancel{3-5x}}{\cancel{3-5x}}}=\boxed{\frac{1}{1}}=\boxed{\mathbf{1}}$$

Example 5.2-22

$$\boxed{\frac{y^2-2y}{y^2+y-6}}=\boxed{\frac{y(y-2)}{(y+3)(y-2)}}=\boxed{\frac{y\cancel{(y-2)}}{(y+3)\cancel{(y-2)}}}=\boxed{\frac{\mathbf{y}}{\mathbf{y+3}}}$$

Example 5.2-23

$$\boxed{\frac{x^2+4x-5}{x^2+2x-15}}=\boxed{\frac{(x-1)(x+5)}{(x+5)(x-3)}}=\boxed{\frac{(x-1)\cancel{(x+5)}}{\cancel{(x+5)}(x-3)}}=\boxed{\frac{\mathbf{x-1}}{\mathbf{x-3}}}$$

Example 5.2-24

$$\boxed{\frac{6a^2-6ab}{a^2-b^2}}=\boxed{\frac{6a(a-b)}{(a-b)(a+b)}}=\boxed{\frac{6a\cancel{(a-b)}}{\cancel{(a-b)}(a+b)}}=\boxed{\frac{\mathbf{6a}}{\mathbf{a+b}}}$$

Example 5.2-25

$$\boxed{\frac{3x^2+9x}{x^3+x^2-6x}}=\boxed{\frac{3x(x+3)}{x\left(x^2+x-6\right)}}=\boxed{\frac{3x(x+3)}{x(x+3)(x-2)}}=\boxed{\frac{3\cancel{x}\cancel{(x+3)}}{\cancel{x}\cancel{(x+3)}(x-2)}}=\boxed{\frac{\mathbf{3}}{\mathbf{x-2}}}$$

Example 5.2-26

$$\boxed{\frac{6x^2+x-1}{3x^2+2x-1}}=\boxed{\frac{(2x+1)(3x-1)}{(3x-1)(x+1)}}=\boxed{\frac{(2x+1)\cancel{(3x-1)}}{\cancel{(3x-1)}(x+1)}}=\boxed{\frac{\mathbf{2x+1}}{\mathbf{x+1}}}$$

Example 5.2-27

$$\boxed{\frac{6x^2+x-2}{3x+2}}=\boxed{\frac{(3x+2)(2x-1)}{3x+2}}=\boxed{\frac{\cancel{(3x+2)}(2x-1)}{\cancel{(3x+2)}}}=\boxed{\frac{2x-1}{1}}=\boxed{\mathbf{2x-1}}$$

Example 5.2-28

$$\boxed{\frac{3y^2+7y+4}{3y^2+y-4}}=\boxed{\frac{(y+1)(3y+4)}{(y-1)(3y+4)}}=\boxed{\frac{(y+1)\cancel{(3y+4)}}{(y-1)\cancel{(3y+4)}}}=\boxed{\frac{\mathbf{y+1}}{\mathbf{y-1}}}$$

Example 5.2-29

$$\boxed{\frac{x^2-y^2}{x^2+4xy+3y^2}}=\boxed{\frac{(x+y)(x-y)}{(x+y)(x+3y)}}=\boxed{\frac{\cancel{(x+y)}(x-y)}{\cancel{(x+y)}(x+3y)}}=\boxed{\frac{\mathbf{x-y}}{\mathbf{x+3y}}}$$

Example 5.2-30

$$\boxed{\frac{2xy+3y^2}{4x^2+4xy-3y^2}}=\boxed{\frac{y(2x+3y)}{(2x+3y)(2x-y)}}=\boxed{\frac{y\cancel{(2x+3y)}}{\cancel{(2x+3y)}(2x-y)}}=\boxed{\mathbf{\frac{y}{2x-y}}}$$

Example 5.2-31

$$\boxed{\frac{5x-5y}{2x^2+xy-3y^2}}=\boxed{\frac{5(x-y)}{(x-y)(2x+3y)}}=\boxed{\frac{5\cancel{(x-y)}}{\cancel{(x-y)}(2x+3y)}}=\boxed{\mathbf{\frac{5}{2x+3y}}}$$

Example 5.2-32

$$\boxed{\frac{25x^2-1}{5x^2-1}}=\boxed{\frac{5^2x^2-1}{\left(\sqrt{5}\right)^2x^2-1}}=\boxed{\frac{(5x)^2-1^2}{\left(\sqrt{5}x\right)^2-1^2}}=\boxed{\mathbf{\frac{(5x+1)(5x-1)}{\left(\sqrt{5}x-1\right)\left(\sqrt{5}x+1\right)}}}$$

Example 5.2-33

$$\boxed{\frac{(u+1)^2}{2u^2+3u+1}}=\boxed{\frac{(u+1)(u+1)}{(u+1)(2u+1)}}=\boxed{\frac{\cancel{(u+1)}(u+1)}{\cancel{(u+1)}(2u+1)}}=\boxed{\mathbf{\frac{u+1}{2u+1}}}$$

Example 5.2-34

$$\boxed{\frac{t^2-9}{t^2-2t-15}}=\boxed{\frac{t^2-3^2}{t^2-2t-15}}=\boxed{\frac{(t-3)(t+3)}{(t+3)(t-5)}}=\boxed{\frac{(t-3)\cancel{(t+3)}}{\cancel{(t+3)}(t-5)}}=\boxed{\mathbf{\frac{t-3}{t-5}}}$$

Example 5.2-35

$$\boxed{\frac{6y^2+7y-3}{-3y^3+y^2}}=\boxed{\frac{(2y+3)(3y-1)}{-y^2(3y-1)}}=\boxed{\frac{(2y+3)\cancel{(3y-1)}}{-y^2\cancel{(3y-1)}}}=\boxed{\frac{2y+3}{-y^2}}=\boxed{\mathbf{-\frac{2y+3}{y^2}}}$$

Practice Problems - Simplifying Algebraic Fractions to Lower Terms

Section 5.2 Practice Problems - Simplify the following algebraic fractions to their lowest terms:

1. $\dfrac{x^2y^2z^5}{-xy^3z^2}=$

2. $-\dfrac{3a^2bc^3}{-9ab^2c}=$

3. $\dfrac{1+2m}{1-2m}=$

4. $\dfrac{2uvw^3}{10u^2v}=$

5. $\dfrac{y^2-4}{y^2-y-6}=$

6. $\dfrac{x^3-3x^2}{x^2-9}=$

7. $\dfrac{x^3+2x^2-15x}{2x^2-5x-3}=$

8. $\dfrac{x^2-2x}{x^3-x^2-2x}=$

9. $\dfrac{x^2+xy-2y^2}{x+2y}=$

10. $\dfrac{6x^2-xy-y^2}{2x^2+xy-y^2}=$

5.3 Math Operations Involving Algebraic Fractions

In this section addition, subtraction, multiplication, and division of algebraic fractions (Cases I through IV) are discussed. In addition, simplification of algebraic expressions involving mixed operations is addressed in Case V.

Case I Addition and Subtraction of Algebraic Fractions with Common Denominators

Algebraic fractions with common denominators are added and subtracted using the following steps:

Step 1 Write the common denominator. Add or subtract the numerators.

Step 2 Simplify the algebraic fraction to its lowest term.

Examples with Steps

The following examples show the steps as to how algebraic fractions with common denominators are added and subtracted:

Example 5.3-1

$$\boxed{\frac{5x}{x+y}-\frac{3x}{x+y}}=$$

Solution:

Step 1

$$\boxed{\frac{5x}{x+y}-\frac{3x}{x+y}}=\boxed{\frac{5x-3x}{x+y}}$$

Step 2

$$\boxed{\frac{5x-3x}{x+y}}=\boxed{\frac{(5-3)x}{x+y}}=\boxed{\mathbf{\frac{2x}{x+y}}}$$

Example 5.3-2

$$\boxed{\frac{2a}{2a+3b}+\frac{3b}{2a+3b}}=$$

Solution:

Step 1

$$\boxed{\frac{2a}{2a+3b}+\frac{3b}{2a+3b}}=\boxed{\frac{2a+3b}{2a+3b}}$$

Step 2

$$\boxed{\frac{2a+3b}{2a+3b}}=\boxed{\frac{\cancel{2a}+\cancel{3b}}{\cancel{2a}+\cancel{3b}}}=\boxed{\frac{1}{1}}=\boxed{\mathbf{1}}$$

Example 5.3-3

$$\boxed{\frac{3x^2+5x+5}{(x+3)(x-1)}-\frac{3x^2+4x+2}{(x+3)(x-1)}}=$$

Solution:

Step 1

$$\boxed{\frac{3x^2+5x+5}{(x+3)(x-1)}-\frac{3x^2+4x+2}{(x+3)(x-1)}}=\boxed{\frac{3x^2+5x+5-\left(3x^2+4x+2\right)}{(x+3)(x-1)}}$$

Step 2

$$\frac{3x^2+5x+5-\left(3x^2+4x+2\right)}{(x+3)(x-1)} = \frac{3x^2+5x+5-3x^2-4x-2}{(x+3)(x-1)}$$

$$= \frac{\left(3x^2-3x^2\right)+(5x-4x)+(5-2)}{(x+3)(x-1)} = \frac{(3-3)x^2+(5-4)x+3}{(x+3)(x-1)} = \frac{0x^2+x+3}{(x+3)(x-1)}$$

$$= \frac{\cancel{(x+3)}}{\cancel{(x+3)}(x-1)} = \boxed{\mathbf{\frac{1}{x-1}}}$$

Example 5.3-4

$$\frac{3a+b}{2a^2b^3}+\frac{2a-b}{2a^2b^3}+\frac{a-2b}{2a^2b^3} =$$

Solution:

Step 1

$$\frac{3a+b}{2a^2b^3}+\frac{2a-b}{2a^2b^3}+\frac{a-2b}{2a^2b^3} = \frac{3a+b+(2a-b)+(a-2b)}{2a^2b^3}$$

Step 2

$$\frac{3a+b+(2a-b)+(a-2b)}{2a^2b^3} = \frac{(3a+2a+a)+(b-b-2b)}{2a^2b^3} = \frac{(3+2+1)a+(1-1-2)b}{2a^2b^3}$$

$$= \frac{6a-2b}{2a^2b^3} = \frac{2(3a-b)}{2a^2b^3} = \boxed{\mathbf{\frac{3a-b}{a^2b^3}}}$$

Example 5.3-5

$$\frac{y^3+3y}{(y-2)(y+3)}-\frac{y^3+2y-3}{(y-2)(y+3)} =$$

Solution:

Step 1

$$\frac{y^3+3y}{(y-2)(y+3)}-\frac{y^3+2y-3}{(y-2)(y+3)} = \frac{y^3+3y-\left(y^3+2y-3\right)}{(y-2)(y+3)}$$

Step 2

$$\frac{y^3+3y-\left(y^3+2y-3\right)}{(y-2)(y+3)} = \frac{y^3+3y-y^3-2y+3}{(y-2)(y+3)} = \frac{\left(y^3-y^3\right)+(3y-2y)+3}{(y-2)(y+3)}$$

$$= \frac{(1-1)y^3+(3-2)y+3}{(y-2)(y+3)} = \frac{0y^3+y+3}{(y-2)(y+3)} = \frac{\cancel{(y+3)}}{(y-2)\cancel{(y+3)}} = \boxed{\mathbf{\frac{1}{y-2}}}$$

Additional Examples - Addition and Subtraction of Algebraic Fractions with Common Denominators

The following examples further illustrate how to add or subtract algebraic fractions with common denominators:

Example 5.3-6

$$\frac{3}{x}+\frac{5}{x}-\frac{2}{x} = \frac{3+5-2}{x} = \boxed{\mathbf{\frac{6}{x}}}$$

Example 5.3-7

$$\boxed{\frac{x^2}{x+1}+\frac{x}{x+1}} = \boxed{\frac{x^2+x}{x+1}} = \boxed{\frac{x\cancel{(x+1)}}{\cancel{(x+1)}}} = \boxed{\frac{x}{1}} = \boxed{x}$$

Example 5.3-8

$$\boxed{\frac{9}{x-3}-\frac{3x}{x-3}} = \boxed{\frac{9-3x}{x-3}} = \boxed{\frac{3(3-x)}{x-3}} = \boxed{\frac{-3\cancel{(x-3)}}{\cancel{(x-3)}}} = \boxed{-\frac{3}{1}} = \boxed{-3}$$

Example 5.3-9

$$\boxed{\frac{4x}{2x-5y}-\frac{10y}{2x-5y}} = \boxed{\frac{4x-10y}{2x-5y}} = \boxed{\frac{2\cancel{(2x-10y)}}{\cancel{(2x-10y)}}} = \boxed{\frac{2}{1}} = \boxed{2}$$

Example 5.3-10

$$\boxed{\frac{x-3}{(x+4)(x-4)}-\frac{x-5}{(x+4)(x-4)}} = \boxed{\frac{x-3-(x-5)}{(x+4)(x-4)}} = \boxed{\frac{x-3-x+5}{(x+4)(x-4)}} = \boxed{\frac{(x-x)+(-3+5)}{(x+4)(x-4)}} = \boxed{\frac{(1-1)x+2}{(x+4)(x-4)}}$$

$$= \boxed{\frac{0x+2}{(x+4)(x-4)}} = \boxed{\mathbf{\frac{2}{(x+4)(x-4)}}}$$

Example 5.3-11

$$\boxed{\frac{x+2}{x-3}+\frac{x+3}{x-3}} = \boxed{\frac{x+2+x+3}{x-3}} = \boxed{\frac{(x+x)+(2+3)}{x-3}} = \boxed{\frac{(1+1)x+5}{x-3}} = \boxed{\mathbf{\frac{2x+5}{x-3}}}$$

Example 5.3-12

$$\boxed{\frac{x^2+3x+2}{(x+1)(x-5)}-\frac{x^2+2x+1}{(x+1)(x-5)}} = \boxed{\frac{x^2+3x+2-\left(x^2+2x+1\right)}{(x+1)(x-5)}} = \boxed{\frac{x^2+3x+2-x^2-2x-1}{(x+1)(x-5)}}$$

$$= \boxed{\frac{\left(x^2-x^2\right)+(3x-2x)+(2-1)}{(x+1)(x-5)}} = \boxed{\frac{(1-1)x^2+(3-2)x+1}{(x+1)(x-5)}} = \boxed{\frac{0x^2+x+1}{(x+1)(x-5)}} = \boxed{\frac{\cancel{(x+1)}}{\cancel{(x+1)}(x-5)}} = \boxed{\mathbf{\frac{1}{x-5}}}$$

Example 5.3-13

$$\boxed{\frac{3x}{x+3}+\frac{2x-1}{x+3}+\frac{3x+2}{x+3}} = \boxed{\frac{3x+2x-1+3x+2}{x+3}} = \boxed{\frac{(3x+2x+3x)+(-1+2)}{x+3}} = \boxed{\frac{(3+2+3)x+1}{x+3}} = \boxed{\mathbf{\frac{8x+1}{x+3}}}$$

Example 5.3-14

$$\boxed{\frac{x^2+2x+4}{x+2}-\frac{x^2-1}{x+2}-\frac{x-3}{x+2}} = \boxed{\frac{x^2+2x+4-\left(x^2-1\right)-(x-3)}{x+2}} = \boxed{\frac{x^2+2x+4-x^2+1-x+3}{x+2}}$$

$$= \boxed{\frac{\left(x^2-x^2\right)+(2x-x)+(4+1+3)}{x+2}} = \boxed{\frac{(1-1)x^2+(2-1)x+8}{x+2}} = \boxed{\frac{0x^2+x+8}{x+2}} = \boxed{\mathbf{\frac{x+8}{x+2}}}$$

Example 5.3-15

$$\boxed{\frac{2x^3+3x^2+5}{(x+1)(x-2)}-\frac{x^3+x^2-4}{(x+1)(x-2)}} = \boxed{\frac{2x^3+3x^2+5-\left(x^3+x^2-4\right)}{(x+1)(x-2)}} = \boxed{\frac{2x^3+3x^2+5-x^3-x^2+4}{(x+1)(x-2)}}$$

$$= \boxed{\frac{\left(2x^3 - x^3\right) + \left(3x^2 - x^2\right) + (5+4)}{(x+1)(x-2)}} = \boxed{\frac{(2-1)x^3 + (3-1)x^2 + 9}{(x+1)(x-2)}} = \boxed{\mathbf{\frac{x^3 + 2x^2 + 9}{(x+1)(x-2)}}}$$

Practice Problems - Addition and Subtraction of Algebraic Fractions with Common Denominators

Section 5.3 Case I Practice Problems - Add or subtract the following algebraic fractions. Reduce the answer to its lowest term.

1. $\frac{x}{7} + \frac{3}{7} =$
2. $\frac{8}{a+b} - \frac{7}{a+b} =$
3. $\frac{3x+1}{2y} + \frac{4x+1}{2y} + \frac{3}{2y} =$
4. $\frac{4x}{x-2} - \frac{8}{x-2} =$
5. $\frac{15a}{5a+b} - \frac{-5b}{5a+b} =$
6. $\frac{6x}{3x^2y^2} + \frac{5y}{3x^2y^2} =$
7. $\frac{x+2}{3x^2+5} - \frac{x-1}{3x^2+5} =$
8. $\frac{x^2+5x+4}{(x+5)(x-2)} - \frac{x^2+3x+1}{(x+5)(x-2)} =$
9. $\frac{y^2}{y-3} - \frac{3y}{y-3} =$
10. $\frac{a^2-2a+1}{(a-1)(a+3)} - \frac{a^2-3a+2}{(a-1)(1+3)} =$

Case II Addition and Subtraction of Algebraic Fractions without Common Denominators

Algebraic fractions without common denominators are solved using the following steps:

Step 1 Obtain a common denominator by multiplying the denominators of the first and second fractions by one another. Cross multiply the numerator of the first fraction with the denominator of the second fraction. Cross multiply the numerator of the second fraction with the denominator of the first fraction. Add or subtract the two products to each other.

Step 2 Simplify the algebraic fraction to its lowest term.

Examples with Steps

The following examples show the steps as to how algebraic fractions without common denominators are added and subtracted:

Example 5.3-16

$$\frac{2}{x-2}+\frac{4}{x+3}=$$

Solution:

Step 1

$$\frac{2}{x-2}+\frac{4}{x+3}=\frac{[2\cdot(x+3)]+[4\cdot(x-2)]}{(x-2)\cdot(x+3)}=\frac{2x+6+4x-8}{(x-2)(x+3)}$$

Step 2

$$\frac{2x+6+4x-8}{(x-2)(x+3)}=\frac{(2x+4x)+(-8+6)}{(x-2)(x+3)}=\frac{6x-2}{(x-2)(x+3)}=\boldsymbol{\frac{2(3x-1)}{(x-2)(x+3)}}$$

Example 5.3-17

$$\frac{5}{x+y}-\frac{3}{x-y}=$$

Solution:

Step 1

$$\frac{5}{x+y}-\frac{3}{x-y}=\frac{[5\cdot(x-y)]-[3\cdot(x+y)]}{(x+y)\cdot(x-y)}=\frac{5x-5y-3x-3y}{(x+y)(x-y)}$$

Step 2

$$\frac{5x-5y-3x-3y}{(x+y)(x-y)}=\frac{(5x-3x)+(-5y-3y)}{(x+y)(x-y)}=\frac{2x-8y}{(x+y)(x-y)}=\boldsymbol{\frac{2(x-4y)}{(x+y)(x-y)}}$$

Example 5.3-18

$$\frac{m+2}{m}-\frac{m}{m-1}=$$

Solution:

Step 1

$$\frac{m+2}{m}-\frac{m}{m-1}=\frac{[(m+2)\cdot(m-1)]-(m\cdot m)}{m\cdot(m-1)}=\frac{m^2-m+2m-2-m^2}{m(m-1)}$$

Step 2

$$\frac{m^2-m+2m-2-m^2}{m(m-1)}=\frac{(m^2-m^2)+(2m-m)-2}{m(m-1)}=\boldsymbol{\frac{m-2}{m(m-1)}}$$

Example 5.3-19

$$\boxed{\frac{1}{x+1}+\left(\frac{2}{x}+\frac{1}{2}\right)}=$$

Solution:

Step 1

$$\boxed{\frac{1}{x+1}+\left(\frac{2}{x}+\frac{1}{2}\right)}=\boxed{\frac{1}{x+1}+\left[\frac{(2\cdot 2)+(1\cdot x)}{2\cdot x}\right]}=\boxed{\frac{1}{x+1}+\frac{4+x}{2x}}=\boxed{\frac{(1\cdot 2x)+[(4+x)\cdot(x+1)]}{(x+1)\cdot 2x}}$$

$$=\boxed{\frac{2x+4x+4+x^2+x}{2x(x+1)}}$$

Step 2

$$\boxed{\frac{2x+4x+4+x^2+x}{2x(x+1)}}=\boxed{\frac{x^2+(4x+2x+x)+4}{2x(x+1)}}=\boxed{\frac{x^2+7x+4}{2x(x+1)}}=\boxed{\mathbf{\frac{(x+0.63)(x+6.37)}{2x(x+1)}}}$$

Example 5.3-20

$$\boxed{\frac{1}{x+1}+\frac{1}{4}+\frac{2}{x}}=$$

Solution:

Step 1

$$\boxed{\frac{1}{x+1}+\frac{1}{4}+\frac{2}{x}}=\boxed{\left(\frac{1}{x+1}+\frac{1}{4}\right)+\frac{2}{x}}=\boxed{\left\{\frac{(1\cdot 4)+[1\cdot(x+1)]}{(x+1)\cdot 4}\right\}+\frac{2}{x}}=\boxed{\left\{\frac{4+x+1}{4(x+1)}\right\}+\frac{2}{x}}$$

$$=\boxed{\frac{x+5}{4(x+1)}+\frac{2}{x}}=\boxed{\frac{[(x+5)\cdot x]+[2\cdot 4(x+1)]}{4(x+1)\cdot x}}=\boxed{\frac{x^2+5x+8x+8}{4x(x+1)}}$$

Step 2

$$\boxed{\frac{x^2+5x+8x+8}{4x(x+1)}}=\boxed{\frac{x^2+(5+8)x+8}{4x(x+1)}}=\boxed{\frac{x^2+13x+8}{4x(x+1)}}=\boxed{\mathbf{\frac{(x+0.65)(x+12.35)}{4x(x+1)}}}$$

Additional Examples - Addition and Subtraction of Algebraic Fractions without Common Denominators

The following examples further illustrate how to add or subtract algebraic fractions without common denominators:

Example 5.3-21

$$\boxed{\frac{3}{x+3}+\frac{5}{x-1}}=\boxed{\frac{[3\cdot(x-1)]+[5\cdot(x+3)]}{(x+3)\cdot(x-1)}}=\boxed{\frac{3x-3+5x+15}{(x+3)(x-1)}}=\boxed{\frac{(3x+5x)+(15-3)}{(x+3)(x-1)}}=\boxed{\frac{(3+5)x+12}{(x+3)(x-1)}}$$

$$=\boxed{\frac{8x+12}{(x+3)(x-1)}}=\boxed{\mathbf{\frac{4(2x+3)}{(x+3)(x-1)}}}$$

Example 5.3-22

$$\boxed{\frac{7}{x^2y^2}-\frac{3}{3xy^2}}=\boxed{\frac{(7\cdot 3xy^2)-(3\cdot x^2y^2)}{x^2y^2\cdot 3xy^2}}=\boxed{\frac{21xy^2-3x^2y^2}{3x^3y^4}}=\boxed{\frac{3xy^2(7-x)}{3x^3y^4}}=\boxed{\frac{\cancel{3}\cancel{x}\cancel{y^2}(7-x)}{\cancel{3}\cancel{x^3}_{x^2}\cancel{y^4}_{y^2}}}=\boxed{\mathbf{\frac{7-x}{x^2y^2}}}$$

Example 5.3-23

$$\frac{x+y}{x-y}+\frac{x}{y} = \frac{[(x+y)\cdot y]+[x\cdot(x-y)]}{y\cdot(x-y)} = \frac{xy+y^2+x^2-xy}{y(x-y)} = \frac{(\cancel{xy}-\cancel{xy})+x^2+y^2}{y(x-y)} = \mathbf{\frac{x^2+y^2}{y(x-y)}}$$

Example 5.3-24

$$\frac{3}{a^2}+\frac{2}{4a} = \frac{(3\cdot 4a)+(2\cdot a^2)}{4a\cdot a^2} = \frac{12a+2a^2}{4a^3} = \frac{2a(6+a)}{4a^3} = \frac{\cancel{2a}(6+a)}{\underset{2a^2}{\cancel{4a^3}}} = \mathbf{\frac{6+a}{2a^2}}$$

Example 5.3-25

$$\frac{x}{yz}+\frac{y}{xz}+\frac{z}{xy} = \left(\frac{x}{yz}+\frac{y}{xz}\right)+\frac{z}{xy} = \left(\frac{(x\cdot xz)+(y\cdot yz)}{yz\cdot xz}\right)+\frac{z}{xy} = \left(\frac{x^2z+y^2z}{xyz^2}\right)+\frac{z}{xy} = \frac{\cancel{z}(x^2+y^2)}{xy\underset{z}{\cancel{z^2}}}+\frac{z}{xy}$$

$$= \frac{x^2+y^2}{xyz}+\frac{z}{xy} = \frac{[xy\cdot(x^2+y^2)]+[(z\cdot xyz)]}{xyz\cdot xy} = \frac{x^3y+xy^3+xyz^2}{x^2y^2z} = \frac{\cancel{xy}(x^2+y^2+z^2)}{\underset{x}{\cancel{x^2}}\,\underset{y}{\cancel{y^2}}\,z} = \mathbf{\frac{x^2+y^2+z^2}{xyz}}$$

Example 5.3-26

$$\frac{y}{x^2-xy}-\frac{x}{xy-y^2} = \frac{[y\cdot(xy-y^2)]-[x\cdot(x^2-xy)]}{(x^2-xy)\cdot(xy-y^2)} = \frac{xy^2-y^3-x^3+x^2y}{x(x-y)\cdot y(x-y)} = \frac{(xy^2-y^3)+(-x^3+x^2y)}{x(x-y)\cdot y(x-y)}$$

$$= \frac{y^2(x-y)-x^2(x-y)}{xy(x-y)^2} = \frac{\cancel{(x-y)}(y^2-x^2)}{xy\underset{(x-y)}{\cancel{(x-y)^2}}} = \frac{y^2-x^2}{xy(x-y)} = \frac{(y-x)(y+x)}{xy(x-y)} = \frac{-\cancel{(x-y)}(y+x)}{xy\cancel{(x-y)}} = \mathbf{-\frac{x+y}{xy}}$$

Example 5.3-27

$$\frac{3a-1}{2a}-\frac{2a-3}{3a} = \frac{[(3a-1)\cdot 3a]-[(2a-3)\cdot 2a]}{2a\cdot 3a} = \frac{9a^2-3a-4a^2+6a}{6a^2} = \frac{(9a^2-4a^2)+(6a-3a)}{6a^2}$$

$$= \frac{5a^2+3a}{6a^2} = \frac{\cancel{a}(5a+3)}{6\underset{a}{\cancel{a^2}}} = \mathbf{\frac{5a+3}{6a}}$$

Example 5.3-28

$$\frac{x-1}{x+2}-\frac{x-2}{x+1} = \frac{[(x-1)\cdot(x+1)]-[(x-2)\cdot(x+2)]}{(x+2)\cdot(x+1)} = \frac{(x^2+\cancel{x}-\cancel{x}-1)-(x^2-\cancel{2x}+\cancel{2x}-4)}{(x+2)(x+1)} = \frac{x^2-1-x^2+4}{(x+2)(x+1)}$$

$$= \frac{(x^2-x^2)+(-1+4)}{(x+2)(x+1)} = \frac{(x^2-x^2)+(-1+4)}{(x+2)(x+1)} = \mathbf{\frac{3}{(x+2)(x+1)}}$$

Example 5.3-29

$$\frac{3}{x}+\frac{2x}{x+y}-4 = \frac{3}{x}+\frac{2x}{x+y}-\frac{4}{1} = \left(\frac{3}{x}+\frac{2x}{x+y}\right)-\frac{4}{1} = \left(\frac{[3\cdot(x+y)]+(2x\cdot x)}{x\cdot(x+y)}\right)-\frac{4}{1} = \left(\frac{3x+3y+2x^2}{x(x+y)}\right)-\frac{4}{1}$$

$$= \boxed{\frac{\left[\left(3x+3y+2x^2\right)\cdot 1\right]-\left[4\cdot x(x+y)\right]}{x(x+y)\cdot 1}} = \boxed{\frac{3x+3y+2x^2-4x(x+y)}{x(x+y)}} = \boxed{\frac{3x+3y+2x^2-4x^2-4xy}{x(x+y)}}$$

$$= \boxed{\frac{\left(2x^2-4x^2\right)+3x+3y-4xy}{x(x+y)}} = \boxed{\frac{\mathbf{-2x^2+3x+3y-4xy}}{\mathbf{x(x+y)}}}$$

Example 5.3-30

$$\boxed{\frac{m}{m-n}-\frac{n}{m+n}} = \boxed{\frac{[m\cdot(m+n)]-[n\cdot(m-n)]}{(m-n)\cdot(m+n)}} = \boxed{\frac{m^2+mn-mn+n^2}{m^2+mn-mn-n^2}} = \boxed{\frac{m^2+\cancel{mn}-\cancel{mn}+n^2}{m^2+\cancel{mn}-\cancel{mn}-n^2}} = \boxed{\frac{\mathbf{m^2+n^2}}{\mathbf{m^2-n^2}}}$$

Practice Problems - Addition and Subtraction of Algebraic Fractions without Common Denominators

Section 5.3 Case II Practice Problems - Add or subtract the following algebraic fractions. Reduce the answer to its lowest term.

1. $\frac{3}{4x^2}+\frac{5}{2x^3}=$
2. $\frac{x}{x+4}-\frac{2}{x-1}=$
3. $\frac{a-b}{a+b}-\frac{a}{b}=$
4. $\frac{x^2}{x+3}+\frac{5x}{x-5}=$
5. $\frac{1}{4x^2y^2z}-\frac{2}{xy^2z}=$
6. $\frac{3}{x+1}+\frac{2}{x-1}-\frac{5}{x}=$
7. $\frac{a}{2a+10}-\frac{3a}{4a+5}=$
8. $5x-\frac{20}{x}=$
9. $\frac{2}{x}-\frac{1}{x+2}-\frac{5}{x-1}=$
10. $\frac{a}{b^2c}-\frac{c}{a^2b}-\frac{b}{ac^2}=$

Case III Multiplication of Algebraic Fractions

To multiply algebraic fractions by one another we first consider simple cases of algebraic fractions where the numerator and the denominator are mostly monomials (Case IIIa). We then consider more difficult algebraic expressions where the terms in the numerator and/or the denominator are mostly polynomials and need to be factored first before reducing the algebraic fraction to lower terms (Case IIIb).

Case IIIa Multiplication of Algebraic Fractions (Simple Cases)

Algebraic fractions are multiplied by one another using the following steps:

Step 1 Write the algebraic expression in fraction form, i.e., write x or $u^2v^2w^3$ as $\frac{x}{1}$ and $\frac{u^2v^2w^3}{1}$, respectively.

Step 2 Multiply the numerator and the denominator of the algebraic fraction terms by one another. Simplify the product to its lowest term.

Examples with Steps

The following examples show the steps as to algebraic fractions are multiplied by one another:

Example 5.3-31

$$\frac{x^2y}{x}\cdot\frac{y}{x^3} =$$

Solution:

Step 1 *Not Applicable*

Step 2

$$\frac{x^2y}{x}\cdot\frac{y}{x^3} = \frac{x^2y\cdot y}{x\cdot x^3} = \frac{x^2y^2}{x^4} = \frac{x^2y^2}{\substack{x^4\\x^2}} = \mathbf{\frac{y^2}{x^2}}$$

Example 5.3-32

$$\frac{a^2b^2c}{a}\cdot\frac{2a}{bc^2} =$$

Solution:

Step 1 *Not Applicable*

Step 2

$$\frac{a^2b^2c}{a}\cdot\frac{2a}{bc^2} = \frac{a^2b^2c\cdot 2a}{a\cdot bc^2} = \frac{2a^3b^2c}{abc^2} = \frac{\substack{a^2\ b\\2\cancel{a^3}\,\cancel{b^2}\,\cancel{c}}}{\substack{\cancel{a}\cancel{b}c^2\\c}} = \mathbf{\frac{2a^2b}{c}}$$

Example 5.3-33

$$u^2v^2w\cdot\frac{1}{uv^2}\cdot\frac{w}{v^2w^3} =$$

Solution:

Step 1 $\boxed{u^2v^2w\cdot\frac{1}{uv^2}\cdot\frac{w}{v^2w^3}} = \boxed{\frac{u^2v^2w}{1}\cdot\frac{1}{uv^2}\cdot\frac{w}{v^2w^3}}$

Step 2 $\boxed{\frac{u^2v^2w}{1}\cdot\frac{1}{uv^2}\cdot\frac{w}{v^2w^3}} = \boxed{\frac{u^2v^2w\cdot 1\cdot w}{1\cdot uv^2\cdot v^2w^3}} = \boxed{\frac{u^2v^2w^2}{uv^4w^3}} = \boxed{\frac{\overset{u}{\cancel{u^2}}\,\cancel{v^2}\,\cancel{w^2}}{\cancel{u}\,\underset{v^2}{\cancel{v^4}}\,\underset{w}{\cancel{w^3}}}} = \boxed{\frac{u}{v^2w}}$

Example 5.3-34

$\boxed{(x-3)\cdot\frac{x}{x^2-9}\cdot\frac{(x+3)}{2x}\cdot 8x^2} =$

Solution:

Step 1 $\boxed{(x-3)\cdot\frac{x}{x^2-9}\cdot\frac{(x+3)}{2x}\cdot 8x^2} = \boxed{\frac{(x-3)}{1}\cdot\frac{x}{x^2-9}\cdot\frac{(x+3)}{2x}\cdot\frac{8x^2}{1}}$

Step 2 $\boxed{\frac{(x-3)}{1}\cdot\frac{x}{x^2-9}\cdot\frac{(x+3)}{2x}\cdot\frac{8x^2}{1}} = \boxed{\frac{(x-3)\cdot x\cdot(x+3)\cdot 8x^2}{1\cdot(x^2-9)\cdot 2x}} = \boxed{\frac{8x^3(x-3)(x+3)}{2x(x-3)(x+3)}}$

$= \boxed{\frac{8x^3\cancel{(x-3)}\cancel{(x+3)}}{2x\cancel{(x-3)}\cancel{(x+3)}}} = \boxed{\frac{8x^3}{2x}} = \boxed{\frac{\overset{4x^2}{\cancel{8x^3}}}{\cancel{2x}}} = \boxed{\frac{4x^2}{1}} = \boxed{4x^2}$

Example 5.3-35

$\boxed{\frac{ab^2}{2}\cdot\left(\frac{1}{a^2b}\cdot\frac{4a^3}{b}\right)} =$

Solution:

Step 1 *Not Applicable*

Step 2 $\boxed{\frac{ab^2}{2}\cdot\left(\frac{1}{a^2b}\cdot\frac{4a^3}{b}\right)} = \boxed{\frac{ab^2}{2}\cdot\left(\frac{1\cdot 4a^3}{a^2b\cdot b}\right)} = \boxed{\frac{ab^2}{2}\cdot\frac{4a^3}{a^2b^2}} = \boxed{\frac{ab^2\cdot 4a^3}{2\cdot a^2b^2}} = \boxed{\frac{4a^4b^2}{2a^2b^2}}$

$= \boxed{\frac{\overset{2a^2}{\cancel{4a^4}}\,\cancel{b^2}}{\cancel{2a^2b^2}}} = \boxed{\frac{2a^2}{1}} = \boxed{2a^2}$

Additional Examples - Multiplication of Algebraic Fractions (Simple Cases)

The following examples further illustrate how to multiply algebraic fractions:

Example 5.3-36

$\boxed{\frac{x^2y^2}{x}\cdot\frac{1}{x^2y}} = \boxed{\frac{x^2y^2\cdot 1}{x\cdot x^2y}} = \boxed{\frac{x^2y^2}{x^3y}} = \boxed{\frac{\cancel{x^2}\,\overset{y}{\cancel{y^2}}}{\underset{x}{\cancel{x^3}}\,\cancel{y}}} = \boxed{\frac{y}{x}}$

Example 5.3-37

$$ab^2c\cdot\frac{a}{b^2c^2}=\frac{ab^2c}{1}\cdot\frac{a}{b^2c^2}=\frac{ab^2c\cdot a}{1\cdot b^2c^2}=\frac{a^2b^2c}{b^2c^2}=\frac{a^2\cancel{b^2}\cancel{c}}{\cancel{b^2}\,\cancel{c^2}_{\,c}}=\boxed{\frac{a^2}{c}}$$

Example 5.3-38

$$\frac{x+1}{2}\cdot\frac{x^2-4}{(x+1)(x-2)}=\frac{x+1}{2}\cdot\frac{(x+2)(x-2)}{(x+1)(x-2)}=\frac{(x+1)\cdot[(x-2)(x+2)]}{2\cdot[(x+1)(x-2)]}=\frac{\cancel{(x+1)}\cdot[\cancel{(x-2)}(x+2)]}{2\cdot[\cancel{(x+1)}\cancel{(x-2)}]}=\boxed{\frac{x+2}{2}}$$

Example 5.3-39

$$\frac{2}{x^2y^2z^2}\cdot\frac{x^2z^3}{4yz}=\frac{2\cdot x^2z^3}{x^2y^2z^2\cdot 4yz}=\frac{2x^2z^3}{4x^2y^3z^3}=\frac{\cancel{2}\cancel{x^2}\cancel{z^3}}{\cancel{4}_{\,2}\cancel{x^2}y^3\cancel{z^3}}=\boxed{\frac{1}{2y^3}}$$

Example 5.3-40

$$\frac{a^2b^3c^3}{abc^2}\cdot\frac{ab^2}{c^2}=\frac{a^2b^3c^3\cdot ab^2}{abc^2\cdot c^2}=\frac{a^3b^5c^3}{abc^4}=\frac{\overset{a^2\,b^4}{\cancel{a^3}\,\cancel{b^5}\,\cancel{c^3}}}{\cancel{a}\cancel{b}\cancel{c^4}_{\,c}}=\boxed{\frac{a^2b^4}{c}}$$

Example 5.3-41

$$\frac{1}{a^2}\cdot\frac{a}{2}\cdot\frac{3a^2}{5}=\frac{1\cdot a\cdot 3a^2}{a^2\cdot 2\cdot 5}=\frac{3a^3}{10a^2}=\frac{3\overset{a}{\cancel{a^3}}}{10\cancel{a^2}}=\boxed{\frac{3a}{10}}$$

Example 5.3-42

$$\frac{uv^2}{w^3}\cdot\frac{w^2}{u^2v^2}\cdot\frac{1}{u}=\frac{uv^2\cdot w^2\cdot 1}{w^3\cdot u^2v^2\cdot u}=\frac{uv^2w^2}{u^3v^2w^3}=\frac{\cancel{u}\cancel{v^2}\cancel{w^2}}{\cancel{u^3}_{\,u^2}\cancel{v^2}\cancel{w^3}_{\,w}}=\boxed{\frac{1}{u^2w}}$$

Example 5.3-43

$$xyz^2\cdot\left(\frac{1}{x^2y^2z^3}\cdot\frac{xy}{z}\right)=\frac{xyz^2}{1}\cdot\left(\frac{1}{x^2y^2z^3}\cdot\frac{xy}{z}\right)=\frac{xyz^2}{1}\cdot\left(\frac{1\cdot xy}{x^2y^2z^3\cdot z}\right)=\frac{xyz^2}{1}\cdot\left(\frac{xy}{x^2y^2z^4}\right)=\frac{xyz^2}{1}\cdot\frac{xy}{x^2y^2z^4}$$

$$=\frac{xyz^2\cdot xy}{1\cdot x^2y^2z^4}=\frac{x^2y^2z^2}{x^2y^2z^4}=\frac{\cancel{x^2}\cancel{y^2}\cancel{z^2}}{\cancel{x^2}\cancel{y^2}\cancel{z^4}_{\,z^2}}=\boxed{\frac{1}{z^2}}$$

Example 5.3-44

$$\left(\frac{ab^2}{c^2}\cdot\frac{1}{b^3}\right)\cdot\left(\frac{ab}{b^2}\cdot\frac{c^3}{a^2}\right)=\left(\frac{ab^2\cdot 1}{c^2\cdot b^3}\right)\cdot\left(\frac{ab\cdot c^3}{b^2\cdot a^2}\right)=\left(\frac{ab^2}{b^3c^2}\right)\cdot\left(\frac{abc^3}{a^2b^2}\right)=\frac{ab^2}{b^3c^2}\cdot\frac{abc^3}{a^2b^2}=\frac{ab^2\cdot abc^3}{b^3c^2\cdot a^2b^2}$$

$$=\frac{a^2b^3c^3}{a^2b^5c^2}=\frac{\cancel{a^2}\cancel{b^3}\overset{c}{\cancel{c^3}}}{\cancel{a^2}\cancel{b^5}_{\,b^2}\cancel{c^2}}=\boxed{\frac{c}{b^2}}$$

Example 5.3-45

$$\boxed{(1+x)\cdot\left(\frac{x^3}{x+1}\cdot\frac{x+2}{x^2}\right)} = \boxed{\frac{(1+x)}{1}\cdot\left(\frac{x^3}{x+1}\cdot\frac{x+2}{x^2}\right)} = \boxed{\frac{(1+x)}{1}\cdot\left[\frac{x^3\cdot(x+2)}{(x+1)\cdot x^2}\right]} = \boxed{\frac{(1+x)\cdot x^3\cdot(x+2)}{1\cdot(x+1)\cdot x^2}}$$

$$= \boxed{\frac{x^3(1+x)(x+2)}{x^2(1+x)}} = \boxed{\frac{\overset{x}{\cancel{x^3}}\cancel{(1+x)}(x+2)}{\cancel{x^2}\cancel{(1+x)}}} = \boxed{\frac{x(x+2)}{1}} = \boxed{\mathbf{x(x+2)}}$$

Practice Problems - Multiplication of Algebraic Fractions (Simple Cases)

Section 5.3 Case IIIa Practice Problems - Multiply the following algebraic fractions by one another. Simplify the answer to its lowest term.

1. $\frac{1}{xy}\cdot\frac{x^2y^2}{2} =$
2. $\frac{2a^2}{a^3}\cdot\frac{1}{8a} =$
3. $xyz\cdot\frac{1}{x^2y^2z^2}\cdot\frac{x^2}{y} =$
4. $\frac{5u^2v^2}{uv}\cdot\frac{uv^3}{15v^2}\cdot\frac{1}{u^4} =$
5. $8x\cdot\frac{2}{x^3}\cdot\frac{1}{4x^2} =$
6. $\frac{x-4}{x+2}\cdot\frac{x^2-4}{2}\cdot\frac{1}{x-4} =$
7. $\frac{5a^2b^2c}{ac}\cdot\frac{bc^2}{a^3}\cdot a^2 =$
8. $\frac{5xyz}{10}\cdot\frac{2x}{y^2z^2} =$
9. $\left(\frac{4w}{w^3}\cdot w^2\right)\cdot\frac{1}{w} =$
10. $\frac{x^2y^2z^2}{x^3}\cdot\left(\frac{y^2}{z^3}\cdot\frac{2z}{x}\right) =$

Case IIIb Multiplication of Algebraic Fractions (More Difficult Cases)

Algebraic fractions are multiplied by one another using the following steps:

Step 1 Use a factoring method to factor the numerators and the denominators completely (see Chapters 3 and 4).

Step 2 Simplify the algebraic expression by canceling the common terms in both the numerator and the denominator. Multiply the numerators and the denominators by one another.

Examples with Steps

The following examples show the steps as to how algebraic fractions are multiplied by one another:

Example 5.3-46

$$\frac{x^2-1}{x^3}\cdot\frac{x^3+2x^2}{x^2+3x+2}=$$

Solution:

Step 1

$$\frac{x^2-1}{x^3}\cdot\frac{x^3+2x^2}{x^2+3x+2}=\frac{(x-1)(x+1)}{x^3}\cdot\frac{x^2(x+2)}{(x+2)(x+1)}$$

Step 2

$$\frac{(x-1)(x+1)}{x^3}\cdot\frac{x^2(x+2)}{(x+2)(x+1)}=\frac{(x-1)\cancel{(x+1)}}{\cancel{x^3}\,x}\cdot\frac{\cancel{x^2}\cancel{(x+2)}}{\cancel{(x+2)}\cancel{(x+1)}}=\frac{x-1}{x}\cdot\frac{1}{1}=\frac{(x-1)\cdot 1}{x\cdot 1}$$

$$=\boxed{\frac{x-1}{x}}$$

Example 5.3-47

$$\frac{4x^2+16x+15}{x+3}\cdot\frac{x^2-9}{2x^2-x-15}=$$

Solution:

Step 1

$$\frac{4x^2+16x+15}{x+3}\cdot\frac{x^2-9}{2x^2-x-15}=\frac{(2x+3)(2x+5)}{x+3}\cdot\frac{(x-3)(x+3)}{(x-3)(2x+5)}$$

Step 2

$$\frac{(2x+3)(2x+5)}{x+3}\cdot\frac{(x-3)(x+3)}{(x-3)(2x+5)}=\frac{(2x+3)\cancel{(2x+5)}}{\cancel{x+3}}\cdot\frac{\cancel{(x-3)}\cancel{(x+3)}}{\cancel{(x-3)}\cancel{(2x+5)}}=\frac{2x+3}{1}\cdot\frac{1}{1}$$

$$=\frac{(2x+3)\cdot 1}{1\cdot 1}=\frac{2x+3}{1}=\boxed{2x+3}$$

Example 5.3-48

$$\frac{x^2-y^2}{3x^2-3xy}\cdot\frac{9x+18y}{x^2+3xy+2y^2}=$$

Solution:

Step 1 $\boxed{\dfrac{x^2-y^2}{3x^2-3xy}\cdot\dfrac{9x+18y}{x^2+3xy+2y^2}} = \boxed{\dfrac{(x+y)(x-y)}{3x(x-y)}\cdot\dfrac{9(x+2y)}{(x+y)(x+2y)}}$

Step 2 $\boxed{\dfrac{(x+y)(x-y)}{3x(x-y)}\cdot\dfrac{9(x+2y)}{(x+y)(x+2y)}} = \boxed{\dfrac{(\cancel{x+y})(\cancel{x-y})}{3x(\cancel{x-y})}\cdot\dfrac{9(\cancel{x+2y})}{(\cancel{x+y})(\cancel{x+2y})}} = \boxed{\dfrac{1}{3x}\cdot\dfrac{9}{1}} = \boxed{\dfrac{1\cdot 9}{3x\cdot 1}}$

$= \boxed{\dfrac{\overset{3}{\cancel{9}}}{\cancel{3}x}} = \boxed{\dfrac{\mathbf{3}}{\boldsymbol{x}}}$

Example 5.3-49

$\boxed{\dfrac{x^2-x-2}{x^2+x-6}\cdot\dfrac{5x^2}{x^2+x}} =$

Solution:

Step 1 $\boxed{\dfrac{x^2-x-2}{x^2+x-6}\cdot\dfrac{5x^2}{x^2+x}} = \boxed{\dfrac{(x-2)(x+1)}{(x-2)(x+3)}\cdot\dfrac{5x^2}{x(x+1)}}$

Step 2 $\boxed{\dfrac{(x-2)(x+1)}{(x-2)(x+3)}\cdot\dfrac{5x^2}{x(x+1)}} = \boxed{\dfrac{(\cancel{x-2})(\cancel{x+1})}{(\cancel{x-2})(x+3)}\cdot\dfrac{5x^2}{x(\cancel{x+1})}} = \boxed{\dfrac{1}{x+3}\cdot\dfrac{5\overset{x}{\cancel{x^2}}}{\cancel{x}}} = \boxed{\dfrac{1}{x+3}\cdot\dfrac{5x}{1}}$

$= \boxed{\dfrac{1\cdot 5x}{(x+3)\cdot 1}} = \boxed{\dfrac{\mathbf{5x}}{\mathbf{x+3}}}$

Example 5.3-50

$\boxed{\dfrac{x-3}{2x^2+3x+1}\cdot\dfrac{x^2+4x+3}{x}\cdot\dfrac{1}{x+3}} =$

Solution:

Step 1 $\boxed{\dfrac{x-3}{2x^2+3x+1}\cdot\dfrac{x^2+4x+3}{x}\cdot\dfrac{1}{x+3}} = \boxed{\dfrac{x-3}{(x+1)(2x+1)}\cdot\dfrac{(x+3)(x+1)}{x}\cdot\dfrac{1}{x+3}}$

Step 2 $\boxed{\dfrac{x-3}{(x+1)(2x+1)}\cdot\dfrac{(x+3)(x+1)}{x}\cdot\dfrac{1}{x+3}} = \boxed{\dfrac{x-3}{(\cancel{x+1})(2x+1)}\cdot\dfrac{(\cancel{x+3})(\cancel{x+1})}{x}\cdot\dfrac{1}{\cancel{x+3}}}$

$= \boxed{\dfrac{x-3}{2x+1}\cdot\dfrac{1}{x}\cdot\dfrac{1}{1}} = \boxed{\dfrac{(x-3)\cdot 1\cdot 1}{(2x+1)\cdot x\cdot 1}} = \boxed{\dfrac{\mathbf{x-3}}{\mathbf{x(2x+1)}}}$

Additional Examples - Multiplication of Algebraic Fractions (More Difficult Cases)

The following examples further illustrate how to multiply algebraic fractions:

Example 5.3-51

$\boxed{\dfrac{x^2+x-6}{x^2-4}\cdot\dfrac{x^2-x-6}{5x-15}} = \boxed{\dfrac{(x+3)(x-2)}{(x-2)(x+2)}\cdot\dfrac{(x+2)(x-3)}{5(x-3)}} = \boxed{\dfrac{(x+3)(\cancel{x-2})}{(\cancel{x-2})(\cancel{x+2})}\cdot\dfrac{(\cancel{x+2})(\cancel{x-3})}{5(\cancel{x-3})}} = \boxed{\dfrac{x+3}{1}\cdot\dfrac{1}{5}} = \boxed{\dfrac{\mathbf{x+3}}{\mathbf{5}}}$

Example 5.3-52

$$\boxed{\frac{x}{2x-4}\cdot\frac{3x-6}{x^2}}=\boxed{\frac{x}{2(x-2)}\cdot\frac{3(x-2)}{x^2}}=\boxed{\frac{\cancel{x}}{2\cancel{(x-2)}}\cdot\frac{3\cancel{(x-2)}}{\underset{x}{\cancel{x^2}}}}=\boxed{\frac{1}{2}\cdot\frac{3}{x}}=\boxed{\frac{1\cdot 3}{2\cdot x}}=\boxed{\mathbf{\frac{3}{2x}}}$$

Example 5.3-53

$$\boxed{\frac{x^2-y^2}{3x^2+xy-4y^2}\cdot\frac{3x^2+10xy+8y^2}{2x+2y}\cdot\frac{1}{x+2y}}=\boxed{\frac{(x-y)(x+y)}{(3x+4y)(x-y)}\cdot\frac{(3x+4y)(x+2y)}{2(x+y)}\cdot\frac{1}{x+2y}}$$

$$=\boxed{\frac{\cancel{(x-y)}\cancel{(x+y)}}{\cancel{(3x+4y)}\cancel{(x-y)}}\cdot\frac{\cancel{(3x+4y)}(x+2y)}{2\cancel{(x+y)}}\cdot\frac{1}{x+2y}}=\boxed{\frac{1}{1}\cdot\frac{\cancel{x+2y}}{2}\cdot\frac{1}{\cancel{x+2y}}}=\boxed{\frac{1\cdot 1\cdot 1}{1\cdot 2\cdot 1}}=\boxed{\mathbf{\frac{1}{2}}}$$

Example 5.3-54

$$\boxed{\frac{(x-2)^2}{(x-3)^2}\cdot\frac{x^2-2x-3}{x+1}\cdot\frac{2}{x}}=\boxed{\frac{(x-2)\cdot(x-2)}{(x-3)\cdot(x-3)}\cdot\frac{(x-3)\cdot(x+1)}{x+1}\cdot\frac{2}{x}}=\boxed{\frac{(x-2)\cdot(x-2)}{(x-3)\cdot\cancel{(x-3)}}\cdot\frac{\cancel{(x-3)}\cdot\cancel{(x+1)}}{\cancel{x+1}}\cdot\frac{2}{x}}$$

$$=\boxed{\frac{(x-2)(x-2)}{(x-3)\cdot 1}\cdot\frac{1}{1}\cdot\frac{2}{x}}=\boxed{\frac{(x-2)(x-2)\cdot 1\cdot 2}{(x-3)\cdot 1\cdot x}}=\boxed{\mathbf{\frac{2(x-2)(x-2)}{x(x-3)}}}$$

Example 5.3-55

$$\boxed{\frac{x^2-x-6}{x-3}\cdot\frac{9x-3}{6x^2+x-1}}=\boxed{\frac{(x-3)(x+2)}{x-3}\cdot\frac{3(3x-1)}{(2x+1)(3x-1)}}=\boxed{\frac{\cancel{(x-3)}(x+2)}{\cancel{x-3}}\cdot\frac{3\cancel{(3x-1)}}{(2x+1)\cancel{(3x-1)}}}=\boxed{\frac{x+2}{1}\cdot\frac{3}{2x+1}}$$

$$=\boxed{\frac{(x+2)\cdot 3}{1\cdot(2x+1)}}=\boxed{\mathbf{\frac{3(x+2)}{2x+1}}}$$

Example 5.3-56

$$\boxed{\frac{a^2+4a-5}{a^2+a-2}\cdot\frac{a^2+5a+6}{a^2+8a+15}}=\boxed{\frac{(a+5)(a-1)}{(a-1)(a+2)}\cdot\frac{(a+2)(a+3)}{(a+5)(a+3)}}=\boxed{\frac{\cancel{(a+5)}\cancel{(a-1)}}{\cancel{(a-1)}\cancel{(a+2)}}\cdot\frac{\cancel{(a+2)}\cancel{(a+3)}}{\cancel{(a+5)}\cancel{(a+3)}}}=\boxed{\frac{1}{1}\cdot\frac{1}{1}}=\boxed{\frac{1}{1}}=\boxed{\mathbf{1}}$$

Example 5.3-57

$$\boxed{\frac{4x^2-4y^2}{x^2+4xy+3y^2}\cdot\frac{3x^2+9xy}{2x^2-2xy}}=\boxed{\frac{4\left(x^2-y^2\right)}{(x+y)(x+3y)}\cdot\frac{3x(x+3y)}{2x(x-y)}}=\boxed{\frac{4(x-y)(x+y)}{(x+y)(x+3y)}\cdot\frac{3x(x+3y)}{2x(x-y)}}$$

$$=\boxed{\frac{4\cancel{(x-y)}\cancel{(x+y)}}{\cancel{(x+y)}\cancel{(x+3y)}}\cdot\frac{3\cancel{x}\cancel{(x+3y)}}{2\cancel{x}\cancel{(x-y)}}}=\boxed{\frac{4}{1}\cdot\frac{3}{2}}=\boxed{\frac{4\cdot 3}{1\cdot 2}}=\boxed{\frac{\overset{6}{\cancel{12}}}{\cancel{2}}}=\boxed{\frac{6}{1}}=\boxed{\mathbf{6}}$$

Example 5.3-58

$$\boxed{\frac{x^2+2x-15}{x^2+x-2}\cdot\frac{x^2+5x+6}{4x+2}\cdot\frac{2x^2-x-1}{x^2-9}}=\boxed{\frac{(x+5)(x-3)}{(x+2)(x-1)}\cdot\frac{(x+3)(x+2)}{2(2x+1)}\cdot\frac{(x-1)(2x+1)}{(x+3)(x-3)}}$$

$$=\boxed{\frac{(x+5)\cancel{(x-3)}}{\cancel{(x+2)}\cancel{(x-1)}}\cdot\frac{\cancel{(x+3)}\cancel{(x+2)}}{2\cancel{(2x+1)}}\cdot\frac{\cancel{(x-1)}\cancel{(2x+1)}}{\cancel{(x+3)}\cancel{(x-3)}}}=\boxed{\frac{x+5}{1}\cdot\frac{1}{2}\cdot\frac{1}{1}}=\boxed{\frac{(x+5)\cdot 1\cdot 1}{1\cdot 2\cdot 1}}=\boxed{\mathbf{\frac{x+5}{2}}}$$

Example 5.3-59

$$\boxed{\frac{3x^2+14xy+8y^2}{2x^2-xy-y^2}\cdot\frac{2x+y}{3x^2+5xy+2y^2}\cdot\frac{x^2-y^2}{2x+8y}} = \boxed{\frac{(3x+2y)(x+4y)}{(2x+y)(x-y)}\cdot\frac{2x+y}{(3x+2y)(x+y)}\cdot\frac{(x+y)(x-y)}{2(x+4y)}}$$

$$= \boxed{\frac{\cancel{(3x+2y)}\cancel{(x+4y)}}{\cancel{(2x+y)}\cancel{(x-y)}}\cdot\frac{\cancel{2x+y}}{\cancel{(3x+2y)}\cancel{(x+y)}}\cdot\frac{\cancel{(x+y)}\cancel{(x-y)}}{2\cancel{(x+4y)}}} = \boxed{\frac{1}{1}\cdot\frac{1}{1}\cdot\frac{1}{2}} = \boxed{\frac{1\cdot1\cdot1}{1\cdot1\cdot2}} = \boxed{\mathbf{\frac{1}{2}}}$$

Example 5.3-60

$$\boxed{\frac{2x^2+x-3}{x^2-1}\cdot\frac{3x+3}{2x^3+3x^2}} = \boxed{\frac{(2x+3)(x-1)}{(x-1)(x+1)}\cdot\frac{3(x+1)}{x^2(2x+3)}} = \boxed{\frac{\cancel{(2x+3)}\cancel{(x-1)}}{\cancel{(x-1)}\cancel{(x+1)}}\cdot\frac{3\cancel{(x+1)}}{x^2\cancel{(2x+3)}}} = \boxed{\frac{1}{1}\cdot\frac{3}{x^2}} = \boxed{\frac{1\cdot3}{1\cdot x^2}}$$

$$= \boxed{\mathbf{\frac{3}{x^2}}}$$

Practice Problems - Multiplication of Algebraic Fractions (More Difficult Cases)

Section 5.3 Case IIIb Practice Problems - Multiply the following algebraic fractions by one another. Simplify the answer to its lowest term.

1. $\frac{x^2+5x-6}{x^2-9}\cdot\frac{x^2-2x-3}{x^2+7x+6} =$

2. $\frac{x^3}{x^2-x-2}\cdot\frac{2x-4}{x} =$

3. $\frac{x^2+xy-6y^2}{x^2-xy-2y^2}\cdot\frac{x^2-y^2}{x+3y} =$

4. $\frac{(x-1)^3}{2x^2+5x-12}\cdot\frac{4x-6}{(x-1)^2} =$

5. $\frac{6x^2+7x+2}{x^2-16}\cdot\frac{x^2+3x-4}{3x^2-x-2} =$

6. $\frac{2x^2+x-3}{x^3+x^2-2x}\cdot\frac{x^2-2x}{4x^2-9} =$

7. $\frac{x^2-2x-15}{2x^2-9x-5}\cdot\frac{2x^2+5x+2}{x^2+5x+6} =$

8. $\frac{3x+2}{6x^2+7x+2}\cdot\frac{2x^2+x}{4} =$

9. $\frac{x^4}{x^2+8x+15}\cdot\frac{x+5}{x^3} =$

10. $\frac{6x^2+17x+5}{3x^2-2x-1}\cdot\frac{x^2-1}{x^3+x^2} =$

Case IV Division of Algebraic Fractions

To divide algebraic fractions by one another we first consider simple cases of algebraic factions where the numerator and the denominator are mostly monomials (Case IVa). We then consider more difficult algebraic expressions where the terms in the numerator and/or the denominator are mostly polynomials and need to be factored first before reducing the algebraic fraction to lower terms (Case IVb).

Case IVa Division of Algebraic Fractions (Simple Cases)

Algebraic fractions are divided by one another using the following steps:

Step 1 Write the algebraic expression in fraction form, i.e., write x or $u^2v^2w^3$ as $\frac{x}{1}$ and $\frac{u^2v^2w^3}{1}$, respectively.

Step 2 Invert the second fraction and change the division sign to a multiplication sign.

Step 3 Multiply the numerator and the denominator of the algebraic fraction terms by one another. Simplify the product to its lowest term, if possible.

Examples with Steps

The following examples show the steps as to algebraic fractions are divided by one another:

Example 5.3-61

$$\frac{1}{x^3y^3z} \div \frac{3z}{x^2y^2} =$$

Solution:

Step 1 *Not Applicable*

Step 2

$$\frac{1}{x^3y^3z} \div \frac{3z}{x^2y^2} = \frac{1}{x^3y^3z} \cdot \frac{x^2y^2}{3z}$$

Step 3

$$\frac{1}{x^3y^3z} \cdot \frac{x^2y^2}{3z} = \frac{1 \cdot x^2y^2}{x^3y^3z \cdot 3z} = \frac{x^2y^2}{3x^3y^3z^2} = \frac{x^2y^2}{3\underset{x}{x^3}\,\underset{y}{y^3}\,z^2} = \mathbf{\frac{1}{3xyz^2}}$$

Example 5.3-62

$$a^2b^2 \div \frac{ab}{3ab^3} =$$

Solution:

Step 1

$$a^2b^2 \div \frac{ab}{3ab^3} = \frac{a^2b^2}{1} \div \frac{ab}{3ab^3}$$

Step 2

$$\frac{a^2b^2}{1} \div \frac{ab}{3ab^3} = \frac{a^2b^2}{1} \cdot \frac{3ab^3}{ab}$$

Step 3 $\boxed{\dfrac{a^2b^2}{1}\cdot\dfrac{3ab^3}{ab}} = \boxed{\dfrac{a^2b^2\cdot 3ab^3}{1\cdot ab}} = \boxed{\dfrac{3a^3b^5}{ab}} = \boxed{\dfrac{\overset{a^2\ b^4}{3\cancel{a^3}\ \cancel{b^5}}}{\cancel{ab}}} = \boxed{\dfrac{3a^2b^4}{1}} = \boxed{\mathbf{3a^2b^4}}$

Example 5.3-63

$\boxed{\dfrac{x^2y^2z}{xyz^2}\div\dfrac{xy^2}{z^3}} =$

Solution:

Step 1 Not Applicable

Step 2 $\boxed{\dfrac{x^2y^2z}{xyz^2}\div\dfrac{xy^2}{z^3}} = \boxed{\dfrac{x^2y^2z}{xyz^2}\cdot\dfrac{z^3}{xy^2}}$

Step 3 $\boxed{\dfrac{x^2y^2z}{xyz^2}\cdot\dfrac{z^3}{xy^2}} = \boxed{\dfrac{x^2y^2z\cdot z^3}{xyz^2\cdot xy^2}} = \boxed{\dfrac{x^2y^2z^4}{x^2y^3z^2}} = \boxed{\dfrac{\cancel{x^2}\ \cancel{y^2}\ \overset{z^2}{\cancel{z^4}}}{\cancel{x^2}\ \underset{y}{\cancel{y^3}}\ \cancel{z^2}}} = \boxed{\dfrac{z^2}{y}}$

Example 5.3-64

$\boxed{\left(\dfrac{1}{xy}\div\dfrac{3x}{y^2}\right)\div\dfrac{y^2}{x^2}} =$

Solution:

Step 1 Not Applicable

Step 2 $\boxed{\left(\dfrac{1}{xy}\div\dfrac{3x}{y^2}\right)\div\dfrac{y^2}{x^2}} = \boxed{\left(\dfrac{1}{xy}\cdot\dfrac{y^2}{3x}\right)\div\dfrac{y^2}{x^2}}$

Step 3 $\boxed{\left(\dfrac{1}{xy}\cdot\dfrac{y^2}{3x}\right)\div\dfrac{y^2}{x^2}} = \boxed{\left(\dfrac{1\cdot y^2}{xy\cdot 3x}\right)\div\dfrac{y^2}{x^2}} = \boxed{\dfrac{y^2}{3x^2y}\div\dfrac{y^2}{x^2}} = \boxed{\dfrac{y^2}{3x^2y}\cdot\dfrac{x^2}{y^2}} = \boxed{\dfrac{y^2\cdot x^2}{3x^2y\cdot y^2}}$

$= \boxed{\dfrac{x^2y^2}{3x^2y^3}} = \boxed{\dfrac{\cancel{x^2}\ \cancel{y^2}}{3\cancel{x^2}\ \underset{y}{\cancel{y^3}}}} = \boxed{\dfrac{\mathbf{1}}{\mathbf{3y}}}$

Example 5.3-65

$\boxed{\dfrac{3x}{x^3}\div\left(\dfrac{1}{x^2}\div 2x^3\right)} =$

Solution:

Step 1 $\boxed{\dfrac{3x}{x^3}\div\left(\dfrac{1}{x^2}\div 2x^3\right)} = \boxed{\dfrac{3x}{x^3}\div\left(\dfrac{1}{x^2}\div\dfrac{2x^3}{1}\right)}$

Step 2 $\boxed{\dfrac{3x}{x^3}\div\left(\dfrac{1}{x^2}\div\dfrac{2x^3}{1}\right)} = \boxed{\dfrac{3x}{x^3}\div\left(\dfrac{1}{x^2}\cdot\dfrac{1}{2x^3}\right)} = \boxed{\dfrac{3x}{x^3}\div\left(\dfrac{1\cdot 1}{x^2\cdot 2x^3}\right)} = \boxed{\dfrac{3x}{x^3}\div\dfrac{1}{2x^5}} = \boxed{\dfrac{3x}{x^3}\cdot\dfrac{2x^5}{1}}$

Step 3
$$\frac{3x}{x^3}\cdot\frac{2x^5}{1} = \frac{3x\cdot 2x^5}{x^3\cdot 1} = \frac{6x^6}{x^3} = \frac{\overset{x^3}{6x^{6}}}{x^{3}} = \frac{6x^3}{1} = \boxed{6x^3}$$

Additional Examples - Division of Algebraic Fractions (Simple Cases)

The following examples further illustrate how to divide algebraic fractions by one another:

Example 5.3-66
$$\frac{x^2y^2z^3}{xy^2}\div\frac{x^2y^2}{yz} = \frac{x^2y^2z^3}{xy^2}\cdot\frac{yz}{x^2y^2} = \frac{x^2y^2z^3\cdot yz}{xy^2\cdot x^2y^2} = \frac{x^2y^3z^4}{x^3y^4} = \frac{x^2y^3z^4}{\underset{x\quad y}{x^3y^4}} = \boxed{\frac{z^4}{xy}}$$

Example 5.3-67
$$\frac{a^2b^2}{bc}\div ab^2 = \frac{a^2b^2}{bc}\div\frac{ab^2}{1} = \frac{a^2b^2}{bc}\cdot\frac{1}{ab^2} = \frac{a^2b^2\cdot 1}{bc\cdot ab^2} = \frac{a^2b^2}{ab^3c} = \frac{\overset{a}{a^2b^2}}{\underset{b}{ab^3c}} = \boxed{\frac{a}{bc}}$$

Example 5.3-68
$$\frac{3}{u^2v^2}\div\frac{9u}{u^3v} = \frac{3}{u^2v^2}\cdot\frac{u^3v}{9u} = \frac{3\cdot u^3v}{u^2v^2\cdot 9u} = \frac{3u^3v}{9u^3v^2} = \frac{3u^3v}{\underset{3\quad\quad v}{9u^3v^2}} = \boxed{\frac{1}{3v}}$$

Example 5.3-69
$$xyz^3\div\frac{x^2y^2z}{2xy} = \frac{xyz^3}{1}\div\frac{x^2y^2z}{2xy} = \frac{xyz^3}{1}\cdot\frac{2xy}{x^2y^2z} = \frac{xyz^3\cdot 2xy}{1\cdot x^2y^2z} = \frac{2x^2y^2z^3}{x^2y^2z} = \frac{\overset{z^2}{2x^2y^2z^3}}{x^2y^2z} = \frac{2z^2}{1} = \boxed{2z^2}$$

Example 5.3-70
$$\left(\frac{xyz}{x^2}\div\frac{x^2y^2}{z^3}\right)\div z^4 = \left(\frac{xyz}{x^2}\cdot\frac{z^3}{x^2y^2}\right)\div z^4 = \left(\frac{xyz\cdot z^3}{x^2\cdot x^2y^2}\right)\div z^4 = \frac{xyz^4}{x^4y^2}\div z^4 = \frac{xyz^4}{x^4y^2}\div\frac{z^4}{1} = \frac{xyz^4}{x^4y^2}\cdot\frac{1}{z^4}$$
$$= \frac{xyz^4\cdot 1}{x^4y^2\cdot z^4} = \frac{xyz^4}{x^4y^2z^4} = \frac{xyz^4}{\underset{x^3\quad y}{x^4y^2z^4}} = \boxed{\frac{1}{x^3y}}$$

Example 5.3-71
$$\frac{u^2v}{v^3}\div\left(u^3\div\frac{u^2}{2v}\right) = \frac{u^2v}{v^3}\div\left(\frac{u^3}{1}\div\frac{u^2}{2v}\right) = \frac{u^2v}{v^3}\div\left(\frac{u^3}{1}\cdot\frac{2v}{u^2}\right) = \frac{u^2v}{v^3}\div\left(\frac{u^3\cdot 2v}{1\cdot u^2}\right) = \frac{u^2v}{v^3}\div\frac{2u^3v}{u^2} = \frac{u^2v}{v^3}\cdot\frac{u^2}{2u^3v}$$
$$= \frac{u^2v\cdot u^2}{v^3\cdot 2u^3v} = \frac{u^4v}{2u^3v^4} = \frac{\overset{u}{u^4v}}{\underset{v^3}{2u^3v^4}} = \boxed{\frac{u}{2v^3}}$$

Example 5.3-72
$$\left(\frac{x^2y^2}{x}\div\frac{xy}{z^2}\right)\div\frac{z^4}{x^2y} = \left(\frac{x^2y^2}{x}\cdot\frac{z^2}{xy}\right)\div\frac{z^4}{x^2y} = \left(\frac{x^2y^2\cdot z^2}{x\cdot xy}\right)\div\frac{z^4}{x^2y} = \frac{x^2y^2z^2}{x^2y}\div\frac{z^4}{x^2y} = \frac{x^2y^2z^2}{x^2y}\cdot\frac{x^2y}{z^4}$$

$$= \boxed{\frac{x^2y^2z^2 \cdot x^2y}{x^2y \cdot z^4}} = \boxed{\frac{x^4y^3z^2}{x^2yz^4}} = \boxed{\frac{\overset{x^2\,y^2}{\cancel{x^4}\,\cancel{y^3}\,\cancel{z^2}}}{\underset{z^2}{\cancel{x^2}\,\cancel{y}\,\cancel{z^4}}}} = \boxed{\mathbf{\frac{x^2y^2}{z^2}}}$$

Example 5.3-73

$$\boxed{\frac{a^3b^2c}{ab^2} \div \frac{a^2b}{c^3}} = \boxed{\frac{a^3b^2c}{ab^2} \cdot \frac{c^3}{a^2b}} = \boxed{\frac{a^3b^2c \cdot c^3}{ab^2 \cdot a^2b}} = \boxed{\frac{a^3b^2c^4}{a^3b^3}} = \boxed{\frac{\cancel{a^3}\,\cancel{b^2}\,c^4}{\underset{b}{\cancel{a^3}\,\cancel{b^3}}}} = \boxed{\mathbf{\frac{c^4}{b}}}$$

Example 5.3-74

$$\boxed{\left(\frac{x^2y}{xy^2} \div \frac{x}{y^2}\right) \div \left(xy \div \frac{x^2y}{xy^3}\right)} = \boxed{\left(\frac{x^2y}{xy^2} \div \frac{x}{y^2}\right) \div \left(\frac{xy}{1} \div \frac{x^2y}{xy^3}\right)} = \boxed{\left(\frac{x^2y}{xy^2} \cdot \frac{y^2}{x}\right) \div \left(\frac{xy}{1} \cdot \frac{xy^3}{x^2y}\right)}$$

$$= \boxed{\left(\frac{x^2y \cdot y^2}{xy^2 \cdot x}\right) \div \left(\frac{xy \cdot xy^3}{1 \cdot x^2y}\right)} = \boxed{\left(\frac{x^2y^3}{x^2y^2}\right) \div \left(\frac{x^2y^4}{x^2y}\right)} = \boxed{\frac{x^2y^3}{x^2y^2} \div \frac{x^2y^4}{x^2y}} = \boxed{\frac{x^2y^3}{x^2y^2} \cdot \frac{x^2y}{x^2y^4}} = \boxed{\frac{x^2y^3 \cdot x^2y}{x^2y^2 \cdot x^2y^4}}$$

$$= \boxed{\frac{x^4y^4}{x^4y^6}} = \boxed{\frac{\cancel{x^4}\,\cancel{y^4}}{\underset{y^2}{\cancel{x^4}\,\cancel{y^6}}}} = \boxed{\mathbf{\frac{1}{y^2}}}$$

Example 5.3-75

$$\boxed{\left(2a^2b \div \frac{1}{ab^2}\right) \div 3a^2b^2} = \boxed{\left(\frac{2a^2b}{1} \div \frac{1}{ab^2}\right) \div \frac{3a^2b^2}{1}} = \boxed{\left(\frac{2a^2b}{1} \cdot \frac{ab^2}{1}\right) \div \frac{3a^2b^2}{1}} = \boxed{\left(\frac{2a^2b \cdot ab^2}{1 \cdot 1}\right) \div \frac{3a^2b^2}{1}}$$

$$= \boxed{\frac{2a^3b^3}{1} \div \frac{3a^2b^2}{1}} = \boxed{\frac{2a^3b^3}{1} \cdot \frac{1}{3a^2b^2}} = \boxed{\frac{2a^3b^3 \cdot 1}{1 \cdot 3a^2b^2}} = \boxed{\frac{2a^3b^3}{3a^2b^2}} = \boxed{\frac{\overset{a\ \ b}{2\cancel{a^3}\,\cancel{b^3}}}{3\cancel{a^2}\,\cancel{b^2}}} = \boxed{\mathbf{\frac{2ab}{3}}}$$

Practice Problems - Division of Algebraic Fractions (Simple Cases)

Section 5.3 Case IVa Practice Problems - Divide the following algebraic fractions. Simplify the answer to its lowest term.

1. $\frac{x^2y}{x^3y^2} \div xy =$
2. $\frac{uv^2w}{vw^2} \div \frac{uv^3}{w} =$
3. $a^2b^2c^4 \div \frac{a^2b}{2ac} =$
4. $\frac{xyz}{x^2z^3} \div \frac{x^2z^2}{yz} =$
5. $\left(\frac{uv^2}{v^3} \div 2u^2\right) \div \frac{uv}{3} =$
6. $\frac{x^2y^2z}{xz} \div \left(x^2y \div \frac{4}{yz^3}\right) =$
7. $a^3b^3 \div \left(\frac{a^2b^2}{b} \div b^3\right) =$
8. $\frac{4xyz}{x^2y^2z^3} \div \frac{8y}{y^2z^3} =$
9. $\left(\frac{3xy}{x^3y^2} \div \frac{2x}{y^2}\right) \div \left(\frac{x}{y^2} \div \frac{x^3}{4y}\right) =$
10. $\left(\frac{m^3n^2}{n^3} \div 3n^2\right) \div m^4 =$

Case IVb Division of Algebraic Fractions (More Difficult Cases)

Algebraic fractions are divided by one another using the following steps:

Step 1 Invert the second fraction and change the division sign to a multiplication sign.

Step 2 Use a factoring method to factor the numerator and the denominator completely (see Chapters 3 and 4).

Step 3 Simplify the algebraic expression by canceling the common terms in both the numerator and the denominator. Multiply the numerators and the denominators by one another.

Examples with Steps

The following examples show the steps as to how algebraic fractions are divided by one another:

Example 5.3-76

$$\frac{x^2+x-2}{x^2-4} \div \frac{x^2-1}{x^2+x-6} =$$

Solution:

Step 1

$$\frac{x^2+x-2}{x^2-4} \div \frac{x^2-1}{x^2+x-6} = \frac{x^2+x-2}{x^2-4} \cdot \frac{x^2+x-6}{x^2-1}$$

Step 2

$$\frac{x^2+x-2}{x^2-4} \cdot \frac{x^2+x-6}{x^2-1} = \frac{(x-1)(x+2)}{(x-2)(x+2)} \cdot \frac{(x-2)(x+3)}{(x-1)(x+1)}$$

Step 3

$$\frac{(x-1)(x+2)}{(x-2)(x+2)} \cdot \frac{(x-2)(x+3)}{(x-1)(x+1)} = \frac{\cancel{(x-1)}\cancel{(x+2)}}{\cancel{(x-2)}\cancel{(x+2)}} \cdot \frac{\cancel{(x-2)}(x+3)}{\cancel{(x-1)}(x+1)} = \frac{1}{1} \cdot \frac{x+3}{x+1} = \frac{1\cdot(x+3)}{1\cdot(x+1)}$$

$$= \boxed{\frac{x+3}{x+1}}$$

Example 5.3-77

$$\frac{6x^2+11x+3}{x-2} \div \frac{3x+1}{x-2} =$$

Solution:

Step 1

$$\frac{6x^2+11x+3}{x-2} \div \frac{3x+1}{x-2} = \frac{6x^2+11x+3}{x-2} \cdot \frac{x-2}{3x+1}$$

Step 2

$$\frac{6x^2+11x+3}{x-2} \cdot \frac{x-2}{3x+1} = \frac{(3x+1)(2x+3)}{x-2} \cdot \frac{x-2}{3x+1}$$

Step 3

$$\frac{(3x+1)(2x+3)}{x-2} \cdot \frac{x-2}{3x+1} = \frac{\cancel{(3x+1)}(2x+3)}{\cancel{(x-2)}} \cdot \frac{\cancel{(x-2)}}{\cancel{(3x+1)}} = \frac{(2x+3)}{1} \cdot \frac{1}{1} = \frac{(2x+3)\cdot 1}{1\cdot 1}$$

$$= \frac{2x+3}{1} = \boxed{2x+3}$$

Example 5.3-78

$$\boxed{x^3+x^2-6x\div\frac{x^2-4}{x^3+2x^2}}=$$

Solution:

Step 1

$$\boxed{x^3+x^2-6x\div\frac{x^2-4}{x^3+2x^2}}=\boxed{\frac{x^3+x^2-6x}{1}\div\frac{x^2-4}{x^3+2x^2}}=\boxed{\frac{x^3+x^2-6x}{1}\cdot\frac{x^3+2x^2}{x^2-4}}$$

Step 2

$$\boxed{\frac{x^3+x^2-6x}{1}\cdot\frac{x^3+2x^2}{x^2-4}}=\boxed{\frac{x\left(x^2+x-6\right)}{1}\cdot\frac{x^2(x+2)}{(x-2)(x+2)}}=\boxed{\frac{x(x-2)(x+3)}{1}\cdot\frac{x^2(x+2)}{(x-2)(x+2)}}$$

Step 3

$$\boxed{\frac{x(x-2)(x+3)}{1}\cdot\frac{x^2(x+2)}{(x-2)(x+2)}}=\boxed{\frac{x(x-2)(x+3)}{1}\cdot\frac{x^2(x+2)}{(x-2)(x+2)}}=\boxed{\frac{x(x+3)}{1}\cdot\frac{x^2}{1}}$$

$$=\boxed{\frac{x(x+3)\cdot x^2}{1\cdot 1}}=\boxed{\frac{x^3(x+3)}{1}}=\boxed{\mathbf{x^3(x+3)}}$$

Example 5.3-79

$$\boxed{\frac{2x^2-3x-2}{2x^2+7x+3}\div\frac{x^2+3x-10}{x^2-25}}=$$

Solution:

Step 1

$$\boxed{\frac{2x^2-3x-2}{2x^2+7x+3}\div\frac{x^2+3x-10}{x^2-25}}=\boxed{\frac{2x^2-3x-2}{2x^2+7x+3}\cdot\frac{x^2-25}{x^2+3x-10}}$$

Step 2

$$\boxed{\frac{2x^2-3x-2}{2x^2+7x+3}\cdot\frac{x^2-25}{x^2+3x-10}}=\boxed{\frac{(x-2)(2x+1)}{(2x+1)(x+3)}\cdot\frac{(x-5)(x+5)}{(x-2)(x+5)}}$$

Step 3

$$\boxed{\frac{(x-2)(2x+1)}{(2x+1)(x+3)}\cdot\frac{(x-5)(x+5)}{(x-2)(x+5)}}=\boxed{\frac{(x-2)(2x+1)}{(2x+1)(x+3)}\cdot\frac{(x-5)(x+5)}{(x-2)(x+5)}}=\boxed{\frac{1}{(x+3)}\cdot\frac{(x-5)}{1}}$$

$$=\boxed{\frac{1\cdot(x-5)}{(x+3)\cdot 1}}=\boxed{\mathbf{\frac{x-5}{x+3}}}$$

Example 5.3-80

$$\boxed{\frac{2x^2+x-3}{x^2+5x+6}\div\frac{x^2-4x+3}{x^2-9}}=$$

Solution:

Step 1

$$\boxed{\frac{2x^2+x-3}{x^2+5x+6}\div\frac{x^2-4x+3}{x^2-9}}=\boxed{\frac{2x^2+x-3}{x^2+5x+6}\cdot\frac{x^2-9}{x^2-4x+3}}$$

Step 2

$$\boxed{\frac{2x^2+x-3}{x^2+5x+6}\cdot\frac{x^2-9}{x^2-4x+3}}=\boxed{\frac{(2x+3)(x-1)}{(x+3)(x+2)}\cdot\frac{(x-3)(x+3)}{(x-3)(x-1)}}$$

Step 3

$$\frac{(2x+3)(x-1)}{(x+3)(x+2)}\cdot\frac{(x-3)(x+3)}{(x-3)(x-1)}=\frac{(2x+3)\cancel{(x-1)}}{\cancel{(x+3)}(x+2)}\cdot\frac{\cancel{(x-3)}\cancel{(x+3)}}{\cancel{(x-3)}\cancel{(x-1)}}=\frac{(2x+3)}{(x+2)}\cdot\frac{1}{1}$$

$$=\frac{(2x+3)\cdot 1}{(x+2)\cdot 1}=\mathbf{\frac{2x+3}{x+2}}$$

Additional Examples - Division of Algebraic Fractions (More Difficult Cases)

The following examples further illustrate how to divide algebraic fractions by one another:

Example 5.3-81

$$\frac{x^2+5x+6}{x^2+5x+4}\div\frac{x+3}{x^2+x}=\frac{x^2+5x+6}{x^2+5x+4}\cdot\frac{x^2+x}{x+3}=\frac{(x+2)(x+3)}{(x+1)(x+4)}\cdot\frac{x(x+1)}{(x+3)}=\frac{(x+2)\cancel{(x+3)}}{\cancel{(x+1)}(x+4)}\cdot\frac{x\cancel{(x+1)}}{\cancel{(x+3)}}$$

$$=\frac{(x+2)}{(x+4)}\cdot\frac{x}{1}=\frac{(x+2)\cdot x}{(x+4)\cdot 1}=\mathbf{\frac{x(x+2)}{x+4}}$$

Example 5.3-82

$$\frac{2x+10}{x^3}\div\frac{x^2-25}{x^2}=\frac{2x+10}{x^3}\cdot\frac{x^2}{x^2-25}=\frac{2(x+5)}{x^3}\cdot\frac{x^2}{(x-5)(x+5)}=\frac{2\cancel{(x+5)}}{\underset{x}{\cancel{x^3}}}\cdot\frac{\cancel{x^2}}{(x-5)\cancel{(x+5)}}=\frac{2}{x}\cdot\frac{1}{(x-5)}$$

$$=\frac{2\cdot 1}{x\cdot(x-5)}=\mathbf{\frac{2}{x(x-5)}}$$

Example 5.3-83

$$\frac{6x^2+5x+1}{2x^2+7x+3}\div 3x^2-2x-1=\frac{6x^2+5x+1}{2x^2+7x+3}\div\frac{3x^2-2x-1}{1}=\frac{6x^2+5x+1}{2x^2+7x+3}\cdot\frac{1}{3x^2-2x-1}$$

$$=\frac{(3x+1)(2x+1)}{(2x+1)(x+3)}\cdot\frac{1}{(3x+1)(x-1)}=\frac{\cancel{(3x+1)}\cancel{(2x+1)}}{\cancel{(2x+1)}(x+3)}\cdot\frac{1}{\cancel{(3x+1)}(x-1)}=\frac{1}{(x+3)}\cdot\frac{1}{(x-1)}=\frac{1\cdot 1}{(x+3)\cdot(x-1)}$$

$$=\mathbf{\frac{1}{(x+3)(x-1)}}$$

Example 5.3-84

$$\frac{x^2-y^2}{5x-5y}\div\frac{3x^2+3xy}{x-y}=\frac{x^2-y^2}{5x-5y}\cdot\frac{x-y}{3x^2+3xy}=\frac{(x-y)(x+y)}{5(x-y)}\cdot\frac{(x-y)}{3x(x+y)}=\frac{(x-y)\cancel{(x+y)}}{5\cancel{(x-y)}}\cdot\frac{\cancel{(x-y)}}{3x\cancel{(x+y)}}$$

$$=\frac{(x-y)}{5}\cdot\frac{1}{3x}=\frac{(x-y)\cdot 1}{5\cdot 3x}=\mathbf{\frac{x-y}{15x}}$$

Example 5.3-85

$$\frac{y-2}{y+2}\div\frac{y^2-y-2}{2y^2+5y+2}=\frac{y-2}{y+2}\cdot\frac{2y^2+5y+2}{y^2-y-2}=\frac{y-2}{y+2}\cdot\frac{(y+2)(2y+1)}{(y-2)(y+1)}=\frac{\cancel{(y-2)}}{\cancel{(y+2)}}\cdot\frac{\cancel{(y+2)}(2y+1)}{\cancel{(y-2)}(y+1)}$$

$$= \boxed{\frac{1}{1}\cdot\frac{(2y+1)}{(y+1)}} = \boxed{\frac{1\cdot(2y+1)}{1\cdot(y+1)}} = \boxed{\frac{\mathbf{2y+1}}{\mathbf{y+1}}}$$

Example 5.3-86

$$\boxed{\frac{a^2b^3c(a+b)}{ab^2c^3(a-b)} \div \frac{a^2-ab-2b^2}{a^2+2ab-3b^2}} = \boxed{\frac{a^2b^3c(a+b)}{ab^2c^3(a-b)}\cdot\frac{a^2+2ab-3b^2}{a^2-ab-2b^2}} = \boxed{\frac{a^2b^3c(a+b)}{ab^2c^3(a-b)}\cdot\frac{(a-b)(a+3b)}{(a+b)(a-2b)}}$$

$$= \boxed{\frac{\overset{a\ \ b}{a^2 b^3} \not{c}(\not{a}+\not{b})}{\not{a}b^2 \underset{c^2}{c^3}(\not{a}-\not{b})}\cdot\frac{(\not{a}-\not{b})(a+3b)}{(\not{a}+\not{b})(a-2b)}} = \boxed{\frac{ab}{c^2}\cdot\frac{(a+3b)}{(a-2b)}} = \boxed{\frac{ab\cdot(a+3b)}{c^2\cdot(a-2b)}} = \boxed{\frac{\mathbf{ab(a+3b)}}{\mathbf{c^2(a-2b)}}}$$

Example 5.3-87

$$\boxed{\frac{x^2-4}{x^2+5x+6} \div \frac{x+2}{x^2-4}} = \boxed{\frac{x^2-4}{x^2+5x+6}\cdot\frac{x^2-4}{x+2}} = \boxed{\frac{(x-2)(x+2)}{(x+3)(x+2)}\cdot\frac{(x-2)(x+2)}{x+2}} = \boxed{\frac{(x-2)(\not{x}+\not{2})}{(x+3)(\not{x}+\not{2})}\cdot\frac{(x-2)(\not{x}+\not{2})}{(\not{x}+\not{2})}}$$

$$= \boxed{\frac{(x-2)}{(x+3)}\cdot\frac{(x-2)}{1}} = \boxed{\frac{(x-2)\cdot(x-2)}{(x+3)\cdot 1}} = \boxed{\frac{(x-2)(x-2)}{x+3}} = \boxed{\frac{\mathbf{(x-2)^2}}{\mathbf{x+3}}}$$

Example 5.3-88

$$\boxed{2-x \div \frac{x^2-x-2}{x^2+4x+3}} = \boxed{\frac{2-x}{1} \div \frac{x^2-x-2}{x^2+4x+3}} = \boxed{\frac{2-x}{1}\cdot\frac{x^2+4x+3}{x^2-x-2}} = \boxed{\frac{2-x}{1}\cdot\frac{(x+3)(x+1)}{(x+1)(x-2)}}$$

$$= \boxed{\frac{-(x-2)}{1}\cdot\frac{(x+3)(x+1)}{(x+1)(x-2)}} = \boxed{-\frac{(\not{x}-\not{2})}{1}\cdot\frac{(x+3)(\not{x}+\not{1})}{(\not{x}+\not{1})(\not{x}-\not{2})}} = \boxed{-\frac{1}{1}\cdot\frac{(x+3)}{1}} = \boxed{-\frac{1\cdot(x+3)}{1\cdot 1}} = \boxed{-\frac{(x+3)}{1}} = \boxed{\mathbf{-x-3}}$$

Example 5.3-89

$$\boxed{\frac{2x^2-x-3}{6x^2+x-1} \div \frac{x^2-1}{9x-3}} = \boxed{\frac{2x^2-x-3}{6x^2+x-1}\cdot\frac{9x-3}{x^2-1}} = \boxed{\frac{(2x-3)(x+1)}{(3x-1)(2x+1)}\cdot\frac{3(3x-1)}{(x-1)(x+1)}} = \boxed{\frac{(2x-3)(\not{x}+\not{1})}{(\not{3}\not{x}-\not{1})(2x+1)}\cdot\frac{3(\not{3}\not{x}-\not{1})}{(x-1)(\not{x}+\not{1})}}$$

$$= \boxed{\frac{(2x-3)}{(2x+1)}\cdot\frac{3}{(x-1)}} = \boxed{\frac{(2x-3)\cdot 3}{(2x+1)\cdot(x-1)}} = \boxed{\frac{\mathbf{3(2x-3)}}{\mathbf{(2x+1)(x-1)}}}$$

Example 5.3-90

$$\boxed{\frac{m^2-1}{6m+3} \div \frac{m^3+m^2}{4m^2+10m+4}} = \boxed{\frac{m^2-1}{6m+3}\cdot\frac{4m^2+10m+4}{m^3+m^2}} = \boxed{\frac{(m-1)(m+1)}{3(2m+1)}\cdot\frac{2(2m^2+5m+2)}{m^2(m+1)}}$$

$$= \boxed{\frac{(m-1)(m+1)}{3(2m+1)}\cdot\frac{2(2m+1)(m+2)}{m^2(m+1)}} = \boxed{\frac{(m-1)(\not{m}+\not{1})}{3(2\not{m}+\not{1})}\cdot\frac{2(2\not{m}+\not{1})(m+2)}{m^2(\not{m}+\not{1})}} = \boxed{\frac{(m-1)}{3}\cdot\frac{2(m+2)}{m^2}}$$

$$= \boxed{\frac{(m-1)\cdot 2(m+2)}{3\cdot m^2}} = \boxed{\frac{\mathbf{2(m-1)(m+2)}}{\mathbf{3m^2}}}$$

Practice Problems - Division of Algebraic Fractions (More Difficult Cases)

Section 5.3 Case IVb Practice Problems - Divide the following algebraic fractions by one another. Simplify the answer to its lowest term.

1. $\dfrac{x^2-2x-15}{x^2-25} \div \dfrac{x^2+3x}{x^2+6x+5} =$

2. $x^3+4x^2-5x \div \dfrac{x^3+5x^2}{x^3} =$

3. $\dfrac{6x^2+7x+2}{3x^2-x-2} \div \dfrac{10x+5}{x^2-2x+1} =$

4. $\dfrac{x^2+x-2}{x^2-x-6} \div \dfrac{x^2+4x-5}{x^2-6x+9} =$

5. $\dfrac{2x^2+7x+3}{3x^4-12x^3} \div \dfrac{2x+1}{x^3-4x^2} =$

6. $\dfrac{32x^3y^3z}{8\left(x^2+5x+6\right)} \div \dfrac{xyz^2}{4x^2(x+3)} =$

7. $\dfrac{2x^2+7x+3}{x^2-1} \div \dfrac{2x^2-3x-2}{x^2-3x+2} =$

8. $\dfrac{x^2+5xy+6y^2}{x^2+5xy+4y^2} \div \dfrac{x^4+3x^3y}{x^2+4xy} =$

9. $\dfrac{6x^2+11x+3}{2x^2+5x+3} \div \dfrac{3x^2-4x+1}{(x-1)^2} =$

10. $\dfrac{a^3b^2c(a-b)^2}{c^2(a+2b)} \div \dfrac{a^2b(a-b)}{c\left(a^2+3ab+2b^2\right)} =$

Case V Mixed Operations Involving Algebraic Fractions

Algebraic fractions are added, subtracted, multiplied, and divided using the following steps:

Step 1 Add, subtract, multiply, and divide the algebraic fractions.

Step 2 Perform additional math operations to reduce the expression to its lowest term.

Examples with Steps

The following examples show the steps as to how math operations involving algebraic fractions are performed:

Example 5.3-91

$$\boxed{\left(\frac{x}{y}\cdot\frac{1}{x^2}\right)\div\frac{x}{y}=}$$

Solution:

Step 1

$$\boxed{\left(\frac{x}{y}\cdot\frac{1}{x^2}\right)\div\frac{x}{y}}=\boxed{\left(\frac{x\cdot 1}{y\cdot x^2}\right)\div\frac{x}{y}}=\boxed{\frac{x}{x^2y}\div\frac{x}{y}}$$

Step 2

$$\boxed{\frac{x}{x^2y}\div\frac{x}{y}}=\boxed{\frac{x}{x^2y}\cdot\frac{y}{x}}=\boxed{\frac{x\cdot y}{x^2y\cdot x}}=\boxed{\frac{xy}{x^3y}}=\boxed{\frac{\cancel{x}\cancel{y}}{\underset{x^2}{\cancel{x^3}}\,\cancel{y}}}=\boxed{\mathbf{\frac{1}{x^2}}}$$

Example 5.3-92

$$\boxed{\left(\frac{1}{b}+1\right)-\left(\frac{1}{2b}+4\right)=}$$

Solution:

Step 1

$$\boxed{\left(\frac{1}{b}+1\right)-\left(\frac{1}{2b}+4\right)}=\boxed{\left(\frac{1}{b}+\frac{1}{1}\right)-\left(\frac{1}{2b}+\frac{4}{1}\right)}=\boxed{\left(\frac{(1\cdot 1)+(1\cdot b)}{b\cdot 1}\right)-\left(\frac{(1\cdot 1)+(4\cdot 2b)}{2b\cdot 1}\right)}$$

$$=\boxed{\left(\frac{1+b}{b}\right)-\left(\frac{1+8b}{2b}\right)}=\boxed{\frac{1+b}{b}-\frac{1+8b}{2b}}$$

Step 2

$$\boxed{\frac{1+b}{b}-\frac{1+8b}{2b}}=\boxed{\frac{[(1+b)\cdot 2b]-[(1+8b)\cdot b]}{b\cdot 2b}}=\boxed{\frac{2b+2b^2-b-8b^2}{2b^2}}$$

$$=\boxed{\frac{(2b-b)+\left(2b^2-8b^2\right)}{2b^2}}=\boxed{\frac{b-6b^2}{2b^2}}=\boxed{\frac{b(1-6b)}{2b^2}}=\boxed{\frac{\cancel{b}(1-6b)}{2\underset{b}{\cancel{b^2}}}}=\boxed{\mathbf{\frac{1-6b}{2b}}}$$

Example 5.3-93

$$\boxed{\left(\frac{uvw}{w^2}\cdot uv\right)+\frac{u}{w^2}=}$$

Solution:

Step 1

$$\boxed{\left(\frac{uvw}{w^2}\cdot uv\right)+\frac{u}{w^2}}=\boxed{\left(\frac{uvw}{w^2}\cdot\frac{uv}{1}\right)+\frac{u}{w^2}}=\boxed{\left(\frac{uvw\cdot uv}{w^2\cdot 1}\right)+\frac{u}{w^2}}=\boxed{\frac{u^2v^2w}{w^2}+\frac{u}{w^2}}$$

Step 2
$$\frac{u^2v^2w}{w^2}+\frac{u}{w^2}=\frac{u^2v^2w+u}{w^2}=\boxed{\frac{u\left(uv^2w+1\right)}{w^2}}$$

Example 5.3-94
$$\left(x-\frac{1}{x}\right)\div x=$$

Solution:

Step 1
$$\left(x-\frac{1}{x}\right)\div x=\left(\frac{x}{1}-\frac{1}{x}\right)\div\frac{x}{1}=\left(\frac{(x\cdot x)-(1\cdot 1)}{1\cdot x}\right)\div\frac{x}{1}=\frac{x^2-1}{x}\div\frac{x}{1}$$

Step 2
$$\frac{x^2-1}{x}\div\frac{x}{1}=\frac{x^2-1}{x}\cdot\frac{1}{x}=\frac{\left(x^2-1\right)\cdot 1}{x\cdot x}=\frac{x^2-1}{x^2}=\boxed{\frac{(x-1)(x+1)}{x^2}}$$

Example 5.3-95
$$\left(y-\frac{4}{y}\right)+\left(y+\frac{2}{y}\right)=$$

Solution:

Step 1
$$\left(y-\frac{4}{y}\right)+\left(y+\frac{2}{y}\right)=\left(\frac{y}{1}-\frac{4}{y}\right)+\left(\frac{y}{1}+\frac{2}{y}\right)=\left(\frac{(y\cdot y)-(4\cdot 1)}{1\cdot y}\right)+\left(\frac{(y\cdot y)+(2\cdot 1)}{1\cdot y}\right)$$
$$=\left(\frac{y^2-4}{y}\right)+\left(\frac{y^2+2}{y}\right)=\frac{y^2-4}{y}+\frac{y^2+2}{y}$$

Step 2
$$\frac{y^2-4}{y}+\frac{y^2+2}{y}=\frac{\left(y^2-4\right)+\left(y^2+2\right)}{y}=\frac{y^2-4+y^2+2}{y}=\frac{2y^2-2}{y}$$
$$=\frac{2\left(y^2-1\right)}{y}=\boxed{\frac{2(y-1)(y+1)}{y}}$$

Additional Examples - Mixed Operations Involving Algebraic Fractions

The following examples further illustrate how to solve math operations involving algebraic fractions:

Example 5.3-96
$$\left(\frac{1}{x}+\frac{1}{y}\right)\cdot\frac{xy}{x^2-y^2}=\left(\frac{(1\cdot y)+(1\cdot x)}{x\cdot y}\right)\cdot\frac{xy}{x^2-y^2}=\left(\frac{y+x}{x\cdot y}\right)\cdot\frac{xy}{x^2-y^2}=\frac{(y+x)\cdot xy}{xy\cdot\left(x^2-y^2\right)}=\frac{(y+x)\cdot xy}{xy\cdot(x-y)(y+x)}$$
$$=\frac{\cancel{(y+x)}\cancel{xy}}{\cancel{xy}(x-y)\cancel{(y+x)}}=\boxed{\frac{1}{x-y}}$$

Example 5.3-97
$$\left(\frac{4}{b}-\frac{2}{b}\right)+\left(\frac{5}{a}-\frac{2}{a}\right)=\left(\frac{4-2}{b}\right)+\left(\frac{5-2}{a}\right)=\frac{2}{b}+\frac{3}{a}=\frac{(2\cdot a)+(3\cdot b)}{b\cdot a}=\boxed{\frac{2a+3b}{ab}}$$

Example 5.3-98

$$\boxed{\left(\frac{x}{y}+\frac{y}{x}\right)-\frac{1}{x}} = \boxed{\left(\frac{(x\cdot x)+(y\cdot y)}{y\cdot x}\right)-\frac{1}{x}} = \boxed{\left(\frac{x^2+y^2}{xy}\right)-\frac{1}{x}} = \boxed{\frac{\left[\left(x^2+y^2\right)\cdot x\right]-(1\cdot xy)}{xy\cdot x}} = \boxed{\frac{x\left(x^2+y^2\right)-xy}{x^2y}}$$

$$= \boxed{\frac{\cancel{x}\left[\left(x^2+y^2\right)-y\right]}{\underset{x}{\cancel{x^2}}\,y}} = \boxed{\mathbf{\frac{x^2+y^2-y}{xy}}}$$

Example 5.3-99

$$\boxed{\left(\frac{2}{a}\div\frac{8}{a^2}\right)\cdot\left(\frac{a}{8}\div a^2\right)} = \boxed{\left(\frac{2}{a}\div\frac{8}{a^2}\right)\cdot\left(\frac{a}{8}\div\frac{a^2}{1}\right)} = \boxed{\left(\frac{2}{a}\cdot\frac{a^2}{8}\right)\cdot\left(\frac{a}{8}\cdot\frac{1}{a^2}\right)} = \boxed{\left(\frac{2\cdot a^2}{a\cdot 8}\right)\cdot\left(\frac{a\cdot 1}{8\cdot a^2}\right)} = \boxed{\frac{2a^2}{8a}\cdot\frac{a}{8a^2}}$$

$$= \boxed{\frac{2a^2\cdot a}{8a\cdot 8a^2}} = \boxed{\frac{2a^3}{64a^3}} = \boxed{\frac{\cancel{2a^3}}{\underset{32}{\cancel{64}}\cancel{a^3}}} = \boxed{\mathbf{\frac{1}{32}}}$$

Example 5.3-100

$$\boxed{\left(\frac{xyz}{x^2y}\div\frac{xy}{y^2z}\right)+\frac{2yz^2}{x^2}} = \boxed{\left(\frac{xyz}{x^2y}\cdot\frac{y^2z}{xy}\right)+\frac{2yz^2}{x^2}} = \boxed{\left(\frac{xyz\cdot y^2z}{x^2y\cdot xy}\right)+\frac{2yz^2}{x^2}} = \boxed{\frac{xy^3z^2}{x^3y^2}+\frac{2yz^2}{x^2}}$$

$$= \boxed{\frac{\left(xy^3z^2\cdot x^2\right)+\left(2yz^2\cdot x^3y^2\right)}{x^3y^2\cdot x^2}} = \boxed{\frac{x^3y^3z^2+2x^3y^3z^2}{x^5y^2}} = \boxed{\frac{3x^3y^3z^2}{x^5y^2}} = \boxed{\frac{3\cancel{x^3}\,\overset{y}{\cancel{y^3}}\,z^2}{\underset{x^2}{\cancel{x^5}}\,\cancel{y^2}}} = \boxed{\mathbf{\frac{3yz^2}{x^2}}}$$

Note that another way of solving this class of problems is by simplifying each fraction term first prior to performing the arithmetic operations as shown below. However, as we have mentioned before, to minimize mistakes, it is good to develop the habit of performing all the math operations first prior to simplifying the fractional expression.

$$\boxed{\left(\frac{xyz}{x^2y}\div\frac{xy}{y^2z}\right)+\frac{2yz^2}{x^2}} = \boxed{\left(\frac{\cancel{xy}z}{\underset{x}{\cancel{x^2}}\,\cancel{y}}\div\frac{\cancel{x}y}{\underset{y}{\cancel{y^2}}\,z}\right)+\frac{2yz^2}{x^2}} = \boxed{\left(\frac{z}{x}\div\frac{x}{yz}\right)+\frac{2yz^2}{x^2}} = \boxed{\left(\frac{z}{x}\cdot\frac{yz}{x}\right)+\frac{2yz^2}{x^2}}$$

$$= \boxed{\left(\frac{z\cdot yz}{x\cdot x}\right)+\frac{2yz^2}{x^2}} = \boxed{\frac{yz^2}{x^2}+\frac{2yz^2}{x^2}} = \boxed{\frac{yz^2+2yz^2}{x^2}} = \boxed{\mathbf{\frac{3yz^2}{x^3}}}$$

Example 5.3-101

$$\boxed{\left(\frac{u}{w}+\frac{v}{w}+\frac{2u}{w}\right)-\left(\frac{3u}{w^2}+\frac{v}{w^2}\right)} = \boxed{\left(\frac{u+v+2u}{w}\right)-\left(\frac{3u+v}{w^2}\right)} = \boxed{\frac{3u+v}{w}-\frac{3u+v}{w^2}} = \boxed{\frac{\left[(3u+v)\cdot w^2\right]-\left[(3u+v)\cdot w\right]}{w\cdot w^2}}$$

$$= \boxed{\frac{(3u+v)w^2-(3u+v)w}{w^3}} = \boxed{\frac{\cancel{w}\left[(3u+v)w-(3u+v)\right]}{\underset{w^2}{\cancel{w^3}}}} = \boxed{\frac{(3u+v)w-(3u+v)}{w^2}} = \boxed{\mathbf{\frac{(3u+v)(w-1)}{w^2}}}$$

Example 5.3-102

$$\left(\frac{abc^2}{ab^2}\cdot\frac{b^3}{a^2}\right)\div\frac{ab^2}{c^3} = \left(\frac{abc^2\cdot b^3}{ab^2\cdot a^2}\right)\div\frac{ab^2}{c^3} = \left(\frac{ab^4c^2}{a^3b^2}\right)\div\frac{ab^2}{c^3} = \frac{\cancel{a}\overset{b^2}{\cancel{b^4}}c^2}{\underset{a^2}{\cancel{a^3}}b^2}\div\frac{ab^2}{c^3} = \frac{b^2c^2}{a^2}\div\frac{ab^2}{c^3}$$

$$= \frac{b^2c^2}{a^2}\cdot\frac{c^3}{ab^2} = \frac{b^2c^2\cdot c^3}{a^2\cdot ab^2} = \frac{b^2c^5}{a^3b^2} = \frac{\cancel{b^2}c^5}{a^3\cancel{b^2}} = \mathbf{\frac{c^5}{a^3}}$$

Example 5.3-103

$$\frac{x}{x-1}+\left(\frac{y}{x+1}-\frac{x}{x+1}\right) = \frac{x}{x-1}+\left(\frac{y-x}{x+1}\right) = \frac{[x\cdot(x+1)]+[(y-x)(x-1)]}{(x-1)\cdot(x+1)} = \frac{x^2+x+xy-y-x^2+x}{(x-1)\cdot(x+1)}$$

$$= \frac{(x^2-x^2)+(x+x)+(xy+y)}{(x-1)(x+1)} = \frac{(1-1)x^2+(1+1)x+(x+1)y}{(x-1)(x+1)} = \frac{0x^2+2x+(x+1)y}{(x-1)(x+1)} = \mathbf{\frac{2x+(x+1)y}{(x-1)(x+1)}}$$

Example 5.3-104

$$\left(\frac{x}{xy}-\frac{1}{y^2}\right)\div\frac{1}{x^2y^3} = \left(\frac{(x\cdot y^2)-(1\cdot xy)}{xy\cdot y^2}\right)\div\frac{1}{x^2y^3} = \left(\frac{xy^2-xy}{xy^3}\right)\div\frac{1}{x^2y^3} = \left(\frac{xy^2-xy}{xy^3}\right)\cdot\frac{x^2y^3}{1}$$

$$= \frac{(xy^2-xy)\cdot x^2y^3}{xy^3\cdot 1} = \frac{(xy^2-xy)\overset{x}{\cancel{x^2}}\cancel{y^3}}{\cancel{x}\cancel{y^3}} = \frac{(xy^2-xy)x}{1} = \frac{x\cdot xy(y-1)}{1} = \frac{x^2y(y-1)}{1} = \mathbf{x^2y(y-1)}$$

Example 5.3-105

$$\left[\left(\frac{a}{b}-\frac{b}{a}\right)+\left(\frac{a}{b}+\frac{b}{a}\right)\right]\div\frac{a^2}{b} = \left[\left(\frac{a\cdot a-b\cdot b}{b\cdot a}\right)+\left(\frac{a\cdot a+b\cdot b}{b\cdot a}\right)\right]\div\frac{a^2}{b} = \left[\left(\frac{a^2-b^2}{ab}\right)+\left(\frac{a^2+b^2}{ab}\right)\right]\div\frac{a^2}{b}$$

$$= \left(\frac{a^2-b^2+a^2+b^2}{ab}\right)\div\frac{a^2}{b} = \left(\frac{a^2+a^2-\cancel{b^2}+\cancel{b^2}}{ab}\right)\div\frac{a^2}{b} = \frac{2a^2}{ab}\div\frac{a^2}{b} = \frac{2a^2}{ab}\cdot\frac{b}{a^2} = \frac{2a^2\cdot b}{ab\cdot a^2} = \frac{2a^2b}{a^3b}$$

$$= \frac{2\cancel{a^2}\cancel{b}}{\underset{a}{\cancel{a^3}}\cancel{b}} = \mathbf{\frac{2}{a}}$$

Example 5.3-106

$$\frac{a^2}{b}\cdot\left[\left(\frac{a}{b}-\frac{b}{a}\right)-\left(\frac{a}{b}+\frac{b}{a}\right)\right] = \frac{a^2}{b}\cdot\left[\left(\frac{a\cdot a-b\cdot b}{b\cdot a}\right)-\left(\frac{a\cdot a+b\cdot b}{b\cdot a}\right)\right] = \frac{a^2}{b}\cdot\left[\left(\frac{a^2-b^2}{ab}\right)-\left(\frac{a^2+b^2}{ab}\right)\right]$$

$$= \frac{a^2}{b}\cdot\left(\frac{a^2-b^2-a^2-b^2}{ab}\right) = \frac{a^2}{b}\cdot\left(\frac{\cancel{a^2}-\cancel{a^2}-b^2-b^2}{ab}\right) = \frac{a^2}{b}\cdot\frac{-2b^2}{ab} = \frac{a^2\cdot -2b^2}{b\cdot ab} = \frac{-2\overset{a}{\cancel{a^2}}\cancel{b^2}}{\cancel{a}\cancel{b^2}} = \mathbf{-2a}$$

Practice Problems - Mixed Operations Involving Algebraic Fractions

Section 5.3 Case V Practice Problems - Solve the following algebraic expressions. Simplify the answer to its lowest term.

1. $\left(\frac{x}{x^3}\cdot\frac{2x}{4}\right)\div x^2 =$

2. $\left(\frac{ab}{b^2}+\frac{a}{c^2}\right)-\frac{1}{b^2c^2} =$

3. $\left(\frac{yz}{z^2}-\frac{y}{y^2}\right)+\left(\frac{1}{z}-\frac{1}{y}\right) =$

4. $\left(\frac{1}{a}\div a^2\right)\cdot\left(\frac{1}{b}\div\frac{b^2}{a^2}\right) =$

5. $\left(\frac{2}{x}-\frac{4}{y}\right)+\left(\frac{1}{x}\div y\right) =$

6. $\left(\frac{u}{v^2}\cdot\frac{v}{u^2}\cdot\frac{1}{v}\right)\div u =$

7. $\left(\frac{a}{b^2}\div\frac{1}{a^2}\right)+\frac{2}{3} =$

8. $\left(1+\frac{2}{x}\right)-\left(\frac{1}{x}+2\right) =$

9. $\left(\frac{x^2y^2z}{z^3}\cdot\frac{z^2}{xy}\right)\div xyz^2 =$

10. $\left[\left(\frac{x}{y^2}+\frac{1}{y^2}\right)-\frac{2}{y}\right]\cdot\frac{y^3}{2x} =$

5.4 Math Operations Involving Complex Algebraic Fractions

A **simple algebraic fraction** is a fraction in which neither the numerator nor the denominator contains a fraction with variables. For example, $\frac{a}{5}$, $\frac{a-1}{a}$, $\frac{2x-1}{3}$, $\frac{3x}{x^2+2x-1}$, and $-\frac{5}{a^2+2a+1}$ are examples of simple algebraic fractions. A **complex algebraic fraction** is a fraction in which either the numerator or the denominator (or both) contains an algebraic fraction. For example, $\frac{1-\frac{1}{w}}{5}$, $\frac{\frac{2x}{5}-\frac{1}{x}}{\frac{2}{x}}$, $\frac{a+\frac{a}{b}}{1-\frac{a}{b}}$, and $\frac{x}{1+\frac{1}{x}}$ are examples of complex algebraic fractions.

Note that an easy way to change complex algebraic fractions to simple algebraic fractions is by multiplying the outer numerator by the outer denominator and the inner denominator by the inner numerator. For example, given the complex fraction $\frac{\frac{x^2}{5}}{\frac{2x}{9}}$; first obtain the numerator of the simple fraction by multiply x^2 , *the outer numerator*, by 9 , *the outer denominator*. Next, obtain the denominator of the simple fraction by multiply 5 , *the inner denominator*, by $2x$, *the inner numerator*. Therefore, the complex fraction $\frac{\frac{x^2}{5}}{\frac{2x}{9}}$ can be written as $\frac{x^2 \cdot 9}{5 \cdot 2x} = \frac{9x^2}{10x} = \frac{9x^{\cancel{2}}}{10\cancel{x}} = \frac{9x}{10}$ which is a simple fraction. In this section addition, subtraction, multiplication, and division of complex algebraic fractions (Cases I through III) are discussed. In addition, simplification of complex algebraic expressions involving mixed operations is addressed in Case IV.

Case I Addition and Subtraction of Complex Algebraic Fractions

Complex algebraic fractions are added or subtracted using the following steps:

Step 1 Add or subtract the algebraic fractions in both the numerator and the denominator. Note that the same process used in simplifying integer (arithmetic) fractions applies to algebraic fractions (See Section 1.2).

Step 2 Change the complex algebraic fraction to a simple fraction . Reduce the algebraic fraction to its lowest term, if possible.

Examples with Steps

The following examples show the steps as to how complex algebraic expressions are added and subtracted:

Example 5.4-1

$$\boxed{\frac{\frac{1}{2x}+\frac{2}{3y}}{\frac{3}{3y}-\frac{5}{2x}}} =$$

Solution:

Step 1

$$\frac{\frac{1}{2x}+\frac{2}{3y}}{\frac{3}{3y}-\frac{5}{2x}}=\frac{\frac{(1\cdot 3y)+(2\cdot 2x)}{2x\cdot 3y}}{\frac{(3\cdot 2x)-(5\cdot 3y)}{3y\cdot 2x}}=\frac{\frac{3y+4x}{6xy}}{\frac{6x-15y}{6xy}}$$

Step 2

$$\frac{\frac{3y+4x}{6xy}}{\frac{6x-15y}{6xy}}=\frac{(3y+4x)\cdot \cancel{6xy}}{\cancel{6xy}\cdot(6x-15y)}=\frac{4x+3y}{6x-15y}=\boldsymbol{\frac{4x+3y}{3(2x-5y)}}$$

Example 5.4-2

$$\frac{\frac{3x^3y^2}{xy}-1}{\frac{x^2y}{xy^2}+1}=$$

Solution:

Step 1

$$\frac{\frac{3x^3y^2}{xy}-1}{\frac{x^2y}{xy^2}+1}=\frac{\frac{3x^3y^2}{xy}-\frac{1}{1}}{\frac{x^2y}{xy^2}+\frac{1}{1}}=\frac{\frac{(3x^3y^2\cdot 1)+(1\cdot xy)}{xy\cdot 1}}{\frac{(x^2y\cdot 1)+(1\cdot xy^2)}{xy^2\cdot 1}}=\frac{\frac{3x^3y^2+xy}{xy}}{\frac{x^2y+xy^2}{xy^2}}$$

Step 2

$$\frac{\frac{3x^3y^2+xy}{xy}}{\frac{x^2y+xy^2}{xy^2}}=\frac{(3x^3y^2+xy)\cdot \cancel{x}\overset{y}{\cancel{y^2}}}{\cancel{xy}\cdot(x^2y+xy^2)}=\frac{(3x^3y^2+xy)y}{x^2y+xy^2}=\frac{\cancel{xy}(3x^2y+1)y}{\cancel{xy}(x+y)}=\boldsymbol{\frac{y(3x^2y+1)}{x+y}}$$

Example 5.4-3

$$\frac{\frac{1}{a}+\frac{1}{b}}{\frac{1}{a}-\frac{1}{b}}-\frac{a}{b}=$$

Solution:

Step 1

$$\frac{\frac{1}{a}+\frac{1}{b}}{\frac{1}{a}-\frac{1}{b}}-\frac{a}{b}=\frac{\frac{(1\cdot b)+(1\cdot a)}{a\cdot b}}{\frac{(1\cdot b)-(1\cdot a)}{a\cdot b}}-\frac{a}{b}=\frac{\frac{b+a}{ab}}{\frac{b-a}{ab}}-\frac{a}{b}$$

Step 2

$$\frac{\frac{b+a}{ab}}{\frac{b-a}{ab}}-\frac{a}{b}=\frac{(b+a)\cdot \cancel{ab}}{\cancel{ab}\cdot(b-a)}-\frac{a}{b}=\frac{b+a}{b-a}-\frac{a}{b}=\frac{[(b+a)\cdot b]-[a\cdot(b-a)]}{(b-a)\cdot b}$$

$$=\frac{b^2+\cancel{ab}-\cancel{ab}+a^2}{b(b-a)}=\boldsymbol{\frac{a^2+b^2}{b(b-a)}}$$

Example 5.4-4

$$\boxed{\frac{\frac{3}{x}-\frac{1}{x-1}}{\frac{2}{x-1}}} =$$

Solution:

Step 1

$$\boxed{\frac{\frac{3}{x}-\frac{1}{x-1}}{\frac{2}{x-1}}} = \boxed{\frac{\frac{3\cdot(x-1)-(1\cdot x)}{x\cdot(x-1)}}{\frac{2}{x-1}}} = \boxed{\frac{\frac{3x-3-x}{x(x-1)}}{\frac{2}{x-1}}} = \boxed{\frac{\frac{(3x-x)-3}{x(x-1)}}{\frac{2}{x-1}}} = \boxed{\frac{\frac{2x-3}{x(x-1)}}{\frac{2}{x-1}}}$$

Step 2

$$\boxed{\frac{\frac{2x-3}{x(x-1)}}{\frac{2}{x-1}}} = \boxed{\frac{(2x-3)\cdot(\cancel{x}-\cancel{1})}{x(\cancel{x}-\cancel{1})\cdot 2}} = \boxed{\mathbf{\frac{2x-3}{2x}}}$$

Example 5.4-5

$$\boxed{\frac{3+\frac{x}{y}}{\frac{xy}{y^2}-4}} =$$

Solution:

Step 1

$$\boxed{\frac{3+\frac{x}{y}}{\frac{xy}{y^2}-4}} = \boxed{\frac{\frac{3}{1}+\frac{x}{y}}{\frac{xy}{y^2}-\frac{4}{1}}} = \boxed{\frac{\frac{(3\cdot y)+(1\cdot x)}{1\cdot y}}{\frac{(xy\cdot 1)-(4\cdot y^2)}{y^2\cdot 1}}} = \boxed{\frac{\frac{3y+x}{y}}{\frac{xy-4y^2}{y^2}}}$$

Step 2

$$\boxed{\frac{\frac{3y+x}{y}}{\frac{xy-4y^2}{y^2}}} = \boxed{\frac{(3y+x)\cdot y^{\cancel{2}}}{\cancel{y}\cdot(xy-4y^2)}} = \boxed{\frac{(3y+x)\cdot y}{xy-4y^2}} = \boxed{\frac{(3y+x)\cdot \cancel{y}}{\cancel{y}\cdot(x-4y)}} = \boxed{\mathbf{\frac{x+3y}{x-4y}}}$$

Additional Examples - Addition and Subtraction of Complex Algebraic Fractions

The following examples further illustrate addition and subtraction of complex algebraic fractions:

Example 5.4-6

$$\boxed{\frac{2-\frac{2}{3x}}{4-\frac{2}{9x}}} = \boxed{\frac{\frac{2}{1}-\frac{2}{3x}}{\frac{4}{1}-\frac{2}{9x}}} = \boxed{\frac{\frac{(2\cdot 3x)-(2\cdot 1)}{1\cdot 3x}}{\frac{(4\cdot 9x)-(2\cdot 1)}{1\cdot 9x}}} = \boxed{\frac{\frac{6x-2}{3x}}{\frac{36x-2}{9x}}} = \boxed{\frac{(6x-2)\cdot 9x}{3x\cdot(36x-2)}} = \boxed{\frac{18x(3x-1)}{6x(18x-1)}} = \boxed{\frac{\overset{3}{\cancel{18}}\cancel{x}(3x-1)}{\cancel{6}\cancel{x}(18x-1)}} = \boxed{\mathbf{\frac{3(3x-1)}{18x-1}}}$$

Example 5.4-7

$$\boxed{\frac{\frac{1}{8}+a}{a+\frac{1}{4}}} = \boxed{\frac{\frac{1}{8}+\frac{a}{1}}{\frac{a}{1}+\frac{1}{4}}} = \boxed{\frac{\frac{(1\cdot 1)+(a\cdot 8)}{8\cdot 1}}{\frac{(a\cdot 4)+(1\cdot 1)}{1\cdot 4}}} = \boxed{\frac{\frac{1+8a}{8}}{\frac{4a+1}{4}}} = \boxed{\frac{(1+8a)\cdot 4}{8\cdot(4a+1)}} = \boxed{\frac{(1+8a)\cdot \cancel{4}}{\underset{2}{\cancel{8}}\cdot(4a+1)}} = \boxed{\mathbf{\frac{1+8a}{2(1+4a)}}}$$

Example 5.4-8

$$\frac{\frac{2x^3y^2z}{x^3yz^2}}{\frac{x^2y}{xz^2}}+\frac{z}{x}=\frac{(2x^3y^2z)\cdot(xz^2)}{(x^3yz^2)\cdot(x^2y)}+\frac{z}{x}=\frac{2\cancel{x^4}\cancel{y^2}\cancel{z^3}^{\,z}}{\cancel{x^5}_{\,x}\cancel{y^2}\cancel{z^2}}+\frac{z}{x}=\frac{2z}{x}+\frac{z}{x}=\frac{2z+z}{x}=\boxed{\mathbf{\frac{3z}{x}}}$$

Example 5.4-9

$$\frac{\frac{8a^3b^2}{4a^2b}}{\frac{6a^3}{3a^2}}+\frac{\frac{ab}{1}}{\frac{1}{a}}=\frac{(8a^3b^2)\cdot(3a^2)}{(4a^2b)\cdot(6a^3)}+\frac{\frac{ab}{1}}{\frac{1}{a}}=\frac{\cancel{24}\cancel{a^5}\cancel{b^2}^{\,b}}{\cancel{24}\cancel{a^5}\cancel{b}}+\frac{ab\cdot a}{1\cdot 1}=\frac{b}{1}+\frac{a^2b}{1}=\frac{b+a^2b}{1}=b+a^2b=\boxed{\mathbf{b(1+a^2)}}$$

Example 5.4-10

$$\frac{\frac{x}{y}+\frac{y}{x}}{\frac{3}{xy}+\frac{5}{xy}}=\frac{\frac{(x\cdot x)+(y\cdot y)}{x\cdot y}}{\frac{3+5}{xy}}=\frac{\frac{x^2+y^2}{xy}}{\frac{8}{xy}}=\frac{(x^2+y^2)\cdot\cancel{xy}}{8\cdot\cancel{xy}}=\boxed{\mathbf{\frac{x^2+y^2}{8}}}$$

Example 5.4-11

$$\frac{\frac{1}{2x}-\frac{1}{3y}}{\frac{2}{3x}+\frac{5}{2y}}-4=\frac{\frac{(1\cdot 3y)-(1\cdot 2x)}{2x\cdot 3y}}{\frac{(2\cdot 2y)+(5\cdot 3x)}{3x\cdot 2y}}-4=\frac{\frac{3y-2x}{6xy}}{\frac{4y+15x}{6xy}}-4=\frac{(3y-2x)\cdot\cancel{6xy}}{\cancel{6xy}\cdot(4y+15x)}-4=\frac{3y-2x}{4y+15x}-4=\frac{3y-2x}{4y+15x}-\frac{4}{1}$$

$$=\frac{(3y-2x)\cdot 1-4\cdot(4y+15x)}{(4y+15x)\cdot 1}=\frac{3y-2x-16y-60x}{4y+15x}=\frac{(3y-16y)+(-60x-2x)}{4y+15x}=\boxed{\mathbf{\frac{-13y-62x}{4y+15x}}}$$

Example 5.4-12

$$\frac{3-\frac{1}{x+1}}{\frac{2}{x+1}+3}=\frac{\frac{3}{1}-\frac{1}{x+1}}{\frac{2}{x+1}+\frac{3}{1}}=\frac{\frac{[3\cdot(x+1)]-(1\cdot 1)}{1\cdot(x+1)}}{\frac{(2\cdot 1)+[3\cdot(x+1)]}{(x+1)\cdot 1}}=\frac{\frac{3x+3-1}{x+1}}{\frac{2+3x+3}{x+1}}=\frac{\frac{3x+2}{x+1}}{\frac{3x+5}{x+1}}=\frac{(3x+2)\cdot\cancel{(x+1)}}{\cancel{(x+1)}\cdot(3x+5)}=\boxed{\mathbf{\frac{3x+2}{3x+5}}}$$

Example 5.4-13

$$\frac{\frac{x}{x+3}+2}{\frac{4}{x+3}-1}=\frac{\frac{x}{x+3}+\frac{2}{1}}{\frac{4}{x+3}-\frac{1}{1}}=\frac{\frac{(x\cdot 1)+[2\cdot(x+3)]}{(x+3)\cdot 1}}{\frac{(4\cdot 1)-[1\cdot(x+3)]}{(x+3)\cdot 1}}=\frac{\frac{x+2x+6}{x+3}}{\frac{4-x-3}{x+3}}=\frac{\frac{3x+6}{x+3}}{\frac{-x+1}{x+3}}=\frac{(3x+6)\cdot\cancel{(x+3)}}{\cancel{(x+3)}\cdot(-x+1)}=\boxed{\mathbf{\frac{3(x+2)}{1-x}}}$$

Example 5.4-14

$$\frac{1+\frac{2}{xy}}{\frac{xy}{x^2y^2}+2}=\frac{\frac{1}{1}+\frac{2}{xy}}{\frac{xy}{x^2y^2}+\frac{2}{1}}=\frac{\frac{(1\cdot xy)+(2\cdot 1)}{1\cdot xy}}{\frac{(xy\cdot 1)+(2\cdot x^2y^2)}{x^2y^2}}=\frac{\frac{xy+2}{xy}}{\frac{xy+2x^2y^2}{x^2y^2}}=\frac{(xy+2)\cdot\cancel{x^2y^2}^{\,xy}}{\cancel{xy}\cdot(xy+2x^2y^2)}=\boxed{\frac{(xy+2)\cdot xy}{xy+2x^2y^2}}$$

$$= \boxed{\frac{(xy+2)\cancel{xy}}{\cancel{xy}(1+2xy)}} = \boxed{\mathbf{\frac{2+xy}{1+2xy}}}$$

Example 5.4-15

$$\boxed{\frac{\frac{2}{a}-\frac{1}{a-1}}{\frac{5}{a-1}}+\frac{1}{a}} = \boxed{\frac{\frac{[2\cdot(a-1)]-(1\cdot a)}{a\cdot(a-1)}}{\frac{5}{a-1}}+\frac{1}{a}} = \boxed{\frac{\frac{2a-2-a}{a(a-1)}}{\frac{5}{a-1}}+\frac{1}{a}} = \boxed{\frac{\frac{a-2}{a(a-1)}}{\frac{5}{a-1}}+\frac{1}{a}} = \boxed{\frac{(a-2)\cdot(a-1)}{[a(a-1)]\cdot 5}+\frac{1}{a}}$$

$$= \boxed{\frac{(a-2)\cdot\cancel{(a-1)}}{5a\cancel{(a-1)}}+\frac{1}{a}} = \boxed{\frac{a-2}{5a}+\frac{1}{a}} = \boxed{\frac{[(a-2)\cdot a]+(1\cdot 5a)}{5a\cdot a}} = \boxed{\frac{a^2-2a+5a}{5a^2}} = \boxed{\frac{a^2+3a}{5a^2}} = \boxed{\frac{\cancel{a}(a+3)}{5a^{\cancel{2}}_{a}}} = \boxed{\mathbf{\frac{a+3}{5a}}}$$

Example 5.4-16

$$\boxed{\frac{\frac{x}{x+2}-\frac{1}{x}}{\frac{1}{x}+\frac{4}{x+2}}+\frac{1}{x}} = \boxed{\frac{\frac{(x\cdot x)-[1\cdot(x+2)]}{(x+2)\cdot x}}{\frac{[1\cdot(x+2)]+(4\cdot x)}{x\cdot(x+2)}}+\frac{1}{x}} = \boxed{\frac{\frac{x^2-x-2}{x(x+2)}}{\frac{x+2+4x}{x(x+2)}}+\frac{1}{x}} = \boxed{\frac{\frac{x^2-x-2}{x(x+2)}}{\frac{5x+2}{x(x+2)}}+\frac{1}{x}} = \boxed{\frac{(x^2-x-2)\cdot x(x+2)}{(5x+2)\cdot x(x+2)}+\frac{1}{x}}$$

$$= \boxed{\frac{(x^2-x-2)\cdot\cancel{x}\cancel{(x+2)}}{(5x+2)\cdot\cancel{x}\cancel{(x+2)}}+\frac{1}{x}} = \boxed{\frac{x^2-x-2}{5x+2}+\frac{1}{x}} = \boxed{\frac{[(x^2-x-2)\cdot x]+[1\cdot(5x+2)]}{(5x+2)\cdot x}} = \boxed{\frac{x^3-x^2-2x+5x+2}{(5x+2)\cdot x}}$$

$$= \boxed{\frac{x^3-x^2+(-2+5)x+2}{x(5x+2)}} = \boxed{\mathbf{\frac{x^3-x^2+3x+2}{x(5x+2)}}}$$

Example 5.4-17

$$\boxed{\frac{a}{\frac{1}{a}-\frac{1}{b}}-a} = \boxed{\frac{\frac{a}{1}}{\frac{1}{a}-\frac{1}{b}}-a} = \boxed{\frac{\frac{a}{1}}{\frac{(1\cdot b)-(1\cdot a)}{a\cdot b}}-a} = \boxed{\frac{\frac{a}{1}}{\frac{b-a}{ab}}-a} = \boxed{\frac{a\cdot ab}{1\cdot(b-a)}-a} = \boxed{\frac{a^2b}{b-a}-\frac{a}{1}} = \boxed{\frac{(a^2b\cdot 1)-[a\cdot(b-a)]}{b-a}}$$

$$= \boxed{\frac{a^2b-ab+a^2}{b-a}} = \boxed{\mathbf{\frac{a(ab-b+a)}{b-a}}}$$

Example 5.4-18

$$\boxed{\frac{\frac{a}{b}}{a+\frac{1}{b}}+\frac{ab}{b^2}} = \boxed{\frac{\frac{a}{b}}{\frac{a}{1}+\frac{1}{b}}+\frac{ab}{b^2}} = \boxed{\frac{\frac{a}{b}}{\frac{(a\cdot b)+(1\cdot 1)}{1\cdot b}}+\frac{ab}{b^2}} = \boxed{\frac{\frac{a}{b}}{\frac{ab+1}{b}}+\frac{ab}{b^2}} = \boxed{\frac{a\cdot b}{b\cdot(ab+1)}+\frac{ab}{b^2}} = \boxed{\frac{a\cancel{b}}{\cancel{b}(ab+1)}+\frac{a\cancel{b}}{b^{\cancel{2}}_{b}}}$$

$$= \boxed{\frac{a}{ab+1}+\frac{a}{b}} = \boxed{\frac{a\cdot b+a\cdot(ab+1)}{(ab+1)\cdot b}} = \boxed{\frac{ab+a^2b+a}{b(ab+1)}} = \boxed{\mathbf{\frac{a(b+ab+1)}{b(ab+1)}}}$$

Example 5.4-19

$$\frac{x+\frac{1}{x+y}}{\frac{1}{x+y}-x} = \frac{\frac{x}{1}+\frac{1}{x+y}}{\frac{1}{x+y}-\frac{x}{1}} = \frac{\frac{[x\cdot(x+y)]+(1\cdot1)}{1\cdot(x+y)}}{\frac{(1\cdot1)-[x\cdot(x+y)]}{(x+y)\cdot1}} = \frac{\frac{x^2+xy+1}{x+y}}{\frac{1-x^2-xy}{x+y}} = \frac{(x^2+xy+1)\cdot(x+y)}{(x+y)\cdot(1-x^2-xy)} = \frac{(x^2+xy+1)\cancel{(x+y)}}{\cancel{(x+y)}(1-x^2-xy)}$$

$$= \frac{x^2+xy+1}{1-x^2-xy} = \boldsymbol{\frac{x^2+xy+1}{-x^2-xy+1}}$$

Example 5.4-20

$$\frac{\frac{x}{y}+2}{1-\frac{x}{y}}-\frac{1}{y} = \frac{\frac{x}{y}+\frac{2}{1}}{\frac{1}{1}-\frac{x}{y}}-\frac{1}{y} = \frac{\frac{(x\cdot1)+(2\cdot y)}{y\cdot1}}{\frac{(1\cdot y)-(x\cdot1)}{1\cdot y}}-\frac{1}{y} = \frac{\frac{x+2y}{y}}{\frac{y-x}{y}}-\frac{1}{y} = \frac{(x+2y)\cdot\cancel{y}}{\cancel{y}\cdot(y-x)}-\frac{1}{y} = \frac{x+2y}{y-x}-\frac{1}{y}$$

$$= \frac{[(x+2y)\cdot y]-[1\cdot(y-x)]}{(y-x)\cdot y} = \frac{xy+2y^2-y+x}{y(y-x)} = \frac{2y^2+xy-y+x}{y(y-x)} = \boldsymbol{\frac{2y^2+(x-1)y+x}{y(y-x)}}$$

Example 5.4-21

$$\frac{\frac{a}{b}+1}{\frac{a}{b}-1}+\frac{a}{b} = \frac{\frac{a}{b}+\frac{1}{1}}{\frac{a}{b}-\frac{1}{1}}+\frac{a}{b} = \frac{\frac{(a\cdot1)+(1\cdot b)}{b\cdot1}}{\frac{(a\cdot1)-(1\cdot b)}{b\cdot1}}+\frac{a}{b} = \frac{\frac{a+b}{b}}{\frac{a-b}{b}}+\frac{a}{b} = \frac{(a+b)\cdot\cancel{b}}{\cancel{b}\cdot(a-b)}+\frac{a}{b} = \frac{a+b}{a-b}+\frac{a}{b}$$

$$= \frac{[(a+b)\cdot b]+[a\cdot(a-b)]}{(a-b)\cdot b} = \frac{ab+b^2+a^2-ab}{b(a-b)} = \frac{\cancel{ab}+b^2+a^2-\cancel{ab}}{b(a-b)} = \boldsymbol{\frac{a^2+b^2}{b(a-b)}}$$

Example 5.4-22

$$\frac{(1+a)-\frac{2}{a}}{\frac{2}{a}+1} = \frac{\frac{(1+a)}{1}-\frac{2}{a}}{\frac{2}{a}+\frac{1}{1}} = \frac{\frac{[a\cdot(1+a)]-(2\cdot1)}{1\cdot a}}{\frac{(2\cdot1)+(1\cdot a)}{a\cdot1}} = \frac{\frac{a+a^2-2}{a}}{\frac{2+a}{a}} = \frac{(a+a^2-2)\cdot\cancel{a}}{(a+2)\cdot\cancel{a}} = \boldsymbol{\frac{a^2+a-2}{a+2}}$$

Example 5.4-23

$$\frac{a+1}{1+\frac{1}{a}}+\frac{1}{a} = \frac{\frac{a+1}{1}}{\frac{1}{1}+\frac{1}{a}}+\frac{1}{a} = \frac{\frac{a+1}{1}}{\frac{(1\cdot a)+(1\cdot1)}{1\cdot a}}+\frac{1}{a} = \frac{\frac{a+1}{1}}{\frac{a+1}{a}}+\frac{1}{a} = \frac{\cancel{(a+1)}\cdot a}{1\cdot\cancel{(a+1)}}+\frac{1}{a} = \frac{a}{1}+\frac{1}{a} = \frac{(a\cdot a)+(1\cdot1)}{1\cdot a} = \boldsymbol{\frac{a^2+1}{a}}$$

Example 5.4-24

$$\frac{a+\frac{1}{a+b}}{ab-\frac{1}{a+b}} = \frac{\frac{a}{1}+\frac{1}{a+b}}{\frac{ab}{1}-\frac{1}{a+b}} = \frac{\frac{[a\cdot(a+b)]+(1\cdot1)}{1\cdot(a+b)}}{\frac{[ab\cdot(a+b)]-(1\cdot1)}{1\cdot(a+b)}} = \frac{\frac{a^2+ab+1}{a+b}}{\frac{a^2b+ab^2-1}{a+b}} = \frac{(a^2+ab+1)\cdot(a+b)}{(a+b)\cdot(a^2b+ab^2-1)}$$

$$= \boxed{\frac{\left(a^2+ab+1\right)\cdot\left(\cancel{a}+\cancel{b}\right)}{\left(\cancel{a}+\cancel{b}\right)\cdot\left(a^2b+ab^2-1\right)}} = \boxed{\mathbf{\frac{a^2+ab+1}{a^2b+ab^2-1}}}$$

Example 5.4-25

$$\boxed{\frac{\frac{3}{x+y}-\frac{2}{x-y}}{\frac{2}{x^2-y^2}}} = \boxed{\frac{\frac{[3\cdot(x-y)]-[2\cdot(x+y)]}{(x+y)(x-y)}}{\frac{2}{x^2-y^2}}} = \boxed{\frac{\frac{3x-3y-2x-2y}{(x+y)(x-y)}}{\frac{2}{(x+y)(x-y)}}} = \boxed{\frac{\frac{x-5y}{(x+y)(x-y)}}{\frac{2}{(x+y)(x-y)}}} = \boxed{\frac{(x-5y)\cdot(x+y)(x-y)}{2\cdot(x+y)(x-y)}}$$

$$= \boxed{\frac{(x-5y)\cdot\left(\cancel{x}+\cancel{y}\right)\left(\cancel{x}-\cancel{y}\right)}{2\cdot\left(\cancel{x}+\cancel{y}\right)\left(\cancel{x}-\cancel{y}\right)}} = \boxed{\mathbf{\frac{x-5y}{2}}}$$

Practice Problems - Addition and Subtraction of Complex Algebraic Fractions

Section 5.4 Case I Practice Problems - Simplify the following complex algebraic fractions. Reduce the answer to its lowest term.

1. $\dfrac{2-\frac{1}{5a}}{3-\frac{2}{15a}} =$

2. $\dfrac{\frac{2x^3y^2z}{4x^2z}-1}{\frac{2x}{xy^2}} =$

3. $\dfrac{\frac{2}{a}+\frac{1}{a^3}}{\frac{2}{a^3}+2} =$

4. $\dfrac{a}{a+\frac{1}{a^2}} =$

5. $\dfrac{\frac{x}{y}-3}{3-\frac{y}{x}} =$

6. $\dfrac{\frac{1}{x}-\frac{1}{x+4}}{\frac{3}{x+4}+1} =$

7. $\dfrac{\frac{w}{w-1}+\frac{1}{w+1}}{1-\frac{1}{w-1}} =$

8. $\dfrac{4-\frac{1}{x+1}}{\frac{2}{x+1}-2} =$

9. $\dfrac{\frac{2}{x^2-1}-\frac{1}{x-1}}{\frac{1}{x-1}+\frac{2}{x+1}} =$

10. $\dfrac{\frac{y+3}{y-1}+\frac{1}{y}}{2-\frac{y+1}{y-1}} =$

Case II Multiplication of Complex Algebraic Fractions

Complex algebraic fractions are multiplied by one another using the following steps:

Step 1 Multiply the numerator and the denominator of the algebraic fraction terms by one another.

Step 2 Change the complex algebraic fraction to a simple fraction. Reduce the algebraic fraction to its lowest term, if possible.

Examples with Steps

The following examples show the steps as to how complex algebraic fractions are multiplied:

Example 5.4-26

$$\boxed{\dfrac{\frac{xy}{x^2}\cdot x}{x^2y^2\cdot\frac{1}{xy}}} =$$

Solution:

Step 1

$$\boxed{\dfrac{\frac{xy}{x^2}\cdot x}{x^2y^2\cdot\frac{1}{xy}}} = \boxed{\dfrac{\frac{xy}{x^2}\cdot\frac{x}{1}}{\frac{x^2y^2}{1}\cdot\frac{1}{xy}}} = \boxed{\dfrac{\frac{xy\cdot x}{x^2\cdot 1}}{\frac{x^2y^2\cdot 1}{1\cdot xy}}} = \boxed{\dfrac{\frac{x^2y}{x^2}}{\frac{x^2y^2}{xy}}}$$

Step 2

$$\boxed{\dfrac{\frac{x^2y}{x^2}}{\frac{x^2y^2}{xy}}} = \boxed{\frac{x^2y\cdot xy}{x^2\cdot x^2y^2}} = \boxed{\frac{x^3y^2}{x^4y^2}} = \boxed{\frac{x^3y^2}{\underset{x}{x^4}\,y^2}} = \boxed{\frac{1}{x}}$$

Example 5.4-27

$$\boxed{\dfrac{\frac{x^2y^2z}{z^2}\cdot\frac{z}{x^3y^3}}{\frac{1}{xy}}} =$$

Solution:

Step 1

$$\boxed{\dfrac{\frac{x^2y^2z}{z^2}\cdot\frac{z}{x^3y^3}}{\frac{1}{xy}}} = \boxed{\dfrac{\frac{x^2y^2z\cdot z}{z^2\cdot x^3y^3}}{\frac{1}{xy}}} = \boxed{\dfrac{\frac{x^2y^2z^2}{x^3y^3z^2}}{\frac{1}{xy}}}$$

Step 2

$$\boxed{\dfrac{\frac{x^2y^2z^2}{x^3y^3z^2}}{\frac{1}{xy}}} = \boxed{\frac{x^2y^2z^2\cdot xy}{x^3y^3z^2\cdot 1}} = \boxed{\frac{x^3y^3z^2}{x^3y^3z^2}} = \boxed{\frac{x^3y^3z^2}{x^3y^3z^2}} = \boxed{\frac{1}{1}} = \boxed{1}$$

Example 5.4-28

$$\frac{a^2b^2\cdot\left(\frac{1}{a^2b^2}\cdot\frac{ab}{a^3}\right)}{\frac{1}{a^2b^3}}=$$

Solution:

Step 1

$$\frac{a^2b^2\cdot\left(\frac{1}{a^2b^2}\cdot\frac{ab}{a^3}\right)}{\frac{1}{a^2b^3}}=\frac{\frac{a^2b^2}{1}\cdot\left(\frac{1}{a^2b^2}\cdot\frac{ab}{a^3}\right)}{\frac{1}{a^2b^3}}=\frac{\frac{a^2b^2}{1}\cdot\left(\frac{1\cdot ab}{a^2b^2\cdot a^3}\right)}{\frac{1}{a^2b^3}}=\frac{\frac{a^2b^2}{1}\cdot\frac{ab}{a^5b^2}}{\frac{1}{a^2b^3}}$$

$$=\frac{\frac{a^2b^2\cdot ab}{1\cdot a^5b^2}}{\frac{1}{a^2b^3}}=\frac{\frac{a^3b^3}{a^5b^2}}{\frac{1}{a^2b^3}}$$

Step 2

$$\frac{\frac{a^3b^3}{a^5b^2}}{\frac{1}{a^2b^3}}=\frac{a^3b^3\cdot a^2b^3}{a^5b^2\cdot 1}=\frac{a^5b^6}{a^5b^2}=\frac{\overset{b^4}{\cancel{a^5}\,\cancel{b^6}}}{\cancel{a^5}\,\cancel{b^2}}=\frac{b^4}{1}=\boxed{\mathbf{b^4}}$$

Example 5.4-29

$$\frac{\frac{u^2v^2}{w}\cdot\frac{w^3}{v^4}}{uw^2}=$$

Solution:

Step 1

$$\frac{\frac{u^2v^2}{w}\cdot\frac{w^3}{v^4}}{uw^2}=\frac{\frac{u^2v^2}{w}\cdot\frac{w^3}{v^4}}{\frac{uw^2}{1}}=\frac{\frac{u^2v^2\cdot w^3}{w\cdot v^4}}{\frac{uw^2}{1}}=\frac{\frac{u^2v^2w^3}{wv^4}}{\frac{uw^2}{1}}$$

Step 2

$$\frac{\frac{u^2v^2w^3}{wv^4}}{\frac{uw^2}{1}}=\frac{u^2v^2w^3\cdot 1}{wv^4\cdot uw^2}=\frac{u^2v^2w^3}{uv^4w^2}=\frac{\overset{u}{\cancel{u^2}}\,\cancel{v^2}\,\cancel{w^3}}{\underset{v^2}{\cancel{u}\,\cancel{v^4}}\,\cancel{w^3}}=\boxed{\mathbf{\frac{u}{v^2}}}$$

Example 5.4-30

$$\frac{5}{\frac{xyz}{x^2y^2z}\cdot\frac{2x}{yz^3}}=$$

Solution:

Step 1

$$\frac{5}{\frac{xyz}{x^2y^2z}\cdot\frac{2x}{yz^3}}=\frac{\frac{5}{1}}{\frac{xyz}{x^2y^2z}\cdot\frac{2x}{yz^3}}=\frac{\frac{5}{1}}{\frac{xyz\cdot 2x}{x^2y^2z\cdot yz^3}}=\frac{\frac{5}{1}}{\frac{2x^2yz}{x^2y^3z^4}}$$

Step 2

$$\frac{\frac{5}{1}}{\frac{2x^2yz}{x^2y^3z^4}} = \frac{5\cdot x^2y^3z^4}{1\cdot 2x^2yz} = \frac{5x^2y^3z^4}{2x^2yz} = \frac{5\cancel{x^2}\overset{y^2}{\cancel{y^3}}\overset{z^3}{\cancel{z^4}}}{2\cancel{x^2}\cancel{y}\cancel{z}} = \frac{5y^2z^3}{2} = \boxed{\mathbf{\frac{5}{2}y^2z^3}}$$

Additional Examples - Multiplication of Complex Algebraic Fractions

The following examples further illustrate how to multiply complex algebraic fractions:

Example 5.4-31

$$\frac{xyz\cdot\frac{2x}{y^2z^2}}{x^2} = \frac{\frac{xyz}{1}\cdot\frac{2x}{y^2z^2}}{x^2} = \frac{\frac{xyz\cdot 2x}{1\cdot y^2z^2}}{x^2} = \frac{\frac{2x^2yz}{y^2z^2}}{x^2} = \frac{\frac{2x^2yz}{y^2z^2}}{\frac{x^2}{1}} = \frac{2x^2yz\cdot 1}{y^2z^2\cdot x^2} = \frac{2x^2yz}{x^2y^2z^2} = \frac{2\cancel{x^2}\cancel{y}\cancel{z}}{\cancel{x^2}\underset{y}{\cancel{y^2}}\underset{z}{\cancel{z^2}}} = \boxed{\mathbf{\frac{2}{yz}}}$$

Example 5.4-32

$$\frac{2}{\frac{x^2y^2}{3x}\cdot\frac{6x}{x^3y^3}} = \frac{2}{\frac{x^2y^2\cdot 6x}{3x\cdot x^3y^3}} = \frac{2}{\frac{6x^3y^2}{3x^4y^3}} = \frac{\frac{2}{1}}{\frac{6x^3y^2}{3x^4y^3}} = \frac{2\cdot 3x^4y^3}{1\cdot 6x^3y^2} = \frac{6x^4y^3}{6x^3y^2} = \frac{\cancel{6}\overset{x}{\cancel{x^4}}\overset{y}{\cancel{y^3}}}{\cancel{6}\cancel{x^3}\cancel{y^2}} = \frac{xy}{1} = \boxed{\mathbf{xy}}$$

Example 5.4-33

$$\frac{ab}{a^3b^2\cdot\frac{3a}{ab^3}} = \frac{ab}{\frac{a^3b^2}{1}\cdot\frac{3a}{ab^3}} = \frac{ab}{\frac{a^3b^2\cdot 3a}{1\cdot ab^3}} = \frac{ab}{\frac{3a^4b^2}{ab^3}} = \frac{\frac{ab}{1}}{\frac{3a^4b^2}{ab^3}} = \frac{ab\cdot ab^3}{1\cdot 3a^4b^2} = \frac{a^2b^4}{3a^4b^2} = \frac{\cancel{a^2}\overset{b^2}{\cancel{b^4}}}{3\underset{a^2}{\cancel{a^4}}\cancel{b^2}} = \boxed{\mathbf{\frac{b^2}{3a^2}}}$$

Example 5.4-34

$$\frac{2x^2y^2z\cdot\frac{1}{x^3y}\cdot\frac{xyz}{4z^2}}{\frac{xy}{z}} = \frac{\frac{2x^2y^2z}{1}\cdot\frac{1}{x^3y}\cdot\frac{xyz}{4z^2}}{\frac{xy}{z}} = \frac{\frac{2x^2y^2z\cdot 1\cdot xyz}{1\cdot x^3y\cdot 4z^2}}{\frac{xy}{z}} = \frac{\frac{2x^3y^3z^2}{4x^3yz^2}}{\frac{xy}{z}} = \frac{2x^3y^3z^2\cdot z}{4x^3yz^2\cdot xy} = \frac{2x^3y^3z^3}{4x^4y^2z^2}$$

$$= \frac{\cancel{2}\cancel{x^3}\overset{y}{\cancel{y^3}}\overset{z}{\cancel{z^3}}}{\underset{2}{\cancel{4}}\underset{x}{\cancel{x^4}}\cancel{y^2}\cancel{z^2}} = \boxed{\mathbf{\frac{yz}{2x}}}$$

Example 5.4-35

$$\frac{\frac{1}{b^2c^2}}{\frac{1}{ab^2c^3}\cdot a^2b\cdot\frac{ab^3c^2}{ab^4c}} = \frac{\frac{1}{b^2c^2}}{\frac{1}{ab^2c^3}\cdot\frac{a^2b}{1}\cdot\frac{ab^3c^2}{ab^4c}} = \frac{\frac{1}{b^2c^2}}{\frac{1\cdot a^2b\cdot ab^3c^2}{ab^2c^3\cdot 1\cdot ab^4c}} = \frac{\frac{1}{b^2c^2}}{\frac{a^3b^4c^2}{a^2b^6c^4}} = \frac{1\cdot a^2b^6c^4}{b^2c^2\cdot a^3b^4c^2} = \frac{a^2b^6c^4}{a^3b^6c^4}$$

$$= \frac{\cancel{a^2}\cancel{b^6}\cancel{c^4}}{\underset{a}{\cancel{a^3}}\cancel{b^6}\cancel{c^4}} = \boxed{\mathbf{\frac{1}{a}}}$$

Example 5.4-36

$$\frac{uv\cdot\frac{1}{u^2v^2}\cdot\frac{2u^3}{v^4}}{\frac{u}{v^3}} = \frac{\frac{uv}{1}\cdot\frac{1}{u^2v^2}\cdot\frac{2u^3}{v^4}}{\frac{u}{v^3}} = \frac{\frac{uv\cdot1\cdot2u^3}{1\cdot u^2v^2\cdot v^4}}{\frac{u}{v^3}} = \frac{\frac{2u^4v}{u^2v^6}}{\frac{u}{v^3}} = \frac{2u^4v\cdot v^3}{u^2v^6\cdot u} = \frac{2u^4v^4}{u^3v^6} = \frac{2\overset{u}{u^4}\,v^4}{u^3\,\underset{v^2}{v^6}} = \boxed{\frac{\mathbf{2u}}{\mathbf{v^2}}}$$

Example 5.4-37

$$\frac{\frac{xyz^2}{xy}\cdot\frac{x^2}{y^3}}{\frac{x}{y}\cdot\frac{y^2}{x^3}} = \frac{\frac{xyz^2\cdot x^2}{xy\cdot y^3}}{\frac{x\cdot y^2}{y\cdot x^3}} = \frac{\frac{x^3yz^2}{xy^4}}{\frac{xy^2}{x^3y}} = \frac{x^3yz^2\cdot x^3y}{xy^4\cdot xy^2} = \frac{x^6y^2z^2}{x^2y^6} = \frac{\overset{x^4}{x^6}\,y^2z^2}{x^2\,\underset{y^4}{y^6}} = \boxed{\frac{\mathbf{x^4z^2}}{\mathbf{y^4}}}$$

Example 5.4-38

$$\frac{\left(\frac{1}{a^3b^2}\cdot\frac{3a^2b}{5a}\right)\cdot\frac{10ab}{b^3}}{\frac{1}{a}} = \frac{\left(\frac{1\cdot3a^2b}{a^3b^2\cdot5a}\right)\cdot\frac{10ab}{b^3}}{\frac{1}{a}} = \frac{\frac{3a^2b}{5a^4b^2}\cdot\frac{10ab}{b^3}}{\frac{1}{a}} = \frac{\frac{3a^2b\cdot10ab}{5a^4b^2\cdot b^3}}{\frac{1}{a}} = \frac{\frac{30a^3b^2}{5a^4b^5}}{\frac{1}{a}} = \frac{30a^3b^2\cdot a}{5a^4b^5\cdot1}$$

$$= \frac{30a^4b^2}{5a^4b^5} = \frac{\overset{6}{30}\,a^4b^2}{5a^4\,\underset{b^3}{b^5}} = \boxed{\frac{\mathbf{6}}{\mathbf{b^3}}}$$

Example 5.4-39

$$\frac{x^2}{\left(x^2y^2\cdot\frac{1}{xy}\right)\cdot\left(\frac{x}{y^2}\cdot\frac{2x^2}{4xy^3}\right)} = \frac{x^2}{\left(\frac{x^2y^2}{1}\cdot\frac{1}{xy}\right)\cdot\left(\frac{x}{y^2}\cdot\frac{2x^2}{4xy^3}\right)} = \frac{x^2}{\left(\frac{x^2y^2\cdot1}{1\cdot xy}\right)\cdot\left(\frac{x\cdot2x^2}{y^2\cdot4xy^3}\right)} = \frac{x^2}{\left(\frac{x^2y^2}{xy}\right)\cdot\left(\frac{2x^3}{4xy^5}\right)}$$

$$= \frac{x^2}{\frac{x^2y^2\cdot2x^3}{xy\cdot4xy^5}} = \frac{x^2}{\frac{2x^5y^2}{4x^2y^6}} = \frac{\frac{x^2}{1}}{\frac{2x^5y^2}{4x^2y^6}} = \frac{x^2\cdot4x^2y^6}{1\cdot2x^5y^2} = \frac{4x^4y^6}{2x^5y^2} = \frac{\overset{2}{4}\,x^4\,\overset{y^4}{y^6}}{2\,\underset{x}{x^5}\,y^2} = \boxed{\frac{\mathbf{2y^4}}{\mathbf{x}}}$$

Example 5.4-40

$$\frac{2}{\frac{2a^3b}{a^2b^3}\cdot\left(\frac{2a}{a^2}\cdot\frac{a^3b^2}{8a}\right)} = \frac{2}{\frac{2a^3b}{a^2b^3}\cdot\left(\frac{2a\cdot a^3b^2}{a^2\cdot8a}\right)} = \frac{2}{\frac{2a^3b}{a^2b^3}\cdot\frac{2a^4b^2}{8a^3}} = \frac{2}{\frac{2a^3b\cdot2a^4b^2}{a^2b^3\cdot8a^3}} = \frac{2}{\frac{4a^7b^3}{8a^5b^3}} = \frac{\frac{2}{1}}{\frac{4a^7b^3}{8a^5b^3}}$$

$$= \frac{2\cdot8a^5b^3}{1\cdot4a^7b^3} = \frac{16a^5b^3}{4a^7b^3} = \frac{\overset{4}{16}\,a^5b^3}{4\,\underset{a^2}{a^7}\,b^3} = \boxed{\frac{\mathbf{4}}{\mathbf{a^2}}}$$

Practice Problems - Multiplication of Complex Algebraic Fractions

Section 5.4 Case II Practice Problems - Multiply the following complex algebraic expressions. Simplify the answer to its lowest term.

1. $\dfrac{y^2}{\dfrac{x^2y^2}{xy}\cdot\dfrac{2x^3y}{x^4}} =$

2. $\dfrac{\dfrac{1}{2x^3y^2}\cdot\dfrac{4x}{y}}{\dfrac{1}{xy^2}} =$

3. $\dfrac{\dfrac{x^2}{x^3}}{2x\cdot\dfrac{3x^2}{18x^5}} =$

4. $\dfrac{\left(x^2yz^2\cdot\dfrac{2x}{y^2}\right)\cdot\dfrac{3x}{x^3z^4}}{\dfrac{1}{y^2z^3}} =$

5. $\dfrac{\left(\dfrac{2a^2b^2}{a^3}\cdot 4ab^3\right)\cdot\dfrac{1}{3ab^4}}{\dfrac{8}{3}} =$

6. $\dfrac{\dfrac{u^3v^3w^2}{w^3}\cdot\dfrac{1}{3w}}{\dfrac{1}{6uv^2}} =$

7. $\dfrac{a^2b^4}{\dfrac{abc^3}{a^2b^2c^2}\cdot\left(abc\cdot\dfrac{2b^3c}{ab^2c^2}\right)} =$

8. $\dfrac{xyz^3\cdot\dfrac{xy}{z^2y^2}}{x^2}\cdot\dfrac{\dfrac{x^2y}{z}}{xy} =$

9. $\dfrac{\dfrac{y^2}{x}\cdot x^3}{\dfrac{x^3y^2}{2xy}\cdot\dfrac{4x^4y^2}{x^2}} =$

10. $\dfrac{\dfrac{xyz}{x^2z^2}\cdot\dfrac{x}{y^3}}{\dfrac{1}{x^2y^2}\cdot\dfrac{xz}{y}} =$

Case III Division of Complex Algebraic Fractions

Complex algebraic fractions are divided by one another using the following steps:

Step 1 Write the algebraic expression in fraction form, i.e., write x^2y^3 as $\frac{x^2y^3}{1}$.

Step 2 a. Invert the second fraction in either the numerator or the denominator, or both.

b. Change the division sign to a multiplication sign.

c. Multiply the numerator and the denominator of the algebraic fraction terms by one another.

Step 3 Change the complex algebraic fraction to a simple fraction. Reduce the algebraic fraction to its lowest term, if possible.

Examples with Steps

The following examples show the steps as to how complex algebraic fractions are divided:

Example 5.4-41

$$\boxed{\frac{\frac{x^3y^3}{x} \div x^2y^2}{x}} =$$

Solution:

Step 1

$$\boxed{\frac{\frac{x^3y^3}{x} \div x^2y^2}{x}} = \boxed{\frac{\frac{x^3y^3}{x} \div \frac{x^2y^2}{1}}{\frac{x}{1}}}$$

Step 2

$$\boxed{\frac{\frac{x^3y^3}{x} \div \frac{x^2y^2}{1}}{\frac{x}{1}}} = \boxed{\frac{\frac{x^3y^3}{x} \cdot \frac{1}{x^2y^2}}{\frac{x}{1}}} = \boxed{\frac{\frac{x^3y^3 \cdot 1}{x \cdot x^2y^2}}{\frac{x}{1}}} = \boxed{\frac{\frac{x^3y^3}{x^3y^2}}{\frac{x}{1}}}$$

Step 3

$$\boxed{\frac{\frac{x^3y^3}{x^3y^2}}{\frac{x}{1}}} = \boxed{\frac{x^3y^3 \cdot 1}{x^3y^2 \cdot x}} = \boxed{\frac{x^3y^3}{x^4y^2}} = \boxed{\frac{\overset{y}{\cancel{x^3}\,\cancel{y^3}}}{\underset{x}{\cancel{x^4}\,\cancel{y^2}}}} = \boxed{\frac{y}{x}}$$

Example 5.4-42

$$\boxed{\frac{\frac{a^3b^2c}{b^2c^3} \div a^3}{\frac{ab}{a^3} \div b^2}} =$$

Solution:

Step 1

$$\boxed{\frac{\frac{a^3b^2c}{b^2c^3} \div a^3}{\frac{ab}{a^3} \div b^2}} = \boxed{\frac{\frac{a^3b^2c}{b^2c^3} \div \frac{a^3}{1}}{\frac{ab}{a^3} \div \frac{b^2}{1}}}$$

Step 2

$$\frac{\frac{a^3b^2c}{b^2c^3} \div \frac{a^3}{1}}{\frac{ab}{a^3} \div \frac{b^2}{1}} = \frac{\frac{a^3b^2c}{b^2c^3} \cdot \frac{1}{a^3}}{\frac{ab}{a^3} \cdot \frac{1}{b^2}} = \frac{\frac{a^3b^2c \cdot 1}{b^2c^3 \cdot a^3}}{\frac{ab \cdot 1}{a^3 \cdot b^2}} = \frac{\frac{a^3b^2c}{a^3b^2c^3}}{\frac{ab}{a^3b^2}}$$

Step 3

$$\frac{\frac{a^3b^2c}{a^3b^2c^3}}{\frac{ab}{a^3b^2}} = \frac{a^3b^2c \cdot a^3b^2}{a^3b^2c^3 \cdot ab} = \frac{a^6b^4c}{a^4b^3c^3} = \frac{\overset{a^2\ b}{\cancel{a^6}\,\cancel{b^4}\,\cancel{c}}}{\underset{c^2}{\cancel{a^4}\,\cancel{b^3}\,\cancel{c^3}}} = \boxed{\mathbf{\frac{a^2b}{c^2}}}$$

Example 5.4-43

$$\frac{\frac{2xy}{x^2y^3} \div \frac{1}{xy}}{\frac{x}{x^3}} =$$

Solution:

Step 1 Not Applicable

Step 2

$$\frac{\frac{2xy}{x^2y^3} \div \frac{1}{xy}}{\frac{x}{x^3}} = \frac{\frac{2xy}{x^2y^3} \cdot \frac{xy}{1}}{\frac{x}{x^3}} = \frac{\frac{2xy \cdot xy}{x^2y^3 \cdot 1}}{\frac{x}{x^3}} = \frac{\frac{2x^2y^2}{x^2y^3}}{\frac{x}{x^3}}$$

Step 3

$$\frac{\frac{2x^2y^2}{x^2y^3}}{\frac{x}{x^3}} = \frac{2x^2y^2 \cdot x^3}{x^2y^3 \cdot x} = \frac{2x^5y^2}{x^3y^3} = \frac{\overset{x^2}{2\,\cancel{x^5}\,\cancel{y^2}}}{\underset{y}{\cancel{x^3}\,\cancel{y^3}}} = \boxed{\mathbf{\frac{2x^2}{y}}}$$

Example 5.4-44

$$\frac{\frac{uv^2w}{w^3} \div \frac{u^2v^3}{u}}{\frac{u}{v}} =$$

Solution:

Step 1 Not Applicable

Step 2

$$\frac{\frac{uv^2w}{w^3} \div \frac{u^2v^3}{u}}{\frac{u}{v}} = \frac{\frac{uv^2w}{w^3} \cdot \frac{u}{u^2v^3}}{\frac{u}{v}} = \frac{\frac{uv^2w \cdot u}{w^3 \cdot u^2v^3}}{\frac{u}{v}} = \frac{\frac{u^2v^2w}{u^2v^3w^3}}{\frac{u}{v}}$$

Step 3

$$\frac{\frac{u^2v^2w}{u^2v^3w^3}}{\frac{u}{v}} = \frac{u^2v^2w \cdot v}{u^2v^3w^3 \cdot u} = \frac{u^2v^3w}{u^3v^3w^3} = \frac{\cancel{u^2}\,\cancel{v^3}\,\cancel{w}}{\underset{u}{\cancel{u^3}}\,\cancel{v^3}\,\underset{w^2}{\cancel{w^3}}} = \boxed{\mathbf{\frac{1}{uw^2}}}$$

Example 5.4-45

$$3xyz \div \frac{\frac{xy}{z} \div z^2}{\frac{xy^2}{xyz^2}} =$$

Solution:

Step 1

$$3xyz \div \frac{\frac{xy}{z} \div z^2}{\frac{xy^2}{xyz^2}} = \frac{3xyz}{1} \div \frac{\frac{xy}{z} \div \frac{z^2}{1}}{\frac{xy^2}{xyz^2}}$$

Step 2

$$\frac{3xyz}{1} \div \frac{\frac{xy}{z} \div \frac{z^2}{1}}{\frac{xy^2}{xyz^2}} = \frac{3xyz}{1} \div \frac{\frac{xy}{z} \cdot \frac{1}{z^2}}{\frac{xy^2}{xyz^2}} = \frac{3xyz}{1} \div \frac{\frac{xy \cdot 1}{z \cdot z^2}}{\frac{xy^2}{xyz^2}} = \frac{3xyz}{1} \div \frac{\frac{xy}{z^3}}{\frac{xy^2}{xyz^2}}$$

Step 3

$$\frac{3xyz}{1} \div \frac{\frac{xy}{z^3}}{\frac{xy^2}{xyz^2}} = \frac{3xyz}{1} \div \frac{xy \cdot xyz^2}{z^3 \cdot xy^2} = \frac{3xyz}{1} \div \frac{x^2y^2z^2}{xy^2z^3} = \frac{3xyz}{1} \cdot \frac{xy^2z^3}{x^2y^2z^2}$$

$$= \frac{3xyz \cdot xy^2z^3}{1 \cdot x^2y^2z^2} = \frac{3x^2y^3z^4}{x^2y^2z^2} = \frac{3\cancel{x^2}\,\overset{y}{\cancel{y^3}}\,\overset{z^2}{\cancel{z^4}}}{\cancel{x^2}\cancel{y^2}\cancel{z^2}} = \frac{3yz^2}{1} = \mathbf{3yz^2}$$

Additional Examples - Division of Complex Algebraic Fractions

The following examples further illustrate how to divide complex algebraic fractions:

Example 5.4-46

$$\frac{xyz^2 \div \frac{x^2z^3}{3x}}{xy} = \frac{\frac{xyz^2}{1} \div \frac{x^2z^3}{3x}}{xy} = \frac{\frac{xyz^2}{1} \cdot \frac{3x}{x^2z^3}}{xy} = \frac{\frac{xyz^2 \cdot 3x}{1 \cdot x^2z^3}}{xy} = \frac{\frac{3x^2yz^2}{x^2z^3}}{xy} = \frac{\frac{3x^2yz^2}{x^2z^3}}{\frac{xy}{1}} = \frac{3x^2yz^2 \cdot 1}{x^2z^3 \cdot xy}$$

$$= \frac{3x^2yz^2}{x^3yz^3} = \frac{3\cancel{x^2}\cancel{y}\cancel{z^2}}{\underset{x}{\cancel{x^3}}\,\cancel{y}\,\underset{z}{\cancel{z^3}}} = \mathbf{\frac{3}{xz}}$$

Example 5.4-47

$$\frac{\frac{2}{x}}{\frac{x^2y}{2x} \div \frac{x^2y^2}{4x^3}} = \frac{\frac{2}{x}}{\frac{x^2y}{2x} \cdot \frac{4x^3}{x^2y^2}} = \frac{\frac{2}{x}}{\frac{x^2y \cdot 4x^3}{2x \cdot x^2y^2}} = \frac{\frac{2}{x}}{\frac{4x^5y}{2x^3y^2}} = \frac{2 \cdot 2x^3y^2}{x \cdot 4x^5y} = \frac{4x^3y^2}{4x^6y} = \frac{\cancel{4x^3}\,\overset{y}{\cancel{y^2}}}{\cancel{4}\,\underset{x^3}{\cancel{x^6}}\,\cancel{y}} = \mathbf{\frac{y}{x^3}}$$

Example 5.4-48

$$\frac{a^2b^2}{ab \div \frac{a^2b}{a^3}} = \frac{a^2b^2}{\frac{ab}{1}\div\frac{a^2b}{a^3}} = \frac{a^2b^2}{\frac{ab}{1}\cdot\frac{a^3}{a^2b}} = \frac{a^2b^2}{\frac{ab\cdot a^3}{1\cdot a^2b}} = \frac{a^2b^2}{\frac{a^4b}{a^2b}} = \frac{\frac{a^2b^2}{1}}{\frac{a^4b}{a^2b}} = \frac{a^2b^2\cdot a^2b}{1\cdot a^4b} = \frac{a^4b^3}{a^4b} = \frac{\overset{b^2}{\cancel{a^4}\,\cancel{b^3}}}{\cancel{a^4}\,\cancel{b}}$$

$$= \frac{b^2}{1} = \boxed{\mathbf{b^2}}$$

Example 5.4-49

$$\frac{\frac{xyz}{2x}\div\frac{x^2y^2}{x^3}}{\frac{x}{y}} = \frac{\frac{xyz}{2x}\cdot\frac{x^3}{x^2y^2}}{\frac{x}{y}} = \frac{\frac{xyz\cdot x^3}{2x\cdot x^2y^2}}{\frac{x}{y}} = \frac{\frac{x^4yz}{2x^3y^2}}{\frac{x}{y}} = \frac{x^4yz\cdot y}{2x^3y^2\cdot x} = \frac{x^4y^2z}{2x^4y^2} = \frac{\cancel{x^4}\,\cancel{y^2}\,z}{2\cancel{x^4}\,\cancel{y^2}} = \boxed{\mathbf{\frac{z}{2}}}$$

Example 5.4-50

$$a^2b^2 \div \frac{\frac{a}{b^3}}{\frac{a^2b}{b}\div\frac{a^3}{b^2}} = a^2b^2\div\frac{\frac{a}{b^3}}{\frac{a^2b}{b}\cdot\frac{b^2}{a^3}} = a^2b^2\div\frac{\frac{a}{b^3}}{\frac{a^2b\cdot b^2}{b\cdot a^3}} = a^2b^2\div\frac{\frac{a}{b^3}}{\frac{a^2b^3}{a^3b}} = a^2b^2\div\frac{a\cdot a^3b}{b^3\cdot a^2b^3}$$

$$= a^2b^2\div\frac{a^4b}{a^2b^6} = a^2b^2\cdot\frac{a^2b^6}{a^4b} = \frac{a^2b^2}{1}\cdot\frac{a^2b^6}{a^4b} = \frac{a^2b^2\cdot a^2b^6}{1\cdot a^4b} = \frac{a^4b^8}{a^4b} = \frac{\overset{b^7}{\cancel{a^4}\,\cancel{b^8}}}{\cancel{a^4}\,\cancel{b}} = \frac{b^7}{1} = \boxed{\mathbf{b^7}}$$

Example 5.4-51

$$\frac{1}{\frac{a^3bc}{b^2c^3}} \div \frac{a}{\frac{a^2b^2}{b}} = \frac{\frac{1}{1}}{\frac{a^3bc}{b^2c^3}} \div \frac{\frac{a}{1}}{\frac{a^2b^2}{b}} = \frac{1\cdot b^2c^3}{1\cdot a^3bc}\div\frac{a\cdot b}{1\cdot a^2b^2} = \frac{b^2c^3}{a^3bc}\div\frac{ab}{a^2b^2} = \frac{b^2c^3}{a^3bc}\cdot\frac{a^2b^2}{ab} = \frac{b^2c^3\cdot a^2b^2}{a^3bc\cdot ab}$$

$$= \frac{a^2b^4c^3}{a^4b^2c} = \frac{\cancel{a^2}\,\overset{b^2}{\cancel{b^4}}\,\overset{c^2}{\cancel{c^3}}}{\underset{a^2}{\cancel{a^4}}\,\cancel{b^2}\,\cancel{c}} = \boxed{\mathbf{\frac{b^2c^2}{a^2}}}$$

Example 5.4-52

$$\frac{\frac{u^2v^2}{uw}\div\frac{uv}{w^2}}{\frac{2uw}{u^2}\div\frac{uv}{w}} = \frac{\frac{u^2v^2}{uw}\cdot\frac{w^2}{uv}}{\frac{2uw}{u^2}\cdot\frac{w}{uv}} = \frac{\frac{u^2v^2\cdot w^2}{uw\cdot uv}}{\frac{2uw\cdot w}{u^2\cdot uv}} = \frac{\frac{u^2v^2w^2}{u^2vw}}{\frac{2uw^2}{u^3v}} = \frac{u^2v^2w^2\cdot u^3v}{u^2vw\cdot 2uw^2} = \frac{u^5v^3w^2}{2u^3vw^3} = \frac{\overset{u^2}{\cancel{u^5}}\,\overset{v^2}{\cancel{v^3}}\,\cancel{w^2}}{2\cancel{u^3}\,\cancel{v}\,\underset{w}{\cancel{w^3}}} = \boxed{\mathbf{\frac{u^2v^2}{2w}}}$$

Example 5.4-53

$$\frac{\frac{2xyz^3}{x^2y^2}\div\frac{z^2}{y^2}}{x^3} = \frac{\frac{2xyz^3}{x^2y^2}\cdot\frac{y^2}{z^2}}{x^3} = \frac{\frac{2xyz^3\cdot y^2}{x^2y^2\cdot z^2}}{x^3} = \frac{\frac{2xy^3z^3}{x^2y^2z^2}}{x^3} = \frac{\frac{2xy^3z^3}{x^2y^2z^2}}{\frac{x^3}{1}} = \frac{2xy^3z^3\cdot 1}{x^2y^2z^2\cdot x^3} = \frac{2xy^3z^3}{x^5y^2z^2}$$

$$= \boxed{\frac{2\cancel{x}\,y^{3}z^{3}}{x^{\cancel{5}}\,y^{2}z^{2}}} = \boxed{\frac{2yz}{x^{4}}}$$

Example 5.4-54

$$\boxed{\frac{3x \div \frac{x^2}{y^2}}{\frac{x}{2} \div y}} = \boxed{\frac{\frac{3x}{1} \div \frac{x^2}{y^2}}{\frac{x}{2} \div \frac{y}{1}}} = \boxed{\frac{\frac{3x}{1} \cdot \frac{y^2}{x^2}}{\frac{x}{2} \cdot \frac{1}{y}}} = \boxed{\frac{\frac{3x \cdot y^2}{1 \cdot x^2}}{\frac{x \cdot 1}{2 \cdot y}}} = \boxed{\frac{\frac{3xy^2}{x^2}}{\frac{x}{2y}}} = \boxed{\frac{3xy^2 \cdot 2y}{x^2 \cdot x}} = \boxed{\frac{6xy^3}{x^3}} = \boxed{\frac{6\cancel{x}y^3}{x^{\cancel{3}}}} = \boxed{\frac{6y^3}{x^2}}$$

Example 5.4-55

$$\boxed{\frac{\frac{2m^2n^2}{n} \div n^3}{\frac{m^2}{n} \div m}} = \boxed{\frac{\frac{2m^2n^2}{n} \div \frac{n^3}{1}}{\frac{m^2}{n} \div \frac{m}{1}}} = \boxed{\frac{\frac{2m^2n^2}{n} \cdot \frac{1}{n^3}}{\frac{m^2}{n} \cdot \frac{1}{m}}} = \boxed{\frac{\frac{2m^2n^2 \cdot 1}{n \cdot n^3}}{\frac{m^2 \cdot 1}{n \cdot m}}} = \boxed{\frac{\frac{2m^2n^2}{n^4}}{\frac{m^2}{mn}}} = \boxed{\frac{2m^2n^2 \cdot mn}{n^4 \cdot m^2}} = \boxed{\frac{2m^3n^3}{m^2n^4}}$$

$$= \boxed{\frac{2m^{\cancel{3}}\,n^{\cancel{3}}}{m^{\cancel{2}}\,n^{\cancel{4}}}} = \boxed{\frac{2m}{n}}$$

Practice Problems - Division of Complex Algebraic Fractions

Section 5.4 Case III Practice Problems - Divide the following complex algebraic fractions. Simplify the answer to its lowest term.

1. $\dfrac{\frac{xy}{x^3} \div \frac{x^2}{xy^2}}{2x} =$

2. $\dfrac{\frac{1}{a} \div \frac{1}{b}}{\frac{a}{ab^2} \div b} =$

3. $xy \div \left(\dfrac{\frac{1}{xy} \div x}{x^3}\right) =$

4. $\dfrac{\frac{u^2vw}{w^2} \div \frac{u^3v}{w}}{w^3} =$

5. $\dfrac{ab^3 \div \frac{a^2b^2}{b}}{a \div \frac{a}{b}} =$

6. $\dfrac{\frac{z^3}{z^2} \div 2z}{z^3 \div \frac{z^2}{2}} =$

7. $\dfrac{\frac{ab^2cd^4}{cd^2} \div \frac{c^2d^3}{b^2}}{a^2b^2} =$

8. $\dfrac{\frac{u^2}{u} \div u^3}{1 \div \frac{1}{u^2}} =$

9. $\dfrac{6x^2y \div \frac{x^2}{y^2}}{\frac{x}{y} \div \frac{x^2}{xy}} =$

10. $\dfrac{m^2 \div \frac{m^3}{n}}{n \div \frac{n}{m}} =$

Case IV Mixed Operations Involving Complex Algebraic Fractions

Complex algebraic fractions are added, subtracted, multiplied, and divided by one another using the following steps:

Step 1 Change the complex algebraic fractions to simple algebraic fractions by performing fractional operations as required.

Step 2 Add, subtract, multiply, or divide the algebraic fractions by each other. Simplify the algebraic fraction to its lowest term, if possible.

Examples with Steps

The following examples show the steps as to how math operations involving complex algebraic fractions are performed:

Example 5.4-56

$$\boxed{\frac{\frac{x}{y}-1}{\frac{x}{y}+1}\div\frac{1}{x+y}} =$$

Solution:

Step 1

$$\boxed{\frac{\frac{x}{y}-1}{\frac{x}{y}+1}\div\frac{1}{x+y}} = \boxed{\frac{\frac{x}{y}-\frac{1}{1}}{\frac{x}{y}+\frac{1}{1}}\div\frac{1}{x+y}} = \boxed{\frac{\frac{(x\cdot 1)-(1\cdot y)}{y\cdot 1}}{\frac{(x\cdot 1)+(1\cdot y)}{y\cdot 1}}\div\frac{1}{x+y}} = \boxed{\frac{\frac{x-y}{y}}{\frac{x+y}{y}}\div\frac{1}{x+y}}$$

$$= \boxed{\frac{(x-y)\cdot y}{y\cdot(x+y)}\div\frac{1}{x+y}} = \boxed{\frac{(x-y)y}{y(x+y)}\div\frac{1}{x+y}}$$

Step 2

$$\boxed{\frac{(x-y)y}{y(x+y)}\div\frac{1}{x+y}} = \boxed{\frac{(x-y)\not{y}}{\not{y}(x+y)}\div\frac{1}{x+y}} = \boxed{\frac{x-y}{x+y}\div\frac{1}{x+y}} = \boxed{\frac{x-y}{x+y}\cdot\frac{x+y}{1}}$$

$$= \boxed{\frac{(x-y)\cdot(x+y)}{(x+y)\cdot 1}} = \boxed{\frac{(x-y)\cdot(\not{x}+\not{y})}{(\not{x}+\not{y})\cdot 1}} = \boxed{\frac{x-y}{1}} = \boxed{x-y}$$

Example 5.4-57

$$\boxed{\frac{\frac{1}{a}-1}{\frac{1}{a}}\cdot\frac{a^2}{1-a}} =$$

Solution:

Step 1

$$\boxed{\frac{\frac{1}{a}-1}{\frac{1}{a}}\cdot\frac{a^2}{1-a}} = \boxed{\frac{\frac{1}{a}-\frac{1}{1}}{\frac{1}{a}}\cdot\frac{a^2}{1-a}} = \boxed{\frac{\frac{(1\cdot 1)-(1\cdot a)}{a\cdot 1}}{\frac{1}{a}}\cdot\frac{a^2}{1-a}} = \boxed{\frac{\frac{1-a}{a}}{\frac{1}{a}}\cdot\frac{a^2}{1-a}}$$

$$= \boxed{\frac{(1-a)\cdot a}{a\cdot 1}\cdot\frac{a^2}{1-a}} = \boxed{\frac{(1-a)a}{a}\cdot\frac{a^2}{1-a}}$$

Step 2

$$\boxed{\frac{(1-a)a}{a}\cdot\frac{a^2}{1-a}}=\boxed{\frac{(1-a)\not{a}}{\not{a}}\cdot\frac{a^2}{1-a}}=\boxed{\frac{1-a}{1}\cdot\frac{a^2}{1-a}}=\boxed{\frac{(1-a)\cdot a^2}{1\cdot(1-a)}}=\boxed{\frac{(\not{1}-\not{a})\cdot a^2}{1\cdot(\not{1}-\not{a})}}=\boxed{\mathbf{a^2}}$$

Example 5.4-58

$$\boxed{\frac{\frac{1}{2x}+\frac{2}{3y}}{\frac{3}{2y}-\frac{1}{3x}}-1}=$$

Solution:

Step 1

$$\boxed{\frac{\frac{1}{2x}+\frac{2}{3y}}{\frac{3}{2y}-\frac{1}{3x}}-1}=\boxed{\frac{\frac{(1\cdot 3y)+(2\cdot 2x)}{2x\cdot 3y}}{\frac{(3\cdot 3x)-(1\cdot 2y)}{2y\cdot 3x}}-1}=\boxed{\frac{\frac{3y+4x}{6xy}}{\frac{9x-2y}{6xy}}-1}=\boxed{\frac{(3y+4x)\cdot 6xy}{6xy\cdot(9x-2y)}-1}$$

Step 2

$$\boxed{\frac{(3y+4x)\cdot 6xy}{6xy\cdot(9x-2y)}-1}=\boxed{\frac{(3y+4x)\not{6xy}}{\not{6xy}(9x-2y)}-1}=\boxed{\frac{3y+4x}{9x-2y}-\frac{1}{1}}=\boxed{\frac{[(3y+4x)\cdot 1]-[1\cdot(9x-2y)]}{(9x-2y)\cdot 1}}$$

$$=\boxed{\frac{3y+4x-9x+2y}{9x-2y}}=\boxed{\frac{(4x-9x)+(3y+2y)}{9x-2y}}=\boxed{\frac{-5x+5y}{9x-2y}}=\boxed{\mathbf{\frac{-5(x-y)}{9x-2y}}}$$

Example 5.4-59

$$\boxed{\frac{\frac{x+1}{x-1}-\frac{x}{x-1}}{x}\cdot\frac{x^2-x}{5}}=$$

Solution:

Step 1

$$\boxed{\frac{\frac{x+1}{x-1}-\frac{x}{x-1}}{x}\cdot\frac{x^2-x}{5}}=\boxed{\frac{\frac{x+1-x}{x-1}}{x}\cdot\frac{x^2-x}{5}}=\boxed{\frac{\frac{\not{x}-\not{x}+1}{x-1}}{x}\cdot\frac{x^2-x}{5}}=\boxed{\frac{\frac{1}{x-1}}{x}\cdot\frac{x^2-x}{5}}$$

$$=\boxed{\frac{\frac{1}{x-1}}{\frac{x}{1}}\cdot\frac{x^2-x}{5}}=\boxed{\frac{1\cdot 1}{(x-1)\cdot x}\cdot\frac{x^2-x}{5}}=\boxed{\frac{1}{x^2-x}\cdot\frac{x^2-x}{5}}$$

Step 2

$$\boxed{\frac{1}{x^2-x}\cdot\frac{x^2-x}{5}}=\boxed{\frac{1\cdot(x^2-\not{x})}{(x^2-\not{x})\cdot 5}}=\boxed{\mathbf{\frac{1}{5}}}$$

Example 5.4-60

$$\boxed{\frac{\frac{x}{y}+2}{1-\frac{x^2}{y^2}}\div\frac{y}{y+x}}=$$

Solution:

Step 1

$$\boxed{\frac{\frac{x}{y}+2}{1-\frac{x^2}{y^2}}\div\frac{y}{y+x}}=\boxed{\frac{\frac{x}{y}+\frac{2}{1}}{\frac{1}{1}-\frac{x^2}{y^2}}\div\frac{y}{y+x}}=\boxed{\frac{\frac{(x\cdot 1)+(2\cdot y)}{y\cdot 1}}{\frac{(1\cdot y^2)-(x^2\cdot 1)}{1\cdot y^2}}\div\frac{y}{y+x}}=\boxed{\frac{\frac{x+2y}{y}}{\frac{y^2-x^2}{y^2}}\div\frac{y}{y+x}}$$

$$= \boxed{\frac{(x+2y)\cdot y^2}{y\cdot\left(y^2-x^2\right)} \div \frac{y}{y+x}}$$

Step 2

$$\boxed{\frac{(x+2y)\cdot y^2}{y\cdot\left(y^2-x^2\right)} \div \frac{y}{y+x}} = \boxed{\frac{(x+2y)\cdot y^2}{y\cdot\left(y^2-x^2\right)} \cdot \frac{y+x}{y}} = \boxed{\frac{y^2(x+2y)\cdot(y+x)}{y\left(y^2-x^2\right)\cdot y}}$$

$$= \boxed{\frac{y^2(x+2y)\cdot(y+x)}{y^2(y-x)(y+x)}} = \boxed{\mathbf{\frac{x+2y}{y-x}}}$$

Additional Examples - Mixed Operations Involving Complex Algebraic Fractions

The following examples further illustrate how to solve math operations involving complex algebraic fractions:

Example 5.4-61

$$\boxed{\frac{\frac{1}{a}+\frac{1}{b}}{a} \cdot \frac{\frac{a}{a+b}}{\frac{b}{a}}} = \boxed{\frac{\frac{1}{a}+\frac{1}{b}}{\frac{a}{1}} \cdot \frac{\frac{a}{a+b}}{\frac{b}{a}}} = \boxed{\frac{\frac{(1\cdot b)+(1\cdot a)}{a\cdot b}}{\frac{a}{1}} \cdot \frac{\frac{a}{a+b}}{\frac{b}{a}}} = \boxed{\frac{\frac{b+a}{ab}}{\frac{a}{1}} \cdot \frac{\frac{a}{a+b}}{\frac{b}{a}}} = \boxed{\frac{(b+a)\cdot 1}{ab\cdot a} \cdot \frac{a\cdot a}{(a+b)\cdot b}}$$

$$= \boxed{\frac{a+b}{a^2b} \cdot \frac{a^2}{b(a+b)}} = \boxed{\frac{(a+b)\cdot a^2}{a^2b\cdot b(a+b)}} = \boxed{\frac{a^2(a+b)}{a^2b^2(a+b)}} = \boxed{\frac{a^2(\not{a}+\not{b})}{a^2b^2(\not{a}+\not{b})}} = \boxed{\mathbf{\frac{1}{b^2}}}$$

Example 5.4-62

$$\boxed{\frac{\frac{1}{2x}+\frac{2}{y}}{\frac{3}{2x}-\frac{1}{y}} - \frac{4}{\frac{3y-2x}{x}}} = \boxed{\frac{\frac{1}{2x}+\frac{2}{y}}{\frac{3}{2x}-\frac{1}{y}} - \frac{\frac{4}{1}}{\frac{3y-2x}{x}}} = \boxed{\frac{\frac{(1\cdot y)+(2\cdot 2x)}{2x\cdot y}}{\frac{(3\cdot y)-(1\cdot 2x)}{2x\cdot y}} - \frac{\frac{4}{1}}{\frac{3y-2x}{x}}} = \boxed{\frac{\frac{y+4x}{2xy}}{\frac{3y-2x}{2xy}} - \frac{\frac{4}{1}}{\frac{3y-2x}{x}}}$$

$$= \boxed{\frac{(y+4x)\cdot 2xy}{2xy\cdot(3y-2x)} - \frac{4\cdot x}{1\cdot(3y-2x)}} = \boxed{\frac{(y+4x)\cdot \not{2xy}}{\not{2xy}\cdot(3y-2x)} - \frac{4x}{(3y-2x)}} = \boxed{\frac{y+4x}{3y-2x} - \frac{4x}{3y-2x}} = \boxed{\frac{y+4x-4x}{3y-2x}}$$

$$= \boxed{\frac{y+\not{4x}-\not{4x}}{3y-2x}} = \boxed{\frac{y}{3y-2x}} = \boxed{\frac{y}{-(2x-3y)}} = \boxed{\mathbf{-\frac{y}{2x-3y}}}$$

Example 5.4-63

$$\boxed{\frac{\frac{x}{y}+\frac{y}{x}}{\frac{1}{xy}} \cdot \frac{4}{x^2+y^2}} = \boxed{\frac{\frac{x\cdot x+y\cdot y}{y\cdot x}}{\frac{1}{xy}} \cdot \frac{4}{x^2+y^2}} = \boxed{\frac{\frac{x^2+y^2}{xy}}{\frac{1}{xy}} \cdot \frac{4}{x^2+y^2}} = \boxed{\frac{\left(x^2+y^2\right)\cdot xy}{xy\cdot 1} \cdot \frac{4}{x^2+y^2}}$$

$$= \boxed{\frac{\left(x^2+y^2\right)\cdot \not{xy}}{\not{xy}} \cdot \frac{4}{x^2+y^2}} = \boxed{\frac{x^2+y^2}{1} \cdot \frac{4}{x^2+y^2}} = \boxed{\frac{\left(x^2+y^2\right)\cdot 4}{1\cdot\left(x^2+y^2\right)}} = \boxed{\frac{4\left(\not{x^2}+\not{y^2}\right)}{\left(\not{x^2}+\not{y^2}\right)}} = \boxed{\frac{4}{1}} = \boxed{\mathbf{4}}$$

Example 5.4-64

$$\frac{\frac{x}{y}-1}{1-\frac{x}{y}}+1=\frac{\frac{x}{y}-\frac{1}{1}}{\frac{1}{1}-\frac{x}{y}}+1=\frac{\frac{(x\cdot1)-(1\cdot y)}{y\cdot1}}{\frac{(1\cdot y)-(x\cdot1)}{1\cdot y}}+1=\frac{\frac{x-y}{y}}{\frac{y-x}{y}}+1=\frac{(x-y)\cdot y}{y\cdot(y-x)}+1=\frac{(x-y)\cdot\cancel{y}}{\cancel{y}\cdot(y-x)}+1=\frac{x-y}{y-x}+1$$

$$=\frac{x-y}{-(x-y)}+1=\frac{(\cancel{x-y})}{-(\cancel{x-y})}+1=\frac{1}{-1}+1=-\frac{1}{1}+1=-1+1=\boxed{0}$$

Example 5.4-65

$$\frac{\frac{1}{a+b}-\frac{1}{a-b}}{\frac{1}{a^2-b^2}}=\frac{\frac{[1\cdot(a-b)]-[1\cdot(a+b)]}{(a+b)\cdot(a-b)}}{\frac{1}{a^2-b^2}}=\frac{\frac{(a-b)-(a+b)}{a^2-ab+ab-b^2}}{\frac{1}{a^2-b^2}}=\frac{\frac{a-b-a-b}{a^2-\cancel{ab}+\cancel{ab}-b^2}}{\frac{1}{a^2-b^2}}=\frac{\frac{\cancel{a}-b-\cancel{a}-b}{a^2-\cancel{ab}+\cancel{ab}-b^2}}{\frac{1}{a^2-b^2}}$$

$$=\frac{\frac{-2b}{a^2-b^2}}{\frac{1}{a^2-b^2}}=\frac{-2b\cdot(a^2-b^2)}{(a^2-b^2)\cdot1}=\frac{-2b\cdot\cancel{(a^2-b^2)}}{\cancel{(a^2-b^2)}}=\frac{-2b}{1}=\boxed{-2b}$$

Example 5.4-66

$$\frac{\frac{x}{y}+1}{1-\frac{x^2}{y^2}}=\frac{\frac{x}{y}+\frac{1}{1}}{\frac{1}{1}-\frac{x^2}{y^2}}=\frac{\frac{(x\cdot1)+(1\cdot y)}{y\cdot1}}{\frac{(1\cdot y^2)-(x^2\cdot1)}{1\cdot y^2}}=\frac{\frac{x+y}{y}}{\frac{y^2-x^2}{y^2}}=\frac{(x+y)\cdot y^2}{y\cdot(y^2-x^2)}=\frac{(x+y)\cdot y^2}{y\cdot(y-x)(y+x)}=\frac{y^2(x+y)}{y(y-x)(x+y)}$$

$$=\frac{\overset{y}{\cancel{y^2}}(\cancel{x+y})}{\cancel{y}(y-x)(\cancel{x+y})}=\boxed{\frac{y}{y-x}}$$

Example 5.4-67

$$\frac{\frac{5}{x+1}-\frac{1}{x+1}}{1+\frac{1}{x+1}}=\frac{\frac{5}{x+1}-\frac{1}{x+1}}{\frac{1}{1}+\frac{1}{x+1}}=\frac{\frac{5-1}{x+1}}{\frac{[1\cdot(x+1)]+(1\cdot1)}{1\cdot(x+1)}}=\frac{\frac{4}{x+1}}{\frac{(x+1)+1}{x+1}}=\frac{\frac{4}{x+1}}{\frac{x+1+1}{x+1}}=\frac{\frac{4}{x+1}}{\frac{x+2}{x+1}}=\frac{4\cdot(x+1)}{(x+1)\cdot(x+2)}$$

$$=\frac{4\cdot\cancel{(x+1)}}{\cancel{(x+1)}\cdot(x+2)}=\boxed{\frac{4}{x+2}}$$

Example 5.4-68

$$\frac{\frac{a}{b}\div\frac{b^2}{a}}{\frac{a}{b^2}}\cdot\frac{\frac{a}{b}}{\frac{1}{ab}}=\frac{\frac{a}{b}\cdot\frac{a}{b^2}}{\frac{a}{b^2}}\cdot\frac{\frac{a}{b}}{\frac{1}{ab}}=\frac{\frac{a\cdot a}{b\cdot b^2}}{\frac{a}{b^2}}\cdot\frac{\frac{a}{b}}{\frac{1}{ab}}=\frac{\frac{a^2}{b^3}}{\frac{a}{b^2}}\cdot\frac{\frac{a}{b}}{\frac{1}{ab}}=\frac{a^2\cdot b^2}{b^3\cdot a}\cdot\frac{a\cdot ab}{b\cdot1}=\frac{a^2b^2}{ab^3}\cdot\frac{a^2b}{b}=\frac{a^2b^2\cdot a^2b}{ab^3\cdot b}$$

$$= \boxed{\frac{a^4b^3}{ab^4}} = \boxed{\frac{\overset{a^3}{\cancel{a^4}}\, \cancel{b^3}}{\cancel{a}\, \underset{b}{\cancel{b^4}}}} = \boxed{\frac{\mathbf{a^3}}{\mathbf{b}}}$$

Example 5.4-69

$$\boxed{\frac{4}{\frac{x^2+x-2}{\frac{1}{x-1}+1}}+1} = \boxed{\frac{4}{\frac{x^2+x-2}{\frac{1}{x-1}+\frac{1}{1}}}+1} = \boxed{\frac{4}{\frac{x^2+x-2}{\frac{(1\cdot 1)+[1\cdot(x-1)]}{(x-1)\cdot 1}}}+1} = \boxed{\frac{4}{\frac{x^2+x-2}{\frac{1+x-1}{x-1}}}+1} = \boxed{\frac{4}{\frac{x^2+x-2}{\frac{\cancel{1}+x-\cancel{1}}{x-1}}}+1} = \boxed{\frac{4}{\frac{x^2+x-2}{\frac{x}{x-1}}}+1}$$

$$= \boxed{\frac{4\cdot(x-1)}{(x^2+x-2)\cdot x}+1} = \boxed{\frac{4(x-1)}{x(x-1)(x+2)}+1} = \boxed{\frac{4(\cancel{x-1})}{x(\cancel{x-1})(x+2)}+1} = \boxed{\frac{4}{x(x+2)}+1} = \boxed{\frac{4}{x(x+2)}+\frac{1}{1}}$$

$$= \boxed{\frac{(4\cdot 1)+[1\cdot x(x+2)]}{x(x+2)\cdot 1}} = \boxed{\frac{4+x^2+2x}{x(x+2)}} = \boxed{\frac{\mathbf{x^2+2x+4}}{\mathbf{x(x+2)}}}$$

Example 5.4-70

$$\boxed{\frac{\frac{1}{u}+\frac{1}{u^3}}{1-\frac{1}{u^3}} \div \frac{\frac{1}{2}}{u^3}} = \boxed{\frac{\frac{1}{u}+\frac{1}{u^3}}{\frac{1}{1}-\frac{1}{u^3}} \div \frac{\frac{1}{2}}{\frac{u^3}{1}}} = \boxed{\frac{\frac{(1\cdot u^3)+(1\cdot u)}{u\cdot u^3}}{\frac{(1\cdot u^3)-(1\cdot 1)}{1\cdot u^3}} \div \frac{1\cdot 1}{2\cdot u^3}} = \boxed{\frac{\frac{u^3+u}{u^4}}{\frac{u^3-1}{u^3}} \div \frac{1}{2u^3}} = \boxed{\frac{(u^3+u)\cdot u^3}{u^4\cdot(u^3-1)} \div \frac{1}{2u^3}}$$

$$= \boxed{\frac{u^4(u^2+1)}{u^4(u^3-1)} \div \frac{1}{2u^3}} = \boxed{\frac{u^2+1}{u^3-1} \div \frac{1}{2u^3}} = \boxed{\frac{u^2+1}{u^3-1}\cdot\frac{2u^3}{1}} = \boxed{\frac{(u^2+1)\cdot 2u^3}{(u^3-1)\cdot 1}} = \boxed{\frac{2u^3(u^2+1)}{u^3-1}} = \boxed{\frac{\mathbf{2u^3(u^2+1)}}{\mathbf{(u-1)(u^2+u+1)}}}$$

Example 5.4-71

$$\boxed{\frac{\frac{a}{b}\cdot\frac{b^2}{a}}{\frac{a}{b^2}} \div \frac{\frac{a}{b}}{\frac{1}{ab}}} = \boxed{\frac{\frac{a\cdot b^2}{b\cdot a}}{\frac{a}{b^2}} \div \frac{\frac{a}{b}}{\frac{1}{ab}}} = \boxed{\frac{\frac{ab^2}{ab}}{\frac{a}{b^2}} \div \frac{\frac{a}{b}}{\frac{1}{ab}}} = \boxed{\frac{ab^2\cdot b^2}{ab\cdot a} \div \frac{a\cdot ab}{b\cdot 1}} = \boxed{\frac{ab^4}{a^2b} \div \frac{a^2b}{b}} = \boxed{\frac{ab^4}{a^2b}\cdot\frac{b}{a^2b}} = \boxed{\frac{ab^4\cdot b}{a^2b\cdot a^2b}}$$

$$= \boxed{\frac{ab^5}{a^4b^2}} = \boxed{\frac{\cancel{a}\,\overset{b^3}{\cancel{b^5}}}{\underset{a^3}{\cancel{a^4}}\,\cancel{b^2}}} = \boxed{\frac{\mathbf{b^3}}{\mathbf{a^3}}}$$

Example 5.4-72

$$\boxed{\frac{\frac{x}{y}+\frac{y}{x}}{\frac{1}{xy}}-1} = \boxed{\frac{\frac{x\cdot x+y\cdot y}{y\cdot x}}{\frac{1}{xy}}-\frac{1}{1}} = \boxed{\frac{\frac{x^2+y^2}{xy}}{\frac{1}{xy}}-\frac{1}{1}} = \boxed{\frac{(x^2+y^2)\cdot xy}{xy\cdot 1}-\frac{1}{1}} = \boxed{\frac{(x^2+y^2)\cdot \cancel{xy}}{\cancel{xy}}-\frac{1}{1}} = \boxed{\frac{x^2+y^2}{1}-\frac{1}{1}}$$

$$= \boxed{\frac{(x^2+y^2)-1}{1}} = \boxed{\mathbf{x^2+y^2-1}}$$

Practice Problems - Mixed Operations Involving Complex Algebraic Fractions

Section 5.4 Case IV Practice Problems - Solve the following complex algebraic fractions. Simplify the answer to its lowest term.

1. $\dfrac{a-\dfrac{2}{3}}{a+\dfrac{4}{5}} =$

2. $\dfrac{1+\dfrac{a}{b}}{1-\dfrac{a}{b}} =$

3. $\dfrac{\dfrac{1}{x}-\dfrac{1}{y}}{\dfrac{1}{x}+\dfrac{1}{y}} \cdot \dfrac{1}{y-x} =$

4. $\dfrac{\dfrac{1}{x^2}+\dfrac{1}{y^2}}{\dfrac{1}{x^2y^2}-1} \div \dfrac{\dfrac{1}{3}}{\dfrac{1-xy}{2}} =$

5. $\dfrac{\dfrac{x}{y}+2}{1-\dfrac{x}{y}} \cdot \left(1-\dfrac{x}{y}\right) =$

6. $\dfrac{\dfrac{1}{a}+\dfrac{1}{b}}{a} \div \dfrac{\dfrac{1}{a}-\dfrac{1}{b}}{b} =$

7. $\dfrac{2-\dfrac{1}{y+1}}{\dfrac{2}{y+1}-2} + \dfrac{1}{2y} =$

8. $\dfrac{\dfrac{1}{2x}+\dfrac{2}{5y}}{\dfrac{2}{5y}-\dfrac{1}{2x}} \div (5y+4x) =$

9. $\dfrac{\dfrac{1}{x^2}-\dfrac{1}{y^2}}{\dfrac{1}{x^2y^2}} \cdot \dfrac{2}{y^2-x^2} =$

10. $\dfrac{\dfrac{1}{x}-\dfrac{1}{y}}{\dfrac{1}{xy}} + x =$

5.5 Solving One Variable Equations Containing Algebraic Fractions

In section 4.5 Case II, solutions to quadratic equations, which are a class of non-linear equations, containing fractional coefficients were addressed. In this section, solutions to one variable linear and non-linear equations containing algebraic fractions are discussed. Again, note that in dealing with fractional equations the solution(s) may not satisfy the original equation. This is because fractions may encounter division by zero which is undefined. Therefore, it is essential that the solution(s) to a linear or non-linear equation be checked by substitution into the original equation in order to ensure division by zero does not occur. One variable equations containing algebraic fractions are solved using the following steps:

Step 1 a. Apply the fraction rules to simplify the algebraic expression.

b. Cross multiply the terms in both sides of the equation.

Step 2 Solve the equation. Check the answer(s) by substituting the solution(s) into the original equation. Disregard any apparent solution(s).

Examples with Steps

The following examples show the steps as to how equations containing algebraic fractions are solved:

Example 5.5-1

$$\boxed{\frac{1}{3}x+\frac{1}{2}x=x+2}$$

Solution:

Step 1

$$\boxed{\frac{1}{3}x+\frac{1}{2}x=x+2}\;;\;\boxed{\frac{x}{3}+\frac{x}{2}=x+2}\;;\;\boxed{\frac{(x\cdot 2)+(x\cdot 3)}{3\cdot 2}=x+2}\;;\;\boxed{\frac{2x+3x}{6}=x+2}$$

$$;\;\boxed{\frac{5x}{6}=\frac{x+2}{1}}\;;\;\boxed{5x\cdot 1=(x+2)\cdot 6}\;;\;\boxed{5x=6x+12}$$

Step 2

$$\boxed{5x=6x+12}\;;\;\boxed{5x-6x=6x-6x+12}\;;\;\boxed{-x=0+12}\;;\;\boxed{-x=12}\;;\;\boxed{\frac{-x}{-1}=\frac{12}{-1}}$$

$$;\;\boxed{x=-\frac{12}{1}}\;;\;\boxed{\mathbf{x=-12}}$$

Check: $\frac{1}{3}\cdot(-12)+\frac{1}{2}\cdot(-12)\overset{?}{=}-12+2\;;\;-\frac{12}{3}-\frac{12}{2}\overset{?}{=}-10\;;\;\frac{(-12\cdot 2)+(-12\cdot 3)}{3\cdot 2}\overset{?}{=}-10\;;\;\frac{-24-36}{6}\overset{?}{=}-10$

$;\;-\frac{\overset{10}{\cancel{60}}}{\cancel{6}}\overset{?}{=}-10\;;\;-\frac{10}{1}\overset{?}{=}-10\;;\;-10=-10$

Example 5.5-2

$$\boxed{\frac{y+1}{3}-\frac{y-1}{5}=1}$$

Solution:

Step 1

$$\boxed{\frac{y+1}{3}-\frac{y-1}{5}=1}\;;\;\boxed{\frac{[(y+1)\cdot 5]-[(y-1)\cdot 3]}{3\cdot 5}=1}\;;\;\boxed{\frac{5y+5-3y+3}{15}=1}$$

$$; \boxed{\frac{(5y-3y)+(3+5)}{15}=1} ; \boxed{\frac{2y+8}{15}=\frac{1}{1}} ; \boxed{(2y+8)\cdot 1 = 15\cdot 1} ; \boxed{2y+8=15}$$

Step 2 $\boxed{2y+8=15}$; $\boxed{2y+8-8=15-8}$; $\boxed{2y+0=7}$; $\boxed{2y=7}$; $\boxed{\frac{\cancel{2}y}{\cancel{2}}=\frac{7}{2}}$; $\boxed{y=\frac{7}{2}}$; $\boxed{y=3\frac{1}{2}}$

; $\boxed{\mathbf{y=3.5}}$

Check: $\frac{3.5+1}{3}-\frac{3.5-1}{5}\overset{?}{=}1$; $\frac{4.5}{3}-\frac{2.5}{5}\overset{?}{=}1$; $\frac{(4.5\cdot 5)-(2.5\cdot 3)}{3\cdot 5}\overset{?}{=}1$; $\frac{22.5-7.5}{15}\overset{?}{=}1$; $\frac{\cancel{15}}{\cancel{15}}\overset{?}{=}1$; $\frac{1}{1}\overset{?}{=}1$; $1=1$

Example 5.5-3

$$\boxed{\frac{2}{2x+5}+\frac{3}{3x+2}=\frac{1}{2}}$$

Solution:

Step 1 $\boxed{\frac{2}{2x+5}+\frac{3}{3x+2}=\frac{1}{2}}$; $\boxed{\frac{[2\cdot(3x+2)]+[3\cdot(2x+5)]}{(2x+5)\cdot(3x+2)}=\frac{1}{2}}$; $\boxed{\frac{6x+4+6x+15}{6x^2+4x+15x+10}=\frac{1}{2}}$

; $\boxed{\frac{(6x+6x)+(15+4)}{6x^2+(4x+15x)+10}=\frac{1}{2}}$; $\boxed{\frac{12x+19}{6x^2+19x+10}=\frac{1}{2}}$; $\boxed{(12x+19)\cdot 2 = 1\cdot\left(6x^2+19x+10\right)}$

; $\boxed{24x+38=6x^2+19x+10}$

Step 2 $\boxed{24x+38=6x^2+19x+10}$; $\boxed{6x^2+(19x-24x)+10-38=0}$; $\boxed{6x^2-5x-28=0}$

; $\boxed{(x-2.62)(x+1.78)=0}$

Therefore, the two apparent solutions are $\boxed{\mathbf{x=2.62}}$ and $\boxed{\mathbf{x=-1.78}}$

Check: 1. $\frac{2}{(2\cdot 2.62)+5}+\frac{3}{(3\cdot 2.62)+2}\overset{?}{=}\frac{1}{2}$; $\frac{2}{5.24+5}+\frac{3}{7.86+2}\overset{?}{=}\frac{1}{2}$; $\frac{2}{10.24}+\frac{3}{9.86}\overset{?}{=}\frac{1}{2}$

; $0.195+0.305\overset{?}{=}0.5$; $0.5=0.5$

2. $\frac{2}{(2\cdot -1.78)+5}+\frac{3}{(3\cdot -1.78)+2}\overset{?}{=}\frac{1}{2}$; $\frac{2}{-3.56+5}+\frac{3}{-5.34+2}\overset{?}{=}\frac{1}{2}$; $\frac{2}{1.44}-\frac{3}{3.34}\overset{?}{=}\frac{1}{2}$

; $1.39-0.89\overset{?}{=}0.5$; $0.5=0.5$

Thus, $\mathbf{x=2.62}$ and $\mathbf{x=-1.78}$ are both solutions to the original equation.

Example 5.5-4

$$\boxed{\frac{y-1}{y}=1+\frac{4}{y-1}}$$

Solution:

Step 1 $\boxed{\frac{y-1}{y}=1+\frac{4}{y-1}}$; $\boxed{\frac{y-1}{y}=\frac{1}{1}+\frac{4}{y-1}}$; $\boxed{\frac{y-1}{y}=\frac{[1\cdot(y-1)]+(4\cdot 1)}{1\cdot(y-1)}}$; $\boxed{\frac{y-1}{y}=\frac{y-1+4}{y-1}}$

; $\boxed{\frac{y-1}{y}=\frac{y+3}{y-1}}$; $\boxed{(y-1)\cdot(y-1)=(y+3)\cdot y}$; $\boxed{y^2-y-y+1=y^2+3y}$

$; \boxed{y^2-2y+1=y^2+3y}$

Step 2 $\boxed{y^2-2y+1=y^2+3y}$; $\boxed{\left(y^2-y^2\right)+(-2y-3y)+1=0}$; $\boxed{-5y+1=0}$; $\boxed{-5y=-1}$

$; \boxed{\frac{-\cancel{5}y}{-\cancel{5}}=\frac{-1}{-5}}$; $\boxed{y=\frac{-1}{-5}}$; $\boxed{y=\frac{1}{5}}$; $\boxed{\mathbf{y=0.2}}$

Check: $\frac{0.2-1}{0.2}\overset{?}{=}1+\frac{4}{0.2-1}$; $\frac{-0.8}{0.2}\overset{?}{=}1+\frac{4}{-0.8}$; $-\frac{0.8}{0.2}\overset{?}{=}1-\frac{4}{0.8}$; $-4\overset{?}{=}1-5$; $-4=-4$

Example 5.5-5

$\boxed{\frac{x+3}{x-1}-\frac{x}{x+2}=\frac{5}{x^2+x-2}}$

Solution:

Step 1 $\boxed{\frac{x+3}{x-1}-\frac{x}{x+2}=\frac{5}{x^2+x-2}}$; $\boxed{\frac{[(x+3)\cdot(x+2)]-[x\cdot(x-1)]}{(x-1)\cdot(x+2)}=\frac{5}{x^2+x-2}}$

$; \boxed{\frac{x^2+2x+3x+6-x^2+x}{x^2+2x-x-2}=\frac{5}{x^2+x-2}}$; $\boxed{\frac{\left(x^2-x^2\right)+(2x+3x+x)+6}{x^2+(2x-x)-2}=\frac{5}{x^2+x-2}}$

$; \boxed{\frac{6x+6}{x^2+x-2}=\frac{5}{x^2+x-2}}$; $\boxed{(6x+6)\cdot\left(x^2+x-2\right)=5\cdot\left(x^2+x-2\right)}$

Step 2 $\boxed{(6x+6)\cdot\left(x^2+x-2\right)=5\cdot\left(x^2+x-2\right)}$; $\boxed{(6x+6)\cdot\left(\cancel{x^2+x-2}\right)=5\cdot\left(\cancel{x^2+x-2}\right)}$

$; \boxed{6x+6=5}$; $\boxed{6x=5-6}$; $\boxed{6x=-1}$; $\boxed{\frac{\cancel{6}x}{\cancel{6}}=\frac{-1}{6}}$; $\boxed{x=-\frac{1}{6}}$; $\boxed{\mathbf{x=-0.167}}$

Check: $\frac{-0.167+3}{-0.167-1}-\frac{-0.167}{-0.167+2}\overset{?}{=}\frac{5}{(-0.167)^2-0.167-2}$; $\frac{2.833}{-1.167}-\frac{-0.167}{1.833}\overset{?}{=}\frac{5}{0.0278-0.167-2}$

$; -\frac{2.833}{1.167}+\frac{0.167}{1.833}\overset{?}{=}\frac{5}{-2.139}$; $-2.428+0.091\overset{?}{=}-2.337$; $-2.337=-2.337$

Additional Examples - Solving One Variable Equations Containing Algebraic Fractions

The following examples further illustrate how to solve equations containing algebraic fractions:

Example 5.5-6

$\boxed{\frac{x}{3}+\frac{x}{2}=5}$; $\boxed{\frac{(2\cdot x)+(3\cdot x)}{3\cdot 2}=5}$; $\boxed{\frac{2x+3x}{6}=5}$; $\boxed{\frac{5x}{6}=\frac{5}{1}}$; $\boxed{5x\cdot 1=5\cdot 6}$; $\boxed{5x=30}$; $\boxed{x=\frac{\overset{6}{\cancel{30}}}{\cancel{5}}}$; $\boxed{x=\frac{6}{1}}$; $\boxed{\mathbf{x=6}}$

Check: $\frac{6}{3}+\frac{6}{2}\overset{?}{=}5$; $\frac{6\cdot 2+6\cdot 3}{3\cdot 2}\overset{?}{=}5$; $\frac{12+18}{6}\overset{?}{=}5$; $\frac{\overset{5}{\cancel{30}}}{\cancel{6}}\overset{?}{=}5$; $\frac{5}{1}\overset{?}{=}5$; $5=5$

Example 5.5-7

$\boxed{\frac{2}{x}=\frac{5}{10}}$; $\boxed{2\cdot 10=5\cdot x}$; $\boxed{20=5x}$; $\boxed{5x=20}$; $\boxed{\frac{\cancel{5}x}{\cancel{5}}=\frac{20}{5}}$; $\boxed{x=\frac{\overset{4}{\cancel{20}}}{\cancel{5}}}$; $\boxed{x=\frac{4}{1}}$; $\boxed{\mathbf{x=4}}$

Check: $\frac{2}{4}\overset{?}{=}\frac{5}{10}$; $\frac{\cancel{2}}{\underset{2}{\cancel{4}}}\overset{?}{=}\frac{\cancel{5}}{\underset{2}{\cancel{10}}}$; $\frac{1}{2}=\frac{1}{2}$

Example 5.5-8

$\boxed{\frac{y-2}{3y+1}=\frac{1}{2}}$; $\boxed{2\cdot(y-2)=1\cdot(3y+1)}$; $\boxed{2y-4=3y+1}$; $\boxed{2y-3y=1+4}$; $\boxed{-y=5}$; $\boxed{\frac{-y}{-1}=\frac{5}{-1}}$; $\boxed{\mathbf{y=-5}}$

Check: $\frac{-5-2}{(3\cdot -5)+1}\overset{?}{=}\frac{1}{2}$; $\frac{-7}{-15+1}\overset{?}{=}\frac{1}{2}$; $\frac{-7}{-14}\overset{?}{=}\frac{1}{2}$; $\frac{7}{14}\overset{?}{=}\frac{1}{2}$; $\frac{\cancel{7}}{\underset{2}{\cancel{14}}}\overset{?}{=}\frac{1}{2}$; $\frac{1}{2}=\frac{1}{2}$

Example 5.5-9

$\boxed{\frac{u}{3}+\frac{u}{5}=2}$; $\boxed{\frac{u}{3}+\frac{u}{5}=\frac{2}{1}}$; $\boxed{\frac{(u\cdot 5)+(u\cdot 3)}{3\cdot 5}=\frac{2}{1}}$; $\boxed{\frac{5u+3u}{15}=\frac{2}{1}}$; $\boxed{\frac{8u}{15}=\frac{2}{1}}$; $\boxed{8u\cdot 1=2\cdot 15}$; $\boxed{8u=30}$; $\boxed{\frac{\cancel{8}u}{\cancel{8}}=\frac{30}{8}}$

; $\boxed{u=\frac{\overset{15}{\cancel{30}}}{\underset{4}{\cancel{8}}}}$; $\boxed{u=\frac{15}{4}}$; $\boxed{u=3\frac{3}{4}}$; $\boxed{\mathbf{u=3.75}}$

Check: $\frac{3.75}{3}+\frac{3.75}{5}\overset{?}{=}2$; $1.25+0.75\overset{?}{=}2$; $2=2$

Example 5.5-10

$\boxed{\frac{2}{x+1}+\frac{3}{x-1}=\frac{4}{x^2-1}}$; $\boxed{\frac{[2\cdot(x-1)]+[3\cdot(x+1)]}{(x+1)\cdot(x-1)}=\frac{4}{x^2-1}}$; $\boxed{\frac{2x-2+3x+3}{x^2-\cancel{x}+\cancel{x}-1}=\frac{4}{x^2-1}}$; $\boxed{\frac{5x+1}{x^2-1}=\frac{4}{x^2-1}}$

; $\boxed{(5x+1)\cdot(x^2-1)=4\cdot(x^2-1)}$; $\boxed{(5x+1)\cdot(\cancel{x^2}-\cancel{1})=4\cdot(\cancel{x^2}-\cancel{1})}$; $\boxed{5x+1=4}$; $\boxed{5x=4-1}$; $\boxed{5x=3}$; $\boxed{\frac{\cancel{5}x}{\cancel{5}}=\frac{3}{5}}$

; $\boxed{x=\frac{3}{5}}$; $\boxed{\mathbf{x=0.6}}$

Check: $\frac{2}{0.6+1}+\frac{3}{0.6-1}\overset{?}{=}\frac{4}{0.6^2-1}$; $\frac{2}{1.6}+\frac{3}{-0.4}\overset{?}{=}\frac{4}{0.36-1}$; $1.25-7.5\overset{?}{=}\frac{4}{-0.64}$; $-6.25\overset{?}{=}-\frac{4}{0.64}$; $-6.25=-6.25$

Example 5.5-11

$\boxed{\frac{u}{u+1}=\frac{2}{3}-\frac{1}{u+1}}$; $\boxed{\frac{u}{u+1}=\frac{2\cdot(u+1)-1\cdot 3}{3\cdot(u+1)}}$; $\boxed{\frac{u}{u+1}=\frac{2u+2-3}{3u+3}}$; $\boxed{\frac{u}{u+1}=\frac{2u-1}{3u+3}}$; $\boxed{u\cdot(3u+3)=(2u-1)\cdot(u+1)}$

; $\boxed{3u^2+3u=2u^2+2u-u-1}$; $\boxed{3u^2-2u^2+3u-2u+u+1=0}$; $\boxed{u^2+2u+1=0}$; $\boxed{(u+1)^2=0}$; $\boxed{(u+1)^2=0}$

; $\boxed{\sqrt{(u+1)^2}=\pm\sqrt{0}}$; $\boxed{u+1=0}$; $\boxed{u=-1}$

A second and perhaps easier way of solving this problem would be as follows:

$\boxed{\frac{u}{u+1}=\frac{2}{3}-\frac{1}{u+1}}$; $\boxed{\frac{u}{u+1}+\frac{1}{u+1}=\frac{2}{3}}$; $\boxed{\frac{u+1}{u+1}=\frac{2}{3}}$; $\boxed{3\cdot(u+1)=2\cdot(u+1)}$; $\boxed{3u+3=2u+2}$; $\boxed{3u-2u=2-3}$

; $\boxed{u=-1}$

Check: $\frac{-1}{-1+1}\overset{?}{=}\frac{2}{3}-\frac{1}{-1+1}$; $\frac{-1}{0}\overset{?}{=}\frac{2}{3}-\frac{1}{0}$; we know that $\frac{1}{0}$ is undefined. Therefore, **the original equation has no solution.**

Example 5.5-12

$\boxed{\frac{1}{x+2}+\frac{1}{x-2}=\frac{1}{x^2-4}}$; $\boxed{\frac{1\cdot(x-2)+1\cdot(x+2)}{(x+2)\cdot(x-2)}=\frac{1}{x^2-4}}$; $\boxed{\frac{x-2+x+2}{x^2-2x+2x-4}=\frac{1}{x^2-4}}$

; $\boxed{\frac{x-\cancel{2}+x+\cancel{2}}{x^2-\cancel{2x}+\cancel{2x}-4}=\frac{1}{x^2-4}}$; $\boxed{\frac{2x}{x^2-4}=\frac{1}{x^2-4}}$; $\boxed{\frac{2x}{x^2-4}-\frac{1}{x^2-4}=0}$; $\boxed{\frac{2x-1}{x^2-4}=0}$; $\boxed{\frac{2x-1}{x^2-4}=\frac{0}{1}}$

; $\boxed{(2x-1)\cdot 1=0\cdot\left(x^2-4\right)}$; $\boxed{2x-1=0}$; $\boxed{2x=1}$; $\boxed{\frac{\cancel{2}x}{\cancel{2}}=\frac{1}{2}}$; $\boxed{x=\frac{1}{2}}$; $\boxed{\mathbf{x=0.5}}$

Check: $\frac{1}{0.5+2}+\frac{1}{0.5-2}\overset{?}{=}\frac{1}{0.5^2-4}$; $\frac{1}{2.5}+\frac{1}{-1.5}\overset{?}{=}\frac{1}{0.25-4}$; $\frac{1}{2.5}-\frac{1}{1.5}\overset{?}{=}-\frac{1}{3.75}$; $\frac{(1\cdot 1.5)-(1\cdot 2.5)}{2.5\cdot 1.5}\overset{?}{=}-\frac{1}{3.75}$

; $\frac{1.5-2.5}{3.75}\overset{?}{=}-\frac{1}{3.75}$; $\frac{-1}{3.75}\overset{?}{=}-\frac{1}{3.75}$; $-\frac{1}{3.75}=-\frac{1}{3.75}$

Example 5.5-13

$\boxed{\frac{a+1}{5}-\frac{a}{7}=\frac{1}{3}}$; $\boxed{\frac{[(a+1)\cdot 7]-(a\cdot 5)}{5\cdot 7}=\frac{1}{3}}$; $\boxed{\frac{7a+7-5a}{35}=\frac{1}{3}}$; $\boxed{\frac{2a+7}{35}=\frac{1}{3}}$; $\boxed{(2a+7)\cdot 3=1\cdot 35}=\boxed{6a+21=35}$

; $\boxed{6a=35-21}$; $\boxed{6a=14}$; $\boxed{\frac{\cancel{6}a}{\cancel{6}}=\frac{14}{6}}$; $\boxed{a=\frac{\overset{7}{\cancel{14}}}{\underset{3}{\cancel{6}}}}$; $\boxed{a=\frac{7}{3}}$; $\boxed{a=2\frac{1}{3}}$; $\boxed{\mathbf{a=2.33}}$

Check: $\frac{2.33+1}{5}-\frac{2.33}{7}\overset{?}{=}\frac{1}{3}$; $\frac{3.33}{5}-\frac{2.33}{7}\overset{?}{=}\frac{1}{3}$; $\frac{(3.33\cdot 7)-(2.33\cdot 5)}{5\cdot 7}\overset{?}{=}\frac{1}{3}$; $\frac{23.31-11.65}{35}\overset{?}{=}\frac{1}{3}$; $\frac{11.66}{35}\overset{?}{=}\frac{1}{3}$

; $0.33=0.33$

Example 5.5-14

$\boxed{\frac{4}{5}=\frac{a}{a-1}}$; $\boxed{4\cdot(a-1)=a\cdot 5}$; $\boxed{4a-4=5a}$; $\boxed{4a-5a=4}$; $\boxed{-a=4}$; $\boxed{\mathbf{a=-4}}$

Check: $\frac{4}{5}\overset{?}{=}\frac{-4}{-4-1}$; $\frac{4}{5}\overset{?}{=}\frac{-4}{-5}$; $\frac{4}{5}=\frac{4}{5}$

Example 5.5-15

$\boxed{\frac{1}{x}-\frac{2}{x}=\frac{3}{x}+1}$; $\boxed{\frac{1}{x}-\frac{2}{x}-\frac{3}{x}=1}$; $\boxed{\frac{1-2-3}{x}=1}$; $\boxed{\frac{1-5}{x}=1}$; $\boxed{\frac{-4}{x}=\frac{1}{1}}$; $\boxed{-4\cdot 1=1\cdot x}$; $\boxed{-4=x}$; $\boxed{\mathbf{x=-4}}$

Check: $\frac{1}{-4}-\frac{2}{-4}\overset{?}{=}\frac{3}{-4}+1$; $-\frac{1}{4}+\frac{2}{4}\overset{?}{=}-\frac{3}{4}+1$; $\frac{-1+2}{4}\overset{?}{=}-\frac{3}{4}+\frac{1}{1}$; $\frac{1}{4}\overset{?}{=}\frac{(-3\cdot 1)+(1\cdot 4)}{4\cdot 1}$; $\frac{1}{4}\overset{?}{=}\frac{-3+4}{4}$; $\frac{1}{4}=\frac{1}{4}$

Example 5.5-16

$\boxed{\frac{1}{x-1}=\frac{2}{x^2-1}}$; $\boxed{\frac{1}{x-1}-\frac{2}{x^2-1}=0}$; $\boxed{\frac{\left[1\cdot\left(x^2-1\right)\right]-\left[2\cdot(x-1)\right]}{(x-1)\cdot\left(x^2-1\right)}=0}$; $\boxed{\frac{x^2-1-2x+2}{(x-1)\cdot\left(x^2-1\right)}=0}$; $\boxed{\frac{x^2-2x+1}{(x-1)\cdot\left(x^2-1\right)}=0}$

; $\boxed{\dfrac{(x-1)^2}{(x-1)\cdot\left(x^2-1\right)}=0}$; $\boxed{\dfrac{(x-1)(x-1)}{(x-1)\cdot\left(x^2-1\right)}=0}$; $\boxed{\dfrac{(x-1)(\cancel{x-1})}{(\cancel{x-1})\cdot\left(x^2-1\right)}=0}$; $\boxed{\dfrac{x-1}{x^2-1}=0}$; $\boxed{\dfrac{x-1}{x^2-1}=\dfrac{0}{1}}$

; $\boxed{(x-1)\cdot 1=0\cdot\left(x^2-1\right)}$; $\boxed{x-1=0}$; $\boxed{x=1}$

Check: $\dfrac{1}{1-1}\overset{?}{=}\dfrac{2}{1^2-1}$; $\dfrac{1}{0}\overset{?}{=}\dfrac{2}{1-1}$; $\dfrac{1}{0}\overset{?}{=}\dfrac{2}{0}$ Since division by zero is undefined, therefore, **the original equation has no solution.**

Example 5.5-17

$\boxed{\dfrac{x+1}{2}-\dfrac{x+2}{3}=2}$; $\boxed{\dfrac{[(x+1)\cdot 3]-[(x+2)\cdot 2]}{2\cdot 3}=2}$; $\boxed{\dfrac{3x+3-2x-4}{6}=2}$; $\boxed{\dfrac{(3x-2x)+(-4+3)}{6}=2}$; $\boxed{\dfrac{x-1}{6}=\dfrac{2}{1}}$

; $\boxed{(x-1)\cdot 1=2\cdot 6}$; $\boxed{x-1=12}$; $\boxed{x=12+1}$; $\boxed{\mathbf{x=13}}$

Check: $\dfrac{13+1}{2}-\dfrac{13+2}{3}\overset{?}{=}2$; $\dfrac{14}{2}-\dfrac{15}{3}\overset{?}{=}2$; $\dfrac{(14\cdot 3)-(15\cdot 2)}{2\cdot 3}\overset{?}{=}2$; $\dfrac{42-30}{6}\overset{?}{=}2$; $\dfrac{12}{6}\overset{?}{=}2$; $\dfrac{\overset{2}{\cancel{12}}}{\cancel{6}}\overset{?}{=}2$; $\dfrac{2}{1}\overset{?}{=}2$; $2=2$

Example 5.5-18

$\boxed{x+\dfrac{2x+1}{3}=5}$; $\boxed{\dfrac{x}{1}+\dfrac{2x+1}{3}=5}$; $\boxed{\dfrac{(x\cdot 3)+[(2x+1)\cdot 1]}{1\cdot 3}=5}$; $\boxed{\dfrac{3x+2x+1}{3}=5}$; $\boxed{\dfrac{5x+1}{3}=\dfrac{5}{1}}$; $\boxed{(5x+1)\cdot 1=5\cdot 3}$

; $\boxed{5x+1=15}$; $\boxed{5x=15-1}$; $\boxed{5x=14}$; $\boxed{\dfrac{\cancel{5}x}{\cancel{5}}=\dfrac{14}{5}}$; $\boxed{x=\dfrac{14}{5}}$; $\boxed{x=2\dfrac{4}{5}}$; $\boxed{\mathbf{x=2.8}}$

Check: $2.8+\dfrac{(2\cdot 2.8)+1}{3}\overset{?}{=}5$; $2.8+\dfrac{5.6+1}{3}\overset{?}{=}5$; $2.8+\dfrac{6.6}{3}\overset{?}{=}5$; $2.8+2.2\overset{?}{=}5$; $5=5$

Example 5.5-19

$\boxed{\dfrac{1}{x}+\dfrac{1}{x+1}=\dfrac{2}{3x}}$; $\boxed{\dfrac{[1\cdot(x+1)]+(1\cdot x)}{x\cdot(x+1)}=\dfrac{2}{3x}}$; $\boxed{\dfrac{x+1+x}{x(x+1)}=\dfrac{2}{3x}}$; $\boxed{\dfrac{2x+1}{x^2+x}=\dfrac{2}{3x}}$; $\boxed{(2x+1)\cdot 3x=2\cdot\left(x^2+x\right)}$

; $\boxed{6x^2+3x=2x^2+2x}$; $\boxed{6x^2-2x^2+3x-2x=0}$; $\boxed{4x^2+x=0}$; $\boxed{x(4x+1)=0}$

Thus, the solutions are $\boxed{x=0}$ and $\boxed{4x+1=0}$; $\boxed{4x=-1}$; $\boxed{\dfrac{4x}{4}=\dfrac{-1}{4}}$; $\boxed{x=-\dfrac{1}{4}}$; $\boxed{\mathbf{x=-0.25}}$

Check: 1. $\dfrac{1}{0}+\dfrac{1}{0+1}\overset{?}{=}\dfrac{2}{3x}$ Substitution of $x=0$ into the original equation result in division by zero which is undefined. **Therefore,** $x=0$ **is not a solution.**

2. $\dfrac{1}{-0.25}+\dfrac{1}{-0.25+1}\overset{?}{=}\dfrac{2}{3\cdot(-0.25)}$; $-\dfrac{1}{0.25}+\dfrac{1}{0.75}\overset{?}{=}\dfrac{2}{-0.75}$; $-4+1.33\overset{?}{=}-2.66$; $-2.66=-2.66$.

Thus, $x=-0.25$ is the solution to the original equation.

Example 5.5-20

$\boxed{\dfrac{2a+1}{3}-\dfrac{a+3}{4}=\dfrac{1}{2}}$; $\boxed{\dfrac{[(2a+1)\cdot 4]-[(a+3)\cdot 3]}{3\cdot 4}=\dfrac{1}{2}}$; $\boxed{\dfrac{8a+4-3a-9}{12}=\dfrac{1}{2}}$; $\boxed{\dfrac{(8a-3a)+(-9+4)}{12}=\dfrac{1}{2}}$

; $\boxed{\frac{5a-5}{12}=\frac{1}{2}}$; $\boxed{(5a-5)\cdot 2=1\cdot 12}$; $\boxed{10a-10=12}$; $\boxed{10a=12+10}$; $\boxed{10a=22}$; $\boxed{a=\frac{\overset{11}{\cancel{22}}}{\underset{5}{\cancel{10}}}}$; $\boxed{a=\frac{11}{5}}$; $\boxed{a=2\frac{1}{5}}$

; $\boxed{\mathbf{a=2.2}}$

Check: $\frac{(2\cdot 2.2)+1}{3}-\frac{2.2+3}{4}\overset{?}{=}\frac{1}{2}$; $\frac{4.4+1}{3}-\frac{5.2}{4}\overset{?}{=}\frac{1}{2}$; $\frac{5.4}{3}-\frac{5.2}{4}\overset{?}{=}\frac{1}{2}$; $\frac{(5.4\cdot 4)-(5.2\cdot 3)}{3\cdot 4}\overset{?}{=}\frac{1}{2}$; $\frac{21.6-15.6}{12}\overset{?}{=}\frac{1}{2}$

; $\frac{6}{12}\overset{?}{=}\frac{1}{2}$; $\frac{\cancel{6}}{\underset{2}{\cancel{12}}}\overset{?}{=}\frac{1}{2}$; $\frac{1}{2}=\frac{1}{2}$

Example 5.5-21

$\boxed{\frac{2}{x}=\frac{5}{x}+1}$; $\boxed{\frac{2}{x}-\frac{5}{x}=1}$; $\boxed{\frac{2-5}{x}=1}$; $\boxed{\frac{-3}{x}=1}$; $\boxed{\frac{-3}{x}=\frac{1}{1}}$; $\boxed{-3\cdot 1=1\cdot x}$; $\boxed{\mathbf{x=-3}}$

Check: $\frac{2}{-3}\overset{?}{=}\frac{5}{-3}+1$; $-\frac{2}{3}\overset{?}{=}-\frac{5}{3}+\frac{1}{1}$; $-\frac{2}{3}\overset{?}{=}\frac{(-5\cdot 1)+(1\cdot 3)}{3\cdot 1}$; $-\frac{2}{3}\overset{?}{=}\frac{-5+3}{3}$; $-\frac{2}{3}=-\frac{2}{3}$

Example 5.5-22

$\boxed{x+\frac{1}{x}=2}$; $\boxed{\frac{x}{1}+\frac{1}{x}=2}$; $\boxed{\frac{(x\cdot x)+(1\cdot 1)}{1\cdot x}=2}$; $\boxed{\frac{x^2+1}{x}=\frac{2}{1}}$; $\boxed{(x^2+1)\cdot 1=2\cdot x}$; $\boxed{x^2+1=2x}$; $\boxed{x^2-2x+1=0}$

; $\boxed{(x-1)^2=0}$; $\boxed{\sqrt{(x-1)^2}=\pm\sqrt{0}}$; $\boxed{x-1=0}$; $\boxed{\mathbf{x=1}}$

Check: $1+\frac{1}{1}\overset{?}{=}2$; $1+1\overset{?}{=}2$; $2=2$

Example 5.5-23

$\boxed{\frac{2x-1}{3}-\frac{2x}{5}=\frac{2}{3}}$; $\boxed{\frac{[(2x-1)\cdot 5]-(2x\cdot 3)}{3\cdot 5}=\frac{2}{3}}$; $\boxed{\frac{10x-5-6x}{15}=\frac{2}{3}}$; $\boxed{\frac{(10x-6x)-5}{15}=\frac{2}{3}}$; $\boxed{\frac{4x-5}{15}=\frac{2}{3}}$

; $\boxed{(4x-5)\cdot 3=2\cdot 15}$; $\boxed{12x-15=30}$; $\boxed{12x=30+15}$; $\boxed{12x=45}$; $\boxed{x=\frac{\overset{15}{\cancel{45}}}{\underset{4}{\cancel{12}}}}$; $\boxed{x=\frac{15}{4}}$; $\boxed{x=3\frac{3}{4}}$; $\boxed{\mathbf{x=3.75}}$

Check: $\frac{(2\cdot 3.75)-1}{3}-\frac{(2\cdot 3.75)}{5}\overset{?}{=}\frac{2}{3}$; $\frac{7.5-1}{3}-\frac{7.5}{5}\overset{?}{=}\frac{2}{3}$; $\frac{6.5}{3}-\frac{7.5}{5}\overset{?}{=}\frac{2}{3}$; $\frac{(6.5\cdot 5)-(7.5\cdot 3)}{3\cdot 5}\overset{?}{=}\frac{2}{3}$

; $\frac{32.5-22.5}{15}\overset{?}{=}\frac{2}{3}$; $\frac{10}{15}\overset{?}{=}\frac{2}{3}$; $0.66=0.66$

Example 5.5-24

$\boxed{\frac{2x-4}{x+1}+\frac{3}{x+1}=4}$; $\boxed{\frac{2x-4}{x+1}+\frac{3}{x+1}=\frac{4}{1}}$; $\boxed{\frac{2x-4+3}{x+1}=\frac{4}{1}}$; $\boxed{\frac{2x-1}{x+1}=\frac{4}{1}}$; $\boxed{(2x-1)\cdot 1=4\cdot(x+1)}$

; $\boxed{2x-1=4x+4}$; $\boxed{2x-4x=4+1}$; $\boxed{-2x=5}$; $\boxed{\frac{-\cancel{2}x}{-\cancel{2}}=\frac{5}{-2}}$; $\boxed{x=-\frac{5}{2}}$; $\boxed{x=-2\frac{1}{2}}$; $\boxed{\mathbf{x=-2.5}}$

Check: $\frac{2\cdot(-2.5)-4}{-2.5+1}+\frac{3}{-2.5+1}\overset{?}{=}4$; $\frac{-5-4}{-1.5}+\frac{3}{-1.5}\overset{?}{=}4$; $\frac{-9}{-1.5}+\frac{3}{-1.5}\overset{?}{=}4$; $\frac{-9+3}{-1.5}\overset{?}{=}4$; $\frac{-6}{-1.5}\overset{?}{=}4$; $\frac{6}{1.5}\overset{?}{=}4$

; $4=4$

Example 5.5-25

$$\boxed{\frac{2x-1}{x+1}=\frac{2x}{3x-1}}\;;\;\boxed{\frac{2x-1}{x+1}-\frac{2x}{3x-1}=0}\;;\;\boxed{\frac{[(2x-1)\cdot(3x-1)]-[2x\cdot(x+1)]}{(x+1)\cdot(3x-1)}=0}\;;\;\boxed{\frac{6x^2-2x-3x+1-2x^2-2x}{3x^2-x+3x-1}=0}$$

$$;\;\boxed{\frac{\left(6x^2-2x^2\right)+(-2x-3x-2x)+1}{3x^2+(-x+3x)-1}=0}\;;\;\boxed{\frac{4x^2-7x+1}{3x^2+2x-1}=\frac{0}{1}}\;;\;\boxed{\left(4x^2-7x+1\right)\cdot 1=0\cdot\left(3x^2+2x-1\right)}$$

$$;\;\boxed{4x^2-7x+1=0}\;;\;\boxed{(x-1.59)(x-0.158)=0}$$

Therefore, the solutions are $\boxed{x=1.59}$ and $\boxed{x=0.158}$

Check: 1. $\frac{(2\cdot 1.59)-1}{1.59+1}\overset{?}{=}\frac{2\cdot 1.59}{(3\cdot 1.59)-1}\;;\;\frac{3.18-1}{2.59}\overset{?}{=}\frac{3.18}{4.77-1}\;;\;\frac{2.18}{2.59}\overset{?}{=}\frac{3.18}{3.77}\;;\;0.84=0.84$

2. $\frac{(2\cdot 0.158)-1}{0.158+1}\overset{?}{=}\frac{2\cdot 0.158}{(3\cdot 0.158)-1}\;;\;\frac{0.316-1}{1.158}\overset{?}{=}\frac{0.316}{0.474-1}\;;\;\frac{-0.684}{1.158}\overset{?}{=}\frac{0.316}{-0.526}\;;\;-0.6=-0.6$

Thus, $x=1.59$ and $x=0.158$ are both solutions to the original equation.

Practice Problems - Solving One Variable Equations Containing Algebraic Fractions

Section 5.5 Practice Problems - Solve the following equations. Check for any apparent solution(s) by substituting the solution(s) into the original equation.

1. $x+\frac{x}{2}=3$
2. $\frac{x-3}{2x-1}=\frac{1}{3}$
3. $\frac{3}{x+5}-\frac{2}{x+5}=2$
4. $\frac{y}{2}-\frac{y}{5}=5$
5. $\frac{4}{x+3}-\frac{3}{x-3}=\frac{5}{x^2-9}$
6. $\frac{1}{x-1}+5=\frac{2}{x-1}$
7. $\frac{b+2}{3}-b=\frac{1}{3}$
8. $\frac{x-1}{x+2}+\frac{4}{x^2+5x+6}=\frac{x+1}{x+3}$
9. $\frac{1}{x+3}+\frac{2x}{x^2-9}=\frac{1}{x-3}$
10. $\frac{x+3}{x^2+2x}=\frac{1}{x+2}+\frac{1}{x}$

Chapter 6
Logarithms

Quick Reference to Chapter 6 Case Problems

$\boxed{\log_3 x + \log_3 6 = \log_2 8 - \log_2 4}$; $\boxed{\ln(x-5) + \ln x = \log_4 2}$; $\boxed{\ln(x+2) + \ln x = 0}$

$\boxed{(0.00425)(0.00035)} =$; $\boxed{\left(\sqrt[3]{125}\right)\left(\sqrt[4]{2^4 \cdot 3}\right)} =$; $\boxed{\dfrac{3.5^3 \cdot \sqrt[3]{2510}}{2.32^2}} =$

$\boxed{\log_{10} \dfrac{1}{\sqrt[3]{(t+2)^2}}} =$; $\boxed{\ln \dfrac{5w^2}{w^2 + 2w - 15}} =$; $\boxed{\log_3 \dfrac{9x^3}{(2x-3)^2}} =$

$\boxed{\log(x+1) + [2\log x - 3\log(x+1)]} =$; $\boxed{2\log(x+1) + \dfrac{1}{5}\log(x-2)} =$; $\boxed{(\log_{10} 5 + \log_{10} x) - \log_{10} 25} =$

$\boxed{y = y_0 \dfrac{1}{1 - e^{ax}}}$; $\boxed{B = B_0\left(1 + \dfrac{m}{n}\right)^{nt}}$; $\boxed{K = K_0 \dfrac{1}{5^x}}$

Chapter 6 - Logarithms

The objective of this chapter is to improve the student's ability to solve and simplify logarithmic expressions. The laws of logarithms are addressed in Section 6.1. Computation of common logarithms, natural logarithms, logarithms other than base 10 and e, and antilogarithms is addressed in Section 6.2. Sections 6.3 and 6.4 show the steps as to how one variable exponential and logarithmic equations are solved. The use of logarithms in solving numerical expressions, expanding logarithmic expressions from a single logarithmic term, and combining logarithmic expressions into a single logarithmic term are addressed in Section 6.5. Advanced logarithmic problems are addressed in Section 6.6. Cases presented in each section are concluded by solving additional examples with practice problems to further enhance the students ability. It is the author's hope that by the conclusion of this final chapter students have considerably improved their math skills and have gained confidence to pursue more advanced math concepts.

6.1 Introduction to Logarithms

In this section students are introduced to logarithmic and exponential expressions (Case I) and learn about the logarithmic laws (Case II).

Case I Logarithmic and Exponential Expressions

The equations $y = log_a x$ and $x = a^y$ are equivalent equations that determine the same functions. This means that exponential expressions can be written in logarithmic form and logarithmic expressions can be written in exponential form. This is because,

$$log_a x = y \ ; \ a^{log_a x} = a^y \ ; \ x = a^y \qquad x > 0, \ a \neq 1$$

and

$$x = a^y \ ; \ log_a x = log_a a^y \ ; \ log_a x = y$$

The similarities between logarithms and exponents can be presented in the following way:

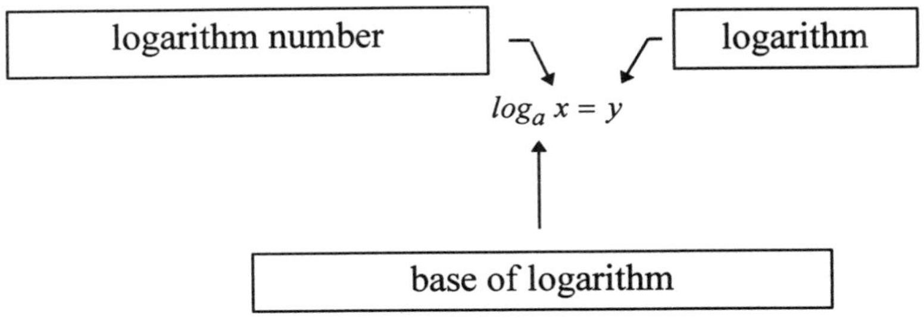

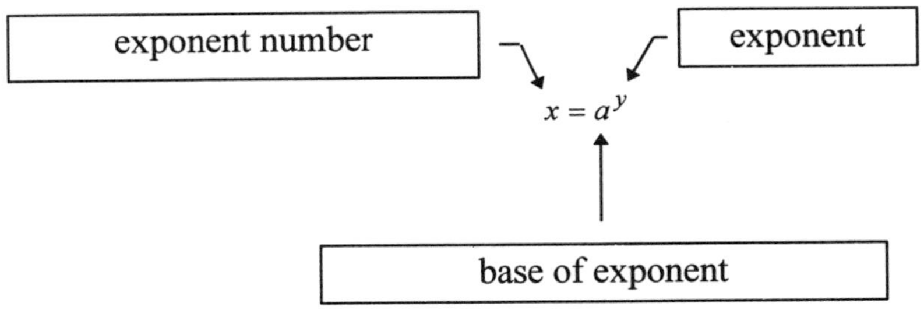

Therefore, we see that logarithms are simply exponents!

Note that the base a in logarithmic and exponential forms is the same. In the logarithmic form this appears as the subscript and in the exponential form it is the number that is raised to a power. Also note that x is a number in both the logarithmic and exponential forms and y is the logarithm in the logarithmic form and the exponent in the exponential form. In this book;

1. Given $y = log_a x$, we refer to a as the *base of the logarithm* and x as the *logarithm number*, and
2. Given $x = a^y$, we refer to a as the *exponent base* and y as the *exponent power*.

Exponential expressions can be written in their equivalent logarithmic form using the following steps:

Step 1 Multiply both sides of the equation by log to the same base as the base of the exponential function.

Step 2 Apply the logarithmic property $log_a a^y = y$.

The following examples show the steps as to how exponential expressions are changed to logarithmic expressions. This is another way of noting that logarithms are simply exponents.

a. $2^3 = 8$; $log_2 2^3 = log_2 8$; $\mathbf{3 = log_2 8}$

b. $3^{-2} = \frac{1}{9}$; $log_3 3^{-2} = log_3 \frac{1}{9}$; $\mathbf{-2 = log_3 \frac{1}{9}}$

c. $5^3 = 125$; $log_5 5^3 = log_5 125$; $\mathbf{3 = log_5 125}$

d. $64^{\frac{1}{3}} = 4$; $log_{64} 64^{\frac{1}{3}} = log_{64} 4$; $\mathbf{\frac{1}{3} = log_{64} 4}$

e. $4^{\frac{1}{2}} = 2$; $log_4 4^{\frac{1}{2}} = log_4 2$; $\mathbf{\frac{1}{2} = log_4 2}$

f. $e^{0.5} = 1.649$; $ln e^{0.5} = ln 1.649$; $\mathbf{0.5 = ln 1.649}$

g. $4^{-\frac{1}{2}} = \frac{1}{2}$; $log_4 4^{-\frac{1}{2}} = log_4 \frac{1}{2}$; $\mathbf{-\frac{1}{2} = log_4 \frac{1}{2}}$

h. $5^0 = 1$; $log_5 5^0 = log_5 1$; $\mathbf{0 = log_5 1}$

Logarithmic expressions can also be written in their equivalent exponential form using the following steps:

Step 1 Raise both sides of the equation to the same base as the base of the logarithm.

Step 2 Apply the logarithmic property $a^{log_a x} = x$.

The following examples show the steps as to how logarithmic expressions are changed to exponential form:

a. $log_{10} 1000 = 3$; $10^{log_{10} 1000} = 10^3$; $\mathbf{1000 = 10^3}$

b. $log_2 8 = 3$; $2^{log_2 8} = 2^3$; $\mathbf{8 = 2^3}$

c. $log_{10} 0.001 = -3$; $10^{log_{10} 0.001} = 10^{-3}$; $\mathbf{0.001 = 10^{-3}}$

d. $log_3 81 = 4$; $3^{log_3 81} = 3^4$; $\mathbf{81 = 3^4}$

e. $log_2 \frac{1}{4} = -2$; $2^{log_2 \frac{1}{4}} = 2^{-2}$; $\mathbf{\frac{1}{4} = 2^{-2}}$

f. $log_{\frac{1}{2}} 4 = -2$; $\frac{1}{2}^{log_{\frac{1}{2}} 4} = \frac{1}{2}^{-2}$; $\mathbf{4 = \frac{1}{2}^{-2}}$

g. $log_2 128 = 7$; $2^{log_2 128} = 2^7$; $\mathbf{128 = 2^7}$

h. $log_{10} 1 = 0$; $10^{log_{10} 1} = 10^0$; $\mathbf{1 = 10^0}$

Note: The subject of exponents is extensively addressed in Chapters 3 and 5 of the *"Mastering Algebra - An Introduction"* book and hence is not addressed in this section. Students are encouraged to review these chapters before pursuing with the remaining sections of this book. The exponent laws are listed below for reference.

Table 6.1-1: Exponent Laws 1 through 7

I. **Multiplication**	$a^m \cdot a^n = a^{m+n}$	When multiplying positive exponential terms, if bases a are the same, add the exponents m and n.
II. **Power of a Power**	$\left(a^m\right)^n = a^{mn}$	When raising an exponential term to a power, multiply the powers (exponents) m and n.
III. **Power of a Product**	$(a \cdot b)^m = a^m \cdot b^m$	When raising a product to a power, raise each factor a and b to the power m.
IV. **Power of a Fraction**	$\left(\frac{a}{b}\right)^m = \frac{a^m}{b^m}$	When raising a fraction to a power, raise the numerator and the denominator to the power m.
V. **Division**	$\frac{a^m}{a^n} = a^m \cdot a^{-n} = a^{m-n}$	When dividing exponential terms, if the bases a are the same, subtract exponents m and n.
VI. **Negative Power**	$a^{-n} = \frac{1}{a^n}$	A non-zero base a raised to the $-n$ power equals 1 divided by the base a to the n power.
VII. **Zero Power**	$a^0 = 1$	A non-zero base a raised to the zero power is always equal to 1.

Practice Problems - Logarithmic and Exponential Expressions

A. Section 6.1 Case I Practice Problems - Write the following exponential expressions in their equivalent logarithmic form:

1. $2^5 = 32$
2. $4^{-3} = \frac{1}{64}$
3. $7^2 = 49$
4. $64^{\frac{1}{2}} = 8$
5. $81^{\frac{1}{4}} = 3$
6. $e^2 = 7.389$
7. $27^{-\frac{1}{3}} = \frac{1}{3}$
8. $100^{-\frac{1}{2}} = \frac{1}{10}$
9. $125^{\frac{1}{3}} = 5$
10. $1000^0 = 1$

B. Section 6.1 Case I Practice Problems - Write the following logarithmic expressions in their equivalent exponential form:

1. $log_{10} 10000 = 4$
2. $log_4 64 = 3$
3. $log_{10} 0.1 = -1$
4. $log_5 625 = 4$
5. $log_2 \frac{1}{32} = -5$
6. $log_{\frac{1}{3}} 27 = -3$
7. $log_3 243 = 5$
8. $log_2 256 = 8$
9. $log_{10} 0.0001 = -4$
10. $log_{\frac{1}{5}} 3125 = -5$

Case II The Laws of Logarithm

As was stated earlier, a logarithm is an exponent. Therefore, the laws of exponents can be translated into the laws of logarithms by using the definition of logarithm:

$$y = log_a x \text{ if and only if } a^y = x$$

There are three logarithmic laws, i.e., the product law, the quotient law, and the power law. These laws are used in order to simplify logarithmic expressions and should be memorized.

Product Law: The product law of logarithms states that the logarithm of the product xy is equal to the sum of the logarithms of x and y.

$$log_a xy = log_a x + log_a y \quad x \rangle 0, y \rangle 0$$

Example 6.1-1: Rewrite each of the following logarithms as the sum of two or more logarithms.

a. $log_4 4 \cdot 3 =$ b. $log_{10} 2 \cdot 3xy =$ c. $log_3 3 \cdot 32$

Solution:

a. $log_4 4 \cdot 3 = log_4 4 + log_4 3 = \mathbf{1 + log_4 3}$ Note: $log_4 4 = 1$

b. $log_{10} 2 \cdot 3xy = \mathbf{log_{10} 2 + log_{10} 3 + log_{10} x + log_{10} y}$

c. $log_3 3 \cdot 32 = log_3 3 + log_3 32 = \mathbf{1 + log_3 32}$ Note: $log_3 3 = 1$

Quotient Law: The quotient law of logarithms states that the logarithm of the quotient $\frac{x}{y}$ is equal to the logarithm of the numerator x minus the logarithm of the denominator y.

$$log_a \frac{x}{y} = log_a x - log_a y \quad x \rangle 0, y \rangle 0$$

Example 6.1-2: Rewrite each of the following logarithms as the difference of two logarithms.

a. $log_4 \frac{23}{6} =$ b. $log_{10} \frac{1000}{x} =$ c. $log_2 \frac{2}{7}$

Solution:

a. $log_4 \frac{23}{6} = \mathbf{log_4 23 - log_4 6}$

b. $log_{10} \frac{1000}{x} = log_{10} 1000 - log_{10} x = \mathbf{3 - log_{10} x}$ Note: $log_{10} 1000 = 3$

c. $log_2 \frac{2}{7} = log_2 2 - log_2 7 = \mathbf{1 - log_2 7}$ Note: $log_2 2 = 1$

Power Law: The power law of logarithms states that the logarithm of a number, or a variable x, raised to a power m, i.e., $log_a x^m$ is equal to the product of the power m multiplied by the logarithm of x.

$$log_a x^m = m \cdot log_a x \quad x \rangle 0, \textit{ and m is a real number}$$

Example 6.1-3: Rewrite each of the following logarithms using the power law of the logarithms.

a. $log_3 5^3 =$ b. $log_{10} \sqrt[3]{4} =$ c. $\log_2 x^{\frac{1}{5}}$

Solution:

a. $log_3 5^3 = \mathbf{3 log_3 5}$ b. $log_{10} \sqrt[3]{4} = \log_{10} 4^{\frac{1}{3}} = \mathbf{\frac{1}{3} log_{10} 4}$ c. $\log_2 x^{\frac{1}{5}} = \mathbf{\frac{1}{5} log_2 x}$

Example 6.1-4: Use laws 1 through 3 to rewrite each of the following logarithms.

a. $log_3 \frac{125}{\sqrt{x}} =$ b. $log_{10} \frac{25x}{y^2} =$ c. $log_5 3x^{\frac{1}{3}} y^2$

Solution:

a. $log_3 \frac{125}{\sqrt{x}} = log_3 125 - log_3 \sqrt{x} = log_3 5^3 - log_3 x^{\frac{1}{2}} = \mathbf{3 log_3 5 - \frac{1}{2} log_3 x}$

b. $log_{10} \frac{25x}{y^2} = log_{10} 25x - log_{10} y^2 = \mathbf{log_{10} 25 + log_{10} x - 2 log_{10} y}$

c. $log_5 3x^{\frac{1}{3}} y^2 = log_5 3 + log_5 x^{\frac{1}{3}} + log_5 y^2 = \mathbf{log_5 3 + \frac{1}{3} log_5 x + 2 log_5 y}$

The following are additional logarithmic rules which should also be memorized:

Rule No. I:

$$log_a 1 = 0$$

Examples:

1. $log_2 1 = \mathbf{0}$ 2. $log_{0.3} 1 = \mathbf{0}$ 3. $log_e 1 = ln 1 = \mathbf{0}$ 4. $log_{\frac{1}{2}} 1 = \mathbf{0}$

5. $log_{20} 1 = \mathbf{0}$ 6. $log_{0.05} 1 = \mathbf{0}$ 7. $log_{\frac{2}{3}} 1 = \mathbf{0}$ 8. $log_{80} 1 = \mathbf{0}$

Rule No. II:

$$log_a a = 1$$

Examples:

1. $log_{10} 10 = \mathbf{1}$ 2. $log_{0.3} 0.3 = \mathbf{1}$ 3. $log_e e = ln e = \mathbf{1}$ 4. $log_{\frac{1}{3}} \frac{1}{3} = \mathbf{1}$

5. $log_{20} 20 = \mathbf{1}$ 6. $log_{0.07} 0.07 = \mathbf{1}$ 7. $log_{\frac{2}{5}} \frac{2}{5} = \mathbf{1}$ 8. $log_{100} 100 = \mathbf{1}$

Rule No. III:

$$log_a x \quad \textit{for } x \le 0 \textit{ is not defined}$$

Examples:

1. $log_{10} 0$ ***is not defined*** 2. $log_2 -2$ ***is not defined*** 3. $log_{10} -5$ ***is not defined***

4. $log_{0.4} 0$ ***is not defined*** 5. $log_e 0 = ln 0$ ***is not defined*** 6. $log_{\frac{2}{3}} 0$ ***is not defined***

Rule No. IV:

$$a^{log_a x} = x \quad \textit{for all } x > 0$$

Examples:

1. $3^{log_3 5} = \mathbf{5}$
2. $\frac{1}{2}^{log_{\frac{1}{2}} 3} = \mathbf{3}$
3. $0.1^{log_{0.1} 36} = \mathbf{36}$
4. $10^{log_{10} 2} = \mathbf{2}$
5. $80^{log_{80} 10} = \mathbf{10}$
6. $0.025^{log_{0.025} 7} = \mathbf{7}$
7. $e^{log_e 2} = e^{ln 2} = \mathbf{2}$
8. $\frac{3}{4}^{log_{\frac{3}{4}} 5} = \mathbf{5}$

Rule No. V:

$$log_a a^x = x\, log_a a = x \times 1 = x \quad \textit{for all real } x$$

Examples:

1. $log_2 2^{x+1} = (x+1) log_2 2 = (x+1) \times 1 = \mathbf{x+1}$
2. $log_{10} 10^u = u\, log_{10} 10 = u \times 1 = \mathbf{u}$
3. $log_{0.1} 0.1^{\sqrt{3}} = \sqrt{3}\, log_{0.1} 0.1 = \sqrt{3} \times 1 = \mathbf{\sqrt{3}}$
4. $log_4 4^{x^2} = x^2 log_4 4 = x^2 \times 1 = \mathbf{x^2}$
5. $log_{\frac{2}{5}}\left(\frac{2}{5}\right)^5 = 5og_{\frac{2}{5}}\left(\frac{2}{5}\right) = 5 \times 1 = \mathbf{5}$
6. $log_5 5^3 = 3 log_5 5 = 3 \times 1 = \mathbf{3}$

Table 6.1-2 summarizes the laws and the rules associated with logarithms:

Table 6.1-2: Summary of Logarithmic Laws and Rules

Product Law:	$log_a xy = log_a x + log_a y \quad x \rangle 0,\ y \rangle 0$		
Quotient Law:	$log_a \frac{x}{y} = log_a x - log_a y \quad x \rangle 0,\ y \rangle 0$		
Power Law:	$log_a x^m = m \cdot log_a x \quad x \rangle 0,$ *and m is a real number*		
Rule No. I:	$log_a 1 = 0$	Rule No. II:	$log_a a = 1$
Rule No. III:	$log_a x \quad$ *for* $x \le 0$ *is not defined*	Rule No. IV:	$a^{log_a x} = x \quad$ *for all* $x \rangle 0$
Rule No. V:	$log_a a^x = x\, log_a a = x \times 1 = x \quad$ *for all real x*		

Practice Problems - The Laws of Logarithm

Section 6.1 Case II Practice Problems - Simplify the following logarithmic expressions:

1. $log_8 8^{u^3} =$
2. $0.02^{log_{0.02} 2} =$
3. $log_{10} \frac{10}{x^2} =$
4. $log_{10} 5 \cdot 2x^2 y^3 =$
5. $log_{25} 25x + log_{25} 1 =$
6. $log_{10} \frac{2}{1000} =$
7. $log_{100} 0 =$
8. $log_{10} \sqrt[5]{3} =$
9. $log_2 \frac{8}{9} =$
10. $log_4 4 \cdot 16 =$

6.2 Computation Involving Logarithms

Most logarithmic expressions are either represented as a logarithm to the base 10, which is referred to as the common logarithm, or as a logarithm to the base e, which is referred to as the natural logarithm. However, logarithmic expressions can be represented in any base. In this section the steps as to how logarithms to the base 10, e, or any other base are solved and simplified are addressed in Cases I, II, and III, respectively. In addition, computation of antilogarithms is addressed in Case IV. Note that logarithmic tables provide the answer to the logarithm of numbers in both base 10 or e. A hand calculator can also provide the answer to logarithm of numbers in both bases. Students are encouraged to learn the steps in using a logarithmic table shown in this section. However, with the accessibility of scientific calculators, the use of logarithmic tables is minimized to cases where no hand calculators are available. Throughout this chapter logarithmic problems are solved using a hand calculator. In addition, note that in this book the numerical values of $log_{10}x$ and lnx are rounded off to four decimal places. This, in many instances, result in obtaining approximate answers to logarithmic expressions.

Case I Computation of Common Logarithms

In this section base 10 logarithms, which are called common logarithms, are discussed. Logarithms to the base 10 are written in the form of $log_{10} x$ or $log x$. Note that since the expression $log_{10} x = b$ is the same as $x = 10^b$, it is easy to calculate the common logarithms of numbers that are to the power of 10 as shown in Table 6.2-1.

Table 6.2-1: The Common Logarithm of Powers of 10

$$log_{10} 0.000001 = log_{10} 1 \times 10^{-6} = log_{10} 1 + log_{10} 10^{-6} = 0 - 6 log_{10} 10 = -6 \times 1 = -6$$

$$log_{10} 0.00001 = log_{10} 1 \times 10^{-5} = log_{10} 1 + log_{10} 10^{-5} = 0 - 5 log_{10} 10 = -5 \times 1 = -5$$

$$log_{10} 0.0001 = log_{10} 1 \times 10^{-4} = log_{10} 1 + log_{10} 10^{-4} = 0 - 4 log_{10} 10 = -4 \times 1 = -4$$

$$log_{10} 0.001 = log_{10} 1 \times 10^{-3} = log_{10} 1 + log_{10} 10^{-3} = 0 - 3 log_{10} 10 = -3 \times 1 = -3$$

$$log_{10} 0.01 = log_{10} 1 \times 10^{-2} = log_{10} 1 + log_{10} 10^{-2} = 0 - 2 log_{10} 10 = -2 \times 1 = -2$$

$$log_{10} 0.1 = log_{10} 1 \times 10^{-1} = log_{10} 1 + log_{10} 10^{-1} = 0 - log_{10} 10 = -1 \times 1 = -1$$

$$log_{10} 1 = log_{10} 1 \times 10^{0} = log_{10} 1 + log_{10} 10^{0} = 0 - 0 log_{10} 10 = -0 \times 1 = 0$$

$$log_{10} 10 = log_{10} 1 \times 10^{1} = log_{10} 1 + log_{10} 10^{1} = 0 + 1 log_{10} 10 = 1 \times 1 = 1$$

$$log_{10} 100 = log_{10} 1 \times 10^{2} = log_{10} 1 + log_{10} 10^{2} = 0 + 2 log_{10} 10 = 2 \times 1 = 2$$

$$log_{10} 1000 = log_{10} 1 \times 10^{3} = log_{10} 1 + log_{10} 10^{3} = 0 + 3 log_{10} 10 = 3 \times 1 = 3$$

Table 6.2-1 (Continued)

$$\log_{10} 10{,}000 = \log_{10} 1\times 10^4 = \log_{10} 1 + \log_{10} 104 = 0 + 4\log_{10} 10 = 4\times 1 = 4$$

$$\log_{10} 100{,}000 = \log_{10} 1\times 10^5 = \log_{10} 1 + \log_{10} 10^5 = 0 + 5\log_{10} 10 = 5\times 1 = 5$$

$$\log_{10} 1{,}000{,}000 = \log_{10} 1\times 10^6 = \log_{10} 1 + \log_{10} 10^6 = 0 + 6\log_{10} 10 = 6\times 1 = 6$$

In cases where numbers are not to the power of 10, we need to either use a table of common logarithms or a calculator in order to find the common logarithm for the numbers. Table 6.2-2 gives the common logarithm of numbers between 1.0 and 9.99 in increments of 0.1. For example, in order to find the common logarithm of a number such as $x = 4.14$ use the following steps:

Step 1 Identify number 4.1 by reading down the left hand column of the table labeled x.

Step 2 Move across the row until the column labeled 4 is reached.

Step 3 Read the value that is given in the intersection of the row with number 4.1 and the column labeled 4, i.e., 0.6170. Therefore, $\log_{10} 4.14 = 0.6170$.

x	0	1	2	3	4	5	6	7	8	9
3.5	0.5441	0.5453	0.5465	0.5478	0.5490	0.5502	0.5514	0.5527	0.5539	0.5551
3.6	0.5563	0.5575	0.5587	0.5599	0.5611	0.5623	0.5635	0.5647	0.5658	0.5670
3.7	0.5682	0.5694	0.5705	0.5717	0.5729	0.5740	0.5752	0.5763	0.5775	0.5786
3.8	0.5798	0.5809	0.5821	0.5832	0.5843	0.5855	0.5866	0.5877	0.5888	0.5899
3.9	0.5911	0.5922	0.5933	0.5944	0.5955	0.5966	0.5977	0.5988	0.5999	0.6010
4.0	0.6021	0.6031	0.6042	0.6053	0.6064	0.6075	0.6085	0.6096	0.6107	0.6117
4.1	0.6128	0.6138	0.6149	0.6160	**0.6170**	0.6180	0.6191	0.6201	0.6212	0.6222
4.2	0.6232	0.6243	0.6253	0.6263	0.6274	0.6284	0.6294	0.6304	0.6314	0.6325
4.3	0.6335	0.6345	0.6355	0.6365	0.6375	0.6385	0.6395	0.6405	0.6415	0.6425
4.4	0.6435	0.6444	0.6454	0.6464	0.6474	0.6484	0.6493	0.6503	0.6513	0.6522
4.5	0.6532	0.6542	0.6551	0.6561	0.6571	0.6580	0.6590	0.6599	0.6609	0.6618
4.6	0.6628	0.6637	0.6646	0.6656	0.6665	0.6675	0.6684	0.6693	0.6702	0.6712
4.7	0.6721	0.6730	0.6739	0.6749	0.6758	0.6767	0.6776	0.6785	0.6794	0.6803
4.8	0.6812	0.6821	0.6830	0.6839	0.6848	0.6857	0.6866	0.6875	0.6884	0.6893
4.9	0.6902	0.6911	0.6920	0.6928	0.6937	0.6946	0.6955	0.6964	0.6972	0.6981
x	0	1	2	3	4	5	6	7	8	9

Using Table 6.2-2 and the above steps one can find the common logarithm of the following numbers:

$\log_{10} 1.12 = 0.0492$ $\log_{10} 2.03 = 0.3075$ $\log_{10} 3.11 = 0.4928$

$\log_{10} 5.07 = 0.7050$ $\log_{10} 1.49 = 0.1732$ $\log_{10} 1.06 = 0.0253$

$\log_{10} 5.57 = 0.7459$ $\log_{10} 2.45 = 0.3892$ $\log_{10} 3.00 = 0.4771$

$\log_{10} 1.29 = 0.1106$ $\log_{10} 8.07 = 0.9069$ $\log_{10} 6.24 = 0.7952$

$\log_{10} 9.26 = 0.9666$ $\log_{10} 2.22 = 0.3464$ $\log_{10} 7.67 = 0.8848$

$\log_{10} 5.21 = 0.7168$ $\log_{10} 4.39 = 0.6425$ $\log_{10} 4.99 = 0.6981$

$\log_{10} 3.47 = 0.5403$ $\log_{10} 9.19 = 0.9633$ $\log_{10} 2.09 = 0.3201$

$\log_{10} 7.28 = 0.8621$ $\log_{10} 5.33 = 0.7267$ $\log_{10} 9.99 = 0.9996$

Practice Problems - Computation of Common Logarithms

A. Section 6.2 Case I Practice Problems - Use the Common Logarithms Table (Table 6.2-2) to find the answer to the following logarithmic expressions:

1. $log_{10} 3.57 =$
2. $log_{10} 3.08 =$
3. $log_{10} 4.53 =$
4. $log_{10} 8.24 =$
5. $log_{10} 7.32 =$
6. $log_{10} 5.55 =$
7. $log_{10} 2.12 =$
8. $log_{10} 9.46 =$
9. $log_{10} 5.29 =$
10. $log_{10} 1.26 =$

Note that in order to find the common logarithm of numbers that are greater than 9.99 use the following steps:

Step 1 Write the logarithm number in scientific notation form, i.e., write $log_{10} 200$ as $log_{10} 2.0 \times 10^2$.

Step 2 Apply the laws of logarithm and simplify the expression (see Table 6.1-2)

Examples with Steps

The following examples show the steps as to how common logarithms are solved:

Example 6.2-1

$$\boxed{log_{10} 42{,}000} =$$

Solution:

Step 1

$$\boxed{log_{10} 42{,}000} = \boxed{log_{10} 4.2 \times 10^4}$$

Step 2

$$\boxed{log_{10} 4.2 \times 10^4} = \boxed{log_{10} 4.2 + log_{10} 10^4} = \boxed{log_{10} 4.2 + 4\, log_{10} 10} = \boxed{0.6232 + 4 \times 1}$$

$$= \boxed{0.6232 + 4} = \boxed{\mathbf{4.6232}}$$

Example 6.2-2

$$\boxed{log_{10} 45^{\frac{2}{3}}} =$$

Solution:

Step 1

$$\boxed{log_{10} 45^{\frac{2}{3}}} = \boxed{\frac{2}{3} log_{10} 45} = \boxed{\frac{2}{3} log_{10} 4.5 \times 10^1}$$

Step 2

$$\boxed{\frac{2}{3} log_{10} 4.5 \times 10^1} = \boxed{\frac{2}{3}\left(log_{10} 4.5 + log_{10} 10^1\right)} = \boxed{\frac{2}{3}(0.6532 + 1)} = \boxed{\frac{2}{3} \times 1.6532} = \boxed{\mathbf{1.1021}}$$

Example 6.2-3

$$\boxed{log_{10} 1000{,}000} =$$

Solution:

Step 1

$$\boxed{log_{10} 1000{,}000} = \boxed{log_{10} 1.0 \times 10^6}$$

Step 2 $\boxed{\log_{10} 1.0 \times 10^6} = \boxed{\log_{10} 1.0 + \log_{10} 10^6} = \boxed{0 + 6\log_{10} 10} = \boxed{6 \times 1} = \boxed{\mathbf{6}}$

Example 6.2-4

$\boxed{\log_{10} 0.000348} =$

Solution:

Step 1 $\boxed{\log_{10} 0.000348} = \boxed{\log_{10} 3.48 \times 10^{-4}}$

Step 2 $\boxed{\log_{10} 3.48 \times 10^{-4}} = \boxed{\log_{10} 3.48 + \log_{10} 10^{-4}} = \boxed{0.5416 - 4\log_{10} 10} = \boxed{0.5416 - 4 \times 1}$

$= \boxed{0.5416 - 4} = \boxed{\mathbf{-3.4584}}$

Example 6.2-5

$\boxed{\log_{10} 4\sqrt{55}} =$

Solution:

Step 1 $\boxed{\log_{10} 4\sqrt{55}} = \boxed{\log_{10} 4 \times 55^{\frac{1}{2}}} = \boxed{\log_{10}\left[\left(4.0 \times 10^0\right) \times \left(5.5 \times 10^1\right)^{\frac{1}{2}}\right]}$

Step 2 $\boxed{\log_{10}\left[\left(4.0 \times 10^0\right) \times (5.5 \times 10)^{\frac{1}{2}}\right]} = \boxed{\log_{10}\left(4.0 \times 10^0\right) + \log_{10}(5.5 \times 10)^{\frac{1}{2}}}$

$= \boxed{\log_{10}(4.0 \times 1) + \frac{1}{2}\log_{10}(5.5 \times 10)} = \boxed{\log_{10} 4.0 + \log_{10} 1 + \frac{1}{2}(\log_{10} 5.5 + \log_{10} 10)}$

$= \boxed{\log_{10} 4.0 + 0 + \frac{1}{2}(\log_{10} 5.5 + 1)} = \boxed{0.6021 + \frac{1}{2}(0.7404 + 1)} = \boxed{0.6021 + \frac{1.7404}{2}}$

$= \boxed{0.6021 + 0.8702} = \boxed{\mathbf{1.4723}}$

Additional Examples - Computation of Common Logarithms

The following examples further illustrate how to solve common logarithms:

Example 6.2-6

$\boxed{\log_{10} 100} = \boxed{\log_{10} 1.0 \times 10^2} = \boxed{\log_{10} 1.0 + \log_{10} 10^2} = \boxed{0 + 2\log_{10} 10} = \boxed{0 + 2} = \boxed{\mathbf{2}}$

Another way of solving the above problem would be:

$\boxed{\log_{10} 100} = \boxed{\log_{10}(10 \cdot 10)} = \boxed{\log_{10} 10 + \log_{10} 10} = \boxed{1 + 1} = \boxed{\mathbf{2}}$ Note: $\log_{10} 10 = 1$

Example 6.2-7

$\boxed{\log_{10} 125} = \boxed{\log_{10} 1.25 \times 10^2} = \boxed{\log_{10} 1.25 + \log_{10} 10^2} = \boxed{0.0969 + 2\log_{10} 10} = \boxed{0.0969 + 2} = \boxed{\mathbf{2.0969}}$

Example 6.2-8

$\boxed{\log_{10} 2{,}350} = \boxed{\log_{10} 2.35 \times 10^3} = \boxed{\log_{10} 2.35 + \log_{10} 10^3} = \boxed{0.3711 + 3\log_{10} 10} = \boxed{0.3711 + 3} = \boxed{\mathbf{3.3711}}$

Example 6.2-9

$\boxed{\log_{10} 5{,}000} = \boxed{\log_{10} 5.0 \times 10^4} = \boxed{\log_{10} 5 + \log_{10} 10^4} = \boxed{0.6989 + 4\log_{10} 10} = \boxed{0.6989 + 4} = \boxed{\mathbf{4.6989}}$

Example 6.2-10

$$\boxed{log_{10} 6} = \boxed{log_{10} 6.0 \times 10^0} = \boxed{log_{10} 6.0 + log_{10} 10^0} = \boxed{0.7782 + 0 \times log_{10} 10} = \boxed{0.7782 + 0} = \boxed{\mathbf{0.7782}}$$

Example 6.2-11

$$\boxed{log_{10} 10} = \boxed{log 1.0 \times 10^1} = \boxed{log_{10} 1 + log_{10} 10^1} = \boxed{0 + 1 \times log_{10} 10} = \boxed{0 + 1} = \boxed{\mathbf{1}}$$

Example 6.2-12

$$\boxed{log_{10} 35} = \boxed{log_{10} 3.5 \times 10^1} = \boxed{log_{10} 3.5 + log_{10} 10^1} = \boxed{0.5441 + 1 \times log_{10} 10} = \boxed{0.5441 + 1} = \boxed{\mathbf{1.5441}}$$

Example 6.2-13

$$\boxed{log_{10} 0.01} = \boxed{log_{10} 1.0 \times 10^{-2}} = \boxed{log_{10} 1.0 + log_{10} 10^{-2}} = \boxed{0 - 2 log_{10} 10} = \boxed{0 - 2} = \boxed{\mathbf{-2}}$$

Example 6.2-14

$$\boxed{log_{10} 0.334} = \boxed{log_{10} 3.34 \times 10^{-1}} = \boxed{log_{10} 3.34 + log_{10} 10^{-1}} = \boxed{0.5237 - 1 \times log_{10} 10} = \boxed{0.5237 - 1} = \boxed{\mathbf{-0.4762}}$$

Example 6.2-15

$$\boxed{log_{10} 0.00007} = \boxed{log_{10} 7.0 \times 10^{-5}} = \boxed{log_{10} 7.0 + log_{10} 10^{-5}} = \boxed{0.8451 - 5 log_{10} 10} = \boxed{0.8451 - 5} = \boxed{\mathbf{-4.1549}}$$

Example 6.2-16

$$\boxed{log_{10} 0.06} = \boxed{log_{10} 6.0 \times 10^{-2}} = \boxed{log_{10} 6 + log_{10} 10^{-2}} = \boxed{0.7782 - 2 log_{10} 10} = \boxed{0.7782 - 2} = \boxed{\mathbf{-1.2218}}$$

Example 6.2-17

$$\boxed{log_{10} 25^{\frac{1}{3}}} = \boxed{\frac{1}{3} log_{10} 25} = \boxed{\frac{1}{3}\left(log_{10} 2.5 \times 10^1\right)} = \boxed{\frac{1}{3}\left[log_{10} 2.5 + log_{10} 10^1\right]} = \boxed{\frac{1}{3}[0.3979 + 1 \times log_{10} 10]}$$

$$= \boxed{\frac{1}{3}[0.3979 + 1]} = \boxed{\frac{1.3979}{3}} = \boxed{\mathbf{0.4659}}$$

Example 6.2-18

$$\boxed{log_{10} 0.000314} = \boxed{log_{10} 3.14 \times 10^{-4}} = \boxed{log_{10} 3.14 + log_{10} 10^{-4}} = \boxed{0.4969 - 4 log_{10} 10} = \boxed{0.4969 - 4} = \boxed{\mathbf{-3.5030}}$$

Example 6.2-19

$$\boxed{log_{10} 43} = \boxed{log_{10} 4.3 \times 10^1} = \boxed{log_{10} 4.3 + log_{10} 10^1} = \boxed{0.6335 + 1 \times log_{10} 10} = \boxed{0.6335 + 1} = \boxed{\mathbf{1.6335}}$$

Example 6.2-20

$$\boxed{log_{10} 0.0515} = \boxed{log_{10} 5.15 \times 10^{-2}} = \boxed{log_{10} 5.15 + log_{10} 10^{-2}} = \boxed{log_{10} 5.15 - 2 log_{10} 10} = \boxed{0.7118 - 2} = \boxed{\mathbf{-1.2882}}$$

Example 6.2-21

$$\boxed{log_{10} 782} = \boxed{log_{10} 7.82 \times 10^2} = \boxed{log_{10} 7.82 + log_{10} 10^2} = \boxed{log_{10} 7.82 + 2 log_{10} 10} = \boxed{0.8932 + 2} = \boxed{\mathbf{2.8932}}$$

Example 6.2-22

$$\boxed{log_{10} 33^{\frac{1}{2}}} = \boxed{\frac{1}{2} log_{10} 33} = \boxed{\frac{1}{2} log_{10} 3.3 \times 10^1} = \boxed{\frac{1}{2}\left[log_{10} 3.3 + log_{10} 10^1\right]} = \boxed{\frac{1}{2}[0.5185 + 1]} = \boxed{\frac{1.5185}{2}} = \boxed{\mathbf{0.7593}}$$

Example 6.2-23

$$\boxed{\log_{10}\sqrt[3]{55}}=\boxed{\log_{10}55^{\frac{1}{3}}}=\boxed{\frac{1}{3}\log_{10}55}=\boxed{\frac{1}{3}\left[\log_{10}5.5\times10^{1}\right]}=\boxed{\frac{1}{3}\left[\log_{10}5.5+\log_{10}10^{1}\right]}=\boxed{\frac{1}{3}[0.7404+1]}$$

$$=\boxed{\frac{1.7404}{3}}=\boxed{\mathbf{0.5801}}$$

Example 6.2-24

$$\boxed{\log_{10}\sqrt[5]{23}}=\boxed{\log_{10}23^{\frac{1}{5}}}=\boxed{\frac{1}{5}\log_{10}23}=\boxed{\frac{1}{5}\left[\log_{10}2.3\times10^{1}\right]}=\boxed{\frac{1}{5}\left[\log_{10}2.3+\log_{10}10^{1}\right]}=\boxed{\frac{1}{5}[0.3617+1]}$$

$$=\boxed{\frac{1.3617}{5}}=\boxed{\mathbf{0.2723}}$$

Example 6.2-25

$$\boxed{\log_{10}\sqrt[5]{4096}}=\boxed{\log_{10}\sqrt[5]{1024\cdot4}}=\boxed{\log_{10}\sqrt[5]{4^{5}\cdot4}}=\boxed{\log_{10}4\sqrt[5]{4}}=\boxed{\log_{10}4\cdot4^{\frac{1}{5}}}=\boxed{\log_{10}4+\log_{10}4^{\frac{1}{5}}}$$

$$=\boxed{0.6021+\frac{1}{5}\log_{10}4}=\boxed{0.6021+\frac{1}{5}\times0.6021}=\boxed{0.6021+0.1204}=\boxed{\mathbf{0.7225}}$$

or we can solve the above problem in the following way:

$$\boxed{\log_{10}\sqrt[5]{4096}}=\boxed{\log_{10}4096^{\frac{1}{5}}}=\boxed{\frac{1}{5}\log_{10}4096}=\boxed{\frac{1}{5}\left[\log_{10}4.096\times10^{3}\right]}=\boxed{\frac{1}{5}\left[\log_{10}4.096+\log_{10}10^{3}\right]}$$

$$=\boxed{\frac{1}{5}[0.6124+3\log_{10}10]}=\boxed{\frac{1}{5}[0.6124+3]}=\boxed{\frac{3.6124}{5}}=\boxed{\mathbf{0.7225}}$$

Example 6.2-26

$$\boxed{\log_{10}\sqrt{147}}=\boxed{\log_{10}\sqrt{49\cdot3}}=\boxed{\log_{10}\sqrt{7^{2}\cdot3}}=\boxed{\log_{10}7\sqrt{3}}=\boxed{\log_{10}7+\log_{10}\sqrt{3}}=\boxed{\log_{10}7+\log_{10}3^{\frac{1}{2}}}$$

$$=\boxed{\log_{10}7+\frac{1}{2}\log_{10}3}=\boxed{0.8451+\frac{1}{2}(0.4771)}=\boxed{0.8451+0.2385}=\boxed{\mathbf{1.0836}}$$

or,

$$\boxed{\log_{10}\sqrt{147}}=\boxed{\log_{10}147^{\frac{1}{2}}}=\boxed{\frac{1}{2}\log_{10}147}=\boxed{\frac{1}{2}\left[\log_{10}1.47\times10^{2}\right]}=\boxed{\frac{1}{2}\left[\log_{10}1.47+\log_{10}10^{2}\right]}$$

$$=\boxed{\frac{1}{2}[0.1673+2\log_{10}10]}=\boxed{\frac{1}{2}[0.1673+2]}=\boxed{\frac{2.1673}{2}}=\boxed{\mathbf{1.0836}}$$

Example 6.2-27

$$\boxed{\log_{10}2\sqrt{250}}=\boxed{\log_{10}2\sqrt{25\cdot10}}=\boxed{\log_{10}2\sqrt{5^{2}\cdot10}}=\boxed{\log_{10}(2\cdot5)\sqrt{10}}=\boxed{\log_{10}10\sqrt{10}}=\boxed{\log_{10}10+\log_{10}\sqrt{10}}$$

$$=\boxed{1+\log_{10}10^{\frac{1}{2}}}=\boxed{1+\frac{1}{2}\log_{10}10}=\boxed{1+\frac{1}{2}}=\boxed{\frac{3}{2}}=\boxed{1\frac{1}{2}}=\boxed{\mathbf{1.5}}$$

or,

$\boxed{log_{10} 2\sqrt{250}} = \boxed{log_{10} 2 + log_{10} \sqrt{250}} = \boxed{log_{10} 2 + log_{10} 250^{\frac{1}{2}}} = \boxed{0.3010 + \frac{1}{2} log_{10} 250}$

$= \boxed{0.3010 + \frac{1}{2}\left[log_{10} 2.50 \times 10^2\right]} = \boxed{0.3010 + \frac{1}{2}\left[log_{10} 2.50 + log_{10} 10^2\right]} = \boxed{0.3010 + \frac{1}{2}[0.3979 + 2\, log_{10} 10]}$

$= \boxed{0.3010 + \frac{1}{2}[0.3979 + 2]} = \boxed{0.3010 + \frac{2.3979}{2}} = \boxed{0.3010 + 1.1989} = \boxed{1.4999} = \boxed{\mathbf{1.5}}$

Example 6.2-28

$\boxed{log_{10} \sqrt[3]{648}} = \boxed{log_{10} \sqrt[3]{216 \cdot 3}} = \boxed{log_{10} \sqrt[3]{6^3 \cdot 3}} = \boxed{log_{10} 6 \cdot \sqrt[3]{3}} = \boxed{log_{10} 6 \cdot 3^{\frac{1}{3}}} = \boxed{log_{10} 6 + log_{10} 3^{\frac{1}{3}}}$

$= \boxed{0.7782 + \frac{1}{3} log_{10} 3} = \boxed{0.7782 + \frac{1}{3}(0.4771)} = \boxed{0.7782 + 0.1590} = \boxed{\mathbf{0.9372}}$

or,

$\boxed{log_{10} \sqrt[3]{648}} = \boxed{log_{10} (648)^{\frac{1}{3}}} = \boxed{\frac{1}{3} log_{10} 648} = \boxed{\frac{1}{3} log_{10} 6.48 \times 10^2} = \boxed{\frac{1}{3}\left[log_{10} 6.48 + log_{10} 10^2\right]}$

$= \boxed{\frac{1}{3}[0.8116 + 2\, log_{10} 10]} = \boxed{\frac{1}{3}[0.8116 + 2]} = \boxed{\frac{2.8116}{3}} = \boxed{\mathbf{0.9372}}$

Example 6.2-29

$\boxed{log_{10} 3\sqrt[4]{324}} = \boxed{log_{10} 3\sqrt[4]{81 \cdot 4}} = \boxed{log_{10} 3\sqrt[4]{3^4 \cdot 4}} = \boxed{log_{10} (3 \cdot 3)\sqrt[4]{4}} = \boxed{log_{10} 9\sqrt[4]{4}} = \boxed{log_{10} 9 + log_{10} \sqrt[4]{4}}$

$= \boxed{log_{10} 9 + log_{10} 4^{\frac{1}{4}}} = \boxed{0.9542 + \frac{1}{4} log_{10} 4} = \boxed{0.9542 + \frac{1}{4}(0.6021)} = \boxed{0.9542 + 0.1505} = \boxed{\mathbf{1.1047}}$

or,

$\boxed{log_{10} 3\sqrt[4]{324}} = \boxed{log_{10} 3(324)^{\frac{1}{4}}} = \boxed{log_{10} 3 + log_{10} 324^{\frac{1}{4}}} = \boxed{0.4771 + \frac{1}{4} log_{10} 324} = \boxed{0.4771 + \frac{1}{4}\left[log_{10} 3.24 \times 10^2\right]}$

$= \boxed{0.4771 + \frac{1}{4}\left[log_{10} 3.24 + log_{10} 10^2\right]} = \boxed{0.4771 + \frac{1}{4}[0.5105 + 2\, log_{10} 10]} = \boxed{0.4771 + \frac{1}{4}[0.5105 + 2]}$

$= \boxed{0.4771 + \frac{2.5105}{4}} = \boxed{0.4771 + 0.6276} = \boxed{\mathbf{1.1047}}$

Example 6.2-30

$\boxed{log_{10} 75{,}300{,}000} = \boxed{log_{10} 7.53 \times 10^7} = \boxed{log_{10} 7.53 + log_{10} 10^7} = \boxed{log_{10} 7.53 + 7\, log_{10} 10} = \boxed{0.8768 + 7} = \boxed{\mathbf{7.8768}}$

Example 6.2-31

$\boxed{log_{10} 2{,}430{,}000} = \boxed{log_{10} 2.43 \times 10^6} = \boxed{log_{10} 2.43 + log_{10} 10^6} = \boxed{log_{10} 2.43 + 6\, log_{10} 10} = \boxed{0.3856 + 6} = \boxed{\mathbf{6.3856}}$

Example 6.2-32

$\boxed{log_{10} 33{,}600{,}000} = \boxed{log_{10} 3.36 \times 10^7} = \boxed{log_{10} 3.36 + log_{10} 10^7} = \boxed{log_{10} 3.36 + 7\, log_{10} 10} = \boxed{0.5263 + 7} = \boxed{\mathbf{7.5263}}$

Example 6.2-33

$\boxed{log_{10} 81{,}200} = \boxed{log_{10} 8.12 \times 10^4} = \boxed{log_{10} 8.12 + log_{10} 10^4} = \boxed{log_{10} 8.12 + 4\,log_{10} 10} = \boxed{0.9096 + 4} = \boxed{\mathbf{4.9096}}$

Example 6.2-34

$\boxed{log_{10} 0.00283} = \boxed{log_{10} 2.83 \times 10^{-3}} = \boxed{log_{10} 2.83 + log_{10} 10^{-3}} = \boxed{log_{10} 2.83 - 3\,log_{10} 10} = \boxed{0.4518 - 3} = \boxed{\mathbf{-2.5482}}$

Example 6.2-35

$\boxed{log_{10} 0.0406} = \boxed{log_{10} 4.06 \times 10^{-2}} = \boxed{log_{10} 4.06 + log_{10} 10^{-2}} = \boxed{log_{10} 4.06 - 2\,log_{10} 10} = \boxed{0.6085 - 2} = \boxed{\mathbf{-1.3915}}$

Example 6.2-36

$\boxed{log_{10} 0.0000453} = \boxed{log_{10} 4.53 \times 10^{-5}} = \boxed{log_{10} 4.53 + log_{10} 10^{-5}} = \boxed{log_{10} 4.53 - 5\,log_{10} 10} = \boxed{0.6561 - 5} = \boxed{\mathbf{-4.3439}}$

Example 6.2-37

$\boxed{log_{10} 0.000472} = \boxed{log_{10} 4.72 \times 10^{-4}} = \boxed{log_{10} 4.72 + log_{10} 10^{-4}} = \boxed{log_{10} 4.72 - 4\,log_{10} 10} = \boxed{0.6739 - 4} = \boxed{\mathbf{-3.3261}}$

Example 6.2-38

$\boxed{log_{10} 82{,}200{,}000} = \boxed{log_{10} 8.22 \times 10^{7}} = \boxed{log_{10} 8.22 + log_{10} 10^{7}} = \boxed{log_{10} 8.22 + 7\,log_{10} 10} = \boxed{0.9149 + 7} = \boxed{\mathbf{7.9149}}$

Example 6.2-39

$\boxed{log_{10} 2{,}000{,}000{,}000} = \boxed{log_{10} 2.0 \times 10^{9}} = \boxed{log_{10} 2.0 + log_{10} 10^{9}} = \boxed{log_{10} 2.0 + 9\,log_{10} 10} = \boxed{0.3010 + 9} = \boxed{\mathbf{9.3010}}$

Example 6.2-40

$\boxed{log_{10} 0.00456} = \boxed{log_{10} 4.56 \times 10^{-3}} = \boxed{log_{10} 4.56 + log_{10} 10^{-3}} = \boxed{log_{10} 4.56 - 3\,log_{10} 10} = \boxed{0.6590 - 3} = \boxed{\mathbf{-2.3410}}$

Practice Problems - Computation of Common Logarithms

B. Section 6.2 Case I Practice Problems - Solve the following common logarithms:

1. $log_{10} 4{,}000 =$
2. $log_{10} 3\sqrt{300} =$
3. $log_{10} 45{,}400{,}000 =$
4. $log_{10} 0.00023 =$
5. $log_{10} \sqrt[3]{5^2} =$
6. $log_{10} 28 =$
7. $log_{10} 2\sqrt[4]{568} =$
8. $log_{10} 0.068 =$
9. $log_{10} 0.00001 =$
10. $log_{10} 450{,}000 =$

Table 6.2-2
Common Logarithms

x	0	1	2	3	4	5	6	7	8	9
1.0	0.0000	0.0043	0.0086	0.0128	0.0170	0.0212	0.0253	0.0294	0.0334	0.0374
1.1	0.0414	0.0453	0.0492	0.0531	0.0569	0.0607	0.0645	0.0682	0.0719	0.0755
1.2	0.0792	0.0828	0.0864	0.0899	0.0934	0.0969	0.1004	0.1038	0.1072	0.1106
1.3	0.1139	0.1173	0.1206	0.1239	0.1271	0.1303	0.1335	0.1367	0.1399	0.1430
1.4	0.1461	0.1492	0.1523	0.1553	0.1584	0.1614	0.1644	0.1673	0.1703	0.1732
1.5	0.1761	0.1790	0.1818	0.1847	0.1875	0.1903	0.1931	0.1959	0.1987	0.2014
1.6	0.2041	0.2068	0.2095	0.2122	0.2148	0.2175	0.2201	0.2227	0.2253	0.2279
1.7	0.2304	0.2330	0.2355	0.2380	0.2405	0.2430	0.2455	0.2480	0.2504	0.9529
1.8	0.2553	0.2577	0.2601	0.2625	0.2648	0.2672	0.2695	0.2718	0.2749	0.2765
1.9	0.2788	0.2810	0.2833	0.2856	0.2878	0.2900	0.2923	0.2945	0.2967	0.2989
2.0	0.3010	0.3032	0.3054	0.3075	0.3096	0.3118	0.3139	0.3160	0.3181	0.3201
2.1	0.3222	0.3243	0.3263	0.3284	0.3304	0.3324	0.3345	0.3365	0.3385	0.3404
2.2	0.3424	0.3444	0.3464	0.3483	0.3502	0.3522	0.3541	0.3560	0.3579	0.3598
2.3	0.3617	0.3636	0.3655	0.3674	0.3692	0.3711	0.3729	0.3747	0.3766	0.3784
2.4	0.3802	0.3820	0.3838	0.3856	0.3874	0.3892	0.3909	0.3927	0.3945	0.3962
2.5	0.3979	0.3997	0.4014	0.4031	0.4048	0.4065	0.4082	0.4099	0.4116	0.4133
2.6	0.4150	0.4166	0.4183	0.4200	0.4216	0.4232	0.4249	0.4265	0.4281	0.4298
2.7	0.4314	0.4330	0.4346	0.4362	0.4378	0.4393	0.4409	0.4425	0.4440	0.4456
2.8	0.4472	0.4487	0.4502	0.4518	0.4533	0.4548	0.4564	0.4579	0.4594	0.4609
2.9	0.4624	0.4639	0.4654	0.4669	0.4683	0.4698	0.4713	0.4728	0.4742	0.4757
3.0	0.4771	0.4786	0.4800	0.4814	0.4829	0.4843	0.4857	0.4871	0.4886	0.4900
3.1	0.4914	0.4928	0.4942	0.4955	0.4969	0.4983	0.4997	0.5011	0.5024	0.5038
3.2	0.5051	0.5065	0.5079	0.5092	0.5105	0.5119	0.5132	0.5145	0.5159	0.5172
3.3	0.5185	0.5198	0.5211	0.5224	0.5237	0.5250	0.5263	0.5276	0.5289	0.5302
3.4	0.5315	0.5328	0.5340	0.5353	0.5366	0.5378	0.5391	0.5403	0.5416	0.5428
3.5	0.5441	0.5453	0.5465	0.5478	0.5490	0.5502	0.5514	0.5527	0.5539	0.5551
3.6	0.5563	0.5575	0.5587	0.5599	0.5611	0.5623	0.5635	0.5647	0.5658	0.5670
3.7	0.5682	0.5694	0.5705	0.5717	0.5729	0.5740	0.5752	0.5763	0.5775	0.5786
3.8	0.5798	0.5809	0.5821	0.5832	0.5843	0.5855	0.5866	0.5877	0.5888	0.5899
3.9	0.5911	0.5922	0.5933	0.5944	0.5955	0.5966	0.5977	0.5988	0.5999	0.6010
4.0	0.6021	0.6031	0.6042	0.6053	0.6064	0.6075	0.6085	0.6096	0.6107	0.6117
4.1	0.6128	0.6138	0.6149	0.6160	0.6170	0.6180	0.6191	0.6201	0.6212	0.6222
4.2	0.6232	0.6243	0.6253	0.6263	0.6274	0.6284	0.6294	0.6304	0.6314	0.6325
4.3	0.6335	0.6345	0.6355	0.6365	0.6375	0.6385	0.6395	0.6405	0.6415	0.6425
4.4	0.6435	0.6444	0.6454	0.6464	0.6474	0.6484	0.6493	0.6503	0.6513	0.6522
4.5	0.6532	0.6542	0.6551	0.6561	0.6571	0.6580	0.6590	0.6599	0.6609	0.6618
4.6	0.6628	0.6637	0.6646	0.6656	0.6665	0.6675	0.6684	0.6693	0.6702	0.6712
4.7	0.6721	0.6730	0.6739	0.6749	0.6758	0.6767	0.6776	0.6785	0.6794	0.6803
4.8	0.6812	0.6821	0.6830	0.6839	0.6848	0.6857	0.6866	0.6875	0.6884	0.6893
4.9	0.6902	0.6911	0.6920	0.6928	0.6937	0.6946	0.6955	0.6964	0.6972	0.6981
5.0	0.6990	0.6998	0.7007	0.7016	0.7024	0.7033	0.7042	0.7050	0.7059	0·7067
5.1	0.7076	0.7084	0.7093	0.7101	0.7110	0.7118	0.7126	0.7135	0.7143	0.7152
5.2	0.7160	0.7168	0.7177	0.7185	0.7193	0.7202	0.7210	0.7218	0.7226	0.7235
5.3	0.7243	0.7251	0.7259	0.7267	0.7275	0.7284	0.7292	0.7300	0.7308	0.7316
5.4	0.7324	0.7332	0.7340	0.7348	0.7356	0.7364	0.7372	0.7380	0.7388	0.7396
x	0	1	2	3	4	5	6	7	8	9

Table 6.2-2 (Continued)
Common Logarithms

x	0	1	2	3	4	5	6	7	8	9
5.5	0.7404	0.7412	0.7419	0.7427	0.7435	0.7443	0.7451	0.7459	0.7466	0.7474
5.6	0.7482	0.7490	0.7497	0.7505	0.7513	0.7520	0.7528	0.7536	0.7543	0.7551
5.7	0.7559	0.7566	0.7574	0.7582	0.7589	0.7597	0.7604	0.7612	0.7619	0.7627
5.8	0.1634	0.7642	0.7649	0.7657	0.7664	0.1672	0.7679	0.7686	0.7694	0.7701
5.9	0.7709	0.7716	0.7123	0.7731	0.7738	0.7745	0.7752	0.7760	0.7767	0.7774
6.0	0.7782	0.7789	0.7796	0.7803	0.7810	0.7818	0.7825	0.7832	0.7839	0.7846
6.1	0.7853	0.7860	0.7868	0.7875	0.7882	0.7889	0.7896	0.7903	0.7910	0.7917
6.2	0.7924	0.7931	0.7938	0.7945	0.7952	0.7959	0.7966	0.7973	0.7980	0.7987
6.3	0.7993	0.8000	0.8007	0.8014	0.8021	0.8028	0.8035	0.8041	0.8048	0.8055
6.4	0.8062	0.8069	0.8075	0.8082	0.8089	0.8096	0.8102	0.8109	0.8116	0.8122
6.5	0.8129	0.8136	0.8142	0.8149	0.8156	0.8162	0.8169	0.8176	0.8182	0.8189
6.6	0.8195	0.8202	0.8209	0.8215	0.8222	0.8228	0.8235	0.8241	0.8248	0.8254
6.7	0.8261	0.8267	0.8274	0.8280	0.8287	0.8293	0.8299	0.8306	0.8312	0.8319
6.8	0.8325	0.8331	0.8338	0.8344	0.8351	0.8357	0.8363	0.8370	0.8376	0.8382
6.9	0.8388	0.8395	0.8401	0.8407	0.8414	0.8420	0.8426	0.8432	0.8439	0.8445
7.0	0.8451	0.8457	0.8463	0.8470	0.8476	0.8482	0.8488	0.8494	0.8500	0.8506
7.1	0.8513	0.8519	0.8525	0.8531	0.8537	0.8543	0.8549	0.8555	0.8561	0.8567
7.2	0.8673	0.8579	0.8585	0.8591	0.8597	0.8603	0.8609	0.8615	0.8621	0.8627
7.3	0.8633	0.8639	0.8645	0.8651	0.8657	0.8663	0.8669	0.8675	0.8681	0.8686
7.4	0.8692	0.8698	0.8704	0.8710	0.8716	0.8722	0.8727	0.8733	0.8739	0.8745
7.5	0.8751	0.8756	0.8762	0.8768	0.8774	0.8779	0.8785	0.8791	0.8797	0.8802
7.6	0.8808	0.8814	0.8820	0.8825	0.8831	0.8837	0.8842	0.8848	0.8854	0.8859
7.7	0.8865	0.8871	0.8876	0.8876	0.8887	0.8893	0.8899	0.8904	0.8910	0.8915
7.8	0.8921	0.8927	0.8932	0.8938	0.8943	0.8949	0.8954	0.8960	0.8965	0.8971
7.9	0.8976	0.8982	0.8987	0.8993	0.8998	0.9004	0.9009	0.9015	0.9020	0.9025
8.0	0.9031	0.9036	0.9042	0.9047	0.9053	0.9058	0.9063	0.9069	0.9074	0.9079
8.1	0.9085	0.9090	0.9096	0.9101	0.9106	0.9112	0.9117	0.9122	0.9128	0.9133
8.2	0.9138	0.9143	0.9149	0.9154	0.9159	0.9165	0.9170	0.9175	0.9180	0.9186
8.3	0.9191	0.9196	0.9201	0.9206	0.9212	0.9217	0.9222	0.9227	0.9232	0.9238
8.4	0.9243	0.9248	0.9253	0.9258	0.9263	0.9269	0.9274	0.9279	0.9284	0.9289
8.5	0.9294	0.9299	0.9304	0.9309	0.9315	0.9320	0.9325	0.9330	0.9335	0.9340
8.6	0.9345	0.9350	0.9355	0.9360	0.9365	0.9370	0.9375	0.9380	0.9385	0.9390
8.7	0.9395	0.9400	0.9405	0.9410	0.9415	0.9420	0.9425	0.9430	0.9435	0.9440
8.8	0.9445	0.9450	0.9455	0.9460	0.9465	0.9469	0.9474	0.9479	0.9484	0.9489
8.9	0.9494	0.9499	0.9504	0.9509	0.9513	0.9518	0.9523	0.9528	0.9533	0.9538
9.0	0.9542	0.9547	0.9552	0.9557	0.9562	0.9566	0.9571	0.9576	0.9581	0.9586
9.1	0.9590	0.9595	0.9600	0.9605	0.9609	0.9614	0.9619	0.9624	0.9628	0.9633
9.2	0.9638	0.9643	0.9647	0.9652	0.9657	0.9661	0.9666	0.9671	0.9675	0.9680
9.3	0.9685	0.9689	0.9694	0.9699	0.9703	0.9708	0.9713	0.9717	0.9722	0.9727
9.4	0.9731	0.9736	0.9741	0.9745	0.9750	0.9754	0.9759	0.9763	0.9768	0.9773
9.5	0.9777	0.9782	0.9786	0.9791	0.9795	0.9800	0.9805	0.9809	0.9814	0.9818
9.6	0.9823	0.9827	0.9832	0.9836	0.9841	0.9845	0.9850	0.9854	0.9859	0.9863
9.7	0.9868	0.9872	0.9877	0.9881	0.9886	0.9890	0.9894	0.9899	0.9903	0.9908
9.8	0.9912	0.9917	0.9921	0.9926	0.9930	0.9934	0.9939	0.9943	0.9948	0.9952
9.9	0.9956	0.9961	0.9965	0.9969	0.9974	0.9978	0.9983	0.9987	0.9991	0.9996
x	0	1	2	3	4	5	6	7	8	9

Case II Computation of Natural Logarithms

In this section base e logarithms, which are also called natural logarithms, are discussed. Logarithmic expressions using base e are written in the form of $log_e x$ or $ln x$. The letter e represents an irrational number and is approximately equal to 2.718282. A calculator or a table of natural logarithms is used in order to find the natural logarithm of numbers. Table 6.2-3 gives the natural logarithm of numbers in various increments. To obtain natural logarithm of numbers other than those indicated in Table 6.2-3 use the following steps and a hand calculator.

Step 1 Write the logarithm number in scientific notation form, i.e., write $ln 250$ as $ln 2.5 \times 10^2$.

Step 2 Apply the laws of logarithm and simplify the expression (see Table 6.1-2).

Examples with Steps

The following examples show the steps as to how natural logarithms are solved:

Example 6.2-41

$\boxed{ln 100} =$ Note: $ln = log_e$

Solution:

Step 1 $\boxed{ln 100} = \boxed{ln 1.0 \times 10^2}$

Step 2 $\boxed{ln 1.0 \times 10^2} = \boxed{ln 1.0 + ln 10^2} = \boxed{0 + 2 ln 10} = \boxed{2 \times 2.3026} = \boxed{\mathbf{4.6052}}$

Example 6.2-42

$\boxed{ln 0.0001} =$

Solution:

Step 1 $\boxed{ln 0.0001} = \boxed{ln 1.0 \times 10^{-4}}$

Step 2 $\boxed{ln 1.0 \times 10^{-4}} = \boxed{ln 1.0 + ln 10^{-4}} = \boxed{0 - 4 ln 10} = \boxed{-4 \times 2.3026} = \boxed{\mathbf{-9.2104}}$

Example 6.2-43

$\boxed{ln \sqrt{e^{\frac{1}{3}}}} =$

Solution:

Step 1 Not Applicable

Step 2 $\boxed{ln \sqrt{e^{\frac{1}{3}}}} = \boxed{ln e^{\frac{\frac{1}{3}}{2}}} = \boxed{ln e^{\frac{1}{3} \times \frac{1}{2}}} = \boxed{ln e^{\frac{1 \times 1}{3 \times 2}}} = \boxed{ln e^{\frac{1}{6}}} = \boxed{\frac{1}{6} ln e} = \boxed{\frac{1}{6} \times 1} = \boxed{\mathbf{\frac{1}{6}}}$

Example 6.2-44

$\boxed{ln \frac{150}{27}} =$

Solution:

Step 1 $\boxed{ln \frac{150}{27}} = \boxed{ln \frac{\overset{50}{\cancel{150}}}{\underset{9}{\cancel{27}}}} = \boxed{ln \frac{50}{9}} = \boxed{ln \frac{5.0 \times 10^1}{9.0 \times 10^0}}$

Step 2 $\boxed{ln\frac{5.0\times10^1}{9.0\times10^0}} = \boxed{ln5.0\times10^1 - ln9.0\times10^0} = \boxed{\left(ln5.0+ln10^1\right)-\left(ln9.0+ln10^0\right)}$

$= \boxed{(1.6094+2.3026)-(2.1972+ln1)} = \boxed{(3.912)-(2.1972+0)} = \boxed{3.912-2.1972} = \boxed{\mathbf{1.7148}}$

Example 6.2-45

$\boxed{5ln4\cdot36^{\frac{1}{2}}} =$

Solution:

Step 1 $\boxed{5ln4\cdot36^{\frac{1}{2}}} = \boxed{5ln4\cdot\left(6^2\right)^{\frac{1}{2}}} = \boxed{5ln4\cdot6^{2\times\frac{1}{2}}} = \boxed{5ln4\cdot6} = \boxed{5ln24} = \boxed{5ln2.4\times10^1}$

Step 2 $\boxed{5ln2.4\times10^1} = \boxed{5\left(ln2.4+ln10^1\right)} = \boxed{5(0.8755+2.3026)} = \boxed{5(3.1781)} = \boxed{\mathbf{15.8905}}$

Additional Examples - Computation of Natural Logarithms

The following examples further illustrate how to solve natural logarithms:

Example 6.2-46

$\boxed{ln200} = \boxed{ln2.0\times10^2} = \boxed{ln2.0+ln10^2} = \boxed{0.6931+2ln10} = \boxed{0.6931+2\times2.3026} = \boxed{0.6931+4.6052}$

$= \boxed{\mathbf{5.2983}}$

Example 6.2-47

$\boxed{ln2{,}450} = \boxed{ln2.45\times10^3} = \boxed{ln2.45+ln10^3} = \boxed{0.8961+3ln10} = \boxed{0.8961+3\times2.3026} = \boxed{0.8961+6.9078}$

$= \boxed{\mathbf{7.8039}}$

Example 6.2-48

$\boxed{ln60{,}000} = \boxed{ln6.0\times10^4} = \boxed{ln6.0+ln10^4} = \boxed{1.7917+4ln10} = \boxed{1.79171+4\times2.3026} = \boxed{1.79171+9.2104}$

$= \boxed{\mathbf{11.0021}}$

Example 6.2-49

$\boxed{ln8} = \boxed{ln8.0\times10^0} = \boxed{ln8.0+ln10^0} = \boxed{2.0794+ln1} = \boxed{2.0794+0} = \boxed{\mathbf{2.0794}}$

Example 6.2-50

$\boxed{ln\sqrt{81}} = \boxed{ln81^{\frac{1}{2}}} = \boxed{\frac{1}{2}ln81} = \boxed{\frac{1}{2}\left(ln8.1\times10^1\right)} = \boxed{\frac{1}{2}\left(ln8.1+ln10^1\right)} = \boxed{\frac{1}{2}(2.0919+2.3026)} = \boxed{\frac{4.3945}{2}} = \boxed{\mathbf{2.1972}}$

or we can solve the above problem in the following way:

$\boxed{ln\sqrt{81}} = \boxed{ln\sqrt{9^2}} = \boxed{ln\left(9^2\right)^{\frac{1}{2}}} = \boxed{ln9^{2\times\frac{1}{2}}} = \boxed{ln9^1} = \boxed{ln9} = \boxed{\mathbf{2.1972}}$

Example 6.2-51

$\boxed{ln\sqrt[3]{54}} = \boxed{ln\sqrt[3]{27\cdot2}} = \boxed{ln\sqrt[3]{3^3\cdot2}} = \boxed{ln3\sqrt[3]{2}} = \boxed{ln3\times2^{\frac{1}{3}}} = \boxed{ln3+ln2^{\frac{1}{3}}} = \boxed{ln3+\frac{1}{3}ln2} = \boxed{1.0986+\frac{1}{3}(0.6931)}$

$= \boxed{1.0986 + 0.2310} = \boxed{\mathbf{1.3296}}$

Example 6.2-52

$\boxed{ln25^{\frac{1}{4}}} = \boxed{\frac{1}{4}ln25} = \boxed{\frac{1}{4}ln2.5 \times 10^1} = \boxed{\frac{1}{4}\left(ln2.5 + ln10^1\right)} = \boxed{\frac{1}{4}(0.9163 + 2.3026)} = \boxed{\frac{3.2189}{4}} = \boxed{\mathbf{0.8047}}$

Example 6.2-53

$\boxed{ln3\sqrt[4]{256}} = \boxed{ln3\sqrt[4]{4^4}} = \boxed{ln3 \cdot 4^{\frac{4}{4}=\frac{1}{1}}} = \boxed{ln3 \cdot 4} = \boxed{ln12} = \boxed{ln1.2 \times 10^1} = \boxed{ln1.2 + ln10} = \boxed{0.1823 + 2.3026}$

$= \boxed{\mathbf{2.4849}}$ or,

$\boxed{ln3\sqrt[4]{256}} = \boxed{ln3\sqrt[4]{4^4}} = \boxed{ln3 \cdot 4^{\frac{4}{4}=\frac{1}{1}}} = \boxed{ln3 \cdot 4} = \boxed{ln3 + ln4} = \boxed{1.0986 + 1.3863} = \boxed{\mathbf{2.4849}}$

Example 6.2-54

$\boxed{ln76{,}000{,}000} = \boxed{ln7.6 \times 10^7} = \boxed{ln7.6 + ln10^7} = \boxed{2.0281 + 7\,ln10} = \boxed{2.0281 + 7 \times 2.3026} = \boxed{2.0281 + 16.1182}$

$= \boxed{\mathbf{18.1463}}$

Example 6.2-55

$\boxed{ln0.00024} = \boxed{ln2.4 \times 10^{-4}} = \boxed{ln2.4 + ln10^{-4}} = \boxed{0.8755 - 4\,ln10} = \boxed{0.8755 - 4 \times 2.3026} = \boxed{0.8755 - 9.2104}$

$= \boxed{\mathbf{-8.3349}}$

Example 6.2-56

$\boxed{ln81{,}000} = \boxed{ln8.1 \times 10^4} = \boxed{ln8.1 + ln10^4} = \boxed{2.0919 + 4\,ln10} = \boxed{2.0919 + 4 \times 2.3026} = \boxed{2.0919 + 9.2104}$

$= \boxed{\mathbf{11.3023}}$

Example 6.2-57

$\boxed{ln2\sqrt{250}} = \boxed{ln2 \times 5\sqrt{10}} = \boxed{ln10\sqrt{10}} = \boxed{ln10 \times 10^{\frac{1}{2}}} = \boxed{ln10 + ln10^{\frac{1}{2}}} = \boxed{ln10 + \frac{1}{2}ln10} = \boxed{2.3026 + \frac{1}{2} \times 2.3026}$

$= \boxed{2.3026 + 1.1513} = \boxed{\mathbf{3.4539}}$

Example 6.2-58

$\boxed{ln\sqrt[3]{44}} = \boxed{ln44^{\frac{1}{3}}} = \boxed{\frac{1}{3}ln44} = \boxed{\frac{1}{3}ln4.4 \times 10^1} = \boxed{\frac{1}{3}\left(ln4.4 + ln10^1\right)} = \boxed{\frac{1}{3}(1.4816 + 2.3026)} = \boxed{\frac{3.7842}{3}} = \boxed{\mathbf{1.2614}}$

Example 6.2-59

$\boxed{ln0.05} = \boxed{ln5.0 \times 10^{-2}} = \boxed{ln5.0 + ln10^{-2}} = \boxed{ln5.0 - 2\,ln10} = \boxed{1.6094 - 2 \times 2.3026} = \boxed{1.6094 - 4.6052}$

$= \boxed{\mathbf{-2.9958}}$

Example 6.2-60

$\boxed{ln3e^2} = \boxed{ln3 + lne^2} = \boxed{ln3 + 2\,lne} = \boxed{ln3 + 2 \times 1} = \boxed{1.0986 + 2} = \boxed{\mathbf{3.0986}}$

Example 6.2-61

$$\boxed{ln\sqrt{e^3}} = \boxed{ln\left(e^3\right)^{\frac{1}{2}}} = \boxed{ln e^{3\times\frac{1}{2}}} = \boxed{ln e^{\frac{3}{2}}} = \boxed{\frac{3}{2} ln e} = \boxed{\frac{3}{2}\times 1} = \boxed{\mathbf{\frac{3}{2}}}$$

Example 6.2-62

$$\boxed{ln\frac{1}{\sqrt{e^3}}} = \boxed{ln\frac{1}{e^{3\frac{1}{2}}}} = \boxed{ln\frac{1}{e^{3\times\frac{1}{2}}}} = \boxed{ln\frac{1}{e^{\frac{3}{2}}}} = \boxed{ln e^{-\frac{3}{2}}} = \boxed{-\frac{3}{2} ln e} = \boxed{-\frac{3}{2}\times 1} = \boxed{\mathbf{-\frac{3}{2}}}$$

Example 6.2-63

$$\boxed{ln\frac{2}{27}} = \boxed{ln\frac{2}{3^3}} = \boxed{ln2 - ln3^3} = \boxed{ln2 - 3ln3} = \boxed{0.6931 - 3\times 1.0986} = \boxed{0.6931 - 3.2958} = \boxed{\mathbf{-2.6027}}$$

Example 6.2-64

$$\boxed{3ln\frac{5}{\sqrt{7}}} = \boxed{3\left(ln5 - ln\sqrt{7}\right)} = \boxed{3\left(ln5 - ln7^{\frac{1}{2}}\right)} = \boxed{3\left(ln5 - \frac{1}{2}ln7\right)} = \boxed{3\left(1.6094 - \frac{1}{2}\times 1.9459\right)} = \boxed{3(1.6094 - 0.9729)}$$

$$= \boxed{3\times 0.6365} = \boxed{\mathbf{1.9095}}$$

Example 6.2-65

$$\boxed{5ln\frac{1}{16}} = \boxed{5(ln1 - ln16)} = \boxed{5(0 - ln16)} = \boxed{-5ln1.6\times 10^1} = \boxed{-5(ln1.6 + ln10)} = \boxed{-5(0.470 + 2.3026)}$$

$$= \boxed{-5\times 2.7726} = \boxed{\mathbf{-13.863}}$$

Example 6.2-66

$$\boxed{ln2\cdot 3^{\frac{1}{4}}} = \boxed{ln2 + ln3^{\frac{1}{4}}} = \boxed{ln2 + \frac{1}{4}ln3} = \boxed{0.6931 + \frac{1}{4}\times 1.0986} = \boxed{0.6931 + \frac{1.0986}{4}} = \boxed{0.6931 + 0.2746}$$

$$= \boxed{\mathbf{0.9677}}$$

Example 6.2-67

$$\boxed{2ln\frac{2^3}{55^0}} = \boxed{2ln\frac{2^3}{1}} = \boxed{2\left(ln2^3 - ln1\right)} = \boxed{2(3ln2 - 0)} = \boxed{2(3\times 0.6931)} = \boxed{2\times 2.0793} = \boxed{\mathbf{4.1586}}$$

Example 6.2-68

$$\boxed{ln\frac{e^{\frac{2}{3}}}{e}} = \boxed{ln e^{\frac{2}{3}} - ln e} = \boxed{\frac{2}{3}ln e - ln e} = \boxed{\frac{2}{3}\times 1 - 1} = \boxed{\frac{2}{3} - \frac{1}{1}} = \boxed{\frac{(2\times 1)-(1\times 3)}{3\times 1}} = \boxed{\frac{2-3}{3}} = \boxed{\mathbf{-\frac{1}{3}}}$$

Another way of solving the above problem is by using the exponent laws as follows:

$$\boxed{ln\frac{e^{\frac{2}{3}}}{e}} = \boxed{ln\frac{e^{\frac{2}{3}}}{e^1}} = \boxed{ln e^{\frac{2}{3}}\cdot e^{-1}} = \boxed{ln e^{\frac{2}{3}}\cdot e^{-\frac{1}{1}}} = \boxed{ln e^{\frac{2}{3}-\frac{1}{1}}} = \boxed{ln e^{\frac{(2\times1)-(1\times3)}{3\times1}}} = \boxed{ln e^{\frac{2-3}{3}}} = \boxed{ln e^{-\frac{1}{3}}} = \boxed{-\frac{1}{3}ln e}$$

$$= \boxed{-\frac{1}{3}\times 1} = \boxed{\mathbf{-\frac{1}{3}}}$$

Example 6.2-69

$$\boxed{ln\frac{\sqrt[3]{3e^2}}{4}} = \boxed{ln\sqrt[3]{3e^2} - ln4} = \boxed{ln\left(3e^2\right)^{\frac{1}{3}} - ln4} = \boxed{\frac{1}{3}ln3e^2 - ln4} = \boxed{\frac{1}{3}\left(ln3 + lne^2\right) - ln4} = \boxed{\frac{1}{3}(ln3 + 2lne) - ln4}$$

$$= \boxed{\frac{1}{3}(1.0986 + 2\times 1) - 1.3863} = \boxed{\frac{1}{3}(1.0986 + 2) - 1.3863} = \boxed{\frac{3.0986}{3} - 1.3863} = \boxed{1.0329 - 1.3863} = \boxed{\mathbf{-0.3534}}$$

Example 6.2-70

$$\boxed{ln\frac{e^{-\frac{2}{3}}\cdot e^3}{e}} = \boxed{ln\frac{e^{-\frac{2}{3}}\cdot e^{\frac{3}{1}}}{e}} = \boxed{ln\frac{e^{-\frac{2}{3}+\frac{3}{1}}}{e}} = \boxed{ln\frac{e^{\frac{(-2\times 1)+(3\times 3)}{3\times 1}}}{e}} = \boxed{ln\frac{e^{\frac{-2+9}{3}}}{e}} = \boxed{ln\frac{e^{\frac{7}{3}}}{e}} = \boxed{lne^{\frac{7}{3}} - lne}$$

$$= \boxed{\frac{7}{3}lne - 1} = \boxed{\frac{7}{3}\times 1 - 1} = \boxed{\frac{7}{3} - \frac{1}{1}} = \boxed{\frac{(7\times 1)-(1\times 3)}{3\times 1}} = \boxed{\frac{7-3}{3}} = \boxed{\frac{4}{3}} = \boxed{\mathbf{1\frac{1}{3}}}$$

Practice Problems - Computation of Natural Logarithms

Section 6.2 Case II Practice Problems - Find the values of the following natural logarithms:

1. $ln38 =$
2. $ln\sqrt{e^3} =$
3. $ln0.0007 =$
4. $ln255{,}000 =$
5. $ln\frac{1}{216} =$
6. $ln\frac{e}{3} =$
7. $ln\sqrt[5]{500} =$
8. $ln\frac{2}{3} =$
9. $500\,lne^{-3} =$
10. $ln49^{\frac{2}{3}} =$

Table 6.2-3
Natural Logarithms

n	$\log_e n$	n	$\log_e n$	n	$\log_e n$	n	$\log_e n$	n	$\log_e n$
		4.5	1.5041	9.0	2.1972	17.0	2.8332	170	5.1358
0.1	-2.3026	4.6	1.5261	9.1	2.2083	17.2	2.8449	175	5.1648
0.2	-1.6094	4.7	1.5476	9.2	2.2192	17.4	2.8565	180	5.1930
0.3	-1.2040	4.8	1.5686	9.3	2.2300	17.6	2.8679	185	5.2204
0.4	-0.9163	4.9	1.5892	9.4	2.2407	17.8	2.8792	190	5.2470
0.5	-0.6931	5.0	1.6094	9.5	2.2513	18.0	2.8904	195	5.2730
0.6	-0.5108	5.1	1.6292	9.6	2.2618	18.2	2.9014	200	5.2983
0.7	-0.3567	5.2	1.6487	9.7	2.2721	18.4	2.9124	250	5.5215
0.8	-0.2231	5.3	1.6677	9.8	2.2824	18.6	2.9232	300	5.7038
0.9	-0.1054	5.4	1.6864	9.9	2.2925	18.8	2.9339	350	5.8579
1.0	0.0000	5.5	1.7047	10.0	2.3026	19.0	2.9444	400	5.9915
1.1	0.0953	5.6	1.7228	10.2	2.3224	19.2	2.9549	450	6.1092
1.2	0.1823	5.7	1.7405	10.4	2.3418	19.4	2.9653	500	6.2146
1.3	0.2624	5.8	1.7579	10.6	2.3609	19.6	2.9755	550	6.3099
1.4	0.3365	5.9	1.7750	10.8	2.3795	19.8	2.9857	600	6.3969
1.5	0.4055	6.0	1.7918	11.0	2.3979	20	2.9957	650	6.4770
1.6	0.4700	6.1	1.8083	11.2	2.4159	25	3.2189	700	6.5511
1.7	0.5306	6.2	1.8245	11.4	2.4336	30	3.4012	750	6.6201
1.8	0.5878	6.3	1.8405	11.6	2.4510	35	3.5553	800	6.6846
1.9	0.6419	6.4	1.8563	11.8	2.4681	40	3.6889	850	6.7452
2.0	0.6931	6.5	1.8718	12.0	2.4849	45	3.8067	900	6.8024
2.1	0.7419	6.6	1.8871	12.2	2.5014	50	3.9120	950	6.8565
2.2	0.7885	6.7	1.9021	12.4	2.5014	55	4.0073	1000	6.9078
2.3	0.8329	6.8	1.9169	12.6	2.5337	60	4.0943	1050	6.9565
2.4	0.8755	6.9	1.9315	12.8	2.5494	65	4.1744	1100	7.0031
2.5	0.9163	7.0	1.9459	13.0	2.5649	70	4.2485	1150	7.0475
2.6	0.9555	7.1	1.9601	13.2	2.5802	75	4.3175	1200	7.0901
2.7	0.9933	7.2	1.9741	13.4	2.5953	80	4.3820	1250	7.1309
2.8	1.0296	7.3	1.9879	13.6	2.6101	85	4.4427	1300	7.1701
2.9	1.0647	7.4	2.0015	13.8	2.6247	90	4.4998	1350	7.2079
3.0	1.0986	7.5	2.0149	14.0	2.6391	95	4.5539	1400	7.2442
3.1	1.1314	7.6	2.0281	14.2	2.6532	100	4.6052	1450	7.2793
3.2	1.1632	7.7	2.04122	14.4	2.6672	105	4.6540	1500	7.3132
3.3	1.1939	7.8	2.0541	14.6	2.6810	110	4.7005	1550	7.3460
3.4	1.2238	7.9	2.0669	14.8	2.6946	115	4.7449	1600	7.3778
3.5	1.2528	8.0	2.0794	15.0	2.7081	120	4.7875	1650	7.4085
3.6	1.2809	8.1	2.0919	15.2	2.7213	125	4.8283	1700	7.4384
3.7	1.3083	8.2	2.1041	15.4	2.7344	130	4.8676	1750	7.4674
3.8	1.3350	8.3	2.1163	15.6	2.7473	135	4.9053	1800	7.4955
3.9	1.3610	8.4	2.1282	15.8	2.7600	140	4.9416	1850	7.5229
4.0	1.3863	8.5	2.1401	16.0	2.7726	145	4.9767	1900	7.5496
4.1	1.4110	8.6	2.1518	16.2	2.7850	150	5.0106	1950	7.5756
4.2	1.4351	8.7	2.1633	16.4	2.7973	155	5.0434	2000	7.6009
4.3	1.4586	8.8	2.1748	16.6	2.8094	160	5.0752	3000	8.0064
4.4	1.4816	8.9	2.1861	16.8	2.8214	165	5.1059	4000	8.2940
x	0	1	2	3	4	5	6	7	8

Case III Computation of Logarithms other than Base 10 or e

Logarithms other than base 10 or e are solved by using the following general rule:

$$log_a b = \frac{log_{10} b}{log_{10} a}$$

For example, $log_3 5 = \frac{log_{10} 5}{log_{10} 3} = \frac{0.6990}{0.4771} = \mathbf{1.465}$; $log_{0.2} 8 = \frac{log_{10} 8}{log_{10} 0.2} = \frac{0.9031}{-0.6989} = \mathbf{-1.292}$

$log_5 2 = \frac{log_{10} 2}{log_{10} 5} = \frac{0.3010}{0.6990} = \mathbf{0.4306}$; $log_8 7 = \frac{log_{10} 7}{log_{10} 8} = \frac{0.8451}{0.9031} = \mathbf{0.9358}$

The steps in solving this class of logarithms are as follows:

Step 1 Change the logarithm number to an exponential number with a base similar to the base of the logarithm, i.e., change $log_3 81$ to $log_3 3^4$.

Note: If a logarithm number can not be changed to an exponential number with a base similar to the base of the logarithm, we then change the given logarithm to a common logarithm by using the general equation $log_a b = \frac{log_{10} b}{log_{10} a}$.

Step 2 Apply the laws of logarithm and solve the logarithmic expression (see Table 6.1-2).

Examples with Steps

The following examples show the steps as to how logarithms other than base 10 or e are solved:

Example 6.2-71

$$\boxed{log_3 \sqrt{243}} =$$

Solution:

I. Step 1

$$\boxed{log_3 \sqrt{243}} = \boxed{log_3 243^{\frac{1}{2}}} = \boxed{\frac{1}{2} log_3 243} = \boxed{\frac{1}{2} log_3 3^5}$$

Step 2

$$\boxed{\frac{1}{2} log_3 3^5} = \boxed{\frac{1}{2} \times 5 log_3 3} = \boxed{\frac{5}{2} log_3 3} = \boxed{\frac{5}{2} \times 1} = \boxed{\frac{5}{2}} = \boxed{\mathbf{2.5}}$$

or we can solve the above example in the following way:

II. Step 1

$$\boxed{log_3 \sqrt{243}} = \boxed{log_3 243^{\frac{1}{2}}} = \boxed{\frac{1}{2} log_3 243}$$

Step 2

$$\boxed{\frac{1}{2} log_3 243} = \boxed{\frac{1}{2} \frac{log_{10} 243}{log_{10} 3}} = \boxed{\frac{1}{2} \frac{log_{10} 2.43 \times 10^2}{log_{10} 3.0 \times 10^0}} = \boxed{\frac{1}{2} \frac{log_{10} 2.43 + log_{10} 10^2}{log_{10} 3 + log_{10} 10^0}}$$

$$= \boxed{\frac{1}{2} \frac{log_{10} 2.43 + 2}{log_{10} 3 + 0}} = \boxed{\frac{1}{2} \times \frac{0.3856 + 2}{0.4771}} = \boxed{\frac{1}{2} \times \frac{2.3856}{0.4771}} = \boxed{\frac{1}{2} \times 5} = \boxed{\frac{5}{2}} = \boxed{\mathbf{2.5}}$$ or,

III. Step 1

$$\boxed{log_3 \sqrt{243}} = \boxed{log_3 15.59}$$

Step 2 $\boxed{log_3 15.59} = \boxed{\frac{log_{10} 15.59}{log_{10} 3}} = \boxed{\frac{log_{10} 1.559 \times 10^1}{log_{10} 3.0 \times 10^0}} = \boxed{\frac{log_{10} 1.559 + log_{10} 10^1}{log_{10} 3 + log_{10} 10^0}} = \boxed{\frac{log_{10} 1.559 + 1}{log_{10} 3 + 0}}$

$= \boxed{\frac{0.193 + 1}{0.4771}} = \boxed{\frac{1.193}{0.4771}} = \boxed{\mathbf{2.5}}$

Example 6.2-72

$\boxed{log_4 \frac{3}{1024}} =$

Solution:

I. Step 1 $\boxed{log_4 \frac{3}{1024}} = \boxed{log_4 3 - log_4 1024} = \boxed{log_4 3 - log_4 4^5}$

Step 2 $\boxed{log_4 3 - log_4 4^5} = \boxed{\frac{log_{10} 3}{log_{10} 4} - 5 log_4 4} = \boxed{\frac{0.4771}{0.6021} - 5 \times 1} = \boxed{0.79 - 5} = \boxed{\mathbf{-4.21}}$ or,

II. Step 1 $\boxed{log_4 \frac{3}{1024}} = \boxed{\log_4 0.00293}$

Step 2 $\boxed{\log_4 0.00293} = \boxed{\frac{\log_{10} 0.00293}{\log_{10} 4}} = \boxed{\frac{\log_{10} 2.93 \times 10^{-3}}{\log_{10} 4}} = \boxed{\frac{0.4668 - 3}{0.6020}} = \boxed{\frac{-2.533}{0.6020}} = \boxed{\mathbf{-4.21}}$

Example 6.2-73

$\boxed{log_{\frac{1}{3}} 27} =$

Solution:

I. Step 1 $\boxed{log_{\frac{1}{3}} 27} = \boxed{log_{\frac{1}{3}} \left(\frac{1}{3}\right)^{-3}}$

Note: $\left(\frac{1}{3}\right)^{-3} = \frac{1}{3^{-3}} = \frac{1}{\frac{1}{3^3}} = \frac{\frac{1}{1}}{\frac{1}{27}} = \frac{1 \times 27}{1 \times 1} = \frac{27}{1} = 27$

Step 2 $\boxed{log_{\frac{1}{3}} \left(\frac{1}{3}\right)^{-3}} = \boxed{-3 log_{\frac{1}{3}} \left(\frac{1}{3}\right)} = \boxed{-3 \times 1} = \boxed{\mathbf{-3}}$

or we can solve the given logarithmic expression in the following way:

II. Step 1 $\boxed{log_{\frac{1}{3}} 27} = \boxed{log_{0.333} 27}$

Step 2 $\boxed{log_{0.333} 27} = \boxed{\frac{log_{10} 27}{log_{10} 0.333}} = \boxed{\frac{log_{10} 2.7 \times 10^1}{log_{10} 3.33 \times 10^{-1}}} = \boxed{\frac{log_{10} 2.7 + log_{10} 10^1}{log_{10} 3.33 + log_{10} 10^{-1}}} = \boxed{\frac{log_{10} 2.7 + 1}{log_{10} 3.33 - 1}}$

$= \boxed{\frac{0.4314 + 1}{0.5224 - 1}} = \boxed{\frac{1.4314}{-0.4776}} = \boxed{\mathbf{-3}}$

Example 6.2-74

$\boxed{log_{0.02} 25} =$

Solution:

Step 1 *Not Applicable*

Step 2 $\boxed{\log_{0.02} 25} = \boxed{\frac{\log_{10} 25}{\log_{10} 0.02}} = \boxed{\frac{\log_{10} 2.5\times 10^1}{\log_{10} 2.0\times 10^{-2}}} = \boxed{\frac{\log_{10} 2.5+\log_{10} 10^1}{\log_{10} 2.0+\log_{10} 10^{-2}}}$

$= \boxed{\frac{\log_{10} 2.5+1}{\log_{10} 2.0-2}} = \boxed{\frac{0.3979+1}{0.3010-2}} = \boxed{\frac{1.3979}{-1.699}} = \boxed{\mathbf{-0.82}}$

Example 6.2-75

$\boxed{\log_{\frac{3}{5}} \frac{625}{81}} =$

Solution:

I. Step 1 $\boxed{\log_{\frac{3}{5}} \frac{625}{81}} = \boxed{\log_{\frac{3}{5}} \left(\frac{3}{5}\right)^{-4}}$

Note: $\left(\frac{3}{5}\right)^{-4} = \frac{1}{\left(\frac{3}{5}\right)^4} = \frac{1}{\frac{81}{625}} = \frac{\frac{1}{1}}{\frac{81}{625}} = \frac{1\times 625}{1\times 81} = \frac{625}{81}$

Step 2 $\boxed{\log_{\frac{3}{5}} \left(\frac{3}{5}\right)^{-4}} = \boxed{-4\log_{\frac{3}{5}} \left(\frac{3}{5}\right)} = \boxed{-4\times 1} = \boxed{\mathbf{-4}}$ or,

II. Step 1 $\boxed{\log_{\frac{3}{5}} \frac{625}{81}} = \boxed{\log_{0.6} \frac{625}{81}}$

Step 2 $\boxed{\log_{0.6} \frac{625}{81}} = \boxed{\log_{0.6} 625 - \log_{0.6} 81} = \boxed{\frac{\log_{10} 625}{\log_{10} 0.6} - \frac{\log_{10} 81}{\log_{10} 0.6}} = \boxed{\frac{\log_{10} 625 - \log_{10} 81}{\log_{10} 0.6}}$

$= \boxed{\frac{\log_{10} 6.25\times 10^2 - \log_{10} 8.1\times 10^1}{\log_{10} 6.0\times 10^{-1}}} = \boxed{\frac{\log_{10} 6.25 + \log_{10} 10^2 - \log_{10} 8.1 - \log_{10} 10^1}{\log_{10} 6.0 + \log_{10} 10^{-1}}}$

$= \boxed{\frac{\log_{10} 6.25 + 2 - \log_{10} 8.1 - 1}{\log_{10} 6.0 - 1}} = \boxed{\frac{0.796+2-0.908-1}{0.778-1}} = \boxed{\frac{0.888}{-0.222}} = \boxed{\mathbf{-4}}$ or,

III. Step 1 $\boxed{\log_{\frac{3}{5}} \frac{625}{81}} = \boxed{\log_{0.6} 7.716}$

Step 2 $\boxed{\log_{0.6} 7.716} = \boxed{\frac{\log_{10} 7.716}{\log_{10} 0.6}} = \boxed{\frac{\log_{10} 7.716\times 10^0}{\log_{10} 6.0\times 10^{-1}}} = \boxed{\frac{\log_{10} 7.716 + \log_{10} 10^0}{\log_{10} 6.0 + \log_{10} 10^{-1}}}$

$= \boxed{\frac{\log_{10} 7.716 + 0}{\log_{10} 6.0 - 1}} = \boxed{\frac{0.888}{0.778-1}} = \boxed{\frac{0.888}{-0.222}} = \boxed{\mathbf{-4}}$

The above examples show that logarithms can be solved in different ways. Students are encouraged to practice solving the additional problems given below in more than one way.

Additional Examples - Computation of Logarithms other than Base 10 or *e*

The following examples further illustrate how to solve logarithms other than base 10 or e:

Example 6.2-76

I. $\boxed{log_2 \frac{16}{64}} = \boxed{log_2 16 - log_2 64} = \boxed{log_2 2^4 - log_2 2^6} = \boxed{4 log_2 2 - 6 log_2 2} = \boxed{(4 \times 1) - (6 \times 1)} = \boxed{4 - 6} = \boxed{\mathbf{-2}}$

or we can also solve the above problem in the following way:

II. $\boxed{log_2 \frac{16}{64}} = \boxed{\log_2 \frac{\cancel{16}}{\underset{4}{\cancel{64}}}} = \boxed{log_2 \frac{1}{4}} = \boxed{log_2 1 - log_2 4} = \boxed{0 - log_2 2^2} = \boxed{0 - 2 log_2 2} = \boxed{-2 \times 1} = \boxed{\mathbf{-2}}$

Example 6.2-77

I. $\boxed{log_5 \frac{1}{125}} = \boxed{log_5 1 - log_5 125} = \boxed{0 - log_5 5^3} = \boxed{-3 log_5 5} = \boxed{-3 \times 1} = \boxed{\mathbf{-3}}$ or,

II. $\boxed{log_5 \frac{1}{125}} = \boxed{log_5 0.008} = \boxed{\frac{log_{10} 0.008}{log_{10} 5}} = \boxed{\frac{log_{10} 8 \times 10^{-3}}{log_{10} 5}} = \boxed{\frac{log_{10} 8 + log_{10} 10^{-3}}{log_{10} 5}} = \boxed{\frac{log_{10} 8 - 3}{log_{10} 5}}$

$= \boxed{\frac{0.9030 - 3}{0.6989}} = \boxed{\frac{-2.097}{0.6989}} = \boxed{\mathbf{-3}}$

Example 6.2-78

$\boxed{log_9 \sqrt[5]{9}} = \boxed{log_9 9^{\frac{1}{5}}} = \boxed{\frac{1}{5} log_9 9} = \boxed{\frac{1}{5} \times 1} = \boxed{\mathbf{\frac{1}{5}}}$

Example 6.2-79

$\boxed{log_8 4} = \boxed{log_8 8^{\frac{2}{3}}} = \boxed{\frac{2}{3} log_8 8} = \boxed{\frac{2}{3} \times 1} = \boxed{\frac{2}{3}} = \boxed{\mathbf{0.667}}$ or, $\boxed{log_8 4} = \boxed{\frac{log_{10} 4}{log_{10} 8}} = \boxed{\frac{0.6021}{0.9031}} = \boxed{\mathbf{0.667}}$

Example 6.2-80

I. $\boxed{log_3 \frac{243}{9}} = \boxed{log_3 243 - log_3 9} = \boxed{log_3 3^5 - log_3 3^2} = \boxed{5 log_3 3 - 2 log_3 3} = \boxed{(5 \times 1) - (2 \times 1)} = \boxed{5 - 2} = \boxed{\mathbf{3}}$

or,

II. $\boxed{log_3 \frac{243}{9}} = \boxed{\log_3 \frac{\overset{27}{\cancel{243}}}{\cancel{9}}} = \boxed{log_3 \frac{27}{1}} = \boxed{log_3 27 - log_3 1} = \boxed{log_3 3^3 - 0} = \boxed{3 log_3 3} = \boxed{3 \times 1} = \boxed{\mathbf{3}}$

Note that we can also solve the above logarithm, after it has been simplified, in the following way:

III. $\boxed{log_3 \frac{243}{9}} = \boxed{\log_3 \frac{\overset{27}{\cancel{243}}}{\cancel{9}}} = \boxed{log_3 \frac{27}{1}} = \boxed{log_3 27} = \boxed{\frac{log_{10} 27}{log_{10} 3}} = \boxed{\frac{1.4313}{0.4771}} = \boxed{\mathbf{3}}$

Example 6.2-81

I. $\boxed{log_2 \sqrt{2048}} = \boxed{log_2 (2048)^{\frac{1}{2}}} = \boxed{\frac{1}{2} log_2 2048} = \boxed{\frac{1}{2} log_2 (512 \times 4)} = \boxed{\frac{1}{2}(log_2 512 + log_2 4)}$

$= \boxed{\frac{1}{2}(log_2 2^9 + log_2 2^2)} = \boxed{\frac{1}{2}(9 log_2 2 + 2 log_2 2)} = \boxed{\frac{1}{2}(9 \times 1 + 2 \times 1)} = \boxed{\frac{1}{2}(9 + 2)} = \boxed{\frac{11}{2}} = \boxed{\mathbf{5.5}}$ or,

II. $\boxed{\log_2 \sqrt{2048}} = \boxed{\log_2 (2048)^{\frac{1}{2}}} = \boxed{\frac{1}{2}\log_2 2048} = \boxed{\frac{1}{2}\,\frac{\log_{10} 2048}{\log_{10} 2}} = \boxed{\frac{1}{2}\,\frac{\log_{10} 2.048 \times 10^3}{\log_{10} 2}}$

$= \boxed{\frac{1}{2}\,\frac{\log_{10} 2.048 + \log_{10} 10^3}{\log_{10} 2}} = \boxed{\frac{1}{2}\,\frac{0.3113 + 3}{0.3010}} = \boxed{\frac{1}{2}\,\frac{3.3113}{0.3010}} = \boxed{\frac{1}{2} \times 11} = \boxed{\frac{11}{2}} = \boxed{\mathbf{5.5}}$ or,

III. $\boxed{\log_2 \sqrt{2048}} = \boxed{\log_2 45.25} = \boxed{\frac{\log_{10} 45.25}{\log_{10} 2}} = \boxed{\frac{\log_{10} 4.525 \times 10^1}{\log_{10} 2}} = \boxed{\frac{\log_{10} 4.525 + \log_{10} 10^1}{\log_{10} 2}} = \boxed{\frac{0.6556 + 1}{0.3010}}$

$= \boxed{\frac{1.6556}{0.3010}} = \boxed{\frac{1.6556}{0.3010}} = \boxed{\mathbf{5.5}}$

Example 6.2-82

I. $\boxed{\log_{0.01} 0.0001} = \boxed{\log_{0.01}(0.01 \times 0.01)} = \boxed{\log_{0.01} 0.01 + \log_{0.01} 0.01} = \boxed{1+1} = \boxed{\mathbf{2}}$ or,

II. $\boxed{\log_{0.01} 0.0001} = \boxed{\log_{0.01} 1 \times 10^{-4}} = \boxed{\log_{0.01} 1 + \log_{0.01} 10^{-4}} = \boxed{0 - 4\log_{0.01} 10} = \boxed{-4\frac{\log_{10} 10}{\log_{10} 0.01}}$

$= \boxed{\frac{-4}{\log_{10} 10^{-2}}} = \boxed{\frac{-4}{-2\log_{10} 10}} = \boxed{\frac{4}{2 \times 1}} = \boxed{\frac{2}{1}} = \boxed{\mathbf{2}}$ or,

III. $\boxed{\log_{0.01} 0.0001} = \boxed{\frac{\log_{10} 0.0001}{\log_{10} 0.01}} = \boxed{\frac{\log_{10} 1 \times 10^{-4}}{\log_{10} 1 \times 10^{-2}}} = \boxed{\frac{\log_{10} 1 + \log_{10} 10^{-4}}{\log_{10} 1 + \log_{10} 10^{-2}}} = \boxed{\frac{0 - 4\log_{10} 10}{0 - 2\log_{10} 10}} = \boxed{\frac{-4 \times 1}{-2 \times 1}}$

$= \boxed{\frac{\overset{2}{\not 4}}{\not 2}} = \boxed{\frac{2}{1}} = \boxed{\mathbf{2}}$

Example 6.2-83

I. $\boxed{\log_{0.1} 0.0001} = \boxed{\log_{0.1}(0.1 \times 0.1 \times 0.1 \times 0.1)} = \boxed{\log_{0.1} 0.1 + \log_{0.1} 0.1 + \log_{0.1} 0.1 + +\log_{0.1} 0.1} = \boxed{1+1+1+1} = \boxed{\mathbf{4}}$

or,

II. $\boxed{\log_{0.1} 0.0001} = \boxed{\frac{\log_{10} 0.0001}{\log_{10} 0.1}} = \boxed{\frac{\log_{10} 1 \times 10^{-4}}{\log_{10} 1 \times 10^{-1}}} = \boxed{\frac{\log_{10} 1 + \log_{10} 10^{-4}}{\log_{10} 1 + \log_{10} 10^{-1}}} = \boxed{\frac{0 - 4\log_{10} 10}{0 - 1\log_{10} 10}} = \boxed{\frac{-4 \times 1}{-1 \times 1}} = \boxed{\frac{4}{1}} = \boxed{\mathbf{4}}$

Example 6.2-84

I. $\boxed{\log_{0.01} 0.001} = \boxed{\log_{0.01}(0.01 \times 0.1)} = \boxed{\log_{0.01} 0.01 + \log_{0.01} 0.1} = \boxed{1 + \log_{0.01} 0.1} = \boxed{1 + \frac{\log_{10} 0.1}{\log_{10} 0.01}}$

$= \boxed{1 + \frac{\log_{10} 1 \times 10^{-1}}{\log_{10} 1 \times 10^{-2}}} = \boxed{1 + \frac{\log_{10} 1 + \log_{10} 10^{-1}}{\log_{10} 1 + \log_{10} 10^{-2}}} = \boxed{1 + \frac{0 - \log_{10} 10}{0 - 2\log_{10} 10}} = \boxed{1 + \frac{-1}{-2}} = \boxed{1 + \frac{1}{2}} = \boxed{\frac{2+1}{2}} = \boxed{\frac{3}{2}} = \boxed{\mathbf{1\frac{1}{2}}}$

or,

II. $\boxed{\log_{0.01} 0.001} = \boxed{\frac{\log_{10} 0.001}{\log_{10} 0.01}} = \boxed{\frac{\log_{10} 1 \times 10^{-3}}{\log_{10} 1 \times 10^{-2}}} = \boxed{\frac{\log_{10} 1 + \log_{10} 10^{-3}}{\log_{10} 1 + \log_{10} 10^{-2}}} = \boxed{\frac{0 - 3\log_{10} 10}{0 - 2\log_{10} 10}} = \boxed{\frac{-3 \times 1}{-2 \times 1}} = \boxed{\frac{3}{2}} = \boxed{\mathbf{1\frac{1}{2}}}$

Example 6.2-85

I. $\boxed{log_{0.25} 16} = \boxed{log_{0.25} 0.25^{-2}} = \boxed{-2\,log_{0.25} 0.25} = \boxed{-2 \times 1} = \boxed{-2}$ Note: $0.25^{-2} = \frac{1}{0.25^2} = \frac{1}{0.0625} = 16$

or,

II. $\boxed{log_{0.25} 16} = \boxed{\frac{log_{10} 16}{log_{10} 0.25}} = \boxed{\frac{1.2041}{-0.6021}} = \boxed{-2}$

Example 6.2-86

$\boxed{log_{\frac{1}{5}} 125} = \boxed{log_{\frac{1}{5}} \left(\frac{1}{5}\right)^{-3}} = \boxed{-3\,log_{\frac{1}{5}} \frac{1}{5}} = \boxed{-3 \times 1} = \boxed{-3}$ Note: $\left(\frac{1}{5}\right)^{-3} = \frac{1}{5^{-3}} = \frac{1}{\frac{1}{5^3}} = \frac{\frac{1}{1}}{\frac{1}{125}} = \frac{1 \times 125}{1 \times 1} = 125$

Example 6.2-87

I. $\boxed{log_{16} 256} = \boxed{log_{16} 16^2} = \boxed{2\,log_{16} 16} = \boxed{2 \times 1} = \boxed{2}$ or,

II. $\boxed{log_{16} 256} = \boxed{\frac{log_{10} 256}{log_{10} 16}} = \boxed{\frac{log_{10} 2.56 \times 10^2}{log_{10} 1.6 \times 10^1}} = \boxed{\frac{log_{10} 2.56 + log_{10} 10^2}{log_{10} 1.6 + log_{10} 10^1}} = \boxed{\frac{0.4082 + 2}{0.2041 + 1}} = \boxed{\frac{2.4082}{1.2041}} = \boxed{2}$

Example 6.2-88

I. $\boxed{log_{16} 768} = \boxed{log_{16}(256 \times 3)} = \boxed{log_{16} 256 + log_{16} 3} = \boxed{log_{16} 16^2 + log_{16} 3} = \boxed{2\,log_{16} 16 + log_{16} 3}$

$= \boxed{2 \times 1 + \frac{log_{10} 3}{log_{10} 16}} = \boxed{2 + \frac{0.4771}{1.2041}} = \boxed{2 + 0.3962} = \boxed{\mathbf{2.3962}}$ or,

II. $\boxed{log_{16} 768} = \boxed{\frac{log_{10} 768}{log_{10} 16}} = \boxed{\frac{log_{10} 7.68 \times 10^2}{log_{10} 1.6 \times 10^1}} = \boxed{\frac{log_{10} 7.68 + log_{10} 10^2}{log_{10} 1.6 + log_{10} 10^1}} = \boxed{\frac{0.8854 + 2}{0.2041 + 1}} = \boxed{\frac{2.8854}{1.2041}} = \boxed{\mathbf{2.3962}}$

Example 6.2-89

I. $\boxed{log_{\frac{1}{2}} 8} = \boxed{log_{\frac{1}{2}} \left(\frac{1}{2}\right)^{-3}} = \boxed{-3\,log_{\frac{1}{2}} \left(\frac{1}{2}\right)} = \boxed{-3 \times 1} = \boxed{-3}$ Note: $\left(\frac{1}{2}\right)^{-3} = \frac{1}{\left(\frac{1}{2}\right)^3} = \frac{1}{\frac{1}{8}} = \frac{\frac{1}{1}}{\frac{1}{8}} = \frac{1 \times 8}{1 \times 1} = \frac{8}{1} = 8$

or,

II. $\boxed{log_{\frac{1}{2}} 8} = \boxed{log_{0.5} 8} = \boxed{\frac{log_{10} 8}{log_{10} 0.5}} = \boxed{\frac{0.9031}{-0.3010}} = \boxed{-3}$

Example 6.2-90

I. $\boxed{log_{0.5} 32} = \boxed{log_{0.5} 8 \cdot 4} = \boxed{log_{0.5} 8 + log_{0.5} 4} = \boxed{log_{0.5} 0.5^{-3} + log_{0.5} 0.5^{-2}} = \boxed{-3\,log_{0.5} 0.5 - 2\,log_{0.5} 0.5}$

$= \boxed{-3 \times 1 - 2 \times 1} = \boxed{-3 - 2} = \boxed{-5}$ or,

II. $\boxed{log_{0.5} 32} = \boxed{\frac{log_{10} 32}{log_{10} 0.5}} = \boxed{\frac{log_{10} 3.2 \times 10^1}{log_{10} 5.0 \times 10^{-1}}} = \boxed{\frac{log_{10} 3.2 + log_{10} 10^1}{log_{10} 5.0 + log_{10} 10^{-1}}} = \boxed{\frac{0.5051 + 1}{0.6990 - 1}} = \boxed{\frac{1.5051}{-0.301}} = \boxed{-5}$

Example 6.2-91

I. $\boxed{log_5 \frac{1}{5}} = \boxed{log_5 1 - log_5 5} = \boxed{0-1} = \boxed{-1}$ or, II. $\boxed{log_5 \frac{1}{5}} = \boxed{log_5 5^{-1}} = \boxed{-1 log_5 5} = \boxed{-1\times 1} = \boxed{-1}$ or,

III. $\boxed{log_5 \frac{1}{5}} = \boxed{log_5 0.2} = \boxed{\frac{log_{10} 0.2}{log_{10} 5}} = \boxed{\frac{-0.6989}{0.6989}} = \boxed{-1}$

Example 6.2-92

$\boxed{log_5 44} = \boxed{\frac{log_{10} 44}{log_{10} 5}} = \boxed{\frac{1.6434}{0.6990}} = \boxed{\mathbf{2.3511}}$

Example 6.2-93

I. $\boxed{log_{\frac{2}{3}} \frac{27}{8}} = \boxed{log_{\frac{2}{3}} \left(\frac{2}{3}\right)^{-3}} = \boxed{-3 log_{\frac{2}{3}} \frac{2}{3}} = \boxed{-3\times 1} = \boxed{-3}$ Note: $\left(\frac{2}{3}\right)^{-3} = \frac{1}{\left(\frac{2}{3}\right)^3} = \frac{1}{\frac{8}{27}} = \frac{\frac{1}{1}}{\frac{8}{27}} = \frac{1\times 27}{1\times 8} = \frac{27}{8}$

or,

II. $\boxed{log_{\frac{2}{3}} \frac{27}{8}} = \boxed{log_{0.6666} 3.375} = \boxed{\frac{log_{10} 3.375}{log_{10} 0.6666}} = \boxed{\frac{0.5283}{-0.1761}} = \boxed{-3}$

Example 6.2-94

I. $\boxed{log_{\frac{1}{3}} 9} = \boxed{log_{\frac{1}{3}} \left(\frac{1}{3}\right)^{-2}} = \boxed{-2 log_{\frac{1}{3}} \left(\frac{1}{3}\right)} = \boxed{-2\times 1} = \boxed{-2}$ Note: $\left(\frac{1}{3}\right)^{-2} = \frac{1}{\left(\frac{1}{3}\right)^2} = \frac{1}{\frac{1}{9}} = \frac{\frac{1}{1}}{\frac{1}{9}} = \frac{1\times 9}{1\times 1} = \frac{9}{1} = 9$

or,

II. $\boxed{log_{\frac{1}{3}} 9} = \boxed{log_{0.3333} 9} = \boxed{\frac{log_{10} 9}{log_{10} 0.3333}} = \boxed{\frac{0.9542}{-0.4772}} = \boxed{-2}$

Example 6.2-95

I. $\boxed{log_{\frac{1}{6}} 1296} = \boxed{log_{\frac{1}{6}} \left(\frac{1}{6}\right)^{-4}} = \boxed{-4 log_{\frac{1}{6}} \frac{1}{6}} = \boxed{-4\times 1} = \boxed{-4}$ Note: $\left(\frac{1}{6}\right)^{-4} = \frac{1}{\left(\frac{1}{6}\right)^4} = \frac{1}{\frac{1}{1296}} = \frac{\frac{1}{1}}{\frac{1}{1296}} = 1296$

or,

II. $\boxed{log_{\frac{1}{6}} 1296} = \boxed{log_{0.1666} 1296} = \boxed{\frac{log_{10} 1296}{log_{10} 0.1666}} = \boxed{log_{10} \frac{1.296\times 10^3}{-0.7783}} = \boxed{\frac{log_{10} 1.296 + log_{10} 10^3}{-0.7783}} = \boxed{\frac{0.1126+3}{-0.7783}} = \boxed{-4}$

Example 6.2-96

I. $\boxed{log_{0.03} 0.003} = \boxed{log_{0.03}(0.03\times 0.1)} = \boxed{log_{0.03} 0.03 + log_{0.03} 0.1} = \boxed{1 + log_{0.03} 0.1} = \boxed{1+\frac{log_{10} 0.1}{log_{10} 0.03}}$

$= \boxed{1+\frac{log_{10} 1\cdot 10^{-1}}{log_{10} 3\cdot 10^{-2}}} = \boxed{1+\frac{log_{10} 1 + log_{10} 10^{-1}}{log_{10} 3 + log_{10} 10^{-2}}} = \boxed{1+\frac{0 - log_{10} 10}{log_{10} 3 - 2 log_{10} 10}} = \boxed{1+\frac{-1}{0.4771-2}} = \boxed{1+\frac{-1}{-1.5229}}$

$= \boxed{1+\frac{1}{1.5229}} = \boxed{1+0.6566} = \boxed{\mathbf{1.6566}}$ or,

II. $\boxed{log_{0.03} 0.003} = \boxed{\dfrac{log_{10} 0.003}{log_{10} 0.03}} = \boxed{\dfrac{-2.5228}{-1.5228}} = \boxed{\dfrac{2.5228}{1.5228}} = \boxed{\mathbf{1.6566}}$

Example 6.2-97

I. $\boxed{log_5 375\sqrt{5}} = \boxed{log_5\left(375 \times \sqrt{5}\right)} = \boxed{log_5 375 + log_5 \sqrt{5}} = \boxed{log_5 (125 \times 3) + log_5 5^{\frac{1}{2}}} = \boxed{\left(log_5 125 + log_5 3\right) + \frac{1}{2} log_5 5}$

$= \boxed{log_5 5^3 + log_5 3 + \frac{1}{2}} = \boxed{3 log_5 5 + \frac{1}{2} + log_5 3} = \boxed{3 + \frac{1}{2} + log_5 3} = \boxed{\frac{7}{2} + log_5 3} = \boxed{3\frac{1}{2} + \frac{log_{10} 3}{log_{10} 5}} = \boxed{3.5 + \frac{0.4771}{0.6990}}$

$= \boxed{3.5 + 0.6825} = \boxed{\mathbf{4.1825}}$ or,

II. $\boxed{log_5 375\sqrt{5}} = \boxed{log_5 375 \times 2 \cdot 236} = \boxed{log_5 838.525} = \boxed{\dfrac{log_{10} 838.525}{log_{10} 5}} = \boxed{\dfrac{2.9235}{0.6989}} = \boxed{\mathbf{4.1825}}$

Example 6.2-98

$\boxed{log_2 48\sqrt[5]{2}} = \boxed{log_2\left(48 \times \sqrt[5]{2}\right)} = \boxed{log_2 48 + log_2 \sqrt[5]{2}} = \boxed{log_2 (16 \times 3) + log_2 2^{\frac{1}{5}}} = \boxed{log_2 16 + log_2 3 + \frac{1}{5} log_2 2}$

$= \boxed{log_2 2^4 + log_2 3 + \frac{1}{5}} = \boxed{4 log_2 2 + log_2 3 + \frac{1}{5}} = \boxed{4 + \frac{1}{5} + log_2 3} = \boxed{\frac{21}{5} + log_2 3} = \boxed{\frac{21}{5} + \frac{log_{10} 3}{log_{10} 2}}$

$= \boxed{\frac{21}{5} + \frac{0.4771}{0.3010}} = \boxed{4.2 + 1.5849} = \boxed{\mathbf{5.7849}}$

Example 6.2-99

$\boxed{log_3 \frac{1}{2000}} = \boxed{log_3 1 - log_3 2000} = \boxed{0 - log_3 2 \times 10^{+3}} = \boxed{-\left(log_3 2 + log_3 10^3\right)} = \boxed{-\left(log_3 2 + 3 log_3 10\right)}$

$= \boxed{-\left[\frac{log_{10} 2}{log_{10} 3} + 3\frac{log_{10} 10}{log_{10} 3}\right]} = \boxed{-\left[\frac{0.3010}{0.4771} + 3\frac{1}{0.4771}\right]} = \boxed{-\left[\frac{0.3010}{0.4771} + \frac{3}{0.4771}\right]} = \boxed{-\frac{3.3010}{0.4771}} = \boxed{\mathbf{-6.9185}}$

Example 6.2-100

I. $\boxed{log_5 \frac{1}{625}} = \boxed{log_5 \frac{1}{5^4}} = \boxed{log_5 1 - log_5 5^4} = \boxed{0 - 4 log_5 5} = \boxed{-4 \times 1} = \boxed{\mathbf{-4}}$ or,

II. $\boxed{log_5 \frac{1}{625}} = \boxed{log_5 1.6 \times 10^{-3}} = \boxed{log_5 1.6 + log_5 10^{-3}} = \boxed{\frac{log_{10} 1.6}{log_{10} 5} - 3 log_5 10} = \boxed{\frac{log_{10} 1.6}{log_{10} 5} - 3\frac{log_{10} 10}{log_{10} 5}}$

$= \boxed{\frac{0.2041}{0.6989} - 3\frac{1}{0.6989}} = \boxed{\frac{0.2041}{0.6989} - \frac{3}{0.6989}} = \boxed{\frac{0.2041 - 3}{0.6989}} = \boxed{\frac{-2.7959}{0.6989}} = \boxed{\mathbf{-4}}$

III. $\boxed{log_5 \frac{1}{625}} = \boxed{log_5 0.0016} = \boxed{\dfrac{log_{10} 0.0016}{log_{10} 5}} = \boxed{\dfrac{-2.7958}{0.6989}} = \boxed{\mathbf{-4}}$

Example 6.2-101

I. $\boxed{log_3 \frac{9}{54}} = \boxed{log_3 9 - log_3 54} = \boxed{log_3 3^2 - log_3 (27 \times 2)} = \boxed{2 log_3 3 - \left[log_3 27 + log_3 2\right]} = \boxed{2 - \left[log_3 3^3 + log_3 2\right]}$

$$= \boxed{2-\left[3\log_3 3+\log_3 2\right]} = \boxed{2-\left[3\times 1-\log_3 2\right]} = \boxed{2-3-\log_3 2} = \boxed{-1-\log_3 2} = \boxed{-1-\frac{\log_{10} 2}{\log_{10} 3}} = \boxed{-1-\frac{0.3010}{0.4771}}$$

$$= \boxed{-1-0.631} = \boxed{\mathbf{-1.631}}$$

II. $\boxed{\log_3 \frac{9}{54}} = \boxed{\log_3 0.1666} = \boxed{\log_{10} \frac{0.1666}{\log_{10} 3}} = \boxed{\frac{-0.7783}{0.4771}} = \boxed{\mathbf{-1.631}}$

Example 6.2-102

$$\boxed{\log_5 33} = \boxed{\frac{\log_{10} 33}{\log_{10} 5}} = \boxed{\frac{1.5185}{0.6989}} = \boxed{\mathbf{2.1726}}$$

Example 6.2-103

I. $\boxed{\log_8 64} = \boxed{\log_8 8^2} = \boxed{2\log_8 8} = \boxed{2\times 1} = \boxed{\mathbf{2}}$ or,

II. $\boxed{\log_8 64} = \boxed{\frac{\log_{10} 64}{\log_{10} 8}} = \boxed{\frac{1.8060}{0.9030}} = \boxed{\mathbf{2}}$

Example 6.2-104

I. $\boxed{\log_e e^2 \cdot e^{\frac{2}{3}}} = \boxed{\log_e e^2 + \log_e e^{\frac{2}{3}}} = \boxed{2\log_e e + \frac{2}{3}\log_e e} = \boxed{2\times 1 + \frac{2}{3}\times 1} = \boxed{2+\frac{2}{3}} = \boxed{\frac{6+2}{3}} = \boxed{\mathbf{\frac{8}{3}}}$ or,

II. $\boxed{\log_e e^2 \cdot e^{\frac{2}{3}}} = \boxed{\log_e e^{\frac{2}{1}} \cdot e^{\frac{2}{3}}} = \boxed{\log_e e^{\frac{2}{1}+\frac{2}{3}}} = \boxed{\log_e e^{\frac{(2\cdot 3)+(1\cdot 2)}{1\cdot 3}}} = \boxed{\log_e e^{\frac{6+2}{3}}} = \boxed{\log_e e^{\frac{8}{3}}} = \boxed{\frac{8}{3}\log_e e} = \boxed{\frac{8}{3}\times 1} = \boxed{\mathbf{\frac{8}{3}}}$

Example 6.2-105

$$\boxed{\log_{\frac{1}{2}} 8} = \boxed{\log_{\frac{1}{2}} \left(\frac{1}{2}\right)^{-3}} = \boxed{-3\log_{\frac{1}{2}} \frac{1}{2}} = \boxed{-3\times 1} = \boxed{\mathbf{-3}}$$

Note: $\left(\frac{1}{2}\right)^{-3} = \frac{1}{\left(\frac{1}{2}\right)^3} = \frac{1}{\frac{1}{8}} = \frac{\frac{1}{1}}{\frac{1}{8}} = \frac{1\times 8}{1\times 1} = \frac{8}{1} = 8$

Example 6.2-106

$$\boxed{\log_{20} 35{,}100} = \boxed{\frac{\log_{10} 35{,}100}{\log_{10} 20}} = \boxed{\frac{\log_{10} 3.51\times 10^4}{\log_{10} 2\times 10^1}} = \boxed{\frac{\log_{10} 3.51+\log_{10} 10^4}{\log_{10} 2+\log_{10} 10^1}} = \boxed{\frac{\log_{10} 3.51+4}{\log_{10} 2+1}} = \boxed{\frac{0.5453+4}{0.3010+1}}$$

$$= \boxed{\frac{4.5453}{1.3010}} = \boxed{\mathbf{3.4936}}$$

Example 6.2-107

$$\boxed{\log_{50} 7000} = \boxed{\frac{\log_{10} 7000}{\log_{10} 50}} = \boxed{\frac{\log_{10} 7\times 10^3}{\log_{10} 5\times 10^1}} = \boxed{\frac{\log_{10} 7+\log_{10} 10^3}{\log_{10} 5+\log_{10} 10^1}} = \boxed{\frac{\log_{10} 7+3}{\log_{10} 5+1}} = \boxed{\frac{0.8451+3}{0.6990+1}} = \boxed{\mathbf{2.2632}}$$

Example 6.2-108

$$\boxed{\log_{12} 43} = \boxed{\frac{\log_{10} 43}{\log_{10} 12}} = \boxed{\frac{\log_{10} 4.3\times 10^1}{\log_{10} 1.2\times 10^1}} = \boxed{\frac{\log_{10} 4.3+\log_{10} 10^1}{\log_{10} 1.2+\log_{10} 10^1}} = \boxed{\frac{\log_{10} 4.3+1\log_{10} 10}{\log_{10} 1.2+1\log_{10} 10}} = \boxed{\frac{\log_{10} 4.3+1}{\log_{10} 1.2+1}}$$

$$= \boxed{\frac{0.6335+1}{0.0792+1}} = \boxed{\frac{1.6335}{1.0792}} = \boxed{\mathbf{1.5136}}$$

Example 6.2-109

$$\log_3\sqrt{3^3} = \log_3\left(3^3\right)^{\frac{1}{2}} = \log_3 3^{3\times\frac{1}{2}} = \log_3 3^{\frac{3}{2}} = \frac{3}{2}\log_3 3 = \frac{3}{2}\times 1 = \boxed{\frac{3}{2}}$$

Example 6.2-110

$$\log_{\frac{1}{3}} 9 = \log_{\frac{1}{3}}\left(\frac{1}{3}\right)^{-2} = -2\log_{\frac{1}{3}}\frac{1}{3} = -2\times 1 = \boxed{-2}$$

Example 6.2-111

I. $$\log_{49} 7 = \log_{49}\sqrt{49} = \log_{49} 49^{\frac{1}{2}} = \frac{1}{2}\log_{49} 49 = \frac{1}{2}\times 1 = \boxed{0.5}$$ or,

II. $$\log_{49} 7 = \frac{\log_{10} 7}{\log_{10} 49} = \frac{\log_{10} 7.0\times 10^0}{\log_{10} 4.9\times 10^1} = \frac{\log_{10} 7.0+\log_{10} 10^0}{\log_{10} 4.9+\log_{10} 10^1} = \frac{0.8451+0}{0.6902+1} = \frac{0.8451}{1.6902} = \boxed{0.5}$$

Example 6.2-112

I. $$\log_{25} 125 = \log_{25} 25\cdot 5 = \log_{25} 25+\log_{25} 5 = 1+\log_{25} 25^{\frac{1}{2}} = 1+\frac{1}{2}\log_{25} 25 = 1+\frac{1}{2} = \frac{2+1}{2} = \frac{3}{2} = \boxed{1.5}$$

or we can solve the problem in the following way:

II. $$\log_{25} 125 = \frac{\log_{10} 125}{\log_{10} 25} = \frac{\log_{10} 1.25\times 10^2}{\log_{10} 2.5\times 10^1} = \frac{\log_{10} 1.25+\log_{10} 10^2}{\log_{10} 2.5+\log_{10} 10^1} = \frac{0.0969+2}{0.3979+1} = \frac{2.0969}{1.3979} = \boxed{1.5}$$

Example 6.2-113

$$\log_{11}\sqrt[3]{11^2} = \log_{11} 11^{\frac{2}{3}} = \frac{2}{3}\log_{11} 11 = \frac{2}{3}\times 1 = \boxed{\frac{2}{3}}$$

Example 6.2-114

I. $$\log_5\sqrt{\frac{1}{25}} = \log_5\left(\frac{1}{25}\right)^{\frac{1}{2}} = \frac{1}{2}\log_5\frac{1}{25} = \frac{1}{2}\log_5\frac{1}{5^2} = \frac{1}{2}\log_5 5^{-2} = -2\times\frac{1}{2}\log_5 5 = -\log_5 5 = \boxed{-1}$$ or,

II. $$\log_5\sqrt{\frac{1}{25}} = \log_5\sqrt{0.04} = \log_5 0.2 = \frac{\log_{10} 2.0\times 10^{-1}}{\log_{10} 5\times 10^0} = \frac{\log_{10} 2+\log_{10} 10^{-1}}{\log_{10} 5+\log_{10} 10^0} = \frac{0.3010-1}{0.699+0}$$

$$= \frac{-0.699}{0.699} = \boxed{-1}$$

Example 6.2-115

I. $$\log_{0.2}\sqrt[3]{\frac{1}{0.008}} = \log_{0.2}\left(\frac{1}{0.008}\right)^{\frac{1}{3}} = \frac{1}{3}\log_{0.2}\frac{1}{0.008} = \frac{1}{3}\log_{0.2}\frac{1}{0.2^3} = \frac{1}{3}\log_{0.2} 0.2^{-3} = -3\times\frac{1}{3}\log_{0.2} 0.2$$

$$= -\log_{0.2} 0.2 = \boxed{-1}$$

or we can solve the above problem in the following way:

II. $\boxed{log_{0.2}\sqrt[3]{\frac{1}{0.008}}} = \boxed{log_{0.2}\sqrt[3]{125}} = \boxed{log_{0.2}125^{\frac{1}{3}}} = \boxed{\frac{1}{3}log_{0.2}125} = \boxed{\frac{1}{3}\times\frac{log_{10}125}{log_{10}0.2}} = \boxed{\frac{1}{3}\times\frac{log_{10}1.25\times10^2}{log_{10}2\times10^{-1}}}$

$= \boxed{\frac{1}{3}\times\frac{log_{10}1.25+log_{10}10^2}{log_{10}2+log_{10}10^{-1}}} = \boxed{\frac{1}{3}\times\frac{0.0969+2}{0.301-1}} = \boxed{\frac{1}{3}\times\frac{2.0969}{-0.699}} = \boxed{\frac{1}{\not{3}}\times-\not{3}} = \boxed{-\frac{1}{1}} = \boxed{-1}$

Practice Problems - Computation of Logarithms other than Base 10 or *e*

Section 6.2 Case III Practice Problems - Solve the following logarithms:

1. $log_4\frac{16}{128} =$
2. $log_3 162 =$
3. $log_{50} 600 =$
4. $log_{\frac{1}{2}} 16 =$
5. $log_3\frac{2}{27} =$
6. $log_{0.04} 0.0004 =$
7. $log_2 32\sqrt[3]{4} =$
8. $log_{\frac{3}{5}}\frac{125}{27} =$
9. $log_{0.2} 50 =$
10. $log_4 30\sqrt[5]{2} =$

Case IV Computing Antilogarithms

The antilogarithm of a number y is simply the logarithm to the base a raised to the number y, i.e., $antilog_a\, y = a^y$. Note that some books use the notation log^{-1} to represent the antilog so that $log_a^{-1}\, y = a^y$. In this book we simply raise the base to the number y. For example, to obtain the antilogarithm of y where $y = log_a x$ we raise both sides of the equation to the base a so that $a^y = a^{log_a x}$ or $a^y = x$. This is a less confusing notation than $log_a^{-1}\, y = log_a^{-1}(log_a x)$ which is the same as $log_a^{-1}\, y = x$. To calculate the antilogarithm of a number use the following steps:

Step 1 Raise both sides of the equation to the same base as the base of the logarithm, i.e., write $log_a x = b$ as $a^{log_a x} = a^b$

Step 2 Apply the logarithmic rule No. IV to solve for the variable x, i.e., change $a^{log_a x} = a^b$ to $x = a^b$. Check the answer by substituting the solution into the original equation.

Examples with Steps

The following examples show the steps in calculating antilogarithms:

Example 6.2-116

$$\boxed{log_3 x = 0.479}$$

Solution:

Step 1 $\boxed{log_3 x = 0.479}$; $\boxed{3^{log_3 x} = 3^{0.479}}$

Step 2 $\boxed{3^{log_3 x} = 3^{0.479}}$; $\boxed{x = 3^{0.479}}$; $\boxed{\mathbf{x = 1.693}}$

Check: $log_3 1.693 \overset{?}{=} 0.479$; $\frac{log_{10} 1.693}{log_{10} 3} \overset{?}{=} 0.479$; $\frac{0.2286}{0.4771} \overset{?}{=} 0.479$; $0.479 = 0.479$

Example 6.2-117

$$\boxed{log_3 x = log_2 8}$$

Solution:

Step 1 $\boxed{log_3 x = log_2 8}$; $\boxed{log_3 x = log_2 2^3}$; $\boxed{log_3 x = 3 log_2 2}$; $\boxed{log_3 x = 3}$; $\boxed{3^{log_3 x} = 3^3}$

Step 2 $\boxed{3^{log_3 x} = 3^3}$; $\boxed{x = 3^3}$; $\boxed{\mathbf{x = 27}}$

Check: $log_3 27 \overset{?}{=} log_2 8$; $\frac{log_{10} 27}{log_{10} 3} \overset{?}{=} log_2 2^3$; $\frac{log_{10} 27}{log_{10} 3} \overset{?}{=} 3 log_2 2$; $\frac{1.4314}{0.4771} \overset{?}{=} 3$; $3 = 3$

Example 6.2-118

$$\boxed{log_5 x = log_4 8}$$

Solution:

Step 1 $\boxed{log_5 x = log_4 8}$; $\boxed{log_5 x = \frac{log_{10} 8}{log_{10} 4}}$; $\boxed{\log_5 x = \frac{0.9031}{0.6021}}$; $\boxed{log_5 x = 1.5}$

Step 2 $\boxed{log_5 x = 1.5}$; $\boxed{5^{log_5 x} = 5^{1.5}}$; $\boxed{x = 5^{1.5}}$; $\boxed{\mathbf{x = 11.2}}$

Check: $log_5 11.2 \stackrel{?}{=} log_4 8$; $\frac{log_{10} 11.2}{log_{10} 5} \stackrel{?}{=} \frac{log_{10} 8}{log_{10} 4}$; $\frac{1.05}{0.699} \stackrel{?}{=} \frac{0.9031}{0.6021}$; $1.5 = 1.5$

Example 6.2-119

$$\boxed{log_{0.4} x = 0.479}$$

Solution:

Step 1 $\boxed{log_{0.4} x = 0.479}$; $\boxed{0.4^{log_{0.4} x} = 0.4^{0.479}}$

Step 2 $\boxed{0.4^{log_{0.4} x} = 0.4^{0.479}}$; $\boxed{x = 0.4^{0.479}}$; $\boxed{\mathbf{x = 0.645}}$

Check: $log_{0.4} 0.645 \stackrel{?}{=} 0.479$; $\frac{log_{10} 0.645}{log_{10} 0.4} \stackrel{?}{=} 0.479$; $\frac{-0.1904}{-0.3979} \stackrel{?}{=} 0.479$; $0.479 = 0.479$

Example 6.2-120

$$\boxed{log_{10} x = log_3 \sqrt{25}}$$

Solution:

Step 1 $\boxed{log_{10} x = log_3 \sqrt{25}}$; $\boxed{\log_{10} x = \log_3 \sqrt{5^2}}$; $\boxed{log_{10} x = log_3 5}$; $\boxed{log_{10} x = \frac{log_{10} 5}{log_{10} 3}}$

; $\boxed{log_{10} x = \frac{0.699}{0.477}}$; $\boxed{log_{10} x = 1.5}$

Step 2 $\boxed{log_{10} x = 1.5}$; $\boxed{10^{log_{10} x} = 10^{1.5}}$; $\boxed{x = 10^{1.5}}$; $\boxed{\mathbf{x = 31.6}}$

Check: $log_{10} 31.6 \stackrel{?}{=} log_3 \sqrt{25}$; $1.5 \stackrel{?}{=} log_3 5$; $1.5 \stackrel{?}{=} \frac{log_{10} 5}{log_{10} 3}$; $1.5 \stackrel{?}{=} \frac{0.6989}{0.4771}$; $1.5 = 1.5$

Additional Examples - Computing Antilogarithms

The following examples further illustrate how to find the antilogarithms of logarithmic expressions:

Example 6.2-121

Given $log_a x = b$ where $a = 10$ *and* $b = 0.2822$ find x, the antilog of b.

$\boxed{log_{10} x = 0.2822}$; $\boxed{10^{log_{10} x} = 10^{0.2822}}$; $\boxed{x = 10^{0.2822}}$; $\boxed{\mathbf{x = 1.915}}$

Check: $log_{10} 1.915 \stackrel{?}{=} 0.2822$; $\frac{log_{10} 1.915}{log_{10} 10} \stackrel{?}{=} 0.2822$; $\frac{0.2822}{1} \stackrel{?}{=} 0.2822$; $0.2822 = 0.2822$

Example 6.2-122

Given $log_a x = b$ where $a = 2$ *and* $b = 0.2822$ find x, the antilog of b.

$\boxed{log_2 x = 0.2822}$; $\boxed{2^{log_{10} x} = 2^{0.2822}}$; $\boxed{x = 2^{0.2822}}$; $\boxed{\mathbf{x = 1.216}}$

Check: $log_2 1.216 \stackrel{?}{=} 0.2822$; $\frac{log_{10} 1.216}{log_{10} 2} \stackrel{?}{=} 0.2822$; $\frac{0.0849}{0.3010} \stackrel{?}{=} 0.2822$; $0.2822 = 0.2822$

Example 6.2-123

Given $log_a x = b$ where $a = 8$ *and* $b = 0.56$ find x, the antilog of b.

$\boxed{log_8 x = 0.56}$; $\boxed{8^{log_8 x} = 8^{0.56}}$; $\boxed{x = 8^{0.56}}$; $\boxed{\mathbf{x = 3.204}}$

Check: $log_8 3.204 \overset{?}{=} 0.56$; $\frac{log_{10} 3.204}{log_{10} 8} \overset{?}{=} 0.56$; $\frac{0.5056}{0.9030} \overset{?}{=} 0.56$; $0.56 = 0.56$

Example 6.2-124

Given $log_a x = b$ where $a = 0.02$ *and* $b = 0.345$ find x, the antilog of b.

$\boxed{log_{0.02} x = 0.345}$; $\boxed{0.02^{log_{0.02} x} = 0.02^{0.345}}$; $\boxed{x = 0.02^{0.345}}$; $\boxed{\mathbf{x = 0.259}}$

Check: $log_{0.02} 0.259 \overset{?}{=} 0.345$; $\frac{log_{10} 0.259}{log_{10} 0.02} \overset{?}{=} 0.345$; $\frac{-0.586}{-1.698} \overset{?}{=} 0.345$; $0.345 = 0.345$

Example 6.2-125

Given $log_a x = b$ where $a = 10$ *and* $b = -0.5801$ find x, the antilog of b.

$\boxed{log_{10} x = -0.5801}$; $\boxed{10^{log_{10} x} = 10^{-0.5801}}$; $\boxed{x = 10^{-0.5801}}$; $\boxed{x = \frac{1}{10^{0.5801}}}$; $\boxed{x = \frac{1}{3.803}}$; $\boxed{\mathbf{x = 0.263}}$

Check: $log_{10} 0.263 \overset{?}{=} -0.5801$; $\frac{log_{10} 0.263}{log_{10} 10} \overset{?}{=} -0.5801$; $\frac{-0.5801}{1} \overset{?}{=} -0.5801$; $-0.5801 = -0.5801$

Example 6.2-126

Given $log_a x = b$ where $a = 0.001$ *and* $b = .075$ find x, the antilog of b.

$\boxed{log_{0.001} x = 0.075}$; $\boxed{0.001^{log_{0.001} x} = 0.001^{0.075}}$; $\boxed{x = 0.001^{0.075}}$; $\boxed{\mathbf{x = 0.596}}$

Check: $log_{0.001} 0.596 \overset{?}{=} 0.075$; $\frac{log_{10} 0.596}{log_{10} 0.001} \overset{?}{=} 0.075$; $\frac{-0.2247}{\log_{10} 10^{-3}} \overset{?}{=} 0.075$; $\frac{-0.2247}{-3} \overset{?}{=} 0.075$; $0.075 = 0.075$

Example 6.2-127

Given $log_a x = b$ where $a = 6$ *and* $b = 0.795$ find x, the antilog of b.

$\boxed{log_6 x = 0.795}$; $\boxed{6^{log_6 x} = 6^{0.795}}$; $\boxed{x = 6^{0.795}}$; $\boxed{\mathbf{x = 4.156}}$

Check: $log_6 4.156 \overset{?}{=} 0.795$; $\frac{log_{10} 4.156}{log_{10} 6} \overset{?}{=} 0.795$; $\frac{0.6186}{0.7782} \overset{?}{=} 0.795$; $0.795 = 0.795$

Example 6.2-128

Given $log_a x = b$ where $a = \sqrt{2}$ *and* $b = -0.346$ find x, the antilog of b.

$\boxed{log_{\sqrt{2}} x = -0.346}$; $\boxed{log_{1.414} x = -0.346}$; $\boxed{1.414^{log_{1.414} x} = 1.414^{-0.346}}$; $\boxed{x = 1.414^{-0.346}}$; $\boxed{x = \frac{1}{1.414^{0.346}}}$; $\boxed{x = \frac{1}{1.127}}$; $\boxed{\mathbf{x = 0.887}}$

Check: $log_{1.414} 0.887 \overset{?}{=} -0.346$; $\frac{log_{10} 0.887}{log_{10} 1.414} \overset{?}{=} -0.346$; $\frac{-0.0521}{0.1504} \overset{?}{=} -0.346$; $-0.346 = -0.346$

Example 6.2-129

Given $log_a x = b$ where $a = 100$ *and* $b = 0.5$ find x, the antilog of b.

$\boxed{log_{100} x = 0.5}$; $\boxed{100^{log_{100} x} = 100^{0.5}}$; $\boxed{x = 100^{0.5}}$; $\boxed{\mathbf{x = 10}}$

Check: $\log_{100} 10 \overset{?}{=} 0.5$; $\dfrac{\log_{10} 10}{\log_{10} 100} \overset{?}{=} 0.5$; $\dfrac{1}{2} \overset{?}{=} 0.5$; $0.5 = 0.5$

Example 6.2-130

Given $\log_a x = b$ where $a = 0.7$ *and* $b = -3$ find x, the antilog of b.

$\boxed{\log_{0.7} x = -3}$; $\boxed{0.7^{\log_{0.7} x} = 0.7^{-3}}$; $\boxed{x = 0.7^{-3}}$; $\boxed{x = \dfrac{1}{0.7^3}}$; $\boxed{x = \dfrac{1}{0.343}}$; $\boxed{\mathbf{x = 2.915}}$

Check: $\log_{0.7} 2.915 \overset{?}{=} -3$; $\dfrac{\log_{10} 2.915}{\log_{10} 0.7} \overset{?}{=} -3$; $\dfrac{0.4646}{-0.1549} \overset{?}{=} -3$; $-3 = -3$

Practice Problems - Computing Antilogarithms

Section 6.2 Case IV Practice Problems - Given $\log_a x = b$ find x, the antilog of b, for the following values of a and b:

1. $a = 10$ *and* $b = 0.453$
2. $a = 0.1$ *and* $b = 0.08$
3. $a = 2$ *and* $b = -0.543$
4. $a = 0.03$ *and* $b = 0.4$
5. $a = \sqrt{3}$ *and* $b = -2$
6. $a = 100$ *and* $b = 0.04$
7. $a = 2$ *and* $b = 3$
8. $a = 10$ *and* $b = -1.35$
9. $a = 4$ *and* $b = -2.3$
10. $a = 1000$ *and* $b = 0.03$

6.3 Solving One Variable Exponential Equations

In this section one variable exponetial expressions with and without similar bases on both sides of the equation (Cases I and II, respectively) are addressed.

Case I Both Sides of the Exponential Equation Have the Same Base

The steps in solving one variable exponential equations when the base on both sides of the equation is the same are as follows:

Step 1 Write both sides of the equation in exponential form with similar bases, i.e., write $5^{x+1} = \frac{1}{625}$ as $5^{x+1} = 5^{-4}$.

Step 2 Equate the exponents on both sides of the exponential expression, i.e., write $5^{x+1} = 5^{-4}$ as $x+1=-4$

Step 3 Solve for the unknown variable x. Check the answer by substituting the solution into the original equation.

Examples with Steps

The following examples show the steps as to how one variable exponential expressions with similar bases on both sides of the equation are solved:

Example 6.3-1

$$\boxed{343^{x+1} = \frac{1}{7}}$$

Solution:

Step 1 $\boxed{343^{x+1} = \frac{1}{7}}$; $\boxed{\left(7^3\right)^{x+1} = 7^{-1}}$; $\boxed{7^{3(x+1)} = 7^{-1}}$; $\boxed{7^{3x+3} = 7^{-1}}$

Step 2 $\boxed{7^{3x+3} = 7^{-1}}$; $\boxed{3x+3=-1}$

Step 3 $\boxed{3x+3=-1}$; $\boxed{3x=-3-1}$; $\boxed{3x=-4}$; $\boxed{\mathbf{x = -\frac{4}{3}}}$

Check: $343^{-\frac{4}{3}+1} \stackrel{?}{=} \frac{1}{7}$; $343^{\frac{-4+3}{3}} \stackrel{?}{=} \frac{1}{7}$; $343^{-\frac{1}{3}} \stackrel{?}{=} \frac{1}{7}$; $\left(7^3\right)^{-\frac{1}{3}} \stackrel{?}{=} \frac{1}{7}$; $7^{3\times-\frac{1}{3}} \stackrel{?}{=} \frac{1}{7}$; $7^{-1} \stackrel{?}{=} \frac{1}{7}$; $\frac{1}{7} = \frac{1}{7}$

Example 6.3-2

$$\boxed{5^{4x+2} = \frac{1}{625}}$$

Solution:

Step 1 $\boxed{5^{4x+2} = \frac{1}{625}}$; $\boxed{5^{4x+2} = 5^{-4}}$; $\boxed{4x+2=-4}$

Step 2 $\boxed{5^{4x+2} = 5^{-4}}$; $\boxed{4x+2=-4}$

Step 3 $\boxed{4x+2=-4}$; $\boxed{4x=-4-2}$; $\boxed{4x=-6}$ = $\boxed{x=-\frac{\overset{3}{\cancel{6}}}{\underset{2}{\cancel{4}}}}$ = $\boxed{x=-\frac{3}{2}}$

Check: $5^{4\times-\frac{3}{2}+2}\overset{?}{=}\frac{1}{625}$; $5^{-\frac{12}{2}+2}\overset{?}{=}\frac{1}{625}$; $5^{-6+2}\overset{?}{=}\frac{1}{625}$; $5^{-4}\overset{?}{=}\frac{1}{625}$; $\frac{1}{5^4}\overset{?}{=}\frac{1}{625}$; $\frac{1}{625}=\frac{1}{625}$

Example 6.3-3

$\boxed{\left(3^2\right)^x=243^{x+1}}$ =

Solution:

Step 1 $\boxed{\left(3^2\right)^x=243^{x+1}}$; $\boxed{3^{2\times x}=\left(3^5\right)^{x+1}}$; $\boxed{3^{2x}=3^{5(x+1)}}$; $\boxed{3^{2x}=3^{5x+5}}$

Step 2 $\boxed{3^{2x}=3^{5x+5}}$; $\boxed{2x=5x+5}$

Step 3 $\boxed{2x=5x+5}$; $\boxed{2x-5x=5}$; $\boxed{-3x=5}$; $\boxed{x=-\frac{5}{3}}$

Check: $\left(3^2\right)^{-\frac{5}{3}}\overset{?}{=}243^{-\frac{5}{3}+1}$; $3^{2\times-\frac{5}{3}}\overset{?}{=}243^{\frac{-5+3}{3}}$; $3^{-\frac{10}{3}}\overset{?}{=}243^{-\frac{2}{3}}$; $3^{-\frac{10}{3}}\overset{?}{=}\left(3^5\right)^{-\frac{2}{3}}$; $3^{-\frac{10}{3}}\overset{?}{=}3^{5\times-\frac{2}{3}}$; $3^{-\frac{10}{3}}=3^{-\frac{10}{3}}$

Example 6.3-4

$\boxed{16^{x+1}=\frac{1}{8}}$

Solution:

Step 1 $\boxed{16^{x+1}=\frac{1}{8}}$; $\boxed{\left(2^4\right)^{x+1}=\frac{1}{2^3}}$; $\boxed{2^{4(x+1)}=2^{-3}}$; $\boxed{2^{4x+4}=2^{-3}}$

Step 2 $\boxed{2^{4x+4}=2^{-3}}$; $\boxed{4x+4=-3}$

Step 3 $\boxed{4x+4=-3}$; $\boxed{4x=-4-3}$; $\boxed{4x=-7}$; $\boxed{x=-\frac{7}{4}}$

Check: $16^{-\frac{7}{4}+1}\overset{?}{=}\frac{1}{8}$; $\left(2^4\right)^{\frac{-7+4}{4}}\overset{?}{=}\frac{1}{8}$; $\left(2^4\right)^{-\frac{3}{4}}\overset{?}{=}\frac{1}{8}$; $2^{4\times-\frac{3}{4}}\overset{?}{=}\frac{1}{8}$; $2^{-\frac{3}{1}}\overset{?}{=}\frac{1}{8}$; $2^{-3}\overset{?}{=}\frac{1}{8}$; $\frac{1}{8}=\frac{1}{8}$

Example 6.3-5

$\boxed{5^x\cdot5^{2x+1}=625}$

Solution:

Step 1 $\boxed{5^x\cdot5^{2x+1}=625}$; $\boxed{5^{x+(2x+1)}=5^4}$; $\boxed{5^{3x+1}=5^4}$

Step 2 $\boxed{5^{3x+1}=5^4}$; $\boxed{3x+1=4}$; $\boxed{3x=4-1}$

Step 3 $\boxed{3x = 4-1}$; $\boxed{3x = 3}$; $\boxed{\mathbf{x = 1}}$

Check: $5^1 \cdot 5^{(2\times 1)+1} \stackrel{?}{=} 625$; $5^1 \cdot 5^3 \stackrel{?}{=} 625$; $5^{1+3} \stackrel{?}{=} 625$; $5^4 \stackrel{?}{=} 625$; $625 = 625$

Additional Examples - Both Sides of the Exponential Equation Have the Same Base

The following examples further illustrate how to solve exponential expressions with similar bases on both sides of the equation:

Example 6.3-6

$\boxed{3^{x-2}3^{x+3} = 27}$; $\boxed{3^{(x-2)+(x+3)} = 3^3}$; $\boxed{3^{2x+1} = 3^3}$; $\boxed{2x+1 = 3}$; $\boxed{2x = 2}$; $\boxed{\mathbf{x = 1}}$

Check: $3^{1-2}3^{1+3} \stackrel{?}{=} 27$; $3^{-1}3^4 \stackrel{?}{=} 27$; $3^{-1+4} \stackrel{?}{=} 27$; $3^3 \stackrel{?}{=} 27$; $27 = 27$

Example 6.3-7

$\boxed{4^{x+5} = 1024}$; $\boxed{4^{x+5} = 4^5}$; $\boxed{x+5 = 5}$; $\boxed{x = 5-5}$; $\boxed{\mathbf{x = 0}}$

Check: $4^{0+5} \stackrel{?}{=} 1024$; $4^5 \stackrel{?}{=} 1024$; $1024 = 1024$

Example 6.3-8

$\boxed{\left(3^2\right)^4 = 81^y}$; $\boxed{3^{2\times 4} = 81^y}$; $\boxed{3^8 = 81^y}$; $\boxed{3^8 = \left(3^4\right)^y}$; $\boxed{3^8 = 3^{4\times y}}$; $\boxed{3^8 = 3^{4y}}$; $\boxed{8 = 4y}$; $\boxed{\mathbf{y = 2}}$

Check: $\left(3^2\right)^4 \stackrel{?}{=} 81^2$; $3^{2\times 4} \stackrel{?}{=} 6561$; $3^8 \stackrel{?}{=} 6561$; $6561 = 6561$

Example 6.3-9

$\boxed{5^x = 25^{x+4}}$; $\boxed{5^x = \left(5^2\right)^{x+4}}$; $\boxed{5^x = 5^{2(x+4)}}$; $\boxed{5^x = 5^{2x+8}}$; $\boxed{x = 2x+8}$; $\boxed{x-2x = 8}$; $\boxed{-x = 8}$; $\boxed{\mathbf{x = -8}}$

Check: $5^{-8} \stackrel{?}{=} 25^{-8+4}$; $5^{-8} \stackrel{?}{=} 25^{-4}$; $5^{-8} \stackrel{?}{=} \left(5^2\right)^{-4}$; $5^{-8} \stackrel{?}{=} 5^{2\times -4}$; $5^{-8} = 5^{-8}$

Example 6.3-10

$\boxed{\left(5^2\right)^{-2} = 125^{t+1}}$; $\boxed{5^{2\times -2} = \left(5^3\right)^{t+1}}$; $\boxed{5^{-4} = 5^{3(t+1)}}$; $\boxed{5^{-4} = 5^{3t+3}}$; $\boxed{-4 = 3t+3}$; $\boxed{-4-3 = 3t}$; $\boxed{-7 = 3t}$

; $\boxed{\mathbf{t = -\frac{7}{3}}}$

Check: $\left(5^2\right)^{-2} \stackrel{?}{=} 125^{-\frac{7}{3}+1}$; $5^{2\times -2} \stackrel{?}{=} 125^{\frac{-7+3}{3}}$; $5^{-4} \stackrel{?}{=} 125^{-\frac{4}{3}}$; $5^{-4} \stackrel{?}{=} \left(5^3\right)^{-\frac{4}{3}}$; $5^{-4} \stackrel{?}{=} 5^{3\times -\frac{4}{3}}$; $5^{-4} = 5^{-4}$

Example 6.3-11

$\boxed{a^{3k+1} \cdot a^2 = a^{k+6}}$; $\boxed{a^{(3k+1)+2} = a^{k+6}}$; $\boxed{a^{3k+3} = a^{k+6}}$; $\boxed{3k+3 = k+6}$; $\boxed{3k-k = 6-3}$; $\boxed{2k = 3}$; $\boxed{\mathbf{k = \frac{3}{2}}}$

Check: $a^{3\times\frac{3}{2}+1} \cdot a^2 \stackrel{?}{=} a^{\frac{3}{2}+6}$; $a^{\frac{9}{2}+1} \cdot a^2 \stackrel{?}{=} a^{\frac{3+12}{2}}$; $a^{\frac{9+2}{2}+2} \stackrel{?}{=} a^{\frac{15}{2}}$; $a^{\frac{11}{2}+2} \stackrel{?}{=} a^{\frac{15}{2}}$; $a^{\frac{11+4}{2}} \stackrel{?}{=} a^{\frac{15}{2}}$; $a^{\frac{15}{2}} = a^{\frac{15}{2}}$

Example 6.3-12

$\boxed{1024^{w}=\left(4^{2}\right)^{3}}$; $\boxed{\left(4^{5}\right)^{w}=4^{2\times3}}$; $\boxed{4^{5\times w}=4^{6}}$; $\boxed{4^{5w}=4^{6}}$; $\boxed{5w=6}$; $\boxed{\mathbf{w=\frac{6}{5}}}$

Check: $1024^{\frac{6}{5}}\overset{?}{=}\left(4^{2}\right)^{3}$; $\left(4^{5}\right)^{\frac{6}{5}}\overset{?}{=}4^{2\times3}$; $4^{\cancel{5}\times\frac{6}{\cancel{5}}}\overset{?}{=}4^{6}$; $4^{6}=4^{6}$

Example 6.3-13

$\boxed{125^{x+2}\cdot5^{x}=\left(5^{2}\right)^{2}\cdot5}$; $\boxed{\left(5^{3}\right)^{x+2}\cdot5^{x}=5^{2\times2}\cdot5^{1}}$; $\boxed{5^{3(x+2)}\cdot5^{x}=5^{4}\cdot5^{1}}$; $\boxed{5^{3x+6}\cdot5^{x}=5^{4+1}}$; $\boxed{5^{(3x+6)+x}=5^{5}}$

; $\boxed{5^{4x+6}=5^{5}}$; $\boxed{4x+6=5}$; $\boxed{4x=5-6}$; $\boxed{4x=-1}$; $\boxed{\mathbf{x=-\frac{1}{4}}}$

Check: $125^{-\frac{1}{4}+2}\cdot5^{-\frac{1}{4}}\overset{?}{=}\left(5^{2}\right)^{2}\cdot5$; $\left(5^{3}\right)^{\frac{-1+8}{4}}\cdot5^{-\frac{1}{4}}\overset{?}{=}5^{2\times2}\cdot5$; $5^{3\times\frac{7}{4}}\cdot5^{-\frac{1}{4}}\overset{?}{=}5^{4}\cdot5^{1}$; $5^{\frac{21}{4}}\cdot5^{-\frac{1}{4}}\overset{?}{=}5^{4+1}$

; $5^{\frac{21}{4}-\frac{1}{4}}\overset{?}{=}5^{5}$; $5^{\frac{21-1}{4}}\overset{?}{=}5^{5}$; $5^{\frac{\cancel{20}}{\cancel{4}}=\frac{5}{1}}\overset{?}{=}5^{5}$; $5^{\frac{5}{1}}\overset{?}{=}5^{5}$; $5^{5}=5^{5}$

Example 6.3-14

$\boxed{5^{2k+3}\cdot125^{k+4}=25}$; $\boxed{5^{2k+3}\cdot\left(5^{3}\right)^{k+4}=25}$; $\boxed{5^{2k+3}\cdot5^{3(k+4)}=5^{2}}$; $\boxed{5^{2k+3}\cdot5^{3k+12}=5^{2}}$; $\boxed{5^{(2k+3)+(3k+12)}=5^{2}}$

; $\boxed{5^{5k+15}=5^{2}}$; $\boxed{5k+15=2}$; $\boxed{5k=2-15}$; $\boxed{5k=-13}$; $\boxed{\mathbf{k=-\frac{13}{5}}}$

Check: $5^{2\times-\frac{13}{5}+3}\cdot125^{-\frac{13}{5}+4}\overset{?}{=}25$; $5^{\frac{-26}{5}+3}\cdot\left(5^{3}\right)^{\frac{-13}{5}+4}\overset{?}{=}25$; $5^{\frac{-26+15}{5}}\cdot\left(5^{3}\right)^{\frac{-13+20}{5}}\overset{?}{=}5^{2}$; $5^{\frac{-11}{5}}\cdot\left(5^{3}\right)^{\frac{7}{5}}\overset{?}{=}5^{2}$

; $5^{\frac{-11}{5}}\cdot5^{3\times\frac{7}{5}}\overset{?}{=}5^{2}$; $5^{\frac{-11}{5}}\cdot5^{\frac{21}{5}}\overset{?}{=}5^{2}$; $5^{\frac{-11+21}{5}}\overset{?}{=}5^{2}$; $5^{\frac{\cancel{10}}{\cancel{5}}=\frac{2}{1}}\overset{?}{=}5^{2}$; $5^{\frac{2}{1}}\overset{?}{=}5^{2}$; $5^{2}=5^{2}$

Example 6.3-15

$\boxed{\left(2^{4}\right)^{3}=64^{x}}$; $\boxed{2^{4\times3}=\left(2^{6}\right)^{x}}$; $\boxed{2^{12}=2^{6\times x}}$; $\boxed{2^{12}=2^{6x}}$; $\boxed{12=6x}$; $\boxed{\mathbf{x=2}}$

Check: $\left(2^{4}\right)^{3}\overset{?}{=}64^{2}$; $2^{4\times3}\overset{?}{=}\left(2^{6}\right)^{2}$; $2^{12}\overset{?}{=}2^{6\times2}$; $2^{12}=2^{12}$

Example 6.3-16

$\boxed{4^{x}\cdot4^{3}=\left(4^{2x}\right)^{3}}$; $\boxed{4^{x+3}=4^{2x\times3}}$; $\boxed{4^{x+3}=4^{6x}}$; $\boxed{x+3=6x}$; $\boxed{3=6x-x}$; $\boxed{3=5x}$; $\boxed{\mathbf{x=\frac{3}{5}}}$

Check: $4^{\frac{3}{5}}\cdot4^{3}\overset{?}{=}\left(4^{2\times\frac{3}{5}}\right)^{3}$; $4^{\frac{3}{5}+3}\overset{?}{=}\left(4^{\frac{6}{5}}\right)^{3}$; $4^{\frac{3+15}{5}}\overset{?}{=}4^{\frac{6}{5}\times3}$; $4^{\frac{18}{5}}\overset{?}{=}4^{\frac{6\times3}{5}}$; $4^{\frac{18}{5}}=4^{\frac{18}{5}}$

Example 6.3-17

$\boxed{27^{3u}=3^{u+4}}$; $\boxed{\left(3^{3}\right)^{3u}=3^{u+4}}$; $\boxed{3^{3\times3u}=3^{u+4}}$; $\boxed{3^{9u}=3^{u+4}}$; $\boxed{9u=u+4}$; $\boxed{9u-u=4}$; $\boxed{8u=4}$; $\boxed{\mathbf{u=\frac{1}{2}}}$

Check: $27^{3\times\frac{1}{2}} \overset{?}{=} 3^{\frac{1}{2}+4}$; $27^{\frac{3}{2}} \overset{?}{=} 3^{\frac{1+8}{2}}$; $\left(3^3\right)^{\frac{3}{2}} \overset{?}{=} 3^{\frac{9}{2}}$; $3^{3\times\frac{3}{2}} \overset{?}{=} 3^{\frac{9}{2}}$; $3^{\frac{3\times3}{2}} \overset{?}{=} 3^{\frac{9}{2}}$; $3^{\frac{9}{2}} = 3^{\frac{9}{2}}$

Example 6.3-18

$\boxed{2^{v+2} = \frac{1}{16}}$; $\boxed{2^{v+2} = 2^{-4}}$; $\boxed{v+2=-4}$; $\boxed{v=-4-2}$; $\boxed{\mathbf{v=-6}}$

Check: $2^{-6+2} \overset{?}{=} \frac{1}{16}$; $2^{-4} \overset{?}{=} \frac{1}{16}$; $\frac{1}{2^4} \overset{?}{=} \frac{1}{16}$; $\frac{1}{16} = \frac{1}{16}$

Example 6.3-19

$\boxed{3^{5q+1}\cdot 3^{q-2} = \frac{1}{27}}$; $\boxed{3^{(5q+1)+(q-2)} = \frac{1}{3^3}}$; $\boxed{3^{6q-1} = 3^{-3}}$; $\boxed{6q-1=-3}$; $\boxed{6q=-3+1}$; $\boxed{6q=-2}$; $\boxed{\mathbf{q=-\frac{1}{3}}}$

Check: $3^{5\times-\frac{1}{3}+1}\cdot 3^{-\frac{1}{3}-2} \overset{?}{=} \frac{1}{27}$; $3^{\frac{-5}{3}+1}\cdot 3^{\frac{-1-6}{3}} \overset{?}{=} \frac{1}{27}$; $3^{\frac{-5+3}{3}}\cdot 3^{\frac{-7}{3}} \overset{?}{=} \frac{1}{27}$; $3^{\frac{-2}{3}}\cdot 3^{\frac{-7}{3}} \overset{?}{=} \frac{1}{27}$; $3^{\frac{-2-7}{3}} \overset{?}{=} \frac{1}{27}$

; $3^{-\frac{9}{3}=-\frac{3}{1}} \overset{?}{=} \frac{1}{27}$; $3^{-3} \overset{?}{=} \frac{1}{27}$; $\frac{1}{3^3} \overset{?}{=} \frac{1}{27}$; $\frac{1}{27} = \frac{1}{27}$

Example 6.3-20

$\boxed{128^{z+1} = \frac{1}{32}}$; $\boxed{\left(2^7\right)^{z+1} = \frac{1}{2^5}}$; $\boxed{2^{7(z+1)} = 2^{-5}}$; $\boxed{2^{7z+7} = 2^{-5}}$; $\boxed{7z+7=-5}$; $\boxed{7z=-5-7}$; $\boxed{7z=-12}$

; $\boxed{\mathbf{z=-\frac{12}{7}}}$

Check: $128^{-\frac{12}{7}+1} \overset{?}{=} \frac{1}{32}$; $128^{\frac{-12+7}{7}} \overset{?}{=} \frac{1}{2^5}$; $128^{-\frac{5}{7}} \overset{?}{=} 2^{-5}$; $\left(2^7\right)^{-\frac{5}{7}} \overset{?}{=} 2^{-5}$; $2^{7\times-\frac{5}{7}} \overset{?}{=} 2^{-5}$; $2^{-\frac{5}{1}} \overset{?}{=} 2^{-5}$

; $2^{-5} = 2^{-5}$

Practice Problems - Both Sides of the Exponential Equation Have the Same Base

Section 6.3 Case I Practice Problems - Solve the following exponential expressions:

1. $3^{x+2}\cdot 3^{3x} = 243$
2. $2^{u-2}2^{u+3} = 2$
3. $125^{2q} = 5^{q+3}$
4. $5^{x+7} = 625$
5. $\left(4^2\right)^{-2} = 256^{t+2}$
6. $2^{3k-1}\cdot 2^2 = 2^{k-4}$
7. $243^w = \left(3^2\right)^2$
8. $4^{5a+1}\cdot 4^{a-3} = \frac{1}{16}$
9. $8^{x+1} = \frac{1}{512}$
10. $\left(3^3\right)^2 = 81^x$

Case II Both Sides of the Exponential Equation do not Have the Same Base

The steps in solving one variable exponential equations when the base on both sides of the equation is not the same are as follows:

Step 1 Multiply both sides of the equation by a logarithm having the same base as the exponent base, i.e., write $5^{x+1} = 3$ as $log_5 5^{x+1} = log_5 3$.

Step 2 a. Apply the logarithmic rules to simplify the equation, i.e., write $log_5 5^{x+1} = log_5 3$ as $x+1 = log_5 3$ (see Section 6.1).

b. Solve for the unknown variable x.

c. Check the answer by substituting the solution into the original equation.

Examples with Steps

The following examples show the steps as to how one variable exponential expressions without similar bases on both sides of the equation are solved:

Example 6.3-21

$$\boxed{3^x = 5}$$

Solution:

Step 1 $\boxed{3^x = 5}$; $\boxed{log_3 3^x = log_3 5}$

Step 2 $\boxed{log_3 3^x = log_3 5}$; $\boxed{x\, log_3 3 = log_3 5}$; $\boxed{x \cdot 1 = log_3 5}$; $\boxed{x = log_3 5}$; $\boxed{x = \frac{log_{10} 5}{log_{10} 3}}$

; $\boxed{x = \frac{0.6989}{0.4771}}$; $\boxed{\mathbf{x = 1.4649}}$

Check: $3^{1.4649} \stackrel{?}{=} 5$; $5 = 5$

Example 6.3-22

$$\boxed{4^{x-2} \cdot 4^{x+5} = 3}$$

Solution:

Step 1 $\boxed{4^{x-2} \cdot 4^{x+5} = 3}$; $\boxed{4^{(x-2)+(x+5)} = 3}$; $\boxed{4^{2x+3} = 3}$; $\boxed{log_4 4^{2x+3} = log_4 3}$

Step 2 $\boxed{log_4 4^{2x+3} = log_4 3}$; $\boxed{(2x+3) log_4 = log_4 3}$; $\boxed{2x+3 = log_4 3}$; $\boxed{2x = \frac{log_{10} 3}{log_{10} 4} - 3}$

; $\boxed{2x = \frac{0.4771}{0.6021} - 3}$; $\boxed{2x = 0.7924 - 3}$; $\boxed{2x = -2.2076}$; $\boxed{x = \frac{-2.2076}{2}}$; $\boxed{\mathbf{x = -1.1038}}$

Check: $4^{-1.1038-2} \cdot 4^{-1.1038+5} \stackrel{?}{=} 3$; $4^{-3.1038} \cdot 4^{3.8962} \stackrel{?}{=} 3$; $\frac{1}{4^{3.1038}} \cdot 221.7 \stackrel{?}{=} 3$; $\frac{221.7}{73.9} \stackrel{?}{=} 3$; $3 = 3$

Example 6.3-23

$$\boxed{2 \cdot 3^x = 3 \cdot 5^{-2x}}$$

Solution:

Step 1 $\boxed{2 \cdot 3^x = 3 \cdot 5^{-2x}}$; $\boxed{log_{10}\left(2 \cdot 3^x\right) = log_{10}\left(3 \cdot 5^{-2x}\right)}$

Step 2 $\boxed{log_{10}\left(2\cdot 3^{x}\right)=log_{10}\left(3\cdot 5^{-2x}\right)}$; $\boxed{log_{10}2+log_{10}3^{x}=log_{10}3+log_{10}5^{-2x}}$

; $\boxed{log_{10}2+x\,log_{10}3=log_{10}3-2x\,log_{10}5}$; $\boxed{0.3010+0.4771x=0.4771-2\times 0.6989x}$

; $\boxed{0.3010-0.4771=-1.3979x-0.4771x}$; $\boxed{-0.1761=-1.875x}$; $\boxed{x=\frac{-0.1761}{-1.875}}$; $\boxed{\mathbf{x=0.094}}$

Check: $2\cdot 3^{0.094}\overset{?}{=}3\cdot 5^{-2\times 0.094}$; $2\cdot 3^{0.094}\overset{?}{=}3\cdot 5^{-0.188}$; $2\times 1.1087\overset{?}{=}3\times\frac{1}{5^{0.188}}$; $2.217\overset{?}{=}\frac{3}{1.353}$

; $2.217=2.217$

Example 6.3-24

$\boxed{log_{3}81^{t-1}=2t}=$

Solution:

Step 1 $\boxed{Not\ Applicable}$

Step 2 $\boxed{log_{3}81^{t-1}=2t}$; $\boxed{log_{3}\left(3^{4}\right)^{t-1}=2t}$; $\boxed{log_{3}3^{4\times(t-1)}=2t}$; $\boxed{log_{3}3^{4t-4}=2t}$; $\boxed{4t-4=2t}$

; $\boxed{4t-2t=4}$; $\boxed{2t=4}$; $\boxed{t=\frac{\overset{2}{\cancel{4}}}{\cancel{2}}}$; $\boxed{t=\frac{2}{1}}$; $\boxed{\mathbf{t=2}}$

Check: $\log_{3}81^{2-1}\overset{?}{=}2\times 2$; $\log_{3}81\overset{?}{=}4$; $\log_{3}3^{4}\overset{?}{=}4$; $4\log_{3}3\overset{?}{=}4$; $4\times 1\overset{?}{=}4$; $4=4$

Example 6.3-25

$\boxed{e^{-3x}=4^{x}\cdot 2^{x-2}}$

Solution:

Step 1 $\boxed{e^{-3x}=4^{x}\cdot 2^{x-2}}$; $\boxed{ln\,e^{-3x}=ln\left(4^{x}\cdot 2^{x-2}\right)}$

Step 2 $\boxed{ln\,e^{-3x}=ln\left(4^{x}\cdot 2^{x-2}\right)}$; $\boxed{-3x\,ln\,e=ln4^{x}+ln2^{x-2}}$; $\boxed{-3x=x\,ln4+(x-2)ln2}$

; $\boxed{-3x=x\,ln4+x\,ln2-2\,ln2}$; $\boxed{-3x=1.386x+0.693x-2\times 0.693}$; $\boxed{-3x=2.079x-1.386}$

; $\boxed{-3x-2.079x=-1.386}$; $\boxed{-5.079x=-1.386}$; $\boxed{x=\frac{-1.386}{-5.079}}$; $\boxed{\mathbf{x=0.273}}$

Check: $e^{-3\times 0.273}\overset{?}{=}4^{0.273}\cdot 2^{0.273-2}$; $e^{-0.819}\overset{?}{=}4^{0.273}\cdot 2^{-1.727}$; $\frac{1}{e^{0.819}}\overset{?}{=}1.46\cdot\frac{1}{2^{1.727}}$; $\frac{1}{2.268}\overset{?}{=}\frac{1.46}{3.310}$

; $0.441=0.441$

Additional Examples - Both Sides of the Exponential Equation do not Have the Same Base

The following examples further illustrate how to solve exponential expressions without similar bases on both sides of the equation:

Example 6.3-26

I. $\boxed{3^{x+1}=5}$; $\boxed{log_{3}3^{x+1}=log_{3}5}$; $\boxed{(x+1)log_{3}3=log_{3}5}$; $\boxed{(x+1)\times 1=log_{3}5}$; $\boxed{x+1=log_{3}5}$; $\boxed{x=log_{3}5-1}$

; $\boxed{x = \frac{log_{10} 5}{log_{10} 3} - 1}$; $\boxed{x = \frac{0.6989}{0.4771} - 1}$; $\boxed{1.4649 - 1}$; $\boxed{\mathbf{x = 0.4649}}$

Check: $3^{0.4649+1} \stackrel{?}{=} 5$; $3^{1.4649} \stackrel{?}{=} 5$; $5 = 5$

Note that we can multiply both sides of the equation by log to any base. For example, let's multiply both sides of the above equation by log_{10}, i.e.,

II. $\boxed{3^{x+1} = 5}$; $\boxed{log_{10} 3^{x+1} = log_{10} 5}$; $\boxed{(x+1) log_{10} 3 = log_{10} 5}$; $\boxed{x + 1 = \frac{log_{10} 5}{log_{10} 3}}$; $\boxed{x = \frac{log_{10} 5}{log_{10} 3} - 1}$

; $\boxed{x = \frac{0.6989}{0.4771} - 1}$; $\boxed{1.4649 - 1}$; $\boxed{\mathbf{x = 0.4649}}$

or, let's multiply both sides of the equation by log_8

III. $\boxed{3^{x+1} = 5}$; $\boxed{log_8 3^{x+1} = log_8 5}$; $\boxed{(x+1) log_8 3 = log_8 5}$; $\boxed{x + 1 = \frac{log_8 5}{log_8 3}}$; $\boxed{x = \frac{log_8 5}{log_8 3} - 1}$; $\boxed{x = \frac{\frac{log_{10} 5}{log_{10} 8}}{\frac{log_{10} 3}{log_{10} 8}} - 1}$

; $\boxed{x = \frac{log_{10} 5 \times log_{10} 8}{log_{10} 8 \times log_{10} 3} - 1}$; $\boxed{x = \frac{log_{10} 5}{log_{10} 3} - 1}$; $\boxed{x = \frac{0.6989}{0.4771} - 1}$; $\boxed{1.4649 - 1}$; $\boxed{\mathbf{x = 0.4649}}$

or, let's multiply both sides of the equation by the natural logarithm $(log_e = ln)$.

IV. $\boxed{3^{x+1} = 5}$; $\boxed{ln 3^{x+1} = ln 5}$; $\boxed{(x+1) ln 3 = ln 5}$; $\boxed{x + 1 = \frac{ln 5}{ln 3}}$; $\boxed{x = \frac{ln 5}{ln 3} - 1}$; $\boxed{x = \frac{1.6094}{1.0986} - 1}$; $\boxed{x = 1.4649 - 1}$

; $\boxed{\mathbf{x = 0.4649}}$

Therefore, irregardless of the logarithm base, where both sides of the equation are multiplied by, the final answer is always the same. In this book, to simplify the process, we multiply both sides of the logarithmic expression by either a logarithm having the same base as the base given in the left hand side of the equation, or by a natural logarithm $(log_e = ln)$. For example, if the problem is stated as:

1. $2^{x+1} \cdot 2^{x+3} = 5$. Since the base on the left hand side is 2 therefore, we multiply both sides by either log_2 or ln.
2. $3^{x+1} = 5$. Since the base on the left hand side is 3 therefore, we multiply both sides by either log_3 or ln.
3. $e^x = 5^{x+2}$. Since the base on the left hand side is e therefore, we multiply both sides by $log_e = ln$.

Example 6.3-27

$\boxed{2^x = 7}$; $\boxed{log_2 2^x = log_2 7}$; $\boxed{x \, log_2 2 = log_2 7}$; $\boxed{x = log_2 7}$; $\boxed{x = \frac{log_{10} 7}{log_{10} 2}}$; $\boxed{x = \frac{0.8451}{0.3010}}$; $\boxed{\mathbf{x = 2.8074}}$

Check: $2^{2.8074} \stackrel{?}{=} 7$; $7 = 7$

Example 6.3-28

$\boxed{2^{x+1} \cdot 2^{x+3} = 5}$; $\boxed{2^{(x+1)+(x+3)} = 5}$; $\boxed{2^{2x+4} = 5}$; $\boxed{log_2 2^{2x+4} = log_2 5}$; $\boxed{(2x+4) log_2 2 = log_2 5}$

; $\boxed{(2x+4)\times 1 = log_2 5}$; $\boxed{2x+4 = log_2 5}$; $\boxed{2x = log_2 5 - 4}$; $\boxed{2x = \frac{log_{10} 5}{log_{10} 2} - 4}$; $\boxed{2x = \frac{0.6989}{0.3010} - 4}$

; $\boxed{2x = 2.3217 - 4}$; $\boxed{2x = -1.6783}$; $\boxed{x = \frac{-1.6783}{2}}$; $\boxed{\mathbf{x = -0.8392}}$

Check: $2^{-0.8392+1} \cdot 2^{-0.8392+3} \overset{?}{=} 5$; $2^{0.1608} \cdot 2^{2.1608} \overset{?}{=} 5$; $2^{0.1608+2.1608} \overset{?}{=} 5$; $2^{2.3216} \overset{?}{=} 5$; $5 = 5$

Example 6.3-29

$\boxed{4 \cdot 3^{-x} = 5^{2x}}$; $\boxed{log_3\left(4 \cdot 3^{-x}\right) = log_3\left(5^{2x}\right)}$; $\boxed{log_3 4 + log_3 3^{-x} = 2x\, log_3 5}$; $\boxed{log_3 4 - x\, log_3 3 = 2x\, log_3 5}$

; $\boxed{log_3 4 - x \times 1 = 2x\, log_3 5}$; $\boxed{log_3 4 - x = 2x\, log_3 5}$; $\boxed{-2x\, log_3 5 - x = -log_3 4}$; $\boxed{2x\, log_3 5 + x = log_3 4}$

; $\boxed{x(2\, log_3 5 + 1) = log_3 4}$; $\boxed{x = \frac{log_3 4}{2\, log_3 5 + 1}}$; $\boxed{x = \frac{log_3 4}{log_3 5^2 + 1}}$; $\boxed{x = \frac{log_3 4}{log_3 25 + 1}}$; $\boxed{x = \frac{\frac{log_{10} 4}{log_{10} 3}}{\frac{log_{10} 25}{log_{10} 3} + 1}}$

; $\boxed{x = \frac{\frac{0.6021}{0.4771}}{\frac{1.3979}{0.4771} + 1}}$; $\boxed{x = \frac{1.2619}{2.9299 + 1}}$; $\boxed{\mathbf{x = 0.3211}}$

We can also solve this problem by multiplying both sides by the natural logarithm, i.e.,

$\boxed{4 \cdot 3^{-x} = 5^{2x}}$; $\boxed{ln\left(4 \cdot 3^{-x}\right) = ln\left(5^{2x}\right)}$; $\boxed{ln4 + ln3^{-x} = 2x\, ln5}$; $\boxed{ln4 - x\, ln3 = 2x\, ln5}$; $\boxed{ln4 - x\, ln3 = 2x\, ln5}$

; $\boxed{-2x\, ln5 - x\, ln3 = -ln4}$; $\boxed{2x\, ln5 + x\, ln3 = ln4}$; $\boxed{x(2\, ln5 + ln3) = ln4}$; $\boxed{x = \frac{ln4}{2\, ln5 + ln3}}$; $\boxed{x = \frac{ln4}{ln5^2 + ln3}}$

; $\boxed{x = \frac{ln4}{ln25 + ln3}}$; $\boxed{x = \frac{1.3863}{3.2189 + 1.0986}}$; $\boxed{x = \frac{1.3863}{4.3175}}$; $\boxed{\mathbf{x = 0.3211}}$

Check: $4 \cdot 3^{-0.3211} \overset{?}{=} 5^{2\times 0.3211}$; $4 \cdot \frac{1}{3^{0.3211}} \overset{?}{=} 5^{0.6422}$; $4 \cdot \frac{1}{1.4229} \overset{?}{=} 2.8111$; $4 \cdot 0.7028 \overset{?}{=} 2.8111$; $2.8111 \overset{?}{=} 2.8111$

Example 6.3-30

$\boxed{e^{-3x} = 2^{x-1}}$; $\boxed{ln\left(e^{-3x}\right) = ln\left(2^{x-1}\right)}$; $\boxed{-3x\, ln e = (x-1)\, ln2}$; $\boxed{-3x \times 1 = x\, ln2 - ln2}$; $\boxed{-3x = x\, ln2 - ln2}$

; $\boxed{-3x - x\, ln2 = -ln2}$; $\boxed{3x + x\, ln2 = ln2}$; $\boxed{x(3 + ln2) = ln2}$; $\boxed{x = \frac{ln2}{3 + ln2}}$; $\boxed{x = \frac{0.6932}{3 + 0.6932}}$; $\boxed{x = \frac{0.6932}{3.6932}}$

; $\boxed{\mathbf{x = 0.1877}}$

Check: $e^{-3(0.1877)} \overset{?}{=} 2^{0.1877-1}$; $e^{-0.5631} \overset{?}{=} 2^{-0.8123}$; $\frac{1}{e^{0.5631}} \overset{?}{=} \frac{1}{2^{0.8123}}$; $\frac{1}{1.756} = \frac{1}{1.756}$

Example 6.3-31

$\boxed{e^{-4x} = 4^{x-3}}$; $\boxed{ln\left(e^{-4x}\right) = ln\left(4^{x-3}\right)}$; $\boxed{-4x\, ln e = (x-3)\, ln4}$; $\boxed{-4x \times 1 = x\, ln4 - 3\, ln4}$; $\boxed{-4x = x\, ln4 - 3\, ln4}$

; $\boxed{-4x - x\,ln4 = -3\,ln4}$; $\boxed{4x + x\,ln4 = 3\,ln4}$; $\boxed{x(4 + ln4) = 3\,ln4}$; $\boxed{x = \frac{3\,ln4}{4 + ln4}}$; $\boxed{x = \frac{ln4^3}{4 + ln4}}$; $\boxed{x = \frac{ln64}{4 + ln4}}$

; $\boxed{x = \frac{4.1589}{4 + 1.3863}}$; $\boxed{x = \frac{4.1589}{5.3863}}$; $\boxed{\mathbf{x = 0.7721}}$

Check: $e^{-4(0.7721)} \stackrel{?}{=} 4^{0.7721-3}$; $e^{-3.0884} \stackrel{?}{=} 4^{-2.2279}$; $\frac{1}{e^{3.0884}} \stackrel{?}{=} \frac{1}{4^{2.2279}}$; $\frac{1}{21.94} = \frac{1}{21.94}$

Example 6.3-32

$\boxed{5 \cdot 3^{3x} = 2 \cdot 4^{-x}}$; $\boxed{ln\left(5 \cdot 3^{3x}\right) = ln\left(2 \cdot 4^{-x}\right)}$; $\boxed{ln5 + ln3^{3x} = ln2 + ln4^{-x}}$; $\boxed{ln5 + 3x\,ln3 = ln2 - x\,ln4}$

; $\boxed{3x\,ln3 + x\,ln4 = ln2 - ln5}$; $\boxed{x(3\,ln3 + ln4) = ln2 - ln5}$; $\boxed{x\left(ln3^3 + ln4\right) = ln2 - ln5}$; $\boxed{x(ln27 + ln4) = ln2 - ln5}$

; $\boxed{x = \frac{ln2 - ln5}{ln27 + ln4}}$; $\boxed{x = \frac{0.6931 - 1.6094}{3.2958 + 1.3863}}$; $\boxed{x = \frac{-0.9163}{4.6821}}$; $\boxed{\mathbf{x = -0.1957}}$

Check: $5 \cdot 3^{3(-0.1957)} \stackrel{?}{=} 2 \cdot 4^{-(-0.1957)}$; $5 \cdot 3^{-0.5871} \stackrel{?}{=} 2 \cdot 4^{0.1957}$; $5 \cdot \frac{1}{3^{0.5871}} \stackrel{?}{=} 2 \cdot 1.3116$; $5 \cdot \frac{1}{1.9059} \stackrel{?}{=} 2.623$

; $\frac{5}{1.9059} \stackrel{?}{=} 2.623$; $2.623 = 2.623$

Example 6.3-33

$\boxed{4^{\frac{x}{2}} \cdot 2^x = 5e^{-x}}$; $\boxed{ln\left(4^{\frac{x}{2}} \cdot 2^x\right) = ln\left(5e^{-x}\right)}$; $\boxed{ln4^{\frac{x}{2}} + ln2^x = ln5 + lne^{-x}}$; $\boxed{\frac{x}{2}ln4 + x\,ln2 = ln5 - x\,lne}$

; $\boxed{\frac{x}{2}ln4 + x\,ln2 + x\,lne = ln5}$; $\boxed{\frac{x}{2}ln4 + x\,ln2 + x = ln5}$; $\boxed{x\left(\frac{1}{2}ln4 + ln2 + 1\right) = ln5}$; $\boxed{x\left(ln4^{\frac{1}{2}} + ln2 + 1\right) = ln5}$

; $\boxed{x = \frac{ln5}{ln4^{\frac{1}{2}} + ln2 + 1}}$; $\boxed{x = \frac{ln5}{ln2 + ln2 + 1}}$; $\boxed{x = \frac{ln5}{2\,ln2 + 1}}$; $\boxed{x = \frac{1.6094}{2 \times 0.6931 + 1}}$; $\boxed{x = \frac{1.6094}{1.3862 + 1}}$; $\boxed{\mathbf{x = 0.6745}}$

Check: $4^{\frac{0.6745}{2}} \cdot 2^{0.6745} \stackrel{?}{=} 5e^{-0.6745}$; $4^{0.33725} \cdot 2^{0.6745} \stackrel{?}{=} 5 \cdot \frac{1}{e^{0.6745}}$; $1.5960 \cdot 1.5960 \stackrel{?}{=} 5 \cdot \frac{1}{1.9631}$; $2.547 = 2.547$

Example 6.3-34

$\boxed{2^x \cdot 3^{x-1} = 4e^{-2x}}$; $\boxed{ln\left(2^x \cdot 3^{x-1}\right) = ln\left(4e^{-2x}\right)}$; $\boxed{ln2^x + ln3^{x-1} = ln4 + lne^{-2x}}$; $\boxed{x\,ln2 + (x-1)ln3 = ln4 - 2x\,lne}$

; $\boxed{x\,ln2 + x\,ln3 - ln3 = ln4 - 2x \times 1}$; $\boxed{x\,ln2 + x\,ln3 - ln3 = ln4 - 2x}$; $\boxed{x\,ln2 + x\,ln3 + 2x = ln4 + ln3}$

; $\boxed{x(ln2 + ln3 + 2) = ln4 + ln3}$; $\boxed{x = \frac{ln4 + ln3}{ln2 + ln3 + 2}}$; $\boxed{x = \frac{ln(4 \cdot 3)}{ln(2 \cdot 3) + 2}}$; $\boxed{x = \frac{ln12}{ln6 + 2}}$; $\boxed{x = \frac{2.485}{1.792 + 2}}$; $\boxed{\mathbf{x = 0.6553}}$

Check: $2^{0.6553} \cdot 3^{0.6553-1} \stackrel{?}{=} 4e^{-2(0.6553)}$; $2^{0.6553} \cdot 3^{-0.3447} \stackrel{?}{=} 4e^{-1.3106}$; $1.5749 \cdot \frac{1}{3^{0.3447}} \stackrel{?}{=} 4 \cdot \frac{1}{e^{1.3106}}$

$; \; 1.5749 \cdot \frac{1}{1.4604} \stackrel{?}{=} 4 \cdot \frac{1}{3.7084} \; ; \; \frac{1.5749}{1.4604} \stackrel{?}{=} \frac{4}{3.7084} \; ; \; 1.078 = 1.078$

Example 6.3-35

$\boxed{e^{\frac{2x}{3}} \cdot 2^{x-2} = 3^{x+4} \cdot e^{x}}$; $\boxed{ln\left(e^{\frac{2x}{3}} \cdot 2^{x-2}\right) = ln\left(3^{x+4} \cdot e^{x}\right)}$; $\boxed{ln e^{\frac{2x}{3}} + ln 2^{x-2} = ln 3^{x+4} + ln e^{x}}$

; $\boxed{\frac{2x}{3} ln e + (x-2) ln 2 = (x+4) ln 3 + x ln e}$; $\boxed{\frac{2x}{3} \times 1 + x ln 2 - 2 ln 2 = x ln 3 + 4 ln 3 + x \times 1}$

; $\boxed{\frac{2x}{3} + x ln 2 - ln 2^{2} = x ln 3 + ln 3^{4} + x}$; $\boxed{\frac{2x}{3} + x ln 2 - x ln 3 - x = ln 3^{4} + ln 2^{2}}$; $\boxed{x\left(\frac{2}{3} + ln 2 - ln 3 - 1\right) = ln 81 + ln 4}$

; $\boxed{x = \frac{ln 81 + ln 4}{\frac{2}{3} - 1 + ln 2 - ln 3}}$; $\boxed{x = \frac{ln(81 \cdot 4)}{\frac{2-3}{3} + ln\frac{2}{3}}}$; $\boxed{x = \frac{ln 324}{-\frac{1}{3} + ln\frac{2}{3}}}$; $\boxed{x = \frac{5.7807}{-0.3333 - 0.4056}}$; $\boxed{\mathbf{x = -7.8234}}$

Check: $e^{\frac{2(-7.8234)}{3}} \cdot 2^{-7.8234-2} \stackrel{?}{=} 3^{-7.8234+4} \cdot e^{-7.8234}$; $e^{-5.2156} \cdot 2^{-9.8234} \stackrel{?}{=} 3^{-3.8234} \cdot e^{-7.8234}$

$; \; \frac{1}{e^{5.2156}} \cdot \frac{1}{2^{9.8234}} \stackrel{?}{=} \frac{1}{3^{3.8234}} \cdot \frac{1}{e^{7.8234}} \; ; \; \frac{1}{184.12} \cdot \frac{1}{906.02} \stackrel{?}{=} \frac{1}{66.72} \cdot \frac{1}{2498.38}$

$; \; 0.00543 \times 0.001103 \stackrel{?}{=} 0.01499 \times 0.0004 \; ; \; 5.99 \times 10^{-6} = 5.99 \times 10^{-6}$

Example 6.3-36

$\boxed{e^{-4x} = 4^{x-3}}$; $\boxed{ln e^{-4x} = ln 4^{x-3}}$; $\boxed{-4x\, ln e = (x-3) ln 4}$; $\boxed{-4x \times 1 = x ln 4 - 3 ln 4}$; $\boxed{-4x = 1.3863x - 3 \times 1.3863}$

; $\boxed{-4x - 1.3863x = -4.1589}$; $\boxed{4x + 1.3863x = 4.1589}$; $\boxed{5.3863x = 4.1589}$; $\boxed{x = \frac{4.1589}{5.3863}}$; $\boxed{\mathbf{x = 0.7721}}$

Check: $e^{-4(0.7721)} \stackrel{?}{=} 4^{0.7721-3}$; $e^{-3.0884} \stackrel{?}{=} 4^{-2.2279}$; $\frac{1}{e^{3.0884}} \stackrel{?}{=} \frac{1}{4^{2.2279}}$; $\frac{1}{21.94} = \frac{1}{21.94}$

Example 6.3-37

$\boxed{5 \cdot 3^{x} = 7 \cdot 4^{\frac{x}{4}}}$; $\boxed{ln\left(5 \cdot 3^{x}\right) = ln\left(7 \cdot 4^{\frac{x}{4}}\right)}$; $\boxed{ln 5 + ln 3^{x} = ln 7 + ln 4^{\frac{x}{4}}}$; $\boxed{ln 5 + x ln 3 = ln 7 + \frac{x}{4} ln 4}$

; $\boxed{1.6094 + 1.0986x = 1.9459 + 1.3863 \times \frac{x}{4}}$ = $\boxed{1.6094 - 1.9459 = 0.3466x - 1.0986x}$; $\boxed{-0.3365 = -0.752x}$

; $\boxed{0.3365 = 0.752x}$; $\boxed{x = \frac{0.3365}{0.752}}$; $\boxed{\mathbf{x = 0.4475}}$

Check: $5 \cdot 3^{0.4475} \stackrel{?}{=} 7 \cdot 4^{\frac{0.4475}{4}}$; $5 \cdot 1.635 \stackrel{?}{=} 7 \cdot 4^{0.1119}$; $8.175 \stackrel{?}{=} 7 \cdot 1.1678$; $8.175 = 8.175$

Example 6.3-38

$\boxed{6\cdot 2^t = 5\cdot 3^{t+1}}$; $\boxed{ln\left(6\cdot 2^t\right) = ln\left(5\cdot 3^{t+1}\right)}$; $\boxed{ln6 + ln2^t = ln5 + ln3^{t+1}}$; $\boxed{ln6 + t\,ln2 = ln5 + (t+1)ln3}$

; $\boxed{1.792 + 0.693t = 1.609 + (t+1)1.099}$; $\boxed{1.792 - 1.609 = (t+1)1.099 - 0.693t}$; $\boxed{0.183 = 1.099t + 1.099 - 0.693t}$

; $\boxed{0.183 - 1.099 = 0.406t}$; $\boxed{-0.916 = 0.406t}$; $\boxed{t = -\frac{0.916}{0.406}}$; $\boxed{\mathbf{t = -2.26}}$

Check: $6\cdot 2^{-2.26} \overset{?}{=} 5\cdot 3^{-2.26+1}$; $6\cdot 2^{-2.26} \overset{?}{=} 5\cdot 3^{-1.26}$; $6\cdot\frac{1}{2^{2.26}} \overset{?}{=} 5\cdot\frac{1}{3^{1.26}}$; $\frac{6}{4.79} \overset{?}{=} \frac{5}{3.99}$; $1.25 = 1.25$

Example 6.3-39

$\boxed{log_8 512^t = t+1}$; $\boxed{log_8\left(8^3\right)^t = t+1}$; $\boxed{log_8 8^{3\times t} = t+1}$; $\boxed{log_8 8^{3t} = t+1}$; $\boxed{3t = t+1}$; $\boxed{3t - t = 1}$; $\boxed{2t = 1}$

; $\boxed{t = \frac{1}{2}}$; $\boxed{\mathbf{t = 0.5}}$

Check: $\log_8 512^{0.5} \overset{?}{=} 0.5 + 1$; $\log_8\left(8^3\right)^{0.5} \overset{?}{=} 1.5$; $\log_8 8^{3\times 0.5} \overset{?}{=} 1.5$; $\log_8 8^{1.5} \overset{?}{=} 1.5$; $1.5 = 1.5$

Example 6.3-40

$\boxed{log_2 32^y = y+2}$; $\boxed{log_2\left(2^5\right)^y = y+2}$; $\boxed{log_2 2^{5\times y} = y+2}$; $\boxed{log_2 2^{5y} = y+2}$; $\boxed{5y = y+2}$; $\boxed{5y - y = 2}$

; $\boxed{4y = 2}$; $\boxed{y = \frac{2}{4}}$; $\boxed{y = \frac{1}{2}}$; $\boxed{\mathbf{y = 0.5}}$

Check: $\log_2 32^{0.5} \overset{?}{=} 0.5 + 2$; $\log_2\left(2^5\right)^{0.5} \overset{?}{=} 2.5$; $\log_2\left(2^{5\times 0.5}\right) \overset{?}{=} 2.5$; $\log_2 2^{2.5} \overset{?}{=} 2.5$; $2.5 = 2.5$

Example 6.3-41

$\boxed{log_x 3 = \frac{2}{3}}$; $\boxed{x^{log_x 3} = x^{\frac{2}{3}}}$; $\boxed{3 = x^{\frac{2}{3}}}$; $\boxed{x^{\frac{2}{3}} = 3}$; $\boxed{x^{\frac{3}{2}\times\frac{2}{3}} = 3^{\frac{3}{2}}}$; $\boxed{x = \sqrt[2]{3^3}}$; $\boxed{x = \sqrt{27}}$; $\boxed{\mathbf{x = 5.1962}}$

Check: $\log_{5.1962} 3 \overset{?}{=} \frac{2}{3}$; $5.1962^{\log_{5.1962} 3} \overset{?}{=} 5.1962^{\frac{2}{3}}$; $3 \overset{?}{=} \sqrt[3]{5.1962^2}$; $3 \overset{?}{=} \sqrt[3]{27}$; $3 \overset{?}{=} \sqrt[3]{3^3}$; $3 = 3$

Example 6.3-42

$\boxed{log_t 0.5 = \frac{1}{3}}$; $\boxed{t^{log_t 0.5} = t^{\frac{1}{3}}}$; $\boxed{0.5 = t^{\frac{1}{3}}}$; $\boxed{0.5^3 = t^{3\times\frac{1}{3}}}$; $\boxed{0.125 = t^{\frac{1}{1}}}$; $\boxed{\mathbf{t = 0.125}}$

Check: $\log_{0.125} 0.5 \overset{?}{=} \frac{1}{3}$; $\frac{\log_{10} 0.5}{\log_{10} 0.125} \overset{?}{=} \frac{1}{3}$; $\frac{-0.3010}{-0.9030} \overset{?}{=} \frac{1}{3}$; $\frac{1}{3} = \frac{1}{3}$

Example 6.3-43

$\boxed{\frac{e^x - e^{-x}}{e^x + e^{-x}} = \frac{1}{4}}$; $\boxed{4\cdot\left(e^x - e^{-x}\right) = 1\cdot\left(e^x + e^{-x}\right)}$; $\boxed{4e^x - 4e^{-x} = e^x + e^{-x}}$; $\boxed{4e^x - e^x = e^{-x} + 4e^{-x}}$

; $\boxed{(4-1)e^x = (1+4)e^{-x}}$; $\boxed{3e^x = 5e^{-x}}$; $\boxed{ln3e^x = ln5e^{-x}}$; $\boxed{ln3 + lne^x = ln5 + lne^{-x}}$; $\boxed{ln3 + x = ln5 - x}$

; $\boxed{x+x=ln5-ln3}$; $\boxed{2x=ln\frac{5}{3}}$; $\boxed{2x=ln1.6667}$; $\boxed{2x=0.5108}$; $\boxed{x=\frac{0.5108}{2}}$; $\boxed{\mathbf{x=0.2554}}$

Check: $\frac{e^{0.2554}-e^{-0.2554}}{e^{0.2554}+e^{-0.2554}}\overset{?}{=}\frac{1}{4}$; $\frac{e^{0.2554}-\frac{1}{e^{0.2554}}}{e^{0.2554}+\frac{1}{e^{0.2554}}}\overset{?}{=}\frac{1}{4}$; $\frac{1.2909-\frac{1}{1.2909}}{1.2909+\frac{1}{1.2909}}\overset{?}{=}\frac{1}{4}$; $\frac{1.2909-0.7746}{1.2909+0.7746}\overset{?}{=}\frac{1}{4}$

; $\frac{0.5163}{2.0655}\overset{?}{=}\frac{1}{4}$; $0.25=0.25$

Example 6.3-44

$\boxed{4=2\left(e^{2t}+ln\sqrt{e}\right)}$; $\boxed{\frac{4}{2}=e^{2t}+lne^{\frac{1}{2}}}$; $\boxed{2=e^{2t}+\frac{1}{2}lne}$; $\boxed{2=e^{2t}+\frac{1}{2}\times 1}$; $\boxed{2=e^{2t}+\frac{1}{2}}$; $\boxed{2-\frac{1}{2}=e^{2t}}$

; $\boxed{e^{2t}=\frac{4-1}{2}}$; $\boxed{e^{2t}=\frac{3}{2}}$; $\boxed{lne^{2t}=ln1.5}$; $\boxed{2t\,lne=0.4055}$; $\boxed{2t=0.4055}$; $\boxed{t=\frac{0.4055}{2}}$; $\boxed{\mathbf{t=0.2027}}$

Check: $4\overset{?}{=}2\left(e^{2\times 0.2027}+\ln\sqrt{e}\right)$; $4\overset{?}{=}2\left(e^{0.4054}+\ln e^{\frac{1}{2}}\right)$; $4\overset{?}{=}2\left(1.5+\frac{1}{2}\ln e\right)$; $4\overset{?}{=}2\left(1.5+\frac{1}{2}\times 1\right)$

; $4\overset{?}{=}2(1.5+0.5)$; $4\overset{?}{=}2\times 2$; $4=4$

Example 6.3-45

$\boxed{log_e e^2+e^{-3x}=3log_e\sqrt[3]{e^4}}$; $\boxed{2log_e e+e^{-3x}=3log_e e^{\frac{4}{3}}}$; $\boxed{2+e^{-3x}=3\times\frac{4}{3}log_e e}$; $\boxed{2+e^{-3x}=\cancel{3}\times\frac{4}{\cancel{3}}}$

; $\boxed{2+e^{-3x}=4}$; $\boxed{e^{-3x}=4-2}$; $\boxed{e^{-3x}=2}$; $\boxed{log_e e^{-3x}=log_e 2}$; $\boxed{-3x=0.693}$; $\boxed{x=-\frac{0.693}{3}}$; $\boxed{\mathbf{x=-0.231}}$

Check: $\log_e e^2+e^{-3\times-0.231}\overset{?}{=}3\log_e\sqrt[3]{e^4}$; $2\log_e e+e^{-3\times-0.231}\overset{?}{=}3\log_e e^{\frac{4}{3}}$; $2\log_e e+e^{0.693}\overset{?}{=}\cancel{3}\times\frac{4}{\cancel{3}}\log_e e$

; $2+e^{0.693}\overset{?}{=}4$; $2+2\overset{?}{=}4$; $4=4$

Practice Problems - Both Sides of the Exponential Equation do not Have the Same Base

Section 6.3 Case II Practice Problems - Solve the following expressions using the logarithmic properties:

1. $5^x=9$
2. $3^x\cdot 3^{x+1}=5$
3. $log_2 32^{x-2}=3x$
4. $e^{-2u}=2^u\cdot 2^{u-2}$
5. $3^{x+1}\cdot 3^{x+2}=6$
6. $2\cdot 2^{-v}=2^{2v}$
7. $2\cdot 3^{2t}=4\cdot 2^{-t}$
8. $2=e^t+ln\sqrt{e}$
9. $2e^{x+2}=2^3$
10. $log_x 2=0.5$

6.4 Solving One Variable Logarithmic Equations

In this section one variable logarithmic expressions are addressed in Cases I and II. Note that the steps in solving one variable logarithmic equations that are in the form of, or can be reduced to the form of, $log_a x = b$ are similar to what we have already learned in Section 6.2 Case IV, which is merely obtaining the antilogarithm of a number.

Case I Solving One Variable Logarithmic Equations (Simple Cases)

One variable logarithmic equations are solved using the following steps:

Step 1 Apply the logarithmic laws to simplify the equation. (In cases were the equation is in the form of $log_a x = b$ raise both sides of the equation to the same base as the base of the logarithm, i.e., write $log_a x = b$ as $a^{log_a x} = a^b$.)

Step 2 Solve for the unknown variable. (In cases were the equation is in the form of $log_a x = b$ the variable x is equal to $x = a^b$.) Check the answer by substituting the solution into the original equation.

Examples with Steps

The following examples show the steps as to how one variable logarithmic equations are solved:

Example 6.4-1

$$\boxed{log_2 u = -2}$$

Solution:

Step 1 $\boxed{log_2 u = -2}$; $\boxed{2^{log_2 u} = 2^{-2}}$

Step 2 $\boxed{2^{log_2 u} = 2^{-2}}$; $\boxed{u = 2^{-2}}$; $\boxed{u = \frac{1}{2^2}}$; $\boxed{\mathbf{u = \frac{1}{4}}}$

Check: $log_2 \frac{1}{4} \stackrel{?}{=} -2$; $log_2 2^{-2} \stackrel{?}{=} -2$; $-2 log_2 2 \stackrel{?}{=} -2$; $-2 \times 1 \stackrel{?}{=} -2$; $-2 = -2$

Example 6.4-2

$$\boxed{log_x 0.00001 = -5}$$

Solution:

Step 1 $\boxed{log_x 0.00001 = -5}$; $\boxed{x^{log_x 0.00001} = x^{-5}}$

Step 2 $\boxed{x^{log_x 0.00001} = x^{-5}}$; $\boxed{0.00001 = x^{-5}}$; $\boxed{10^{-5} = x^{-5}}$; $\boxed{\mathbf{x = 10}}$

Check: $log_{10} 0.00001 \stackrel{?}{=} -5$; $log_{10} 10^{-5} \stackrel{?}{=} -5$; $-5 log_{10} 10 \stackrel{?}{=} -5$; $-5 \times 1 \stackrel{?}{=} -5$; $-5 = -5$

Example 6.4-3

$$\boxed{log_{\frac{1}{5}} (x-1) = -1}$$

Solution:

Step 1 $\boxed{log_{\frac{1}{5}} (x-1) = -1}$; $\boxed{\frac{1}{5}^{log_{\frac{1}{5}}(x-1)} = \frac{1}{5}^{-1}}$

Step 2 $\boxed{\frac{1}{5}^{\log_{\frac{1}{5}}(x-1)} = \frac{1}{5}^{-1}}$; $\boxed{x-1 = \frac{1}{5}^{-1}}$; $\boxed{x-1 = \frac{1}{\frac{1}{5}}}$; $\boxed{x-1 = \frac{\frac{1}{1}}{\frac{1}{5}}}$; $\boxed{x-1 = \frac{1\times 5}{1\times 1}}$; $\boxed{x-1 = \frac{5}{1}}$

; $\boxed{x = 5+1}$; $\boxed{x = 6}$

Check: $\log_{\frac{1}{5}}(6-1) \overset{?}{=} -1$; $\log_{\frac{1}{5}} 5 \overset{?}{=} -1$; $\log_{\frac{1}{5}}\left(\frac{1}{5}\right)^{-1} \overset{?}{=} -1$; $-1\times\log_{\frac{1}{5}}\left(\frac{1}{5}\right) \overset{?}{=} -1$; $-1\times 1 \overset{?}{=} -1$; $-1 = -1$

Example 6.4-4

$\boxed{\log_u \frac{1}{5} = -3}$

Solution:

Step 1 $\boxed{\log_u \frac{1}{5} = -3}$; $\boxed{u^{\log_u \frac{1}{5}} = u^{-3}}$

Step 2 $\boxed{u^{\log_u \frac{1}{5}} = u^{-3}}$; $\boxed{\frac{1}{5} = u^{-3}}$; $\boxed{\left(\frac{1}{5}\right)^{-\frac{1}{3}} = u^{-3\times -\frac{1}{3}}}$; $\boxed{\frac{1}{\left(\frac{1}{5}\right)^{\frac{1}{3}}} = u^{1}}$; $\boxed{\frac{1}{\frac{1}{1.709}} = u}$; $\boxed{\frac{\frac{1}{1}}{\frac{1}{1.709}} = u}$

; $\boxed{u = \frac{1\times 1.709}{1\times 1}}$; $\boxed{u = \frac{1.709}{1}}$; $\boxed{u = 1.709}$

Check: $\log_{1.709}\frac{1}{5} \overset{?}{=} -3$; $\log_{1.709} 5^{-1} \overset{?}{=} -3$; $\log_{1.709} 1.709^{-3} \overset{?}{=} -3$; $-3\log_{1.709} 1.709 \overset{?}{=} -3$; $-3\times 1 \overset{?}{=} -3$; $-3 = -3$

Example 6.4-5

$\boxed{\log_8 2 = x+5}$

Solution:

Step 1 $\boxed{\log_8 2 = x+5}$; $\boxed{8^{\log_8 2} = 8^{x+5}}$

Step 2 $\boxed{8^{\log_8 2} = 8^{x+5}}$; $\boxed{2 = 8^{x+5}}$; $\boxed{2 = \left(2^3\right)^{x+5}}$; $\boxed{2 = 2^{3(x+5)}}$; $\boxed{2^1 = 2^{3x+15}}$; $\boxed{1 = 3x+15}$

; $\boxed{-3x = 15-1}$; $\boxed{-3x = 14}$; $\boxed{x = \frac{14}{-3}}$; $\boxed{x = -4.667}$

Check: $\log_8 2 \overset{?}{=} -4.667+5$; $\log_8 8^{\frac{1}{3}} \overset{?}{=} 0.333$; $\frac{1}{3}\log_8 8 \overset{?}{=} 0.333$; $\frac{1}{3}\times 1 \overset{?}{=} 0.333$; $\frac{1}{3} \overset{?}{=} 0.333$; $0.333 = 0.333$

Additional Examples - Solving One Variable Logarithmic Equations (Simple Cases)

The following examples further illustrate how to solve one variable logarithmic equations:

Example 6.4-6

$\boxed{\log_6 x = 3}$; $\boxed{6^{\log_6 x} = 6^3}$; $\boxed{x = 6^3}$; $\boxed{x = 216}$

Check: $log_6 216 \overset{?}{=} 3$; $log_6 6^3 \overset{?}{=} 3$; $3log_6 6 \overset{?}{=} 3$; $3 \times 1 \overset{?}{=} 3$; $3 = 3$

Example 6.4-7

$\boxed{log_2 u = 2}$; $\boxed{2^{log_2 u} = 2^2}$; $\boxed{u = 2^2}$; $\boxed{\mathbf{u = 4}}$

Check: $log_2 4 \overset{?}{=} 2$; $log_2 2^2 \overset{?}{=} 2$; $2log_2 2 \overset{?}{=} 2$; $2 \times 1 \overset{?}{=} 2$; $2 = 2$

Example 6.4-8

$\boxed{log_{\frac{1}{2}} y = -5}$; $\boxed{\left(\frac{1}{2}\right)^{log_{\frac{1}{2}} y} = \left(\frac{1}{2}\right)^{-5}}$; $\boxed{y = \left(\frac{1}{2}\right)^{-5}}$; $\boxed{y = \frac{1}{\left(\frac{1}{2}\right)^5}}$; $\boxed{y = \frac{1}{\frac{1}{32}}}$; $\boxed{y = \frac{\frac{1}{1}}{\frac{1}{32}}}$; $\boxed{y = \frac{1 \times 32}{1 \times 1}}$; $\boxed{y = \frac{32}{1}}$

; $\boxed{\mathbf{y = 32}}$

Check: $log_{\frac{1}{2}} 32 \overset{?}{=} -5$; $log_{\frac{1}{2}} \left(\frac{1}{2}\right)^{-5} \overset{?}{=} -5$; $-5log_{\frac{1}{2}} \frac{1}{2} \overset{?}{=} -5$; $-5 \times 1 \overset{?}{=} -5$; $-5 = -5$

Example 6.4-9

I. $\boxed{log_2 16 = w}$; $\boxed{2^{log_2 16} = 2^w}$; $\boxed{16 = 2^w}$; $\boxed{2^4 = 2^w}$; $\boxed{\mathbf{w = 4}}$ or,

II. $\boxed{log_2 16 = w}$; $\boxed{log_2 2^4 = w}$; $\boxed{4log_2 2 = w}$; $\boxed{4 \times 1 = w}$; $\boxed{\mathbf{w = 4}}$

Check: $log_2 16 \overset{?}{=} 4$; $log_2 2^4 \overset{?}{=} 4$; $4log_2 2 \overset{?}{=} 4$; $4 \times 1 \overset{?}{=} 4$; $4 = 4$

Example 6.4-10

$\boxed{log_{\frac{1}{4}} x = -3}$; $\boxed{\left(\frac{1}{4}\right)^{log_{\frac{1}{4}} x} = \left(\frac{1}{4}\right)^{-3}}$; $\boxed{x = \left(\frac{1}{4}\right)^{-3}}$; $\boxed{x = \frac{1}{\left(\frac{1}{4}\right)^3}}$; $\boxed{x = \frac{\frac{1}{1}}{\frac{1}{64}}}$; $\boxed{x = \frac{1 \times 64}{1 \times 1}}$; $\boxed{x = \frac{64}{1}}$; $\boxed{\mathbf{x = 64}}$

Check: $log_{\frac{1}{4}} 64 \overset{?}{=} -3$; $log_{\frac{1}{4}} \left(\frac{1}{4}\right)^{-3} \overset{?}{=} -3$; $-3log_{\frac{1}{4}} \frac{1}{4} \overset{?}{=} -3$; $-3 \times 1 \overset{?}{=} -3$; $-3 = -3$

Example 6.4-11

$\boxed{log_{\frac{1}{6}} v = 2}$; $\boxed{\left(\frac{1}{6}\right)^{log_{\frac{1}{6}} v} = \left(\frac{1}{6}\right)^2}$; $\boxed{v = \left(\frac{1}{6}\right)^2}$; $\boxed{\mathbf{v = \frac{1}{36}}}$

Check: $log_{\frac{1}{6}} \frac{1}{36} \overset{?}{=} 2$; $log_{\frac{1}{6}} \left(\frac{1}{6}\right)^2 \overset{?}{=} 2$; $2log_{\frac{1}{6}} \frac{1}{6} \overset{?}{=} 2$; $2 \times 1 \overset{?}{=} 2$; $2 = 2$

Example 6.4-12

$\boxed{log_{0.125} u = -1}$; $\boxed{0.125^{log_{0.125} u} = 0.125^{-1}}$; $\boxed{u = 0.125^{-1}}$; $\boxed{u = \frac{1}{0.125}}$; $\boxed{u = \frac{1}{\frac{125}{1000}}}$; $\boxed{u = \frac{\frac{1}{1}}{\frac{125}{1000}}}$; $\boxed{u = \frac{1 \times 1000}{1 \times 125}}$

; $\boxed{u = \frac{1000}{125}}$; $\boxed{\mathbf{u = 8}}$

Check: $log_{0.125} 8 \overset{?}{=} -1$; $log_{0.125} \frac{1}{0.125} \overset{?}{=} -1$; $log_{0.125} 1 - log_{0.125} 0.125 \overset{?}{=} -1$; $0 - 1 \overset{?}{=} -1$; $-1 = -1$

Example 6.4-13

$\boxed{log_{0.0625}\, x = 2}$; $\boxed{0.0625^{log_{0.0625}\, x} = 2}$; $\boxed{x = 0.0625^2}$; $\boxed{\mathbf{x = 0.0039}}$

Check: $log_{0.0625}\, 0.0039 \overset{?}{=} 2$; $log_{0.0625}\, 0.0625^2 \overset{?}{=} 2$; $2\, log_{0.0625}\, 0.0625 \overset{?}{=} 2$; $2 \times 1 \overset{?}{=} 2$; $2 = 2$

Example 6.4-14

$\boxed{log_{\frac{1}{2}}(x+1) = -2}$; $\boxed{\left(\frac{1}{2}\right)^{log_{\frac{1}{2}}(x+1)} = \left(\frac{1}{2}\right)^{-2}}$; $\boxed{x+1 = \left(\frac{1}{2}\right)^{-2}}$; $\boxed{x+1 = \frac{1}{\left(\frac{1}{2}\right)^2}}$; $\boxed{x+1 = \frac{1}{\frac{1}{4}}}$; $\boxed{x+1 = \frac{\frac{1}{1}}{\frac{1}{4}}}$

; $\boxed{x+1 = \frac{1 \times 4}{1 \times 1}}$; $\boxed{x+1 = \frac{4}{1}}$; $\boxed{x+1 = 4}$; $\boxed{x = 4-1}$; $\boxed{\mathbf{x = 3}}$

Check: $log_{\frac{1}{2}}(3+1) \overset{?}{=} -2$; $log_{\frac{1}{2}}\, 4 \overset{?}{=} -2$; $log_{\frac{1}{2}}\left(\frac{1}{2}\right)^{-2} \overset{?}{=} -2$; $-2\, log_{\frac{1}{2}}\, \frac{1}{2} \overset{?}{=} -2$; $-2 \times 1 \overset{?}{=} -2$; $-2 = -2$

Example 6.4-15

$\boxed{log_x\, 2 = \frac{1}{4}}$; $\boxed{x^{log_x\, 2} = x^{\frac{1}{4}}}$; $\boxed{2 = x^{\frac{1}{4}}}$; $\boxed{2^4 = x^{4 \times \frac{1}{4}}}$; $\boxed{2^4 = x^{\frac{4}{4} = \frac{1}{1}}}$; $\boxed{2^4 = x^{\frac{1}{1}}}$; $\boxed{2^4 = x}$; $\boxed{\mathbf{x = 16}}$

Check: $log_{16}\, 2 \overset{?}{=} \frac{1}{4}$; $\frac{log_{10}\, 2}{log_{10}\, 16} \overset{?}{=} \frac{1}{4}$; $\frac{0.3010}{1.2040} \overset{?}{=} 0.25$; $0.25 = 0.25$

Example 6.4-16

$\boxed{log_y\, 3 = -\frac{1}{2}}$; $\boxed{y^{log_y\, 3} = y^{-\frac{1}{2}}}$; $\boxed{3 = y^{-\frac{1}{2}}}$; $\boxed{3^{-2} = y^{-2 \times -\frac{1}{2}}}$; $\boxed{3^{-2} = y^{\frac{2}{2} = \frac{1}{1}}}$; $\boxed{3^{-2} = y^{\frac{1}{1}}}$; $\boxed{3^{-2} = y}$; $\boxed{y = 3^{-2}}$

; $\boxed{y = \frac{1}{3^2}}$; $\boxed{\mathbf{y = \frac{1}{9}}}$

Note: $\left(\frac{1}{9}\right)^{-\frac{1}{2}} = \frac{1}{\left(\frac{1}{9}\right)^{\frac{1}{2}}} = \frac{1}{\frac{1}{\sqrt{9}}} = \frac{1}{\frac{1}{3}} = \frac{\frac{1}{1}}{\frac{1}{3}} = \frac{1 \times 3}{1 \times 1} = \frac{3}{1} = 3$

Check: $log_{\frac{1}{9}}\, 3 \overset{?}{=} -\frac{1}{2}$; $log_{\frac{1}{9}}\left(\frac{1}{9}\right)^{-\frac{1}{2}} \overset{?}{=} -\frac{1}{2}$; $-\frac{1}{2} log_{\frac{1}{9}}\, \frac{1}{9} \overset{?}{=} -\frac{1}{2}$; $-\frac{1}{2} \times 1 \overset{?}{=} -\frac{1}{2}$; $-\frac{1}{2} = -\frac{1}{2}$

Example 6.4-17

$\boxed{log_u\, 2 = 0.1}$; $\boxed{u^{log_u\, 2} = u^{0.1}}$; $\boxed{2 = u^{0.1}}$; $\boxed{2 = u^{\frac{1}{10}}}$; $\boxed{2^{10} = u^{10 \times \frac{1}{10}}}$; $\boxed{2^{10} = u^{\frac{10}{10} = \frac{1}{1}}}$; $\boxed{2^{10} = u^{\frac{1}{1}}}$; $\boxed{2^{10} = u}$

; $\boxed{u = 2^{10}}$; $\boxed{\mathbf{u = 1024}}$

Check: $log_{1024}\, 2 \overset{?}{=} 0.1$; $\frac{log_{10}\, 2}{log_{10}\, 1024} \overset{?}{=} 0.1$; $\frac{0.30103}{3.0103} \overset{?}{=} 0.1$; $\frac{1}{10} \overset{?}{=} 0.1$; $0.1 = 0.1$

Example 6.4-18

$\boxed{log_x\, 0.001 = -3}$; $\boxed{x^{log_x\, 0.001} = x^{-3}}$; $\boxed{0.001 = x^{-3}}$; $\boxed{10^{-3} = x^{-3}}$; $\boxed{\mathbf{x = 10}}$

Check: $log_{10}\, 0.001 \overset{?}{=} -3$; $log_{10}\, 10^{-3} \overset{?}{=} -3$; $-3\, log_{10}\, 10 \overset{?}{=} -3$; $-3 \times 1 \overset{?}{=} -3$; $-3 = -3$

Example 6.4-19

$\boxed{\log_z 0.2 = 2}$; $\boxed{z^{\log_z 0.2} = z^2}$; $\boxed{0.2 = z^2}$; $\boxed{0.2^{\frac{1}{2}} = z^{2\times\frac{1}{2}}}$; $\boxed{0.2^{\frac{1}{2}} = z^{\frac{2}{2}=\frac{1}{1}}}$; $\boxed{0.2^{\frac{1}{2}} = z^{\frac{1}{1}}}$; $\boxed{0.2^{\frac{1}{2}} = z}$; $\boxed{\mathbf{z = 0.4472}}$

Check: $\log_{0.4472} 0.2 \stackrel{?}{=} 2$; $\frac{\log_{10} 0.2}{\log_{10} 0.4472} \stackrel{?}{=} 2$; $\frac{-0.69897}{-0.34949} \stackrel{?}{=} 2$; $2 = 2$

Example 6.4-20

$\boxed{\log_w 0.1 = -1}$; $\boxed{w^{\log_w 0.1} = w^{-1}}$; $\boxed{0.1 = w^{-1}}$; $\boxed{10^{-1} = w^{-1}}$; $\boxed{\mathbf{w = 10}}$

Check: $\log_{10} 0.1 \stackrel{?}{=} -1$; $\log_{10} 10^{-1} \stackrel{?}{=} -1$; $-1\times\log_{10} 10 \stackrel{?}{=} -1$; $-1\times 1 \stackrel{?}{=} -1$; $-1 = -1$

Example 6.4-21

$\boxed{\log_u \frac{3}{4} = -\frac{1}{2}}$; $\boxed{u^{\log_u \frac{3}{4}} = u^{-\frac{1}{2}}}$; $\boxed{\frac{3}{4} = u^{-\frac{1}{2}}}$; $\boxed{\left(\frac{3}{4}\right)^{-2} = u^{-2\times-\frac{1}{2}}}$; $\boxed{\frac{1}{\left(\frac{3}{4}\right)^2} = u^{\frac{2}{2}=\frac{1}{1}}}$; $\boxed{\frac{1}{\frac{9}{16}} = u^1}$; $\boxed{\frac{1}{\frac{9}{16}} = u}$

; $\boxed{\frac{1\times 16}{1\times 9} = u}$; $\boxed{\mathbf{u = \frac{16}{9}}}$

Note: $\left(\frac{16}{9}\right)^{-\frac{1}{2}} = \frac{1}{\left(\frac{16}{9}\right)^{\frac{1}{2}}} = \frac{1}{\sqrt{\frac{16}{9}}} = \frac{1}{\frac{4}{3}} = \frac{\frac{1}{1}}{\frac{4}{3}} = \frac{1\times 3}{4\times 1} = \frac{3}{4}$

Check: $\log_{\frac{16}{9}} \frac{3}{4} \stackrel{?}{=} -\frac{1}{2}$; $\log_{\frac{16}{9}} \left(\frac{16}{9}\right)^{-\frac{1}{2}} \stackrel{?}{=} -\frac{1}{2}$; $-\frac{1}{2}\log_{\frac{16}{9}} \frac{16}{9} \stackrel{?}{=} -\frac{1}{2}$; $-\frac{1}{2}\times 1 \stackrel{?}{=} -\frac{1}{2}$; $-\frac{1}{2} = -\frac{1}{2}$

Example 6.4-22

$\boxed{\log_x \frac{1}{4} = -2}$; $\boxed{x^{\log_x \frac{1}{4}} = x^{-2}}$; $\boxed{\frac{1}{4} = x^{-2}}$; $\boxed{\left(\frac{1}{4}\right)^{-\frac{1}{2}} = x^{-2\times-\frac{1}{2}}}$; $\boxed{\frac{1}{\left(\frac{1}{4}\right)^{\frac{1}{2}}} = x^{\frac{2}{2}=\frac{1}{1}}}$; $\boxed{\frac{1}{\sqrt{\frac{1}{4}}} = x^1}$; $\boxed{\frac{1}{\frac{1}{2}} = x}$; $\boxed{\frac{\frac{1}{1}}{\frac{1}{2}} = x}$

; $\boxed{\frac{1\times 2}{1\times 1} = x}$; $\boxed{x = \frac{2}{1}}$; $\boxed{\mathbf{x = 2}}$

Check: $\log_2 \frac{1}{4} \stackrel{?}{=} -2$; $\log_2 2^{-2} \stackrel{?}{=} -2$; $-2\log_2 2 \stackrel{?}{=} -2$; $-2\times 1 \stackrel{?}{=} -2$; $-2 = -2$

Example 6.4-23

$\boxed{\log_3 \frac{27}{t} = 5}$; $\boxed{\log_3 27 - \log_3 t = 5}$; $\boxed{-\log_3 t = -\log_3 27 + 5}$; $\boxed{\log_3 t = \log_3 27 - 5}$; $\boxed{\log_3 t = \log_3 3^3 - 5}$

; $\boxed{\log_3 t = 3\log_3 3 - 5}$; $\boxed{\log_3 t = 3\times 1 - 5}$; $\boxed{\log_3 t = 3 - 5}$; $\boxed{\log_3 t = -2}$; $\boxed{3^{\log_3 t} = 3^{-2}}$; $\boxed{t = 3^{-2}}$; $\boxed{t = \frac{1}{3^2}}$; $\boxed{\mathbf{t = \frac{1}{9}}}$

Check: $\log_3 \frac{27}{\frac{1}{9}} \stackrel{?}{=} 5$; $\log_3 \frac{27\times 9}{1} \stackrel{?}{=} 5$; $\log_3 \frac{243}{1} \stackrel{?}{=} 5$; $\log_3 243 \stackrel{?}{=} 5$; $\log_3 3^5 \stackrel{?}{=} 5$; $5\log_3 3 \stackrel{?}{=} 5$; $5 = 5$ or,

Check: $\log_3 \frac{27}{3^{-2}} \stackrel{?}{=} 5$; $\log_3 \frac{3^3}{3^{-2}} \stackrel{?}{=} 5$; $\log_3 \frac{3^3\times 3^2}{1} \stackrel{?}{=} 5$; $\log_3 3^{3+2} \stackrel{?}{=} 5$; $\log_3 3^5 \stackrel{?}{=} 5$; $5\log_3 3 \stackrel{?}{=} 5$; $5 = 5$

Example 6.4-24

$\boxed{log_5 1 = u}$; $\boxed{5^{log_5 1} = 5^u}$; $\boxed{1 = 5^u}$; $\boxed{5^0 = 5^u}$; $\boxed{\mathbf{u = 0}}$

Check: $log_5 1 \overset{?}{=} 0$; $log_5 10^0 \overset{?}{=} 0$; $0 \times log_5 10 \overset{?}{=} 0$; $0 = 0$

Note: $log_a 1$ is always equal to zero

Example 6.4-25

$\boxed{ln 1 = x}$; $\boxed{log_e 1 = x}$; $\boxed{e^{log_e 1} = e^x}$; $\boxed{1 = e^x}$; $\boxed{e^0 = e^x}$; $\boxed{\mathbf{x = 0}}$

Check: $log_e 1 \overset{?}{=} 0$; $log_e 10^0 \overset{?}{=} 0$; $0 \times log_e 10 \overset{?}{=} 0$; $0 = 0$

Note: $10^0 = 1$

Example 6.4-26

$\boxed{log_{\frac{1}{8}} 1 = w}$; $\boxed{\left(\frac{1}{8}\right)^{log_{\frac{1}{8}} 1} = \left(\frac{1}{8}\right)^w}$; $\boxed{1 = \left(\frac{1}{8}\right)^w}$; $\boxed{\left(\frac{1}{8}\right)^0 = \left(\frac{1}{8}\right)^w}$; $\boxed{\mathbf{w = 0}}$

Check: $log_{\frac{1}{8}} 1 \overset{?}{=} 0$; $log_{\frac{1}{8}} 10^0 \overset{?}{=} 0$; $0 \times log_{\frac{1}{8}} 10 \overset{?}{=} 0$; $0 = 0$

Example 6.4-27

I. $\boxed{log_8 0.125 = u}$; $\boxed{8^{log_8 0.125} = 8^u}$; $\boxed{0.125 = 8^u}$; $\boxed{8^{-1} = 8^u}$; $\boxed{\mathbf{u = -1}}$ or,

II. $\boxed{log_8 0.125 = u}$; $\boxed{log_8 8^{-1} = u}$; $\boxed{-1 \times log_8 8 = u}$; $\boxed{-1 \times 1 = u}$; $\boxed{\mathbf{u = -1}}$

Check: $log_8 0.125 \overset{?}{=} -1$; $\frac{log_{10} 0.125}{log_{10} 8} \overset{?}{=} -1$; $\frac{-0.9031}{0.9031} \overset{?}{=} -1$; $-1 = -1$

Example 6.4-28

I. $\boxed{log_4 0.5 = x}$; $\boxed{4^{log_4 0.5} = 4^x}$; $\boxed{0.5 = 4^x}$; $\boxed{4^{-\frac{1}{2}} = 4^x}$; $\boxed{-\frac{1}{2} = x}$; $\boxed{\mathbf{x = -0.5}}$ or,

Note: $4^{-\frac{1}{2}} = \frac{1}{4^{\frac{1}{2}}} = \frac{1}{\sqrt{4}} = \frac{1}{\sqrt{2^2}} = \frac{1}{2} = 0.5$

II. $\boxed{log_4 0.5 = x}$; $\boxed{\frac{log_{10} 0.5}{log_{10} 4} = x}$; $\boxed{\frac{-0.3010}{0.6021} = x}$; $\boxed{\mathbf{x = -0.5}}$ or,

III. $\boxed{log_4 0.5 = x}$; $\boxed{log_4 (0.25 \times 2) = x}$; $\boxed{log_4 0.25 + log_4 2 = x}$; $\boxed{log_4 4^{-1} + log_4 2 = x}$

; $\boxed{-1 \times log_4 4 + \frac{log_{10} 2}{log_{10} 4} = x}$; $\boxed{-1 \times 1 + \frac{0.3010}{0.6021} = x}$; $\boxed{-1 + 0.5 = x}$; $\boxed{\mathbf{x = -0.5}}$

Check: $log_4 0.5 \overset{?}{=} -0.5$; $\frac{log_{10} 0.5}{log_{10} 4} \overset{?}{=} -0.5$; $\frac{-0.3010}{0.6021} \overset{?}{=} -0.5$; $-0.5 = -0.5$

Example 6.4-29

$\boxed{log_{16} 4 = t}$; $\boxed{16^{log_{16} 4} = 16^t}$; $\boxed{4 = 16^t}$; $\boxed{4 = \left(4^2\right)^t}$; $\boxed{4 = 4^{2t}}$; $\boxed{4^1 = 4^{2t}}$; $\boxed{1 = 2t}$; $\boxed{\mathbf{t = \frac{1}{2}}}$

Check: $log_{16} 4 \overset{?}{=} \frac{1}{2}$; $log_{16} 16^{\frac{1}{2}} \overset{?}{=} \frac{1}{2}$; $\frac{1}{2} \times log_{16} 16 \overset{?}{=} \frac{1}{2}$; $\frac{1}{2} \times 1 \overset{?}{=} \frac{1}{2}$; $\frac{1}{2} = \frac{1}{2}$ or,

Check: $log_{16} 4 \overset{?}{=} \frac{1}{2}$; $\frac{log_{10} 4}{log_{10} 16} \overset{?}{=} \frac{1}{2}$; $\frac{0.6021}{1.2042} \overset{?}{=} \frac{1}{2}$; $\frac{1}{2} = \frac{1}{2}$

Example 6.4-30

$\boxed{log_{125}5=x}$; $\boxed{125^{log_{125}5}=125^{x}}$; $\boxed{5=125^{x}}$; $\boxed{5=\left(5^{3}\right)^{x}}$; $\boxed{5=5^{3x}}$; $\boxed{5^{1}=5^{3x}}$; $\boxed{1=3x}$; $\boxed{\mathbf{x=\frac{1}{3}}}$

Check: $log_{125}5\overset{?}{=}\frac{1}{3}$; $log_{125}125^{\frac{1}{3}}\overset{?}{=}\frac{1}{3}$; $\frac{1}{3}\times log_{125}125\overset{?}{=}\frac{1}{3}$; $\frac{1}{3}\times 1\overset{?}{=}\frac{1}{3}$; $\frac{1}{3}=\frac{1}{3}$ or,

Check: $log_{125}5\overset{?}{=}\frac{1}{3}$; $\frac{log_{10}5}{log_{10}125}\overset{?}{=}\frac{1}{3}$; $\frac{0.6989}{2.0969}\overset{?}{=}\frac{1}{3}$; $\frac{1}{3}=\frac{1}{3}$

Example 6.4-31

$\boxed{log_{81}3=x+2}$; $\boxed{81^{log_{81}3}=81^{x+2}}$; $\boxed{3=81^{x+2}}$; $\boxed{3=\left(3^{4}\right)^{x+2}}$; $\boxed{3=3^{4x+8}}$; $\boxed{3^{1}=3^{4x+8}}$; $\boxed{1=4x+8}$

; $\boxed{\mathbf{x=-\frac{7}{4}}}$

Check: $log_{81}3\overset{?}{=}-\frac{7}{4}+2$; $log_{81}81^{\frac{1}{4}}\overset{?}{=}\frac{-7}{4}+\frac{2}{1}$; $\frac{1}{4}\overset{?}{=}\frac{(-7\times 1)+(2\times 4)}{4\times 1}$; $\frac{1}{4}\overset{?}{=}\frac{-7+8}{4}$; $\frac{1}{4}=\frac{1}{4}$

Example 6.4-32

I. $\boxed{x=log_{2}\frac{1}{16}}$; $\boxed{2^{x}=2^{log_{2}\frac{1}{16}}}$; $\boxed{2^{x}=\frac{1}{16}}$; $\boxed{2^{x}=\frac{1}{2^{4}}}$; $\boxed{2^{x}=2^{-4}}$; $\boxed{\mathbf{x=-4}}$ or,

II. $\boxed{x=log_{2}\frac{1}{16}}$; $\boxed{x=log_{2}\frac{1}{2^{4}}}$; $\boxed{x=log_{2}2^{-4}}$; $\boxed{x=-4\times log_{2}2}$; $\boxed{x=-4\times 1}$; $\boxed{\mathbf{x=-4}}$

Check: $-4\overset{?}{=}log_{2}\frac{1}{16}$; $-4\overset{?}{=}log_{2}1-log_{2}16$; $-4\overset{?}{=}0-log_{2}2^{4}$; $-4\overset{?}{=}-4\,log_{2}2$; $-4\overset{?}{=}-4\times 1$; $-4=-4$

Example 6.4-33

I. $\boxed{y=log_{5}\frac{1}{125}}$; $\boxed{5^{y}=5^{log_{5}\frac{1}{125}}}$; $\boxed{5^{y}=\frac{1}{125}}$; $\boxed{5^{y}=\frac{1}{5^{3}}}$; $\boxed{5^{y}=5^{-3}}$; $\boxed{\mathbf{y=-3}}$ or,

II. $\boxed{y=log_{5}\frac{1}{125}}$; $\boxed{y=\log_{5}\frac{1}{5^{3}}}$; $\boxed{y=log_{5}5^{-3}}$; $\boxed{y=-3\times log_{5}5}$; $\boxed{y=-3\times 1}$; $\boxed{\mathbf{y=-3}}$

Check: $-3\overset{?}{=}log_{5}\frac{1}{125}$; $-3\overset{?}{=}log_{5}1-log_{5}125$; $-3\overset{?}{=}0-log_{5}5^{3}$; $-3\overset{?}{=}-3\,log_{5}5$; $-3\overset{?}{=}-3\times 1$; $-3=-3$

Example 6.4-34

I. $\boxed{u+2=log_{3}\frac{1}{243}}$; $\boxed{3^{u+2}=3^{log_{3}\frac{1}{243}}}$; $\boxed{3^{u+2}=\frac{1}{243}}$; $\boxed{3^{u+2}=\frac{1}{3^{5}}}$; $\boxed{3^{u+2}=3^{-5}}$; $\boxed{u+2=-5}$; $\boxed{\mathbf{u=-7}}$ or,

II. $\boxed{u+2=log_{3}\frac{1}{243}}$; $\boxed{u+2=log_{3}\frac{1}{3^{5}}}$; $\boxed{u+2=log_{3}3^{-5}}$; $\boxed{u+2=-5\times log_{3}3}$; $\boxed{u+2=-5\times 1}$; $\boxed{u+2=-5}$

; $\boxed{\mathbf{u=-7}}$

Check: $-7+2\overset{?}{=}log_{3}\frac{1}{243}$; $-5\overset{?}{=}log_{3}1-log_{3}243$; $-5\overset{?}{=}0-log_{3}3^{5}$; $-5\overset{?}{=}-5\,log_{3}5$; $-5\overset{?}{=}-5\times 1$; $-5=-5$

Example 6.4-35

I. $\boxed{x-1=log_{10}0.01}$; $\boxed{10^{x-1}=10^{log_{10}0.01}}$; $\boxed{10^{x-1}=0.01}$; $\boxed{10^{x-1}=10^{-2}}$; $\boxed{x-1=-2}$; $\boxed{\mathbf{x=-1}}$ or,

II. $\boxed{x-1=log_{10}0.01}$; $\boxed{x-1=log_{10}10^{-2}}$; $\boxed{x-1=-2\times log_{10}10}$; $\boxed{x-1=-2\times 1}$; $\boxed{x-1=-2}$; $\boxed{\mathbf{x=-1}}$

Check: $-1-1\overset{?}{=}log_{10}0.01$; $-2\overset{?}{=}log_{10}10^{-2}$; $-2\overset{?}{=}-2\,log_{10}10$; $-2\overset{?}{=}-2\times 1$; $-2=-2$

Example 6.4-36

I. $\boxed{w-10=ln\frac{1}{e^{10}}}$; $\boxed{e^{w-10}=e^{ln\frac{1}{e^{10}}}}$; $\boxed{e^{w-10}=\frac{1}{e^{10}}}$; $\boxed{e^{w-10}=e^{-10}}$; $\boxed{w-10=-10}$; $\boxed{\mathbf{w=0}}$ or,

II. $\boxed{w-10=ln\frac{1}{e^{10}}}$; $\boxed{w-10=lne^{-10}}$; $\boxed{w-10=-10\times lne}$; $\boxed{w-10=-10\times 1}$; $\boxed{w=-10+10}$; $\boxed{\mathbf{w=0}}$

Check: $0-10\overset{?}{=}ln\frac{1}{e^{10}}$; $-10\overset{?}{=}lne^{-10}$; $-10\overset{?}{=}-10\,lne$; $-10\overset{?}{=}-10\times 1$; $-10=-10$

Example 6.4-37

I. $\boxed{v=ln\frac{1}{e^{-2}}+lne^{3}}$; $\boxed{v=lne^{2}+lne^{3}}$; $\boxed{v=ln\left(e^{2}\cdot e^{3}\right)}$; $\boxed{v=ln\,e^{5}}$; $\boxed{v=5ln\,e}$; $\boxed{v=5\times 1}$; $\boxed{\mathbf{v=5}}$ or,

II. $\boxed{v=ln\frac{1}{e^{-2}}+lne^{3}}$; $\boxed{v=lne^{2}+lne^{3}}$; $\boxed{v=2\times lne+3\times lne}$; $\boxed{v=(2\times 1)+(3\times 1)}$; $\boxed{v=2+3}$; $\boxed{\mathbf{v=5}}$

Check: $5\overset{?}{=}ln\frac{1}{e^{-2}}+lne^{3}$; $5\overset{?}{=}lne^{2}+lne^{3}$; $5\overset{?}{=}2lne+3lne$; $5\overset{?}{=}(2\times 1)+(3\times 1)$; $5=5$

Example 6.4-38

$\boxed{log_4(x-1)=2}$; $\boxed{4^{log_4(x-1)}=4^{2}}$; $\boxed{x-1=4^{2}}$; $\boxed{x-1=16}$; $\boxed{\mathbf{x=17}}$

Check: $log_4(17-1)\overset{?}{=}2$; $log_4 16\overset{?}{=}2$; $log_4 4^{2}\overset{?}{=}2$; $2\,log_4 4\overset{?}{=}2$; $2\overset{?}{=}2\times 1$; $2=2$

Example 6.4-39

$\boxed{log_8(x+2)=\sqrt{16}}$; $\boxed{log_8(x+2)=\sqrt{4^{2}}}$; $\boxed{log_8(x+2)=4}$; $\boxed{8^{log_8(x+2)}=8^{4}}$; $\boxed{x+2=4096}$; $\boxed{\mathbf{x=4094}}$

Check: $log_8(4094+2)\overset{?}{=}\sqrt{16}$; $log_8(4096)\overset{?}{=}4$; $log_8 8^{4}\overset{?}{=}4$; $4\,log_8 8\overset{?}{=}4$; $4\times 1\overset{?}{=}4$; $4=4$

Example 6.4-40

$\boxed{log_2(u-1)=log_3 16}$; $\boxed{log_2(u-1)=\frac{log_{10}16}{log_{10}3}}$; $\boxed{log_2(u-1)=\frac{1.2041}{0.4771}}$; $\boxed{log_2(u-1)=2.524}$; $\boxed{log_2(u-1)=2.524}$

; $\boxed{2^{log_2(u-1)}=2^{2.524}}$; $\boxed{u-1=5.75}$; $\boxed{\mathbf{u=6.75}}$

Check: $log_2(6.75-1)\overset{?}{=}log_3 16$; $log_2 5.75\overset{?}{=}2.524$; $\frac{log_{10}5.75}{log_{10}2}\overset{?}{=}2.524$; $\frac{0.7597}{0.3010}\overset{?}{=}2.524$; $2.524=2.524$

Example 6.4-41

$\boxed{log_5(w+1)=log_5 625}$; $\boxed{5^{log_5(w+1)}=5^{log_5 625}}$; $\boxed{w+1=625}$; $\boxed{w=625-1}$; $\boxed{\mathbf{w=624}}$

Check: $log_5(624+1)\overset{?}{=}log_5 625$; $log_5 625 = log_5 625$

Example 6.4-42

$\boxed{log_4 y = \frac{1}{2} + log_{10} 1000}$; $\boxed{log_4 y = \frac{1}{2} + log_{10} 10^3}$; $\boxed{log_4 y = \frac{1}{2} + 3 log_{10} 10}$; $\boxed{log_4 y = \frac{1}{2} + 3 \times 1}$; $\boxed{log_4 y = \frac{1}{2} + 3}$

; $\boxed{log_4 y = \frac{1}{2} + \frac{3}{1}}$; $\boxed{log_4 y = \frac{(1\cdot 1)+(3\cdot 2)}{2\cdot 1}}$; $\boxed{log_4 y = \frac{1+6}{2}}$; $\boxed{log_4 y = \frac{7}{2}}$; $\boxed{4^{log_4 y} = 4^{\frac{7}{2}}}$; $\boxed{y = \sqrt[2]{4^7}}$

; $\boxed{y = \sqrt{16384}}$; $\boxed{\mathbf{y = 128}}$

Check: $log_4 128 \overset{?}{=} \frac{1}{2} + log_{10} 1000$; $\frac{log_{10} 128}{log_{10} 4} \overset{?}{=} \frac{1}{2} + 3$; $\frac{2.107}{0.6020} \overset{?}{=} \frac{7}{2}$; $\frac{7}{2} = \frac{7}{2}$

Example 6.4-43

$\boxed{log_{27} x = \frac{1}{3}}$; $\boxed{27^{log_{27} x} = 27^{\frac{1}{3}}}$; $\boxed{x = \sqrt[3]{27^1}}$; $\boxed{x = \sqrt[3]{3^3}}$; $\boxed{\mathbf{x = 3}}$

Check: $log_{27} 3 \overset{?}{=} \frac{1}{3}$; $\frac{log_{10} 3}{log_{10} 27} \overset{?}{=} \frac{1}{3}$; $\frac{0.477}{1.431} \overset{?}{=} \frac{1}{3}$; $\frac{1}{3} = \frac{1}{3}$

Example 6.4-44

$\boxed{3^{5x+3} = \frac{1}{27}}$; $\boxed{3^{5x+3} = 3^{-3}}$; $\boxed{5x+3 = -3}$; $\boxed{5x = -3-3}$; $\boxed{5x = -6}$; $\boxed{\mathbf{x = -\frac{6}{5}}}$

Check: $3^{5\times -\frac{6}{5}+3} \overset{?}{=} \frac{1}{27}$; $3^{\cancel{5}\times -\frac{6}{\cancel{5}}+3} \overset{?}{=} 3^{-3}$; $3^{-6+3} \overset{?}{=} 3^{-3}$; $3^{-3} = 3^{-3}$

Example 6.4-45

$\boxed{64^{x+4} = \frac{1}{512}}$; $\boxed{\left(8^2\right)^{x+4} = 8^{-3}}$; $\boxed{8^{2x+8} = 8^{-3}}$; $\boxed{2x+8 = -3}$; $\boxed{2x = -8-3}$; $\boxed{2x = -11}$; $\boxed{\mathbf{x = -\frac{11}{2}}}$

Check: $64^{-\frac{11}{2}+4} \overset{?}{=} \frac{1}{512}$; $64^{\frac{-11+8}{2}} \overset{?}{=} 8^{-3}$; $64^{-\frac{3}{2}} \overset{?}{=} 8^{-3}$; $\left(8^2\right)^{-\frac{3}{2}} \overset{?}{=} 8^{-3}$; $8^{\cancel{2}\times -\frac{3}{\cancel{2}}} \overset{?}{=} 8^{-3}$; $8^{-3} = 8^{-3}$

Example 6.4-46

$\boxed{log_8 t = -\frac{1}{3}}$; $\boxed{8^{log_8 t} = 8^{-\frac{1}{3}}}$; $\boxed{t = 8^{-\frac{1}{3}}}$; $\boxed{t = \frac{1}{8^{\frac{1}{3}}}}$; $\boxed{t = \frac{1}{\left(2^3\right)^{\frac{1}{3}}}}$; $\boxed{t = \frac{1}{2^{3\times\frac{1}{3}}}}$; $\boxed{t = \frac{1}{2^{\cancel{3}\times\frac{1}{\cancel{3}}}}}$; $\boxed{t = \frac{1}{2^1}}$; $\boxed{\mathbf{t = \frac{1}{2}}}$

Check: $log_8 \frac{1}{2} \overset{?}{=} -\frac{1}{3}$; $log_8 8^{-\frac{1}{3}} \overset{?}{=} -\frac{1}{3}$; $-\frac{1}{3} log_8 8 \overset{?}{=} -\frac{1}{3}$; $-\frac{1}{3} \times 1 \overset{?}{=} -\frac{1}{3}$; $-\frac{1}{3} = -\frac{1}{3}$

Example 6.4-47 Given $5 = 2^{log_b 5}$ solve for b.

Solution: Note that $5 = 2^{log_b 5} = b^{log_b 5}$. Since $b^{log_b 5} = 2^{log_b 5}$, we observe that by equating the terms on both sides of the equality $\boxed{\mathbf{b = 2}}$.

Check: $5 \overset{?}{=} 2^{log_2 5}$; $5 = 5$

Example 6.4-48

$\boxed{\log_y 10{,}000 = 4}$; $\boxed{y^{\log_y 10{,}000} = y^4}$; $\boxed{10{,}000 = y^4}$; $\boxed{10^4 = y^4}$; $\boxed{\mathbf{y = 10}}$

Check: $\log_{10} 10{,}000 \overset{?}{=} 4$; $\log_{10} 10^4 \overset{?}{=} 4$; $4\log_{10} 10 \overset{?}{=} 4$; $4 \times 1 \overset{?}{=} 4$; $4 = 4$

Example 6.4-49 Show that:

$\boxed{\log_2 128 + \log_2 \frac{1}{2} \overset{?}{=} \log_2 64}$; $\boxed{\log_2 64 \cdot 2 + \log_2 2^{-1} \overset{?}{=} \log_2 64}$; $\boxed{\log_2 64 + \log_2 2 - \log_2 2 \overset{?}{=} \log_2 64}$

; $\boxed{\log_2 64 + 1 - 1 \overset{?}{=} \log_2 64}$; $\boxed{\mathbf{\log_2 64 = \log_2 64}}$

Example 6.4-50 Show that:

$\boxed{\log_6 1296 + \log_6 \frac{1}{36} \overset{?}{=} 2}$; $\boxed{\log_6 6^4 + \log_6 \frac{1}{6^2} \overset{?}{=} 2}$; $\boxed{4\log_6 6 + \log_6 6^{-2} \overset{?}{=} 2}$; $\boxed{4\log_6 6 - 2\log_6 6 \overset{?}{=} 2}$

; $\boxed{4 \times 1 - 2 \times 1 \overset{?}{=} 2}$; $\boxed{4 - 2 \overset{?}{=} 2}$; $\boxed{\mathbf{2 = 2}}$

Practice Problems - Solving One Variable Logarithmic Equations (Simple Cases)

Section 6.4 Case I Practice Problems - Solve the following logarithmic equations:

1. $x + 4 = \log_4 \frac{1}{64}$
2. $y - 5 = \ln \sqrt[3]{e^2}$
3. $x = \log_2 \frac{1}{3^{-2}} + \log_2 8$
4. $\log_3 (x + 2) = 5$
5. $\log_4 x = 3 + \log_3 5$
6. $\log_a 1000 = 3$
7. $27^{x+3} = \frac{1}{243}$
8. $\log_3 t = -\frac{1}{5}$
9. $\log_{25} x = \frac{1}{2}$
10. $x + 3 = \log_2 7$

Case II Solving One Variable Logarithmic Equations (More Difficult Cases)

The more difficult class of one variable loagrithmic equations are solved using the following steps:

Step 1 Apply the logarithmic laws to reduce the equation to the form of $log_a x = b$.

Step 2 Solve for the unknown variable x. Check the answer by substituting the solution into the original equation.

Examples with Steps

The following examples show the steps as to how one variable logarithmic equations are solved:

Example 6.4-51

$$\boxed{log_3 x + log_3 6 = log_2 8 - log_2 4}$$

Solution:

Step 1 $\boxed{log_3 x + log_3 6 = log_2 8 - log_2 4}$; $\boxed{log_3 6x = log_2 \frac{8}{4}}$; $\boxed{log_3 6x = log_2 2}$; $\boxed{log_3 6x = 1}$

Step 2 $\boxed{log_3 6x = 1}$; $\boxed{3^{log_3 6x} = 3^1}$; $\boxed{6x = 3}$; $\boxed{x = \frac{\cancel{3}}{\cancel{6}_2}}$; $\boxed{x = \frac{1}{2}}$; $\boxed{\mathbf{x = 0.5}}$

Check: $log_3 0.5 + log_3 6 \stackrel{?}{=} log_2 8 - log_2 4$; $log_3 0.5 \times 6 \stackrel{?}{=} log_2 \frac{8}{4}$; $log_3 3 \stackrel{?}{=} log_2 2$; $1 = 1$

Example 6.4-52

$$\boxed{ln(x-5) + ln x = log_4 2}$$

Solution:

Step 1 $\boxed{ln(x-5) + ln x = log_4 2}$; $\boxed{ln x(x-5) = \frac{log_{10} 2}{log_{10} 4}}$; $\boxed{ln x(x-5) = \frac{0.3010}{0.6021}}$; $\boxed{ln x(x-5) = 0.5}$

Step 2 $\boxed{ln x(x-5) = 0.5}$; $\boxed{e^{ln x(x-5)} = e^{0.5}}$; $\boxed{x(x-5) = 1.65}$; $\boxed{x^2 - 5x = 1.65}$; $\boxed{x^2 - 5x - 1.65 = 0}$

; $\boxed{x - \left(\frac{5 + \sqrt{31.6}}{2}\right) = 0 \; or \; x - \left(\frac{5 - \sqrt{31.6}}{2}\right) = 0}$; $\boxed{x - \left(\frac{5 + 5.62}{2}\right) = 0 \; or \; x - \left(\frac{5 - 5.62}{2}\right) = 0}$

; $\boxed{(x - 5.31) = 0 \; or \; (x + 0.31) = 0}$; $\boxed{\mathbf{x = 5.31}}$

or $\boxed{x = -0.31}$

Check: 1. Substitute $x = 5.31$ into the original equation. Then, $ln(5.31 - 5) + ln 5.31 \stackrel{?}{=} log_4 2$; $ln 0.31 + ln 5.31 \stackrel{?}{=} \frac{log_{10} 2}{log_{10} 4}$; $ln 5.31 \times 0.31 \stackrel{?}{=} \frac{0.3010}{0.6020}$; $ln 1.6461 \stackrel{?}{=} 0.5$; $0.5 = 0.5$. Therefore, $x = 5.31$ is the solution.

2. Substitute $x = -0.31$ into the original equation: Then, $ln(-0.31 - 5) + ln(-0.31) \stackrel{?}{=} log_4 2$; $ln(-5.31) + ln(-0.31) \stackrel{?}{=} log_4 2$. The ln of negative numbers is not defined. Thus, $x = -0.31$ is not a solution.

Example 6.4-53

$$\boxed{\log_8 x + \log_8 7 = 2}$$

Solution:

Step 1 $\boxed{\log_8 x + \log_8 7 = 2}$; $\boxed{\log_8 7x = 2}$

Step 2 $\boxed{\log_8 7x = 2}$; $\boxed{8^{\log_8 7x} = 8^2}$; $\boxed{7x = 64}$; $\boxed{x = \frac{64}{7}}$; $\boxed{\mathbf{x = 9.14}}$

Check: $\log_8 9.14 + \log_8 7 \overset{?}{=} 2$; $\log_8 7 \times 9.14 \overset{?}{=} 2$; $\log_8 64 \overset{?}{=} 2$; $\log_8 8^2 \overset{?}{=} 2$; $2\log_8 8 \overset{?}{=} 2$; $2 \times 1 \overset{?}{=} 2$; $2 = 2$

Example 6.4-54

$$\boxed{\log_5 x = \log_{0.2} 5}$$

Solution:

Step 1 $\boxed{\log_5 x = \log_{0.2} 5}$; $\boxed{\log_5 x = \frac{\log_{10} 5}{\log_{10} 0.2}}$; $\boxed{\log_5 x = \frac{0.6989}{-0.6989}}$; $\boxed{\log_5 x = -1}$

Step 2 $\boxed{\log_5 x = -1}$; $\boxed{5^{\log_5 x} = 5^{-1}}$; $\boxed{x = \frac{1}{5}}$; $\boxed{\mathbf{x = 0.2}}$

Check: $\log_5 0.2 \overset{?}{=} \log_{0.2} 5$; $\frac{\log_{10} 0.2}{\log_{10} 5} \overset{?}{=} \frac{\log_{10} 5}{\log_{10} 0.2}$; $\frac{-0.6989}{0.6989} \overset{?}{=} \frac{0.6989}{-0.6989}$; $-1 = -1$

Example 6.4-55

$$\boxed{\ln x - \ln 10 = \ln 5 + \ln 3}$$

Solution:

Step 1 $\boxed{\ln x - \ln 10 = \ln 5 + \ln 3}$; $\boxed{\ln x = \ln 5 + \ln 3 + \ln 10}$; $\boxed{\ln x = \ln 5 \cdot 3 \cdot 10}$; $\boxed{\ln x = \ln 150}$

Step 2 $\boxed{\ln x = \ln 150}$; $\boxed{e^{\ln x} = e^{\ln 150}}$; $\boxed{\mathbf{x = 150}}$

Check: $\ln 150 - \ln 10 \overset{?}{=} \ln 5 + \ln 3$; $\ln \frac{150}{10} \overset{?}{=} \ln 5 \cdot 3$; $\ln \frac{150}{10} \overset{?}{=} \ln 15$; $\ln \frac{15}{1} \overset{?}{=} \ln 15$; $\ln 15 = \ln 15$

Additional Examples - Solving One Variable Logarithmic Equations (More Difficult Cases)

The following examples further illustrate how to solve one variable logarithmic equations:

Example 6.4-56

$\boxed{\ln(x+2) + \ln x = 0}$; $\boxed{\ln x(x+2) = 0}$; $\boxed{e^{\ln x(x+2)} = e^0}$; $\boxed{x(x+2) = 1}$; $\boxed{x^2 + 2x = 1}$; $\boxed{x^2 + 2x - 1 = 0}$

; $\boxed{x - \left(-1 + \sqrt{2}\right) = 0 \text{ or } x - \left(-1 - \sqrt{2}\right) = 0}$; $\boxed{x - (-1 + 1.414) = 0 \text{ or } x - (-1 - 1.414) = 0}$

; $\boxed{(x - 0.414) = 0 \text{ or } (x + 2.414) = 0}$; $\boxed{\mathbf{x = 0.414}}$ or $\boxed{x = -2.414}$

Check: 1. Substitute $x = 0.414$ into the original equation. Then, $\ln(0.414 + 2) - \ln 0.414 \overset{?}{=} 0$

; $ln2.414 - ln0.414 \stackrel{?}{=} 0$; $0.881 - 0.881 \stackrel{?}{=} 0$; $0 = 0$. Therefore, $x = 0.414$ is the solution.

2. Substitute $x = -2.414$ into the original equation: Then, $ln(-2.414+2) - ln -2.414 \stackrel{?}{=} 0$; $ln(-0.414) - ln2.414 \stackrel{?}{=} 0$. The ln of negative numbers is not defined. Thus, $x = -2.414$ is not a solution.

Example 6.4-57

$\boxed{ln(x+2) - ln x = 2}$; $\boxed{\ln\frac{x+2}{x} = 2}$; $\boxed{e^{\ln\frac{x+2}{x}} = e^2}$; $\boxed{\frac{x+2}{x} = e^2}$; $\boxed{x+2 = e^2 x}$; $\boxed{x - e^2 x = -2}$; $\boxed{x\left(1-e^2\right) = -2}$

; $\boxed{x = -\frac{2}{1-e^2}}$; $\boxed{x = -\frac{2}{1-7.3891}}$; $\boxed{x = -\frac{2}{-6.3891}}$; $\boxed{x = +\frac{2}{6.3891}}$; $\boxed{\mathbf{x = 0.3130}}$

Check: $\ln(0.3130+2) - \ln 0.3130 \stackrel{?}{=} 2$; $0.8385 - (-1.1615) \stackrel{?}{=} 2$; $0.8385 + 1.1615 \stackrel{?}{=} 2$; $2 = 2$

Example 6.4-58

$\boxed{ln(2-x) - ln(3+x) = ln x}$; $\boxed{ln\frac{2-x}{3+x} = ln x}$; $\boxed{e^{ln\frac{2-x}{3+x}} = e^{ln x}}$; $\boxed{\frac{2-x}{3+x} = x}$; $\boxed{2-x = x(3+x)}$; $\boxed{2-x = 3x + x^2}$

; $\boxed{x^2 + 4x - 2 = 0}$; $\boxed{x - \left(-2+\sqrt{6}\right) = 0 \text{ or } x - \left(-2-\sqrt{6}\right) = 0}$; $\boxed{x - (-2+2.449) = 0 \text{ or } x - (-2-2.449) = 0}$

; $\boxed{(x-0.449) = 0 \text{ or } (x+4.449) = 0}$; $\boxed{\mathbf{x = 0.449}}$ or $\boxed{x = -4.449}$

Check: 1. Substitute $x = 0.449$ into the original equation. Then, $ln(2-0.449) - ln(3+0.449) \stackrel{?}{=} ln0.449$; $ln1.551 - ln3.449 \stackrel{?}{=} ln0.449$; $0.4389 - 1.238 \stackrel{?}{=} -0.8$; $-0.8 = -0.8$. Thus, $x = 0.449$ is the solution.

2. Substitute $x = -4.449$ into the original equation: Then, $ln(2+4.449) - ln(3-4.449) \stackrel{?}{=} ln4.449$; $ln6.449 - ln(-1.449) \stackrel{?}{=} ln4.449$. The ln of negative numbers is not defined. Therefore, $x = -4.449$ is not a solution.

Example 6.4-59

$\boxed{ln(x+1) = ln5}$; $\boxed{e^{ln(x+1)} = e^{ln5}}$; $\boxed{x+1 = 5}$; $\boxed{x = 5-1}$; $\boxed{\mathbf{x = 4}}$

Check: $ln(4+1) \stackrel{?}{=} ln5$; $ln5 = ln5$; $1.609 = 1.609$

Example 6.4-60

$\boxed{log 20 = ln x}$; $\boxed{1.3010 = ln x}$; $\boxed{e^{1.3010} = e^{ln x}}$; $\boxed{e^{1.3010} = x}$; $\boxed{\mathbf{x = 3.6729}}$

Check: $log 20 \stackrel{?}{=} ln3.6729$; $1.301 = 1.301$

Example 6.4-61

$\boxed{log_2(x+1) + log_3 243 = log_2 16}$; $\boxed{log_2(x+1) + log_3 3^5 = log_2 2^4}$; $\boxed{log_2(x+1) + 5 = 4}$; $\boxed{log_2(x+1) = 4-5}$

; $\boxed{log_2(x+1) = -1}$; $\boxed{2^{log_2(x+1)} = 2^{-1}}$; $\boxed{x+1 = \frac{1}{2}}$; $\boxed{x = \frac{1}{2} - 1}$; $\boxed{x = \frac{1-2}{2}}$; $\boxed{x = -\frac{1}{2}}$; $\boxed{\mathbf{x = -0.5}}$

Check: $log_2(-0.5+1)+log_3 243 \stackrel{?}{=} log_2 16$; $log_2 0.5 + log_3 3^5 \stackrel{?}{=} log_2 2^4$; $\frac{log_{10} 0.5}{log_{10} 2} + 5 \stackrel{?}{=} 4$; $\frac{-0.3010}{0.3010} + 5 \stackrel{?}{=} 4$

; $-1+5 \stackrel{?}{=} 4$; $4 = 4$

Example 6.4-62

$\boxed{log_5(x+1) = log_5 x + 2}$; $\boxed{log_5(x+1) - log_5 x = 2}$; $\boxed{log_5 \frac{x+1}{x} = 2}$; $\boxed{5^{log_5 \frac{x+1}{x}} = 5^2}$; $\boxed{\frac{x+1}{x} = 25}$; $\boxed{x+1 = 25x}$

; $\boxed{25x - x = 1}$; $\boxed{24x = 1}$; $\boxed{x = \frac{1}{24}}$; $\boxed{\mathbf{x = 0.0417}}$

Check: $\ln(0.3130+2) - \ln 0.3130 \stackrel{?}{=} 2$; $0.8385 - (-1.1615) \stackrel{?}{=} 2$; $0.8385 + 1.1615 \stackrel{?}{=} 2$; $2 = 2$

Example 6.4-63

$\boxed{log_{10}(x+1) - log_{10} x^2 = 1}$; $\boxed{log_{10} \frac{x+1}{x^2} = 1}$; $\boxed{10^{log_{10} \frac{x+1}{x^2}} = 10^1}$; $\boxed{\frac{x+1}{x^2} = 10}$; $\boxed{x+1 = 10x^2}$; $\boxed{10x^2 - x - 1 = 0}$

; $\boxed{x - \left(\frac{1+\sqrt{41}}{20}\right) = 0 \text{ or } x - \left(\frac{1-\sqrt{41}}{20}\right) = 0}$; $\boxed{x - \left(\frac{1+6.4}{20}\right) = 0 \text{ or } x - \left(\frac{1-6.4}{20}\right) = 0}$; $\boxed{x - 0.37 = 0 \text{ or } x + 0.27 = 0}$

; $\boxed{\mathbf{x = 0.37}}$ or $\boxed{x = -0.27}$

Check: 1. Substitute $x = 0.37$ into the original equation. Then, $log_{10}(0.37+1) - log_{10} 0.37^2 \stackrel{?}{=} 1$

; $log_{10} 1.37 - log_{10} 0.1369 \stackrel{?}{=} 1$; $0.137 - (-0.863) \stackrel{?}{=} 1$; $0.137 + 0.863 \stackrel{?}{=} 1$; $1 = 1$. Therefore, $x = 0.37$ is the solution.

2. Substitute $x = -0.27$ into the original equation: Then, $log_{10}(-0.27+1) - log_{10}(-0.27)^2 \stackrel{?}{=} 1$

; $log_{10} 0.73 - 2 log_{10}(-0.27) \stackrel{?}{=} 1$. The logarithm of negative numbers is not defined. Therefore, $x = -0.27$ is not a solution.

Example 6.4-64

$\boxed{log_3(x+1) + log_3 27 = 2}$; $\boxed{log_3 27(x+1) = 2}$; $\boxed{3^{log_3 27(x+1)} = 3^2}$; $\boxed{27(x+1) = 9}$; $\boxed{27x + 27 = 9}$

; $\boxed{27x = 9 - 27}$; $\boxed{27x = -18}$; $\boxed{x = -\frac{18}{27}}$; $\boxed{\mathbf{x = -0.6667}}$

Check: $\ln(0.3130+2) - \ln 0.3130 \stackrel{?}{=} 2$; $0.8385 - (-1.1615) \stackrel{?}{=} 2$; $0.8385 + 1.1615 \stackrel{?}{=} 2$; $2 = 2$

Example 6.4-65

$\boxed{log_2(x+10) = log_2(x^2 - 2)}$; $\boxed{2^{log_2(x+10)} = 2^{log_2(x^2-2)}}$; $\boxed{x + 10 = x^2 - 2}$; $\boxed{x^2 - x - 10 - 2 = 0}$

; $\boxed{x^2 - x - 12 = 0}$; $\boxed{x - \left(\frac{1+\sqrt{49}}{2}\right) = 0 \text{ or } x - \left(\frac{1-\sqrt{49}}{2}\right) = 0}$; $\boxed{x - \left(\frac{1+7}{2}\right) = 0 \text{ or } x - \left(\frac{1-7}{2}\right) = 0}$

; $\boxed{x - 4 = 0 \text{ or } x + 3 = 0}$; $\boxed{\mathbf{x = 4}}$ or $\boxed{\mathbf{x = -3}}$

Check: 1. Substitute $x=4$ into the original equation. Then, $log_2(4+10)\overset{?}{=}log_2(4^2-2)$

; $log_2 14\overset{?}{=}log_2(16-2)$; $log_2 14 = log_2 14$. Therefore, $x=4$ is the first solution.

2. Substitute $x=-3$ into the original equation: Then, $log_2(-3+10)\overset{?}{=}log_2\left[(-3)^2-2\right]$

; $log_2 7\overset{?}{=}log_2(9-2)$; $log_2 7 = log_2 7$. Therefore, $x=-3$ is the second solution.

Example 6.4-66

$\boxed{log_{0.03}(x+5)=log_{0.03}3}$; $\boxed{0.03^{log_{0.03}(x+5)}=0.03^{log_{0.03}3}}$; $\boxed{x+5=3}$; $\boxed{x=3-5}$; $\boxed{\mathbf{x=-2}}$

Check: $log_{0.03}(-2+5)\overset{?}{=}log_{0.03}3$; $log_{0.03}3=log_{0.03}3$

Example 6.4-67

$\boxed{ln(10-x)=ln(x+2)}$; $\boxed{e^{ln(10-x)}=e^{ln(x+2)}}$; $\boxed{10-x=x+2}$; $\boxed{-x-x=2-10}$; $\boxed{-2x=-8}$; $\boxed{x=\frac{-8}{-2}}$; $\boxed{\mathbf{x=4}}$

Check: $ln(10-4)\overset{?}{=}ln(4+2)$; $ln6=ln6$; $1.7918=1.7918$

Example 6.4-68

$\boxed{ln(x+1)-lnx=lnx}$; $\boxed{ln\frac{x+1}{x}=lnx}$; $\boxed{e^{ln\frac{x+1}{x}}=e^{lnx}}$; $\boxed{\frac{x+1}{x}=x}$; $\boxed{x+1=x^2}$; $\boxed{x^2-x-1=0}$

; $\boxed{x-\left(\frac{1+\sqrt{5}}{2}\right)=0 \; or \; x-\left(\frac{1-\sqrt{5}}{2}\right)=0}$; $\boxed{x-\left(\frac{1+2.24}{2}\right)=0 \; or \; x-\left(\frac{1-2.24}{2}\right)=0}$; $\boxed{x-\frac{3.24}{2}=0 \; or \; x+\frac{1.24}{2}=0}$

; $\boxed{x-1.62=0 \; or \; x+0.62=0}$; $\boxed{\mathbf{x=1.62}}$ or $\boxed{x=-0.62}$

Check: 1. Substitute $x=1.62$ into the original equation. Then, $ln(1.62+1)-ln1.62\overset{?}{=}ln1.62$

; $ln2.62-ln1.62\overset{?}{=}ln1.62$; $ln\frac{2.62}{1.62}\overset{?}{=}ln1.62$; $ln1.62=ln1.62$. Therefore, $x=1.62$ is the solution.

2. Substitute $x=-0.62$ into the original equation: Then, $ln(-0.62+1)-ln(-0.62)\overset{?}{=}ln(-0.62)$

; $ln0.38-ln(-0.62)\overset{?}{=}ln(-0.62)$. The ln of negative numbers is not defined. Therefore, $x=-0.62$ is not a solution.

Example 6.4-69

$\boxed{log_4 x+log_4 5=5}$; $\boxed{log_4 5x=5}$; $\boxed{4^{log_4 5x}=4^5}$; $\boxed{5x=1024}$; $\boxed{x=\frac{1024}{5}}$; $\boxed{\mathbf{x=204.8}}$

Check: $log_4 204.8+log_4 5\overset{?}{=}5$; $\frac{log_{10}204.8}{log_{10}4}+\frac{log_{10}5}{log_{10}4}\overset{?}{=}5$; $\frac{2.3113}{0.6021}+\frac{0.6989}{0.6021}\overset{?}{=}5$; $3.8+1.2\overset{?}{=}5$; $5=5$

Note that we can also check the answer in the following way:

Check: $log_4 204.8+log_4 5\overset{?}{=}5$; $log_4(5\times 204.8)\overset{?}{=}5$; $log_4 1024\overset{?}{=}5$; $\frac{log_{10}1024}{log_{10}4}\overset{?}{=}5$; $\frac{3.0103}{0.6021}\overset{?}{=}5$; $5=5$

Example 6.4-70

$\boxed{\ln x - \ln x^3 = \ln 25}$; $\boxed{\ln \frac{x}{x^3} = \ln 25}$; $\boxed{\ln \frac{\not x}{\not x^3} = \ln 25}$; $\boxed{\ln \frac{1}{x^2} = \ln 25}$; $\boxed{e^{\ln \frac{1}{x^2}} = e^{\ln 25}}$; $\boxed{\frac{1}{x^2} = 25}$; $\boxed{25x^2 = 1}$

; $\boxed{x^2 = \frac{1}{25}}$; $\boxed{x^2 = \pm\sqrt{\frac{1}{25}}}$; $\boxed{x = \pm\sqrt{\frac{1}{5^2}}}$; $\boxed{x = \pm\frac{1}{5}}$; $\boxed{\mathbf{x = 0.2}}$ and $\boxed{x = -0.2}$

Check: 1. $\ln 0.2 - \ln 0.2^3 \stackrel{?}{=} \ln 25$; $\ln 0.2 - \ln 0.008 \stackrel{?}{=} \ln 25$; $\ln \frac{0.2}{0.008} \stackrel{?}{=} \ln 25$; $\ln 25 = \ln 25$

2. $\ln(-0.2) - \ln(-0.2)^3 \stackrel{?}{=} \ln 25$. The $\ln$ of negative numbers is not defined. Therefore, $x = -0.2$ is not a solution.

Example 6.4-71

$\boxed{\log_3(x+2) + \log_3 4 = \log_3 50}$; $\boxed{\log_3 4(x+2) = \log_3 50}$; $\boxed{\log_3 4(x+2) = \frac{\log_{10} 50}{\log_{10} 3}}$; $\boxed{\log_3 4(x+2) = \frac{1.6989}{0.4771}}$

; $\boxed{\log_3 4(x+2) = 3.5609}$; $\boxed{3^{\log_3 4(x+2)} = 3^{3.5609}}$; $\boxed{4(x+2) = 50}$; $\boxed{4x + 8 = 50}$; $\boxed{4x = 50 - 8}$; $\boxed{4x = 42}$

; $\boxed{x = \frac{42}{4}}$; $\boxed{\mathbf{x = 10.5}}$

Check: $\log_3(10.5+2) + \log_3 4 \stackrel{?}{=} \log_3 50$; $\log_3 12.5 + \log_3 4 \stackrel{?}{=} \log_3 50$; $\log_3 4 \times 12.5 \stackrel{?}{=} \log_3 50$; $\log_3 50 = \log_3 50$

Example 6.4-72

$\boxed{\log(x-1) + 2\log x = \log x}$; $\boxed{\log(x-1) + \log x^2 = \log x}$; $\boxed{\log x^2(x-1) = \log x}$; $\boxed{10^{\log x^2(x-1)} = 10^{\log x}}$

; $\boxed{x^2(x-1) = x}$; $\boxed{x^3 - x^2 = x}$; $\boxed{x^3 - x^2 - x = 0}$; $\boxed{x(x^2 - x - 1) = 0}$; $\boxed{x = 0}$ and $\boxed{x^2 - x - 1 = 0}$

; $\boxed{x^2 - x - 1 = 0}$; $\boxed{x - \left(\frac{1+\sqrt{5}}{2}\right) = 0 \; or \; x - \left(\frac{1-\sqrt{5}}{2}\right) = 0}$; $\boxed{x - \left(\frac{1+2.236}{2}\right) = 0 \; or \; x - \left(\frac{1-2.236}{2}\right) = 0}$

; $\boxed{x - 1.618 = 0 \; or \; x + 0.618 = 0}$; $\boxed{\mathbf{x = 1.618}}$ or $\boxed{x = -0.618}$

Check: 1. Substitute $x = 0$ into the original equation. Then, $\log(0-1) + 2\log 0 \stackrel{?}{=} \log 0$. The $\ln$ of zero and negative numbers is not defined. Thus, $x = 0$ is not a solution.

2. Substitute $x = 1.618$ into the original equation. Then, $\log(1.618-1) + 2\log 1.618 \stackrel{?}{=} \log 1.618$; $\log 0.618 + \log 1.618^2 \stackrel{?}{=} \log 1.618$; $\log 0.618 + \log 2.618 \stackrel{?}{=} \log 1.618$; $\log 0.618 \times 2.618 \stackrel{?}{=} \log 1.618$; $\log 1.618 = \log 1.618$. Thus, $x = 1.618$ is the solution.

3. Substitute $x = -0.618$ into the original equation: Then, $\log(-0.618-1) + 2\log(-0.618) \stackrel{?}{=}$; $\log(-0.618)$. Since the $\ln$ of negative numbers is not defined therefore, $x = -0.618$ is not a solution.

Example 6.4-73

$\boxed{ln(3x+1)-ln(x+3)=lne}$; $\boxed{ln\frac{3x+1}{x+3}=1}$; $\boxed{e^{ln\frac{3x+1}{x+3}}=e^{1}}$; $\boxed{\frac{3x+1}{x+3}=2.718}$; $\boxed{3x+1=2.718(x+3)}$

; $\boxed{3x+1=2.718x+8.154}$; $\boxed{3x-2.718x=8.154-1}$; $\boxed{0.282x=7.154}$; $\boxed{x=\frac{7.154}{0.282}}$; $\boxed{\mathbf{x=25.4}}$

Check: $ln(3\times 25.4+1)-ln(25.4+3)\overset{?}{=}lne$; $ln77.2-ln28.4\overset{?}{=}1$; $4.35-3.35\overset{?}{=}1$; $1=1$

Example 6.4-74

$\boxed{log_6(8-x)=log_6(x+1)}$; $\boxed{6^{log_6(8-x)}=6^{log_6(x+1)}}$; $\boxed{8-x=x+1}$; $\boxed{8-1=x+x}$; $\boxed{2x=7}$; $\boxed{x=\frac{7}{2}}$; $\boxed{\mathbf{x=3.5}}$

Check: $log_6(8-3.5)\overset{?}{=}log_6(3.5+1)$; $log_6 4.5=log_6 4.5$

Example 6.4-75

$\boxed{lnx+ln5=\ln 8}$; $\boxed{ln5x=ln8}$; $\boxed{e^{ln5x}=e^{ln8}}$; $\boxed{5x=8}$; $\boxed{x=\frac{8}{5}}$; $\boxed{\mathbf{x=1.6}}$

Check: $ln1.6+ln15\overset{?}{=}ln8$; $ln5\times 1.6\overset{?}{=}ln8$; $ln8=ln8$

Example 6.4-76

$\boxed{log_2 6-log_2 x=log_2 4}$; $\boxed{log_2\frac{6}{x}=log_2 4}$; $\boxed{2^{log_2\frac{6}{x}}=2^{log_2 4}}$; $\boxed{\frac{6}{x}=4}$; $\boxed{4x=6}$; $\boxed{x=\frac{\overset{3}{\cancel{6}}}{\underset{2}{\cancel{4}}}}$; $\boxed{x=\frac{3}{2}}$; $\boxed{\mathbf{x=1.5}}$

Check: $log_2 6-log_2 1.5\overset{?}{=}log_2 4$; $log_2\frac{6}{1.5}\overset{?}{=}log_2 4$; $log_2 4=log_2 4$

Example 6.4-77

$\boxed{log_2 5=lnx}$; $\boxed{\frac{log_{10}5}{log_{10}2}=lnx}$; $\boxed{\frac{0.6989}{0.3010}=lnx}$; $\boxed{2.3219=lnx}$; $\boxed{e^{2.3219}=e^{lnx}}$; $\boxed{e^{2.3219}=x}$; $\boxed{\mathbf{x=10.195}}$

Check: $log_2 5\overset{?}{=}ln10.195$; $\frac{log_{10}5}{log_{10}2}\overset{?}{=}ln10.195$; $\frac{0.6989}{0.3010}\overset{?}{=}2.3219$; $2.3219=2.3219$

Example 6.4-78

$\boxed{log_3 x=log_{0.4}7}$; $\boxed{log_3 x=\frac{log_{10}7}{log_{10}0.4}}$; $\boxed{log_3 x=\frac{0.8451}{-0.3979}}$; $\boxed{log_3 x=-2.1239}$; $\boxed{3^{log_3 x}=3^{-2.1239}}$

; $\boxed{x=3^{-2.1239}}$; $\boxed{x=\frac{1}{3^{2.1239}}}$; $\boxed{x=\frac{1}{10.3124}}$; $\boxed{\mathbf{x=0.0969}}$

Check: $log_3 0.0969\overset{?}{=}log_{0.4}7$; $\frac{log_{10}0.0969}{log_{10}3}\overset{?}{=}\frac{log_{10}7}{log_{10}0.4}$; $\frac{-1.0137}{0.4771}\overset{?}{=}\frac{0.8451}{-0.3979}$; $-2.124=-2.124$

Example 6.4-79

$\boxed{log_5 x+log_5 6=log_3 81-log_3 9}$; $\boxed{\log_5 6x=\log_3\frac{\overset{9}{\cancel{81}}}{\cancel{9}}}$; $\boxed{log_5 6x=log_3 9}$; $\boxed{log_5 6x=log_3 3^2}$

; $\boxed{log_5 6x=2log_3 3}$; $\boxed{log_5 6x=2}$; $\boxed{5^{log_5 6x}=5^2}$; $\boxed{6x=25}$; $\boxed{x=\frac{25}{6}}$; $\boxed{\mathbf{x=4.1667}}$

Check: $log_5 4.1667 + log_5 6 \stackrel{?}{=} log_3 81 - log_3 9$; $log_5 6 \times 4.1667 \stackrel{?}{=} log_3 \frac{81}{9}$; $log_5 25 \stackrel{?}{=} log_3 9$; $log_5 5^2 \stackrel{?}{=} log_3 3^2$

; $2 log_5 5 \stackrel{?}{=} 2 log_3 3$; $2 \times 1 = 2 \times 1$; $2 = 2$

Example 6.4-80

$\boxed{log_5(x-1) - log_5 25 = log_5 625}$; $\boxed{log_5(x-1) - log_5 5^2 = log_5 5^4}$; $\boxed{log_5(x-1) - 2 = 4}$; $\boxed{log_5(x-1) = 4 + 2}$

; $\boxed{log_5(x-1) = 6}$; $\boxed{5^{log_5(x-1)} = 5^6}$; $\boxed{x - 1 = 15625}$; $\boxed{\mathbf{x = 15626}}$

Check: $log_5(15626-1) - log_5 25 \stackrel{?}{=} log_5 625$; $log_5 5^6 - log_5 5^2 \stackrel{?}{=} log_5 5^4$; $log_5 \frac{5^6}{5^2} \stackrel{?}{=} log_5 5^4$

; $log_5 5^6 \cdot 5^2 \stackrel{?}{=} log_5 5^4$; $log_5 5^{6-2} \stackrel{?}{=} log_5 5^4$; $log_5 5^4 = log_5 5^4$; $4 = 4$

Practice Problems - Solving One Variable Logarithmic Equations (More Difficult Cases)

Section 6.4 Case II Practice Problems - Solve the following equations using the logarithmic properties:

1. $log_3 6 - log_3 u = log_3 27$
2. $log_5 x = log_{0.5} 9$
3. $log_5(x+2) - log_5 4 = log_5 50$
4. $log(x+2) + log x = log 10$
5. $ln(x+2) = ln(5-x)$
6. $log_3(x+1) + log_3 x = log_3 9$
7. $log(x+2) - log(x+3) = log 10$
8. $log_8(x+2) = log_8(10-x)$
9. $log_2(x+2) + log_5 5 = log_{0.3} 10$
10. $log_2(x+8) - log_2 x = log_2 16$

6.5 Use of Logarithms in Solving Math Operations and Algebraic Expressions

One of the advantages of working with logarithms is that numerical expressions that are being added, subtracted, multiplied, or divided by one another can be solved using the logarithmic laws without the use of hand calculators, as shown in Case I. In addition, using the laws of logarithm, logarithmic expressions can either be expanded to simpler logarithmic terms (Case II) or be combined to a single term (Case III).

Case I Solving Numerical Expressions Using Logarithms

In section 6.2 Case IV we learned how to find the antilogarithm of numbers. In this section we use the logarithmic laws and the approach we learned in computing antilogarithms to solve for numerical expressions such as radicals, decimals, or exponents that are either added, subtracted, multiplied, or divided by one another. The following show the steps as to how math operations involving numerical expressions are solved using the laws of logarithm.

Step 1 a. Let a variable such as x be equal to the given numerical expression.

b. Multiply both sides of the equation by a logarithm to any base. (To simplify the process select log to the base 10.)

Step 2 Apply the laws of logarithm to reduce the equation to the form of $log_{10} x = b$.

Step 3 Solve for the variable x by taking the antilogarithm of the number.

Note: Since the answers to logarithm numbers are rounded to the first four digits therefore, the computation of numerical numbers, in many instances, result in approximate values.

Examples with Steps

The following examples show the steps as to how numerical expressions are solved using the logarithmic properties:

Example 6.5-1

$$\boxed{(0.00425)(0.00035)} =$$

Solution:

Step 1 $\boxed{Let\ x = (0.00425)(0.00035)}$ then; $\boxed{log\, x = log(0.00425)(0.00035)}$

Step 2 $\boxed{log\, x = log(0.00425)(0.00035)}$; $\boxed{log\, x = log\, 0.00425 + log\, 0.00035}$

; $\boxed{log\, x = log\, 4.25 \times 10^{-3} + log\, 3.5 \times 10^{-4}}$; $\boxed{log\, x = log\, 4.25 + log\, 10^{-3} + log\, 3.5 + log\, 10^{-4}}$

; $\boxed{log\, x = log\, 4.25 - 3\, log\, 10 + log\, 3.5 - 4\, log\, 10}$; $\boxed{log\, x = 0.6284 - 3 + 0.5441 - 4}$

; $\boxed{log\, x = -5.8275}$

Step 3 $\boxed{log\, x = -5.8275}$; $\boxed{10^{log\, x} = 10^{-5.8275}}$; $\boxed{x = \frac{1}{10^{5.8275}}}$; $\boxed{\mathbf{x = 1.4876 \times 10^{-6}}}$

Example 6.5-2

$\sqrt{0.319} =$

Solution:

Step 1 $Let\ x = \sqrt{0.319}$ then; $log_{10} x = log_{10} \sqrt{0.319}$

Step 2 $log_{10} x = log_{10} \sqrt{0.319}$; $log_{10} x = log_{10} 0.319^{\frac{1}{2}}$; $log_{10} x = \frac{1}{2} log_{10} 0.319$

; $log_{10} x = \frac{1}{2} log_{10} 3.19 \times 10^{-1}$; $log_{10} x = \frac{1}{2}\left(log_{10} 3.19 + log_{10} 10^{-1}\right)$

; $log_{10} x = \frac{1}{2}(0.5038 - 1)$; $log_{10} x = \frac{-0.4962}{2}$; $log_{10} x = -0.2481$

Step 3 $log_{10} x = -0.2481$; $10^{log_{10} x} = 10^{-0.2481}$; $x = \frac{1}{10^{0.2481}}$; $x = \frac{1}{1.7705}$; $\mathbf{x = 0.5648}$

Example 6.5-3

$66^{2.5} =$

Solution:

Step 1 $Let\ x = 66^{2.5}$ then; $log_{10} x = log_{10} 66^{2.5}$

Step 2 $log_{10} x = log_{10} 66^{2.5}$; $log_{10} x = 2.5 log_{10} 66$; $log_{10} x = 2.5 log_{10} 6.6 \times 10^{1}$

; $log_{10} x = 2.5\left[log_{10} 6.6 + log_{10} 10^{1}\right]$; $log_{10} x = 2.5[0.8195 + 1]$; $log_{10} x = 2.5 \times 1.8195$

; $\log_{10} x = 4.5488$

Step 3 $\log_{10} x = 4.5488$; $10^{\log_{10} x} = 10^{4.5488}$; $x = 10^{4.5488}$; $\mathbf{x = 3.5383 \times 10^{4}}$

Example 6.5-4

$325^{\frac{2}{3}} =$

Solution:

Step 1 $Let\ x = 325^{\frac{2}{3}}$ then; $log_{10} x = log_{10} 325^{\frac{2}{3}}$

Step 2 $log_{10} x = log_{10} 325^{\frac{2}{3}}$; $log_{10} x = \frac{2}{3} log_{10} 325$; $log_{10} x = \frac{2}{3} log_{10} 3.25 \times 10^{2}$

; $log_{10} x = \frac{2}{3}\left[log_{10} 3.25 + log_{10} 10^{2}\right]$; $log_{10} x = \frac{2}{3}[0.5119 + 2]$; $log_{10} x = 1.6746$

Step 3 $log_{10} x = 1.6746$; $10^{log_{10} x} = 10^{1.6746}$; $x = 10^{1.6746}$; $\mathbf{x = 47.2715}$

Example 6.5-5

$\left(\sqrt[3]{125}\right)\left(\sqrt[4]{2^4}\cdot 3\right) =$

Solution:

Step 1

$\boxed{Let\ x = \left(\sqrt[3]{125}\right)\left(\sqrt[4]{2^4}\cdot 3\right)}$ then; $\boxed{log_{10}\, x = log_{10}\left(\sqrt[3]{125}\right)\left(\sqrt[4]{2^4}\cdot 3\right)}$

Step 2

$\boxed{log_{10}\, x = log_{10}\left(\sqrt[3]{125}\right)\left(\sqrt[4]{2^4}\cdot 3\right)}$; $\boxed{log_{10}\, x = log_{10}\, 125^{\frac{1}{3}}\cdot 2\cdot 3^{\frac{1}{4}}}$

; $\boxed{log_{10}\, x = log_{10}\, 125^{\frac{1}{3}} + log_{10}\, 2 + log_{10}\, 3^{\frac{1}{4}}}$; $\boxed{log_{10}\, x = \frac{1}{3} log_{10}\, 125 + log_{10}\, 2 + \frac{1}{4} log_{10}\, 3}$

; $\boxed{log_{10}\, x = \frac{1}{3}\times 2.0969 + 0.3010 + \frac{1}{4}\times 0.47712}$; $\boxed{log_{10}\, x = 0.6989 + 0.3010 + 0.1193}$

; $\boxed{log_{10}\, x = 1.1192}$

Step 3

$\boxed{log_{10}\, x = 1.1192}$; $\boxed{10^{log_{10}\, x} = 10^{1.1192}}$; $\boxed{x = 10^{1.1192}}$; $\boxed{\mathbf{x = 13.1583}}$

Additional Examples - Solving Numerical Expressions Using Logarithms

Use the logarithmic laws to compute the following numerical expressions:

Example 6.5-6

$\boxed{(38.5)(0.0002)}$; $\boxed{Let\ x = (38.5)(0.0002)}$ then; $\boxed{log_{10}\, x = log_{10}(38.5)(0.0002)}$; $\boxed{log_{10}\, x = log_{10}\, 38.5 + log_{10}\, 0.0002}$

; $\boxed{log_{10}\, x = log_{10}\, 3.85\times 10^1 + log_{10}\, 2.0\times 10^{-4}}$; $\boxed{log_{10}\, x = log_{10}\, 3.85 + log_{10}\, 10^1 + log_{10}\, 2.0 + log_{10}\, 10^{-4}}$

; $\boxed{log_{10}\, x = log_{10}\, 3.85 + log_{10}\, 10 + log_{10}\, 2.0 - 4\, log_{10}\, 10}$; $\boxed{log_{10}\, x = 0.5855 + 1 + 0.3010 - 4}$; $\boxed{log_{10}\, x = -2.1135}$

; $\boxed{10^{log_{10}\, x} = 10^{-2.1135}}$; $\boxed{x = 10^{-2.1135}}$; $\boxed{x = \frac{1}{10^{2.1135}}}$; $\boxed{x = \frac{1}{129.867}}$; $\boxed{\mathbf{x = 0.0077}}$

As was stated earlier, these types of problems can be solved using *log* to any base. Let's solve the above problem using log_2.

$\boxed{(38.5)(0.0002)}$; $\boxed{Let\ x = (38.5)(0.0002)}$ then; $\boxed{log_2\, x = log_2(38.5)(0.0002)}$; $\boxed{log_2\, x = log_2\, 38.5 + log_2\, 0.0002}$

; $\boxed{log_2\, x = log_2\, 3.85\times 10^1 + log_2\, 2.0\times 10^{-4}}$; $\boxed{log_2\, x = log_2\, 3.85 + log_2\, 10^1 + log_2\, 2.0 + log_2\, 10^{-4}}$

; $\boxed{log_2\, x = log_2\, 3.85 + log_2\, 10 + log_2\, 2.0 - 4\, log_2\, 10}$; $\boxed{log_2\, x = \frac{log_{10}\, 3.85}{log_{10}\, 2} + \frac{log_{10}\, 10}{log_{10}\, 2} + \frac{log_{10}\, 2.0}{log_{10}\, 2} - 4\frac{log_{10}\, 10}{log_{10}\, 2}}$

; $\boxed{log_2\, x = \frac{log_{10}\, 3.85 + log_{10}\, 10 + log_{10}\, 2.0 - 4\, log_{10}\, 10}{log_{10}\, 2}}$; $\boxed{log_2\, x = \frac{0.5855 + 1 + 0.3010 - 4}{0.3010}}$; $\boxed{log_2\, x = -\frac{2.1135}{0.3010}}$

; $\boxed{log_2 x = -7.0216}$; $\boxed{2^{log_2 x} = 2^{-7.0216}}$; $\boxed{x = 2^{-7.0216}}$; $\boxed{x = \frac{1}{2^{7.0216}}}$; $\boxed{x = \frac{1}{129.931}}$; $\boxed{\mathbf{x = 0.0077}}$

Note that the final answer using either log_{10} or log_2 is the same - as it should be. To simplify the process, the remaining examples are solved using log_{10}. However, students are encouraged to practice solving the remaining problems using logarithm to bases such as log_3, log_8, and $log_{0.2}$. Your answers should be the same as the answers shown in examples 6.5-7 through 6.5-25.

Example 6.5-7

$\boxed{(30,000)(1,700)}$; $\boxed{Let\ x = (30,000)(1,700)}$ then; $\boxed{log x = log(30,000)(1,700)}$; $\boxed{log x = log 30,000 + log 1,700}$

; $\boxed{log x = log 3.0 \times 10^4 + log 1.7 \times 10^3}$; $\boxed{log x = log 3.0 + log 10^4 + log 1.7 + log 10^3}$

; $\boxed{log x = log 3.0 + 4 log 10 + log 1.7 + 3 log 10}$; $\boxed{log x = 0.4771 + 4 + 0.2304 + 3}$; $\boxed{log x = 7.7075}$

; $\boxed{10^{log x} = 10^{7.7075}}$; $\boxed{x = 10^{7.7075}}$; $\boxed{\mathbf{x = 5.0992 \times 10^7}}$

Example 6.5-8

$\boxed{(8,250)(0.025)}$; $\boxed{Let\ x = (8,250)(0.025)}$ then; $\boxed{log x = log(8,250)(0.025)}$; $\boxed{log x = log 8,250 + log 0.025}$

; $\boxed{log x = log 8.25 \times 10^3 + log 2.5 \times 10^{-2}}$; $\boxed{log x = log 8.25 + log 10^3 + log 2.5 + log 10^{-2}}$

; $\boxed{log x = log 8.25 + 3 log 10 + log 2.5 - 2 log 10}$; $\boxed{log x = 0.9165 + 3 + 0.3979 - 2}$; $\boxed{log x = 2.3144}$

; $\boxed{10^{log x} = 10^{2.3144}}$; $\boxed{x = 10^{2.3144}}$; $\boxed{\mathbf{x = 206.25}}$

Example 6.5-9

$\boxed{(850,000)(12.5)}$; $\boxed{Let\ x = (850,000)(12.5)}$ then; $\boxed{log x = log(850,000)(12.5)}$; $\boxed{log x = log 850,000 + log 12.5}$

; $\boxed{log x = log 8.5 \times 10^5 + log 1.25 \times 10^1}$; $\boxed{log x = log 8.5 + log 10^5 + log 1.25 + log 10}$

; $\boxed{log x = log 8.5 + 5 log 10 + log 1.25 + log 10}$; $\boxed{log x = 0.9294 + 5 + 0.0969 + 1}$; $\boxed{log x = 7.0263}$

; $\boxed{10^{log x} = 10^{7.0263}}$; $\boxed{x = 10^{7.0263}}$; $\boxed{\mathbf{x = 1.0624 \times 10^7}}$

Example 6.5-10

$\boxed{26^{0.05}}$; $\boxed{Let\ x = 26^{0.05}}$ then; $\boxed{log_{10} x = log_{10} 26^{0.05}}$; $\boxed{log_{10} x = 0.05 log_{10} 26}$; $\boxed{log_{10} x = 0.05 log_{10} 2.6 \times 10^1}$

; $\boxed{log_{10} x = 0.05\left[log_{10} 2.6 + log_{10} 10^1\right]}$; $\boxed{log_{10} x = 0.05[0.4149 + 1]}$; $\boxed{log_{10} x = 0.05 \times 1.4149}$; $\boxed{log_{10} x = 0.0707}$

; $\boxed{10^{log_{10} x} = 10^{0.0707}}$; $\boxed{x = 10^{0.0707}}$; $\boxed{\mathbf{x = 1.1768}}$

Example 6.5-11

$\boxed{\sqrt{32.5}} = \boxed{32.5^{\frac{1}{2}}}$; $\boxed{Let\ x = 32.5^{\frac{1}{2}}}$ then; $\boxed{log x = log 32.5^{\frac{1}{2}}}$; $\boxed{log x = \frac{1}{2} log 32.5}$; $\boxed{log x = \frac{1}{2} log 3.25 \times 10^1}$

; $\boxed{log\,x = \frac{1}{2}\left[log\,3.25 + log\,10^1\right]}$; $\boxed{log\,x = \frac{1}{2}[0.5119 + 1]}$; $\boxed{log\,x = \frac{1}{2} \times 1.5119}$; $\boxed{log\,x = 0.7559}$

; $\boxed{10^{log_{10} x} = 10^{0.7559}}$; $\boxed{x = 10^{0.7559}}$; $\boxed{\mathbf{x = 5.7}}$

Example 6.5-12

$\boxed{\sqrt[3]{2.45}} = \boxed{2.45^{\frac{1}{3}}}$; $\boxed{Let\ x = 2.45^{\frac{1}{3}}}$ then; $\boxed{log_{10}\,x = log_{10}\,2.45^{\frac{1}{3}}}$; $\boxed{log_{10}\,x = \frac{1}{3} log_{10}\,2.45}$; $\boxed{log_{10}\,x = \frac{1}{3} \times 0.3892}$

; $\boxed{log_{10}\,x = 0.12973}$; $\boxed{10^{log_{10} x} = 10^{0.12973}}$; $\boxed{x = 10^{0.12973}}$; $\boxed{\mathbf{x = 1.3481}}$

Example 6.5-13

$\boxed{\sqrt[4]{32.5}} = \boxed{32.5^{\frac{1}{4}}}$; $\boxed{Let\ x = 32.5^{\frac{1}{4}}}$ then; $\boxed{\log x = \log 32.5^{\frac{1}{4}}}$; $\boxed{\log x = \frac{1}{4} \log 32.5}$; $\boxed{\log x = \frac{1}{4} \log 3.25 \times 10^1}$

; $\boxed{\log x = \frac{1}{4}\left[\log 3.25 + \log 10^1\right]}$; $\boxed{\log x = \frac{1}{4}[0.5119 + 1]}$; $\boxed{\log x = \frac{1}{4} \times 1.5119}$; $\boxed{\log x = 0.3779}$

; $\boxed{10^{\log_{10} x} = 10^{0.3779}}$; $\boxed{x = 10^{0.3779}}$; $\boxed{\mathbf{x = 2.3873}}$

Example 6.5-14

$\boxed{4.03^{1.2}}$; $\boxed{Let\ x = 4.03^{1.2}}$ then; $\boxed{log_{10}\,x = log_{10}\,4.03^{1.2}}$; $\boxed{log_{10}\,x = 1.2\,log_{10}\,4.03}$; $\boxed{log_{10}\,x = 1.2 \times 0.6053}$

; $\boxed{log_{10}\,x = 0.7264}$; $\boxed{10^{log_{10} x} = 10^{0.7264}}$; $\boxed{x = 10^{0.7264}}$; $\boxed{\mathbf{x = 5.3259}}$

Example 6.5-15

$\boxed{\sqrt[3]{22.8^{-\frac{1}{2}}}} = \boxed{\left(22.8^{-\frac{1}{2}}\right)^{\frac{1}{3}}} = \boxed{22.8^{-\frac{1}{2} \times \frac{1}{3}}} = \boxed{22.8^{-\frac{1}{6}}}$; $\boxed{Let\ x = 22.8^{-\frac{1}{6}}}$ then; $\boxed{log_{10}\,x = log\,22.8^{-\frac{1}{6}}}$

; $\boxed{log_{10}\,x = -\frac{1}{6} log_{10}\,22.8}$; $\boxed{log_{10}\,x = -\frac{1}{6} \times 1.3579}$; $\boxed{log_{10}\,x = -0.2263}$; $\boxed{10^{log_{10} x} = 10^{-0.2263}}$; $\boxed{x = 10^{-0.2263}}$

; $\boxed{x = \frac{1}{10^{0.2263}}}$; $\boxed{x = \frac{1}{1.6838}}$; $\boxed{\mathbf{x = 0.5938}}$

Example 6.5-16

$\boxed{55^{-\frac{1}{3}}}$; $\boxed{Let\ x = 55^{-\frac{1}{3}}}$ then; $\boxed{log_{10}\,x = log_{10}\,55^{-\frac{1}{3}}}$; $\boxed{log_{10}\,x = -\frac{1}{3} log_{10}\,55}$; $\boxed{log_{10}\,x = -\frac{1}{3} log_{10}\,5.5 \times 10^1}$

; $\boxed{log_{10}\,x = -\frac{1}{3}\left[log_{10}\,5.5 + log_{10}\,10^1\right]}$; $\boxed{log_{10}\,x = -\frac{1}{3}[0.7404 + 1]}$; $\boxed{log_{10}\,x = -\frac{1}{3} \times 1.7404}$; $\boxed{log_{10}\,x = -0.5801}$

; $\boxed{10^{log_{10} x} = 10^{-0.5801}}$; $\boxed{x = 10^{-0.5801}}$; $\boxed{x = \frac{1}{10^{0.5801}}}$; $\boxed{x = \frac{1}{3.8027}}$; $\boxed{\mathbf{x = 0.2629}}$

Example 6.5-17

$\boxed{0.6^{-0.02}}$; $\boxed{Let\ x = 0.6^{-0.02}}$ then; $\boxed{log\,x = log\,0.6^{-0.02}}$; $\boxed{log\,x = -0.02\,log\,0.6}$; $\boxed{log\,x = -0.02\times -0.2218}$

; $\boxed{log\,x = 0.0044}$; $\boxed{10^{log_{10}x} = 10^{0.0044}}$; $\boxed{x = 10^{0.0044}}$; $\boxed{\mathbf{x = 1.0102}}$

Example 6.5-18

$\boxed{\left(\sqrt[5]{3^5\cdot 2}\right)\left(\sqrt[3]{2^3\cdot 3}\right)} = \boxed{\left(3^5\cdot 2\right)^{\frac{1}{5}}\left(2^3\cdot 3\right)^{\frac{1}{3}}} = \boxed{3^{\not 5\times\frac{1}{\not 5}}\cdot 2^{\frac{1}{5}}\cdot 2^{\not 3\times\frac{1}{\not 3}}\cdot 3^{\frac{1}{3}}} = \boxed{3^1\cdot 2^{\frac{1}{5}}\cdot 2^1\cdot 3^{\frac{1}{3}}} = \boxed{3\cdot 2^{\frac{1}{5}}\cdot 2\cdot 3^{\frac{1}{3}}}$

$= \boxed{6\cdot 2^{\frac{1}{5}}\cdot 3^{\frac{1}{3}}}$; $\boxed{Let\ x = 6\cdot 2^{\frac{1}{5}}\cdot 3^{\frac{1}{3}}}$ then; $\boxed{log\,x = log\,6\cdot 2^{\frac{1}{5}}\cdot 3^{\frac{1}{3}}}$; $\boxed{log\,x = log\,6 + log\,2^{\frac{1}{5}} + log\,3^{\frac{1}{3}}}$

; $\boxed{log\,x = log\,6 + \frac{1}{5}log\,2 + \frac{1}{3}log\,3}$; $\boxed{log\,x = 0.7782 + \frac{1}{5}\times 0.3010 + \frac{1}{3}\times 0.4771}$; $\boxed{log\,x = 0.7782 + 0.0602 + 0.1590}$

; $\boxed{log\,x = 0.9974}$; $\boxed{10^{log_{10}x} = 10^{0.9974}}$; $\boxed{x = 10^{0.9974}}$; $\boxed{\mathbf{x = 9.9403}}$

Example 6.5-19

$\boxed{\frac{0.025}{5650}}$; $\boxed{Let\ x = \frac{0.025}{5650}}$ then; $\boxed{log_{10}\,x = log_{10}\frac{0.025}{5650}}$; $\boxed{log_{10}\,x = log_{10}\,0.025 - log_{10}\,5650}$

; $\boxed{log_{10}\,x = log_{10}\,2.5\times 10^{-2} - log_{10}\,5.65\times 10^3}$; $\boxed{log_{10}\,x = log_{10}\,2.5 + log_{10}\,10^{-2} - \left(log_{10}\,5.65 + log_{10}\,10^3\right)}$

; $\boxed{log_{10}\,x = log_{10}\,2.5 - 2\,log_{10}\,10 - log_{10}\,5.65 - 3\,log_{10}\,10}$; $\boxed{log_{10}\,x = 0.3979 - 2 - 0.7520 - 3}$; $\boxed{log_{10}\,x = -5.3541}$

; $\boxed{10^{log_{10}x} = 10^{-5.3541}}$; $\boxed{x = 10^{-5.3541}}$; $\boxed{x = \frac{1}{10^{5.3541}}}$; $\boxed{x = \frac{1}{2.2599\times 10^5}}$; $\boxed{\mathbf{x = 4.4248\times 10^{-6}}}$

Example 6.5-20

$\boxed{\frac{555}{0.285^{-\frac{2}{3}}}}$; $\boxed{Let\ x = \frac{555}{0.285^{-\frac{2}{3}}}}$ then; $\boxed{log_{10}\,x = log_{10}\frac{555}{0.285^{-\frac{2}{3}}}}$; $\boxed{log_{10}\,x = log_{10}\,555 - log_{10}\,0.285^{-\frac{2}{3}}}$

; $\boxed{\log_{10} x = \log_{10} 5.55\times 10^2 + \frac{2}{3}\log_{10} 0.285}$; $\boxed{\log_{10} x = \log_{10} 5.55\times 10^2 + \frac{2}{3}\log_{10} 2.85\times 10^{-1}}$

; $\boxed{\log_{10} x = \left(\log_{10} 5.55 + \log_{10} 10^2\right) + \frac{2}{3}\left(\log_{10} 2.85 - \log_{10} 10^1\right)}$; $\boxed{\log_{10} x = (0.7443 + 2) + \frac{2}{3}(0.4548 - 1)}$

; $\boxed{\log_{10} x = 2.7443 - 0.3635}$; $\boxed{\log_{10} x = 2.3809}$; $\boxed{10^{\log_{10} x} = 10^{2.3809}}$; $\boxed{x = 10^{2.3809}}$; $\boxed{\mathbf{x = 240.380}}$

Example 6.5-21

$\boxed{\frac{0.00028}{\sqrt[3]{2^3\,3}}} = \boxed{\frac{0.00028}{2\cdot 3^{\frac{1}{3}}}}$; $\boxed{Let\ x = \frac{0.00028}{2\cdot 3^{\frac{1}{3}}}}$ then; $\boxed{log_{10}\,x = log_{10}\frac{0.00028}{2\cdot 3^{\frac{1}{3}}}}$; $\boxed{log_{10}\,x = log_{10}\,0.00028 - log_{10}\,2\cdot 3^{\frac{1}{3}}}$

; $\boxed{\log_{10} x = \log_{10} 2.8 \times 10^{-4} - \left(\log_{10} 2 + \log_{10} 3^{\frac{1}{3}}\right)}$; $\boxed{\log_{10} x = \log_{10} 2.8 + \log_{10} 10^{-4} - \left(\log_{10} 2 + \frac{1}{3}\log_{10} 3\right)}$

; $\boxed{\log_{10} x = \log_{10} 2.8 - 4\log_{10} 10 - \log_{10} 2 - \frac{1}{3}\log_{10} 3}$; $\boxed{\log_{10} x = 0.4471 - 4 - 0.3010 - \frac{1}{3} \times 0.4771}$

; $\boxed{\log_{10} x = -4.0129}$; $\boxed{10^{\log_{10} x} = 10^{-4.0129}}$; $\boxed{x = 10^{-4.0129}}$; $\boxed{x = \frac{1}{10^{4.0129}}}$; $\boxed{\mathbf{x = 9.7073 \times 10^{-5}}}$

Example 6.5-22

$\boxed{\frac{\sqrt[3]{25}}{\sqrt[5]{125}}} = \boxed{\frac{25^{\frac{1}{3}}}{125^{\frac{1}{5}}}}$; $\boxed{Let\ x = \frac{25^{\frac{1}{3}}}{125^{\frac{1}{5}}}}$ then; $\boxed{\log_{10} x = \log_{10} \frac{25^{\frac{1}{3}}}{125^{\frac{1}{5}}}}$; $\boxed{\log_{10} x = \log_{10} 25^{\frac{1}{3}} - \log_{10} 125^{\frac{1}{5}}}$

; $\boxed{\log_{10} x = \frac{1}{3}\log_{10} 25 - \frac{1}{5}\log_{10} 125}$; $\boxed{\log_{10} x = \frac{1}{3}\log_{10} 2.5 \times 10^1 - \frac{1}{5}\log_{10} 1.25 \times 10^2}$

; $\boxed{\log_{10} x = \frac{1}{3}\left(\log_{10} 2.5 + \log_{10} 10^1\right) - \frac{1}{5}\left(\log_{10} 1.25 + \log_{10} 10^2\right)}$; $\boxed{\log_{10} x = \frac{1}{3}\left(\log_{10} 2.5 + 1\right) - \frac{1}{5}\left(\log_{10} 1.25 + 2\right)}$

; $\boxed{\log_{10} x = \frac{1}{3}(0.3979 + 1) - \frac{1}{5}(0.0969 + 2)}$; $\boxed{\log_{10} x = \frac{1.3979}{3} - \frac{2.0969}{5}}$; $\boxed{\log_{10} x = 0.4659 - 0.4194}$

; $\boxed{\log_{10} x = 0.0465}$; $\boxed{10^{\log_{10} x} = 10^{0.0465}}$; $\boxed{x = 10^{0.0465}}$; $\boxed{\mathbf{x = 1.1130}}$

Example 6.5-23

$\boxed{\frac{0.0005^{-1}}{5.23^{\frac{1}{3}}}}$; $\boxed{Let\ x = \frac{0.0005^{-1}}{5.23^{\frac{1}{3}}}}$ then ; $\boxed{\log_{10} x = \log_{10} \frac{0.0005^{-1}}{5.23^{\frac{1}{3}}}}$; $\boxed{\log_{10} x = \log_{10} 0.0005^{-1} - \log_{10} 5.23^{\frac{1}{3}}}$

; $\boxed{\log_{10} x = -\log_{10} 0.0005 - \frac{1}{3}\log_{10} 5.23}$; $\boxed{\log_{10} x = -\log_{10} 5.0 \times 10^{-4} - \frac{1}{3} \times 0.7185}$

; $\boxed{\log_{10} x = -\left(\log_{10} 5.0 + \log_{10} 10^{-4}\right) - 0.2395}$; $\boxed{\log_{10} x = -\left(\log_{10} 5.0 - 4\log_{10} 10\right) - 0.2395}$

; $\boxed{\log_{10} x = -(0.6990 - 4) - 0.2395}$; $\boxed{\log_{10} x = 3.0615}$; $\boxed{10^{\log_{10} x} = 10^{3.0615}}$; $\boxed{x = 10^{3.0615}}$; $\boxed{\mathbf{x = 1152.13}}$

Example 6.5-24

$\boxed{\frac{2.57^{-\frac{1}{2}} \cdot 1.25^{\frac{2}{3}}}{\sqrt[5]{243} \cdot \sqrt[3]{27}}} = \boxed{\frac{2.57^{-\frac{1}{2}} \cdot 1.25^{\frac{2}{3}}}{\sqrt[5]{3^5} \cdot \sqrt[3]{3^3}}} = \boxed{\frac{2.57^{-\frac{1}{2}} \cdot 1.25^{\frac{2}{3}}}{3 \cdot 3}} = \boxed{\frac{2.57^{-\frac{1}{2}} \cdot 1.25^{\frac{2}{3}}}{9}}$; $\boxed{Let\ x = \frac{2.57^{-\frac{1}{2}} \cdot 1.25^{\frac{2}{3}}}{9}}$ then,

; $\boxed{\log_{10} x = \log_{10} \frac{2.57^{-\frac{1}{2}} \cdot 1.25^{\frac{2}{3}}}{9}}$; $\boxed{\log_{10} x = \log_{10} 2.57^{-\frac{1}{2}} \cdot 1.25^{\frac{2}{3}} - \log_{10} 9}$

; $\boxed{log_{10}\, x = log_{10}\, 2.57^{-\frac{1}{2}} + log_{10}\, 1.25^{\frac{2}{3}} - log_{10}\, 9}$; $\boxed{log_{10}\, x = -\frac{1}{2} log_{10}\, 2.57 + \frac{2}{3} log_{10}\, 1.25 - 0.9542}$

; $\boxed{log_{10}\, x = -\frac{1}{2} \times 0.4099 + \frac{2}{3} \times 0.0969 - 0.9542}$; $\boxed{log_{10}\, x = -0.2049 + 0.0646 - 0.9542}$

; $\boxed{log_{10}\, x = -1.0945}$; $\boxed{10^{log_{10}\, x} = 10^{-1.0945}}$; $\boxed{x = 10^{-1.0945}}$; $\boxed{x = \frac{1}{10^{1.0945}}}$; $\boxed{x = \frac{1}{12.4308}}$; $\boxed{\mathbf{x = 0.0804}}$

Example 6.5-25

$\boxed{\frac{3.5^3 \cdot \sqrt[3]{2510}}{2.32^2}} = \boxed{\frac{3.5^3 \cdot 2510^{\frac{1}{3}}}{2.32^2}}$; $\boxed{Let\ x = \frac{3.5^3 \cdot 2510^{\frac{1}{3}}}{2.32^2}}$ then; $\boxed{log_{10}\, x = log_{10}\, \frac{3.5^3 \cdot 2510^{\frac{1}{3}}}{2.32^2}}$

; $\boxed{log_{10}\, x = log_{10}\, 3.5^3 \cdot 2510^{\frac{1}{3}} - log_{10}\, 2.32^2}$; $\boxed{log_{10}\, x = log_{10}\, 3.5^3 + log_{10}\, 2510^{\frac{1}{3}} - log_{10}\, 2.32^2}$

; $\boxed{log_{10}\, x = 3\, log_{10}\, 3.5 + \frac{1}{3} log_{10}\, 2510 - 2\, log_{10}\, 2.32}$; $\boxed{log_{10}\, x = 3 \times 0.5441 + \frac{1}{3} log_{10}\, 2.51 \times 10^3 - 2 \times 0.3655}$

; $\boxed{log_{10}\, x = 1.6323 + \frac{1}{3}\left(log_{10}\, 2.51 + log_{10}\, 10^3\right) - 0.7310}$; $\boxed{log_{10}\, x = 1.6323 + \frac{1}{3} \times 0.3997 + \frac{1}{3} \times 3 - 0.7310}$

; $\boxed{log_{10}\, x = 1.6323 + 0.1332 + 1 - 0.7310}$; $\boxed{log_{10}\, x = 2.0345}$; $\boxed{10^{log_{10}\, x} = 10^{2.0345}}$; $\boxed{x = 10^{2.0345}}$; $\boxed{\mathbf{x = 108.26}}$

Practice Problems - Solving Numerical Expressions Using Logarithms

Section 6.5 Case I Practice Problems - Use the properties of logarithm to solve each expression:

1. $(0.00025)(12{,}000{,}000) =$
2. $(8755)(0.000165) =$
3. $\sqrt[5]{0.35} =$
4. $\frac{3650}{2.25} =$
5. $\sqrt[5]{5.09^3} =$
6. $0.983^{5.6} =$
7. $\frac{0.00057^{0.05}}{5554^{0.002}} =$
8. $\frac{0.148^{2.5}}{33.5^{1.2}} =$
9. $\frac{2355^{\frac{1}{2}}}{0.235^2} =$
10. $\frac{28^{0.05}\sqrt[3]{2.3^2}}{(0.00008)\sqrt{3.05^3}} =$

Case II Expanding Logarithmic Expressions From a Single Term

Logarithmic expressions are expanded to single logarithmic terms using the following steps:

Step 1 Simplify the logarithmic expression by factoring the numerator and/or the denominator, if necessary.

Step 2 Apply the properties of logarithm and simplify further.

Examples with Steps

The following examples show the steps as to how logarithmic expressions are expanded:

Example 6.5-26

$$\boxed{log_{10}\frac{1}{\sqrt[3]{(t+2)^2}}} =$$

Solution:

Step 1 $\boxed{\textit{Not Applicable}}$

Step 2

$$\boxed{log_{10}\frac{1}{\sqrt[3]{(t+2)^2}}} = \boxed{log_{10}\frac{1}{(t+2)^{\frac{2}{3}}}} = \boxed{log_{10}1 - log_{10}(t+2)^{\frac{2}{3}}} = \boxed{0 - \frac{2}{3}log_{10}(t+2)}$$

$$= \boxed{\mathbf{-\frac{2}{3}log_{10}(t+2)}}$$

Example 6.5-27

$$\boxed{ln\frac{5w^2}{w^2+2w-15}} =$$

Solution:

Step 1

$$\boxed{ln\frac{5w^2}{w^2+2w-15}} = \boxed{ln\frac{5w^2}{(w-3)(w+5)}}$$

Step 2

$$\boxed{ln\frac{5w^2}{(w-3)(w+5)}} = \boxed{ln5w^2 - ln[(w-3)(w+5)]} = \boxed{ln5 + ln\,w^2 - [ln(w-3) + ln(w+5)]}$$

$$= \boxed{\mathbf{ln5 + 2\,ln\,w - ln(w-3) - ln(w+5)}}$$

Example 6.5-28

$$\boxed{ln\frac{2x(x+2)}{x^2-1}} =$$

Solution:

Step 1

$$\boxed{ln\frac{2x(x+2)}{x^2-1}} = \boxed{ln\frac{2x(x+2)}{(x-1)(x+1)}}$$

Step 2 $\boxed{ln\frac{2x(x+2)}{(x-1)(x+1)}} = \boxed{ln2x(x+2) - log(x-1)(x+1)} = \boxed{ln2x + ln(x+2) - [ln(x-1) + ln(x+1)]}$

$= \boxed{ln2x + ln(x+2) - [ln(x-1) + ln(x+1)]} = \boxed{\mathbf{ln2 + lnx + ln(x+2) - ln(x-1) - ln(x+1)}}$

Example 6.5-29

$\boxed{log_3 \frac{9x^3}{(2x-3)^2} - log_3 27} =$

Solution:

Step 1 $\boxed{Not\ Applicable}$

Step 2 $\boxed{log_3 \frac{9x^3}{(2x-3)^2} - log_3 27} = \boxed{log_3 \frac{9x^3}{(2x-3)^2} - log_3 3^3} = \boxed{log_3 9x^3 - log_3 (2x-3)^2 - 3\,log_3 3}$

$= \boxed{log_3 9 + log_3 x^3 - 2\,log_3(2x-3) - 3 \times 1} = \boxed{log_3 3^2 + 3\,log_3 x - 2\,log_3(2x-3) - 3}$

$= \boxed{2\,log_3 3 + 3\,log_3 x - 2\,log_3(2x-3) - 3} = \boxed{2 \times 1 + 3\,log_3 x - 2\,log_3(2x-3) - 3}$

$= \boxed{3\,log_3 x - 2\,log_3(2x-3) + (2-3)} = \boxed{\mathbf{3\,log_3 x - 2\,log_3(2x-3) - 1}}$

Example 6.5-30

$\boxed{log_8 \frac{2x^2 - x - 3}{64x^3(x+6)} + log_8 \frac{1}{512}} =$

Solution:

Step 1 $\boxed{log_8 \frac{2x^2 - x - 3}{64x^3(x+6)} + log_8 \frac{1}{512}} = \boxed{log_8 \frac{(2x-3)(x+1)}{8^2 x^3 (x+6)} + log_8 512^{-1}}$

$= \boxed{log_8 \frac{(2x-3)(x+1)}{8^2 x^3 (x+6)} + log_8 \left(8^3\right)^{-1}} = \boxed{log_8 \frac{(2x-3)(x+1)}{8^2 x^3 (x+6)} + log_8 8^{-3}}$

Step 2 $\boxed{log_8 \frac{(2x-3)(x+1)}{8^2 x^3 (x+6)} + log_8 8^{-3}} = \boxed{log_8[(2x-3)(x+1)] - log_8\left[8^2 x^3 (x+6)\right] + log_8 8^{-3}}$

$= \boxed{log_8(2x-3) + log_8(x+1) - \left[2\,log_8 8 + log_8 x^3 + log_8(x+6)\right] - 3\,log_8 8}$

$= \boxed{log_8(2x-3) + log_8(x+1) - \left[2 \times 1 + 3\,log_8 x + log_8(x+6)\right] - 3 \times 1}$

$= \boxed{log_8(2x-3) + log_8(x+1) - 2 - 3\,log_8 x - log_8(x+6) - 3}$

$= \boxed{\mathbf{log_8(2x-3) + log_8(x+1) - 3\,log_8 x - log_8(x+6) - 5}}$

Additional Examples - Expanding Logarithmic Expressions From a Single Term

The following examples further illustrate how to expand logarithmic expressions using the properties of logarithms:

Example 6.5-31

$$\boxed{ln\left[\frac{(x^2+1)^3(x-1)}{x(x+1)^2}\right]} = \boxed{ln\left[(x^2+1)^3(x-1)\right]-ln\,x(x+1)^2} = \boxed{ln(x^2+1)^3+ln(x-1)-\left[ln\,x+ln(x+1)^2\right]}$$

$$= \boxed{3\,ln(x^2+1)+ln(x-1)-\left[ln\,x+2\,ln(x+1)\right]} = \boxed{\mathbf{3\,ln(x^2+1)+ln(x-1)-ln\,x-2\,ln(x+1)}}$$

Note: $log(x+y)\neq log\,x+log\,y$ because $log\,x+log\,y=log\,xy$. The logarithmic expression $log(x+y)$ can not be simplified any further. For example, $ln(x^2+1)\neq ln\,x^2+ln\,1$ or $log_3(3y+5)\neq log_3 3y+log_3 5$

Example 6.5-32

$$\boxed{ln\frac{\sqrt{x^2+2}}{(x+1)^3x^2}} = \boxed{ln\sqrt{x^2+2}-ln\left[(x+1)^3x^2\right]} = \boxed{ln(x^2+2)^{\frac{1}{2}}-ln\left[(x+1)^3x^2\right]} = \boxed{\frac{1}{2}ln(x^2+2)-\left[ln(x+1)^3+ln\,x^2\right]}$$

$$= \boxed{\frac{1}{2}ln(x^2+2)-\left[3\,ln(x+1)+2\,ln\,x\right]} = \boxed{\mathbf{\frac{1}{2}ln(x^2+2)-3\,ln(x+1)-2\,ln\,x}}$$

Example 6.5-33

$$\boxed{ln\frac{y^3(y+1)^{\frac{1}{2}}}{(y-1)^4}} = \boxed{ln\left[y^3(y+1)^{\frac{1}{2}}\right]-ln(y-1)^4} = \boxed{ln\,y^3+ln(y+1)^{\frac{1}{2}}-4\,ln(y-1)} = \boxed{\mathbf{3\,ln\,y+\frac{1}{2}ln(y+1)-4\,ln(y-1)}}$$

Example 6.5-34

$$\boxed{ln\frac{\sqrt{x^2+1}(2x+1)}{x^3(x+1)}} = \boxed{ln\left[\sqrt{x^2+1}(2x+1)\right]-ln\left[x^3(x+1)\right]} = \boxed{ln\sqrt{x^2+1}+ln(2x+1)-\left[ln\,x^3+ln(x+1)\right]}$$

$$= \boxed{ln(x^2+1)^{\frac{1}{2}}+ln(2x+1)-\left[3\,ln\,x+ln(x+1)\right]} = \boxed{\mathbf{\frac{1}{2}ln(x^2+1)+ln(2x+1)-3\,ln\,x-ln(x+1)}}$$

Example 6.5-35

$$\boxed{ln\frac{\sqrt[4]{t+1}(t+2)^2}{\sqrt[3]{t+3}}-ln\sqrt{t^3}} = \boxed{ln\left[\sqrt[4]{t+1}(t+2)^2\right]-ln\left(\sqrt[3]{t+3}\right)-ln\,t^{\frac{3}{2}}} = \boxed{ln\sqrt[4]{t+1}+ln(t+2)^2-ln(t+3)^{\frac{1}{3}}-ln\,t^{\frac{3}{2}}}$$

$$= \boxed{ln(t+1)^{\frac{1}{4}}+2\,ln(t+2)-\frac{1}{3}ln(t+3)-\frac{3}{2}ln\,t} = \boxed{\mathbf{\frac{1}{4}ln(t+1)+2\,ln(t+2)-\frac{1}{3}ln(t+3)-\frac{3}{2}ln\,t}}$$

Example 6.5-36

$$\boxed{ln\frac{5z^3(z+1)^4}{2z-1}+ln\,z^3} = \boxed{ln\left[5z^3(z+1)^4\right]-ln(2z-1)+ln\,z^3} = \boxed{ln\,5+ln\,z^3+ln(z+1)^4-ln(2z-1)+ln\,z^3}$$

$= \boxed{ln5 + 2\,ln\,z^3 + 4\,ln(z+1) - ln(2z-1)} = \boxed{\mathbf{ln\,5 + 6\,ln\,z + 4\,ln(z+1) - ln(2z-1)}}$

Example 6.5-37

$\boxed{log_{10}\frac{x\left(\sqrt{x+1}\right)(2x+1)^3}{10}} = \boxed{log_{10}\left[x\left(\sqrt{x+1}\right)(2x+1)^3\right] - log_{10}10} = \boxed{log_{10}x + log_{10}\left(\sqrt{x+1}\right) + log_{10}(2x+1)^3 - 1}$

$= \boxed{log_{10}x + log_{10}(x+1)^{\frac{1}{2}} + 3\,log_{10}(2x+1) - 1} = \boxed{\mathbf{log_{10}\,x + \frac{1}{2}log_{10}(x+1) \;\; + 3\,log_{10}(2x+1) - 1}}$

Example 6.5-38

$\boxed{ln\frac{\sqrt{w^5-1}}{4w(w-1)} + ln\,e^2} = \boxed{ln\left(\sqrt{w^5-1}\right) - ln[4w(w-1)] + 2\,ln\,e} = \boxed{ln\left(w^5-1\right)^{\frac{1}{2}} - [ln\,4w + ln(w-1)] + 2\times 1}$

$= \boxed{\frac{1}{2}ln\left(w^5-1\right) - [ln\,4 + ln\,w + ln(w-1)] + 2} = \boxed{\mathbf{\frac{1}{2}ln\left(w^5-1\right) - ln\,4 - ln\,w - ln(w-1) + 2}}$

Example 6.5-39

$\boxed{log\left(x^2+x-6\right) - log\,2x} = \boxed{log[(x+3)(x-2)] - [log\,2 + log\,x]} = \boxed{\mathbf{log(x+3) + log(x-2) - log\,2 - log\,x}}$

Example 6.5-40

$\boxed{log\frac{t-3}{t^2-4}} = \boxed{log(t-3) - log\left(t^2-4\right)} = \boxed{log(t-3) - [log(t-2)(t+2)]} = \boxed{\mathbf{log(t-3) - log(t-2) - log(t+2)}}$

Example 6.5-41

$\boxed{log\frac{u^3-2u^2}{3(u+2)}} = \boxed{log\frac{u^2(u-2)}{3(u+2)}} = \boxed{log\,u^2(u-2) - log\,3(u+2)} = \boxed{\left[log\,u^2 + log(u-2)\right] - [log\,3 + log(u+2)]}$

$= \boxed{\mathbf{2\,log\,u + log(u-2) - log\,3 - log(u+2)}}$

Example 6.5-42

$\boxed{log\,5x\sqrt[5]{(x-1)^2}} = \boxed{log\,5x(x-1)^{\frac{2}{5}}} = \boxed{log\,5x + log(x-1)^{\frac{2}{5}}} = \boxed{\mathbf{log\,5 + log\,x + \frac{2}{5}log(x-1)}}$

Example 6.5-43

$\boxed{log\frac{x^2+x-2}{(x-3)}} = \boxed{log\frac{(x-1)(x+2)}{(x-3)}} = \boxed{log[(x-1)(x+2)] - log(x-3)} = \boxed{\mathbf{log(x-1) + log(x+2) - log(x-3)}}$

Example 6.5-44

$\boxed{log\left(\frac{(3y+1)^{\frac{2}{3}}}{y^2}\right)^3} = \boxed{log\frac{(3y+1)^{\cancel{3}\times\frac{2}{\cancel{3}}}}{y^{2\times 3}}} = \boxed{log\frac{(3y+1)^2}{y^6}} = \boxed{log(3y+1)^2 - log\,y^6} = \boxed{\mathbf{2\,log(3y+1) - 6\,log\,y}}$

Example 6.5-45

$$\boxed{ln\sqrt{\frac{x+2}{(x+3)^2}}} = \boxed{ln\left(\frac{x+2}{(x+3)^2}\right)^{\frac{1}{2}}} = \boxed{ln\frac{(x+2)^{1\times\frac{1}{2}}}{(x+3)^{2\times\frac{1}{2}}}} = \boxed{ln\frac{(x+2)^{\frac{1}{2}}}{(x+3)^{\frac{1}{1}}}} = \boxed{ln\frac{(x+2)^{\frac{1}{2}}}{(x+3)}} = \boxed{ln(x+2)^{\frac{1}{2}} - ln(x+3)}$$

$$= \boxed{\frac{1}{2}ln(x+2) - ln(x+3)}$$

or, we can solve the above problem in the following way:

$$\boxed{ln\sqrt{\frac{x+2}{(x+3)^2}}} = \boxed{ln\left(\frac{x+2}{(x+3)^2}\right)^{\frac{1}{2}}} = \boxed{\frac{1}{2}ln\frac{x+2}{(x+3)^2}} = \boxed{\frac{1}{2}\left[ln(x+2) - ln(x+3)^2\right]} = \boxed{\frac{1}{2}\left[ln(x+2) - 2\,ln(x+3)\right]}$$

$$= \boxed{\frac{1}{2}ln(x+2) - 2\times\frac{1}{2}ln(x+3)} = \boxed{\frac{1}{2}ln(x+2) - ln(x+3)}$$

Example 6.5-46

$$\boxed{\log\left[\sqrt[5]{\frac{x^2}{4x-3}}\cdot\left(\frac{3x+1}{x}\right)^{\frac{1}{3}}\right]} = \boxed{\log\left[\left(\frac{x^2}{4x-3}\right)^{\frac{1}{5}}\cdot\left(\frac{3x+1}{x}\right)^{\frac{1}{3}}\right]} = \boxed{\log\left(\frac{x^2}{4x-3}\right)^{\frac{1}{5}} + \log\left(\frac{3x+1}{x}\right)^{\frac{1}{3}}}$$

$$= \boxed{\frac{1}{5}\log\frac{x^2}{4x-3} + \frac{1}{3}\log\frac{3x+1}{x}} = \boxed{\frac{1}{5}\left[\log x^2 - \log(4x-3)\right] + \frac{1}{3}\left[\log(3x+1) - \log x\right]}$$

$$= \boxed{\frac{1}{5}\left[2\log x - \log(4x-3)\right] + \frac{1}{3}\left[\log(3x+1) - \log x\right]} = \boxed{\frac{2}{5}log\,x - \frac{1}{5}log(4x-3) + \frac{1}{3}log(3x+1) - \frac{1}{3}log\,x}$$

Example 6.5-47

$$\boxed{log_{10}\frac{100(x-2)^2}{x^2-3}} = \boxed{log_{10}100 + log_{10}(x-2)^2 - log_{10}\left(x^2-3\right)} = \boxed{2 + 2\,log_{10}(x-2) - log_{10}\left(x-\sqrt{3}\right)\left(x+\sqrt{3}\right)}$$

$$= \boxed{2 + 2\,log_{10}(x-2) - \left[log_{10}\left(x-\sqrt{3}\right) + log_{10}\left(x+\sqrt{3}\right)\right]} = \boxed{2 + 2\,log_{10}(x-2) - log_{10}\left(x-\sqrt{3}\right) - log_{10}\left(x+\sqrt{3}\right)}$$

Example 6.5-48

$$\boxed{ln\frac{(x+3)^{\frac{2}{3}}x^2}{x^{-\frac{1}{4}}}} = \boxed{ln\left[(x+3)^{\frac{2}{3}}x^2\right] - ln\,x^{-\frac{1}{4}}} = \boxed{ln(x+3)^{\frac{2}{3}} + ln\,x^2 + \frac{1}{4}ln\,x} = \boxed{\frac{2}{3}ln(x+3) + ln\,x^2 + \frac{1}{4}ln\,x}$$

$$= \boxed{\frac{2}{3}ln(x+3) + 2\,ln\,x + \frac{1}{4}ln\,x} = \boxed{\frac{2}{3}ln(x+3) + \left(2+\frac{1}{4}\right)ln\,x} = \boxed{\frac{2}{3}ln(x+3) + \left(\frac{8+1}{4}\right)ln\,x} = \boxed{\frac{2}{3}ln(x+3) + \frac{9}{4}ln\,x}$$

We can also solve the above problem in the following way:

$$\boxed{ln\frac{(x+3)^{\frac{2}{3}}x^2}{x^{-\frac{1}{4}}}} = \boxed{ln\frac{(x+3)^{\frac{2}{3}}x^2x^{+\frac{1}{4}}}{1}} = \boxed{ln\left[(x+3)^{\frac{2}{3}}x^2x^{\frac{1}{4}}\right]} = \boxed{ln\left[(x+3)^{\frac{2}{3}}x^{2+\frac{1}{4}}\right]} = \boxed{ln\left[(x+3)^{\frac{2}{3}}x^{\frac{8+1}{4}}\right]}$$

$$= \boxed{ln\left[(x+3)^{\frac{2}{3}}x^{\frac{9}{4}}\right]} = \boxed{ln\left[(x+3)^{\frac{2}{3}}x^{\frac{9}{4}}\right]} = \boxed{ln(x+3)^{\frac{2}{3}} + ln\,x^{\frac{9}{4}}} = \boxed{\mathbf{\frac{2}{3}ln(x+3)+\frac{9}{4}ln\,x}}$$

Example 6.5-49

$$\boxed{log\frac{x^2-2x-15}{x^2-16}} = \boxed{log\frac{(x+3)(x-5)}{(x-4)(x+4)}} = \boxed{log[(x+3)(x-5)] - log[(x-4)(x+4)]}$$

$$= \boxed{log(x+3)+log(x-5)-[log(x-4)+log(x+4)]} = \boxed{\mathbf{log(x+3)+log(x-5)-log(x-4)-log(x+4)}}$$

Example 6.5-50

$$\boxed{ln\frac{3x(x-1)^{\frac{1}{3}}}{x^2-2}} = \boxed{ln\,3x(x-1)^{\frac{1}{3}} - ln(x^2-2)} = \boxed{ln\,3+ln\,x+ln(x-1)^{\frac{1}{3}} - ln(x-\sqrt{2})(x+\sqrt{2})}$$

$$= \boxed{ln\,3+ln\,x+\frac{1}{3}ln(x-1)-\left[ln(x-\sqrt{2})+ln(x+\sqrt{2})\right]} = \boxed{\mathbf{ln\,3+ln\,x+\frac{1}{3}ln(x-1)-ln(x-\sqrt{2})-ln(x+\sqrt{2})}}$$

Example 6.5-51

$$\boxed{log_{10}\sqrt[3]{\frac{v^2+v-2}{v^2+v-6}}} = \boxed{log_{10}\left(\frac{v^2+v-2}{v^2+v-6}\right)^{\frac{1}{3}}} = \boxed{\frac{1}{3}log_{10}\frac{v^2+v-2}{v^2+v-6}} = \boxed{\frac{1}{3}log_{10}\frac{(v-1)(v+2)}{(v+3)(v-2)}}$$

$$= \boxed{\frac{1}{3}\left[log_{10}(v-1)(v+2)-log_{10}(v+3)(v-2)\right]} = \boxed{\mathbf{\frac{1}{3}\left[log_{10}(v-1)+log_{10}(v+2)-log_{10}(v+3)-log_{10}(v-2)\right]}}$$

Example 6.5-52

$$\boxed{log_{10}\left(\sqrt{\frac{u^5}{2u-3}}\cdot 3u^2\right)} = \boxed{log_{10}\left(\frac{u^5}{2u-3}\right)^{\frac{1}{2}} + log\,3u^2} = \boxed{\frac{1}{2}\left[log_{10}\frac{u^5}{2u-3}\right] + log_{10}3 + log_{10}u^2}$$

$$= \boxed{\frac{1}{2}\left[5\,log_{10}u - log_{10}(2u-3)\right] + log_{10}3 + 2\,log_{10}u} = \boxed{\frac{5}{2}log_{10}u - \frac{1}{2}log_{10}(2u-3) + log_{10}3 + 2log_{10}u}$$

$$= \boxed{\left(\frac{5}{2}+2\right)log_{10}u - \frac{1}{2}log_{10}(2u-3) + log_{10}u} = \boxed{\mathbf{\frac{9}{2}log_{10}u - \frac{1}{2}log_{10}(2u-3) + log_{10}u}}$$

Example 6.5-53

$$\boxed{log_2\frac{6t^2-t-1}{t^2\sqrt{t+1}}} = \boxed{log_2\frac{(3t+1)(2t-1)}{t^2\sqrt{t+1}}} = \boxed{log_2(3t+1)(2t-1) - log_2 t^2\sqrt{t+1}}$$

$$= \boxed{log_2(3t+1)+log_2(2t-1)-\left[log_2 t^2 + log_2(t+1)^{\frac{1}{2}}\right]} = \boxed{\mathbf{log_2(3t+1)+log_2(2t-1)-2\,log_2\,t-\frac{1}{2}log_2(t+1)}}$$

Example 6.5-54

$$\boxed{log_{10}\frac{y^3-3y^2}{\sqrt{y^3}}} = \boxed{log_{10}\frac{y^2(y-3)}{\sqrt{y^3}}} = \boxed{log_{10}\,y^2(y-3)-log_{10}\sqrt{y^3}} = \boxed{log_{10}\,y^2(y-3)-log_{10}\,y^{\frac{3}{2}}}$$

$$= \boxed{log_{10}\,y^2 + log_{10}(y-3)-\frac{3}{2}log_{10}\,y} = \boxed{2\,log_{10}\,y + log_{10}(y-3)-\frac{3}{2}log_{10}\,y} = \boxed{\left(2-\frac{3}{2}\right)log_{10}\,y + log_{10}(y-3)}$$

$$= \boxed{\left(\frac{4-3}{2}\right)log_{10}\,y + log_{10}(y-3)} = \boxed{\mathbf{\frac{1}{2}log_{10}\,y + log_{10}(y-3)}}$$

We can also solve the above problem in the following way:

$$\boxed{log_{10}\frac{y^3-3y^2}{\sqrt{y^3}}} = \boxed{log_{10}\frac{y^2(y-3)}{\left(y^3\right)^{\frac{1}{2}}}} = \boxed{log_{10}\frac{y^2(y-3)}{y^{\frac{3}{2}}}} = \boxed{log_{10}\frac{y^2y^{-\frac{3}{2}}(y-3)}{1}} = \boxed{log_{10}\,y^{2-\frac{3}{2}}(y-3)}$$

$$= \boxed{log_{10}\,y^{\frac{4-3}{2}}(y-3)} = \boxed{log_{10}\,y^{\frac{1}{2}}(y-3)} = \boxed{log_{10}\,y^{\frac{1}{2}} + log_{10}(y-3)} = \boxed{\mathbf{\frac{1}{2}log_{10}\,y + log_{10}(y-3)}}$$

Example 6.5-55

$$\boxed{log_{10}\frac{2v-3}{v^3\sqrt{1+2v}}} = \boxed{log_{10}(2v-3)-log_{10}\left(v^3\sqrt{1+2v}\right)} = \boxed{log_{10}(2v-3)-\left[log_{10}\,v^3 + log_{10}\sqrt{1+2v}\right]}$$

$$= \boxed{log_{10}(2v-3)-3\,log_{10}\,v-log_{10}(1+2v)^{\frac{1}{2}}} = \boxed{\mathbf{log_{10}(2v-3)-3\,log_{10}\,v-\frac{1}{2}log_{10}(1+2v)}}$$

Practice Problems - Expanding Logarithmic Expressions From a Single Term

Section 6.5 Case II Practice Problems - Expand the following logarithmic expressions:

1. $log\frac{3x}{2x^2+x-1} =$
2. $ln\sqrt{\frac{(1-x)^2}{(1+x)}} =$
3. $log_3\frac{3x\sqrt{(x+1)^3}}{(x-1)^2} =$
4. $log_8\frac{64x^3}{\sqrt{(x-1)^4}} =$
5. $ln\frac{6x^2+7x-3}{x^2+3x+2} =$
6. $log_5\frac{125y^5\left(\sqrt[5]{y}+2\right)}{8(y-1)^{\frac{1}{2}}} =$
7. $log_5\frac{125z^2}{\sqrt{8}(z-1)^2} =$
8. $log_{10}\frac{3w}{11\sqrt{w^2+1}} =$
9. $log_3\frac{27x^3}{2(x-1)^{\frac{1}{2}}} =$
10. $log_4\frac{64t^2(t+1)}{\sqrt[3]{7}} =$

Case III Combining Logarithmic Expressions into a Single Term

Logarithmic expressions are combined into a single term using the following steps:

Step 1 Apply the logarithmic rules No. 1 through 5, as appropriate (see Section 6.1, Table 6.1-2).

Step 2 Apply the laws of logarithm such as the product and quotient laws to combine the logarithmic expression into a single logarithm.

Examples with Steps

The following examples show the steps as to how logarithmic expressions are combined into a single logarithmic term:

Example 6.5-56

$$\boxed{(2\log_{10}5+3\log_{10}u)-\log_{10}10}=$$

Solution:

Step 1

$$\boxed{(2\log_{10}5+3\log_{10}u)-\log_{10}10}=\boxed{\left(\log_{10}5^2+\log_{10}u^3\right)-\log_{10}10}$$

$$=\boxed{\left(\log_{10}25+\log_{10}u^3\right)-\log_{10}10}$$

Step 2

$$\boxed{\left(\log_{10}25+\log_{10}u^3\right)-\log_{10}10}=\boxed{\log_{10}25u^3-\log_{10}10}=\boxed{\log_{10}\frac{\overset{5}{\cancel{25}}u^3}{\underset{2}{\cancel{10}}}}=\boxed{\mathbf{\log_{10}\frac{5u^3}{2}}}$$

Example 6.5-57

$$\boxed{\log(x+1)+[2\log x-3\log(x+1)]}=$$

Solution:

Step 1

$$\boxed{\log(x+1)+[2\log x-3\log(x+1)]}=\boxed{\log(x+1)+\left[\log x^2-\log(x+1)^3\right]}$$

Step 2

$$\boxed{\log(x+1)+\left[\log x^2-\log(x+1)^3\right]}=\boxed{\log(x+1)+\log\frac{x^2}{(x+1)^3}}=\boxed{\log\left[\cancel{(x+1)}\cdot\frac{x^2}{(x+1)^{\cancel{3}=2}}\right]}$$

$$=\boxed{\log\frac{x^2}{(x+1)^2}}=\boxed{\mathbf{\log\frac{x^2}{x^2+2x+1}}}$$

Example 6.5-58

$$\boxed{\left[2\log(x+1)+\frac{1}{5}\log(x-2)\right]-\log(x+1)}=$$

Solution:

Step 1

$$\boxed{\left[2\log(x+1)+\frac{1}{5}\log(x-2)\right]-\log(x+1)}=\boxed{\left[\log(x+1)^2+\log(x-2)^{\frac{1}{5}}\right]-\log(x+1)}$$

Step 2 $\left[\log(x+1)^2+\log(x-2)^{\frac{1}{5}}\right]-\log(x+1)$ $=$ $\log\left[(x+1)^2\cdot(x-2)^{\frac{1}{5}}\right]-\log(x+1)$

$=$ $\log\frac{(x+1)^{2=1}\cdot(x-2)^{\frac{1}{5}}}{(x+1)}$ $=$ $\log\frac{(x+1)(x-2)^{\frac{1}{5}}}{1}$ $=$ $\boldsymbol{\log\left[(x+1)\sqrt[5]{(x-2)}\right]}$

Example 6.5-59

$(4\log_2 2+\log_2 x)-(2\log_2 3+3\log_2 x)$ $=$

Solution:

Step 1 $(4\log_2 2+\log_2 x)-(2\log_2 3+3\log_2 x)$ $=$ $\left(\log_2 2^4+\log_2 x\right)-\left(\log_2 3^2+\log_2 x^3\right)$

$=$ $\left(\log_2 16+\log_2 x\right)-\left(\log_2 9+\log_2 x^3\right)$

Step 2 $\left(\log_2 16+\log_2 x\right)-\left(\log_2 9+\log_2 x^3\right)$ $=$ $\left(\log_2 16\cdot x\right)-\left(\log_2 9\cdot x^3\right)$

$=$ $\log_2 16x-\log_2 9x^3$ $=$ $\log_2\frac{16\not{x}}{9x^{\not{3}}}$ (with x^2 remaining in the denominator) $=$ $\boldsymbol{\log_2\frac{16}{9x^2}}$

Example 6.5-60

$(\log_{10}4+\log_{10}Q)+(2\log_{10}Q+2\log_{10}2)$ $=$

Solution:

Step 1 $(\log_{10}4+\log_{10}Q)+(2\log_{10}Q+2\log_{10}2)$ $=$ $(\log_{10}4+\log_{10}Q)+\left(\log_{10}Q^2+\log_{10}2^2\right)$

$=$ $(\log_{10}4+\log_{10}Q)+\left(\log_{10}Q^2+\log_{10}4\right)$

Step 2 $(\log_{10}4+\log_{10}Q)+\left(\log_{10}Q^2+\log_{10}4\right)$ $=$ $\log_{10}(4\cdot Q)+\log_{10}\left(4\cdot Q^2\right)$

$=$ $\log_{10}\left(4Q\cdot 4Q^2\right)$ $=$ $\log_{10}16Q^{1+2}$ $=$ $\boldsymbol{\log_{10}16Q^3}$

Additional Examples - Combining Logarithmic Expressions into a Single Term

The following examples further illustrate how to combine logarithmic expressions into a single logarithmic term:

Example 6.5-61

$(\log_3 5+\log_3 x)-(3\log_3 x+2\log_3 5)$ $=$ $\log_3 5x-\left(\log_3 x^3+\log_3 5^2\right)$ $=$ $\log_3 5x-\left(\log_3 x^3+\log_3 25\right)$

$=$ $\log_3 5x-\log_3 25x^3$ $=$ $\log_3\frac{\not{5}x}{\not{25}x^{\not{3}}}$ (with $5\ x^2$ remaining in the denominator) $=$ $\boldsymbol{\log_3\frac{1}{5x^2}}$

Example 6.5-62

$$\boxed{(\log_2 2-\log_2 u)+(2\log_2 u-\log_2 8)} = \boxed{\log_2\frac{2}{u}+\left(\log_2 u^2-\log_2 8\right)} = \boxed{\log_2\frac{2}{u}+\log_2\frac{u^2}{8}} = \boxed{\log_2\left(\frac{2}{u}\cdot\frac{u^2}{8}\right)}$$

$$= \boxed{\log_2\frac{\overset{u}{2u^2}}{\underset{4}{8u}}} = \boxed{\mathbf{\log_2\frac{u}{4}}}$$

Example 6.5-63

$$\boxed{(\log_{10} 5+\log_{10} x)-\log_{10} 25} = \boxed{\log_{10} 5x-\log_{10} 25} = \boxed{\log_{10}\frac{5x}{\underset{5}{25}}} = \boxed{\mathbf{\log_{10}\frac{x}{5}}}$$

Example 6.5-64

$$\boxed{(2\log_{10} x+\log_{10} 5)+(3\log_{10} x+\log_{10} 3)} = \boxed{\left(\log_{10} x^2+\log_{10} 5\right)+\left(\log_{10} x^3+\log_{10} 3\right)} = \boxed{\log_{10} 5x^2+\log_{10} 3x^3}$$

$$= \boxed{\log_{10}\left(5x^2\cdot 3x^3\right)} = \boxed{\log_{10} 15x^{2+3}} = \boxed{\mathbf{\log_{10} 15x^5}}$$

Example 6.5-65

$$\boxed{(2\log_5 5+\log_5 w)-\frac{1}{2}\log_5 13} = \boxed{\left(\log_5 5^2+\log_5 w\right)-\log_5 13^{\frac{1}{2}}} = \boxed{(\log_5 25+\log_5 w)-\log_5\sqrt{13}}$$

$$= \boxed{\log_5 25w-\log_5\sqrt{13}} = \boxed{\mathbf{\log_5\frac{25w}{\sqrt{13}}}}$$

Example 6.5-66

$$\boxed{(5\log_2 w-\log_2 5)+\left(2\log_2 w-\frac{1}{2}\log_2 3\right)} = \boxed{\left(\log_2 w^5-\log_2 5\right)+\left(\log_2 w^2-\log_2 3^{\frac{1}{2}}\right)} = \boxed{\log_2\frac{w^5}{5}+\log_2\frac{w^2}{\sqrt{3}}}$$

$$= \boxed{\log_2\left(\frac{w^5}{5}\cdot\frac{w^2}{\sqrt{3}}\right)} = \boxed{\log_2\frac{w^5\cdot w^2}{5\cdot\sqrt{3}}} = \boxed{\log_2\frac{w^{5+2}}{5\sqrt{3}}} = \boxed{\mathbf{\log_2\frac{w^7}{5\sqrt{3}}}}$$

Example 6.5-67

$$\boxed{\left[2\log(x-1)+\frac{1}{3}\log(x+2)\right]-\log x} = \boxed{\left[\log(x-1)^2+\log(x+2)^{\frac{1}{3}}\right]-\log x} = \boxed{\log\left[(x-1)^2\cdot(x+2)^{\frac{1}{3}}\right]-\log x}$$

$$= \boxed{\mathbf{\log\frac{(x-1)(x-1)(x+2)^{\frac{1}{3}}}{x}}}$$

Example 6.5-68

$$\boxed{\left[2\log_{10} u-3\log_{10}(u+1)\right]+\left[2\log_{10}(u+1)-2\log_{10} 5\right]} = \boxed{\left[\log_{10} u^2-\log_{10}(u+1)^3\right]+\left[\log_{10}(u+1)^2-\log_{10} 5^2\right]}$$

$$= \boxed{\log_{10}\frac{u^2}{(u+1)^3}+\log_{10}\frac{(u+1)^2}{25}} = \boxed{\log_{10}\left[\frac{u^2}{(u+1)^3}\cdot\frac{(u+1)^2}{25}\right]} = \boxed{\log_{10}\frac{u^2\cdot(u+1)^2}{25\cdot(u+1)^3}} = \boxed{\mathbf{\log_{10}\frac{u^2}{25(u+1)}}}$$

Example 6.5-69

$$\boxed{5\log x-\left[\log x-\log(x-1)\right]}=\boxed{\log x^5-\left[\log x-\log(x-1)\right]}=\boxed{\log x^5-\log\frac{x}{x-1}}=\boxed{\log\frac{\frac{x^5}{x}}{x-1}}=\boxed{\log\frac{\frac{x^5}{1}}{\frac{x}{x-1}}}$$

$$\boxed{\log\frac{x^5\cdot(x-1)}{1\cdot x}}=\boxed{\log\frac{x^4\,\cancel{x^5}(x-1)}{\cancel{x}}}=\boxed{\mathbf{\log x^4(x-1)}}$$

Example 6.5-70

$$\boxed{\left(\log_6 15-\frac{1}{2}\log_6 3\right)+2\log_6 y}=\boxed{\left(\log_6 15-\log_6 3^{\frac{1}{2}}\right)+\log_6 y^2}=\boxed{\left(\log_6 15-\log_6\sqrt{3}\right)+\log_6 y^2}$$

$$=\boxed{\log_6\frac{15}{\sqrt{3}}+\log_6 y^2}=\boxed{\log_6\left(\frac{15}{\sqrt{3}}\cdot y^2\right)}=\boxed{\mathbf{\log_6\frac{15}{\sqrt{3}}y^2}}$$

Practice Problems - Combining Logarithmic Expressions into a Single Term

Section 6.5 Case III Practice Problems - Combine the following logarithmic expressions into a single logarithmic term:

1. $2\log_8 5+4\log_8 u+\log_8 10=$
2. $(4\log_3 2-\log_3 u)+(2\log_3 3-3\log_3 u)=$
3. $3\log_{10} x+\left[\log_{10}(x-2)-\log_{10}(x+1)\right]=$
4. $\left[\log(x-1)-3\log(x+2)\right]-\log x=$
5. $\left(\log_2 7-\frac{1}{3}\log_2 3\right)+3\log_2 x=$
6. $\log_4 3-(\log_4 3+2\log_4 x)=$
7. $\left[2\log x+3\log(x-1)\right]-\left[2\log(x-1)+\log 2\right]=$
8. $(\log_{10} 12-2\log_{10} 3)+2\log_{10} x=$
9. $(\log_{10} 12-3\log_{10} 2)-(\log_{10} 3-2\log_{10} w)=$
10. $\log(x+1)-\left[3\log(x+2)-3\log x\right]=$

6.6 Advanced Logarithmic Problems

This chapter is concluded by addressing logarithmic and exponential problems that have practical application in physics, chemistry, and applied mathematics. Note that this class of problems is solved for a specific number assigned to each variable. For further practice, students are encouraged to solve and check the answer by selecting other values. Exponential equations are solved using the following steps:

Step 1 Substitute the constant values into the original equation and simplify the equation as appropriate.

Step 2 a. Multiply both sides of the equation by either log_{10} or a logarithm having the same base as the base of the exponential expression.

b. Solve for the unknown variable by applying the logarithmic laws. Check the answer by substituting the solution into the original equation.

Examples with Steps

The following examples show the steps as to how exponential equations are solved using the logarthmic properties:

Example 6.6-1

Given $K = K_0(1+r)^t$ solve for t if $K = 4000$, $K_0 = 1000$, and $r = 0.04$.

Solution:

Step 1 $\boxed{K = K_0(1+r)^t}$; $\boxed{4000 = 1000(1+0.04)^t}$; $\boxed{\frac{\overset{4}{\cancel{4000}}}{\cancel{1000}} = (1+0.04)^t}$; $\boxed{4 = (1+0.04)^t}$

Step 2 $\boxed{4 = (1+0.04)^t}$; $\boxed{log_{10} 4 = log_{10}(1+0.04)^t}$; $\boxed{log_{10} 4 = t\, log_{10}(1+0.04)}$

; $\boxed{0.6021 = t \times 0.017}$; $\boxed{0.6021 = 0.017t}$; $\boxed{t = \frac{0.6021}{0.017}}$; $\boxed{\mathbf{t = 35.42}}$

Check: $4000 \overset{?}{=} 1000(1+0.04)^{35.42}$; $4000 \overset{?}{=} 1000 \times 1.04^{35.42}$; $4000 \overset{?}{=} 1000 \times 4.0$; $4000 = 4000$

Example 6.6-2

Given $M = M_0\left(1-e^{-kt}\right)$ solve for t if $M = 30$, $M_0 = 40$, and $k = 10$.

Solution:

Step 1 $\boxed{M = M_0\left(1-e^{-kt}\right)}$; $\boxed{30 = 40\left(1-e^{-10t}\right)}$; $\boxed{\frac{\overset{3}{\cancel{30}}}{\underset{4}{\cancel{40}}} = \left(1-e^{-10t}\right)}$; $\boxed{\frac{3}{4} = 1-e^{-10t}}$

; $\boxed{\frac{3}{4} - 1 = -e^{-10t}}$; $\boxed{\frac{3}{4} - \frac{1}{1} = -e^{-10t}}$; $\boxed{\frac{3-4}{4} = -e^{-10t}}$; $\boxed{-\frac{1}{4} = -e^{-10t}}$; $\boxed{e^{-10t} = \frac{1}{4}}$

Step 2 $\boxed{e^{-10t} = \frac{1}{4}}$; $\boxed{log_e e^{-10t} = log_e \frac{1}{4}}$; $\boxed{-10t = log_e 1 - log_e 4}$; $\boxed{-10t = 0 - log_e 4}$

; $\boxed{10t = \log_e 4}$; $\boxed{t = \frac{\log_e 4}{10}}$; $\boxed{t = \frac{1.3863}{10}}$; $\boxed{\mathbf{t = 0.1386}}$

Check: $30 \stackrel{?}{=} 40\left(1 - e^{-10\times 0.1386}\right)$; $30 \stackrel{?}{=} 40\left(1 - e^{-1.386}\right)$; $30 \stackrel{?}{=} 40 - 40e^{-1.386}$; $30 \stackrel{?}{=} 40 - \frac{40}{e^{1.386}}$

; $30 \stackrel{?}{=} 40 - \frac{40}{4}$; $30 \stackrel{?}{=} 40 - 10$; $30 = 30$

Example 6.6-3

Given $B = B_0\left(1 + \frac{m}{n}\right)^{nt}$ solve for t if $B = 300$, $B_0 = 100$, $m = 5$, and $n = 2$.

Solution:

Step 1 $\boxed{B = B_0\left(1 + \frac{m}{n}\right)^{nt}}$; $\boxed{300 = 100\left(1 + \frac{5}{2}\right)^{2t}}$; $\boxed{\frac{\overset{3}{\cancel{300}}}{\cancel{100}} = \left(\frac{2+5}{2}\right)^{2t}}$; $\boxed{\frac{3}{1} = \frac{7}{2}^{2t}}$; $\boxed{3 = \frac{7}{2}^{2t}}$

Step 2 $\boxed{3 = \frac{7}{2}^{2t}}$; $\boxed{\log_{10} 3 = \log_{10} \frac{7}{2}^{2t}}$; $\boxed{\log_{10} 3 = 2t \log_{10} \frac{7}{2}}$; $\boxed{\log_{10} 3 = 2t \log_{10} 3.5}$

; $\boxed{0.4771 = 2t \times 0.5441}$; $\boxed{0.4771 = 1.0882t}$; $\boxed{t = \frac{0.4771}{1.0882}}$; $\boxed{\mathbf{t = 0.4384}}$

Check: $300 \stackrel{?}{=} 100\left(1 + \frac{5}{2}\right)^{2\times 0.4384}$; $300 \stackrel{?}{=} 100\left(1 + \frac{5}{2}\right)^{0.8768}$; $300 \stackrel{?}{=} 100\left(\frac{2+5}{2}\right)^{0.8768}$; $300 \stackrel{?}{=} 100 \times \frac{7}{2}^{0.8768}$

; $300 \stackrel{?}{=} 100 \times 3.5^{0.8768}$; $300 \stackrel{?}{=} 100 \times 3$; $300 = 300$

Example 6.6-4

Given $Y = Y_0 \cdot 4^{-\frac{t}{2000}}$ solve for t if $Y = 10$ and $Y_0 = 5Y$.

Solution:

Step 1 $\boxed{Y = Y_0 \cdot 4^{-\frac{t}{2000}}}$; $\boxed{10 = 50 \cdot 4^{-\frac{t}{2000}}}$; $\boxed{\frac{\cancel{10}}{\underset{5}{\cancel{50}}} = 4^{-\frac{t}{2000}}}$; $\boxed{\frac{1}{5} = 4^{-\frac{t}{2000}}}$; $\boxed{0.2 = 4^{-\frac{t}{2000}}}$

Step 2 $\boxed{0.2 = 4^{-\frac{t}{2000}}}$; $\boxed{\log_4 0.2 = \log_4 4^{-\frac{t}{2000}}}$; $\boxed{\log_4 0.2 = -\frac{t}{2000}\log_4 4}$; $\boxed{\frac{\log_{10} 0.2}{\log_{10} 4} = -\frac{t}{2000}}$

; $\boxed{\frac{-0.6989}{0.6021} = -\frac{t}{2000}}$; $\boxed{1.1608 = \frac{t}{2000}}$; $\boxed{t = 1.1608 \times 2000}$; $\boxed{\mathbf{t = 2321.6}}$

Check: $10 \stackrel{?}{=} 50 \cdot 4^{-\frac{2321.6}{2000}}$; $10 \stackrel{?}{=} 50 \cdot 4^{-\frac{2321.6}{2000}}$; $10 \stackrel{?}{=} 50 \cdot 4^{-1.1608}$; $10 \stackrel{?}{=} 50 \cdot \frac{1}{4^{1.1608}}$; $10 \stackrel{?}{=} 50 \cdot \frac{1}{5}$; $10 = 10$

Example 6.6-5

Given $Q = Q_0 e^{-kt}$ solve for t if $Q = 20$, $Q_0 = 50$, and $k = 0.002$.

Solution:

Step 1 $\boxed{Q = Q_0 e^{-kt}}$; $\boxed{20 = 50e^{-0.002t}}$; $\boxed{\frac{\overset{2}{\cancel{20}}}{\underset{5}{\cancel{50}}} = e^{-0.002t}}$; $\boxed{\frac{2}{5} = e^{-0.002t}}$; $\boxed{0.4 = e^{-0.002t}}$

Step 2 $\boxed{0.4 = e^{-0.002t}}$; $\boxed{log_e 0.4 = log_e e^{-0.002t}}$; $\boxed{-0.9163 = -0.002t\, log_e e}$; $\boxed{-0.9163 = -0.002t}$

; $\boxed{t = \frac{0.9163}{0.002}}$; $\boxed{\mathbf{t = 458.15}}$

Check: $20 \stackrel{?}{=} 50e^{-0.002\times 458.15}$; $20 \stackrel{?}{=} 50e^{-0.9163}$; $20 \stackrel{?}{=} 50 \times \frac{1}{e^{0.9163}}$; $20 \stackrel{?}{=} 50 \times \frac{1}{2.5}$; $20 \stackrel{?}{=} 50 \times 0.4$

; $20 = 20$

Additional Examples - Advanced Logarithmic Problems

The following examples further illustrate how to solve exponential equations using the logarithmic properties:

Example 6.6-6

Given $y = y_0\left(1 - e^{-kt}\right)$ solve for t if $y = 5$, $y_0 = 30$, and $k = 2$.

$\boxed{y = y_0\left(1 - e^{-at}\right)}$; $\boxed{5 = 30\left(1 - e^{-2t}\right)}$; $\boxed{\frac{\not{5}}{\not{30}_{\,6}} = \left(1 - e^{-2t}\right)}$; $\boxed{\frac{1}{6} = \left(1 - e^{-2t}\right)}$; $\boxed{0.1667 - 1 = -e^{-2t}}$; $\boxed{-0.8333 = -e^{-2t}}$

; $\boxed{0.8333 = e^{-2t}}$; $\boxed{log_e 0.8333 = log_e e^{-2t}}$; $\boxed{-0.1824 = -2t\, log_e e}$; $\boxed{-0.1824 = -2t}$; $\boxed{t = \frac{0.1824}{2}}$; $\boxed{\mathbf{t = 0.0911}}$

Check: $5 \stackrel{?}{=} 30\left(1 - e^{-2\times 0.0911}\right)$; $5 \stackrel{?}{=} 30\left(1 - e^{-0.1822}\right)$; $5 \stackrel{?}{=} 30\left(1 - \frac{1}{e^{0.1822}}\right)$; $5 \stackrel{?}{=} 30\left(1 - \frac{1}{1.1998}\right)$

; $5 \stackrel{?}{=} 30(1 - 0.8334)$; $5 \stackrel{?}{=} 30 \times 0.1667$; $5 = 5$

Example 6.6-7

Given $A = A_0 2^{t-1}$ solve for t if $A = 2^5$ and $A_0 = 4$.

$\boxed{A = A_0 2^{t-1}}$; $\boxed{2^5 = 4 \cdot 2^{t-1}}$; $\boxed{2^5 = 2^2 \cdot 2^{t-1}}$; $\boxed{2^5 \cdot 2^{-2} = 2^{t-1}}$; $\boxed{2^{5-2} = 2^{t-1}}$; $\boxed{2^3 = 2^{t-1}}$; $\boxed{t - 1 = 3}$; $\boxed{\mathbf{t = 4}}$

Check: $2^5 \stackrel{?}{=} 4 \cdot 2^{4-1}$; $2^5 \stackrel{?}{=} 4 \cdot 2^3$; $2^5 \stackrel{?}{=} 2^2 \cdot 2^3$; $2^5 \stackrel{?}{=} 2^{2+3}$; $2^5 = 2^5$

Example 6.6-8

Given $K = K_0 \frac{1}{5^x}$ solve for x if $K = 125$ and $K_0 = 5$.

$\boxed{K = K_0 \frac{1}{5^x}}$; $\boxed{125 = 5\frac{1}{5^x}}$; $\boxed{\frac{\overset{25}{\not{125}}}{\not{5}} = \frac{1}{5^x}}$; $\boxed{\frac{25}{1} = \frac{1}{5^x}}$; $\boxed{\frac{5^2}{1} = \frac{1}{5^x}}$; $\boxed{\frac{1}{5^{-2}} = \frac{1}{5^x}}$; $\boxed{\mathbf{x = -2}}$

Another way of solving the above problem would be:

$\boxed{K = K_0 \frac{1}{5^x}}$; $\boxed{125 = 5\frac{1}{5^x}}$; $\boxed{\frac{\overset{5}{\not{125}}}{\not{5}} = \frac{1}{5^x}}$; $\boxed{\frac{25}{1} = \frac{1}{5^x}}$; $\boxed{25 \cdot 5^x = 1 \cdot 1}$; $\boxed{5^x = \frac{1}{25}}$; $\boxed{log_5 5^x = log_5 \frac{1}{25}}$

; $\boxed{x = log_5 1 - log_5 25}$; $\boxed{x = 0 - log_5 5^2}$; $\boxed{x = -2\, log_5 5}$; $\boxed{x = -2 \times 1}$; $\boxed{\mathbf{x = -2}}$

Check: $125 \stackrel{?}{=} 5\frac{1}{5^{-2}}$; $125 \stackrel{?}{=} 5 \cdot 5^2$; $125 \stackrel{?}{=} 5 \cdot 25$; $125 = 125$

Example 6.6-9

Given $y = y_0 \frac{1}{1-e^{ax}}$ solve for x if $y = 40$, $y_0 = 20$, and $a = 0.5$.

$\boxed{y = y_0 \frac{1}{1-e^{ax}}}$; $\boxed{40 = 20\frac{1}{1-e^{0.5x}}}$; $\boxed{\frac{\overset{2}{\cancel{40}}}{\cancel{20}} = \frac{1}{1-e^{0.5x}}}$; $\boxed{\frac{2}{1} = \frac{1}{1-e^{0.5x}}}$; $\boxed{log_e 2 = log_e \frac{1}{1-e^{0.5x}}}$

; $\boxed{log_e 2 = log_e 1 - log_e\left(1-e^{0.5x}\right)}$; $\boxed{0.6931 = 0 - log_e\left(1-e^{0.5x}\right)}$; $\boxed{0.6931 = -log_e\left(1-e^{0.5x}\right)}$

; $\boxed{-0.6931 = log_e\left(1-e^{0.5x}\right)}$; $\boxed{e^{-0.6931} = e^{log_e\left(1-e^{0.5x}\right)}}$; $\boxed{\frac{1}{e^{0.6931}} = 1-e^{0.5x}}$; $\boxed{\frac{1}{2} = 1-e^{0.5x}}$; $\boxed{0.5-1 = -e^{0.5x}}$

; $\boxed{0.5 = e^{0.5x}}$; $\boxed{log_e 0.5 = log_e e^{0.5x}}$; $\boxed{log_e 0.5 = 0.5x}$; $\boxed{-0.6931 = 0.5x}$; $\boxed{x = \frac{-0.6931}{0.5}}$; $\boxed{\mathbf{x = -1.3862}}$

Check: $40 \overset{?}{=} 20\frac{1}{1-e^{0.5\times -1.3862}}$; $40 \overset{?}{=} \frac{20}{1-e^{-0.6931}}$; $40 \overset{?}{=} \frac{20}{1-e^{-0.6931}}$; $40 \overset{?}{=} \frac{20}{1-0.5}$; $40 \overset{?}{=} \frac{20}{0.5}$; $40 = 40$

Example 6.6-10

Given $N = N_0 e^{kt}$ solve for t if $N = 4N_0$ and $k = 0.05$.

$\boxed{N = N_0 e^{kt}}$; $\boxed{4N_0 = N_0 e^{0.05t}}$; $\boxed{\frac{4\cancel{N_0}}{\cancel{N_0}} = e^{0.05t}}$; $\boxed{\frac{4}{1} = e^{0.05t}}$; $\boxed{4 = e^{0.05t}}$; $\boxed{ln 4 = ln e^{0.05t}}$; $\boxed{ln 4 = 0.05t}$

; $\boxed{t = \frac{ln 4}{0.05}}$; $\boxed{t = \frac{1.3863}{0.05}}$; $\boxed{\mathbf{t = 27.726}}$

Check: $4N_0 \overset{?}{=} N_0 e^{0.05\times 27.726}$; $4N_0 \overset{?}{=} N_0 e^{1.3863}$; $4N_0 \overset{?}{=} N_0 e^{1.3863}$; $4N_0 = 4N_0$

Example 6.6-11

Given $W = W_0\left(4-e^{4t-2}\right)$ solve for t if $W = 20$ and $W_0 = 1000$.

$\boxed{W = W_0\left(4-e^{4t-2}\right)}$; $\boxed{20 = 1000\left(4-e^{4t-2}\right)}$; $\boxed{\frac{\cancel{20}}{\underset{50}{\cancel{1000}}} = \left(4-e^{4t-2}\right)}$; $\boxed{\frac{1}{50} = 4-e^{4t-2}}$; $\boxed{0.02-4 = -e^{4t-2}}$

; $\boxed{-3.98 = -e^{4t-2}}$; $\boxed{3.98 = e^{4t-2}}$; $\boxed{log_e 3.98 = log_e e^{4t-2}}$; $\boxed{1.3813 = 4t-2}$; $\boxed{1.3813+2 = 4t}$; $\boxed{3.3813 = 4t}$

; $\boxed{t = \frac{3.3813}{4}}$; $\boxed{\mathbf{t = 0.8453}}$

Check: $20 \overset{?}{=} 1000\left(4-e^{4\times 0.8453-2}\right)$; $20 \overset{?}{=} 1000\left(4-e^{3.3812-2}\right)$; $20 \overset{?}{=} 1000\left(4-e^{1.3812}\right)$; $20 \overset{?}{=} 1000(4-3.9797)$

; $20 \overset{?}{=} 1000\times 0.020$; $20 = 20$

Example 6.6-12

Given $N = N_0 e^{-k(t-1)}$ solve for t if $N = 5$, $N_0 = 25$, and $k = 0.05$.

$\boxed{N = N_0 e^{-k(t-1)}}$; $\boxed{5 = 25e^{-0.05(t-1)}}$; $\boxed{\frac{\cancel{5}}{\underset{5}{\cancel{25}}} = e^{-0.05t+0.05}}$; $\boxed{\frac{1}{5} = e^{-0.05t+0.05}}$; $\boxed{0.2 = e^{-0.05t+0.05}}$

; $\boxed{log_e\, 0.2 = log_e\, e^{-0.05t+0.05}}$; $\boxed{log_e\, 0.2 = -0.05t + 0.05}$; $\boxed{-1.6094 = -0.05t + 0.05}$; $\boxed{-1.6094 - 0.05 = -0.05t}$

; $\boxed{-1.6594 = -0.05t}$; $\boxed{t = \frac{1.6594}{0.05}}$; $\boxed{\mathbf{t = 33.188}}$

Check: $5 \overset{?}{=} 25e^{-0.05(33.188-1)}$; $5 \overset{?}{=} 25e^{-0.05\times 32.188}$; $5 \overset{?}{=} 25e^{-1.6094}$; $5 \overset{?}{=} 25\frac{1}{e^{1.6094}}$; $5 \overset{?}{=} \frac{25}{5}$; $5 = 5$

Example 6.6-13

Given $B = B_0 e^{-kx}$ solve for x if $B = 10$, $B_0 = 2$, and $k = 0.01$.

$\boxed{B = B_0 e^{-kx}}$; $\boxed{10 = 2e^{-0.01x}}$; $\boxed{\frac{\overset{5}{\cancel{10}}}{\cancel{2}} = e^{-0.01x}}$; $\boxed{5 = e^{-0.01x}}$; $\boxed{log_e\, 5 = log_e\, e^{-0.01x}}$; $\boxed{log_e\, 5 = -0.01x}$

; $\boxed{1.6094 = -0.01x}$; $\boxed{-\frac{1.6094}{0.01} = x}$; $\boxed{\mathbf{x = -160.94}}$

Check: $10 \overset{?}{=} 2e^{-0.01\times -160.94}$; $10 \overset{?}{=} 2e^{1.6094}$; $10 \overset{?}{=} 2\times 5$; $10 = 10$

Example 6.6-14

Given $F = F_0\left(1 - e^{kt}\right)$ solve for t if $F = -2F_0$, and $k = 2$.

$\boxed{F = F_0\left(1 - e^{kt}\right)}$; $\boxed{-2F_0 = F_0\left(1 - e^{2t}\right)}$; $\boxed{\frac{-2\cancel{F_0}}{\cancel{F_0}} = \left(1 - e^{2t}\right)}$; $\boxed{-2 = 1 - e^{2t}}$; $\boxed{-2 - 1 = -e^{2t}}$; $\boxed{-3 = -e^{2t}}$

; $\boxed{3 = e^{2t}}$; $\boxed{log_e\, 3 = log_e\, e^{2t}}$; $\boxed{1.0986 = 2t}$; $\boxed{t = \frac{1.0986}{2}}$; $\boxed{\mathbf{t = 0.5493}}$

Check: $-2F_0 \overset{?}{=} F_0\left(1 - e^{2\times 0.5493}\right)$; $-2F_0 \overset{?}{=} F_0\left(1 - e^{1.0986}\right)$; $-2F_0 \overset{?}{=} F_0(1-3)$; $-2F_0 = -2F_0$

Example 6.6-15

Given $K + 10 = K_0 \cdot 2^{-\frac{a}{5000}}$ solve for a if $K = 25$ and $K_0 = 10$.

$\boxed{K + 10 = K_0 \cdot 2^{-\frac{a}{5000}}}$; $\boxed{25 + 10 = 10 \cdot 2^{-\frac{a}{5000}}}$; $\boxed{\frac{35}{10} = 2^{-\frac{a}{5000}}}$; $\boxed{3.5 = 2^{-\frac{a}{5000}}}$; $\boxed{log_2\, 3.5 = log_2\, 2^{-\frac{a}{5000}}}$

; $\boxed{\frac{log_{10}\, 3.5}{log_{10}\, 2} = -\frac{a}{5000} log_2\, 2}$; $\boxed{\frac{0.5441}{0.3010} = -\frac{a}{5000}}$; $\boxed{1.8075 = -\frac{a}{5000}}$; $\boxed{\mathbf{a = -9037.5}}$

Check: $25 + 10 \overset{?}{=} 10 \cdot 2^{-\frac{-9037.5}{5000}}$; $35 \overset{?}{=} 10 \cdot 2^{1.8075}$; $35 \overset{?}{=} 10 \cdot 3.5$; $35 = 35$

Example 6.6-16

Given $P = S\left[\frac{r}{1-(1+r)^{-n}}\right]$ solve for n if $P = 500$, $S = 2000$, and $r = 0.12$.

$\boxed{P = S\left[\frac{r}{1-(1+r)^{-n}}\right]}$; $\boxed{500 = 2000\left[\frac{0.12}{1-(1+0.12)^{-n}}\right]}$; $\boxed{\frac{\cancel{500}}{\underset{4}{\cancel{2000}}} = \frac{0.12}{1-1.12^{-n}}}$; $\boxed{\frac{1}{4} = \frac{0.12}{1-1.12^{-n}}}$

; $\boxed{1\times\left[1-1.12^{-n}\right]=4\times 0.12}$; $\boxed{1-1.12^{-n}=0.48}$; $\boxed{-1.12^{-n}=0.48-1}$; $\boxed{1.12^{-n}=0.52}$; $\boxed{log_{1.12}1.12^{-n}=log_{1.12}0.52}$

; $\boxed{-n=log_{1.12}0.52}$; $\boxed{-n=\frac{log_{10}0.52}{log_{10}1.12}}$; $\boxed{-n=\frac{-0.2839}{0.0492}}$; $\boxed{-n=-5.7703}$; $\boxed{\frac{-n}{-1}=\frac{-5.7703}{-1}}$; $\boxed{\mathbf{n=5.7703}}$

Check: $500\overset{?}{=}2000\left[\frac{0.12}{1-(1+0.12)^{-5.7703}}\right]$; $500\overset{?}{=}2000\left[\frac{0.12}{1-1.12^{-5.7703}}\right]$; $500\overset{?}{=}2000\left[\frac{0.12}{1-0.5199}\right]$

; $500\overset{?}{=}2000\left[\frac{0.12}{0.4801}\right]$; $500\overset{?}{=}2000\times 0.25$; $500=500$

Example 6.6-17

Given $L=10log_{10}\frac{I}{I_0}$ solve for I if $I_0=10^{-8}$ and $L=100$.

$\boxed{L=10log_{10}\frac{I}{I_0}}$; $\boxed{100=10log_{10}\frac{I}{10^{-8}}}$; $\boxed{\frac{10}{\cancel{100}}{\cancel{10}}=log_{10}\frac{I}{10^{-8}}}$; $\boxed{\frac{10}{1}=log_{10}I-log_{10}10^{-8}}$; $\boxed{10=log_{10}I+8}$

; $\boxed{10-8=log_{10}I}$; $\boxed{2=log_{10}I}$; $\boxed{10^2=10^{log_{10}I}}$; $\boxed{10^2=I}$; $\boxed{\mathbf{I=100}}$

Check: $100\overset{?}{=}10\log_{10}\frac{100}{10^{-8}}$; $100\overset{?}{=}10\log_{10}100-10\log_{10}10^{-8}$; $100\overset{?}{=}10\times 2+80\log_{10}10$; $100\overset{?}{=}20+80$

; $100=100$

Example 6.6-18

Given $V=P\left(1+\frac{r}{100n}\right)^{nk}$ solve for r if $V=4$, $P=2$, $n=1$, and $k=12$.

$\boxed{V=P\left(1+\frac{r}{100n}\right)^{nk}}$; $\boxed{4=2\left(1+\frac{r}{100\cdot 1}\right)^{1\cdot 12}}$; $\boxed{\frac{\overset{2}{\cancel{4}}}{\cancel{2}}=\left(1+\frac{r}{100}\right)^{12}}$; $\boxed{2=\left(1+\frac{r}{100}\right)^{12}}$; $\boxed{log_{10}2=log_{10}\left(1+\frac{r}{100}\right)^{12}}$

; $\boxed{0.3010=12log_{10}\left(1+\frac{r}{100}\right)}$; $\boxed{\frac{0.3010}{12}=log_{10}\left(1+\frac{r}{100}\right)}$; $\boxed{0.02508=log_{10}\left(1+\frac{r}{100}\right)}$; $\boxed{10^{0.02508}=10^{log_{10}\left(1+\frac{r}{100}\right)}}$

; $\boxed{1.05944=1+\frac{r}{100}}$ = $\boxed{1.05944-1=\frac{r}{100}}$; $\boxed{0.05944=\frac{r}{100}}$; $\boxed{\mathbf{r=5.944}}$

Check: $4\overset{?}{=}2\left(1+\frac{5.944}{100\cdot 1}\right)^{1\cdot 12}$; $4\overset{?}{=}2(1+0.05944)^{12}$; $4\overset{?}{=}2(1.05944)^{12}$; $4\overset{?}{=}2\times 1.9998$; $4=4$

Example 6.6-19

Given $W=-nRT\,ln\left(\frac{v_f}{v_o}\right)$ solve for v_f if $W=5000$, $T=373$, $R=10$, $n=2$, and $v_0=10$.

$\boxed{W=-nRT\,ln\left(\frac{v_f}{v_o}\right)}$; $\boxed{5000=-2\cdot 10\cdot 373\cdot ln\left(\frac{v_f}{10}\right)}$; $\boxed{5000=-7460\left(ln\,v_f-ln10\right)}$; $\boxed{-\frac{5000}{7460}=\left(ln\,v_f-ln10\right)}$

; $\boxed{-0.6702=ln\,v_f-2.3026}$; $\boxed{-0.6702+2.3026=ln\,v_f}$; $\boxed{1.6324=ln\,v_f}$; $\boxed{e^{1.6324}=e^{ln\,v_f}}$; $\boxed{\mathbf{v_f=5.1161}}$

Check: $5000\stackrel{?}{=}-2\cdot 10\cdot 373\cdot \ln\left(\frac{5.1161}{10}\right)$; $5000\stackrel{?}{=}-7460\cdot \ln 0.5116$; $5000\stackrel{?}{=}-7460\cdot -0.6702$; $5000=5000$

Example 6.6-20

Given $A=A_0\cdot 4^{4t+0.2}$ solve for t if $A=1000$ and $A_0=100$.

$\boxed{A=A_0\cdot 4^{4t+0.2}}$; $\boxed{1000=100\cdot 4^{4t+0.2}}$; $\boxed{\frac{\overset{10}{\cancel{1000}}}{\cancel{100}}=4^{4t+0.2}}$; $\boxed{\frac{10}{1}=4^{4t+0.2}}$; $\boxed{10=4^{4t+0.2}}$; $\boxed{\log_4 10=\log_4 4^{4t+0.2}}$

; $\boxed{\log_4 10=4t+0.2}$; $\boxed{\frac{\log_{10}10}{\log_{10}4}-0.2=4t}$; $\boxed{\frac{1}{0.6021}-0.2=4t}$; $\boxed{1.6608-0.2=4t}$; $\boxed{t=\frac{1.4608}{4}}$; $\boxed{\mathbf{t=0.3652}}$

Check: $1000\stackrel{?}{=}100\cdot 4^{4\times 0.3652+0.2}$; $1000\stackrel{?}{=}100\cdot 4^{1.4608+0.2}$; $1000\stackrel{?}{=}100\cdot 4^{1.6608}$; $1000\stackrel{?}{=}100\cdot 10$; $1000=1000$

Practice Problems - Advanced Logarithmic Problems

Section 6.6 Practice Problems - Solve the following equations using the logarithmic properties:

1. Given $P=P_0\left(1+\frac{r}{k}\right)^{kt}$ solve for t if $P=20$, $P_0=10$, $r=1$, and $k=2$.
2. Given $A=A_0+ke^{-t}$ solve for t if $A=100$, $A_0=10$, and $k=2$.
3. Given $B=B_0+ke^{-(t+a)}$ solve for t if $B=200$, $B_0=25$, $k=5$, and $a=0.01$.
4. Given $N=N_0e^{-\frac{t}{25000}}$ solve for t if $N=2$ and $N_0=5N$.
5. Given $Q=Q_0(1-m)^{0.05t}$ solve for t if $Q=0.4$, $Q_0=40$, and $m=0.5$.
6. Given $K=K_0 4^{-\frac{kr}{200}}$ solve for r if $K=5K_0$, $K_0=0.1$, and $k=1$.
7. Given $M=M_0e^{-k(t-4)}$ solve for t if $M=5$, $M_0=500$, and $k=0.1$.
8. Given $Y=Y_0(1+r)^n$ solve for n if $Y=10$, $Y_0=2$, and $r=0.25$.
9. Given $A=A_0\left(2^{-5t}\right)+\ln A_0$ solve for t if $A=10$ and $A_0=0.02$.
10. Given $U=U_0e^{4t-1}+2$ solve for t if $U=8$ and $U_0=4$.

Appendix - Exercise Solutions

Chapter 1 Solutions:

Section 1.1a Solutions - Signed Numbers

1. $\frac{-95}{-5} = \frac{95}{5} = \mathbf{19}$

2. $(-20) \times (-8) = +160 = \mathbf{160}$

3. $(-33) + (-14) = -33 - 14 = \mathbf{-47}$

4. $(-18) - (-5) = (-18) + (5) = -18 + 5 = \mathbf{-13}$

5. $(-20) + 8 = -20 + 8 = \mathbf{-12}$

6. $\frac{48}{-4} = -\frac{48}{4} = \mathbf{-12}$

Section 1.1b Solutions - Using Parentheses and Brackets in Mixed Operations

1. $(28 \div 4) \times 3 = (7) \times 3 = 7 \times 3 = \mathbf{21}$
2. $250 + (15 \div 3) = 250 + (5) = 250 + 5 = \mathbf{255}$
3. $28 \div [(23+5) \times 8] = 28 \div [(28) \times 8] = 28 \div [28 \times 8] = 28 \div [224] = 28 \div 224 = \mathbf{0.125}$
4. $[(255-15) \div 20] + 8 = [(240) \div 20] + 8 = [240 \div 20] + 8 = [12] + 8 = 12 + 8 = \mathbf{20}$
5. $[230 \div (15 \times 2)] + 12 = [230 \div (30)] + 12 = [230 \div 30] + 12 = [7.67] + 12 = 7.67 + 12 = \mathbf{19.67}$
6. $55 \times [(28+2) \div 3] = 55 \times [(30) \div 3] = 55 \times [30 \div 3] = 55 \times [10] = 55 \times 10 = \mathbf{550}$

Section 1.2a Solutions - Simplifying Integer Fractions

1. $\frac{60}{150} = \frac{60 \div 30}{150 \div 30} = \mathbf{\frac{2}{5}}$

2. $\frac{8}{18} = \frac{8 \div 2}{18 \div 2} = \mathbf{\frac{4}{9}}$

3. $\frac{355}{15} = \frac{355 \div 5}{15 \div 5} = \frac{71}{3} = \mathbf{23\frac{2}{3}}$

4. $\mathbf{\frac{3}{8}}$ **is in its lowest term.**

5. $\frac{27}{6} = \frac{27 \div 3}{6 \div 3} = \frac{9}{2} = \mathbf{4\frac{1}{2}}$

6. $\frac{33}{6} = \frac{33 \div 3}{6 \div 3} = \frac{11}{2} = \mathbf{5\frac{1}{2}}$

Section 1.2b Cases I and II Solutions - Adding Integer Fractions with or without a Common Denominator

1. $\frac{4}{9} + \frac{2}{9} = \frac{4+2}{9} = \frac{\overset{2}{\cancel{6}}}{\underset{3}{\cancel{9}}} = \mathbf{\frac{2}{3}}$

2. $\frac{3}{8} + \frac{2}{5} = \frac{3}{8} + \frac{2}{5} = \frac{(3 \times 5) + (2 \times 8)}{8 \times 5} = \frac{15+16}{40} = \mathbf{\frac{31}{40}}$

3. $\frac{3}{8} + \frac{2}{4} + \frac{5}{6} = \left(\frac{3}{8} + \frac{2}{4}\right) + \frac{5}{6} = \left(\frac{(3 \times 4) + (2 \times 8)}{8 \times 4}\right) + \frac{5}{6} = \left(\frac{12+16}{32}\right) + \frac{5}{6} = \left(\frac{28}{32}\right) + \frac{5}{6} = \frac{28}{32} + \frac{5}{6} = \frac{(28 \times 6) + (5 \times 32)}{32 \times 6}$

$= \frac{168+160}{192} = \frac{\overset{41}{\cancel{328}}}{\underset{24}{\cancel{192}}} = \frac{41}{24} = \mathbf{1\frac{17}{24}}$

4. $\frac{4}{5}+\frac{2}{5}+\frac{3}{5} = \frac{4+2+3}{5} = \frac{9}{5} = \mathbf{1\frac{4}{5}}$

5. $5+\frac{0}{10}+\frac{6}{1}+\frac{4}{8} = \left(\frac{5}{1}+\frac{0}{10}\right)+\left(\frac{6}{1}+\frac{4}{8}\right) = \left(\frac{5}{1}+0\right)+\left(\frac{(6\times 8)+(4\times 1)}{1\times 8}\right) = \left(\frac{5}{1}\right)+\left(\frac{48+4}{8}\right) = \frac{5}{1}+\left(\frac{52}{8}\right) = \frac{5}{1}+\frac{52}{8}$

$= \frac{(5\times 8)+(52\times 1)}{1\times 8} = \frac{40+52}{8} = \frac{\overset{23}{\cancel{92}}}{\underset{2}{\cancel{8}}} = \frac{23}{2} = \mathbf{11\frac{1}{2}}$

6. $\left(\frac{3}{16}+\frac{1}{8}\right)+\frac{1}{6} = \left(\frac{(3\times 8)+(1\times 16)}{16\times 8}\right)+\frac{1}{6} = \left(\frac{24+16}{128}\right)+\frac{1}{6} = \left(\frac{40}{128}\right)+\frac{1}{6} = \frac{\overset{5}{\cancel{40}}}{\underset{16}{\cancel{128}}}+\frac{1}{6} = \frac{5}{16}+\frac{1}{6} = \frac{(5\times 6)+(1\times 16)}{16\times 6}$

$= \frac{30+16}{96} = \frac{\overset{23}{\cancel{46}}}{\underset{48}{\cancel{96}}} = \mathbf{\frac{23}{48}}$

Section 1.2b Cases III and IV Solutions - Subtracting Integer Fractions with or without a Common Denominator

1. $\frac{3}{5}-\frac{2}{5} = \frac{3-2}{5} = \mathbf{\frac{1}{5}}$

2. $\frac{2}{5}-\frac{3}{4} = \frac{(2\times 4)-(3\times 5)}{5\times 4} = \frac{8-15}{20} = \mathbf{-\frac{7}{20}}$

3. $\frac{12}{15}-\frac{3}{15}-\frac{6}{15} = \frac{12-3-6}{15} = \frac{\overset{1}{\cancel{3}}}{\underset{5}{\cancel{15}}} = \mathbf{\frac{1}{5}}$

4. $\frac{5}{8}-\frac{3}{4}-\frac{1}{3} = \left(\frac{5}{8}-\frac{3}{4}\right)-\frac{1}{3} = \left(\frac{(5\times 4)-(3\times 8)}{8\times 4}\right)-\frac{1}{3} = \left(\frac{20-24}{32}\right)-\frac{1}{3} = \left(\frac{-4}{32}\right)-\frac{1}{3} = \frac{\overset{-1}{\cancel{-4}}}{\underset{8}{\cancel{32}}}-\frac{1}{3} = \frac{-1}{8}-\frac{1}{3}$

$= \frac{(-1\times 3)-(1\times 8)}{8\times 3} = \frac{-3-8}{24} = \mathbf{-\frac{11}{24}}$

5. $\left(\frac{2}{8}-\frac{1}{6}\right)-\frac{2}{5} = \left(\frac{(2\times 6)-(1\times 8)}{8\times 6}\right)-\frac{2}{5} = \left(\frac{12-8}{48}\right)-\frac{2}{5} = \left(\frac{4}{48}\right)-\frac{2}{5} = \frac{\overset{1}{\cancel{4}}}{\underset{12}{\cancel{48}}}-\frac{2}{5} = \frac{1}{12}-\frac{2}{5} = \frac{(1\times 5)-(2\times 12)}{12\times 5} = \mathbf{-\frac{19}{60}}$

6. $28-\left(\frac{1}{8}-\frac{2}{3}\right) = \frac{28}{1}-\left(\frac{(1\times 3)-(2\times 8)}{8\times 3}\right) = \frac{28}{1}-\left(\frac{3-16}{24}\right) = \frac{28}{1}-\left(\frac{-13}{24}\right) = \frac{28}{1}-\frac{-13}{24} = \frac{28}{1}+\frac{13}{24} = \frac{(28\times 24)+(13\times 1)}{1\times 24}$

$= \frac{672+13}{24} = \frac{685}{24} = \mathbf{28\frac{13}{24}}$

Section 1.2b Case V Solutions - Multiplying Integer Fractions with or without a Common Denominator

1. $\frac{4}{8}\times\frac{3}{5} = \frac{\overset{1}{\cancel{4}}\times 3}{\underset{2}{\cancel{8}}\times 5} = \frac{1\times 3}{2\times 5} = \mathbf{\frac{3}{10}}$

2. $\frac{4}{8}\times\frac{5}{6}\times 100 = \frac{4}{8}\times\frac{5}{6}\times\frac{100}{1} = \frac{4\times 5\times 100}{8\times 6\times 1} = \frac{\overset{125}{\cancel{2000}}}{\underset{3}{\cancel{48}}} = \frac{125}{3} = \mathbf{41\frac{2}{3}}$

3. $\frac{7}{3}\times\frac{9}{4}\times\frac{6}{3} = \frac{7\times\overset{3}{\cancel{9}}\times\overset{2}{\cancel{6}}}{\underset{1}{\cancel{3}}\times 4\times\underset{1}{\cancel{3}}} = \frac{7\times 3\times 2}{1\times 4\times 1} = \frac{7\times 3\times\overset{1}{\cancel{2}}}{1\times\underset{2}{\cancel{4}}\times 1} = \frac{7\times 3\times 1}{1\times 2\times 1} = \frac{21}{2} = \mathbf{10\frac{1}{2}}$

4. $34\times\frac{1}{5}\times\frac{3}{17}\times\frac{1}{8}\times 20 = \frac{34}{1}\times\frac{1}{5}\times\frac{3}{17}\times\frac{1}{8}\times\frac{20}{1} = \frac{\overset{2}{\cancel{34}}\times 1\times 3\times 1\times\overset{4}{\cancel{20}}}{1\times\underset{1}{\cancel{5}}\times\underset{1}{\cancel{17}}\times 8\times 1} = \frac{2\times 1\times 3\times 1\times 4}{1\times 1\times 1\times 8\times 1} = \frac{\overset{3}{\cancel{24}}}{\underset{1}{\cancel{8}}} = \frac{3}{1} = \mathbf{3}$

5. $\left(\frac{2}{55}\times 3\right)\times\left(\frac{4}{5}\times\frac{25}{8}\right) = \left(\frac{2}{55}\times\frac{3}{1}\right)\times\left(\frac{\overset{1}{\cancel{4}}\times\overset{5}{\cancel{25}}}{\underset{1}{\cancel{5}}\times\underset{2}{\cancel{8}}}\right) = \left(\frac{2\times 3}{55\times 1}\right)\times\left(\frac{1\times 5}{1\times 2}\right) = \left(\frac{6}{55}\right)\times\left(\frac{5}{2}\right) = \frac{6}{55}\times\frac{5}{2} = \frac{\overset{3}{\cancel{6}}\times\overset{1}{\cancel{5}}}{\underset{11}{\cancel{55}}\times\underset{1}{\cancel{2}}} = \frac{3\times 1}{11\times 1} = \mathbf{\frac{3}{11}}$

6. $\left(1000\times\frac{1}{5}\right)\times\left(\frac{25}{5}\times\frac{1}{8}\right)\times\frac{0}{100} = \mathbf{0}$

Section 1.2b Case VI Solutions - Dividing Integer Fractions with or without a Common Denominator

1. $\frac{8}{10}\div\frac{4}{30} = \frac{8}{10}\times\frac{30}{4} = \frac{\overset{2}{\cancel{8}}\times\overset{3}{\cancel{30}}}{\underset{1}{\cancel{10}}\times\underset{1}{\cancel{4}}} = \frac{2\times 3}{1\times 1} = \frac{6}{1} = \mathbf{6}$

2. $\left(\frac{3}{8}\div\frac{12}{16}\right)\div\frac{4}{8} = \left(\frac{3}{8}\times\frac{16}{12}\right)\div\frac{4}{8} = \left(\frac{\overset{1}{\cancel{3}}\times\overset{2}{\cancel{16}}}{\underset{1}{\cancel{8}}\times\underset{4}{\cancel{12}}}\right)\div\frac{\overset{1}{\cancel{4}}}{\underset{2}{\cancel{8}}} = \left(\frac{1\times 2}{1\times 4}\right)\div\frac{1}{2} = \left(\frac{2}{4}\right)\div\frac{1}{2} = \frac{2}{4}\div\frac{1}{2} = \frac{2}{4}\times\frac{2}{1} = \frac{2\times 2}{4\times 1} = \frac{\overset{1}{\cancel{4}}}{\underset{1}{\cancel{4}}} = \frac{1}{1} = \mathbf{1}$

3. $\left(\frac{4}{16}\div\frac{1}{32}\right)\div 8 = \left(\frac{4}{16}\times\frac{32}{1}\right)\div\frac{8}{1} = \left(\frac{4\times\overset{2}{\cancel{32}}}{\underset{1}{\cancel{16}}\times 1}\right)\div\frac{8}{1} = \left(\frac{4\times 2}{1\times 1}\right)\div\frac{8}{1} = \frac{8}{1}\div\frac{8}{1} = \frac{8}{1}\times\frac{1}{8} = \frac{\overset{1}{\cancel{8}}\times 1}{1\times\underset{1}{\cancel{8}}} = \frac{1\times 1}{1\times 1} = \frac{1}{1} = \mathbf{1}$

4. $12\div\left(\frac{9}{8}\div\frac{27}{16}\right) = \frac{12}{1}\div\left(\frac{9}{8}\times\frac{16}{27}\right) = \frac{12}{1}\div\left(\frac{\overset{1}{\cancel{9}}\times\overset{2}{\cancel{16}}}{\underset{1}{\cancel{8}}\times\underset{3}{\cancel{27}}}\right) = \frac{12}{1}\div\left(\frac{1\times 2}{1\times 3}\right) = \frac{12}{1}\div\left(\frac{2}{3}\right) = \frac{12}{1}\div\frac{2}{3} = \frac{12}{1}\times\frac{3}{2} = \frac{\overset{6}{\cancel{12}}\times 3}{1\times\underset{1}{\cancel{2}}} = \mathbf{18}$

5. $\left(\frac{2}{20}\div\frac{4}{5}\right)\div 2 = \left(\frac{2}{20}\times\frac{5}{4}\right)\div\frac{2}{1} = \left(\frac{\overset{1}{\cancel{2}}\times\overset{1}{\cancel{5}}}{\underset{4}{\cancel{20}}\times\underset{2}{\cancel{4}}}\right)\div\frac{2}{1} = \left(\frac{1\times 1}{4\times 2}\right)\div\frac{2}{1} = \left(\frac{1}{8}\right)\div\frac{2}{1} = \frac{1}{8}\div\frac{2}{1} = \frac{1}{8}\times\frac{1}{2} = \frac{1\times 1}{8\times 2} = \mathbf{\frac{1}{16}}$

6. $\left(\frac{4}{15}\div\frac{8}{30}\right)\div\left(\frac{1}{5}\div\frac{4}{35}\right) = \left(\frac{4}{15}\times\frac{30}{8}\right)\div\left(\frac{1}{5}\times\frac{35}{4}\right) = \left(\frac{\overset{1}{\cancel{4}}\times\overset{2}{\cancel{30}}}{\underset{1}{\cancel{15}}\times\underset{2}{\cancel{8}}}\right)\div\left(\frac{1\times\overset{7}{\cancel{35}}}{\underset{1}{\cancel{5}}\times 4}\right) = \left(\frac{1\times 2}{1\times 2}\right)\div\left(\frac{1\times 7}{1\times 4}\right) = \left(\frac{2}{2}\right)\div\left(\frac{7}{4}\right) = \frac{2}{2}\div\frac{7}{4}$

$= \frac{2}{2}\times\frac{4}{7} = \frac{2\times\overset{2}{\cancel{4}}}{\underset{1}{\cancel{2}}\times 7} = \frac{2\times 2}{1\times 7} = \mathbf{\frac{4}{7}}$

Section 1.2 Appendix Solutions - Changing Improper Fractions to Mixed Fractions

1. $\frac{83}{4} = \mathbf{20\frac{3}{4}}$
2. $\frac{13}{3} = \mathbf{4\frac{1}{3}}$
3. $-\frac{26}{5} = -\left(\mathbf{5\frac{1}{5}}\right)$
4. $\frac{67}{10} = \mathbf{6\frac{7}{10}}$
5. $\frac{9}{2} = \mathbf{4\frac{1}{2}}$
6. $-\frac{332}{113} = -\left(\mathbf{2\frac{106}{113}}\right)$

Section 1.3a Case I Solutions - Real Numbers Raised to Positive Integer Exponents

1. $4^3 = 4\cdot 4\cdot 4 = \mathbf{64}$
2. $(-10)^4 = -10\cdot -10\cdot -10\cdot -10 = \mathbf{+10000}$
3. $0.25^3 = 0.25\cdot 0.25\cdot 0.25 = \mathbf{0.0156}$
4. $12^5 = 12\cdot 12\cdot 12\cdot 12\cdot 12 = \mathbf{248832}$
5. $-(3)^5 = -(3\cdot 3\cdot 3\cdot 3\cdot 3) = \mathbf{-243}$
6. $489^0 = \mathbf{1}$

Section 1.3a Case II Solutions - Real Numbers Raised to Negative Integer Exponents

1. $4^{-3} = \frac{1}{4^3} = \frac{1}{4\cdot 4\cdot 4} = \mathbf{\frac{1}{64}}$

2. $(-5)^{-4} = \frac{1}{(-5)^4} = \frac{1}{(-5)\cdot(-5)\cdot(-5)\cdot(-5)} = \mathbf{\frac{1}{625}}$

3. $0.25^{-3} = \frac{1}{0.25^3} = \frac{1}{(0.25)\cdot(0.25)\cdot(0.25)} = \mathbf{\frac{1}{0.0156}}$

4. $12^{-5} = \frac{1}{12^5} = \frac{1}{12\cdot 12\cdot 12\cdot 12\cdot 12} = \mathbf{\frac{1}{248832}}$

5. $-(3)^{-4} = -\frac{1}{3^4} = -\frac{1}{3\cdot 3\cdot 3\cdot 3} = \mathbf{-\frac{1}{81}}$

6. $48^{-2} = \frac{1}{48^2} = \frac{1}{48\cdot 48} = \mathbf{\frac{1}{2304}}$

Section 1.3b Case I Solutions - Multiplying Positive Integer Exponents

1. $x^2\cdot x^3\cdot x = x^2\cdot x^3\cdot x^1 = x^{2+3+1} = \mathbf{x^6}$

2. $2\cdot a^2\cdot b^0\cdot a^3\cdot b^2 = 2\cdot\left(a^2\cdot a^3\right)\cdot\left(b^0\cdot b^2\right) = 2\cdot\left(a^{2+3}\right)\cdot\left(b^{0+2}\right) = \mathbf{2a^5b^2}$

3. $\frac{4}{-6}a^2b^3ab^4b^5 = -\frac{\overset{2}{\cancel{4}}}{\underset{3}{\cancel{6}}}a^2b^3a^1b^4b^5 = -\frac{2}{3}\left(a^2a^1\right)\cdot\left(b^3b^4b^5\right) = -\frac{2}{3}\left(a^{2+1}\right)\cdot\left(b^{3+4+5}\right) = \mathbf{-\frac{2}{3}a^3b^{12}}$

4. $2^3\cdot 2^2\cdot x^{2a}\cdot x^{3a}\cdot x^a = \left(2^3\cdot 2^2\right)\cdot\left(x^{2a}\cdot x^{3a}\cdot x^a\right) = \left(2^{3+2}\right)\cdot\left(x^{2a+3a+a}\right) = 2^5\cdot x^{6a} = \mathbf{32x^{6a}}$

5. $\left(x\cdot y^2\cdot z^3\right)^0\cdot w^2z^3zw^4z^2 = 1\cdot w^2z^3z^1w^4z^2 = \left(w^2w^4\right)\cdot\left(z^3z^1z^2\right) = \left(w^{2+4}\right)\cdot\left(z^{3+1+2}\right) = \mathbf{w^6z^6}$

6. $2^0\cdot 4^2\cdot 4^2\cdot 2^2\cdot 4^1 = \left(2^0\cdot 2^2\right)\cdot\left(4^2\cdot 4^2\cdot 4^1\right) = \left(2^{0+2}\right)\cdot\left(4^{2+2+1}\right) = 2^2\cdot 4^5 = 4\cdot 1024 = \mathbf{4096}$

Section 1.3b Case II Solutions - Dividing Positive Integer Exponents

1. $\frac{x^5}{x^3} = \frac{x^5x^{-3}}{1} = \frac{x^{5-3}}{1} = \frac{x^2}{1} = \mathbf{x^2}$

2. $\frac{a^2b^3}{a} = \frac{a^2b^3}{a^1} = \frac{\left(a^2a^{-1}\right)b^3}{1} = \frac{a^{2-1}b^3}{1} = \frac{a^1b^3}{1} = \frac{ab^3}{1} = \mathbf{ab^3}$

3. $\frac{a^3b^3c^2}{a^2b^6c} = \frac{a^3b^3c^2}{a^2b^6c^1} = \frac{\left(a^3a^{-2}\right)\cdot\left(c^2c^{-1}\right)}{b^6b^{-3}} = \frac{\left(a^{3-2}\right)\cdot\left(c^{2-1}\right)}{b^{6-3}} = \frac{a^1\cdot c^1}{b^3} = \mathbf{\frac{ac}{b^3}}$

4. $\frac{3^2\cdot\left(rs^2\right)}{(2rs)\cdot r^3} = \frac{9\cdot\left(rs^2\right)}{\left(2r^1s\right)\cdot r^3} = \frac{9rs^2}{2\left(r^3r^1\right)\cdot s} = \frac{9}{2}\frac{rs^2}{r^{3+1}s} = \frac{9}{2}\frac{r^1s^2}{r^4s^1} = \frac{9}{2}\frac{s^2s^{-1}}{r^4r^{-1}} = \frac{9}{2}\frac{s^{2-1}}{r^{4-1}} = \frac{9}{2}\frac{s^1}{r^3} = \mathbf{4\frac{1}{2}\left(\frac{s}{r^3}\right)}$

5. $\frac{2p^2q^3pr^4}{-6p^4q^2r} = -\frac{\overset{1}{\cancel{2}}}{\underset{3}{\cancel{6}}}\frac{p^2q^3p^1r^4}{p^4q^2r^1} = -\frac{1}{3}\frac{\left(q^3q^{-2}\right)\cdot\left(r^4r^{-1}\right)}{p^4p^{-2}p^{-1}} = -\frac{1}{3}\frac{\left(q^{3-2}\right)\cdot\left(r^{4-1}\right)}{p^{4-2-1}} = -\frac{1}{3}\frac{q^1r^3}{p^1} = \mathbf{-\frac{1}{3}\left(\frac{qr^3}{p}\right)}$

6. $\frac{\left(k^2l^3\right)\cdot\left(kl^2m^0\right)}{k^4l^3m^5} = \frac{\left(k^2l^3\right)\cdot\left(kl^2\cdot 1\right)}{k^4l^3m^5} = \frac{\left(k^2l^3\right)\cdot\left(kl^2\right)}{k^4l^3m^5} = \frac{k^2l^3k^1l^2}{k^4l^3m^5} = \frac{l^3l^{-3}l^2}{\left(k^4k^{-2}k^{-1}\right)m^5} = \frac{l^{3-3+2}}{\left(k^{4-2-1}\right)m^5} = \frac{l^2}{k^1m^5} = \mathbf{\frac{l^2}{km^5}}$

Section 1.3b Case III Solutions - Adding and Subtracting Positive Integer Exponents

1. $x^2+4xy-2x^2-2xy+z^3 = \left(x^2-2x^2\right)+(4xy-2xy)+z^3 = (1-2)x^2+(4-2)xy+z^3 = \mathbf{-x^2+2xy+z^3}$

2. $\left(a^3+2a^2+4^3\right)-\left(4a^3+20\right) = \left(a^3+2a^2+4^3\right)+\left(-4a^3-20\right) = a^3+2a^2+64-4a^3-20 = \left(a^3-4a^3\right)+2a^2+(64-20)$
$= (1-4)a^3+2a^2+44 = \mathbf{-3a^3+2a^2+44}$

3. $3x^4+2x^2+2x^4-\left(x^4-2x^2+3\right) = 3x^4+2x^2+2x^4+\left(-x^4+2x^2-3\right) = 3x^4+2x^2+2x^4-x^4+2x^2-3$
$= \left(3x^4+2x^4-x^4\right)+\left(2x^2+2x^2\right)-3 = (3+2-1)x^4+(2+2)x^2-3 = \mathbf{4x^4+4x^2-3}$

4. $-\left(-2l^3a^3+2l^2a^2-5^3\right)-\left(4l^3a^3-20\right) = \left(+2l^3a^3-2l^2a^2+5^3\right)+\left(-4l^3a^3+20\right) = 2l^3a^3-2l^2a^2+125-4l^3a^3+20$
$= \left(2l^3a^3-4l^3a^3\right)-2l^2a^2+(125+20) = (2-4)l^3a^3-2l^2a^2+145 = \mathbf{-2l^3a^3-2l^2a^2+145}$

5. $\left(m^{3n}-4m^{2n}\right)-\left(2m^{3n}+3m^{2n}\right)+5m = \left(m^{3n}-4m^{2n}\right)+\left(-2m^{3n}-3m^{2n}\right)+5m = m^{3n}-4m^{2n}-2m^{3n}-3m^{2n}+5m$
$= \left(m^{3n}-2m^{3n}\right)+\left(-4m^{2n}-3m^{2n}\right)+5m = (1-2)m^{3n}+(-4-3)m^{2n}+5m = \mathbf{-m^{3n}-7m^{2n}+5m}$

6. $\left(-7z^3+3z-5\right)-\left(-3z^3+z-4\right)+5z+20 = \left(-7z^3+3z-5\right)+\left(3z^3-z+4\right)+5z+20 = -7z^3+3z-5+3z^3-z+4+5z+20$
$= \left(-7z^3+3z^3\right)+(3z-z+5z)+(-5+4+20) = (-7+3)z^3+(3-1+5)z+19 = \mathbf{-4z^3+7z+19}$

Section 1.3c Case I Solutions - Multiplying Negative Integer Exponents

1. $\left(3^{-3}\cdot 2^{-1}\right)\cdot\left(2^{-3}\cdot 3^{-2}\cdot 2\right) = 3^{-3}\cdot 2^{-1}\cdot 2^{-3}\cdot 3^{-2}\cdot 2^1 = \left(3^{-3}\cdot 3^{-2}\right)\cdot\left(2^{-1}\cdot 2^{-3}\cdot 2^1\right) = \left(3^{-3-2}\right)\cdot\left(2^{-1-3+1}\right) = 3^{-5}\cdot 2^{-3}$
$= \frac{1}{3^5\cdot 2^3} = \frac{1}{243\cdot 8} = \mathbf{\frac{1}{1944}}$

2. $a^{-6}\cdot b^{-4}\cdot a^{-1}\cdot b^{-2}\cdot a^0 = \left(a^{-6}a^{-1}a^0\right)\cdot\left(b^{-2}b^{-4}\right) = \left(a^{-6-1+0}\right)\cdot\left(b^{-2-4}\right) = a^{-7}b^{-6} = \mathbf{\frac{1}{a^7b^6}}$

3. $\left(a^{-2}\cdot b^{-3}\right)^2\cdot\left(a\cdot b^{-2}\right) = \left(a^{-2\times 2}\cdot b^{-3\times 2}\right)\cdot\left(a\cdot b^{-2}\right) = \left(a^{-4}\cdot b^{-6}\right)\cdot\left(a\cdot b^{-2}\right) = a^{-4}\cdot b^{-6}\cdot a\cdot b^{-2} = \left(a^{-4}\cdot a^1\right)\cdot\left(b^{-6}\cdot b^{-2}\right)$
$= \left(a^{-4+1}\right)\cdot\left(b^{-6-2}\right) = a^{-3}\cdot b^{-8} = \mathbf{\frac{1}{a^3b^8}}$

4. $(-2)^{-4}\left(r^{-2}s^2t\right)\cdot\left(r^3st^{-2}s^{-1}\right) = \frac{1}{(-2)^4}r^{-2}s^2t^1r^3s^1t^{-2}s^{-1} = \frac{1}{(-2\cdot-2\cdot-2\cdot-2)}\left(r^{-2}r^3\right)\cdot\left(s^2s^1s^{-1}\right)\cdot\left(t^1t^{-2}\right)$
$= \frac{1}{+16}\left(r^{-2+3}\right)\cdot\left(s^{2+1-1}\right)\cdot\left(t^{1-2}\right) = \frac{1}{16}r^1\cdot s^2\cdot t^{-1} = \frac{1}{16}\cdot\frac{rs^2}{t^1} = \mathbf{\frac{1}{16}\left(\frac{rs^2}{t}\right)}$

5. $\left(\frac{4}{5}\right)^{-4}2^2v^{-5}2^{-4}v^3v^{-2} = \left(\frac{4^{1\times-4}}{5^{1\times-4}}\right)\cdot\left(2^22^{-4}\right)\cdot\left(v^{-5}v^3v^{-2}\right) = \left(\frac{4^{-4}}{5^{-4}}\right)\cdot\left(2^{2-4}\right)\cdot\left(v^{-5+3-2}\right) = \left(\frac{5^4}{4^4}\right)\cdot 2^{-2}\cdot v^{-4}$
$= \left(\frac{625}{256}\right)\cdot\frac{1}{2^2v^4} = \left(\frac{625}{256}\right)\cdot\frac{1}{4\cdot v^4} = \left(\frac{625}{256\cdot 4}\right)\cdot\frac{1}{v^4} = \left(\frac{625}{1024}\right)\cdot\frac{1}{v^4} = \mathbf{\frac{625}{1024v^4}}$

6. $2^{-1}\cdot 3^2\cdot 3^{-5}\cdot 2^2\cdot 2^0 = \left(2^{-1}\cdot 2^2\cdot 2^0\right)\cdot\left(3^2\cdot 3^{-5}\right) = \left(2^{-1+2+0}\right)\cdot\left(3^{2-5}\right) = 2^1\cdot 3^{-3} = 2\cdot 3^{-3} = \frac{2}{3^3} = \mathbf{\frac{2}{27}}$

Section 1.3c Case II Solutions - Dividing Negative Integer Exponents

1. $\frac{x^{-2}x}{x^3x^0} = \frac{x^{-2}x^1}{x^3x^0} = \frac{x^{-2+1}}{x^{3+0}} = \frac{x^{-1}}{x^3} = \frac{1}{x^3x^1} = \frac{1}{x^{3+1}} = \mathbf{\frac{1}{x^4}}$

2. $\frac{-2a^{-2}b^3}{-6a^{-1}b^{-2}} = +\frac{\overset{1}{\cancel{2}}a^{-2}b^3}{\underset{3}{\cancel{6}}a^{-1}b^{-2}} = \frac{b^3b^2}{3a^2a^{-1}} = \frac{b^{3+2}}{3a^{2-1}} = \frac{b^5}{3a^1} = \mathbf{\frac{b^5}{3a}}$

3. $\frac{-(-3)^{-4}}{3\cdot(-3)^{-3}} = -\frac{(-3)^3}{3\cdot(-3)^4} = -\frac{-3\cdot-3\cdot-3}{3\cdot(-3\cdot-3\cdot-3\cdot-3)} = -\frac{-27}{3\cdot(+81)} = +\frac{27}{3\cdot81} = \frac{27}{243} = \mathbf{\frac{1}{9}}$

4. $\frac{-3^3y^{-3}yw}{(-3)^{-2}y^2w^{-3}} = -\frac{3^3y^{-3}y^1w^1}{(-3)^{-2}y^2w^{-3}} = -\frac{27\cdot(-3)^2w^1w^3}{y^2y^3y^{-1}} = -\frac{27\cdot(-3\cdot-3)w^{1+3}}{y^{2+3-1}} = -\frac{(27\cdot9)w^4}{y^4} = -\mathbf{\frac{243w^4}{y^4}}$

5. $\frac{a^{-2}b^2a^{-5}y^{-2}}{a^{-3}y} = \frac{a^{-2}b^2a^{-5}y^{-2}}{a^{-3}y^1} = \frac{b^2}{(a^{-3}a^2a^5)\cdot(y^2y^1)} = \frac{b^2}{a^{-3+2+5}\cdot y^{2+1}} = \frac{b^2}{a^4\cdot y^3} = \mathbf{\frac{b^2}{a^4y^3}}$

6. $\frac{(x\cdot y\cdot z)^0\cdot yx^{-2}}{x^{-4}y^{-1}} = \frac{1\cdot yx^{-2}}{x^{-4}y^{-1}} = \frac{y^1x^{-2}}{x^{-4}y^{-1}} = \frac{(x^4x^{-2})\cdot(y^1y^1)}{1} = \frac{(x^{4-2})\cdot(y^{1+1})}{1} = \frac{x^2\cdot y^2}{1} = \mathbf{x^2y^2}$

Section 1.3c Case III Solutions - Adding and Subtracting Negative Integer Exponents

1. $x^{-1}+2x^{-2}+3x^{-1}-6x^{-2} = (2x^{-2}-6x^{-2})+(x^{-1}+3x^{-1}) = (2-6)x^{-2}+(1+3)x^{-1} = -4x^{-2}+4x^{-1} = \frac{-4}{x^2}+\frac{4}{x^1}$

$= -\frac{4}{x^2}+\frac{4}{x} = \frac{(-4\cdot x)+(4\cdot x^2)}{x^2\cdot x} = \frac{4x^2-4x}{x^3} = \frac{4x(x-1)}{x^3} = \frac{4(x-1)}{x^3\cdot x^{-1}} = \frac{4(x-1)}{x^{3-1}} = \mathbf{\frac{4(x-1)}{x^2}}$

2. $(3a^{-4}-b^{-2})+(-2a^{-4}+3b^{-2}) = 3a^{-4}-b^{-2}-2a^{-4}+3b^{-2} = (3a^{-4}-2a^{-4})+(-b^{-2}+3b^{-2}) = (3-2)a^{-4}+(-1+3)b^{-2}$

$= a^{-4}+2b^{-2} = \frac{1}{a^4}+\frac{2}{b^2} = \frac{(b^2\cdot1)+(2\cdot a^4)}{a^4\cdot b^2} = \mathbf{\frac{b^2+2a^4}{a^4b^2}}$

3. $(xy)^{-1}+y^{-2}+4(xy)^{-1}-3y^{-2}+2^{-3} = \left[(xy)^{-1}+4(xy)^{-1}\right]+(y^{-2}-3y^{-2})+2^{-3} = [1+4](xy)^{-1}+(1-3)y^{-2}+2^{-3}$

$= 5(xy)^{-1}-2y^{-2}+2^{-3} = \frac{5}{xy}-\frac{2}{y^2}+\frac{1}{2^3} = \left(\frac{5}{xy}-\frac{2}{y^2}\right)+\frac{1}{8} = \left(\frac{(5\cdot y^2)-(2\cdot xy)}{xy\cdot y^2}\right)+\frac{1}{8} = \left(\frac{5y^2-2xy}{xy^3}\right)+\frac{1}{8}$

$= \frac{[8\cdot(5y^2-2xy)]+(1\cdot xy^3)}{8\cdot xy^3} = \frac{40y^2-16xy+xy^3}{8xy^3} = \frac{xy^3+40y^2-16xy}{8xy^3} = \frac{y(xy^2+40y-16x)}{8xy^3} = \frac{xy^2+40y-16x}{8xy^3y^{-1}}$

$= \frac{xy^2+40y-16x}{8xy^{3-1}} = \mathbf{\frac{xy^2+40y-16x}{8xy^2}}$

4. $4x^{-1}+y^{-3}+5y^{-3} = 4x^{-1}+(y^{-3}+5y^{-3}) = 4x^{-1}+(1+5)y^{-3} = 4x^{-1}+6y^{-3} = \frac{4}{x}+\frac{6}{y^3} = \frac{(4\cdot y^3)+(6\cdot x)}{x\cdot y^3} = \mathbf{\frac{4y^3+6x}{xy^3}}$

5. $m^{-5}-(m^{-2}-3m^{-5}+m^0)+3m^{-2} = m^{-5}-(m^{-2}-3m^{-5}+1)+3m^{-2} = m^{-5}+(-m^{-2}+3m^{-5}-1)+3m^{-2}$

$= m^{-5}-m^{-2}+3m^{-5}-1+3m^{-2} = (m^{-5}+3m^{-5})+(-m^{-2}+3m^{-2})-1 = (1+3)m^{-5}+(-1+3)m^{-2}-1 = 4m^{-5}+2m^{-2}-1$

$= \frac{4}{m^5}+\frac{2}{m^2}-1 = \left(\frac{4}{m^5}+\frac{2}{m^2}\right)-1 = \left(\frac{(4\cdot m^2)+(2\cdot m^5)}{m^5\cdot m^2}\right)-1 = \left(\frac{4m^2+2m^5}{m^{5+2}}\right)-1 = \frac{4m^2+2m^5}{m^7}-\frac{1}{1}$

$= \frac{[1\cdot(4m^2+2m^5)]-(1\cdot m^7)}{m^7\cdot1} = \frac{4m^2+2m^5-m^7}{m^7} = \frac{m^2(4+2m^3-m^5)}{m^7} = \frac{4+2m^3-m^5}{m^7m^{-2}} = \frac{4+2m^3-m^5}{m^{7-2}} = \mathbf{\frac{-m^5+2m^3+4}{m^5}}$

6. $\left(a^3\right)^{-2}+\left(a^{-2}b\right)^2-6a^{-6}+3a^{-4}b^2 = \left(a^{3\times-2}\right)+\left(a^{-2\times2}b^{1\times2}\right)-6a^{-6}+3a^{-4}b^2 = a^{-6}+a^{-4}b^2-6a^{-6}+3a^{-4}b^2$

$= \left(a^{-6}-6a^{-6}\right)+\left(a^{-4}b^2+3a^{-4}b^2\right) = (1-6)a^{-6}+(1+3)a^{-4}b^2 = -5a^{-6}+4a^{-4}b^2 = -\frac{5}{a^6}+\frac{4b^2}{a^4} = \frac{-\left(5\cdot a^4\right)+\left(a^6\cdot 4b^2\right)}{a^6\cdot a^4}$

$= \frac{-5a^4+4a^6b^2}{a^{6+4}} = \frac{a^4\left(-5+4a^2b^2\right)}{a^{10}} = \frac{-5+4a^2b^2}{a^{10}a^{-4}} = \frac{-5+4a^2b^2}{a^{10-4}} = \mathbf{\frac{4a^2b^2-5}{a^6}}$

Section 1.4a Case I Solutions - Roots and Radical Expressions

1. $\sqrt[2]{98} = \sqrt{98} = \sqrt{49\cdot2} = \sqrt{7^2\cdot2} = \mathbf{7\sqrt{2}}$

2. $3\sqrt{75} = 3\sqrt{25\cdot3} = 3\sqrt{5^2\cdot3} = (3\cdot5)\sqrt{3} = \mathbf{15\sqrt{3}}$

3. $\sqrt[3]{125} = \sqrt[3]{5^3} = \mathbf{5}$

4. $\sqrt[5]{3125} = \sqrt[5]{5^5} = \mathbf{5}$

5. $\sqrt[4]{162} = \sqrt[4]{81\cdot2} = \sqrt[4]{3^4\cdot2} = \mathbf{3\sqrt[4]{2}}$

6. $\sqrt[2]{192} = \sqrt{192} = \sqrt{64\cdot3} = \sqrt{8^2\cdot3} = \mathbf{8\sqrt{3}}$

Section 1.4a Case II Solutions - Rational, Irrational, Real, and Imaginary Numbers

1. $\frac{5}{8}$; is a rational and real number

2. $\sqrt{45} = \sqrt{9\cdot5} = 3\sqrt{5}$; is an irrational and real number

3. 450 ; is a rational and real number

4. $-\frac{2}{\sqrt{10}}$; is an irrational and real number

5. $-\sqrt{-5}$; is not a real number

6. $\frac{\sqrt{5}}{-2}$; is an irrational and real number

Section 1.4a Case III Solutions - Simplifying Radical Expressions with Real Numbers as a Radicand

1. $-\sqrt{49} = -\sqrt{7\cdot7} = -\sqrt{7\cdot7} = -\sqrt{7^1\cdot7^1} = -\sqrt{7^{1+1}} = -\sqrt{7^2} = \mathbf{-7}$

2. $\sqrt{54} = \sqrt{9\cdot5} = \sqrt{(3\cdot3)\cdot5} = \sqrt{\left(3^1\cdot3^1\right)\cdot5} = \sqrt{3^{1+1}\cdot5} = \sqrt{3^2\cdot5} = \mathbf{3\sqrt{5}}$

3. $-\sqrt{500} = -\sqrt{100\cdot5} = -\sqrt{(10\cdot10)\cdot5} = -\sqrt{\left(10^1\cdot10^1\right)\cdot5} = -\sqrt{\left(10^{1+1}\right)\cdot5} = -\sqrt{10^2\cdot5} = \mathbf{-10\sqrt{5}}$

4. $\sqrt[5]{3^5\cdot5} = \mathbf{3\sqrt[5]{5}}$

5. $\sqrt[2]{216} = \sqrt{216} = \sqrt{36\cdot6} = \sqrt{(6\cdot6)\cdot6} = \sqrt{\left(6^1\cdot6^1\right)\cdot6} = \sqrt{6^{1+1}\cdot6} = \sqrt{6^2\cdot6} = \mathbf{6\sqrt{6}}$

6. $-\frac{1}{4}\sqrt[4]{4^5\cdot2} = -\frac{1}{4}\sqrt[4]{4^{4+1}\cdot2} = -\frac{1}{4}\sqrt[4]{\left(4^4\cdot4^1\right)\cdot2} = -\frac{1}{4}\cdot4\sqrt[4]{(4\cdot2)} = -\frac{\frac{1}{4}}{\frac{4}{1}}\sqrt[4]{8} = \mathbf{-\sqrt[4]{8}}$

Section 1.4b Case I Solutions - Multiplying Monomial Expressions in Radical Form, with Real Numbers

1. $\sqrt{72}\cdot\sqrt{75} = \sqrt{36\cdot2}\cdot\sqrt{25\cdot3} = \sqrt{6^2\cdot2}\cdot\sqrt{5^2\cdot3} = 6\sqrt{2}\cdot5\sqrt{3} = (6\cdot5)\sqrt{2\cdot3} = \mathbf{30\sqrt{6}}$

2. $-3\sqrt{20}\cdot2\sqrt{32} = -3\sqrt{4\cdot5}\cdot2\sqrt{16\cdot2} = -3\sqrt{2^2\cdot5}\cdot2\sqrt{4^2\cdot2} = -(3\cdot2)\sqrt{5}\cdot(2\cdot4)\sqrt{2} = -6\sqrt{5}\cdot8\sqrt{2} = -(6\cdot8)\sqrt{5\cdot2} = \mathbf{-48\sqrt{10}}$

3. $\sqrt[2]{16}\cdot\sqrt[2]{27} = \sqrt{16}\cdot\sqrt{27} = \sqrt{4^2}\cdot\sqrt{9\cdot 3} = 4\cdot\sqrt{3^2\cdot 3} = (4\cdot 3)\sqrt{3} = \mathbf{12\sqrt{3}}$

4. $\sqrt{64}\cdot\sqrt{100}\cdot\sqrt{54} = \sqrt{8^2}\cdot\sqrt{10^2}\cdot\sqrt{9\cdot 6} = (8\cdot 10)\cdot\sqrt{3^2\cdot 6} = (80\cdot 3)\sqrt{6} = \mathbf{240\sqrt{6}}$

5. $-\sqrt{125}\cdot -2\sqrt{98} = +2\sqrt{25\cdot 5}\cdot\sqrt{49\cdot 2} = 2\sqrt{5^2\cdot 5}\cdot\sqrt{7^2\cdot 2} = (2\cdot 5)\sqrt{5}\cdot 7\sqrt{2} = (10\cdot 7)\sqrt{5\cdot 2} = \mathbf{70\sqrt{10}}$

6. $\sqrt[4]{625}\cdot\sqrt[4]{324}\cdot\sqrt[4]{48} = \sqrt[4]{5^4}\cdot\sqrt[4]{81\cdot 4}\cdot\sqrt[4]{16\cdot 3} = 5\cdot\sqrt[4]{3^4\cdot 4}\cdot\sqrt[4]{2^4\cdot 3} = 5\cdot 3\sqrt[4]{4}\cdot 2\sqrt[4]{3} = (5\cdot 3\cdot 2)\sqrt[4]{4\cdot 3} = \mathbf{30\sqrt[4]{12}}$

Section 1.4b Case II Solutions - Multiplying Binomial Expressions in Radical Form, with Real Numbers

1. $\left(2\sqrt{3}+1\right)\cdot\left(2+\sqrt{2}\right) = (2\cdot 2)\sqrt{3} + \left(2\sqrt{3}\cdot\sqrt{2}\right) + (1\cdot 2) + \left(1\cdot\sqrt{2}\right) = 4\sqrt{3} + 2\sqrt{3\cdot 2} + 2 + \sqrt{2} = \mathbf{4\sqrt{3} + 2\sqrt{6} + \sqrt{2} + 2}$

2. $\left(1+\sqrt{5}\right)\cdot\left(\sqrt{8}+\sqrt{5}\right) = \left(1+\sqrt{5}\right)\cdot\left(\sqrt{4\cdot 2}+\sqrt{5}\right) = \left(1+\sqrt{5}\right)\cdot\left(\sqrt{2^2\cdot 2}+\sqrt{5}\right) = \left(1+\sqrt{5}\right)\cdot\left(2\sqrt{2}+\sqrt{5}\right)$

$= \left(1\cdot 2\sqrt{2}\right) + \left(1\cdot\sqrt{5}\right) + \left(2\sqrt{2}\cdot\sqrt{5}\right) + \left(\sqrt{5}\cdot\sqrt{5}\right) = 2\sqrt{2} + \sqrt{5} + 2\sqrt{2\cdot 5} + \sqrt{5\cdot 5} = 2\sqrt{2} + \sqrt{5} + 2\sqrt{10} + \sqrt{5^2}$

$= \mathbf{2\sqrt{2} + \sqrt{5} + 2\sqrt{10} + 5}$

3. $\left(2-\sqrt{2}\right)\cdot\left(3+\sqrt{2}\right) = (2\cdot 3) + \left(2\cdot\sqrt{2}\right) - \left(3\cdot\sqrt{2}\right) - \left(\sqrt{2}\cdot\sqrt{2}\right) = 6 + 2\sqrt{2} - 3\sqrt{2} - \sqrt{2\cdot 2} = 6 + (2-3)\sqrt{2} - \sqrt{2^2}$

$= 6 - \sqrt{2} - 2 = (6-2) - \sqrt{2} = \mathbf{4 - \sqrt{2}}$

4. $\left(5+\sqrt{5}\right)\cdot\left(5-\sqrt{5^3}\right) = \left(5+\sqrt{5}\right)\cdot\left(5-\sqrt{5^{2+1}}\right) = \left(5+\sqrt{5}\right)\cdot\left(5-\sqrt{5^2\cdot 5^1}\right) = \left(5+\sqrt{5}\right)\cdot\left(5-5\sqrt{5}\right)$

$= (5\cdot 5) - (5\cdot 5)\sqrt{5} + \left(5\cdot\sqrt{5}\right) - \left(5\sqrt{5}\cdot\sqrt{5}\right) = 25 - 25\sqrt{5} + 5\sqrt{5} - 5\sqrt{5\cdot 5} = 25 + (-25+5)\sqrt{5} - 5\sqrt{5^2}$

$= 25 - 20\sqrt{5} - 5\cdot 5 = 25 - 20\sqrt{5} - 25 = (25-25) - 20\sqrt{5} = \mathbf{-20\sqrt{5}}$

5. $\left(2+\sqrt{6}\right)\cdot\left(\sqrt[4]{16}-\sqrt{18}\right) = \left(2+\sqrt{6}\right)\cdot\left(\sqrt[4]{2^4}-\sqrt{9\cdot 2}\right) = \left(2+\sqrt{6}\right)\cdot\left(2-\sqrt{3^2\cdot 2}\right) = \left(2+\sqrt{6}\right)\cdot\left(2-3\sqrt{2}\right)$

$= (2\cdot 2) - (2\cdot 3)\sqrt{2} + \left(2\cdot\sqrt{6}\right) - \left(3\sqrt{2}\cdot\sqrt{6}\right) = 4 - 6\sqrt{2} + 2\sqrt{6} - 3\sqrt{2\cdot 6} = 4 - 6\sqrt{2} + 2\sqrt{6} - 3\sqrt{12}$

$= 4 - 6\sqrt{2} + 2\sqrt{6} - 3\sqrt{4\cdot 3} = 4 - 6\sqrt{2} + 2\sqrt{6} - 3\sqrt{2^2\cdot 3} = 4 - 6\sqrt{2} + 2\sqrt{6} - (3\cdot 2)\sqrt{3} = \mathbf{4 - 6\sqrt{2} + 2\sqrt{6} - 6\sqrt{3}}$

6. $\left(2-\sqrt{5}\right)\cdot\left(\sqrt{45}+\sqrt[4]{81}\right) = \left(2-\sqrt{5}\right)\cdot\left(\sqrt{9\cdot 5}+\sqrt[4]{3^4}\right) = \left(2-\sqrt{5}\right)\cdot\left(\sqrt{3^2\cdot 5}+3\right) = \left(2-\sqrt{5}\right)\cdot\left(3\sqrt{5}+3\right)$

$= (2\cdot 3)\sqrt{5} + (2\cdot 3) - \left(3\sqrt{5}\cdot\sqrt{5}\right) - \left(3\cdot\sqrt{5}\right) = 6\sqrt{5} + 6 - 3\sqrt{5\cdot 5} - 3\sqrt{5} = 6\sqrt{5} + 6 - 3\sqrt{5^2} - 3\sqrt{5} = 6\sqrt{5} + 6 - (3\cdot 5) - 3\sqrt{5}$

$= 6\sqrt{5} - 3\sqrt{5} + 6 - 15 = (6-3)\sqrt{5} - 9 = 3\sqrt{5} - 9 = \mathbf{3\left(\sqrt{5} - 3\right)}$

Section 1.4b Case III Solutions - Multiplying Monomial and Binomial Expressions in Radical Form, with Real Numbers

1. $2\sqrt{3}\cdot\left(2+\sqrt{2}\right) = (2\cdot 2)\sqrt{3}+\left(2\sqrt{3}\cdot\sqrt{2}\right) = 4\sqrt{3}+2\sqrt{3\cdot 2} = 4\sqrt{3}+2\sqrt{6} = \mathbf{2\left(2\sqrt{3}+\sqrt{6}\right)}$

2. $\sqrt{5}\cdot\left(\sqrt{8}+\sqrt{5}\right) = \left(\sqrt{5}\cdot\sqrt{8}\right)+\left(\sqrt{5}\cdot\sqrt{5}\right) = \left(\sqrt{5\cdot 8}\right)+\left(\sqrt{5\cdot 5}\right) = \sqrt{40}+\sqrt{5^2} = \sqrt{4\cdot 10}+5 = \sqrt{2^2\cdot 10}+5 = \mathbf{5+2\sqrt{10}}$

3. $-\sqrt{8}\cdot\left(3-\sqrt{3}\right) = -\sqrt{4\cdot 2}\cdot\left(3-\sqrt{3}\right) = -\sqrt{2^2\cdot 2}\cdot\left(3-\sqrt{3}\right) = -2\sqrt{2}\cdot\left(3-\sqrt{3}\right) = \left(-(2\cdot 3)\cdot\sqrt{2}\right)+\left(2\sqrt{2}\cdot\sqrt{3}\right)$

$= -6\sqrt{2}+2\sqrt{2\cdot 3} = -6\sqrt{2}+2\sqrt{6} = \mathbf{2\left(\sqrt{6}-3\sqrt{2}\right)}$

4. $4\sqrt{98}\cdot\left(3-\sqrt{2^3}\right) = 4\sqrt{49\cdot 2}\cdot\left(3-\sqrt{2^{2+1}}\right) = 4\sqrt{7^2\cdot 2}\cdot\left(3-\sqrt{2^2\cdot 2^1}\right) = (4\cdot 7)\sqrt{2}\cdot\left(3-2\sqrt{2}\right) = 28\sqrt{2}\cdot\left(3-2\sqrt{2}\right)$

$= (28\cdot 3)\sqrt{2}-(28\cdot 2)\cdot\left(\sqrt{2}\cdot\sqrt{2}\right) = 84\sqrt{2}-56\left(\sqrt{2\cdot 2}\right) = 84\sqrt{2}-56\sqrt{2^2} = 84\sqrt{2}-(56\cdot 2) = 84\sqrt{2}-112 = \mathbf{4\left(21\sqrt{2}-28\right)}$

5. $\sqrt[4]{48}\cdot\left(\sqrt[4]{324}+\sqrt[4]{32}\right) = \sqrt[4]{16\cdot 3}\cdot\left(\sqrt[4]{81\cdot 4}+\sqrt[4]{16\cdot 2}\right) = \sqrt[4]{2^4\cdot 3}\cdot\left(\sqrt[4]{3^4\cdot 4}+\sqrt[4]{2^4\cdot 2}\right) = 2\sqrt[4]{3}\cdot\left(3\sqrt[4]{4}+2\sqrt[4]{2}\right)$

$= (2\cdot 3)\cdot\left(\sqrt[4]{3}\cdot\sqrt[4]{4}\right)+(2\cdot 2)\cdot\left(\sqrt[4]{3}\cdot\sqrt[4]{2}\right) = 6\cdot\left(\sqrt[4]{3\cdot 4}\right)+4\cdot\left(\sqrt[4]{3\cdot 2}\right) = 6\sqrt[4]{12}+4\sqrt[4]{6} = \mathbf{2\left(3\sqrt[4]{12}+2\sqrt[4]{6}\right)}$

6. $2\sqrt{5}\cdot\left(\sqrt{45}+\sqrt[4]{81}\right) = 2\sqrt{5}\cdot\left(\sqrt{9\cdot 5}+\sqrt[4]{3^4}\right) = 2\sqrt{5}\cdot\left(\sqrt{3^2\cdot 5}+3\right) = 2\sqrt{5}\cdot\left(3\sqrt{5}+3\right) = (2\cdot 3)\left(\sqrt{5}\cdot\sqrt{5}\right)+(2\cdot 3)\sqrt{5}$

$= 6\left(\sqrt{5\cdot 5}\right)+6\sqrt{5} = 6\sqrt{5^2}+6\sqrt{5} = (6\cdot 5)+6\sqrt{5} = 30+6\sqrt{5} = \mathbf{6\left(5+\sqrt{5}\right)}$

Section 1.4b Case IV Solutions - Rationalizing Radical Expressions - Monomial Denominators with Real Numbers

1. $\sqrt{\frac{1}{8}} = \sqrt{\frac{1}{4\cdot 2}} = \sqrt{\frac{1}{2^2\cdot 2}} = \frac{1}{2}\cdot\sqrt{\frac{1}{2}} = \frac{1}{2}\cdot\frac{\sqrt{1}}{\sqrt{2}} = \frac{1}{2}\cdot\frac{1}{\sqrt{2}} = \frac{1}{2}\left(\frac{1}{\sqrt{2}}\times\frac{\sqrt{2}}{\sqrt{2}}\right) = \frac{1}{2}\left(\frac{1\times\sqrt{2}}{\sqrt{2}\times\sqrt{2}}\right) = \frac{1}{2}\left(\frac{\sqrt{2}}{\sqrt{2\cdot 2}}\right) = \frac{1}{2}\cdot\frac{\sqrt{2}}{\sqrt{2^2}}$

$= \frac{1}{2}\cdot\frac{\sqrt{2}}{2} = \frac{1\times\sqrt{2}}{2\cdot 2} = \mathbf{\frac{\sqrt{2}}{4}}$

2. $\sqrt[2]{\frac{50}{7}} = \sqrt{\frac{50}{7}} = \sqrt{\frac{25\cdot 2}{7}} = \sqrt{\frac{5^2\cdot 2}{7}} = 5\sqrt{\frac{2}{7}} = 5\frac{\sqrt{2}}{\sqrt{7}} = 5\frac{\sqrt{2}}{\sqrt{7}}\times\frac{\sqrt{7}}{\sqrt{7}} = 5\frac{\sqrt{2}\times\sqrt{7}}{\sqrt{7}\times\sqrt{7}} = 5\frac{\sqrt{2\cdot 7}}{\sqrt{7\cdot 7}} = 5\frac{\sqrt{14}}{\sqrt{7^2}} = \mathbf{5\frac{\sqrt{14}}{7}}$

3. $\frac{\sqrt{75}}{-5} = -\frac{\sqrt{25\cdot 3}}{5} = -\frac{\sqrt{5^2\cdot 3}}{5} = -\frac{\overset{1}{\cancel{5}}\sqrt{3}}{\underset{1}{\cancel{5}}} = -\frac{\sqrt{3}}{1} = \mathbf{-\sqrt{3}}$

4. $\sqrt[3]{\frac{25}{16}} = \sqrt[3]{\frac{25}{8\cdot 2}} = \sqrt[3]{\frac{25}{2^3\cdot 2}} = \frac{1}{2}\sqrt[3]{\frac{25}{2}} = \frac{1}{2}\frac{\sqrt[3]{25}}{\sqrt[3]{2^1}} = \frac{1}{2}\left(\frac{\sqrt[3]{25}}{\sqrt[3]{2^1}}\times\frac{\sqrt[3]{2^2}}{\sqrt[3]{2^2}}\right) = \frac{1}{2}\left(\frac{\sqrt[3]{25}\times\sqrt[3]{4}}{\sqrt[3]{2^1}\times\sqrt[3]{2^2}}\right) = \frac{1}{2}\left(\frac{\sqrt[3]{25\cdot 4}}{\sqrt[3]{2^1\cdot 2^2}}\right) = \frac{1}{2}\cdot\frac{\sqrt[3]{100}}{\sqrt[3]{2^{1+2}}}$

$= \frac{1}{2}\cdot\frac{\sqrt[3]{100}}{\sqrt[3]{2^3}} = \frac{1}{2}\cdot\frac{\sqrt[3]{100}}{2} = \frac{1\cdot\sqrt[3]{100}}{2\cdot 2} = \mathbf{\frac{\sqrt[3]{100}}{4}}$

5. $\sqrt[5]{\frac{32}{8}} = \sqrt[5]{\frac{2^5}{2^3}} = 2\sqrt[5]{\frac{1}{2^3}} = 2\frac{1}{\sqrt[5]{2^3}} = 2\frac{1}{\sqrt[5]{2^3}}\times\frac{\sqrt[5]{2^2}}{\sqrt[5]{2^2}} = 2\frac{1\times\sqrt[5]{2^2}}{\sqrt[5]{2^3}\times\sqrt[5]{2^2}} = 2\frac{\sqrt[5]{2^2}}{\sqrt[5]{2^3\cdot 2^2}} = 2\frac{\sqrt[5]{4}}{\sqrt[5]{2^{3+2}}} = 2\frac{\sqrt[5]{4}}{\sqrt[5]{2^5}} = \frac{\overset{1}{\cancel{2}}}{1}\cdot\frac{\sqrt[5]{4}}{\underset{1}{\cancel{2}}}$

$= \frac{1}{1}\cdot\frac{\sqrt[5]{4}}{1} = \frac{1\cdot\sqrt[5]{4}}{1\cdot 1} = \frac{\sqrt[5]{4}}{1} = \mathbf{\sqrt[5]{4}}$

The following are two other ways to solve this problem:

5. $\sqrt[5]{\frac{32}{8}} = \sqrt[5]{\frac{2^5}{2^3}} = \sqrt[5]{2^5\cdot 2^{-3}} = \sqrt[5]{2^{5-3}} = \sqrt[5]{2^2} = \mathbf{\sqrt[5]{4}}$ or, $\sqrt[5]{\frac{32}{8}} = \sqrt[5]{\frac{\overset{4}{\cancel{32}}}{\underset{1}{\cancel{8}}}} = \sqrt[5]{\frac{4}{1}} = \mathbf{\sqrt[5]{4}}$

6. $\dfrac{-3\sqrt{100}}{-5\sqrt{3000}} = +\dfrac{3\sqrt{10^2}}{5\sqrt{100\cdot 30}} = \dfrac{3\cdot 10}{5\sqrt{10^2\cdot 30}} = \dfrac{30}{(5\cdot 10)\sqrt{30}} = \dfrac{\overset{3}{\cancel{30}}}{\underset{5}{\cancel{50}}\sqrt{30}} = \dfrac{3}{5\sqrt{30}} = \dfrac{3}{5}\cdot\dfrac{1}{\sqrt{30}} = \dfrac{3}{5}\left(\dfrac{1}{\sqrt{30}}\times\dfrac{\sqrt{30}}{\sqrt{30}}\right)$

$= \dfrac{3}{5}\left(\dfrac{1\times\sqrt{30}}{\sqrt{30}\times\sqrt{30}}\right) = \dfrac{3}{5}\left(\dfrac{\sqrt{30}}{\sqrt{30\cdot 30}}\right) = \dfrac{3}{5}\cdot\dfrac{\sqrt{30}}{\sqrt{30^2}} = \dfrac{3}{5}\cdot\dfrac{\sqrt{30}}{30} = \dfrac{\overset{1}{\cancel{3}}\cdot\sqrt{30}}{5\cdot\underset{10}{\cancel{30}}} = \dfrac{1\cdot\sqrt{30}}{5\cdot 10} = \mathbf{\dfrac{\sqrt{30}}{50}}$

Section 1.4b Case V Solutions - Rationalizing Radical Expressions - Binomial Denominators with Real Numbers

1. $\dfrac{7}{1+\sqrt{7}} = \dfrac{7}{1+\sqrt{7}}\times\dfrac{1-\sqrt{7}}{1-\sqrt{7}} = \dfrac{7\times\left(1-\sqrt{7}\right)}{\left(1+\sqrt{7}\right)\times\left(1-\sqrt{7}\right)} = \dfrac{7\left(1-\sqrt{7}\right)}{(1\cdot 1)+\left(1\cdot\sqrt{7}\right)-\left(1\cdot\sqrt{7}\right)-\left(\sqrt{7}\cdot\sqrt{7}\right)} = \dfrac{7\left(1-\sqrt{7}\right)}{1+\sqrt{7}-\sqrt{7}-\sqrt{7\cdot 7}}$

$= \dfrac{7\left(1-\sqrt{7}\right)}{1-\sqrt{7^2}} = \dfrac{7\left(1-\sqrt{7}\right)}{1-7} = \dfrac{7\left(1-\sqrt{7}\right)}{-6} = \mathbf{-\dfrac{7\left(1-\sqrt{7}\right)}{6}}$

2. $\dfrac{1-\sqrt{18}}{2+\sqrt{18}} = \dfrac{1-\sqrt{9\cdot 2}}{2+\sqrt{9\cdot 2}} = \dfrac{1-\sqrt{3^2\cdot 2}}{2+\sqrt{3^2\cdot 2}} = \dfrac{1-3\sqrt{2}}{2+3\sqrt{2}} = \dfrac{1-3\sqrt{2}}{2+3\sqrt{2}}\times\dfrac{2-3\sqrt{2}}{2-3\sqrt{2}} = \dfrac{\left(1-3\sqrt{2}\right)\times\left(2-3\sqrt{2}\right)}{\left(2+3\sqrt{2}\right)\times\left(2-3\sqrt{2}\right)}$

$= \dfrac{(1\cdot 2)-(1\cdot 3)\sqrt{2}-(2\cdot 3)\sqrt{2}+(3\cdot 3)\cdot\left(\sqrt{2}\cdot\sqrt{2}\right)}{(2\cdot 2)-(2\cdot 3)\sqrt{2}+(2\cdot 3)\sqrt{2}-(3\cdot 3)\cdot\left(\sqrt{2}\cdot\sqrt{2}\right)} = \dfrac{2-3\sqrt{2}-6\sqrt{2}+9\sqrt{2\cdot 2}}{4-6\sqrt{2}+6\sqrt{2}-9\sqrt{2\cdot 2}} = \dfrac{2-(3+6)\sqrt{2}+9\sqrt{2^2}}{4-9\sqrt{2^2}} = \dfrac{2-9\sqrt{2}+(9\cdot 2)}{4-(9\cdot 2)}$

$= \dfrac{2-9\sqrt{2}+18}{4-18} = \dfrac{(2+18)-9\sqrt{2}}{-14} = \mathbf{-\dfrac{20-9\sqrt{2}}{14}}$

3. $\dfrac{\sqrt{5}}{\sqrt{5}+\sqrt{2}} = \dfrac{\sqrt{5}}{\sqrt{5}+\sqrt{2}}\times\dfrac{\sqrt{5}-\sqrt{2}}{\sqrt{5}-\sqrt{2}} = \dfrac{\sqrt{5}\times\left(\sqrt{5}-\sqrt{2}\right)}{\left(\sqrt{5}+\sqrt{2}\right)\times\left(\sqrt{5}-\sqrt{2}\right)} = \dfrac{\left(\sqrt{5}\cdot\sqrt{5}\right)-\left(\sqrt{5}\cdot\sqrt{2}\right)}{\left(\sqrt{5}\cdot\sqrt{5}\right)-\left(\sqrt{5}\cdot\sqrt{2}\right)+\left(\sqrt{2}\cdot\sqrt{5}\right)-\left(\sqrt{2}\cdot\sqrt{2}\right)}$

$= \dfrac{\sqrt{5\cdot 5}-\sqrt{5\cdot 2}}{\sqrt{5\cdot 5}-\sqrt{5\cdot 2}+\sqrt{2\cdot 5}-\sqrt{2\cdot 2}} = \dfrac{\sqrt{5^2}-\sqrt{10}}{\sqrt{5^2}-\sqrt{10}+\sqrt{10}-\sqrt{2^2}} = \dfrac{5-\sqrt{10}}{5-2} = \mathbf{\dfrac{5-\sqrt{10}}{3}}$

4. $\dfrac{3-\sqrt{5}}{\sqrt{7}-\sqrt{4}} = \dfrac{3-\sqrt{5}}{\sqrt{7}-\sqrt{2^2}} = \dfrac{3-\sqrt{5}}{\sqrt{7}-2} = \dfrac{3-\sqrt{5}}{\sqrt{7}-2}\times\dfrac{\sqrt{7}+2}{\sqrt{7}+2} = \dfrac{\left(3-\sqrt{5}\right)\times\left(\sqrt{7}+2\right)}{\left(\sqrt{7}-2\right)\times\left(\sqrt{7}+2\right)} = \dfrac{\left(3\cdot\sqrt{7}\right)+(3\cdot 2)-\left(\sqrt{5}\cdot\sqrt{7}\right)-\left(2\cdot\sqrt{5}\right)}{\left(\sqrt{7}\cdot\sqrt{7}\right)+\left(2\cdot\sqrt{7}\right)-\left(2\cdot\sqrt{7}\right)-(2\cdot 2)}$

$= \dfrac{3\sqrt{7}+6-\sqrt{5\cdot 7}-2\sqrt{5}}{\sqrt{7\cdot 7}+2\sqrt{7}-2\sqrt{7}-4} = \dfrac{3\sqrt{7}+6-\sqrt{35}-2\sqrt{5}}{\sqrt{7^2}-4} = \dfrac{3\sqrt{7}+6-\sqrt{35}-2\sqrt{5}}{7-4} = \mathbf{\dfrac{3\sqrt{7}-\sqrt{35}-2\sqrt{5}+6}{3}}$

5. $\dfrac{-3+\sqrt{3}}{4+\sqrt{5}} = \dfrac{-3+\sqrt{3}}{4+\sqrt{5}}\times\dfrac{4-\sqrt{5}}{4-\sqrt{5}} = \dfrac{\left(-3+\sqrt{3}\right)\times\left(4-\sqrt{5}\right)}{\left(4+\sqrt{5}\right)\times\left(4-\sqrt{5}\right)} = \dfrac{-(3\cdot 4)+\left(3\cdot\sqrt{5}\right)+\left(4\cdot\sqrt{3}\right)-\left(\sqrt{3}\cdot\sqrt{5}\right)}{(4\cdot 4)-\left(4\cdot\sqrt{5}\right)+\left(4\cdot\sqrt{5}\right)-\left(\sqrt{5}\cdot\sqrt{5}\right)}$

$= \dfrac{-12+3\sqrt{5}+4\sqrt{3}-\sqrt{3\cdot 5}}{16-4\sqrt{5}+4\sqrt{5}-\sqrt{5\cdot 5}} = \dfrac{-12+3\sqrt{5}+4\sqrt{3}-\sqrt{15}}{16-\sqrt{5^2}} = \dfrac{3\sqrt{5}+4\sqrt{3}-\sqrt{15}-12}{16-5} = \mathbf{\dfrac{3\sqrt{5}+4\sqrt{3}-\sqrt{15}-12}{11}}$

6. $\dfrac{3-\sqrt{3}}{3+\sqrt{3}} = \dfrac{3-\sqrt{3}}{3+\sqrt{3}}\times\dfrac{3-\sqrt{3}}{3-\sqrt{3}} = \dfrac{\left(3-\sqrt{3}\right)\times\left(3-\sqrt{3}\right)}{\left(3+\sqrt{3}\right)\times\left(3-\sqrt{3}\right)} = \dfrac{(3\cdot 3)-\left(3\cdot\sqrt{3}\right)-\left(3\cdot\sqrt{3}\right)+\left(\sqrt{3}\cdot\sqrt{3}\right)}{(3\cdot 3)-\left(3\cdot\sqrt{3}\right)+\left(3\cdot\sqrt{3}\right)-\left(\sqrt{3}\cdot\sqrt{3}\right)} = \dfrac{9-3\sqrt{3}-3\sqrt{3}+\sqrt{3\cdot 3}}{9-3\sqrt{3}+3\sqrt{3}-\sqrt{3\cdot 3}}$

$= \dfrac{9-(3+3)\sqrt{3}+\sqrt{3^2}}{9-\sqrt{3^2}} = \dfrac{9-6\sqrt{3}+3}{9-3} = \dfrac{(9+3)-6\sqrt{3}}{6} = \dfrac{12-6\sqrt{3}}{6} = \dfrac{\cancel{6}\left(2-\sqrt{3}\right)}{\cancel{6}} = \dfrac{2-\sqrt{3}}{1} = \mathbf{2-\sqrt{3}}$

Section 1.4b Case VI Solutions - Adding and Subtracting Radical Terms

1. $5\sqrt{3}+8\sqrt{3} = (5+8)\sqrt{3} = \mathbf{13\sqrt{3}}$

2. $2\sqrt[3]{3}-4\sqrt[3]{3} = (2-4)\sqrt[3]{3} = \mathbf{-2\sqrt[3]{3}}$

3. $12\sqrt[4]{5}+8\sqrt[4]{5}+2\sqrt[4]{3} = (12+8+2)\sqrt[4]{5} = \mathbf{22\sqrt[4]{5}}$

4. $a\sqrt{ab}-b\sqrt{ab}+c\sqrt{ab} = (\boldsymbol{a-b+c})\sqrt{\boldsymbol{ab}}$

5. $3x\sqrt[3]{x}-2x\sqrt[3]{x}+4x\sqrt[3]{x^2} = (3x-2x)\sqrt[3]{x}+4x\sqrt[3]{x^2} = \boldsymbol{x\sqrt[3]{x}+4x\sqrt[3]{x^2}}$

6. $5\sqrt[3]{2}+8\sqrt[3]{5}$ ***can not be simplified***

Section 1.5a Case I Solutions - Classification of Polynomials

Polynomials	Standard Form	Type	Degree	No. of Terms
1. $3x+2x^3-6$	$2x^3+3x-6$	trinomial	3	3
2. $-6y^8+2$	$-6y^8+2$	binomial	8	2
3. $2w+6w^2+8w^5$	$8w^5+6w^2+2w$	trinomial	5	3
4. $6y$	$6y$	monomial	1	1
5. $\sqrt{72}$	$\sqrt{72}$	monomial	0	1
6. $-16+2x^4$	$2x^4-16$	binomial	4	2

Section 1.5a Case II Solutions - Simplifying Polynomials

1. $-x^3+4x-8x^2+3x-5x^3-5x = \left(-x^3-5x^3\right)+(4x+3x-5x)-8x^2 = (-1-5)x^3+(4+3-5)x-8x^2$
$= -6x^3+2x-8x^2 = \mathbf{-6x^3-8x^2+2x}$

2. $2y+2y^3-5+4y-5y^3+1+y = (2y+4y+y)+\left(2y^3-5y^3\right)+(-5+1) = (2+4+1)y+(2-5)y^3-4 = 7y-3y^3-4$
$= \mathbf{-3y^3+7y-4}$

3. $2a^5+2a^2-3+4a^5+a^2 = \left(2a^5+4a^5\right)+\left(2a^2+a^2\right)-3 = (2+4)a^5+(2+1)a^2-3 = \mathbf{6a^5+3a^2-3}$

4. $3x+2x^4+2x^3-7x-5x^4 = (3x-7x)+\left(2x^4-5x^4\right)+2x^3 = (3-7)x+(2-5)x^4+2x^3 = -4x-3x^4+2x^3$
$= \mathbf{-3x^4+2x^3-4x}$

5. $2rs+4r^3s^3-20+2rs-5r^3s^3-3 = (2rs+2rs)+\left(4r^3s^3-5r^3s^3\right)+(-20-3) = (2+2)rs+(4-5)r^3s^3-23$
$= 4rs-r^3s^3-23 = \mathbf{-r^3s^3+4rs-23}$

6. $2xyz+2x^3y^3z^3+10-4xyz-4 = (2xyz-4xyz)+2x^3y^3z^3+(10-4) = (2-4)xyz+2x^3y^3z^3+6 = -2xyz+2x^3y^3z^3+6$
$= \mathbf{2x^3y^3z^3-2xyz+6}$

Section 1.5b Case I a Solutions - Multiplying Monomials by Monomials

1. $(2ax)\cdot\left(3a^2x^2\right) = (2\cdot 3)\cdot\left(a\cdot a^2\right)\cdot\left(x\cdot x^2\right) = 6\cdot a^{1+2}\cdot x^{1+2} = \mathbf{6a^3x^3}$

2. $\left(5x^2y^2\right)\cdot(2x)\cdot(4y) = (5\cdot 2\cdot 4)\cdot\left(x^2\cdot x\right)\cdot\left(y^2\cdot y\right) = 40\cdot x^{2+1}\cdot y^{2+1} = \mathbf{40x^3y^3}$

3. $\left(6x^2\right)^0\cdot\left(3x^2\right)\cdot(-2x) = 1\cdot\left(3x^2\right)\cdot(-2x) = -(3\cdot 2)\cdot\left(x^2\cdot x\right) = -6\cdot x^{2+1} = \mathbf{-6x^3}$

4. $\left(x^2y\right)\cdot(3xy)\cdot\left(4x^3y^2\right) = (3\cdot 4)\left(x^2\cdot x\cdot x^3\right)\cdot\left(y\cdot y\cdot y^2\right) = 12\left(x^{2+1+3}\right)\cdot\left(y^{1+1+2}\right) = \mathbf{12x^6y^4}$

5. $\left(3x^2y^2\right)\cdot\left(2xy^0\right)\cdot\left(5x^0y\right) = \left(3x^2y^2\right)\cdot(2x)\cdot(5y) = (3\cdot 2\cdot 5)\left(x^2\cdot x\right)\cdot\left(y^2\cdot y\right) = 30x^{2+1}\cdot y^{2+1} = \mathbf{30x^3y^3}$

6. $\left(8a^2b^2\right)\cdot(2a)\cdot\left(3a^2b^3\right) = (8\cdot 2\cdot 3)\left(a^2\cdot a\cdot a^2\right)\cdot\left(b^2\cdot b^3\right) = 48a^{2+1+2}\cdot b^{2+3} = \mathbf{48a^5b^5}$

Section 1.5b Case I b Solutions - Multiplying Polynomials by Monomials

1. $2\cdot\left(5x^2+6x-2x^2-x+5\right) = 2\cdot\left[\left(5x^2-2x^2\right)+(6x-x)+5\right] = 2\cdot\left[(5-2)x^2+(6-1)x+5\right] = 2\cdot\left[3x^2+5x+5\right]$
$= (2\cdot 3)x^2+(2\cdot 5)x+(2\cdot 5) = \mathbf{6x^2+10x+10}$

2. $\left(2x^2y-5y^2+3x^2y-2y^2+3\right)\cdot\left(3x^2y^2\right) = \left[\left(2x^2y+3x^2y\right)+\left(-5y^2-2y^2\right)+3\right]\cdot\left(3x^2y^2\right)$
$= \left[(2+3)x^2y+(-5-2)y^2+3\right]\cdot\left(3x^2y^2\right) = \left[5x^2y-7y^2+3\right]\cdot\left(3x^2y^2\right)$
$= (5\cdot 3)\cdot\left(x^2\cdot x^2\right)\cdot\left(y\cdot y^2\right)-(7\cdot 3)\cdot x^2\cdot\left(y^2\cdot y^2\right)+(3\cdot 3)x^2y^2 = 15\cdot x^{2+2}\cdot y^{1+2}-21\cdot x^2\cdot y^{2+2}+9x^2y^2$
$= \mathbf{15x^4y^3-21x^2y^4+9x^2y^2}$

3. $\left(5x^3+2x^2-5+3x-2x^3\right)\cdot(-2x)^2 = \left[\left(5x^3-2x^3\right)+2x^2-5+3x\right]\cdot 4x^2 = \left[(5-2)x^3+2x^2-5+3x\right]\cdot 4x^2$
$= \left[3x^3+2x^2+3x-5\right]\cdot 4x^2 = (3\cdot 4)\cdot\left(x^3\cdot x^2\right)+(2\cdot 4)\cdot\left(x^2\cdot x^2\right)+(3\cdot 4)\cdot\left(x\cdot x^2\right)-(5\cdot 4)x^2$
$= 12\cdot x^{3+2}+8\cdot x^{2+2}+12\cdot x^{1+2}-20x^2 = \mathbf{12x^5+8x^4+12x^3-20x^2}$

4. $6w\cdot\left(4w+2w^2+2-3w+w^2\right) = 6w\cdot\left[(4w-3w)+\left(2w^2+w^2\right)+2\right] = 6w\cdot\left[(4-3)w+(2+1)w^2+2\right] = 6w\cdot\left[w+3w^2+2\right]$
$= 6w\cdot\left[3w^2+w+2\right] = (6\cdot 3)\cdot\left(w^2\cdot w\right)+6(w\cdot w)+(2\cdot 6)w = 18w^{2+1}+6w^2+12w = \mathbf{18w^3+6w^2+12w}$

5. $2x\cdot\left(2x^2\right)^2\cdot\left(5x^2+3x-2x^2+x-2\right) = 2x\cdot 4x^4\cdot\left[\left(5x^2-2x^2\right)+(3x+x)-2\right] = (2\cdot 4)\left(x\cdot x^4\right)\cdot\left[(5-2)x^2+(3+1)x-2\right]$
$= 8x^5\cdot\left[3x^2+4x-2\right] = (8\cdot 3)\cdot\left(x^5\cdot x^2\right)+(8\cdot 4)\cdot\left(x^5\cdot x\right)-(8\cdot 2)x^5 = 24\cdot x^{5+2}+32\cdot x^{5+1}-16x^5 = \mathbf{24x^7+32x^6-16x^5}$

6. $\left(\sqrt{162}+\sqrt{9}x-2x^2+\sqrt{16}x^3\right)\cdot\left(2x^3\right) = \left(\sqrt{81\cdot 2}+\sqrt{3^2}x-2x^2+\sqrt{4^2}x^3\right)\cdot\left(2x^3\right) = \left(\sqrt{9^2\cdot 2}+3x-2x^2+4x^3\right)\cdot\left(2x^3\right)$
$= \left(9\sqrt{2}+3x-2x^2+4x^3\right)\cdot\left(2x^3\right) = (9\cdot 2)\sqrt{2}x^3+(3\cdot 2)\cdot\left(x\cdot x^3\right)-(2\cdot 2)\cdot\left(x^2\cdot x^3\right)+(4\cdot 2)\cdot\left(x^3\cdot x^3\right)$
$= 18\sqrt{2}x^3+6x^{1+3}-4x^{2+3}+8x^{3+3} = 18\sqrt{2}x^3+6x^4-4x^5+8x^6 = \mathbf{8x^6-4x^5+6x^4+18\sqrt{2}x^3}$

Section 1.5b Case II Solutions - Multiplying Binomials by Binomials

1. $(x+3)(x-2) = (x\cdot x)-(2\cdot x)+(3\cdot x)-(2\cdot 3) = x^2-2x+3x-6 = x^2+(-2x+3x)-6 = x^2+(-2+3)x-6$
$= \mathbf{x^2+x-6}$

2. $(-y+8)(y-6) = -(y\cdot y)+(6\cdot y)+(8\cdot y)-(8\cdot 6) = -y^2+6y+8y-48 = -y^2+(6y+8y)-48 = -y^2+(6+8)y-48$

$= \mathbf{-y^2 + 14y - 48}$

3. $(x^2 - 2xy)(-y^2 + 2xy) = -(x^2 \cdot y^2) + 2(x^2 \cdot x)y + 2x(y \cdot y^2) - (2 \cdot 2)(x \cdot x)(y \cdot y) = -x^2y^2 + 2x^3y + 2xy^3 - 4x^2y^2$

$= (-x^2y^2 - 4x^2y^2) + 2x^3y + 2xy^3 = (-1-4)x^2y^2 + 2x^3y + 2xy^3 = -5x^2y^2 + 2x^3y + 2xy^3 = \mathbf{2x^3y - 5x^2y^2 + 2xy^3}$

4. $(a^3 - a^2)(a - 6) = (a^3 \cdot a) - (6 \cdot a^3) - (a^2 \cdot a) + (6 \cdot a^2) = a^4 - 6a^3 - a^3 + 6a^2 = a^4 + (-6a^3 - a^3) + 6a^2$

$= a^4 + (-6-1)a^3 + 6a^2 = \mathbf{a^4 - 7a^3 + 6a^2}$

5. $(\sqrt{x^3} - 2x\sqrt{x^5})(\sqrt{x} - 4) = (\sqrt{x^{2+1}} - 2x\sqrt{x^{2+2+1}})(\sqrt{x} - 4) = (\sqrt{x^2 \cdot x^1} - 2x\sqrt{x^2 \cdot x^2 \cdot x^1})(\sqrt{x} - 4)$

$= [x\sqrt{x} - 2x(x \cdot x)\sqrt{x}](\sqrt{x} - 4) = [x\sqrt{x} - 2(x \cdot x^2)\sqrt{x}](\sqrt{x} - 4) = [x\sqrt{x} - 2x^3\sqrt{x}](\sqrt{x} - 4)$

$= x(\sqrt{x} \cdot \sqrt{x}) - (4 \cdot x\sqrt{x}) - (2x^3 \cdot \sqrt{x} \cdot \sqrt{x}) + (2 \cdot 4)x^3 \cdot \sqrt{x} = x(\sqrt{x \cdot x}) - 4x\sqrt{x} - 2x^3\sqrt{x \cdot x} + 8x^3\sqrt{x}$

$= x\sqrt{x^2} - 4x\sqrt{x} - 2x^3\sqrt{x^2} + 8x^3\sqrt{x} = x \cdot x - 4x\sqrt{x} - 2x^3 \cdot x + 8x^3\sqrt{x} = x^2 - 4x\sqrt{x} - 2x^4 + 8x^3\sqrt{x}$

$= \mathbf{-2x^4 + 8x^3\sqrt{x} + x^2 - 4x\sqrt{x}}$

6. $(\sqrt[3]{y^5} - \sqrt[3]{y^2})(\sqrt[3]{y^7} - \sqrt[3]{y}) = (\sqrt[3]{y^{3+2}} - \sqrt[3]{y^2})(\sqrt[3]{y^{3+3+1}} - \sqrt[3]{y}) = (\sqrt[3]{y^3 \cdot y^2} - \sqrt[3]{y^2})(\sqrt[3]{y^3 \cdot y^3 \cdot y^1} - \sqrt[3]{y})$

$= (y\sqrt[3]{y^2} - \sqrt[3]{y^2})[(y \cdot y)\sqrt[3]{y} - \sqrt[3]{y}] = (y\sqrt[3]{y^2} - \sqrt[3]{y^2})[y^2\sqrt[3]{y} - \sqrt[3]{y}]$

$= (y \cdot y^2)(\sqrt[3]{y^2} \cdot \sqrt[3]{y}) - y(\sqrt[3]{y^2} \cdot \sqrt[3]{y}) - y^2(\sqrt[3]{y^2} \cdot \sqrt[3]{y}) + (\sqrt[3]{y^2} \cdot \sqrt[3]{y}) = y^3\sqrt[3]{y^2 \cdot y} - y\sqrt[3]{y^2 \cdot y} - y^2\sqrt[3]{y^2 \cdot y} + \sqrt[3]{y^2 \cdot y}$

$= y^3\sqrt[3]{y^3} - y\sqrt[3]{y^3} - y^2\sqrt[3]{y^3} + \sqrt[3]{y^3} = (y^3 \cdot y) - (y \cdot y) - (y^2 \cdot y) + y = y^4 - y^2 - y^3 + y = \mathbf{y^4 - y^3 - y^2 + y}$

Section 1.5b Case IIIa Solutions - Dividing Monomials by Monomials

1. $\dfrac{-4xyz}{-8xyz} = +\dfrac{4xyz}{8xyz} = \dfrac{\overset{1}{\cancel{4}}\cancel{xyz}}{\underset{2}{\cancel{8}}\cancel{xyz}} = \mathbf{\dfrac{1}{2}}$

2. $\dfrac{u^2v^3w}{-uw^4} = -\dfrac{u^2v^3w^1}{u^1w^4} = -\dfrac{(u^2u^{-1})v^3}{w^4w^{-1}} = -\dfrac{(u^{2-1})v^3}{w^{4-1}} = -\dfrac{u^1v^3}{w^3} = \mathbf{-\dfrac{uv^3}{w^3}}$

3. $\dfrac{\sqrt{72x^2y^4}}{-12xy^2} = -\dfrac{x\sqrt{(36 \cdot 2)y^4}}{12xy^2} = -\dfrac{\cancel{x}\sqrt{(6^2 \cdot 2)y^{2+2}}}{12\cancel{x}y^2} = -\dfrac{6\sqrt{2(y^2 \cdot y^2)}}{12y^2} = -\dfrac{6(y \cdot y)\sqrt{2}}{12y^2} = -\dfrac{6y^2\sqrt{2}}{12y^2} = -\dfrac{\overset{1}{\cancel{6}}\sqrt{2}}{\underset{2}{\cancel{12}}(y^2y^{-2})}$

$= -\dfrac{\sqrt{2}}{2y^{2-2}} = -\dfrac{\sqrt{2}}{2y^0} = \mathbf{-\dfrac{\sqrt{2}}{2}}$

4. $\dfrac{-36x^3y^3z^4}{-\sqrt{25}xyz^3} = +\dfrac{36x^3y^3z^4}{\sqrt{5^2}xyz^3} = \dfrac{36x^3y^3z^4}{5x^1y^1z^3} = \dfrac{36(x^3x^{-1}) \cdot (y^3y^{-1}) \cdot (z^4z^{-3})}{5} = \dfrac{36}{5}\dfrac{(x^{3-1}) \cdot (y^{3-1}) \cdot (z^{4-3})}{1} = \mathbf{7\frac{1}{5}(x^2y^2z)}$

5. $\frac{-9a^2b^2c^3}{\sqrt[3]{27a^6b^3c^3}} = -\frac{9a^2b^2c^3}{bc\sqrt[3]{3^3a^6}} = -\frac{\overset{3}{\cancel{9}}a^2b^2c^3}{\underset{1}{\cancel{3}}bc\sqrt[3]{a^{3+3}}} = -\frac{3a^2b^2c^3}{b^1c^1\sqrt[3]{a^3\cdot a^3}} = -\frac{a^2\left(b^2b^{-1}\right)\cdot\left(c^3c^{-1}\right)}{a\cdot a} = -\frac{3a^2b^{2-1}c^{3-1}}{a^2} = -\frac{3b^1c^2}{a^2a^{-2}}$

$= -\frac{3bc^2}{a^{2-2}} = -\frac{3bc^2}{a^0} = -\frac{3bc^2}{1} = \mathbf{-3bc^2}$

6. $\frac{-24lm^3n^2}{12l^2mn} = -\frac{\overset{2}{\cancel{24}}lm^3n^2}{\underset{1}{\cancel{12}}l^2mn} = -\frac{2l^1m^3n^2}{l^2m^1n^1} = -\frac{2\left(m^3m^{-1}\right)\cdot\left(n^2n^{-1}\right)}{l^2l^{-1}} = -\frac{2\left(m^{3-1}\right)\cdot\left(n^{2-1}\right)}{l^{2-1}} = -\frac{2m^2n^1}{l^1} = \mathbf{-\frac{2m^2n}{l}}$

Section 1.5b Case IIIb Solutions - Dividing Binomials by Monomials

1. $\frac{98-46}{-12} = -\frac{\overset{13}{\cancel{52}}}{\underset{3}{\cancel{12}}} = -\frac{13}{3} = \mathbf{-4\frac{1}{3}}$

2. $\frac{x^3y^3z+4x^2y^2}{-2xy^2z} = \frac{x^3y^3z}{-2xy^2z} + \frac{4x^2y^2}{-2xy^2z} = -\frac{x^3y^3z^1}{2x^1y^2z^1} - \frac{\overset{2}{\cancel{4}}x^2y^2}{\underset{1}{\cancel{2}}x^1y^2z} = -\frac{\left(x^3x^{-1}\right)\cdot\left(y^3y^{-2}\right)}{2\left(z^1z^{-1}\right)} - \frac{2\left(x^2x^{-1}\right)}{\left(y^2y^{-2}\right)z}$

$= -\frac{x^{3-1}\cdot y^{3-2}}{2z^{1-1}} - \frac{2x^{2-1}}{y^{2-2}z} = -\frac{x^2\cdot y^1}{2z^0} - \frac{2x^1}{y^0z} = \mathbf{-\frac{x^2y}{2} - \frac{2x}{z}}$

3. $\frac{-a^3b^3c+a^2bc^2}{-a^2b^2c^2} = \frac{-a^3b^3c}{-a^2b^2c^2} + \frac{a^2bc^2}{-a^2b^2c^2} = +\frac{a^3b^3c^1}{a^2b^2c^2} - \frac{a^2b^1c^2}{a^2b^2c^2} = \frac{\left(a^3a^{-2}\right)\cdot\left(b^3b^{-2}\right)}{c^2c^{-1}} - \frac{\left(a^2a^{-2}\right)\cdot\left(c^2c^{-2}\right)}{b^2b^{-1}}$

$= \frac{a^{3-2}\cdot b^{3-2}}{c^{2-1}} - \frac{a^{2-2}\cdot c^{2-2}}{b^{2-1}} = \frac{a^1\cdot b^1}{c^1} - \frac{a^0\cdot c^0}{b^1} = \frac{ab}{c} - \frac{1\cdot 1}{b} = \mathbf{\frac{ab}{c} - \frac{1}{b}}$

4. $\frac{\sqrt[4]{a^5b^4c^3}-\sqrt[3]{a^3b^6c}}{\sqrt{a^2b^4c^6}} = \frac{b\sqrt[4]{a^{4+1}c^3}-a\sqrt[3]{b^{3+3}c}}{a\sqrt{b^{2+2}c^{2+2+2}}} = \frac{b\sqrt[4]{\left(a^4\cdot a^1\right)\cdot c^3}-a\sqrt[3]{\left(b^3\cdot b^3\right)\cdot c}}{a\sqrt{\left(b^2\cdot b^2\right)\cdot\left(c^2\cdot c^2\cdot c^2\right)}} = \frac{ab\sqrt[4]{ac^3}-a(b\cdot b)\sqrt[3]{c}}{a(b\cdot b)(c\cdot c\cdot c)}$

$= \frac{ab\sqrt[4]{ac^3}-ab^2\sqrt[3]{c}}{ab^2c^3} = \frac{ab\sqrt[4]{ac^3}}{ab^2c^3} + \frac{-ab^2\sqrt[3]{c}}{ab^2c^3} = \frac{\cancel{a}b\sqrt[4]{ac^3}}{\cancel{a}b^2c^3} - \frac{\cancel{a}b^2\sqrt[3]{c}}{\cancel{a}b^2c^3} = \frac{\sqrt[4]{ac^3}}{\left(b^2b^{-1}\right)\cdot c^3} - \frac{\sqrt[3]{c}}{\left(b^2b^{-2}\right)\cdot c^3} = \frac{\sqrt[4]{ac^3}}{b^{2-1}\cdot c^3} - \frac{\sqrt[3]{c}}{b^{2-2}\cdot c^3}$

$= \frac{\sqrt[4]{ac^3}}{b^1c^3} - \frac{\sqrt[3]{c}}{b^0c^3} = \mathbf{\frac{\sqrt[4]{ac^3}}{bc^3} - \frac{\sqrt[3]{c}}{c^3}}$

5. $\frac{m^3n^2l+ml^2}{mnl} = \frac{m^3n^2l}{mnl} + \frac{ml^2}{mnl} = \frac{m^3n^2l}{mnl} + \frac{\cancel{m}l^2}{\cancel{m}nl} = \frac{m^3n^2}{m^1n^1} + \frac{l^2}{nl^1} = \frac{\left(m^3m^{-1}\right)\cdot\left(n^2n^{-1}\right)}{1} + \frac{l^2l^{-1}}{n} = \frac{m^{3-1}n^{2-1}}{1} + \frac{l^{2-1}}{n}$

$= \frac{m^2n^1}{1} + \frac{l^1}{n} = \mathbf{m^2n + \frac{l}{n}}$

6. $\frac{36y^2-18y^3}{-9y} = \frac{36y^2}{-9y} + \frac{-18y^3}{-9y} = -\frac{\overset{4}{\cancel{36}}y^2}{\underset{1}{\cancel{9}}y} + \frac{\overset{2}{\cancel{18}}y^3}{\underset{1}{\cancel{9}}y} = -\frac{4y^2}{y^1} + \frac{2y^3}{y^1} = -\frac{4y^2y^{-1}}{1} + \frac{2y^3y^{-1}}{1} = -\frac{4y^{2-1}}{1} + \frac{2y^{3-1}}{1}$

$= -\frac{4y^1}{1} + \frac{2y^2}{1} = \mathbf{2y^2 - 4y}$

Section 1.5b Case IIIc Solutions - Dividing Polynomials by Polynomials

1. Divide $3x^2 + 10x + 7$ by $x + 3$.

$$
\begin{array}{r l}
 & 3x + 1 \\
x + 3 & \big) +3x^2 + 10x + 7 \\
 & \overset{-}{+}3x^2 \overset{-}{+} 9x \\
\hline
 & + x + 7 \\
 & \overset{-}{+} x \overset{-}{+} 3 \\
\hline
 & + 4
\end{array}
$$

The answer is $\mathbf{3x+1}$ with remainder of $+\mathbf{4}$, or

$$\mathbf{3x + 1 + \frac{4}{x+3}}.$$

2. Divide $x^4 + 7x^3 + 13x^2 + 17x + 10$ by $x + 5$.

$$
\begin{array}{r l}
 & x^3 + 2x^2 + 3x + 2 \\
x + 5 & \big) +x^4 + 7x^3 + 13x^2 + 17x + 10 \\
 & \overset{-}{+}x^4 \overset{-}{+} 5x^3 \\
\hline
 & +2x^3 + 13x^2 \\
 & \overset{-}{+}2x^3 \overset{-}{+} 10x^2 \\
\hline
 & + 3x^2 + 17x \\
 & \overset{-}{+} 3x^2 \overset{-}{+} 15x \\
\hline
 & + 2x + 10 \\
 & \overset{-}{+} 2x \overset{-}{+} 10 \\
\hline
 & 0
\end{array}
$$

The answer is $\mathbf{x^3 + 2x^2 + 3x + 2}$ with remainder of zero.

3. Divide $x^6 - x^5 - 2x^4 - x^3 + 2x^2 + 5x - 10$ by $x - 2$.

$$
\begin{array}{r l}
 & x^5 + x^4 - x^2 + 5 \\
x - 2 & \big) +x^6 - x^5 - 2x^4 - x^3 + 2x^2 + 5x - 10 \\
 & \overset{-}{+}x^6 \pm 2x^5 \\
\hline
 & + x^5 - 2x^4 \\
 & \overset{-}{+} x^5 \pm 2x^4 \\
\hline
 & -x^3 + 2x^2 \\
 & \pm x^3 \overset{-}{+} 2x^2 \\
\hline
 & +5x - 10 \\
 & \overset{-}{+}5x \pm 10 \\
\hline
 & 0
\end{array}
$$

The answer is $\mathbf{x^5 + x^4 - x^2 + 5}$ with remainder of zero.

4. Divide $-2x^4 + 5x^3 - 4x^2 + 16x - 15$ by $-2x + 5$.

$$
\begin{array}{r l}
 & x^3 + 2x - 3 \\
-2x + 5 & \big) -2x^4 + 5x^3 - 4x^2 + 16x - 15 \\
 & \pm 2x^4 \overset{-}{+} 5x^3 \\
\hline
 & -4x^2 + 16x \\
 & \pm 4x^2 \overset{-}{+} 10x \\
\hline
 & + 6x - 15 \\
 & \overset{-}{+} 6x \pm 15 \\
\hline
 & 0
\end{array}
$$

The answer is $\mathbf{x^3 + 2x - 3}$ with remainder of zero.

5. Divide $2x^4 - 13x^3 + 13x^2 + 15x - 35$ by $x - 5$.

$$
\begin{array}{r l}
 & 2x^3 - 3x^2 - 2x + 5 \\
x - 5 & \big) +2x^4 - 13x^3 + 13x^2 + 15x - 35 \\
 & \overset{-}{+}2x^4 \pm 10x^3 \\
\hline
 & - 3x^3 + 13x^2 \\
 & \pm 3x^3 \overset{-}{+} 15x^2 \\
\hline
 & -2x^2 + 15x \\
 & \pm 2x^2 \overset{-}{+} 10x \\
\hline
 & + 5x - 35 \\
 & \overset{-}{+} 5x \pm 25 \\
\hline
 & - 10
\end{array}
$$

The answer is $\mathbf{2x^3 - 3x^2 - 2x + 5}$ with remainder of $-\mathbf{10}$, or

$$\mathbf{2x^3 - 3x^2 - 2x + 5 - \frac{10}{x-5}}.$$

6. Divide $-2x^4 + 7x^3 - 6x^2 - 2x + 3$ by $-2x + 3$.

$$
\begin{array}{r l}
 & x^3 - 2x^2 + 1 \\
-2x + 3 & \big) -2x^4 + 7x^3 - 6x^2 - 2x + 3 \\
 & \pm 2x^4 \overset{-}{+} 3x^3 \\
\hline
 & +4x^3 - 6x^2 \\
 & \overset{-}{+}4x^3 \pm 6x^2 \\
\hline
 & -2x + 3 \\
 & \pm 2x \overset{-}{+} 3 \\
\hline
 & 0
\end{array}
$$

The answer is $\mathbf{x^3 - 2x^2 + 1}$ with remainder of zero.

Section 1.5b Case IV Solutions - Adding and Subtracting Polynomials Horizontally

1. $\left(x^3+2x^5-3x+2\right)+\left(3x^3+x-x^5\right) = \left(2x^5+x^3-3x+2\right)+\left(-x^5+3x^3+x\right) = \left(2x^5-x^5\right)+\left(x^3+3x^3\right)+\left(-3x+x\right)+2$

$= (2-1)x^5+(1+3)x^3+(-3+1)x+2 = \mathbf{x^5+4x^3-2x+2}$

2. $\left(y-y^2+2y^4+3y^2-3\right)+\left(2y^4+y^3+5-y^2\right) = \left(2y^4+3y^2-y^2+y-3\right)+\left(2y^4+y^3-y^2+5\right)$

$= \left(2y^4+2y^4\right)+y^3+\left(3y^2-y^2-y^2\right)+y+(-3+5) = (2+2)y^4+y^3+(3-1-1)y^2+y+2 = \mathbf{4y^4+y^3+y^2+y+2}$

3. $\left(3x-3x^2+5x-3\right)-\left(-2x+5-x^2+2\right) = \left(3x-3x^2+5x-3\right)+\left(2x-5+x^2-2\right) = \left(-3x^2+3x+5x-3\right)+\left(x^2+2x-5-2\right)$

$= \left(-3x^2+x^2\right)+(3x+5x+2x)+(-3-5-2) = (-3+1)x^2+(3+5+2)x-10 = \mathbf{-2x^2+10x-10}$

4. $\left(xyz+2x^2yz+4xyz\right)+\left(4x^2yz-x^2yz+2xyz\right) = \left(2x^2yz+xyz+4xyz\right)+\left(4x^2yz-x^2yz+2xyz\right)$

$= \left(2x^2yz+4x^2yz-x^2yz\right)+(xyz+4xyz+2xyz) = (2+4-1)x^2yz+(1+4+2)xyz = \mathbf{5x^2yz+7xyz}$

5. $\left(-2ab-3+2a^2b^2\right)+\left(-3ab+a^2b^2+2(ab)^0\right) = \left(-2ab-3+2a^2b^2\right)+\left(-3ab+a^2b^2+2\right)$

$= \left(2a^2b^2-2ab-3\right)+\left(a^2b^2-3ab+2\right) = \left(2a^2b^2+a^2b^2\right)+(-2ab-3ab)+(-3+2) = (2+1)a^2b^2+(-2-3)ab-1$

$= \mathbf{3a^2b^2-5ab-1}$

6. $\left(5x^6-x^5-4x^4+3x+x^2\right)-\left(x-3x^2+x^4-3x^6\right) = \left(5x^6-x^5-4x^4+3x+x^2\right)+\left(-x+3x^2-x^4+3x^6\right)$

$= \left(5x^6+3x^6\right)-x^5+\left(-4x^4-x^4\right)+\left(x^2+3x^2\right)+(3x-x) = (5+3)x^6-x^5+(-4-1)x^4+(1+3)x^2+(3-1)x$

$= \mathbf{8x^6-x^5-5x^4+4x^2+2x}$

Section 1.5b Case V Solutions - Adding and Subtracting Polynomials Vertically

1. $\left(x^2+2x+x^3\right)+\left(3x-2x^3\right) = \left(x^3+x^2+2x\right)+\left(-2x^3+3x\right) =$

$$\begin{array}{rrrr} + & x^3 & +x^2 & +2x \\ & -2x^3 & & +3x \\ \hline - & \mathbf{x^3} & \mathbf{+x^2} & \mathbf{+5x} \end{array}$$

2. $\left(y+y^2+3y^3+4\right)+\left(-2+y^2+3y^2+2y\right) = \left(3y^3+y^2+y+4\right)+\left[\left(y^2+3y^2\right)+2y-2\right]$

$= \left(3y^3+y^2+y+4\right)+\left[(1+3)y^2+2y-2\right] = \left(3y^3+y^2+y+4\right)+\left(4y^2+2y-2\right) =$

$$\begin{array}{rrrr} 3y^3 & +y^2 & +y & +4 \\ & +4y^2 & +2y & -2 \\ \hline \mathbf{3y^3} & \mathbf{+5y^2} & \mathbf{+3y} & \mathbf{+2} \end{array}$$

3. $\left(x^3+x^2-3+3x^2\right)-\left(-2x^3-5x+5\right) = \left[x^3+\left(x^2+3x^2\right)-3\right]+\left(2x^3+5x-5\right) = \left[x^3+(1+3)x^2-3\right]+\left(2x^3+5x-5\right)$

$= \left(x^3+4x^2-3\right)+\left(2x^3+5x-5\right) =$

$$\begin{array}{rrrr} x^3 & +4x^2 & & -3 \\ 2x^3 & & +5x & -5 \\ \hline \mathbf{3x^3} & \mathbf{+4x^2} & \mathbf{+5x} & \mathbf{-8} \end{array}$$

4. $\left(z^5+3z^2+z-2z^2-4z+2\right)+\left(z^2+4z^5+z^0\right)=\left[z^5+\left(3z^2-2z^2\right)+(z-4z)+2\right]+\left(4z^5+z^2+1\right)$

$=\left[z^5+(3-2)z^2+(1-4)z+2\right]+\left(4z^5+z^2+1\right)=\left(z^5+z^2-3z+2\right)+\left(4z^5+z^2+1\right)=$

$$\begin{array}{r} z^5+z^2-3z+2 \\ 4z^5+z^2\quad\;\;+1 \\ \hline \mathbf{5z^5+2z^2-3z+3} \end{array}$$

5. $-\left(a^3-2a+a+2-3a^3\right)+\left(-2a^3-4a-3\right)=\left(-a^3+2a-a-2+3a^3\right)+\left(-2a^3-4a-3\right)$

$=\left[\left(3a^3-a^3\right)+(2a-a)-2\right]+\left(-2a^3-4a-3\right)=\left[(3-1)a^3+(2-1)a-2\right]+\left(-2a^3-4a-3\right)$

$=\left(2a^3+a-2\right)+\left(-2a^3-4a-3\right)=$

$$\begin{array}{r} +2a^3+\;\;a-2 \\ -2a^3-4a-3 \\ \hline \mathbf{0a^3-3a-5} \end{array}$$

6. $\left(u^2+2u+u+5\right)+\left(-2u^2-3-5u-8\right)=\left[u^2+(2u+u)+5\right]+\left[-2u^2-5u+(-3-8)\right]=\left[u^2+(2+1)u+5\right]+\left[-2u^2-5u-11\right]$

$=\left(u^2+3u+5\right)+\left(-2u^2-5u-11\right)=$

$$\begin{array}{r} +\;u^2+3u+5 \\ -2u^2-5u-11 \\ \hline \mathbf{-\;u^2-2u-6} \end{array}$$

Chapter 2 Solutions:

Section 2.1 Solutions - Introduction to Linear Equations

1. Determine whether 2 is the solution to each of the following equations:
   a. $3\cdot 2-2\stackrel{?}{=}10$; $6-2\stackrel{?}{=}10$; $4\neq 10$. Therefore, 2 is not the solution to $3x-2=10$.
   b. $-2\cdot 2+3\stackrel{?}{=}2$; $-4+3\stackrel{?}{=}2$; $-1\neq 2$. Therefore, 2 is not the solution to $-2x+3=x$.
   c. $6-2\stackrel{?}{=}2\cdot 2+1$; $4\stackrel{?}{=}4+1$; $4\neq 5$. Therefore, 2 is not the solution to $6-x=2x+1$.
   d. $2\cdot 2-8\stackrel{?}{=}-3\cdot 2+2$; $4-8\stackrel{?}{=}-6+2$; $-4=-4$. **Therefore, 2 is the solution to $2x-8=-3x+2$.**
2. Determine if $y=-2$ is the solution to the following equations:
   a. $-2+3\stackrel{?}{=}-2\cdot 2$; $1\neq -4$. Therefore, $y=-2$ is not the solution to $y+3=-2y$.
   b. $(6\cdot -2)-2\stackrel{?}{=}(8\cdot -2)+2$; $-12-2\stackrel{?}{=}-16+2$; $-14=-14$. **Therefore, $y=-2$ is the solution to $6y+y=8y+2$.**
   c. $6+(3\cdot -2)\stackrel{?}{=}0$; $6-6\stackrel{?}{=}0$; $0=0$. **Therefore, $y=-2$ is the solution to $6+3y=0$.**
   d. $3\cdot -2\stackrel{?}{=}5-(-2)$; $-6\stackrel{?}{=}5+2$; $-6\neq 7$. Therefore, $y=-2$ is not the solution to $3y=5-y$.
3. Given the algebraic equation $2x-8=(x-5)+3$, does $x=0$, $x=-1$, and $x=6$ satisfy the original equation?
   a. Let $x=0$ in $2x-8=(x-5)+3$. Then, $2\cdot 0-8\stackrel{?}{=}(0-5)+3$; $0-8\stackrel{?}{=}-5+3$; $-8\neq -2$. Therefore, $x=0$ does not satisfy $2x-8=(x-5)+3$.
   b. Let $x=-1$ in $2x-8=(x-5)+3$. Then, $(2\cdot -1)-8\stackrel{?}{=}(-1-5)+3$; $-2-8\stackrel{?}{=}-6+3$; $-10\neq -3$. Therefore, $x=-1$ does not satisfy $2x-8=(x-5)+3$.
   c. Let $x=6$ in $2x-8=(x-5)+3$. Then, $2\cdot 6-8\stackrel{?}{=}(6-5)+3$; $12-8\stackrel{?}{=}1+3$; $4=4$. **Therefore, $x=6$ satisfies $2x-8=(x-5)+3$.**
4. Does $a=2$ satisfy any of the following equations?
   a. $3\cdot 2+2\stackrel{?}{=}4\cdot 2$; $6+2\stackrel{?}{=}8$; $8=8$. **Therefore, $a=2$ is the solution to $3a+2=4a$.**
   b. $3+7\cdot 2\stackrel{?}{=}18$; $3+14\stackrel{?}{=}18$; $17\neq 18$. Therefore, $a=2$ is not the solution to $3+7a=18$.
   c. $-5\cdot 2+3\stackrel{?}{=}-3\cdot 2-1$; $-10+3\stackrel{?}{=}-6-1$; $-7=-7$. **Therefore, $a=2$ is the solution to $-5a+3=-3a-1$.**
   d. $8\stackrel{?}{=}2+3$; $8\neq 5$. Therefore, $a=2$ is not the solution to $8=a+3$.

Section 2.2 Case I Solutions - Addition and Subtraction of Linear Equations

1. $x-13=12$; $x-13+13=12+13$; $x+0=25$; $\mathbf{x=25}$
2. $8+h=20$; $8-8+h=20-8$; $0+h=12$; $\mathbf{h=12}$
3. $5=x-3$; $5+3=x-3+3$; $8=x+0$; $8=x$; $\mathbf{x=8}$
4. $-3=u-5$; $-3+5=u-5+5$; $2=u+0$; $2=u$; $\mathbf{u=2}$
5. $2.8+x=-3.7$; $2.8-2.8+x=-3.7-2.8$; $0+x=-6.5$; $\mathbf{x=-6.5}$
6. $x-\frac{3}{8}=2\frac{3}{8}$; $x-\frac{3}{8}+\frac{3}{8}=2\frac{3}{8}+\frac{3}{8}$; $x+0=\frac{(2\cdot 8)+3}{8}+\frac{3}{8}$; $x=\frac{16+3}{8}+\frac{3}{8}$; $x=\frac{19}{8}+\frac{3}{8}$; $x=\frac{19+3}{8}$; $x=\frac{22}{8}$; $\mathbf{x=2.75}$

7. $4.9+x=1\frac{2}{3}$; $4.9-4.9+x=1\frac{2}{3}-4.9$; $0+x=\frac{(1\cdot3)+2}{3}-4.9$; $x=\frac{3+2}{3}-4.9$; $x=\frac{5}{3}-4.9$; $x=1.67-4.9$; $\boldsymbol{x=-3.23}$

8. $u+2\frac{1}{3}=-2\frac{3}{5}$; $u+2\frac{1}{3}-2\frac{1}{3}=-2\frac{3}{5}-2\frac{1}{3}$; $u+0=-2\frac{3}{5}-2\frac{1}{3}$; $u=-\frac{(2\cdot5)+3}{5}-\frac{(2\cdot3)+1}{3}$; $u=-\frac{10+3}{5}-\frac{6+1}{3}$

; $u=-\frac{13}{5}-\frac{7}{3}$; $u=\frac{(-13\cdot3)+(-7\cdot5)}{5\cdot3}$; $u=\frac{-39-35}{15}$; $u=-\frac{74}{15}$; $\boldsymbol{u=-4.93}$

9. $6\frac{2}{3}=y-2\frac{4}{5}$; $6\frac{2}{3}+2\frac{4}{5}=y-2\frac{4}{5}+2\frac{4}{5}$; $\frac{(6\cdot3)+2}{3}+\frac{(2\cdot5)+4}{5}=y+0$; $\frac{18+2}{3}+\frac{10+4}{5}=y+0$; $\frac{20}{3}+\frac{14}{5}=y$

; $\frac{(20\cdot5)+(14\cdot3)}{3\cdot5}=y$; $\frac{100+42}{15}=y$; $\frac{142}{15}=y$; $9.46=y$; $\boldsymbol{y=9.46}$

10. $y-2.38=-3\frac{2}{5}$; $y-2.38+2.38=-3\frac{2}{5}+2.38$; $y+0=-\frac{(3\cdot5)+2}{5}+2.38$; $y=-\frac{15+2}{5}+2.38$; $y=-\frac{17}{5}+2.38$

; $y=-3.4+2.38$; $\boldsymbol{y=-1.02}$

Section 2.2 Case II Solutions - Multiplication and Division of Linear Equations

1. $3y=-\frac{2}{3}$; $3y\cdot\frac{1}{3}=-\frac{2}{3}\cdot\frac{1}{3}$; $y=-\frac{2\cdot1}{3\cdot3}$; $\boldsymbol{y=-\frac{2}{9}}$

2. $-\frac{1}{2}x=1\frac{2}{3}$; $-\frac{x}{2}=\frac{(1\cdot3)+2}{3}$; $-\frac{x}{2}=\frac{3+2}{3}$; $-\frac{x}{2}=\frac{5}{3}$; $-\frac{x}{2}\cdot2=\frac{5}{3}\cdot2$; $-x=\frac{10}{3}$; $-x=3.33$; $\frac{-x}{-1}=\frac{3.33}{-1}$; $\boldsymbol{x=-3.33}$

3. $\frac{3}{8}=-2h$; $\frac{3}{8}\cdot\frac{1}{2}=-2h\cdot\frac{1}{2}$; $\frac{3\cdot1}{8\cdot2}=-h$; $\frac{3}{16}=-h$; $0.187=-h$; $\frac{0.187}{-1}=\frac{-h}{-1}$; $-0.187=h$; $\boldsymbol{h=-0.187}$

1. Note that in cases where the variable, in this case h, is in the right hand side of the equation we can solve for h without isolating the variable to the left hand side by applying the multiplication or division rules. However, in the very last step we should move the variable h to the left hand side of the equation and the solution to the right hand side of the equation.

2. Another method of solving for h is by applying the addition or subtraction rules as shown below:

$\frac{3}{8}=-2h$; $\frac{3}{8}+2h=-2h+2h$; $\frac{3}{8}+2h=0$; $\frac{3}{8}-\frac{3}{8}+2h=0-\frac{3}{8}$; $0+2h=-\frac{3}{8}$; $2h=-\frac{3}{8}$; $2h\cdot\frac{1}{2}=-\frac{3}{8}\cdot\frac{1}{2}$

; $h=-\frac{3\cdot1}{8\cdot2}$; $h=-\frac{3}{16}$; $\boldsymbol{h=-0.187}$

4. $\frac{x}{8}=-2$; $\frac{x}{8}\cdot8=-2\cdot8$; $\boldsymbol{x=-16}$

5. $-x=-35$; $\frac{-x}{-1}=\frac{-35}{-1}$; $\boldsymbol{x=35}$

6. $2\frac{1}{8}u=-1\frac{1}{2}$; $\frac{(2\cdot8)+1}{8}u=-\frac{(1\cdot2)+1}{2}$; $\frac{16+1}{8}u=-\frac{2+1}{2}$; $\frac{17}{8}u=-\frac{3}{2}$; $\frac{8}{17}\cdot\frac{17}{8}u=-\frac{3}{2}\cdot\frac{8}{17}$; $u=-\frac{3\cdot8}{2\cdot17}$; $u=-\frac{3\cdot4}{1\cdot17}$

; $u=-\frac{12}{17}$; $\boldsymbol{u=-0.71}$

7. $-w=1\frac{4}{5}$; $-w=\frac{(1\cdot5)+4}{5}$; $-w=\frac{5+4}{5}$; $-w=\frac{9}{5}$; $-w=1.8$; $\frac{-w}{-1}=\frac{1.8}{-1}$; $\boldsymbol{w=-1.8}$

8. $-\frac{1}{2}y=-12$; $-\frac{1}{2}y\cdot2=-12\cdot2$; $-y=-24$; $\frac{-y}{-1}=\frac{-24}{-1}$; $\boldsymbol{y=24}$

9. $2.8x=-1.4$; $\frac{2.8x}{2.8}=\frac{-1.4}{2.8}$; $x=-\frac{\frac{14}{10}}{\frac{28}{10}}$; $x=-\frac{14\cdot10}{10\cdot28}$; $x=-\frac{140}{280}$; $x=-\frac{14}{28}$; $x=-\frac{1}{2}$; $\boldsymbol{x=-0.5}$

10. $-2\frac{3}{5}x=-4.3$; $-\frac{(2\cdot5)+3}{5}x=-4.3$; $-\frac{10+3}{5}x=-4.3$; $-\frac{13}{5}x=-\frac{4.3}{1}$; $-\frac{13}{5}x\cdot-\frac{5}{13}=-\frac{4.3}{1}\cdot-\frac{5}{13}$; $x=\frac{-4.3\cdot-5}{1\cdot13}$; $x=\frac{21.5}{13}$; $\mathbf{x=1.654}$

Section 2.2 Case III Solutions - Mixed Operations Involving Linear Equations

1. $3x-20=5x-8$; $3x-20+20=5x-8+20$; $3x+0=5x+12$; $3x=5x+12$; $3x-5x=5x-5x+12$; $-2x=0+12$; $-2x=12$; $\frac{-2x}{-2}=\frac{12}{-2}$; $x=-\frac{12}{2}$; $\mathbf{x=-6}$

2. $-6y+2=-3+10y$; $-6y-10y+2=-3+10y-10y$; $-16y+2=-3+0$; $-16y+2=-3$; $-16y+2-2=-3-2$; $-16y+0=-5$; $-16y=-5$; $\frac{-16y}{-16}=\frac{-5}{-16}$; $y=\frac{5}{16}$; $\mathbf{y=0.313}$

3. $\frac{x}{-2}+3=5$; $\frac{x}{-2}+3-3=5-3$; $\frac{x}{-2}+0=2$; $\frac{x}{-2}=2$; $-2\cdot\frac{x}{-2}=-2\cdot2$; $\mathbf{x=-4}$

4. $5x-3=-15$; $5x-3+3=-15+3$; $5x+0=-12$; $5x=-12$; $\frac{5x}{5}=\frac{-12}{5}$; $x=-\frac{12}{5}$; $\mathbf{x=-2.4}$

5. $\frac{y}{4}+4=-3$; $\frac{y}{4}+4-4=-3-4$; $\frac{y}{4}+0=-7$; $\frac{y}{4}=-7$; $\frac{y}{4}\cdot4=-7\cdot4$; $\mathbf{y=-28}$

6. $5+\frac{w}{2}=10$; $5-5+\frac{w}{2}=10-5$; $0+\frac{w}{2}=5$; $\frac{w}{2}=5$; $2\cdot\frac{w}{2}=5\cdot2$; $\mathbf{w=10}$

7. $25-3y=2y$; $25-3y-2y=2y-2y$; $25-5y=0$; $25-25-5y=0-25$; $0-5y=-25$; $-5y=-25$; $\frac{-5y}{-5}=\frac{-25}{-5}$; $y=\frac{25}{5}$; $\mathbf{y=5}$

8. $10y+2=8y$; $10y+2-2=8y-2$; $10y+0=8y-2$; $10y=8y-2$; $10y-8y=8y-8y-2$; $2y=0-2$; $2y=-2$; $\frac{2y}{2}=\frac{-2}{2}$; $y=-\frac{2}{2}$; $\mathbf{y=-1}$

9. $\frac{2}{3}x+5=12$; $\frac{2}{3}x+5-5=12-5$; $\frac{2}{3}x+0=12-5$; $\frac{2}{3}x=7$; $\frac{3}{2}\cdot\frac{2}{3}x=\frac{3}{2}\cdot7$; $x=\frac{3}{2}\cdot\frac{7}{1}$; $x=\frac{3\cdot7}{2\cdot1}$; $x=\frac{21}{2}$; $\mathbf{x=10.5}$

10. $m+\frac{1}{2}=4m-\frac{2}{3}$; $m-4m+\frac{1}{2}=4m-4m-\frac{2}{3}$; $-3m+\frac{1}{2}=0-\frac{2}{3}$; $-3m+\frac{1}{2}=-\frac{2}{3}$; $-3m+\frac{1}{2}-\frac{1}{2}=-\frac{2}{3}-\frac{1}{2}$; $-3m+0=-\frac{2}{3}-\frac{1}{2}$; $-3m=\frac{(-2\cdot2)+(-1\cdot3)}{3\cdot2}$; $-3m=\frac{-4-3}{6}$; $-3m=\frac{-7}{6}$; $-\frac{1}{3}\cdot-3m=-\frac{7}{6}\cdot-\frac{1}{3}$; $m=\frac{7\cdot1}{6\cdot3}$; $m=\frac{7}{18}$; $\mathbf{m=0.388}$

Section 2.3 Case I Solutions - Solving Linear Equations Containing Parentheses and Brackets

1. $x-(2x+3)=3$; $x-2x-3=3$; $-x-3=3$; $-x-3+3=3+3$; $-x+0=6$; $-x=6$; $\frac{-x}{-1}=\frac{6}{-1}$; $\mathbf{x=-6}$

2. $2+3(x-1)=-3-(x+5)$; $2+3x-3=-3-x-5$; $(2-3)+3x=(-3-5)-x$; $-1+3x=-8-x$; $-1+1+3x=-8+1-x$; $0+3x=-7-x$; $3x=-7-x$; $3x+x=-7-x+x$; $4x=-7+0$; $4x=-7$; $\frac{4x}{4}=\frac{-7}{4}$; $\mathbf{x=-1.75}$

3. $2-3(x-1)+5x=0$; $2-3x+3+5x=0$; $(2+3)+(-3x+5x)=0$; $5+2x=0$; $+5-5+2x=0-5$; $0+2x=-5$

$; \ 2x=-5 \ ; \ \frac{2x}{2}=\frac{-5}{2} \ ; \ x=-\frac{5}{2} \ ; \ \mathbf{x=-2.5}$

4. $-4(-x+1)-3x=2(x-1) \ ; \ 4x-4-3x=2x-2 \ ; \ (4x-3x)-4=2x-2 \ ; \ x-4=2x-2 \ ; \ x-2x-4=2x-2x-2$

$; \ -x-4=0-2 \ ; \ -x-4=-2 \ ; \ -x-4+4=-2+4 \ ; \ -x+0=2 \ ; \ -x=2 \ ; \ \frac{-x}{-1}=\frac{2}{-1} \ ; \ \mathbf{x=-2}$

5. $2[5-(x-2)]-(x-3)=0 \ ; \ 2[5-x+2]-x+3=0 \ ; \ 2[7-x]-x+3=0 \ ; \ 14-2x-x+3=0 \ ; \ (14+3)+(-2x-x)=0$

$; \ 17-3x=0 \ ; \ 17-17-3x=0-17 \ ; \ 0-3x=-17 \ ; \ -3x=-17 \ ; \ \frac{-3x}{-3}=\frac{-17}{-3} \ ; \ x=\frac{17}{3} \ ; \ \mathbf{x=5.67}$

6. $(x-5)-[3(x-1)+2]=2 \ ; \ (x-5)-[3x-3+2]=2 \ ; \ x-5-[3x-1]=2 \ ; \ x-5-3x+1=2 \ ; \ (x-3x)+(-5+1)=2$

$; \ -2x-4=2 \ ; \ -2x-4+4=2+4 \ ; \ -2x+0=6 \ ; \ -2x=6 \ ; \ \frac{-2x}{-2}=\frac{6}{-2} \ ; \ x=-\frac{6}{2} \ ; \ \mathbf{x=-3}$

7. $3-[(-x+1)+2]=3x-5 \ ; \ 3-[-x+1+2]=3x-5 \ ; \ 3-[-x+3]=3x-5 \ ; \ 3+x-3=3x-5 \ ; \ (3-3)+x=3x-5$

$; \ 0+x=3x-5 \ ; \ x=3x-5 \ ; \ x-3x=3x-3x-5 \ ; \ -2x=0-5 \ ; \ -2x=-5 \ ; \ \frac{-2x}{-2}=\frac{-5}{-2} \ ; \ x=\frac{5}{2} \ ; \ \mathbf{x=2.5}$

8. $-[(5-x)+(3-4x)]=8 \ ; \ -[5-x+3-4x]=8 \ ; \ -[(5+3)+(-x-4x)]=8 \ ; \ -[8-5x]=8 \ ; \ -8+5x=8$

$; \ -8+8+5x=8+8 \ ; \ 0+5x=16 \ ; \ 5x=16 \ ; \ \frac{5x}{5}=\frac{16}{5} \ ; \ \mathbf{x=3.2}$

9. $3+(2x+5)-4x=3-3x \ ; \ 3+2x+5-4x=3-3x \ ; \ (3+5)+(2x-4x)=3-3x \ ; \ 8-2x=3-3x \ ; \ 8-8-2x=3-8-3x$

$; \ 0-2x=-5-3x \ ; \ -2x=-5-3x \ ; \ -2x+3x=-5-3x+3x \ ; \ x=-5+0 \ ; \ \mathbf{x=-5}$

10. $6(x-2)-2(x+1)=3(x+2) \ ; \ 6x-12-2x-2=3x+6 \ ; \ (6x-2x)+(-12-2)=3x+6 \ ; \ 4x-14=3x+6$

$; \ 4x-3x-14=3x-3x+6 \ ; \ x-14=0+6 \ ; \ x-14=6 \ ; \ x-14+14=6+14 \ ; \ x+0=20 \ ; \ \mathbf{x=20}$

Section 2.3 Case II Solutions - Solving Linear Equations Containing Integer Fractions

1. $\frac{1}{2}y=\frac{1}{5}y+5 \ ; \ \frac{1}{2}y-\frac{1}{5}y=\frac{1}{5}y-\frac{1}{5}y+5 \ ; \ \frac{y}{2}-\frac{y}{5}=0+5 \ ; \ \frac{(y\cdot 5)-(2\cdot y)}{2\cdot 5}=5 \ ; \ \frac{5y-2y}{10}=5 \ ; \ \frac{3y}{10}=5 \ ; \ \frac{3y}{10}=\frac{5}{1}$

$; \ 3y\cdot 1=5\cdot 10 \ ; \ 3y=50 \ ; \ \frac{3y}{3}=\frac{50}{3} \ ; \ y=\frac{50}{3} \ ; \ \mathbf{y=16.67}$

A second way to solve this problem is as follows:

$\frac{1}{2}y=\frac{1}{5}y+5 \ ; \ \frac{1}{2}y-\frac{1}{5}y=\frac{1}{5}y-\frac{1}{5}y+5 \ ; \ \frac{y}{2}-\frac{y}{5}=0+5 \ ; \ \left(\frac{1}{2}-\frac{1}{5}\right)y=5 \ ; \ \left(\frac{(1\cdot 5)-(2\cdot 1)}{2\cdot 5}\right)y=5 \ ; \ \left(\frac{5-2}{10}\right)y=5$

$; \ \frac{3}{10}y=5 \ ; \ \frac{3y}{10}-5=5-5 \ ; \ \frac{3y}{10}-\frac{5}{1}=0 \ ; \ \frac{(3y\cdot 1)-(5\cdot 10)}{10\cdot 1}=0 \ ; \ \frac{3y-50}{10}=0 \ ; \ \frac{3y-50}{10}=\frac{0}{1} \ ; \ (3y-50)\cdot 1=10\cdot 0$

$; \ 3y-50=0 \ ; \ 3y-50+50=0+50 \ ; \ 3y+0=50 \ ; \ 3y=50 \ ; \ \frac{3y}{3}=\frac{50}{3} \ ; \ y=\frac{50}{3} \ ; \ \mathbf{y=16.67}$

2. $x=3-\frac{x}{2} \ ; \ x+\frac{x}{2}=3-\frac{x}{2}+\frac{x}{2} \ ; \ \frac{x}{1}+\frac{x}{2}=3+0 \ ; \ \frac{(x\cdot 2)+(1\cdot x)}{1\cdot 2}=3 \ ; \ \frac{2x+x}{2}=3 \ ; \ \frac{3x}{2}=3 \ ; \ \frac{3}{2}x=3 \ ; \ \frac{2}{3}\cdot\frac{3}{2}x=3\cdot\frac{2}{3} \ ; \ \mathbf{x=2}$

3. $y+\frac{2}{3}y=1\frac{2}{3} \ ; \ \left(1+\frac{2}{3}\right)y=\frac{1\cdot 3+2}{3} \ ; \ \left(\frac{1}{1}+\frac{2}{3}\right)y=\frac{3+2}{3} \ ; \ \left(\frac{(1\cdot 3)+(1\cdot 2)}{3}\right)y=\frac{5}{3} \ ; \ \left(\frac{3+2}{3}\right)y=\frac{5}{3} \ ; \ \frac{5}{3}y=\frac{5}{3}$

$; \frac{\cancel{3}}{\cancel{5}}\cdot\frac{\cancel{5}}{\cancel{3}}y=\frac{\cancel{5}}{\cancel{3}}\cdot\frac{\cancel{3}}{\cancel{5}} \; ; \; \mathbf{y=1}$

4. $u-\frac{1}{3}u=6 \; ; \; \frac{u}{1}-\frac{u}{3}=6 \; ; \; \frac{(u\cdot 3)-(1\cdot u)}{1\cdot 3}=6 \; ; \; \frac{3u-u}{3}=6 \; ; \; \frac{2u}{3}=\frac{6}{1} \; ; \; 2u\cdot 1=6\cdot 3 \; ; \; 2u=18 \; ; \; \frac{\cancel{2}u}{\cancel{2}}=\frac{\cancel{18}}{\cancel{2}} \; ; \; u=\frac{9}{1} \; ; \; \mathbf{u=9}$

5. $s+1\frac{2}{3}=2\frac{3}{5}s \; ; \; s+\frac{(1\cdot 3)+2}{3}=\frac{(2\cdot 5)+3}{5}s \; ; \; s+\frac{3+2}{3}=\frac{10+3}{5}s \; ; \; s+\frac{5}{3}=\frac{13}{5}s \; ; \; s+\frac{5}{3}-\frac{5}{3}=\frac{13}{5}s-\frac{5}{3} \; ; \; s+0=\frac{13}{5}s-\frac{5}{3}$

$; \; s=\frac{13}{5}s-\frac{5}{3} \; ; \; s-\frac{13}{5}s=\frac{13}{5}s-\frac{13}{5}s-\frac{5}{3} \; ; \; s-\frac{13}{5}s=0-\frac{5}{3} \; ; \; \frac{s}{1}-\frac{13s}{5}=-\frac{5}{3} \; ; \; \frac{(5\cdot s)-(13s\cdot 1)}{1\cdot 5}=-\frac{5}{3} \; ; \; \frac{5s-13s}{5}=-\frac{5}{3}$

$; \; \frac{-8s}{5}=-\frac{5}{3} \; ; \; -8s\cdot 3=-5\cdot 5 \; ; \; -24s=-25 \; ; \; \frac{\cancel{-24}s}{\cancel{-24}}=+\frac{25}{24} \; ; \; s=\frac{25}{24} \; ; \; \mathbf{s=1.04}$

6. $\frac{w}{3}-1\frac{2}{3}w=4 \; ; \; \frac{w}{3}-\frac{(1\cdot 3)+2}{3}w=4 \; ; \; \frac{w}{3}-\frac{3+2}{3}w=4 \; ; \; \frac{w}{3}-\frac{5w}{3}=4 \; ; \; \frac{w-5w}{3}=4 \; ; \; \frac{-4w}{3}=\frac{4}{1} \; ; \; -4w\cdot 1=3\cdot 4$

$; \; -4w=12 \; ; \; \frac{\cancel{-4}w}{\cancel{-4}}=\frac{12}{-4} \; ; \; w=-\frac{\cancel{12}}{\cancel{4}} \; ; \; w=-\frac{3}{1} \; ; \; \mathbf{w=-3}$

7. $x-\frac{2}{3}=1\frac{1}{4}x \; ; \; x-\frac{2}{3}=\frac{(1\cdot 4)+1}{4}x \; ; \; x-\frac{2}{3}=\frac{4+1}{4}x \; ; \; x-\frac{2}{3}=\frac{5}{4}x \; ; \; x-\frac{5}{4}x-\frac{2}{3}=\frac{5}{4}x-\frac{5}{4}x \; ; \; x-\frac{5}{4}x-\frac{2}{3}=0$

$; \; x-\frac{5}{4}x-\frac{2}{3}+\frac{2}{3}=0+\frac{2}{3} \; ; \; x-\frac{5}{4}x+0=\frac{2}{3} \; ; \; x-\frac{5}{4}x=\frac{2}{3} \; ; \; \left(1-\frac{5}{4}\right)x=\frac{2}{3} \; ; \; \left(\frac{1}{1}-\frac{5}{4}\right)x=\frac{2}{3} \; ; \; \left(\frac{(1\cdot 4)-(1\cdot 5)}{1\cdot 4}\right)x=\frac{2}{3}$

$; \; \left(\frac{4-5}{4}\right)x=\frac{2}{3} \; ; \; -\frac{1}{4}x=\frac{2}{3} \; ; \; -\frac{x}{\cancel{4}}\cdot\cancel{-4}=\frac{2}{3}\cdot -4 \; ; \; x=-\frac{8}{3} \; ; \; x=-2\frac{2}{3} \; ; \; \mathbf{x=-2.67}$

8. $t=5-\frac{2}{3}t \; ; \; t+\frac{2}{3}t=5-\frac{2}{3}t+\frac{2}{3}t \; ; \; t+\frac{2}{3}t=5+0 \; ; \; \left(1+\frac{2}{3}\right)t=5 \; ; \; \left(\frac{1}{1}+\frac{2}{3}\right)t=5 \; ; \; \left(\frac{(1\cdot 3)+(1\cdot 2)}{1\cdot 3}\right)t=5 \; ; \; \frac{3+2}{3}t=5$

$; \; \frac{5}{3}t=5 \; ; \; \frac{\cancel{3}}{\cancel{5}}\cdot\frac{\cancel{5}}{\cancel{3}}t=\cancel{5}\cdot\frac{3}{\cancel{5}} \; ; \; t=\frac{3}{1} \; ; \; \mathbf{t=3}$

9. $4\frac{2}{3}z-2\frac{3}{5}z=1\frac{1}{4} \; ; \; \frac{(4\cdot 3)+2}{3}z-\frac{(2\cdot 5)+3}{5}z=\frac{(1\cdot 4)+1}{4} \; ; \; \frac{12+2}{3}z-\frac{10+3}{5}z=\frac{4+1}{4} \; ; \; \frac{14}{3}z-\frac{13}{5}z=\frac{5}{4} \; ; \; \left(\frac{14}{3}-\frac{13}{5}\right)z=\frac{5}{4}$

$; \; \left(\frac{(14\cdot 5)-(13\cdot 3)}{3\cdot 5}\right)z=\frac{5}{4} \; ; \; \left(\frac{70-39}{15}\right)z=\frac{5}{4} \; ; \; \frac{31}{15}z=\frac{5}{4} \; ; \; \frac{\cancel{15}}{\cancel{31}}\cdot\frac{\cancel{31}}{\cancel{15}}z=\frac{5}{4}\cdot\frac{15}{31} \; ; \; z=\frac{5\cdot 15}{4\cdot 31} \; ; \; z=\frac{75}{124} \; ; \; \mathbf{z=0.605}$

10. $6+\frac{1}{2}t=t-1\frac{2}{5}t \; ; \; 6+\frac{1}{2}t=t-\frac{(1\cdot 5)+2}{5}t \; ; \; 6+\frac{1}{2}t=t-\frac{5+2}{5}t \; ; \; 6+\frac{1}{2}t=t-\frac{7}{5}t \; ; \; 6+\frac{1}{2}t=\left(1-\frac{7}{5}\right)t \; ; \; 6+\frac{1}{2}t=\left(\frac{1}{1}-\frac{7}{5}\right)t$

$6+\frac{1}{2}t=\left(\frac{(1\cdot 5)-(1\cdot 7)}{1\cdot 5}\right)t \; ; \; 6+\frac{1}{2}t=\left(\frac{5-7}{5}\right)t \; ; \; 6+\frac{1}{2}t=-\frac{2}{5}t \; ; \; 6+\frac{1}{2}t+\frac{2}{5}t=-\frac{2}{5}t+\frac{2}{5}t \; ; \; 6+\frac{1}{2}t+\frac{2}{5}t=0$

$; \; 6-6+\frac{1}{2}t+\frac{2}{5}t=0-6 \; ; \; 0+\frac{1}{2}t+\frac{2}{5}t=-6 \; ; \; \frac{1}{2}t+\frac{2}{5}t=-6 \; ; \; \left(\frac{1}{2}+\frac{2}{5}\right)t=-6 \; ; \; \left(\frac{(1\cdot 5)+(2\cdot 2)}{2\cdot 5}\right)t=-6 \; ; \; \left(\frac{5+4}{10}\right)t=-6$

$; \; \frac{9}{10}t=-6 \; ; \; \frac{\cancel{10}}{\cancel{9}}\cdot\frac{\cancel{9}}{\cancel{10}}t=-6\cdot\frac{10}{9} \; ; \; t=-\frac{6}{1}\cdot\frac{10}{9} \; ; \; t=-\frac{6\cdot 10}{1\cdot 9} \; ; \; t=-\frac{60}{9} \; ; \; \mathbf{t=-6.67}$

A second way to solve this problem is as follows:

$6+\frac{1}{2}t=t-1\frac{2}{5}t \; ; \; 6+\frac{1}{2}t=t-\frac{(1\cdot 5)+2}{5}t \; ; \; 6+\frac{1}{2}t=t-\frac{5+2}{5}t \; ; \; 6+\frac{1}{2}t=t-\frac{7}{5}t \; ; \; 6+\frac{1}{2}t=\left(1-\frac{7}{5}\right)t \; ; \; 6+0.5t=(1-1.4)t$

$; 6+0.5t=-0.4t \; ; \; 6+0.5t+0.4t=-0.4t+0.4t \; ; \; 6+0.9t=0 \; ; \; 6-6+0.9t=0-6 \; ; \; 0+0.9t=-6 \; ; \; 0.9t=-6 \; ; \; \frac{\cancel{0.9}t}{\cancel{0.9}}=\frac{-6}{0.9}$

$; t=-\frac{6}{0.9}=t=-\frac{\frac{6}{1}}{\frac{9}{10}} \; ; \; t=-\frac{6\cdot10}{1\cdot9} \; ; \; t=-\frac{60}{9} \; ; \; \mathbf{t=-6.67}$

Section 2.3 Case III Solutions - Solving Linear Equations Containing Decimals

1. $0.35-0.2x=0.5+0.1x \; ; \; 0.35-0.2x-0.1x=0.5+0.1x-0.1x \; ; \; 0.35-0.3x=0.5+0 \; ; \; 0.35-0.3x=0.5$

$; 0.35-0.35-0.3x=0.5-0.35 \; ; \; 0.0-0.3x=0.15 \; ; \; -0.3x=0.15 \; ; \; \frac{\cancel{-0.3}x}{\cancel{-0.3}}=\frac{0.15}{-0.3} \; ; \; x=-\frac{\frac{15}{100}}{\frac{3}{10}} \; ; \; x=-\frac{\overset{5}{\cancel{15}}\cdot\cancel{10}}{\underset{10}{\cancel{100}}\cdot\cancel{3}} \; ; \; x=-\frac{\cancel{5}}{\underset{2}{\cancel{10}}}$

$; x=-\frac{1}{2} \; ; \; \mathbf{x=-0.5}$

2. $5.2x+0.1(x-0.25)=0.2x \; ; \; 5.2x+0.1x-0.025=0.2x \; ; \; 5.3x-0.025=0.2x \; ; \; 5.3x-0.2x-0.025=0.2x-0.2x$

$; 5.1x-0.025=0 \; ; \; 5.1x-0.025+0.025=0+0.025 \; ; \; 5.1x+0=0.025 \; ; \; 5.1x=0.025 \; ; \; \frac{\cancel{5.1}x}{\cancel{5.1}}=\frac{0.025}{5.1} \; ; \; x=\frac{\frac{25}{1000}}{\frac{51}{10}}$

$; x=\frac{25\cdot\cancel{10}}{\underset{100}{\cancel{1000}}\cdot51} \; ; \; x=\frac{25}{5100} \; ; \; \mathbf{x=0.0049}$

3. $0.4(x-2)-0.2(x-1)=0.25 \; ; \; 0.4x-0.8-0.2x+0.2=0.25 \; ; \; [0.4x-0.2x]+(0.2-0.8)=0.25 \; ; \; 0.2x-0.6=0.25$

$; 0.2x-0.6+0.6=0.25+0.6 \; ; \; 0.2x+0=0.85 \; ; \; 0.2x=0.85 \; ; \; \frac{\cancel{0.2}x}{\cancel{0.2}}=\frac{0.85}{0.2} \; ; \; x=\frac{\frac{85}{100}}{\frac{2}{10}} \; ; \; x=\frac{85\cdot\cancel{10}}{\underset{10}{\cancel{100}}\cdot2} \; ; \; x=\frac{85}{20} \; ; \; \mathbf{x=4.25}$

4. $1.2x+0.56-0.6x=1.25x \; ; \; (1.2x-0.6x)+0.56=1.25x \; ; \; 0.6x+0.56=1.25x \; ; \; 0.6x-1.25x+0.56=1.25x-1.25x$

$; -0.65x+0.56=0 \; ; \; -0.65x+0.56-0.56=0-0.56 \; ; \; -0.65x+0=-0.56 \; ; \; -0.65x=-0.56 \; ; \; \frac{\cancel{-0.65}x}{\cancel{-0.65}}=\frac{-0.56}{-0.65} \; ; \; x=+\frac{\frac{56}{100}}{\frac{65}{100}}$

$; x=\frac{56\cdot\cancel{100}}{\cancel{100}\cdot65} \; ; \; x=\frac{56}{65} \; ; \; \mathbf{x=0.862}$

5. $(x-0.5)-[(x+0.1)-3x]=-x \; ; \; x-0.5-[x+0.1-3x]=-x \; ; \; x-0.5-[0.1-2x]=-x \; ; \; x-0.5-0.1+2x=-x$

$; (x+2x)+(-0.5-0.1)=-x \; ; \; 3x-0.6=-x \; ; \; 3x+x-0.6=-x+x \; ; \; 4x-0.6=0 \; ; \; 4x-0.6+0.6=0+0.6 \; ; \; 4x+0=0.6$

$; 4x=0.6 \; ; \; \frac{\cancel{4}x}{\cancel{4}}=\frac{0.6}{4} \; ; \; x=\frac{\frac{6}{10}}{\frac{4}{1}} \; ; \; x=\frac{\overset{3}{\cancel{6}}\cdot1}{10\cdot\underset{2}{\cancel{4}}} \; ; \; x=\frac{3}{20} \; ; \; \mathbf{x=0.15}$

6. $5(0.02x+0.002)-0.5x=1.25 \; ; \; 0.1x+0.01-0.5x=1.25 \; ; \; (0.1x-0.5x)+0.01=1.25 \; ; \; -0.4x+0.01=1.25$

$; -0.4x+0.01-0.01=1.25-0.01 \; ; \; -0.4x+0=1.24 \; ; \; -0.4x=1.24 \; ; \; \frac{\cancel{-0.4}x}{\cancel{-0.4}}=\frac{1.24}{-0.4} \; ; \; x=-\frac{\frac{124}{100}}{\frac{4}{10}} \; ; \; x=-\frac{\overset{31}{\cancel{124}}\cdot\cancel{10}}{\underset{10}{\cancel{100}}\cdot\cancel{4}} \; ; \; x=-\frac{31}{10}$

$; x=-3\frac{1}{10} \; ; \; \mathbf{x=-3.1}$

7. $0.5x=-(2-2.5x)+2.8 \; ; \; 0.5x=-2+2.5x+2.8 \; ; \; 0.5x=(-2+2.8)+2.5x \; ; \; 0.5x=0.8+2.5x \; ; \; 0.5x-2.5x=0.8+2.5x-2.5x$

$; -2x = 0.8 + 0 \; ; \; -2x = 0.8 \; ; \; \frac{-2x}{-2} = \frac{0.8}{-2} \; ; \; x = -\frac{\frac{8}{10}}{\frac{2}{1}} \; ; \; x = -\frac{\overset{4}{8} \cdot 1}{10 \cdot 2} \; ; \; x = -\frac{4}{10} \; ; \; \mathbf{x = -0.4}$

8. $1.35 - 0.5(x + 0.2) = 0 \; ; \; 1.35 - 0.5x - 0.1 = 0 \; ; \; -0.5x + (1.35 - 0.1) = 0 \; ; \; -0.5x + 1.25 = 0 \; ; \; -0.5x + 1.25 - 1.25 = 0 - 1.25$

$; -0.5x + 0 = -1.25 \; ; \; -0.5x = -1.25 \; ; \; \frac{-0.5x}{-0.5} = \frac{-1.25}{-0.5} \; ; \; x = +\frac{1.25}{0.5} \; ; \; x = \frac{\frac{125}{100}}{\frac{5}{10}} \; ; \; x = \frac{\overset{25}{125} \cdot 10}{\underset{10}{100} \cdot 5} \; ; \; x = \frac{25}{10} \; ; \; \mathbf{x = 2.5}$

9. $0.5[-(0.8x - 0.2) - 5] = 2.2x \; ; \; 0.5[-0.8x + 0.2 - 5] = 2.2x \; ; \; 0.5[-0.8x - 4.8] = 2.2x \; ; \; -0.4x - 2.4 = 2.2x$

$; -0.4x - 2.2x - 2.4 = 2.2x - 2.2x \; ; \; -2.6x - 2.4 = 0 \; ; \; -2.6x - 2.4 + 2.4 = 0 + 2.4 \; ; \; -2.6x + 0 = 2.4 \; ; \; -2.6x = 2.4$

$; \frac{-2.6x}{-2.6} = \frac{2.4}{-2.6} \; ; \; x = -\frac{\frac{24}{10}}{\frac{26}{10}} \; ; \; x = -\frac{\overset{12}{24} \cdot 10}{10 \cdot \underset{13}{26}} \; ; \; x = -\frac{12}{13} \; ; \; \mathbf{x = -0.923}$

10. $0.25x - 1.3 + (1.2x - 1.7) = -2.8 \; ; \; 0.25x - 1.3 + 1.2x - 1.7 = -2.8 \; ; \; (0.25x + 1.2x) + (-1.7 - 1.3) = -2.8 \; ; \; 1.45x - 3 = -2.8$

$; 1.45x - 3 + 3 = -2.8 + 3 \; ; \; 1.45x + 0 = 0.2 \; ; \; 1.45x = 0.2 \; ; \; \frac{1.45x}{1.45} = \frac{0.2}{1.45} \; ; \; x = \frac{\frac{2}{10}}{\frac{145}{100}} \; ; \; x = \frac{2 \cdot \overset{10}{100}}{10 \cdot 145} \; ; \; x = \frac{20}{145} \; ; \; \mathbf{x = 0.138}$

Section 2.4 Solutions - Formulas

1. I. $V = \pi r^2 h \; ; \; V = \pi h r^2 \; ; \; \frac{V}{\pi h} = \frac{\pi h r^2}{\pi h} \; ; \; \frac{V}{\pi h} = r^2 \; ; \; r^2 = \frac{V}{\pi h} \; ; \; \sqrt{r^2} = \pm\sqrt{\frac{V}{\pi h}} \; ; \; \mathbf{r = \pm\sqrt{\frac{V}{\pi h}}}$

   II. $V = \pi r^2 h \; ; \; V = \pi r^2 h \; ; \frac{V}{\pi r^2} = \frac{\pi r^2 h}{\pi r^2} \; ; \; \frac{V}{\pi r^2} = h \; ; \; \mathbf{h = \frac{V}{\pi r^2}}$

2. I. $2x + 2y = 3(x + y) - 5 \; ; \; 2x + 2y = 3x + 3y - 5 \; ; \; 2x - 3x + 2y = (3x - 3x) + 3y - 5 \; ; \; -x + 2y = 0 + 3y - 5$

   $; -x + 2y = 3y - 5 \; ; \; -x + (2y - 2y) = 3y - 2y - 5 \; ; \; -x + 0 = y - 5 \; ; \; -x = y - 5 \; ; \; \frac{-x}{-1} = \frac{y-5}{-1} \; ; \; x = -\frac{y-5}{1} \; ; \; \mathbf{x = -y + 5}$

   II. $2x + 2y = 3(x + y) - 5 \; ; \; (2x - 2x) + 2y = (3x - 2x) + 3y - 5 \; ; \; 0 + 2y = x + 3y - 5 \; ; \; 2y = x + 3y - 5$

   $; 2y - 3y = x + (3y - 3y) - 5 \; ; \; -y = x + 0 - 5 \; ; \; -y = x - 5 \; ; \; \frac{-y}{-1} = \frac{x-5}{-1} \; ; \; y = -\frac{x-5}{1} \; ; \; \mathbf{y = -x + 5}$

3. $C = 2\pi r \; ; \; \frac{C}{2\pi} = \frac{2\pi r}{2\pi} \; ; \; \frac{C}{2\pi} = r \; ; \; \mathbf{r = \frac{C}{2\pi}}$

4. I. $d = rt \; ; \; \frac{d}{r} = \frac{rt}{r} \; ; \; \frac{d}{r} = t \; ; \; \mathbf{t = \frac{d}{r}}$

   II. $d = rt \; ; \; \frac{d}{t} = \frac{rt}{t} \; ; \; \frac{d}{t} = r \; ; \; \mathbf{r = \frac{d}{t}}$

5. I. $y - b = \frac{1}{3}x + \frac{2}{3}b \; ; \; y - b - \frac{2}{3}b = \frac{1}{3}x + \frac{2}{3}b - \frac{2}{3}b \; ; \; y - b - \frac{2}{3}b = \frac{1}{3}x + 0 \; ; \; y - b - 0.67b = \frac{1}{3}x \; ; \; y - 1.67b = \frac{1}{3}x$

   $; 3 \cdot (y - 1.67b) = 3 \cdot \frac{1}{3}x \; ; \; 3y - 5.01b = x \; ; \; \mathbf{x = 3y - 5.01b}$

   II. $y - b = \frac{1}{3}x + \frac{2}{3}b \; ; \; y - b - \frac{2}{3}b = \frac{1}{3}x + \frac{2}{3}b - \frac{2}{3}b \; ; \; y - b - \frac{2}{3}b = \frac{1}{3}x + 0 \; ; \; y - b - 0.67b = \frac{1}{3}x \; ; \; y - 1.67b = \frac{1}{3}x$

   $; y - y - 1.67b = \frac{1}{3}x - y \; ; \; 0 - 1.67b = \frac{1}{3}x - y \; ; \; -1.67b = 0.33x - y \; ; \; \frac{-1.67b}{-1.67} = \frac{0.33x - y}{-1.67} \; ; \; \mathbf{b = -\frac{0.33x - y}{1.67}}$

6. $y=\frac{a-b-c}{3}$; $\frac{y}{1}=\frac{a-b-c}{3}$; $3\cdot y=1\cdot(a-b-c)$; $3y=a-b-c$; $3y-a=(a-a)-b-c$; $3y-a=0-b-c$

; $3y-a=-b-c$; $3y-a+b=(-b+b)-c$; $3y-a+b=0-c$; $3y-a+b=-c$; $\frac{3y-a+b}{-1}=\frac{-c}{-1}$

; $-(3y-a+b)=c$; $-3y+a-b=c$; $\mathbf{c=-3y+a-b}$

7. I. $m=\frac{y-b}{x}$; $\frac{m}{1}=\frac{y-b}{x}$; $m\cdot x=(y-b)\cdot 1$; $mx=y-b$; $mx+b=y-b+b$; $mx+b=y$; $\mathbf{y=mx+b}$

II. $m=\frac{y-b}{x}$; $\frac{m}{1}=\frac{y-b}{x}$; $m\cdot x=(y-b)\cdot 1$; $mx=y-b$; $mx-y=y-y-b$; $mx-y=0-b$; $mx-y=-b$

; $\frac{mx-y}{-1}=\frac{-b}{-1}$; $-\frac{mx-y}{1}=\frac{b}{1}$; $-mx+y=b$; $\mathbf{b=-mx+y}$

8. I. $V=\frac{1}{3}\pi r^2h$; $V=\frac{\pi r^2h}{3}$; $\frac{V}{1}=\frac{\pi r^2h}{3}$; $3\cdot V=1\cdot\left(\pi r^2h\right)$; $3V=\pi r^2h$; $\frac{3V}{r^2h}=\frac{\pi \cancel{r^2h}}{\cancel{r^2h}}$; $\frac{3V}{r^2h}=\pi$; $\boldsymbol{\pi=\frac{3V}{r^2h}}$

II. $V=\frac{1}{3}\pi r^2h$; $V=\frac{\pi r^2h}{3}$; $\frac{V}{1}=\frac{\pi r^2h}{3}$; $3\cdot V=1\cdot\left(\pi r^2h\right)$; $3V=\pi r^2h$; $3V=\pi hr^2$; $\frac{3V}{\pi h}=\frac{\cancel{\pi h}r^2}{\cancel{\pi h}}$; $\frac{3V}{\pi h}=r^2$

; $r^2=\frac{3V}{\pi h}$; $\sqrt{r^2}=\pm\sqrt{\frac{3V}{\pi h}}$; $\boldsymbol{r=\pm\sqrt{\frac{3V}{\pi h}}}$

III. $V=\frac{1}{3}\pi r^2h$; $V=\frac{\pi r^2h}{3}$; $\frac{V}{1}=\frac{\pi r^2h}{3}$; $3\cdot V=1\cdot\left(\pi r^2h\right)$; $3V=\pi r^2h$; $\frac{3V}{\pi r^2}=\frac{\cancel{\pi r^2}h}{\cancel{\pi r^2}}$; $\frac{3V}{\pi r^2}=h$; $\boldsymbol{h=\frac{3V}{\pi r^2}}$

9. I. $E=mc^2$; $\frac{E}{m}=\frac{\cancel{m}c^2}{\cancel{m}}$; $\frac{E}{m}=c^2$; $c^2=\frac{E}{m}$; $\sqrt{c^2}=\pm\sqrt{\frac{E}{m}}$; $\boldsymbol{c=\pm\sqrt{\frac{E}{m}}}$

II. $E=mc^2$; $\frac{E}{c^2}=\frac{mc^2}{c^2}$; $\frac{E}{c^2}=m$; $\boldsymbol{m=\frac{E}{c^2}}$

10. I. $y-(2x-3y)+3=5y-x$; $y-2x+3y+3=5y-x$; $(y+3y)-2x+3=5y-x$; $4y-2x+3=5y-x$

; $(4y-4y)-2x+3=5y-4y-x$; $0-2x+3=y-x$; $-2x+3=y-x$; $-2x+3-3=y-x-3$; $-2x+0=y-x-3$

; $-2x=y-x-3$; $-2x+x=y-\cancel{x}+\cancel{x}-3$; $-x=y-3$; $\frac{-x}{-1}=\frac{y-3}{-1}$; $x=-(y-3)$; $\mathbf{x=3-y}$

II. $y-(2x-3y)+3=5y-x$; $y-2x+3y+3=5y-x$; $(y+3y)-2x+3=5y-x$; $4y-2x+3=5y-x$

; $(4y-5y)-2x+3=5y-5y-x$; $-y-2x+3=0-x$; $-y-2x+3=-x$; $-y+(-2x+2x)+3=-x+2x$

; $-y+0+3=x$; $-y+3=x$; $-y+3-3=x-3$; $-y+0=x-3$; $-y=x-3$; $\frac{-y}{-1}=\frac{x-3}{-1}$; $y=-\frac{x-3}{1}$; $\mathbf{y=3-x}$

Section 2.5 Case I Solutions - Addition and Subtraction of Linear Inequalities

1. $x-10>12$; $x-10+10>12+10$; $x+0>12+10$; $\mathbf{x>22}$

2. $-3\le -u+8$; $-3+u\le -u+u+8$; $-3+u\le 0+8$; $-3+u\le 8$; $-3+3+u\le 8+3$; $0+u\le 11$; $\mathbf{u\le 11}$

3. $8<-x+5$; $8+x<-x+x+5$; $8+x<0+5$; $8+x<5$; $8-8+x<5-8$; $0+x<-3$; $\mathbf{x<-3}$

4. $3.2+w\ge -2.8$; $3.2-3.2+w\ge -2.8-3.2$; $0+w\ge -6$; $\mathbf{w\ge -6}$

5. $0.65+t > 2\frac{2}{3}$; $0.65-0.65+t > \frac{(2\cdot 3)+2}{3}-0.65$; $0+t > \frac{6+2}{3}-0.65$; $t > \frac{8}{3}-0.65$; $t > 2.67-0.65$; $\boldsymbol{t > 2.02}$

6. $s-\frac{3}{5} < 1\frac{3}{5}$; $s-\frac{3}{5}+\frac{3}{5} < \frac{(1\cdot 5)+3}{5}+\frac{3}{5}$; $s+0 < \frac{5+3}{5}+\frac{3}{5}$; $s < \frac{8}{5}+\frac{3}{5}$; $s < \frac{8+3}{5}$; $s < \frac{11}{5}$; $\boldsymbol{s < 2.2}$

7. $0.8+w \geq 1\frac{1}{3}+0.9$; $0.8-0.8+w \geq \frac{(1\cdot 3)+1}{3}+0.9-0.8$; $0+w \geq \frac{3+1}{3}+0.1$; $w \geq \frac{4}{3}+0.1$; $w \geq 1.33+0.1$; $\boldsymbol{w \geq 1.43}$

8. $-1\frac{2}{7} < -h+2\frac{3}{8}$; $-\frac{(1\cdot 7)+2}{7} < -h+\frac{(2\cdot 8)+3}{8}$; $-\frac{7+2}{7} < -h+\frac{16+3}{8}$; $-\frac{9}{7} < -h+\frac{19}{8}$; $-\frac{9}{7}+h < -h+h+\frac{19}{8}$; $-\frac{9}{7}+h < 0+\frac{19}{8}$; $-\frac{9}{7}+h < \frac{19}{8}$; $-\frac{9}{7}+\frac{9}{7}+h < \frac{19}{8}+\frac{9}{7}$; $h < \frac{(19\cdot 7)+(9\cdot 8)}{8\cdot 7}$; $h < \frac{133+72}{56}$; $h < \frac{205}{56}$; $\boldsymbol{h < 3.66}$

9. $y-1.25 \leq -2\frac{3}{4}$; $y-1.25+1.25 \leq -\frac{(2\cdot 4)+3}{4}+1.25$; $y+0 \leq -\frac{8+3}{4}+1.25$; $y \leq -\frac{11}{4}+1.25$; $y \leq -2.75+1.25$; $\boldsymbol{y \leq -1.5}$

10. $x+1\frac{2}{3} < 2\frac{2}{5}-\frac{2}{7}$; $x+\frac{(1\cdot 3)+2}{3} < \frac{(2\cdot 5)+2}{5}-\frac{2}{7}$; $x+\frac{3+2}{3} < \frac{10+2}{5}-\frac{2}{7}$; $x+\frac{5}{3} < \frac{12}{5}-\frac{2}{7}$; $x+\frac{5}{3} < \frac{(12\cdot 7)-(2\cdot 5)}{5\cdot 7}$; $x+\frac{5}{3} < \frac{84-10}{35}$; $x+\frac{5}{3} < \frac{74}{35}$; $x+\frac{5}{3}-\frac{5}{3} < \frac{74}{35}-\frac{5}{3}$; $x+0 < \frac{(74\cdot 3)-(5\cdot 35)}{35\cdot 3}$; $x < \frac{222-175}{105}$; $x < \frac{47}{105}$; $\boldsymbol{x < 0.45}$

Section 2.5 Case II Solutions - Multiplication and Division of Linear Inequalities

1. $4y > -\frac{2}{3}$; $\frac{1}{4}\cdot 4y > -\frac{2}{3}\cdot\frac{1}{4}$; $y > -\frac{2\cdot 1}{3\cdot 4}$; $y > -\frac{2}{12}$; $\boldsymbol{y > -0.167}$

2. $-\frac{2}{3}x \leq 1\frac{2}{3}$; $-\frac{2}{3}x \leq \frac{(1\cdot 3)+2}{3}$; $-\frac{2}{3}x \leq \frac{3+2}{3}$; $-\frac{2}{3}x \leq \frac{5}{3}$; $-\frac{3}{2}\cdot-\frac{2}{3}x \geq \frac{5}{3}\cdot-\frac{3}{2}$; $x \geq -\frac{5\cdot 3}{3\cdot 2}$; $x \geq -\frac{5}{2}$; $\boldsymbol{x \geq -2.5}$

3. $\frac{2}{5} < -2h$; $\frac{2}{5}+2h < -2h+2h$; $\frac{2}{5}+2h < 0$; $\frac{2}{5}-\frac{2}{5}+2h < 0-\frac{2}{5}$; $0+2h < -\frac{2}{5}$; $2h < -\frac{2}{5}$; $\frac{1}{2}\cdot 2h < -\frac{2}{5}\cdot\frac{1}{2}$; $h < -\frac{2\cdot 1}{5\cdot 2}$; $\boldsymbol{h < -\frac{1}{5}}$ Another way of solving the problem is: $\frac{2}{5} < -2h$; $\frac{2}{5}\cdot\frac{1}{-2} > \frac{-2h}{-2}$; $-\frac{2}{10} > h$; $-\frac{1}{5} > h$ or $\boldsymbol{h < -\frac{1}{5}}$

4. $\frac{w}{7} \geq -3$; $7\cdot\frac{w}{7} \geq -3\cdot 7$; $\frac{w}{1} \geq -21$; $\boldsymbol{w \geq -21}$

5. $\frac{w}{-2} < -5$; $-2\cdot\frac{w}{-2} > -5\cdot -2$; $\frac{w}{1} > +10$; $\boldsymbol{w > 10}$

6. $2\frac{1}{4}u > -1\frac{1}{5}$; $\frac{(2\cdot 4)+1}{4}u > -\frac{(1\cdot 5)+1}{5}$; $\frac{8+1}{4}u > -\frac{5+1}{5}$; $\frac{9}{4}u > -\frac{6}{5}$; $\frac{4}{9}\cdot\frac{9}{4}u > -\frac{6}{5}\cdot\frac{4}{9}$; $u > -\frac{6\cdot 4}{5\cdot 9}$; $u > -\frac{24}{45}$; $\boldsymbol{u > -0.53}$

7. $-2x \leq 2\frac{3}{4}+1$; $-2x \leq \frac{(2\cdot 4)+3}{4}+\frac{1}{1}$; $-2x \leq \frac{8+3}{4}+\frac{1}{1}$; $-2x \leq \frac{11}{4}+\frac{1}{1}$; $-2x \leq \frac{(11\cdot 1)+(1\cdot 4)}{4\cdot 1}$; $-2x \leq \frac{11+4}{4}$; $-2x \leq \frac{15}{4}$; $-\frac{1}{2}\cdot-2x \geq \frac{15}{4}\cdot-\frac{1}{2}$; $x \geq \frac{15}{4}\cdot-\frac{1}{2}$; $x \geq -\frac{15\cdot 1}{4\cdot 2}$; $x \geq -\frac{15}{8}$; $\boldsymbol{x \geq -1.875}$

8. $3.28x \geq 2.4$; $\frac{3.28x}{3.28} \geq \frac{2.4}{3.28}$; $x \geq \frac{2.4}{3.28}$; $x \geq \frac{\frac{24}{10}}{\frac{328}{100}}$; $x \geq \frac{24\cdot 100}{10\cdot 328}$; $x \geq \frac{2400}{3280}$; $\boldsymbol{x \geq 0.732}$

9. $-\frac{1}{4}y > -2$; $-\frac{1}{4}\cdot -4y < -2\cdot -4$; $+y < +8$; $\boldsymbol{y < 8}$

10. $5\frac{1}{3} < -2\frac{3}{4}x \; ; \; \frac{(5\cdot3)+1}{3} < -\frac{(2\cdot4)+3}{4}x \; ; \; \frac{15+1}{3} < -\frac{8+3}{4}x \; ; \; \frac{16}{3} < -\frac{11}{4}x \; ; \; \frac{16}{3}+\frac{11}{4}x < -\frac{11}{4}x+\frac{11}{4}x \; ; \; \frac{16}{3}+\frac{11}{4}x < 0$

$; \; \frac{16}{3}-\frac{16}{3}+\frac{11}{4}x < 0-\frac{16}{3} \; ; \; 0+\frac{11}{4}x < -\frac{16}{3} \; ; \; \frac{11}{4}x < -\frac{16}{3} \; ; \; \frac{4}{11}\cdot\frac{11}{4}x < -\frac{16}{3}\cdot\frac{4}{11} \; ; \; x < -\frac{16\cdot4}{3\cdot11} \; ; \; x < -\frac{64}{33} \; ; \; \mathbf{x < -1.94}$

Section 2.5 Case III Solutions - Mixed Operations Involving Linear Inequalities

1. $-2x-9 > 9x-20 \; ; \; -2x-9x-9 > 9x-9x-20 \; ; \; -11x-9 > 0-20 \; ; \; -11x-9 > -20 \; ; \; -11x-9+9 > -20+9$

$; \; -11x+0 > -11 \; ; \; -11x > -11 \; ; \; \frac{-11x}{-11} < \frac{-11}{-11} \; ; \; \mathbf{x < 1}$

2. $15x+3 \le 20x \; ; \; 15x-20x+3 \le 20x-20x \; ; \; -5x+3 \le 0 \; ; \; -5x+3-3 \le 0-3 \; ; \; -5x+0 \le -3 \; ; \; -5x \le -3 \; ; \; \frac{-5x}{-5} \ge \frac{-3}{-5}$

$; \; x \ge +\frac{3}{5} \; ; \; \mathbf{x \ge 0.6}$

3. $-4x+5 < 10 \; ; \; -4x+5-5 < 10-5 \; ; \; -4x+0 < 5 \; ; \; -4x < 5 \; ; \; \frac{-4x}{-4} > \frac{5}{-4} \; ; \; x > -\frac{5}{4} \; ; \; x > -1\frac{1}{4} \; ; \; \mathbf{x > -1.25}$

4. $-12t+4 > 4t-8 \; ; \; -12t-4t+4 > 4t-4t-8 \; ; \; -16t+4 > 0-8 \; ; \; -16t+4 > -8 \; ; \; -16t+4-4 > -8-4 \; ; \; -16t+0 > -12$

$; \; -16t > -12 \; ; \; \frac{-16t}{-16} < \frac{-12}{-16} \; ; \; t < +\frac{12}{16} \; ; \; t < \frac{3}{4} \; ; \; \mathbf{t < 0.75}$

5. $-4w-5 \ge 8w+17 \; ; \; -4w-8w-5 \ge 8w-8w+17 \; ; \; -12w-5 \ge 0+17 \; ; \; -12w-5 \ge 17 \; ; \; -12w-5+5 \ge 17+5$

$; \; -12w+0 \ge 22 \; ; \; -12w \ge 22 \; ; \; \frac{-12w}{-12} \le \frac{22}{-12} \; ; \; w \le -\frac{22}{12} \; ; \; w \le -\frac{11}{6} \; ; \; w \le -1\frac{5}{6} \; ; \; \mathbf{w \le -1.83}$

6. $10y-4 < 4y-12 \; ; \; 10y-4y-4 < 4y-4y-12 \; ; \; 6y-4 < 0-12 \; ; \; 6y-4 < -12 \; ; \; 6y-4+4 < -12+4 \; ; \; 6y+0 < -8$

$; \; 6y < -8 \; ; \; \frac{6y}{6} < -\frac{8}{6} \; ; \; y < -\frac{4}{3} \; ; \; y < -1\frac{1}{3} \; ; \; \mathbf{y < -1.33}$

7. $\frac{y}{3}-5\frac{2}{3} > 1\frac{3}{5} \; ; \; \frac{y}{3}-\frac{(5\cdot3)+2}{3} > \frac{(1\cdot5)+3}{5} \; ; \; \frac{y}{3}-\frac{15+2}{3} > \frac{5+3}{5} \; ; \; \frac{y}{3}-\frac{17}{3} > \frac{8}{5} \; ; \; \frac{y}{3}-\frac{17}{3}+\frac{17}{3} > \frac{8}{5}+\frac{17}{3} \; ; \; \frac{y}{3}+0 > \frac{(8\cdot3)+(17\cdot5)}{5\cdot3}$

$; \; \frac{y}{3} > \frac{24+85}{15} \; ; \; \frac{y}{3} > \frac{109}{15} \; ; \; \frac{y}{3} > 7.27 \; ; \; 3\cdot\frac{y}{3} > 3\cdot7.27 \; ; \; \mathbf{y > 21.81}$

8. $3\frac{4}{5} < t+2\frac{1}{3} \; ; \; \frac{(3\cdot5)+4}{5} < t+\frac{(2\cdot3)+1}{3} \; ; \; \frac{15+4}{5} < t+\frac{6+1}{3} \; ; \; \frac{19}{5} < t+\frac{7}{3} \; ; \; \frac{19}{5}-\frac{7}{3} < t+\frac{7}{3}-\frac{7}{3} \; ; \; \frac{19}{5}-\frac{7}{3} < t+0 \; ; \; \frac{19}{5}-\frac{7}{3} < t$

$; \; \frac{(19\cdot3)-(7\cdot5)}{5\cdot3} < t \; ; \; \frac{57-35}{15} < t \; ; \; \frac{22}{15} < t \; ; \; 1\frac{7}{15} < t \; ; \; 1.47 < t$ or $\mathbf{t > 1.47}$. A second way to solve this problem is as follows:

$3\frac{4}{5} < t+2\frac{1}{3} \; ; \; 3\frac{4}{5}-t < t-t+2\frac{1}{3} \; ; \; 3\frac{4}{5}-t < 0+2\frac{1}{3} \; ; \; 3\frac{4}{5}-t < 2\frac{1}{3} \; ; \; \frac{(3\cdot5)+4}{5}-t < \frac{(2\cdot3)+1}{3} \; ; \; \frac{15+4}{5}-t < \frac{6+1}{3}$

$; \; \frac{19}{5}-t < \frac{7}{3} \; ; \; \frac{19}{5}-\frac{19}{5}-t < \frac{7}{3}-\frac{19}{5} \; ; \; 0-t < \frac{7}{3}-\frac{19}{5} \; ; \; -t < \frac{(7\cdot5)-(19\cdot3)}{3\cdot5} \; ; \; -t < \frac{35-57}{15} \; ; \; -t < -\frac{22}{15} \; ; \; t > \frac{22}{15} \; ; \; \mathbf{t > 1.47}$

9. $-3.4 \ge w-2\frac{3}{5} \; ; \; -3.4-w \ge w-w-\frac{(2\cdot5)+3}{5} \; ; \; -3.4-w \ge +0-\frac{10+3}{5} \; ; \; -3.4-w \ge -\frac{13}{5} \; ; \; -3.4-w \ge -2.6$

$; \; -3.4+3.4-w \ge -2.6+3.4 \; ; \; 0-w \ge 0.8 \; ; \; -w \ge 0.8 \; ; \; \frac{-w}{-1} \le \frac{0.8}{-1} \; ; \; \mathbf{w \le -0.8}$

10. $0.48x+2.5 < 1.5x-0.35 \; ; \; 0.48x-1.5x+2.5 < 1.5x-1.5x-0.35 \; ; \; -1.02x+2.5 < 0-0.35 \; ; \; -1.02x+2.5 < -0.35$

$; \; -1.02x+2.5+0.35 < -0.35+0.35 \; ; \; -1.02x+2.85 < 0 \; ; \; -1.02x+2.85-2.85 < 0-2.85 \; ; \; -1.02x+0 < -2.85$

$; \; -1.02x < -2.85 \; ; \; \frac{-1.02}{-1.02}x > \frac{-2.85}{-1.02} \; ; \; x > +\frac{2.85}{1.02} \; ; \; x > \frac{\frac{285}{100}}{\frac{102}{100}} \; ; \; x > \frac{285\cdot100}{100\cdot102} \; ; \; x > \frac{285}{102} \; ; \; x > 2\frac{81}{102} \; ; \; \mathbf{x > 2.794}$

Chapter 3 Solutions:

Section 3.1 Case I Solutions - Factoring the Greatest Common Factor to Monomial Terms

1. a. $5x^3 = 5 \cdot x \cdot x^2 = 5 \cdot x \cdot x \cdot x$
   b. $15x = 3 \cdot 5 \cdot x$
   Therefore, the common terms are 5 and x. Thus, G.C.F. = $5 \cdot x$ = $\mathbf{5x}$
2. a. $18x^2y^3z^4 = 2 \cdot 9 \cdot x \cdot x \cdot y \cdot y^2 \cdot z^2 \cdot z^2 = 2 \cdot 3 \cdot 3 \cdot x \cdot x \cdot y \cdot y \cdot y \cdot z \cdot z \cdot z \cdot z$
   b. $24xy^4z^5 = 8 \cdot 3 \cdot x \cdot y^2 \cdot y^2 \cdot z^2 \cdot z^3 = 2 \cdot 4 \cdot 3 \cdot x \cdot y \cdot y \cdot y \cdot y \cdot z \cdot z \cdot z \cdot z^2 = 2 \cdot 2 \cdot 2 \cdot 3 \cdot x \cdot y \cdot y \cdot y \cdot y \cdot z \cdot z \cdot z \cdot z \cdot z$
   Therefore, the common terms are 2, 3, x, y, y, y, z, z, z, z, and z. Thus, G.C.F. = $2 \cdot 3 \cdot x \cdot y \cdot y \cdot y \cdot z \cdot z \cdot z \cdot z$
   $= \mathbf{6xy^3z^4}$
3. a. $16a^2bc^3 = 2 \cdot 8 \cdot a \cdot a \cdot b \cdot c \cdot c^2 = 2 \cdot 2 \cdot 4 \cdot a \cdot a \cdot b \cdot c \cdot c \cdot c = 2 \cdot 2 \cdot 2 \cdot 2 \cdot a \cdot a \cdot b \cdot c \cdot c \cdot c$
   b. $38ab^4c^2 = 2 \cdot 19 \cdot a \cdot b^2 \cdot b^2 \cdot c^2 = 2 \cdot 19 \cdot a \cdot b \cdot b \cdot b \cdot b \cdot c \cdot c$
   c. $6a^3bc = 2 \cdot 3 \cdot a \cdot a^2 \cdot b \cdot c = 2 \cdot 3 \cdot a \cdot a \cdot a \cdot b \cdot c$
   Therefore, the common terms are 2, a, b, and c. Thus, G.C.F. = $2 \cdot a \cdot b \cdot c$ = $\mathbf{2abc}$
4. a. $r^5s^4 = r^2 \cdot r^3 \cdot s^2 \cdot s^2 = r \cdot r \cdot r \cdot r^2 \cdot s \cdot s \cdot s \cdot s = r \cdot r \cdot r \cdot r \cdot r \cdot s \cdot s \cdot s \cdot s$
   b. $4r^3s^2 = 2 \cdot 2 \cdot r \cdot r^2 \cdot s \cdot s = 2 \cdot 2 \cdot r \cdot r \cdot r \cdot s \cdot s$
   c. $3rs = 3 \cdot r \cdot s$
   Therefore, the common terms are r and s. Thus, G.C.F. = $r \cdot s$ = $\mathbf{rs}$
5. a. $10u^2vw^3 = 2 \cdot 5 \cdot u \cdot u \cdot w \cdot w^2 = 2 \cdot 5 \cdot u \cdot u \cdot w \cdot w \cdot w$
   b. $2uv^3w^2 = 2 \cdot u \cdot v \cdot v^2 \cdot w \cdot w = 2 \cdot u \cdot v \cdot v \cdot v \cdot w \cdot w$
   c. $uv^2 = u \cdot v \cdot v$
   Therefore, the common terms are u and v. Thus, G.C.F. = $u \cdot v$ = $\mathbf{uv}$
6. a. $19a^3b^3 = 19 \cdot a \cdot a^2 \cdot b \cdot b^2 = 19 \cdot a \cdot a \cdot a \cdot b \cdot b \cdot b$
   b. $12ab^2 = 2 \cdot 6 \cdot a \cdot b \cdot b = 2 \cdot 2 \cdot 3 \cdot a \cdot b \cdot b$
   c. $6ab = 2 \cdot 3 \cdot a \cdot b$
   Therefore, the common terms are a and b. Thus, G.C.F. = $a \cdot b$ = $\mathbf{ab}$
7. a. $30xyz = 2 \cdot 15 \cdot x \cdot y \cdot z$
   b. $2x = 2 \cdot x$
   c. z
   There is no common terms among the three monomials.
8. a. $25p^2q^7 = 5 \cdot 5 \cdot p \cdot p \cdot q^2 \cdot q^5 = 5 \cdot 5 \cdot p \cdot p \cdot q \cdot q \cdot q^2 \cdot q^3 = 5 \cdot 5 \cdot p \cdot p \cdot q \cdot q \cdot q \cdot q \cdot q \cdot q^2 = 5 \cdot 5 \cdot p \cdot p \cdot q \cdot q \cdot q \cdot q \cdot q \cdot q \cdot q$
   b. $5pq = 5 \cdot p \cdot q$
   c. $p^3 = p \cdot p^2 = p \cdot p \cdot p$
   Therefore, the common term is p. Thus, G.C.F. = $\mathbf{p}$
9. a. $a^3b^3c = a \cdot a^2 \cdot b \cdot b^2 \cdot c = a \cdot a \cdot a \cdot b \cdot b \cdot b \cdot c$
   b. $a^7b^7 = a^2 \cdot a^5 \cdot b^2 \cdot b^5 = a \cdot a \cdot a \cdot a^2 \cdot a^2 \cdot b \cdot b \cdot b \cdot b^2 \cdot b^2 = a \cdot a \cdot a \cdot a \cdot a \cdot a \cdot a \cdot b \cdot b \cdot b \cdot b \cdot b \cdot b \cdot b$
   Therefore, the common terms are a, a, a, b, b, and b. Thus, G.C.F. = $a \cdot a \cdot a \cdot b \cdot b \cdot b$ = $\mathbf{a^3b^3}$
10. a. $z^5 = z \cdot z^4 = z \cdot z^2 \cdot z^2 = z \cdot z \cdot z \cdot z \cdot z$
    b. $xy^3z = x \cdot y \cdot y^2 \cdot z = x \cdot y \cdot y \cdot y \cdot z$
    c. $3x^2y = 3 \cdot x \cdot x \cdot y$
    There is no common terms among the three monomials.

Section 3.1 Case II Solutions - Factoring the Greatest Common Factor to Binomial and Polynomial Terms

1. a. $18x^3y^3 = 2 \cdot 9 \cdot x \cdot x^2 \cdot y \cdot y^2 = 2 \cdot 3 \cdot 3 \cdot x \cdot x \cdot x \cdot y \cdot y \cdot y$
   b. $12x^2y = 2 \cdot 6 \cdot x \cdot x \cdot y = 2 \cdot 2 \cdot 3 \cdot x \cdot x \cdot y$
   Therefore, the common terms are 2, 3, x, x, and y. This implies that G.C.F. = $2 \cdot 3 \cdot x \cdot x \cdot y = 6x^2y$. Thus,

$18x^3y^3 - 12x^2y = \mathbf{6x^2y(3xy^2 - 2)}$

2. a. $3a^2b^3c = 3 \cdot a \cdot a \cdot b \cdot b^2 \cdot c = 3 \cdot a \cdot a \cdot b \cdot b \cdot b \cdot c$

b. $15ab^2c^3 = 3 \cdot 5 \cdot a \cdot b \cdot b \cdot c \cdot c^2 = 3 \cdot 5 \cdot a \cdot b \cdot b \cdot c \cdot c \cdot c$

Therefore, the common terms are 3, a, b, b, and c. This implies that G.C.F. = $3 \cdot a \cdot b \cdot b \cdot c = 3ab^2c$. Thus, $3a^2b^3c + 15ab^2c^3 = \mathbf{3ab^2c(ab + 5c^2)}$

3. a. $xyz^3 = x \cdot y \cdot z \cdot z^2 = x \cdot y \cdot z \cdot z \cdot z$

b. $4x^2y^2z^5 = 2 \cdot 2 \cdot x \cdot x \cdot y \cdot y \cdot z^2 \cdot z^3 = 2 \cdot 2 \cdot x \cdot x \cdot y \cdot y \cdot z \cdot z \cdot z \cdot z^2 = 2 \cdot 2 \cdot x \cdot x \cdot y \cdot y \cdot z \cdot z \cdot z \cdot z \cdot z$

Therefore, the common terms are x, y, z, z, and z. This implies that G.C.F. = $x \cdot y \cdot z \cdot z \cdot z = xyz^3$. Thus, $xyz^3 + 4x^2y^2z^5 = \mathbf{xyz^3(1 + 4xyz^2)}$

4. a. $25p^3 = 5 \cdot 5 \cdot p \cdot p^2 = 5 \cdot 5 \cdot p \cdot p \cdot p$

b. $5p^2q^3 = 5 \cdot p \cdot p \cdot q \cdot q^2 = 5 \cdot p \cdot p \cdot q \cdot q \cdot q$

c. $pq = p \cdot q$

Therefore, the common term is p. This implies that G.C.F. = p. Thus, $25p^3 + 5p^2q^3 + pq = \mathbf{p(25p^2 + 5pq^3 + q)}$

5. a. $r^2s^2t = r \cdot r \cdot s \cdot s \cdot t$

b. $5rst^2 = 5 \cdot r \cdot s \cdot t \cdot t$

Therefore, the common terms are r, s, and t. This implies that G.C.F. = $r \cdot s \cdot t = rst$. Thus, $r^2s^2t - 5rst^2 = \mathbf{rst(rs - 5t)}$

6. a. $36x^3yz^3 = 2 \cdot 18 \cdot x \cdot x^2 \cdot y \cdot z \cdot z^2 = 2 \cdot 2 \cdot 9 \cdot x \cdot x \cdot x \cdot y \cdot z \cdot z \cdot z = 2 \cdot 2 \cdot 3 \cdot 3 \cdot x \cdot x \cdot x \cdot y \cdot z \cdot z \cdot z$

b. $4xy^2z^4 = 2 \cdot 2 \cdot x \cdot y \cdot y \cdot z^2 \cdot z^2 = 2 \cdot 2 \cdot x \cdot y \cdot y \cdot z \cdot z \cdot z \cdot z$

c. $12x^3y^3z = 2 \cdot 6 \cdot x \cdot x^2 \cdot y \cdot y^2 \cdot z = 2 \cdot 2 \cdot 3 \cdot x \cdot x \cdot x \cdot y \cdot y \cdot y \cdot z$

Therefore, the common terms are 2, 2, x, y, and z. This implies that G.C.F. = $2 \cdot 2 \cdot x \cdot y \cdot z = 4xyz$. Thus, $36x^3yz^3 + 4xy^2z^4 - 12x^3y^3z = \mathbf{4xyz(9x^2z^2 + yz^3 - 3x^2y^2)}$

7. a. $17ab^2c^2d^3 = 17 \cdot a \cdot b \cdot b \cdot c \cdot c \cdot d \cdot d^2 = 17 \cdot a \cdot b \cdot b \cdot c \cdot c \cdot d \cdot d \cdot d$

b. $7a^3bc = 7 \cdot a \cdot a^2 \cdot b \cdot c = 7 \cdot a \cdot a \cdot a \cdot b \cdot c$

c. $8a^2cd^2 = 2 \cdot 4 \cdot a \cdot a \cdot c \cdot d \cdot d = 2 \cdot 2 \cdot 2 \cdot a \cdot a \cdot c \cdot d \cdot d$

Therefore, the common terms are a, and c. This implies that G.C.F. = $a \cdot c = ac$. Thus, $17ab^2c^2d^3 - 7a^3bc + 8a^2cd^2 = \mathbf{ac(17b^2cd^3 - 7a^2b + 8ad^2)}$

8. a. $5pq^2r = 5 \cdot p \cdot q \cdot q \cdot r$

b. $30p^2q^2 = 6 \cdot 5 \cdot p \cdot p \cdot q \cdot q = 2 \cdot 3 \cdot 5 \cdot p \cdot p \cdot q \cdot q$

c. $20p^3qr^3 = 4 \cdot 5 \cdot p \cdot p^2 \cdot q \cdot r \cdot r^2 = 2 \cdot 2 \cdot 5 \cdot p \cdot p \cdot p \cdot q \cdot r \cdot r \cdot r$

Therefore, the common terms are 5, p, and q. This implies that G.C.F. = $5 \cdot p \cdot q = 5pq$. Thus, $5pq^2r + 30p^2q^2 - 20p^3qr^3 = \mathbf{5pq(qr + 6pq - 4p^2r^3)}$

9. a. $9x^2y^2z = 3 \cdot 3 \cdot x \cdot x \cdot y \cdot y \cdot z$

b. $3xyz = 3 \cdot x \cdot y \cdot z$

Therefore, the common terms are 3, x, y, and z. This implies that G.C.F. = $3 \cdot x \cdot y \cdot z = 3xyz$. Thus, $9x^2y^2z - 3xyz = \mathbf{3xyz(3xy - 1)}$

10. a. $7abc^2d = 7 \cdot a \cdot b \cdot c \cdot c \cdot d$

b. $49a^3b^3 = 7 \cdot 7 \cdot a \cdot a^2 \cdot b \cdot b^2 = 7 \cdot 7 \cdot a \cdot a \cdot a \cdot b \cdot b \cdot b$

c. $14a^2b^3d = 2 \cdot 7 \cdot a \cdot a \cdot b \cdot b^2 \cdot d = 2 \cdot 7 \cdot a \cdot a \cdot b \cdot b \cdot b \cdot d$

Therefore, the common terms are 7, a, and b. This implies that G.C.F. = $7 \cdot a \cdot b = 7ab$. Thus, $7abc^2d + 49a^3b^3 - 14a^2b^3d = \mathbf{7ab(c^2d + 7a^2b^2 - 2ab^2d)}$

Section 3.2 Solutions - Factoring Polynomials Using the Grouping Method

1. $2ab-5b-6a+15 = b(2a-5)-3(2a-5) = \mathbf{(2a-5)(b-3)}$
2. $y^3+4y^2+y+4 = y^2(y+4)+(y+4) = \mathbf{(y+4)(y^2+1)}$
3. $42x^2y+21xy-70x-35 = 21xy(2x+1)-35(2x+1) = \mathbf{(2x+1)(21xy-35)}$
4. $(x+y)^3+(x+y)^2+x+y = (x+y)^3+(x+y)^2+(x+y) = (x+y)\left[(x+y)^2+(x+y)+1\right] = \mathbf{(x+y)\{(x+y)[(x+y)+1]+1\}}$
5. $4(a+b)^2+32a+32b = 4(a+b)^2+32(a+b) = 4(a+b)[(a+b)+8] = \mathbf{4(a+b)[a+b+8]}$
6. $36r^3s-6r^2s+18r-3 = 6r^2s(6r-1)+3(6r-1) = (6r-1)(6r^2s+3) = \mathbf{3(6r-1)(2r^2s+1)}$
7. $3u^2+7u+2 = 3u^2+6u+u+2 = 3u(u+2)+u+2 = 3u(u+2)+(u+2) = \mathbf{(u+2)(3u+1)}$
8. $25(p+q)^4+5(p+q)^2+2p+2q = 25(p+q)^4+5(p+q)^2+2(p+q) = (p+q)\left[25(p+q)^3+5(p+q)+2\right]$
   $= \mathbf{(p+q)\left\{5(p+q)\left[5(p+q)^2+1\right]+2\right\}}$
9. $ax^3+bx^3-ax^2y^2-bx^2y^2 = x^3(a+b)-x^2y^2(a+b) = (a+b)(x^3-x^2y^2) = \mathbf{x^2(a+b)(x-y^2)}$
10. $6r^3s^2+6r+9r^2s^2+9 = 6r(r^2s^2+1)+9(r^2s^2+1) = (r^2s^2+1)(6r+9) = \mathbf{3(r^2s^2+1)(2r+3)}$

Section 3.3 Case I Solutions - Factoring Trinomials of the Form ax^2+bx+c where $a=1$

1. $x^2-2x-15 = \mathbf{(x+3)(x-5)}$
2. $y^2-9y+8 = \mathbf{(y-1)(y-8)}$
3. $t^2+2t-15 = \mathbf{(t-3)(t+5)}$
4. $y^2-2y+11$ **is prime**
5. $x^2+10x+21 = \mathbf{(x+3)(x+7)}$
6. $u^2+4u-32 = \mathbf{(u-4)(u+8)}$
7. $a^2+9a+18 = \mathbf{(a+3)(a+6)}$
8. $w^2-11w+30 = \mathbf{(w-6)(w-5)}$
9. $x^2-8x-20 = \mathbf{(x+2)(x-10)}$
10. $v^2+120v+2000 = \mathbf{(v+20)(v+100)}$

Section 3.3 Case II Solutions - Factoring Trinomials of the Form ax^2+bx+c where $a \rangle 1$

1. $10x^2+11x-35 = 10x^2+(25-14)x-35 = 10x^2+25x-14x-35 = 5x(2x+5)-7(2x+5) = \mathbf{(2x+5)(5x-7)}$
2. $6x^2-x-12 = 6x^2+(-9+8)x-12 = 6x^2-9x+8x-12 = 3x(2x-3)+4(2x-3) = \mathbf{(2x-3)(3x+4)}$
3. $-7x^2+46x+21 = -7x^2+(49-3)x+21 = -7x^2+49x-3x+21 = 7x(-x+7)+3(-x+7) = \mathbf{(-x+7)(7x+3)}$
4. $6x^2-11xy+3y^2 = 3y^2+(-11x)y+6x^2 = 3y^2+(-9-2)xy+6x^2 = 3y^2-9xy-2xy+6x^2 = 3y(y-3x)-2x(y-3x)$
   $= \mathbf{(y-3x)(3y-2x)}$
5. $6x^2+x-40 = 6x^2+(16-15)x-40 = 6x^2+16x-15x-40 = 2x(3x+8)-5(3x+8) = \mathbf{(3x+8)(2x-5)}$
6. $2x^2+3x-27 = 2x^2+(9-6)x-27 = 2x^2+9x-6x-27 = x(2x+9)-3(2x+9) = \mathbf{(2x+9)(x-3)}$
7. $12x^2+10y^2-23xy = 12x^2+(-23y)x+10y^2 = 12x^2+(-8y-15y)x+10y^2 = 12x^2-8yx-15yx+10y^2$
   $= 4x(3x-2y)-5y(3x-2y) = \mathbf{(3x-2y)(4x-5y)}$

8. $5x^2 - 17x + 14 = 5x^2 + (-10 - 7)x + 14 = 5x^2 - 10x - 7x + 14 = 5x(x-2) - 7(x-2) = \mathbf{(x-2)(5x-7)}$

9. $18x^2 + 9x - 20 = 18x^2 + (24 - 15)x - 20 = 18x^2 + 24x - 15x - 20 = 6x(3x+4) - 5(3x+4) = \mathbf{(3x+4)(6x-5)}$

10. $27x^2 + 42x + 16 = 27x^2 + (18 + 24)x + 16 = 27x^2 + 18x + 24x + 16 = 9x(3x+2) + 8(3x+2) = \mathbf{(3x+2)(9x+8)}$

Section 3.4 Case I Solutions - Factoring Polynomials Using the Difference of Two Squares Method

1. $x^3 - 16x = x(x^2 - 16) = x(x^2 - 4^2) = \mathbf{x(x-4)(x+4)}$

2. $(x+1)^2 - (y+3)^2 = [(x+1) - (y+3)][(x+1) + (y+3)] = (x + 1 - y - 3)(x + 1 + y + 3) = \mathbf{(x - y - 2)(x + y + 4)}$

3. $t^5 - 81t = t(t^4 - 81) = t\left(t^{2^2} - 9^2\right) = t(t^2 - 9)(t^2 + 9) = t(t^2 - 3^2)(t^2 + 9) = \mathbf{t(t-3)(t+3)(t^2+9)}$

4. $(x^2 + 10x + 25) - y^2 = (x+5)^2 - y^2 = [(x+5) - y][(x+5) + y] = (x + 5 - y)(x + 5 + y) = \mathbf{(x - y + 5)(x + y + 5)}$

5. $c^4 - 9c^2 = c^2(c^2 - 9) = c^2(c^2 - 3^2) = \mathbf{c^2(c-3)(c+3)}$

6. $p^2 - q^2 - 4q - 4 = p^2 - (q^2 + 4q + 4) = p^2 - (q+2)^2 = [p - (q+2)][p + (q+2)] = \mathbf{(p - q - 2)(p + q + 2)}$

7. $x^2 - y^2 + 6y - 9 = x^2 - (y^2 - 6y + 9) = x^2 - (y-3)^2 = [x - (y-3)][x + (y-3)] = \mathbf{(x - y + 3)(x + y - 3)}$

8. $p^2 + q^2 + 4q + 4 = p^2 + (q^2 + 4q + 4) = p^2 + (q+2)^2$

Note that the answer is in the same form as $a^2 + b^2$ which **is a prime polynomial** and can not be factored.

9. $m^{16} - 256 = m^{8^2} - 16^2 = (m^8 - 16)(m^8 + 16) = \left(m^{4^2} - 4^2\right)(m^8 + 16) = (m^4 - 4)(m^4 + 4)(m^8 + 16)$

$= \left(m^{2^2} - 2^2\right)(m^4 + 4)(m^8 + 16) = \mathbf{(m^2 - 2)(m^2 + 2)(m^4 + 4)(m^8 + 16)}$

10. $r^2 - (s+7)^2 = [r - (s+7)][r + (s+7)] = \mathbf{(r - s - 7)(r + s + 7)}$

Section 3.4 Case II Solutions - Factoring Polynomials Using the Sum and Difference of Two Cubes Method

1. $4x^6 + 4 = 4(x^6 + 1) = 4\left(x^{2^3} + 1^3\right) = 4\left[(x^2)^3 + 1^3\right] = 4(x^2 + 1)\left[(x^2)^2 - x^2 \cdot 1 + 1^2\right] = \mathbf{4(x^2 + 1)(x^4 - x^2 + 1)}$

2. $x^6y^6 + 8 = x^{2^3}y^{2^3} + 2^3 = (x^2y^2)^3 + 2^3 = (x^2y^2 + 2)\left[(x^2y^2)^2 - 2 \cdot x^2y^2 + 2^2\right] = \mathbf{(x^2y^2 + 2)(x^4y^4 - 2x^2y^2 + 4)}$

3. $(x+2)^3 - y^3 = (x+2)^3 - y^3 = [(x+2) - y]\left[(x+2)^2 + (x+2) \cdot y + y^2\right] = \mathbf{(x - y + 2)\left[(x+2)^2 + (x+2)y + y^2\right]}$

4. $2r^6 - 128 = 2(r^6 - 64) = 2\left(r^{2^3} - 4^3\right) = 2\left[(r^2)^3 - 4^3\right] = 2(r^2 - 4)\left[(r^2)^2 + 4 \cdot r^2 + 4^2\right] = \mathbf{2(r^2 - 4)(r^4 + 4r^2 + 16)}$

5. $(x-7)^3 + y^3 = (x-7)^3 + y^3 = [(x-7) + y]\left[(x-7)^2 - (x-7) \cdot y + y^2\right] = \mathbf{(x + y - 7)\left[(x-7)^2 - (x-7)y + y^2\right]}$

6. $x^6y^5 + x^3y^2 = x^3y^2(x^3y^3 + 1) = x^3y^2\left[(xy)^3 + 1^3\right] = x^3y^2\left\{[(xy) + 1]\left[(xy)^2 - (xy) \cdot 1 + 1^2\right]\right\}$

$= \mathbf{x^3y^2\left\{(xy + 1)\left[(xy)^2 - xy + 1\right]\right\}}$

7. $27a^4 - 125a = a(27a^3 - 125) = a(3^3a^3 - 5^3) = a\left[(3a)^3 - 5^3\right] = a\left\{(3a - 5)\left[(3a)^2 + 3a \cdot 5 + 5^2\right]\right\}$

$= \mathbf{a(3a - 5)(9a^2 + 15a + 25)}$

8. $x^5y+64x^2y^4 = x^2y\left(x^3+64y^3\right) = x^2y\left(x^3+4^3y^3\right) = x^2y\left[x^3+(4y)^3\right] = x^2y\left\{(x+4y)\left[x^2-x\cdot 4y+(4y)^2\right]\right\}$

$= \mathbf{x^2y\left[(x+4y)\left(x^2-4xy+16y^2\right)\right]}$

9. $u^3+(v+1)^3 = u^3+(v+1)^3 = [u+(v+1)]\left[u^2-u\cdot(v+1)+(v+1)^2\right] = \mathbf{(u+v+1)\left[u^2-u(v+1)+(v+1)^2\right]}$

10. $a^3-(b+7)^3 = a^3-(b+7)^3 = [a-(b+7)]\left[a^2+a\cdot(b+7)+(b+7)^2\right] = \mathbf{(a-b-7)\left[a^2+a(b+7)+(b+7)^2\right]}$

Section 3.5 Case I Solutions - Factoring Perfect Square Trinomials

1. $x^2+18x+81 = x^2+18x+9^2 = x^2+2\cdot(x\cdot 9)+9^2 = \mathbf{(x+9)^2}$

2. $9+64p^2-48p = 64p^2-48p+9 = 8^2p^2-48p+3^2 = (8p)^2-2\cdot(8p\cdot 3)+3^2 = \mathbf{(8p-3)^2}$

3. $9w^2+25+30w = 9w^2+30w+25 = 3^2w^2+30w+5^2 = (3w)^2+2\cdot(3w\cdot 5)+5^2 = \mathbf{(3w+5)^2}$

4. $25+k^2-10k = k^2-10k+25 = k^2-2\cdot(k\cdot 5)+5^2 = \mathbf{(k-5)^2}$

5. $49x^2-84x+36 = 7^2x^2-84x+6^2 = (7x)^2-2\cdot(7x\cdot 6)+6^2 = \mathbf{(7x-6)^2}$

6. $1+16z+64z^2 = 64z^2+16z+1 = 8^2z^2+16z+1^2 = (8z)^2+2\cdot(8z\cdot 1)+1^2 = \mathbf{(8z+1)^2}$

7. $100u^4-40u^2v^2+4v^4 = 10^2u^{2^2}-40u^2v^2+2^2v^{2^2} = \left(10u^2\right)^2-2\cdot\left(10u^2\cdot 2v^2\right)+\left(2v^2\right)^2 = \mathbf{\left(10u^2-2v^2\right)^2}$

8. $49p^2-126pq+81q^2 = 7^2p^2-126pq+9^2q^2 = (7p)^2-2\cdot(7p\cdot 9q)+(9q)^2 = \mathbf{(7p-9q)^2}$

9. $25x^4-30x^2y^2+9y^4 = 5^2x^{2^2}-30x^2y^2+3^2y^{2^2} = \left(5x^2\right)^2-2\cdot\left(5x^2\cdot 3y^2\right)+\left(3y^2\right)^2 = \mathbf{\left(5x^2-3y^2\right)^2}$

10. $9x^2+12xy+4y^2 = 3^2x^2+12xy+2^2y^2 = (3x)^2+2\cdot(3x\cdot 2y)+(2y)^2 = \mathbf{(3x+2y)^2}$

Section 3.5 Case II Solutions - Factoring Other Types of Polynomials

1. $a^3+7a^2-9a-63 = a^2(a+7)-9(a+7) = (a+7)\left(a^2-9\right) = (a+7)\left(a^2-3^2\right) = \mathbf{(a+7)(a-3)(a+3)}$

2. $2x^2+16x-40 = 2\left(x^2+8x-20\right) = \mathbf{2(x+10)(x-2)}$

3. $a^2-b^2+9a+9b = \left(a^2-b^2\right)+9(a+b) = (a-b)(a+b)+9(a+b) = \mathbf{(a+b)[(a-b)+9]}$

4. $6y^2+39y+60 = 3\left(2y^2+13y+20\right) = 3\left(2y^2+13y+20\right) = 3\left[2y^2+(8+5)y+20\right] = \mathbf{3\left[2y^2+8y+5y+20\right]}$

$= 3[2y(y+4)+5(y+4)] = 3[(y+4)(2y+5)] = \mathbf{3(y+4)(2y+5)}$

5. $2w^3-4w^2-16w = 2w\left(w^2-2w-8\right) = \mathbf{2w(w-4)(w+2)}$

6. $-25x^4+70x^2y-49y^2 = -\left(25x^4-70x^2y+49y^2\right) = -\left(5^2x^{2^2}-70x^2y+7^2y^2\right) = -\left[\left(5x^2\right)^2-2\cdot\left(5x^2\cdot 7y\right)+(7y)^2\right]$

$= \mathbf{-\left(5x^2-7y\right)^2}$

7. $12y^3+26y^2+10y = 2y\left(6y^2+13y+5\right) = 2y\left[6y^2+(3+10)y+5\right] = 2y\left[6y^2+3y+10y+5\right] = 2y[3y(2y+1)+5(2y+1)]$

$= 2y[(2y+1)(3y+5)] = \mathbf{2y(2y+1)(3y+5)}$

8. $24x^3 + 74x^2 - 35x = x(24x^2 + 74x - 35) = x[24x^2 + (84-10)x - 35] = x[24x^2 + 84x - 10x - 35]$

$= x[12x(2x+7) - 5(2x+7)] = x[(2x+7)(12x-5)] = \mathbf{x(2x+7)(12x-5)}$

9. $24x^2 + 78x + 45 = 3(8x^2 + 26x + 15) = 3[8x^2 + (20+6)x + 15] = 3[8x^2 + 20x + 6x + 15] = 3[4x(2x+5) + 3(2x+5)]$

$= 3[(2x+5)(4x+3)] = \mathbf{3(2x+5)(4x+3)}$

10. $16u^8 - 256u^4 = 16u^4(u^4 - 16) = 16u^4\left(u^{2^2} - 4^2\right) = 16u^4(u^2-4)(u^2+4) = 16u^4(u^2-2^2)(u^2+4)$

$= \mathbf{16u^4(u-2)(u+2)(u^2+4)}$

Chapter 4 Solutions:

Section 4.1 Solutions - Quadratic Equations and the Quadratic Formula

1. First - Write the quadratic equation $3x = -5 + 2x^2$ in standard form $ax^2 + bx + c = 0$.

   $3x = -5 + 2x^2$; $-2x^2 + 3x = -5 + 2x^2 - 2x^2$; $-2x^2 + 3x = -5 + 0$; $-2x^2 + 3x = -5$; $-2x^2 + 3x + 5 = -5 + 5$; $-2x^2 + 3x + 5 = 0$

   Second - Equate the a, b, and c coefficients with the coefficients of the given quadratic equation.

   Thus, $\boldsymbol{a = -2}$, $\boldsymbol{b = 3}$, and $\boldsymbol{c = 5}$

2. First - Write the quadratic equation $2x^2 = 5$ in standard form $ax^2 + bx + c = 0$.

   $2x^2 = 5$; $2x^2 - 5 = 5 - 5$; $2x^2 - 5 = 0$ which is the same as $2x^2 + 0x - 5 = 0$

   Second - Equate the a, b, and c coefficients with the coefficients of the given quadratic equation.

   Thus, $\boldsymbol{a = 2}$, $\boldsymbol{b = 0}$, and $\boldsymbol{c = -5}$

3. First - Write the quadratic equation $3w^2 - 5w = 2$ in standard form $aw^2 + bw + c = 0$.

   $3w^2 - 5w = 2$; $3w^2 - 5w - 2 = 2 - 2$; $3w^2 - 5w - 2 = 0$

   Second - Equate the a, b, and c coefficients with the coefficients of the given quadratic equation.

   Thus, $\boldsymbol{a = 3}$, $\boldsymbol{b = -5}$, and $\boldsymbol{c = -2}$

4. First - Write the quadratic equation $15 = -y^2 - 3$ in standard form $ay^2 + by + c = 0$.

   $15 = -y^2 - 3$; $y^2 + 15 = -y^2 + y^2 - 3$; $y^2 + 15 = 0 - 3$; $y^2 + 15 = -3$; $y^2 + 15 + 3 = -3 + 3$; $y^2 + 18 = 0$; which is the same as $y^2 + 0y + 18 = 0$

   Second - Equate the a, b, and c coefficients with the coefficients of the given quadratic equation.

   Thus, $\boldsymbol{a = 1}$, $\boldsymbol{b = 0}$, and $\boldsymbol{c = 18}$

5. First - Write the quadratic equation $x^2 + 3 = 5x$ in standard form $ax^2 + bx + c = 0$.

   $x^2 + 3 = 5x$; $x^2 - 5x + 3 = 5x - 5x$; $x^2 - 5x + 3 = 0$

   Second - Equate the a, b, and c coefficients with the coefficients of the given quadratic equation.

   Thus, $\boldsymbol{a = 1}$, $\boldsymbol{b = -5}$, and $\boldsymbol{c = 3}$

6. First - Write the quadratic equation $-u^2 + 2 = 3u$ in standard form $au^2 + bu + c = 0$.

   $-u^2 + 2 = 3u$; $-u^2 - 3u + 2 = 3u - 3u$; $-u^2 - 3u + 2 = 0$

   Second - Equate the a, b, and c coefficients with the coefficients of the given quadratic equation.

   Thus, $\boldsymbol{a = -1}$, $\boldsymbol{b = -3}$, and $\boldsymbol{c = 2}$

7. The quadratic equation $y^2 + 5y - 2 = 0$ is already in standard form $ay^2 + by + c = 0$. Therefore, simply equate the a, b, and c coefficients with the coefficients of the given quadratic equation to obtain $\boldsymbol{a = 1}$, $\boldsymbol{b = 5}$, and $\boldsymbol{c = -2}$.

8. First - Write the quadratic equation $-3x^2 = 2x - 1$ in standard form $ax^2 + bx + c = 0$.

   $-3x^2 = 2x - 1$; $-3x^2 - 2x = 2x - 2x - 1$; $-3x^2 - 2x = 0 - 1$; $-3x^2 - 2x = -1$; $-3x^2 - 2x + 1 = -1 + 1$; $-3x^2 - 2x + 1 = 0$

   Second - Equate the a, b, and c coefficients with the coefficients of the given quadratic equation.

   Thus, $\boldsymbol{a = -3}$, $\boldsymbol{b = -2}$, and $\boldsymbol{c = 1}$

9. First - Write the quadratic equation $p^2 = p - 1$ in standard form $ap^2 + bp + c = 0$.

   $p^2 = p - 1$; $p^2 - p = p - p - 1$; $p^2 - p = 0 - 1$; $p^2 - p = -1$; $p^2 - p + 1 = -1 + 1$; $p^2 - p + 1 = 0$

Second - Equate the a, b, and c coefficients with the coefficients of the given quadratic equation.

Thus, $\mathbf{a=1}$, $\mathbf{b=-1}$, and $\mathbf{c=1}$

10. First - Write the quadratic equation $3x-2=x^2$ in standard form $ax^2+bx+c=0$.

$$3x-2=x^2 \; ; \; -x^2+3x-2=x^2-x^2 \; ; \; -x^2+3x-2=0$$

Second - Equate the a, b, and c coefficients with the coefficients of the given quadratic equation.

Thus, $\mathbf{a=-1}$, $\mathbf{b=3}$, and $\mathbf{c=-2}$

Section 4.2 Case I Solutions - Solving Quadratic Equations of the Form ax^2+bx+c where $a=1$

1. $x^2=-5x-6$ Write the equation in standard form, i.e., $x^2+5x+6=0$.

Let: $a=1$, $b=5$, and $c=6$. Then,

Given: $x=\frac{-b\pm\sqrt{b^2-4ac}}{2a}$; $x=\frac{-5\pm\sqrt{5^2-4\times1\times6}}{2\times1}$; $x=\frac{-5\pm\sqrt{25-24}}{2}$; $x=\frac{-5\pm\sqrt{1}}{2}$; $x=\frac{-5\pm1}{2}$ therefore,

I. $x=\frac{-5+1}{2}$; $x=-\frac{\overset{2}{\cancel{4}}}{\cancel{2}}$; $x=-\frac{2}{1}$; $\mathbf{x=-2}$ and

II. $x=\frac{-5-1}{2}$; $x=-\frac{\overset{3}{\cancel{6}}}{\cancel{2}}$; $x=-\frac{3}{1}$; $\mathbf{x=-3}$

Check: I. *Let* $x=-2$ *in* $x^2=-5x-6$; $(-2)^2\overset{?}{=}(-5\times-2)-6$; $4\overset{?}{=}10-6$; $4=4$

II. *Let* $x=-3$ *in* $x^2=-5x-6$; $(-3)^2\overset{?}{=}(-5\times-3)-6$; $9\overset{?}{=}15-6$; $9=9$

Therefore, the equation $x^2+5x+6=0$ can be factored to $\mathbf{(x+2)(x+3)=0}$.

2. $y^2-40y=-300$ Write the equation in standard form, i.e., $y^2-40y+300=0$.

Let: $a=1$, $b=-40$, and $c=300$. Then,

Given: $y=\frac{-b\pm\sqrt{b^2-4ac}}{2a}$; $y=\frac{-(-40)\pm\sqrt{(-40)^2-4\times1\times300}}{2\times1}$; $y=\frac{40\pm\sqrt{1600-1200}}{2}$; $y=\frac{40\pm\sqrt{400}}{2}$

; $y=\frac{40\pm\sqrt{20^2}}{2}$; $y=\frac{40\pm20}{2}$ therefore, I. $y=\frac{40+20}{2}$; $y=\frac{\overset{30}{\cancel{60}}}{\cancel{2}}$; $y=\frac{30}{1}$; $\mathbf{y=30}$ and

II. $y=\frac{40-20}{2}$; $y=\frac{\overset{10}{\cancel{20}}}{\cancel{2}}$; $y=\frac{10}{1}$; $\mathbf{y=10}$

Check: I. *Let* $y=30$ *in* $y^2-40y=-300$; $(30)^2-40\cdot30\overset{?}{=}-300$; $900-1200\overset{?}{=}-300$; $-300=-300$

II. *Let* $y=10$ *in* $y^2-40y=-300$; $(10)^2-40\cdot10\overset{?}{=}-300$; $100-400\overset{?}{=}-300$; $-300=-300$

Therefore, the equation $y^2-40y+300=0$ can be factored to $\mathbf{(y-30)(y-10)=0}$.

3. $-x=-x^2+20$ Write the equation in standard form, i.e., $x^2-x-20=0$.

Let: $a=1$, $b=-1$, and $c=-20$. Then,

Given: $x=\frac{-b\pm\sqrt{b^2-4ac}}{2a}$; $x=\frac{-(-1)\pm\sqrt{(-1)^2-4\times1\times-20}}{2\times1}$; $x=\frac{1\pm\sqrt{1+80}}{2}$; $x=\frac{1\pm\sqrt{81}}{2}$; $x=\frac{1\pm\sqrt{9^2}}{2}$

$; \; x=\frac{1\pm 9}{2}$ therefore, I. $x=\frac{1+9}{2} \; ; \; x=\frac{\overset{5}{\cancel{10}}}{\cancel{2}} \; ; \; x=\frac{5}{1} \; ; \; \mathbf{x=5}$ and

II. $x=\frac{1-9}{2} \; ; \; x=-\frac{\overset{4}{\cancel{8}}}{\cancel{2}} \; ; \; x=-\frac{4}{1} \; ; \; \mathbf{x=-4}$

Check: I. *Let* $x=5$ *in* $-x=-x^2+20 \; ; \; -5\overset{?}{=}-5^2+20 \; ; \; -5\overset{?}{=}-25+20 \; ; \; -5=-5$

II. *Let* $x=-4$ *in* $-x=-x^2+20 \; ; \; -(-4)\overset{?}{=}-(-4)^2+20 \; ; \; 4\overset{?}{=}-16+20 \; ; \; 4=4$

Therefore, the equation $x^2-x-20=0$ can be factored to $\mathbf{(x-5)(x+4)=0}$.

4. $x^2+3x+4=0$ The equation is already in standard form.

Let: $a=1$, $b=3$, and $c=4$. Then,

Given: $x=\frac{-b\pm\sqrt{b^2-4ac}}{2a} \; ; \; x=\frac{-3\pm\sqrt{3^2-4\times 1\times 4}}{2\times 1} \; ; \; x=\frac{-3\pm\sqrt{9-16}}{2} \; ; \; x=\frac{-3\pm\sqrt{-7}}{2}$

Since the number under the radical is negative (an imaginary number), the given equation is not factorable.

5. $x^2-80-2x=0$ Write the equation in standard form, i.e., $x^2-2x-80=0$.

Let: $a=1$, $b=-2$, and $c=-80$. Then,

Given: $x=\frac{-b\pm\sqrt{b^2-4ac}}{2a} \; ; \; x=\frac{-(-2)\pm\sqrt{(-2)^2-4\times 1\times -80}}{2\times 1} \; ; \; x=\frac{2\pm\sqrt{4+320}}{2} \; ; \; x=\frac{2\pm\sqrt{324}}{2} \; ; \; x=\frac{2\pm\sqrt{18^2}}{2}$

$; \; x=\frac{2\pm 18}{2}$ therefore, I. $x=\frac{2+18}{2} \; ; \; x=\frac{\overset{10}{\cancel{20}}}{\cancel{2}} \; ; \; x=\frac{10}{1} \; ; \; \mathbf{x=10}$ and

II. $x=\frac{2-18}{2} \; ; \; x=-\frac{\overset{8}{\cancel{16}}}{\cancel{2}} \; ; \; x=-\frac{8}{1} \; ; \; \mathbf{x=-8}$

Check: I. *Let* $x=10$ *in* $x^2-80-2x=0 \; ; \; 10^2-80-2\cdot 10\overset{?}{=}0 \; ; \; 100-80-20\overset{?}{=}0 \; ; \; 100-100\overset{?}{=}0 \; ; \; 0=0$

II. *Let* $x=-8$ *in* $x^2-80-2x=0 \; ; \; (-8)^2-80-2\cdot(-8)\overset{?}{=}0 \; ; \; 64-80+16\overset{?}{=}0 \; ; \; 80-80\overset{?}{=}0 \; ; \; 0=0$

Therefore, the equation $x^2-2x-80=0$ can be factored to $\mathbf{(x-10)(x+8)=0}$.

6. $x^2+4x+4=0$ The equation is already in standard form.

Let: $a=1$, $b=4$, and $c=4$. Then,

Given: $x=\frac{-b\pm\sqrt{b^2-4ac}}{2a} \; ; \; x=\frac{-4\pm\sqrt{4^2-4\times 1\times 4}}{2\times 1} \; ; \; x=\frac{-4\pm\sqrt{16-16}}{2} \; ; \; x=\frac{-4\pm\sqrt{0}}{2} \; ; \; x=\frac{-4\pm 0}{2}$

$; \; x=-\frac{\overset{2}{\cancel{4}}}{\cancel{2}} \; ; \; x=-\frac{2}{1} \; ; \; x=-2$. In this case the equation has one repeated solution, i.e., $\mathbf{x=-2}$ and $\mathbf{x=-2}$.

Check: *Let* $x=-2$ *in* $x^2+4x+4=0 \; ; \; (-2)^2+4\cdot(-2)+4\overset{?}{=}0 \; ; \; 4-8+4\overset{?}{=}0 \; ; \; 8-8\overset{?}{=}0 \; ; \; 0=0$

Therefore, the equation $x^2+4x+4=0$ can be factored to $\mathbf{(x+2)(x+2)=0}$.

7. $-6=-w^2+w$ Write the equation in standard form, i.e., $w^2-w-6=0$

Let: $a=1$, $b=-1$, and $c=-6$. Then,

Given: $w = \frac{-b \pm \sqrt{b^2 - 4ac}}{2a}$; $w = \frac{-(-1) \pm \sqrt{(-1)^2 - 4 \times 1 \times -6}}{2 \times 1}$; $w = \frac{1 \pm \sqrt{1+24}}{2}$; $w = \frac{1 \pm \sqrt{25}}{2}$; $w = \frac{1 \pm \sqrt{5^2}}{2}$

; $w = \frac{1 \pm 5}{2}$ therefore, I. $w = \frac{1+5}{2}$; $w = \frac{\overset{3}{\cancel{6}}}{\cancel{2}}$; $w = \frac{3}{1}$; $\mathbf{w = 3}$ and

II. $w = \frac{1-5}{2}$; $w = -\frac{\overset{2}{\cancel{4}}}{\cancel{2}}$; $w = -\frac{2}{1}$; $\mathbf{w = -2}$

Check: I. *Let* $w = 3$ *in* $-6 = -w^2 + w$; $-6 \overset{?}{=} -(3^2) + 3$; $-6 \overset{?}{=} -9 + 3$; $-6 = -6$

II. *Let* $w = -2$ *in* $-6 = -w^2 + w$; $-6 \overset{?}{=} -(-2)^2 + (-2)$; $-6 \overset{?}{=} -4 - 2$; $-6 = -6$

Therefore, the equation $w^2 - w - 6 = 0$ can be factored to $\mathbf{(w-3)(w+2) = 0}$.

8. $4x = x^2$ Write the equation in standard form, i.e., $x^2 - 4x = 0$.

Let: $a = 1$, $b = -4$, and $c = 0$. Then,

Given: $x = \frac{-b \pm \sqrt{b^2 - 4ac}}{2a}$; $x = \frac{-(-4) \pm \sqrt{(-4)^2 - 4 \times 1 \times 0}}{2 \times 1}$; $x = \frac{4 \pm \sqrt{16 - 0}}{2}$; $x = \frac{4 \pm \sqrt{16}}{2}$; $x = \frac{4 \pm \sqrt{4^2}}{2}$

; $x = \frac{4 \pm 4}{2}$; therefore, I. $x = \frac{4-4}{2}$; $x = \frac{0}{2}$; $x = 0$ and

II. $x = \frac{4+4}{2}$; $x = \frac{\overset{4}{\cancel{8}}}{\cancel{2}}$; $x = \frac{4}{1}$; $x = 4$

Check: I. *Let* $x = 0$ *in* $4x = x^2$; $4 \cdot 0 = 0^2$; $0 = 0$

II. *Let* $x = 4$ *in* $4x = x^2$; $4 \cdot 4 = 4^2$; $16 = 16$

Therefore, the equation $x^2 - 4x = 0$ can be factored to $(x+0)(x-4) = 0$ which is the same as $x(x-4) = 0$.

9. $z^2 - 37z - 120 = 0$ The equation is already in standard form

Let: $a = 1$, $b = -37$, and $c = -120$. Then,

Given: $z = \frac{-b \pm \sqrt{b^2 - 4ac}}{2a}$; $z = \frac{-(-37) \pm \sqrt{(-37)^2 - 4 \times 1 \times -120}}{2 \times 1}$; $z = \frac{37 \pm \sqrt{1369 + 480}}{2}$; $z = \frac{37 \pm \sqrt{1849}}{2}$

; $z = \frac{37 \pm \sqrt{43^2}}{2}$; $z = \frac{37 \pm 43}{2}$ therefore, I. $z = \frac{37+43}{2}$; $z = \frac{\overset{40}{\cancel{80}}}{\cancel{2}}$; $z = \frac{40}{1}$; $\mathbf{z = 40}$ and

II. $z = \frac{37-43}{2}$; $z = -\frac{\overset{3}{\cancel{6}}}{\cancel{2}}$; $z = -\frac{3}{1}$; $\mathbf{z = -3}$

Check: I. *Let* $z = 40$ *in* $z^2 - 37z - 120 = 0$; $40^2 - 37 \cdot 40 - 120 \overset{?}{=} 0$; $1600 - 1480 - 120 \overset{?}{=} 0$; $1600 - 1600 \overset{?}{=} 0$; $0 = 0$

II. *Let* $z = -3$ *in* $z^2 - 37z - 120 = 0$; $(-3)^2 - 37 \cdot (-3) - 120 \overset{?}{=} 0$; $9 + 111 - 120 \overset{?}{=} 0$; $120 - 120 \overset{?}{=} 0$; $0 = 0$

Therefore, the equation $z^2 - 37z - 120 = 0$ can be factored to $\mathbf{(z-40)(z+3) = 0}$.

10. $x^2 - 20 = -8x$ Write the equation in standard form, i.e., $x^2 + 8x - 20 = 0$

Let: $a = 1$, $b = 8$, and $c = -20$. Then,

Given: $x = \frac{-b \pm \sqrt{b^2 - 4ac}}{2a}$; $x = \frac{-8 \pm \sqrt{8^2 - 4 \times 1 \times -20}}{2 \times 1}$; $x = \frac{-8 \pm \sqrt{64 + 80}}{2}$; $x = \frac{-8 \pm \sqrt{144}}{2}$; $x = \frac{-8 \pm \sqrt{12^2}}{2}$

$; \; x=\frac{-8\pm 12}{2}$ therefore, I. $x=\frac{-8+12}{2} \; ; \; x=\frac{\overset{2}{\not 4}}{\not 2} \; ; \; x=\frac{2}{1} \; ; \; \mathbf{x=2}$ and

II. $x=\frac{-8-12}{2} \; ; \; x=-\frac{\overset{10}{\not 20}}{\not 2} \; ; \; x=-\frac{10}{1} \; ; \; \mathbf{x=-10}$

Check: I. *Let* $x=2$ *in* $x^2-20=-8x \; ; \; 2^2-20\overset{?}{=}-8\cdot 2 \; ; \; 4-20\overset{?}{=}-16 \; ; \; -16=-16$

II. *Let* $x=-10$ *in* $x^2-20=-8x \; ; \; (-10)^2-20\overset{?}{=}-8\cdot(-10) \; ; \; 100-20\overset{?}{=}+80 \; ; \; 80=80$

Therefore, the equation $x^2+8x-20=0$ can be factored to $\mathbf{(x-2)(x+10)=0}$.

Section 4.2 Case II Solutions - Solving Quadratic Equations of the Form ax^2+bx+c where $a\rangle 1$

1. $4u^2+6u+1=0$ The quadratic equation is already in standard form.

Let: $a=4$, $b=6$, and $c=1$. Then,

Given: $u=\frac{-b\pm\sqrt{b^2-4ac}}{2a} \; ; \; u=\frac{-6\pm\sqrt{6^2-4\times 4\times 1}}{2\times 4} \; ; \; u=\frac{-6\pm\sqrt{36-16}}{8} \; ; \; u=\frac{-6\pm\sqrt{20}}{8} \; ; \; u=\frac{-6\pm 4.47}{8}$

therefore, I. $u=\frac{-6+4.47}{8} \; ; \; u=-\frac{1.53}{8} \; ; \; \mathbf{u=-0.19}$ and

II. $u=\frac{-6-4.47}{8} \; ; \; u=-\frac{10.47}{8} \; ; \; \mathbf{u=-1.31}$

The solution set is $\{\mathbf{-1.31, -0.9}\}$.

Check: I. *Let* $u=-0.19$ *in* $4u^2+6u+1=0 \; ; \; 4\cdot(-0.19)^2+6\cdot -0.19+1\overset{?}{=}0 \; ; \; 4\cdot 0.036-1.14+1\overset{?}{=}0 \; ; \; 0.14-1.14+1\overset{?}{=}0$

$; \; 1.14-1.14\overset{?}{=}0 \; ; \; 0=0$

II. *Let* $u=-1.31$ *in* $4u^2+6u+1=0 \; ; \; 4\cdot(-1.31)^2+6\cdot -1.31+1\overset{?}{=}0 \; ; \; 4\cdot 1.716-7.86+1\overset{?}{=}0 \; ; \; 6.86-7.86+1\overset{?}{=}0$

$; \; 7.86-7.86\overset{?}{=}0 \; ; \; 0=0$

Therefore, the equation $4u^2+6u+1=0$ can be factored to $\mathbf{(u+0.19)(u+1.31)=0}$.

2. $4w^2+10w=-3$ Write the equation in standard form, i.e., $4w^2+10w+3=0$.

Let: $a=4$, $b=10$, and $c=3$. Then,

Given: $w=\frac{-b\pm\sqrt{b^2-4ac}}{2a} \; ; \; w=\frac{-10\pm\sqrt{10^2-4\times 4\times 3}}{2\times 4} \; ; \; w=\frac{-10\pm\sqrt{100-48}}{8} \; ; \; w=\frac{-10\pm\sqrt{52}}{8}$

$; \; w=\frac{-10\pm 7.2}{8}$ therefore, I. $w=\frac{-10+7.2}{8} \; ; \; w=-\frac{2.8}{8} \; ; \; \mathbf{w=-0.35}$ and

II. $w=\frac{-10-7.2}{8} \; ; \; w=-\frac{17.2}{8} \; ; \; \mathbf{w=-2.15}$

The solution set is $\{\mathbf{-2.15, -0.35}\}$.

Check: I. *Let* $w=-0.35$ *in* $4w^2+10w=-3 \; ; \; 4\cdot(-0.35)^2+10\cdot -0.35\overset{?}{=}-3 \; ; \; 4\cdot 0.123-3.5\overset{?}{=}-3 \; ; \; 0.5-3.5\overset{?}{=}-3$

$; \; -3=-3$

II. *Let* $w=-2.15$ *in* $4w^2+10w=-3 \; ; \; 4\cdot(-2.15)^2+10\cdot -2.15\overset{?}{=}-3 \; ; \; 4\cdot 4.62-21.5\overset{?}{=}-3 \; ; \; 18.5-21.5\overset{?}{=}-3$

$; \; -3=-3$

Therefore, the equation $4w^2+10w+3=0$ can be factored to $\mathbf{(w+0.35)(w+2.15)=0}$.

3. $6x^2+4x-2=0$ The quadratic equation is already in standard form.

Let: $a = 6$, $b = 4$, and $c = -2$. Then,

Given: $x = \frac{-b \pm \sqrt{b^2 - 4ac}}{2a}$; $x = \frac{-4 \pm \sqrt{4^2 - 4 \times 6 \times -2}}{2 \times 6}$; $x = \frac{-4 \pm \sqrt{16 + 48}}{12}$; $x = \frac{-4 \pm \sqrt{64}}{12}$; $x = \frac{-4 \pm \sqrt{8^2}}{12}$

; $x = \frac{-4 \pm 8}{12}$ therefore, I. $x = \frac{-4 + 8}{12}$; $x = \frac{\cancel{4}}{\underset{3}{\cancel{12}}}$; $x = \frac{1}{3}$; $\mathbf{x = 0.33}$ and

II. $x = \frac{-4 - 8}{12}$; $x = -\frac{\cancel{12}}{\cancel{12}}$; $x = -\frac{1}{1}$; $\mathbf{x = -1}$

The solution set is $\{\mathbf{-1, 0.33}\}$.

Check: I. *Let* $x = 0.33$ *in* $6x^2 + 4x - 2 = 0$; $6 \cdot 0.33^2 + 4 \cdot 0.33 - 2 \overset{?}{=} 0$; $6 \cdot 0.111 + 1.32 - 2 \overset{?}{=} 0$; $0.67 + 1.32 - 2 \overset{?}{=} 0$; $2 - 2 \overset{?}{=} 0$; $0 = 0$

II. *Let* $x = -1$ *in* $6x^2 + 4x - 2 = 0$; $6 \cdot (-1)^2 + 4 \cdot -1 - 2 \overset{?}{=} 0$; $6 \cdot 1 - 4 - 2 \overset{?}{=} 0$; $6 - 6 \overset{?}{=} 0$; $0 = 0$

Therefore, the equation $6x^2 + 4x - 2 = 0$ can be factored to $\mathbf{(x - 0.33)(x + 1) = 0}$.

4. $15y^2 + 3 = -14y$ Write the equation in standard form, i.e., $15y^2 + 14y + 3 = 0$.

Let: $a = 15$, $b = 14$, and $c = 3$. Then,

Given: $y = \frac{-b \pm \sqrt{b^2 - 4ac}}{2a}$; $y = \frac{-14 \pm \sqrt{(-14)^2 - 4 \times 15 \times 3}}{2 \times 15}$; $y = \frac{-14 \pm \sqrt{196 - 180}}{30}$; $y = \frac{-14 \pm \sqrt{16}}{30}$; $y = \frac{-14 \pm \sqrt{4^2}}{30}$

; $y = \frac{-14 \pm 4}{30}$ therefore, I. $y = \frac{-14 + 4}{30}$; $y = -\frac{\cancel{10}}{\underset{3}{\cancel{30}}}$; $y = -\frac{1}{3}$; $\mathbf{y = -0.33}$ and

II. $y = \frac{-14 - 4}{30}$; $y = -\frac{\overset{3}{\cancel{18}}}{\underset{5}{\cancel{30}}}$; $y = -\frac{3}{5}$; $\mathbf{y = -0.6}$

The solution set is $\{\mathbf{-0.6, -0.33}\}$.

Check I. *Let* $y = -0.33$ *in* $15y^2 + 3 = -14y$; $15 \cdot (-0.33)^2 + 3 \overset{?}{=} -14 \cdot -0.33$; $15 \cdot 0.108 + 3 \overset{?}{=} 4.62$; $1.62 + 3 \overset{?}{=} 4.62$; $4.62 = 4.62$

II. *Let* $y = -0.6$ *in* $15y^2 + 3 = -14y$; $15 \cdot (-0.6)^2 + 3 \overset{?}{=} -14 \cdot -0.6$; $15 \cdot 0.36 + 3 \overset{?}{=} 8.4$; $5.4 + 3 \overset{?}{=} 8.4$; $8.4 = 8.4$

Therefore, the equation $15y^2 + 3 = -14y$ can be factored to $\mathbf{(y + 0.6)(w + 0.33) = 0}$.

5. $2x^2 - 5x + 3 = 0$ The equation is in standard form.

Let: $a = 2$, $b = -5$, and $c = 3$. Then,

Given: $x = \frac{-b \pm \sqrt{b^2 - 4ac}}{2a}$; $x = \frac{-(-5) \pm \sqrt{(-5)^2 - 4 \times 2 \times 3}}{2 \times 2}$; $x = \frac{5 \pm \sqrt{25 - 24}}{4}$; $x = \frac{5 \pm \sqrt{1}}{4}$; $x = \frac{5 \pm 1}{4}$

therefore, I. $x = \frac{5 + 1}{4}$; $x = \frac{\overset{3}{\cancel{6}}}{\underset{2}{\cancel{4}}}$; $x = \frac{3}{2}$; $\mathbf{x = 1.5}$ and

II. $x = \frac{5 - 1}{4}$; $x = \frac{\cancel{4}}{\cancel{4}}$; $x = \frac{1}{1}$; $\mathbf{x = 1}$

The solution set is $\{\mathbf{1, 1.5}\}$.

Check I. *Let* $x = 1$ *in* $2x^2 - 5x + 3 = 0$; $2 \cdot 1^2 - 5 \cdot 1 + 3 \overset{?}{=} 0$; $2 \cdot 1 - 5 + 3 \overset{?}{=} 0$; $2 - 5 + 3 \overset{?}{=} 0$; $5 - 5 \overset{?}{=} 0$; $0 = 0$

II. *Let* $x = 1.5$ *in* $2x^2 - 5x + 3 = 0$; $2 \cdot 1.5^2 - 5 \cdot 1.5 + 3 \overset{?}{=} 0$; $2 \cdot 2.25 - 7.5 + 3 \overset{?}{=} 0$; $4.5 - 7.5 + 3 \overset{?}{=} 0$

$; 7.5 - 7.5 \stackrel{?}{=} 0 \; ; \; 0 = 0$

Therefore, the equation $2x^2 - 5x + 3 = 0$ can be factored to $\mathbf{(x-1)(x-1.5)=0}$.

6. $2x^2 + xy - y^2 = 0$ $\quad$ x is varaible $\quad$ Write the equation in standard form, i.e., $2x^2 + yx - y^2 = 0$.

Let: $a = 2$, $b = y$, and $c = -y^2$. Then,

Given: $x = \dfrac{-b \pm \sqrt{b^2 - 4ac}}{2a}$; $x = \dfrac{-y \pm \sqrt{y^2 - 4 \times 2 \times -y^2}}{2 \times 2}$; $x = \dfrac{-y \pm \sqrt{y^2 + 8y^2}}{4}$; $x = \dfrac{-y \pm \sqrt{9y^2}}{4}$; $x = \dfrac{-y \pm 3y}{4}$

therefore, I. $x = \dfrac{-y + 3y}{4}$; $x = \dfrac{\overset{1}{\cancel{2}}y}{\underset{2}{\cancel{4}}}$; $x = \dfrac{1}{2}y$; $\mathbf{x = 0.5y}$ and

II. $x = \dfrac{-y - 3y}{4}$; $x = \dfrac{-4y}{4}$; $x = -\dfrac{\cancel{4}}{\cancel{4}}y$; $\mathbf{x = -y}$

The solution set is $\{\mathbf{-y, 0.5y}\}$.

I. *Let* $x = 0.5y$ *in* $2x^2 + xy - y^2 = 0$; $2 \cdot (0.5y)^2 + (0.5y) \cdot y - y^2 \stackrel{?}{=} 0$; $2 \cdot 0.25y^2 + 0.5y^2 - y^2 \stackrel{?}{=} 0$

$; 0.5y^2 + 0.5y^2 - y^2 \stackrel{?}{=} 0 \; ; \; y^2 - y^2 \stackrel{?}{=} 0 \; ; \; 0 = 0$

II. *Let* $x = -y$ *in* $2x^2 + xy - y^2 = 0$; $2 \cdot (-y)^2 + (-y) \cdot y - y^2 \stackrel{?}{=} 0$; $2y^2 - y^2 - y^2 \stackrel{?}{=} 0$; $2y^2 - 2y^2 \stackrel{?}{=} 0$; $0 = 0$

Therefore, the equation $2x^2 + xy - y^2 = 0$ can be factored to $\mathbf{(x+y)(x-0.5y)=0}$.

7. $6x^2 + 7x - 3 = 0$ $\quad$ The equation is already in standard form.

Let: $a = 6$, $b = 7$, and $c = -3$. Then,

Given: $x = \dfrac{-b \pm \sqrt{b^2 - 4ac}}{2a}$; $x = \dfrac{-7 \pm \sqrt{7^2 - 4 \times 6 \times -3}}{2 \times 6}$; $x = \dfrac{-7 \pm \sqrt{49 + 72}}{12}$; $x = \dfrac{-7 \pm \sqrt{121}}{12}$; $x = \dfrac{-7 \pm \sqrt{11^2}}{12}$

; $x = \dfrac{-7 \pm 11}{12}$ therefore, I. $x = \dfrac{-7 + 11}{12}$; $x = \dfrac{\cancel{4}}{\underset{3}{\cancel{12}}}$; $x = \dfrac{1}{3}$; $\mathbf{x = 0.33}$ and

II. $x = \dfrac{-7 - 11}{12}$; $x = -\dfrac{\overset{3}{\cancel{18}}}{\underset{2}{\cancel{12}}}$; $x = -\dfrac{3}{2}$; $\mathbf{x = -1.5}$

The solution set is $\{\mathbf{-1.5, 0.33}\}$.

Check I. *Let* $x = 0.33$ *in* $6x^2 + 7x - 3 = 0$; $6 \cdot (0.33)^2 + 7 \cdot 0.33 - 3 \stackrel{?}{=} 0$; $6 \cdot 0.11 + 2.31 - 3 \stackrel{?}{=} 0$; $0.66 + 2.31 - 3 \stackrel{?}{=} 0$

$; 3 - 3 \stackrel{?}{=} 0 \; ; \; 0 = 0$

II. *Let* $x = -1.5$ *in* $6x^2 + 7x - 3 = 0$; $6 \cdot (-1.5)^2 + 7 \cdot -1.5 - 3 \stackrel{?}{=} 0$; $6 \cdot 2.25 - 10.5 - 3 \stackrel{?}{=} 0$; $13.5 - 10.5 - 3 \stackrel{?}{=} 0$

$; 13.5 - 13.5 \stackrel{?}{=} 0 \; ; \; 0 = 0$

Therefore, the equation $6x^2 + 7x - 3 = 0$ can be factored to $\mathbf{(x+1.5)(x-0.33)=0}$.

8. $5x^2 = -3x$ $\quad$ Write the equation in standard form, i.e., $5x^2 + 3x = 0$.

Let: $a = 5$, $b = 3$, and $c = 0$. Then,

Given: $x = \dfrac{-b \pm \sqrt{b^2 - 4ac}}{2a}$; $x = \dfrac{-3 \pm \sqrt{3^2 - 4 \times 5 \times 0}}{2 \times 5}$; $x = \dfrac{-3 \pm \sqrt{9 - 0}}{10}$; $x = \dfrac{-3 \pm \sqrt{9}}{10}$; $x = \dfrac{-3 \pm \sqrt{3^2}}{10}$

; $x = \dfrac{-3 \pm 3}{10}$ therefore, I. $x = \dfrac{-3 + 3}{10}$; $x = \dfrac{0}{10}$; $\mathbf{x = 0}$ and

$$\text{II.}\quad x=\frac{-3-3}{10}\ ;\ x=-\frac{\overset{3}{\cancel{6}}}{\underset{5}{\cancel{10}}}\ ;\ \boldsymbol{x=-\frac{3}{5}}$$

Check: I. $Let\ x=0\ in\quad 5x^2=-3x\ ;\ 5\cdot 0^2\overset{?}{=}-3\cdot 0\ ;\ 5\cdot 0\overset{?}{=}-3\cdot 0\ ;\ 0=0$

II. $Let\ x=-\frac{3}{5}\ in\quad 5x^2=-3x\ ;\ 5\cdot\left(-\frac{3}{5}\right)^2\overset{?}{=}-3\cdot-\frac{3}{5}\ ;\ \cancel{5}\cdot\frac{9}{\underset{5}{\cancel{25}}}\overset{?}{=}+\frac{9}{5}\ ;\ \frac{9}{5}=\frac{9}{5}$

The solution set is $\left\{\boldsymbol{-\frac{3}{5},0}\right\}$.

Therefore, the equation $5x^2+3x=0$ can be factored to $(x+0)\left(x+\frac{3}{5}\right)=0$ which is the same as $\boldsymbol{x\left(x+\frac{3}{5}\right)=0}$.

Note that this equation can further be simplified in order to obtain the original form of the quadratic equation as follows:

$$;\ x\left(\frac{x}{1}+\frac{3}{5}\right)=0\ ;\ x\left(\frac{(5\cdot x)+(1\cdot 3)}{1\cdot 5}\right)=0\ ;\ x\left(\frac{5x+3}{5}\right)=0\ ;\ \frac{5x^2+3x}{5}=0\ ;\ \frac{5x^2+3x}{5}=\frac{0}{1}\ ;\ \left(5x^2+3x\right)\cdot 1=0\cdot 5$$

$$;\ 5x^2+3x=0$$

9. $3x^2+4x+5=0$ The equation is already in standard form

Let: $a=3$, $b=4$, and $c=5$. Then,

Given: $x=\frac{-b\pm\sqrt{b^2-4ac}}{2a}\ ;\ x=\frac{-4\pm\sqrt{4^2-4\times 3\times 5}}{2\times 3}\ ;\ x=\frac{-4\pm\sqrt{16-60}}{6}\ ;\ x=\frac{-4\pm\sqrt{-44}}{6}$

Since the number under the radical is a negative number (an imaginary number) therefore, **the equation $\boldsymbol{3x^2+4x+5=0}$ has no real solutions.**

10. $-3y^2+13y+10=0$ The equation is in standard form.

Let: $a=-3$, $b=13$, and $c=10$. Then,

Given: $y=\frac{-b\pm\sqrt{b^2-4ac}}{2a}\ ;\ y=\frac{-13\pm\sqrt{13^2-4\times-3\times 10}}{2\times-3}\ ;\ y=\frac{-13\pm\sqrt{169+120}}{-6}\ ;\ y=\frac{-13\pm\sqrt{289}}{-6}$

$;\ y=\frac{-13\pm\sqrt{17^2}}{-6}\ ;\ y=\frac{-13\pm 17}{-6}$ therefore, I. $y=\frac{-13+17}{-6}\ ;\ y=\frac{\overset{2}{\cancel{4}}}{-\underset{3}{\cancel{6}}}\ ;\ y=-\frac{2}{3}\ ;\ \boldsymbol{y=-0.66}$ and

II. $y=\frac{-13-17}{-6}\ ;\ y=\frac{-\overset{5}{\cancel{30}}}{-\cancel{6}}\ ;\ y=\frac{5}{1}\ ;\ \boldsymbol{y=5}$

The solution set is $\{\boldsymbol{-0.66, 5}\}$.

Check I. $Let\ y=5\ in\quad -3y^2+13y+10=0\ ;\ -3\cdot 5^2+13\cdot 5+10\overset{?}{=}0\ ;\ -3\cdot 25+65+10\overset{?}{=}0\ ;\ -75+65+10\overset{?}{=}0$

$;\ -75+75\overset{?}{=}0\ ;\ 0=0$

II. $Let\ y=-0.66\ in\quad -3y^2+13y+10=0\ ;\ -3\cdot(-0.66)^2+13\cdot-0.66+10\overset{?}{=}0\ ;\ -3\cdot 0.436-8.58+10\overset{?}{=}0$

$;\ -1.32-8.58+10\overset{?}{=}0\ ;\ -10+10\overset{?}{=}0\ ;\ 0=0$

Therefore, the equation $-3y^2+13y+10=0$ can be factored to $\boldsymbol{(y+0.66)(y-5)=0}$.

Section 4.3 Solutions - Solving Quadratic Equations Using the Square Root Property Method

1. First - Take the square root of both sides of the equation $(2y+5)^2=36$, i.e., $\sqrt{(2y+5)^2}=\pm\sqrt{36}$

Second - Simplify the terms on both sides to obtain the solutions, i.e., $\sqrt{(2y+5)^2}=\pm\sqrt{36}\ ;\ 2y+5=\pm 6$

Therefore the two solutions are: I. $2y+5=-6$; $2y=-6-5$; $2y=-11$; $\frac{2y}{2}=-\frac{11}{2}$; $y=-\frac{11}{2}$; $\mathbf{y=-5.5}$ and

II. $2y+5=+6$; $2y=6-5$; $2y=1$; $\frac{2y}{2}=\frac{1}{2}$; $y=\frac{1}{2}$; $\mathbf{y=0.5}$

Thus, the solution set is $\{\mathbf{-5.5, 0.5}\}$ and the equation $(2y+5)^2=36$ can be factored to $\mathbf{(y+5.5)(y-0.5)=0}$.

Check: I. *Let* $y=-5.5$ *in* $(2y+5)^2=36$; $(2\cdot -5.5+5)^2\stackrel{?}{=}36$; $(-11+5)^2\stackrel{?}{=}36$; $(-6)^2\stackrel{?}{=}36$; $6^2\stackrel{?}{=}36$; $36=36$

II. *Let* $y=0.5$ *in* $(2y+5)^2=36$; $(2\cdot 0.5+5)^2\stackrel{?}{=}36$; $(1+5)^2\stackrel{?}{=}36$; $6^2\stackrel{?}{=}36$; $36=36$

2. First - Take the square root of both sides of the equation $(x+1)^2=7$, i.e., $\sqrt{(x+1)^2}=\pm\sqrt{7}$

Second - Simplify the terms on both sides to obtain the solutions, i.e., $\sqrt{(x+1)^2}=\pm\sqrt{7}$; $x+1=\pm 2.65$

Therefore the two solutions are: I. $x+1=-2.65$; $x=-2.65-1$; $\mathbf{x=-3.65}$ and

II. $x+1=+2.65$; $x=2.65-1$; $\mathbf{x=1.65}$

Thus, the solution set is $\{\mathbf{-3.65, 1.65}\}$ and the equation $(x+1)^2=7$ can be factored to $\mathbf{(x+3.65)(x-1.65)=0}$.

Check: I. *Let* $x=-3.65$ *in* $(x+1)^2=7$; $(-3.65+1)^2\stackrel{?}{=}7$; $(-2.65)^2\stackrel{?}{=}7$; $7=7$

II. *Let* $x=1.65$ *in* $(x+1)^2=7$; $(1.65+1)^2\stackrel{?}{=}7$; $2.65^2\stackrel{?}{=}7$; $7=7$

3. First - Take the square root of both sides of the equation $(2x-3)^2=1$, i.e., $\sqrt{(2x-3)^2}=\pm\sqrt{1}$

Second - Simplify the terms on both sides to obtain the solutions, i.e., $\sqrt{(2x-3)^2}=\pm\sqrt{1}$; $2x-3=\pm 1$

Therefore the two solutions are: I. $2x-3=-1$; $2x=-1+3$; $2x=2$; $\frac{2x}{2}=\frac{2}{2}$; $x=\frac{1}{1}$; $\mathbf{x=1}$ and

II. $2x-3=+1$; $2x=1+3$; $2x=4$; $\frac{2x}{2}=\frac{\overset{2}{4}}{2}$; $x=\frac{2}{1}$; $\mathbf{x=2}$

Thus, the solution set is $\{\mathbf{1, 2}\}$ and the equation $(2x-3)^2=1$ can be factored to $\mathbf{(x-1)(x-2)=0}$.

Check: I. *Let* $x=1$ *in* $(2x-3)^2=1$; $(2\cdot 1-3)^2\stackrel{?}{=}1$; $(2-3)^2\stackrel{?}{=}1$; $(-1)^2\stackrel{?}{=}1$; $1=1$

II. *Let* $x=2$ *in* $(2x-3)^2=1$; $(2\cdot 2-3)^2\stackrel{?}{=}1$; $(4-3)^2\stackrel{?}{=}1$; $1^2\stackrel{?}{=}1$; $1=1$

4. First - Write the equation $x^2+3=0$ in the form of $x^2=b$, i.e., $x^2=-3$

Second - Take the square root of both sides of the equation, i.e., $\sqrt{x^2}=\pm\sqrt{-3}$

Since the number under the radical is a negative number (an imaginary number) therefore, **the equation $x^2+3=0$ has no real solutions.**

5. First - Take the square root of both sides of the equation $(y-5)^2=5$, i.e., $\sqrt{(y-5)^2}=\pm\sqrt{5}$

Second - Simplify the terms on both sides to obtain the solutions, i.e., $\sqrt{(y-5)^2}=\pm\sqrt{5}$; $y-5=\pm 2.24$

Therefore the two solutions are: I. $y-5=-2.24$; $y=-2.24+5$; $\mathbf{y=2.76}$ and

II. $y-5=+2.24$; $y=2.24+5$; $\mathbf{y=7.24}$

Thus, the solution set is $\{\mathbf{-2.76, 7.24}\}$ and the equation $(y-5)^2=5$ can be factored to $\mathbf{(y-2.76)(y-7.24)=0}$.

Check: I. *Let* $y=2.76$ *in* $(y-5)^2=5$; $(2.76-5)^2\stackrel{?}{=}5$; $(-2.24)^2\stackrel{?}{=}5$; $5=5$

II. *Let* $y=7.24$ *in* $(y-5)^2=5$; $(7.24-5)^2\stackrel{?}{=}5$; $(2.24)^2\stackrel{?}{=}5$; $5=5$

6. First - Write the equation $16x^2-25=0$ in the form of $ax^2=b$, i.e., $16x^2=25$

Second - Divide both sides of the equation $16x^2 = 25$ by the coefficient of x, i.e., $\frac{\cancel{16}x^2}{\cancel{16}} = \frac{25}{16}$; $x^2 = \frac{25}{16}$

Third - Take the square root of both sides of the equation, i.e., $\sqrt{x^2} = \pm\sqrt{\frac{25}{16}}$

Fourth - Simplify the terms on both sides to obtain the solutions, i.e., $\mathbf{x = \pm\frac{5}{4}}$

Therefore, the solution set is $\left\{-\frac{5}{4}, \frac{5}{4}\right\}$ and the equation $16x^2 - 25 = 0$ can be factored to $\left(x - \frac{5}{4}\right)\left(x + \frac{5}{4}\right) = 0$ which is the same as $\mathbf{(4x-5)(4x+5) = 0}$.

Check: I. *Let* $x = -\frac{5}{4}$ *in* $16x^2 - 25 = 0$; $16 \cdot \left(-\frac{5}{4}\right)^2 - 25 \stackrel{?}{=} 0$; $\cancel{16} \cdot \frac{25}{\cancel{16}} - 25 \stackrel{?}{=} 0$; $25 - 25 \stackrel{?}{=} 0$; $0 = 0$

II. *Let* $x = -\frac{5}{4}$ *in* $16x^2 - 25 = 0$; $16 \cdot \left(\frac{5}{4}\right)^2 - 25 \stackrel{?}{=} 0$; $\cancel{16} \cdot \frac{25}{\cancel{16}} - 25 \stackrel{?}{=} 0$; $25 - 25 \stackrel{?}{=} 0$; $0 = 0$

7. First - Write the equation $x^2 - 49 = 0$ in the form of $x^2 = b$, i.e., $x^2 = 49$

Second - Take the square root of both sides of the equation, i.e., $\sqrt{x^2} = \pm\sqrt{49}$

Third - Simplify the terms on both sides to obtain the solutions, i.e., $\mathbf{x = \pm 7}$

Therefore, the solution set is $\{\mathbf{-7, 7}\}$ and the equation $x^2 = 49$ can be factored to $\mathbf{(x-7)(x+7) = 0}$.

Check: I. *Let* $x = -7$ *in* $x^2 - 49 = 0$; $(-7)^2 - 49 \stackrel{?}{=} 0$; $49 - 49 \stackrel{?}{=} 0$; $0 = 0$

II. *Let* $x = 7$ *in* $x^2 - 49 = 0$; $7^2 - 49 \stackrel{?}{=} 0$; $49 - 49 \stackrel{?}{=} 0$; $0 = 0$

8. First - Take the square root of both sides of the equation $(3x-1)^2 = 25$, i.e., $\sqrt{(3x-1)^2} = \pm\sqrt{25}$

Second - Simplify the terms on both sides to obtain the solutions, i.e., $\sqrt{(3x-1)^2} = \pm\sqrt{25}$; $3x - 1 = \pm 5$

Therefore, the two solutions are: I. $3x - 1 = -5$; $3x = -5 + 1$; $3x = -4$; $\frac{\cancel{3}x}{\cancel{3}} = -\frac{4}{3}$; $\mathbf{x = -1.33}$ and

II. $3x - 1 = +5$; $3x = 5 + 1$; $3x = 6$; $\frac{\cancel{3}x}{\cancel{3}} = \frac{\overset{2}{\cancel{6}}}{\cancel{3}}$; $x = \frac{2}{1}$; $\mathbf{x = 2}$

Thus, the solution set is $\{\mathbf{-1.33, 2}\}$ and the equation $(3x-1)^2 = 25$ can be factored to $\mathbf{(x+1.33)(x-2) = 0}$.

Check: I. *Let* $x = -1.33$ *in* $(3x-1)^2 = 25$; $(3 \cdot -1.33 - 1)^2 \stackrel{?}{=} 25$; $(-4-1)^2 \stackrel{?}{=} 25$; $(-5)^2 \stackrel{?}{=} 25$; $5^2 \stackrel{?}{=} 25$; $25 = 25$

II. *Let* $x = 2$ *in* $(3x-1)^2 = 25$; $(3 \cdot 2 - 1)^2 \stackrel{?}{=} 25$; $(6-1)^2 \stackrel{?}{=} 25$; $5^2 \stackrel{?}{=} 25$; $25 = 25$

9. First - Take the square root of both sides of the equation $(x-2)^2 = -7$, i.e., $\sqrt{(x-2)^2} = \pm\sqrt{-7}$

Since the number under the radical is a negative number (an imaginary number) therefore, **the equation $\mathbf{(x-2)^2 = -7}$ has no real solutions.**

10. First - Take the square root of both sides of the equation $\left(x - \frac{1}{3}\right)^2 = \frac{1}{9}$, i.e., $\sqrt{\left(x - \frac{1}{3}\right)^2} = \pm\sqrt{\frac{1}{9}}$

Second - Simplify the terms on both sides to obtain the solutions, i.e., $\sqrt{\left(x - \frac{1}{3}\right)^2} = \pm\sqrt{\frac{1}{9}}$; $x - \frac{1}{3} = \pm\frac{1}{3}$

Therefore the two solutions are: I. $x - \frac{1}{3} = -\frac{1}{3}$; $x = -\frac{1}{3} + \frac{1}{3}$; $x = \frac{-1+1}{3}$; $x = \frac{0}{3}$; $\mathbf{x = 0}$ and

II. $x - \frac{1}{3} = +\frac{1}{3}$; $x = \frac{1}{3} + \frac{1}{3}$; $x = \frac{1+1}{3}$; $\mathbf{x = \frac{2}{3}}$

Thus, the solution set is $\left\{\mathbf{0, \frac{2}{3}}\right\}$ and the equation $\left(x-\frac{1}{3}\right)^2=\frac{1}{9}$ can be factored to $(x+0)\left(x-\frac{2}{3}\right)=0$ which is the same as $x\left(x-\frac{2}{3}\right)=0$ or $\boldsymbol{x(3x-2)=0}$.

Check: I. *Let* $x=0$ *in* $\left(x-\frac{1}{3}\right)^2=\frac{1}{9}$; $\left(0-\frac{1}{3}\right)^2\overset{?}{=}\frac{1}{9}$; $\left(-\frac{1}{3}\right)^2\overset{?}{=}\frac{1}{9}$; $\frac{1}{9}=\frac{1}{9}$

II. *Let* $x=\frac{2}{3}$ *in* $\left(x-\frac{1}{3}\right)^2=\frac{1}{9}$; $\left(\frac{2}{3}-\frac{1}{3}\right)^2\overset{?}{=}\frac{1}{9}$; $\left(\frac{2-1}{3}\right)^2\overset{?}{=}\frac{1}{9}$; $\left(\frac{1}{3}\right)^2\overset{?}{=}\frac{1}{9}$; $\frac{1}{9}=\frac{1}{9}$

Section 4.4 Case I Solutions - Solving Quadratic Equations of the Form $ax^2+bx+c=0$, where $a=1$, by Completing the Square

1. First - Write the equation $x^2+10x-2=0$ in the form of $x^2+bx=-c$, i.e., $x^2+10x=2$.

   Second - Complete the square and simplify. $x^2+10x=2$; $x^2+10x+\left(\frac{\overset{5}{\cancel{10}}}{\cancel{2}}\right)^2=2+\left(\frac{\overset{5}{\cancel{10}}}{\cancel{2}}\right)^2$; $x^2+10x+5^2=2+5^2$

   ; $x^2+10x+25=2+25$; $x^2+10x+25=27$; $(x+5)^2=27$

   Third - Take the square root of both sides of the equation and solve for x.

   $(x+5)^2=27$; $\sqrt{(x+5)^2}=\pm\sqrt{27}$; $x+5=\pm5.19$. Therefore,

   I. $x+5=+5.19$; $x=5.19-5$; $\boldsymbol{x=0.19}$ and II. $x+5=-5.19$; $x=-5.19-5$; $\boldsymbol{x=-10.19}$

   The solution set is $\{\mathbf{-10.19, 0.19}\}$.

   Fourth - Check the answers and write the quadratic equation in its factored form.

   I. *Let* $x=0.19$ *in* $x^2+10x-2=0$; $0.19^2+10\cdot0.19-2\overset{?}{=}0$; $0.036+1.9-2\overset{?}{=}0$; $2-2\overset{?}{=}0$; $0=0$

   II. *Let* $x=-10.19$ *in* $x^2+10x-2=0$; $(-10.19)^2+10\cdot-10.19-2\overset{?}{=}0$; $103.8-101.9-2\overset{?}{=}0$; $101.9-101.9\overset{?}{=}0$; $0=0$

   Therefore, the equation $x^2+10x-2=0$ can be factored to $\boldsymbol{(x+10.19)(x-0.19)=0}$.

2. First - Write the equation $x^2-x-1=0$ in the form of $x^2+bx=-c$, i.e., $x^2-x=1$.

   Second - Complete the square and simplify. $x^2-x=1$; $x^2-x+\left(-\frac{1}{2}\right)^2=1+\left(-\frac{1}{2}\right)^2$; $x^2-x+\frac{1}{4}=1+\frac{1}{4}$

   ; $\left(x-\frac{1}{2}\right)^2=1.25$

   Third - Take the square root of both sides of the equation and solve for x.

   $\left(x-\frac{1}{2}\right)^2=1.25$; $\sqrt{\left(x-\frac{1}{2}\right)^2}=\pm\sqrt{1.25}$; $x-\frac{1}{2}=\pm1.118$; $x-0.5=\pm1.118$. Therefore,

   I. $x-0.5=+1.118$; $x=1.118+0.5$; $\boldsymbol{x=1.618}$ and II. $x-0.5=-1.118$; $x=-1.118+0.5$; $\boldsymbol{x=-0.618}$

   The solution set is $\{\mathbf{-0.618, 1.618}\}$.

   Fourth - Check the answers and write the quadratic equation in its factored form.

   I. *Let* $x=1.618$ *in* $x^2-x-1=0$; $1.618^2-1.618-1\overset{?}{=}0$; $2.618-1.618-1\overset{?}{=}0$; $2.618-2.618\overset{?}{=}0$; $0=0$

   II. *Let* $x=-0.618$ *in* $x^2-x-1=0$; $(-0.618)^2-(-0.618)-1\overset{?}{=}0$; $0.381+0.618-1\overset{?}{=}0$; $-0.618+0.618\overset{?}{=}0$; $0=0$

   Therefore, the equation $x^2-x-1=0$ can be factored to $\boldsymbol{(x+0.618)(x-1.618)=0}$.

3. First - Write the equation $x(x+2)=80$ in the form of $x^2+bx=-c$, i.e., $x^2+2x=80$.

Second - Complete the square and simplify. $x^2+2x=80$; $x^2+2x+\left(\frac{2}{2}\right)^2=80+\left(\frac{2}{2}\right)^2$; $x^2+2x+1=80+1$

; $(x+1)^2=81$

Third - Take the square root of both sides of the equation and solve for x.

$(x+1)^2=81$; $\sqrt{(x+1)^2}=\pm\sqrt{81}$; $x+1=\pm 9$. Therefore,

I. $x+1=+9$; $x=9-1$; $\mathbf{x=8}$ and II. $x+1=-9$; $x=-9-1$; $\mathbf{x=-10}$

The solution set is $\{\mathbf{-10, 8}\}$.

Fourth - Check the answers and write the quadratic equation in its factored form.

I. *Let* $x=8$ *in* $x(x+2)=80$; $8(8+2)\overset{?}{=}80$; $8\cdot 10\overset{?}{=}80$; $80=80$

II. *Let* $x=-10$ *in* $x(x+2)=80$; $-10(-10+2)\overset{?}{=}80$; $-10\cdot -8\overset{?}{=}80$; $80=80$

Therefore, the equation $x(x+2)=80$ can be factored to $\mathbf{(x+10)(x-8)=0}$.

4. First - Write the equation $y^2-10y+5=0$ in the form of $y^2+by=-c$, i.e., $y^2-10y=-5$.

Second - Complete the square and simplify. $y^2-10y=-5$; $y^2-10y+\left(-\frac{\overset{5}{\cancel{10}}}{\cancel{2}}\right)^2=-5+\left(-\frac{\overset{5}{\cancel{10}}}{\cancel{2}}\right)^2$; $y^2-10y+5^2=-5+5^2$

; $y^2-10y+25=-5+25$; $y^2-10y+25=20$; $(y-5)^2=20$

Third - Take the square root of both sides of the equation and solve for y.

$(y-5)^2=20$; $\sqrt{(y-5)^2}=\pm\sqrt{20}$; $y-5=\pm 4.47$. Therefore,

I. $y-5=+4.47$; $y=4.47+5$; $\mathbf{y=9.47}$ and II. $y-5=-4.47$; $y=-4.47+5$; $\mathbf{y=0.53}$

The solution set is $\{\mathbf{0.53, 9.47}\}$.

Fourth - Check the answers and write the quadratic equation in its factored form.

I. *Let* $y=0.53$ *in* $y^2-10y+5=0$; $0.53^2-10\cdot 0.53+5\overset{?}{=}0$; $0.3-5.3+5\overset{?}{=}0$; $5.3-5.3\overset{?}{=}0$; $0=0$

II. *Let* $y=9.47$ *in* $y^2-10y+5=0$; $9.47^2-10\cdot 9.47+5\overset{?}{=}0$; $89.7-94.7+5\overset{?}{=}0$; $94.7-94.7\overset{?}{=}0$; $0=0$

Therefore, the equation $y^2-10y+5=0$ can be factored to $\mathbf{(y-0.53)(y-9.47)=0}$.

5. First - Write the equation $x^2+4x-5=0$ in the form of $x^2+bx=-c$, i.e., $x^2+4x=5$.

Second - Complete the square and simplify. $x^2+4x=5$; $x^2+4x+\left(\frac{\overset{2}{\cancel{4}}}{\cancel{2}}\right)^2=5+\left(\frac{\overset{2}{\cancel{4}}}{\cancel{2}}\right)^2$; $x^2+4x+2^2=5+2^2$

; $x^2+4x+4=5+4$; $x^2+4x+4=9$; $(x+2)^2=9$

Third - Take the square root of both sides of the equation and solve for x.

$(x+2)^2=9$; $\sqrt{(x+2)^2}=\pm\sqrt{9}$; $x+2=\pm 3$. Therefore,

I. $x+2=+3$; $x=3-2$; $\mathbf{x=1}$ and II. $x+2=-3$; $x=-2-3$; $\mathbf{x=-5}$

The solution set is $\{\mathbf{-5, 1}\}$.

Fourth - Check the answers and write the quadratic equation in its factored form.

I. *Let* $x=1$ *in* $x^2+4x-5=0$; $1^2+4\cdot 1-5\overset{?}{=}0$; $1+4-5\overset{?}{=}0$; $5-5\overset{?}{=}0$; $0=0$

II. *Let* $x=-5$ *in* $x^2+4x-5=0$; $(-5)^2+4\cdot(-5)-5\overset{?}{=}0$; $25-20-5\overset{?}{=}0$; $25-25\overset{?}{=}0$; $0=0$

Therefore, the equation $x^2+4x-5=0$ can be factored to $\mathbf{(x+5)(x-1)=0}$.

6. The equation $y^2+4y=14$ is already in the form of $y^2+by=-c$.

First - Complete the square and simplify. $y^2+4y=14$; $y^2+4y+\left(\dfrac{4}{2}\right)^2=14+\left(\dfrac{4}{2}\right)^2$; $y^2+4y+2^2=14+2^2$

; $y^2+4y+4=14+4$; $y^2+4y+4=18$; $(y+2)^2=18$

Second - Take the square root of both sides of the equation and solve for y.

$(y+2)^2=8$; $\sqrt{(y+2)^2}=\pm\sqrt{18}$; $y+2=\pm 4.24$. Therefore,

I. $y+2=+4.24$; $y=4.24-2$; $\mathbf{y=2.24}$ and II. $y+2=-4.24$; $y=-4.24-2$; $\mathbf{y=-6.24}$

The solution set is $\{\mathbf{-6.24, 2.24}\}$.

Third - Check the answers and write the quadratic equation in its factored form.

I. *Let* $y=2.24$ *in* $y^2+4y=14$; $2.24^2+4\cdot 2.24\overset{?}{=}14$; $5+9\overset{?}{=}14$; $14=14$

II. *Let* $y=-6.24$ *in* $y^2+4y=14$; $(-6.24)^2+4\cdot -6.24\overset{?}{=}14$; $39-25\overset{?}{=}14$; $14=14$

Therefore, the equation $y^2+4y=14$ can be factored to $\mathbf{(y+6.24)(y-2.24)=0}$.

7. First - Write the equation $w^2+\frac{1}{3}w-\frac{1}{2}=0$ in the form of $w^2+bw=-c$, i.e., $w^2+\frac{1}{3}w=\frac{1}{2}$.

Second - Complete the square and simplify. $w^2+\frac{1}{3}w=\frac{1}{2}$; $w^2+\frac{1}{3}w+\left(\frac{1}{6}\right)^2=\frac{1}{2}+\left(\frac{1}{6}\right)^2$; $w^2+\frac{1}{3}w+\frac{1}{36}=\frac{1}{2}+\frac{1}{36}$

; $w^2+\frac{1}{3}w+\frac{1}{36}=\dfrac{(1\cdot 36)+(1\cdot 2)}{2\cdot 36}$; $w^2+\frac{1}{3}w+\frac{1}{36}=\frac{36+2}{72}$; $w^2+\frac{1}{3}w+\frac{1}{36}=\frac{38}{72}$; $\left(w+\frac{1}{6}\right)^2=\frac{38}{72}$

Third - Take the square root of both sides of the equation and solve for w.

$\left(w+\frac{1}{6}\right)^2=\frac{38}{72}$; $\sqrt{\left(w+\frac{1}{6}\right)^2}=\pm\sqrt{\frac{38}{72}}$; $w+\frac{1}{6}=\pm\sqrt{0.527}$; $w+0.167=\pm 0.726$. Therefore,

I. $w+0.167=+0.726$; $w=0.726-0.167$; $\mathbf{w=0.56}$ and II. $w+0.167=-0.726$; $w=-0.726-0.167$

; $\mathbf{w=-0.89}$ The solution set is $\{\mathbf{-0.89, 0.56}\}$.

Fourth - Check the answers and write the quadratic equation in its factored form.

I. *Let* $w=0.56$ *in* $w^2+\frac{1}{3}w-\frac{1}{2}=0$; $0.56^2+\frac{1}{3}\cdot 0.56-\frac{1}{2}\overset{?}{=}0$; $0.31+0.19-0.5\overset{?}{=}0$; $0.5-0.5\overset{?}{=}0$; $0=0$

II. *Let* $w=-0.89$ *in* $w^2+\frac{1}{3}w-\frac{1}{2}=0$; $(-0.89)^2+\frac{1}{3}\cdot(-0.89)-\frac{1}{2}\overset{?}{=}0$; $0.79-0.29-0.5\overset{?}{=}0$; $0.79-0.79\overset{?}{=}0$

; $0=0$

Therefore, the equation $w^2+\frac{1}{3}w-\frac{1}{2}=0$ can be factored to $\mathbf{(w+0.89)(w-0.56)=0}$.

8. The equation $z^2+3z=-\frac{1}{4}$ is already in the form of $z^2+bz=-c$.

First - Complete the square and simplify. $z^2+3z=-\frac{1}{4}$; $z^2+3z+\left(\frac{3}{2}\right)^2=-\frac{1}{4}+\left(\frac{3}{2}\right)^2$; $z^2+3z+\frac{9}{4}=-\frac{1}{4}+\frac{9}{4}$

; $z^2+3z+\frac{9}{4}=\frac{-1+9}{4}$; $z^2+3z+\frac{9}{4}=\frac{8}{4}$; $\left(z+\frac{3}{2}\right)^2=2$

Second - Take the square root of both sides of the equation and solve for z.

$\left(z+\frac{3}{2}\right)^2=2$; $\sqrt{\left(z+\frac{3}{2}\right)^2}=\pm\sqrt{2}$; $z+\frac{3}{2}=\pm 1.414$. Therefore,

I. $z+\frac{3}{2}=+1.414$; $z=-\frac{3}{2}+1.414$; $z=-1.5+1.414$; $z=\mathbf{-0.086}$ and II. $z+\frac{3}{2}=-1.414$; $z=-\frac{3}{2}-1.414$; ; $z=-1.5-1.414$; $z=\mathbf{-2.914}$ The solution set is $\{\mathbf{-2.914, -0.086}\}$.

Third - Check the answers and write the quadratic equation in its factored form.

I. *Let* $z=-0.1$ *in* $z^2+3z=-\frac{1}{4}$; $(-0.086)^2+3\cdot(-0.086)\overset{?}{=}-0.25$; $0.008-0.258\overset{?}{=}-0.25$; $-0.25=-0.25$

II. *Let* $z=-2.914$ *in* $z^2+3z=-\frac{1}{4}$; $(-2.914)^2+3\cdot(-2.914)\overset{?}{=}-0.25$; $8.49-8.74\overset{?}{=}-0.25$; $-0.25=-0.25$

Therefore, the equation $z^2+3z=-\frac{1}{4}$ can be factored to $\mathbf{(z+2.914)(z+0.086)=0}$.

9. First - Write the equation $z^2+\frac{5}{3}z-\frac{1}{2}=0$ in the form of $z^2+bz=-c$, i.e., $z^2+\frac{5}{3}z=\frac{1}{2}$.

Second - Complete the square and simplify. $z^2+\frac{5}{3}z=\frac{1}{2}$; $z^2+\frac{5}{3}z+\left(\frac{5}{6}\right)^2=\frac{1}{2}+\left(\frac{5}{6}\right)^2$; $z^2+\frac{5}{3}z+\frac{25}{36}=\frac{1}{2}+\frac{25}{36}$

; $z^2+\frac{5}{3}z+\frac{25}{36}=\frac{(1\cdot 36)+(25\cdot 2)}{2\cdot 36}$; $z^2+\frac{5}{3}z+\frac{25}{36}=\frac{36+50}{72}$; $z^2+\frac{5}{3}z+\frac{25}{36}=\frac{86}{72}$; $\left(z+\frac{5}{6}\right)^2=\frac{86}{72}$

Third - Take the square root of both sides of the equation and solve for z.

$\left(z+\frac{5}{6}\right)^2=\frac{86}{72}$; $\sqrt{\left(z+\frac{5}{6}\right)^2}=\pm\sqrt{\frac{86}{72}}$; $z+\frac{5}{6}=\pm\sqrt{1.194}$; $z+0.83=\pm 1.09$. Therefore,

I. $z+0.83=+1.09$; $z=1.09-0.83$; $z=\mathbf{0.26}$ and II. $z+0.83=-1.09$; $z=-1.09-0.83$; $z=\mathbf{-1.92}$

The solution set is $\{\mathbf{-1.92, 0.26}\}$.

Fourth - Check the answers and write the quadratic equation in its factored form.

I. *Let* $z=0.26$ *in* $z^2+\frac{5}{3}z-\frac{1}{2}=0$; $0.26^2+\frac{5}{3}\cdot 0.26-\frac{1}{2}\overset{?}{=}0$; $0.07+0.43-0.5\overset{?}{=}0$; $0.5-0.5\overset{?}{=}0$; $0=0$

II. *Let* $z=-1.92$ *in* $z^2+\frac{5}{3}z-\frac{1}{2}=0$; $(-1.92)^2+\frac{5}{3}\cdot(-1.92)-\frac{1}{2}\overset{?}{=}0$; $3.7-3.2-0.5\overset{?}{=}0$; $3.7-3.7\overset{?}{=}0$; $0=0$

Therefore, the equation $z^2+\frac{5}{3}z-\frac{1}{2}=0$ can be factored to $\mathbf{(z+1.92)(z-0.26)=0}$.

10. The equation $x^2-6x=-4$ is already in the form of $x^2+bx=-c$.

First - Complete the square and simplify. $x^2-6x=-4$; $x^2-6x+\left(-\frac{\overset{3}{\cancel{6}}}{\cancel{2}}\right)^2=-4+\left(-\frac{\overset{3}{\cancel{6}}}{\cancel{2}}\right)^2$; $x^2-6x+3^2=-4+3^2$

; $x^2-6x+9=-4+9$; $x^2-6x+9=5$; $(x-3)^2=5$

Second - Take the square root of both sides of the equation and solve for x.

$(x-3)^2=5$; $\sqrt{(x-3)^2}=\pm\sqrt{5}$; $x-3=\pm 2.24$. Therefore,

I. $x-3=+2.24$; $x=2.24+3$; $\mathbf{x=5.24}$ and II. $x-3=-2.24$; $x=-2.24+3$; $\mathbf{x=0.76}$

The solution set is $\{\mathbf{0.76, 5.24}\}$.

Third - Check the answers and write the quadratic equation in its factored form.

I. *Let* $x=0.76$ *in* $x^2-6x=-4$; $0.76^2-6\cdot 0.76\overset{?}{=}-4$; $0.57-4.57\overset{?}{=}-4$; $-4=-4$

II. *Let* $x=5.24$ *in* $x^2-6x=-4$; $5.24^2-6\cdot 5.24\overset{?}{=}-4$; $27.45-31.45\overset{?}{=}-4$; $-4=-4$

Therefore, the equation $x^2-6x=-4$ can be factored to $\mathbf{(x-0.76)(x-5.24)=0}$.

Section 4.4 Case II Solutions - Solving Quadratic Equations of the Form $ax^2+bx+c=0$, where $a\rangle 1$, by Completing the Square

1. First - Write the equation $4u^2+6u+1=0$ in the form of $au^2+bu=-c$, i.e., $4u^2+6u=-1$.

   Second - Divide both sides of the equation by the coefficient of u^2, i.e., $\frac{\cancel{4}}{\cancel{4}}u^2+\frac{\overset{3}{\cancel{6}}}{\underset{2}{\cancel{4}}}u=-\frac{1}{4}$; $u^2+\frac{3}{2}u=-\frac{1}{4}$

   Third - Complete the square and simplify. $u^2+\frac{3}{2}u=-\frac{1}{4}$; $u^2+\frac{3}{2}u+\left(\frac{3}{4}\right)^2=-\frac{1}{4}+\left(\frac{3}{4}\right)^2$; $u^2+\frac{3}{2}u+\frac{9}{16}=-\frac{1}{4}+\frac{9}{16}$

   ; $\left(u+\frac{3}{4}\right)^2=\frac{(-1\cdot16)+(9\cdot4)}{4\cdot16}$; $\left(u+\frac{3}{4}\right)^2=\frac{-16+36}{64}$; $\left(u+\frac{3}{4}\right)^2=\frac{\overset{5}{\cancel{20}}}{\underset{16}{\cancel{64}}}$; $\left(u+\frac{3}{4}\right)^2=\frac{5}{16}$

   Fourth - Take the square root of both sides of the equation and solve for u.

   $\left(u+\frac{3}{4}\right)^2=\frac{5}{16}$; $\sqrt{\left(u+\frac{3}{4}\right)^2}=\pm\sqrt{\frac{5}{16}}$; $u+\frac{3}{4}=\pm\sqrt{0.313}$; $u+0.75=\pm0.56$. Therefore,

   I. $u+0.75=+0.56$; $u=0.56-0.75$; $\boldsymbol{u=-0.19}$ and II. $u+0.75=-0.56$; $u=-0.56-0.75$; $\boldsymbol{u=-1.31}$

   The solution set is $\{\mathbf{-1.31,\ -0.19}\}$.

   Fifth - Check the answers and write the quadratic equation in its factored form.

   I. *Let* $u=-0.19$ *in* $4u^2+6u+1=0$; $4\cdot(-0.19)^2+6\cdot-0.19+1\overset{?}{=}0$; $4\cdot0.036-1.14+1\overset{?}{=}0$; $0.14-1.14+1\overset{?}{=}0$; $1.14-1.14\overset{?}{=}0$; $0=0$

   II. *Let* $u=-1.31$ *in* $4u^2+6u+1=0$; $4\cdot(-1.31)^2+6\cdot-1.31+1\overset{?}{=}0$; $4\cdot1.716-7.86+1\overset{?}{=}0$; $6.86-7.86+1\overset{?}{=}0$; $7.86-7.86\overset{?}{=}0$; $0=0$

   Therefore, the equation $4u^2+6u+1=0$ can be factored to $\mathbf{(u+1.31)(u+0.19)=0}$.

2. The equation $4w^2+10w=-3$ is already in standard form of $aw^2+bw=-c$.

   First - Divide both sides of the equation by the coefficient of w^2, i.e., $\frac{\cancel{4}}{\cancel{4}}w^2+\frac{\overset{5}{\cancel{10}}}{\underset{2}{\cancel{4}}}w=-\frac{3}{4}$; $w^2+\frac{5}{2}w=-\frac{3}{4}$

   Second - Complete the square and simplify. $w^2+\frac{5}{2}w=-\frac{3}{4}$; $w^2+\frac{5}{2}w+\left(\frac{5}{4}\right)^2=-\frac{3}{4}+\left(\frac{5}{4}\right)^2$; $w^2+\frac{5}{2}w+\frac{25}{16}=-\frac{3}{4}+\frac{25}{16}$

   ; $\left(w+\frac{5}{4}\right)^2=\frac{(-3\cdot16)+(25\cdot4)}{4\cdot16}$; $\left(w+\frac{5}{4}\right)^2=\frac{-48+100}{64}$; $\left(w+\frac{5}{4}\right)^2=\frac{\overset{13}{\cancel{52}}}{\underset{16}{\cancel{64}}}$; $\left(w+\frac{5}{4}\right)^2=\frac{13}{16}$

   Third - Take the square root of both sides of the equation and solve for w.

   $\left(w+\frac{5}{4}\right)^2=\frac{13}{16}$; $\sqrt{\left(w+\frac{5}{4}\right)^2}=\pm\sqrt{\frac{13}{16}}$; $w+\frac{5}{4}=\pm\sqrt{0.813}$; $w+1.25=\pm0.9$. Therefore,

   I. $w+1.25=+0.9$; $w=0.9-1.25$; $\boldsymbol{w=-0.35}$ and II. $w+1.25=-0.9$; $w=-0.9-1.25$; $\boldsymbol{w=-2.15}$

   The solution set is $\{\mathbf{-2.15,\ -0.35}\}$.

   Fourth - Check the answers and write the quadratic equation in its factored form.

   I. *Let* $w=-0.35$ *in* $4w^2+10w=-3$; $4\cdot(-0.35)^2+10\cdot-0.35\overset{?}{=}-3$; $4\cdot0.123-3.5\overset{?}{=}-3$; $0.5-3.5\overset{?}{=}-3$; $-3=-3$

   II. *Let* $w=-2.15$ *in* $4w^2+10w=-3$; $4\cdot(-2.15)^2+10\cdot-2.15\overset{?}{=}-3$; $4\cdot4.62-21.5\overset{?}{=}-3$; $18.5-21.5\overset{?}{=}-3$; $-3=-3$

   Therefore, the equation $4w^2+10w=-3$ can be factored to $\mathbf{(w+2.15)(w+0.35)=0}$.

3. First - Write the equation $6x^2+4x-2=0$ in the form of $ax^2+bx=-c$, i.e., $6x^2+4x=2$.

Second - Divide both sides of the equation by the coefficient of x^2, i.e., $\frac{\cancel{6}}{\cancel{6}}x^2+\frac{\cancel{4}^{\,2}}{\cancel{6}_{\,3}}x=\frac{\cancel{2}}{\cancel{6}_{\,3}}$; $x^2+\frac{2}{3}x=\frac{1}{3}$

Third - Complete the square and simplify. $x^2+\frac{2}{3}x=\frac{1}{3}$; $x^2+\frac{2}{3}x+\left(\frac{2}{\cancel{6}_{\,3}}\right)^2=\frac{1}{3}+\left(\frac{2}{\cancel{6}_{\,3}}\right)^2$; $x^2+\frac{2}{3}x+\left(\frac{1}{3}\right)^2=\frac{1}{3}+\left(\frac{1}{3}\right)^2$

; $x^2+\frac{2}{3}x+\frac{1}{9}=\frac{1}{3}+\frac{1}{9}$; $\left(x+\frac{1}{3}\right)^2=\frac{(1\cdot 9)+(1\cdot 3)}{3\cdot 9}$; $\left(x+\frac{1}{3}\right)^2=\frac{9+3}{27}$; $\left(x+\frac{1}{3}\right)^2=\frac{12}{27}$

Fourth - Take the square root of both sides of the equation and solve for x.

$\left(x+\frac{1}{3}\right)^2=\frac{12}{27}$; $\sqrt{\left(x+\frac{1}{3}\right)^2}=\pm\sqrt{\frac{12}{27}}$; $x+\frac{1}{3}=\pm\sqrt{0.44}$; $x+0.33=\pm 0.66$. Therefore,

I. $x+0.33=+0.66$; $x=0.66-0.33$; $\mathbf{x=0.33}$ and II. $x+0.33=-0.66$; $x=-0.66-0.33$; $\mathbf{x=-1}$

The solution set is $\{\mathbf{-1, 0.33}\}$.

Fifth - Check the answers and write the quadratic equation in its factored form.

I. *Let* $x=0.33$ *in* $6x^2+4x-2=0$; $6\cdot(0.33)^2+4\cdot 0.33-2\overset{?}{=}0$; $6\cdot 0.11+1.32-2\overset{?}{=}0$; $0.66+1.32-2\overset{?}{=}0$

; $2-2\overset{?}{=}0$; $0=0$

II. *Let* $x=-1$ *in* $6x^2+4x-2=0$; $6\cdot(-1)^2+4\cdot -1-2\overset{?}{=}0$; $6\cdot 1-4-2\overset{?}{=}0$; $6-4-2\overset{?}{=}0$; $6-6\overset{?}{=}0$; $0=0$

Therefore, the equation $6x^2+4x-2=0$ can be factored to $\mathbf{(x-0.33)(x+1)=0}$.

4. First - Write the equation $15y^2+3=-14y$ in the form of $ay^2+by=-c$, i.e., $15y^2+14y=-3$.

Second - Divide both sides of the equation by the coefficient of y^2, i.e., $\frac{\cancel{15}}{\cancel{15}}y^2+\frac{14}{15}y=-\frac{\cancel{3}}{\cancel{15}_{\,5}}$; $y^2+\frac{14}{15}y=-\frac{1}{5}$

Third - Complete the square. $y^2+\frac{14}{15}y=-\frac{1}{5}$; $y^2+\frac{14}{15}y+\left(\frac{14}{30}\right)^2=-\frac{1}{5}+\left(\frac{14}{30}\right)^2$; $y^2+\frac{14}{15}y+\frac{196}{900}=-\frac{1}{5}+\frac{196}{900}$

; $\left(y+\frac{\cancel{14}^{\,7}}{\cancel{30}_{\,15}}\right)^2=\frac{(-1\cdot 900)+(196\cdot 5)}{5\cdot 900}$; $\left(y+\frac{7}{15}\right)^2=\frac{-900+980}{4500}$; $\left(y+\frac{7}{15}\right)^2=\frac{\cancel{80}^{\,4}}{\cancel{4500}_{\,225}}$; $\left(y+\frac{7}{15}\right)^2=\frac{4}{225}$

Fourth - Take the square root of both sides of the equation and solve for y.

$\left(y+\frac{7}{15}\right)^2=\frac{4}{225}$; $\sqrt{\left(y+\frac{7}{15}\right)^2}=\pm\sqrt{\frac{4}{225}}$; $y+\frac{7}{15}=\pm\sqrt{0.02}$; $y+0.46=\pm 0.13$. Therefore,

I. $y+0.46=+0.13$; $y=0.13-0.46$; $\mathbf{y=-0.33}$ and II. $y+0.46=-0.13$; $y=-0.13-0.46$; $\mathbf{y=-0.59}$

The solution set is $\{\mathbf{-0.59, -0.33}\}$.

Fifth - Check the answers and write the quadratic equation in its factored form.

I. *Let* $y=-0.33$ *in* $15y^2+3=-14y$; $15\cdot(-0.33)^2+3\overset{?}{=}-14\cdot -0.33$; $15\cdot 0.108+3\overset{?}{=}4.62$; $1.62+3\overset{?}{=}4.62$

; $4.62=4.62$

II. *Let* $y=-0.59$ *in* $15y^2+3=-14y$; $15\cdot(-0.59)^2+3\overset{?}{=}-14\cdot -0.59$; $15\cdot 0.348+3\overset{?}{=}8.26$; $5.23+3\overset{?}{=}8.26$

; $8.26=8.26$

Therefore, the equation $15y^2+3=-14y$ can be factored to $\mathbf{(y+0.59)(w+0.33)=0}$.

5. First - Write the equation $2x^2-5x+3=0$ in the form of $ax^2+bx=-c$, i.e., $2x^2-5x=-3$.

Second - Divide both sides of the equation by the coefficient of x^2, i.e., $\frac{\cancel{2}}{\cancel{2}}x^2-\frac{5}{2}x=-\frac{3}{2}$; $x^2-\frac{5}{2}x=-\frac{3}{2}$

Third - Complete the square and simplify. $x^2 - \frac{5}{2}x = -\frac{3}{2}$; $x^2 - \frac{5}{2}x + \left(-\frac{5}{4}\right)^2 = -\frac{3}{2} + \left(-\frac{5}{4}\right)^2$; $x^2 - \frac{5}{2}x + \frac{25}{16} = -\frac{3}{2} + \frac{25}{16}$

; $\left(x - \frac{5}{4}\right)^2 = \frac{(-3 \cdot 16) + (2 \cdot 25)}{2 \cdot 16}$; $\left(x - \frac{5}{4}\right)^2 = \frac{-48 + 50}{32}$; $\left(x - \frac{5}{4}\right)^2 = \frac{\cancel{2}}{\underset{16}{\cancel{32}}}$; $\left(x - \frac{5}{4}\right)^2 = \frac{1}{16}$

Fourth - Take the square root of both sides of the equation and solve for x.

$\left(x - \frac{5}{4}\right)^2 = \frac{1}{16}$; $\sqrt{\left(x - \frac{5}{4}\right)^2} = \pm\sqrt{\frac{1}{16}}$; $x - \frac{5}{4} = \pm\frac{1}{4}$; $x - 1.25 = \pm 0.25$. Therefore,

I. $x - 1.25 = +0.25$; $x = 0.25 + 1.25$; $\mathbf{x = 1.5}$ and II. $x - 1.25 = -0.25$; $x = -0.25 + 1.25$; $\mathbf{x = 1}$

The solution set is $\{\mathbf{1, 1.5}\}$.

Fifth - Check the answers and write the quadratic equation in its factored form.

I. *Let* $x = 1$ *in* $2x^2 - 5x + 3 = 0$; $2 \cdot 1^2 - 5 \cdot 1 + 3 \overset{?}{=} 0$; $2 \cdot 1 - 5 + 3 \overset{?}{=} 0$; $2 - 5 + 3 \overset{?}{=} 0$; $5 - 5 \overset{?}{=} 0$; $0 = 0$

II. *Let* $x = 1.5$ *in* $2x^2 - 5x + 3 = 0$; $2 \cdot 1.5^2 - 5 \cdot 1.5 + 3 \overset{?}{=} 0$; $2 \cdot 2.25 - 7.5 + 3 \overset{?}{=} 0$; $4.5 - 7.5 + 3 \overset{?}{=} 0$; $7.5 - 7.5 \overset{?}{=} 0$; $0 = 0$

Therefore, the equation $2x^2 - 5x + 3 = 0$ can be factored to $\mathbf{(x-1)(x-1.5) = 0}$.

6. First - Write the equation $2x^2 + xy - y^2 = 0$, where x is variable, in the form of $ax^2 + bx = -c$, i.e., $2x^2 + yx = y^2$.

Second - Divide both sides of the equation by the coefficient of x^2, i.e., $\frac{\cancel{2}}{\cancel{2}}x^2 + \frac{y}{2}x = \frac{y^2}{2}$; $x^2 + \frac{y}{2}x = \frac{y^2}{2}$

Third - Complete the square and simplify. $x^2 + \frac{y}{2}x = \frac{y^2}{2}$; $x^2 + \frac{y}{2}x + \left(\frac{y}{4}\right)^2 = \frac{y^2}{2} + \left(\frac{y}{4}\right)^2$; $x^2 + \frac{y}{2}x + \left(\frac{y}{4}\right)^2 = \frac{y^2}{2} + \frac{y^2}{16}$

; $\left(x + \frac{y}{4}\right)^2 = \frac{(y^2 \cdot 16) + (y^2 \cdot 2)}{2 \cdot 16}$; $\left(x + \frac{y}{4}\right)^2 = \frac{16y^2 + 2y^2}{32}$; $\left(x + \frac{y}{4}\right)^2 = \frac{\overset{9}{\cancel{18}}}{\underset{16}{\cancel{32}}}y^2$; $\left(x + \frac{y}{4}\right)^2 = \frac{9}{16}y^2$

Fourth - Take the square root of both sides of the equation and solve for x.

$\left(x + \frac{y}{4}\right)^2 = \frac{9}{16}y^2$; $\sqrt{\left(x + \frac{y}{4}\right)^2} = \pm\sqrt{\frac{9}{16}y^2}$; $x + \frac{y}{4} = \pm\frac{3}{4}y$; $x + 0.25y = \pm 0.75y$. Therefore,

I. $x + 0.25y = +0.75y$; $x = 0.75y - 0.25y$; $\mathbf{x = 0.5y}$ and II. $x + 0.25y = -0.75y$; $x = -0.75y - 0.25y$; $\mathbf{x = -y}$

The solution set is $\{\mathbf{-y, 0.5y}\}$.

Fifth - Check the answers and write the quadratic equation in its factored form.

I. *Let* $x = 0.5y$ *in* $2x^2 + xy - y^2 = 0$; $2 \cdot (0.5y)^2 + (0.5y) \cdot y - y^2 \overset{?}{=} 0$; $2 \cdot 0.25y^2 + 0.5y^2 - y^2 \overset{?}{=} 0$

; $0.5y^2 + 0.5y^2 - y^2 \overset{?}{=} 0$; $y^2 - y^2 \overset{?}{=} 0$; $0 = 0$

II. *Let* $x = -y$ *in* $2x^2 + xy - y^2 = 0$; $2 \cdot (-y)^2 + (-y) \cdot y - y^2 \overset{?}{=} 0$; $2y^2 - y^2 - y^2 \overset{?}{=} 0$; $2y^2 - 2y^2 \overset{?}{=} 0$; $0 = 0$

Therefore, the equation $2x^2 + xy - y^2 = 0$ can be factored to $\mathbf{(x+y)(x-0.5y) = 0}$.

7. First - Write the equation $6x^2 + 7x - 3 = 0$ in the form of $ax^2 + bx = -c$, i.e., $6x^2 + 7x = 3$.

Second - Divide both sides of the equation by the coefficient of x^2, i.e., $\frac{\cancel{6}}{\cancel{6}}x^2 + \frac{7}{6}x = \frac{\overset{1}{\cancel{3}}}{\underset{2}{\cancel{6}}}$; $x^2 + \frac{7}{6}x = \frac{1}{2}$

Third - Complete the square and simplify. $x^2 + \frac{7}{6}x = \frac{1}{2}$; $x^2 + \frac{7}{6}x + \left(\frac{7}{12}\right)^2 = \frac{1}{2} + \left(\frac{7}{12}\right)^2$; $x^2 + \frac{7}{6}x + \frac{49}{144} = \frac{1}{2} + \frac{49}{144}$

; $\left(x + \frac{7}{12}\right)^2 = \frac{(1 \cdot 144) + (49 \cdot 2)}{2 \cdot 144}$; $\left(x + \frac{7}{12}\right)^2 = \frac{144 + 98}{288}$; $\left(x + \frac{7}{12}\right)^2 = \frac{242}{288}$

Fourth - Take the square root of both sides of the equation and solve for x.

$\left(x+\frac{7}{12}\right)^2=\frac{242}{288}$; $\sqrt{\left(x+\frac{7}{12}\right)^2}=\pm\sqrt{\frac{242}{288}}$; $x+\frac{7}{12}=\pm\sqrt{0.84}$; $x+0.58=\pm0.92$. Therefore,

I. $x+0.58=+0.92$; $x=0.92-0.58$; $\boldsymbol{x=0.34}$ and II. $x+0.58=-0.92$; $x=-0.92-0.58$; $\boldsymbol{x=-1.5}$

The solution set is $\{\mathbf{-1.5, 0.34}\}$.

Fifth - Check the answers and write the quadratic equation in its factored form.

I. *Let* $x=0.34$ *in* $6x^2+7x-3=0$; $6\cdot(0.34)^2+7\cdot0.34-3\overset{?}{=}0$; $6\cdot0.115+2.38-3\overset{?}{=}0$; $0.69+2.38-3\overset{?}{=}0$; $3-3\overset{?}{=}0$; $0=0$

II. *Let* $x=-1.5$ *in* $6x^2+7x-3=0$; $6\cdot(-1.5)^2+7\cdot-1.5-3\overset{?}{=}0$; $6\cdot2.25-10.5-3\overset{?}{=}0$; $13.5-10.5-3\overset{?}{=}0$; $13.5-13.5\overset{?}{=}0$; $0=0$

Therefore, the equation $6x^2+7x-3=0$ can be factored to $\boldsymbol{(x+1.5)(x-0.34)=0}$.

8. First - Write the equation $5x^2=-3x$ in the form of $ax^2+bx=-c$, i.e., $5x^2+3x=0$.

Second - Divide both sides of the equation by the coefficient of x^2, i.e., $\frac{\cancel{5}}{\cancel{5}}x^2+\frac{3}{5}x=\frac{0}{5}$; $x^2+\frac{3}{5}x=0$

Third - Complete the square and simplify. $x^2+\frac{3}{5}x=0$; $x^2+\frac{3}{5}x+\left(\frac{3}{10}\right)^2=0+\left(\frac{3}{10}\right)^2$; $x^2+\frac{3}{5}u+\frac{9}{100}=\frac{9}{100}$; $\left(x+\frac{3}{10}\right)^2=\frac{9}{100}$

Fourth - Take the square root of both sides of the equation and solve for x.

$\left(x+\frac{3}{10}\right)^2=\frac{9}{100}$; $\sqrt{\left(x+\frac{3}{10}\right)^2}=\pm\sqrt{\frac{9}{100}}$; $x+\frac{3}{10}=\pm\frac{3}{10}$. Therefore,

I. $x+\frac{3}{10}=+\frac{3}{10}$; $x=\frac{3}{10}-\frac{3}{10}$; $\boldsymbol{x=0}$ and

II. $x+\frac{3}{10}=-\frac{3}{10}$; $x=-\frac{3}{10}-\frac{3}{10}$; $x=\frac{-3-3}{10}$; $x=-\frac{\overset{3}{\cancel{6}}}{\underset{5}{\cancel{10}}}$; $x=-\frac{3}{5}$; $\boldsymbol{x=-\frac{3}{5}}$

The solution set is $\left\{\mathbf{-\frac{3}{5}, 0}\right\}$.

Fifth - Check the answers and write the quadratic equation in its factored form.

I. *Let* $x=0$ *in* $5x^2=-3x$; $5\cdot0^2\overset{?}{=}-3\cdot0$; $5\cdot0\overset{?}{=}-3\cdot0$; $0=0$

II. *Let* $x=-\frac{3}{5}$ *in* $4x^2=-3x$; $5\cdot\left(-\frac{3}{5}\right)^2\overset{?}{=}-3\cdot-\frac{3}{5}$; $\cancel{5}\cdot\frac{9}{\underset{5}{\cancel{25}}}\overset{?}{=}\frac{9}{5}$; $\frac{9}{5}=\frac{9}{5}$

Therefore, the equation $5x^2=-3x$ can be factored to $(x+0)\left(x+\frac{3}{5}\right)=0$ which is the same as $\boldsymbol{x(5x+3)=0}$.

9. First - Write the equation $3x^2+4x+5=0$ in the form of $ax^2+bx=-c$, i.e., $3x^2+4x=-5$.

Second - Divide both sides of the equation by the coefficient of x^2, i.e., $\frac{\cancel{3}}{\cancel{3}}x^2+\frac{4}{3}x=-\frac{5}{3}$; $x^2+\frac{4}{3}x=-\frac{5}{3}$

Third - Complete the square and simplify. $x^2+\frac{4}{3}x=-\frac{5}{3}$; $x^2+\frac{4}{3}x+\left(\frac{\overset{2}{\cancel{4}}}{\underset{3}{\cancel{6}}}\right)^2=-\frac{5}{3}+\left(\frac{\overset{2}{\cancel{4}}}{\underset{3}{\cancel{6}}}\right)^2$; $x^2+\frac{4}{3}x+\left(\frac{2}{3}\right)^2=-\frac{5}{3}+\left(\frac{2}{3}\right)^2$

; $x^2+\frac{4}{3}x+\frac{4}{9}=-\frac{5}{3}+\frac{4}{9}$; $\left(x+\frac{2}{3}\right)^2=\frac{(-5\cdot9)+(4\cdot3)}{3\cdot9}$; $\left(x+\frac{2}{3}\right)^2=\frac{-45+12}{27}$; $\left(x+\frac{2}{3}\right)^2=-\frac{\overset{11}{\cancel{33}}}{\underset{9}{\cancel{27}}}$; $\left(x+\frac{2}{3}\right)^2=-\frac{11}{9}$

Fourth - Take the square root of both sides of the equation and solve for x.

$$\left(x+\frac{2}{3}\right)^2=-\frac{11}{9} \; ; \; \sqrt{\left(x+\frac{2}{3}\right)^2}=\pm\sqrt{-\frac{11}{9}} \; ; \; x+\frac{2}{3}=\pm\sqrt{-1.22}$$

Since the number under the radical is a negative number (an imaginary number) therefore, **the equation $3x^2+4x+5=0$ has no real solutions.**

10. First - Write the equation $-3y^2+13y+10=0$ in the form of $ay^2+by=-c$, i.e., $-3y^2+13y=-10$.

Second - Divide both sides of the equation by the coefficient of y^2, i.e., $\frac{-3}{-3}y^2+\frac{13}{-3}y=\frac{-10}{-3}$; $y^2-\frac{13}{3}y=\frac{10}{3}$

Third - Complete the square and simplify. $y^2-\frac{13}{3}y=\frac{10}{3}$; $y^2-\frac{13}{3}y+\left(-\frac{13}{6}\right)^2=\frac{10}{3}+\left(-\frac{13}{6}\right)^2$

; $y^2-\frac{13}{3}y+\frac{169}{36}=\frac{10}{3}+\frac{169}{36}$; $\left(y-\frac{13}{6}\right)^2=\frac{(10\cdot 36)+(169\cdot 3)}{3\cdot 36}$; $\left(y-\frac{13}{6}\right)^2=\frac{360+507}{108}$; $\left(y-\frac{13}{6}\right)^2=\frac{867}{108}$

Fourth - Take the square root of both sides of the equation and solve for y.

$\left(y-\frac{13}{6}\right)^2=\frac{867}{108}$; $\sqrt{\left(y-\frac{13}{6}\right)^2}=\pm\sqrt{\frac{867}{108}}$; $y-\frac{13}{6}=\pm\sqrt{8.03}$; $y-2.17=\pm 2.83$. Therefore,

I. $y-2.17=+2.83$; $y=2.83+2.17$; $\mathbf{y=5}$ and II. $y-2.17=-2.83$; $y=-2.83+2.17$; $\mathbf{y=-0.66}$

The solution set is $\{\mathbf{-0.66, 5}\}$.

Fifth - Check the answers and write the quadratic equation in its factored form.

I. *Let* $y=5$ *in* $-3y^2+13y+10=0$; $-3\cdot 5^2+13\cdot 5+10\overset{?}{=}0$; $-3\cdot 25+65+10\overset{?}{=}0$; $-75+65+10\overset{?}{=}0$; $-75+75\overset{?}{=}0$; $0=0$

II. *Let* $y=-0.66$ *in* $-3y^2+13y+10=0$; $-3\cdot(-0.66)^2+13\cdot -0.66+10\overset{?}{=}0$; $-3\cdot 0.436-8.58+10\overset{?}{=}0$; $-1.32-8.58+10\overset{?}{=}0$; $-10+10\overset{?}{=}0$; $0=0$

Therefore, the equation $-3y^2+13y+10=0$ can be factored to $\mathbf{(y+0.66)(y-5)=0}$.

Section 4.5 Case I Solutions - Solving Quadratic Equations Containing Radicals

1. First - Move $-y+2$ terms of the equation $\sqrt{-9y+28}-y+2=0$ to the right hand side of the equation to obtain $\sqrt{-9y+28}=y-2$

Second - Square both sides of the equation $\sqrt{-9y+28}=y-2$; $\left(\sqrt{-9y+28}\right)^2=(y-2)^2$; $-9y+28=(y-2)^2$

Third - Complete the square on the right hand side of the equation and simplify. $-9y+28=(y-2)^2$; $-9y+28=y^2+4-4y$; $0=y^2+(4-28)+(9y-4y)$; $0=y^2-24+5y$

Fourth - Write the quadratic equation $0=y^2-24+5y$ in standard form, i.e., $y^2+5y-24=0$

Fifth - Solve the quadratic equation by choosing a solution method. $y^2+5y-24=0$; $(y-3)(y+8)=0$.

Therefore, the two apparent solutions are: $y-3=0$; $y=3$ and $y+8=0$; $y=-8$

Sixth - Check the answers by substituting the y values into the original equation.

I. *Let* $y=3$ *in* $\sqrt{-9y+28}-y+2=0$; $\sqrt{-9\cdot 3+28}-3+2\overset{?}{=}0$; $\sqrt{-27+28}-1\overset{?}{=}0$; $\sqrt{1}-1\overset{?}{=}0$; $1-1\overset{?}{=}0$; $0=0$

II. *Let* $y=-8$ *in* $\sqrt{-9y+28}-y+2=0$; $\sqrt{-9\cdot(-8)+28}-(-8)+2\overset{?}{=}0$; $\sqrt{72+28}+8+2\overset{?}{=}0$; $\sqrt{100}+10\overset{?}{=}0$; $\sqrt{10^2}+10\overset{?}{=}0$; $10+10\overset{?}{=}0$; $20\neq 0$

Therefore, $\mathbf{y=3}$ is the only real solution to the equation $\sqrt{-9y+28}-y+2=0$.

2. First - Square both sides of the equation. $2x = \sqrt{9x+3}$; $(2x)^2 = \left(\sqrt{9x+3}\right)^2$; $4x^2 = 9x + 3$

Second - Write the quadratic equation $4x^2 = 9x + 3$ in standard form, i.e., $4x^2 - 9x - 3 = 0$

Third - Solve the quadratic equation by choosing a solution method. $4x^2 - 9x - 3 = 0$; $(x+0.3)(x-2.55) = 0$.

Therefore, the two apparent solutions are: $x + 0.3 = 0$; $x = -0.3$ and $x - 2.55 = 0$; $x = 2.55$

Fourth - Check the answers by substituting the x values into the original equation.

I. *Let* $x = -0.3$ *in* $2x = \sqrt{9x+3}$; $2\cdot(-0.3) \overset{?}{=} \sqrt{9\cdot(-0.3)+3}$; $-0.6 \overset{?}{=} \sqrt{-2.7+3}$; $-0.6 \overset{?}{=} \sqrt{0.3}$; $-0.6 \neq 0.547$

II. *Let* $x = 2.55$ *in* $2x = \sqrt{9x+3}$; $2\cdot 2.55 \overset{?}{=} \sqrt{9\cdot 2.55+3}$; $5.1 \overset{?}{=} \sqrt{22.95+3}$; $5.1 \overset{?}{=} \sqrt{25.95}$; $5.1 = 5.1$

Therefore, the equation $2x = \sqrt{9x+3}$ has one real solution, i.e., $\mathbf{x = 2.55}$.

3. First - Write the quadratic equation $t^2 = -\sqrt{5}t$ in standard form, i.e., $t^2 + \sqrt{5}t = 0$

Second - Solve the quadratic equation by choosing a solution method. $t^2 + \sqrt{5}t = 0$; $t\left(t+\sqrt{5}\right) = 0$.

Therefore, the two apparent solutions are: $t = 0$ and $t + \sqrt{5} = 0$; $t = -\sqrt{5}$

Third - Check the answers by substituting the t values into the original equation.

I. *Let* $t = 0$ *in* $t^2 = -\sqrt{5}t$; $0^2 \overset{?}{=} -\sqrt{5}\cdot 0$; $0 = 0$

II. *Let* $t = 0$ *in* $t^2 = -\sqrt{5}t$; $\left(-\sqrt{5}\right)^2 \overset{?}{=} -\sqrt{5}\cdot -\sqrt{5}$; $+5 \overset{?}{=} +\sqrt{5\cdot 5}$; $5 \overset{?}{=} \sqrt{5^2}$; $5 = 5$

Therefore, $\mathbf{t = 0}$ and $\mathbf{t = -\sqrt{5}}$ are the real solutions to $t^2 = -\sqrt{5}t$. Furthermore, the equation $t^2 = -\sqrt{5}t$ can be factored to $\mathbf{t\left(t+\sqrt{5}\right) = 0}$.

4. First - Write the quadratic equation $y^2 - \sqrt{8}y = 7$ in standard form. $y^2 - \sqrt{8}y - 7 = 0$

Second - Solve the quadratic equation by choosing a solution method. $y^2 - \sqrt{8}y - 7 = 0$; $(y+1.6)(y-4.4) = 0$.

Therefore, the two apparent solutions are: $y + 1.6 = 0$; $y = -1.6$ and $y - 4.4 = 0$; $y = 4.4$

Third - Check the answers by substituting the y values into the original equation.

I. *Let* $y = -1.6$ *in* $y^2 - \sqrt{8}y = 7$; $(-1.6)^2 - \sqrt{8}\cdot(-1.6) \overset{?}{=} 7$; $2.56 - 2.83\cdot(-1.6) \overset{?}{=} 7$; $2.56 + 4.53 \overset{?}{=} 7$; $7 = 7$

II. *Let* $y = 4.4$ *in* $y^2 - \sqrt{8}y = 7$; $4.4^2 - \sqrt{8}\cdot 4.4 \overset{?}{=} 7$; $19.36 - 2.83\cdot 4.4 \overset{?}{=} 7$; $19.4 - 12.4 \overset{?}{=} 7$; $7 = 7$

Therefore, $\mathbf{y = -1.6}$ and $\mathbf{y = 4.4}$ are the real solutions to $y^2 - \sqrt{8}y = 7$. Furthermore, the equation $y^2 - \sqrt{8}y = 7$ can be factored to $\mathbf{(y+1.6)(y-4.4) = 0}$.

5. First - Write the quadratic equation $\sqrt{5}x = 2x^2$ in standard form, i.e., $2x^2 - \sqrt{5}x = 0$

Second - Solve the quadratic equation by choosing a solution method. $2x^2 - \sqrt{5}x = 0$; $x\left(2x-\sqrt{5}\right) = 0$.

Therefore, the two apparent solutions are: $x = 0$ and $2x - \sqrt{5} = 0$; $2x = \sqrt{5}$; $x = \frac{\sqrt{5}}{2}$

Third - Check the answers by substituting the x values into the original equation.

I. *Let* $x = 0$ *in* $\sqrt{5}x = 2x^2$; $\sqrt{5}\cdot 0 \overset{?}{=} 2\cdot 0^2$; $0 = 0$

II. *Let* $x = \frac{\sqrt{5}}{2}$ *in* $\sqrt{5}x = 2x^2$; $\sqrt{5}\cdot\frac{\sqrt{5}}{2} \overset{?}{=} 2\cdot\left(\frac{\sqrt{5}}{2}\right)^2$; $\frac{\sqrt{5\cdot 5}}{2} \overset{?}{=} 2\cdot\frac{5}{4}$; $\frac{\sqrt{5^2}}{2} \overset{?}{=} \frac{5}{2}$; $\frac{5}{2} = \frac{5}{2}$

Therefore, $\mathbf{x = 0}$ and $\mathbf{x = \frac{\sqrt{5}}{2}}$ are the real solutions to $\sqrt{5}x = 2x^2$. Furthermore, the equation $\sqrt{5}x = 2x^2$ can be factored to $x\left(2x-\sqrt{5}\right) = 0$ which is the same as $\mathbf{x\left(x-\frac{\sqrt{5}}{2}\right) = 0}$.

6. First - Square both sides of the equation. $\sqrt{x^2-12}=2$; $\left(\sqrt{x^2-12}\right)^2=2^2$; $x^2-12=4$

Second - Write the quadratic equation $x^2-12=4$ in standard form, i.e., $x^2-12-4=0$; $x^2-16=0$

Third - Solve the quadratic equation by choosing a solution method. $x^2-16=0$; $(x-4)(x+4)=0$.

Therefore, the two apparent solutions are: $x-4=0$; $x=+4$ and $x-4=0$; $x=-4$

Fourth - Check the answers by substituting the x values into the original equation.

I. *Let* $x=-4$ *in* $\sqrt{x^2-12}=2$; $\sqrt{(-4)^2-12}\overset{?}{=}2$; $\sqrt{16-12}\overset{?}{=}2$; $\sqrt{4}\overset{?}{=}2$; $\sqrt{2^2}\overset{?}{=}2$; $2=2$

II. *Let* $x=4$ *in* $\sqrt{x^2-12}=2$; $\sqrt{4^2-12}\overset{?}{=}2$; $\sqrt{16-12}\overset{?}{=}2$; $\sqrt{4}\overset{?}{=}2$; $\sqrt{2^2}\overset{?}{=}2$; $2=2$

Therefore, $\boldsymbol{x=4}$ and $\boldsymbol{x=-4}$ are the real solutions to $\sqrt{x^2-12}=2$. Furthermore, the equation $\sqrt{x^2-12}=2$ can be factored to $\boldsymbol{(x+4)(x-4)=0}$.

7. First - Square both sides of the equation. $\sqrt{-8x-4}=2x+1$; $\left(\sqrt{-8x-4}\right)^2=(2x+1)^2$; $-8x-4=(2x+1)^2$

Second - Complete the square on the right hand side of the equation and simplify. $-8x-4=(2x+1)^2$

; $-8x-4=4x^2+1+4x$; $-8x-4x-4-1=4x^2$; $-12x-5=4x^2$

Third - Write the quadratic equation $-12x-5=4x^2$ in standard form, i.e., $4x^2+12x+5=0$

Fourth - Solve the quadratic equation by choosing a solution method. $4x^2+12x+5=0$; $\left(x+\frac{1}{2}\right)\left(x+\frac{5}{2}\right)=0$.

Therefore, the two apparent solutions are: $x+\frac{1}{2}=0$; $x=-\frac{1}{2}$ and $x+\frac{5}{2}=0$; $x=-\frac{5}{2}$

Fifth - Check the answers by substituting the x values into the original equation.

I. *Let* $x=-\frac{1}{2}$ *in* $\sqrt{-8x-4}=2x+1$; $\sqrt{-\overset{4}{\cancel{8}}\cdot\left(-\frac{1}{\cancel{2}}\right)-4}\overset{?}{=}2\cdot\left(-\frac{1}{2}\right)+1$; $\sqrt{4-4}\overset{?}{=}-1+1$; $\sqrt{0}\overset{?}{=}0$; $0=0$

II. *Let* $x=-\frac{5}{2}$ *in* $\sqrt{-8x-4}=2x+1$; $\sqrt{-\overset{4}{\cancel{8}}\cdot\left(-\frac{5}{\cancel{2}}\right)-4}\overset{?}{=}2\cdot\left(-\frac{5}{2}\right)+1$; $\sqrt{20-4}\overset{?}{=}-5+1$; $\sqrt{16}\overset{?}{=}-4$; $4\neq-4$

Therefore, the equation $\sqrt{-8x-4}=2x+1$ has one real solution, i.e., $\boldsymbol{x=-\frac{1}{2}}$.

8. First - Square both sides of the equation. $x=\sqrt{-x+2}$; $x^2=\left(\sqrt{-x+2}\right)^2$; $x^2=-x+2$

Second - Write the quadratic equation $x^2=-x+2$ in standard form, i.e., $x^2+x-2=0$

Third - Solve the quadratic equation by choosing a solution method. $x^2+x-2=0$; $(x-1)(x+2)=0$.

Therefore, the two apparent solutions are: $x-1=0$; $x=1$ and $x+2=0$; $x=-2$

Fourth - Check the answers by substituting the x values into the original equation.

I. *Let* $x=1$ *in* $x=\sqrt{-x+2}$; $1\overset{?}{=}\sqrt{-1+2}$; $1\overset{?}{=}\sqrt{1}$; $1=1$

II. *Let* $x=-2$ *in* $x=\sqrt{-x+2}$; $-2\overset{?}{=}\sqrt{-(-2)+2}$; $-2\overset{?}{=}\sqrt{2+2}$; $-2\overset{?}{=}\sqrt{4}$; $-2\overset{?}{=}\sqrt{2^2}$; $-2\neq2$

Therefore, the equation $x=\sqrt{-x+2}$ has one real solution, i.e., $\boldsymbol{x=1}$.

9. First - Square both sides of the equation. $x=\sqrt{-2x+3}$; $x^2=\left(\sqrt{-2x+3}\right)^2$; $x^2=-2x+3$

Second - Write the quadratic equation $x^2=-2x+3$ in standard form, i.e., $x^2+2x-3=0$

Third - Solve the quadratic equation by choosing a solution method. $x^2+2x-3=0$; $(x-1)(x+3)=0$.

Therefore, the two apparent solutions are: $x-1=0$; $x=1$ and $x+3=0$; $x=-3$

Fourth - Check the answers by substituting the x values into the original equation.

I. *Let* $x = 1$ *in* $x = \sqrt{-2x+3}$; $1 \overset{?}{=} \sqrt{-2 \cdot 1 + 3}$; $1 \overset{?}{=} \sqrt{-2+3}$; $1 \overset{?}{=} \sqrt{1}$; $1 = 1$

II. *Let* $x = -3$ *in* $x = \sqrt{-2x+3}$; $-3 \overset{?}{=} \sqrt{-2 \cdot -3 + 3}$; $-3 \overset{?}{=} \sqrt{6+3}$; $-3 \overset{?}{=} \sqrt{9}$; $-3 \overset{?}{=} \sqrt{3^2}$; $-3 \neq 3$

Therefore, the equation $x = \sqrt{-2x+3}$ has one real solution, i.e., $\mathbf{x = 1}$.

10. First - Square both sides of the equation. $\sqrt{x^2+3} = x+1$; $\left(\sqrt{x^2+3}\right)^2 = (x+1)^2$; $x^2+3 = (x+1)^2$

Second - Complete the square on the right hand side of the equation and simplify. $x^2+3 = (x+1)^2$; $x^2+3 = x^2+1+2x$; $x^2 - x^2 + 3 - 1 - 2x = 0$; $x^2 - x^2 + 2 - 2x = 0$; $2 - 2x = 0$

Third - Solve the equation, i.e., $2 - 2x = 0$; $2(1-x) = 0$.

Therefore, the apparent solution is: $1 - x = 0$; $-x = -1$; $x = 1$

Fourth - Check the answer by substituting the x value into the original equation.

Let $x = 1$ *in* $\sqrt{x^2+3} = x+1$; $\sqrt{1^2+3} \overset{?}{=} 1+1$; $\sqrt{1+3} \overset{?}{=} 2$; $\sqrt{4} \overset{?}{=} 2$; $\sqrt{2^2} \overset{?}{=} 2$; $2 = 2$

Therefore, $\mathbf{x = 1}$ is the real solution to $\sqrt{x^2+3} = x+1$.

Section 4.5 Case II Solutions - Solving Quadratic Equations Containing Fractions

1. $\frac{8}{y+1} = y-1$; $\frac{8}{y+1} = \frac{y-1}{1}$; $8 \cdot 1 = (y-1)\cdot(y+1)$; $8 = y^2 + y - y - 1$; $8 = y^2 - 1$; $8+1 = y^2$; $y^2 = 9$; $\sqrt{y^2} = \pm\sqrt{9}$; $y = \pm\sqrt{3^2}$; $y = \pm 3$

Therefore, the two solutions are $\mathbf{y = +3}$ and $\mathbf{y = -3}$. In addition, the fractional equation $\frac{8}{y+1} = y-1$ can be expressed in factored form as $\mathbf{(y-3)(y+3) = 0}$.

Check: I. *Let* $y = +3$ *in* $\frac{8}{y+1} = y-1$; $\frac{8}{3+1} \overset{?}{=} 3-1$; $\frac{\overset{2}{\cancel{8}}}{\cancel{4}} \overset{?}{=} 2$; $\frac{2}{1} \overset{?}{=} 2$; $2 = 2$

II. *Let* $y = -3$ *in* $\frac{8}{y+1} = y-1$; $\frac{8}{-3+1} \overset{?}{=} -3-1$; $-\frac{\overset{4}{\cancel{8}}}{\cancel{2}} \overset{?}{=} -4$; $-\frac{4}{1} \overset{?}{=} -4$; $-4 = -4$

2. $\frac{11x+15}{x} = -2x$; $\frac{11x+15}{x} = -\frac{2x}{1}$; $(11x+15)\cdot 1 = 2x \cdot x$; $11x + 15 = 2x^2$; $2x^2 - 11x - 15 = 0$; $\mathbf{(x+3)(2x+5) = 0}$

Therefore, the two solutions are $x + 3 = 0$; $\mathbf{x = -3}$ and $2x + 5 = 0$; $2x = -5$; $x = -\frac{5}{2}$; $\mathbf{x = -2.5}$.

Check: I. *Let* $x = -3$ *in* $\frac{11x+15}{x} = -2x$; $\frac{11\cdot(-3)+15}{-3} \overset{?}{=} -2\cdot(-3)$; $\frac{-33+15}{-3} \overset{?}{=} 6$; $\frac{-18}{-3} \overset{?}{=} 6$; $\frac{\overset{6}{\cancel{18}}}{\cancel{3}} \overset{?}{=} 6$; $\frac{6}{1} \overset{?}{=} 6$; $6 = 6$

II. *Let* $x = -2.5$ *in* $\frac{11x+15}{x} = -2x$; $\frac{11\cdot(-2.5)+15}{-2.5} \overset{?}{=} -2\cdot(-2.5)$; $\frac{-27.5+15}{-2.5} \overset{?}{=} 5$; $\frac{-12.5}{-2.5} \overset{?}{=} 5$; $\frac{\overset{5}{\cancel{12.5}}}{\cancel{2.5}} \overset{?}{=} 5$; $\frac{5}{1} \overset{?}{=} 5$; $5 = 5$

3. $\frac{x^2}{x+3} = \frac{1}{x+3}$; $\frac{x^2}{x+3} - \frac{1}{x+3} = 0$; $\frac{x^2-1}{x+3} = 0$; $\frac{x^2-1}{x+3} = \frac{0}{1}$; $(x^2-1)\cdot 1 = 0\cdot(x+3)$; $x^2 - 1 = 0$; $x^2 = 1$; $\sqrt{x^2} = \pm\sqrt{1}$; $x = \pm 1$. Therefore, the two solutions are $\mathbf{x = +1}$ and $\mathbf{x = -1}$. In addition, the fractional equation $\frac{x^2}{x+3} = \frac{1}{x+3}$ can be expressed in factored form as $\mathbf{(x+1)(x-1) = 0}$.

Check: I. *Let* $x = +1$ *in* $\frac{x^2}{x+3} = \frac{1}{x+3}$; $\frac{1^2}{1+3} \overset{?}{=} \frac{1}{1+3}$; $\frac{1}{4} = \frac{1}{4}$

II. *Let* $x = -1$ *in* $\frac{x^2}{x+3} = \frac{1}{x+3}$; $\frac{(-1)^2}{-1+3} \overset{?}{=} \frac{1}{-1+3}$; $\frac{1}{2} = \frac{1}{2}$

4. $\frac{1-2u}{u}=-u$; $\frac{1-2u}{u}=-\frac{u}{1}$; $(1-2u)\cdot 1=-u\cdot u$; $1-2u=-u^2$; $u^2-2u+1=0$; $(u-1)^2=0$; $\sqrt{(u-1)^2}=\pm\sqrt{0}$; $u-1=\pm 0$; $u=1$

Therefore, there are two repeated solution of $\boldsymbol{u=1}$. In addition, the fractional equation $\frac{1-2u}{u}=-u$ can be expressed in factored form as $(\boldsymbol{u-1})\cdot(\boldsymbol{u-1})=\boldsymbol{0}$.

Check: *Let* $u=+1$ *in* $\frac{1-2u}{u}=-u$; $\frac{1-(2\cdot 1)}{1}\overset{?}{=}-1$; $\frac{1-2}{1}\overset{?}{=}-1$; $-\frac{1}{1}\overset{?}{=}-1$; $-1=-1$

5. $x=\frac{3}{x}-2$; $x+2=\frac{3}{x}$; $\frac{x+2}{1}=\frac{3}{x}$; $(x+2)\cdot x=1\cdot 3$; $x^2+2x=3$; $x^2+2x-3=0$; $(\boldsymbol{x+3})(\boldsymbol{x-1})=\boldsymbol{0}$

Therefore, the two solutions are $x+3=0$; $\boldsymbol{x=-3}$ and $x-1=0$; $\boldsymbol{x=1}$.

Check: I. *Let* $x=-3$ *in* $x=\frac{3}{x}-2$; $-3\overset{?}{=}\frac{3}{-3}-2$; $-3\overset{?}{=}-1-2$; $-3=-3$

II. *Let* $x=1$ *in* $x=\frac{3}{x}-2$; $1\overset{?}{=}\frac{3}{1}-2$; $1\overset{?}{=}3-2$; $1=1$

6. $\frac{3x-10}{x}=-x$; $\frac{3x-10}{x}=-\frac{x}{1}$; $1\cdot(3x-10)=-x\cdot x$; $3x-10=-x^2$; $x^2+3x-10=0$; $(\boldsymbol{x+5})(\boldsymbol{x-2})=\boldsymbol{0}$

Therefore, the two solutions are $x+5=0$; $\boldsymbol{x=-5}$ and $x-2=0$; $\boldsymbol{x=2}$.

Check: I. *Let* $x=-5$ *in* $\frac{3x-10}{x}=-x$; $\frac{3\cdot(-5)-10}{-5}\overset{?}{=}-(-5)$; $\frac{-15-10}{-5}\overset{?}{=}5$; $\frac{-25}{-5}\overset{?}{=}5$; $\frac{\overset{5}{\cancel{25}}}{\cancel{5}}\overset{?}{=}5$; $\frac{5}{1}\overset{?}{=}5$; $5=5$

II. *Let* $x=2$ *in* $\frac{3x-10}{x}=-x$; $\frac{3\cdot 2-10}{2}\overset{?}{=}-2$; $\frac{6-10}{2}\overset{?}{=}-2$; $-\frac{\overset{2}{\cancel{4}}}{\cancel{2}}\overset{?}{=}-2$; $-\frac{2}{1}\overset{?}{=}-2$; $-2=-2$

7. $u=\frac{49}{u}$; $\frac{u}{1}=\frac{49}{u}$; $u\cdot u=49\cdot 1$; $u^2=49$; $\sqrt{u^2}=\pm\sqrt{49}$; $u=\pm\sqrt{7^2}$; $u=\pm 7$

Therefore, the two solutions are $\boldsymbol{u=+7}$ and $\boldsymbol{u=-7}$. In addition, the fractional equation $u=\frac{49}{u}$ can be expressed in factored form as $(\boldsymbol{u-7})(\boldsymbol{u+7})=\boldsymbol{0}$.

Check: I. *Let* $u=+7$ *in* $u=\frac{49}{u}$; $7\overset{?}{=}\frac{\overset{7}{\cancel{49}}}{\cancel{7}}$; $7\overset{?}{=}\frac{7}{1}$; $7=7$

II. *Let* $u=-7$ *in* $u=\frac{49}{u}$; $-7\overset{?}{=}\frac{49}{-7}$; $\frac{-7}{1}\overset{?}{=}\frac{49}{-7}$; $-7\cdot -7\overset{?}{=}49\cdot 1$; $49=49$

8. $6x+17=-\frac{5}{x}$; $\frac{6x+17}{1}=-\frac{5}{x}$; $(6x+17)\cdot x=-5\cdot 1$; $6x^2+17x=-5$; $6x^2+17x+5=0$; $(\boldsymbol{2x+5})(\boldsymbol{3x+1})=\boldsymbol{0}$

Therefore, the two solutions are $2x+5=0$; $2x=-5$; $x=-\frac{5}{2}$; $\boldsymbol{x=-2.5}$ and $3x+1=0$; $3x=-1$; $x=-\frac{1}{3}$; $\boldsymbol{x=-0.333}$.

Check: I. *Let* $x=-2.5$ *in* $6x+17=-\frac{5}{x}$; $(6\cdot -2.5)+17\overset{?}{=}-\frac{5}{-2.5}$; $-15+17\overset{?}{=}\frac{5}{2.5}$; $2=2$

II. *Let* $x=-0.333$ *in* $6x+17=-\frac{5}{x}$; $(6\cdot -0.333)+17\overset{?}{=}-\frac{5}{-0.333}$; $-1.98+17\overset{?}{=}\frac{5}{0.333}$; $15.02=15.02$

9. $y+4=-\frac{3}{y}$; $\frac{y+4}{1}=-\frac{3}{y}$; $(y+4)\cdot y=-3\cdot 1$; $y^2+4y=-3$; $y^2+4y+3=0$; $(\boldsymbol{y+3})(\boldsymbol{y+1})=\boldsymbol{0}$

Therefore, the two solutions are $y+3=0$; $\boldsymbol{y=-3}$, and $y+1=0$; $\boldsymbol{y=-1}$.

Check: I. *Let* $y=-3$ *in* $y+4=-\frac{3}{y}$; $-3+4\overset{?}{=}\frac{-3}{-3}$; $1\overset{?}{=}\frac{\cancel{3}}{\cancel{3}}$; $1\overset{?}{=}\frac{1}{1}$; $1=1$

II. *Let* $y=-1$ *in* $y+4=-\frac{3}{y}$; $-1+4\overset{?}{=}\frac{-3}{-1}$; $3\overset{?}{=}\frac{3}{1}$; $3=3$

10. $3x = \frac{-5x-2}{x}$; $\frac{3x}{1} = \frac{-5x-2}{x}$; $3x \cdot x = (-5x-2) \cdot 1$; $3x^2 = -5x-2$; $3x^2 + 5x + 2 = 0$; $\mathbf{(3x+2)(x+1) = 0}$

Therefore, the two solutions are $3x + 2 = 0$; $3x = -2$; $x = -\frac{2}{3}$; $\mathbf{x = -0.67}$ and $x + 1 = 0$; $\mathbf{x = -1}$.

Check: I. *Let* $x = -0.67$ *in* $3x = \frac{-5x-2}{x}$; $3 \times -0.67 \stackrel{?}{=} \frac{(-5 \times -0.67) - 2}{-0.67}$; $-0.21 \stackrel{?}{=} \frac{3.35 - 2}{-0.67}$; $-0.21 \stackrel{?}{=} -\frac{1.35}{0.67}$; $-0.21 = -0.21$

II. *Let* $x = -1$ *in* $3x = \frac{-5x-2}{x}$; $3 \times -1 \stackrel{?}{=} \frac{(-5 \times -1) - 2}{-1}$; $-3 \stackrel{?}{=} \frac{5-2}{-1}$; $-3 \stackrel{?}{=} -\frac{3}{1}$; $-3 = -3$

Section 4.6 Solutions - How to Choose the Best Factoring or Solution Method

1. **First Method:** (The Trial and Error Method)

Write the equation $x^2 = 16$ in the standard quadratic equation form $ax^2 + bx + c = 0$, i.e., write $x^2 = 16$ as $x^2 + 0x - 16 = 0$. Consider the left hand side of the equation which is a polynomial. To factor the given polynomial we need to obtain two numbers whose sum is 0 and whose product is -16. Let's construct a table as follows:

Sum	*Product*
$1 - 1 = 0$	$1 \cdot (-1) = -1$
$2 - 2 = 0$	$2 \cdot (-2) = -4$
$3 - 3 = 0$	$3 \cdot (-3) = -9$
$\mathbf{4 - 4 = 0}$	$\mathbf{4 \cdot (-4) = -16}$

The last line contains the sum and the product of the two numbers that we need. Thus, $x^2 = 16$ or $x^2 + 0x - 16 = 0$ can be factored to $\mathbf{(x-4)(x+4) = 0}$

Second Method: (The Quadratic Formula Method)

First, write the equation in the standard quadratic equation form $ax^2 + bx + c = 0$, i.e., write $x^2 = 16$ as $x^2 + 0x - 16 = 0$. Second, equate the coefficients of $x^2 + 0x - 16 = 0$ with the standard quadratic equation by letting $a = 1$, $b = 0$, and $c = -16$. Then,

Given: $x = \frac{-b \pm \sqrt{b^2 - 4ac}}{2a}$; $x = \frac{-0 \pm \sqrt{0^2 - (4 \times 1 \times -16)}}{2 \times 1}$; $x = \frac{\pm\sqrt{0 + 64}}{2}$; $x = \frac{\pm\sqrt{64}}{2}$; $x = \pm\frac{\sqrt{8^2}}{2}$; $x = \pm\frac{8}{2}$.

Therefore, the two solutions are $\mathbf{x = -4}$ and $\mathbf{x = 4}$ and the equation $x^2 + 0x - 16 = 0$ can be factored to $\mathbf{(x+4)(x-4) = 0}$.

Third Method: (The Square Root Property Method)

Take the square root of both sides of the equation, i.e., write $x^2 = 16$ as $\sqrt{x^2} = \pm\sqrt{16}$; $x = \pm\sqrt{4^2}$; $x = \pm 4$. Thus, $\mathbf{x = +4}$ and $\mathbf{x = -4}$ are the solution sets to the equation $x^2 = 16$ which can be represented in its factorable form as $\mathbf{(x+4)(x-4) = 0}$.

Check: $(x-4)(x+4) = 0$; $x \cdot x + 4 \cdot x - 4 \cdot x + 4 \cdot (-4) = 0$; $x^2 + 4x - 4x - 16 = 0$; $x^2 + (4-4)x - 16 = 0$; $x^2 + 0x - 16 = 0$

From the above three methods using the Square Root Property method is the easiest method to use. The Trial and Error method is the second easiest method to use. Followed by the Quadratic Formula method which is the longest and somewhat a more difficult way of obtaining the factored terms.

2. **First Method:** (The Trial and Error Method)

Consider the left hand side of the equation which is a polynomial. To factor the polynomial $x^2 + 7x + 3$ we need to obtain two numbers whose sum is 7 and whose product is 3. However, after few trials, it becomes clear that such a combination of integer numbers is not possible to obtain. Therefore, **the given equation is not factorable and is referred to as PRIME**.

Second Method: (The Quadratic Formula Method)

Given the standard quadratic equation $ax^2 + bx + c = 0$, equate the coefficients of $x^2 + 7x + 3 = 0$ with the standard quadratic equation by letting $a = 1$, $b = 7$, and $c = 3$. Then,

Given: $x = \frac{-b \pm \sqrt{b^2 - 4ac}}{2a}$; $x = \frac{-7 \pm \sqrt{7^2 - (4 \times 1 \times 3)}}{2 \times 1}$; $x = \frac{-7 \pm \sqrt{49 - 12}}{2}$; $x = \frac{-7 \pm \sqrt{37}}{2}$; $x = \frac{-7 \pm 6.08}{2}$.

Therefore, the two solutions are $\mathbf{x = -6.54}$ and $\mathbf{x = -0.46}$ and the equation $x^2 + 7x + 3 = 0$ can be factored to $\mathbf{(x + 6.54)(x + 0.46) = 0}$.

Third Method: (Completing the Square Method)

$x^2 + 7x + 3 = 0$; $x^2 + 7x = -3$; $x^2 + 7x + \left(\frac{7}{2}\right)^2 = -3 + \left(\frac{7}{2}\right)^2$; $x^2 + 7x + \frac{49}{4} = -3 + \frac{49}{4}$; $\left(x + \frac{7}{2}\right)^2 = -\frac{3}{1} + \frac{49}{4}$

; $\left(x + \frac{7}{2}\right)^2 = \frac{(-3 \cdot 4) + (1 \cdot 49)}{1 \cdot 4}$; $\left(x + \frac{7}{2}\right)^2 = \frac{-12 + 49}{4}$; $\left(x + \frac{7}{2}\right)^2 = \frac{37}{4}$; $x + \frac{7}{2} = \pm\sqrt{\frac{37}{4}}$; $x + \frac{7}{2} = \pm\frac{\sqrt{37}}{2}$. Therefore, the two solutions are $\mathbf{x = -6.54}$ and $\mathbf{x = -0.46}$ and the equation $x^2 + 7x + 3 = 0$ can be factored to $\mathbf{(x + 6.54)(x + 0.46) = 0}$.

Check: I. *Let* $x = -0.46$ *in* $x^2 + 7x + 3 = 0$; $(-0.46)^2 + 7 \cdot (-0.46) + 3 \overset{?}{=} 0$; $0.2 - 3.2 + 3 \overset{?}{=} 0$; $-3 + 3 \overset{?}{=} 0$; $0 = 0$

II. *Let* $x = -6.54$ *in* $x^2 + 7x + 3 = 0$; $(-6.54)^2 + 7 \cdot (-6.54) + 3 \overset{?}{=} 0$; $42.8 - 45.8 + 3 \overset{?}{=} 0$; $42.8 - 42.8 \overset{?}{=} 0$; $0 = 0$

Therefore, the equation $x^2 + 7x + 3 = 0$ can be factored to $(x + 0.46)(x + 6.54) = 0$.

From the above three methods using the Quadratic Formula method may be the faster method than Completing the Square method.

3. **First Method:** (The Square Root Property Method)

$(3x + 4)^2 = 36$; $\sqrt{(3x + 4)^2} = \pm\sqrt{36}$; $3x + 4 = \pm 6$; $3x = \pm 6 - 4$; $x = \frac{\pm 6 - 4}{3}$. Thus, the two solutions are $x = \frac{6 - 4}{3}$; $\mathbf{x = \frac{2}{3}}$; and $x = \frac{-6 - 4}{3}$; $\mathbf{x = -\frac{10}{3}}$ and the equation $(3x + 4)^2 = 36$ can be factored to $\left(x - \frac{2}{3}\right)\left(x + \frac{10}{3}\right) = 0$ which is the same as $\mathbf{(3x - 2)(3x + 10) = 0}$.

Second Method: (The Quadratic Formula Method)

Complete the square term on the left hand side and write the equation in standard form, i.e., $(3x + 4)^2 = 36$; $9x^2 + 24x + 16 = 36$; $9x^2 + 24x + 16 - 36 = 36 - 36$; $9x^2 + 24x - 20 = 0$.

Given the standard quadratic equation $ax^2 + bx + c = 0$, equate the coefficients of $9x^2 + 24x - 20 = 0$ with the standard quadratic equation by letting $a = 9$, $b = 24$, and $c = -20$. Then,

Given: $x = \frac{-b \pm \sqrt{b^2 - 4ac}}{2a}$; $x = \frac{-24 \pm \sqrt{24^2 - (4 \times 9 \times -20)}}{2 \times 9}$; $x = \frac{-24 \pm \sqrt{576 + 720}}{18}$; $x = \frac{-24 \pm \sqrt{1296}}{18}$

; $x = \frac{-24 \pm \sqrt{36^2}}{18}$; $x = \frac{-24 \pm 36}{18}$. Therefore, the two solutions are $x = \frac{-24 - 36}{18}$; $x = -\frac{\overset{10}{\cancel{60}}}{\underset{3}{\cancel{18}}}$; $\mathbf{x = -\frac{10}{3}}$; and

$x = \frac{-24 + 36}{18}$; $x = \frac{\overset{2}{\cancel{12}}}{\underset{3}{\cancel{18}}}$; $\mathbf{x = \frac{2}{3}}$ and the equation $(3x + 4)^2 = 36$ can be factored to $\left(x - \frac{2}{3}\right)\left(x + \frac{10}{3}\right) = 0$ which is the same as $\mathbf{(3x - 2)(3x + 10) = 0}$.

Third Method: (Completing-the-Square Method)

First complete the square term on the left hand side and simplify the equation, i.e., $(3x + 4)^2 = 36$; $9x^2 + 24x + 16 = 36$; $9x^2 + 24x + 16 - 16 = 36 - -16$; $9x^2 + 24x = 20$; $\frac{\cancel{9}}{\cancel{9}}x^2 + \frac{24}{9}x = \frac{20}{9}$; $x^2 + \frac{24}{9}x = \frac{20}{9}$.

Then, complete the square in the following way:

$$x^2+\frac{24}{9}x=\frac{20}{9}\ ;\ x^2+\frac{24}{9}x+\left(\frac{\overset{4}{\cancel{24}}}{\underset{3}{\cancel{18}}}\right)^2=\frac{20}{9}+\left(\frac{\overset{4}{\cancel{24}}}{\underset{3}{\cancel{18}}}\right)^2\ ;\ x^2+\frac{24}{9}x+\left(\frac{4}{3}\right)^2=\frac{20}{9}+\left(\frac{4}{3}\right)^2\ ;\ x^2+\frac{24}{9}x+\frac{16}{9}=\frac{20}{9}+\frac{16}{9}$$

$$;\ \left(x+\frac{4}{3}\right)^2=\frac{20+16}{9}\ ;\ \left(x+\frac{4}{3}\right)^2=\frac{36}{9}\ ;\ \sqrt{\left(x+\frac{4}{3}\right)^2}=\pm\sqrt{\frac{36}{9}}\ ;\ x+\frac{4}{3}=\pm\sqrt{\frac{6^2}{3^2}}\ ;\ x+\frac{4}{3}=\pm\frac{6}{3}\ ;\ x=-\frac{4}{3}\pm\frac{6}{3}\ ;\ x=-\frac{4\pm 6}{3}.$$

Therefore, the two solutions are $x=\frac{-4-6}{3}$; $\mathbf{x=-\frac{10}{3}}$; and $x=\frac{-4+6}{3}$; $\mathbf{x=\frac{2}{3}}$. In addition, the equation $(3x+4)^2=36$ can be factored to $\left(x-\frac{2}{3}\right)\left(x+\frac{10}{3}\right)=0$ which is the same as $\mathbf{(3x-2)(3x+10)=0}$.

Check: $(3x-2)(3x+10)=0\ ;\ 3x\cdot 3x+10\cdot 3x-2\cdot 3x-2\cdot 10=0\ ;\ 9x^2+30x-6x-20=0\ ;\ 9x^2+(30-6)x-20=0$

$;\ 9x^2+24x-20=0$ which is the same as $(3x+4)^2=36$.

From the above three methods the Square Root Property method is the easiest method in factoring the quadratic equation, followed by the Quadratic Formula method and Completing the Square method.

4. **First Method:** (The Trial and Error Method)

Consider the left hand side of the equation which is a polynomial. To factor the polynomial $x^2+11x+30$ we need to obtain two numbers whose sum is 11 and whose product is 30. Let's construct a table as follows:

Sum	***Product***
$1+10=11$	$1\cdot 10=10$
$2+9=11$	$2\cdot 9=18$
$3+8=11$	$3\cdot 8=24$
$4+7=11$	$4\cdot 7=28$
$\mathbf{5+6=11}$	$\mathbf{5\cdot 6=30}$

The last line contains the sum and the product of the two numbers that we need. Thus, $x^2+11x+30=0$ can be factored to $\mathbf{(x+5)(x+6)=0}$

Second Method: (The Quadratic Formula Method)

Given the standard quadratic equation $ax^2+bx+c=0$, equate the coefficients of $x^2+11x+30=0$ with the standard quadratic equation by letting $a=1$, $b=11$, and $c=30$. Then,

Given: $x=\frac{-b\pm\sqrt{b^2-4ac}}{2a}\ ;\ x=\frac{-11\pm\sqrt{11^2-(4\times 1\times 30)}}{2\times 1}\ ;\ x=\frac{-11\pm\sqrt{121-120}}{2}\ ;\ x=\frac{-11\pm\sqrt{1}}{2}\ ;\ x=\frac{-11\pm 1}{2}.$

Therefore, the two solutions are $x=\frac{-11+1}{2}$; $x=-\frac{\overset{5}{\cancel{10}}}{\cancel{2}}$; $\mathbf{x=-5}$ and $x=\frac{-11-1}{2}$; $x=-\frac{\overset{6}{\cancel{12}}}{\cancel{2}}$; $\mathbf{x=-6}$ and the equation $x^2+11x+30=0$ can be factored to $\mathbf{(x+5)(x+6)=0}$.

Third Method: (Completing-the-Square Method)

$$x^2+11x+30=0\ ;\ x^2+11x=-30\ ;\ x^2+11x+\left(\frac{11}{2}\right)^2=-30+\left(\frac{11}{2}\right)^2\ ;\ x^2+11x+\frac{121}{4}=-30+\frac{121}{4}$$

$$;\ \left(x+\frac{11}{2}\right)^2=-\frac{30}{1}+\frac{121}{4}\ ;\ \left(x+\frac{11}{2}\right)^2=\frac{(-30\cdot 4)+(1\cdot 121)}{1\cdot 4}\ ;\ \left(x+\frac{11}{2}\right)^2=\frac{-120+121}{4}\ ;\ \left(x+\frac{11}{2}\right)^2=\frac{1}{4}\ ;\ x+\frac{11}{2}=\pm\sqrt{\frac{1}{4}}$$

$;\ x+\frac{11}{2}=\pm\frac{1}{2}$. Therefore, the two solutions are $\mathbf{x=-5}$ and $\mathbf{x=-6}$ and the equation $x^2+11x+30=0$ can be factored to $\mathbf{(x+5)(x+6)=0}$.

Check: $(x+5)(x+6)=0\ ;\ x\cdot x+6\cdot x+5\cdot x+5\cdot 6=0\ ;\ x^2+6x+5x+30=0\ ;\ x^2+(6+5)x+30=0\ ;\ x^2+11x+30=0$

From the above three methods using the Trial and Error method is the easiest method to obtain the factored terms. Completing the Square method is the second easiest method to use, followed by the Quadratic Formula method which is the longest and perhaps the most difficult way of obtaining the factored terms.

5. **First Method:** (The Trial and Error Method)

Consider the left hand side of the equation which is a polynomial. To factor the polynomial $5t^2 + 4t - 1$ we need to obtain two numbers whose sum is 4 and whose product is $5 \cdot -1 = -5$. Let's construct a table as follows:

Sum	***Product***
$8 - 4 = 4$	$8 \cdot (-4) = -32$
$7 - 3 = 4$	$7 \cdot (-3) = -21$
$6 - 2 = 4$	$6 \cdot (-2) = -12$
$\mathbf{5 - 1 = 4}$	$\mathbf{5 \cdot (-1) = -5}$

The last line contains the sum and the product of the two numbers that we need. Thus, $5t^2 + 4t - 1 = 0$; $5t^2 + (5-1)t - 1 = 0$; $5t^2 + 5t - t - 1 = 0$; $5t(t+1) - (t+1) = 0$; $\mathbf{(t+1)(5t-1) = 0}$

Second Method: (The Quadratic Formula Method)

Given the standard quadratic equation $at^2 + bt + c = 0$, equate the coefficients of $5t^2 + 4t - 1 = 0$ with the standard quadratic equation by letting $a = 5$, $b = 4$, and $c = -1$. Then,

Given: $t = \frac{-b \pm \sqrt{b^2 - 4ac}}{2a}$; $t = \frac{-4 \pm \sqrt{4^2 - (4 \times 5 \times -1)}}{2 \times 5}$; $t = \frac{-4 \pm \sqrt{16 + 20}}{10}$; $t = \frac{-4 \pm \sqrt{36}}{10}$; $t = \frac{-4 \pm \sqrt{6^2}}{6}10$; $t = \frac{-4 \pm 6}{10}$. Therefore, the two solutions are $\mathbf{t = -1}$ and $\mathbf{t = \frac{1}{5}}$ and the equation $5t^2 + 4t - 1 = 0$ can be factored to $(t+1)\left(t - \frac{1}{5}\right) = 0$ which is the same as $\mathbf{(t+1)(5t-1) = 0}$.

Third Method: (Completing-the-Square Method)

$5t^2 + 4t - 1 = 0$; $5t^2 + 4t = 1$; $\frac{\cancel{5}}{\cancel{5}}t^2 + \frac{4}{5}t = \frac{1}{5}$; $t^2 + \frac{4}{5}t = \frac{1}{5}$; $t^2 + \frac{4}{5}t + \left(\frac{\overset{2}{\cancel{4}}}{\underset{5}{\cancel{10}}}\right)^2 = \frac{1}{5} + \left(\frac{\overset{2}{\cancel{4}}}{\underset{5}{\cancel{10}}}\right)^2$; $t^2 + \frac{4}{5}t + \left(\frac{2}{5}\right)^2 = \frac{1}{5} + \left(\frac{2}{5}\right)^2$; $t^2 + \frac{4}{5}t + \frac{4}{25} = \frac{1}{5} + \frac{4}{25}$; $\left(t + \frac{2}{5}\right)^2 = \frac{(1 \cdot 25) + (4 \cdot 5)}{5 \cdot 25}$; $\left(t + \frac{2}{5}\right)^2 = \frac{25 + 20}{125}$; $\left(t + \frac{2}{5}\right)^2 = \frac{\overset{9}{\cancel{45}}}{\underset{25}{\cancel{125}}}$; $\left(t + \frac{2}{5}\right)^2 = \frac{9}{25}$; $\sqrt{\left(t + \frac{2}{5}\right)^2} = \pm\sqrt{\frac{9}{25}}$; $t + \frac{2}{5} = \pm\sqrt{\frac{3^2}{5^2}}$; $t + \frac{2}{5} = \pm\frac{3}{5}$; $t = -\frac{2}{5} \pm \frac{3}{5}$. Therefore, the two solutions are $t = -\frac{2}{5} - \frac{3}{5}$; $t = \frac{-2-3}{5}$; $t = -\frac{5}{5}$; $\mathbf{t = -1}$ and $t = -\frac{2}{5} + \frac{3}{5}$; $t = \frac{-2+3}{5}$; $\mathbf{t = \frac{1}{5}}$ and the equation $5t^2 + 4t - 1 = 0$ can be factored to $(t+1)\left(t - \frac{1}{5}\right) = 0$ which is the same as $\mathbf{(t+1)(5t-1) = 0}$.

Check: I. *Let* $t = \frac{1}{5}$ *in* $5t^2 + 4t - 1 = 0$; $5 \cdot \left(\frac{1}{5}\right)^2 + 4 \cdot \left(\frac{1}{5}\right) - 1 \overset{?}{=} 0$; $\frac{\cancel{5}}{\underset{5}{\cancel{25}}} + \frac{4}{5} - 1 \overset{?}{=} 0$; $\frac{1}{5} + \frac{4}{5} - 1 \overset{?}{=} 0$; $\frac{1+4}{5} - 1 \overset{?}{=} 0$; $\frac{\cancel{5}}{\cancel{5}} - 1 \overset{?}{=} 0$; $1 - 1 \overset{?}{=} 0$; $0 = 0$

II. *Let* $t = -1$ *in* $5t^2 + 4t - 1 = 0$; $5 \cdot (-1)^2 + 4 \cdot (-1) - 1 \overset{?}{=} 0$; $5 - 4 - 1 \overset{?}{=} 0$; $5 - 5 \overset{?}{=} 0$; $0 = 0$

From the above three methods using the Trial and Error method is the easiest method to obtain the factored terms. The Quadratic Formula method is the second easiest method to use, followed by Completing-the-Square method.

6. **First Method:** (The Trial and Error Method)

To apply the Trial and Error method to the equation $(2x+6)^2 = 36$ we need to complete and simplify the square in the left hand side of the equation, i.e., $(2x+6)^2 = 36$; $4x^2 + 36 + 24x = 36$; $4x^2 + 24x + 36 - 36 = 36 - 36$; $4x^2 + 24x + 0 = 0$

; $\frac{\overset{6}{\cancel{24}}}{\cancel{4}}$ … $\frac{\cancel{4}}{\cancel{4}}x^2 + \frac{\overset{6}{\cancel{24}}}{\cancel{4}}x + 0 = 0$; $x^2 + 6x + 0 = 0$. Consider the left hand side of the equation which is a polynomial. To factor the polynomial $x^2 + 6x + 0$ we need to obtain two numbers whose sum is 6 and whose product is $6 \cdot 0 = 0$. Let's construct a table as follows:

Sum	***Product***
$4+2=6$	$4 \cdot 2 = 8$
$5+1=6$	$5 \cdot 1 = 5$
$\mathbf{6+0=6}$	$\mathbf{6 \cdot 0 = 0}$

The last line contains the sum and the product of the two numbers that we need. Thus, $(2x+6)^2 = 36$ can be factored to $(x+0)(x+6) = 0$ which is the same as $\mathbf{x(x+6)=0}$

Second Method: (The Greatest Common Factoring Method)

First complete the square term on the left hand side and simplify the equation:

$(2x+6)^2 = 36$; $4x^2 + 24x + 36 = 36$; $4x^2 + 24x = 36 - 36$; $4x^2 + 24x = 0$; $\frac{\cancel{4}}{\cancel{4}}x^2 + \frac{\overset{6}{\cancel{24}}}{\cancel{4}}x = \frac{0}{4}$; $x^2 + 6x = 0$

Then, Factor out the greatest common monomial term x.

$x^2 + 6x = 0$; $x(x+6) = 0$. Thus, the two solution to the equation are: $\mathbf{x=0}$ and $x+6=0$; $\mathbf{x=-6}$

Hence, the equation $(2x+6)^2 = 36$ can be factored to $(x+0)(x+6) = 0$ which is the same as $\mathbf{x(x+6)=0}$.

Third Method: (The Square Root Property Method)

$(2x+6)^2 = 36$; $\sqrt{(2x+6)^2} = \pm\sqrt{36}$; $2x+6 = \pm 6$; $2x = -6 \pm 6$. Thus, the two solutions are $2x = -6+6$; $2x = 0$; $\frac{\cancel{2}}{\cancel{2}}x = \frac{0}{2}$; $\mathbf{x=0}$ and $2x = -6-6$; $2x = -12$; $\frac{\cancel{2}}{\cancel{2}}x = -\frac{\overset{6}{\cancel{12}}}{\cancel{2}}$; $\mathbf{x=-6}$ and the equation $(2x+6)^2 = 36$ can be factored to $(x+0)(x+6) = 0$ which is the same as $\mathbf{x(x+6)=0}$.

Check: I. *Let* $x = 0$ *in* $(2x+6)^2 = 36$; $[(2 \cdot 0)+6]^2 \overset{?}{=} 36$; $6^2 \overset{?}{=} 36$; $36 = 36$

II. *Let* $x = -6$ *in* $(2x+6)^2 = 36$; $[(2 \cdot -6)+6]^2 \overset{?}{=} 36$; $(-12+6)^2 \overset{?}{=} 36$; $(-6)^2 \overset{?}{=} 36$; $36 = 36$

From the above three methods the Greatest Common Factoring method is the easiest method. The Square Root Property method is the second easiest, followed by the Trail and Error method.

7. **First Method:** (The Trial and Error Method)

Consider the left hand side of the equation which is a polynomial. To factor the polynomial $y^2 - 8y + 15$ we need to obtain two numbers whose sum is -8 and whose product is 15. Let's construct a table as follows:

Sum	***Product***
$-4-4=-8$	$-4 \cdot -4 = 16$
$\mathbf{-5-3=-8}$	$\mathbf{-5 \cdot -3 = 15}$
$-6-2=-8$	$-6 \cdot -2 = 12$
$-7-1=-8$	$-7 \cdot -1 = 7$

The second line contains the sum and the product of the two numbers that we need. Thus, $y^2 - 8y + 15 = 0$ can be factored to $\mathbf{(y-3)(y-5)=0}$.

Second Method: (The Quadratic Formula Method)

Given the standard quadratic equation $ax^2 + bx + c = 0$, equate the coefficients of $y^2 - 8y + 15 = 0$ with the standard quadratic equation by letting $a = 1$, $b = -8$, and $c = 15$. Then,

Given: $y = \frac{-b \pm \sqrt{b^2 - 4ac}}{2a}$; $y = \frac{-(-8) \pm \sqrt{(-8)^2 - (4 \times 1 \times 15)}}{2 \times 1}$; $y = \frac{8 \pm \sqrt{64 - 60}}{2 \times 1}$; $y = \frac{8 \pm \sqrt{4}}{2}$; $y = \frac{8 \pm \sqrt{2^2}}{2}$

; $y=\frac{8\pm 2}{2}$. Therefore, the two solutions are $y=\frac{8-2}{2}$; $y=\frac{\cancel{6}}{\cancel{2}}$; $\mathbf{y=3}$ and $y=\frac{8+2}{2}$; $y=\frac{\overset{5}{\cancel{10}}}{\cancel{2}}$; $\mathbf{y=5}$ and the equation $y^2-8y+15=0$ can be factored to $\mathbf{(y-3)(y-5)=0}$.

Third Method: (Completing-the-Square Method)

$y^2-8y+15=0$; $y^2-8y=-15$; $y^2-8y+\left(-\frac{\overset{4}{\cancel{8}}}{\cancel{2}}\right)^2=-15+\left(-\frac{\overset{4}{\cancel{8}}}{\cancel{2}}\right)^2$; $y^2-8y+\left(-\frac{4}{1}\right)^2=-15+\left(-\frac{4}{1}\right)^2$

; $y^2-8y+16=-15+16$; $y^2-8y+16=1$; $(y-4)^2=1$; $\sqrt{(y-4)^2}=\pm\sqrt{1}$; $y-4=\pm 1$; $y=4\pm 1$. Therefore, the two solutions are $\mathbf{y=3}$ and $\mathbf{y=5}$ and the equation $y^2-8y+15=0$ can be factored to $\mathbf{(y-3)(y-5)=0}$.

Check: $(y-3)(y-5)=0$; $y\cdot y-5\cdot y-3\cdot y+(-3)\cdot(-5)=0$; $y^2-5y-3y+15=0$; $y^2+(-5-3)y+15=0$; $y^2-8y+15=0$

From the above three methods using the Trial and Error method is the easiest method to obtain the factored terms. Completing-the-Square method is the second easiest method to use, followed by the Quadratic Formula method which is the longest and perhaps the most difficult way of obtaining the factored terms.

8. **First Method:** (The Trial and Error Method)

Write the equation $w^2=-7$ in the standard quadratic equation form $aw^2+bw+c=0$, i.e., write $w^2=-7$ as $w^2+0w+7=0$. Consider the left hand side of the equation which is a polynomial. To factor the polynomial w^2+0w+7 we need to obtain two numbers whose sum is 0 and whose product is 7. Let's construct a table as follows:

Sum	***Product***
$1-1=0$	$1\cdot(-1)=-1$
$2-2=0$	$2\cdot(-2)=-4$
$3-3=0$	$3\cdot(-3)=-9$
$4-4=0$	$4\cdot(-4)=-16$

After several trials it becomes clear that the given equation can not be simplified using the Trail and Error method. Therefore, **the given equation is not factorable and is referred to as PRIME.**

Second Method: (The Quadratic Formula Method)

First, write the equation in the standard quadratic equation form $aw^2+bw+c=0$, i.e., write $w^2=-7$ as $w^2+0w+7=0$. Second, equate the coefficients of $w^2+0w+7=0$ with the standard quadratic equation by letting $a=1$,, $b=0$, and $c=7$. Then,

Given: $w=\frac{-b\pm\sqrt{b^2-4ac}}{2a}$; $w=\frac{-0\pm\sqrt{0^2-(4\times 1\times 7)}}{2\times 1}$; $w=\frac{\pm\sqrt{0-28}}{2}$; $w=\frac{\pm\sqrt{-28}}{2}$. **However, since the number under the radical is a negative number, the given equation has no real solution and can not be factored.**

Third Method: (The Square Root Property Method)

Take the square root of both sides of the equation, i.e., write $w^2=-7$ as $\sqrt{w^2}=\pm\sqrt{-7}$; $w=\pm\sqrt{-7}$

Again, since the number under the radical is negative, the given equation has no real solution and can not be factored.

9. **First Method:** (The Trial and Error Method)

Consider the left hand side of the equation which is a polynomial. To factor the polynomial $6x^2+x-1$ we need to obtain two numbers whose sum is 1 and whose product is $6\cdot -1=-6$. Let's construct a table as follows:

Sum	***Product***
$5-4=1$	$5\cdot -4=-20$
$4-3=1$	$4\cdot -3=-12$
$\mathbf{3-2=1}$	$\mathbf{2\cdot -3=-6}$
$2-1=1$	$2\cdot -1=-2$

The third line contains the sum and the product of the two numbers that we need. Therefore, $6x^2 + x - 1 = 0$; $6x^2 + (3-2)x - 1 = 0$; $6x^2 + 3x - 2x - 1 = 0$; $3x(2x+1) - (2x+1) = 0$; $\mathbf{(2x+1)(3x-1) = 0}$

Second Method: (The Quadratic Formula Method)

Given the standard quadratic equation $ax^2 + bx + c = 0$, equate the coefficients of $6x^2 + x - 1 = 0$ with the standard quadratic equation by letting $a = 6$, $b = 1$, and $c = -1$. Then,

Given: $x = \frac{-b \pm \sqrt{b^2 - 4ac}}{2a}$; $x = \frac{-1 \pm \sqrt{1^2 - (4 \times 6 \times -1)}}{2 \times 6}$; $x = \frac{-1 \pm \sqrt{1+24}}{12}$; $x = \frac{-1 \pm \sqrt{25}}{12}$; $x = \frac{-1 \pm \sqrt{5^2}}{12}$

; $x = \frac{-1 \pm 5}{12}$. Therefore, the two solutions are $x = \frac{-1-5}{12}$; $x = -\frac{\cancel{6}}{\cancel{12}_{2}}$; $\mathbf{x = -\frac{1}{2}}$ and $x = \frac{-1+5}{12}$; $x = \frac{\cancel{4}}{\cancel{12}_{3}}$; $\mathbf{x = \frac{1}{3}}$ and

the equation $6x^2 + x - 1 = 0$ can be factored to $\left(x + \frac{1}{2}\right)\left(x - \frac{1}{3}\right) = 0$ which is the same as $\mathbf{(2x+1)(3x-1) = 0}$.

Third Method: (Completing-the-Square Method)

$6x^2 + x - 1 = 0$; $6x^2 + x = 1$; $\frac{\cancel{6}}{\cancel{6}}x^2 + \frac{1}{6}x = \frac{1}{6}$; $x^2 + \frac{1}{6}x = \frac{1}{6}$; $x^2 + \frac{1}{6}x + \left(\frac{1}{12}\right)^2 = \frac{1}{6} + \left(\frac{1}{12}\right)^2$

; $x^2 + \frac{1}{6}x + \frac{1}{144} = \frac{1}{6} + \frac{1}{144}$; $\left(x + \frac{1}{12}\right)^2 = \frac{(1 \cdot 144) + (1 \cdot 6)}{6 \cdot 144}$; $\left(x + \frac{1}{12}\right)^2 = \frac{144+6}{864}$; $\left(x + \frac{1}{12}\right)^2 = \frac{\cancel{150}^{75}}{\cancel{864}_{432}}$; $\left(x + \frac{1}{12}\right)^2 = \frac{75}{432}$

; $\sqrt{\left(x + \frac{1}{12}\right)^2} = \pm\sqrt{\frac{75}{432}}$; $x + \frac{1}{12} = \pm\sqrt{0.174}$; $x + 0.083 = \pm 0.417$; $x = -0.083 \pm 0.417$.

Therefore, the two solutions are $x = -0.083 - 0.417$; $x = -0.5$; $\mathbf{x = -\frac{1}{2}}$ and $x = -0.083 + 0.417$; $x = 0.33$; $\mathbf{x = \frac{1}{3}}$ and

the equation $6x^2 + x - 1 = 0$ can be factored to $\left(x + \frac{1}{2}\right)\left(x - \frac{1}{3}\right) = 0$ which is the same as $\mathbf{(2x+1)(3x-1) = 0}$.

Check: I. *Let* $t = \frac{1}{3}$ *in* $6x^2 + x - 1 = 0$; $6 \cdot \left(\frac{1}{3}\right)^2 + \frac{1}{3} - 1 \stackrel{?}{=} 0$; $\cancel{6}^{2} \cdot \frac{1}{\cancel{9}_{3}} + \frac{1}{3} - 1 \stackrel{?}{=} 0$; $\frac{2}{3} + \frac{1}{3} - 1 \stackrel{?}{=} 0$; $\frac{2+1}{3} - 1 \stackrel{?}{=} 0$; $\frac{3}{3} - 1 \stackrel{?}{=} 0$

; $1 - 1 \stackrel{?}{=} 0$; $0 = 0$

II. *Let* $t = -\frac{1}{2}$ *in* $6x^2 + x - 1 = 0$; $6 \cdot \left(-\frac{1}{2}\right)^2 - \frac{1}{2} - 1 \stackrel{?}{=} 0$; $\cancel{6}^{3} \cdot \frac{1}{\cancel{4}_{2}} - \frac{1}{2} - 1 \stackrel{?}{=} 0$; $\frac{3}{2} - \frac{1}{2} - 1 \stackrel{?}{=} 0$; $\frac{3-1}{2} - 1 \stackrel{?}{=} 0$

; $\frac{2}{2} - 1 \stackrel{?}{=} 0$; $1 - 1 \stackrel{?}{=} 0$; $0 = 0$

From the above three methods using the Quadratic Formula method is the easiest method to obtain the factored terms. The Trial and Error method is the second easiest method to use, followed by Completing-the-Square method.

10. **First Method:** (The Perfect Square Approach)

First - Check and see if the coefficients of the first and the last term are perfect squares, i.e., $x^2 - 4x + 4$; $x^2 - 4x + 2^2$

Second - Check and see if by multiplying the base of the last term by $-2x$ we can obtain the middle term, i.e., $2 \cdot -2x = -4x$.

Third - Since $-4x$ is the same as the middle term of the given polynomial therefore, the trinomial can be written as:

$x^2 - 4x + 2^2 = (x-2)^2 = \mathbf{(x-2)(x-2)}$

Second Method: (The Quadratic Formula Method)

Based on our earlier stated assumption, we write the given trinomial as a quadratic equation, i.e., $x^2 - 4x + 4 = 0$ and apply the Quadratic Formula by letting $a = 1$, $b = -4$, and $c = 4$. Then,

Given: $x = \frac{-b \pm \sqrt{b^2 - 4ac}}{2a}$; $x = \frac{-(-4) \pm \sqrt{(-4)^2 - (4 \times 1 \times 4)}}{2 \times 1}$; $x = \frac{4 \pm \sqrt{16 - 16}}{2}$; $x = \frac{4 \pm \sqrt{0}}{2}$; $x = \frac{4 \pm 0}{2}$; $x = \frac{\overset{2}{\cancel{4}}}{\cancel{2}}$;

$x = 2$. Therefore, the quadratic equation has two repeated solution, i.e., $\mathbf{x = 2}$ and $\mathbf{x = 2}$ and the equation $x^2 - 4x + 4 = 0$ can be factored to $\mathbf{(x-2)(x-2) = 0}$.

Third Method: (Trial and Error Method)

Consider the left hand side of the equation which is a polynomial. To factor the polynomial $x^2 - 4x + 4$ we need to obtain two numbers whose sum is -4 and whose product is 4. Let's construct a table as follows:

Sum	***Product***
$-6 + 2 = -4$	$-6 \cdot 2 = -12$
$-5 + 1 = -4$	$-5 \cdot 1 = -5$
$-3 - 1 = -4$	$-3 \cdot -1 = 3$
$\mathbf{-2 - 2 = -4}$	$\mathbf{-2 \cdot -2 = 4}$

The last line contains the sum and the product of the two numbers that we need. Thus, $x^2 - 4x + 4 = 0$ can be factored to $\mathbf{(x-2)(x-2) = 0}$.

Check: $(x-2)(x-2) = 0$; $x \cdot x - 2 \cdot x - 2 \cdot x + (-2) \cdot (-2) = 0$; $x^2 - 2x - 2x + 4 = 0$; $x^2 + (-2-2)x + 4 = 0$; $x^2 - 4x + 4 = 0$

From the above three methods the Perfect Square method is the easiest method in factoring the trinomial followed by the Trial and Error method and the Quadratic Formula method.

Chapter 5 Solutions:

Section 5.1 Solutions - Introduction to Algebraic Fractions

A. Write the correct sign for the following fractions:

1. $-\frac{2}{-5} = +\frac{\mathbf{2}}{\mathbf{5}}$
2. $+\frac{-3}{-6} = +\frac{3}{6} = +\frac{\overset{1}{\cancel{3}}}{\underset{2}{\cancel{6}}} = +\frac{\mathbf{1}}{\mathbf{2}}$
3. $-\frac{-8}{-4} = -\frac{8}{4} = -\frac{\overset{2}{\cancel{8}}}{4} = -\frac{2}{1} = \mathbf{-2}$
4. $\frac{10}{-2} = -\frac{10}{2} = -\frac{\overset{5}{\cancel{10}}}{\cancel{2}} = -\frac{5}{1} = \mathbf{-5}$
5. $-\frac{5}{-15} = +\frac{5}{15} = +\frac{\cancel{5}}{\underset{3}{\cancel{15}}} = +\frac{\mathbf{1}}{\mathbf{3}}$
6. $+\frac{-8}{6} = -\frac{8}{6} = -\frac{\overset{4}{\cancel{8}}}{\underset{3}{\cancel{6}}} = -\frac{\mathbf{4}}{\mathbf{3}}$
7. $-\frac{+3}{+15} = -\frac{3}{15} = -\frac{\cancel{3}}{\underset{5}{\cancel{15}}} = -\frac{\mathbf{1}}{\mathbf{5}}$
8. $-\frac{2}{+12} = -\frac{2}{12} = -\frac{\cancel{2}}{\underset{6}{\cancel{12}}} = -\frac{\mathbf{1}}{\mathbf{6}}$
9. $-\frac{-6}{3} = +\frac{6}{3} = +\frac{\overset{2}{\cancel{6}}}{\cancel{3}} = +\frac{2}{1} = \mathbf{+2}$
10. $+\frac{2}{-3} = -\frac{\mathbf{2}}{\mathbf{3}}$

B. State the value(s) of the variable for which the following fractions are not defined.

1. $\frac{3}{x-1}$ is not defined when $\boldsymbol{x=1}$
2. $\frac{x-5}{5-x}$ is not defined when $\boldsymbol{x=5}$
3. $-\frac{1}{x}$ is not defined when $\boldsymbol{x=0}$
4. $\frac{x}{x+10}$ is not defined when $\boldsymbol{x=-10}$
5. $\frac{2x}{3x-5}$ is not defined when $\boldsymbol{x=\frac{5}{3}}$
6. $\frac{5x-2}{x-7}$ is not defined when $\boldsymbol{x=7}$
7. $\frac{4}{x^3-x} = \frac{4}{x\left(x^2-1\right)} = \frac{4}{x(x-1)(x+1)}$ is not defined when $\boldsymbol{x=0}$, $\boldsymbol{x=1}$, and $\boldsymbol{x=-1}$
8. $\frac{x+1}{x^2+5x+4} = \frac{x+1}{(x+1)(x+4)}$ is not defined when $\boldsymbol{x=-1}$, and $\boldsymbol{x=-4}$
9. $\frac{3x}{2x^2-5x-3} = \frac{3x}{(2x+1)(x-3)}$ is not defined when $\boldsymbol{x=-\frac{1}{2}}$, and $\boldsymbol{x=3}$
10. $\frac{x-3}{x^2-2x-3} = \frac{x-3}{(x+1)(x-3)}$ is not defined when $\boldsymbol{x=-1}$, and $\boldsymbol{x=3}$

C. State which of the following algebraic fractions are equivalent fractions.

1. $\frac{2x}{3y}$ and $\frac{4x}{9y}$ are **not equivalent** fractions.
2. $\frac{3x+1}{2x}$ and $\frac{9x+3}{6x}$ are **equivalent** fractions.
3. $\frac{2}{a-b}$ and $-\frac{2}{b-a}$ are **equivalent** fractions.
4. $\frac{x-5}{x+1}$ and $\frac{5-x}{x-1}$ are **not equivalent** fractions.
5. $-\frac{a}{a-1}$ and $\frac{a}{1-a}$ are **equivalent** fractions.
6. $\frac{3-x}{-x}$ and $\frac{x+3}{x}$ are **not equivalent** fractions.
7. $\frac{u-w}{w}$ and $\frac{-2u+2w}{2w}$ are **not equivalent** fractions.
8. $\frac{a^2}{1-a^2}$ and $\frac{-a^2}{a^2-1}$ are **equivalent** fractions.
9. $\frac{2-x}{-3}$ and $\frac{x}{3}-\frac{2}{3}$ are **equivalent** fractions.
10. $-\frac{x+2}{-2}$ and $\frac{-x-2}{2}$ are **not equivalent** fractions.

Section 5.2 Solutions - Simplifying Algebraic Fractions to Lower Terms

1. $\frac{x^2y^2z^5}{-xy^3z^2} = -\frac{\overset{x}{\cancel{x^2}}\,\cancel{y^2}\,\overset{z^3}{\cancel{z^5}}}{\cancel{x}\,\underset{y}{\cancel{y^3}}\,\cancel{z^2}} = -\frac{\boldsymbol{xz^3}}{\boldsymbol{y}}$
2. $-\frac{3a^2bc^3}{-9ab^2c} = +\frac{\cancel{3}\,\overset{a}{\cancel{a^2}}\,\cancel{b}\,\overset{c^2}{\cancel{c^3}}}{\underset{3}{\cancel{9}}\,\cancel{a}\,\underset{b}{\cancel{b^2}}\,\cancel{c}} = \frac{\boldsymbol{ac^2}}{\boldsymbol{3b}}$

3. $\frac{1+2m}{1-2m}$ **Can not be simplified.**

4. $\frac{2uvw^3}{10u^2v} = \frac{\cancel{2}\cancel{u}\cancel{v}w^3}{\underset{5}{\cancel{10}}\underset{u}{\cancel{u^2}}\cancel{v}} = \mathbf{\frac{w^3}{5u}}$

5. $\frac{y^2-4}{y^2-y-6} = \frac{y^2-2^2}{(y+2)(y-3)} = \frac{(y-2)(y+2)}{(y+2)(y-3)} = \frac{(y-2)\cancel{(y+2)}}{\cancel{(y+2)}(y-3)} = \mathbf{\frac{y-2}{y-3}}$

6. $\frac{x^3-3x^2}{x^2-9} = \frac{x^2(x-3)}{x^2-3^2} = \frac{x^2(x-3)}{(x-3)(x+3)} = \frac{x^2\cancel{(x-3)}}{\cancel{(x-3)}(x+3)} = \mathbf{\frac{x^2}{x+3}}$

7. $\frac{x^3+2x^2-15x}{2x^2-5x-3} = \frac{x(x^2+2x-15)}{2x^2-5x-3} = \frac{x(x+5)(x-3)}{(x-3)(2x+1)} = \frac{x(x+5)\cancel{(x-3)}}{\cancel{(x-3)}(2x+1)} = \mathbf{\frac{x(x+5)}{2x+1}}$

8. $\frac{x^2-2x}{x^3-x^2-2x} = \frac{x(x-2)}{x(x^2-x-2)} = \frac{x(x-2)}{x(x+1)(x-2)} = \frac{\cancel{x}\cancel{(x-2)}}{\cancel{x}(x+1)\cancel{(x-2)}} = \mathbf{\frac{1}{x+1}}$

9. $\frac{x^2+xy-2y^2}{x+2y} = \frac{(x+2y)(x-y)}{x+2y} = \frac{\cancel{(x+2y)}(x-y)}{\cancel{(x+2y)}} = \frac{x-y}{1} = \mathbf{x-y}$

10. $\frac{6x^2-xy-y^2}{2x^2+xy-y^2} = \frac{(3x+y)(2x-y)}{(2x-y)(x+y)} = \frac{(3x+y)\cancel{(2x-y)}}{\cancel{(2x-y)}(x+y)} = \mathbf{\frac{3x+y}{x+y}}$

Section 5.3 Case I Solutions - Addition amd Subtraction of Algebraic Fractions with Common Denominators

1. $\frac{x}{7}+\frac{3}{7} = \mathbf{\frac{x+3}{7}}$

2. $\frac{8}{a+b}-\frac{7}{a+b} = \frac{8-7}{a+b} = \mathbf{\frac{1}{a+b}}$

3. $\frac{3x+1}{2y}+\frac{4x+1}{2y}+\frac{3}{2y} = \frac{3x+1+4x+1+3}{2y} = \frac{(3x+4x)+(1+1+3)}{2y} = \frac{(3+4)x+5}{2y} = \mathbf{\frac{7x+5}{2y}}$

4. $\frac{4x}{x-2}-\frac{8}{x-2} = \frac{4x-8}{x-2} = \frac{4\cancel{(x-2)}}{\cancel{(x-2)}} = \frac{4}{1} = \mathbf{4}$

5. $\frac{15a}{5a+b}-\frac{-5b}{5a+b} = \frac{15a-(-5b)}{5a+b} = \frac{15a+5b}{5a+b} = \mathbf{\frac{5(3a+b)}{5a+b}}$ The answer is in its lowest terms.

6. $\frac{6x}{3x^2y^2}+\frac{5y}{3x^2y^2} = \mathbf{\frac{6x+5y}{3x^2y^2}}$ The answer is in its lowest terms.

7. $\frac{x+2}{3x^2+5}-\frac{x-1}{3x^2+5} = \frac{x+2-(x-1)}{3x^2+5} = \frac{x+2-x+1}{3x^2+5} = \frac{(x-x)+(2+1)}{3x^2+5} = \frac{(1-1)x+3}{3x^2+5} = \frac{0x+3}{3x^2+5} = \mathbf{\frac{3}{3x^2+5}}$

8. $\frac{x^2+5x+4}{(x+5)(x-2)}-\frac{x^2+3x+1}{(x+5)(x-2)} = \frac{x^2+5x+4-(x^2+3x+1)}{(x+5)(x-2)} = \frac{x^2+5x+4-x^2-3x-1}{(x+5)(x-2)} = \frac{(x^2-x^2)+(5x-3x)+(4-1)}{(x+5)(x-2)}$

$= \frac{(1-1)x^2+(5-3)x+3}{(x+5)(x-2)} = \frac{0x^2+2x+3}{(x+5)(x-2)} = \mathbf{\frac{2x+3}{(x+5)(x-2)}}$

9. $\frac{y^2}{y-3}-\frac{3y}{y-3} = \frac{y^2-3y}{y-3} = \frac{y\cancel{(y-3)}}{\cancel{(y-3)}} = \frac{y}{1} = \mathbf{y}$

10. $\frac{a^2-2a+1}{(a-1)(a+3)}-\frac{a^2-3a+2}{(a-1)(1+3)} = \frac{a^2-2a+1-(a^2-3a+2)}{(a-1)(a+3)} = \frac{a^2-2a+1-a^2+3a-2}{(a-1)(a+3)} = \frac{(a^2-a^2)+(3a-2a)+(1-2)}{(a-1)(a+3)}$

$= \frac{(1-1)a^2+(3-2)a-1}{(a-1)(a+3)} = \frac{0a^2+a-1}{(a-1)(a+3)} = \frac{\cancel{(a-1)}}{\cancel{(a-1)}(a+3)} = \mathbf{\frac{1}{a+3}}$

Section 5.3 Case II Solutions - Addition and Subtraction of Algebraic Fractions without Common Denominators

1. $\frac{3}{4x^2}+\frac{5}{2x^3}=\frac{\left(3\cdot 2x^3\right)+\left(5\cdot 4x^2\right)}{4x^2\cdot 2x^3}=\frac{6x^3+20x^2}{8x^5}=\frac{\cancel{2x^2}(3x+10)}{\underset{4}{\cancel{8}}\underset{x^3}{\cancel{x^5}}}=\mathbf{\frac{3x+10}{4x^3}}$

2. $\frac{x}{x+4}-\frac{2}{x-1}=\frac{[x\cdot(x-1)]-[2\cdot(x+4)]}{(x+4)\cdot(x-1)}=\frac{x^2-x-(2x+8)}{(x+4)(x-1)}=\frac{x^2-x-2x-8}{(x+4)(x-1)}=\mathbf{\frac{x^2-3x-8}{(x+4)(x-1)}}$

3. $\frac{a-b}{a+b}-\frac{a}{b}=\frac{[b\cdot(a-b)]-[a\cdot(a+b)]}{b\cdot(a+b)}=\frac{\left(ab-b^2\right)-\left(a^2+ab\right)}{b(a+b)}=\frac{\cancel{ab}-b^2-a^2-\cancel{ab}}{b(a+b)}=\frac{-b^2-a^2}{b(a+b)}=\mathbf{\frac{-\left(a^2+b^2\right)}{b(a+b)}}$

4. $\frac{x^2}{x+3}+\frac{5x}{x-5}=\frac{\left[x^2\cdot(x-5)\right]+[5x\cdot(x+3)]}{(x+3)\cdot(x-5)}=\frac{\left(x^3-5x^2\right)+\left(5x^2+15x\right)}{(x+3)(x-5)}=\frac{x^3-\cancel{5x^2}+\cancel{5x^2}+15x}{(x+3)(x-5)}=\frac{x^3+15x}{(x+3)(x-5)}$

$=\mathbf{\frac{x\left(x^2+15\right)}{(x+3)(x-5)}}$

5. $\frac{1}{4x^2y^2z}-\frac{2}{xy^2z}=\frac{\left(1\cdot xy^2z\right)-\left(2\cdot 4x^2y^2z\right)}{4x^2y^2z\cdot xy^2z}=\frac{xy^2z-8x^2y^2z}{4x^3y^4z^2}=\frac{\cancel{xy^2z}(y-8x)}{4\underset{x^2}{\cancel{x^3}}\underset{y^2}{\cancel{y^4}}\underset{z}{\cancel{z^2}}}=\mathbf{\frac{1-8x}{4x^2y^2z}}$

6. $\frac{3}{x+1}+\frac{2}{x-1}-\frac{5}{x}=\left(\frac{3}{x+1}+\frac{2}{x-1}\right)-\frac{5}{x}=\left(\frac{[3\cdot(x-1)]+[2\cdot(x+1)]}{(x+1)\cdot(x-1)}\right)-\frac{5}{x}=\left(\frac{3x-3+2x+2}{(x+1)(x-1)}\right)-\frac{5}{x}=\left(\frac{5x-1}{(x+1)(x-1)}\right)-\frac{5}{x}$

$=\frac{[x\cdot(5x-1)]-\{5\cdot[(x+1)(x-1)]\}}{x\cdot[(x+1)(x-1)]}=\frac{\left(5x^2-x\right)-\left[5\cdot\left(x^2-\cancel{x}+\cancel{x}-1\right)\right]}{x(x+1)(x-1)}=\frac{\left(5x^2-x\right)-\left[5\cdot\left(x^2-1\right)\right]}{x(x+1)(x-1)}=\frac{\left(5x^2-x\right)-\left(5x^2-5\right)}{x(x+1)(x-1)}$

$=\frac{\cancel{5x^2}-x-\cancel{5x^2}+5}{x(x+1)(x-1)}=\frac{-x+5}{x(x+1)(x-1)}=\mathbf{-\frac{(x-5)}{x(x+1)(x-1)}}$

7. $\frac{a}{2a+10}-\frac{3a}{4a+5}=\frac{[a\cdot(4a+5)]-[3a\cdot(2a+10)]}{(2a+10)\cdot(4a+5)}=\frac{\left(4a^2+5a\right)-\left(6a^2+30a\right)}{2(a+5)(4a+5)}=\frac{4a^2+5a-6a^2-30a}{2(a+5)(4a+5)}$

$=\frac{\left(4a^2-6a^2\right)+(5a-30a)}{2(a+5)(4a+5)}=\frac{-2a^2-25a}{2(a+5)(4a+5)}=\frac{-a(2a+25)}{2(a+5)(4a+5)}=\mathbf{-\frac{a(2a+25)}{2(a+5)(4a+5)}}$

8. $5x-\frac{20}{x}=\frac{5x}{1}-\frac{20}{x}=\frac{(5x\cdot x)-(20\cdot 1)}{1\cdot x}=\frac{5x^2-20}{x}=\frac{5\left(x^2-4\right)}{x}=\frac{5\left(x^2-2^2\right)}{x}=\mathbf{\frac{5(x-2)(x+2)}{x}}$

9. $\frac{2}{x}-\frac{1}{x+2}-\frac{5}{x-1}=\left(\frac{2}{x}-\frac{1}{x+2}\right)-\frac{5}{x-1}=\left(\frac{2\cdot(x+2)-(1\cdot x)}{x\cdot(x+2)}\right)-\frac{5}{x-1}=\left(\frac{2x+4-x}{x(x+2)}\right)-\frac{5}{x-1}=\frac{x+4}{x(x+2)}-\frac{5}{x-1}$

$=\frac{[(x+4)\cdot(x-1)]-[5\cdot x(x+2)]}{x(x+2)\cdot(x-1)}=\frac{\left(x^2-x+4x-4\right)-\left(5x^2+10x\right)}{x(x+2)(x-1)}=\frac{x^2+3x-4-5x^2-10x}{x(x+2)(x-1)}=\frac{\left(x^2-5x^2\right)+(3x-10x)-4}{x(x+2)(x-1)}$

$=\frac{-4x^2-7x-4}{x(x+2)(x-1)}=\mathbf{\frac{-4x^2-7x-4}{x(x+2)(x-1)}}$

10. $\frac{a}{b^2c}-\frac{c}{a^2b}-\frac{b}{ac^2}=\left(\frac{a}{b^2c}-\frac{c}{a^2b}\right)-\frac{b}{ac^2}=\left(\frac{\left(a\cdot a^2b\right)-\left(c\cdot b^2c\right)}{b^2c\cdot a^2b}\right)-\frac{b}{ac^2}=\left(\frac{a^3b-b^2c^2}{a^2b^3c}\right)-\frac{b}{ac^2}$

$$= \frac{\left[\left(a^3b - b^2c^2\right)\cdot ac^2\right] - \left[b \cdot a^2b^3c\right]}{a^2b^3c \cdot ac^2} = \frac{a^4bc^2 - ab^2c^4 - a^2b^4c}{a^3b^3c^3} = \frac{\cancel{abc}\left(a^3c - bc^3 - ab^3\right)}{\underset{a^2\,b^2\,c^2}{\cancel{a^3}\,\cancel{b^3}\,\cancel{c^3}}} = \mathbf{\frac{a^3c - bc^3 - ab^3}{a^2b^2c^2}}$$

Section 5.3 Case IIIa Solutions - Multiplication of Algebraic Fractions (Simple Cases)

1. $\frac{1}{xy}\cdot\frac{x^2y^2}{2} = \frac{1\cdot x^2y^2}{xy\cdot 2} = \frac{x^2y^2}{2xy} = \frac{\overset{x}{\cancel{x^2}}\,\overset{y}{\cancel{y^2}}}{2\cancel{xy}} = \mathbf{\frac{xy}{2}}$

2. $\frac{2a^2}{a^3}\cdot\frac{1}{8a} = \frac{2a^2\cdot 1}{a^3\cdot 8a} = \frac{2a^2}{8a^4} = \frac{\cancel{2}\cancel{a^2}}{\underset{4\;a^2}{\cancel{8}\,\cancel{a^4}}} = \mathbf{\frac{1}{4a^2}}$

3. $xyz\cdot\frac{1}{x^2y^2z^2}\cdot\frac{x^2}{y} = \frac{xyz}{1}\cdot\frac{1}{x^2y^2z^2}\cdot\frac{x^2}{y} = \frac{xyz\cdot 1\cdot x^2}{1\cdot x^2y^2z^2\cdot y} = \frac{x^3yz}{x^2y^3z^2} = \frac{\overset{x}{\cancel{x^3}}\,\cancel{y}\cancel{z}}{\cancel{x^2}\,\underset{y^2}{\cancel{y^3}}\,\underset{z}{\cancel{z^2}}} = \mathbf{\frac{x}{y^2z}}$

4. $\frac{5u^2v^2}{uv}\cdot\frac{uv^3}{15v^2}\cdot\frac{1}{u^4} = \frac{5u^2v^2\cdot uv^3\cdot 1}{uv\cdot 15v^2\cdot u^4} = \frac{5u^3v^5}{15u^5v^3} = \frac{\cancel{5}\cancel{u^3}\,\overset{v^2}{\cancel{v^5}}}{\underset{3\;u^2}{\cancel{15}\,\cancel{u^5}}\,\cancel{v^3}} = \mathbf{\frac{v^2}{3u^2}}$

5. $8x\cdot\frac{2}{x^3}\cdot\frac{1}{4x^2} = \frac{8x}{1}\cdot\frac{2}{x^3}\cdot\frac{1}{4x^2} = \frac{8x\cdot 2\cdot 1}{1\cdot x^3\cdot 4x^2} = \frac{16x}{4x^5} = \frac{\overset{4}{\cancel{16}}\cancel{x}}{\cancel{4}\,\underset{x^4}{\cancel{x^5}}} = \mathbf{\frac{4}{x^4}}$

6. $\frac{x-4}{x+2}\cdot\frac{x^2-4}{2}\cdot\frac{1}{x-4} = \frac{(x-4)\cdot\left(x^2-4\right)\cdot 1}{(x+2)\cdot 2\cdot (x-4)} = \frac{(x-4)(x-2)(x+2)}{2(x+2)(x-4)} = \frac{\cancel{(x-4)}(x-2)\cancel{(x+2)}}{2\cancel{(x+2)}\cancel{(x-4)}} = \mathbf{\frac{x-2}{2}}$

7. $\frac{5a^2b^2c}{ac}\cdot\frac{bc^2}{a^3}\cdot a^2 = \frac{5a^2b^2c}{ac}\cdot\frac{bc^2}{a^3}\cdot\frac{a^2}{1} = \frac{5a^2b^2c\cdot bc^2\cdot a^2}{ac\cdot a^3\cdot 1} = \frac{5a^4b^3c^3}{a^4c} = \frac{5\cancel{a^4}b^3\,\overset{c^2}{\cancel{c^3}}}{\cancel{a^4}\cancel{c}} = \frac{5b^3c^2}{1} = \mathbf{5b^3c^2}$

8. $\frac{5xyz}{10}\cdot\frac{2x}{y^2z^2} = \frac{5xyz\cdot 2x}{10\cdot y^2z^2} = \frac{10x^2yz}{10y^2z^2} = \frac{\cancel{10}x^2\cancel{y}\cancel{z}}{\cancel{10}\,\underset{y}{\cancel{y^2}}\,\underset{z}{\cancel{z^2}}} = \mathbf{\frac{x^2}{yz}}$

9. $\left(\frac{4w}{w^3}\cdot w^2\right)\cdot\frac{1}{w} = \left(\frac{4w}{w^3}\cdot\frac{w^2}{1}\right)\cdot\frac{1}{w} = \left(\frac{4w\cdot w^2}{w^3\cdot 1}\right)\cdot\frac{1}{w} = \left(\frac{4w^3}{w^3}\right)\cdot\frac{1}{w} = \frac{4\cancel{w^3}}{\cancel{w^3}}\cdot\frac{1}{w} = \frac{4}{1}\cdot\frac{1}{w} = \frac{4\cdot 1}{1\cdot w} = \mathbf{\frac{4}{w}}$

10. $\frac{x^2y^2z^2}{x^3}\cdot\left(\frac{y^2}{z^3}\cdot\frac{2z}{x}\right) = \frac{x^2y^2z^2}{x^3}\cdot\left(\frac{y^2\cdot 2z}{z^3\cdot x}\right) = \frac{x^2y^2z^2}{x^3}\cdot\left(\frac{2y^2z}{xz^3}\right) = \frac{x^2y^2z^2}{x^3}\cdot\frac{2y^2z}{xz^3} = \frac{x^2y^2z^2\cdot 2y^2z}{x^3\cdot xz^3} = \frac{2x^2y^4z^3}{x^4z^3}$

$= \frac{2\cancel{x^2}y^4\cancel{z^3}}{\underset{x^2}{\cancel{x^4}}\,\cancel{z^3}} = \mathbf{\frac{2y^4}{x^2}}$

Section 5.3 Case IIIb Solutions - Multiplication of Algebraic Fractions (More Difficult Cases)

1. $\frac{x^2+5x-6}{x^2-9}\cdot\frac{x^2-2x-3}{x^2+7x+6} = \frac{(x-1)(x+6)}{(x-3)(x+3)}\cdot\frac{(x+1)(x-3)}{(x+6)(x+1)} = \frac{(x-1)\cancel{(x+6)}}{\cancel{(x-3)}(x+3)}\cdot\frac{\cancel{(x+1)}\cancel{(x-3)}}{\cancel{(x+6)}\cancel{(x+1)}} = \mathbf{\frac{x-1}{x+3}}$

2. $\frac{x^3}{x^2-x-2}\cdot\frac{2x-4}{x} = \frac{x^3}{(x-2)(x+1)}\cdot\frac{2(x-2)}{x} = \frac{\overset{x^2}{\cancel{x^3}}}{\cancel{(x-2)}(x+1)}\cdot\frac{2\cancel{(x-2)}}{\cancel{x}} = \frac{x^2}{x+1}\cdot\frac{2}{1} = \mathbf{\frac{2x^2}{x+1}}$

3. $\frac{x^2+xy-6y^2}{x^2-xy-2y^2}\cdot\frac{x^2-y^2}{x+3y} = \frac{(x-2y)(x+3y)}{(x-2y)(x+y)}\cdot\frac{(x-y)(x+y)}{(x+3y)} = \frac{\cancel{(x-2y)}\cancel{(x+3y)}}{\cancel{(x-2y)}\cancel{(x+y)}}\cdot\frac{(x-y)\cancel{(x+y)}}{\cancel{(x+3y)}} = \frac{x-y}{1} = \mathbf{x-y}$

4. $\frac{(x-1)^3}{2x^2+5x-12}\cdot\frac{4x-6}{(x-1)^2} = \frac{(x-1)(x-1)^2}{(2x-3)(x+4)}\cdot\frac{2(2x-3)}{(x-1)^2} = \frac{(x-1)\cancel{(x-1)^2}}{\cancel{(2x-3)}(x+4)}\cdot\frac{2\cancel{(2x-3)}}{\cancel{(x-1)^2}} = \frac{x-1}{x+4}\cdot\frac{2}{1} = \mathbf{\frac{2(x-1)}{x+4}}$

5. $\frac{6x^2+7x+2}{x^2-16}\cdot\frac{x^2+3x-4}{3x^2-x-2} = \frac{(2x+1)(3x+2)}{(x-4)(x+4)}\cdot\frac{(x-1)(x+4)}{(3x+2)(x-1)} = \frac{(2x+1)\cancel{(3x+2)}}{(x-4)\cancel{(x+4)}}\cdot\frac{\cancel{(x-1)}\cancel{(x+4)}}{\cancel{(3x+2)}\cancel{(x-1)}} = \mathbf{\frac{2x+1}{x-4}}$

6. $\frac{2x^2+x-3}{x^3+x^2-2x}\cdot\frac{x^2-2x}{4x^2-9} = \frac{(x-1)(2x+3)}{x\left(x^2+x-2\right)}\cdot\frac{x(x-2)}{(2x-3)(2x+3)} = \frac{\cancel{(x-1)}\cancel{(2x+3)}}{\cancel{x}\cancel{(x-1)}(x+2)}\cdot\frac{\cancel{x}(x-2)}{(2x-3)\cancel{(2x+3)}} = \mathbf{\frac{x-2}{(x+2)(2x-3)}}$

7. $\frac{x^2-2x-15}{2x^2-9x-5}\cdot\frac{2x^2+5x+2}{x^2+5x+6} = \frac{(x-5)(x+3)}{(2x+1)(x-5)}\cdot\frac{(x+2)(2x+1)}{(x+3)(x+2)} = \frac{\cancel{(x-5)}\cancel{(x+3)}}{\cancel{(2x+1)}\cancel{(x-5)}}\cdot\frac{\cancel{(x+2)}\cancel{(2x+1)}}{\cancel{(x+3)}\cancel{(x+2)}} = \mathbf{1}$

8. $\frac{3x+2}{6x^2+7x+2}\cdot\frac{2x^2+x}{4} = \frac{3x+2}{(2x+1)(3x+2)}\cdot\frac{x(2x+1)}{4} = \frac{\cancel{(3x+2)}}{\cancel{(2x+1)}\cancel{(3x+2)}}\cdot\frac{x\cancel{(2x+1)}}{4} = \mathbf{\frac{x}{4}}$

9. $\frac{x^4}{x^2+8x+15}\cdot\frac{x+5}{x^3} = \frac{x^4}{(x+3)(x+5)}\cdot\frac{x+5}{x^3} = \frac{\overset{x}{\cancel{x^4}}}{(x+3)\cancel{(x+5)}}\cdot\frac{\cancel{(x+5)}}{\cancel{x^3}} = \mathbf{\frac{x}{x+3}}$

10. $\frac{6x^2+17x+5}{3x^2-2x-1}\cdot\frac{x^2-1}{x^3+x^2} = \frac{(2x+5)(3x+1)}{(3x+1)(x-1)}\cdot\frac{(x-1)(x+1)}{x^2(x+1)} = \frac{(2x+5)\cancel{(3x+1)}}{\cancel{(3x+1)}\cancel{(x-1)}}\cdot\frac{\cancel{(x-1)}\cancel{(x+1)}}{x^2\cancel{(x+1)}} = \mathbf{\frac{2x+5}{x^2}}$

Section 5.3 Case IVa Solutions - Division of Algebraic Fractions (Simple Cases)

1. $\frac{x^2y}{x^3y^2}\div xy = \frac{x^2y}{x^3y^2}\div\frac{xy}{1} = \frac{x^2y}{x^3y^2}\cdot\frac{1}{xy} = \frac{x^2y\cdot 1}{x^3y^2\cdot xy} = \frac{x^2y}{x^4y^3} = \frac{\cancel{x^2}\cancel{y}}{\underset{x^2}{\cancel{x^4}}\,\underset{y^2}{\cancel{y^3}}} = \mathbf{\frac{1}{x^2y^2}}$

2. $\frac{uv^2w}{vw^2}\div\frac{uv^3}{w} = \frac{uv^2w}{vw^2}\cdot\frac{w}{uv^3} = \frac{uv^2w\cdot w}{vw^2\cdot uv^3} = \frac{uv^2w^2}{uv^4w^2} = \frac{\cancel{u}\cancel{v^2}\cancel{w^2}}{\cancel{u}\,\underset{v^2}{\cancel{v^4}}\,\cancel{w^2}} = \mathbf{\frac{1}{v^2}}$

3. $a^2b^2c^4\div\frac{a^2b}{2ac} = \frac{a^2b^2c^4}{1}\div\frac{a^2b}{2ac} = \frac{a^2b^2c^4}{1}\cdot\frac{2ac}{a^2b} = \frac{a^2b^2c^4\cdot 2ac}{1\cdot a^2b} = \frac{2a^3b^2c^5}{a^2b} = \frac{2\,\overset{a}{\cancel{a^3}}\,\overset{b}{\cancel{b^2}}\,c^5}{\cancel{a^2}\cancel{b}} = \frac{2abc^5}{1} = \mathbf{2abc^5}$

4. $\frac{xyz}{x^2z^3}\div\frac{x^2z^2}{yz} = \frac{xyz}{x^2z^3}\cdot\frac{yz}{x^2z^2} = \frac{xyz\cdot yz}{x^2z^3\cdot x^2z^2} = \frac{xy^2z^2}{x^4z^5} = \frac{\cancel{x}y^2\cancel{z^2}}{\underset{x^3}{\cancel{x^4}}\,\underset{z^3}{\cancel{z^5}}} = \mathbf{\frac{y^2}{x^3z^3}}$

5. $\left(\frac{uv^2}{v^3}\div 2u^2\right)\div\frac{uv}{3} = \left(\frac{uv^2}{v^3}\div\frac{2u^2}{1}\right)\div\frac{uv}{3} = \left(\frac{uv^2}{v^3}\cdot\frac{1}{2u^2}\right)\div\frac{uv}{3} = \left(\frac{uv^2\cdot 1}{v^3\cdot 2u^2}\right)\div\frac{uv}{3} = \frac{uv^2}{2u^2v^3}\div\frac{uv}{3} = \frac{uv^2}{2u^2v^3}\cdot\frac{3}{uv}$

$= \frac{uv^2\cdot 3}{2u^2v^3\cdot uv} = \frac{3uv^2}{2u^3v^4} = \frac{3\cancel{u}\cancel{v^2}}{2\,\underset{u^2}{\cancel{u^3}}\,\underset{v^2}{\cancel{v^4}}} = \mathbf{\frac{3}{2u^2v^2}}$

6. $\frac{x^2y^2z}{xz}\div\left(x^2y\div\frac{4}{yz^3}\right) = \frac{x^2y^2z}{xz}\div\left(\frac{x^2y}{1}\div\frac{4}{yz^3}\right) = \frac{x^2y^2z}{xz}\div\left(\frac{x^2y}{1}\cdot\frac{yz^3}{4}\right) = \frac{x^2y^2z}{xz}\div\left(\frac{x^2y\cdot yz^3}{1\cdot 4}\right) = \frac{x^2y^2z}{xz}\div\frac{x^2y^2z^3}{4}$

$= \frac{x^2y^2z}{xz}\cdot\frac{4}{x^2y^2z^3} = \frac{x^2y^2z\cdot 4}{xz\cdot x^2y^2z^3} = \frac{4x^2y^2z}{x^3y^2z^4} = \frac{4\cancel{x^2}\cancel{y^2}\cancel{z}}{\underset{x}{\cancel{x^3}}\,\cancel{y^2}\,\underset{z^3}{\cancel{z^4}}} = \mathbf{\frac{4}{xz^3}}$

7. $a^3b^3\div\left(\frac{a^2b^2}{b}\div b^3\right) = \frac{a^3b^3}{1}\div\left(\frac{a^2b^2}{b}\div\frac{b^3}{1}\right) = \frac{a^3b^3}{1}\div\left(\frac{a^2b^2}{b}\cdot\frac{1}{b^3}\right) = \frac{a^3b^3}{1}\div\left(\frac{a^2b^2\cdot 1}{b\cdot b^3}\right) = \frac{a^3b^3}{1}\div\frac{a^2b^2}{b^4} = \frac{a^3b^3}{1}\cdot\frac{b^4}{a^2b^2}$

$$= \frac{a^3b^3 \cdot b^4}{1 \cdot a^2b^2} = \frac{a^3b^7}{a^2b^2} = \frac{\overset{a}{\cancel{a^3}}\ \overset{b^5}{\cancel{b^7}}}{\cancel{a^2}\cancel{b^2}} = \frac{ab^5}{1} = \mathbf{ab^5}$$

8. $$\frac{4xyz}{x^2y^2z^3} \div \frac{8y}{y^2z^3} = \frac{4xyz}{x^2y^2z^3} \cdot \frac{y^2z^3}{8y} = \frac{4xyz \cdot y^2z^3}{x^2y^2z^3 \cdot 8y} = \frac{4xy^3z^4}{8x^2y^3z^3} = \frac{\cancel{4}\cancel{x}\cancel{y^3}\ \overset{z}{\cancel{z^4}}}{\underset{2}{\cancel{8}}\underset{x}{\cancel{x^2}}\ \cancel{y^3}\cancel{z^3}} = \frac{z}{\mathbf{2x}}$$

9. $$\left(\frac{3xy}{x^3y^2} \div \frac{2x}{y^2}\right) \div \left(\frac{x}{y^2} \div \frac{x^3}{4y}\right) = \left(\frac{3xy}{x^3y^2} \cdot \frac{y^2}{2x}\right) \div \left(\frac{x}{y^2} \cdot \frac{4y}{x^3}\right) = \left(\frac{3xy \cdot y^2}{x^3y^2 \cdot 2x}\right) \div \left(\frac{x \cdot 4y}{y^2 \cdot x^3}\right) = \frac{3xy^3}{2x^4y^2} \div \frac{4xy}{x^3y^2} = \frac{3xy^3}{2x^4y^2} \cdot \frac{x^3y^2}{4xy}$$

$$= \frac{3xy^3 \cdot x^3y^2}{2x^4y^2 \cdot 4xy} = \frac{3x^4y^5}{8x^5y^3} = \frac{3\cancel{x^4}\ \overset{y^2}{\cancel{y^5}}}{8\underset{x}{\cancel{x^5}}\ \cancel{y^3}} = \frac{\mathbf{3y^2}}{\mathbf{8x}}$$

10. $$\left(\frac{m^3n^2}{n^3} \div 3n^2\right) \div m^4 = \left(\frac{m^3n^2}{n^3} \div \frac{3n^2}{1}\right) \div \frac{m^4}{1} = \left(\frac{m^3n^2}{n^3} \cdot \frac{1}{3n^2}\right) \div \frac{m^4}{1} = \left(\frac{m^3n^2 \cdot 1}{n^3 \cdot 3n^2}\right) \div \frac{m^4}{1} = \frac{m^3n^2}{3n^5} \div \frac{m^4}{1} = \frac{m^3n^2}{3n^5} \cdot \frac{1}{m^4}$$

$$= \frac{m^3n^2 \cdot 1}{3n^5 \cdot m^4} = \frac{m^3n^2}{3m^4n^5} = \frac{\cancel{m^3}\cancel{n^2}}{3\underset{m}{\cancel{m^4}}\ \underset{n^3}{\cancel{n^5}}} = \frac{\mathbf{1}}{\mathbf{3mn^3}}$$

Section 5.3 Case IVb Solutions - Division of Algebraic Fractions (More Difficult Cases)

1. $$\frac{x^2-2x-15}{x^2-25} \div \frac{x^2+3x}{x^2+6x+5} = \frac{x^2-2x-15}{x^2-25} \cdot \frac{x^2+6x+5}{x^2+3x} = \frac{(x+3)(x-5)}{(x-5)(x+5)} \cdot \frac{(x+5)(x+1)}{x(x+3)} = \frac{\cancel{(x+3)}\cancel{(x-5)}}{\cancel{(x-5)}\cancel{(x+5)}} \cdot \frac{\cancel{(x+5)}(x+1)}{x\cancel{(x+3)}}$$

$$= \frac{1}{1} \cdot \frac{x+1}{x} = \frac{\mathbf{x+1}}{\mathbf{x}}$$

2. $$x^3+4x^2-5x \div \frac{x^3+5x^2}{x^3} = \frac{x^3+4x^2-5x}{1} \cdot \frac{x^3}{x^3+5x^2} = \frac{x\left(x^2+4x-5\right)}{1} \cdot \frac{x^3}{x^2(x+5)} = \frac{x(x+5)(x-1)}{1} \cdot \frac{x^3}{x^2(x+5)}$$

$$= \frac{x\cancel{(x+5)}(x-1)}{1} \cdot \frac{\overset{x}{\cancel{x^3}}}{\cancel{x^2}\cancel{(x+5)}} = \frac{x(x-1)}{1} \cdot \frac{x}{1} = \frac{x^2(x-1)}{1} = \mathbf{x^2(x-1)}$$

3. $$\frac{6x^2+7x+2}{3x^2-x-2} \div \frac{10x+5}{x^2-2x+1} = \frac{6x^2+7x+2}{3x^2-x-2} \cdot \frac{x^2-2x+1}{10x+5} = \frac{(2x+1)(3x+2)}{(3x+2)(x-1)} \cdot \frac{(x-1)^2}{5(2x+1)} = \frac{\cancel{(2x+1)}\cancel{(3x+2)}}{\cancel{(3x+2)}\cancel{(x-1)}} \cdot \frac{(x-1)^2}{5\cancel{(2x+1)}}$$

$$= \frac{1}{1} \cdot \frac{x-1}{5} = \frac{\mathbf{x-1}}{\mathbf{5}}$$

4. $$\frac{x^2+x-2}{x^2-x-6} \div \frac{x^2+4x-5}{x^2-6x+9} = \frac{x^2+x-2}{x^2-x-6} \cdot \frac{x^2-6x+9}{x^2+4x-5} = \frac{(x+2)(x-1)}{(x-3)(x+2)} \cdot \frac{(x-3)^2}{(x+5)(x-1)} = \frac{\cancel{(x+2)}\cancel{(x-1)}}{\cancel{(x-3)}\cancel{(x+2)}} \cdot \frac{(x-3)^2}{(x+5)\cancel{(x-1)}}$$

$$= \frac{1}{1} \cdot \frac{x-3}{x+5} = \frac{\mathbf{x-3}}{\mathbf{x+5}}$$

5. $$\frac{2x^2+7x+3}{3x^4-12x^3} \div \frac{2x+1}{x^3-4x^2} = \frac{2x^2+7x+3}{3x^4-12x^3} \cdot \frac{x^3-4x^2}{2x+1} = \frac{(2x+1)(x+3)}{3x^3(x-4)} \cdot \frac{x^2(x-4)}{2x+1} = \frac{\cancel{(2x+1)}(x+3)}{3\underset{x}{\cancel{x^3}}\cancel{(x-4)}} \cdot \frac{\cancel{x^2}\cancel{(x-4)}}{\cancel{(2x+1)}}$$

$$= \frac{x+3}{3x} \cdot \frac{1}{1} = \frac{\mathbf{x+3}}{\mathbf{3x}}$$

6. $$\frac{32x^3y^3z}{8\left(x^2+5x+6\right)} \div \frac{xyz^2}{4x^2(x+3)} = \frac{32x^3y^3z}{8\left(x^2+5x+6\right)} \cdot \frac{4x^2(x+3)}{xyz^2} = \frac{32x^3y^3z}{8(x+2)(x+3)} \cdot \frac{4x^2(x+3)}{xyz^2} = \frac{\overset{4}{\cancel{32}}\overset{x^2}{\cancel{x^3}}\ \overset{y^2}{\cancel{y^3}}\ \cancel{z}}{\cancel{8}(x+2)\cancel{(x+3)}} \cdot \frac{4x^2\cancel{(x+3)}}{\cancel{x}\cancel{y}\ \underset{z}{\cancel{z^2}}}$$

$$= \frac{4x^2y^2}{x+2}\cdot\frac{4x^2}{z} = \mathbf{\frac{16x^4y^2}{z(x+2)}}$$

7. $$\frac{2x^2+7x+3}{x^2-1} \div \frac{2x^2-3x-2}{x^2-3x+2} = \frac{2x^2+7x+3}{x^2-1}\cdot\frac{x^2-3x+2}{2x^2-3x-2} = \frac{(x+3)(2x+1)}{(x-1)(x+1)}\cdot\frac{(x-1)(x-2)}{(2x+1)(x-2)}$$

$$= \frac{(x+3)\cancel{(2x+1)}}{\cancel{(x-1)}(x+1)}\cdot\frac{\cancel{(x-1)}\cancel{(x-2)}}{\cancel{(2x+1)}\cancel{(x-2)}} = \frac{x+3}{x+1}\cdot\frac{1}{1} = \mathbf{\frac{x+3}{x+1}}$$

8. $$\frac{x^2+5xy+6y^2}{x^2+5xy+4y^2} \div \frac{x^4+3x^3y}{x^2+4xy} = \frac{x^2+5xy+6y^2}{x^2+5xy+4y^2}\cdot\frac{x^2+4xy}{x^4+3x^3y} = \frac{(x+3y)(x+2y)}{(x+y)(x+4y)}\cdot\frac{x(x+4y)}{x^3(x+3y)} = \frac{\cancel{(x+3y)}(x+2y)}{(x+y)\cancel{(x+4y)}}\cdot\frac{\cancel{x}\cancel{(x+4y)}}{\underset{x^2}{\cancel{x^3}}\cancel{(x+3y)}}$$

$$= \frac{x+2y}{x+y}\cdot\frac{1}{x^2} = \mathbf{\frac{x+2y}{x^2(x+y)}}$$

9. $$\frac{6x^2+11x+3}{2x^2+5x+3} \div \frac{3x^2-4x+1}{(x-1)^2} = \frac{6x^2+11x+3}{2x^2+5x+3}\cdot\frac{(x-1)^2}{3x^2-4x+1} = \frac{(3x+1)(2x+3)}{(2x+3)(x+1)}\cdot\frac{(x-1)^2}{(3x-1)(x-1)}$$

$$= \frac{(3x+1)\cancel{(2x+3)}}{\cancel{(2x+3)}(x+1)}\cdot\frac{\overset{(x-1)}{\cancel{(x-1)^2}}}{(3x-1)\cancel{(x-1)}} = \frac{3x+1}{x+1}\cdot\frac{x-1}{3x-1} = \mathbf{\frac{(3x+1)(x-1)}{(x+1)(3x-1)}}$$

10. $$\frac{a^3b^2c(a-b)^2}{c^2(a+2b)} \div \frac{a^2b(a-b)}{c(a^2+3ab+2b^2)} = \frac{a^3b^2c(a-b)^2}{c^2(a+2b)}\cdot\frac{c(a^2+3ab+2b^2)}{a^2b(a-b)} = \frac{a^3b^2c(a-b)^2}{c^2(a+2b)}\cdot\frac{c(a+2b)(a+b)}{a^2b(a-b)}$$

$$= \frac{a^3b^2c\overset{(a-b)}{\cancel{(a-b)^2}}}{c^2\cancel{(a+2b)}}\cdot\frac{c\cancel{(a+2b)}(a+b)}{a^2b\cancel{(a-b)}} = \frac{a^3b^2c(a-b)}{c^2}\cdot\frac{c(a+b)}{a^2b} = \frac{a^3b^2c^2(a-b)(a+b)}{a^2bc^2} = \frac{\overset{a}{\cancel{a^3}}\,\overset{b}{\cancel{b^2}}\,\cancel{c^2}(a-b)(a+b)}{\cancel{a^2}\cancel{b}\cancel{c^2}}$$

$$= \frac{ab(a-b)(a+b)}{1} = \mathbf{ab(a-b)(a+b)}$$

Section 5.3 Case V Solutions - Mixed Operations Involving Algebraic Fractions

1. $$\left(\frac{x}{x^3}\cdot\frac{2x}{4}\right)\div x^2 = \left(\frac{x\cdot 2x}{x^3\cdot 4}\right)\div\frac{x^2}{1} = \frac{2x^2}{4x^3}\div\frac{x^2}{1} = \frac{2x^2}{4x^3}\cdot\frac{1}{x^2} = \frac{2x^2\cdot 1}{4x^3\cdot x^2} = \frac{\cancel{2x^2}}{\underset{2}{\cancel{4}}x^3\cdot\cancel{x^2}} = \mathbf{\frac{1}{2x^3}}$$

2. $$\left(\frac{ab}{b^2}+\frac{a}{c^2}\right)-\frac{1}{b^2c^2} = \left(\frac{(ab\cdot c^2)+(a\cdot b^2)}{b^2\cdot c^2}\right)-\frac{1}{b^2c^2} = \frac{abc^2+ab^2}{b^2c^2}-\frac{1}{b^2c^2} = \frac{abc^2+ab^2-1}{b^2c^2} = \mathbf{\frac{ab(c^2+b)-1}{b^2c^2}}$$

3. $$\left(\frac{yz}{z^2}-\frac{y}{y^2}\right)+\left(\frac{1}{z}-\frac{1}{y}\right) = \left(\frac{(yz\cdot y^2)-(y\cdot z^2)}{z^2\cdot y^2}\right)+\left(\frac{(1\cdot y)-(1\cdot z)}{z\cdot y}\right) = \left(\frac{y^3z-yz^2}{z^2y^2}\right)+\left(\frac{y-z}{zy}\right) = \frac{yz(y^2-z)}{z^2y^2}+\frac{y-z}{zy}$$

$$= \frac{\cancel{y}\cancel{z}(y^2-z)}{\underset{z}{\cancel{z^2}}\,\underset{y}{\cancel{y^2}}}+\frac{y-z}{zy} = \frac{y^2-z}{zy}+\frac{y-z}{zy} = \frac{y^2-z+y+z}{zy} = \frac{y^2-\cancel{z}+y+\cancel{z}}{zy} = \frac{y^2+y}{zy} = \frac{\cancel{y}(y+1)}{z\cancel{y}} = \mathbf{\frac{y+1}{z}}$$

4. $$\left(\frac{1}{a}\div a^2\right)\cdot\left(\frac{1}{b}\div\frac{b^2}{a^2}\right) = \left(\frac{1}{a}\div\frac{a^2}{1}\right)\cdot\left(\frac{1}{b}\div\frac{b^2}{a^2}\right) = \left(\frac{1}{a}\cdot\frac{1}{a^2}\right)\cdot\left(\frac{1}{b}\cdot\frac{a^2}{b^2}\right) = \left(\frac{1\cdot 1}{a\cdot a^2}\right)\cdot\left(\frac{1\cdot a^2}{b\cdot b^2}\right) = \frac{1}{a^3}\cdot\frac{a^2}{b^3} = \frac{1\cdot a^2}{a^3\cdot b^3}$$

$$= \frac{a^2}{a^3b^3} = \frac{\cancel{a^2}}{\underset{a}{\cancel{a^3}}\,b^3} = \mathbf{\frac{1}{ab^3}}$$

5. $\left(\frac{2}{x}-\frac{4}{y}\right)+\left(\frac{1}{x}\div y\right) = \left(\frac{2}{x}-\frac{4}{y}\right)+\left(\frac{1}{x}\div\frac{y}{1}\right) = \left(\frac{(2\cdot y)-(4\cdot x)}{x\cdot y}\right)+\left(\frac{1}{x}\cdot\frac{1}{y}\right) = \left(\frac{2y-4x}{xy}\right)+\left(\frac{1\cdot 1}{x\cdot y}\right) = \frac{2y-4x}{xy}+\frac{1}{xy}$

$= \frac{2y-4x+1}{xy} = \mathbf{\frac{2(y-2x)+1}{xy}}$

6. $\left(\frac{u}{v^2}\cdot\frac{v}{u^2}\cdot\frac{1}{v}\right)\div u = \left(\frac{u}{v^2}\cdot\frac{v}{u^2}\cdot\frac{1}{v}\right)\div\frac{u}{1} = \left(\frac{u\cdot v\cdot 1}{v^2\cdot u^2\cdot v}\right)\div\frac{u}{1} = \frac{uv}{u^2v^3}\div\frac{u}{1} = \frac{uv}{u^2v^3}\cdot\frac{1}{u} = \frac{uv\cdot 1}{u^2v^3\cdot u} = \frac{uv}{u^3v^3} = \frac{\cancel{uv}}{\underset{u^2\,v^2}{\cancel{u^3}\,\cancel{v^3}}} = \mathbf{\frac{1}{u^2v^2}}$

7. $\left(\frac{a}{b^2}\div\frac{1}{a^2}\right)+\frac{2}{3} = \left(\frac{a}{b^2}\cdot\frac{a^2}{1}\right)+\frac{2}{3} = \left(\frac{a\cdot a^2}{b^2\cdot 1}\right)+\frac{2}{3} = \frac{a^3}{b^2}+\frac{2}{3} = \frac{(3\cdot a^3)+(2\cdot b^2)}{3\cdot b^2} = \mathbf{\frac{3a^3+2b^2}{3b^2}}$

8. $\left(1+\frac{2}{x}\right)-\left(\frac{1}{x}+2\right) = \left(\frac{1}{1}+\frac{2}{x}\right)-\left(\frac{1}{x}+\frac{2}{1}\right) = \left(\frac{(1\cdot x)+(2\cdot 1)}{1\cdot x}\right)-\left(\frac{(1\cdot 1)+(2\cdot x)}{x\cdot 1}\right) = \left(\frac{x+2}{x}\right)-\left(\frac{1+2x}{x}\right) = \frac{x+2-(1+2x)}{x}$

$= \frac{x+2-1-2x}{x} = \frac{(x-2x)+(2-1)}{x} = \frac{-x+1}{x} = \mathbf{\frac{1-x}{x}}$

9. $\left(\frac{x^2y^2z}{z^3}\cdot\frac{z^2}{xy}\right)\div xyz^2 = \left(\frac{x^2y^2z}{z^3}\cdot\frac{z^2}{xy}\right)\div\frac{xyz^2}{1} = \left(\frac{x^2y^2z\cdot z^2}{z^3\cdot xy}\right)\div\frac{xyz^2}{1} = \frac{x^2y^2z^3}{xyz^3}\div\frac{xyz^2}{1} = \frac{x^2y^2z^3}{xyz^3}\cdot\frac{1}{xyz^2}$

$= \frac{x^2y^2z^3\cdot 1}{xyz^3\cdot xyz^2} = \frac{x^2y^2z^3}{x^2y^2z^5} = \frac{\cancel{x^2y^2z^3}}{\underset{z^2}{\cancel{x^2y^2z^5}}} = \mathbf{\frac{1}{z^2}}$

10. $\left[\left(\frac{x}{y^2}+\frac{1}{y^2}\right)-\frac{2}{y}\right]\cdot\frac{y^3}{2x} = \left[\left(\frac{x+1}{y^2}\right)-\frac{2}{y}\right]\cdot\frac{y^3}{2x} = \left[\frac{(x+1)\cdot y-(2\cdot y^2)}{y^2\cdot y}\right]\cdot\frac{y^3}{2x} = \left[\frac{xy+y-2y^2}{y^3}\right]\cdot\frac{y^3}{2x} = \frac{(xy+y-2y^2)\cdot y^3}{y^3\cdot 2x}$

$= \frac{(xy+y-2y^2)y^3}{2xy^3} = \frac{xy+y-2y^2}{2x} = \frac{y(x+1-2y)}{2x} = \mathbf{\frac{y(x-2y+1)}{2x}}$

Section 5.4 Case I Solutions - Addition and Subtraction of Complex Algebraic Fractions

1. $\frac{2-\frac{1}{5a}}{3-\frac{2}{15a}} = \frac{\frac{2}{1}-\frac{1}{5a}}{\frac{3}{1}-\frac{2}{15a}} = \frac{\frac{(2\cdot 5a)-(1\cdot 1)}{1\cdot 5a}}{\frac{(3\cdot 15a)-(2\cdot 1)}{1\cdot 15a}} = \frac{\frac{10a-1}{5a}}{\frac{45a-2}{15a}} = \frac{(10a-1)\cdot\overset{3}{\cancel{15a}}}{\cancel{5a}\cdot(45a-2)} = \frac{3(10a-1)}{45a-2} = \mathbf{\frac{30a-3}{45a-2}}$

2. $\frac{\frac{2x^3y^2z}{4x^2z}}{\frac{2x}{xy^2}}-1 = \frac{(2x^3y^2z)\cdot(xy^2)}{(4x^2z)\cdot(2x)}-1 = \frac{2x^4y^4z}{8x^3z}-1 = \frac{\cancel{2}\overset{x}{\cancel{x^4}}\,y^4\cancel{z}}{\underset{4}{\cancel{8}}\cancel{x^3}\cancel{z}}-1 = \frac{xy^4}{4}-1 = \frac{xy^4}{4}-\frac{1}{1} = \frac{(xy^4\cdot 1)-(1\cdot 4)}{4\cdot 1} = \mathbf{\frac{xy^4-4}{4}}$

3. $\frac{\frac{2}{a}+\frac{1}{a^3}}{\frac{2}{a^3}+2} = \frac{\frac{2}{a}+\frac{1}{a^3}}{\frac{2}{a^3}+\frac{2}{1}} = \frac{\frac{(2\cdot a^3)+(1\cdot a)}{a\cdot a^3}}{\frac{(2\cdot 1)+(2\cdot a^3)}{a^3\cdot 1}} = \frac{\frac{2a^3+a}{a^4}}{\frac{2+2a^3}{a^3}} = \frac{(2a^3+a)\cdot a^3}{a^4\cdot(2+2a^3)} = \frac{\cancel{a}(2a^2+1)}{2\cancel{a}(1+a^3)} = \frac{2a^2+1}{2(1+a^3)} = \mathbf{\frac{2a^2+1}{2(1+a)(1-a+a^2)}}$

4. $\frac{a}{a+\frac{1}{a^2}} = \frac{\frac{a}{1}}{\frac{a}{1}+\frac{1}{a^2}} = \frac{\frac{a}{1}}{\frac{(a\cdot a^2)+(1\cdot 1)}{1\cdot a^2}} = \frac{\frac{a}{1}}{\frac{a^3+1}{a^2}} = \frac{a\cdot a^2}{(a^3+1)\cdot 1} = \frac{a^3}{a^3+1} = \mathbf{\frac{a^3}{(a+1)(a^2-a+1)}}$

5. $$\frac{\frac{x}{y}-3}{3-\frac{y}{x}} = \frac{\frac{x}{y}-\frac{3}{1}}{\frac{3}{1}-\frac{y}{x}} = \frac{\frac{(x\cdot1)-(3\cdot y)}{y\cdot1}}{\frac{(3\cdot x)-(y\cdot1)}{1\cdot x}} = \frac{\frac{x-3y}{y}}{\frac{3x-y}{x}} = \frac{x\cdot(x-3y)}{y\cdot(3x-y)} = \mathbf{\frac{x(x-3y)}{y(3x-y)}}$$

6. $$\frac{\frac{1}{x}-\frac{1}{x+4}}{\frac{3}{x+4}+1} = \frac{\frac{1}{x}-\frac{1}{x+4}}{\frac{3}{x+4}+\frac{1}{1}} = \frac{\frac{[1\cdot(x+4)]-(1\cdot x)}{x\cdot(x+4)}}{\frac{(3\cdot1)+[1\cdot(x+4)]}{(x+4)\cdot1}} = \frac{\frac{\cancel{x}+4-\cancel{x}}{x\cdot(x+4)}}{\frac{3+x+4}{x+4}} = \frac{\frac{4}{x\cdot(x+4)}}{\frac{x+7}{x+4}} = \frac{4\cdot(\cancel{x+4})}{x\cdot(\cancel{x+4})\cdot(x+7)} = \mathbf{\frac{4}{x(x+7)}}$$

7. $$\frac{\frac{w}{w-1}+\frac{1}{w+1}}{1-\frac{1}{w-1}} = \frac{\frac{w}{w-1}+\frac{1}{w+1}}{\frac{1}{1}-\frac{1}{w-1}} = \frac{\frac{[w\cdot(w+1)]+[1\cdot(w-1)]}{(w-1)(w+1)}}{\frac{[1\cdot(w-1)]-(1\cdot1)}{1\cdot(w-1)}} = \frac{\frac{w^2+w+w-1}{(w-1)(w+1)}}{\frac{w-1-1}{w-1}} = \frac{\frac{w^2+2w-1}{(w-1)(w+1)}}{\frac{w-2}{w-1}}$$

$$= \frac{(w^2+2w-1)\cdot(\cancel{w-1})}{[(\cancel{w-1})(w+1)]\cdot(w-2)} = \mathbf{\frac{w^2+2w-1}{(w+1)(w-2)}}$$

8. $$\frac{4-\frac{1}{x+1}}{\frac{2}{x+1}-2} = \frac{\frac{4}{1}-\frac{1}{x+1}}{\frac{2}{x+1}-\frac{2}{1}} = \frac{\frac{[4\cdot(x+1)]-(1\cdot1)}{1\cdot(x+1)}}{\frac{(2\cdot1)-[2\cdot(x+1)]}{(x+1)\cdot1}} = \frac{\frac{4x+4-1}{x+1}}{\frac{2-(2x+2)}{x+1}} = \frac{\frac{4x+3}{x+1}}{\frac{2-2x-2}{x+1}} = \frac{\frac{4x+3}{x+1}}{\frac{-2x}{x+1}} = \frac{(4x+3)\cdot(\cancel{x+1})}{-2x\cdot(\cancel{x+1})} = \mathbf{\frac{4x+3}{-2x}}$$

9. $$\frac{\frac{2}{x^2-1}-\frac{1}{x-1}}{\frac{1}{x-1}+\frac{2}{x+1}} = \frac{\frac{[2\cdot(x-1)]-[1\cdot(x^2-1)]}{(x^2-1)\cdot(x-1)}}{\frac{[1\cdot(x+1)]+[2\cdot(x-1)]}{(x-1)\cdot(x+1)}} = \frac{\frac{2x-2-x^2+1}{[(x-1)\cdot(x+1)]\cdot(x-1)}}{\frac{x+1+2x-2}{(x-1)\cdot(x+1)}} = \frac{\frac{-x^2+2x-1}{[(x-1)\cdot(x+1)]\cdot(x-1)}}{\frac{3x-1}{(x-1)\cdot(x+1)}}$$

$$= \frac{(-x^2+2x-1)\cdot(\cancel{x-1})\cdot(\cancel{x+1})}{[(x-1)(\cancel{x+1})]\cdot(\cancel{x-1})\cdot(3x-1)} = \frac{-x^2+2x-1}{(x-1)\cdot(3x-1)} = \frac{-(x^2-2x+1)}{(x-1)\cdot(3x-1)} = \frac{\overset{(x-1)}{-(x-1)^2}}{(\cancel{x-1})\cdot(3x-1)} = \frac{-(x-1)}{3x-1} = \mathbf{-\frac{x-1}{3x-1}}$$

10. $$\frac{\frac{y+3}{y-1}+\frac{1}{y}}{2-\frac{y+1}{y-1}} = \frac{\frac{y+3}{y-1}+\frac{1}{y}}{\frac{2}{1}-\frac{y+1}{y-1}} = \frac{\frac{[(y+3)\cdot y]+[1\cdot(y-1)]}{(y-1)\cdot y}}{\frac{[2\cdot(y-1)]-[(y+1)\cdot1]}{1\cdot(y-1)}} = \frac{\frac{y^2+3y+y-1}{(y-1)\cdot y}}{\frac{2y-2-y-1}{(y-1)}} = \frac{\frac{y^2+4y-1}{(y-1)\cdot y}}{\frac{y-3}{(y-1)}} = \frac{(y^2+4y-1)\cdot(\cancel{y-1})}{[(\cancel{y-1})\cdot y]\cdot(y-3)}$$

$$= \mathbf{\frac{y^2+4y-1}{y(y-3)}}$$

Section 5.4 Case II Solutions - Multiplication of Complex Algebraic Fractions

1. $$\frac{y^2}{\frac{x^2y^2}{xy}\cdot\frac{2x^3y}{x^4}} = \frac{y^2}{\frac{x^2y^2\cdot2x^3y}{xy\cdot x^4}} = \frac{y^2}{\frac{2x^5y^3}{x^5y}} = \frac{\frac{y^2}{1}}{\frac{2x^5y^3}{x^5y}} = \frac{y^2\cdot x^5y}{1\cdot2x^5y^3} = \frac{x^5y^3}{2x^5y^3} = \frac{\cancel{x^5}\cancel{y^3}}{2\cancel{x^5}\cancel{y^3}} = \mathbf{\frac{1}{2}}$$

2. $$\frac{\frac{1}{2x^3y^2}\cdot\frac{4x}{y}}{\frac{1}{xy^2}} = \frac{\frac{1\cdot4x}{2x^3y^2\cdot y}}{\frac{1}{xy^2}} = \frac{\frac{4x}{2x^3y^3}}{\frac{1}{xy^2}} = \frac{4x\cdot xy^2}{2x^3y^3\cdot1} = \frac{4x^2y^2}{2x^3y^3} = \frac{\overset{2}{\cancel{4}}\cancel{x^2}\cancel{y^2}}{\cancel{2}\underset{x}{\cancel{x^3}}\underset{y}{\cancel{y^3}}} = \mathbf{\frac{2}{xy}}$$

3. $$\frac{\frac{x^2}{x^3}}{2x\cdot\frac{3x^2}{18x^5}} = \frac{\frac{x^2}{x^3}}{\frac{2x}{1}\cdot\frac{3x^2}{18x^5}} = \frac{\frac{x^2}{x^3}}{\frac{2x\cdot 3x^2}{1\cdot 18x^5}} = \frac{\frac{x^2}{x^3}}{\frac{6x^3}{18x^5}} = \frac{x^2\cdot 18x^5}{x^3\cdot 6x^3} = \frac{18x^7}{6x^6} = \frac{\overset{3}{\cancel{18}}\,\overset{x}{\cancel{x^7}}}{\cancel{6}\,\cancel{x^6}} = \frac{3x}{1} = \mathbf{3x}$$

4. $$\frac{\left(x^2yz^2\cdot\frac{2x}{y^2}\right)\cdot\frac{3x}{x^3z^4}}{\frac{1}{y^2z^3}} = \frac{\left(\frac{x^2yz^2}{1}\cdot\frac{2x}{y^2}\right)\cdot\frac{3x}{x^3z^4}}{\frac{1}{y^2z^3}} = \frac{\left(\frac{x^2yz^2\cdot 2x}{1\cdot y^2}\right)\cdot\frac{3x}{x^3z^4}}{\frac{1}{y^2z^3}} = \frac{\frac{2x^3yz^2}{y^2}\cdot\frac{3x}{x^3z^4}}{\frac{1}{y^2z^3}} = \frac{\frac{2x^3yz^2\cdot 3x}{y^2\cdot x^3z^4}}{\frac{1}{y^2z^3}} = \frac{\frac{6x^4yz^2}{x^3y^2z^4}}{\frac{1}{y^2z^3}}$$

$$= \frac{6x^4yz^2\cdot y^2z^3}{x^3y^2z^4\cdot 1} = \frac{6x^4y^3z^5}{x^3y^2z^4} = \frac{6\overset{x}{\cancel{x^4}}\,\overset{y}{\cancel{y^3}}\,\overset{z}{\cancel{z^5}}}{\cancel{x^3}\,\cancel{y^2}\,\cancel{z^4}} = \frac{6xyz}{1} = \mathbf{6xyz}$$

5. $$\frac{\left(\frac{2a^2b^2}{a^3}\cdot 4ab^3\right)\cdot\frac{1}{3ab^4}}{\frac{8}{3}} = \frac{\left(\frac{2a^2b^2}{a^3}\cdot\frac{4ab^3}{1}\right)\cdot\frac{1}{3ab^4}}{\frac{8}{3}} = \frac{\left(\frac{2a^2b^2\cdot 4ab^3}{a^3\cdot 1}\right)\cdot\frac{1}{3ab^4}}{\frac{8}{3}} = \frac{\frac{8a^3b^5}{a^3}\cdot\frac{1}{3ab^4}}{\frac{8}{3}} = \frac{\frac{8a^3b^5\cdot 1}{a^3\cdot 3ab^4}}{\frac{8}{3}} = \frac{\frac{8a^3b^5}{3a^4b^4}}{\frac{8}{3}}$$

$$= \frac{8a^3b^5\cdot 3}{3a^4b^4\cdot 8} = \frac{24a^3b^5}{24a^4b^4} = \frac{\cancel{24}\,\cancel{a^3}\,\overset{b}{\cancel{b^5}}}{\cancel{24}\,\underset{a}{\cancel{a^4}}\,\cancel{b^4}} = \mathbf{\frac{b}{a}}$$

6. $$\frac{\frac{u^3v^3w^2}{w^3}\cdot\frac{1}{3w}}{\frac{1}{6uv^2}} = \frac{\frac{u^3v^3w^2\cdot 1}{w^3\cdot 3w}}{\frac{1}{6uv^2}} = \frac{\frac{u^3v^3w^2}{3w^4}}{\frac{1}{6uv^2}} = \frac{u^3v^3w^2\cdot 6uv^2}{3w^4\cdot 1} = \frac{6u^4v^5w^2}{3w^4} = \frac{\overset{2}{\cancel{6}}\,u^4v^5\cancel{w^2}}{\cancel{3}\,\underset{w^2}{\cancel{w^4}}} = \mathbf{\frac{2u^4v^5}{w^2}}$$

7. $$\frac{a^2b^4}{\frac{abc^3}{a^2b^2c^2}\cdot\left(abc\cdot\frac{2b^3c}{ab^2c^2}\right)} = \frac{a^2b^4}{\frac{abc^3}{a^2b^2c^2}\cdot\left(\frac{abc}{1}\cdot\frac{2b^3c}{ab^2c^2}\right)} = \frac{a^2b^4}{\frac{abc^3}{a^2b^2c^2}\cdot\left(\frac{abc\cdot 2b^3c}{1\cdot ab^2c^2}\right)} = \frac{a^2b^4}{\frac{abc^3}{a^2b^2c^2}\cdot\frac{2ab^4c^2}{ab^2c^2}} = \frac{a^2b^4}{\frac{abc^3\cdot 2ab^4c^2}{a^2b^2c^2\cdot ab^2c^2}}$$

$$= \frac{a^2b^4}{\frac{2a^2b^5c^5}{a^3b^4c^4}} = \frac{\frac{a^2b^4}{1}}{\frac{2a^2b^5c^5}{a^3b^4c^4}} = \frac{a^2b^4\cdot a^3b^4c^4}{1\cdot 2a^2b^5c^5} = \frac{a^5b^8c^4}{2a^2b^5c^5} = \frac{\overset{a^3}{\cancel{a^5}}\,\overset{b^3}{\cancel{b^8}}\,\cancel{c^4}}{2\cancel{a^2}\,\cancel{b^5}\,\underset{c}{\cancel{c^5}}} = \mathbf{\frac{a^3b^3}{2c}}$$

8. $$\frac{xyz^3\cdot\frac{xy}{z^2y^2}}{x^2}\cdot\frac{\frac{x^2y}{z}}{xy} = \frac{\frac{xyz^3}{1}\cdot\frac{xy}{z^2y^2}}{x^2}\cdot\frac{\frac{x^2y}{z}}{xy} = \frac{\frac{xyz^3\cdot xy}{1\cdot z^2y^2}}{x^2}\cdot\frac{\frac{x^2y}{z}}{xy} = \frac{\frac{x^2y^2z^3}{z^2y^2}}{x^2}\cdot\frac{\frac{x^2y}{z}}{xy} = \frac{\frac{x^2y^2z^3}{z^2y^2}}{\frac{x^2}{1}}\cdot\frac{\frac{x^2y}{z}}{\frac{xy}{1}}$$

$$= \frac{x^2y^2z^3\cdot 1}{z^2y^2\cdot x^2}\cdot\frac{x^2y\cdot 1}{z\cdot xy} = \frac{x^2y^2z^3}{x^2y^2z^2}\cdot\frac{x^2y}{xyz} = \frac{x^2y^2z^3\cdot x^2y}{x^2y^2z^2\cdot xyz} = \frac{x^4y^3z^3}{x^3y^3z^3} = \frac{\overset{x}{\cancel{x^4}}\,\cancel{y^3}\,\cancel{z^3}}{\cancel{x^3}\,\cancel{y^3}\,\cancel{z^3}} = \frac{x}{1} = \mathbf{x}$$

9. $$\frac{\frac{y^2}{x}\cdot x^3}{\frac{x^3y^2}{2xy}\cdot\frac{4x^4y^2}{x^2}} = \frac{\frac{y^2}{x}\cdot\frac{x^3}{1}}{\frac{x^3y^2}{2xy}\cdot\frac{4x^4y^2}{x^2}} = \frac{\frac{y^2\cdot x^3}{x\cdot 1}}{\frac{x^3y^2\cdot 4x^4y^2}{2xy\cdot x^2}} = \frac{\frac{x^3y^2}{x}}{\frac{4x^7y^4}{2x^3y}} = \frac{x^3y^2\cdot 2x^3y}{x\cdot 4x^7y^4} = \frac{2x^6y^3}{4x^8y^4} = \frac{\cancel{2}\,\cancel{x^6}\,\cancel{y^3}}{\underset{2}{\cancel{4}}\,\underset{x^2}{\cancel{x^8}}\,\underset{y}{\cancel{y^4}}} = \mathbf{\frac{1}{2x^2y}}$$

10. $$\frac{\frac{xyz}{x^2z^2}\cdot\frac{x}{y^3}}{\frac{1}{x^2y^2}\cdot\frac{xz}{y}} = \frac{\frac{xyz\cdot x}{x^2z^2\cdot y^3}}{\frac{1\cdot xz}{x^2y^2\cdot y}} = \frac{\frac{x^2yz}{x^2y^3z^2}}{\frac{xz}{x^2y^3}} = \frac{x^2yz\cdot x^2y^3}{x^2y^3z^2\cdot xz} = \frac{x^4y^4z}{x^3y^3z^3} = \frac{\overset{x}{\cancel{x^4}}\,\overset{y}{\cancel{y^4}}\,\cancel{z}}{\cancel{x^3}\,\cancel{y^3}\,\underset{z^2}{\cancel{z^3}}} = \mathbf{\frac{xy}{z^2}}$$

Section 5.4 Case III Solutions - Division of Complex Algebraic Fractions

1. $$\frac{\frac{xy}{x^3}\div\frac{x^2}{xy^2}}{2x} = \frac{\frac{xy}{x^3}\cdot\frac{xy^2}{x^2}}{2x} = \frac{\frac{xy\cdot xy^2}{x^3\cdot x^2}}{2x} = \frac{\frac{x^2y^3}{x^5}}{2x} = \frac{\frac{x^2y^3}{x^5}}{\frac{2x}{1}} = \frac{x^2y^3\cdot 1}{x^5\cdot 2x} = \frac{x^2y^3}{2x^6} = \frac{\cancel{x^2}y^3}{2\underset{x^4}{\cancel{x^6}}} = \mathbf{\frac{y^3}{2x^4}}$$

2. $$\frac{\frac{1}{a}\div\frac{1}{b}}{\frac{a}{ab^2}\div b} = \frac{\frac{1}{a}\div\frac{1}{b}}{\frac{a}{ab^2}\div\frac{b}{1}} = \frac{\frac{1}{a}\cdot\frac{b}{1}}{\frac{a}{ab^2}\cdot\frac{1}{b}} = \frac{\frac{1\cdot b}{a\cdot 1}}{\frac{a\cdot 1}{ab^2\cdot b}} = \frac{\frac{b}{a}}{\frac{a}{ab^3}} = \frac{b\cdot ab^3}{a\cdot a} = \frac{ab^4}{a^2} = \frac{\cancel{a}b^4}{\underset{a}{\cancel{a^2}}} = \mathbf{\frac{b^4}{a}}$$

3. $$xy\div\left(\frac{\frac{1}{xy}\div x}{x^3}\right) = xy\div\left(\frac{\frac{1}{xy}\div\frac{x}{1}}{x^3}\right) = xy\div\left(\frac{\frac{1}{xy}\cdot\frac{1}{x}}{x^3}\right) = xy\div\left(\frac{\frac{1\cdot 1}{xy\cdot x}}{x^3}\right) = xy\div\frac{\frac{1}{x^2y}}{x^3} = xy\div\frac{\frac{1}{x^2y}}{\frac{x^3}{1}} = xy\div\frac{1\cdot 1}{x^2y\cdot x^3}$$

$$= xy\div\frac{1}{x^5y} = xy\cdot\frac{x^5y}{1} = \frac{xy}{1}\cdot\frac{x^5y}{1} = \frac{xy\cdot x^5y}{1\cdot 1} = \frac{x^6y^2}{1} = \mathbf{x^6y^2}$$

4. $$\frac{\frac{u^2vw}{w^2}\div\frac{u^3v}{w}}{w^3} = \frac{\frac{u^2vw}{w^2}\cdot\frac{w}{u^3v}}{w^3} = \frac{\frac{u^2vw\cdot w}{w^2\cdot u^3v}}{w^3} = \frac{\frac{u^2vw^2}{u^3vw^2}}{w^3} = \frac{\frac{u^2vw^2}{u^3vw^2}}{\frac{w^3}{1}} = \frac{u^2vw^2\cdot 1}{u^3vw^2\cdot w^3} = \frac{u^2vw^2}{u^3vw^5} = \frac{\cancel{u^2}\cancel{v}\cancel{w^2}}{\underset{u}{\cancel{u^3}}\cancel{v}\underset{w^3}{\cancel{w^5}}} = \mathbf{\frac{1}{uw^3}}$$

5. $$\frac{ab^3\div\frac{a^2b^2}{b}}{a\div\frac{a}{b}} = \frac{\frac{ab^3}{1}\div\frac{a^2b^2}{b}}{\frac{a}{1}\div\frac{a}{b}} = \frac{\frac{ab^3}{1}\cdot\frac{b}{a^2b^2}}{\frac{a}{1}\cdot\frac{b}{a}} = \frac{\frac{ab^3\cdot b}{1\cdot a^2b^2}}{\frac{a\cdot b}{1\cdot a}} = \frac{\frac{ab^4}{a^2b^2}}{\frac{ab}{a}} = \frac{ab^4\cdot a}{a^2b^2\cdot ab} = \frac{a^2b^4}{a^3b^3} = \frac{\cancel{a^2}\overset{b}{\cancel{b^4}}}{\underset{a}{\cancel{a^3}}\cancel{b^3}} = \mathbf{\frac{b}{a}}$$

6. $$\frac{\frac{z^3}{z^2}\div 2z}{z^3\div\frac{z^2}{2}} = \frac{\frac{z^3}{z^2}\div\frac{2z}{1}}{\frac{z^3}{1}\div\frac{z^2}{2}} = \frac{\frac{z^3}{z^2}\cdot\frac{1}{2z}}{\frac{z^3}{1}\cdot\frac{2}{z^2}} = \frac{\frac{z^3\cdot 1}{z^2\cdot 2z}}{\frac{z^3\cdot 2}{1\cdot z^2}} = \frac{\frac{z^3}{2z^3}}{\frac{2z^3}{z^2}} = \frac{z^3\cdot z^2}{2z^3\cdot 2z^3} = \frac{z^5}{4z^6} = \frac{\cancel{z^5}}{4\underset{z}{\cancel{z^6}}} = \mathbf{\frac{1}{4z}}$$

7. $$\frac{\frac{ab^2cd^4}{cd^2}\div\frac{c^2d^3}{b^2}}{a^2b^2} = \frac{\frac{ab^2cd^4}{cd^2}\cdot\frac{b^2}{c^2d^3}}{a^2b^2} = \frac{\frac{ab^2cd^4\cdot b^2}{cd^2\cdot c^2d^3}}{a^2b^2} = \frac{\frac{ab^4cd^4}{c^3d^5}}{a^2b^2} = \frac{\frac{ab^4cd^4}{c^3d^5}}{\frac{a^2b^2}{1}} = \frac{ab^4cd^4\cdot 1}{c^3d^5\cdot a^2b^2} = \frac{ab^4cd^4}{a^2b^2c^3d^5}$$

$$= \frac{\cancel{a}\overset{b^2}{\cancel{b^4}}\cancel{c}\cancel{d^4}}{\underset{a}{\cancel{a^2}}\cancel{b^2}\underset{c^2}{\cancel{c^3}}\underset{d}{\cancel{d^5}}} = \mathbf{\frac{b^2}{ac^2d}}$$

8. $$\frac{\frac{u^2}{u}\div u^3}{1\div\frac{1}{u^2}} = \frac{\frac{u^2}{u}\div\frac{u^3}{1}}{\frac{1}{1}\div\frac{1}{u^2}} = \frac{\frac{u^2}{u}\cdot\frac{1}{u^3}}{\frac{1}{1}\cdot\frac{u^2}{1}} = \frac{\frac{u^2\cdot 1}{u\cdot u^3}}{\frac{1\cdot u^2}{1\cdot 1}} = \frac{\frac{u^2}{u^4}}{\frac{u^2}{1}} = \frac{u^2\cdot 1}{u^4\cdot u^2} = \frac{u^2}{u^6} = \frac{\cancel{u^2}}{\underset{u^4}{\cancel{u^6}}} = \mathbf{\frac{1}{u^4}}$$

9. $$\frac{6x^2y\div\frac{x^2}{y^2}}{\frac{x}{y}\div\frac{x^2}{xy}} = \frac{\frac{6x^2y}{1}\div\frac{x^2}{y^2}}{\frac{x}{y}\div\frac{x^2}{xy}} = \frac{\frac{6x^2y}{1}\cdot\frac{y^2}{x^2}}{\frac{x}{y}\cdot\frac{xy}{x^2}} = \frac{\frac{6x^2y\cdot y^2}{1\cdot x^2}}{\frac{x\cdot xy}{y\cdot x^2}} = \frac{\frac{6x^2y^3}{x^2}}{\frac{x^2y}{x^2y}} = \frac{6x^2y^3\cdot x^2y}{x^2\cdot x^2y} = \frac{6x^4y^4}{x^4y} = \frac{6\cancel{x^4}\overset{y^3}{\cancel{y^4}}}{\cancel{x^4}\cancel{y}} = \frac{6y^3}{1} = \mathbf{6y^3}$$

10. $$\frac{m^2\div\frac{m^3}{n}}{n\div\frac{n}{m}} = \frac{\frac{m^2}{1}\div\frac{m^3}{n}}{\frac{n}{1}\div\frac{n}{m}} = \frac{\frac{m^2}{1}\cdot\frac{n}{m^3}}{\frac{n}{1}\cdot\frac{m}{n}} = \frac{\frac{m^2\cdot n}{1\cdot m^3}}{\frac{n\cdot m}{1\cdot n}} = \frac{\frac{m^2n}{m^3}}{\frac{mn}{n}} = \frac{m^2n\cdot n}{m^3\cdot mn} = \frac{m^2n^2}{m^4n} = \frac{\cancel{m^2}\overset{n}{\cancel{n^2}}}{\underset{m^2}{\cancel{m^4}}\cancel{n}} = \mathbf{\frac{n}{m^2}}$$

Section 5.4 Case IV Solutions - Mixed Operations Involving Complex Algebraic Fractions

1. $$\frac{a-\frac{2}{3}}{a+\frac{4}{5}} = \frac{\frac{a}{1}-\frac{2}{3}}{\frac{a}{1}+\frac{4}{5}} = \frac{\frac{(a\cdot 3)-(2\cdot 1)}{1\cdot 3}}{\frac{(a\cdot 5)+(4\cdot 1)}{1\cdot 5}} = \frac{\frac{3a-2}{3}}{\frac{5a+4}{5}} = \frac{(3a-2)\cdot 5}{3\cdot(5a+4)} = \mathbf{\frac{5(3a-2)}{3(5a+4)}}$$

2. $$\frac{1+\frac{a}{b}}{1-\frac{a}{b}} = \frac{1+\frac{a}{b}}{1-\frac{a}{b}} = \frac{\frac{1}{1}+\frac{a}{b}}{\frac{1}{1}-\frac{a}{b}} = \frac{\frac{(1\cdot b)+(a\cdot 1)}{1\cdot b}}{\frac{(1\cdot b)-(a\cdot 1)}{1\cdot b}} = \frac{\frac{b+a}{b}}{\frac{b-a}{b}} = \frac{b\cdot(b+a)}{b\cdot(b-a)} = \frac{\cancel{b}(b+a)}{\cancel{b}(b-a)} = \mathbf{\frac{b+a}{b-a}}$$

3. $$\frac{\frac{1}{x}-\frac{1}{y}}{\frac{1}{x}+\frac{1}{y}}\cdot\frac{1}{y-x} = \frac{\frac{(1\cdot y)-(1\cdot x)}{x\cdot y}}{\frac{(1\cdot y)+(1\cdot x)}{x\cdot y}}\cdot\frac{1}{y-x} = \frac{\frac{y-x}{xy}}{\frac{y+x}{xy}}\cdot\frac{1}{y-x} = \frac{xy(y-x)}{xy(x+y)}\cdot\frac{1}{y-x} = \frac{\cancel{xy}(y-x)}{\cancel{xy}(x+y)}\cdot\frac{1}{y-x} = \frac{y-x}{x+y}\cdot\frac{1}{y-x}$$

$$= \frac{(y-x)\cdot 1}{(x+y)\cdot(y-x)} = \frac{\cancel{(y-x)}}{(x+y)\cancel{(y-x)}} = \mathbf{\frac{1}{x+y}}$$

4. $$\frac{\frac{1}{x^2}+\frac{1}{y^2}}{\frac{1}{x^2y^2}-1}\div\frac{\frac{1}{3}}{\frac{1-xy}{2}} = \frac{\frac{1}{x^2}+\frac{1}{y^2}}{\frac{1}{x^2y^2}-\frac{1}{1}}\div\frac{\frac{1}{3}}{\frac{1-xy}{2}} = \frac{\frac{(1\cdot y^2)+(1\cdot x^2)}{x^2\cdot y^2}}{\frac{(1\cdot 1)-(1\cdot x^2y^2)}{x^2y^2\cdot 1}}\div\frac{\frac{1}{3}}{\frac{1-xy}{2}} = \frac{\frac{y^2+x^2}{x^2y^2}}{\frac{1-x^2y^2}{x^2y^2}}\div\frac{\frac{1}{3}}{\frac{1-xy}{2}}$$

$$= \frac{x^2y^2(y^2+x^2)}{x^2y^2(1-x^2y^2)}\div\frac{1\cdot 2}{3\cdot(1-xy)} = \frac{\cancel{x^2y^2}(y^2+x^2)}{\cancel{x^2y^2}(1-x^2y^2)}\div\frac{2}{3(1-xy)} = \frac{x^2+y^2}{1-x^2y^2}\div\frac{2}{3(1-xy)} = \frac{x^2+y^2}{1-x^2y^2}\cdot\frac{3(1-xy)}{2}$$

$$= \frac{x^2+y^2}{(1-xy)(1+xy)}\cdot\frac{3(1-xy)}{2} = \frac{x^2+y^2}{\cancel{(1-xy)}(1+xy)}\cdot\frac{3\cancel{(1-xy)}}{2} = \frac{x^2+y^2}{1+xy}\cdot\frac{3}{2} = \frac{(x^2+y^2)\cdot 3}{(1+xy)\cdot 2} = \mathbf{\frac{3(x^2+y^2)}{2(1+xy)}}$$

5. $$\frac{\frac{x}{y}+2}{1-\frac{x}{y}}\cdot\left(1-\frac{x}{y}\right) = \frac{\frac{x}{y}+\frac{2}{1}}{\frac{1}{1}-\frac{x}{y}}\cdot\left(\frac{1}{1}-\frac{x}{y}\right) = \frac{\frac{(x\cdot 1)+(2\cdot y)}{y\cdot 1}}{\frac{(1\cdot y)-(x\cdot 1)}{1\cdot y}}\cdot\left(\frac{(1\cdot y)-(x\cdot 1)}{1\cdot y}\right) = \frac{\frac{x+2y}{y}}{\frac{y-x}{y}}\cdot\left(\frac{y-x}{y}\right) = \frac{y\cdot(x+2y)}{y\cdot(y-x)}\cdot\left(\frac{y-x}{y}\right)$$

$$= \frac{\cancel{y}(x+2y)}{\cancel{y}\cancel{(y-x)}}\cdot\frac{\cancel{(y-x)}}{y} = \frac{x+2y}{1}\cdot\frac{1}{y} = \frac{(x+2y)\cdot 1}{1\cdot y} = \mathbf{\frac{x+2y}{y}}$$

6. $$\frac{\frac{1}{a}+\frac{1}{b}}{a}\div\frac{\frac{1}{a}-\frac{1}{b}}{b} = \frac{\frac{1}{a}+\frac{1}{b}}{\frac{a}{1}}\div\frac{\frac{1}{a}-\frac{1}{b}}{\frac{b}{1}} = \frac{\frac{(1\cdot b)+(1\cdot a)}{a\cdot b}}{\frac{a}{1}}\div\frac{\frac{(1\cdot b)-(1\cdot a)}{a\cdot b}}{\frac{b}{1}} = \frac{\frac{b+a}{ab}}{\frac{a}{1}}\div\frac{\frac{b-a}{ab}}{\frac{b}{1}} = \frac{(b+a)\cdot 1}{ab\cdot a}\div\frac{(b-a)\cdot 1}{ab\cdot b}$$

$$= \frac{b+a}{a^2b}\div\frac{b-a}{ab^2} = \frac{b+a}{a^2b}\cdot\frac{ab^2}{b-a} = \frac{(b+a)\cdot ab^2}{a^2b\cdot(b-a)} = \frac{\cancel{a}\,\overset{b}{\cancel{b^2}}(b+a)}{\underset{a}{\cancel{a^2}}\,\cancel{b}(b-a)} = \mathbf{\frac{b(b+a)}{a(b-a)}}$$

7. $$\frac{2-\frac{1}{y+1}}{\frac{2}{y+1}-2}+\frac{1}{2y} = \frac{\frac{2}{1}-\frac{1}{y+1}}{\frac{2}{y+1}-\frac{2}{1}}+\frac{1}{2y} = \frac{\frac{[2\cdot(y+1)]-(1\cdot 1)}{1\cdot(y+1)}}{\frac{(2\cdot 1)-[2\cdot(y+1)]}{(y+1)\cdot 1}}+\frac{1}{2y} = \frac{\frac{2y+2-1}{y+1}}{\frac{2-2y-2}{y+1}}+\frac{1}{2y} = \frac{\frac{2y+1}{y+1}}{\frac{-2y}{y+1}}+\frac{1}{2y}$$

$= \frac{(2y+1)\cdot(y+1)}{(y+1)\cdot -2y} + \frac{1}{2y} = \frac{(2y+1)(y+1)}{-2y(y+1)} + \frac{1}{2y} = \frac{2y+1}{-2y} + \frac{1}{2y} = -\frac{2y+1}{2y} + \frac{1}{2y} = \frac{-(2y+1)+1}{2y} = \frac{-2y-1+1}{2y}$

$= \frac{-2y-1+1}{2y} = \frac{-2y}{2y} = -\frac{2y}{2y} = -\frac{1}{1} = \mathbf{-1}$

8. $\frac{\frac{1}{2x}+\frac{2}{5y}}{\frac{2}{5y}-\frac{1}{2x}} \div (5y+4x) = \frac{\frac{1}{2x}+\frac{2}{5y}}{\frac{2}{5y}-\frac{1}{2x}} \div \frac{(5y+4x)}{1} = \frac{\frac{(1\cdot 5y)+(2\cdot 2x)}{2x\cdot 5y}}{\frac{(2\cdot 2x)-(1\cdot 5y)}{5y\cdot 2x}} \div \frac{(5y+4x)}{1} = \frac{\frac{5y+4x}{10xy}}{\frac{4x-5y}{10xy}} \div \frac{(5y+4x)}{1}$

$= \frac{(5y+4x)\cdot 10xy}{10xy\cdot(4x-5y)} \div \frac{(5y+4x)}{1} = \frac{(5y+4x)10xy}{10xy(4x-5y)} \div \frac{(5y+4x)}{1} = \frac{(5y+4x)}{(4x-5y)}\cdot\frac{1}{(5y+4x)} = \frac{(5y+4x)}{(4x-5y)}\cdot\frac{1}{(5y+4x)}$

$= \frac{1}{(4x-5y)}\cdot\frac{1}{1} = \frac{1\cdot 1}{(4x-5y)\cdot 1} = \mathbf{\frac{1}{4x-5y}}$

9. $\frac{\frac{1}{x^2}-\frac{1}{y^2}}{\frac{1}{x^2y^2}}\cdot\frac{2}{y^2-x^2} = \frac{\frac{(1\cdot y^2)-(1\cdot x^2)}{x^2\cdot y^2}}{\frac{1}{x^2y^2}}\cdot\frac{2}{y^2-x^2} = \frac{\frac{y^2-x^2}{x^2y^2}}{\frac{1}{x^2y^2}}\cdot\frac{2}{y^2-x^2} = \frac{(y^2-x^2)\cdot x^2y^2}{x^2y^2\cdot 1}\cdot\frac{2}{y^2-x^2}$

$= \frac{(y^2-x^2)x^2y^2}{x^2y^2}\cdot\frac{2}{y^2-x^2} = \frac{y^2-x^2}{1}\cdot\frac{2}{y^2-x^2} = \frac{(y^2-x^2)\cdot 2}{1\cdot(y^2-x^2)} = \frac{2(y^2-x^2)}{(y^2-x^2)} = \frac{2}{1} = \mathbf{2}$

10. $\frac{\frac{1}{x}-\frac{1}{y}}{\frac{1}{xy}} + x = \frac{\frac{1}{x}-\frac{1}{y}}{\frac{1}{xy}} + \frac{x}{1} = \frac{\frac{(1\cdot y)-(1\cdot x)}{x\cdot y}}{\frac{1}{xy}} + \frac{x}{1} = \frac{\frac{y-x}{xy}}{\frac{1}{xy}} + \frac{x}{1} = \frac{(y-x)\cdot xy}{xy\cdot 1} + \frac{x}{1} = \frac{(y-x)xy}{xy} + \frac{x}{1} = \frac{y-x}{1} + \frac{x}{1}$

$= \frac{y-x+x}{1} = \frac{y-x+x}{1} = \frac{y}{1} = \mathbf{y}$

Section 5.5 Solutions - Solving One Variable Equations Containing Algebraic Fractions

1. $x+\frac{x}{2}=3$; $\frac{x}{1}+\frac{x}{2}=3$; $\frac{(x\cdot 2)+(x\cdot 1)}{1\cdot 2}=3$; $\frac{2x+x}{2}=3$; $\frac{3x}{2}=\frac{3}{1}$; $3x\cdot 1=3\cdot 2$; $3x=6$; $\frac{3x}{3}=\frac{6}{3}$ (reduced to $\frac{2}{1}$) ; $x=\frac{2}{1}$; $\mathbf{x=2}$

2. $\frac{x-3}{2x-1}=\frac{1}{3}$; $(x-3)\cdot 3=1\cdot(2x-1)$; $3x-9=2x-1$; $3x-2x-9=2x-2x-1$; $x-9=0-1$; $x-9=-1$; $x-9+9=-1+9$; $x-0=8$; $\mathbf{x=8}$

3. $\frac{3}{x+5}-\frac{2}{x+5}=2$; $\frac{3-2}{x+5}=2$; $\frac{1}{x+5}=\frac{2}{1}$; $1\cdot 1=2\cdot(x+5)$; $1=2x+10$; $1-2x=2x-2x+10$; $1-2x=0+10$; $1-2x=10$; $1-1-2x=10-1$; $0-2x=9$; $-2x=9$; $\frac{-2x}{-2}=\frac{9}{-2}$; $x=-\frac{9}{2}$; $\mathbf{x=-4\frac{1}{2}}$

4. $\frac{y}{2}-\frac{y}{5}=5$; $\frac{(y\cdot 5)-(y\cdot 2)}{2\cdot 5}=5$; $\frac{5y-2y}{10}=5$; $\frac{3y}{10}=\frac{5}{1}$; $3y\cdot 1=5\cdot 10$; $3y=50$; $\frac{3y}{3}=\frac{50}{3}$; $\mathbf{y=16\frac{2}{3}}$

5. $\frac{4}{x+3}-\frac{3}{x-3}=\frac{5}{x^2-9}$; $\frac{[4\cdot(x-3)]-[3\cdot(x+3)]}{(x+3)\cdot(x-3)}=\frac{5}{x^2-9}$; $\frac{4x-12-(3x+9)}{(x+3)\cdot(x-3)}=\frac{5}{x^2-9}$; $\frac{4x-12-3x-9}{x^2-3x+3x-3\cdot 3}=\frac{5}{x^2-9}$

$; \frac{(4x-3x)+(-12-9)}{x^2-9}=\frac{5}{x^2-9} \; ; \; \frac{x-21}{x^2-9}=\frac{5}{x^2-9} \; ; \; (x-21)\cdot\left(\cancel{x^2-9}\right)=5\cdot\left(\cancel{x^2-9}\right) \; ; \; x-21=5 \; ; \; x-21+21=5+21$

$; \; x+0=26 \; ; \; \mathbf{x=26}$

6. $\frac{1}{x-1}+5=\frac{2}{x-1} \; ; \; \frac{1}{x-1}+\frac{5}{1}=\frac{2}{x-1} \; ; \; \frac{(1\cdot1)+[5\cdot(x-1)]}{(x-1)\cdot1}=\frac{2}{x-1} \; ; \; \frac{1+5x-5}{x-1}=\frac{2}{x-1} \; ; \; \frac{5x-4}{x-1}=\frac{2}{x-1}$

$; \; (5x-4)\cdot(x-1)=2\cdot(x-1) \; ; \; (5x-4)(\cancel{x-1})=2(\cancel{x-1}) \; ; \; 5x-4=2 \; ; \; 5x-4+4=2+4 \; ; \; 5x+0=6 \; ; \; 5x=6 \; ; \; \frac{\cancel{5}x}{\cancel{5}}=\frac{6}{5}$

$; \; x=\frac{6}{5} \; ; \; \mathbf{x=1\frac{1}{5}}$

7. $\frac{b+2}{3}-b=\frac{1}{3} \; ; \; \frac{b+2}{3}-\frac{b}{1}=\frac{1}{3} \; ; \; \frac{[(b+2)\cdot1]-(b\cdot3)}{3\cdot1}=\frac{1}{3} \; ; \; \frac{b+2-3b}{3}=\frac{1}{3} \; ; \; \frac{2-2b}{3}=\frac{1}{3} \; ; \; (2-2b)\cdot3=1\cdot3 \; ; \; \cancel{3}(2-2b)=\cancel{3}$

$; \; 2-2b=1 \; ; \; 2-2-2b=1-2 \; ; \; 0-2b=-1 \; ; \; -2b=-1 \; ; \; \frac{\cancel{-2}b}{\cancel{-2}}=\frac{-1}{-2} \; ; \; \mathbf{b=\frac{1}{2}}$

8. $\frac{x-1}{x+2}+\frac{4}{x^2+5x+6}=\frac{x+1}{x+3} \; ; \; \frac{x-1}{x+2}+\frac{4}{(x+3)(x+2)}=\frac{x+1}{x+3} \; ; \; \frac{x-1}{x+2}-\frac{x+1}{x+3}+\frac{4}{(x+3)(x+2)}=\frac{x+1}{x+3}-\frac{x+1}{x+3}$

$; \; \frac{x-1}{x+2}-\frac{x+1}{x+3}+\frac{4}{(x+3)(x+2)}=0 \; ; \; \frac{[(x-1)(x+3)]-[(x+1)(x+2)]}{(x+2)(x+3)}+\frac{4}{(x+3)(x+2)}=0$

$; \; \frac{(x^2+3x-x-3)-(x^2+2x+x+2)}{(x+2)(x+3)}+\frac{4}{(x+3)(x+2)}=0 \; ; \; \frac{(x^2+2x-3)-(x^2+3x+2)}{(x+2)(x+3)}+\frac{4}{(x+3)(x+2)}=0$

$; \; \frac{x^2+2x-3-x^2-3x-2}{(x+2)(x+3)}+\frac{4}{(x+3)(x+2)}=0 \; ; \; \frac{(x^2-x^2)+(2x-3x)+(-3-2)}{(x+2)(x+3)}+\frac{4}{(x+3)(x+2)}=0$

$; \; \frac{-x-5}{(x+2)(x+3)}+\frac{4}{(x+3)(x+2)}=0 \; ; \; \frac{-x-5+4}{(x+2)(x+3)}=0 \; ; \; \frac{-x-1}{(x+2)(x+3)}=\frac{0}{1} \; ; \; (-x-1)\cdot1=0\cdot[(x+2)(x+3)] \; ; \; -x-1=0$

$; \; -x-1+1=0+1 \; ; \; -x+0=1 \; ; \; -x=1 \; ; \; \frac{-x}{-1}=\frac{1}{-1} \; ; \; \mathbf{x=-1}$

9. $\frac{1}{x+3}+\frac{2x}{x^2-9}=\frac{1}{x-3} \; ; \; \frac{1}{x+3}-\frac{1}{x-3}+\frac{2x}{x^2-9}=\frac{1}{x-3}-\frac{1}{x-3} \; ; \; \frac{1}{x+3}-\frac{1}{x-3}+\frac{2x}{x^2-9}=0$

$; \; \frac{[1\cdot(x-3)]-[1\cdot(x+3)]}{(x+3)(x-3)}+\frac{2x}{x^2-9}=0 \; ; \; \frac{\cancel{x}-3-\cancel{x}-3}{x^2-\cancel{3x}+\cancel{3x}-3\cdot3}+\frac{2x}{x^2-9}=0 \; ; \; \frac{-6}{x^2-9}+\frac{2x}{x^2-9}=0 \; ; \; \frac{-6+2x}{x^2-9}=0$

$; \; \frac{-6+2x}{x^2-9}=\frac{0}{1} \; ; \; (-6+2x)\cdot1=0\cdot(x^2-9) \; ; \; -6+2x=0 \; ; \; -6+6+2x=0+6 \; ; \; 0+2x=6 \; ; \; 2x=6 \; ; \; \frac{\cancel{2}x}{\cancel{2}}=\frac{\overset{3}{\cancel{6}}}{\cancel{2}} \; ; \; x=\frac{3}{1}$

$; \; x=3$ Note that $x=3$ is an apparent solution. Substitution of $x=3$ into the original equation results in division by zero which is not defined.

10. $\frac{x+3}{x^2+2x}=\frac{1}{x+2}+\frac{1}{x} \; ; \; \frac{x+3}{x^2+2x}=\frac{(1\cdot x)+[1\cdot(x+2)]}{(x+2)\cdot x}+\frac{1}{x} \; ; \; \frac{x+3}{x^2+2x}=\frac{x+x+2}{x(x+2)} \; ; \; \frac{x+3}{x^2+2x}=\frac{2x+2}{x^2+2x}$

$; \; (x+3)\cdot(x^2+2x)=(2x+2)\cdot(x^2+2x) \; ; \; (x+3)(\cancel{x^2+2x})=(2x+2)(\cancel{x^2+2x}) \; ; \; x+3=2x+2 \; ; \; x-2x+3=2x-2x+2$

$; \; -x+3=0+2 \; ; \; -x+3=2 \; ; \; -x+3-3=2-3 \; ; \; -x+0=-1 \; ; \; -x=-1 \; ; \; \frac{-x}{-1}=\frac{-\cancel{1}}{-\cancel{1}} \; ; \; \mathbf{x=1}$

Chapter 6 Solutions:

Section 6.1 Case I Solutions - Logarithmic and Exponential Expressions

A. Write the following exponential expressions in their equivalent logarithmic form:

1. $2^5 = 32$; $log_2 2^5 = log_2 32$; $\mathbf{5 = log_2 32}$

2. $4^{-3} = \frac{1}{64}$; $log_4 4^{-3} = log_4 \frac{1}{64}$; $\mathbf{-3 = log_4 \frac{1}{64}}$

3. $7^2 = 49$; $log_7 7^2 = log_7 49$; $\mathbf{2 = log_7 49}$

4. $64^{\frac{1}{2}} = 8$; $log_{64} 64^{\frac{1}{2}} = log_{64} 8$; $\mathbf{\frac{1}{2} = log_{64} 8}$

5. $81^{\frac{1}{4}} = 3$; $log_{81} 81^{\frac{1}{4}} = log_{81} 3$; $\mathbf{\frac{1}{4} = log_{81} 3}$

6. $e^2 = 7.389$; $lne^2 = ln 7.389$; $\mathbf{2 = ln 7.389}$

7. $27^{-\frac{1}{3}} = \frac{1}{3}$; $log_{27} 27^{-\frac{1}{3}} = log_{27} \frac{1}{3}$; $\mathbf{-\frac{1}{3} = log_{27} \frac{1}{3}}$

8. $100^{-\frac{1}{2}} = \frac{1}{10}$; $log_{100} 100^{-\frac{1}{2}} = log_{100} \frac{1}{10}$; $\mathbf{-\frac{1}{2} = log_{100} \frac{1}{10}}$

9. $125^{\frac{1}{3}} = 5$; $log_{125} 125^{\frac{1}{3}} = log_{125} 5$; $\mathbf{\frac{1}{3} = log_{125} 5}$

10. $1000^0 = 1$; $log_{1000} 1000^0 = log_{1000} 1$; $\mathbf{0 = log_{1000} 1}$

B. Write the following logarithmic expressions in their equivalent exponential form:

1. $log_{10} 10000 = 4$; $10^{log_{10} 10000} = 10^4$; $\mathbf{10000 = 10^4}$

2. $log_4 64 = 3$; $4^{log_4 64} = 4^3$; $\mathbf{64 = 4^3}$

3. $log_{10} 0.1 = -1$; $10^{log_{10} 0.1} = 10^{-1}$; $\mathbf{0.1 = 10^{-1}}$

4. $log_5 625 = 4$; $5^{log_5 625} = 5^4$; $\mathbf{625 = 5^4}$

5. $log_2 \frac{1}{32} = -5$; $2^{log_2 \frac{1}{32}} = 2^{-5}$; $\mathbf{\frac{1}{32} = 2^{-5}}$

6. $log_{\frac{1}{3}} 27 = -3$; $\frac{1}{3}^{log_{\frac{1}{3}} 27} = \frac{1}{3}^{-3}$; $\mathbf{27 = \frac{1}{3}^{-3}}$

7. $log_3 243 = 5$; $3^{log_3 243} = 3^5$; $\mathbf{243 = 3^5}$

8. $log_2 256 = 8$; $2^{log_2 256} = 2^8$; $\mathbf{256 = 2^8}$

9. $log_{10} 0.0001 = -4$; $10^{log_{10} 0.0001} = 10^{-4}$; $\mathbf{0.0001 = 10^{-4}}$

10. $log_{\frac{1}{5}} 3125 = -5$; $\frac{1}{5}^{log_{\frac{1}{5}} 3125} = \frac{1}{5}^{-5}$; $\mathbf{3125 = \frac{1}{5}^{-5}}$

Section 6.1 Case II Solutions - The Laws of Logarithm

1. $log_8 8^{u^3} = u^3 log_8 8 = u^3 \times 1 = \mathbf{u^3}$
2. $0.02^{log_{0.02} 2} = \mathbf{2}$
3. $log_{10} \frac{10}{x^2} = log_{10} 10 - log_{10} x^2 = \mathbf{1 - 2 log_{10} x}$
4. $log_{10} 5 \cdot 2x^2 y^3 = log_{10} 5 + log_{10} 2 + log_{10} x^2 + log_{10} y^3 = \mathbf{log_{10} 5 + log_{10} 2 + 2 log_{10} x + 3 log_{10} y}$
5. $log_{25} 25x + log_{25} 1 = log_{25} 25 + log_{25} x + 0 = \mathbf{1 + log_{25} x}$
6. $log_{10} \frac{2}{1000} = log_{10} 2 - log_{10} 1000 = \mathbf{log_{10} 2 - 3}$
7. $\mathbf{log_{100} 0}$ ***is not defined***
8. $log_{10} \sqrt[5]{3} = log_{10} 3^{\frac{1}{5}} = \mathbf{\frac{1}{5} log_{10} 3}$
9. $log_2 \frac{8}{9} = log_2 8 - log_2 9 = log_2 2^3 - log_2 9 = 3 log_2 2 - log_2 9 = 3 \times 1 - log_2 9 = \mathbf{3 - log_2 9}$
10. $log_4 4 \cdot 16 = log_4 4 + log_4 16 = 1 + log_4 4^2 = 1 + 2 log_4 4 = 1 + 2 \times 1 = 1 + 2 = \mathbf{3}$

Section 6.2 Case I Solutions - Computation of Common Logarithms

A. Use the Common Logarithms Table to solve the following logarithmic expresions:

1. $log_{10} 3.57 = \mathbf{0.5527}$
2. $log_{10} 3.08 = \mathbf{0.4886}$
3. $log_{10} 4.53 = \mathbf{0.6561}$
4. $log_{10} 8.24 = \mathbf{0.9159}$
5. $log_{10} 7.32 = \mathbf{0.8645}$
6. $log_{10} 5.55 = \mathbf{0.7443}$
7. $log_{10} 2.12 = \mathbf{0.3263}$
8. $log_{10} 9.46 = \mathbf{0.9759}$
9. $log_{10} 5.29 = \mathbf{0.7235}$
10. $log_{10} 1.26 = \mathbf{0.1004}$

B. Solve the following common logarithms:

1. $log_{10} 4{,}000 = log_{10} 4.0 \times 10^3 = log_{10} 4.0 + log_{10} 10^3 = log_{10} 4.0 + 3\,log_{10} 10 = 0.6021 + 3 = \mathbf{3.6021}$

2. $log_{10} 3\sqrt{300} = log_{10} 3 + log_{10} \sqrt{300} = log_{10} 3 + log_{10} 300^{\frac{1}{2}} = log_{10} 3 + \frac{1}{2} log_{10} 300 = log_{10} 3 + \frac{1}{2} log_{10} 3.0 \times 10^2$

$= log_{10} 3 + \frac{1}{2}\left(log_{10} 3.0 + log_{10} 10^2\right) = log_{10} 3 + \frac{1}{2}\left(log_{10} 3.0 + 2\,log_{10} 10\right) = 0.4771 + \frac{1}{2}(0.4771 + 2) = 0.4771 + \frac{1}{2} \times 2.4771$

$= 0.4771 + 1.2386 = \mathbf{1.7157}$

3. $log_{10} 45{,}400{,}000 = log_{10} 4.54 \times 10^7 = log_{10} 4.54 + log_{10} 10^7 = log_{10} 4.54 + 7\,log_{10} 10 = 0.6571 + 7 = \mathbf{7.6571}$

4. $log_{10} 0.00023 = log_{10} 2.3 \times 10^{-4} = log_{10} 2.3 + log_{10} 10^{-4} = log_{10} 2.3 - 4\,log_{10} 10 = 0.3617 - 4 = \mathbf{-3.6383}$

5. $log_{10} \sqrt[3]{5^2} = log_{10} 5^{\frac{2}{3}} = \frac{2}{3} log_{10} 5 = \frac{2}{3} \times 0.6989 = \frac{1.3978}{3} = \mathbf{0.4659}$

6. $log_{10} 28 = log_{10} 2.8 \times 10^1 = log_{10} 2.8 + log_{10} 10^1 = 0.4472 + log_{10} 10 = 0.4472 + 1 = \mathbf{1.4472}$

7. $log_{10} 2\sqrt[4]{568} = log_{10} 2 + log_{10} \sqrt[4]{568} = log_{10} 2 + log_{10} 568^{\frac{1}{4}} = log_{10} 2 + \frac{1}{4} log_{10} 568 = log_{10} 2 + \frac{1}{4}\left(log_{10} 5.68 \times 10^2\right)$

$= log_{10} 2 + \frac{1}{4}\left(log_{10} 5.68 + log_{10} 10^2\right) = log_{10} 2 + \frac{1}{4}\left(log_{10} 5.68 + 2\,log_{10} 10\right) = 0.3010 + \frac{1}{4}(0.7543 + 2) = 0.3010 + \frac{2.7543}{4}$

$= 0.3010 + 0.6886 = \mathbf{0.9896}$

8. $log_{10} 0.068 = log_{10} 6.8 \times 10^{-2} = log_{10} 6.8 + log_{10} 10^{-2} = log_{10} 6.8 - 2\,log_{10} 10 = 0.8325 - 2 = \mathbf{-1.1675}$

9. $log_{10} 0.00001 = log_{10} 1.0 \times 10^{-5} = log_{10} 1.0 + log_{10} 10^{-5} = log_{10} 1.0 - 5\,log_{10} 10 = 0 - 5 = \mathbf{-5}$

10. $log_{10} 450{,}000 = log_{10} 4.5 \times 10^5 = log_{10} 4.5 + log_{10} 10^5 = log_{10} 4.5 + 5\,log_{10} 10 = 0.6532 + 5 = \mathbf{5.6532}$

Section 6.2 Case II Solutions - Computation of Natural Logarithms

1. $ln 38 = ln 3.8 \times 10^1 = ln 3.8 + ln 10 = 1.3350 + 2.3026 = \mathbf{3.6376}$

2. $ln\sqrt{e^3} = ln\left(e^3\right)^{\frac{1}{2}} = ln\, e^{3 \times \frac{1}{2}} = ln\, e^{\frac{3}{2}} = \frac{3}{2} ln\, e = \frac{3}{2} \times 1 = \frac{3}{2} = \mathbf{1.5}$

3. $ln 0.0007 = ln 7.0 \times 10^{-4} = ln 7.0 + ln 10^{-4} = ln 7.0 - 4\,ln 10 = 1.9459 - 4 \times 2.3026 = 1.9459 - 9.2104 = \mathbf{-7.2645}$

4. $ln 255{,}000 = ln 2.55 \times 10^5 = ln 2.55 + ln 10^5 = ln 2.55 + 5\,ln 10 = 0.9361 + 5 \times 2.3026 = 0.9361 + 11.513 = \mathbf{12.4491}$

5. $ln\frac{1}{216} = ln 1 - ln 216 = 0 - ln 2.16 \times 10^2 = -\left(ln 2.16 + ln 10^2\right) = -ln 2.16 - 2\,ln 10 = -0.7701 - 2 \times 2.3026$

$= -0.7701 - 4.6052 = \mathbf{-5.3753}$

6. $ln\frac{e}{3} = lne - ln3 = 1 - 1.0986 = \mathbf{-0.0986}$

7. $ln\sqrt[5]{500} = ln500^{\frac{1}{5}} = \frac{1}{5}ln500 = \frac{1}{5}ln\left(5.0\times10^2\right) = \frac{1}{5}\left(ln5.0 + ln10^2\right) = \frac{1}{5}(ln5.0 + 2ln10) = \frac{1}{5}(1.6094 + 2\times2.3026)$

$= \frac{1}{5}(1.6094 + 4.6052) = \frac{1}{5}\times6.2146 = \mathbf{1.2429}$

8. $ln\frac{2}{3} = ln2 - ln3 = 0.6932 - 1.0986 = \mathbf{-0.4054}$

9. $500\,lne^{-3} = 500\times-3\,lne = 500(-3\times1) = 500\times-3 = \mathbf{-1500}$

10. $ln49^{\frac{2}{3}} = \frac{2}{3}ln49 = \frac{2}{3}ln4.9\times10^1 = \frac{2}{3}(ln4.9 + ln10) = \frac{2}{3}(1.5892 + 2.3026) = \frac{2}{3}\times3.8918 = \frac{7.7836}{3} = \mathbf{2.5945}$

Section 6.2 Case III Solutions - Computation of Logarithms other than Base 10 or e

1. $log_4\frac{16}{128} = log_4\frac{\overset{1}{\cancel{16}}}{\underset{8}{\cancel{128}}} = log_4\frac{1}{8} = log_4 8^{-1} = -log_4 8 = -\frac{log_{10}8}{log_{10}4} = -\frac{0.9031}{0.6021} = \mathbf{-1.4999}$

2. $log_3 162 = log_3 2\cdot81 = log_3 2 + log_3 81 = log_3 2 + log_3 3^4 = log_3 2 + 4log_3 3 = \frac{log_{10}2}{log_{10}3} + 4\times1 = \frac{0.3010}{0.4771} + 4$

$= 0.6309 + 4 = \mathbf{4.6309}$ or,

$log_3 162 = \frac{log_{10}162}{log_{10}3} = \frac{log1.62\times10^2}{log3} = \frac{log1.62 + log10^2}{log3} = \frac{log1.62 + 2log10}{log3} = \frac{0.2094 + 2}{0.4771} = \frac{2.2094}{0.4771} = \mathbf{4.6309}$

3. $log_{50}600 = \frac{log_{10}600}{log_{10}50} = \frac{log6.0\times10^2}{log5.0\times10^1} = \frac{log6.0 + log10^2}{log5.0 + log10^1} = \frac{log6.0 + 2log10}{log5.0 + log10} = \frac{0.7782 + 2}{0.6989 + 1} = \frac{2.7782}{1.6989} = \mathbf{1.6353}$

4. $log_{\frac{1}{2}}16 = log_{0.5}16 = \frac{log_{10}16}{log_{10}0.5} = \frac{1.2041}{-0.3010} = \mathbf{-4}$ or,

$log_{\frac{1}{2}}16 = log_{\frac{1}{2}}\left(\frac{1}{2}\right)^{-4} = -4log_{\frac{1}{2}}\frac{1}{2} = -4\times1 = \mathbf{-4}$ Note: $\left(\frac{1}{2}\right)^{-4} = \frac{1}{\left(\frac{1}{2}\right)^4} = \frac{1}{\frac{1}{16}} = \frac{\frac{1}{1}}{\frac{1}{16}} = \frac{1\times16}{1\times1} = \frac{16}{1} = 16$

5. $log_3\frac{2}{27} = log_3 2 - log_3 27 = log_3 2 - log_3 3^3 = log_3 2 - 3log_3 3 = log_3 2 - 3\times1 = \frac{log_{10}2}{log_{10}3} - 3 = \frac{0.3010}{0.4771} - 3$

$= 0.6309 - 3 = \mathbf{-2.3691}$ or;

$log_3\frac{2}{27} = log_3 2 - log_3 27 = \frac{log_{10}2}{log_{10}3} - \frac{log_{10}27}{log_{10}3} = \frac{log2 - log27}{log3} = \frac{log2 - log2.7\times10^1}{log3} = \frac{log2 - (log2.7 + log10)}{log3}$

$= \frac{log2 - log2.7 - 1}{log3} = \frac{0.3010 - 0.4313 - 1}{0.4771} = \frac{0.3010 - 1.4313}{0.4771} = \frac{-1.1303}{0.4771} = \mathbf{-2.3691}$

6. $log_{0.04}0.0004 = \frac{log_{10}0.0004}{log_{10}0.04} = \frac{log_{10}4.0\times10^{-4}}{log_{10}4.0\times10^{-2}} = \frac{log_{10}4.0 + log_{10}10^{-4}}{log_{10}4.0 + log_{10}10^{-2}} = \frac{log_{10}4.0 - 4log_{10}10}{log_{10}4.0 - 2log_{10}10} = \frac{log_{10}4.0 - 4}{log_{10}4.0 - 2}$

$= \frac{0.6021 - 4}{0.6021 - 2} = \frac{-3.3979}{-1.3979} = \mathbf{2.431}$

7. $log_2 32\sqrt[3]{4} = log_2 32 + log_2\sqrt[3]{4} = log_2 32 + log_2 4^{\frac{1}{3}} = log_2 2^5 + \frac{1}{3}log_2 4 = 5log_2 2 + \frac{1}{3}log_2 2^2 = 5log_2 2 + \frac{2}{3}log_2 2$

$= 5 \times 1 + \frac{2}{3} \times 1 = 5 + \frac{2}{3} = 5 + 0.6667 = \mathbf{5.6667}$

8. $log_{\frac{3}{5}} \frac{125}{27} = log_{\frac{3}{5}} \left(\frac{3}{5}\right)^{-3} = -3 log_{\frac{3}{5}} \frac{3}{5} = -3 \times 1 = \mathbf{-3}$ Note: $\left(\frac{3}{5}\right)^{-3} = \frac{1}{\left(\frac{3}{5}\right)^3} = \frac{1}{\frac{27}{125}} = \frac{\frac{1}{1}}{\frac{27}{125}} = \frac{1 \times 125}{1 \times 27} = \frac{125}{27}$

9. $log_{0.2} 50 = \frac{log_{10} 50}{log_{10} 0.2} = \frac{log_{10} 5.0 \times 10^1}{log_{10} 2 \times 10^{-1}} = \frac{log_{10} 5.0 + log_{10} 10^1}{log_{10} 2 + log_{10} 10^{-1}} = \frac{log_{10} 5.0 + log_{10} 10}{log_{10} 2 - log_{10} 10} = \frac{0.6989 + 1}{0.3010 - 1} = \frac{1.6989}{-0.699} = \mathbf{-2.43}$

10. $log_4 30\sqrt[5]{2} = log_4 30 + log_4 \sqrt[5]{2} = log_4 30 + log_4 2^{\frac{1}{5}} = log_4 30 + \frac{1}{5} log_4 2 = \frac{log_{10} 30}{log_{10} 4} + \frac{1}{5}\frac{log_{10} 2}{log_{10} 4} = \frac{1.4771}{0.6021} + \frac{1}{5}\frac{0.3010}{0.6021}$

$= 2.4533 + \frac{1}{5} \times 0.5 = 2.4533 + 0.1 = \mathbf{2.5533}$

Section 6.2 Case IV Solutions - Computing Antilogarithms

1. $log_{10} x = 0.453$; $10^{log_{10} x} = 10^{0.453}$; $x = 10^{0.453}$; $\mathbf{x = 2.8379}$

Check: $log_{10} 2.8379 \stackrel{?}{=} 0.453$; $0.453 = 0.453$

2. $log_{0.1} x = 0.08$; $0.1^{log_{0.1} x} = 0.1^{0.08}$; $x = 0.1^{0.08}$; $\mathbf{x = 0.8318}$

Check: $log_{0.1} 0.8318 \stackrel{?}{=} 0.08$; $\frac{log_{10} 0.8318}{log_{10} 0.1} \stackrel{?}{=} 0.08$; $\frac{-0.08}{-1} \stackrel{?}{=} 0.08$; $0.08 = 0.08$

3. $log_2 x = -0.543$; $2^{log_2 x} = 2^{-0.543}$; $x = 2^{-0.543}$; $x = \frac{1}{2^{0.543}}$; $x = \frac{1}{1.4569}$; $\mathbf{x = 0.6863}$

Check: $log_2 0.6863 \stackrel{?}{=} -0.543$; $\frac{log_{10} 0.6863}{log_{10} 2} \stackrel{?}{=} -0.543$; $\frac{-0.1635}{0.3010} \stackrel{?}{=} -0.543$; $-0.543 = -0.543$

4. $log_{0.03} x = 0.4$; $0.03^{log_{0.03} x} = 0.03^{0.4}$; $x = 0.03^{0.4}$; $\mathbf{x = 0.2459}$

Check: $log_{0.03} 0.2459 \stackrel{?}{=} 0.4$; $\frac{log_{10} 0.2459}{log_{10} 0.03} \stackrel{?}{=} 0.4$; $\frac{-0.6092}{-1.5229} \stackrel{?}{=} 0.4$; $0.4 = 0.4$

5. $log_{\sqrt{3}} x = -2$; $log_{1.732} x = -2$; $1.732^{log_{1.732} x} = 1.732^{-2}$; $x = 1.732^{-2}$; $x = \frac{1}{1.732^2}$; $\mathbf{x = \frac{1}{3}}$

Check: $log_{\sqrt{3}} \frac{1}{3} \stackrel{?}{=} -2$; $log_{\sqrt{3}} \left(\sqrt{3}\right)^{-2} \stackrel{?}{=} -2$; $-2 log_{\sqrt{3}} \sqrt{3} \stackrel{?}{=} -2$; $-2 \times 1 \stackrel{?}{=} -2$; $-2 = -2$ Note: $\left(\sqrt{3}\right)^{-2} = 3^{\frac{1}{2} \times -2} = 3^{-1} = \frac{1}{3}$

6. $log_{100} x = 0.04$; $100^{log_{100} x} = 100^{0.04}$; $x = 100^{0.04}$; $\mathbf{x = 1.2023}$

Check: $log_{100} 1.2023 \stackrel{?}{=} 0.04$; $\frac{log_{10} 1.2023}{log_{10} 100} \stackrel{?}{=} 0.04$; $\frac{0.08}{2} \stackrel{?}{=} 0.04$; $0.04 = 0.04$

7. $log_2 x = 3$; $2^{log_2 x} = 2^3$; $x = 2^3$; $\mathbf{x = 8}$

Check: $log_2 8 \stackrel{?}{=} 3$; $log_2 2^3 \stackrel{?}{=} 3$; $3 log_2 2 \stackrel{?}{=} 3$; $3 \times 1 \stackrel{?}{=} 3$; $3 = 3$

8. $log_{10} x = -1.35$; $10^{log_{10} x} = 10^{-1.35}$; $x = 10^{-1.35}$; $x = \frac{1}{10^{1.35}}$; $x = \frac{1}{22.387}$; $\mathbf{x = 0.0447}$

Check: $log_{10} 0.0447 \stackrel{?}{=} -1.35$; $\frac{log_{10} 0.0447}{log_{10} 10} \stackrel{?}{=} -1.35$; $\frac{-1.35}{1} \stackrel{?}{=} -1.35$; $-1.35 = -1.35$

9. $log_4 x = -2.3$; $4^{log_4 x} = 4^{-2.3}$; $x = 4^{-2.3}$; $x = \frac{1}{4^{2.3}}$; $x = \frac{1}{24.25}$; $\mathbf{x = 0.0412}$

Check: $log_4 0.0412 \stackrel{?}{=} -2.3$; $\frac{log_{10} 0.0412}{log_{10} 4} \stackrel{?}{=} -2.3$; $\frac{-1.3851}{0.6021} \stackrel{?}{=} -2.3$; $-2.3 = -2.3$

10. $log_{1000} x = 0.03$; $1000^{log_{1000} x} = 1000^{0.03}$; $x = 1000^{0.03}$; $\mathbf{x = 1.2303}$

Check: $log_{1000} 1.2303 \stackrel{?}{=} 0.03$; $\frac{log_{10} 1.2303}{log_{10} 1000} \stackrel{?}{=} 0.03$; $\frac{0.09}{3} \stackrel{?}{=} 0.03$; $0.03 = 0.03$

Section 6.3 Case I Solutions - Both Sides of the Exponential Equation Have the Same Base

1. $3^{x+2} \cdot 3^{3x} = 243$; $3^{x+2} \cdot 3^{3x} = 3^5$; $3^{x+3x+2} = 3^5$; $3^{4x+2} = 3^5$; $4x + 2 = 5$; $4x = 5 - 2$; $x = \frac{3}{4}$; $\mathbf{x = 0.75}$

Check: $3^{0.75+2} \cdot 3^{3\times 0.75} \stackrel{?}{=} 243$; $3^{2.75} \cdot 3^{2.25} \stackrel{?}{=} 243$; $3^{2.75+2.25} \stackrel{?}{=} 243$; $3^5 \stackrel{?}{=} 243$; $243 = 243$

2. $2^{u-2} 2^{u+3} = 2$; $2^{u-2+u+3} = 2$; $2^{2u+1} = 2^1$; $2u + 1 = 1$; $2u = 1 - 1$; $2u = 0$; $u = \frac{0}{2}$; $\mathbf{u = 0}$

Check: $2^{0-2} 2^{0+3} \stackrel{?}{=} 2$; $2^{3-2} \stackrel{?}{=} 2$; $2^1 \stackrel{?}{=} 2$; $2 = 2$

3. $125^{2q} = 5^{q+3}$; $\left(5^3\right)^{2q} = 5^{q+3}$; $5^{3\times 2q} = 5^{q+3}$; $5^{6q} = 5^{q+3}$; $6q = q + 3$; $6q - q = 3$; $5q = 3$; $q = \frac{3}{5}$; $\mathbf{q = 0.6}$

Check: $125^{2\times 0.6} \stackrel{?}{=} 5^{0.6+3}$; $125^{1.2} \stackrel{?}{=} 5^{3.6}$; $\left(5^3\right)^{1.2} \stackrel{?}{=} 5^{3.6}$; $5^{3\times 1.2} \stackrel{?}{=} 5^{3.6}$; $5^{3.6} = 5^{3.6}$

4. $5^{x+7} = 625$; $5^{x+7} = 5^4$; $x + 7 = 4$; $x = 4 - 7$; $\mathbf{x = -3}$

Check: $5^{-3+7} \stackrel{?}{=} 625$; $5^4 \stackrel{?}{=} 625$; $625 = 625$

5. $\left(4^2\right)^{-2} = 256^{t+2}$; $4^{2\times -2} = \left(4^4\right)^{t+2}$; $4^{-4} = 4^{4(t+2)}$; $4^{-4} = 4^{4t+8}$; $-4 = 4t + 8$; $-4 - 8 = 4t$; $t = -\frac{12}{4}$; $\mathbf{t = -3}$

Check: $\left(4^2\right)^{-2} \stackrel{?}{=} 256^{-3+2}$; $4^{2\times -2} \stackrel{?}{=} 256^{-1}$; $4^{-4} \stackrel{?}{=} 256^{-1}$; $4^{-4} \stackrel{?}{=} \left(4^4\right)^{-1}$; $4^{-4} \stackrel{?}{=} 4^{4\times -1}$; $4^{-4} = 4^{-4}$

6. $2^{3k-1} \cdot 2^2 = 2^{k-4}$; $2^{3k-1+2} = 2^{k-4}$; $2^{3k+1} = 2^{k-4}$; $3k + 1 = k - 4$; $3k - k = -4 - 1$; $2k = -5$; $k = -\frac{5}{2}$; $\mathbf{k = -2.5}$

Check: $2^{3\times -2.5-1} \cdot 2^2 \stackrel{?}{=} 2^{-2.5-4}$; $2^{-7.5-1} \cdot 2^2 \stackrel{?}{=} 2^{-6.5}$; $2^{-8.5} \cdot 2^2 \stackrel{?}{=} 2^{-6.5}$; $2^{-8.5+2} \stackrel{?}{=} 2^{-6.5}$; $2^{-6.5} = 2^{-6.5}$

7. $243^w = \left(3^2\right)^2$; $\left(3^5\right)^w = 3^{2\times 2}$; $3^{5\times w} = 3^4$; $3^{5w} = 3^4$; $5w = 4$; $w = \frac{4}{5}$; $\mathbf{w = 0.8}$

Check: $243^{0.8} \stackrel{?}{=} \left(3^2\right)^2$; $81 \stackrel{?}{=} 3^{2\times 2}$; $81 \stackrel{?}{=} 3^4$; $81 = 81$

8. $4^{5a+1} \cdot 4^{a-3} = \frac{1}{16}$; $4^{5a+1+a-3} = \frac{1}{4^2}$; $4^{6a-2} = 4^{-2}$; $6a - 2 = -2$; $6a = 2 - 2$; $6a = 0$; $a = \frac{0}{6}$; $\mathbf{a = 0}$

Check: $4^{5\times 0+1} \cdot 4^{0-3} \stackrel{?}{=} \frac{1}{16}$; $4^1 \cdot 4^{-3} \stackrel{?}{=} \frac{1}{16}$; $4^{1-3} \stackrel{?}{=} \frac{1}{16}$; $4^{-2} \stackrel{?}{=} \frac{1}{16}$; $\frac{1}{4^2} \stackrel{?}{=} \frac{1}{16}$; $\frac{1}{16} = \frac{1}{16}$

9. $8^{x+1} = \frac{1}{512}$; $8^{x+1} = \frac{1}{8^3}$; $8^{x+1} = 8^{-3}$; $x + 1 = -3$; $x = -3 - 1$; $\mathbf{x = -4}$

Check: $8^{-4+1} \stackrel{?}{=} \frac{1}{512}$; $8^{-3} \stackrel{?}{=} \frac{1}{512}$; $\frac{1}{8^3} \stackrel{?}{=} \frac{1}{512}$; $\frac{1}{512} = \frac{1}{512}$

10. $\left(3^3\right)^2 = 81^x$; $3^{3\times 2} = \left(3^4\right)^x$; $3^6 = 3^{4x}$; $6 = 4x$; $x = \frac{6}{4}$; $x = \frac{3}{2}$; $\mathbf{x = 1.5}$

Check: $\left(3^3\right)^2 \stackrel{?}{=} 81^{1.5}$; $3^{3\times 2} \stackrel{?}{=} 729$; $3^6 \stackrel{?}{=} 729$; $729 = 729$

Section 6.3 Case II Solutions - Both Sides of the Exponential Equation do not Have the Same Base

1. $5^x = 9$; $log_5 5^x = log_5 9$; $x = log_5 9$; $x = \frac{log_{10} 9}{log_{10} 5}$; $x = \frac{0.9542}{0.6989}$; $\mathbf{x = 1.3652}$

   Check: $5^{1.3652} \stackrel{?}{=} 9$; $9 = 9$

2. $3^x \cdot 3^{x+1} = 5$; $3^{x+x+1} = 5$; $3^{2x+1} = 5$; $log_3 3^{2x+1} = log_3 5$; $2x+1 = log_3 5$; $2x+1 = \frac{0.6989}{0.4771}$; $2x+1 = 1.4649$; $2x = 1.4649 - 1$; $2x = 0.4649$; $x = \frac{0.4649}{2}$; $\mathbf{x = 0.2325}$

   Check: $3^{0.2325} \cdot 3^{0.2325+1} \stackrel{?}{=} 5$; $3^{0.2325} \cdot 3^{1.2325} \stackrel{?}{=} 5$; $3^{0.2325+1.2325} \stackrel{?}{=} 5$; $3^{1.465} \stackrel{?}{=} 5$; $5 = 5$

3. $log_2 32^{x-2} = 3x$; $log_2 \left(2^5\right)^{x-2} = 3x$; $log_2 2^{5(x-2)} = 3x$; $log_2 2^{5x-10} = 3x$; $5x - 10 = 3x$; $5x - 3x = 10$; $2x = 10$; $x = \frac{10}{2}$; $x = \frac{5}{1}$; $\mathbf{x = 5}$

   Check: $log_2 32^{5-2} \stackrel{?}{=} 3 \times 5$; $log_2 32^3 \stackrel{?}{=} 15$; $log_2 \left(2^5\right)^3 \stackrel{?}{=} 15$; $log_2 2^{5\times3} \stackrel{?}{=} 15$; $log_2 2^{15} \stackrel{?}{=} 15$; $15 log_2 2 \stackrel{?}{=} 15$; $15 = 15$

4. $e^{-2u} = 2^u \cdot 2^{u-2}$; $e^{-2u} = 2^{u+u-2}$; $e^{-2u} = 2^{2u-2}$; $log_e e^{-2u} = log_e 2^{2u-2}$; $-2u = (2u-2) log_e 2$; $-2u = (2u-2)\frac{log_{10} 2}{log_{10} e}$; $-2u = (2u-2)\frac{log_{10} 2}{log_{10} 2.7183}$; $-2u = (2u-2)\frac{0.3010}{0.4343}$; $-2u = 0.693(2u-2)$; $-2u = 1.386u - 1.386$; $-2u - 1.386u = -1.386$; $-3.386u = -1.386$; $u = \frac{-1.386}{-3.386}$; $\mathbf{u = 0.4093}$

   Check: $e^{-2\times0.4093} \stackrel{?}{=} 2^{0.4093} \cdot 2^{0.4093-2}$; $e^{-0.8186} \stackrel{?}{=} 2^{0.4093} \cdot 2^{-1.5907}$; $e^{-0.8186} \stackrel{?}{=} 2^{0.4093-1.5907}$; $e^{-0.8186} \stackrel{?}{=} 2^{-1.1814}$; $\frac{1}{e^{0.8186}} \stackrel{?}{=} \frac{1}{2^{1.1814}}$; $\frac{1}{2.267} = \frac{1}{2.267}$

5. $3^{x+1} \cdot 3^{x+2} = 6$; $3^{x+1+x+2} = 6$; $3^{2x+3} = 6$; $log_3 3^{2x+3} = log_3 6$; $2x+3 = log_3 6$; $2x+3 = \frac{log_{10} 6}{log_{10} 3}$; $2x+3 = \frac{0.7782}{0.4771}$; $2x+3 = 1.6311$; $2x = 1.6311 - 3$; $2x = -1.3689$; $x = -\frac{1.3689}{2}$; $\mathbf{x = -0.6845}$

   Check: $3^{-0.6845+1} \cdot 3^{-0.6845+2} \stackrel{?}{=} 6$; $3^{0.3155} \cdot 3^{1.3155} \stackrel{?}{=} 6$; $3^{0.3155+1.3155} \stackrel{?}{=} 6$; $3^{1.631} \stackrel{?}{=} 6$; $6 = 6$

6. $2 \cdot 2^{-v} = 2^{2v}$; $2^1 \cdot 2^{-v} = 2^{2v}$; $2^{1-v} = 2^{2v}$; $1 - v = 2v$; $-v - 2v = -1$; $-3v = -1$; $v = \frac{-1}{-3}$; $v = \frac{1}{3}$; $\mathbf{v = 0.333}$

   Check: $2 \cdot 2^{-0.333} \stackrel{?}{=} 2^{2\times0.333}$; $2^1 \cdot 2^{-0.333} \stackrel{?}{=} 2^{0.666}$; $2^{1-0.333} \stackrel{?}{=} 2^{0.666}$; $2^{0.666} = 2^{0.666}$

7. $2 \cdot 3^{2t} = 4 \cdot 2^{-t}$; $3^{2t} = \frac{4}{2} \cdot 2^{-t}$; $3^{2t} = 2 \cdot 2^{-t}$; $3^{2t} = 2^1 \cdot 2^{-t}$; $3^{2t} = 2^{1-t}$; $log_3 3^{2t} = log_3 2^{1-t}$; $2t = (1-t) log_3 2$; $2t = (1-t)\frac{log_{10} 2}{log_{10} 3}$; $2t = (1-t)\frac{0.3010}{0.4771}$; $2t = 0.6309(1-t)$; $2t = 0.6309 - 0.6309t$; $2t + 0.6309t = 0.6309$; $2.6309t = 0.6309$; $t = \frac{0.6309}{2.6309}$; $\mathbf{t = 0.2398}$

   Check: $2 \cdot 3^{2\times0.2398} \stackrel{?}{=} 4 \cdot 2^{-0.2398}$; $2 \times 3^{0.4796} \stackrel{?}{=} 4 \times \frac{1}{2^{0.2398}}$; $2 \times 1.6937 \stackrel{?}{=} 4 \times \frac{1}{1.1808}$; $3.387 \stackrel{?}{=} 4 \times 0.8468$; $3.387 = 3.387$

8. $2 = e^t + ln\sqrt{e}$; $2 = e^t + ln e^{\frac{1}{2}}$; $2 = e^t + \frac{1}{2} ln e$; $2 = e^t + \frac{1}{2}$; $2 - \frac{1}{2} = e^t$; $\frac{4-1}{2} = e^t$; $e^t = \frac{3}{2}$; $e^t = 1.5$

; $log_e e^t = log_e 1.5$; $t = ln 1.5$; $\mathbf{t = 0.4055}$

Check: $2 \overset{?}{=} e^{0.4055} + ln\sqrt{e}$; $2 \overset{?}{=} e^{0.4055} + ln e^{\frac{1}{2}}$; $2 \overset{?}{=} 1.5 + \frac{1}{2} ln e$; $2 \overset{?}{=} 1.5 + \frac{1}{2} \times 1$; $2 \overset{?}{=} 1.5 + \frac{1}{2}$; $2 \overset{?}{=} 1.5 + 0.5$; $2 = 2$

9. $2e^{x+2} = 2^3$; $2e^{x+2} = 8$; $e^{x+2} = \frac{8}{2}$; $ln e^{x+2} = ln 4$; $x + 2 = ln 4$; $x + 2 = 1.3863$; $x = 1.3863 - 2$; $\mathbf{x = -0.6137}$

Check: $2e^{-0.6137+2} \overset{?}{=} 2^3$; $2e^{1.3863} \overset{?}{=} 8$; $2 \times 4 \overset{?}{=} 8$; $8 = 8$

10. $log_x 2 = 0.5$; $x^{log_x 2} = x^{0.5}$; $2 = x^{0.5}$; $2 = x^{\frac{1}{2}}$; $2^2 = x^{2\times\frac{1}{2}}$; $2^2 = x^{\frac{1}{1}}$; $\mathbf{x = 4}$

Check: $log_4 2 \overset{?}{=} 0.5$; $log_4 4^{\frac{1}{2}} \overset{?}{=} 0.5$; $\frac{1}{2} log_4 4 \overset{?}{=} 0.5$; $\frac{1}{2} \times 1 \overset{?}{=} 0.5$; $\frac{1}{2} \overset{?}{=} 0.5$; $0.5 = 0.5$

Section 6.4 Case I Solutions - Solving One Variable Logarithmic Equations (Simple Cases)

1. $x + 4 = log_4 \frac{1}{64}$; $x + 4 = log_4 64^{-1}$; $x + 4 = log_4 \left(4^3\right)^{-1}$; $x + 4 = log_4 4^{3\times -1}$; $x + 4 = log_4 4^{-3}$; $x + 4 = -3 log_4 4$; $x + 4 = -3$; $x = -3 - 4$; $\mathbf{x = -7}$

Check: $-7 + 4 \overset{?}{=} log_4 \frac{1}{64}$; $-3 \overset{?}{=} log_4 64^{-1}$; $-3 \overset{?}{=} log_4 \left(4^3\right)^{-1}$; $-3 \overset{?}{=} log_4 4^{-3}$; $-3 \overset{?}{=} -3 log_4 4$; $-3 \overset{?}{=} -3 \times 1$; $-3 = -3$

2. $y - 5 = ln\sqrt[3]{e^2}$; $y - 5 = ln e^{\frac{2}{3}}$; $y - 5 = \frac{2}{3} ln e$; $y - 5 = \frac{2}{3} \times 1$; $y - 5 = 0.6667$; $y = 5 + 0.6667$; $\mathbf{y = 5.6667}$

Check: $5.6667 - 5 \overset{?}{=} ln\sqrt[3]{e^2}$; $0.6667 \overset{?}{=} ln e^{\frac{2}{3}}$; $0.6667 \overset{?}{=} \frac{2}{3} ln e$; $0.6667 \overset{?}{=} \frac{2}{3} \times 1$; $0.6667 \overset{?}{=} \frac{2}{3}$; $0.6667 = 0.6667$

3. $x = log_2 \frac{1}{3^{-2}} + log_2 8$; $x = log_2 3^2 + log_2 2^3$; $x = 2 log_2 3 + 3 log_2 2$; $x = 2 log_2 3 + 3$; $x = 2\frac{log_{10} 3}{log_{10} 2} + 3$; $x = 2 \times \frac{0.4771}{0.3010} + 3$; $x = 2 \times 1.585 + 3$; $x = 3.17 + 3$; $\mathbf{x = 6.17}$

Check: $6.17 \overset{?}{=} log_2 \frac{1}{3^{-2}} + log_2 8$; $6.17 \overset{?}{=} log_2 3^2 + log_2 8$; $6.17 \overset{?}{=} log_2 9 + log_2 8$; $6.17 \overset{?}{=} log_2 9 \cdot 8$; $6.17 \overset{?}{=} log_2 72$; $6.17 \overset{?}{=} \frac{log_{10} 72}{log_{10} 2}$; $6.17 \overset{?}{=} \frac{1.8573}{0.3010}$; $6.17 = 6.17$

4. $log_3(x + 2) = 5$; $3^{log_3(x+2)} = 3^5$; $x + 2 = 243$; $x = 243 - 2$; $\mathbf{x = 241}$

Check: $log_3(241 + 2) \overset{?}{=} 5$; $log_3 243 \overset{?}{=} 5$; $log_3 3^5 \overset{?}{=} 5$; $5 log_3 3 \overset{?}{=} 5$; $5 \times 1 \overset{?}{=} 5$; $5 = 5$

5. $log_4 x = 3 + log_3 5$; $log_4 x = 3 + \frac{log_{10} 5}{log_{10} 3}$; $log_4 x = 3 + \frac{0.6989}{0.4771}$; $log_4 x = 3 + 1.4649$; $log_4 x = 4.4649$; $4^{log_4 x} = 4^{4.4649}$; $x = 4^{4.4649}$; $\mathbf{x = 487.68}$

Check: $log_4 487.68 \overset{?}{=} 3 + log_3 5$; $\frac{log_{10} 487.68}{log_{10} 4} \overset{?}{=} 3 + \frac{log_{10} 5}{log_{10} 3}$; $\frac{2.6881}{0.6021} \overset{?}{=} 3 + \frac{0.6989}{0.4771}$; $4.465 \overset{?}{=} 3 + 1.465$; $4.465 = 4.465$

6. $log_a 1000 = 3$; $a^{log_a 1000} = a^3$; $1000 = a^3$; $10^3 = a^3$; $\mathbf{a = 10}$

Check: $log_{10} 1000 \overset{?}{=} 3$; $log_{10} 1 \times 10^3 \overset{?}{=} 3$; $log_{10} 1 + log_{10} 10^3 \overset{?}{=} 3$; $0 + 3 log_{10} 10 \overset{?}{=} 3$; $3 \times 1 \overset{?}{=} 3$; $3 = 3$

7. $27^{x+3} = \frac{1}{243}$; $27^{x+3} = 243^{-1}$; $log_{27} 27^{x+3} = log_{27} 243^{-1}$; $x + 3 = -log_{27} 243$; $x + 3 = -\frac{log_{10} 243}{log_{10} 27}$; $x + 3 = -\frac{2.3856}{1.4314}$

$; \; x+3=-1.6667 \; ; \; x=-3-1.6667 \; ; \; \mathbf{x=-4.6667}$

Check: $27^{-4.6667+3} \overset{?}{=} \frac{1}{243} \; ; \; 27^{-1.6667} \overset{?}{=} \frac{1}{243} \; ; \; \frac{1}{27^{1.6667}} \overset{?}{=} \frac{1}{243} \; ; \; \frac{1}{243} = \frac{1}{243}$

8. $log_3 t = -\frac{1}{5} \; ; \; log_3 t = -0.2 \; ; \; 3^{log_3 t} = 3^{-0.2} \; ; \; t = 3^{-0.2} \; ; \; t = \frac{1}{3^{0.2}} \; ; \; t = \frac{1}{1.2457} \; ; \; \mathbf{t = 0.8027}$

Check: $log_3 0.8027 \overset{?}{=} -\frac{1}{5} \; ; \; \frac{log_{10} 0.8027}{log_{10} 3} \overset{?}{=} -0.2 \; ; \; \frac{-0.0955}{0.4771} \overset{?}{=} -0.2 \; ; \; -0.2 = -0.2$

9. $log_{25} x = \frac{1}{2} \; ; \; 25^{log_{25} x} = 25^{\frac{1}{2}} \; ; \; x = 25^{\frac{1}{2}} \; ; \; x = \sqrt{25} \; ; \; x = \sqrt{5^2} \; ; \; \mathbf{x = 5}$

Check: $log_{25} 5 \overset{?}{=} \frac{1}{2} \; ; \; log_{25} 25^{\frac{1}{2}} \overset{?}{=} \frac{1}{2} \; ; \; \frac{1}{2} log_{25} 25 \overset{?}{=} \frac{1}{2} \; ; \; \frac{1}{2} \times 1 \overset{?}{=} \frac{1}{2} \; ; \; \frac{1}{2} = \frac{1}{2}$

10. $x+3 = log_2 7 \; ; \; x+3 = \frac{log_{10} 7}{log_{10} 2} \; ; \; x+3 = \frac{0.8451}{0.3010} \; ; \; x+3 = 2.8076 \; ; \; x = 2.8076-3 \; ; \; \mathbf{x = -0.1924}$

Check: $-0.1924+3 \overset{?}{=} log_2 7 \; ; \; 2.8076 \overset{?}{=} \frac{log_{10} 7}{log_{10} 2} \; ; \; 2.8076 \overset{?}{=} \frac{0.8451}{0.3010} \; ; \; 2.8076 = 2.8076$

Section 6.4 Case II Solutions - Solving One Variable Logarithmic Equations (More Difficult Cases)

1. $log_3 6 - log_3 u = log_3 27 \; ; \; log_3 \frac{6}{u} = log_3 27 \; ; \; 3^{log_3 \frac{6}{u}} = 3^{log_3 27} \; ; \; \frac{6}{u} = 27 \; ; \; \frac{6}{u} = \frac{27}{1} \; ; \; 6 \cdot 1 = 27 \cdot u \; ; \; 6 = 27u \; ; \; \frac{6}{27} = \frac{\mathbf{27u}}{\mathbf{27}}$

$; \; \frac{6}{27} = u \; ; \; u = \frac{6}{27} \; ; \; \mathbf{u = 0.2222}$

Check: $log_3 6 - log_3 0.2222 \overset{?}{=} log_3 27 \; ; \; log_3 \frac{6}{0.2222} \overset{?}{=} log_3 27 \; ; \; log_3 27 = log_3 27$

2. $log_5 x = log_{0.5} 9 \; ; \; log_5 x = \frac{log_{10} 9}{log_{10} 0.5} \; ; \; log_5 x = \frac{0.9542}{-0.3010} \; ; \; log_5 x = -3.1698 \; ; \; 5^{log_5 x} = 5^{-3.1698} \; ; \; x = \frac{1}{5^{3.1698}}$

$; \; x = \frac{1}{164.284} \; ; \; \mathbf{x = 0.0061}$

Check: $log_5 0.0061 \overset{?}{=} log_{0.5} 9 \; ; \; \frac{log_{10} 0.0061}{log_{10} 5} \overset{?}{=} \frac{log_{10} 9}{log_{10} 0.5} \; ; \; \frac{-2.2147}{0.6989} \overset{?}{=} \frac{0.9542}{-0.3010} \; ; \; -3.17 = -3.17$

3. $log_5(x+2) - log_5 4 = log_5 50 \; ; \; log_5 \frac{x+2}{4} = log_5 50 \; ; \; 5^{log_5 \frac{x+2}{4}} = 5^{log_5 50} \; ; \; \frac{x+2}{4} = 50 \; ; \; \frac{x+2}{4} = \frac{50}{1} \; ; \; (x+2) \cdot 1 = 50 \cdot 4$

$; \; x+2 = 200 \; ; \; x+2-2 = 200-2 \; ; \; x+0 = 188 \; ; \; \mathbf{x = 188}$

Check: $log_5(188+2) - log_5 4 \overset{?}{=} log_5 50 \; ; \; log_5 200 - log_5 4 \overset{?}{=} log_5 50 \; ; \; log_5 \frac{200}{4} \overset{?}{=} log_5 50 \; ; \; log_5 50 = log_5 50$

4. $log(x+2) + log\, x = log 10 \; ; \; log[(x+2) \cdot x] = log 10 \; ; \; log(x^2+2x) = 1 \; ; \; 10^{log(x^2+2x)} = 10^1 \; ; \; x^2+2x = 10$

$; \; x^2+2x-10 = 0 \; ; \; x - \left(\frac{-2+\sqrt{44}}{2}\right) = 0 \; or \; x - \left(\frac{-2-\sqrt{44}}{2}\right) = 0 \; ; \; x - \left(\frac{-2+6.6332}{2}\right) = 0 \; or \; x - \left(\frac{-2-6.6332}{2}\right) = 0$

$; \; x - \left(\frac{4.6332}{2}\right) = 0 \; or \; x - \left(\frac{-8.6332}{2}\right) = 0 \; ; \; x - 2.3166 = 0 \; or \; x + 4.3166 = 0 \; ; \; \boxed{\mathbf{x = 2.3166}}$ or $\boxed{x = -4.3166}$

Check: 1. Substitute $x = 2.3116$ in the original equation. Then, $log(2.3166+2) + log 2.3166 \overset{?}{=} log 10$

$; \; log 4.3166 + log 2.3166 \overset{?}{=} 1 \; ; \; 0.635 + 0.365 \overset{?}{=} 1 \; ; \; 1 = 1$

2. Substitute $x = -4.3166$ in the original equation. Then, $log(-4.3166+2) + log -4.3166 \overset{?}{=} log 10$

; $log -2.3166 + log -4.3166 \overset{?}{=} 1$. Since the log of negative numbers is not defined thus, $x = -4.3166$ is not a solution.

5. $ln(x+2) = ln(5-x)$; $e^{ln(x+2)} = e^{ln(5-x)}$; $x+2 = 5-x$; $x+x+2 = 5-x+x$; $2x+2 = 5+0$; $2x+2 = 5$

; $2x+2-2 = 5-2$; $2x+0 = 3$; $2x = 3$; $\frac{2x}{2} = \frac{3}{2}$; $x = \frac{3}{2}$; $\mathbf{x = 1.5}$

Check: $ln(1.5+2) \overset{?}{=} ln(5-1.5)$; $ln 3.5 = ln 3.5$

6. $log_3(x+1) + log_3 x = log_3 9$; $log_3[(x+1)\cdot x] = log_3 3^2$; $log_3(x^2+x) = 2\,log_3 3$; $log_3(x^2+x) = 2 \times 1$

; $log_3(x^2+x) = 2$; $3^{log_3(x^2+x)} = 3^2$; $x^2+x = 9$; $x^2+x-9 = 0$; $x - \left(\frac{-1+\sqrt{37}}{2}\right) = 0$ or $x - \left(\frac{-1-\sqrt{37}}{2}\right) = 0$

; $x - \left(\frac{-1+6.0827}{2}\right) = 0$ or $x - \left(\frac{-1-6.0827}{2}\right) = 0$; $x - \left(\frac{5.0827}{2}\right) = 0$ or $x - \left(\frac{-7.0827}{2}\right) = 0$

; $x - 2.5413 = 0$ or $x + 3.5413 = 0$; $\boxed{\mathbf{x = 2.5413}}$ or $\boxed{x = -3.5413}$

Check: 1. Substitute $x = 2.5413$ in the original equation. Then, $log_3(2.5413+1) + log_3 2.5413 \overset{?}{=} log_3 9$

; $log_3 3.5413 + log_3 2.5413 \overset{?}{=} log_3 9$; $log_3(3.5413 \times 2.5413) \overset{?}{=} log_3 9$; $log_3 9 = log_3 9$

2. Substitute $x = -3.5413$ in the original equation. Then, $log_3(-3.5413+1) + log_3 -3.5413 \overset{?}{=} log_3 9$

; $log_3 -2.5413 + log_3 -3.5413 \overset{?}{=} log_3 9$. Since the log of negative numbers is not defined thus, $x = -3.5413$ is not a solution.

7. $log(x+2) - log(x+3) = log 10$; $log\frac{x+2}{x+3} = 1$; $10^{log\frac{x+2}{x+3}} = 10^1$; $\frac{x+2}{x+3} = 10$; $\frac{x+2}{x+3} = \frac{10}{1}$; $(x+2)\cdot 1 = 10\cdot(x+3)$

; $x+2 = 10x+30$; $x - 10x + 2 = 10x - 10x + 30$; $-9x+2 = 0+30$; $-9x+2 = 30$; $-9x+2-2 = 30-2$; $-9x+0 = 28$

; $-9x = 28$; $\frac{-9x}{-9} = \frac{28}{-9}$; $x = -\frac{28}{9}$; $x = -3\frac{1}{9}$; $\mathbf{x = -3.111}$

Check: $log(-3.111+2) - log(-3.111+3) \overset{?}{=} log 10$; $log(-1.111) - log(-0.111) \overset{?}{=} log 10$; $log\frac{-1.111}{-0.111} \overset{?}{=} log 10$; $log 10 = log 10$

8. $log_8(x+2) = log_8(10-x)$; $8^{log_8(x+2)} = 8^{log_8(10-x)}$; $x+2 = 10-x$; $x+x+2 = 10-x+x$; $2x+2 = 10+0$; $2x+2 = 10$

; $2x+2-2 = 10-2$; $2x+0 = 8$; $2x = 8$; $\frac{2x}{2} = \frac{8}{2}$; $x = \frac{4}{1}$; $\mathbf{x = 4}$

Check: $log_8(4+2) \overset{?}{=} log_8(10-4)$; $log_8 6 = log_8 6$

9. $log_2(x+2) + log_5 5 = log_{0.3} 10$; $log_2(x+2) + 1 = \frac{log_{10} 10}{log_{10} 0.3}$; $log_2(x+2) + 1 = \frac{1}{-0.5229}$; $log_2(x+2) + 1 = -1.9125$

; $log_2(x+2) + 1 - 1 = -1.9125 - 1$; $log_2(x+2) + 0 = -2.9125$; $log_2(x+2) = -2.9125$; $2^{log_2(x+2)} = 2^{-2.9125}$

; $x+2 = \frac{1}{2^{2.9125}}$; $x+2 = \frac{1}{7.5292}$; $x+2 = 0.1328$; $x+2-2 = 0.1328-2$; $x+0 = -1.867$; $\mathbf{x = -1.867}$

Check: $log_2(-1.867+2) + log_5 5 \overset{?}{=} log_{0.3} 10$; $log_2 0.133 + 1 \overset{?}{=} log_{0.3} 10$; $\frac{log_{10} 0.133}{log_{10} 2} + 1 \overset{?}{=} \frac{log_{10} 10}{log_{10} 0.3}$; $\frac{-0.8761}{0.3010} + 1 \overset{?}{=} \frac{1}{-0.5229}$

; $-2.91 + 1 \overset{?}{=} -1.91$; $-1.91 = -1.91$

10. $log_2(x+8) - log_2 x = log_2 16$; $log_2 \frac{x+8}{x} = log_2 16$; $2^{log_2 \frac{x+8}{x}} = 2^{log_2 16}$; $\frac{x+8}{x} = 16$; $\frac{x+8}{x} = \frac{16}{1}$; $(x+8)\cdot 1 = 16 \cdot x$

; $x+8 = 16x$; $x - 16x + 8 = 16x - 16x$; $-15x + 8 = 0$; $-15x + 8 - 8 = 0 - 8$; $-15x + 0 = -8$; $-15x = -8$

; $\frac{-15x}{-15} = \frac{-8}{-15}$; $x = +\frac{8}{15}$; $\mathbf{x = 0.5333}$

Check: $log_2(0.5333+8) - log_2 0.5333 \stackrel{?}{=} log_2 16$; $log_2 8.5333 - log_2 0.5333 \stackrel{?}{=} log_2 16$; $log_2 \frac{8.5333}{0.5333} \stackrel{?}{=} log_2 16$

; $log_2 16 = log_2 16$

Section 6.5 Case I Solutions - Solving Numerical Expressions Using Logarithms

1. $Let\ x = (0.00025)(12{,}000{,}000)$ then, $log\,x = log(0.00025)(12{,}000{,}000)$; $log\,x = log(0.00025) + log(12{,}000{,}000)$

; $log\,x = log\,2.5 \times 10^{-4} + log\,1.2 \times 10^{7}$; $log\,x = log\,2.5 + log\,10^{-4} + log\,1.2 + log\,10^{7}$

; $log\,x = log\,2.5 - 4\,log\,10 + log\,1.2 + 7\,log\,10$; $log\,x = log\,2.5 - 4 + log\,1.2 + 7$; $log\,x = 0.3979 - 4 + 0.0792 + 7$

; $log\,x = 0.3979 + 0.0792 + 3$; $log\,x = 3.4771$; $10^{log\,x} = 10^{3.4771}$; $x = 10^{3.4771}$; $\mathbf{x = 2999.85}$

2. $Let\ x = (8755)(0.000165)$ then, $log\,x = log(8755)(0.000165)$; $log\,x = log(8755) + log(0.000165)$

; $log\,x = log\,8.755 \times 10^{3} + log\,1.65 \times 10^{-4}$; $log\,x = log\,8.755 + log\,10^{3} + log\,1.65 + log\,10^{-4}$

; $log\,x = log\,8.755 + 3\,log\,10 + log\,1.65 - 4\,log\,10$; $log\,x = log\,8.755 + 3 + log\,1.65 - 4$; $log\,x = 0.9423 + 0.2175 - 1$

; $log\,x = 0.1598$; $10^{log\,x} = 10^{0.1598}$; $x = 10^{0.1598}$; $\mathbf{x = 1.4447}$

3. First Method: $Let\ x = \sqrt[5]{0.35}$ then, $log_{10} x = log_{10} \sqrt[5]{0.35}$: $log_{10} x = log_{10} 0.35^{\frac{1}{5}}$; $log_{10} x = log_{10} 0.35^{0.2}$

; $log_{10} x = 0.2\,log_{10} 0.35$; $log_{10} x = 0.2 \times -0.4559$; $log_{10} x = -0.09118$; $10^{log_{10} x} = 10^{-0.09118}$

; $x = 10^{-0.09118}$; $x = \frac{1}{10^{0.09118}}$; $x = \frac{1}{1.2336}$; $\mathbf{x = 0.8106}$

Second Method: $Let\ x = \sqrt[5]{0.35}$ then, $log_2 x = log_2 \sqrt[5]{0.35}$: $log_2 x = log_2 0.35^{\frac{1}{5}}$; $log_2 x = log_2 0.35^{0.2}$

; $log_2 x = 0.2\,log_2 0.35$; $log_2 x = 0.2 \times \frac{log_{10} 0.35}{log_{10} 2}$; $log_2 x = 0.2 \times \frac{-0.4559}{0.3010}$; $log_2 x = -0.3029$; $2^{log_2 x} = 2^{-0.3029}$

; $x = 2^{-0.3029}$; $x = \frac{1}{2^{0.3029}}$; $x = \frac{1}{1.2336}$; $\mathbf{x = 0.8106}$

4. $Let\ x = \frac{3650}{2.25}$ then, $log\,x = log\frac{3650}{2.25}$; $log\,x = log\,3650 - log\,2.25$; $log\,x = log\,3.65 \times 10^{3} - log\,2.25$

; $log\,x = log\,3.65 + log\,10^{3} - log\,2.25$; $log\,x = log\,3.65 + 3\,log\,10 - log\,2.25$; $log\,x = log\,3.65 + 3 - log\,2.25$

; $log\,x = 0.5623 + 3 - 0.3522$; $log\,x = 3.2101$; $10^{log\,x} = 10^{3.2101}$; $x = 10^{3.2101}$; $\mathbf{x = 1622.2}$

5. $Let\ x = \sqrt[5]{5.09^3}$ then, $log\,x = log\sqrt[5]{5.09^3}$; $log\,x = log\,5.09^{\frac{3}{5}}$; $log\,x = \frac{3}{5}log\,5.09$; $log\,x = \frac{3}{5} \times 0.7067$; $log\,x = 0.4240$

; $10^{log\,x} = 10^{0.4240}$; $x = 10^{0.4240}$; $\mathbf{x = 2.6546}$

6. $Let\ x = 0.983^{5.6}$ then, $log\,x = log\,0.983^{5.6}$; $log\,x = 5.6\,log\,0.983$; $log\,x = 5.6 \times -0.00745$; $log\,x = -0.0417$

; $10^{log\,x} = 10^{-0.0417}$; $x = \frac{1}{10^{0.0417}}$; $x = \frac{1}{1.1008}$; $\mathbf{x = 0.9084}$

7. $Let\ x = \frac{0.00057^{0.05}}{5554^{0.002}}$ then, $log\,x = log\frac{0.00057^{0.05}}{5554^{0.002}}$; $log\,x = log\,0.00057^{0.05} - log\,5554^{0.002}$

$; \log x = 0.05 \log 0.00057 - 0.002 \log 5554 \;;\; \log x = 0.05\left(\log 5.7 \times 10^{-4}\right) - 0.002\left(\log 5.554 \times 10^{3}\right)$

$; \log x = 0.05\left(\log 5.7 + \log 10^{-4}\right) - 0.002\left(\log 5.554 + \log 10^{3}\right) \;;\; \log x = 0.05(\log 5.7 + -4 \log 10) - 0.002(\log 5.554 + 3 \log 10)$

$; \log x = 0.05(\log 5.7 - 4) - 0.002(\log 5.554 + 3) \;;\; \log x = 0.05(0.7558 - 4) - 0.002(0.7446 + 3)$

$; \log x = (0.05 \times -3.2442) - (0.002 \times 3.7446) \;;\; \log x = -0.1622 - 0.0075 \;;\; \log x = -0.1697 \;;\; 10^{\log x} = 10^{-0.1697}$

$; x = 10^{-0.1697} \;;\; x = \frac{1}{10^{0.1697}} \;;\; x = \frac{1}{1.4780} \;;\; \mathbf{x = 0.676}$

8. $Let\ x = \frac{0.148^{2.5}}{33.5^{1.2}}$ then, $\log x = \log \frac{0.148^{2.5}}{33.5^{1.2}} \;;\; \log x = \log 0.148^{2.5} - \log 33.5^{1.2} \;;\; \log x = 2.5 \log 0.148 - 1.2 \log 33.5$

$; \log x = 2.5 \log 1.48 \times 10^{-1} - 1.2 \log 3.35 \times 10^{1} \;;\; \log x = 2.5\left(\log 1.48 + \log 10^{-1}\right) - 1.2\left(\log 3.35 + \log 10^{1}\right)$

$; \log x = 2.5(\log 1.48 - \log 10) - 1.2(\log 3.35 + \log 10) \;;\; \log x = 2.5(\log 1.48 - 1) - 1.2(\log 3.35 + 1)$

$; \log x = 2.5(0.1703 - 1) - 1.2(0.5250 + 1) \;;\; \log x = 2.5 \times -0.8297 - 1.2 \times 1.525 \;;\; \log x = -2.0743 - 1.83 \;;\; \log x = -3.9043$

$; 10^{\log x} = 10^{-3.9043} \;;\; x = 10^{-3.9043} \;;\; x = \frac{1}{10^{3.9043}} \;;\; x = \frac{1}{8022.32} \;;\; \mathbf{x = 0.000124}$

9. $Let\ x = \frac{2355^{\frac{1}{2}}}{0.235^{2}}$ then, $\log x = \log \frac{2355^{\frac{1}{2}}}{0.235^{2}} \;;\; \log x = \log 2355^{\frac{1}{2}} - \log 0.235^{2} \;;\; \log x = \frac{1}{2} \log 2355 - 2 \log 0.235$

$; \log x = 0.5 \log 2.355 \times 10^{3} - 2 \log 2.35 \times 10^{-1} \;;\; \log x = 0.5\left(\log 2.355 + \log 10^{3}\right) - 2\left(\log 2.35 + \log 10^{-1}\right)$

$; \log x = 0.5(\log 2.355 + 3 \log 10) - 2(\log 2.35 - \log 10) \;;\; \log x = 0.5(\log 2.355 + 3) - 2(\log 2.35 - 1)$

$; \log x = 0.5(0.3719 + 3) - 2(0.3711 - 1) \;;\; \log x = 0.5 \times 3.3719 - 2 \times -0.6289 \;;\; \log x = 1.686 + 1.2579 \;;\; \log x = 2.9439$

$; 10^{\log x} = 10^{2.9439} \;;\; \mathbf{x = 878.8}$

10. $Let\ x = \frac{28^{0.05} \sqrt[3]{2.3^{2}}}{(0.00008)\sqrt{3.05^{3}}}$ then, $\log x = \log \frac{28^{0.05} \sqrt[3]{2.3^{2}}}{(0.00008)\sqrt{3.05^{3}}} \;;\; \log x = \log\left(28^{0.05} \sqrt[3]{2.3^{2}}\right) - \log\left[(0.00008)\sqrt{3.05^{3}}\right]$

$; \log x = \log 28^{0.05} + \log \sqrt[3]{2.3^{2}} - \log 0.00008 - \log \sqrt{3.05^{3}} \;;\; \log x = 0.05 \log 28 + \log 2.3^{\frac{2}{3}} - \log 0.00008 - \log 3.05^{\frac{3}{2}}$

$; \log x = 0.05 \log 2.8 \times 10^{1} + \frac{2}{3} \log 2.3 - \log 8 \times 10^{-5} - \frac{3}{2} \log 3.05 \;;\; \log x = 0.05(\log 2.8 + 1) + \frac{2}{3} \log 2.3 - (\log 8 - 5) - \frac{3}{2} \log 3.05$

$; \log x = 0.05(0.4471 + 1) + \frac{2}{3} \times 0.3617 - (0.9031 - 5) - \frac{3}{2} \times 0.4843 \;;\; \log x = 0.0724 + 0.2411 + 4.0969 - 0.7265 \;;\; \log x = 3.684$

$; 10^{\log x} = 10^{3.684} \;;\; x = 10^{3.684} \;;\; \mathbf{x = 4830.5}$

Section 6.5 Case II Solutions - Expanding Logarithmic Expressions from a Single Term

1. $\log \frac{3x}{2x^{2} + x - 1} = \log \frac{3x}{(2x - 1)(x + 1)} = \log 3x - \log[(2x - 1)(x + 1)] = \log 3 + \log x - [\log(2x - 1) + \log(x + 1)]$

$= \mathbf{\log 3 + \log x - \log(2x - 1) - \log(x + 1)}$

2. $\ln \sqrt{\frac{(1 - x)^{2}}{(1 + x)}} = \ln\left(\frac{(1 - x)^{2}}{1 + x}\right)^{\frac{1}{2}} = \frac{1}{2} \ln \frac{(1 - x)^{2}}{1 + x} = \frac{1}{2}\left[\ln(1 - x)^{2} - \ln(1 + x)\right] = \frac{1}{2}[2 \ln(1 - x) - \ln(1 + x)]$

$= \frac{2}{2} \ln(1 - x) - \frac{1}{2} \ln(1 + x) = \mathbf{\ln(1 - x) - \frac{1}{2} \ln(1 + x)}$

3. $log_3 \frac{3x\sqrt{(x+1)^3}}{(x-1)^2} = log_3 \frac{3x\left[(x+1)^3\right]^{\frac{1}{2}}}{(x-1)^2} = log_3 \frac{3x(x+1)^{3\times\frac{1}{2}}}{(x-1)^2} = log_3 \frac{3x(x+1)^{\frac{3}{2}}}{(x-1)^2} = log_3\left[3x(x+1)^{\frac{3}{2}}\right] - log_3(x-1)^2$

$= log_3 3x + log_3(x+1)^{\frac{3}{2}} - log_3(x-1)^2 = log_3 3 + log_3 x + \frac{3}{2}log_3(x+1) - 2\,log_3(x-1)$

$= \mathbf{1 + log_3\, x + \frac{3}{2}log_3(x+1) - 2\,log_3(x-1)}$

4. $log_8 \frac{64x^3}{\sqrt{(x-1)^4}} = log_8 64x^3 - log_8\sqrt{(x-1)^4} = log_8 64 + log_8 x^3 - log_8\left[(x-1)^4\right]^{\frac{1}{2}} = log_8 8^2 + 3\,log_8 x - \frac{1}{2}log_8(x-1)^4$

$= 2\,log_8 8 + 3\,log_8 x - \frac{1}{2}\times \overset{2}{\cancel{4}}\left[log_8(x-1)\right] = 2\times 1 + 3\,log_8 x - 2\,log_8(x-1) = \mathbf{2 + 3\,log_8\, x - 2\,log_8(x-1)}$

5. $ln\frac{6x^2+7x-3}{x^2+3x+2} = ln\frac{(2x+3)(3x-1)}{(x+2)(x+1)} = ln[(2x+3)(3x-1)] - ln[(x+2)(x+1)] = ln(2x+3) + ln(3x-1) - [ln(x+2) + ln(x+1)]$

$= \mathbf{ln(2x+3) + ln(3x-1) - ln(x+2) - ln(x+1)}$

6. $log_5\frac{125y^5\left(\sqrt[5]{y+2}\right)}{8(y-1)^{\frac{1}{2}}} = log_5\left[125y^5\left(\sqrt[5]{y+2}\right)\right] - log_5 8(y-1)^{\frac{1}{2}} = log_5 125 + log_5 y^5 + log_5\sqrt[5]{y+2} - \left[log_5 8 + log_5(y-1)^{\frac{1}{2}}\right]$

$= log_5 5^3 + 5\,log_5 y + log_5(y+2)^{\frac{1}{5}} - \left[log_5 8 + \frac{1}{2}log_5(y-1)\right] = 3\,log_5 5 + 5\,log_5 y + \frac{1}{5}log_5(y+2) - log_5 8 - \frac{1}{2}log_5(y-1)$

$= \mathbf{3 + 5\,log_5\, y + \frac{1}{5}log_5(y+2) - log_5 8 - \frac{1}{2}log_5(y-1)}$

7. $log_5\frac{125z^2}{\sqrt{8}(z-1)^2} = log_5 125z^2 - log_5\left[\sqrt{8}(z-1)^2\right] = log_5 125 + log_5 z^2 - \left[log_5\sqrt{8} + log_5(z-1)^2\right]$

$= log_5 5^3 + 2\,log_5 z - log_5\sqrt{8} - log_5(z-1)^2 = 3\,log_5 5 + 2\,log_5 z - log_5 8^{\frac{1}{2}} - 2\,log_5(z-1)$

$= 3\times 1 + 2\,log_5 z - \frac{1}{2}log_5 8 - 2\,log_5(z-1) = \mathbf{3 + 2\,log_5\, z - \frac{1}{2}log_5 8 - 2\,log_5(z-1)}$

8. $log_{10}\frac{3w}{11\sqrt{w^2+1}} = log_{10} 3w - log_{10} 11\sqrt{w^2+1} = log_{10} 3 + log_{10} w - \left(log_{10} 11 + log_{10}\sqrt{w^2+1}\right)$

$= log_{10} 3 + log_{10} w - log_{10} 11 - log_{10}\left(w^2+1\right)^{\frac{1}{2}} = \mathbf{log_{10} 3 - log_{10} 11 + log_{10}\, w - \frac{1}{2}log_{10}\left(w^2+1\right)}$

9. $log_3\frac{27x^3}{2(x-1)^{\frac{1}{2}}} = log_3 27x^3 - log_3 2(x-1)^{\frac{1}{2}} = log_3 27 + log_3 x^3 - \left[log_3 2 + log_3(x-1)^{\frac{1}{2}}\right]$

$= log_3 3^3 + log_3 x^3 - log_3 2 - \frac{1}{2}log_3(x-1) = 3\,log_3 3 + 3\,log_3 x - log_3 2 - \frac{1}{2}log_3(x-1) = 3\times 1 + 3\,log_3 x - log_3 2 - \frac{1}{2}log_3(x-1)$

$= \mathbf{3 + 3\,log_3\, x - log_3 2 - \frac{1}{2}log_3(x-1)}$

10. $log_4\frac{64t^2(t+1)}{\sqrt[3]{7}} = log_4\left[64t^2(t+1)\right] - log_4\sqrt[3]{7} = log_4 64 + log_4 t^2 + log_4(t+1) - log_4 7^{\frac{1}{3}}$

$= log_4 4^3 + 2\,log_4 t + log_4(t+1) - \frac{1}{3}log_4 7 = 3\,log_4 4 + 2\,log_4 t + log_4(t+1) - \frac{1}{3}log_4 7 = 3\times 1 + 2\,log_4 t + log_4(t+1) - \frac{1}{3}log_4 7$

$= \mathbf{3 + 2\,log_4\, t + log_4(t+1) - \frac{1}{3}log_4 7}$

Section 6.5 Case III Solutions - Combining Logarithmic Expressions into a Single Term

1. $2\log_8 5+4\log_8 u+\log_8 10 = \log_8 5^2+\log_8 u^4+\log_8 10 = \log_8 25+\log_8 u^4+\log_8 10 = \log_8 25\times 10\times u^4 = \boldsymbol{\log_8 250u^4}$

2. $(4\log_3 2-\log_3 u)+(2\log_3 3-3\log_3 u) = \left(\log_3 2^4-\log_3 u\right)+\left(\log_3 3^2-\log_3 u^3\right) = \log_3\frac{2^4}{u}+\log_3\frac{3^2}{u^3} = \log_3\frac{16}{u}+\log_3\frac{9}{u^3}$

$= \log_3\frac{16}{u}\times\frac{9}{u^3} = \log_3\frac{16\times 9}{u\times u^3} = \boldsymbol{\log_3\frac{144}{u^4}}$

3. $3\log_{10}x+\left[\log_{10}(x-2)-\log_{10}(x+1)\right] = \log_{10}x^3+\log_{10}\left[\frac{x-2}{x+1}\right] = \log_{10}x^3\times\frac{x-2}{x+1} = \boldsymbol{\log_{10}\frac{x^3(x-2)}{x+1}}$

4. $\left[\log(x-1)-3\log(x+2)\right]-\log x = \left[\log(x-1)-\log(x+2)^3\right]-\log x = \left[\log\frac{x-1}{(x+2)^3}\right]-\log x = \log\frac{\frac{x-1}{(x+2)^3}}{x} = \boldsymbol{\log\frac{x-1}{x(x+2)^3}}$

5. $\left(\log_2 7-\frac{1}{3}\log_2 27\right)+3\log_2 x = \left(\log_2 7-\log_2 27^{\frac{1}{3}}\right)+\log_2 x^3 = \left(\log_2 7-\log_2\sqrt[3]{27}\right)+\log_2 x^3 = \log_2\frac{7}{\sqrt[3]{27}}+\log_2 x^3$

$= \log_2\frac{7}{\sqrt[3]{3^3}}+\log_2 x^3 = \log_2\frac{7}{3}+\log_2 x^3 = \log_2\frac{7}{3}\times x^3 = \boldsymbol{\log_2\frac{7x^3}{3}}$

6. $\log_4 3-(\log_4 3+2\log_4 x) = \log_4 3-\left(\log_4 3+\log_4 x^2\right) = \log_4 3-\log_4 3x^2 = \log_4\frac{3}{3x^2} = \boldsymbol{\log_4\frac{1}{x^2}}$

7. $\left[2\log x+3\log(x-1)\right]-\left[2\log(x-1)+\log 2\right] = \left[\log x^2+\log(x-1)^3\right]-\left[\log(x-1)^2+\log 2\right] = \left[\log x^2(x-1)^3\right]-\left[\log 2(x-1)^2\right]$

$= \log\frac{x^2(x-1)^3}{2(x-1)^2} = \boldsymbol{\log\frac{x^2(x-1)}{2}}$

8. $(\log_{10}12-2\log_{10}3)+2\log_{10}x = \left(\log_{10}12-\log_{10}3^2\right)+\log_{10}x^2 = (\log_{10}12-\log_{10}9)+\log_{10}x^2 = \log_{10}\frac{12}{9}+\log_{10}x^2$

$= \log_{10}\frac{4}{3}+\log_{10}x^2 = \log_{10}\frac{4}{3}\times x^2 = \boldsymbol{\log_{10}\frac{4x^2}{3}}$

9. $(\log_{10}12-3\log_{10}2)-(\log_{10}3-2\log_{10}w) = \left(\log_{10}12-\log_{10}2^3\right)-\left(\log_{10}3-\log_{10}w^2\right) = \left(\log_{10}\frac{12}{8}\right)-\left(\log_{10}\frac{3}{w^2}\right)$

$= \left(\log_{10}\frac{3}{2}\right)-\left(\log_{10}\frac{3}{w^2}\right) = \log_{10}\frac{\frac{3}{2}}{\frac{3}{w^2}} = \log_{10}\frac{3\times w^2}{3\times 2} = \boldsymbol{\log_{10}\frac{w^2}{2}}$

10. $\log(x+1)-\left[3\log(x+2)-3\log x\right] = \log(x+1)-\left[\log(x+2)^3-\log x^3\right] = \log(x+1)-\left[\log\frac{(x+2)^3}{x^3}\right] = \log\frac{(x+1)}{\frac{(x+2)^3}{x^3}}$

$= \log\frac{x^3\times(x+1)}{(x+2)^3} = \boldsymbol{\log\frac{x^3(x+1)}{(x+2)^3}}$

Section 6.6 Solutions - Advanced Logarithmic Problems

1. Given $P=P_0\left(1+\frac{r}{k}\right)^{kt}$ solve for t if $P=20$, $P_0=10$, $r=1$, and $k=2$.

$P=P_0\left(1+\frac{r}{k}\right)^{kt}$; $20=10\left(1+\frac{1}{2}\right)^{2\times t}$; $\frac{20}{10}=\left(\frac{2+1}{2}\right)^{2t}$; $\frac{2}{1}=\left(\frac{3}{2}\right)^{2t}$; $2=1.5^{2t}$; $\log_{1.5}2=\log_{1.5}1.5^{2t}$; $\log_{1.5}2=2t$

$; \; 2t = \frac{log_{10} 2}{log_{10} 1.5} \; ; \; 2t = \frac{0.3010}{0.1761} \; ; \; 2t = 1.7093 \; ; \; t = \frac{1.7093}{2} \; ; \; \mathbf{t = 0.8547}$

Check: $20 \stackrel{?}{=} 10\left(1+\frac{1}{2}\right)^{2\times 0.8547} \; ; \; 20 \stackrel{?}{=} 10\left(\frac{2+1}{2}\right)^{1.7093} \; ; \; 20 \stackrel{?}{=} 10\cdot\left(\frac{3}{2}\right)^{1.7093} \; ; \; 20 \stackrel{?}{=} 10\times 1.5^{1.7093} \; ; \; 20 \stackrel{?}{=} 10\times 2 \; ; \; 20 = 20$

2. Given $A = A_0 + ke^{-t}$ solve for t if $A = 100$, $A_0 = 10$, and $k = 2$.

$A = A_0 + ke^{-t} \; ; \; 100 = 10 + 2e^{-t} \; ; \; 100 - 10 = 2e^{-t} \; ; \; 90 = 2e^{-t} \; ; \; \frac{90}{2} = e^{-t} \; ; \; 45 = e^{-t} \; ; \; log_e 45 = log_e e^{-t} \; ; \; log_e 45 = -t$
$; \; t = -log_e 45 \; ; \; \mathbf{t = -3.8067}$

Check: $100 \stackrel{?}{=} 10 + 2e^{-(-3.8067)} \; ; \; 100 \stackrel{?}{=} 10 + 2e^{3.8067} \; ; \; 100 \stackrel{?}{=} 10 + 2\times 45 \; ; \; 100 \stackrel{?}{=} 10 + 90 \; ; \; 100 = 100$

3. Given $B = B_0 + ke^{-(t+a)}$ solve for t if $B = 200$, $B_0 = 25$, $k = 5$, and $a = 0.01$.

$B = B_0 + ke^{-(t+a)} \; ; \; 200 = 25 + 5e^{-(t+0.01)} \; ; \; 200 - 25 = 5e^{-(t+0.01)} \; ; \; 175 = 5e^{-(t+0.01)} \; ; \; \frac{175}{5} = e^{-(t+0.01)} \; ; \; 35 = e^{-(t+0.01)}$
$; \; ln35 = lne^{-(t+0.01)} \; ; \; ln35 = -(t+0.01) \; ; \; ln35 = -t - 0.01 \; ; \; t = -ln35 - 0.01 \; ; \; t = -3.5554 - 0.01 \; ; \; \mathbf{t = -3.5654}$

Check: $200 \stackrel{?}{=} 25 + 5e^{-(-3.5654+0.01)} \; ; \; 200 \stackrel{?}{=} 25 + 5e^{3.5654-0.01} \; ; \; 200 \stackrel{?}{=} 25 + 5e^{3.5554} \; ; \; 200 \stackrel{?}{=} 25 + 5\times 35 \; ; \; 200 \stackrel{?}{=} 25 + 175$
$; \; 200 = 200$

4. Given $N = N_0 e^{-\frac{t}{1000}}$ solve for t if $N = 2$ and $N_0 = 5N$.

$N = N_0 e^{-\frac{t}{1000}} \; ; \; 2 = 10e^{-\frac{t}{1000}} \; ; \; \frac{2}{10} = e^{-\frac{t}{1000}} \; ; \; \frac{1}{5} = e^{-\frac{t}{1000}} \; ; \; 0.2 = e^{-\frac{t}{1000}} \; ; \; ln0.2 = lne^{-\frac{t}{1000}} \; ; \; ln0.2 = -\frac{t}{1000}$
$; \; 1000\times ln0.2 = -t \; ; \; t = -1000\times ln0.2 \; ; \; t = -1000\times -1.6094 \; ; \; \mathbf{t = 1609.4}$

Check: $2 \stackrel{?}{=} 10e^{-\frac{1609.4}{1000}} \; ; \; 2 \stackrel{?}{=} 10e^{-1.6094} \; ; \; 2 \stackrel{?}{=} 10\cdot\frac{1}{e^{1.6094}} \; ; \; 2 \stackrel{?}{=} 10\cdot\frac{1}{5} \; ; \; 2 \stackrel{?}{=} 10\times 0.2 \; ; \; 2 = 2$

5. Given $Q = Q_0(1-m)^{0.05t}$ solve for t if $Q = 0.4$, $Q_0 = 40$, and $m = 0.5$.

$Q = Q_0(1-m)^{0.05t} \; ; \; 0.4 = 40(1-0.5)^{0.05t} \; ; \; \frac{0.4}{40} = 0.5^{0.05t} \; ; \; 0.01 = 0.5^{0.05t} \; ; \; log_{0.5} 0.01 = log_{0.5} 0.5^{0.05t} \; ; \; log_{0.5} 0.01 = 0.05t$
$; \; 0.05t = \frac{log_{10} 0.01}{log_{10} 0.5} \; ; \; 0.05t = \frac{-2}{-0.3010} \; ; \; 0.05t = 6.6439 \; ; \; t = \frac{6.6439}{0.05} \; ; \; \mathbf{t = 132.88}$

Check: $0.4 \stackrel{?}{=} 40(1-0.5)^{0.05\times 132.88} \; ; \; 0.4 \stackrel{?}{=} 40\times 0.5^{6.644} \; ; \; 0.4 \stackrel{?}{=} 40\times 0.01 \; ; \; 0.4 = 0.4$

6. Given $K = K_0 4^{-\frac{kr}{200}}$ solve for r if $K = 5K_0$, $K_0 = 0.1$, and $k = 1$.

$K = K_0 4^{-\frac{kr}{200}} \; ; \; 5\times 0.1 = 0.1\times 4^{-\frac{1\times r}{200}} \; ; \; \frac{0.5}{0.1} = 4^{-\frac{r}{200}} \; ; \; 5 = 4^{-0.005r} \; ; \; log_4 5 = log_4 4^{-0.005r} \; ; \; log_4 5 = -0.005r$
$; \; -0.005r = \frac{log_{10} 5}{log_{10} 4} \; ; \; -0.005r = \frac{0.6989}{0.6021} = -0.005r = 1.1608 \; ; \; r = \frac{1.1608}{-0.005} \; ; \; \mathbf{r = -232.16}$

Check: $0.5 \stackrel{?}{=} 0.1\times 4^{-\frac{1\times -232.16}{200}} \; ; \; 0.5 \stackrel{?}{=} 0.1\times 4^{1.1608} \; ; \; 0.5 \stackrel{?}{=} 0.1\times 5 \; ; \; 0.5 = 0.5$

7. Given $M = M_0 e^{-k(t-4)}$ solve for t if $M = 5$, $M_0 = 500$, and $k = 0.1$.

$M = M_0 e^{-k(t-4)} \; ; \; 5 = 500e^{-0.1(t-4)} \; ; \; \frac{5}{500} = e^{-0.1t+0.4} \; ; \; \frac{1}{100} = e^{-0.1t+0.4} \; ; \; 0.01 = e^{-0.1t+0.4} \; ; \; ln0.01 = lne^{-0.1t+0.4}$
$; \; -4.6052 = -0.1t + 0.4 \; ; \; -4.6052 - 0.4 = -0.1t \; ; \; -0.1t = -5.0052 \; ; \; t = \frac{5.0052}{0.1} \; ; \; \mathbf{t = 50.052}$

Check: $5 \stackrel{?}{=} 500e^{-0.1(50.052-4)} \; ; \; 5 \stackrel{?}{=} 500e^{-0.1(46.052)} \; ; \; 5 \stackrel{?}{=} 500e^{-4.6052} \; ; \; 5 \stackrel{?}{=} 500\times\frac{1}{e^{4.6052}} \; ; \; 5 \stackrel{?}{=} 500\times\frac{1}{100} \; ; \; 5 \stackrel{?}{=} 500\times 0.01$
$; \; 5 = 5$

8. Given $Y = Y_0(1+r)^n$ solve for n if $Y = 10$, $Y_0 = 2$, and $r = 0.25$.

$Y = Y_0(1+r)^n$; $10 = 2(1+0.25)^n$; $\frac{10}{2} = 1.25^n$; $5 = 1.25^n$; $log_{1.25} 5 = log_{1.25} 1.25^n$; $log_{1.25} 5 = n$; $n = \frac{log_{10} 5}{log_{10} 1.25}$

; $n = \frac{0.6989}{0.0969}$; $\mathbf{n = 7.2126}$

Check: $10 \overset{?}{=} 2(1+0.25)^{7.2126}$; $\frac{10}{2} \overset{?}{=} 1.25^{7.2126}$; $\frac{5}{1} \overset{?}{=} 1.25^{7.2126}$; $5 = 5$

9. Given $A = A_0\left(2^{-5t}\right) + ln\, A_0$ solve for t if $A = 10$ and $A_0 = 0.02$.

$A = A_0\left(2^{-5t}\right) + ln\, A_0$; $10 = 0.02\left(2^{-5t}\right) + ln\, 0.02$; $10 = 0.02\left(2^{-5t}\right) - 3.912$; $10 + 3.912 = 0.02\left(2^{-5t}\right)$; $\frac{13.912}{0.02} = 2^{-5t}$

; $695.6 = 2^{-5t}$; $log_2 695.6 = log_2 2^{-5t}$; $log_2 695.6 = -5t$; $-5t = \frac{log_{10} 695.6}{log_{10} 2}$; $-5t = \frac{2.8424}{0.3010}$; $-5t = 9.4422$

; $t = \frac{9.4422}{-5}$; $\mathbf{t = -1.8884}$

Check: $10 \overset{?}{=} 0.02\left(2^{-5 \times -1.8884}\right) + ln\, 0.02$; $10 \overset{?}{=} 0.02\left(2^{9.442}\right) - 3.912$; $10 \overset{?}{=} 0.02 \times 695.55 - 3.912$; $10 \overset{?}{=} 13.912 - 3.912$

; $10 = 10$

10. Given $U = U_0 e^{4t-1} + 2$ solve for t if $U = 8$ and $U_0 = 4$.

$U = U_0 e^{4t-1} + 2$; $8 = 4e^{4t-1} + 2$; $8 = 2\left(2e^{4t-1} + 1\right)$; $\frac{8}{2} = \left(2e^{4t-1} + 1\right)$; $\frac{4}{1} = 2e^{4t-1} + 1$; $4 - 1 = 2e^{4t-1}$ 1; $\frac{3}{2} = e^{4t-1}$

; $1.5 = e^{4t-1}$; $ln 1.5 = ln\, e^{4t-1}$; $0.4055 = 4t - 1$; $0.4055 + 1 = 4t$; $4t = 1.4055$; $t = \frac{1.4055}{4}$; $\mathbf{t = 0.3514}$

Check: $8 \overset{?}{=} 4e^{4 \times 0.3514 - 1} + 2$; $8 \overset{?}{=} 4e^{1.4056-1} + 2$; $8 \overset{?}{=} 4e^{0.4056} + 2$; $8 \overset{?}{=} 4 \times 1.5 + 2$; $8 \overset{?}{=} 6 + 2$; $8 = 8$

Glossary

The following glossary terms are used throughout this book:

Absolute value - The numerical value or magnitude of a quantity, as of a negative number, without regard to its sign. The symbol for absolute value is two parallel lines "| |". For instance, $|-2| = |2| = 2$, $|-35| = |35| = 35$, $|-0.23| = |0.23| = 0.23$, and $|-5.13| = |5.13| = 5.13$ are some examples of how absolute value is used.

Accurately - Without any error; exactly.

Addend - Any of a set of numbers to be added.

Addition - The process of adding two or more numbers to get a number called the sum.

Adequate - To consider or treat as equal. To make or set equal.

Advanced - Having gone beyond an initial or elementary stage; progressive.

Algebra - A branch of mathematics that deals with the relations between numbers.

Algebraic approach - An approach in which only numbers, letters, and arithmetic operations are used.

Algebraic expression - Designating an expression, equation, or function in which only numbers, letters, and arithmetic operations are contained or used.

Algebraic fractions - A fraction having variables in either the numerator or the denominator or both.

Antilogarithm - The number corresponding to a logarithm.

Apparent - Appearing to the eye or to the judgment; seeming, often in distinction to real; obvious.

Application - The act of applying or putting to use.

Apply - To put on. To put to or adapt for particular use. To use.

Approximation - An amount or estimate nearly exact or correct.

Area - The amount or size of a surface: The area of a floor 10 feet by 10 feet is a 100 square feet.

Arithmetic fractions - A fraction having positive or negative whole numbers in the numerator and the denominator; an integer fraction.

Associative - Pertaining to an operation in which the result is the same regardless of the way the elements are grouped, as, in addition, $2+(4+5) = (2+4)+5 = 11$ and, in multiplication, $2 \times (4 \times 5) = (2 \times 4) \times 5 = 40$.

Assumption - The act of assuming; supposition; the act of taking for granted.

Base - *a.* The number on which a system of numeration is based. For example, the base of the decimal system is 10. Computers use the binary system, which has the base 2. *b.* A number that is to be multiplied by itself the number of times indicated by an exponent or logarithm. For example, in 2^5, 2 is the base and 5 is the exponent.

Binomial - An expression consisting of two terms connected by a plus or minus sign. For example, $a+b$, $\sqrt{x^3}-\sqrt{y}$, x^3+3x, and a^2b^3-3ab are referred to as binomials.

Brackets [ ] - A pair of symbols used to enclose a mathematical expression.

Cancel - To cross out or mark. To subtract, a common term, from both sides of an equation or inequality in the same way.

Case - Supporting facts offered in justification of a statement.

Centigrade - Of or indicating the temperature scale used for ordinary purposes throughout the world. In this system water freezes at 0 degrees and boils at 100 degrees.

Change - To replace by another; alter; transform.

Check - To test or examine, as for accuracy or completeness.

Circle - A plane curve all of whose points are equally distant from a point in the plane, called the center.

Class - A group of persons or things that have something in common, a set, collection, group.

Classification - The act, process, or result of classifying.

Classify - To put or divide into classes or groups.

Coefficient - A number placed in front of an algebraic expression and multiplying it; factor. For example, in the expression $3x^2+5x=2$, 3 is the coefficient of x^2, and 5 is the coefficient of x.

Combine - To bring together; unite; join; merge.

Common - Belonging equally to two or more quantities.

Common denominator - A common multiple of the denominators of two or more fractions. For example, 10 is a common denominator of $\frac{1}{2}$ and $\frac{3}{5}$.

Common divisor - A number or quantity that can evenly divide two or more other numbers or quantities. For example, 4 is a common divisor of 12 and 20.

Common factor - Another name for common divisor.

Common fraction - A fraction whose numerator and denominator are both integers (whole numbers).

Common logarithm - Logarithms using 10 as a base.

Commutative - Pertaining to an operation in which the order of the elements does not affect the result, as, in addition, $5+3 = 3+5$ and, in multiplication, $5\times3 = 3\times5$.

Complex fractions - A fraction in which either the numerator or the denominator or both contain a fraction.

Computation - The act or method of computing; calculation.

Compute - To figure by using mathematics; calculate.

Concept - A general idea or notion.

Conjugate - Inversely related to one of a group of otherwise identical properties.

Constant - Remaining the same; not changing. A number or other thing that never changes.

Construct - To make by putting parts together; build.

Conversion - A change in the form of a quantity or an expression without a change in the value.

Convert - To change from one form or use to another; transform.

Correspond - To be in agreement with each other.

Cube - The third power of a number or quantity.

Cube root $\left(\sqrt[3]{\ }\right)$ - A number which, cubed, equals the number given. For example, the cube root of 216 is 6.

Decimal number - Any number written using base 10; a number containing a decimal point.

Decimal point - A period placed to the left of a decimal.

Decrease - Reduce; make less; lessen usually refers to decrease in numbers.

Degree - The greatest sum of the exponents of the variables in a term of a polynomial or polynomial equation. For example, the polynomial $w^3 + 3w + 5$ is a third degree polynomial.

Denominator - The term below the line in a fraction; the divisor of the numerator. For example, in the fraction $\frac{3}{5}$, 5 is the denominator.

Descend - To move from a higher to a lower place. To go down.

Descending order - Decreasing order.

Difference - The amount by which one quantity differs from another; remainder left after subtraction.

Digit - Any of the numerals from 0 through 9 - in the base-ten system.

Distributive - Of the principle in multiplication that allows the multiplier to be used separately with each term of the multiplicand.

Dividend - A quantity to be divided. For example, in the problem $14 \div 2$, 14 is called the dividend.

Division - The process of finding how many times a number (the divisor) is contained in another number (the dividend). The number of times equals the quotient.

Divisor - The quantity by which another quantity, the dividend, is to be divided. For example, in the problem $14 \div 2$, 2 is called the divisor.

Enhance - To add to; to increase or make greater.

Equal - Exactly the same. Of the same quantity, size, number, value, degree, intensity, or quality.

Equality - The condition or quality of being equal.

Equate - To make or set equal. To put in the form of an equation.

Equation - A mathematical expression involving the use of an equal sign. For example, $x^3 + 3x^2 + 5x = 3$ is referred to as an equation.

Equivalent algebraic fractions - Algebraic fractions that are the same.

Even number - A number which is exactly divisible by two; not odd. For example, $(0, 2, 4, 6, 8, 10, \ldots)$ are even numbers.

Exact order - Not deviating in form or content; precise.

Example - One that is representative of a group as a whole; a sample.

Expanded form - To write, a quantity, as a sum of terms, as a continued product, or as another extended form.

Exponent - A number placed as a superscript to show how many times another number is to be placed as a factor. For example, in the problem $5^3 = 5 \times 5 \times 5 = 125$, 3 is an exponent.

Exponential - Containing, involving, or expressed as an exponent.

Exponential notation - A way of expressing a number as the product of the factor and 10 raised to some power. The factor is either a whole number or a decimal number. For example, the exponential notation form of 0.0353, 0.048, 489, 3987 are 35.3×10^{-3}, 48×10^{-3}, 48.9×10^{1}, and 398.7×10^{1}, respectively.

Express - To tell in words; state. Given in direct terms; explicit; precise; plain.

Expression - A designation of any symbolic mathematical form, such as an equation. The means by which something is expressed.

Extend - To stretch out. To make greater or broader; increase.

Factor - One of two or more quantities having a designated product. For example, 3 and 5 are factors of 15.

Factorize - Resolve into factors.

Fahrenheit - Of or indicating the temperature scale used for ordinary purposes in the U.S. and Great Britain. In this system water freezes at 32 degrees and boils at 212 degrees.

Familiar - Well aquatinted, as through experience or study.

Form - A specific type; kind.

Formula - A statement of some fact or relationship in mathematical terms.

Fraction - A number which indicates the ratio between two quantities in the form of $\frac{a}{b}$ such that a is any real number and b is any real number not equal to zero.

Fractional - Having to do with or making up a fraction.

General - Not precise or detailed. Not limited to one class of things. Relating to all.

Greater than $(\rangle)$ - A symbol used to compare two numbers with the greater number given first. For example, $5 \rangle 2$, $23 \rangle 20$, $50 \rangle 10$.

Greatest common factor - A greatest number that divides two or more numbers without a remainder. For example, 6 is the greatest common factor among 6, 12, and 36.

Group - An assemblage of objects or numbers.

Horizontal - Flat. Parallel to the horizon. Something that is horizontal, as a line, plane, or bar.

Identical - Exactly alike; the very same.

Identify - To recognize. To establish the identity of.

Illustrate - To explain or make clear, as by examples or comparisons.

Imaginary number - The positive square root of a negative number. For example, $\sqrt{-5}$, $\sqrt{-3}$, and $\sqrt{-1}$ are imaginary numbers. Not real number.

Improper fraction - A fraction in which the numerator is larger than or equal to the denominator. For example, $\frac{6}{5}$, $\frac{10}{9}$, and $\frac{23}{7}$ are improper fractions.

Increment - An increase or addition. The amount by which a quantity increases.

Index - A number or symbol, often written as a subscript or superscript to a mathematical expression, that indicates an operation to be performed on. For example, in the problem $\sqrt[3]{x^2}$, 3 is referred to as an index.

Indicate - To point out; to suggest; to show or signify.

Inequality $(\neq)$ - A relation indicating that the two numbers are not the same.

Instance - A case or example given as an illustration or proof.

Integer fraction - A fraction having positive or negative whole numbers in the numerator and the denominator.

Integer number - Any member of the set of positive whole numbers $(1, 2, 3, 4, \ldots)$, negative whole numbers $(-1, -2, -3, -4, \ldots)$, and zero is an integer number.

Introduction - To inform of something for the first time. The act of introducing.

Invert - To turn upside down. To reverse the order of.

Irrational number - A number not capable of being expressed by an integer (a whole number) or an integer fraction (quotient of an integer). For example, $\sqrt{3}$, π, and $\sqrt[4]{7}$ are irrational numbers.

Law - A general principle or rule that is obeyed in all cases to which it is applicable.

Less than $(\langle)$ - A symbol used to compare two numbers with the lesser number given first. For example, $5 \langle 8$, $23 \langle 30$, $12 \langle 25$.

Lessen - To make smaller or less; to decrease or diminish.

Like terms - Similar terms.

Linear - Of or having to do with a line or lines. Of the first degree, as an equation.

Linear equation - An algebraic equation in which variables are used as factors no more than once in each term. For example, $3x + 5y = 10$ is a linear equation.

Logarithms - The exponent or power to which a fixed number, called the base, must be raised in order to produce a given number, called the antilogarithm.

Lowest term - Smallest value.

Match - A person or thing that is exactly like another, counterpart.

Mathematical operation - The process of performing addition, subtraction, multiplication, and division in a specified sequence.

Method - A way of doing or accomplishing something.

Minimize - To reduce to the least possible amount; reduce to a minimum.

Mixed fraction - A fraction made up of a positive or negative whole number and an integer fraction.

Mixed operation - Combining addition, subtraction, multiplication, and division in a math process is defined as a mixed operation.

Monomial - An expression consisting of only one term. Being a simple algebraic term. For example, 5, $\sqrt{xy}$, x^3, and $2ab$ are referred to as monomials.

Multiplicand - The number that is or is to be multiplied by another.

Multiplication - The process of finding the number obtained by repeated additions of a number a specified number of times: Multiplication is symbolized in various ways, i.e., $3 \times 4 = 12$ or $3 \cdot 4 = 12$, which means $3 + 3 + 3 + 3 = 12$, to add the number three together four times.

Multiplier - The number by which the multiplicand is multiplied. For example, if 3 is multiplied by 4, 3 is the multiplicand, 4 is the multiplier, and 12 is the product.

Natural logarithm - Logarithms using e as a base. Also called Napierian logarithm.

Negative number - A quantity less than zero.

Not Applicable - In this book *Not Applicable* pertains to a *step* that can not be put to a specific use. A *step* that is not relevant.

Not real number - Imaginary number.

Numerator - The term above the line in a fraction. For example, in the fraction $\frac{3}{5}$, 3 is the numerator.

Numerical coefficient - Coefficients represented by numbers rather than letters.

Objective - An end toward which efforts are directed; a goal or end.

Observe - To look on without actively participating; to regard attentively; to see or notice.

Obtain - To gain possession of, especially by effort; get; acquire.

Odd number - A number having a remainder of one when divided by two; not even. For example, $(1, 3, 5, 7, 9, 11, \ldots)$ are even numbers.

Operation - A process or action, such as addition, subtraction, multiplication, or division, performed in a specified sequence and in accordance with specific rules of procedure.

Original - Belonging to or pertaining to the origin, source, or beginning of something.

Parentheses ( ) - A pair of symbols used to enclose a sum, product, or other mathematical expressions.

Perform - To accomplish; to do; to execute.

Polynomial - An algebraic function of two or more summed terms, each term consisting of a constant multiplier and one or more variables raised to a power. For example, the general form of a polynomial of degree n in a single real variable x is $P(x) = a_n x^n + a_{n-1} x^{n-1} + a_{n-2} x^{n-2} + \cdots + a_0$.

Positive number - A quantity greater than zero.

Power - An exponent. The result of a number multiplied by itself a given number of times. For example, the third power of 3 is 27.

Practical - Pertaining or relating to practice or action.

Practice - To exercise or perform repeatedly in order to acquire or polish a skill.

Primary - Something that is first in degree, quality, or importance. Occurring first in time or sequence. Original.

Prime factorization - A factorization that shows only prime factors. For example, $21 = 1 \times 3 \times 7$.

Prime number - A number that has itself and unity as its only factors. For example, 2, 3, 5, 7, and 11 are prime numbers since they have no common divisor except unity.

Principal - First, highest, or foremost in importance.

Problem - Something to be done or solved.

Proceed - To go on. To continue. To begin and carry on an activity.

Process - A series of operations or a method for producing something. A series of actions, changes, or functions that bring about an end or result.

Product - The quantity obtained by multiplying two or more quantities together.

Proficiency - The state of being proficient; very skilled; expert.

Proper - Strictly belonging or suitable; fitting.

Proper fraction - A fraction in which the numerator is smaller than the denominator.

Quadratic equations - Indicating a mathematical expression or equation of the second degree. For example, $x^2+3x-2=0$ is a quadratic equation.

Quality - That which makes something the way it is; distinctive feature or characteristic.

Quantity - An amount or number.

Quotient - The quantity resulting from division of one quantity by another.

Radical - The root of a quantity as indicated by the radical sign. Indicating or having to do with a square root or cube root.

Radical expression - A mathematical expression or form in which radical signs appear.

Radical sign $(\sqrt{\ })$ - A sign that indicates a specified root of the number written under it. For example, $\sqrt[3]{27}$ = the cube root of 27, which is, 3.

Radicand - The quantity under a radical sign. For example, 27 is the radicand of $\sqrt[3]{27}$.

Radius - A straight line from the center of a circle or sphere to the circumference or surface.

Rational number - A number that can be represented as an integer (a whole number) or an integer fraction (quotient of integers). For example, $\frac{1}{5}$, $-\frac{2}{15}$, $12=\frac{12}{1}$, $-230=-\frac{230}{1}=\frac{230}{-1}=\ldots$, $-10=-\frac{10}{1}=-\frac{100}{10}=-\frac{50}{5}=\frac{350}{-35}=\ldots$, and $0.13=\frac{13}{100}=\frac{130}{1000}=\frac{26}{200}=\ldots$ are rational numbers.

Rationalization - The act, process, or practice of rationalizing.

Rationalize - To remove radicals without changing the value of an expression or roots of an equation.

Real number - A number that is either a rational number or an irrational number. For example, $\frac{3}{5}$, $-\frac{4}{13}$, -23, 0.13, $\sqrt{5}$, and π are real numbers.

Reduce - To make less in size or amount. To diminish in size, quantity, or value.

Reference - The directing of attention to a person or thing.

Re-group - A repeated assemblage of objects or numbers.

Remainder - *a.* What is left when a smaller number is subtracted from a larger number. *b.* What is left undivided when one number is divided by another that is not one of its factors.

Represent - Stand for; to depict; portray; to describe.

Require - To have need of. To order or insist upon.

Respectively - In their respective order; individually in their given order.

Result - To end in a particular way. The consequence of a particular action. An outcome.

Resultant - That which results. Consequence.

Review - To go over or examine again. To think back on.

Revise - To change or modify. To read carefully so as to correct errors or make improvements and changes.

Revision - The result of revising. Something that has been revised.

Root - A quantity that, multiplied by itself a specified number of times, produces a given quantity. For example, 5 is the square root (5×5) of 25 and the cube root $(5 \times 5 \times 5)$ of 125.

Round number - A number that is revised or rounded to the nearest unit, as ten, hundred or thousand. For example, 200 is a round number for 199 or 201.

Rounded off - To make into a round number.

Rule - A method or procedure prescribed for computing or solving a problem.

Scientific notation - A way of expressing a number as the product of the factor and 10 raised to some power. The factor is always of the form where the decimal point is to the right of the first non-zero digit. For example, the scientific notation form of 0.0353, 0.048, 489, 3987 are 3.53×10^{-2}, 4.8×10^{-2}, 4.89×10^{2}, 3.987×10^{3}, respectively.

Section - One of several component parts of something; piece; portion.

Sequence - The order in which one thing comes after another. A number of things following each other; series.

Show - Demonstrate; to point out; indicate.

Sign - A mark or symbol having an accepted and specific meaning. For example, the sign $+$ implies addition.

Signed number - A number which can have a positive or negative value as designated by $+$ or $-$ symbol. A signed number with no accompanying symbol is understood to be positive.

Similar - Alike but not completely the same.

Similar radicals - Radical expressions with the same index and the same radicand. For example, $\sqrt[3]{x^2}$, $5\sqrt[3]{x^2}$, and $3\sqrt[3]{x^2}$ are referred to as similar radicals.

Simplify - Make easier; less complex.

Solution - The act, method, or process of solving a problem. The answer to a problem.

Solution set - The set of all the values that satisfy an equation or inequality.

Solve - To find a solution to; answer.

Special - Exceptional. Surpassing what is common or usual.

Specific example - An example that is precise and explicit.

Square - To find the equivalent of in square measure; to multiply, as a number or quantity, by itself.

Square root $(\sqrt{\ })$ - The factor of a number which, multiplied by itself, gives the original number. For example, the square root of 36 is 6.

Standard - Any type, model, or example for comparison. Serving as a gauge or model.

Step - One of a series of actions or measures taken toward some end.

Sub-group - A distinct group within a group.

Subject - A topic discussed in writing.

Subscript - A number, letter, or a symbol, written below and to the right or left of a character. For example, 2 is the subscript in x_2.

Substitute - To put in the place of another; to put in exchange.

Subtraction - The mathematical process of finding the difference between two numbers.

Sum - The amount obtained as a result of adding two or more numbers together.

Summary - Reduced into few words; concise.

Superscript - A number, letter, or a symbol, written above a character. For example, 5 is the superscript in y^5.

Symbol - A sign used to represent a mathematical operation.

Technique - A special method of doing something. The systematic procedure by which a complex or scientific task is accomplished.

Term - The parts of a mathematical expression that are added or subtracted. For example, in the equation $ax^3 + bx^2 + cx - d$, ax^3, bx^2, cx, and d are referred to as terms.

Trinomial - An expression consisting of three terms connected by a plus or minus sign. For example, $a^2 + a + 3$, $\sqrt[3]{x^2} + \sqrt[3]{x} - 5$, and $x^3 + 3x^2 + 2$ are referred to as trinomials.

Type - An example or model; kind.

Variable - A quantity capable of assuming any of a set of values. Having no fixed quantitative value.

Vertical - Upright. At right angles to the horizon. Straight up and down.

Whole number - A whole number is defined as an integer number.

With - Having as a possession, attribute, or characteristic.

Without - In the absence of; with no or none of.

Zero - The symbol or numeral 0. The point, marked 0, from which positive or negative quantities are reckoned on a graduated scale.

The following references were used in developing this glossary:

1) *The Webster's New World Dictionary of American English, Victoria E. Neufeldt, editor in chief, third college edition, 1995.*

2) *The American Heritage Dictionary of the English Language, William Morris, editor, third edition, 1994.*

3) *HBJ School Dictionary, Harcourt Brace Jovanovich publishing, fourth edition, 1985.*

Index

D

M

About the Author

Said Hamilton received his B.S. degree in Electrical Engineering from Oklahoma State University and Master's degree, also in Electrical Engineering from the University of Texas at Austin. He has taught a number of math and engineering courses as a visiting lecturer at the University of Oklahoma, Department of Mathematics, and as a faculty member at Rose State College, Department of Engineering Technology, at Midwest City, Oklahoma. He is currently working in the field of aerospace technology and has published numerous technical papers.

About the Editor

Pat Eblen received his Bachelor of Science degree in Electrical Engineering from the University of Kentucky where he was a member of Eta Kappa Nu Electrical Engineering Honor Society. He has worked in the aerospace industry for nearly twenty years where he has received numerous awards for contributions to spacecraft technology programs. Mr. Eblen enjoys studying mathematical theories in probability and quantum mechanics and has developed several original concepts in these fields.

Order Form

call 1-800-209-8186 to order

Or: Use the order form

Last Name ______________________ First Name ______________ M.I. ____

Address ____

City ________________ State __________ Zip Code __________________

No. of Books	Book Price ($49.95)	Total Price
		$
	Subtotal	$
	Shipping and Handling (Add $4.50 for the first book $3.00 for each additional book)	$
	Sales Tax Va. residents add 4.5% ($49.95 × 0.045 = $2.25)	$
	Total Payment	$

Enclosed is:

A check ☐ Master Card ☐

VISA ☐ American Express ☐

Account Number ______________________ Expiration Date ________________

Signature __________________________

Make checks payable to Hamilton Education Guides.

Please send completed form to:

Hamilton Education Guides

P.O. Box 681

Vienna, Va. 22183